Electronic Instrument Handbook

Other McGraw-Hill Books of Interest

ALI • *Digital Switching Systems*
ASH • *Dynamic Routing in Telecommunications Networks*
BEDELL • *Cellular/PCS Management*
BENNER • *Fibre Channel*
BEST • *Phase-Locked Loops,* Third Edition
BILLINGS • *Switchmade Power Supply Handbook,* Second Edition
CLAYTON • *McGraw-Hill Illustrated Telecom Dictionary*
DAVIS • *ATM For Public Networks*
FAYNBERG • *Intelligent Network Standards*
FEIT • *TCP/IP,* Second Edition
GALLAGHER • *Mobile Telecommunications Networking with IS-41*
GORALSKI • *Introduction to ATM Networking*
HARTE • *Cellular and PCS: The Big Picture*
HARTE • *GSM Superphones*
HELDMAN • *Competitive Telecommunications*
KESSLER • *ISDN,* Third Edition
KURUPPILLAI • *Wireless PCS*
LACHS • *Fiber Optic Communications*
LEE • *Mobile Cellular Telecommunications,* Second Edition
LEE • *Mobile Communications Engineering,* Second Edition
LOGSON • *Mobile Communication Satellites*
MACARIO • *Cellular Radio,* Second Edition
MULLER • *Desktop Encyclopedia of Telecommunications*
NAVABI • *Verilog Digital System Design*
PECAR • *Telecommunications Factbook*
RODDY • *Satellite Communications,* Second Edition
ROHDE et al. • *Communication Receivers,* Second Edition
RUSSELL • *Signaling System #7,* Second Edition
RUSSELL • *Telecommunications Protocols,* Second Edition
SIMON et al. • *Spread Spectrum Communications Handbook*
SMITH • *Cellular Design and Optimization*
SMITH • *Practical Cellular and PCS Design*
TSAKALAKIS • *PCS Network Deployment*
TURIN • *Digital Transmission Systems*
WINCH • *Telecommunication Transmission Systems,* Second Edition

Electronic Instrument Handbook

Clyde F. Coombs Jr. *Editor-in-Chief*

Third Edition

McGraw-Hill

New York San Francisco Washington, D.C. Auckland Bogotá
Caracas Lisbon London Madrid Mexico City Milan
Montreal New Delhi San Juan Singapore
Sydney Tokyo Toronto

Library of Congress Cataloging-in-Publication Data

Electronic instrument handbook / Clyde F. Coombs, Jr., editor-in-chief.—3rd ed.
p. cm.
ISBN 0-07-012618-6
1. Electronic instruments. I. Coombs, Clyde F.
TK7878.4.E55 1999
621.3815′4—dc21 99-35923
CIP

McGraw-Hill 

A Division of The **McGraw·Hill** *Companies*

1 2 3 4 5 6 7 8 9 0 DOC/DOC 9 0 4 3 2 1 0 9

ISBN 0-07-012618-6

The sponsoring editor for this book was Stephen S. Chapman and the production supervisor was Sherri Souffrance. It was set in Century Schoolbook by The PRD Group, Inc.

Printed and bound by R. R. Donnelley & Sons Company.

This book is printed on acid-free paper.

Contents in Brief

Part 8 Microwave Passive Devices

Part 9 Using Electronic Instruments

Part 10 Instruments in Systems

Part 11 Software and Connectivity for Instrumentation

Part 12 Distributed and Networked Instrumentation

Contents

Part 7 Digital Domain Instruments

Chapter 29. Logic Analyzers *Steve Warntjes* 29.1

Chapter 30. Protocol Analyzers *Steve Witt* 30.1

Chapter 31. Bit Error Rate Measuring Instruments: Pattern Generators and Error Detectors *Hugh Walker* 31.1

Part 8 Microwave Passive Devices

Chapter 32. Microwave Passive Devices *Frank K. David* 32.1

Part 9 Using Electronic Instruments

Chapter 33. Impedance Considerations *K. D. Baker and D. A. Burt* 33.1

For Ann

Contributors

K. D. Baker *Department of Electrical Engineering, Utah State University, Logan, Utah* (CHAPS. 33–37)

D. A. Burt *Department of Electrical Engineering, Utah State University, Logan, Utah* (CHAPS. 33–37)

Rex Chappell *Agilent Technologies, Santa Clara, Calif.* (CHAP. 19)

Phillip J. Christ *Agilent Technologies, Loveland, Colo.* (CHAP. 41)

John J. Corcoran *Agilent Technologies, Palo Alto, Calif.* (CHAP. 6)

Fred Cruger *Agilent Technologies, Lake Stevens, Washington* (CHAP. 4)

Leonard S. Cutler *Agilent Technologies, Palo Alto, Calif.* (CHAP. 20)

Frank R. David *Agilent Technologies, Santa Rosa, Calif.* (CHAP. 32)

Alan J. De Vilbiss *Agilent Technologies, Colorado Springs, Colo.* (CHAP. 14)

J. Flemming Dias *Agilent Technologies, Palo Alto, Calif.* (CHAP. 5)

Calvin Erickson *Agilent Technologies, Loveland, Colo.* (CHAP. 39)

James S. Gallo *Agilent Technologies, Rockaway, N.J.* (CHAP. 11)

Geri Georg *Agilent Technologies, Loveland, Colo.* (CHAP. 46)

John H. Guilford *Agilent Technologies, Lake Stevens, Wash.* (CHAP. 9)

Daniel R. Harkins *Agilent Technologies, Santa Rosa, Calif.* (CHAP. 28)

William Heinz *Agilent Technologies, Santa Clara, Calif.* (CHAPS. 8, 18)

James L. Hook *Agilent Technologies, Santa Clara, Calif.* (CHAP. 27)

Waguih Ishak *Agilent Technologies, Palo Alto, Calif.* (CHAPS. 22–25)

Charles Kingsford-Smith *Agilent Technologies, Lake Stevens, Wash.* (CHAPS. 7, 16)

Darren Kwock *Agilent Technologies, Loveland, Colo.* (CHAP. 45)

Kang Lee *National Institute of Standards and Technology, Gaithersburg, Maryland* (CHAP. 47)

James M. McGillivary *Agilent Technologies, Rockaway, N.J.* (CHAP. 38)

Tim Mikkelsen *Agilent Technologies, Loveland, Colo.* (CHAP. 10)

Joseph E. Mueller *Agilent Technologies, Loveland, Colo.* (CHAP. 43)

Yoh Narimatsu *Yokogawa-Hewlett-Packard Ltd., Kobe, Japan* (CHAP. 26)

Andreas Pfaff *Agilent Technologies, Boeblingen, Germany* (CHAP. 17)

Ronald E. Pratt *Agilent Technologies, Santa Clara, Calif.* (CHAP. 15)

David B. Richey *Agilent Technologies, Loveland, Colo.* (CHAP. 40)

Jan Ryles *Agilent Technologies, Loveland, Colo.* (CHAPS. 12, 44)

Alan W. Schmidt *Agilent Technologies, Santa Rosa, Calif.* (CHAP. 21)

Bonnie Stahlin *Agilent Technologies, Loveland, Colo.* (CHAPS. 1, 41)

Scott Stever *Agilent Technologies, Loveland, Colo.* (CHAP. 13)

Tim Tillson *Agilent Technologies, Loveland, Colo.* (CHAP. 42)

Hugh Walker *Agilent Technologies, South Queensferry, Scotland* (CHAP. 31)
Steven B. Warntjes *Agilent Technologies, Colorado Springs, Colo.* (CHAP. 29)
Stephen Witt *Agilent Technologies, Colorado Springs, Colo.* (CHAP. 30)
David R. Workman *Consultant, Littleton, Colo.* (CHAPS. 2,3)

Foreword*

The advancement of science and technology is matched by a parallel progress in the art of measurement. It can, in fact, be said that the quickest way to assess the state of a nation's science and technology is to examine the measurements that are being made and the way in which the data accumulated by measurements are utilized.

The reasons for this are simple. As science and technology move ahead, phenomena and relations are discovered that make new types of measurements desirable. Concurrently, advances in science and technology provide means of making new kinds of measurements that add to understanding. This in turn leads to discoveries that make still more measurements both possible and desirable.

It is thus axiomatic that sophisticated science and technology are associated with sophisticated measurements, while simple-minded science is associated with only elementary measuring techniques.

As the art of measurement has advanced, the technology of making measurements has increasingly relied on electrical and electronic methods. This comes about for two reasons. First, once information is transformed into electrical form, it can be readily processed in ways that will meet the needs of a great variety of individual situations. Second, most phenomena, such as temperature, speed, distance, light, sound, and pressure, can be readily transformed into electrical indications for processing and interpretation.

The result has been that during the last 30 years, there has developed a remarkable world of instruments based on electronics, which both supports and feeds on the ever-advancing frontiers of knowledge, and concurrently makes it possible to carry on the old tasks more easily and with greater accuracy.

Modern electronic instruments are typically direct-reading, making it unnecessary to resort to calibration curves. Increasingly, their outputs are available in digital form, which eliminates the necessity of even reading the indication of a needle or the scale of a cathode-ray tube. Moreover, data in digital form can be processed through a computer that can instantly perform necessary ancillary calculations; this eliminates possibilities of error and saves time of high-priced personnel. Through the use of recorders and cathode-ray oscilloscopes, it is now even possible to draw the final results in the form of plotted curves, thereby further speeding up the entire process of gathering and analyzing data.

---

* Reprinted from the First Edition.

A third of a century ago, most electronic measurements were made with instruments which the experimenter had constructed with his own hands. More often than not these early instruments were not only inconvenient but also useless unless operated by highly skilled personnel, preferably the men who had built them.

This situation has now changed completely. Today one can usually buy a much better instrument than he can build, and one does not have to possess expert knowledge about a particular instrument in order to keep it in adjustment and functioning properly.

At the same time, even with the marvelous array of professionally made instruments that are listed in catalogs today, the user must provide an input of his own in order to take full advantage of the opportunities available to him. He must know what the instruments he uses, or is considering purchasing, will and will not measure; types of difficulties that can arise in making measurements under special or unusual conditions; possibilities and limitations; and the errors that can be introduced by distortions in waveform, by noise, by stray electric currents, etc. Today's user of instruments must also consider the characteristics of what he wishes to measure, and then relate these characteristics to the properties, possibilities, and limitations of the measuring instrument he plans to use.

It is the purpose of this book to help a worker in some field of science and technology match his needs with those of the world of instruments, in situations in which he is a nonexpert "consumer" of the fruits of instrumentation technology.

Frederick E. Terman

Preface

This is a book about electronic instruments. In this new edition, we have included descriptions and discussions on how software, computers, and networks have come together, with the standard hardware to add new dimensions to the concept of electronic instruments. These include distributed measurement, software-defined instruments, and virtual instruments, in various forms. At the same time, these technologies have combined with the standard hardware to perform general-purpose as well as complex instrument functions. Naturally, this new material is in addition to the comprehensive information on the more traditional instruments, their building block elements, and how they can work together in systems. It provides a one-volume source of information on all aspects of electronic instruments as they are actually defined and used.

A major objective of the book is to provide descriptions of the technology and functions of electronic instrumentation that will demystify these important devices. The instrument chapters describe what they are, what they do, how they do it, and how to get the most from them. As instruments have become more capable, they also have become more complex and less intuitive in their use. Although the fundamental purposes and functions of electronic instruments have not changed significantly over the years, the instruments themselves and how much information—or even involvement in the solution—they can provide has seen a revolution. It is critical that all those who use these devices understand and feel comfortable with the tools of their trade. This is an especially important issue in many fields such as medicine, chemistry, and communications where electronic instrumentation, used by technologists not necessarily trained in electronics, forms the basis of measurement and control.

This book is designed to be a resource for all who use electronic instrumentation. This includes those just entering the field of electronic measurement, as well as experienced professionals. The information available here ranges from very basic to very technical. The chapters, therefore, are organized to provide a progression from the general nature of the elements of signal generation, sensing, and processing, along with interface technology and products, to the combination of these elements into instruments. It proceeds on the organization of instruments into specific systems, through to distributed test systems, using networks, and includes the various aspects of what has come to be called the "virtual instrument."

In this book we recognize that software and computers have become full partners in defining the function and construction of instruments. In addition, combining these with either Local Area Networks (LANs) or Wide Area

Networks (WANs), such as the Internet, adds a totally new dimension to electronic test and measurement technology. This ability has become an important, often critical, part of the technologists' capability to get data on remote or dispersed elements, process data into information, or use information to provide solutions to the total problem at hand. We describe information on these elements, what they are, how they work, how they are integrated into electronic instrumentation, how they are connected, and how they create these capabilities that has not been available before.

Although the act of measurement itself is the result of using instruments, this is not a "measurements" book. The great number of possible measurements makes their definition in this book unfeasible and potentially confusing. As a result, specific measurements are discussed only as examples of applications of the instruments themselves. It is felt that with a clear understanding of what the instruments do and how they do it, the reader is in the best position of define the solution to their specific measurement problem.

The international team of authors provide unequalled expertise in the function, design, and application of the instruments they discuss. They each represent years of experience working with the theory as well as the use of the devices. In addition, they are familiar with a wide range of instruments and instrumentation and how they work together to achieve user objectives.

The early encouragement of Bill Hewlett, Bill Terry, John Doyle, and Ned Barnholdt from the management team at Hewlett-Packard and Agilent Technologies is gratefully acknowledged. In addition, the work of Chuck Gustafson, Bob Allen, and John Page in clarifying the Table of Contents and potential contributors was invaluable. I would like to emphasize my appreciation to Chuck Gustafson, as without his help, this book would not have been able to meet the objectives we set for it.

Clyde F. Coombs, Jr.

Chapter

1

Introduction to Electronic Instruments and Measurements

Bonnie Stahlin*
Agilent Technologies
Loveland, Colorado

1.1 Introduction

This chapter provides an overview of both the software and hardware components of instruments and instrument systems. It introduces the principles of electronic instrumentation, the basic building blocks of instruments, and the way that software ties these blocks together to create a solution. This chapter introduces practical aspects of the design and the implementation of instruments and systems.

Instruments and systems participate in environments and topologies that range from the very simple to the extremely complex. These include applications as diverse as:

- Design verification at an engineer's workbench
- Testing components in the expanding semiconductor industry
- Monitoring and testing of multinational telecom networks

1.2 Instrument Software

Hardware and software work in concert to meet these diverse applications. Instrument software includes the firmware or embedded software in instru-

---

* Additional material adapted from "Introduction to Electronic Instruments" by Randy Coverstone, Electronic Instrument Handbook 2nd edition, Chapter 4, McGraw-Hill, 1995, and Joe Mueller, Hewlett-Packard Co., Loveland.

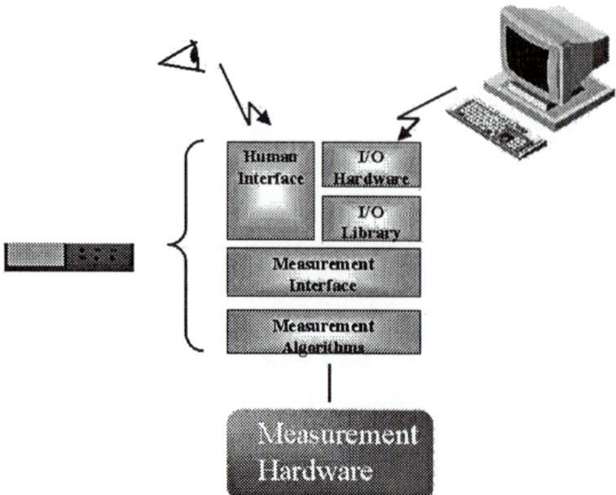

Figure 1.1 Instrument embedded software.

ments that integrates the internal hardware blocks into a subystem that performs a useful measurement. Instrument software also includes system software that integrates multiple instruments into a single system. These systems are able to perform more extensive analysis than an individual instrument or combine several instruments to perform a task that requires capabilities not included in a single instrument. For example, a particular application might require both a source and a measuring instrument.

1.2.1 Instrument embedded software

Figure 1.1 shows a block diagram of the embedded software layers of an instrument. The I/O hardware provides the physical interface between the computer and the instrument. The I/O software delivers the messages to and from the computer to the instrument interface software. The measurement interface software translates requests from the computer or the human into the fundamental capabilities implemented by the instrument. The measurement algorithms work in conjunction with the instrument hardware to actually sense physical parameters or generate signals.

The embedded software simplifies the instrument design by:

- Orchestrating the discrete hardware components to perform a complete measurement or sourcing function.
- Providing the computer interaction. This includes the I/O protocols, parsing the input, and formatting responses.
- Providing a friendly human interface that allows the user to enter numeric values in whatever units are convenient and generally interface to the instrument in a way that the user naturally expects.
- Performing instrument calibration.

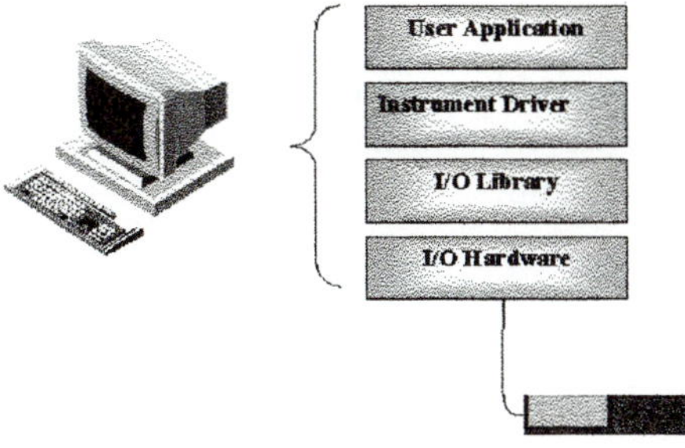

Figure 1.2 Software layers on the host side for instrument to computer connection.

1.2.2 System software

Figure 1.2 shows the key software layers required on the host side for instrument systems. Systems typically take instruments with generic capabilities and provide some specific function. For instance, an oscilloscope and a function generator can be put together in a system to measure transistor gain. The exact same system with different software could be used to test the fuel injector from a diesel engine.

Generally, the system itself:

- Automates a task that would be either complex or repetitive if performed manually.
- Can perform more complex analysis or capture trends that would be impractical with a single instrument.
- Is specific to a particular application.
- Can integrate the results of the test into a broader application. For instance, the system test could run in a manufacturing setting where the system is also responsible for handling the devices being tested as they come off the production line.

Please refer to Part 11 of this handbook for an in-depth discussion of instrument software.

1.3 Instruments

In test and measurement applications, it is commonplace to refer to the part of the real or physical world that is of interest as the *device under test* (*DUT*). A measurement instrument is used to determine the value or magnitude of a physical variable of the DUT. A source instrument generates some sort of stimulus that is used to stimulate the DUT. Although a tremendous variety of instruments exist, all share some basic principles. This section introduces these basic principles of the function and design of electronic instruments.

1.3.1 Performance attributes of measurements

The essential purpose of instruments is to sense or source things in the physical world. The performance of an instrument can thus be understood and characterized by the following concepts:

- *Connection* to the variable of interest. The inability to make a suitable connection could stem from physical requirements, difficulty of probing a silicon wafer, or from safety considerations (the object of interest or its environment might be hazardous).
- *Sensitivity* refers to the smallest value of the physical property that is detectable. For example, humans can smell sulfur if its concentration in air is a few parts per million. However, even a few parts per billion are sufficient to corrode electronic circuits. Gas chromatographs are sensitive enough to detect such weak concentrations.
- *Resolution* specifies the smallest change in a physical property that causes a change in the measurement or sourced quantity. For example, humans can detect loudness variations of about 1 dB, but a sound level meter may detect changes as small as 0.001 dB.
- *Dynamic Range* refers to the span from the smallest to the largest value of detectable stimuli. For instance, a voltmeter can be capable of registering input from 10 microvolts to 1 kilovolt.
- *Linearity* specifies how the output changes with respect to the input. The output of perfectly linear device will always increase in direct proportion to an increase in its input. For instance, a perfectly linear source would increase its output by exactly 1 millivolt if it were adjusted from 2 to 3 millivolts. Also, its output would increase by exactly 1 millivolt if it were adjusted from 10.000 to 10.001 volts.
- *Accuracy* refers to the degree to which a measurement corresponds to the true value of the physical input.
- *Lag and Settling Time* refer to the amount of time that lapses between requesting measurement or output and the result being achieved.
- *Sample Rate* is the time between successive measurements. The sample rate can be limited by either the acquisition time (the time it takes to determine the magnitude of the physical variable of interest) or the output rate (the amount of time required to report the result).

1.3.2 Ideal instruments

As shown in Fig. 1.3, the role of an instrument is as a transducer, relating properties of the physical world to information. The transducer has two primary interfaces; the input is connected to the physical world (DUT) and the output is information communicated to the operator. (For stimulus instruments, the roles of input and output are reversed—that is, the input is the

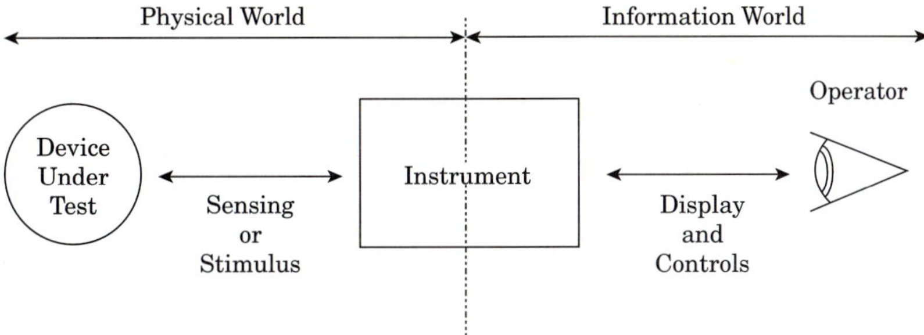

Figure 1.3 Ideal instruments.

information and the output is the physical stimulus of the DUT.) The behavior of the instrument as a transducer can be characterized in terms of its transfer function—the ratio of the output to the input. Ideally, the transfer function of the instrument would be simply a unit conversion. For example, a voltmeter's transfer function could be "X degrees of movement in the display meter per electrical volt at the DUT."

A simple instrument example. A common example of an instrument is the mercury-bulb thermometer (Fig. 1.4). Since materials expand with increasing temperature, a thermometer can be constructed by bringing a reservoir of mercury into thermal contact with the device under test. The resultant volume of mercury is thus related to the temperature of the DUT. When a small capillary is connected to the mercury reservoir, the volume of mercury can be detected by the height that the mercury rises in the capillary column. Ideally, the length of the mercury in the capillary is directly proportional to the temperature of the reservoir. (The transfer function would be X inches of mercury in the column per degree.) Markings along the length of the column can be calibrated to indicate the temperature of the DUT.

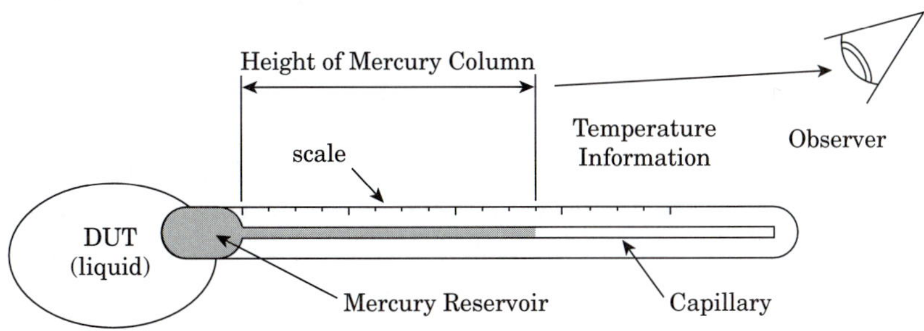

Figure 1.4 A mercury-bulb thermometer.

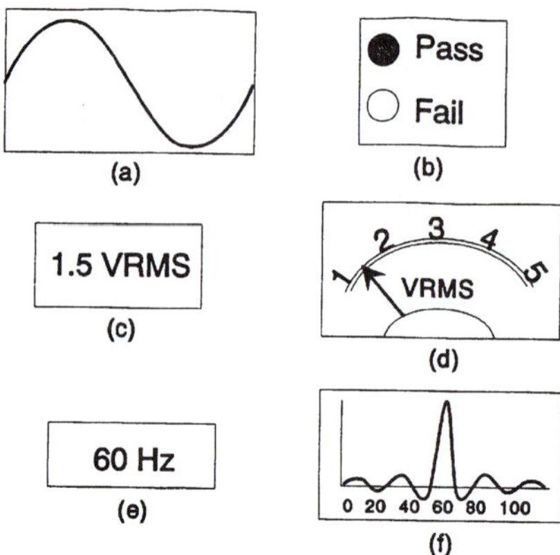

Figure 1.5 Some alternate information displays for an electrical signal.

1.3.3 Types of instruments

Although all instruments share the same basic role, there is a large variety of instruments. As mentioned previously, some instruments are used for measurements, while others are designed to provide stimulus. Figure 1.3 illustrates three primary elements of instruments that can be used to describe variations among instruments.

1. The interface to the DUT depends on the nature of the physical property to be measured (e.g., temperature, pressure, voltage, mass, time) and the type of connection to the instrument. Different instruments are used to measure different things.
2. The operator interface is determined by the kind of information desired about the physical property, and the means by which the information is communicated. For example, the user of an instrument that detects electrical voltage may desire different information about the electrical signal (e.g., rms voltage, peak voltage, waveform shape, frequency, etc.), depending upon the application. The interface to the instrument may be a colorful graphic display for a human, or it may be an interface to a computer. Figure 1.5 illustrates several possible information displays for the same electrical signal.
3. The fidelity of the transformation that takes place within the instrument itself—the extent to which the actual instrument behaves like an ideal instrument—is the third element that differentiates instruments. The same limitations of human perception described in the introduction apply to the behavior of instruments. The degree to which the instrument overcomes these limitations (for example, the accuracy, sensitivity, and sample rate) is the primary differentiator between instruments of similar function.

1.3.4 Electronic instruments

Electronic instruments have several advantages over purely mechanical ones, including:

- Electronic instruments are a natural choice when measuring electrical devices.
- The sophistication of electronics allows for improved signal and information processing within the instrument. Electronic instruments can make sophisticated measurements, incorporate calibration routines within the instrument, and present the information to the user in a variety of formats.
- Electronic instruments enable the use of computers as controllers of the instruments for fully automated measurement systems.

1.4 The Signal Flow of Electronic Instruments

Although the design of individual instruments varies greatly, there are common building blocks. Figure 1.6 illustrates a generic design of a digital electronic instrument. The figure depicts a chain of signal processing elements, each converting information to a form required for input to the next block. In the past, most instruments were purely analog, with analog data being fed directly to analog displays. Currently, however, most instruments being developed contain a digital information processing stage as shown in Fig. 1.6.

1.4.1 Device under Test (DUT) connections

Beginning at the bottom of Fig. 1.6 is the device under test (DUT). As the primary purpose of the instrument is to gain information about some physical property of the DUT, a connection must be made between the instrument and the DUT. This requirement often imposes design constraints on the instrument. For example, the instrument may need to be portable, or the connection to the DUT may require a special probe. The design of the thermometer in the earlier example assumes that the mercury reservoir can be immersed into the DUT that is, presumably, a fluid. It also assumes that the fluid's temperature is considerably lower than the melting point of glass.

1.4.2 Sensor or actuator

Continuing up from the DUT in Fig. 1.6 is the first transducer in the signal flow of the instrument—the sensor. This is the element that is in physical (not necessarily mechanical) contact with the DUT. The sensor must respond to the physical variable of interest and convert the physical information into an electrical signal. Often, the physical variable of interest is itself an electri-

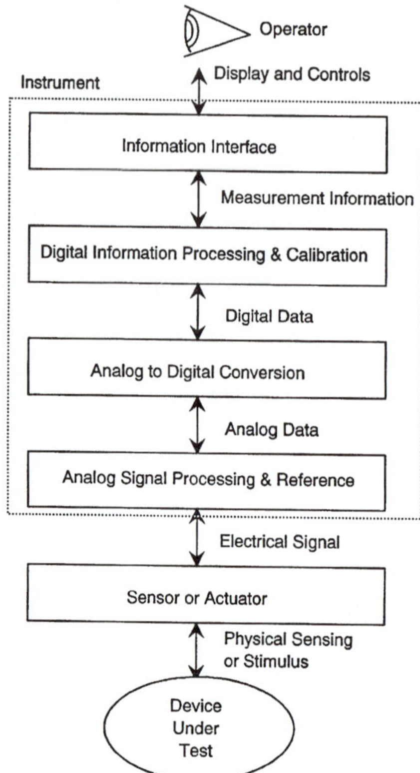

Figure 1.6 The signal flow diagram.

cal signal. In that case, the "sensor" is simply an electrical connection. In other cases, however, the physical variable of interest is not electrical. Examples of sensors include a piezoelectric crystal that converts pressure to voltage, or a thermocouple that converts temperature into a voltage. The advantage of such sensors is that, by converting the physical phenomenon of interest into an electrical signal, the rest of the signal chain can be implemented with a general-purpose electronic instrument.

An ideal sensor would be unobtrusive to the DUT; that is, its presence would not affect the state or behavior of the device under test. To make a measurement, some energy must flow between the DUT and the instrument. If the act of measurement is to have minimal impact on the property being measured, then the amount of energy that the sensor takes from the DUT must be minimized. In the thermometer example, the introduction of the mercury bulb must not appreciably cool the fluid being tested if an accurate temperature reading is desired. Attempting to measure the temperature of a single snowflake with a mercury-bulb thermometer is hopeless.

The sensor should be sensitive to the physical parameter of interest while remaining unresponsive to other effects. For instance, a pressure transducer should not be affected by the temperature of the DUT. The output of a sensor is usually a voltage, resistance, or electric current that is proportional to the magnitude of the physical variable of interest.

In the case of a stimulus instrument, the role of this stage is to convert an electrical signal into a physical stimulus of the DUT. In this case, some form of actuator is used. Examples of actuators are solenoids and motors to convert electrical signals into mechanical motion, loudspeakers to convert electrical signals into sound, and heaters to convert electrical signals into thermal energy.

1.4.3 Analog signal processing and reference

Analog signal processing. The next stage in the signal flow shown in Fig. 1.6 is the analog signal conditioning within the instrument. This stage often contains circuitry that is quite specific to the particular type of instrument. Functions of this stage may include amplification of very low voltage signals coming from the sensor, filtering of noise, mixing of the sensor's signal with a reference signal (to convert the frequency of the signal, for instance), or special circuitry to detect specific features in the input waveform. A key operation in this stage is the comparison of the analog signal with a reference value.

Analog reference. Ultimately, the value of a measurement depends upon its accuracy, that is, the extent to which the information corresponds to the true value of the property being measured. The information created by a measurement is a comparison of the unknown physical variable of the DUT with a reference, or known value. This requires the use of a physical *standard* or physical quantity whose value is known. A consequence of this is that each instrument must have its own internal reference standard as an integral part of the design if it is to be capable of making a measurement. For example, an instrument designed to measure the time between events or the frequency of a signal must have some form of clock as part of the instrument. Similarly, an instrument that needs to determine the magnitude of an electrical signal must have some form of internal voltage direct reference. The quality of this internal standard imposes limitations on the obtainable precision and accuracy of the measurement.

In the mercury-bulb thermometer example, the internal reference is not a fixed temperature. Rather, the internal reference is a fixed, or known, amount of mercury. In this case, the reference serves as an indirect reference, relying on a well-understood relationship between temperature and volume of mercury. The output of the analog processing section is a voltage or current that is scaled in both amplitude and frequency to be suitable for input to the next stage of the instrument.

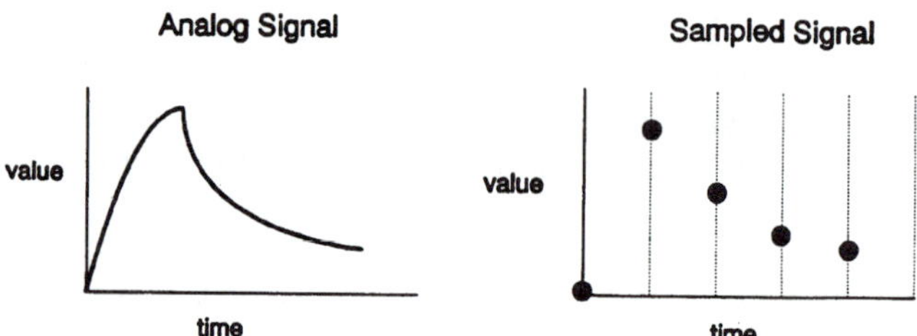

Figure 1.7 A comparison of analog and sampled signals.

1.4.4 Analog-to-digital conversion

For many instruments, the data typically undergo some form of analog-to-digital conversion. The purpose of this stage is to convert the continuously varying analog signal into a series of numbers that can be processed digitally. This is accomplished in two basic steps: (1) the signal is sampled, and (2) the signal is *quantized,* or digitized.

Sampling is the process of converting a signal that is continuously varying over time to a series of values that are representative of the signal at discrete points in time. Figure 1.7 illustrates an analog signal and the resulting sampled signal. The time between samples is the measure of the sample rate of the conversion. In order to represent an analog signal accurately, the sample rate must be high enough that the analog signal does not change appreciably between samples. Put another way: Given a sequence of numbers representing an analog signal, the maximum frequency that can be detected is proportional to the sample rate of the analog-to-digital conversion.

The second step of analog-to-digital conversion, quantization, is illustrated in Fig. 1.8. As shown in the figure, the principal effect of quantization is to round off the signal amplitude to limited precision. While this is not particularly desirable, some amount of quantization is inevitable since digital computation cannot deal with infinite precision arithmetic. The precision of the quantization is usually measured by the number of bits required by a digital representation of the largest possible signal. If N is the number of bits, then the number of output values possible is 2**N. The output range is from a

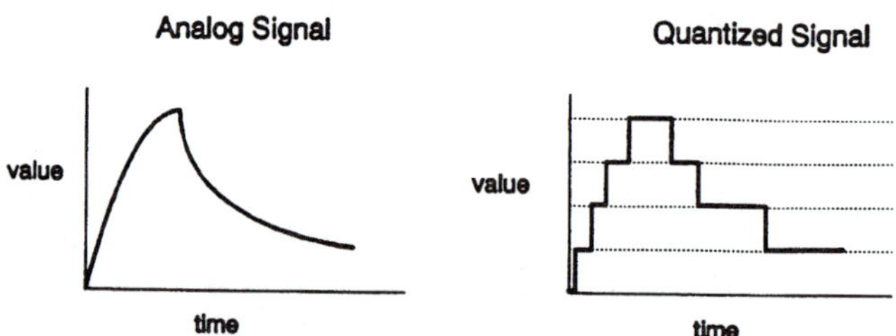

Figure 1.8 A comparison of analog and quantized signals.

smallest output of zero to a maximum value of 2**N−1. For example, an 8-bit analog-to-digital converter (ADC) could output 2**8, or 256 possible discrete values. The output range would be from 0 to 255. If the input range of the converter is 0 to 10 V, then the precision of the converter would be (10 − 0)/255, or 0.039 V. This quantization effect imposes a tradeoff between the range and precision of the measurement. In practice, the precision of the quantization is a cost and accuracy tradeoff made by the instrument designer, but the phenomenon must be understood by the user when selecting the most appropriate instrument for a given application.

The output of the analog-to-digital conversion stage is thus a succession of numbers. Numbers appear at the output at the sample rate, and their precision is determined by the design of the ADC. These digital data are fed into the next stage of the instrument, the digital processing stage. [For a stimulus instrument, the flow of information is reversed—a succession of numbers from the digital processing stage is fed into a digital-to-analog converter (DAC) that converts them into a continuous analog voltage. The analog voltage is then fed into the analog signal processing block.]

1.4.5 Digital information processing and calibration

Digital processing. The digital processing stage is essentially a dedicated computer that is optimized for the control and computational requirements of the instrument. It usually contains one or more microprocessors and/or digital-signal-processor circuits that are used to perform calculations on the raw data that come from the ADC. The data are converted into measurement information. Conversions performed by the digital processing stage include:

- Extracting information—for example, calculating the rise time or range of the signal represented by the data.
- Converting them to a more meaningful form—for example, performing a discrete Fourier transform to convert time-domain to frequency-domain data.
- Combining them with other relevant information—for example, an instrument that provides both stimulus of the DUT and response measurements may take the ratio of the response to the stimulus level to determine the transfer function of the DUT.
- Formatting the information for communication via the information interface—for example, three-dimensional data may be illustrated by two dimensions plus color.

Another function of processing at this stage is the application of calibration factors to the data. The sophistication of digital processing and its relatively low cost have allowed instrument designers to incorporate more complete error compensation and calibration factors into the information, thereby improving the accuracy, linearity, and precision of the measurements.

Calibration. External reference standards are used to check the overall accuracy of instruments. When an instrument is used to measure the value of a standard DUT, the instrument's reading can be compared with the known true value, with the difference being a measure of the instrument's error. For example, the thermometer's accuracy may be tested by measuring the temperature of water that is boiling or freezing, since the temperature at which these phase changes occur is defined to be 100°C and 0°C, respectively.

The source of the error may be due to differences between the instrument's internal reference and the standard DUT or may be introduced by other elements of the signal flow of the instrument. Discrepancies in the instrument's internal reference or nonlinearities in the instrument's signal chain may introduce errors that are repeatable, or *systematic*. When systematic errors are understood and predictable, a *calibration* technique can be used to adjust the output of the instrument to more nearly correspond to the true value. For example, if it is known that the markings on the thermometer are off by a fixed distance (determined by measuring the temperature of a reference DUT whose temperature has been accurately determined by independent means), then the indicated temperature can be adjusted by subtracting the known offset before reporting the temperature result. Unknown systematic errors, however, are particularly dangerous, since the erroneous results may be misinterpreted as being correct. These may be minimized by careful experiment design. In critical applications, an attempt is made to duplicate the results via independent experiments.

In many cases the errors are purely random and thus limit the measurement precision of the instrument. In these cases, the measurement results can often be improved by taking multiple readings and performing statistical analysis on the set of results to yield a more accurate estimate of the desired variable's value. The statistical compensation approach assumes that something is known about the nature of the errors. When all understood and repeatable error mechanisms have been compensated, the remaining errors are expressed as a measurement uncertainty in terms of accuracy or precision of the readings.

Besides performing the digital processing of the measurement information, the digital processing stage often controls the analog circuitry, the user interface, and an input/output (I/O) channel to an external computer.

1.4.6 Information interface

When a measurement is made of the DUT, the instrument must communicate that information if it is to be of any real use. The final stage in the signal flow diagram (Fig. 1.6) is the presentation of the measurement results through the information interface. This is usually accomplished by having the microprocessor either control various display transducers to convey information to the instrument's operator or communicate directly with an external computer.

Whether it is to a human operator or a computer, similar considerations apply to the design of the information interface.

Interfaces to human operators. In this case, the displays (e.g., meters and gauges) and controls (e.g., dials and buttons) must be a good match to human sensory capabilities. The readouts must be easy to see and the controls easy to manipulate. This provides an appropriate physical connection to the user. Beyond this, however, the information must be presented in a form that is meaningful to the user. For example, text must be in the appropriate language, and the values must be presented with corresponding units (e.g., volts or degrees) and in an appropriate format (e.g., text or graphics). Finally, if information is to be obtained and communicated accurately, the operator interface should be easy to learn and use properly. Otherwise the interface may lead the operator to make inaccurate measurements or to misinterpret the information obtained from the instrument.

Computer interfaces. The same considerations used for human interfaces apply in an analogous manner to computer interfaces. The interface must be a good match to the computer. This requirement applies to the transmission of signals between the instrument and the computer. This means that both devices must conform to the same interface standards that determine the size and shape of the connectors, the voltage levels on the wires, and the manner in which the signals on the wires are manipulated to transfer information. Common examples of computer interfaces are RS-232 (serial), Centronics (parallel), SCSI, or LAN. Some special instrumentation interfaces (GPIB, VXI, and MMS) are often used in measurement systems. (These are described later in this chapter and in other chapters of this book.)

The communication between the instrument and computer must use a form that is meaningful to each. This consideration applies to the format of the information, the language of commands, and the data structures employed. Again, there are a variety of standards to choose from, including Standard Commands for Programmable Instruments (SCPI) and IEEE standards for communicating text and numbers.

The ease of learning requirement applies primarily to the job of the system developer or programmer. This means that the documentation for the instrument must be complete and comprehensible, and that the developer must have access to the programming tools needed to develop the computer applications that interact with the instrument. Finally, the ease of use requirement relates to the style of interaction between the computer and the instrument. For example, is the computer blocked from doing other tasks while the instrument is making a measurement? Does the instrument need to be able to interrupt the computer while it is doing some other task? If so, the interface and the operating system of the computer must be designed to respond to the interrupt in a timely manner.

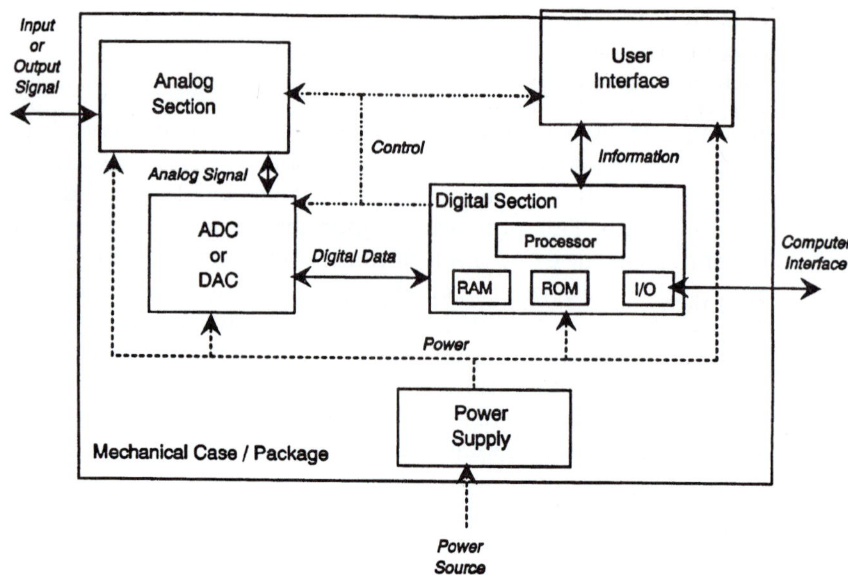

Figure 1.9 An instrument block diagram.

1.5 The Instrument Block Diagram

While the design of the signal flow elements focuses on measurement performance, the physical components chosen and their methods of assembly will determine several important specifications of the instrument, namely, its cost, weight, size, and power consumption. In addition, the instrument designer must consider the compatibility of the instrument with its environment. Environmental specifications include ranges of temperature, humidity, vibration, shock, chemicals, and pressure. These are often specified at two levels: The first is the range over which the instrument can be expected to operate within specifications, and the second (larger) is the range that will not cause permanent damage to the instrument.

In order to build an instrument that implements a signal flow like that of Fig. 1.6, additional elements such as a mechanical case and power supply are required. A common design of an instrument that implements the signal flow path discussed above is illustrated in Fig. 1.9. As shown in the figure, the building blocks of the signal flow path are present as physical devices in the instrument. In addition, there are two additional support elements, the mechanical case and package and the power supply.

1.5.1 Mechanical case and package

The most visible component of instruments is the mechanical package, or case. The case must provide support of the various electronic components, ensuring their electrical, thermal, electromagnetic, and physical containment and protection. The case is often designed to fit into a standard 19-in-wide rack, or it may provide carrying handles if the instrument is designed to be portable. The case supports a number of connectors that are used to interface the instrument with its environment. The connections illustrated in Fig. 1.9 include a power cable, the input connections for the sensor, a computer interface, and the front panel for the user interface. The case must also protect the electrical environment of the instrument. The instrument usually contains a lot of very sensitive circuitry. Thus it is important for the case to protect the circuitry from stray electromagnetic fields (such as radio waves). It is likewise important that electromagnetic emissions created by the instrument itself are not sent into the environment where they could interfere with other electronic devices.

Similarly, the package must provide for adequate cooling of the contents. This may not be a concern if the other elements of the instrument do not generate much heat and are not adversely affected by the external temperature of the instrument's environment within the range of intended use. However, most instruments are cooled by designing some form of natural or forced convection (airflow) through the instrument. This requires careful consideration of the space surrounding the instrument to ensure that adequate airflow is possible and that the heat discharged by the instrument will not adversely affect adjacent devices. Airflow through the case may cause electromagnetic shielding problems by providing a path for radiated energy to enter or leave the instrument. In addition, if a fan is designed into the instrument to increase the amount of cooling airflow, the fan itself may be a source of electromagnetic disturbances.

1.5.2 Power supply

Figure 1.9 also illustrates a power supply within the instrument. The purpose of the power supply is to convert the voltages and frequencies of an external power source (such as 110 V ac, 60 Hz) into the levels required by the other elements of the instrument. Most digital circuitry requires 5 V dc, while analog circuitry has varying voltage requirements (typically, ± 12 V dc, although some elements such as CRTs may have much higher voltage requirements).

The power supply design also plays a major role in providing the proper electrical isolation of various elements, both internal and external to the instrument. Internally, it is necessary to make sure that the power supplied to the analog signal conditioning circuitry, for instance, is not corrupted by spurious signals introduced by the digital processing section. Externally, it is important for the power supply to isolate the instrument from voltage and frequency fluctuations present on the external power grid, and to shield the external power source from conducted emissions that may be generated internal to the instrument.

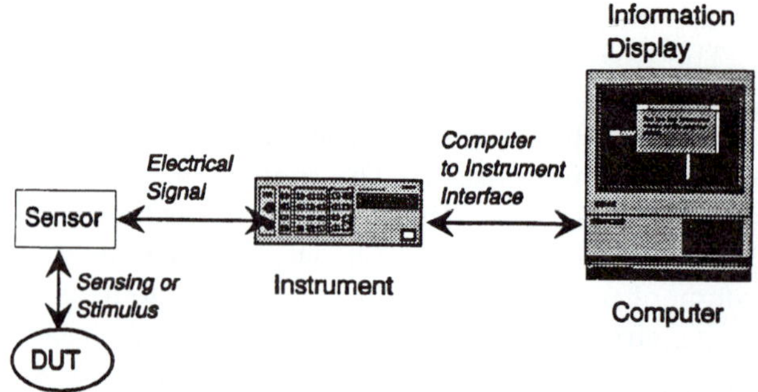

Figure 1.10 A simple measurement system.

1.6 Measurement Systems

One of the advantages of electronic instruments is their suitability for incorporation into measurement systems. A measurement system is built by connecting one or more instruments together, usually with one or more computers. Figure 1.10 illustrates a simple example of such a system.

1.6.1 Distributing the "instrument"

When a measurement system is constructed by connecting an instrument with a computer, the functionality is essentially the same as described in the signal flow diagram (Fig. 1.6), although it is distributed between the two hardware components as illustrated in Fig. 1.11. Comparison of the signal flow diagram for a computer-controlled instrument (Fig. 1.11) with that of a stand-alone instrument (Fig. 1.6) shows the addition of a second stage of digital information processing and an interface connection between the computer and the instrument. These additions constitute the two primary advantages of such a system.

Digital processing in the computer. The digital information processing capabilities of the computer can be used to automate the operation of the instrument. This capability is important when control of the instrument needs to be faster than human capabilities allow, when a sequence of operations is to be repeated accurately, or when unattended operation is desired.

Beyond mere automation of the instrument, the computer can run a special-purpose program to customize the measurements being made to a specific application. One such specific application would be to perform the calculations necessary to compute a value of interest based on indirect measurements. For example, the moisture content of snow is measured indirectly by weighing a known volume of snow. In this case, the instrument makes the weight measurement and the computer can perform the calculation that determines the density of the snow and converts the density information into moisture content.

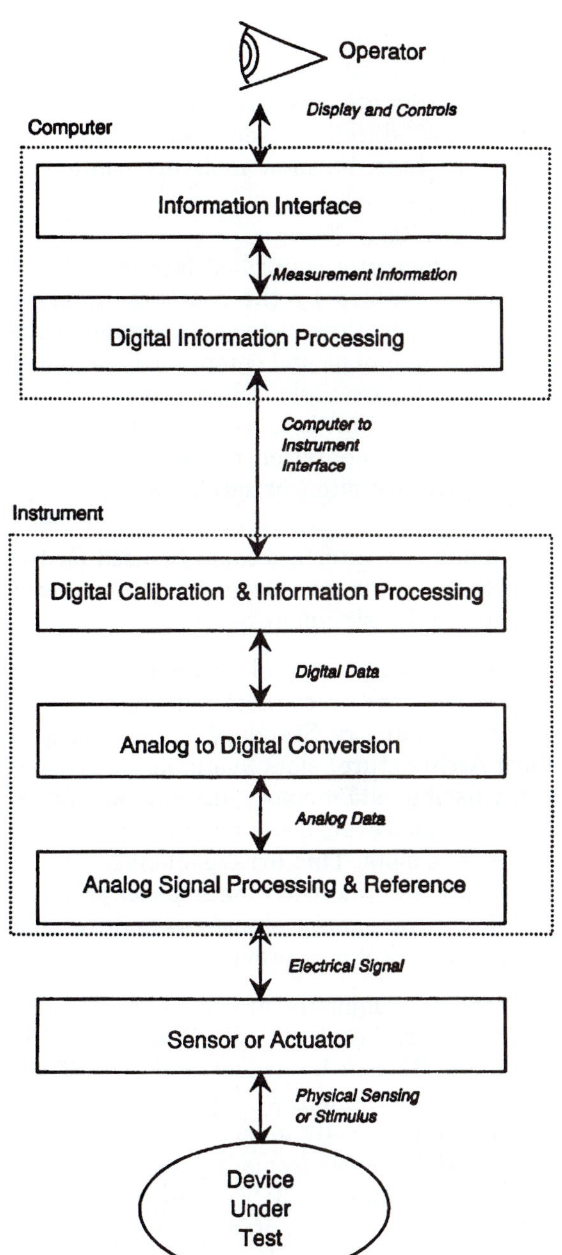

Figure 1.11 The signal flow diagram for a computer-controlled instrument.

Finally, the computer can generate a new interface for the user that displays snow moisture content rather than the raw weight measurement made by the instrument. The software running on the computer in this example is often referred to as a "virtual instrument," since it presents an interface that is equivalent to an instrument—in this case, a "snow moisture content instrument."

Remote instruments. A second use of a computer-controlled instrument is to exploit the distribution of functionality enabled by the computer interface connection to the instrument. The communications between instrument and computer over this interface allow the instrument and computer to be placed in different locations. This is desirable, for example, when the instrument must accompany the DUT in an environment that is inhospitable to the operator. Some examples of this would be instrumentation placed into environmental chambers, wind tunnels, explosive test sites, or satellites.

Computer-instrument interfaces. Although any interface could be used for this purpose, a few standards are most common for measurement systems: computer backplanes, computer interfaces, and instrument buses.

Computer backplanes. These buses are used for internal expansion of a computer. (Buses are interfaces that are designed to connect multiple devices.) The most common of these is the ISA (Industry Standard Architecture) or EISA (Extended Industry Standard Architecture) slots available in personal computers. These buses are typically used to add memory, display controllers, or interface cards to the host computer. However, some instruments are designed to plug directly into computer bus slots. This arrangement is usually the lowest-cost option, but the performance of the instrument is compromised by the physical lack of space, lack of electromagnetic shielding, and the relatively noisy power supply that computer backplanes provide.

Computer interfaces. These interfaces are commonly provided by computer manufacturers to connect computers to peripherals or other computers. The most common of these interfaces are RS-232, SCSI, parallel, and LAN. These interfaces have several advantages for measurement systems over computer buses, including: (1) The instruments are physically independent of the computer, so their design can be optimized for measurement performance. (2) The instruments can be remote from the computer. (3) The interfaces, being standard for the computer, are supported by the computer operating systems and a wide variety of software tools. Despite these advantages, these interfaces have limitations in measurement systems applications, particularly when the application requires tight timing synchronization among multiple instruments.

Instrument buses. These interfaces, developed by instrument manufacturers, have been optimized for measurement applications. The most common of

Figure 1.12 VXI instruments.

these special instrument interfaces is the General Purpose Interface Bus, GPIB, also known as IEEE-488. GPIB is a parallel bus designed to connect standalone instruments to a computer, as shown in Fig. 1.10. In this case, a GPIB interface is added to the computer, usually by installing an interface card in the computer's expansion bus. Other instrument bus standards are VXI (VMEbus Extended for Instrumentation) and MMS (Modular Measurement System). VXI and MMS are cardcage designs that support the mechanical, electrical, and communication requirements of demanding instrumentation applications. Figure 1.12 is a photograph of VXI instrumentation. Note that in the case of VXI instruments, the instruments themselves have no user interface, as they are designed solely for incorporation into computer-controlled measurement systems.

Cardcage systems use a mainframe that provides a common power supply, cooling, mechanical case, and communication bus. The system developer can then select a variety of instrument modules to be assembled into the mainframe. Figure 1.13 illustrates the block diagram for a cardcage-based instrument system. Comparison with the block diagram for a single instrument (Fig. 1.9) shows that the cardcage system has the same elements but there are multiple signal-flow elements sharing common support blocks such as the power supply, case, and the data and control bus. One additional element is the computer interface adapter. This element serves as a bridge between the control and data buses of the mainframe and an interface to an external com-

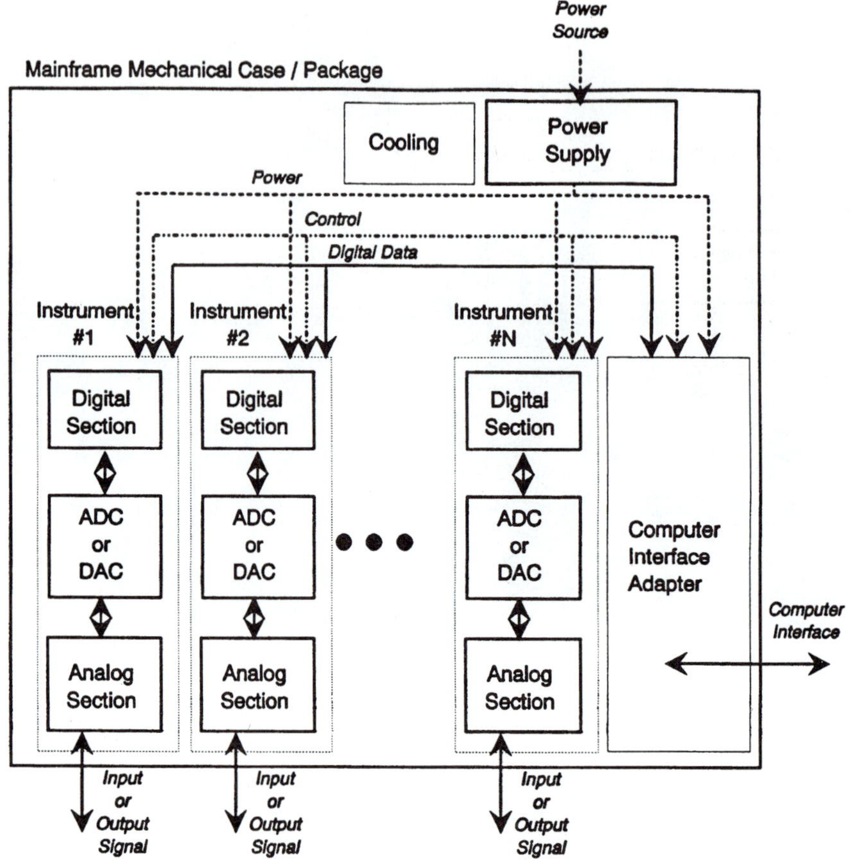

Figure 1.13 The block diagram for a cardcage instrument system.

puter. (In some cases, the computer may be inserted or embedded in the mainframe next to the instruments where it interfaces directly to the control and data buses of the cardcage.)

1.6.2 Multiple instruments in a measurement system

A common element in the design of each of the instrument buses is the provision to connect multiple instruments together, all controlled by a single computer as shown in Fig. 1.14. A typical example of such a system configuration is composed of several independent instruments all mounted in a 19-in-wide rack, all connected to the computer via GPIB (commonly referred to as a "rack and stack" system).

Multiple instruments may be used when several measurements are required on the same DUT. In some cases, a variety of measurements must be made concurrently on a DUT; for example, a power measurement can be

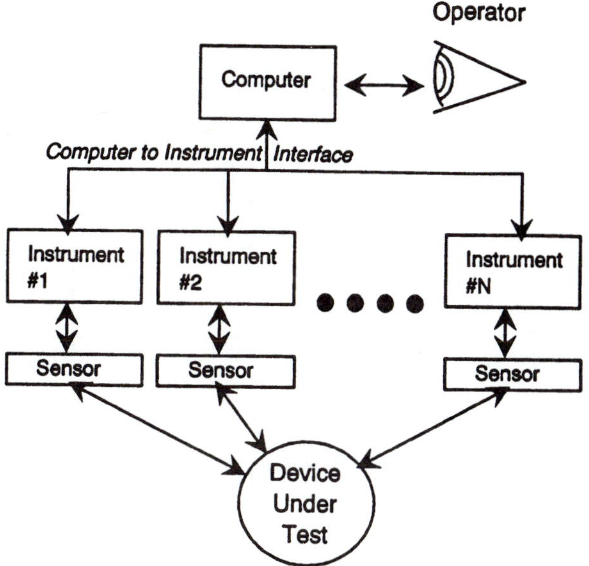

Figure 1.14 A measurement system with multiple instruments.

made by simultaneously measuring a voltage and a current. In other cases, a large number of similar measurements requires duplication of instruments in the system. This is particularly common when testing complex DUTs such as integrated circuits or printed circuit boards, where hundreds of connections are made to the DUT.

Multiple instruments are also used when making stimulus-response measurements. In this case, one of the instruments does not make a measurement but rather provides a signal that is used to stimulate the DUT in a controlled manner. The other instruments measure the response of the DUT to the applied stimulus. This technique is useful to characterize the behavior of the DUT. A variant of this configuration is the use of instruments as surrogates for the DUT's expected environment. For example, if only part of a device is to be tested, the instruments may be used to simulate the missing pieces, providing the inputs that the part being tested would expect to see in normal operation.

Another use of multiple instruments is to measure several DUTs simultaneously with the information being consolidated by the computer. This allows simultaneous testing (batch testing) of a group of DUTs for greater testing throughput in a production environment. Note that this could also be accomplished by simply duplicating the measurement system used to test a single DUT. However, using multiple instruments connected to a single computer not only saves money on computers, it also provides a mechanism for the centralized control of the tests and the consolidation of test results from the different DUTs.

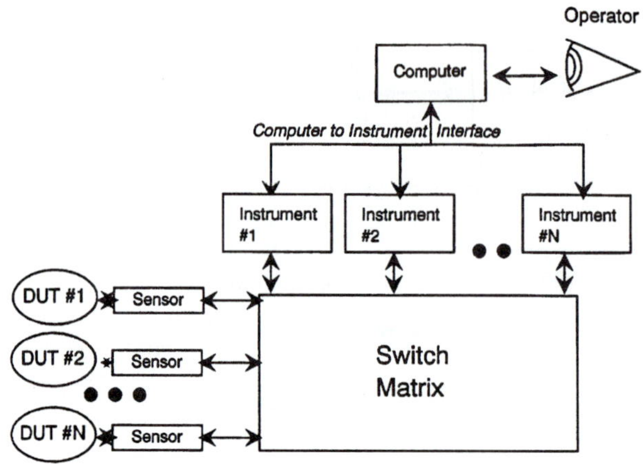

Figure 1.15 Using a switch matrix for batch testing.

Economies can be realized if the various measurements made on multiple DUTs do not need to be made simultaneously. In this case, a single set of instruments can be used to measure several DUTs by connecting all the instruments and DUTs to a switch matrix, as illustrated in Fig. 1.15. Once these connections are made, the instruments may be used to measure any selected DUT by programmatic control of the switch matrix. This approach may be used, for example, when a batch of DUTs is subjected to a long-term test with periodic measurements made on each.

1.6.3 Multiple computers in a measurement system

As the information processing needs increase, the number of computers required in the measurement system also increases. Lower cost and improved networking have improved the cost-effectiveness of multicomputer configurations.

Real time. Some measurement systems add a second computer to handle special real-time requirements. There are several types of real-time needs that may be relevant, depending on the application:

- *Not real time.* The completion of a measurement or calculation can take as long as necessary. Most information processing falls into this category, where the value of the result does not depend on the amount of time that it takes to complete the task. Consequently, most general-purpose computers are developed to take advantage of this characteristic—when the task becomes more difficult, the computer simply spends more time on it.
- *"Soft" real time.* The task must complete within a deadline if the result is to be useful. In this case, any computer will suffice as long as it is fast

enough. However, since most modern operating systems are multitasking, they cannot in general guarantee that each given task will be completed by a specified time or even that any particular task will be completed in the same amount of time if the task is repeated.

- *"Hard" real time.* The result of a task is incorrect if the task is not performed at a specific time. For example, an instrument that is required to sample an input signal 100 times in a second must perform the measurements at rigidly controlled times. It is not satisfactory if the measurements take longer than 1 s or even if all 100 samples are made within 1 s. Each sample must be taken at precisely 1/100-s intervals. Hard real-time requirements may specify the precise start time, stop time, and duration of a task. The results of a poorly timed measurement are not simply late, they're wrong.

Since the physical world (the world of DUTs) operates in real time, the timing requirements of measurement systems become more acute as the elements get closer to the DUT. Usually, the hard real-time requirements of the system are handled completely within the instruments themselves. This requires that the digital processing section of the instrument be designed to handle its firmware tasks in real time.

In some cases, it is important for multiple instruments to be coordinated or for certain information processing tasks to be completed in hard real time. For example, an industrial process control application may have a safety requirement that certain machines be shut down within a specified time after a measurement reaches a predetermined value. Figure 1.16 illustrates a measurement system that has a computer dedicated to real-time instrument control and measurement processing. In this case, the real-time computer is embedded in an instrument mainframe (such as a VXI cardcage) where it interfaces directly with the instrument data and control bus. A second interface on the real-time computer is used to connect to a general-purpose computer that provides for the non-real-time information-processing tasks and the user interface.

A further variant of the system illustrated in Fig. 1.16 is the incorporation of multiple real-time computers. Although each instrument typically performs real-time processing, the system designer may augment the digital processing capabilities of the instruments by adding multiple real-time processors. This would be necessary, in particular, when several additional information-processing tasks must be executed simultaneously.

Multiple consumers of measurement results. A more common requirement than the support of real-time processes is simply the need to communicate measurement results to several general-purpose computers. Figure 1.17 illustrates one possible configuration of such a system. As shown in Fig. 1.17,

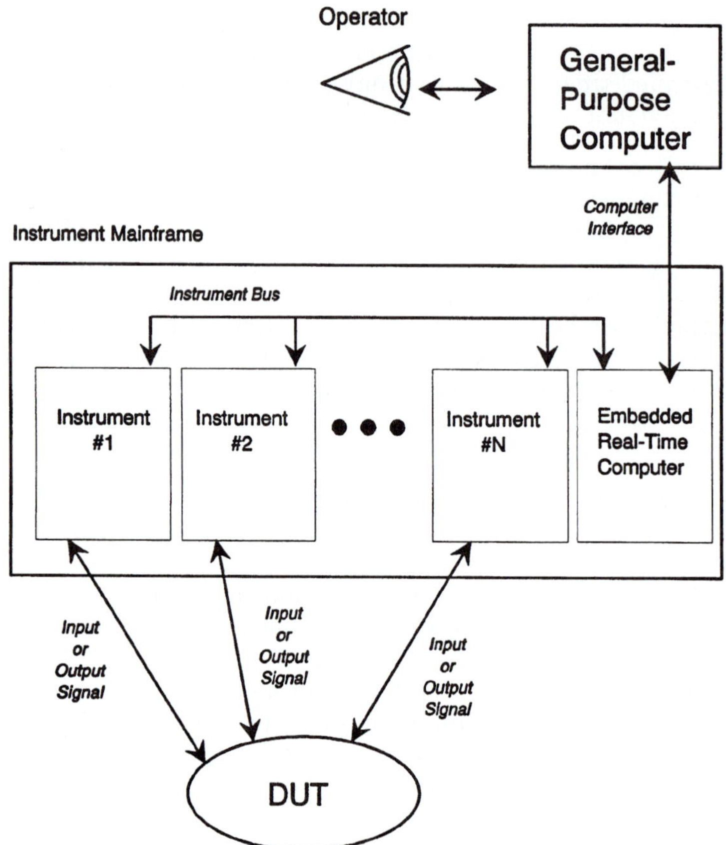

Figure 1.16 A system with an embedded real-time computer for measurement.

several operations may run on different computers yet require interaction with measurements, such as:

- *Analysis and presentation of measurement results.* There may be several different operator interfaces at different locations in the system. For example, a person designing the DUT at a workstation may desire to compare the performance of a DUT with the expected results derived by running a simulation on the model of the DUT.
- *Test coordination.* A single computer may be used to schedule and coordinate the operation of several different instrument subsystems.
- *System development and administration.* A separate computer may be used to develop new test and measurement software routines or to monitor the operation of the system.

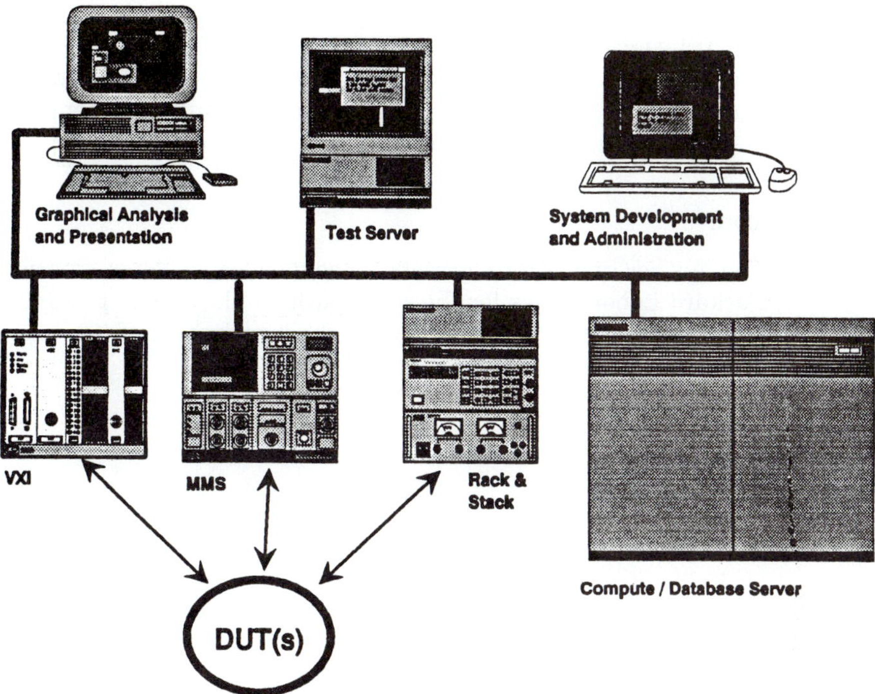

Figure 1.17 A networked measurement system.

- *Database.* Measurement results may be communicated to or retrieved from a centralized database.
- *Other measurement subsystems.* The information from measurements taken at one location may need to be incorporated or integrated with the operation of a measurement subsystem at another location. For example, a manufacturing process control system often requires that measurements taken at one point in the process are used to adjust the stimulus at another location in the process.

1.7 Summary

All instruments and measurement systems share the same basic purpose, namely, the connection of information and the physical world. The variety of physical properties of interest and the diversity of information requirements for different applications give rise to the abundance of available instruments. However, these instruments all share the same basic performance attributes. Given the physical and information interface requirements, the keys to the design of these systems are the creation of suitable basic building blocks and the arrangement and interconnection of these blocks. The design of these basic building blocks and their supporting elements determines the fidelity of the transformation between physical properties and information. The arrangement and connection of these blocks allow the creation of systems that range from a single compact instrument to a measurement and information system that spans the globe.

Acknowledgment

The author wishes to gratefully acknowledge Peter Robrish of Hewlett-Packard Laboratories for the contribution of key ideas presented here as well as his consultations and reviews in the preparation of this chapter.

Chapter

2

Calibration, Traceability, and Standards

David R. Workman
Consultant, Littleton, Colorado

2.1 Metrology and Metrologists

The accepted name for the field of calibration is "metrology," one definition for which is "the science that deals with measurement."[1] Calibration facilities are commonly called metrology laboratories. Individuals who are primarily engaged in calibration services are called "metrologists," a title used to describe both technicians and engineers. Where subcategorization is required, they are referred to as "metrology technicians" or "metrology engineers."

All technical, engineering, and scientific disciplines utilize measurement technology in their fields of endeavor. Metrology concentrates on the fundamental scientific concepts of measurement support for all disciplines. In addition to calibration, the requirements for a competent metrologist include detailed knowledge regarding contractual quality assurance requirements, test methodology and system design, instrument specifications, and performance analysis. Metrologists commonly perform consultation services for company programs that relate to their areas of expertise.

2.2 Definitions for Fundamental Calibration Terms

The definitions of related terms are a foundation for understanding calibration. Many of the following definitions are taken directly or paraphrased for added clarity from the contents of MIL-STD-45662A.[2] Other definitions are based on accepted industrial usage. Additional commentary is provided where deemed necessary.

2.2.2 Calibration

Calibration is the comparison of measurement and test equipment (M&TE) or a measurement standard of unknown accuracy to a measurement standard of known accuracy to detect, correlate, report, or eliminate by adjustment any variation in the accuracy of the instrument being compared. In other words, calibration is the process of comparing a measurement device whose accuracy is unknown or has not been verified to one with known characteristics. The purposes of a calibration are to ensure that a measurement device is functioning within the limit tolerances that are specified by its manufacturer, characterize its performance, or ensure that it has the accuracy required to perform its intended task.

2.2.2 Measurement and test equipment (M&TE)

M&TE are all devices used to measure, gauge, test, inspect, or otherwise determine compliance with prescribed technical requirements.

2.2.3 Measurement standards

Measurement standards are the devices used to calibrate M&TE or other measurement standards and provide traceability to accepted references. Measurement standards are M&TE to which an instrument requiring calibration is compared, whose application and control set them apart from other M&TE.

2.2.4 Reference standard

A reference standard is the highest level of measurement standard available in a calibration facility for a particular measurement function. The term usually refers to standards calibrated by an outside agency. Reference standards for some measurement disciplines are capable of being verified locally or do not require verification (i.e., cesium beam frequency standard). Applications of most reference standards are limited to the highest local levels of calibration. While it is not accepted terminology, owing to confusion with national standards, some organizations call them "primary standards."

2.2.5 Transfer or working standards

Transfer standards, sometimes called working standards, are measurement standards whose characteristics are determined by direct comparison or through a chain of calibrations against reference standards. Meter calibrators are a common type of transfer standards.

2.2.6 Artifact standards

Artifact standards are measurement standards that are represented by a physical embodiment. Common examples of artifacts include resistance, capacitance, inductance, and voltage standards.

2.2.7 Intrinsic standards

Intrinsic standards are measurement standards that require no external calibration services. Examples of intrinsic standards include the Josephson array voltage standard, iodine-stabilized helium-neon laser length standard, and cesium beam frequency standard. While these are standalone capabilities, they are commonly not accepted as being properly traceable without some form of intercomparison program against other reference sources.

2.2.8 Consensus and industry accepted standards

A consensus standard is an artifact or process that is used as a de facto standard by the contractor and its customer when no recognized U.S. national standard is available. In other terms, consensus standard refers to an artifact or process that has no clear-cut traceability to fundamental measurement units but is accepted methodology. Industry accepted standards are consensus standards that have received overall acceptance by the industrial community. A good example of an industry accepted standard is the metal blocks used for verifying material hardness testers. In some cases, consensus standards are prototypes of an original product to which subsequent products are compared.

2.2.9 Standard reference materials (SRMs)

Standard reference materials (SRMs) are materials, chemical compounds, or gases that are used to set up or verify performance of M&TE. SRMs are purchased directly from NIST or other quality approved sources. Examples of SRMs include pure metal samples used to establish temperature freezing points and radioactive materials with known or determined characteristics. SRMs are often used as consensus standards.

2.3 Traceability

As illustrated in Fig. 2.1, traceability is the ability to relate individual measurement results through an unbroken chain of calibrations to one or more of the following:

1. U.S. national standards that are maintained by the National Institute of Standards and Technology (NIST) and U.S. Naval Observatory
2. Fundamental or natural physical constants with values assigned or accepted by NIST
3. National standards of other countries that are correlated with U.S. national standards
4. Ratio types of calibrations
5. Consensus standards
6. Standard reference materials (SRMs)

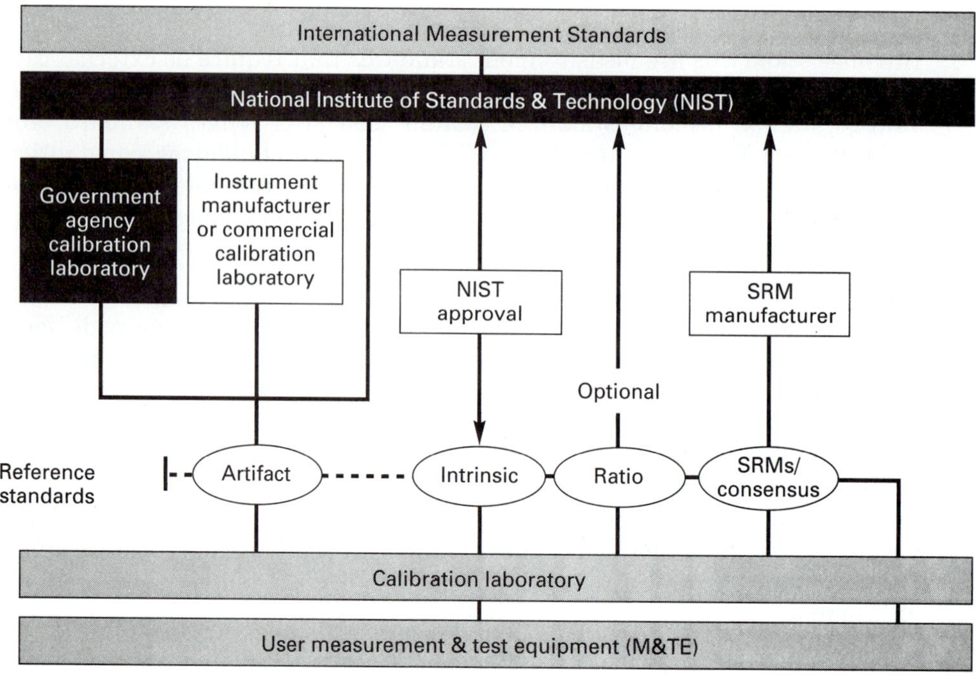

Figure 2.1 Measurement traceability to national standards.

In other words, traceability is the process of ensuring that the accuracy of measurements can be traced to an accepted measurement reference source.

2.4 Calibration Types

There are two fundamental types of calibrations, report and limit tolerance.

2.4.1 Report calibration

A report calibration is the type issued by NIST when it tests a customer's instrument. It provides the results of measurements and a statement of measurement uncertainty. Report calibrations are also issued by nongovernment calibration laboratories. A report provides no guarantee of performance beyond the time when the data were taken. To obtain knowledge of the device's change with time, or other characteristics, the equipment owners must perform their own evaluation of data obtained from several calibrations.

2.4.2 Limit tolerance calibration

Limit tolerance calibrations are the type most commonly used in industry. The purpose of a limit calibration is to compare an instrument's measured performance against nominal performance specifications. If an instrument submitted for calibration does not conform to required specifications, it is considered to be received "out of tolerance." It is then repaired or adjusted to correct the out-of-tolerance condition, retested, and returned to its owner. A calibration label is applied to indicate when the calibration was performed and when the next service will be due. This type of calibration is a certification that guarantees performance for a given period.

A limit tolerance calibration consists of three steps:

1. Calibrate the item and determine performance data as received for calibration (as found).
2. If found to be out of tolerance, perform necessary repairs or adjustments to bring it within tolerance.
3. Recalibrate the item and determine performance data before it is returned to the customer (as left).

If the item meets specification requirements in step 1, no further service is required and the "as found" and "as left" data are the same. Procedures that give step-by-step instructions on how to adjust a device to obtain proper performance, but do not include tests for verification of performance before and after adjustment, are not definitive calibrations.

Often the acceptance (as found) and adjustment (as left) tolerances are different. When an item has predictable drift characteristics and is specified to have a nominal accuracy for a given period of time, the acceptance tolerance includes an allowance for drift. The adjustment tolerance is often tighter. Before return to the user, a device must be adjusted for conformance with adjustment tolerances to ensure that normal drift does not cause it to exceed specifications before its next calibration.

2.5 Calibration Requirements

When users require a level of confidence in data taken with a measurement device, an instrument's calibration should be considered important. Individuals who lack knowledge of calibration concepts often believe that the operator functions performed during setup are a calibration. In addition, confusion between the primary functional operations of an instrument and its accuracy is common.

"Functionality" is apparent to an operator but "measurement accuracy" is invisible. An operator can determine if equipment is working correctly but has no means for determining the magnitude of errors in performed measurements. Typically, measurement accuracy is either ignored, taken on faith, or guaranteed with a calibration that is performed by someone other than the user. These are the fundamental reasons why instruments require periodic calibrations.

Manufacturers normally perform the initial calibrations on measurement devices. Subsequent calibrations are obtained by either returning the equipment to its manufacturer or obtaining the required service from a calibration laboratory that has the needed capabilities.

Subsequent calibrations must be performed when a measurement instrument's performance characteristics change with time. There are many accepted reasons why these changes occur. The more common reasons for change include:

1. Mechanical wear
2. Electrical component aging
3. Operator abuse
4. Unauthorized adjustment

Most measurement devices require both initial and periodic calibrations to maintain the required accuracy. Many manufacturers of electronic test equipment provide recommendations for the time intervals between calibrations that are necessary to maintain specified performance capabilities.

While the theoretical reasons for maintaining instruments on a periodic calibration cycle are evident, in practice many measurement equipment owners and users submit equipment for subsequent calibrations only when they are required to or a malfunction is evident and repair is required. Reasons for this practice are typically the inconvenience of losing the use of an instrument and calibration cost avoidance.

Routine periodic calibrations are usually performed only where mandated requirements exist to have them done. Because of user reluctance to obtain calibration services, surveillance and enforcement systems are often required to ensure that mandates are enforced.

2.6 Check Standards and Cross-Checks

In almost all situations, users must rely on a calibration service to ascertain accuracy and adjust measurement instrumentation for specified performance. While it is possible for users to maintain measurement standards for self-calibration, this is not an accepted or economical practice. The lack of specific knowledge and necessary standards makes it difficult for equipment users to maintain their own calibration efforts. Because of the involved costs, duplication of calibration efforts at each equipment location is usually cost-prohibitive. Users normally rely on a dedicated calibration facility to perform needed services.

A primary user need is to ensure that, after being calibrated, an instrument's performance does not change significantly before its next calibration. This is a vital need if the possibility exists that products will have to be recalled because of bad measurements. There is a growing acceptance for the use of "cross-checks" and "check standards" to verify that instruments or measurement systems do not change to the point of invalidating performance requirements.

2.6.1 Cross-check

A cross-check is a test performed by the equipment user to ensure that the performance of a measurement device or system does not change significantly with time or use. It does not verify the absolute accuracy of the device or system.

2.6.2 Check standard

A check standard can be anything the measurement device or system will measure. The only requirement is that it will not change significantly with time. As an example, an uncalibrated piece of metal can be designated a "check standard" for verifying the performance of micrometers after they have been calibrated. The block of metal designated to be a check standard must be measured before the micrometer is put into use, with measured data recorded. If the process is repeated each day and noted values are essentially the same, one is assured that the micrometer's performance has not changed. If data suddenly change, it indicates that the device's characteristics have changed and it should be submitted for repair or readjustment.

The problem with instruments on a periodic recall cycle is that if it is determined to be out of tolerance when received for calibration, without a cross-check program one has no way of knowing when the device went bad. If the calibration recall cycle was 12 months, potentially up to 12 months of products could be suspect and subject to recall. If a cross-check program is used, suspect products can be limited to a single period between checks.

2.7 Calibration Methodology

Calibration is the process of comparing a known device, which will be called a standard instrument, to an unknown device, which will be referred to as the test instrument. There are two fundamental methodologies for accomplishing comparisons:

1. Direct comparisons
2. Indirect comparisons

2.7.1 Direct comparison calibration

The basic direct comparison calibration setups are shown in Fig. 2.2. Where meters or generators are the test instruments, the required standards are opposite. If the test instrument is a meter, a standard generator is required. If the test instrument is a generator, a standard meter is required. A transducer calibration requires both generator and meter standards.

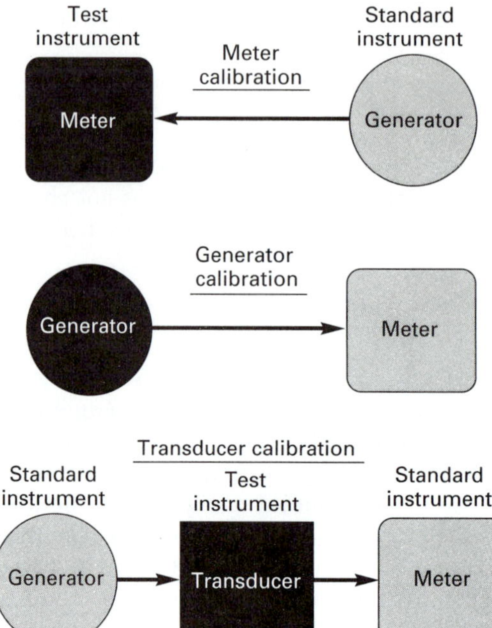

Figure 2.2 Direct comparison calibration setups.

When the test instrument is a meter, the generator applies a known stimulus to the meter. The ratio of meter indication to known generator level quantifies the meter's error. The simplified uncertainty of the measurement is the certainty of the standard value plus the resolution and repeatability of the test instrument. If a limit tolerance is being verified, the noted deviation from nominal is compared to the allowable performance limit. If the noted deviation exceeds the allowance, the instrument is considered to be out of tolerance. The same principles apply in reverse when the test instrument is a generator and the standard is a meter.

Transducer characteristics are expressed as a ratio between the device's output to its input, in appropriate input and output measurement units. As an example, a pressure transducer that has a voltage output proportional to a psi input would have an output expressed in volts or millivolts per psi. If the transducer is a voltage amplifier, the output is expressed in volts per volt, or a simple numerical ratio. In simplest terms, the measurement uncertainty is the additive uncertainties of the standard generator and meter.

2.7.2 Indirect comparisons

Indirect comparisons (Fig. 2.3) are calibration methods where a standard is compared to a like test instrument. In other words, a standard meter is compared to a test meter, standard generator to test generator, and standard transducer to test transducer.

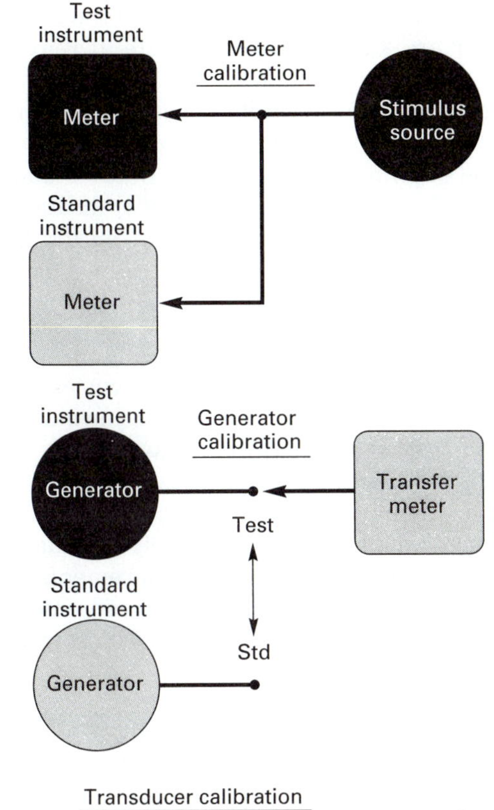

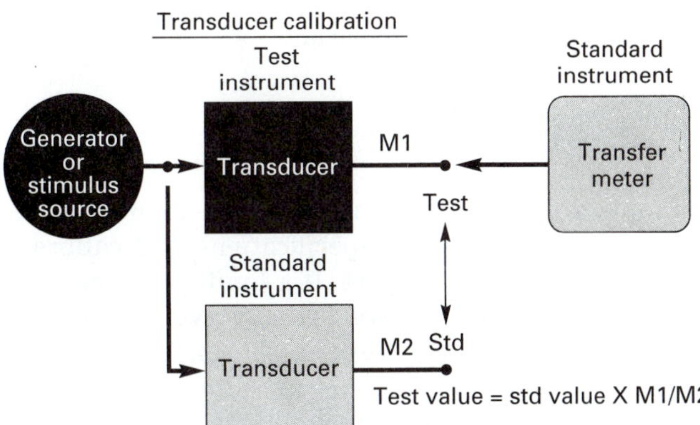

Figure 2.3 Indirect comparison calibration setups.

If the test instrument is a meter, the same stimulus is applied simultaneously to both the test and standard meters. The calibration test consists of a comparison of the test instrument's indication to the standard's indication. With the exception that the source stimulus must have the required level of stability during the comparison process, its actual magnitude is unimportant.

With outputs set to the same nominal levels, a "transfer meter" is used to measure both the standard and test generators. If the resolution and linearity of the transfer meter are known to be adequate, it requires no further calibration.

Indirect comparison transducer calibrations are similar to generator calibrations, except that an approximate level of stimulus source is required. Given the same level of stimulus, the outputs of the standard and test transducers are measured. The sensitivity of the test transducer is found by multiplying the determined ratio of the two outputs by the known sensitivity of the standard.

2.7.3 Ratiometer comparisons

A special type of measurement instrument known as a ratiometer is commonly used in calibration applications for comparing standard and test generators. These types of instruments typically have a high resolution capability for determining ratios and minute differences between standard and test instruments. When the standard and test devices have nominally the same values, the error of the ratiometer can be effectively eliminated by interchanging the two units and taking a second set of measurements. The addition of the two measurements divided by the number of measurements effectively eliminates the error contribution of the ratiometer. Common types of ratiometers include:

1. Kelvin ratio bridge—resistor comparisons
2. Transformer test sets—transformer ratio testing
3. Analytical balances—mass comparisons

2.8 Instrument Specifications and Calibration Tests

The tools available to a user for determining the accuracy of performed measurements are the manufacturer's specifications and calibration, if such is presented in the form of a data report. If a limit tolerance calibration is performed, instrument specifications are the only available tool for uncertainty analysis. Where calibration is concerned, the instrument specifications are the basis of test requirements.

Because of the virtually infinite combination of indicatable measurement values on even the simplest instrument, there has probably never been a calibration performed that could scientifically claim to verify all manufacturer's specifications. As an example, if a dc instrument has one range and three digits of resolution, it would require one thousand tests to verify all possible readings. If the instrument has multiple ranges and broad-range ac measurement capabilities, billions of tests and many years could be required to verify all possible combinations. If such were the case, calibration would not be possible and instruments would be worn out before they reached the hands of a user. Another important consideration is the fact that, in many situations, standards are simply not available to test all possible measurement capabilities. As an example, an rf impedance meter may have the capability to measure ac resistance at frequencies up to 1 GHz. The highest-frequency test performed by NIST is at 250 MHz. Calibration tests are actually selected to typify the performance of a measurement device on the basis of logical requirements, standards availability, and economics.

Another factor to consider is that not all specifications require verification on a periodic basis. Instrument specifications can be separated into four categories:

1. Academic—specifications that are of interest to the purchaser but have no bearing on measurement capability (example: instrument dimensions and weight)
2. Evaluation—specifications that may require a one-time validation before purchase (example: temperature and humidity performance characteristics)
3. Soft—specifications that are measurable but are not deemed critical or are verified indirectly by performance tests on other functions (example: input impedance)
4. Hard—specifications that require periodic verification (examples: accuracy, linearity, drift, etc.)

Even when hard specifications exist, if it is known that equipment users do not use a particular function, a calibrating organization may justifiably elect not to perform calibration tests on that particular operating feature. Primary quality control efforts for a calibration program require:

1. Documentation must exist to delineate what tests are performed, with specified limit tolerances, against specified standards.
2. Equipment users are made aware of any performance features that are not verified by the calibration process.

While variations exist between types of measurement equipment, calibration performance tests usually consist of the following:

1. Basic accuracy or sensitivity (unit to unit) at a level approximating full scale of the device for each operating range and function of the device. If the device is an ac measurement instrument, the tests are performed at a selected reference frequency.
2. Linearity tests referenced to full scale on at least one range. The number of linearity tests required depends on the instrument type, limit tolerance, and device application.
3. Frequency response tests referenced to the basic accuracy and sensitivity tests over the frequency range of use (ac measurement instruments).

A more detailed discussion and information regarding the determination of requirements and preparation of calibration procedures can be found in the NCSL publication RP-3.[3]

2.9 Calibration Standard Requirements

The fundamental requirement of a calibration standard that is used to perform a limit tolerance calibration is that it must have better accuracy than the device it calibrates. The mathematical relationship between the standard and test device accuracy, called the accuracy ratio, can be expressed in either percent or a numerical ratio. The importance of this relationship is best understood when the fundamental equation for worst case measurement accuracy (Eq. 2.1*a*,*b*) is examined. As shown, the uncertainty of any single measurement is the additive effect of many factors, among which is the accuracy of the device that calibrated the instrument.

$$U_t = \pm \sum_{1}^{5} U_n \tag{2.1a}$$

or

$$U_t = \pm(U_1 + U_2 + U_3 + U_4 + U_5) \tag{2.1b}$$

where U_t = Total uncertainty, percent or parts per million (ppm) of indicated value
U_1 = Indicated value tolerance
U_2 = Range full-scale tolerance
U_3 = Time stability tolerance (change per time period times number of periods since calibration)
U_4 = Temperature tolerance (change per degree times difference between operating and calibration temperatures)
U_5 = Calibration standard uncertainty

$$U_{\mathrm{RRS}} = \pm \left[\left(\sum_{1}^{5} U_n^2 \right) \Big/ 5 \right] \tag{2.2}$$

where U_{RRS} = root sum of the squares uncertainty, percent or ppm of indication.

Because errors are vector quantities, statistics indicate that when multiple factors are involved, the uncertainties are not directly additive. A common method for indicating simple statistical uncertainty is to calculate the root sum of the square, or RSS, as it is more commonly known (Eq. 2.2). Use of this formula provides the rationale for a 25 percent or 4 : 1 accuracy ratio requirement. In simple terms, if the standard is four times better than the

TABLE 2.1 Accuracy Ratio Chain Example

Requirement	Required accuracy ratio	Percent accuracy ratio	Approximate accuracy chain requirement
National standard requirement	Legal value	Legal value	±0.000016% (0.16 ppm)
Reference calibration laboratory	4:1	25%	±0.000063% (0.63 ppm)
Reference standard	4:1	25%	±0.00025% (2.5 ppm)
Working standard	4:1	25%	±0.001% (10 ppm)
Product test instrument	10:1	10%	±0.004% (40 ppm)
Product test requirement	Prescribed requirement	Prescribed requirement	±0.04% (400 ppm)

device it tests, it will not make a significant addition to the resultant uncertainty if an RSS analysis is used.

When limit tolerances are being verified, a methodology called the measurement analysis or subtraction tolerance analysis provides a technique that is highly effective in determining if a device is in or out of tolerance. Three fundamental relationships form the basis of the methodology:

1. If a noted deviation is less than the measurement device tolerance minus the standard uncertainty ($T_1 = T_{\text{test}} - U_{\text{std}}$), the test instrument is within tolerance.

2. If a noted deviation is greater than the measurement device tolerance plus the standard uncertainty ($T_h = T_{\text{test}} + U_{\text{std}}$), the test instrument is out of tolerance.

3. If the magnitude of a noted deviation falls between the high and low limits ($>T_1$ but $<T_h$), there is no way of knowing whether the device is in or out of tolerance. To make this determination, it must be retested with a higher-accuracy standard.

Accuracy ratios are not applicable when performing a report type of calibration. The accuracy of a comparison between standards equals the reference standard uncertainty plus the uncertainty associated with the transfer methodology. In many cases, the transfer uncertainty can be 10 or more times better than that of the reference standard. Given adequate methodology, it is possible to transfer reference standards through several echelons of calibration laboratories without significant accuracy degradation.

Understanding the real requirements of accuracy ratio is extremely important for both technical adequacy and economics of test operations. An example of an accuracy ratio chain is shown in Table 2.1. As indicated, if 4:1 ratios were applied at all levels of calibration service, a national standard capability of approximately ± 0.000016 percent would be required to support a ±0.04 percent product test requirement. By contrast, using report methodology (see

TABLE 2.2 Report and Accuracy Ratio Chain Example

Requirement	Required accuracy ratio	Percent accuracy ratio	Approximate accuracy chain requirement
National standard requirement	Legal value	Legal value	±0.0007% (7 ppm)
Reference calibration laboratory	Report	Report	±0.0008% (8 ppm)
Reference standard	Report	Report	±0.0009% (9 ppm)
Working standard	4:1	25%	±0.0010% (10 ppm)
Product test instrument	10:1	10%	±0.004% (40 ppm)
Product test requirement	Prescribed requirement	Prescribed requirement	±0.04% (400 ppm)

Table 2.2) with a transfer capability 10 times better than the required uncertainty, a national standard uncertainty of ±0.0007 percent would support the same requirement. Two important principles are illustrated by this hypothetical example:

1. Product test requirements determine needed instrumentation and the levels of calibration efforts required to support them. Unrealistic requirements can significantly increase costs of both test programs and calibration support.

2. When calibration efforts by organizations other than NIST are based solely on a chain of accuracy ratios, requirements may be indicated for NIST capabilities that are unobtainable. Appropriate standards measurement and support technology is required at all levels to avoid what is commonly known as the "accuracy crunch."

References

1. "The American Heritage Dictionary," 2d ed., Houghton Mifflin Company, Boston, Mass.
2. "Calibration System Requirements," MIL-STD-45662A, Aug. 1, 1988, Commander, U.S. Army Missile Command, Attn.: AMSMI-RD-SE-TD-ST, Redstone Arsenal, AL 35898-5270.
3. "Recommended Practice Calibration Procedures," NCSL Information Manual RP-3, Aug. 19, 1988, National Conference of Standards Laboratories, 1800-30th St., Suite 305B, Boulder, CO 80301.

Bibliography

"Quality Program Requirements," MIL-Q-9858A, Mar. 8, 1985, HQ USAF/RDCM, Washington, DC 20330.

"Inspection System Requirements," MIL-I-45028A, Dec. 16, 1963, Commander, U.S. Army Natick Research and Development Laboratories, Attn.: DRDNA-ES, Natick, MA 01760.

"Recommended Practice Laboratory Design," NCSL Information Manual RP 7, July 10, 1986, National Conference of Standards Laboratories, 1800-30th St., Suite 305B, Boulder, CO 80301.

Quality Standards Series, ISO 9000 Series (9000-9004), International Organization for Standardization, Case postale 56, CH-1211, Genève 20, Switzerland.

Quality Standards Series, ANSI/ASQC Q Series (90-94), American Society for Quality Control, 310 West Wisconsin Ave., Milwaukee, WI 53203.

"General Requirements for the Competence of Calibration and Testing Laboratories," Guide 25, International Organization for Standardization, Case postale 56, CH-1211, Genève 20, Switzerland.

National Voluntary Laboratory Accreditation Program (NVLAP), NIST/NVLAP, Bldg. 411, Room 162, Gaithersburg, MD 20899.

"Calibration Status—A Key Element of Measurement Systems Management," D. R. Workman, 1993 National Conference of Standards Laboratories (NCSL) Symposium.

Chapter

3

Basic Electronic Standards

David R. Workman*

Consultant, Littleton, Colorado

3.1 International System of Measurement Units

This section follows today's concepts and practices in the use of measurement systems. It is based upon the Système International (SI), or International System of Units, and more specifically upon the meter-kilogram-second-ampere system (mkas) of relationships of units of electrical quantities.

In the genealogy of measurement quantities we picture a lineage of derived quantities that are related to the base quantities. This is shown in Fig. 3.1. In this delineation the derived quantities are limited to those associated with the electrical measurement system, plus several others that are of general interest. About thirty such derived quantities are recognized today.

Although various physical quantities could be considered as basic in a measurement system, at least academically, the SI is based upon those of length, mass, time, electric current, temperature, and luminous intensity. The quantities of length come to be called mechanical quantities and units by common usage. Some systems, such as the centimeter-gram-second system (cgs) or the meter-kilogram-second system (mks) for electromagnetic quantities, recognize only three base units. Both these systems are coupled to the metric system of units. In the older cgs system (actually two, the electrostatic and the electromagnetic) the base units are the centimeter, gram, and second. In the mks system the base units are the meter, kilogram, and second. The latter

* Based upon material developed by Wilbert F. Snyder for "Basic Electronic Handbook," 1st ed., McGraw-Hill, New York, 1972.

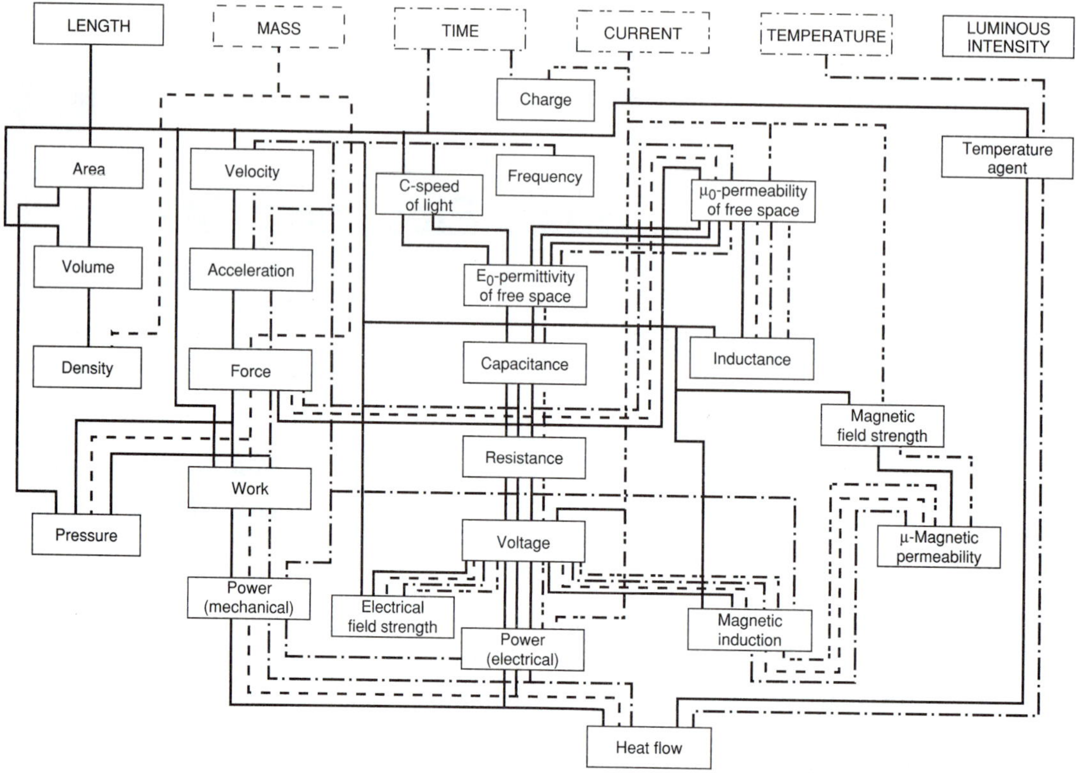

Figure 3.1 Relation of derived quantities to base quantities. For example, time and current yield a charge, or length and temperature yield gradient, or current and resistance yield voltage.

system is the foundation of the now universally accepted International System of Units. The National Bureau of Standards [currently National Institute of Standards and Technology (NIST)] adopted the International System of Units in 1964.[1,2]

1. *Metre (m), or meter—length.* The meter is the length equal to 1,650,763.73 wavelengths in vacuum of radiation corresponding to the transition between the levels 2_{p10} and 5_{d5} (orange-red line) of the krypton-86 atom (excited at the triple point* of nitrogen, 63.15 kelvins).
2. *Kilogram (kg)—mass.* The kilogram is the unit of mass; it is equal to the mass of the international prototype of the kilogram.
3. *Second (s)—time.* The second[3] is the duration of 9,192,631,770 periods of

* Triple point is the temperature of equilibrium between the solid, liquid, and vapor states.

the radiation corresponding to the transition between the two hyperfine levels of the ground state of the cesium-133 atom.

4. *Ampere (A)—electric current.* The ampere is the constant current which, if maintained in two straight parallel conductors of infinite length, of negligible circular cross section, and placed 1 meter apart in vacuum, would produce between these conductors a force equal to 2×10^{-7} newton per meter of length.

5. *Kelvin (K)—temperature.* The kelvin,* unit of thermodynamic temperature, is the fraction 1/273.16 of the thermodynamic temperature of the triple point of water. [The International Practical Temperature Scale of 1968 (IPTS 68)[4] and the International Practical Celsius Temperature Scale are referenced to the triple point of water and at least five other reference points.]

6. *Candela (cd)—luminous intensity.* The candela is the luminous intensity, in the perpendicular direction, of a surface of 1/600,000 square meter of a blackbody at the tempeature of freezing platinum under a pressure of 101,325 newtons per square meter. (In the SI the candela is considered to be a base unit of luminous intensity, although it is not completely a physical unit, as it involves the wavelength sensitivity of the average human eye.)

   To the six SI base units have been added two supplemental units, the radian for plane angles and the steradian for solid angles.

7. *Radian (rad).* Plane angle subtended by an arc of a circle equal in length to the radius of the circle.

8. *Steradian (sr).* Solid angle subtended at the center of a sphere by a portion of the surface whose area is equal to the square of the radius of the sphere.

Of interest to physicists, and particularly to electrical engineers, are the derived units that express the magnitude of the various electrical and magnetic quantities. Approximately thirty of these are considered as SI units, including the units of the rather simple quantities of area, volume, and frequency (see Sec. 3.1.2). The subject of derived quantities and the associated units is an extensive one in physics. It is a field in which many ideas have been expressed, with the unfortunate result that much confusion exists in the understanding of the various systems that have evolved. This has been particularly true in the area of electricity. In each of the various electrical systems one starts by expressing electrical quantities with fundamental and

* The name *kelvin* (symbol: °K) of the unit of thermodynamic temperature was changed from *degree kelvin* (symbol: K) by action of the 13th General Conference on Weights and Measures, Paris, October 1967. See *NBS Tech. News Bull.*, vol. 53, p. 12, January 1968.

relatively simple equations, which relate the electrical quantities to mechanical quantities, such as force, work (energy), and power. The latter, in turn, are expressible in terms of length, mass, and time. The mathematical process known as dimensional analysis is used as an aid to keep the physical and mathematical steps logical in using the equations. There is also the process known as "rationalization" (also "subrationalization") that is applied to the equations of some of the electrical measurement systems. The process treats the factor 4π (associated with spherical symmetry and commonly found in many equations) in various ways. This too has caused considerable confusion in its application. A detailed listing of the derived quantities is given in Sec. 3.1.2.

Figure 3.1 shows the relationships of the derived quantities to the basic quantities. A more detailed chart could show this relationship in dimensional terminology. Even greater detailing could show the relationships by definition and physical equations. Lineage of the derived quantities is indicated by hatched lines representing the several fundamental quantities. Each derived quantity (outlined in a block) is related to one or more base quantities and in some cases, such as electric power, through a chain of derived quantities. These relationships are expressed by physical equations and can be verified by expressing the relationships in dimensional terminology. Two of these chains, namely, mechanical and electrical quantities, have a common equivalency in the quantity of power. In turn, they have a common "sink," depicted by heat flow, mechanical power as the mechanical equivalent of heat, and electrical power as completely dissipated into heat.

Of interest in this genealogy of the International System is the role played by the permeability of free space.* Also of interest is the role played by the permittivity of free space.† By definition, current is considered to be a base quantity, although it is not independent of length, mass, and time. It is related, by definition, to these "mechanical" quantities by assigning a value of 4×10^{-7} henry/meter to the permeability of free space.‡ Experimentally, the relationship of current to the mechanical quantities is established by means of the current balance or the Pellat-type dynamometer (see Sec. 3.1.1). Volt-

---

* Permeability of free space μ_0 is a derived quantity and is expressed as the ratio of magnetic flux density (induction) B_0 to magnetic field intensity H_0 in free space. In the mksa system it is assigned the value of $4\pi \times 10^{-7}$ henry/meter. The quantity can also be expressed in terms of Maxwell's equations.

† Permittivity of free space ε_0 is a derived quantity and is expressed as the ratio of electric flux density (displacement) D_0 to electric field strength E_0 in free space. In the mksa system ε_0 has a value of 8.8542×10^{-12} farad-meter. The quantity can also be expressed in terms of Maxwell's equations.

‡ This rather confusing subject of the definition of the electrical units in the mksa system and the relation of the electrical units to the mechanical units is discussed and clarified by Page. See C. H. Page, Definition of "Ampere" and "Magnetic Constant," *Proc. IEEE*, vol. 53, no. 1, pp. 100–101, January 1965. Also, refer to Secs. 3.1.1 and 3.1.2.

age is derived from current and resistance through the relationship of Ohm's law. In early work the unit of resistance was established from the reactance of a self-inductor or mutual inductor. More recently, greater accuracy has been attained by using a computable capacitor.*

The value of permittivity of free space ε_0 is derived from the defined value of the permeability of free space μ_0 and the best value of the observed speed of light c, the relationship being $\varepsilon_0 = 1/\mu_0 c^2$. A presently used value of the permittivity of free space is 8.8542×10^{-12} farad/meter.[5] Derivation of resistance by the capacitor method is indicated in the genealogy diagram.

3.1.1 Development of the electrical systems of units

The development of electrical systems of units has progressed over a period of nearly one and a half centuries. During this time no fewer than eight recognized systems have evolved.[6] In consequence, much confusion has existed because of varied terminology, multiplicity of concepts and methods of approach, complexity of the subject, lack of standardization, and lack of understanding of various viewpoints. The electrical system of units as we know them today had its origin with the expression of Ohm's law, $E = IR$, in 1827. In 1833, Gauss first related magnetic quantities to mechanical units by experimental methods.† At later dates, Weber developed methods of measuring electric current and resistance in terms of the mechanical units of length, mass, and time. Such methods of relating the electrical quantities (or heat, light, etc.) to the so-called mechanical units of length, mass, and time are known as "absolute" methods,‡ with no special reason except that early workers in the field chose this terminology.

Formed in 1861, and continuing its influence for many years, was the Com-

---

* W. K. Clothier, A Calculable Standard of Capacitance, *Metrologia*, vol. 1, pp. 36–55, 1965. A capacitor designed upon a theorem that permits a small capacitance (approximately 1 picofarad) to be calculated directly in terms of unit lengths of cylindrical solid rods placed in various geometrical configurations, with an uncertainty of about 1 part in 107.

† A paper entitled On the Intensity of the Earth's Magnetic Field Expressed in Absolute Measure, written by Gauss in 1833, gave the first example of determining magnetic and electrical quantities by measurements in terms of length, mass, and time. Gauss measured the horizontal component of the earth's magnetic field by using a deflection magnetometer and a vibration magnetometer in combination.

‡ The word "absolute" can be confusing. Its long-time usage in electrical measurements has been that of relating electrical units to the mechanical units of length, mass, and time. However, sometimes it is given the connotation of a state of perfection; that is, no errors, deviations, or residuals exist in the system of measurement units.

Other connotations have been given to the expression "absolute" method of measurement, for example, "bootstrap" techniques that are employed for the calibration of devices used to determine the magnitude of dimensionless quantities such as attenuation. In this usage the measurement is not dependent upon any other device or upon measured characteristics of the device itself. Another example of an "absolute" method of measurement is the calibration of an "ideal" rotary-vane attenuator by measuring the angles of rotation of various settings of the revolving section of the attenuator.

mittee on Electrical Standards[7] appointed by the British Association for the Advancement of Science. This committee, under the chairmanship of Clerk Maxwell in its early years, established the centimeter-gram-second (cgs) system of electrical units in both the electrostatic and electromagnetic relationship of units, and also the practical units, all expressed in terms of the metric system. The cgs system of electrical units, by its very name, was directly related to the mechanical units and thus was an "absolute" system. The magnitudes of the practical units were selected to be of greater convenience to engineering applications than some of the more extreme values of the electrostatic and electromagnetic units.

During a period of many years, after the establishment of the cgs system, the basic units of the volt, ohm, and ampere became embodied in such standards as the stand cell* (culminating in the Weston saturated cell), the mercury ohm,† and the silver voltameter‡ (also known as the coulometer). The electrical units in terms of these standards became known as the international units and existed as such until 1948. In some countries, including the United States, these units were known as the "legal" units.

During the long period of use of the cgs system highly precise methods of determining the volt, ohm, and ampere in terms of the mechanical units were carried out by the national laboratories of England, Germany, and the United States. Eventually, unit values of the ohm and volt became embodied in the standard 1-Ω resistor of the Thomas or similar type and in the Weston saturated cell (very nearly 1 V), and they remain such today. As a result of these measurements (using the current balance and the self- or mutual-inductance method of measuring resistance), beginning Jan. 1, 1948, a new set of values known as the absolute volt and absolute ohm were assigned to national standards of voltage and resistance.[8] During the transition period it was neces-

---

* The early Clark cell, developed in 1873, was followed by the Weston cell as a standard of voltage. The Weston standard cell has a positive electrode of mercury, a negative electrode of a cadmium amalgam (mercury alloy), and a solution of cadmium sulfate as the electrolyte. Two forms of the cell are used, the unsaturated (electrolyte) and the saturated (electrolyte). The saturated cell is more stable but has a higher tempeature coefficient than the unsaturated cell and thus must be carefully temperature-controlled. These stand cells operate at approximately 1.02 V. In recent years specially selected zener diodes have come into common use as voltage standards.

† The mercury ohm, which had its origin in Germany, became a considerably less than satisfactory international unit of resistance. In its final development it was defined as the resistance of a column of pure mercury 106.3 cm long and having a mass of 14.4521 g (approximately 1 mm^2 in cross section). It went into general disuse during the early 1900s, although used for occasional reference as late as the 1930s.

‡ The silver voltameter served for many years as the standard for electric current. The international ampere was defined as the current which when passed through a solution of silver nitrate deposits silver on a platinum cup (cathode) at the rate of 0.001118 g/s under specified conditions. Because of lack of agreement between different forms of the voltameter and of the more fundamental approach in determining the ampere by the current balance (see the following paragraph in the text), the silver voltameter has fallen into disuse.

TABLE 3.1 International (SI) System of Units

Basic units			Derived units			
Quantity	Name	Symbol	Quantity	Name	Symbol	Formula
Length	Meter	m	Charge	Coulomb	C	A × s
Mass	Kilogram	kg	Capacitance	Farad	F	a × s/V
Time	Second	s	Inductance	Henry	H	V × s/A
Electrical current	Ampere	A	Potential	Volt	V	W/A
Temperature	Kelvin	K	Resistance	Ohm	Ω	V/A
Luminous intensity	Candela	cd	Energy (work/heat)	Joule	J	N × M
Amount of substance	Mole	mol	Force	Newton	N	kg × m/s^2
			Frequency	Hertz	Hz	c/s
			Illuminance	Lux	lx	lm/m^2
Supplementary units			Luminous flux	Lumen	lm	cd × sr
Plane angle	Radian	rad	Magnetic flux	Weber	Wb	V × s
Solid angle	Steradian	sr	Magnetic flux density	Tesla	T	Wb/m^2
			Power	Watt	W	J/s
			Pressure	Pascal	Pa	n/m^2

sary to state their relationship to the international units that had been used for over a half century. Although these values remain essentially the same today as in 1948, they are subject to slight changes with more accurate methods of determination. Such a change brought a change to the legal U.S. volt[9] on Jan. 1, 1969, because of the development of the computable-capacitor method of determining the ohm,[10] and a more precise current balance[11] and the Pellat-type dynamometer methods[12] of determining the ampere.

In 1954, the 10th General Conference on Weights and Measures established the "Système International" (SI), or International System of Units, based upon the meter, kilogram, second, ampere, degree Kelvin (now kelvin), and candela.

The electrical units of the International System of Units are based on the mksa system adopted by the International Electrotechnical Commission (IEC) as a development of Giorgi's[13] 1901 proposal for a four-dimensional mks system.* The system includes the ampere and is therefore known as the meter-kilogram-second-ampere system.

3.1.2 The international system of derived units

Table 3.1 lists the derived units of the International System of Units. Table 3.2 lists the prefixes for decimal multiples and submultiples of SI units.

* L. H. A. Carr, The M.K.S. or Giorgi System of Units, *Proc. Inst. Elec. Engrs. (London)*, vol. 97, pt. 1, pp. 235–240, 1950. This paper was one of four papers of a symposium on the mks system of units. The four papers and an extensive discussion are found in this referenced publication. A. E. Kennelly, The M.K.S. System of Units. *J. Inst. Elec. Engrs. (London)*, vol. 78, pp. 235–244, 1936. A. E. Kennelly, I.E.C. Adopts M.K.S. system of Units, *Trans. Am. Inst. Elec. Engrs.*, vol. 54, pp. 1373–1384, December 1935. The last two references are papers that are quite similar and contain a very extensive list of references on systems of electrical units.

TABLE 3.2 Multiple and Submultiple Prefixes

Symbol	Prefix	Multiple	Symbol	Prefix	Multiple
T	tera	10^{12}	c	centi	10^{-2}
G	giga	10^{9}	m	milli	10^{-3}
M	mega	10^{6}	μ	micro	10^{-6}
k	kilo	10^{3}	n	nano	10^{-9}
h	hecto	10^{2}	p	pico	10^{-12}
da	deca	10^{1}	f	femto	10^{-15}
d	deci	10^{-1}	a	atto	10^{-18}

3.1.3 Internationalization of electrical units and standards

The almost universal use of the metric system has probably been a determining force to engender international agreements on electrical units and standards. The agreement process is rather involved. It begins with the International Advisory Committee on Electricity that passes its recommendations on to the International Committee on Weights and Measures. Final action is taken by the (International) General Conference on Weights and Measures. The latter is a meeting of delegates from countries that hold to the Treaty of the Meter. The General Conference convenes every few years.

The International Bureau of Weights and Measures serves as the laboratory facility for this international system that has evolved over nearly a 100-year period. It is located at Sèvres, France, in suburban Paris. Cooperating in the international program are a number of national standardizing laboratories such as the National Bureau of Standards (USA); the National Physical Laboratory (England), formerly the Physikalisch-Technische Bundesanstalt (West Germany), formerly the Physikalisch-Technische Reichsanstalt (Germany).

3.1.4 National Institute of Standards and Technology (NIST)

The NIST, founded in 1901, is the national standardizing laboratory of the United States. Responsibility for a national system of electromagnetic measurements in the frequency range of zero (direct current) to above 300 GHz is with the Electricity Division (Electromagnetics Division, Quantum Electronics Division) and the Time and Frequency Division.

The Institute's mission is to develop and maintain the national standards of measurements and furnish essential services that lead to accurate and uniform physical measurements throughout the nation. The mission provides the central basis for a complete and consistent system of physical measurements that is coordinated with those of other nations.

The NIST Reference Standards and the NIST Working Standards maintained by the Institute for Basic Standards for electrical and radio-frequency quantities are used in Echelon I.

3.1.5 An echelon of standards

It is natural that an echelon or hierarchy of standards will evolve if a measurement system is to show lineage or traceability to a common source. Conversely, this common source provides the point from which a chain of measurement leads to the ultimate user of a system. The term base is used to designate the standards at the source of a measurement system. However, the term "prototype" standard[14] is frequently used, particularly if it is of an arbitrary type such as the international 1-kilogram mass. National standardizing laboratories have replicas of the international kilogram; in the United States the NIST has Prototype Kilogram No. 20. Atomic standards of length have supplanted the former meter bar, and in the SI system atomic standards of time (interval as differentiated from epic) have supplanted the second as determined from the rotation of the earth on its axis or around the sun. Probably the ultimate of all basic standards is atomic[15] in nature rather than macroscopic.

An echelon of standards of a measurement system is shown in Fig. 3.2. Such an echelon is fairly common in practice, although it is somewhat idealized as shown in this format. The numbering of the echelons is based upon a standardizing program for electronic measurements prepared by an IEEE committee* on basic standards and calibration methods. The general concept of this format comes from the work of a committee within the Interservice (Army, Navy, Air Force) Calibration Conference of 1960, with the purpose of attaining a more uniform nomenclature for classes or echelons of laboratory standards.

In brief, the echelon structure as shown by Echelon I in Fig. 3.2 is typical of chains of standards as they exist within the National Bureau of Standards. Much of the terminology used in this presentation is after McNish.[16] The system of measurement quantities in Echelon I consists of the base units that are embodied in prototype standards such as standards of length and mass, and of the many derived units that are expressed by their mathematical relationships to the base units. Many of these derived units can be embodied in standards; for example, the saturated Weston-type cell serves as the standard for the volt. Whether standards exist singly, such as Prototype Kilogram No. 20, or uniquely as the group of 40 saturated standard cells, they are classed as the national standards. The national standards serve to establish and stabilize the units of measurement for the United States.

---

* A Program to Provide Information on the Accuracy of Electromagnetic Measurements, *Report* 62 IRE 20, TR2; also *Proc. IEEE*, vol. 51, no. 4, pp. 569–574. April 1963: also, see any subsequent IEEE Standards Report on State of the Art of Measuring (specific electromagnetic quantity) prepared by the various subcommittees of the Electromagnetic Measurement State-of-the-Art Committee of the IEEE Instrumentation and Measurement Group.

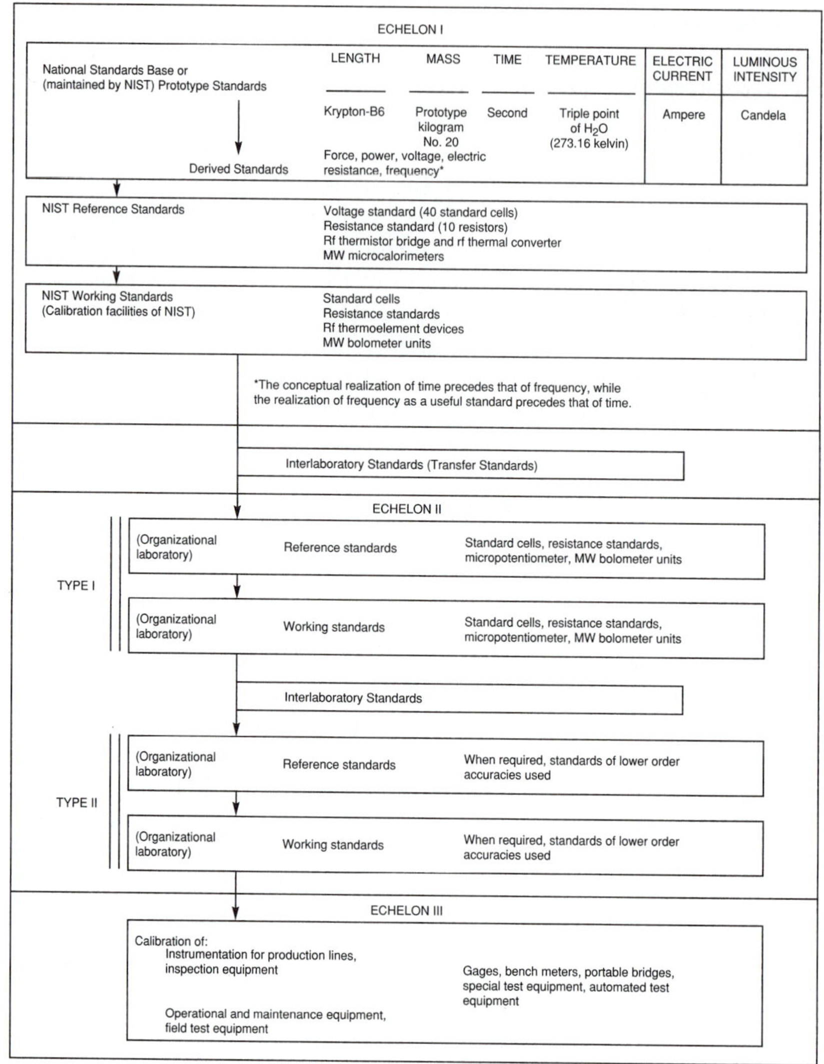

Figure 3.2 An echelon of standards of a measurement system.

3.2 Traceability of Standards

3.2.1 Standard capabilities and traceability

In the years that have intervened since the publication of the First Edition of the "Electronic Instrumentation Handbook" many measurement capabilities have changed dramatically, while others have remained relatively the same. A major change has occurred in the recognition of items called intrinsic standards, which are standards based on defined values and natural physical law. Because these standards produce a defined output, they theoretically do not require calibration by NIST to determine their values or uncertainty. Over the years many standards have been developed and used that fall into this category. Historical intrinsic standards include ice and metal freezing point temperature standards. Where economically justified, the cesium beam frequency standard has replaced the traditional crystal oscillator compared to NIST frequency transmissions. The most recent players in the intrinsic standard area include the Josephson junction dc voltage reference and Hall-effect resistance standards. If an organization can afford the procurement and maintenance costs of intrinsic standards, as justified by their requirements, they can effectively have the same capabilities as NIST. Of note is the fact that the existence of capabilities identical to NIST does not in itself evidence measurement capabilities or required traceability to national standards. Proof of capabilities requires some form of measurement assurance program that relates performed measurements to NIST capabilities.

The most notable change in dc voltage standards, at levels below NIST, is the replacement of the traditional standard cell with a zener reference. Where resistance measurements were traditionally tied to 1-ohm standards, there is currently increased use of 10-k (10,000) ohm standards as a primary reference point. The use of traditional supply/voltage divider/potentiometer calibration setups has almost universally been replaced by multifunction ac/dc calibrators, many of which are relatively self-calibrating against as few as three fundamental voltage and resistance standards (1 Ω, 10 kΩ, and 10 V dc zener reference).

An almost universal change that has occurred is the replacement of analog-type meter instrumentation with digital readouts. In the areas of basic electrical unit calibration and measurement (dc and ac voltage and current, and resistance) high-resolution digital meters have, in many cases, eliminated classical potentiometer, voltage divider, bridge, and ac/dc transfer instrumentation in both calibration and use applications. This quiet revolution also extends to rf technology and other areas of measurement science. While the faces of transfer instrumentation have changed, the basic methods for transferring measurement capabilities from the NIST to users remain essentially the same.

Another notable change is increased use of computer controlled automated test systems at all levels of effort. Automated test systems have allowed for levels of testing and statistical analysis of data that were only dreamed of a few years ago.

3.2.2 Traceability echelons

The definitive echelons of traceability are:

Echelon I (I)	Standards and measurement techniques used by NIST or a comparable national laboratory

Interlaboratory Standards (IS)	Transfer standards
Echelon II (II)	Standards and measurements used by industrial and government standards laboratories
Echelon III (III)	Instruments used for production work and quality control in manufacturing, for maintenance, and for general measurement purposes

To obtain information on current capabilities and costs of NIST services, a "NIST Calibration Services Users Guide—Special Publication 250" is required. A copy can be obtained from:

Physical Measurements Services Program
National Institute of Standards and Technology
Gaithersburg, MD 20899

3.3 Standards Maintained by NIST

The following is a list of standards maintained by NIST and available for comparison for calibration of instrumentation:

- Resistance (dc and low-frequency)
- Voltage (dc)
- DC ratio
- AC-DC transfer (low-frequency voltage or current)
- AC ratio (low-frequency)
- Capacitance and inductance (low-frequency)
- Power and energy (dc and low-frequency)
- Voltage
- Current
- Power, CW, and pulse
- Impedance (immittance)
- Phase shift
- Attenuation
- Noise
- Power 1 to 40 GHz (rectangular waveguide, also coaxial line)
- Impedance to 40 GHz (rectangular waveguide)
- Phase shift
- Attenuation 1 to 40 GHz (rectangular waveguide, also coaxial line)

- Noise 1 to 40 GHz (rectangular waveguide) (measurements at NIST limited to certain frequencies)
- Frequency

References

1. NBS Tech. News Bull., vol. 48, pp. 61–62. April 1964.
2. Mechtly, S. A.: "The International System of Units—Physical Constants and Conversion Factors," NASA SP-7012, National Aeronautics and Space Administration, Washington, D.C., 1964.
3. Atomic Second Adopted as International Unit of Time, NBS Tech. News Bull., vol. 52, p. 12, January 1968.
4. New Definitions Authorized for SI Base Units, NBS Tech. News Bull., vol. 53, pp. 12–13. January 1969.
5. Electric Engineering Units and Constants, NBS Tech. News Bull., vol. 49, no. 5, p. 75, May 1965.
6. Varner, W. R.: "The Fourteen Systems of Units," Vantage Press, Inc., New York, 1961. F. B. Silsbee, Systems of Electrical Units, NBS Monograph 56, September 1962.
7. Reports of the Committee on Electrical Standards of the British Association for Advancement of Science, Cambridge University Press, 1913.
8. Announcement of Changes in Electrical and Photometric Units, NBS Circ. 459, May 1947.
9. Reference Base of the Volt to Be Changed, NBS Tech. News Bull., vol. 52, pp. 204–206, September 1968. Also, 1968 actions, International Committee of Weights and Measures, NBS Tech. News Bull., vol. 53, pp. 12–13, January 1969.
10. Cutkosky, R. D.: Evaluation of the NBS Unit of Resistance Based on a Computable Capacitor, J. Res. NBS, vol. 65A, pp. 147–158, June 1961.
11. Driscoll, R. L., and R. D. Cutkosky: Measurement of Current with the National Bureau of Standards Current Balance, J. Res. NBS, vol. 60, pp. 297–305, April 1958.
12. Driscoll, R. L.: Measurement of Current with the Pellat Electrodynamometer, J. Res. NBS, vol. 60, pp. 287–296, April 1959.
13. Giorgi, G.: Rational Units of Electromagnetism, Atti dell' Associoz Elettr Italiene, 1901, p. 201.
14. McNish, A. G.: Classification and Nomenclature for Standards of Measurements, IRE Trans. Instr., vol. I-7, pp. 371–378, December 1958.
15. Huntoon, R. D., and U. Fano: Atomic Definition of Primary Standards, Nature, vol. 116, pp. 167–168, July 29, 1950.
16. McNish, A. G.: Classification and Nomenclature for Standards of Measurements, IRE Trans. Instr., vol. I-7, pp. 371–378, December 1958.

Bibliography

"ASTM Metric Practice Guide," National Bureau of Standards Handbook 102, Mar. 10, 1967.

Chertov, A. G.: "Units of Measures of Physical Quantities," Hayden Book Company, Inc., New York, 1964.

Harris F. K.: Units and Standards of Electrical Measure, Electron. World, vol. 72, pp. 29–32, 55, August 1964.

Harris, F. K.: Electrical Units, 19th Ann. ISA Conf. Proc., vol. 19, pt. 1, paper 12, p. 1–1–64.

Huntoon, R. D.: The Measurement System of the United States, NBS Misc. Publ. 291, pp. 89–98, July 1967.

Huntoon, R. D.: Concept of the National Measurement System, Science, vol. 158, no. 3797, pp. 67–71, Oct. 6, 1967.

McNish, A. G.: Dimensions, Units and Standards, Physics Today, vol. 10, pp. 12–25, April 1957.

McNish, A. G.: The International System of Units (SI). Materials Research and Standards, vol. 5, pp. 528–532, October 1965.

NBS Interprets Policy on SI Units, NBS Tech. News Bull., vol. 52, pp. 121, 124, June 1968.

Pontius, P. E.: Measurement Philosophy of the Pilot Program for Mass Calibration, NBS Tech. Note 288, May 6, 1966.

Pontius, P. E., and J. M. Cameron: Realistic Uncertainties and the Mass Measurement Process, an Illustrated Review, NBS Monograph 103, Aug. 15, 1967.
Silsbee, F. B.: Establishment and Maintenance of the Electrical Units, NBS Circ. 475, June 1949.
Silsbee, F. B.: Extension and Dissemination of the Electric and Magnetic Units, NBS Circ. 531, July 1952.
Silsbee, F. G.: Systems of Electrical Units, J. Res. NBS, vol. 66C, pp. 137–178, April–June 1962; reprinted in NBS Monograph 56, September 1962.
Varner, W. R.: "The Fourteen Systems of Units," Vantage Press, Inc., New York, 1961.
Young, L.: "Systems of Units in Electricity and Magnetism," Oliver & Boyd, Ltd., Edinburgh, 1967.

Chapter

4

Data-Acquisition Systems

Fred Cruger
Agilent Technologies
Lake Stevens, Washington

4.1 Introduction to Data-Acquisition Systems

In the context of this chapter, *data-acquisition systems* are generally those products designed to capture basic electro-mechanical phenomena by measuring the electrical output from a variety of transducers. The dominant measurements include temperature, pressure, strain, flow, vibration, voltage, current, etc. These parameters represent the "real" (physical) world. In specialized applications, more esoteric measurements (field strength, chemical content, etc.) can become a necessary piece of a data-acquisition system; such special requirements often affect the final choice of system architecture. This chapter, therefore, presents the fundamental architectural choices and the factors that help to differentiate the alternatives.

4.1.1 Development of data acquisition

Various forms of data acquisition have been commercially available for decades, but technologies and techniques continue to evolve. For example, voltmeters that periodically print results, tape recorders that are optimized for multi-channel recording, and strip chart recorders that make hard-copy graphs have been used in everything from measuring the durability of mechanical devices during life test, to in-flight recording of airframe structural movements, to measuring and recording changes in barometric pressure, river levels, or seismic events. These have evolved into complex, distributed, computer controlled, and reconfigurable systems.

Sensors, the devices that are actually used to measure physical phenomena, typically require some sort of "signal conditioning" to be compatible with measurement hardware. Such conditioning consists of amplifiers (for sensors generating very low-level signals), filters (to limit the amount of "noise" on the signal), isolation (to protect the sensors from interacting with one another and/or to protect the measurement system from possible damaging inputs), and whatever other circuitry is necessary to adapt the "sensor" to the measurement hardware. The adoption of digital technology in measurement operations and the invention of new sensors has led to the development of corresponding signal conditioning. The packaging of sensors and signal conditioners has also grown progressively smaller through the years.

Types of data acquisition systems. A variety of choices in packaging and architecture have evolved, and each alternative offers some unique advantages while presenting some unique costs and challenges. These include:

- Data loggers
- Computer backplanes
- PC plug-ins
- Network-based systems

Data loggers. Originally, simple electronic instruments, such as voltmeters, were equipped with signal conditioning that enabled them to measure the output from devices, such as thermocouples, and deliver "scaled" answers in the appropriate engineering units. Some units even included strip-chart recorders to provide a permanent record. By the 1970s, such instruments ("data loggers," Fig. 4.1a) were equipped with digital interfaces that enabled engineers to control the measurements and collect the data with computers. By 1975, a number of competing companies had taken a major step in measurement system architecture by cooperating in the development of an interface standard (IEEE-488) that allowed instruments from different companies to be combined into more complex computer-controlled systems. With the advent of such standards, it became apparent that, in many cases, the technical challenges associated with making measurement could be less focused on the measurements themselves, and more focused on handling the measurement data and information. However, the "data logger" still just acquires and records the output of sensing devices. It does not develop information from the data, or inherently, provide the capability to control the function and uses of the instrumentation involved.

Computer backplane. In the 1980s, a number of companies cooperated to create the next measurement architecture by starting with an information system architecture (i.e., a popular computer backplane) and modifying it to satisfy some of the subtleties of making good measurements (shielding, cooling, triggering). Starting with the popular VME bus (see Fig. 4.1b and refer to

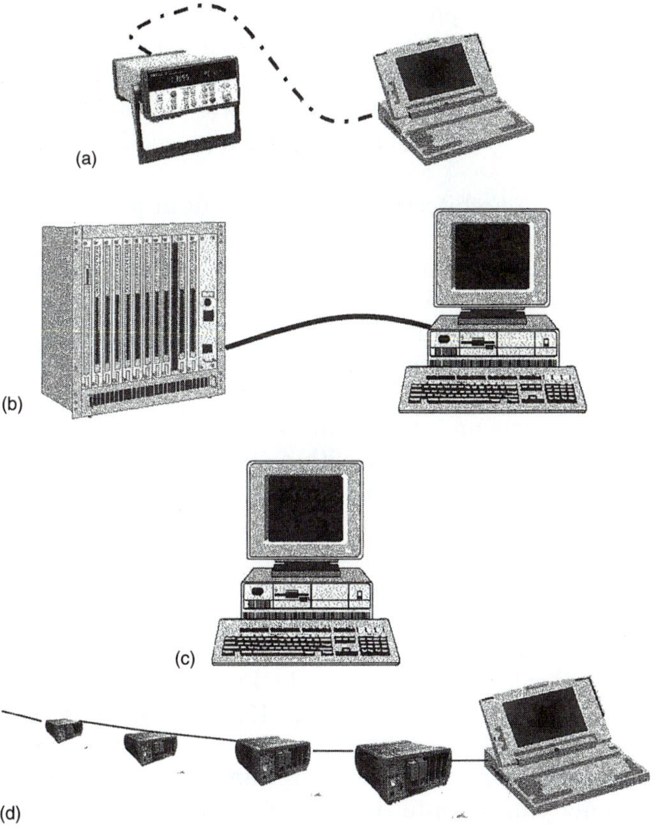

Figure 4.1 A data-logger solution is often a stand-alone instrument, but could include a low-cost interface to a personal computer (a). A computer backplane solution makes use of a separate mainframe holding multiple modules, connected to a personal computer with a high-speed interface (b). A PC plug-in solution generally uses available card slots in the computer to hold measurement modules (c). A network-based solution can be physically distributed, inheriting the advantages of LAN technology (d).

Chapter 40 for a more-detailed description of VME), they called the new standard *VME eXtended for Instrumentation* (*VXI*).

The strength of such a design is that large amounts of data can be moved at high speed on a parallel digital backplane, providing a very high "information bandwidth." Innovative uses of "local buses" and "front-panel data ports" make it possible to move up to 100 MB/s between modules without impacting the performance of the VME/VXI bus. Hundreds of channels, each sampling at 10s of Ks/s, can be accommodated, simultaneously making measurements and recording the data digitally. These multi-channel architectures enable designers and manufacturers of airplanes, automobiles, trains, ships, engines, and other complex physical structures to characterize and perfect their designs. The aerospace, defense, and transportation industries take advantage of the "standard" by mixing and matching technologies from a variety of vendors. Large, long-lived systems can "evolve" as technology progresses in different areas, with system modules being replaced individually, as needed.

PC plug-in. VXI-like architectures were overkill in performance for those users who needed only a few channels of measurement, and whose signal set required not millions of samples per second, but perhaps only hundreds or thousands of samples per second. For those users, a different measurement architecture became dominant: the PC plug-in card.

As the PC gained in popularity, many companies provided "accessory" boards that plugged directly into the computer. Some of these boards were measurement products and they enabled users to build limited-channel-count measurement systems physically within their PC (See Fig. 4.1c and refer to Chapter 40 for more-detailed information). Although the total information bandwidth was lower than that of VXI/VME and the measurement "purity" was limited by the rather unfriendly electronic environment of the PC, such systems satisfied the needs of many users. The total data-acquisition dollars spent on straightforward physical measurements (temperature, strain, and pressure) began to shift from the high-performance mainframe solutions to the lower-cost PC solutions. However, the PC business evolved far faster than the measurement business. By the late 1990s, it became apparent that the popular PCs were rapidly becoming laptops, with no plug-in slots. That forced the PC plug-in suppliers to create "mainframes" designed to accept the cards that once simply plugged into the PCs. That, in turn, made the architecture more expensive and less able to move from one generation to the next. It also encouraged a proliferation of alternate proprietary packaging approaches, although most of them still relied on the PC as a controller.

Network-based systems. The biggest drawbacks to the generations based on computers were the cost and complexity of mainframes and the need for specialized I/O to connect the mainframes to the controller of choice (typically a PC). If the mainframe could be eliminated, so would its cost. Similarly, if the specialized I/O could be eliminated, so would its cost and maintenance. Therefore, the next step in data-acquisition development was to eliminate the parallel backplane, which, by its physical nature, carried a large cost burden, and substitute a serial backplane, essentially a single cable or fiber (see Fig. 4.1d).

Instead of creating the need for specialized I/O, the serial standard chosen can be used to ensure lasting value, evolutionary improvement in performance, and unmatched flexibility. Using such standards as Ethernet, small-measurement modules can be connected to PCs, sometimes at great distances. This new architecture provides an information bandwidth greater than that typically available from a standalone data logger, but less than

TABLE 4.1 The Relatively New LAN-Based Architectures

	Strengths	Weaknesses
Data logger	Low cost per input channel. Ease of installation. Ease of operation.	Low channel count. Limited measurement types. Limited overall information bandwidth.
PC plug-in	Moderate cost per input channel. Tight link between measurement and PC. Large variety of measurement functions. Excellent software support tools. Large number of system integrators.	Unfriendly EMI environment for some measurements. Limited channel count. PC plug-in slots disappearing. Some user-programming generally required.
Computer-backplane	Very high channel counts possible. High information bandwidth. Large variety of measurement functions. International system integrators for very complex applications.	Relatively high cost. Learning curve for new users. Some user-programming generally required (turnkey software growing).
Network-based	Low cost per input channel. Dispersed measurements very easy. Usable with Web-browser access. "Inherit" LAN benefits (wireless, speed improvements, etc.).	Limited information bandwidth. Limited measurement types (growing).

that available from a computer parallel-backplane architecture, such as VXI. It encourages the design of systems that digitize the measurement data as close to the point of measurement as possible, then bring the data back to the controller digitally. It clearly, therefore, revolutionizes the way that users think about cabling issues. It has led to the development of "smart sensors," those that can be plugged directly into such a "measurement network" (much as a telephone extension can be plugged into the wall at home).

Architecture comparisons. There are potentially many "hybrid" architectures, but each type has specific advantages and disadvantages. Combinations of these technologies can be created to address specific data-acquisition problems, but to outline the considerations taken by system designers, it is necessary to cover the four fundamental measurement architectures of interest to data acquisition: data loggers (standalone boxes, scanning voltmeters), PC plug-in cards (including proprietary modules, etc.), computer-backplane cards (VME, VXI, PXI, etc.), and network-based systems (particularly Ethernet). The advantages and disadvantages of each are summarized in Table 4.1.

Data logger. Standalone data loggers with multiplexed front-ends are available, providing channel counts close to 100. Their fundamental strength comes from their user friendliness and the pleasant "out-of-box" experience provided by limiting their functionality and providing a simple user interface. They are often provided with software drivers that enable them to be controlled by popular PC packages or by user-written software.

PC plug-in cards. PC plug-in cards are available in almost infinite variety, with input channel counts from one to dozens of channels per card. Their fundamental strength comes from eliminating the need for a separate measurement mainframe and from the tight data link between the measurement cards and the PC controller. Proprietary modular systems typically share this tight link to a PC, although they might be packaged separately to achieve specific goals for durability, ruggedness, battery operation, or other specialized needs.

Computer backplane cards. VME/VXI data-acquisition cards are available from many suppliers, as are a variety of mainframes to hold them, and computer interfaces to connect them to a PC. Their fundamental strength comes from being able to accommodate large channel counts, provide an extremely wide range of measurements, and record/analyze the resultant data at high real-time speeds.

Network-based systems. The relatively new LAN-based architectures (Table 4.1) are usable as benchtop systems, as well as distributed systems, and their physical deployment flexibility is enhanced by the rapid advancement in computer technologies. Their fundamental strength comes from the serial digital transmission of data from the measurement modules to the PC controller (minimizing analog cabling), but the problems they can address must fit within the data rates of the LAN link.

4.2 Information Rate and Data Rate

The choice of any given data-acquisition architecture depends more on the intended information flow within a system than on any other issue. As a result, it is necessary to differentiate between data bandwidth and information bandwidth. *Data* represents the instantaneous status of a given sensor, and *information* is the meaning of this status. For example, many data-acquisition systems are used to monitor vibration in equipment to ensure that nothing is degraded or broken. Although any individual vibration channel might be sampled at a rate of 50 kilosamples per second (100 kilobytes per second), the system user might only want to know the peak and RMS value (4 bytes) of vibration once every second or so. In such a case, the data rate on a single channel is 100,000 bytes/sec, but the information rate is 4 bytes/sec. A more extreme example, far more specialized in nature, might be the monitoring of RF field strength. The input signal bandwidth might be many MHz wide, but the information bandwidth a mere few Hz wide. Data-acquisition modules are generally selected according to the data rates and signal conditioning required for specific input channels, but the overall architecture is determined more by the information bandwidth required of the entire system. The system designer must consider the current needs for information, potential future system growth, and the flexibility to accommodate various measurement types, to ascertain at what point in the system the information

TABLE 4.2 Typical Data-Acquisition Channel Counts and Sample Rates

Measurement parameter	Elec. motor	Elec. pump	Transmission	Auto engine	Jet engine	Sample rate (Sa/s) on each channel
Temperature	8	8	1	8	32	1–10
Pressure		1	4	4	16	500–2000
Voltage/Current	4	2				1–100
Flow		2				1000–1500
Position			5			100–1500
RPM	1	2	5	1	2	1000–2000
Vibration	2	2			32	1000–25000
Torque	1		4	4		500–2000
Total # Channels	16	18	19	33	82	

bottlenecks are likely to occur. Table 4.2 offers typical data-acquisition channel counts and sample rates.

4.2.1 Consuming bandwidths

The typical data rate required by an individual input channel is described in Table 4.2. It is strictly a function of the physical parameter being measured. If the channel count is limited, all data can all be moved real-time to the PC for recording. In such a multi-channel system, the total data rate between the measurement modules and the PC is simply the sum of all the individual channels. For many R&D applications wherein a device or design is being characterized, all the data samples are recorded for analysis—even when making noise and vibration measurements (e.g., wide bandwidth, high sample rates). As mentioned earlier, they can be moved on high-speed data paths to digital recording/processing modules within the VME/VXI mainframe.

4.2.2 Establishing requirements

To design or specify a data-acquisition system, it is absolutely necessary that the user understand the phenomena being measured and analyzed. Table 4.3 offers questions that must be answered in order to know what architecture to choose.

Data logger. A simple data-logger represents no problem to the user, as long as it can measure the input signal(s) at the appropriate rate. Depending on how it stores data (memory, tape, chart paper, etc.), the user simply needs to ensure that enough storage is available to accommodate the measurement interval. Any digital electronic link to a PC is likely to have more than enough bandwidth to pass all the data while it is being collected, the limita-

TABLE 4.3 Data-Acquisition System Design Questions

For any measurement channel the user must ask:

1. What is the data bandwidth of the analog signal? In other words, how fast must it be sampled by the system to be sure that nothing is missed?
2. Must the system process the real-time data in some fashion, or does it pass "raw" data to the PC?
3. Is there some way to "compress" the amount of data without eliminating the "information" sought?
4. How often is information needed regarding any individual channel? Can the data be averaged?

For the overall measurement system, the user must ask:

5. How many channels must be monitored?
6. Can they be sequenced (i.e. multiplex in time) or must they be sampled simultaneously?
7. If some input signal types require a variety of sample rates, must everything run at the highest sample rate, or can the system handle mixed sample rates?
8. What/where are the data bottlenecks likely to be?

tion then becoming one of file size and storage on the PC. The digital link might become more challenging if the user decides to use multiple data loggers to achieve a higher channel count than is possible with just one. Then the links to the PC must be replicated (or multiplexed) so that the possibility exists that the user could create a situation in which the digital links are a bottleneck or an implementation problem. The strength of the standalone data logger comes from the simplicity of a single measurement instrument (with a single link to a PC).

Computer backplane. As the channel count increases, particularly in systems with a large number of wide bandwidth channels, the aggregate data rate of all channels might exceed the capacity of the digital link between the measurement modules and the PC. In such cases, it is often possible to either "compress" the data by processing or averaging the data within the measurement module or to record the raw data within the measurement mainframe for later download to the PC. Both capabilities are available from VXI providers. High-speed data paths exist in VME/VXI to permit real-time data recording/processing at data rates far greater than the data could be moved to a PC. Of course, the strength of a VME/VXI architecture is its ability to deal with such high channel counts and wide data bandwidths; this is reflected in the cost of the measurement mainframe and the links to a PC controller.

PC plug-in card. In a PC plug-in or VME/VXI design, the user generally has to "configure" each data-acquisition card by specifying the sample rate or setting up the multiplexing scheme. For reasonable channel counts, it is possible to simplify the system design by passing raw measurement data to the PC—even on input channels that measure noise and vibration (typically, wide

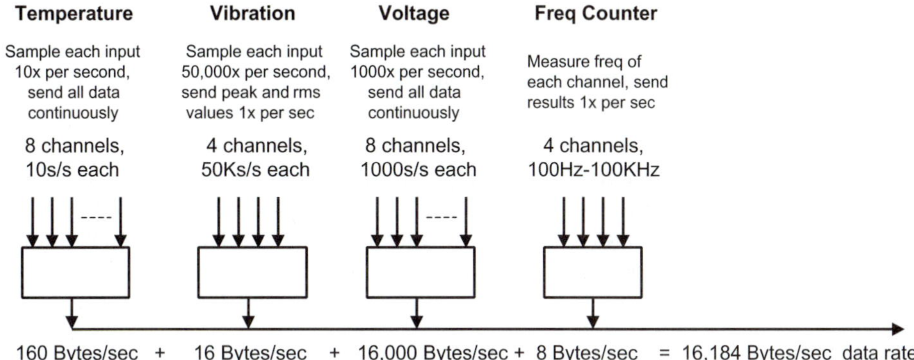

Figure 4.2 Block diagram showing representative sample and data rates in multi-channel system.

bandwidths), and allowing the PC to do the signal processing (FFTs, averaging, etc.). Depending on the cards chosen, it might be necessary to sample at the rate of the highest channel required, essentially generating more data than is necessary for the other channels, but ensuring that the highest bandwidth channel is satisfied. Should this be the situation, the total number of channels might be limited by the speed of the PC, rather than the space inside the PC. Still, the tight fast link between the measurement cards and the PC are a fundamental strength of a PC plug-in architecture.

Network-based system. If, however, a networked architecture has appeal (perhaps for cost or remote monitoring reasons), it becomes important not to "clog" the data and control path with unnecessary data samples. It is extremely important in such cases to perform the signal processing (averaging, FFT, etc.) in the measurement module to avoid unnecessarily consuming the information bandwidth of the LAN (see Fig. 4.2).

4.3 Data and Control Connections

The cost and complexity of cabling is often the largest challenge in multichannel data acquisition systems. Low-level analog signals often need to be shielded from external influences (electromagnetic effects, electrical noise, etc.). Computer interface cables need to be chosen according to whether data and control is passed in parallel or serial form, and whether the distances involved significantly impact the speed and robustness of the digital transmissions.

4.3.1 Signal cabling

Clearly, the number of measurement channels required, the types of measurements being made, and the physical distance between the measurement points and the measurement system are the primary factors that determine the actual signal cabling efforts and expense. For single-ended voltage measurements, that means a single connection, for differential measurements a double connection, or for sensors (such as strain gauges), perhaps as many as six to eight connections per measurement channel.

As cable lengths get to 10 meters or more, and as channel counts increase beyond 30 to 40 channels, the cost of installing and maintaining the signal cable can easily exceed the purchase price of the measurement instrument. This is a consideration for all data loggers, PC plug-ins, and computer backplane systems, unless external remote signal conditioning and multiplexing is available.

Data logger. In a data-logger application, the number of signals is limited and analog signals are routed to the test system using a single wire (or a single set of wires) for each signal. The problem is inherently manageable because of the relatively low channel count and simple measurements provided by data loggers.

Computer backplane. When a computer-backplane architecture (such as VXI) is required to maintain adequate information bandwidth and/or channel count, straightforward analog connections can result in immense cable bundles. Strain-gauge measurements on a large commercial airframe, for example, can involve thousands of measurement points, hence tens of thousands of wire connections. The sheer size and weight of the cable bundle can affect the physical response of the device being measured! In such cases, multiplexing becomes very attractive, in order to reduce cabling. Manufacturers of VXI data-acquisition systems often offer "external" signal conditioning that can be placed close to the points of measurement. In some cases, the signal conditioning is intended to amplify and buffer the raw signals, making measurements more accurate and less susceptible to noise and interference. In other cases, the primary purpose is to multiplex signals together onto a single connection. The number of long-run cables that return to the measurement mainframe are thereby minimized, and the user can minimize the expense of installing and managing huge cable bundles. In large installations, this can cut total system installation costs by as much as 40 percent.

Figure 4.3 depicts the difference between a nonmultiplexed strain-gauge measurement system and a multiplexed system. Without multiplexing near the point of measurement, one six-conductor wire must be strung from each strain gauge to the measurement hardware. A 1000-channel system that requires 1000 cables in a large bundle is connected to 1000 separate input channels (multiple input cards). By multiplexing near the strain gauges, however, the long-run cables can be reduced by a factor of 32, and can be connected to a single input card, saving significant cabling and measurement hardware expense.

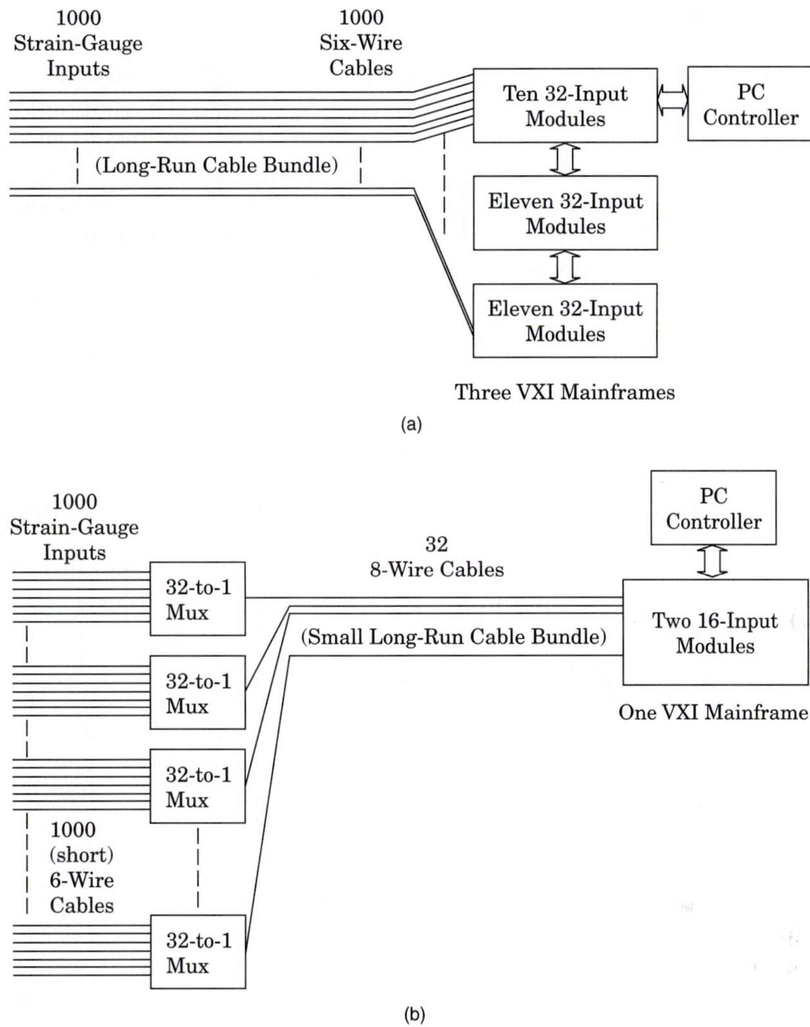

Figure 4.3 The difference between a nonmultiplexed strain-gauge measurement system and a multiplexed system: Without multiplexing, a 1000-channel strain-gauge measurement could require three mainframes and 32 input modules. Cabling costs could be large (a). By multiplexing the input channels close to the point of measurement, large cable bundles, measurement hardware (input modules and mainframes), and overall costs can be reduced. Of course, the structure being measured must be suitable for data bandwidths low enough to tolerate the multiplexing speed (b).

PC plug-in cards. The same cabling techniques used for "data loggers" are normally used for connection to PC plug-in cards, where the channel count is not excessive, and the distances allow for adequate analog connections. Some proprietary modular systems have an advantage, in that small modules can be located closer to the point of measurement, reducing cable lengths, and thereby reducing the effects of noise pickup, crosstalk, and other factors that can degrade low-level analog signals. Connection to the measurement system controller must then be made by a digital link that can cover the required distance, typically a low-cost link (such as RS-232 or FireWire). Of course, the user must consider the speed of that digital link because it is the "bottleneck" in moving data from the real world to the PC controller.

Network-based systems. The network-based architectures represent the ultimate in multiplexing, albeit digital instead of analog. Because the measurement modules can be placed close to the point of measurement, analog connection lengths are minimized. Digitization of all the data occurs in the modules, so data brought back to the PC controller is all in digital form. The multiplexing and demultiplexing happens transparently as part of the Ethernet information protocols. As long as the aggregate data rates do not exceed the available information bandwidth of the LAN, the user is "protected" from having to manage the detailed multiplexing/demultiplexing task. Distance limitations are mitigated because the digital medium can be extended far more easily without measurement risk than can analog cable bundles. In fact, if spatial requirements mandate a wireless connection, it can be implemented easily using off-the-shelf LAN hardware without impact on the measurements being made.

4.3.2 Control and I/O cabling

Connections between the measurement hardware and the PC controller have a major impact on the overall capability and cost of the complete system. It is impractical and unwise, however, to assume that the advertised theoretical data-handling speeds of any given interface will be reached during any practical application. Although the user must evaluate alternatives that provide the necessary speed (i.e., information bandwidth) for any particular application, shown in Table 4.4 are benchmarks run on actual interfaces using realistic data-acquisition hardware. The numbers shown are intended only as examples to illustrate the wide differences in performance under directly comparable conditions. The performance comparisons are for interfaces connected to a VXI mainframe containing a "typical" digitizer and a 64-channel data-acquisition card. The interfaces compared are a GPIB (IEEE-488), FireWire® (IEEE-1366), MXI-2 (unique to VXI), and an "embedded" controller (i.e., a PC/controller placed directly on the computer backplane of the measurement system).

TABLE 4.4 The Numbers Shown Are Intended Only as Examples to Illustrate the Wide Differences in Performance Under Directly Comparable Conditions

	GPIB (IEEE-488)	FireWire® (IEEE-1366)	MXI-2	Embedded Controller
Theoretical Max. Rate	1 MB/s	7 MB/s	17 MB/s	22 MB/s
I/O Transfer Time to Collect Digitized Blocks of Data, For Various Block Sizes:	143 ms (14 KB/s) 355 ms 639 ms 1207 ms 73130 ms (27 KB/s)	53 ms (38 KB/s) 90 ms 139 ms 239 ms 12743 ms (0.16 MB/s)	48 ms (42 KB/s) 83 ms 130 ms 224 ms 12106 ms (0.17 MB/s)	47 ms (43 KB/s) 79 ms 130 ms 209 ms 11033 ms (0.18 MB/s)
I/O Transfer Time for Small and Large Data Blocks from 64-channel Data Acquisition Card:				
64 Readings	205 ms	1.27 ms	0.34 ms	.43 ms
50 K Readings	2445 ms	122 ms	87 ms	47 ms

Data loggers. In the case of data loggers or scanning voltmeters, focus is generally on the cost and ease of use for a low-channel-count application. Many data-logger applications have no need for connection to a PC, so the friendly standalone user interface provides a very high value to the user. If a PC connection is required, low-cost, low-data-rate connections (such as RS-232) are very popular and easy to implement because virtually all PCs are equipped with such a serial interface. For short distances, higher data rates can be handled with IEEE-488 connections, albeit at a somewhat higher price. As more PCs have included USB and FireWire interfaces as standard components, more measurement devices that use such links have appeared on the market. In all cases, the information bandwidth from a data logger is unlikely to be further hampered by the performance of the PC I/O used because the digital transmission speeds are typically far greater than the bandwidths of the signals being measured. Both control information and data flows across the same digital connection without adversely impacting overall performance.

Computer backplane. VME/VXI/PXI have a broad spectrum of connection possibilities, each with advantages and disadvantages. For instance, it is possible to use an "embedded" controller, which means that the PC physically plugs into one of the slots in the mainframe. This allows for direct data transfer over the VME/VXI/PCI backplane between the measurement modules and the PC, and eliminates any need for a special I/O connection between the PC and the mainframe. However, the downside to this is that embedded controllers often cost two or three times more than their equivalent-performance counterparts that are not designed to plug into the mainframe. The rapid evolution of PC technology makes the desktop or laptop PC much less expensive and often far more powerful than embedded machines of the same vintage. So, users often choose to use an external (to the measurement mainframe) controller, which requires the use of specialized I/O. For VXI, these connections are available in a variety of prices and performances, including IEEE-488, MXI, and FireWire interfaces.

Interfaces must be chosen carefully, with insight into the continuous data rate required and the maximum latency that can be tolerated. For example, if a system is intended to be set up once, then run continuously for long periods of data collection, the user would be interested primarily in the efficiency and speed of transferring large blocks of data from the mainframe to the PC. On the other hand, if a system is designed to be set up repeatedly, armed, and triggered on short-duration repetitive events, the user might be more interested in the efficiency and speed of transferring very small blocks of data and/or control information. The "latency" of responding to a single-byte transfer might actually make one particular type of interface unusable, and the total speed for large block transfers might make another interface inadequate to sustain continuous data collection. To some extent, this highlights an inherent difference between "control" and "data" transfers. Control tends to occur intermittently, while data is more of a continuous requirement. For that reason, VME/VXI designers provide alternate paths for data transmission, often at speeds far greater than might be required for mere control and set-up purposes. These alternate paths are typically used to move data between measurement modules or from measurement to recording modules, rather than from modules to the PC.

In VXI, the "local bus" was defined to allow such flexibility, and commercial implementations demonstrate data rates up to 100 Mb/sec between VXI modules. In data-acquisition applications, this is particularly valuable in moving very high-channel-count data from the measurement modules to digital recording modules for post-measurement analysis. In VME, the "front-panel data port" can provide similar capability. Such alternate data paths allow continuous data movement without using any of the bandwidth of the VME/VXI backplane, which is thereby available for more "mundane" data or control tasks, or for moving summary data to the PC. For systems with very high channel count and information bandwidths, such alternate paths are an absolute must. Figure 4.4 is a block diagram of an example system.

PC plug-in cards. In the case of PC plug-in cards, the need for special I/O is gone because the boards already reside in the PC. That is the greatest fundamental strength of the PC plug-in architecture. Again, both control and data are moved across the internal PC backplane, which provides high data rates up to the limit of the number of cards that can be plugged into the PC physically. For proprietary modular systems, both control information and data typically occur over the same low-cost digital links mentioned earlier (RS-232, USB, and FireWire), so the system design must accommodate the combined information bandwidth.

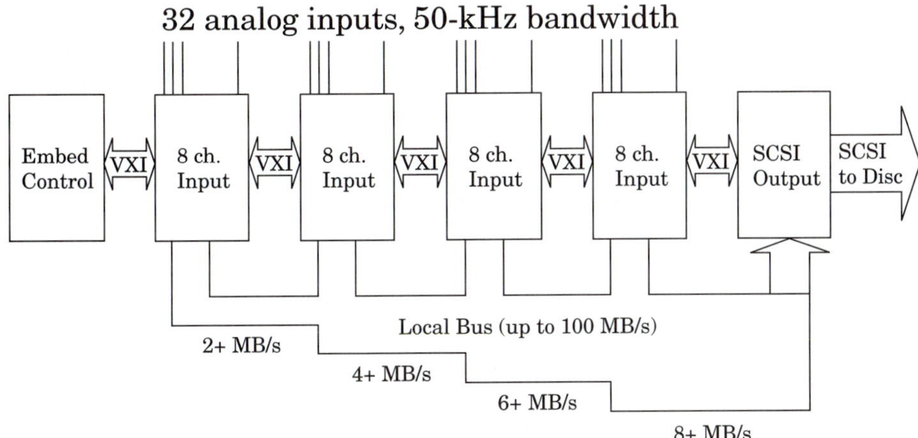

Figure 4.4 A block diagram that shows a VXI data-acquisition system utilizing local bus links and disc recording. In a wide information bandwidth system, such as the one shown, a continuous stream of data can be recorded (at rates greater than 8 Mb/s), and the embedded controller can easily display any eight vibration spectrums while also checking all channels for over-limit conditions. This is made possible by the computer backplane.

Network-based systems. Clearly, a LAN-based architecture uses the same data path for control and for data as the computer-based architecture. The fact that it is a serial link allows for maximum flexibility. The connection works as well over long distances as it does over short ones, allowing the measurement modules to be used in benchtop or in remote measurement applications. Advances in LAN technology can be "inherited" by the data-acquisition system, permitting connections to existing LAN structures, or even wireless connections. However, the system designer must give more thought to the total information bandwidth required because all data flows across the same serial connection.

If the measurement modules are connected to an existing LAN, the user must realize that LAN availability will affect system performance, and that other LAN transactions can drastically reduce the information bandwidth of the data-acquisition system. Therefore, the concept of performing data compression, averaging, scaling, and units conversion in the measurement module is an important issue to address as the total potential channel count of the system increases. As more "smart sensors" are created, providing their own calibrated digitized output data, the flexibility and utility of LAN architectures will be further enhanced (for more detailed information on smart sensors, see Chapter 47).

4.4 Software Alternatives

Software choices play a large part in selecting a data-acquisition system. The software required to configure the system, to operate the system, and to analyze the data can easily become the determining factor when choosing between two otherwise equivalent hardware alternatives. In extremely large high-performance data-acquisition systems, the choice of hardware is often left to the system integrator. High-speed acquisition, recording, real-time analysis, and system control can be so complex that only the system integrator truly understands the detailed inner workings of the system.

Once the end-user selects the software with the user interface and data analysis capabilities required, the hardware choice is limited to those supported by the software supplier. In simpler applications, however, the choice of hardware and software might become iterative. That is, when an end-user or small-system integrator must create a unique solution, a variety of hardware/software combinations might need to be considered, each with some advantages and each with some unique costs. In all cases, software choices are biased according to the dominant operating systems and "standard" application packages with which the user is familiar.

From a user perspective, software alternatives should minimize the need to learn new user interfaces, enable simple data transfers between collection and analysis and reporting packages, and they should avoid the need for custom programming whenever possible.

4.4.1 Data logger

Although standalone data loggers can be used in "turn-key" form, requiring no software, those that are to be used in conjunction with a PC generally must satisfy the software constraints imposed by the user.

At a minimum, the user might want to set up the instrument for taking measurements and then collect the data over a link, such as RS-232. In addition, the user might be predisposed to use common software packages for data display and analysis. In most cases, the ultimate result from a data-collection task is a report, followed by a decision.

Because the entire value statement of a data logger is steeped in "ease of use," the software associated with such a product is typically expected to operate in straightforward "spreadsheet" fashion—a style that is familiar and comfortable to PC users. Software utilities to change data from one file format to another have been created for many measurement devices. The need for such utilities, however, has been eliminated by the industry-wide convergence on software structures and techniques dominated by PC software developers. Users expect to be able to move the data from the measurement world into whatever PC standard package is to be used for display and/or analysis. The ability to "drag and drop" in typical PC style enhances the simplicity of a data-logger solution. Any additional data-handling complexity (file translation or programming) detracts from the simplicity.

4.4.2 Computer backplane

As the test requirement moves higher in channel count, it becomes less likely that the ultimate user of the equipment is prepared to create the entire test solution. Complex structural test systems, noise/vibration/harshness test systems, or drive-by acoustic test systems, are typically left to system integrators, who specialize in the various applications. They invest man-years into coordinating the behavior of measurement modules with recording and display devices, and more man-years into multi-channel computations and analysis packages. Multi-dimensional real-time displays, comparative analyses between live and archived data, and links to computer-aided design systems are data-acquisition system features offered by such system integrators. Therefore, the most-common software alternatives in high-end VME/VXI systems (and even in high-performance PC plug-in solutions) are typically measurement and analysis packages offered by specialized system integrators.

Those users who need the information bandwidth of a computer backplane solution, but whose problem is unique enough that no pre-packaged solution is available to them, or for whom an available solution is significant overkill (and, therefore, too costly), face a software challenge. However, the convergence of users on the PC as the controller and analysis tool of choice has resulted in software products specifically targeted at the needs of these users. Measurement hardware providers have standardized on a set of "plug-and-play" attributes expected to be met by driver software. At an even higher level, measurement-configuration and data-collection software has been created to provide a "spreadsheet look and feel," guiding the user through setup and operation, and transferring the data transparently to the PC report/analysis package of choice. For many applications, such products can provide complete data-acquisition functionality without requiring any programming! The unique dominance of particular PC operating systems, spreadsheets, word-processing tools, and math libraries has made possible such focused data-acquisition software. In doing so, it has become possible to maintain a consistent look and feel across data-acquisition systems ranging from data loggers all the way up to multi-channel VME/VXI systems.

4.4.3 PC plug-in cards

PC plug-in systems, on the other hand, are not available in a "standalone, turnkey" form. Users expect to configure the cards (setting voltage ranges, multiplexing schemes, filtering, etc.) and to be able to flexibly interconnect multiple cards to achieve multi-channel solutions. The most popular software products for such tasks are based on test-oriented graphical user interfaces, allowing users to create "virtual" instruments (see Chapter 46) and systems on-screen, to create measurement sequences that satisfy complex measurement procedures, and to report the data in graphical or tabular form. These software products do require the user to do programming, but operate with special languages that are very oriented toward test and measurement. They provide the user a very high degree of flexibility in configuring and reconfiguring test hardware to perform multiple tasks. They rely upon the hardware suppliers to provide "drivers" that are compatible with the higher-level software tools, in accordance with published standards. Virtually all hardware suppliers consider the "driver" to be an inherent part of their measurement product offering.

4.4.4 Network-based systems

For classic "test-and-measurement" applications, the programming-free software products designed for data loggers, PC plug-ins, and computer backplane systems can also be used with the network-based systems. The user paradigm for such software is to set up the necessary measurements, make the measurements on a device under test, then analyze the results. Within the constraints of information bandwidth, the various data-acquisition architectures can all be used for such tests. However, as network-based data-acquisition systems grow in popularity, more opportunities are being uncovered for the creative application of "data acquisition." The measurement modules used in some LAN-based systems can be accessed with Web browser tools that have become common among PC users. This enhances the usability of a network-based architecture because it makes obvious the ease with which "remote" measurements can be performed. Only the network-based architecture can take advantage of low-cost networking (including wireless connections) and the ease of use of a Web browser to gain access to remote measurement points. Applications that require "measurements," but are not directed specifically at a "device under test," include environmental monitoring, machine condition monitoring, industrial automation, utilities monitoring, and (potentially) the control of widely dispersed systems. Web browser access is not as fast as access provided by specialized data-acquisition software, but if it is fast enough for the task at hand, it eliminates the learning curve associated with specialized software tools. The exchange of data between a Web browser and other common PC packages (spreadsheets and word processors) is straightforward, allowing a user to go from a widely dispersed measurement problem to a concise summary report with tremendous flexibility.

Chapter

5

Transducers

J. Fleming Dias
Agilent Technologies
Palo Alto, California

5.1 Introduction

In general terms, the transduction process involves the transformation of one form of energy into another form. This process consists of sensing with specificity the input energy from the measurand by means of a "sensing element" and then transforming it into another form by a "transduction element." The sensor-transduction element combination shown in Fig. 5.1 will henceforth be referred to as the "transducer." Measurand relates to the quantity, property, or state that the transducer seeks to translate into an electrical output.

As an example, consider a "walkie-talkie" intercom set where the loudspeaker also functions as a microphone. At the input end, the loudspeaker functions as an acoustoelectric transducer and at the output end as an electroacoustic transducer. Moreover, in the reverse direction, the functions of the loudspeakers are interchanged, and for this reason we say that the loudspeaker is a bidirectional transducer and the transduction process is reversible.

Another example of reversible transduction is seen in piezoelectric materials; when an electric voltage is applied to the faces of a piezoelectric substrate, it produces a change in its physical dimensions; and conversely, when the material is physically deformed, an electric charge is generated on these faces. In this transducer, the sensing and transduction functions cannot be separated as easily, and it represents a good example of a practical transducer used in the field of nondestructive testing (NDT) of materials and in medical ultrasound imaging of body tissues and organs. This is a bidirectional transducer, but most practical transducers are not bidirectional.

Transducers may be classified as self-generating or externally powered.

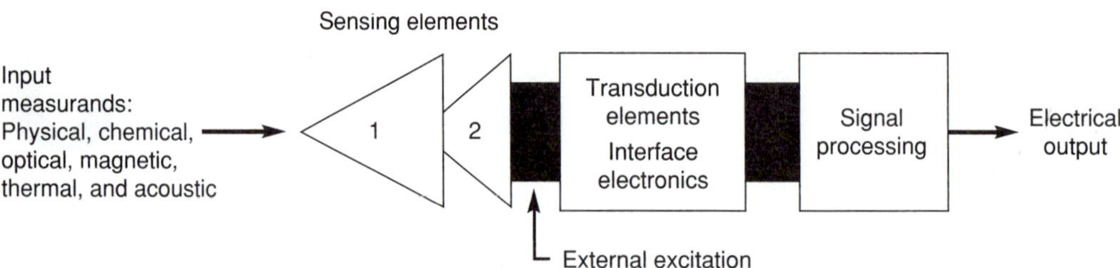

Figure 5.1 A transducer model showing multiple sensing elements, often found in some chemical and biochemical transducers. The output of the transduction element is interfaced with suitable electronic circuits to provide usable analog or digital signal output.

Self-generating transducers develop their own voltage or current and in the process absorb all the energy needed from the measurand. Externally powered transducers, as the name implies, must have power supplied from an external source, though they may absorb some energy from the measurand. The Hall-effect transducer in Sec. 5.9 and the integrated-circuit temperature transducer in Sec. 5.13.4 are examples of externally powered transducers, whereas the loudspeaker and the piezoelectric substrate are self-generating transducers.

5.2 Transduction Mechanisms and Measurands

The operation of a transducer is tightly coupled to one or more electrical phenomena or electrical effects.[1–3,12] These effects are listed below. Some relate to more advanced concepts for transducers that are leaving the research and development laboratories and making an entry into the commercial world. In addition, the most useful and important measurands are also listed.

5.2.1 Transduction mechanisms

Capacitive

Inductive and electromagnetic

Resistive and thermoresistive

Piezoresistive effect

Hall effect

Lateral effect

Extrinsic, interferometric, and evanescent effects in optical fibers

Magnetoresistive effect

Piezoelectric effect

Tunneling effect

Thermoelectric effects (Seebeck and Peltier)

Ionization effects

Photoelectric effect

Photoresistive effect

Photovoltaic effect

Acoustooptic effect

Fluorescence and fluorescence quenching effect

Field effect

Doppler effect

5.2.2 Measurands

Displacement
Position
Velocity
Acceleration
Force and load
Strain
Rotation and encoding
Vibrations
Flow
Temperature
Pressure
Vacuum
Relative humidity
Atomic and surface profiles
Gas concentration and pH
pH and partial pressures of O_2 and CO_2 in blood
Infrared radiation
Torque
Magnetic fields
Acoustic fields
Medical imaging
Nondestructive testing
Audio fields and noise
Rotation and guidance

5.3 Classification of Transducers

We have tabulated the class of transducers and some important measurands in Table 5.1. The name of each class is keyed to the transduction mechanism, but its explanation is given along with the description of the transducer. Table 5.1 should allow the instrumentation technologist to select the transducer most suitable for the measurand of choice. Moreover, cross indexing to the literature in the reference section is provided, wherever required, to the transducer class-measurand combination.

In addition, the table also includes several other transducers such as fiber-optic transducers, surface-profiling transducers (SPT), wave-propagation transducers (WPT), intravascular imaging and Doppler transducers, surface acoustic wave (SAW) transducers, acoustooptic (AO) transducers, Hall-effect transducers, and ChemFET transducers.

5.4 Selection of Transducers

Research and development in the transducer industry has traditionally been very productive. Many new forms and rapid improvements of old forms are continuously reported. One of the most successful improvements in transducers is the incorporation of integrated circuits for signal conditioning, with the basic transducer unit. These are known as smart transducers.[11] When selecting a transducer, in addition to the question of cost, careful attention must be given to the following:[1,4]

Sensitivity
Range
Physical properties
Output impedance
Power requirements
Noise

TABLE 5.1 Classification of Transducers and Some Important Measurands

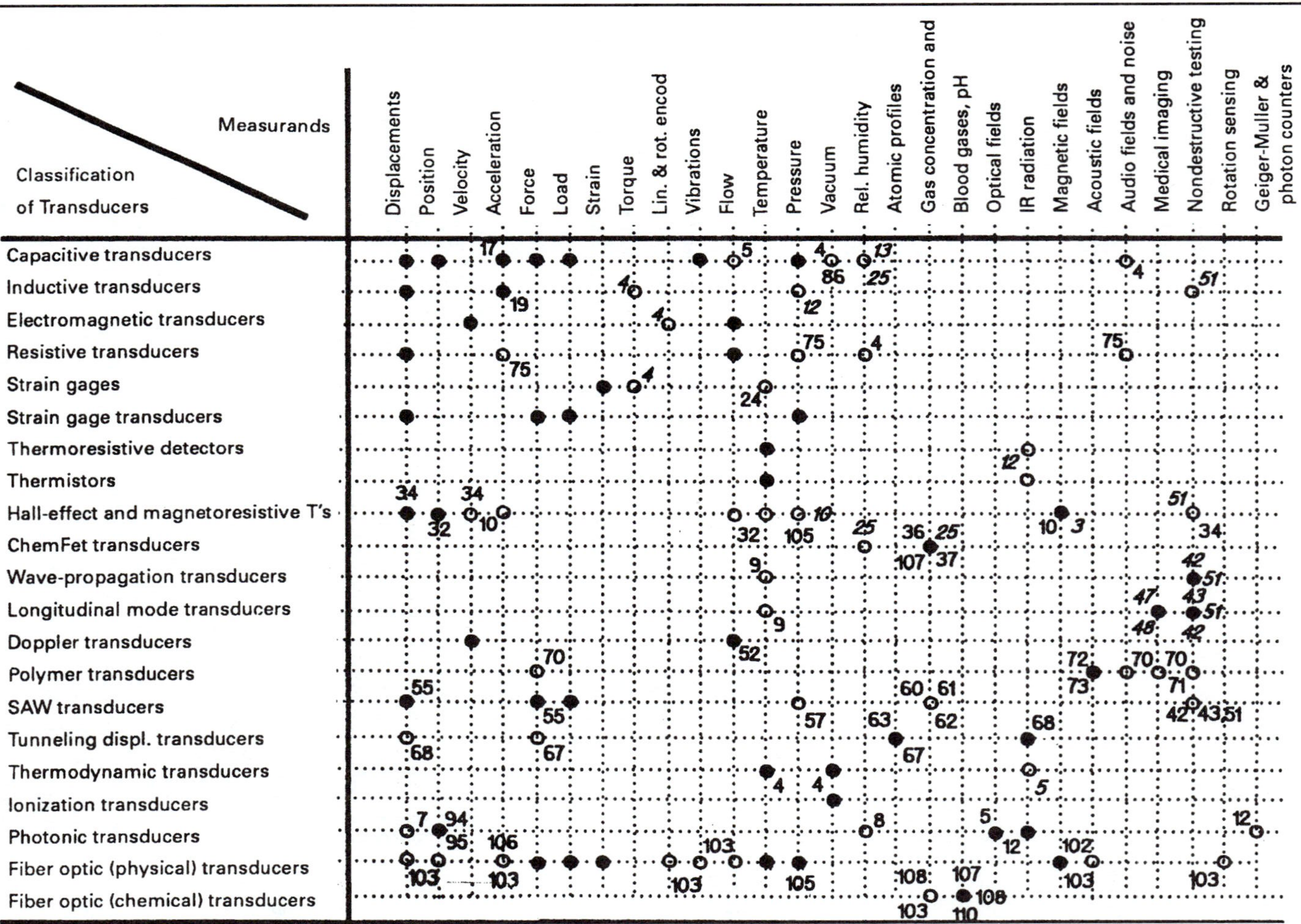

Classification of Transducers \ Measurands	Displacements	Position	Velocity	Acceleration	Force	Load	Strain	Torque	Lin. & rot. encod	Vibrations	Flow	Temperature	Pressure	Vacuum	Rel. humidity	Atomic profiles	Gas concentration and	Blood gases, pH	Optical fields	IR radiation	Magnetic fields	Acoustic fields	Audio fields and noise	Medical imaging	Nondestructive testing	Rotation sensing	Geiger-Muller & photon counters
Capacitive transducers	●	●		● 17	●	●				●	○ 5		●	○ 4, 86	○ 13, 25								○ 4				
Inductive transducers	●			● 19				○ 4					○ 12												○ 51		
Electromagnetic transducers			●						○ 4		●																
Resistive transducers	●			○ 75							●		○ 75		○ 4								○ 75				
Strain gages							●	○ 4				○ 24															
Strain gage transducers	●				●	●							●														
Thermoresistive detectors												●								○ 12							
Thermistors												●								○							
Hall-effect and magnetoresistive T's	● 34	● 32	○ 34	○ 10							○	○ 32	○ 10, 105								● 10, 3				○ 51, 34		
ChemFet transducers															○ 25		● 36, 25, 107, 37										
Wave-propagation transducers												○ 9													● 42, 51		
Longitudinal mode transducers												○ 9												● 47, 48	● 43, 51, 42		
Doppler transducers			●								● 52																
Polymer transducers					○ 70																	● 72, 73	○ 70	○ 70, 71	○		
SAW transducers	● 55				●	● 55							○ 57				○ 60, 61, 62								○ 42, 43, 51		
Tunneling displ. transducers	○ 68				○ 67											● 63, 67				● 68							
Thermodynamic transducers												● 4		● 4						○ 5							
Ionization transducers														●													
Photonic transducers	○ 7	● 94, 95													○ 8				● 5, 12	●							○ 12
Fiber optic (physical) transducers	○ 103	○		○ 106, 103	●	●	●		○ 103	○ 103	○	●	● 105								●	○ 102, 103				○ 103	
Fiber optic (chemical) transducers																	○ 108, 103	● 107, 108, 110									

NOTES: Open circles indicate transducers not covered in this chapter. Numerals indicate reference numbers in Sec. 5.17.

Loading effects and distortion	Error or accuracy
Frequency response	Calibration
Electrical output format	Environment

5.5 Capacitive Transducers

The change in capacitance in response to a measurand has many applications in physical transducers. Displacements, velocity, acceleration, force, pressure, vacuum, flow, fluid level, audio sound field, and relative humidity can be measured using capacitive transducers.

The capacitance between parallel conducting plates with a dielectric material between them is given by

$$C_s = \frac{\varepsilon_0 \varepsilon_p A(n-1)}{d} \tag{5.1}$$

where ε_0 = permittivity of free space
= 8.854×10^{-12} farad/m
ε_p = relative permittivity or dielectric constant of the material, the value usually tabulated in handbooks ($\varepsilon_p = 1$ for air)
A = area that is common to both plates (overlap area, m^2)
d = separation between plates, m
n = number of plates
C_s = capacitance, farads

Equation 5.1 indicates that the capacitance varies linearly with the area A and the dielectric constant of the material, but it varies inversely with the separation between plates. Any changes in the above-mentioned parameters caused by a measurand, and taken one at a time, provide practical transduction mechanisms.

Figure 5.2 shows some of the configurations where the changes in C_s are used to measure physical measurands. The first two lend themselves to the measurement of displacement, force, flow, vacuum, and pressure, and the third configuration could be used to measure the changes in the dielectric constant brought about by the absorption of moisture[13] or a chemical reaction with the dielectric material.

Figure 5.3 shows the diametrical cross section of a pressure transducer constructed totally from fused quartz, which has a very small temperature coefficient of expansion. The transducer consists of a circular diaphragm rigidly clamped by brazing it to a fused quartz body. A very shallow cavity has been sputter etched in the body to provide the separation Z_0 between the capacitor plates. This cavity is vented to the atmosphere by a small hole in the body of the structure, allowing the transducer to measure gage pressure.

The diaphragm has an annular electrode of metalized chrome and gold on

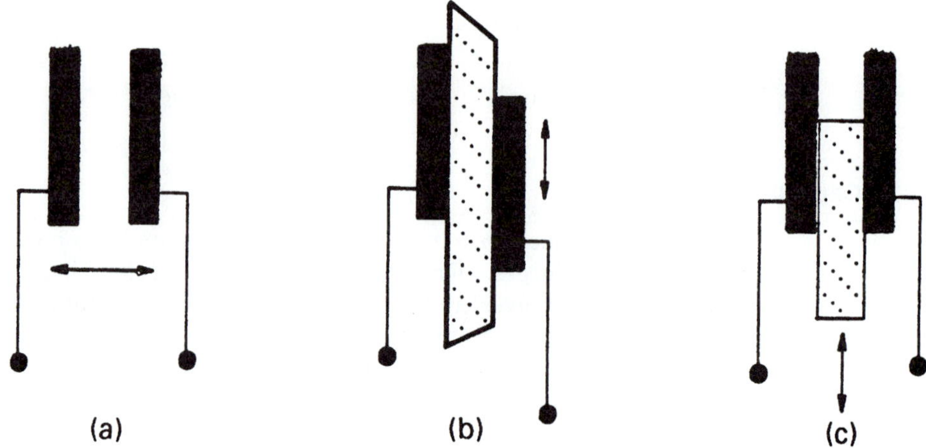

Figure 5.2 Capacitive displacement transducers. When used with a force summing device, they can provide a measure of several physical measurands.

the inside face, and a common electrode is deposited on the bottom of the etched cavity. The capacitance of this transducer as a function of the applied pressure is given by[15]

$$C_R = \frac{8.85\pi K_2}{2Z_0} \log_e \left\{ \frac{K_2 - (a^2 - b_2^2)[K_2 + (a^2 - b_1^2)]}{K_2 + (a^2 - b_2^2)[K_2 - (a^2 - b_1^2)]} \right\} \tag{5.2}$$

where

$$K_2 = \left\{ \frac{Z_0 h^3}{3/16[(1-\mu^2)/E]P} \right\}^{1/2}$$

and C_R = capacitance, pf
μ = Poisson's ratio (0.17 for fused quartz)
E = Young's modulus (745×10^3 kg/cm^3 for fused quartz)
a = radius of etched cavity, m
h = thickness of diaphragm, m
Z_0 = depth of etched cavity, m
P = applied pressure, kg/cm^2

The units of b_1 and b_2 are meters. Using Eq. 5.2, the capacitance of a circular electrode transducer in the center of the diaphragm can be obtained by setting b_1 equal to 0 and b_2 equal to the desired radius. This construction is

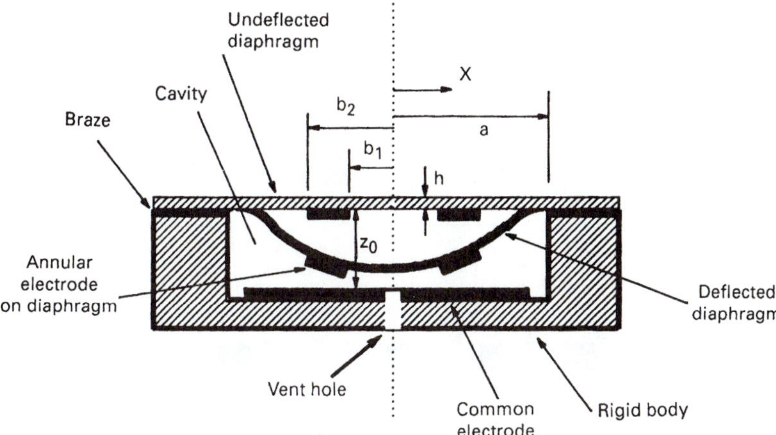

Figure 5.3 Capacitive pressure transducer. A complete fused quartz construction results in a rugged transducer with excellent thermal and elastic stability. (*From Dias et al.*[15] *Reprinted by permission. Copyright © 1980, Instrument Society of America from "ISATransactions," vol. 19, no. 3*)

currently used in an invasive blood pressure transducer[15,16] shown in Fig. 5.4. It has a circular sensing electrode in the center of the diaphragm, an annular reference electrode very close to the clamped edge of the diaphragm, and a full electrode at the bottom of the etched cavity. The reference capacitor changes very little with pressure.

A monolithic capacitor-type accelerometer made from silicon by microma-

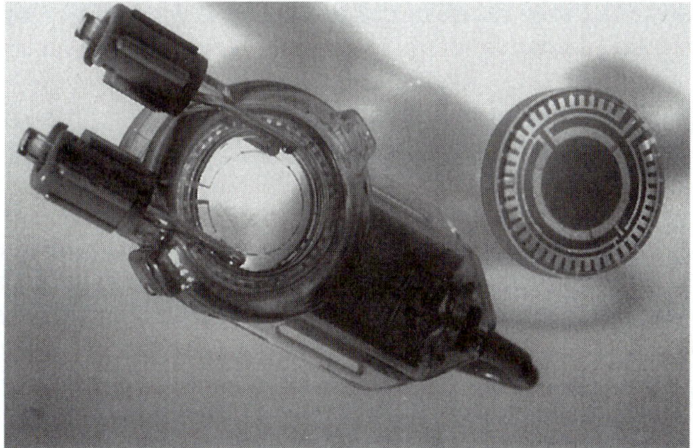

Figure 5.4 Capacitive blood pressure transducer. Measures a full-scale pressure of 300 mm Hg. The thermal drift is 0.01 percent FS/°C. Linearity and hysteresis is less than 0.5 percent FS. (*Courtesy of Hewlett-Packard Company, Palo Alto, Calif.*)

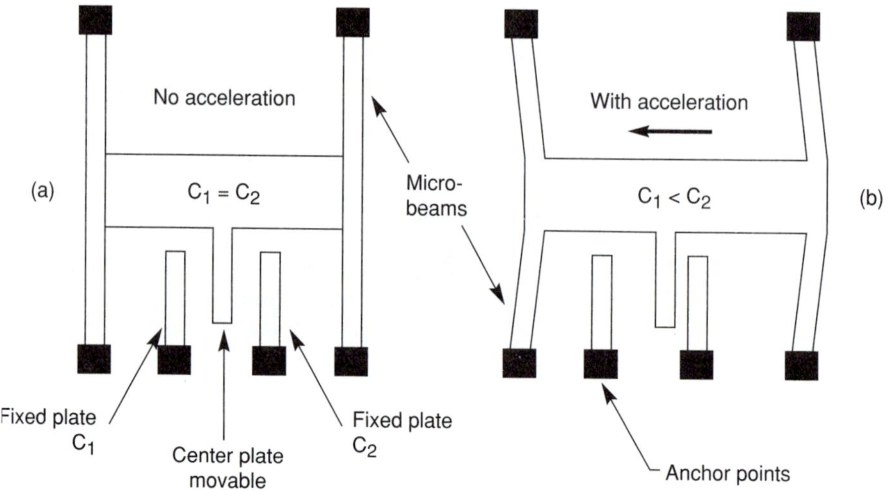

Figure 5.5 Capacitive micromachined accelerometer. (*From Ref. 17. Courtesy of Analog Devices, Norwood, Mass.*)

chining techniques[17] is shown in Fig. 5.5*a*. It is made up of several differential capacitors, and each capacitor section consists of two fixed outer plates and a center plate that is movable. Fig. 5.5*b* shows the deflected position of the center plate when the transducer experiences an acceleration. The deflection, and consequently the capacitance change, is proportional to the acceleration.

In many applications, capacitive transducers are connected in an ac bridge circuit to obtain an electrical output proportional to the measurand. In others it can be made a part of an *LC* oscillator, where the oscillator frequency is proportional to the magnitude of the measurand. When differential sensing is used, the sensitivity can be doubled and the temperature sensitivity reduced.

5.6 Inductive Transducers

In these transducers, the transduction mechanism is one where the self-inductance of a single coil or the mutual inductance between two coils is changed by a measurand. In general, the measurand could be a linear or rotary displacement, pressure, force, torque, vibration velocity, and acceleration. The inductance changes are brought about by the movement of a concentric magnetic core.

The inductance of a single coil increases as the core is inserted in the coil and reaches a maximum value when it is centered on the coil length. Similarly, two separate coils L_1 and L_2 wound on the same bobbin can also be used as a displacement transducer. Any measurand that moves the core directly through a summing device will produce a change in the impedance of the coils that is proportional to the magnitude of the measurand. The coils can

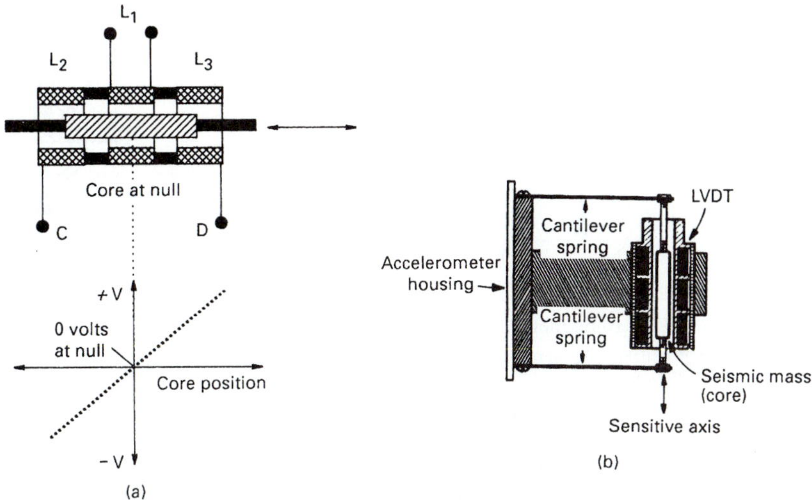

Figure 5.6 Inductive transducers. (*a*) Linear variable differential transformer (LVDT) measures linear displacements. With suitable summing devices it will measure force, load, pressure, and torque. (*b*) LVDT accelerometer. The core acts as the seismic mass and the leaf springs provide the restoring force. This is an open-loop accelerometer. (*From Ref. 19. Courtesy of Lucas Schaevitz Engineering, Camden, New Jersey.*)

be used as the adjacent arms of an impedance bridge. Since L_1 increases by the same amount that L_2 decreases, or vice versa, the bridge output will be doubled.

A variation of the inductive transducer, shown schematically in Fig. 5.6*a*, is known as the linear variable differential transformer (LVDT).[18,19] This transducer consists of a primary coil L_1, two interconnected coils L_2, L_3, and a common magnetic core. The coils are wound on a hollow nonmagnetic glass-filled nylon tube and the core slides coaxially inside the tube. The excitation frequency for coil L_1 ranges from 1 to 10 kHz. Coils L_2 and L_3 are wound in phase opposition in a way that the voltages induced in them by coil L_1 are 180° out of phase. Consequently, the voltage at terminals *c-d* is zero when the core is centered inside the tube between coils L_2 and L_3. When the core is moved away from the null position, the voltage at terminals *c-d* changes in amplitude and phase (polarity). This change, when brought about by a measurand, is proportional to the magnitude of the measurand. LVDTs are available in linear stroke lengths of ±1 to ±300 mm and sensitivities of 1.7 to 250 mV/V/mm, depending on the stroke length.

The same transduction mechanism is also used in a rotary variable differential transformer (RVDT) to measure angular displacements and torque.[20] To achieve good linearity, the angle of rotation is limited to ±40°. The LVDT can be used with Bourdon tubes, bellows, and proving rings to measure force and pressure.

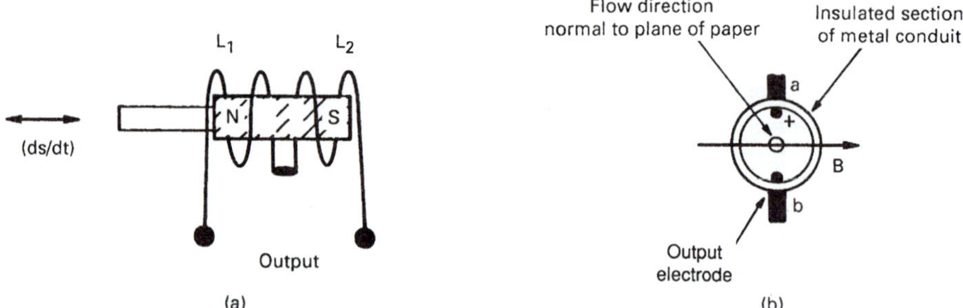

Figure 5.7 Electromagnetic Transducers. (*a*) Linear velocity transducer (LVT). (*b*) Flow velocity transducer.

5.7 Electromagnetic Transducers

When a moving conductor of length l or a single-turn coil of the same length moves with a velocity ds/dt across and perpendicular to the lines of magnetic flux of density B, an emf is generated in the conductor (coil) which is given by Faraday's law as $e = Bl\, ds/dt$. Now $l\, ds$ represents an area through which the flux lines cross during the time dt, and $Bl\, ds$ is the corresponding differential flux $d\phi$ through that area. The emf generated corresponding to N turns is

$$e = N\frac{d\phi}{dt} \tag{5.3}$$

Accordingly, the magnitude of the emf depends on the rate of crossing the lines of magnetic flux. This transduction mechanism is utilized in a velocity-measuring transducer. Figure 5.7*a* shows two coils L_1 and L_2, which are connected in phase opposition. The measurand is linked to the permanent magnet, which slides freely in the coils. The rate of movement of the magnet determines the velocity of the measurand. The output voltage is proportional to the velocity for all positions of the magnet. These transducers are known as linear velocity transducers (LVTs).[19]

Another application of this transduction mechanism is in sensing the flow velocity V of an electrically conducting fluid as shown in Fig. 5.7*b*. The flow is into the plane of the paper. The magnetic field is normal to the flow. The emf is generated along the diameter *a-b* normal to the flow and the magnetic field B. The voltage across the electrodes inserted into the flow at a and b is proportional to the flow velocity.

5.8 Resistive Transducers

Resistive transducers have many and varied applications in the transduction of measurands such as displacements, mechanical strain, pressure, force and load, temperature, and fluid velocity into electrical outputs. The transduction

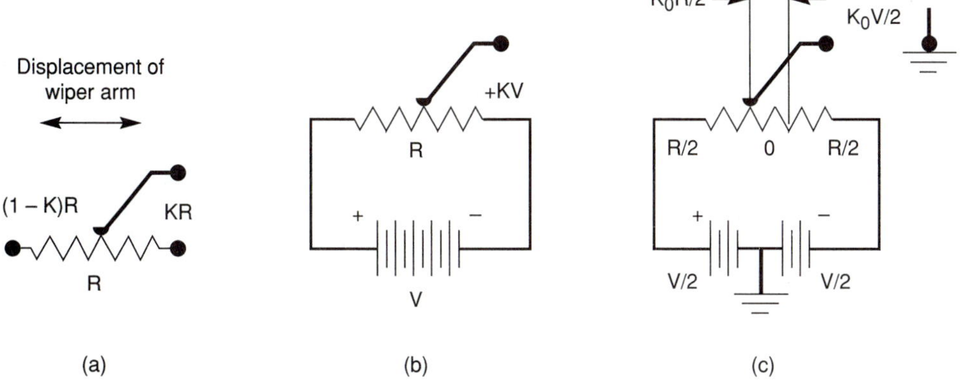

Figure 5.8 Potentiometric displacement transducers. (*a*) Resistance proportional to displacement or position. (*b*) Voltage proportional to the same measurands. (*c*) Displacement is measured around a null position and the output voltage is $\pm K_oV_o$. K_o is referenced to the center of the resistor.

mechanisms are based on the change in resistance brought about by the measurands.

5.8.1 Potentiometric transducers

A potentiometric transducer is a mechanically driven variable resistor. It consists of a wire-wound fixed resistor and a wiper arm that slides over it and in so doing taps a different segment of the resistor, as shown diagrammatically in Fig. 5.8*a* and *b*, where K represents a fraction of the resistor that is tapped. The displacement to be measured is linked by a shaft to the wiper arm, and a measure of the displacement is the fractional resistance KR or the fractional voltage KV. This is the transduction mechanism.

The resolution that one can achieve with this transducer depends on the gage of the nickel alloy or platinum wire used. For extremely fine resolution, the wire is replaced by a metallized ceramic or a film resistor. If the resistance wire is wound on a doughnut-shaped tube, the wiper will measure angular displacements. The output voltage corresponding to a displacement, force, or pressure is a fraction of the external voltage V, and therefore it does not need any amplification to activate external circuitry.

5.8.2 Resistance strain gages

Carbon granules, packed in a small volume in the shape of a button and connected in series with a voltage source and a load resistor, have been used in the past as microphones. Using this transduction mechanism, carbon-strip strain gages were developed in the early 1930s. These were, in turn, followed by unbonded and bonded wire strain gages, foil strain gages, and semiconductor strain gages.[21,24]

Wire strain gages. The resistance of a metallic wire is a function of the material resistivity ρ, its length L, and its cross-sectional area A and is given by

$$R = \rho \frac{L}{A} \tag{5.4}$$

When the wire is stretched, its length increases by ΔL, its diameter decreases by Δd, and its resistance increases by ΔR, assuming that the resistivity remains constant. Substituting these variations in Eq. 5.4 and neglecting higher-order terms, the following expression is obtained:[21–25]

$$\frac{\Delta R}{R} = \frac{(1 + 2\mu)\Delta L}{L} \tag{5.5}$$

where μ is Poisson's ratio, i.e., the ratio of transverse strain to the axial strain. The strain sensitivity is defined as

$$\frac{\Delta R/R}{\Delta L/L} = 1 + 2\mu \tag{5.6}$$

In unbonded strain gages, as the name implies, the resistance wire is strung between insulated posts. One post is attached to a fixed frame and the other to a frame constrained to move in a fixed direction. Movements that place the wire in tension are measured by the corresponding changes in resistance. This arrangement finds use in certain force and acceleration transducers.

In bonded strain gages, a meandering grid of fine resistance wire is sandwiched between two thin layers of paper and is then impregnated with a resin to provide the necessary strength. The gage is then bonded to the structural members for the determination of the strain at the desired location. The strain sensitivity of an "encapsulated and bonded" gage is defined by the gage factor

$$GF = \frac{\Delta R/R}{\Delta L/L} \tag{5.7}$$

where ΔR = resistance change, Ω
R = initial resistance of the gage, Ω
ΔL = change in length, mm
L = initial length of grid, mm

Wire strain gages are rapidly being replaced by foil strain gages, which can be mass-produced using standard photolithographic techniques.

Foil strain gages. Foil gages are fabricated from constantan or nickel-chromium alloy[23] sheet material that is reduced to 0.0025 to 0.0005 mm in thick-

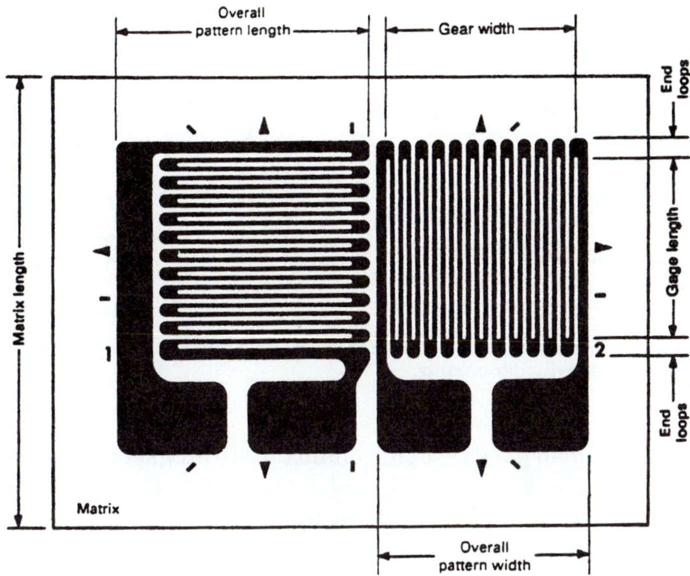

Figure 5.9 A two-element 90° rosette-type foil gage. Many gage configurations are used in stress analysis of mechanical structures and in strain gage transducers.[21–23] (*Courtesy of the Micro-Measurements Division of Measurements Group, Inc., Raleigh, North Carolina.*)

ness. This foil is then laminated to a backing material and coated with photoresist. Using a multiple-image negative of a gage, the composite gage pattern is transferred to the photoresist on the foil using photolithographic techniques. After developing and chemical etching processes, the gages are completely defined and can be isolated for lead-wire attachment, encapsulation, and establishing the gage factor for the batch of gages. The gage factor for foil gages is also defined by Eq. 5.7.

Figure 5.9 shows a gage pattern of a two-element 90° rosette.[23] The axis of the gage on the right is aligned with the principal axis. The gage on the left is automatically aligned to measure the transverse strain due to Poisson's ratio. The gage length is an important parameter in gage selection, as it is the effective length, excluding the end loops, over which the transduction takes place. The grid width together with the gage length determines the area of the strain field. For example, when measuring strains in a reinforced-concrete beam in the presence of aggregate and cement-sand mix, one is interested in the average strain, and for this reason long-gage-length gages should be used.

Transverse sensitivity is exhibited by foil gages owing to a strain field normal to the principal axial field. Since the grid lines of a foil gage are wide, the transverse strain couple through the backing into the grid, and this causes a change in the resistance of the grid.

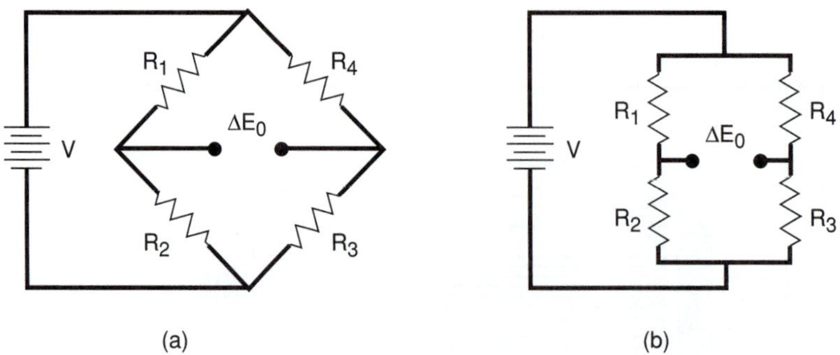

Figure 5.10 (*a*) Wheatstone bridge circuit. (*b*) The Wheatstone bridge circuit redrawn as two voltage-divider circuits.

Semiconductor strain gages. Semiconductor strain gages operate on the transduction mechanism known as piezoresistive effect. It is defined as the change in electrical resistivity brought about by an elastic strain field. In some semiconductors the piezoresistive effect is quite large. Both *p*-type and *n*-type silicon are used in the fabrication of these gages. When a semiconductor gage is strained, the distribution of the number of charge carriers and their mobility changes and consequently the resistivity changes. Semiconductor gages also have a gage factor as their figure of merit,[24] which is given by

$$GF = \frac{\Delta R/R}{\Delta L/L} = 1 + 2\mu + \pi_{\mathrm{L}} Y \tag{5.8}$$

where μ = Poisson's ratio
π_L = longitudinal piezoresistive coefficient
Y = Young's modulus

The first two terms of Eq. 5.8 correspond to dimensional changes similar to wire and foil gages, but the third term is due to piezoresistivity. Also, π_L is larger than $(1 + 2\mu)$ by a factor of about 100. The magnitude of the gage factor also depends on the direction along which the stress is applied to the semiconductor.[25] Silicon diaphragms with diffused strain gages are used in miniature pressure transducers.[24,25]

Basic measuring circuits. The Wheatstone bridge circuit shown in Fig. 5.10*a* can be redrawn as in Fig. 5.10*b*, which shows that the full bridge is made up of two voltage-divider circuits. These circuits are commonly used for static and dynamic strain measurements.

The strain-induced incremental output voltage ΔE_0, corresponding to changes in resistance of the four arms, is given by Ref. 21 as

$$\Delta E_0 = \frac{V_a}{(1-a)^2}\left(\frac{\Delta R_1}{R_1} - \frac{\Delta R_2}{R_2} + \frac{\Delta R_3}{R_3} - \frac{\Delta R_4}{R_4}\right)(1-n) \tag{5.9}$$

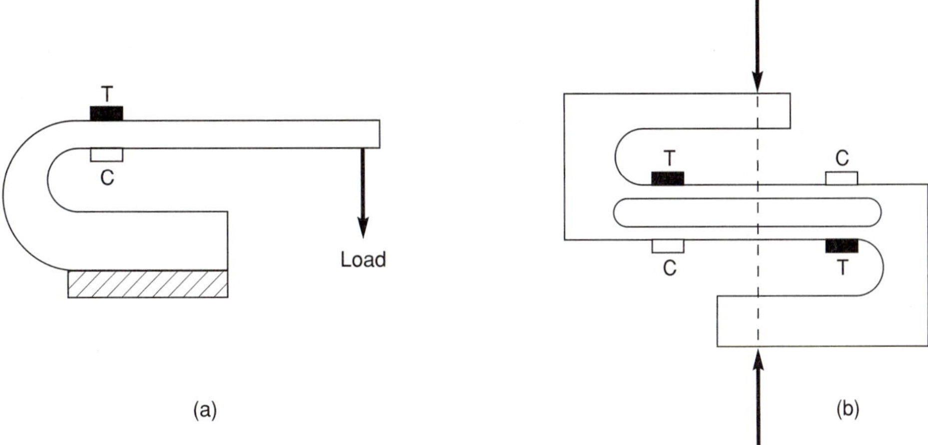

Figure 5.11 (*a*) A simple cantilever-type summing device. Any displacement of the free end of the cantilever induces equal but opposite strains at locations *T* and *C*. When strain gages are bonded at these locations and connected to a Wheatstone bridge, its output will determine the value of the load placed at the free end. (*b*) A complex summing device. Equal and opposite strains are induced at two locations. A complete Wheatstone bridge can be formed by using strain gages bonded at locations *T* and *C* and used in the opposite arms of the bridge. The output of the bridge is four times that of a single active gage, and it is fully temperature compensated. (*Redrawn with permission from Strain Gage Users Handbook, Elsevier, Bethel, Conn.*)

where $R_1/R_2 = a = R_4/R_3$ and n is a nonlinearity factor,[21] which can be neglected for wire and foil gages.

For an equal-arm bridge, $a = 1$. In this bridge, if only R_1 is an active gage and R_2, R_3, and R_4 are fixed resistances not affected by the strain field, then

$$\Delta E_o = \frac{1}{4}\left(V \times GF \times \frac{\Delta L}{L}\right) \tag{5.10}$$

since

$$\frac{\Delta R_1}{R_1} = GF \times \frac{\Delta L}{L}$$

If, instead, R_3 and R_4 are not affected by any strain field, then ΔR_3 and ΔR_4 are each equal to zero. In this case, the bridge circuit reduces to the voltage-divider circuit, where R_3 in series with R_4 is across the voltage supply V and

$$\Delta E_o = \frac{V}{4}\left(\frac{\Delta R_1}{R_1} - \frac{\Delta R_2}{R_2}\right)(1 - n) \tag{5.11}$$

When $R_1 = R_2$ and if these gages sense equal but opposite strains, as in the case of a cantilever beam, shown in Fig. 5.11*a*, the incremental bridge output ΔE_0 will double since $\Delta R_2 = -\Delta R_1$. For this situation, n becomes zero.

Similarly, if the four arms have identical resistances, and R_1, R_2 and R_3, R_4 sense equal but opposite strain fields, as in the case of a shaft in torsion or a load cell shown in Fig. 5.11*b*, the incremental output will quadruple and the nonlinearity term will again be zero.[21,23]

$$\Delta E_o = \left(V \times GF \times \frac{\Delta L}{L} \right) \tag{5.12}$$

5.8.3 Physical strain gage transducers

Strain gages are used in physical transducers to measure the strain induced in a summing device by a displacement, force and load, pressure, or torque. Figure 5.11*a* shows a simple cantilever.

Figure 5.11*b* shows a type of spring element used in load cells, that is, transducers for measuring loads. In each of these, the bending of a composite beam is utilized to obtain strains of opposite signs designated as T for tensile and C for compressive. When strain gages are fixed at those locations, the bridge output could be doubled or quadrupled if two or four active gages are used. Semiconductor gages can also be used with various types of spring elements to measure displacements, pressure, temperature, and force.[24]

5.8.4 Thermoresistive detectors

Metals and semiconductors experience an increase in their electrical resistivity when heated, and consequently their resistance increases with temperature. This transduction mechanism, in metals like nickel, nichrome, tungsten, copper, and platinum, is used in resistance temperature detectors (RTDs).[26] Platinum resistance temperature detectors (PRTDs) yield a reproducible resistance-temperature relationship, and their resistance varies quite linearly with temperature.

The relationship between resistance and temperature for a platinum wire RTD is given by the Callender-Van Dusen equation[1,4]

$$R_t = R_0 \left\{ 1 + \alpha \left[t + \delta \left(1 - \frac{t}{100} \right) \frac{t}{100} + \beta \left(1 - \frac{t}{100} \right) \left(\frac{t}{100} \right)^3 \right] \right\} \tag{5.13}$$

where R_t = resistance at some temperature t
R_0 = ice point resistance (0.01°C or 273.16 K)
α = temperature coefficient of resistance near 0°C
β = Van Dusen constant
t = temperature, degrees Celsius

Typical values are $\alpha = 0.003926$, $\delta = 1.491$, and $\beta = 0$ when $t > 0$ and 0.1103 when $t < 0$. The exact values of α, δ, β, and R_0 are obtained by measuring the resistance of the detector at four temperatures including $t = 0$°C and solving the resulting equations.

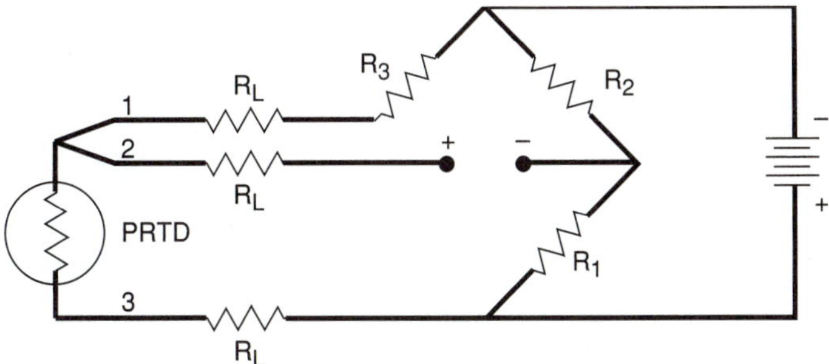

Figure 5.12 A PRTD in a three-wire configuration. The effect of the R_L's is canceled since they are in the opposite arms of the bridge, in series with R_3 and the PRTD. In a four-wire arrangement (not shown), the PRTD is powered by a current source and the voltage drop across it is measured with the remaining two wires.

The PRTD assembly is composed of a resistance element made from 99.9 percent pure platinum wire, a sheath that encloses the element, and lead wires that connect the element to the external measuring circuitry. RTDs are also made from platinum metal film with a laser trimming system and are bonded directly to the surface under test. Owing to the intimate thermal contact, the self-heating is minimal and therefore they can be operated at a higher excitation voltage.

In most measuring circuits, the PRTD forms an arm of a dc Wheatstone bridge circuit. Since the RTD is connected by wires to the bridge, there is a need to compensate for the wire resistance R_L. For this reason, the RTD is supplied in a three-wire (as shown in Fig. 5.12) and a four-wire configuration.

5.8.5 Thermistors

The thermistor is a thermally sensitive resistor, but unlike the RTD, it exhibits a correspondingly large change in resistance. Its resistance, in general, decreases as the temperature increases. Thermistors are made by sintering a combination of oxides into plain beads or beads in a glass rod.[27] Oxides of manganese, nickel, cobalt, copper, iron, and titanium are commonly used.

The resistance vs. temperature relationship for a thermistor can be represented by a third-degree polynomial,[27,28] as shown below.

$$\log_e R_T = a_0 + \frac{a_1}{T} + \frac{a_2}{T^2} + \frac{a_3}{T^3} \tag{5.14}$$

where R_T = resistance, Ω, at T/K

a_0, a_1, a_2, and a_3 = unique numerical constants

T = (°C + 273.15)K

Since there are four unknowns, at least four calibration points are required to solve four simultaneous equations to obtain the values of the constants.

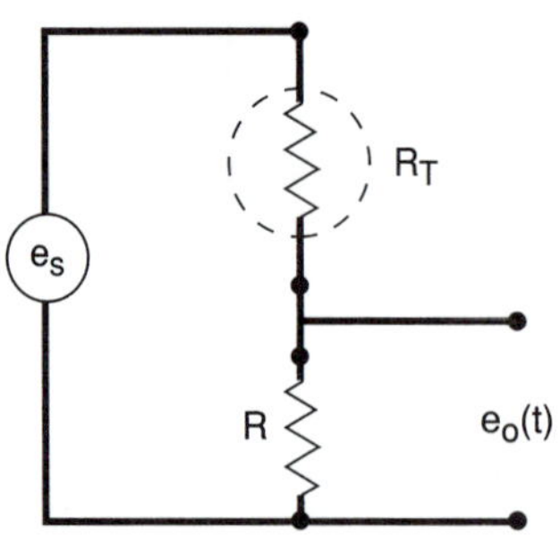

Figure 5.13 A voltage-divider circuit for a thermistor.

Figure 5.14 Hot-wire anemometer. (*Courtesy of Dantec Electronics, Inc., Mahwah, N.J.*)

The Wheatstone bridge circuit configurations used with strain gages and RTDs are also used with thermistors. In many cases, a single voltage-divider circuit is also used as shown in Fig. 5.13. This circuit has the advantage of providing an output voltage $e_o(t)$ that increases as the temperature increases,

$$e_o(t) = e_s\left(\frac{R}{R + R_T}\right) \tag{5.15}$$

By selecting an appropriate value of R, it is possible to operate in the quasilinear region of the e_o vs. t plot.[27]

5.8.6 Hot-wire anemometer

The hot-wire (h-w) or hot-film (h-f) anemometer is a thermoresistive transducer used in the microstructure analysis of gas and liquid flows. Consequently it is useful in the study of flow problems related to the design of airplane wings, propellers, ventilation systems, and blood-velocity measurements for medical research.[30] The anemometers have a very small sensing element, which accounts for their high spatial resolution and fast response.

The transduction mechanism in these transducers is the change in resistance of the sensor element, brought about by the convective heat lost because of the velocity of the fluid. The changes in resistance are caused by the thermoresistive effect and are measured with a Wheatstone bridge circuit. The output of the bridge can be adequately expressed by King's law,[29]

$$e_o^2 = A + BV^{0.5} + CV \tag{5.16}$$

where e_o is the output voltage of the bridge setup, V is the velocity of the fluid, and A, B, and C are constants.

These transducers consist of a thermoresistive sensing element suspended between two support prongs, a transducer handle, and electrical connections to the support prongs. The h-w transducer most commonly consists of a 5-μm-diameter platinum-plated tungsten wire as shown in Fig. 5.14.[30] The hot-film sensing element consists of a thin nickel film deposited on a 70-μm-diameter quartz fiber, which is suspended between the two prongs. Typical sensor element resistance for the h-w transducer is 3.5 Ω, and the velocity measuring range is from 0.2 to 500 m/s.

5.9 Hall-Effect Transducers

The material used in the manufacture of Hall-effect devices is a p-type or an n-type semiconductor. Typical examples are indium arsenide, indium arsenide phosphide, and doped silicon. Figure 5.15 shows a section of a p-doped semiconductor subjected to a magnetic field B_z in the z direction and an electric field E_x in the x direction. A current I_x flows in the x direction.

The holes move in the x direction across the magnetic field and experience an upward magnetic force, which results in the accumulation of holes on the top surface and electrons on the bottom face, as indicated in Fig. 5.15. An electric field E_y, known as the Hall field, is set up as a consequence, and this is known as the Hall effect.[31] The corresponding Hall voltage $V_H = E_y t$. Since there is no flow of current in the y direction, the magnetic force equilibrates with the electric force that is exerted on the holes, and as a result the Hall voltage can be expressed as

$$V_H = \frac{R_H(I_x B_z)}{h} \tag{5.17}$$

where R_H is the Hall coefficient. R_H is negative for an n-type semiconductor. h is the dimension parallel in B_Z. Equation 5.17 represents the transduction mechanism.

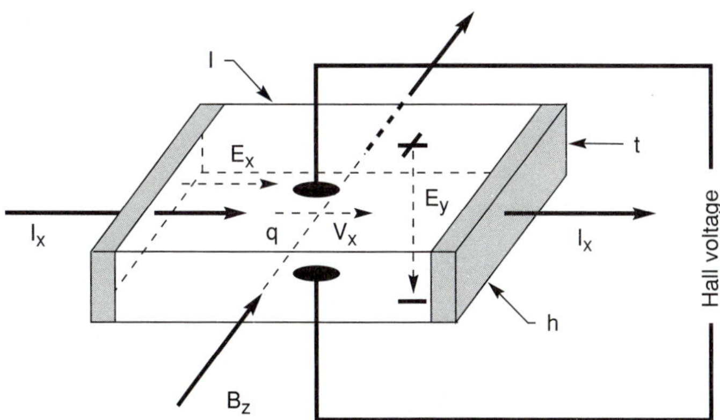

Figure 5.15 Hall-effect device. The Hall-effect transducer (HET) consists of the semiconductor device, a differential amplifier, and a voltage regulator to maintains I_x constant.

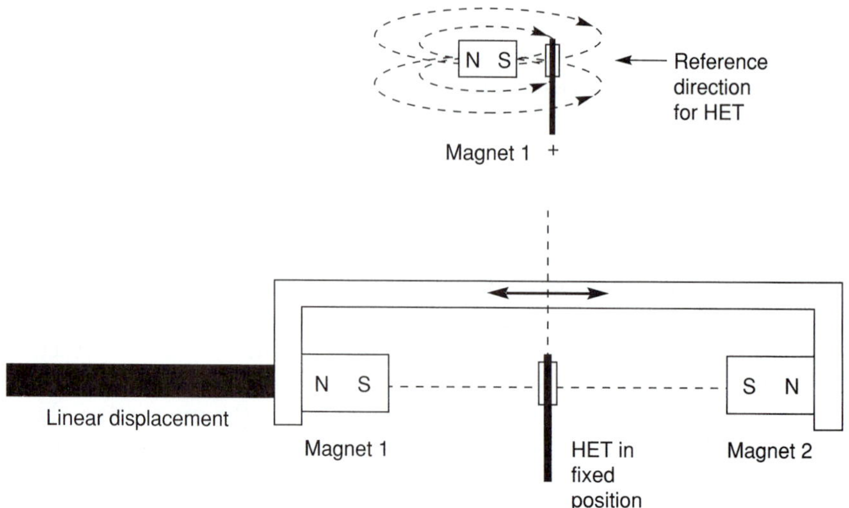

Figure 5.16 Hall-effect displacement transducer. The insert shows the magnetic field reference direction for the HET. When the flux lines enter the transducer in the reference direction, the output voltage is positive. It is negative when the lines enter against the reference direction. (*From Ref. 32. Courtesy of Micro Switch, a Division of Honeywell, Inc., Freeport, Ill.*)

Figure 5.16 shows the HET being used to measure small displacements. If the displacement brings magnet 1 close to the transducer, the output voltage will be increasingly positive, and it will be incresingly negative if magnet 2 moves closer to the HET.[32]

The magnetoresistance effect is closely associated with Hall-effect transducers.[33] If the length l of the device in Fig. 5.15 is made much shorter than its width h, the Hall voltage can be almost short-circuited. As a consequence, the charge carriers move at the Hall angle to the x direction. The increase in path length causes an increase in resistance of the device, and this is known as the geometrical magnetoresistance effect.[33] Transducers for measuring angular velocity of ferrous gear wheels have been developed based on this effect.[10,34]

5.10 Chemfet Transducers

The operation of the chemical field-effect transistor (ChemFET) transducer is similar to the operation of a standard metal-oxide-semiconductor field-effect transistor (MOSFET). ChemFETs are miniature transducers that are used for the measurement of the concentration of certain gases and for the determination of hydrogen ion concentrations (pH).

Figure 5.17 shows a cross section of a conventional MOSFET device.[35] It consists of a source, a drain, and a channel that enables the flow of current I_D from the source to the drain. An aluminum gate electrode on the oxide

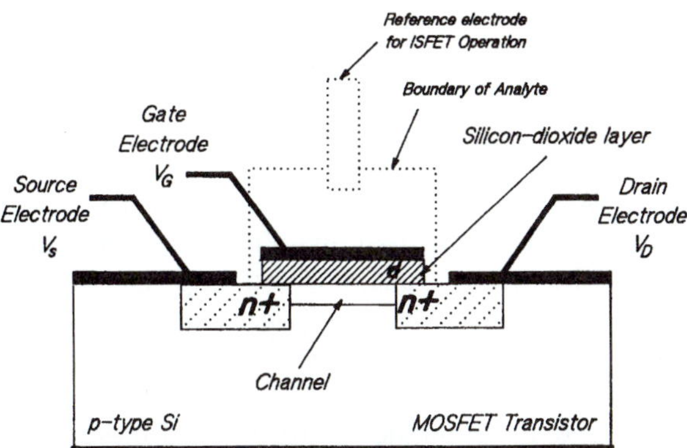

Figure 5.17 Metal-oxide-semiconductor field-effect transistor (MOSFET) used as a ChemFET transducer. In ChemFET transducers the conventional aluminum gate electrode is not used and the region above the silicon dioxide layer is exposed to the gas or liquid to be analyzed. For gas analysis, metals like palladium and platinum are used as gate electrodes. For liquids, the SiO_2 layer is exposed directly, without any gate electrode.

layer d above the channel modulates I_D as a function of the gate voltage V_G. The V_D vs. I_D characteristics of a MOSFET as a function of V_G display a linear region,[35] and for small values of V_D and I_D, the channel behaves as a resistor. The transduction mechanism in a MOSFET is basically to sense the charge on the gate electrode and then use that charge to modulate the flow of charges in the channel between the source and the drain.

The ChemFET uses a modification of the MOSFET transduction mechanism.[37] The palladium (Pd) gate MOSFET for hydrogen gas sensing and the ion-sensitive FET (ISFET) are two important examples of ChemFET transducers. In the Pd gate FET transducer the aluminum gate electrode is replaced by a Pd gate. Molecular hydrogen (measurand), in the air or by itself, is absorbed at the Pd surface, where it undergoes a catalytic dissociation into atomic hydrogen (H_a). The atomic hydrogen then diffuses through the bulk of the Pd electrode and forms a dipole layer at the Pd-SiO_2 interface.[36] The polarization caused by the dipoles modulates the channel current I_D in direct proportion to the hydrogen ion concentration. In another ChemFET, a 10-nm-thick platinum film evaporated on top of the Pd gate electrode enables the measurement of ammonia (NH_3) concentraton. If, instead, the gate electrode is a perforated film of platinum, the ChemFET will measure carbon monoxide.

The ISFET does not have a gate electrode and the SiO_2 gate dielectric is exposed directly to the aqueous solution or the analyte whose pH is to be determined.[37] For proper operation, an electrode is placed in the analyte and referenced to the bulk semiconductor. The transduction mechanism here is the formation of a charge at the analyte-oxide interface which is proportional to the pH of the analyte. This charge then modulates the channel current

I_D, and the electrical output of the ISFET is proportional to the pH of the analyte.

5.11 Piezoelectric Wave-Propagation Transducers

Piezoelectricity derives its name from the Greek word "piezein," to press. When a piezoelectric crystal is strained by an applied stress, an electric polarization is produced within the material which is proportional to the magnitude and sign of the strain—this is the direct piezoelectric effect. The converse effect takes place when a polarizing electric field produces an elastic strain in the same material.

Typical materials used in transducers are crystalline quartz, lithium niobate, several compositions of ferroelectric ceramics, ferroelectric polymers,[39] and evaporated or sputtered films of cadmium sulfide and zinc oxide. The ferroelectric ceramics commonly used are PZT4, PZT5A, PZT5H, and PZT7. The elastic, dielectric, and piezoelectric properties of these materials can be found in Refs. 38, 39, and 40. In addition, using these PZT materials, new and more efficient piezocomposites have been developed.[45]

Ferroelectric ceramics are not piezoelectric when they are manufactured and therefore have to be poled. Poling is done by applying a high dc voltage to the electrode faces normal to the thickness direction, while maintaining the ceramic in a high-temperature environment.[38] Ceramic transducers are generally rectangular, circular with parallel faces, or have a spherical curvature.

The conversion efficiency of a piezoelectric transducer is defined by the electromechanical coupling factor K.

$$K^2 = \frac{\text{mechanical energy converted to electrical energy}}{\text{input mechanical energy}}$$

for the direct piezoelectric effect, and

$$K^2 = \frac{\text{electrical energy converted to mechanical energy}}{\text{input electrical energy}}$$

for the converse piezoelectric effect.

K^2 is always less than unity. Typical values of K are 0.1 for quartz, 0.5 to 0.7 for ferroelectric ceramics, and 0.2 to 0.3 for copolymer materials like P(VDF-TrFE).

Another quantity that is necessary in characterizing a transducer is its characteristic impedance Z_0 which is analogous to the characteristic impedance of an electrical transmission line

$$Z_0 = V_\rho$$

where V = velocity of the acoustic wave in the material
ρ = density of the material

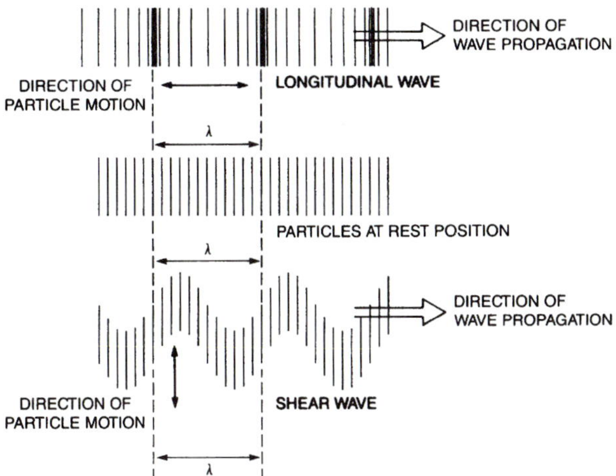

Figure 5.18 Longitudinal and shear waves. (*From Ref. 43. Coutresy of Panametrics, Inc., Waltham, Mass.*)

The unit of acoustic impedance is $kg/m^2 \cdot s$, and it is expressed in Rayls. A megaRayl (MRayl) is equal to 10^6 Rayls. Typical values of Z are 33.7 MRayl for quartz, 40.6 MRayl for brass, 1.5 MRayl for water, and 0.00043 MRayl for air.

The piezoelectric transducers that are most commonly used generate longitudinal and shear waves which propagate with velocities V_L and V_S, respectively. Figure 5.18 shows characteristics of these waves. The longitudinal mode is a compression wave, and its particle motion is in the direction of wave propagation. On the other hand, the particle motion in a shear wave is normal to the direction of wave propagation.

In some chemical, environmental, and physical transducers, surface acoustic waves (SAWs) are used. These waves travel on the surface, where most of their energy is confined. SAWs propagate with a velocity V_R that is less than V_L and V_S. If the medium on which they propagate is piezoelectric, then the electric field associated with the SAW can be used to detect the wave with interdigital transducers (IDTs).[41]

In wave-propagation transducers, maximum energy is transfered to another medium, when its impedance Z_M is equal to the impedance Z_0 of the transducer. For example, the energy could be efficiently transferred from a PZT5A longitudinal mode transducer into a brass rod, whereas the transfer into water would be quite inefficient.

To optimize that transfer, a quarter-wave matching layer is commonly used. The impedance Z_{ML} of the matching layer should be approximately $\sqrt{Z_0 Z_M}$ and its thickness should be a quarter wavelength, evaluated using V_L for that material. Several matching layers can be used for further optimiza-

TABLE 5.2 Acoustic and Piezoelectric Properties of Materials Used in Longitudinal and Shear Wave Transducers

Material	V_L, m/s	V_S, m/s	Density, 10^3 kg/m^3	Z_L, 10^6 kg/m·s	Z_S, 10^6 kg/m·s	Piezoelectric coupling coefficient K_t
PZT4	4600	2630	7.5	34.5	19.7	0.51
PZT5A	4350	2260	7.75	33.7	17.5	0.49
PZT5H	4560	2375	7.5	34.2	17.8	0.51
Lithium niobate ($LiNbO_3$)	7360 (36° Y-cut)	4440 (163° Y-cut)	4.64	34.2	20.6	0.49 (63° Y-cut) 0.55 (163° Y-cut)
Zinc oxide (ZnO)	6330 (Z-cut)	2720 (X-cut)	5.68	36.0	15.5	0.28 (Z-cut)
Cadmium sulfide (CdS)	4460 (Z-cut)	1760 (X-cut)	4.82	21.5	8.48	0.32 (X-cut)
Quartz X-cut	5740	3800	2.65	15.2	10.1	0.093
Copolymer film P(VDFTrFE)	2400	—	1.88	4.51	—	0.2–0.3 max.
Fused quartz	5960	3760	2.20	13.1	8.3	
Brass	4700	2100	8.64	40.6	18.1	
Hysol epoxy C-9-4183/3561	2160	—	3.27	7.04		
Water (body tissue)	1500	—	1.00	1.50		

tion.[41] Table 5.2 shows acoustic parameters of materials suitable for generating longitudinal and shear waves.

5.11.1 Longitudinal mode transducers

Figure 5.19 shows a piezoelectric transducer with two electroded faces normal to the thickness direction, which is also the poling direction. When a signal generator at f_0 is connected to these electrodes, longitudinal waves with a velocity V_L are generated within the material and, consequently, two ultrasonic waves propagate into the adjoining medium. If the adjoining medium is air ($Z = 0.00043$ MRayl), most of the energy is trapped as it would be in a resonator. Maximum excitation occurs when the thickness t is equal to one-half wavelength in the material according to the formula

$$t = \frac{V_L}{2f_0} \tag{5.18}$$

However, maximum energy will be transferred to the adjoining medium when its acoustic impedance Z_M equals the impedance Z_0 of the transducer. If not, an impedance matching layer can be interposed between them.

Conversely, when ultrasonic waves impinge on one or both faces, a voltage is developed across these electrodes. As before, the voltage is maximum when the transducer thickness satisfies Eq. 5.18.

The combination of piezoelectric transducers with matching layers and acoustic absorbers, interfacing with adjoining media at the acoustic ports, can be modeled by the Mason or KLM electrical equivalent circuits. These are

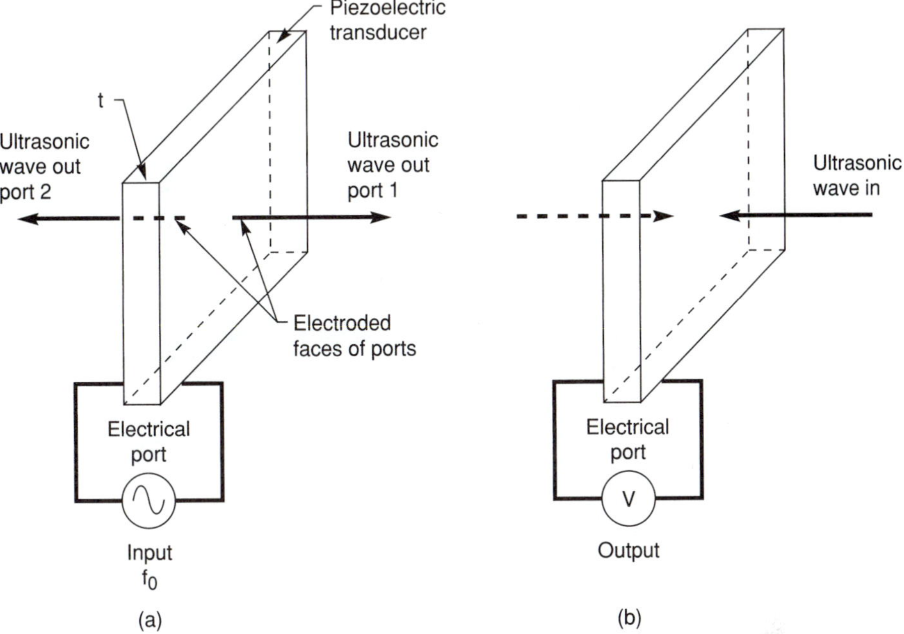

Figure 5.19 Longitudinal mode transducer, illustrating the piezoelectric effect. (*a*) Converse piezoelectric effect. (*b*) Direct piezoelectric effect.

useful in developing matching networks for maximum power transfer into and out of the electrical port.[41]

The shear wave-propagation transducers are quite similar to these transducers, except that the poling direction lies in the plane perpendicular to the propagation direction. The velocity of propagation V_S and the acoustic impedance Z_S is almost half of the value for longitudinal mode (V_L) transducers. These transducers are used mostly in nondestructive testing (NDT).

Single-element transducers. The transducers that we consider here are those connected with NDT of structural materials and medical ultrasound. In many ways the transducers used in these fields are quite similar, and there has been an explosion in their variety and availability in the last decade.[42,43] The frequencies employed in NDT and medical ultrasound cover the range from 100 kHz to 50 MHz.

A single-element transducer is shown in Fig. 5.20. The active element, a piezoelectric ceramic, is poled in the thickness direction. The backing has an impedance Z_B, and absorbs some or all of the energy from one face. The wear plate sometimes includes a quarter-wave matching layer to optimize the energy transfer of the propagating wave into the adjoining medium.

Figure 5.21 shows the on-axis (dashed line) and the transverse ultrasonic fields of a circular piston transducer. The on-axis acoustic pressure field is

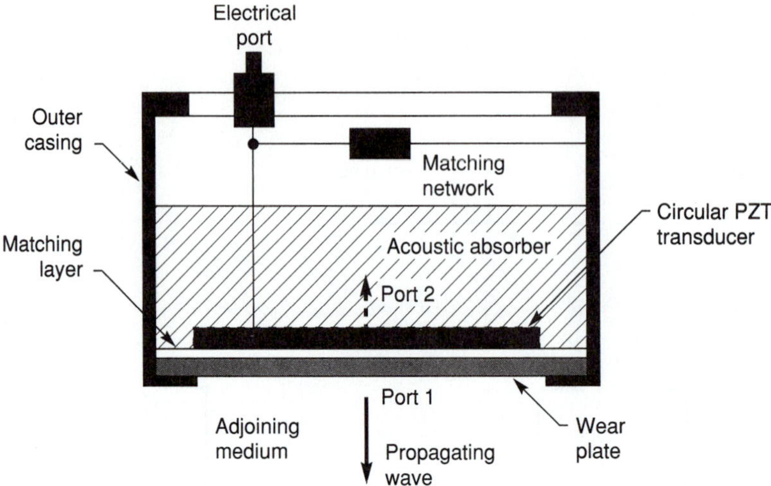

Figure 5.20 A single-element longitudinal mode transducer.

divided into two regions, the near field, or Fresnel region and the far field, or Fraunhofer region. The extent of the Fresnel region is indicated by N. In this region the field goes through a series of maxima and minima and ends with a last maximum, which is considered as the effective focus of the transducer.[44]

The distance N is given by

$$N = \left(\frac{D^2}{4\lambda} - \frac{\lambda}{4}\right) \tag{5.19}$$

with D = diameter of transducer element
λ = wavelength in propagation medium ($= V_M/f_0$)

where f_0 = excitation frequency
V_M = propagation velocity in the medium

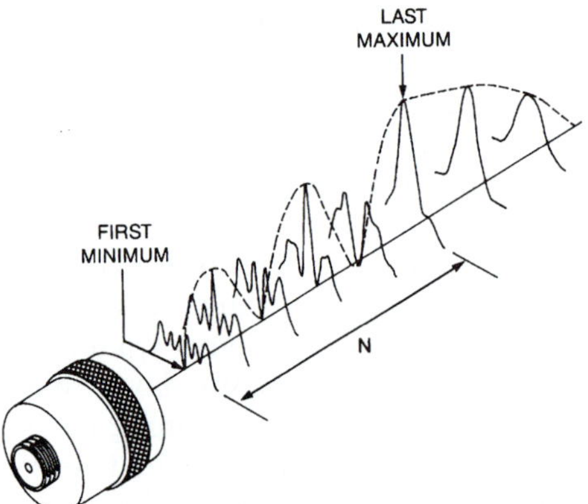

Figure 5.21 The on-axis acoustic pressure field of a single-element transducer (*From Ref. 43. Courtesy of Panametrics, Inc., Waltham, Mass.*)

In the transverse direction, the acoustic beam spreads as one moves away from the transducer face and its intensity drops. Focusing decreases the beam diameter and increases the intensity of the beam. Focusing can be achieved by bonding an acoustic lens to the PZT transducer or by using a spherically shaped PZT transducer. This increase augments the sensitivity of the transducer to locate small targets. In the pulse-echo mode, the −6-dB beam diameter at the effective focus is

$$\text{Beam diameter} = \frac{1.028FV_M}{f_0 D} \tag{5.20}$$

where F is the effective focal length in the medium.[44]

Single-element rectangular transducers can be assembled as a linear phased array transducer. In this configuration, the acoustic beam can be deflected in a sector scan format and focused by using delayed excitation signals.[41,42,46]

Mechanical imaging transducers. Mechanical imaging transducers are used to visualize and display the location of impedance discontinuities in a medium through which a short ultrasonic pulse is made to propagate.[47,48] This method of echo location is known as the pulse-echo technique.[49]

Figure 5.22 illustrates the fundamental basis for the pulse-echo technique. An ultrasonic transmit/receive (T/R) transducer is excited by an electrical pulse, and a corresponding stress pulse propagates into the medium. This stress pulse travels with a velocity V, and it encounters an acoustic impedance discontinuity at a distance d. At this discontinuity, some of the pulse energy is reflected back, and the remaining pulse energy propagates farther into the medium. The first reflected pulse travels back to the same transducer, and since the T/R transducer is bidirectional, it registers the arrival of that pulse by generating an electrical signal at its output. This represents the transduction mechanism in the pulse-echo mode. The total distance traveled by the initial pulse is $2d$. Assuming that the velocity V is constant, the total round-trip time t is $2d/V$. Now, t can be determined from an oscilloscope display, and consequently the distance d of the discontinuity from the transducer face is $Vt/2$. Other discontinuities can also be located in this manner.

Figure 5.23 shows the cross section of a blood vessel,[50] which could very well represent a cylindrical structure like a cladded metal pipe. In the center is a small single-element transducer which rotates when driven by a flexible shaft connected to a motor. In medical practice, the flexible shaft and transducer reside inside a catheter, which is a polyethylene tubing. The catheter is not shown, for clarity. The space around the transducer is filled with an acoustic coupling fluid. The catheter (Sonicath*) is inserted into the blood

---

* Sonicath is a trademark of Boston Scientific Corporation.

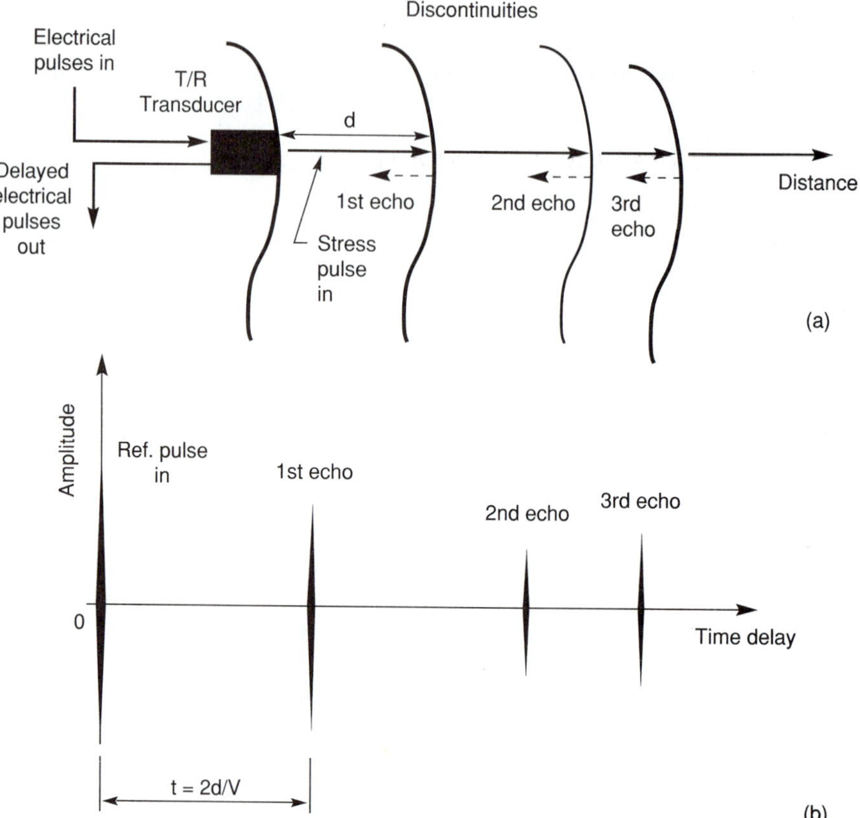

Figure 5.22 The fundamental basis of the pulse-echo technique used in NDT and medical imaging. Ultrasound is reflected whenever there is a change in the acoustic impedance of the medium.

vessel by making a small incision and is then advanced to the desired location. When the transducer is excited, the ultrasonic stress pulse propagates through the fluid, catheter wall, and surrounding blood to reach the linings of the artery wall including the intima, media, and adventitia.[47]

For a given direction, the transducer sends an ultrasonic pulse toward the artery walls and remains in that position long enough to receive all the echoes from the impedance discontinuities. The transducer then rotates, points in another direction, and receives all the echoes. The process is repeated for a 360° circular scan. All the echoes are stored in memory. The image processing circuitry then uses these signals to modulate the intensity of a CRT which displays the cross-sectional image of the blood vessel as shown in Fig. 5.23.

Angle-beam transducers. Angle-beam transducers are used in nondestructive evaluation of castings and riveted steel connections and in the inspection of

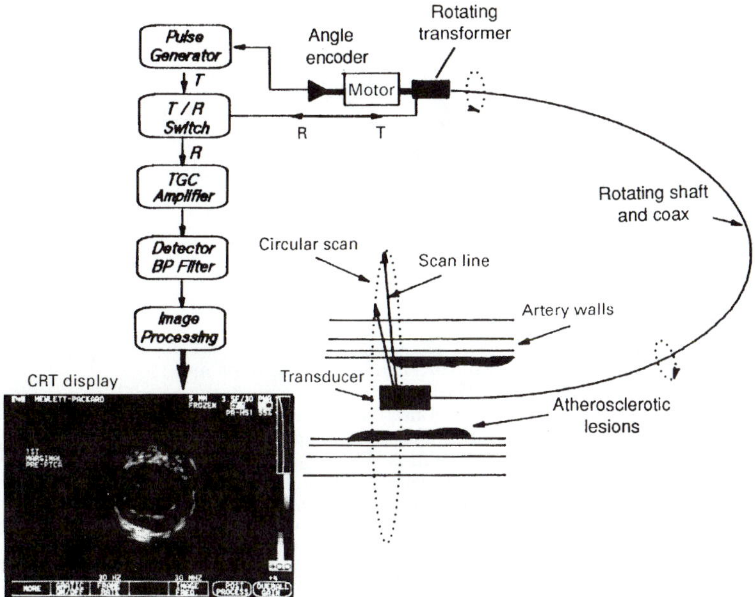

Figure 5.23 Intravascular imaging transducer and processing circuits used to image the inside wall of an artery. The displayed image shows in cross section the eccentric plaque buildup on the wall. (*Reproduced courtesy of Hewlett-Packard Co., Andover, Mass.*)

welded structural elements by the pulse-echo technique. This technique requires the ultrasonic beam to travel at a small angle to the surface of the structure.[42,43,51]

Angle-beam transducers are based on the principle that a longitudinal wave incident on a solid 1–solid 2 interface is mode converted into a refracted shear wave and a refracted longitudinal wave, propagating in solid 2 as shown in Fig. 5.24. The directions of the refracted waves are dictated by Snell's law.[51] These waves are used to selectively investigate welded joints, cracks, and other structural faults. According to Snell's law, as θ_{L1} is increased, θ_{L2} and θ_{S2} also increase. Corresponding to θ_{L1} (crit.), θ_{L2} becomes 90° and V_{L2} ceases to exist, and only V_{S2} propagates in solid 2. If θ_{L1} is increased much beyond θ_{L1} (crit.), V_{S2} also disappears. In this situation a SAW propagates on the surface of solid 2. SAWs are used in NDT to detect surface cracks.

Doppler effect transducers. When a sound wave at a given frequency is reflected from a moving target, the frequency of the reflected or backscattered sound is different. The shift in frequency is caused by the Doppler effect. The frequency is up-shifted if the target is moving toward the observer and down-

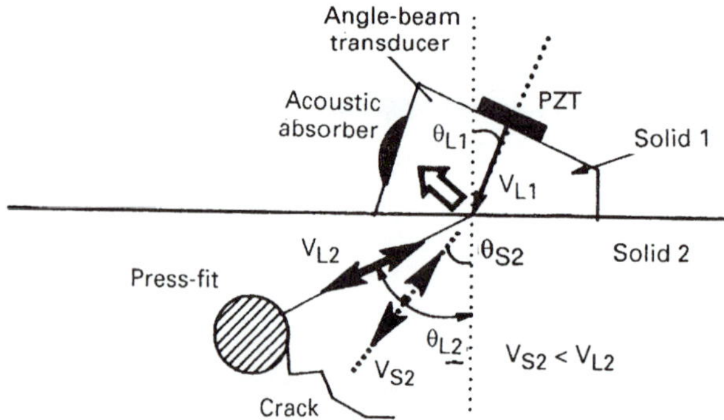

Figure 5.24 Angle-beam transducer. The beam directions are dictated by Snell's law, $\sin \theta_{L1}/V_{L1} = \sin \theta_{L2}/V_{L2} = \sin \theta_{S2}/V_{S2}$. If solid 1 is plexiglass ($V_{L1} = 2750$ m/s) and solid 2 is rolled aluminum ($V_{L2} = 6420$ m/s and $V_{S2} = 3040$ m/s), then at $\theta_{L1} = 25.4°$, V_{L2} ceases to exist in solid 2. When $\theta_{L1} = 64.8°$, the shear wave also ceases to exist. Beyond that angle a surface acoustic wave is generated which propagates on the surface of solid 2.

shifted if it's moving away. The Doppler shift in frequency is proportional to the velocity of the moving target and is given by

$$f_D = 2 f_0 \frac{V \cos \theta}{C} \tag{5.21}$$

where f_D = Doppler shift in frequency
f_0 = transmitted ultrasound frequency
C = velocity of ultrasound in the moving fluid
θ = angle between velocity vector V and the transmitted ultrasonic beam

Two techniques are used in the measurement of fluid flow velocity by the Doppler technique. The continuous-wave (CW) method, as used to determine the flow velocity of slurries in a pipe, is illustrated by Fig. 5.25*a*.[52] The trans-

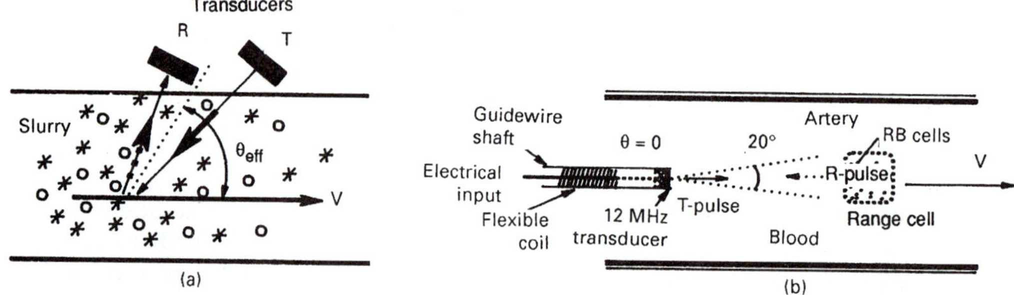

Figure 5.25 Measurement of velocity by Doppler shift. (*a*) Continuous-wave (CW) method. (*b*) Pulse-wave (PW) method gives the peak velocity. (*Part (b) from Ref. 53. Courtesy of Cardiometrics, Mountain View, Calif.*)

TABLE 5.3 Piezoelectric and Physical Properties of Organic Polymers

Property	PVDF	P(VDF-TrFE)	P(VDF-TeFE)	P(VDCN-VAc)
Dielectric constant ϵ_r at 1–10 MHz	6.0	5.0	5.5	6.0
Dielectric loss tangent, tan δ_e at 1–10 MHz	0.25	0.12	0.20	N/A
Mechanical Q, 1/tan δ_m	10	25	15	N/A
Stiffness constant c_{33}^D (10^9 N/m^2)	9.1	11.3	9.2	8.2
Electromechanical coupling, k_t	0.20	0.30	0.21	0.22
Density (10^3 kg/m^3)	1.78	1.88	1.90	1.20
Sound velocity, m/s	2200	2400	2200	2620
Acoustic impedance, Z_0 (10^6 kg/m^{2}s)	3.9	4.5	4.2	3.1
Piezoelectric h_{33} (10^9 V/m)	−2.6	−4.7	−2.9	−2.6

SOURCE: From Brown.[71] Courtesy of IEEE Ultrasonics Symposium, 1992.

mitted CW signal is partly reflected by the suspended particles or gas bubbles in the slurry. This backscattered signal is received by a second transducer and its output is compared with the transmitted signal. The Doppler shifted signal f_D is given by Eq. 5.21. Knowing θ and f_0, the velocity V can be obtained.

The second method, as used in a medical diagnostic application, is illustrated by Fig. 5.25*b*. A pulsed-wave (PW) signal[47,53] is used to measure the blood flow velocity in a small blood sample volume or range cell localized in the bloodstream of a coronary artery. The device is constructed from a 0.45-mm-diameter, flexible and steerable guide wire with a 12-MHz transducer integrated into its tip. The transducer transmits a sequence of 0.83-μs-duration pulses at a pulse repetition frequency of 40 kHz into the bloodstream. The range cell is located by time (range) gating the Doppler shifted backscattered signal generated by the red blood cells and received by the same transducer. This signal is compared with the transmitted signal, where as before the velocity V can be calculated using Eq. 5.21. In this case, cos $\theta = 1$.

Polymer transducers. In 1969, Kawai[69] discovered that polyvinylidene difluoride (PVDF, PVF_2), which is an organic ferroelectric polymer, displayed strong piezoelectric properties. In the last decade, an organic copolymer known as polyvinylidene difluoride trifluoroethylene P(VDF-TrFE) with a higher coupling coefficient has become available. These polymers are used in hydrophones, audio microphones, and robotic tactile transducers. In addition, they find application in wave-propagation transducers for NDT, medical imaging, hi-fi stereophones, and tweeters.[70] Table 5.3 shows the piezoelectric and related properties of these copolymers.[71]

A hydrophone is a very sensitive pressure transducer that is used to map the temporal and spatial acoustic pressure field of another transducer that is propagating acoustic energy through a fluid. Hydrophones have to satisfy many requirements: low acoustic impedance to optimally match the fluid impedance and cause the least disturbance to the field being measured, small sensing spot size to obtain good spatial resolution, large bandwidth and flat frequency response to respond to harmonics of the measured signal, and good

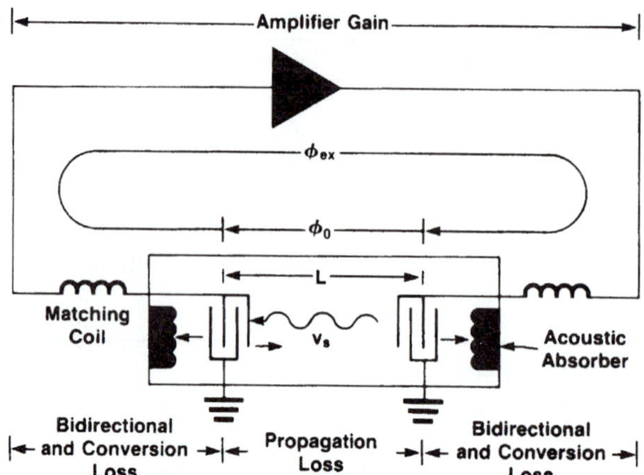

Figure 5.26 SAW oscillator. (*From Ref. 58. Copyright © 1981, Hewlett-Packard Co. Reproduced with permission.*)

linearity to handle the wide dynamic range of pressures. Two types of hydrophones most commonly used are the membrane hydrophone[72] and the Lewin-type needle hydrophone.[73]

In the membrane hydrophone, a metallized 250- to 500-μm-diameter circular dot is vacuum-deposited on either side of a poled copolymer film using a simple mask. This results in a single-element dot transducer. Suitable connections are made to the dots by vacuum-deposited metallic traces. The copolymer is then stretched and held in position over a circular hoop. The pressure of the acoustic field is measured as the electrical output of the dot transducer. The needle hydrophone consists of a poled, 0.5- to 1.0-mm-diameter, PVDF or copolymer film transducer which is bonded to the flattened end of a hypodermic needle but electrically insulated from it. The electrical output of the transducer gives a measure of the acoustic field pressure.

5.11.2 Surface acoustic wave (SAW) transducers

SAW transducers are used to measure physical, chemical, and environmental measurands. These transducers consist of a SAW delay line with two IDTs, inserted in the feedback loop of an amplifier[54,55] as shown in Fig. 5.26. The SAW, in traveling between the IDTs, undergoes a phase shift ϕ_0, and when the total phase shift ϕ_T around the loop is $2n\pi$ and the total attenuation is less than the amplifier loop gain, the circuit goes into oscillation around the center frequency f_0 of the IDT, which is given by nV_S/L. The total phase shift is

$$\phi_T = \phi_0 \pm \Delta\phi_0 \pm \phi_{\mathrm{EXT}} \tag{5.22}$$

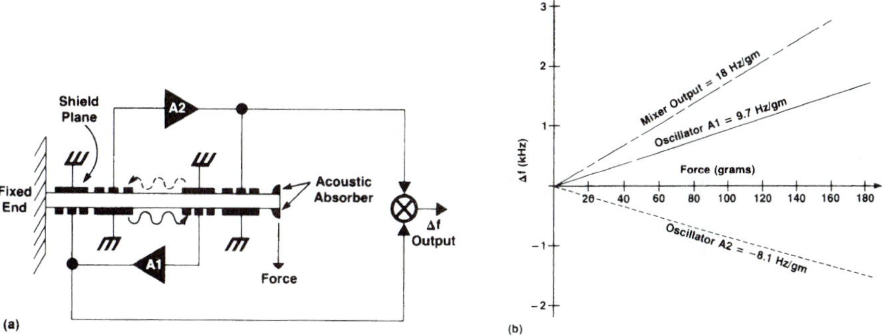

Figure 5.27 Dual SAW-oscillator force transducer. (*a*) Oscillators on opposite faces of quartz substrate achieve temperature compensation. (*b*) Change in oscillator frequency as a function of the force. Mixer output shows almost double the sensitivity. (*From Ref. 58. Copyright © 1981, Hewlett-Packard Co. Reproduced with permission.*)

where $\phi_0 = \omega T = 2\pi f_0 L/V_S$ and V_S is the SAW velocity. $\Delta\phi_0$ is the incremental phase shift caused by straining or changing the temperature of the substrate or by a change in V_S within the length L. $\Delta\phi_{\text{EXT}}$ is the phase shift caused by temperature-sensitive circuit elements.[56] n is an integer.

Any perturbation in the loop phase ϕ_T forces the system to alter the oscillator frequency by adjusting the total phase shift to be again a multiple of 2π. The oscillator frequency changes when L and/or V_S change, and this represents the transduction mechanism, where the output is a variable-frequency voltage proportional to the measurand. Similar results are obtained with SAW resonators.[57]

Physical transducers. In transducers used for force, acceleration, and temperature measurements, the length L is modified by the measurands. In crystalline quartz devices, the fractional change in SAW velocity is small compared with the surface strain caused by a change in L. We can therefore expect a decrease in oscillator frequency when L is increased by direct axial tension or by an increase in temperature.[55,56] So, in transducers that measure force, the temperature must be maintained constant. Figure 5.27 shows a dual oscillator configuration where the temperature effects are reduced and the force sensitivity is doubled. For a force acting downward, the A1 oscillator increases its frequency and the A2 oscillator decreases its frequency. The difference in frequency is a measure of the force.[55,58]

Chemical and environmental transducers. In chemical SAW transducers, the pairs of IDTs are formed on the same side of the substrate as shown in Fig. 5.28,[59–61] in contrast to the physical SAW transducers described earlier. In the intervening space L of the sensing oscillator, a (bio)chemical interface layer is deposited. The second oscillator serves as a reference. Mass loading caused by the interface layer will change the sensing oscillator frequency. Tempera-

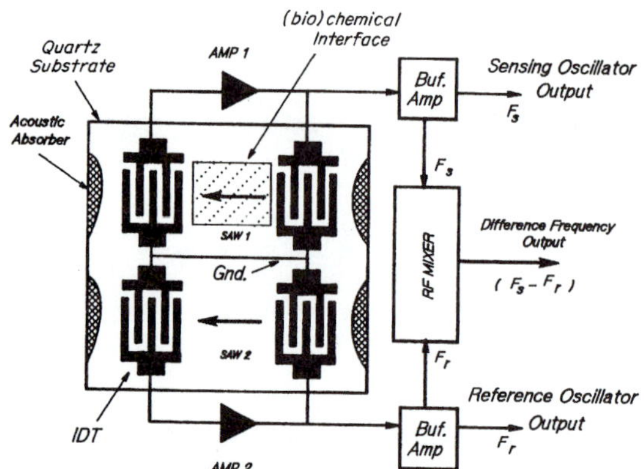

Figure 5.28 Dual SAW oscillator used as a chemical transducer.

ture and stress affect both oscillators equally, and they represent common mode signals. The difference in frequency ($F_S - F_R$) will then be proportional to the magnitude of the measurand, to the extent that it perturbs the interface layer. The transduction mechanisms are otherwise identical to those described for physical transducers.

The magnitude of the change in frequency caused by mass loading of a polymeric interface layer is given by[61]

$$\Delta f = (K_1 + K_2) f_0^2 (h\rho) \tag{5.23}$$

where K_1 and K_2 are material constants of the piezoelectric substrate, h is the thickness, and ρ is the density of the interface layer. The product $h\rho$ is the mass per unit area of the interface layer.

In chemical, environmental, and biochemical transducers, the changes in the frequency F_S are primarily brought about by the mass loading effect of the interface layer. A biochemical transducer uses the selectivity of enzymes and antibodies to mass load the layer, whereas a chemical sensor depends on the adsorption and chemisorption of the analyte gases.[60] These changes alter the SAW velocity. The transduction takes place when the velocity changes cause changes in the total electrical phase shift ϕ_T around the oscillator loop and corresponding changes in the oscillator frequency proportional to the mass loading of the interface layer.

In addition to frequency changes caused by the interface layer, the SAW attenuates as it propagates between the IDTs. Viscoelastic polymer interface layers, on absorption of volatile organic species, become soft because of plasticity effects and the SAW propagation losses increase because of softening of the polymer. The attenuation of the SAW represents yet another transduction mechanism in a chemical sensor which has been used to identify several chemical species for a particular chemical environment.[62]

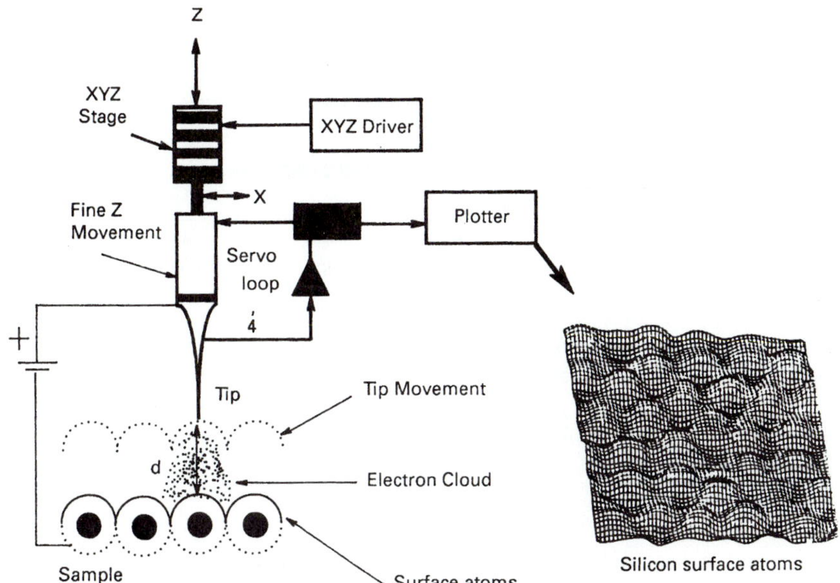

Figure 5.29 Transduction mechanism of the tunneling displacement transducer. (*From Ref. 63. Insert courtesy of Scientific American, New York, New York.*)

5.12 Tunneling Displacement Transducers

In 1986, Binnig and Rohrer were awarded the Nobel Prize in Physics for the scanning tunneling microscope (STM). The physical basis for the surface profiling transducer (SPT) used in the STM is the phenomenon of electron tunneling. It represents the flow of electrons between two conducting surfaces under the influence of a bias voltage. The resulting tunneling current or the tunneling effect[63] is a measure of the separation between these conductors.

Figure 5.29 illustrates the transduction mechanism of the SPT. In practice, one of the conductors is replaced by an extremely fine tip and the other conductor is the surface to be profiled. When the clouds of electrons surrounding the tip and the sample surface are made to overlap and a bias potential is applied between them, a tunneling current I_t is established. This current[64] can be represented by

$$I_t \propto e^{-2d/S_0} \tag{5.24}$$

where d is the distance between the tip and the sample surface and S_0 is a constant for a given work function. I_t is an extremely sensitive function of the distance d. In practical terms, when the tunneling current is well established, a change in d equal to one atomic diameter will change I_t by a factor of 1000. The SPT utilizes this sensitivity in a STM to profile the surfaces of materials at the atomic level. The scanning can be done in vacuum, liquids, or gases.[63,65]

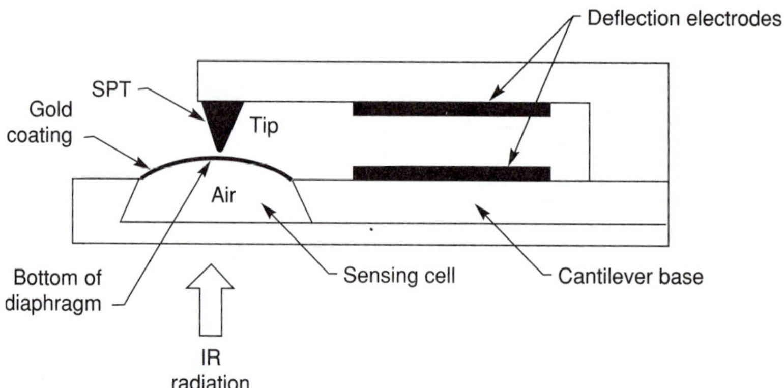

Figure 5.30 Infrared radiation transducer. (*From Ref. 68. Courtesy of American Institute of Physics, Woodbury, New York.*)

The angstrom movements in an STM are achieved with Inchworm* motors and a scanning tube.[65,66] The image shown is of atoms on the surface of a silicon substrate.[63] Based on similar principles, but on a transduction mechanism that involves repulsive forces, an atomic force microscope (AFM) has been developed for surface profiling.[67]

The SPT has also been used in an infrared radiation detector[68] as shown in Fig. 5.30. The tip can be lowered and raised by means of an electrostatic force between two electroded surfaces of the cantilever. The infrared sensor consists of a gold-coated membrane that traps a small volume of air or helium between itself and the base of the transducer. The radiation that is absorbed by the gold coating causes the gas to expand and deflect the membrane. The deflection, which is measured by the SPT, is a measure of the infrared radiation.

5.13 Thermodynamic Transducers

In thermodynamic systems, heat flows from one system to another whenever there is a temperature difference between them. Heat flow takes place by heat conduction, heat convection, and thermal radiation.

The calibration of platinum resistance thermometers, thermocouples, and the measurement of heat flow require the precise determination of temperature. In 1989, the International Committee on Weights and Measures adopted the International Temperature Scale of 1990 (ITS-90). The unit of the fundamental physical quantity known as thermodynamic temperature, symbol T_{90}, is the kelvin, symbol K. The relationship between the Interna-

* Inchworm is a registered trademark of Burleigh Instruments, Inc.

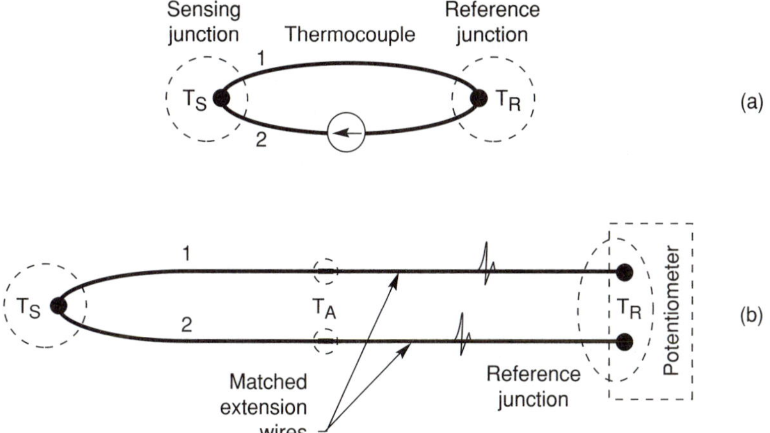

Figure 5.31 Thermocouples. (*a*) Seebeck effect. (*b*) Measurement setup for remote sensing.

tional Celsius Temperature, symbol t_{90}, and T_{90} is $t_{90} = T_{90}/\mathrm{K} - 273.15$. ITS-90 also defines specific fixed-point temperatures corresponding to various states of substances, like the triple point (TP) temperature of water (0.01°C, or 273.16 kelvins). A complete list of these fixed points is given in Ref. 76. These fixed points are used in the calibration of platinum resistance thermometers and thermocouples by evaluating the coefficients of the polynomials that represent their outputs. According to ITS-90, a platinum resistance thermometer, in the range from 0°C to the freezing point of aluminum, would be calibrated at the TP of water (0.01°C) and the freezing points of tin (231.928°C), zinc (419.527°C), and aluminum (660.323°C or 933.473 kelvins).

5.13.1 Thermocouples

When two dissimilar metal wires are joined at the ends and are kept at different temperature T_R and T_S, a continuous thermoelectric current flows owing to the Seebeck effect as shown in Fig. 5.31*a*. If the wires are cut, an open-circuit Seebeck voltage is measured which is proportional to the temperature difference. If T_R is constant, the Seeback voltage is proportional to T_S. The associated Peltier effect is the increase or decrease of the junction temperature when an external current flows in the thermocouple (TC) wires. Reference 77 gives a description of these effects and their interrelationships.

Figure 5.31*b* shows an arrangement when the measuring equipment is remote from the sensing TC. Matched extension TC wires are used and the reference junctions are formed at the measuring equipment. The reference junctions must be kept at a constant temperature, mostly at ambient room temperature T_A. This requirement can also be satisfied by electronically providing a reference voltage compensation.[78]

TABLE 5.4 Properties of Commonly Used Thermocouples

Type	Composition	Temp. range max, °C	Environment (bare wire)	Output, mV at 500°C
B	Platinum—30% rhodium vs. platinum–6% rhodium	0–1700	Oxidizing or inert; don't insert in metal tubes; used in glass industry	1.241
E	Nickel-chromium (Chromel) vs. copper-nickel (Constantan)	−200–900	Oxidizing or inert; limited use in vacuum or reducing	36.99
J	Iron vs. copper-nickel	0–750	Reducing, vacuum, inert; limited use in oxidizing at high temp.	27.388
K	Nickel-chromium (Chromel) vs. nickel-aluminum (Alumel)	−200–1250	Clean oxidizing and inert; use in vacuum or reducing; wide temp. range	20.64
R	Platinum–13% rhodium vs. platinum	0–1450	Oxidizing or inert; don't insert in metal tubes	4.71
S	Platinum–10% rhodium vs. platinum	0–1450	Same as R type	4.234
T	Copper vs. copper-nickel (Constantan)	−200–350	Mild oxidizing, vacuum or inert; low temp. or cryogenic applications	20.869 mV at 400°C

SOURCE: Data from Ref. 76, courtesy of Omega Engineering, Inc.

Thermocouples used in industry are made from several combinations of metals. Table 5.4 summarizes the properties of TCs that are commonly used, and a complete set of thermocouple reference tables is given in Ref. 76. The basic measuring instrument used with these thermocouples is the potentiometer, but direct-reading, analog, and digital meters of many kinds are also available from manufacturers.

5.13.2 Thermocouple vacuum gages

A thermocouple gage is used for monitoring pressures below atmospheric pressure (vacuum) in the range from 2 to 10^{-3} torr. A gage consists of a filament heated by a constant-current source and a thermocouple attached to it to monitor its temperature. The filament and the thermocouple are housed in an enclosure with an opening connected to the vacuum chamber whose gas pressure is to be measured. A schematic diagram of the TC gage is shown in Fig. 5.32.

The basic transduction mechanism can be described as the loss of heat caused when the surrounding gas molecules impinge on the heated filament and take heat away from it, resulting in a decrease in its temperature.[79] As the pressure decreases, a smaller number of molecules impinge on the filament, and consequently its temperature rises. This rise in temperature is indicated by the electrical output of the thermocouple and is proportional to the pressure (vacuum). In order to improve the lower limit of 10^{-3} torr, several TCs are connected in series as is done in a thermopile.

The Pirani gage also operates on the same basic transduction mechanism. The changes in temperature of its tungsten filament are measured as changes in its resistance with a Wheatstone bridge. The out-of-balance voltage is pro-

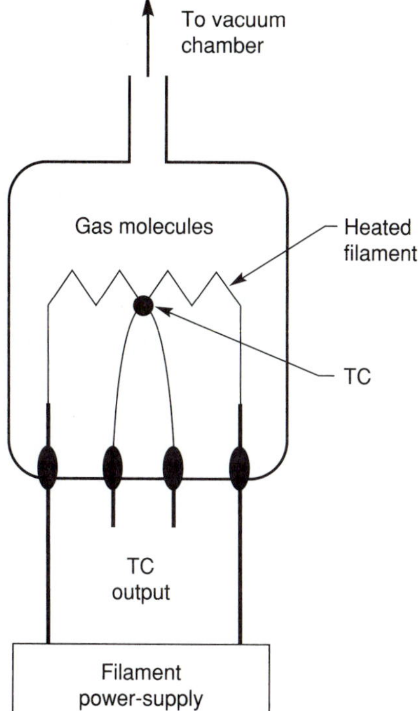

Figure 5.32 Thermocouple vacuum gauge.

portional to the vacuum level. The range of the Pirani gage is around 10^{-3} to 100 torr.[86] The Convectron is a Pirani gage with improvements in its accuracy, repeatability, and response time. The upper limit extends to 1000 torr.

5.13.3 Crystalline quartz thermometers

A quartz crystal, when coupled to conventional oscillator circuitry by replacing its resonant tank circuit, results in an oscillator whose frequency is controlled by the crystal as shown schematically in Fig. 5.33. The precise control depends on the crystal geometry and orientation with respect to the crystallographic axis referred to as a cut. The AT cut, for example, is used in very stable frequency oscillators.[80]

In a quartz thermometer, the transduction mechanism consists of the changes in the elastic and piezoelectric properties of the crystal as a function of temperature. These changes result in corresponding changes of the oscillator frequency. The frequency of an oscillator can be represented by a third-order polynomial in temperature t[80] and is expressed with reference to 25°C, as shown below.

$$f(t) = f_{25}[1 + A(t - 25) + B(t - 25)^2 + C(t - 25)^3] \tag{5.25}$$

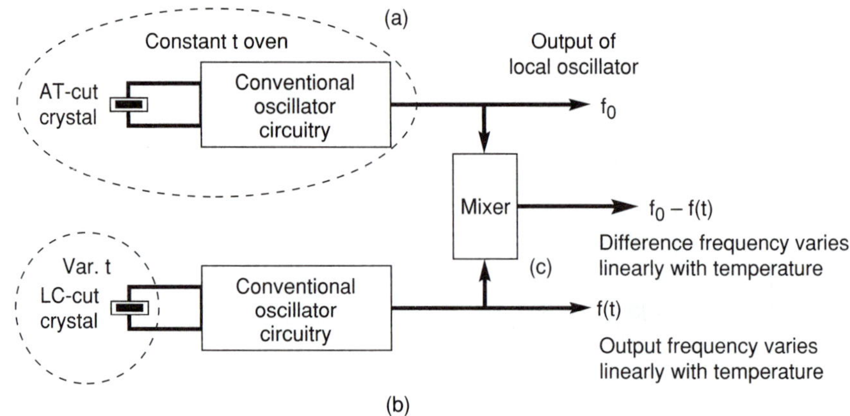

Figure 5.33 Crystalline quartz thermometer.

where *A, B, C* are the temperature coefficients of frequency, f_{25} is the oscillator frequency at 25°C, and *t* is the temperature to be sensed. Hammond[81] discovered a crystal cut where *A* was large and *B, C* were simultaneously zero, designated as the linear coefficient (LC) cut. This cut is used as a transduction element in quartz thermometers. Figure 5.33 shows an oscillator using the LC cut. Its output is a voltage whose frequency changes linearly with temperature.

The outputs from the two oscillators are mixed, and the output of the mixer is a much lower frequency but retains the linear relationship between frequency and temperature. The quartz thermometer measures temperatures in the range −80 to 250°C with an accuracy of ±0.075°C. The sensitivity is 1000 Hz/°C and the corresponding resolution is 0.0001°C.

5.13.4 IC temperature transducers

The IC temperature (ICT) transducer is a two-terminal monolithic integrated circuit whose output is a current or a voltage that is directly proportional to temperature *T/K.* The transduction mechanism utilized in this transducer is the dependence of the base-emitter voltage V_{be} of a silicon transistor, on the ambient temperature *T/K.*

If two well-matched IC transistors are connected as shown in Fig. 5.34*a*, then the difference between the V_{be}'s[82] of the transistors Q_1 and Q_2 is proportional to *T*, if the ratio of the respective emitter areas Ω is constant. It is made constant by designing the emitter areas of Q_1 and Q_2 to be in a fixed ratio and operating them at equal collector currents.

When the temperature being monitored changes, the collector currents of Q_1 and Q_2 also change. The control circuit as drawn in Fig. 5.34*a*[82,83] forces

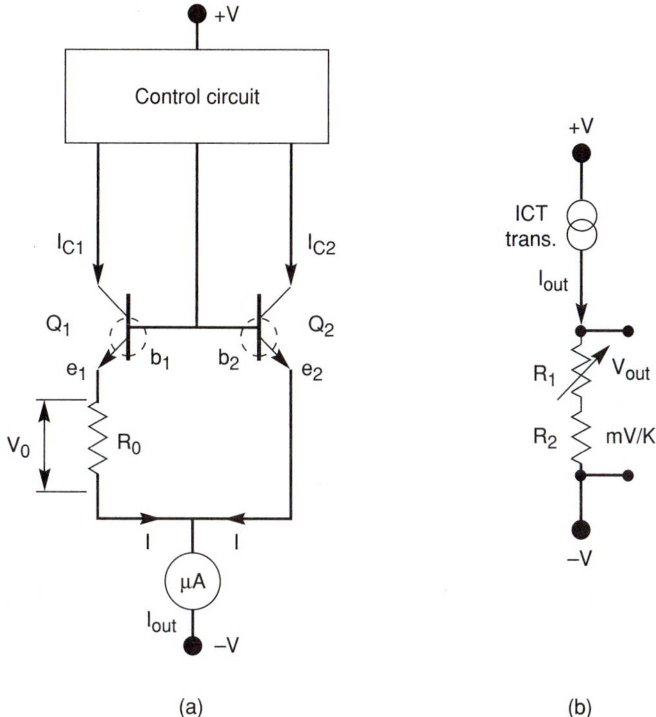

Figure 5.34 Integrated-circuit temperature transducer. (*From Ref. 82. Courtesy of Analog Devices, Norwood, Mass.*)

the collector currents to equalize, and V_0 is then proportional to the temperature being measured. The total output current I_{OUT} through Q_1 and Q_2 is

$$I_{OUT} = \left(\frac{\text{constant}}{R_0}\right) T \tag{5.26}$$

I_{OUT} is then proportional to T/K if R_0 is constant. In practice, R_0 has a very small temperature coefficient of resistance. In Fig. 5.34*b*, the transducer is connected in series with R_1 and R_2 to produce a voltage drop proportional to T/K. The ICT transducer operates in the range -55 to $+150°C$ with very good linearity.

5.14 Ionization Transducers

Ionization transducers (vacuum gages) are used to measure low pressures (vacuum levels) below atmospheric pressure (760 torr). The basic transduction mechanism in these gages is the generation of positive ions from the gas molecules that are present in the chambers to be evacuated. The ions are

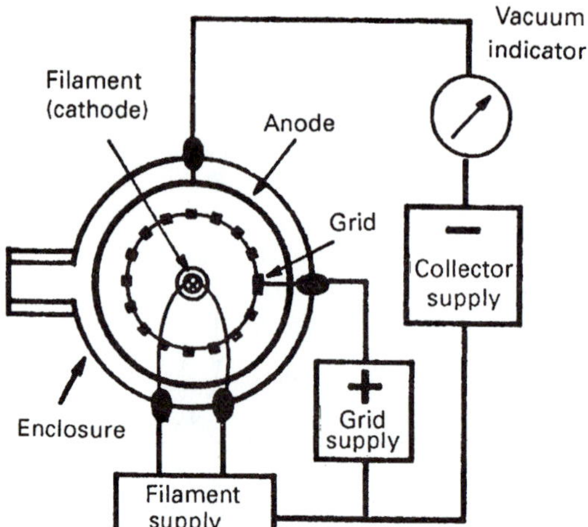

Figure 5.35 Hot-cathode vacuum gage.

generated as a result of collisions between the gas molecules and the high-energy electrons that are generated specifically for that purpose. The resulting ion current is proportional to the gas pressure in the chamber. A complete description of several vacuum gages is given in Refs. 84 and 85. Practical details about these gages are given in Refs. 86 and 87.

5.14.1 Hot-cathode vacuum gage

The triode or thermionic emission ionization transducer is a hot-cathode vacuum gage. The top cross-sectional view is shown diagrammatically in Fig. 5.35. It consists of a tungsten filament and cathode in the center. Concentric with the cathode is the grid, which is made from a fine nickel wire helix, held in position by upright supports. Surrounding the grid is the external nickel plate electrode, which is concentric with the grid. The gage is housed in an enclosure with an opening that is connected to the chamber whose pressure is being measured.

In this configuration, the electrons are accelerated from the cathode to the grid (+180 V), but most electrons not collected by the fine helix move toward the plate (−20 V). As they move in the vicinity of the plate, the electrons get repelled by the negative potential of the plate. These electrons undergo several oscillations between the grid and the plate, and during this process they collide with the gas molecules, creating positive ions. The ions are attracted toward the negatively biased plate collector, and an ion current flows in the plate circuit. This ion current, within a certain pressure range, is proportional to the total pressure (vacuum) in the chamber. The effective range for the gage is 10^{-8} to 10^{-3} torr.

The Bayard Alpert vacuum transducer is an example of a hot-cathode ultra-high-vacuum gage. The collector is in the center and consists of a fine

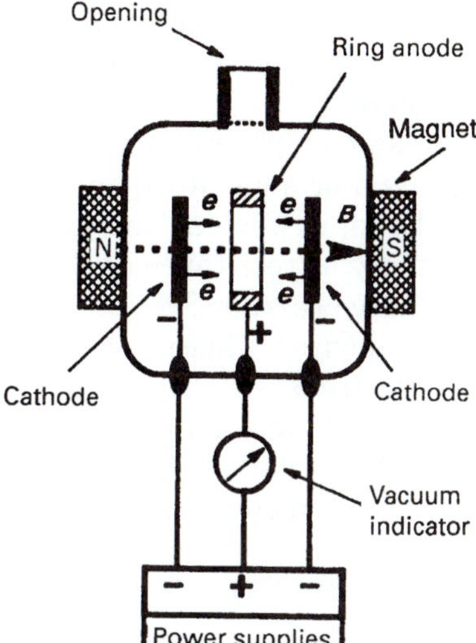

Figure 5.36 Cold-cathode vacuum gage.

nickel wire. The filament and cathode located outside the grid. Its operation is similar to that of the triode gage. The ion current is proportional to the gas pressure down to 10^{-10} torr, because the undesired x-rays generated by the grid are minimally intercepted by the wire collector. Its operating range is from 4×10^{-10} to 5×10^{-2} torr. To measure pressures lower than 10^{-10} torr, the glass envelope is not used. The operating range of the nude gage is 2×10^{-11} to 1×10^{-3} torr.

5.14.2 Cold-cathode vacuum gage

In this gage, the electrons are generated by a large electric field instead of a hot cathode. The gage consists of two plate cathodes made from zirconium or thorium. In the space between these cathodes is placed a ring-shaped anode. A magnet provides a magnetic field normal to the plane of the cathode plates, as shown diagrammatically in Fig. 5.36. It is also known as the Penning or Philips gage. A high voltage is applied between the anode and the cathodes, producing electrons that leave the cold cathode surface. These electrons, in the presence of the magnetic field B, travel in spiral paths toward the positively biased ring anode. Only a few electrons are captured by the anode, but the rest make several passes through the ring anode, and in the process ionize the gas. The positive ions collected by the negatively biased cathodes produce the usual ion current proportional to the gas pressure. The effective range of this gage is 1×10^{-8} to 2×10^{-2} torr.

5.15 Photonic Transducers

Photonic transducers measure the light intensity by measuring the impulses of finite energy content, generally known as light quanta or photons. In these transducers, the energy of the photons is converted into a proportional electrical output by means of several transduction mechanisms. In this section, in addition to others, we consider the following:

Photoemissive detectors, where the energy of a photon removes an electron from a metal surface placed in a vacuum or a gas-filled environment.[35,88,89]

Photoconductive detectors, where the input photons energy creates electron-hole pairs which change the conductivity or resistance of a semiconductor by increasing the number of available charge carriers.[35,90]

Photovoltaic detectors, p-n junction devices where the input photons also generate electron-hole pairs. The electrons and holes provide additional mechanisms for current conduction. These detectors can be operated in the photovoltaic mode or the photoconductive mode.[35,91–93]

The energy of a photon, when expressed in electron volts, is given by

$$E_{ph} = (4.13 \times 10^{-15}) f \tag{5.27}$$

where E_{ph} is proportional to the light frequency f and inversely proportional to its wavelength λ. Consequently, there is a long wavelength limit to λ beyond which the detector will be cut off when E_{ph} is barely less than the work function E_w in photoemissive detectors[88] and the energy bandgap in semiconductor detectors.[35]

5.15.1 Photoelectric detectors

Photoelectric or photoemissive detectors operate on the photoelectric effect, which is a process in which electrons are liberated from a material surface after absorbing energy from a photon. All the energy from a photon is transferred to a single electron in a metal, and that electron, having acquired the energy in excess of E_w, will emerge from the metal surface with a kinetic energy given by Einstein's photoelectric equation[88]

$$\frac{1}{2} m v^2 = E_{ph} - E_w \tag{5.28}$$

where v is the velocity of the emerging electron, m is the mass of an electron, and E_w is the work function. E_w is the energy required by an electron at the Fermi level to leave the metal surface with the highest kinetic energy.

A photoemissive detector is the basic component of vacuum or gas phototubes and photomultiplier tubes. A phototube consists of a semicircular photoemissive cathode concentric with a central rodlike anode in a special glass enclosure. The incoming photons impinge on the photocathode and generate photoelectrons which leave the cathode surface, and because of the electric field between the anode and cathode, these electrons are collected by the anode. This results in an anode current that is proportional to the number of photons impinging on the cathode.

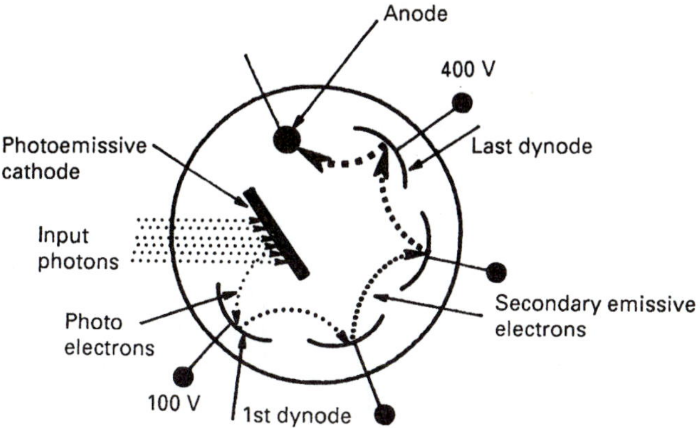

Figure 5.37 Photomultiplier tube. (*From Ref. 89. Courtesy of Hamamatsu Corporation, Bridgewater, New Jersey.*)

The photomultiplier tube (PMT) operates on the same basic transduction mechanism as the phototube, but the PMT has provisions for realizing high current gain by current multiplication. The components of the PMT are shown schematically in Fig. 5.37. Photoelectrons are generated as in a phototube. These photoelectrons are focused onto the first element of the PMT chain, known as a dynode.

The photoelectrons from the first dynode are directed in succession to other dynodes, which are at progressively higher potential. The electrons accelerate toward these dynodes, and on striking them, they generate many more electrons by a process known as secondary emission. As an example, if there are 10 stages and if each primary electron produced 4 secondary electrons, then the total number of electrons at the end of the dynode chain would be 4^{10} or approximately 10^6 electrons. The electrons from the last dynode stage are collected by the anode. The resulting amplified current is proportional to the number of input photons.

5.15.2 Photoconductive detectors

Photoconductive detectors decrease their terminal resistance when exposed to light. Figure 5.38*a* shows a practical representation of a photoconductive detector. It consists of a thin-film trace of a semiconductor such as cadmium sulfide deposited on a ceramic substrate. The length of the meandering trace is *L,* the width is *W,* and *t* is the thickness. There are two metallizations on either side of the trace to which electrical contacts can be made. When a voltage *V* is applied, a current flows across the width of the trace, all along the trace. The resistance of the trace at any given light input is $R_0 = \rho_R W/L$ and the corresponding current is $I_0 = (V/\rho_R)(L/W)$, where ρ_R is the sheet resistivity of the semiconductor trace and $1/\rho_R$ is the sheet conductivity ρ_C. The

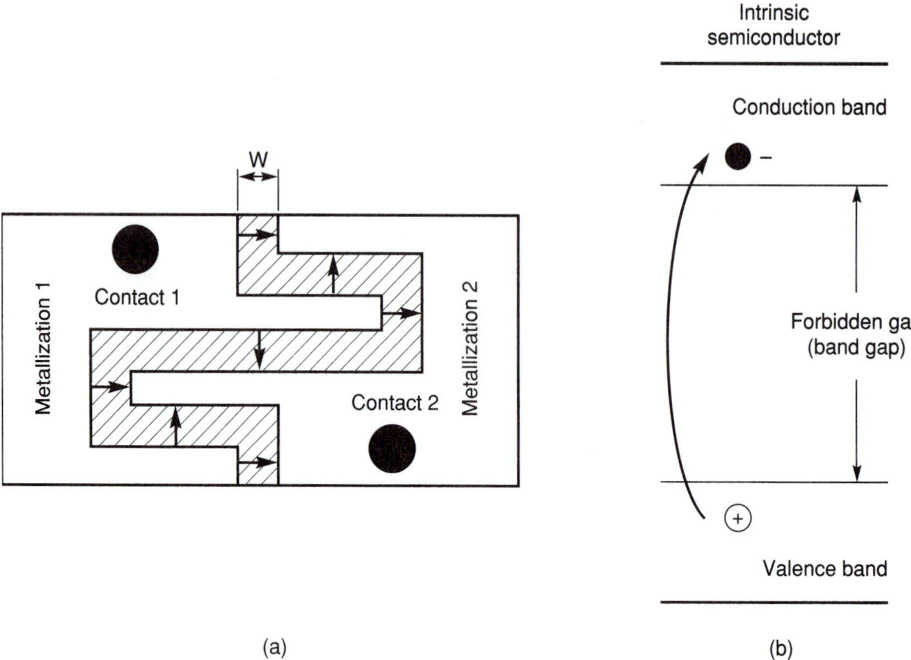

Figure 5.38 (*a*) Photoconductive detector. (*b*) Energy level diagram for an intrinsic semiconductor.

light input modulates ρ_C and consequently the resistance and the current across the trace.[90]

The transduction mechanism can be qualitatively explained with the aid of the energy level diagram for an intrinsic semiconductor shown in Fig. 5.38*b*. An intrinsic semiconductor absorbs a photon when its energy barely exceeds the energy of the forbidden gap and simultaneously creates an electron-hole pair. The electron makes a single transition across the bandgap into the conduction band. This electron and the corresponding hole in the valence band contribute to the increase in total current. As a consequence, the conductivity increases and the resistance of the photoconductor decreases with increased photon input. In extrinsic semiconductors, owing to the added impurities, donor and acceptor ionized states are created in the forbidden gap, which also contribute additional carriers to the conduction and valence band, which increase the conductivity even further. The addition of impurities increases the quantum efficiency and the long wavelength cutoff[35] of photoconductive detectors.

Photoconductors made from lead sulfide (PbS) and lead selenide (PbSe) are used as infrared detectors in the range of optical wavelengths from 1 to 3 μm and 1 to 6 μm, respectively. Cadmium sulfide (CdS) cells are used in exposure meters, light dimmers, photoelectric relays, and other consumer-type applications. Cadmium sulfide cells can be optimized in the range of 515 to 730 nm, which is close to the spectral response of the human eye.

5.15.3 Photovoltaic detectors

Photovoltaic detectors are junction-based semiconductor devices. Photovoltaic detectors can be further classified as operating in the photovoltaic mode or the photoconductive mode. When a *p-n* junction is formed, a carrier concentration gradient exists because of the majority carriers in the *p*-type and *n*-type regions. Excess electrons from the *n*-type region diffuse into the *p*-type region and the excess holes diffuse in the opposite direction in order to equalize this gradient. The movement of charges across the junction generates an electric field E within the photodiode as indicated in Fig. 5.39*a*, which prevents any further diffusion of carriers.

When the energy of a photon that impinges on the photodiode is equal to or greater than the energy bandgap, electron-hole pairs are generated in the regions shown in Fig. 5.39*a*. In the depletion region, the electrons move toward the *n* side and the holes toward the *p* side under the influence of the electric field E. On the *p* side, only the electrons move toward the *n* side and the hole is prevented from moving by the field. On the *n* side, the situation is reversed. The accumulation of charges on either side of the junction results in an open-circuit voltage V_{OC} that is proportional to the photon input.

Figure 5.39*b* shows the *I-V* characteristics of a photovoltaic detector. Without any photon input, the *I-V* plot is identical to that of a rectifying diode. As the photon input increases, the plot is successively displaced downward. For any photon input level, the open-circuit voltage V_{oc} is proportional to that input. If the output terminals are shorted, the short circuit current I_{sc} is also proportional to the photon input. This corresponds to operation in the photovoltaic mode, as indicated by the fourth quadrant of the *I-V* plot.

When a reverse bias is applied to the photovoltaic detector, the load line is displaced to the third quadrant by an amount equal to the reverse bias. In this quadrant, the photovoltaic detector behaves as a current source and is said to operate in the photoconductive mode.

5.15.4 Position sensing transducers

Structurally modified photodiodes are used in the measurement of small angles, distances, and machine vibrations, which are related to linear position sensing.[94] The lateral effect PIN photodiode is generally used for these measurements, and its cross section is shown in Fig. 5.40*a*. Photons from a laser source impinge on the *p*-type front surface of the PIN diode and are readily absorbed within the depletion region, where electron-hole pairs are generated. The electrons move toward the *n*-type region and generate a photocurrent I_0.

The operation of the lateral-effect photodiode can be represented by a variable contact resistor, as shown in Fig. 5.40*b*. The location of the laser spot

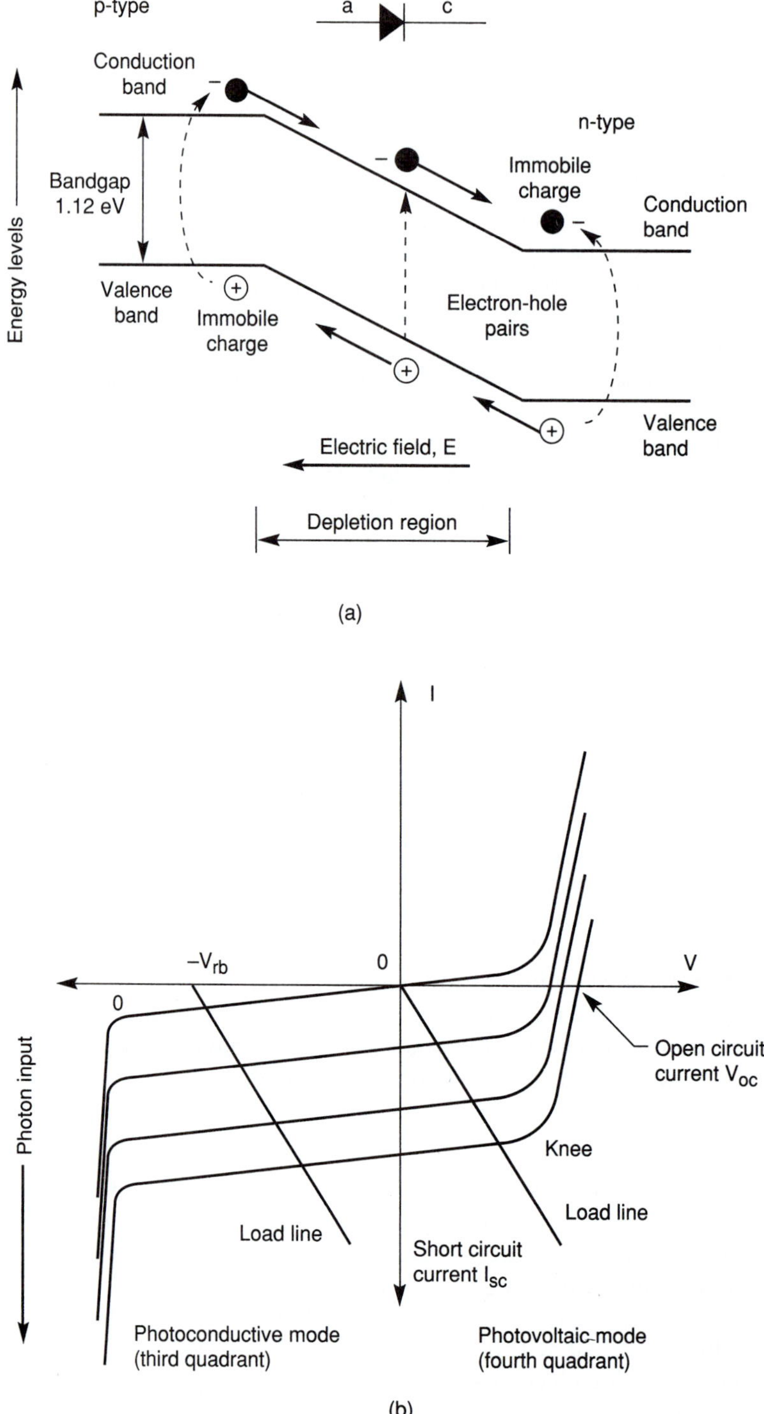

Figure 5.39 Photovoltaic detectors. (*a*) Energy-level diagram for a *p-n* photodiode showing the transduction mechanisms. (*b*) *I-V* characteristics of photovoltaic detectors.

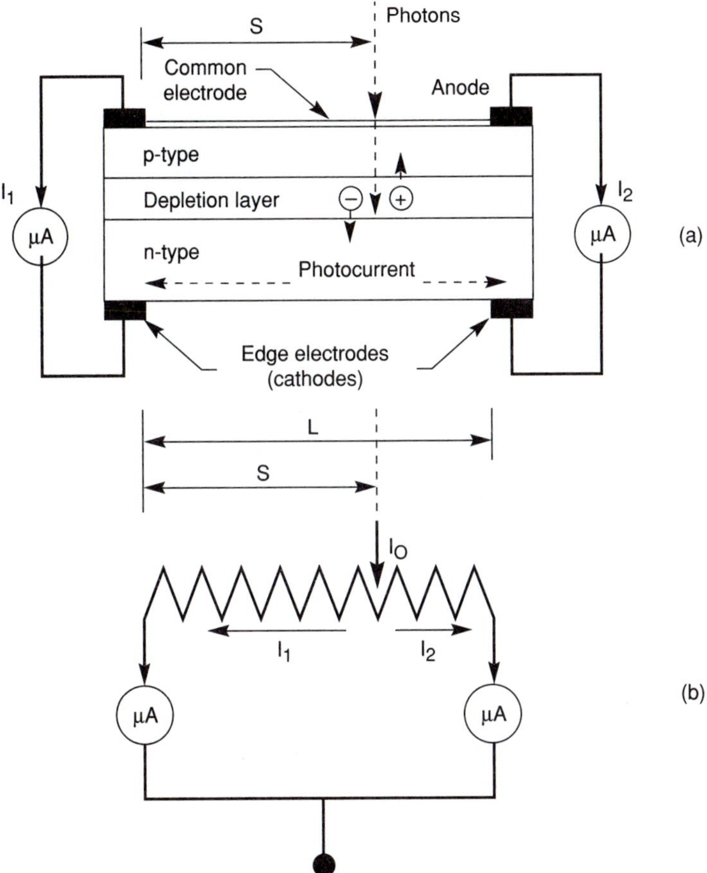

Figure 5.40 Position sensing transducer.

defines that variable contact. The photocurrent I_0 divides between the paths to the two edge electrodes in a ratio given by[94,95]

$$\frac{I_1 - I_2}{I_1 + I_2} = 1 - \frac{2S}{L} \tag{5.29}$$

where L is the distance between the electrodes, S is the distance of the laser spot from one of the electrodes, and $I_0 = I_1 + I_2$. Using the above equation, S can be determined by measuring I_1 and I_2. This is the transduction mechanism of the position sensing transducer. It measures displacements along a single axis.

5.15.5 Acousto-optic deflection transducers

The acousto-optic (AO) deflection transducer, aside from other applications,[96] is used to detect physical vibrations by converting them into a corresponding

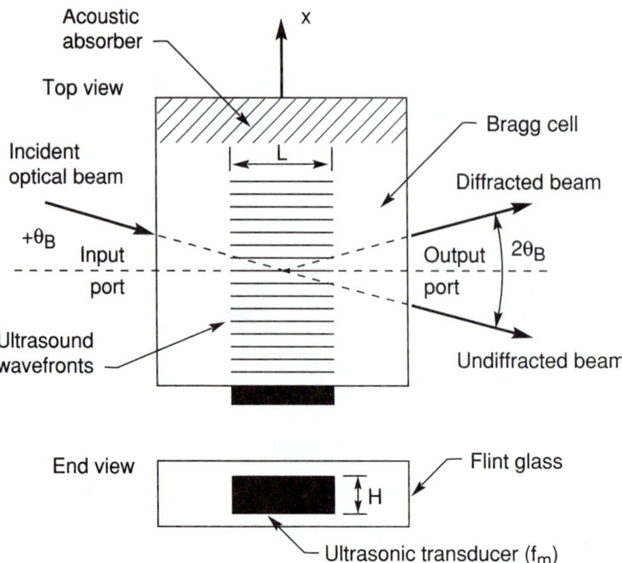

Figure 5.41 Acoustooptic deflection transducer.

phase-modulated electronic signal. The primary mechanism is the interaction of an optical wave (photons) with an acoustic wave (phonons), to produce a new optical wave (photons).[97]

Figure 5.41*a* shows diagrammatically an acoustooptic interaction cell, also known as the Bragg cell. It consists of a solid piece of optical-quality flint glass which acts as the interaction medium. It has optical windows on two opposite sides which serve as ports for the entry and exit of the optical beam. A high-frequency ultrasound transducer is bonded to one of the two remaining sides. The ultrasound transducer launches longitudinal waves in the interaction region and produces regions of compression and rarefaction, which are regions of varying refractive index. These regions are equivalent to a three-dimensional diffraction grating.

An optical (laser) beam entering the Bragg cell at an angle θ and passing through the length L of the grating experiences the periodic changes in the refractive index at the excitation frequency ω_m of the ultrasonic transducer. Consequently, the optical beam leaving the cell is phase-modulated and consists of an optical carrier at ω_c and pairs of optical sidebands around ω_c, at ω_c, $\pm N\omega_m$, where N is an integer corresponding to the order of the sidebands. There is a unique value of $\theta = \theta_B$ corresponding to the optical wavelength λ and the acoustic wavelength Λ, when only one of the first-order sidebands grows by constructive interference and the other, along with all the higher sidebands, diminishes owing to destructive interference.[96] This is known as

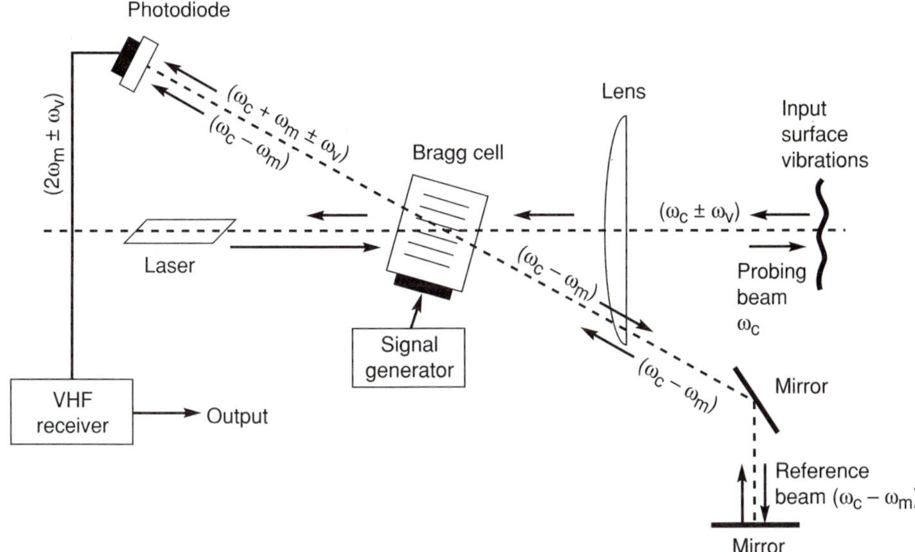

Figure 5.42 Noncontact vibration-measuring setup using a Bragg cell. (*From Ref. 98. Courtesy of Optical Society of America, Washington, D.C.*)

the Bragg effect, and θ_B is called the Bragg angle. The angle is given by the Bragg diffraction equation

$$\sin \theta_B = \frac{\lambda}{2\Lambda} = \frac{\lambda}{2v_m} f_m \tag{5.30}$$

where v_m is the ultrasound velocity in the medium and f_m is the ultrasonic modulation frequency of the Δn wave. The angle between the Bragg deflected beam and the carrier is $2\theta_B$.

Figure 5.42 shows an application where the Bragg cell is used as a noncontact vibration-measuring transducer.[98] A He-Ne laser is used as the source of the optical beam. The Bragg cell is used to measure the surface vibrations ω_v. On the face of the photodiode, the demodulation of the optical signals takes place.[98] The output current of the photodiode has a component at $(2\omega_m + \omega_v)$. The VHF receiver measures the sideband amplitude, which is proportional to the amplitude of the surface vibrations. A minimum peak displacement of 2×10^{-12} m, with 300 μW of power at the photodiode, was measured using this system. Similar arrangements are currently used in commercial systems.[99,100]

5.16 Fiber-Optic Transducers

The role of optical fibers in telecommunication systems is now well established. The recognition of their role in transducers, about two decades ago, led to an explosive number of applications in physical, chemical, and biochemical

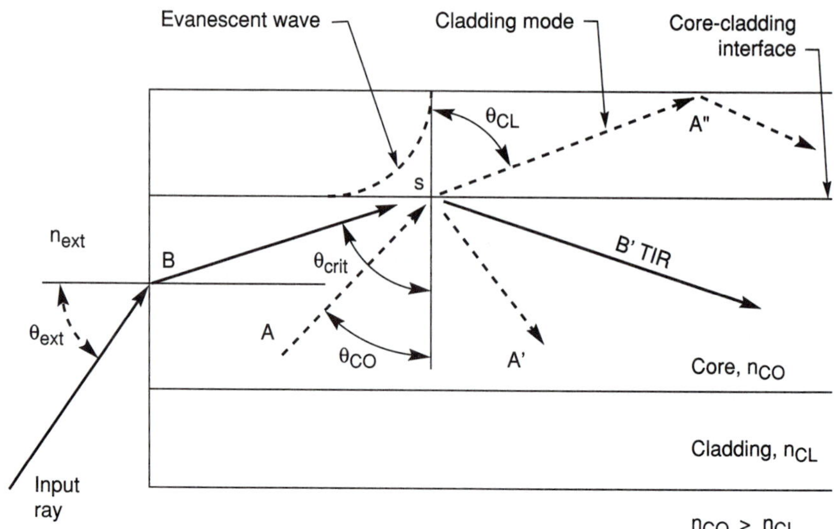

Figure 5.43 Total internal reflection in optical fibers.

transducers.[101–103] These transducers can be separated into extrinsic and intrinsic types. In extrinsic fiber-optic transducers, the transduction process takes place external to the fiber and the fiber itself plays a passive role as a light conduit. In intrinsic transducers, however, the transduction process takes place within the optical fiber. The measurand modulates a parameter of the fiber, such as the refractive index, and the fiber in turn modulates the light beam propagating through its core.

The optical fiber used in these transducers is of the cladded-core type used in communication links,[104] as shown schematically in Fig. 5.43. The core is made of silica glass, and surrounding the core is a concentric silica cladding. The light beam in an optical fiber propagates mainly in the core by total internal reflection (TIR), as shown in Fig. 5.43. A light ray, striking the core-cladding interface at S, is partly reflected (A') and partly refracted (A'') into the cladding. By Snell's law, $n_{CO} \sin \theta_{CO} = n_{CL} \sin \theta_{CL}$. As θ_{CO} is increased, θ_{CL} ultimately becomes 90° at $\theta_{CO} = \theta_{\text{crit.}}$ where $\sin \theta_{\text{crit.}} = n_{CL}/n_{CO}$ and the incident ray (B) is totally reflected (B') within the core by TIR. In addition, there is a nonpropagating wave with an exponentially decaying field normal to the core-cladding interface as a consequence of the boundary conditions. This is known as the evanescent wave, and it forms part of the transduction mechanism of some physical and biomedical transducers.[102,108]

A ray from a laser diode propagating through an external medium of refractive index n_{ext} and striking the fiber end at an angle θ_{ext} has to satisfy Snell's law at that interface, and maintain TIR at the core-cladding interface. Consequently,

$$n_{\text{ext}} \sin \theta_{\text{ext}} = n_{co} \sin (90 - \theta_{\text{crit.}}) \tag{5.31}$$

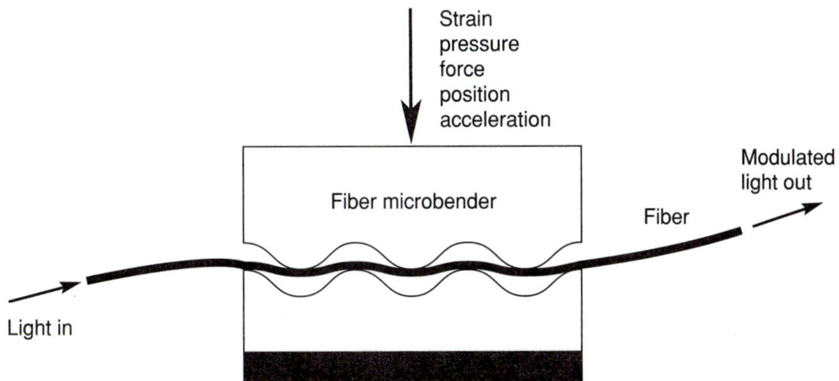

Figure 5.44 Fiber-optic microbend transducer.

and the above equation reduces to

$$n_{\text{ext}} \sin \theta_{\text{ext}} = (n_{co}^2 - n_{ct}^2)^{1/2} \tag{5.32}$$

where $n_{\text{ext}} \sin \theta_{\text{ext}}$ is known as the numerical aperture (NA) of the fiber. θ_{ext} sets a limit on the cone angle that the light source can have for maximum coupling of optical power into the sensing fiber.

5.16.1 Physical transducers

The most direct application of an optical fiber to measure physical measurands is the microbend transducer, as shown in Fig. 5.44.[102,103,105] The fiber is deformed by a measurand such as displacement, strain, pressure, force, acceleration,[106] or temperature. In each case, the transduction mechanism is the decrease in light intensity of the beam through the fiber core as a result of the deformation, which causes the core modes to lose energy to the cladding modes. A photodetector is used to provide a corresponding electrical output. This is an intrinsic-type transducer.

Figure 5.45 shows a pressure transducer and a temperature transducer

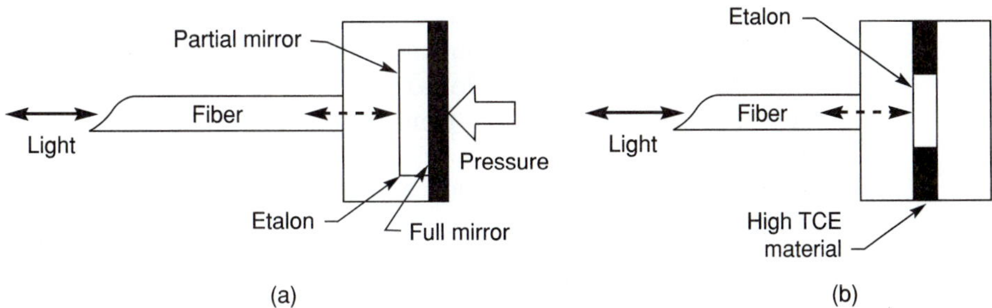

Figure 5.45 Fiber-optic temperature and pressure transducers. (*From Gordon Mitchell, of Future Focus, Woodinville, Wash.*[103] *Courtesy of John Wiley & Sons.*)

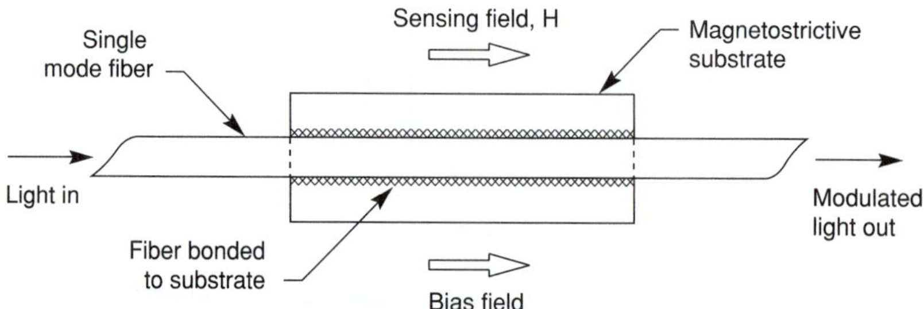

Figure 5.46 Fiber-optic magnetic field transducer.

configured as a Fabry-Perot etalon or interferometer.[103] The spacing between mirrors is critical to the operation of the etalon as a physical transducer. Any change in that spacing, brought about by a measurand, will cause the transmittance of the etalon to change in proportion to the measurand. These extrinsic transducers are small in size and can be constructed by using integrated circuit techniques.

Optical fibers are also used to measure magnetic and acoustic fields.[102,103] In a magnetic field transducer as shown in Fig. 5.46, the transduction mechanism consists of the longitudinal strain in the magnetostrictive material caused by the magnetic field and a corresponding strain in the core. This strain modulates the refractive index of the core and produces a field-dependent phase shift in the output optical beam. An acoustic field transducer can also be made along these lines by replacing the magnetostrictive substrate with an acoustically sensitive coating around the fiber cladding. Consequently, this produces an acoustic field-dependent phase shift of the optical beam. A Mach-Zhender interferometer is used to obtain an electrical output proportional to the measurands.[103]

5.16.2 Fluorescence chemical transducers

The knowledge of the partial pressure of oxygen (PO_2), the pH, and the partial pressure of carbon dioxide (PCO_2) in arterial blood is very important in the management of critically ill patients. Sensing elements with specific reagent chemistries for each of the above measurands are attached to the tips of each of three optical fibers. When used as intravascular transducers, the three fibers are enclosed in a single catheter which is then inserted into the artery through an incision for real-time continuous monitoring of the blood gases and pH.[107]

Light from a laser diode source at a specific wavelength enters the fiber at one end (proximal), propagates through the core, and irradiates the sensing element at the distal end inside the artery. The sensing element contains a dye with fluorescent molecules which, on absorption of the photons, are excited to a higher-energy state. In that state, the molecule loses some of its

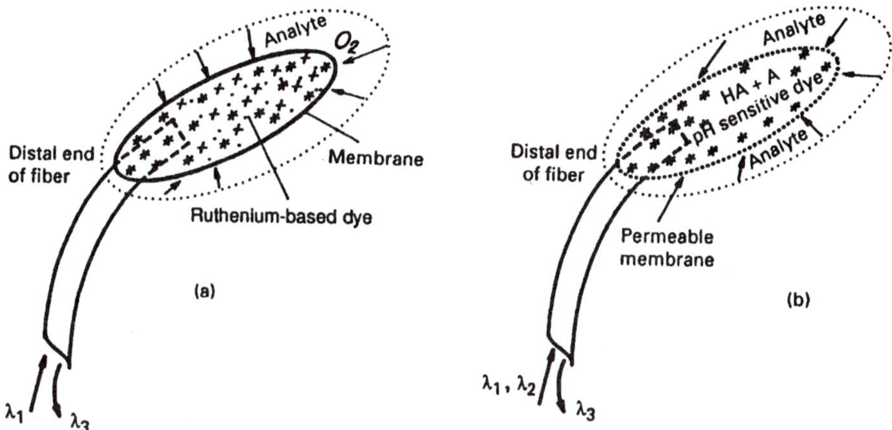

Figure 5.47 Fiber-optic fluorescence chemical transducers. (*a*) Oxygen transducer based on the fluorescence quenching effect. (*b*) pH transducer.

energy to the lattice structure. As it returns with lower energy from an intermediate state of excitation, it fluoresces at a different wavelength.[108,111] Since the energy of a photon is inversely proportional to the wavelength, the energy in the emitted light is of a longer wavelength. The presence of an analyte (measurand) modifies the intensity of the reemitted (fluorescent) light. It can be measured using wavelength-selective devices and photonic detectors, whose output is proportional to the concentration of the measurand. An analyte such as oxygen quenches the fluorescence intensity. This is the basic transduction mechanism in PO_2 transducers.

Based on the fluorescence quenching effect,[108–110] intravascular PO_2 transducers have been built as shown in Fig. 5.47*a*.[111] The relationship between PO_2 (measured) and the fluorescence characteristics of the dye molecule[109] is given by the rearranged form of the Stern-Voltmer equation

$$PO_2 = \left(\frac{I_0}{I} - 1\right)K \tag{5.33}$$

where I is the fluorescence intensity after quenching by oxygen and I_0 is the intensity in the absence of oxygen. K is the quenching constant, and it represents the slope of the PO_2 vs. I_0/I plot, which characterizes the performance of the oxygen transducer.[108] By measuring I_0 at the excitation wavelength λ_1 and I at the fluorescence wavelength λ_3, the partial pressure of oxygen can be determined.

The sensing element in a pH transducer is a pH-sensitive dye, whose acid from (HA) and the base form (A^-) fluoresce at different excitation wavelengths, but the longer wavelength of fluorescence emission is the same for both forms, as indicated in Fig. 5.47*b*.[111] This is the basic transduction mechanism. In this case, the pH can be expressed as[108]

$$\mathrm{pH} = pK_a - \log\left(\frac{I_{\mathrm{HA}}}{\mathrm{I}_{\mathrm{A}^-}}\right) a \tag{5.34}$$

where I_{HA} is the fluorescence intensity corresponding to the acid form, I_{A^-} is the intensity for the base form, a is a constant, and K_{a} is the indicator constant. By determining the ratio of the fluorescence intensities, the pH of the analyte (measurand) can be determined from Eq. 5.34.

A PCO_2 transducer is essentially a pH transducer that selectively measures the changes in pH of a buffer solution brought about by changes in CO_2 concentration. The sensing element consists of a selective membrane that is impermeable to hydrogen ions (H^+) and water (H_2O), but it is highly permeable to CO_2. Enclosed within this membrane is a pH transducer immersed in a buffer solution of sodium bicarbonate ($NaCO_3$) in water. When CO_2 diffuses through the membrane and mixes with the water from the buffer solution, it produces by dissociation H^+ ions. Consequently, the changes in pH of the buffer solution are a measure of the CO_2 concentration.

References

1. Coombs, Clyde F., Jr., Ed., *Basic Electronic Instrument Handbook*, McGraw-Hill, New York, 1972.
2. Oliver, Frank J., *Practical Instrumentation Transducers, Transducer Fundamentals and Physical Effects*, chap. 1, Hayden, New York, 1971.
3. Mazda, F. F., *Electronic Instruments and Measurement Techniques*, Cambridge University Press, Cambridge, 1987.
4. Norton, H. N., *Handbook of Transducers*, Prentice-Hall, Englewood Cliffs, N.J., 1989.
5. Usher, M. J., *Sensors and Transducers*, Macmillan, London, 1985.
6. *Magnetic Sensors Data Book*, Siemens Components, Inc., Iselin, N.J., 1989.
7. Dukes, J. N., and G. B. Gordon, "A Two-Hundred-Foot Yardstick with Graduations Every Microinch," *Hewlett-Packard Journal*, August 1970.
8. Cole, Wayne, "Advanced Hygrometer for Dew/Frost Point Hygrometry," *Sensors*, July 1992, pp. 11–40.
9. Lynworth, L. C., "Temperature Profiling Using Multizone Ultrasonic Waveguides," James F. Schooley, Ed., *Temperature: Its Measurement and Control in Science and Industry,* vol. 5, part 2, pp. 1181–1190, American Institute of Physics, 1982. See also Tasman et al., "Ultrasonic Thin-Wire Thermometry for Nuclear Applications" (same volume).
10. *The Magnetoresistive Sensor,* Philips Technical Publication 268, Amperex Electronics Corporation, Hicksville, Long Island, N.Y., 1988.
11. Isotron Accelerometer, Model 63–100, Endevco Corporation, San Juan Capistrano, Calif., 1992.
12. Diefenderfer, A. J., *Principles of Electronic Instrumentation*, Saunders, New York, 1979.
13. Capacitive Humidity Sensor, Series 691, Philips Components, Mineral Wells, Tex., 1987, pp. 684–712.
14. Levine, Robert J., "Capacitance Transducer Theory, Design Formulae and Associated Circuitry," 20th Annual ISA Conference and Exhibit, Los Angeles, Calif., Oct. 4–7, 1965.
15. Dias, J. F., H. E. Karrer, and A. Tykulsky, "Capacitive Blood Pressure Transducer," *ISA Transactions*, vol. 19, no. 3, pp. 19–23, 1980.
16. Dias, J. F., H. E. Karrer, and A. Tykulsky, High Fidelity Pressure Transducer, U.S. Patent 4,064,550, Dec. 20, 1977.

17. Monolithic Accelerometer with Signal Conditioning, Product Specifications, ADXL50, Analog Devices, Norwood, Mass., 1993.
18. *The Pressure, Strain and Force Handbook*, vol. 28, LVDT Section J, Omega Engineering, Inc., Stamford, Conn., 1992.
19. Herceg, E. E., *Handbook of Measurement and Control, Schaevitz Engineering*, rev. edition, Pennsauken, N.J., March 1989.
20. Schaevitz Linear/Rotary Position Sensors, Catalog SCH-101, Lucas Schaevitz, Pennsauken, N.J., 1992.
21. Hannah, R. L., and S. A. Reed, Eds., *Strain Gage Users Handbook*, Elsevier and Society for Experimental Mechanics, Bethel, Conn., 1992.
22. "Strain Gages—Section E" and "Load Cells—Section F," *The Pressure, Strain and Force Handbook*, vol. 28, Omega Engineering, Inc., Stamford, Conn., 1992.
23. *Strain Gage Selection*, TN-505-2, Measurements Group, Inc., Raleigh, N.C., 1989.
24. *Kulite Semiconductor Strain Gage Manual*, KSGM-3, Kulite Semiconductor Products, Inc., Leonia, N.J.
25. Middelhoek, S., and S. A. Audet, *Silicon Sensors*, chap. 3, Academic Press, New York, 1989.
26. King, David J., "Resistance Elements and RTD's," *The Temperature Handbook*, vol. 28, Omega Engineering, Inc., 1992.
27. *Thermistor Sensor Handbook*, Thermometrics, Edison, N.J., 1987.
28. Siwek, W. R., *Thermometrics*, Edison, N.J., private communication, 1993.
29. Lomas, Charles G., *Fundamentals of Hot Wire Anemometry*, Cambridge University Press, Cambridge, 1986.
30. Constant Temperature Anemometer, Equipment Catalogue, Dantec Electronics, Inc., Mahwah, N.J., 1992.
31. Sze, S. M., *Semiconductor Devices, Physics and Technology*, pp. 36–40, Wiley, New York, 1985.
32. "Hall Effect Transducers," Microswitch, Honeywell Division, Freeport, Ill., 1982.
33. Middelhoek, S., and S. A. Audet, *Silicon Sensors*, chap. 5, Academic Press, New York, 1989.
34. *Magnetic Sensors Data Book*, Siemens, Iselin, N. J., 1989.
35. Sze, S. M., *Physics of Semiconductor Devices*, chap. 8 (MOSFET), Wiley, New York, 1981.
36. Edmonds, T. E., Ed., *Chemical Sensors*, chap. 10, Chapman-Hall, New York, 1988.
37. Madou, Marc, and S. Roy Martin, "Chemical Sensing with Solid State Devices," chap. 8, *Principles of ChemFET Operation*, Academic Press, New York, 1991.
38. Jaffe, B., W. R. Cook, Jr., and H. Jaffe, *Piezoelectric Ceramics*, Academic Press, New York, 1971.
39. Newnham, R. E., L. J. Bowen, K. A. Klicker, and L. E. Cross, "Composite Piezoelectronic Transducers," *Materials in Engineering*, vol. 2, pp. 93–106, 1980.
40. O. E. Mattiat, Ed., *Ultrasonic Transducer Materials*, Plenum Press, New York, 1971.
41. Kino, G. S., *Acoustic Waves, Devices, Imaging and Analog Signal Processing*, Prentice-Hall, Englewood Cliffs, N.J., 1987.
42. *Aerotech Transducers*, Krautkramer Branson, Lewistown, Pa., 1990.
43. *Ultrasonic Transducers for Non-Destructive Testing*, Panametrics, Inc., Waltham, Mass., 1992.
44. Fowler, K. A., F. H. C. Hotchkiss, and T. V. Yamartino, *Important Characteristics of Ultrasonic Transducers and Factors Related to Their Application*, Panametrics NDT Division, Waltham, Mass. See also ASTM E1065-90, October 1990.
45. Smith, W. A., "New Opportunities in Ultrasonic Transducers Emerging from Innovation in Piezoelectric Materials," SPIE International Symposium, July 19, 1992.
46. Dias, J. F., *Ultrasonic Imaging*, vol. 3, no. 4, 1981, pp. 352–368. See also Karrer et al., IEEE Ultrasonics Symposium Proceedings, 1980.
47. Born, N., and J. Roelandt, Eds., *Intravascular Ultrasound*, Kluwer Academic Publishers, 1989.
48. Pandian, Natesa G., Ed., "Intravascular and Intracardiac Ultrasound Imaging: Current Research and Future Directions," *A Journal of Cardiovascular Ultrasound and Allied Techniques*, vol. 7, no. 4, July 1990.
49. Wells, Peter N. T., *Physical Principles of Ultrasonic Diagnosis*, chap. 4 (Pulse Echo Techniques), Academic Press, New York, 1969.
50. Tykulsky, A., Hewlett-Packard Co. ISY Division, Andover, Mass., private communication, 1993. R. Pittaro, Hewlett-Packard Labs, Palo Alto, Calif., private communication, 1993.
51. Bray, D. E., and R. K. Stanley, *Non-Destructive Evaluation*, McGraw-Hill, New York, 1989.

52. *The Flow and Level Handbook*, Omega Engineering, Inc., Stamford, Conn., 1992.
53. Doucette, J. W., P. D. Corl, H. M. Payne, A. E. Flynn, M. Goto, M. Nassi, and J. Segal, "Validation of a Doppler Guide Wire for Intravascular Measurement of Coronary Artery Flow Velocity," *Circulation*, vol. 85, no. 5, pp. 1899–1911, May 1992.
54. Maines, J. D., G. S. Paige, A. F. Saunders, and A. S. Young, "Determination of Delay Time Variations in Acoustic Surface-Wave Structures," *Electronics Letters*, vol. 5, no. 26, pp. 678–680, Dec. 27, 1969.
55. Dias, J. F., and H. E. Karrer, "Stress Effects in Acoustic Surface-Wave Circuits and Applications to Pressure and Force Transducers," *IEEE International Solid State Circuits Conference Proceedings*, 1974, pp. 166–167.
56. Dias, J. F., H. E., Karrer, J. A. Kusters, J. H. Matsinger, and H. B. Schulz, *IEEE Transactions on Sonics and Ultrasonics*, vol. SU-22, no. 1, pp. 46–50, January 1975.
57. Adams, C. A., J. F. Dias, H. E. Karrer, and J. A. Kusters, "Frequency and Stress Sensitivity of SAW Resonators," *Electronics Letters*, vol. 12, no. 22, pp. 580–582, Oct. 28, 1976.
58. Dias, J. F., "Physical Sensors Using SAW Devices," *Hewlett-Packard Journal*, December 1981, pp. 18–20.
59. Wohltjen, H., "Surface Acoustic Wave Microsensors," Transducers 87, *Proceedings of the International Conference on Solid State Sensors, and Actuators*, Tokyo, Japan, June 1987.
60. Nieuwenhuizen, M. S., and A. Venema, "Surface Acoustic Wave Chemical Sensors," *Sensors and Activators*, 5, pp. 261–300, 1989.
61. Wohltjen, H., "Mechanism of Operation and Design Consideration for Surface Acoustic Wave Device Vapour Sensors," *Sensors and Activators*, 5, pp. 307–325, 1984.
62. Frye, G. C., S. J. Martin, R. W. Cernosek, and K. B. Pfeifer, "Portable Acoustic Wave Sensor Systems for On-Line Monitoring of Volatile Organics," *International Journal of Environmentally Conscious Manufacturing*, vol. 1, no. 1, pp. 37–45, 1992.
63. Binnig, G., and Heinrich Rohrer, "The Scanning Tunneling Microscope," *Scientific American*, August 1985, pp. 50–55.
64. Sonnenfeld, R., J. Schneir, and P. K. Hansma, "Scanning Tunneling Microscopy," R. E. White, J. O'M. Bokris, and B. E. Conway, Eds., *Modern Aspects of Electrochemistry,* no. 21, Plenum Publishing Company, pp. 1–28, 1990.
65. *The Scanning Probe Microscope Book*, Burleigh Instruments, Inc., Fishers, N.Y.
66. *The Piezo Book*, Burleigh Instruments, Inc., Fishers, N.Y.
67. Binnig, G., C. F. Quate, and Ch. Gerber, "Atomic Force Microscope," *Physical Review Letters*, vol. 56, pp. 930–933, 1986.
68. Kenny, T. W., W. J. Kaiser, S. B. Waltman, and J. K. Reynolds, "Novel Infrared Detector Based on a Tunneling Displacement Transducer," *Applied Physics Letters*, vol. 59, no. 15, 7, pp. 1820–1822, October 1991.
69. Kawai, H., "The Piezoelectricity of Poly (Vinylidene Fluoride)," *Japan Journal of Applied Physics*, vol. 8, pp. 975–976, 1969.
70. Wang, T. T., J. M. Herbert, and A. M. Glass, Eds., *The Applications of Ferroelectric Polymers*, Chapman Hall, 1988.
71. Brown, L. F., "Ferroelectric Polymers," *IEEE Ultrasonics Symposium*, October 1992.
72. DeReggi, A. S., S. C. Roth, J. M. Keeney, S. Edelman, and G. R. Harris, "Piezoelectric Polymer Probe for Ultrasonic Applications," *Journal of the Acoustical Society of America*, vol. 69, no. 3, March 1983.
73. Lewin, P. A., "Miniature Piezoelectric Polymer Hydrophone Problems," *Ultrasonics*, vol. 19, pp. 213–216, 1981.
74. *Dynamics Instrumentation Catalog*, Endevco, San Juan Capistrano, Calif., 1992.
75. General catalog, K2.006, 6/91, Kistler Instruments Corporation, Amhurst, N.Y., 1991.
76. "The International Temperature Scale of 1990," *The Temperature Handbook*, Omega Engineering, Inc., Stamford, Conn., 1992.
77. Zemansky, Mark W., *Heat and Thermodynamics*, chap. 13, "Thermoelectric Phenomena," McGraw-Hill, New York, 1968.
78. Sheingold, D. H., Ed., *Transducer Interfacing Handbook*, Analog Devices, Inc., Norwood, Mass., 1981.
79. Guthrie, A., *Vacuum Technology*, chap. 6, "Measurement of Pressure," Wiley, New York, 1963.
80. Karrer, H. E., "The Piezoelectric Resonator as a Temperature Sensor," ISA-71, Oct. 4–7, 1971.

81. Hammond, D. L., and A. Benjaminson, "The Linear Quartz Thermometer," *Hewlett-Packard Journal*, March 1965.
82. *Special Linear Reference Manual*, Analog Devices, Norwood, Mass., June 1992, pp. 9-7 to 9-15.
83. Yeung, King-Wah W., Hewlett-Packard Laboratories, Palo Alto, Calif., private communication, 1993.
84. Guthrie, Andrew, *Vacuum Technology*, Wiley, New York, 1963.
85. Roth, A., *Vacuum Technology*, Elsevier, New York, 1982.
86. *Varian Vacuum Products 1993/1994 Catalog*, Varian Associates, Inc., Lexington, Mass.
87. *Ionization Gage Tube Catalog*, Granville Phillips Company, Boulder, Colo., 1991.
88. Semat, H., *Introduction to Atomic and Nuclear Physics*, 5th ed., Holt, Rinehart and Winston, New York, 1972.
89. *Photomultiplier Tubes*, Hamamatsu Photonics, Bridgewater, N.J., October 1990 catalog.
90. *CdS Photoconductive Cells*, EG&G Vactec Optoelectronics, St. Louis, Mo., June 1990.
91. *Silicon Photodiodes*, EG&G Vactec Optoelectronics, St. Louis, Mo., October 1991.
92. *Photodiodes*, Hamamatsu Photonics, Bridgewater, N.J., June 1991.
93. *Optoelectronics Components Catalog*, UDT Sensors, Inc., Hawthorne, Calif., 1990.
94. Edwards, I., "Using Photodetectors for Position Sensing," *Sensors*, December 1988.
95. *Position-Sensitive Detectors*, Hamamatsu Photonics, Bridgewater, N.J., October 1988.
96. Adler, Robert, "Interaction between Light and Sound," *IEEE Spectrum*, May 1967, pp. 42–54.
97. Yariv, Ammon, *Introduction to Optical Electronics*, 2d ed., pp. 337–351, Holt, Rinehart and Winston, New York, 1976.
98. Whitman, R. L., and A. Korpel, "Probing of Acoustic Surface Perturbations by Coherent Light," *Applied Optics*, vol. 8, no. 8, pp. 1567–1576, 1969.
99. *Vibrometer*, Publication 9608E, Dantec Electronics, Inc., Allendale, N.J., 1990.
100. Application Note OP-35-0, Ultra Optec, Inc., Boucherville, Quebec, Canada, 1991.
101. F. P. Milanovich and A. Katzir, Eds., "Fiber Optic Sensors in Medical Diagnostics," *SPIE*, vol. 1886, Jan. 21, 1993.
102. Giallorenzi, T. G., J. A. Bucaro, A. Dandridge, G. H. Siegel, J. H. Cole, and S. C. Rashleigh, "Optical Fiber Sensor Technology," *IEEE Transactions on Microwave Theory and Techniques*, vol. MTT-30, no. 4, pp. 472–511, April 1982.
103. Udd, Eric, Ed., *Fiber Optic Sensors*, Wiley, New York, 1991.
104. Barnoski, M. K., *Fundamentals of Optical Fiber Communications*, Academic Press, New York, 1976.
105. Fields, J. H., and J. H. Cole, "Fiber Microbend Acoustic Sensor," *Applied Optics*, vol. 19, pp. 3265–3267, 1980.
106. Miers, D. R., D. Raj, and J. W. Berthold, "Design and Characterization of Fiber Optic Accelerometers," *Proceedings SPIE* 838, pp. 314–317, 1987.
107. Foster-Smith, R., and W. Marshall, *Intra-Arterial Blood Gas Monitoring—Part 1: Optical Sensor Technology,* Puritan-Bennett, Carlsbad, Calif., 1992. See also *Introducing the CDI 2000 Blood Gas Monitoring System,* 3M Health Care, Tustin, Calif., 1992.
108. Wise, D. L., and L. B. Wingard, Eds., *Biosensors with Fiberoptics,* Humana Press, New Jersey, 1991.
109. Mauze, G. R., and R. R. Holloway, "A Noncrosslinked Organosilicon Matrix for Luminescence Quenching-Based Fiber Optic Oxygen Sensor," *SPIE,* vol. 1886, January 1993.
110. Gehrich, J. L., D. W. Lubbers, N. Opitz, D. R. Hansman, W. W. Miller, J. K. Tusa, and M. Yafuso, "Optical Fluorescence and Its Application to an Intervascular Blood Gas Monitoring System," *IEEE Transactions on Biomedical Engineering,* vol. BME-33, no. 2, pp. 117–132, February 1986.
111. Mauze, G. R., Hewlett-Packard Laboratories, Palo Alto, Calif., private communication, 1993.

Chapter

6

Analog-to-Digital Converters

John J. Corcoran
Agilent Technologies
Palo Alto, California

6.1 Introduction

Analog-to-digital converters, often referred to as A/D converters or simply ADCs, have played an increasingly important role in instrumentation in recent years. Once central to the function of only basic instruments like digital voltmeters, ADCs are now at the heart of more complex instruments like oscilloscopes and spectrum analyzers. In many cases, external instrument specifications are limited by the performance of the internal A/D converter.

The increasing importance of ADCs to instruments has been driven by the development of high-performance integrated circuit (IC) technology. This has enabled higher-speed and higher-resolution ADCs to be designed, manufactured, and sold at reasonable cost. Equally important, advanced IC technology has led to the microprocessor and fast digital signal processing capability, which are essential in providing a low-cost transformation from the raw data generated by the ADC to the measurement results sought by the user (Fig. 6.1).

To understand modern instrument performance it is important to understand the basics of ADCs. This chapter first defines an analog-to-digital converter and describes typical ADC building blocks. The three most common ADC architectures are then described in some detail. The chapter concludes with a discussion of A/D converter specifications and testing techniques.

6.2 What Is An Analog-to-Digital Converter?

The basic function of an A/D converter is to convert an analog value (typically represented by a voltage) into binary bits that give a "good" approximation to

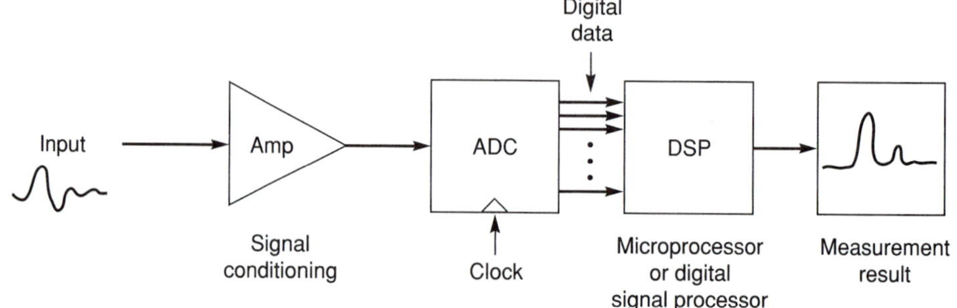

Figure 6.1 A generic ADC-based measurement system. Digital signal processing transforms the raw ADC data into a more usable result.

that analog value. Conceptually (if not physically), this process can be viewed as forming a ratio between the input signal and a known reference voltage V_{ref}, and then rounding the result to the nearest n-bit binary integer (Fig. 6.2). Mathematically, this process can be represented by

$$D = \text{rnd}\left(\frac{V_{in}}{V_{ref}} 2^n\right) \tag{6.1}$$

where V_{in} is the analog value (assumed here to have an allowed range of 0 to V_{ref}), D is the data output word, and n is the resolution of the converter (the number of bits in D). The "rnd" function represents a rounding of the term in parentheses to the nearest integer.

The reference voltage is typically a precise value generated internally by commercial converters. In some cases it may be externally supplied. In either case the reference voltage essentially sets the full-scale input range of the converter.

6.2.1 Resolution

Clearly, the quality of the converter's approximation of the input will improve when the converter resolution n is larger, since the rounding error is smaller. This rounding error, called the quantization error, is an inevitable property of even an otherwise perfect ADC. The quantization error has a range of ±½

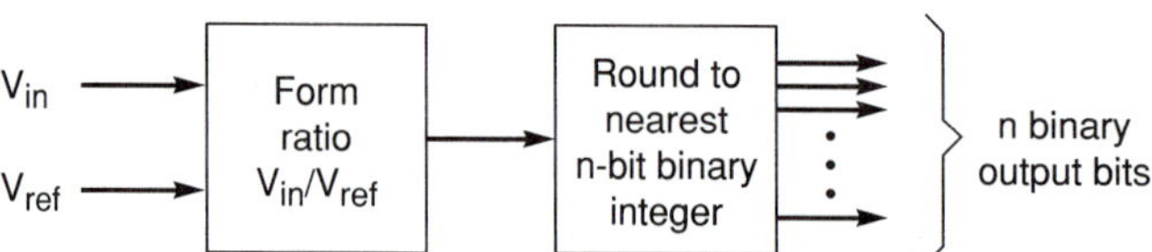

Figure 6.2 Conceptual view of the function provided by an ADC: Form the ratio between V_{in} and V_{ref}, then round to the nearest n-bit binary integer.

TABLE 6.1 Binary versus Decimal Representation

ADC resolution in binary bits	Approximate equivalent in decimal digits
6	2
8	2.5
10	3
12	3.5
16	5
20	6

LSB (least significant bit), where one LSB = $V_{ref}/2^n$. Resolutions are typically expressed in binary bits. Corresponding resolutions in decimal digits for a few common cases are shown in Table 6.1.

6.2.2 Sample rate

By their nature, ADCs are discrete time devices, delivering new output results only periodically at a frequency called the sample rate. In some cases the input may be essentially instantaneously sampled, and in that case each ADC output code can be associated quite directly with a particular input sampling time (Fig. 6.3*a*). In other cases the input may be intentionally averaged over a long period to reduce noise, and each ADC output code is then associated with one of these periods (Fig. 6.3*b*). In either case the rate of output code occurrence is an important property of the converter and, along with the resolution, defines the basic suitability of the converter for a particular application.

6.2.3 Errors

Various problems nearly always occur that make the errors of an ADC larger than the quantization error. Offset and gain errors can be introduced by circuits which process the input and reference. Noise can add a statistical variation in output code even if the ADC input is not changing. And linearity errors, which represent deviations from the linear correspondence between V_{in} and D in Eq. 6.1, are quite common. ADC errors and testing techniques are described more completely in Secs. 6.7 and 6.8.

6.2.4 Building blocks of analog-to-digital converters

A/D converters are typically built from a few common electronic building blocks. In the analog half of the converter these include sample and hold circuits, comparators, integrators, and digital-to-analog (D/A) converters. For the reader unfamiliar with these analog components, their functions are briefly described below. In the digital half of the converter, conventional logic gates and latches are used.

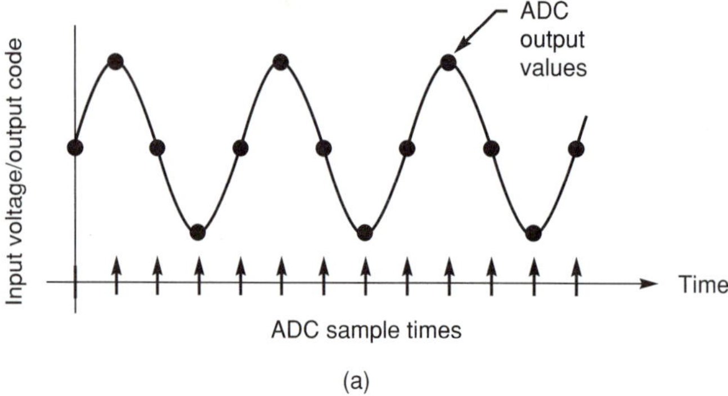

(a)

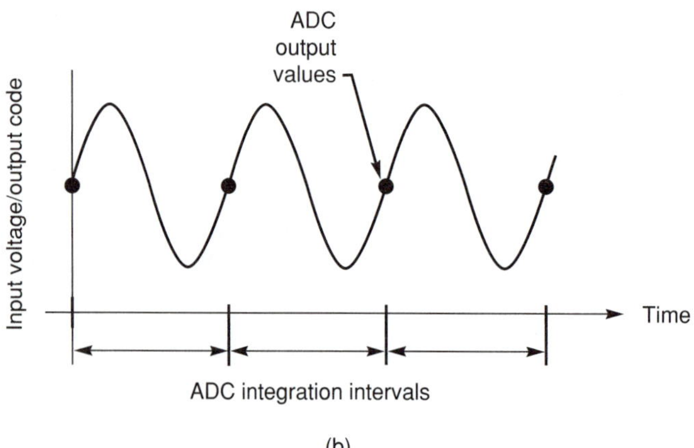

(b)

Figure 6.3 (*a*) Example of a sampling ADC converting a sine wave into a succession of values which approximately reproduce the sine-wave shape. (*b*) Example of an integrating ADC averaging a sine wave over its period. The output codes then represent the average value of the sine wave.

Sample and hold circuits are used to take a sample of the input voltage and hold it essentially constant while the A/D conversion is taking place. This function is usually implemented with a switch and capacitor (Fig. 6.4). The switch can be implemented in various electronic ways, with perhaps the most obvious being a single MOS transistor.

Comparators convert the difference between their two analog inputs into a logic zero or one, depending on the sign of that difference. Unclocked comparators (Fig. 6.5*a*) often consist of simply a very high gain amplifier with an output swing limited to the two logic state voltages. But the output of such an amplifier will be logically ambiguous if the two inputs are very close to-

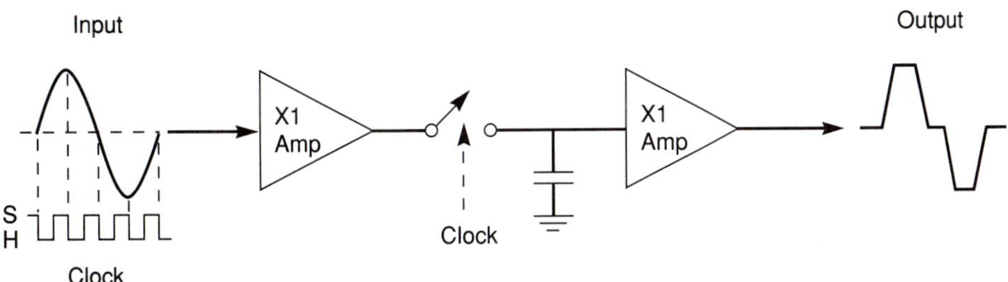

Figure 6.4 Simplified sample and hold circuit. When the switch is opened by the clock signal, a sample of the input is captured on the hold capacitor.

gether. For this reason, clocked comparators are nearly always used in A/D converters.

A clocked comparator (Fig. 6.5*b*) consists of an amplifier and a switchable positive feedback element. When the clock occurs, the positive feedback is turned on and the gain of the amplifier is thus made essentially infinite.

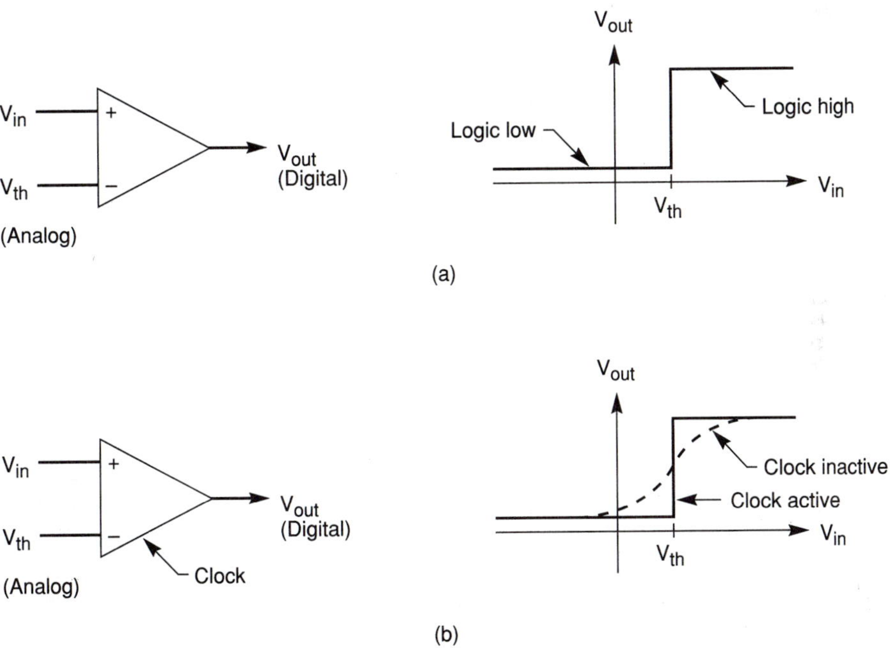

Figure 6.5 (*a*) An unclocked comparator can be made from a high-gain differential amplifier. The output is a logic high if the input exceeds the threshold V_{th}. Unclocked comparators are not usually used in ADCs. (*b*) A clocked comparator typically has low gain when the clock is inactive, and infinite gain (from positive feedback) when the clock becomes active. The output goes to a logic high or low depending on the difference between V_{in} and V_{th} when the clock occurs.

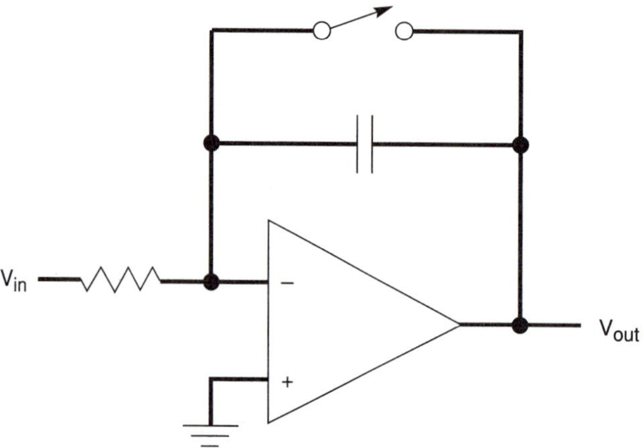

Figure 6.6 A simple integrator circuit. The switch clears charge from the capacitor before each new integration.

Although ambiguous outputs still can occur, the likelihood can be made very low.

Clocked comparators also lend themselves well to use in A/D converters since they essentially "sample" their input at the time of the clock. This allows properly synchronized operation of the comparators within the overall converter algorithm. In some cases clocked comparators can also eliminate the need for a separate sample and hold circuit.

Integrators (Fig. 6.6) in ADCs are generally implemented with the familiar

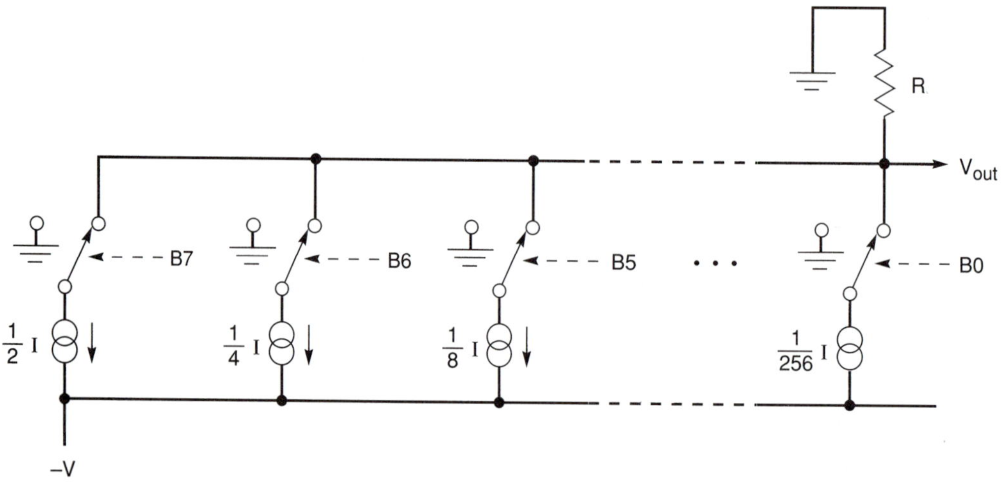

Figure 6.7 Simplified block diagram of an 8-bit digital-to-analog converter (DAC). Binary weighted currents are switched to the output resistor based on the input bits B0–B7.

op-amp and capacitor configuration. The average of the input signal over the integration time is stored in the capacitor and appears with negative sign at the amplifier output. Memory of the last value integrated must be removed before each new integration time can begin. Not surprisingly, integrators are a key element in the integrating ADC, as described below.

Digital-to-analog (D/A) converters are of course the inverse of analog-to-digital converters. DACs are used as building blocks for ADCs (instead of the reverse) since DACs are typically much simpler elements. A block diagram for a basic D/A is shown in Fig. 6.7. Binary weighted currents are switched (or not switched) to an output summing node based on the digital input code. The selected currents develop a voltage across a resistor, which is the desired analog output.

6.3 Types of Analog-to-Digital Converters

There are three types of A/D converters in common use in instruments today, described here as integrating, parallel, and multistep. A fourth type, the oversampled converter,[1–3] is not widely used in instrumentation and therefore is not covered here. Its low noise and low distortion, however, may make it well suited to certain moderate sample rate instrument applications in the future.

6.3.1 Integrating ADCs

Integrating ADCs function by integrating (averaging) the input signal over a fixed time, in order to reduce noise and eliminate interfering signals. This makes them most suitable for digitizing signals that are not very rapidly changing, as in the signals most commonly applied to digital voltmeters. They may also find applications where the signal is changing to some extent but the desired result is in fact a time average of the signal.

The integration time is typically set to one or more periods of the local ac power line in order to eliminate noise from that source. With 60-Hz power, as in the United States, this would mean an integration time that is a multiple of 16.67 ms. Many periods of the line might be selected as the averaging time when it is also desired to average random noise present on the signal and thus increase the resolution of the measurement.

In general, integrating converters are chosen for applications (like digital voltmeters) where high resolution and accuracy are important but where extraordinarily high sample rates are not. Resolutions can exceed 28 bits (8½ decimal digits) at a few samples/s, and 16 bits at 100K samples/s.[4] Integrating ADCs are described in more detail in Sec. 6.4.

6.3.2 Parallel ADCs

Parallel ADCs are used in applications where very high bandwidth is required but moderate resolution (6 to 10 bits) is acceptable, as in digitizing oscilloscopes. These applications require essentially instantaneous sampling of the input signal and high sample rates to achieve their high bandwidth.

The parallel architecture (described in more detail in Sec. 6.5) suits this need since it essentially consists of an array of comparators, all of which are clocked to sample the input simultaneously. Since comparators can be clocked at very high rates, the A/D converter sample rate can be correspondingly high. The price paid for this speed is the large number of comparators, exponentially increasing with resolution, and the associated high power and input capacitance. Sample rates in excess of 1 GHz have been achieved with parallel converters or arrays of interleaved parallel converters.[5–7]

6.3.3 Multistep ADCs

Multistep A/D converters convert the input in a series of steps instead of "all at once" as in a parallel converter. Multistep converters are generally built from building blocks consisting of parallel ADCs and digital-to-analog converters (DACs). Some speed is sacrificed owing to the sequential operation of these components. Furthermore, a sample and hold circuit (see Secs. 6.2.4 and 6.5.3) is required to capture the changing input signal and hold it constant through the sequential operation of the converter.

Multistep converters are typically used in moderate sample rate (1- to 100-MHz) applications. The multistep architecture offers higher resolution or lower power than parallel converters operating at the same rate. Spectrum analyzers, for example, typically require 12 to 14 bits of resolution (for about –80 dB distortion) and gain benefits in bandwidth by using the highest sample rate available. These resolutions are not very practical with parallel converters but are well within the range of the multistep architecture. Multistep converters are described in more detail in Sec. 6.6.

6.4 Integrating Analog-to-Digital Converters

Integrating ADCs are used where very high resolution at comparatively low sample rate is desired. They function by integrating (averaging) an input signal over a selected period of time and are therefore usually used for measuring dc voltages. The averaging has the effect of reducing the noise on the input. If the averaging time is chosen to be one or more power-line cycles, power-line interference is largely removed from the measurement.

The most common application is in digital voltmeters, which take advantage of the exceptional resolution, linearity, stability, and noise rejection typical of the integrating architecture.

6.4.1 The dual slope architecture

The dual slope approach is perhaps the most commonly used integrating A/D architecture (Fig. 6.8). There are two half cycles, referred to here as the up slope and the down slope. The input signal is integrated during the up slope

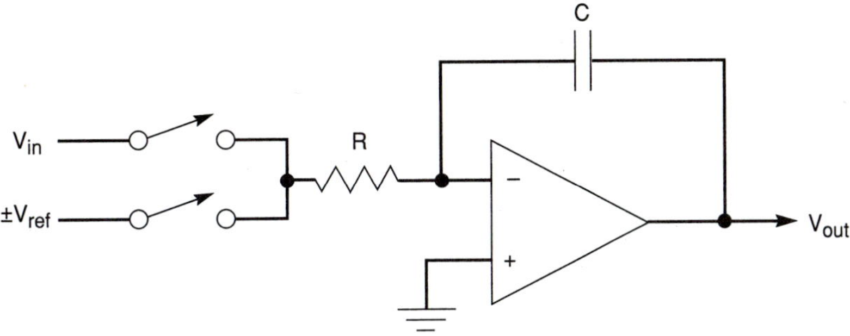

Figure 6.8 Simplified dual slope ADC block diagram.

for a fixed time (Fig. 6.9). Then a reference of opposite sign is integrated during the down slope to return the integrator output to zero. The time required for the down slope is proportional to the value of the input and is the "output" of the ADC.

The up slope cycle can be described mathematically as follows:

$$V_p = -\frac{T_{up} V_{in}}{RC} \tag{6.2}$$

where V_p is the peak value reached at the integrator output during the up slope, T_{up} is the known up slope integration time, V_{in} is the input signal, and R and C are the integrator component values.

The down slope can be similarly described by

$$V_p = \frac{T_{dn} V_{ref}}{RC} \tag{6.3}$$

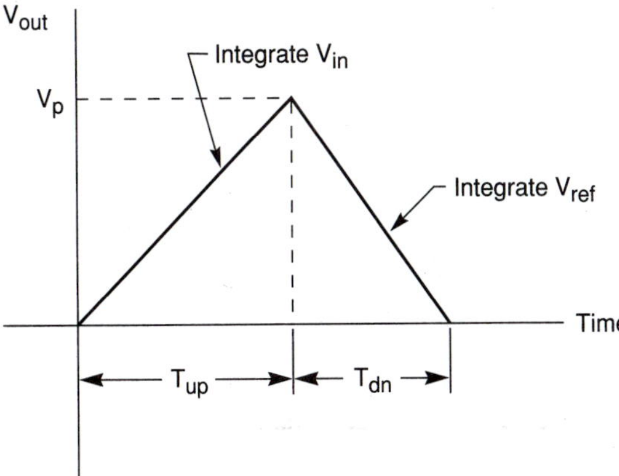

Figure 6.9 Typical dual slope ADC waveform. The input is integrated first for a fixed time T_{up}. The reference is integrated second, and the time T_{dn} is measured.

where T_{dn} is the unknown time for the down slope, and V_{ref} is the known reference. Equating 6.2 and 6.3 and solving for T_{dn}, the output of the ADC:

$$T_{dn} = -\frac{T_{up} V_{in}}{V_{ref}} \tag{6.4}$$

It should be noted here that V_{in} and V_{ref} will always be of opposite sign (to assure a return to zero in the integrator), so that T_{dn} will always be positive.

It can be immediately seen in Eq. 6.4 that the values of R and C do not appear in T_{dn}, so that their values are not critical. This is a result of the same components having been used for both the up and down slopes. Similarly, if the times T_{up} and T_{dn} are defined by counting periods of a single clock, the exact period of that clock will not affect the accuracy of the ADC. Restating the output in terms of the number of periods of the clock:

$$N_{dn} = -\frac{N_{up} V_{in}}{V_{ref}} \tag{6.5}$$

where N_{up} is the fixed number of clock periods used in the up slope and N_{dn} is the number of clock periods required to return the integrator output to zero.

Potential error sources. It is clear from Eq. 6.5 that N_{dn}, the numeric output of the ADC, is dependent only on the input, the reference, and the known value N_{up}. Errors in V_{ref} will affect the gain accuracy of the ADC, but this is implicit in any converter.

Offset errors can enter if the voltage at the start of the up slope is different from the voltage at the end of the down slope. If a single comparator on the integrator output is used to define the zero crossing time in both slopes, its offset will not be important. However, offset errors can also occur because of charge injection from the switches that select the input and reference. In very high accuracy voltmeter applications, these offsets are usually compensated by an auto-zero cycle.[8]

Linearity of the converter can be affected by "memory" effects in the integrator capacitor. This can be caused by a phenomenon called dielectric absorption, in which charge is effectively absorbed by the capacitor dielectric during long exposure to one voltage and then returned to the plates of the capacitor later when another voltage is applied. Choice of a dielectric material with very low absorption can be used to minimize this effect.

Resolution-speed tradeoff. The up slope integration time can be set to exactly N_{up} periods of the clock. However, the time to return the integrator output to zero will not be exactly an integer number of clock periods, since V_{in} can assume any value. In fact, there will always be an ambiguity of ± 1 count in how well N_{dn} describes V_{in}.

The resolution of a dual slope ADC is thus essentially one count in N_{max}, where N_{max} is the number of counts accumulated in the down slope after integrating a full-scale input $V_{in} = V_{fs}$ (assumed to be fixed). Referring to Eq. 6.5,

$$N_{max} = -\frac{N_{up}V_{fs}}{V_{ref}} \tag{6.6}$$

To improve resolution, N_{max} must be increased. This can be done by increasing N_{up}, which has the effect of linearly increasing the time required for both the up slope and the down slope. Or V_{ref} could be lowered, so that the up slope time is constant but the down slope time is increased linearly. In either case (in the limit), the increased resolution requires a linear increase in the number of clock periods in the conversion. Assuming a practical limit to the minimum clock period, increased resolution comes at the direct expense of conversion time. This problem can be alleviated significantly by the use of the multislope architecture.

6.4.2 The multislope architecture

A block diagram of a typical multislope ADC[4] is shown in Fig. 6.10. It differs from the dual slope approach in that there are separate up and down integration resistors, and furthermore, there are multiple values for the down slope integration resistors.

Using different resistors for the up and down slope portions introduces the possibility of errors due to resistor mismatch. The dual slope is immune to this problem since only one resistor is used. However, high-quality resistor networks with good temperature tracking and linearity can overcome this disadvantage.[4]

The advantage of the multislope architecture is a decrease in conversion time or an increase in resolution. A significant reduction in conversion time can be obtained first of all by making R_{up} (connected to V_{in}) considerably smaller. The integrator charging current will be increased, utilizing the full dynamic range of the integrator in less time.

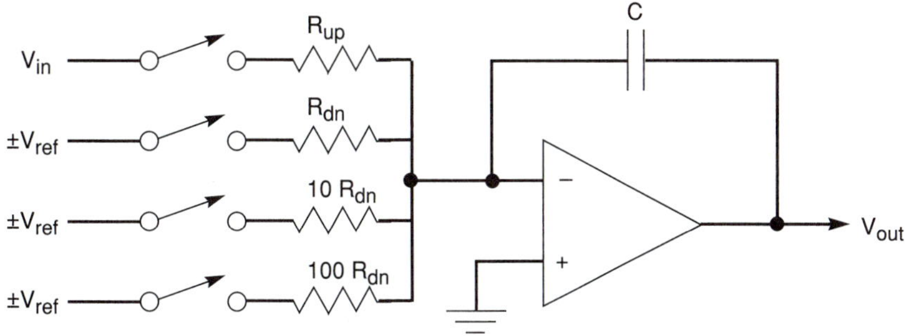

Figure 6.10 Multislope ADC block diagram. Separate up and down integration resistors (and multiple down resistor values) increase conversion speed.

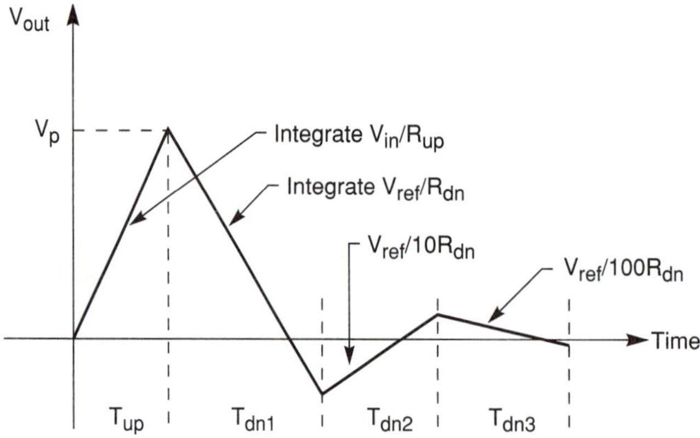

Figure 6.11 Typical multislope ADC waveform. The input is integrated first through resistor R_{up}. References of alternating sign are then integrated through resistors of increasing value.

Next, the time required for the down slope at a given resolution can be reduced by operating multiple "down" slopes, each at successively lower currents (Fig. 6.11). In the example of Fig. 6.11, the first down current is opposite in sign to the input, and sufficiently large that the integrator will cross zero in less than 10 counts.

When the integrator output crosses zero, the current is turned off at the next clock transition. The amount by which the integrator overshoots zero depends on the exact input voltage. To digitize this "residue" accurately, a second, 10 times lower, opposite sign down slope current is selected. Zero is crossed again but from the opposite direction, and with 10 times lower slope. The overshoot is once again proportional to the exact input but will now be 10 times lower in amplitude owing to the lower slope. The counts accumulated in this down slope phase are accorded 10 times lower significance.

An indefinite number of these down slopes can be successively applied, each adding (in this example) a decade to the resolution but adding very little percentage increase to the overall conversion time. The multislope approach can be implemented with decade steps in the down slopes as described here, or with other ratios. Even further increases in resolution can be achieved by applying a multislope up period, in which both the input and an offsetting reference current are applied.[4,8] In summary the multislope approach offers dramatic improvements in resolution-speed tradeoff compared with the simple dual slope architecture, at the expense of more complexity and the need for well-matched resistors.

6.5 Parallel Analog-to-Digital Converters

As mentioned in Sec. 6.3, parallel ADCs are used in applications where very high bandwidth and sample rates are required, and moderate resolution is

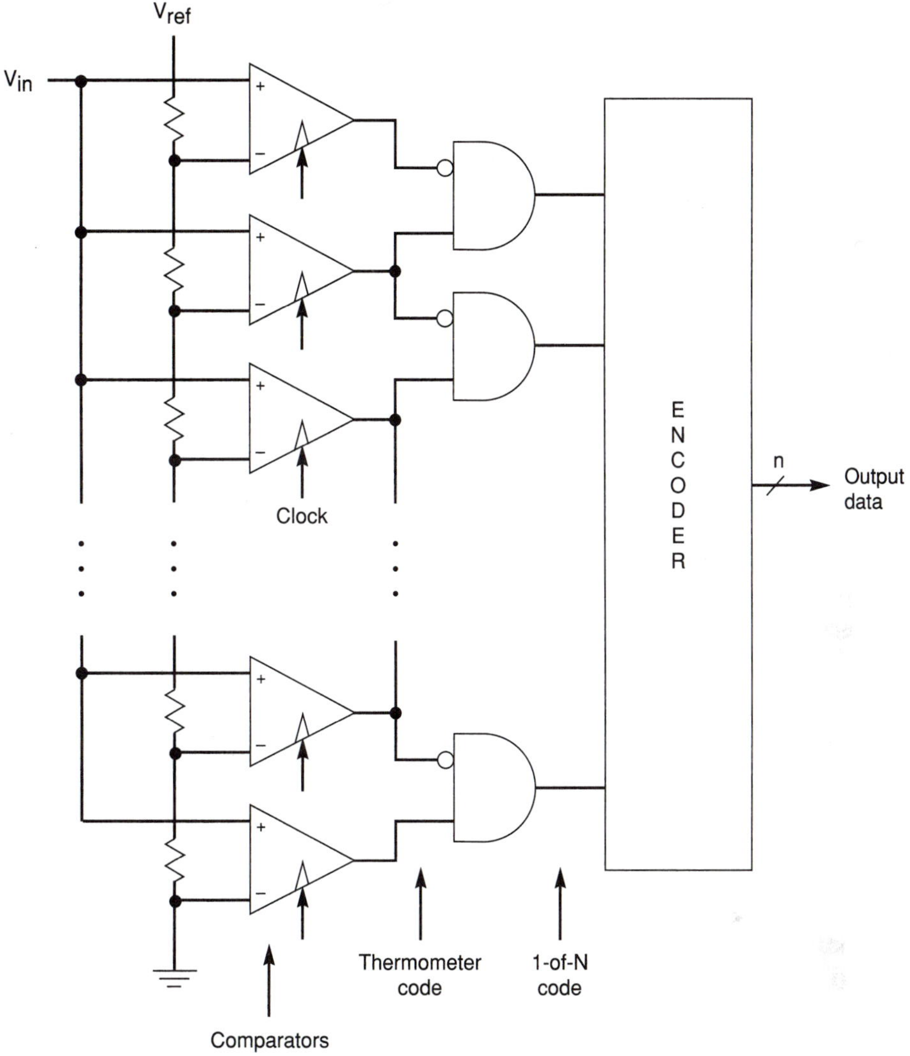

Figure 6.12 Block diagram of a flash A/D converter. The clocked comparators sample the input and form a thermometer code, which is converted to binary.

acceptable. A typical application is real-time digitizing oscilloscopes, which can capture all the information in a signal in a single occurrence. ADCs are also used in repetitive digitizing oscilloscopes, but high real-time sample rates are not required (see Chap. 14 for more detail).

6.5.1 Flash converters

The most familiar of the parallel class of A/D converters is the "flash" converter (Fig. 6.12), so called because 2^n clocked comparators simultaneously

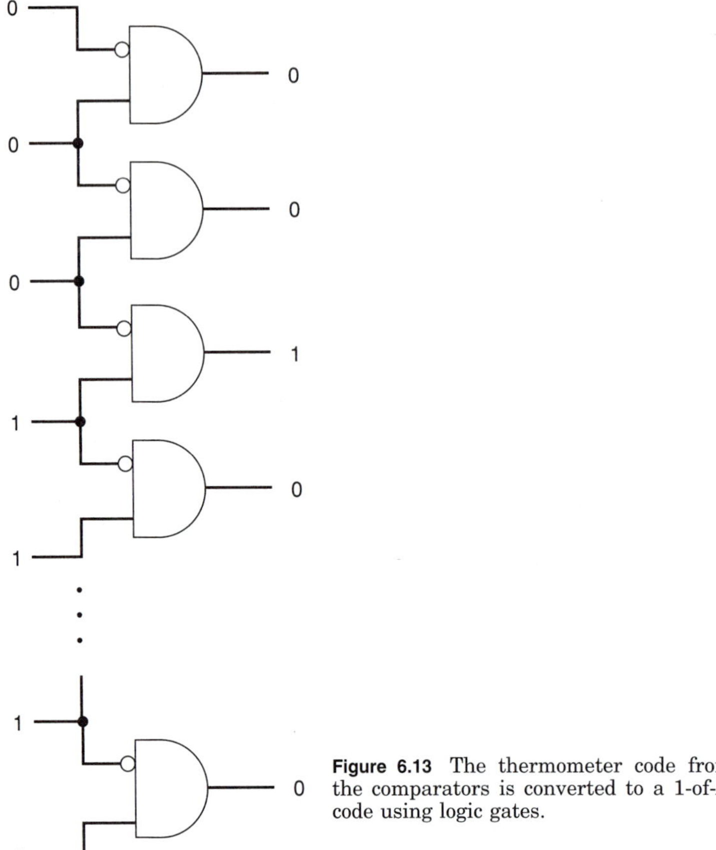

Figure 6.13 The thermometer code from the comparators is converted to a 1-of-N code using logic gates.

sample the waveform (where n is the converter resolution). Each comparator is supplied with a distinct threshold voltage, generated by resistor dividers from the main converter reference voltage. These thresholds taken together span the input range of the converter. The output bits from these comparators form a thermometer code, so called because it can be displayed as a column of continuous ones below a similar sequence of zeros (Fig. 6.13). The transition from ones to zeros in the sequence indicates the value of the sampled input signal. This transition is detectable with a simple logic gate, resulting in a "1-of-N" code (where $N = 2^n$), since only one bit is a one. The 1-of-N code can then be encoded further with straightforward logic to a binary code of n bits, the desired output of the converter.

Flash converters are very fast, since the speed of clocked comparators and logic can be quite high. This makes them well suited to real-time oscilloscope applications. However, numerous disadvantages also exist. The complexity of the circuits grows rapidly as resolution is increased, since there are 2^n clocked comparators. Furthermore, the power, input capacitance, clock capacitance, and physical extent of the comparator array on the integrated circuit all increase directly with the number of comparators.

The size of the comparator array is important since the flash converter typically samples rapidly changing input signals. If all comparators do not sample the input at the same point on the waveform, errors can result. Furthermore, propagation delays of the signal to all comparators are difficult to match as the array size increases. This is one reason that flash converters are typically used in conjunction with a sample and hold circuit (Secs. 6.2.4 and 6.5.3), which samples the input and ideally provides an unchanging signal to all comparators at the time of clocking.

Modifications to the flash architecture can be used to reduce the cost of higher resolution. These techniques, which include analog encoding,[9–11] folding,[12,13] and interpolating,[14–16] can reduce input capacitance and the size of the comparator array considerably.

6.5.2 Dynamic errors in parallel ADCs

The flash A/D architecture and its variants all suffer to some extent from dynamic errors if no sample and hold is used. Dynamic errors are defined here as those which result when high-frequency input signals are applied to the ADC. A common dynamic error is that due to the large nonlinear (voltage-dependent) input capacitance of the ADC. This capacitance is nonlinear since it consists largely of semiconductor junctions. When this input capacitance is driven from a finite impedance source, distortion can occur at high frequencies.

Another class of dynamic errors occurs if the input and clock signal are not delivered simultaneously to all the comparators in the ADC. Even in a monolithic implementation, the physical separation of the comparators can be large enough to make this difficult for very high frequency inputs. For a 1-GHz sine wave at the zero crossing, the rate of change is so high that in 10 ps the signal changes by 3 percent of full scale. To accurately digitize this signal, all comparators must be driven by the same point on the signal when the clock occurs. If there are mismatches in the delays in the clock or signal distribution to the comparators of only 10 ps, there will be a difference of 3 percent in the value of the signal perceived by different comparators. The resulting set of comparator outputs, after interpretation by the encoder which follows, could result in large output code errors.

Both of these errors tend to get worse as the converter resolution is increased, since the input capacitance and size of the comparator array both grow. This may limit practically achievable resolution before power and complexity constraints enter in. Sample and hold circuits are typically used with parallel ADCs to relieve these problems.

6.5.3 Sample and hold circuits

Sample and hold circuits eliminate dynamic errors from parallel ADCs by assuring that the comparator input signal is not changing when the compara-

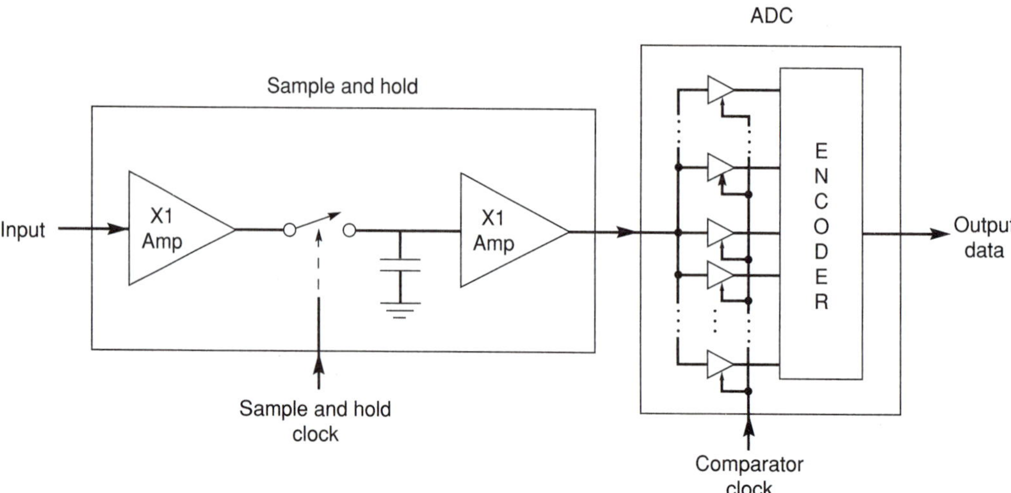

Figure 6.14 Sample and hold circuit driving parallel ADC. When the switch opens, the input value is held on the capacitor. When the ADC input line has settled to a stable value, the comparators are clocked.

tor clock occurs. A conceptual model of a sample and hold driving an ADC is shown in Fig. 6.14. When the switch is closed, the voltage on the hold capacitor tracks the input signal. When the switch is opened, the capacitor holds the input value at that instant. This value is applied to the ADC input through the amplifier, and after settling, a stable value is available for the comparators. Only then are the comparators clocked, eliminating the signal distribution problem referred to above and all other dynamic errors associated with the comparators.

Of course, there are limits to the dynamic performance of the sample and hold circuit as well. To the extent that it has nonlinear input capacitance, the same high-frequency distortion referred to above (Sec. 6.5.2) will occur. However, this effect will typically be much lower, since sample and hold input capacitance is typically much lower than that of a parallel converter.

Another common sample and hold dynamic problem is aperture distortion. This refers to distortion introduced by the nonzero turn-off time of the sampling circuit in the system. This can introduce distortion when sampling a high-frequency signal, since the effective sampling point on the signal can be a function of the signal rate of change (slew rate) and direction. For this reason, much attention is paid to the design of the switch used in the sample and hold circuit. MOS transistors can be used directly as sampling switches, and improvements in transistor speed lead to better sample and hold performance.

Another high-performance sampler configuration often used is the diode bridge, shown in Fig. 6.15. With the currents flowing in the direction shown, the switch is on. The input signal is effectively connected to the hold capacitor

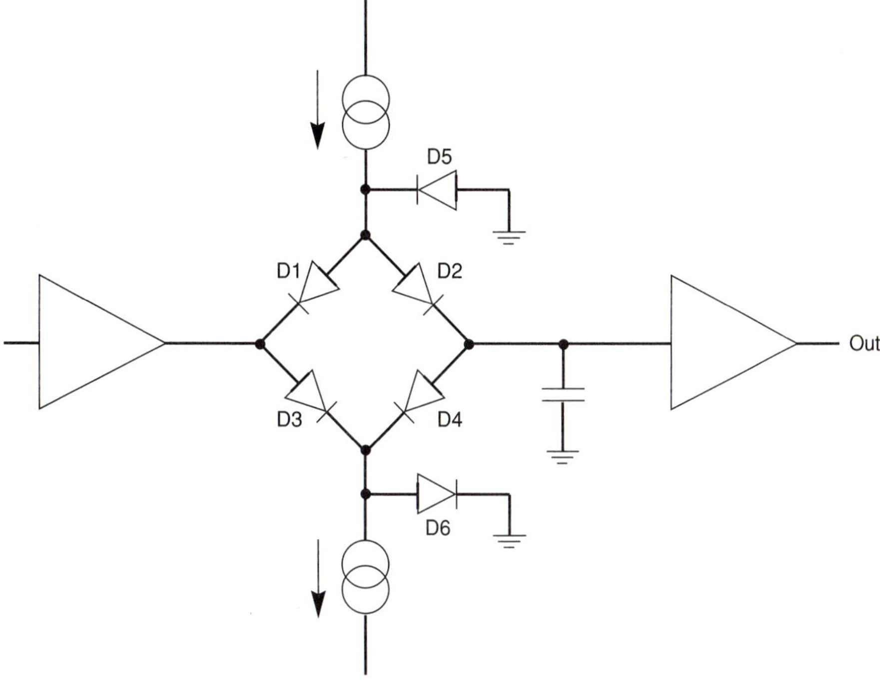

Figure 6.15 Diode bridge circuit for use as sampling switch. The switch is on with the currents flowing in the direction shown. It is turned off by reversing the direction of current flow.

through the conducting diodes D1 to D4. Diodes D5 and D6 are off. To turn the switch off, the currents are reversed. Diodes D5 and D6 now conduct, and all the other diodes are off. The input signal is isolated from the hold capacitor by the OFF series diodes D1 to D4 and the ON shunt diodes D5 and D6.

Diode bridge samplers are often built from Schottky diodes, which have the advantage of no stored charge. They can thus be turned off quickly, offering low aperture distortion. Very high performance sample and hold circuits have been built using this approach.[5,17]

6.5.4 Interleaving ADCs

Regardless of the sample rate of the A/D converters available, higher sample rates are always desired. This is especially true in real-time oscilloscope applications where the realizable bandwidth is directly proportional to the sample rate. To obtain higher sample rates, arrays of converters are often interleaved. For example, four 1-GHz converters, driven by a single input signal, might be operated with their clocks spaced at 90° intervals. This creates an aggregate input sample rate of 4 GHz, raising the realizable bandwidth from a typical 250 MHz to 1 GHz. (Of course, to achieve this 1-GHz bandwidth, the sampling circuit in the ADC must also have 1-GHz bandwidth).

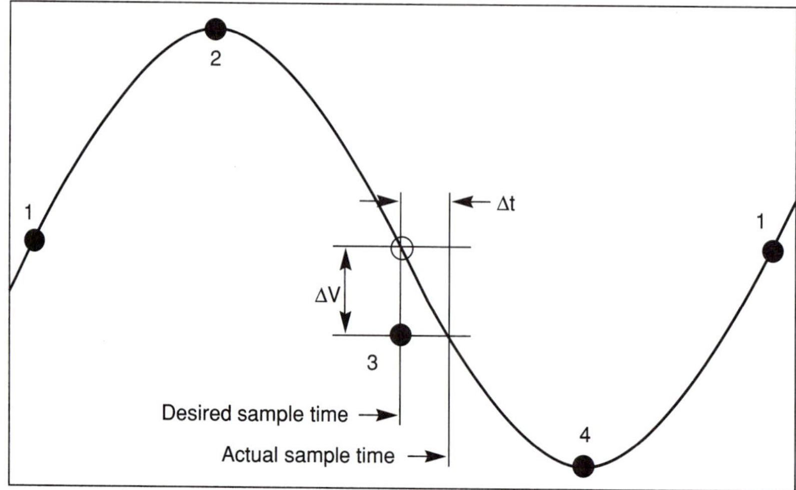

Figure 6.16 The effect of timing errors in an interleaved ADC system. The clock to ADC number 3 is misphased in this example, resulting in a significant voltage error for high-frequency inputs.

But interleaving often introduces errors due to mismatches in the characteristics of the individual ADCs. Offset and gain errors in a single noninterleaved ADC may produce relatively innocuous errors not important to the application. In an interleaved system, differences in the offset and gain of individual ADCs can be translated to spurious frequency components at submultiples of the sample rate. These would be especially undesirable if the spectrum of the signal were of interest.

Fortunately, offset and gain errors in interleaved ADC systems can generally be calibrated out. More difficult is to remove the effect of dynamic mismatches in the ADCs. These have two sources: inaccurate phasing of the clocks that interleave the ADCs, and different bandwidths in the sampler circuits that precede the ADCs.

The effect of clock phase errors is illustrated in Fig. 6.16, which shows the effect of one misphased converter clock in a four-way interleaved ADC system. For a 1-GHz input signal, a 10-ps clock phase error results in a 3 percent error in the value of the sample taken. This is a direct result of the high slew rate of the signal to be digitized. Misphased clocks in interleaved ADC systems can produce spurious frequency components and changes in shape or timing in a reconstructed waveform. A two-rank sample and hold circuit[5] which samples the input with only one sampler can essentially eliminate this problem. Calibration procedures that adjust the phases of the clocks can also help to reduce these effects.

The effect of bandwidth mismatch is similar to that of timing mismatch. Calibration to reduce the effect is more difficult, however, requiring adjustment of the frequency response of an analog circuit rather than merely adjusting the delay of a digital one.

Section 6.8.4 describes a test that can measure the timing errors due to use of interleaving in an ADC system.

6.6 Multistep Analog-to-Digital Converters

Multistep converters are often used when the resolution requirements of the application exceed the resolution available in parallel converters. A typical application for multistep converters is in direct digitizing spectrum analyzers, where 12-bit resolution is typically required at the highest sample rate available. Direct digitizing spectrum analyzers are defined here as those that use a Fourier transform of an ADC output record to compute a spectrum. They typically offer higher measurement throughput than classic analog spectrum analyzers with a swept oscillator and mixer architecture. "Multistep" as used here includes a wide variety of ADC architectures, including the two-step, ripple-through, and pipeline architectures described below. Multistep also includes the simple but lower-speed successive approximation architecture.

6.6.1 Two-step analog-to-digital converters

A simple example of a multistep ADC is a two-step converter with 12-bit resolution (Fig. 6.17). The input signal is captured by a sample and hold and digitized by a parallel converter with 6-bit resolution. The digital result is then converted by a digital-to-analog converter (DAC) to analog form and subtracted from the input. The resulting small "residue" (the difference between the input and the nearest one of the 64 ADC "rounding" levels) is amplified by 64 and then digitized by another parallel 6-bit ADC. The two

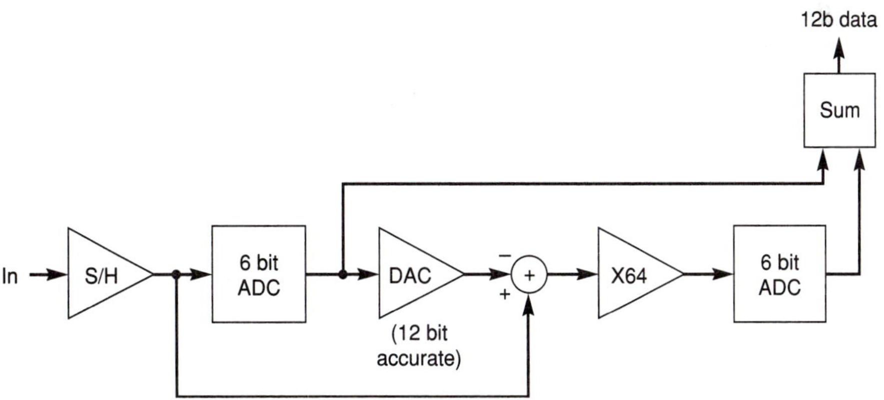

Figure 6.17 Block diagram of a 12-bit two-step ADC. The two 6-bit ADC output codes are summed to give the 12-bit result.

6-bit results are summed with appropriate weights to yield the 12-bit output code.

The advantage of the two-step architecture is clear from this example. The signal has been resolved to 12 bits, but only 128 (2×64) comparators are required. A fully parallel converter would have required 4096 comparators. A two-step converter offers lower power, complexity, and input capacitance than a parallel converter of the same speed.

The price paid is the addition of a sample and hold circuit and DAC. The sample and hold circuit is needed to capture a sample of the input and hold it constant across the sequential operation of the two parallel converters and the DAC. (This sampler can be avoided in parallel converters, where the clocked comparators function intrinsically as samplers). The DAC must be precise to the desired output resolution of the converter (12 bits in the above example).

6.6.2 Ripple-through analog-to-digital converters

The two-step architecture makes a substantial reduction in the number of comparators when compared with the parallel architecture. However, 128 comparators are still required in the 12-bit example of Fig. 6.17. Further reductions are possible by using more stages in the conversion process, with fewer bits per stage and correspondingly lower gain in the residue amplifiers. A three-stage converter resolving four bits per stage would require 48 comparators. A 12-stage converter resolving one bit per stage would require only 12 comparators. Converters of this type (with more than two ADC stages but only one sample and hold) are often called "ripple-through" converters.[18,19] A one bit per stage ripple-through architecture is shown in Fig. 6.18. Each stage includes a single comparator, a 1-bit DAC, a subtractor, and an amplifier with a gain of 2. In each stage, one bit is resolved and a residue is passed on to the next stage. Each stage's comparator is clocked when the prior stage's output has settled, leading to a "ripple" of activity down the converter.

The one bit per stage architecture minimizes comparator count, but it does require more amplifier and DAC stages than a two-step converter. However, these are very simple stages, and overall component count is generally lower in ripple-through converters than in two-step converters. On the other hand, the sample rates of one bit per stage converters tend to be lower than two-step converters within a given technology. This is largely a result of the larger number of sequential operations required. To increase speeds further, pipelining is employed.

6.6.3 Pipelined analog-to-digital converters

A pipelined converter increases speed relative to other multistage converters by simultaneous instead of sequential operation of the comparators, DACs, and amplifiers in the circuit. This is achieved by interposing sample and hold circuits between the stages.

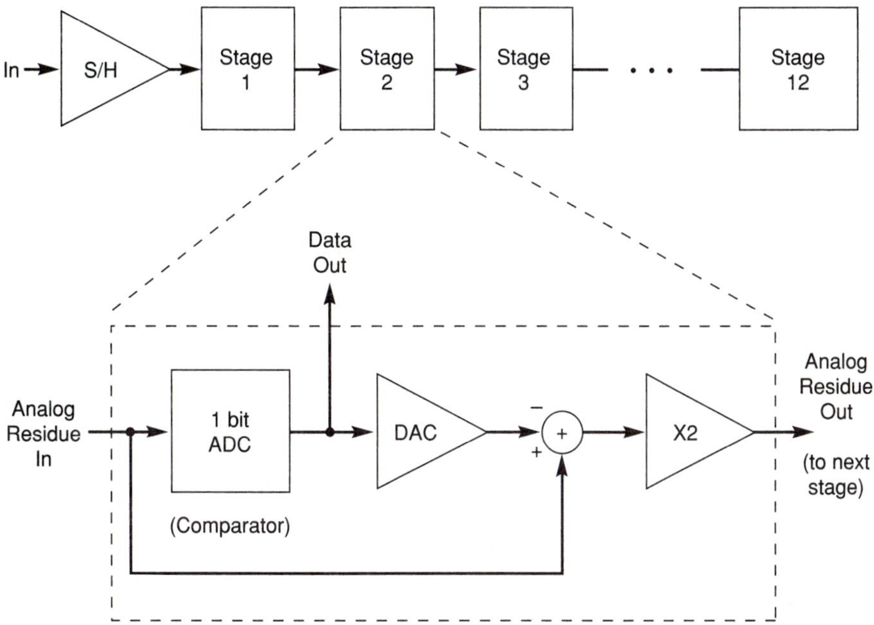

Figure 6.18 Block diagram of a 12-bit, one bit per stage ripple-through converter. Each stage includes a 1 bit ADC, DAC, and amplifier with a gain of 2.

Figure 6.19 is a block diagram for a one bit per stage pipeline converter. It is similar to the architecture of Fig. 6.18, with the addition of sample and hold circuits. Each sample and hold circuit holds the residue from the prior stage. On every clock period, a new sample of the input is taken, and the amplified residues advance one stage down the "pipeline" (really an analog shift register). At any time, there are m input samples being processed in the converter, where m is the number of stages in the pipeline. At every clock occurrence, m data bits are generated. Since these data bits apply to m different input samples, they must be collected into words that properly represent each sample. A tapered length digital shift register can be used for this operation.

In exact analogy to the ripple-through architecture above, pipeline ADCs can be optimized by varying the number of stages used and the number of bits per stage.[20–22] For example, three bits per stage could be employed in conjunction with gain of eight residue amplifiers. Four of these stages would produce a 12-bit converter. Comparator count is minimized by using fewer bits per stage, at the expense of more stages and sample and hold circuits. The optimum configuration depends on whether the goal is to minimize power, complexity, or conversion time, and on the details of the circuit elements employed.

Overall, the pipeline architecture can be very efficient in component count

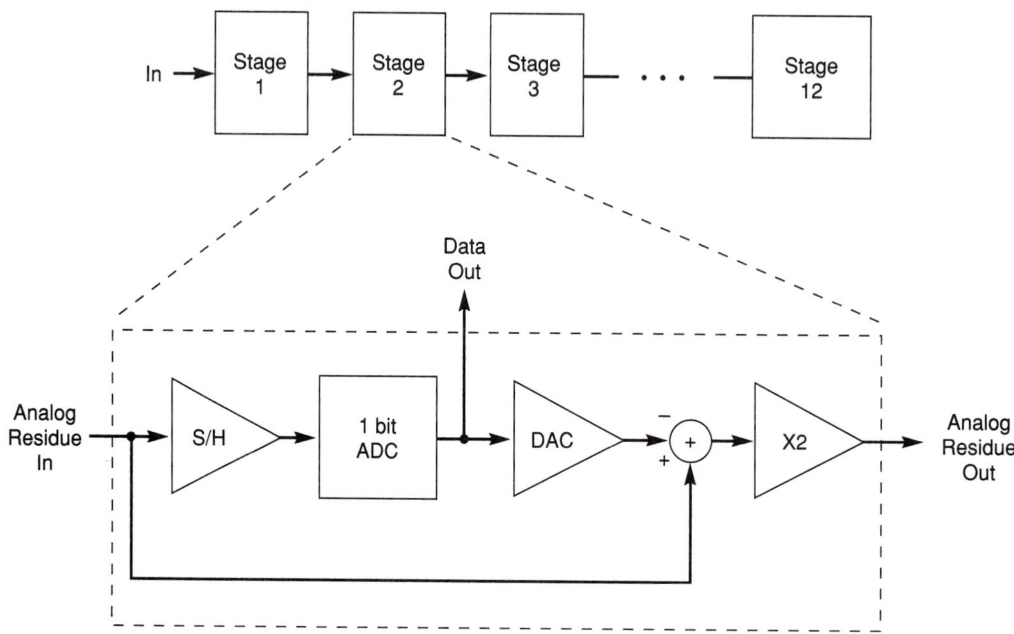

Figure 6.19 Block diagram of a one bit per stage pipeline ADC. Each stage includes a sample and hold circuit, allowing simultaneous processing of multiple input samples and higher conversion rates than the ripple-through architecture.

and power consumption for high-resolution ADCs. It is typically faster than the two-step, ripple-through, and other nonpipelined multistep converters owing to its simultaneous processing of multiple input samples.

6.7 Static ADC Errors and Testing Techniques

Numerous specifications are in common use to describe ADC errors. These errors can be grouped into two classes: static errors and dynamic errors. Static errors are those errors which occur with dc or very low frequency input signals. Dynamic errors are additional errors which occur with high-frequency input signals. Note that "input" here means the analog input, not the clock, which is assumed to be at the normal specified value in both cases.

Which errors are most relevant depends on the instrument application. Static errors are most important for digital voltmeters, but dynamic errors are typically the limiting factor in digitizing oscilloscopes and spectrum analyzers. In this section, common static ADC errors are defined and methods for measuring these errors are described. Section 6.8 covers dynamic errors. In both sections, errors particularly relevant to specific instrument applications are noted.

Code transition	Code bin	Code bin width
	$2^N - 1$	
$T[2^N - 1]$		
	$2^N - 2$	$W[2^N - 2]$
$T[2^N - 2]$		
•	•	•
•	•	•
•	•	•
T[K + 2]		
	K + 1	W[K + 1]
T[K + 1]		
	K	W[K]
T[K]		
•	•	•
•	•	•
•	•	•
T[2]		
	1	W[1}
T[1]		
	0	

Figure 6.20 Relationship between ADC code bins, code transition voltages $T(k)$, and code bin widths $W(k)$.

6.7.1 ADC output codes and code transition levels

To describe ADC errors concisely, it is necessary to first establish a nomenclature that defines the relationship between the input voltage and output code of an ADC. The conventions used in IEEE Standard 1057, "Trial Use Standard for Digitizing Waveform Recorders,"[23] are followed below.

An ideal ADC will produce the same output code across a small range of input voltages. This range is set by the resolution of the ADC and is often called the bin width. Given an input voltage, it is possible to predict what output code would be generated by an ideal ADC. However, the converse is not true. Given an output code, the input might have been any of the voltages within the bin width.

However, the transitions between codes do have a single corresponding input voltage. The transition between one code and another can be loosely defined as the input voltage where, in the presence of some noise, the ADC output is equally likely to be either one of the two adjacent codes. Figure 6.20 defines these transitions between codes as $T(k)$. That is, $T(k)$ is the precise input voltage required to produce the transition between code k and code $k - 1$ at the output of the converter.

Likewise, Fig. 6.20 defines the bin width for each code of the converter. The bin width $W(k)$ is defined as the difference between the two code transitions that bound code k. That is,

$$W(k) = T(k + 1) - T(k) \tag{6.7}$$

In an ideal converter, every bin width would have the value Q, the quantization value of the converter, Q is equal to the full-scale range of the converter divided by 2^n, where n is the converter resolution.

6.7.2 Offset, gain, and linearity errors

In what follows, ADC offset, gain, and linearity errors are defined and test techniques for them are described.

Definitions. Offset and gain errors in ADCs are analogous to the same errors in amplifiers. If an ADC has an offset error, there will be a systematic shift in the values of the transition voltages $T(k)$ from their nominal values. It is possible to estimate the offset error from a single transition voltage measurement at the middle of the converter's range. But since this measurement may include nonlinearity and gain errors, it is an imprecise estimate of offset. A more meaningful measure can be obtained by a least mean squares fit of the complete set of transition values $T(k)$ to a straight line through the ideal $T(k)$. The offset required to produce the best fit of the actual values to this straight line is the offset of the converter.

Likewise, a gain error manifests itself as a scaling of all the transition voltages to higher or lower absolute values. Equivalently, a gain error exists if the average code bin width is higher or lower than the nominal value Q. Once again, the gain error can be obtained by noting the gain term required to produce the best straight-line fit of the $T(k)$ to their ideal values.

Linearity errors are traditionally defined by integral nonlinearity (INL) and differential nonlinearity (DNL). The integral nonlinearity describes the deviation of the transition levels $T(k)$ from their nominal values after a best straight-line fit has been performed to remove offset and gain errors. The differential nonlinearity represents the deviations of the bin widths $W(k)$ from their nominal value Q, again assuming at least a gain error correction has been performed.

INL and DNL errors are usually described in units of least significant bits (LSBs), where one LSB $= Q$ in the nomenclature used here. The integral nonlinearity error in terms of LSBs is

$$\mathrm{INL}(k) = \frac{T(k) - (k-1)Q}{Q} \qquad \text{for } k = 2 \text{ to } 2^n - 1 \tag{6.8}$$

where it is assumed that offset and gain errors have been removed and $T(1) = 0$. Likewise the differential nonlinearity error in LSBs is

$$\mathrm{DNL}(k) = \frac{W(k) - Q}{Q} \qquad \text{for } k = 1 \text{ to } 2^n - 2 \tag{6.9}$$

Clearly, INL and DNL errors are related. In fact, DNL is the first difference of INL, i.e.,

$$\mathrm{DNL}(k) = \mathrm{INL}(k+1) - \mathrm{INL}(k) \tag{6.10}$$

This statement can be readily verified by substituting Eq. 6.8 in Eq. 6.10 and noting that $T(k + 1) - T(k) = W(k)$.

Two qualitative measures of ADC performance related to INL and DNL are missing codes and monotonicity. If a converter has some codes that never appear at the output, the converter is said to have missing codes. This is equivalent to a bin width $W(k) = 0$ for those codes, and a significant associated DNL error. Monotonicity refers to a converter whose output codes continuously increase (decrease) when the input signal is continuously increased (decreased). When testing for monotonicity, the effects of noise must be averaged out.

Offset, gain, and linearity errors are dc measures of performance and as such put an upper bound on overall accuracy for any converter. They are especially critical in digital voltmeter applications, however, where absolute accuracy in making dc measurements is the task at hand. Integrating ADCs are especially well suited to these applications because of their exceptional linearity and monotonicity (see Sec. 6.4).

Testing techniques. To measure offset, gain, and linearity errors, several techniques can be used. The purpose is generally to accurately locate the $T(k)$. Once this is done, the offset, gain, and linearity errors can be quickly calculated. Two locating techniques will be described, one using a digital-to-analog converter (DAC), the other a device called a tracking loop.

The DAC approach requires a D/A converter with accuracy and resolution substantially better than the ADC to be measured. The technique involves moving the output of the DAC slowly across the input range of the ADC and recording the code required at the DAC input to locate each transition voltage $T(k)$. Clearly this requires a way of judging the statistics of the decisions at the ADC output to find the 50 percent likelihood condition. This is best done by storing ADC conversion values in memory and analyzing them by computer.

Another technique (Fig. 6.21) makes use of a "tracking loop" to locate the $T(k)$. The computer controlling the process requests the loop to find a particular $T(k)$ by sending code k to a digital magnitude comparator, which compares k with the output of the ADC. If the ADC output is lower, the integrator output is ramped up, increasing the ADC input. When the ADC outputs code k or higher, the integrator ramps down. Eventually the ADC alternates between code k and code $k - 1$ (assuming a well-behaved converter with low noise). The average value of the input signal to the ADC under this balanced condition is the desired value $T(k)$. The measurement of $T(k)$ is performed at the ADC input using a digital voltmeter (DVM).

A few precautions must be taken to ensure an accurate measurement. The

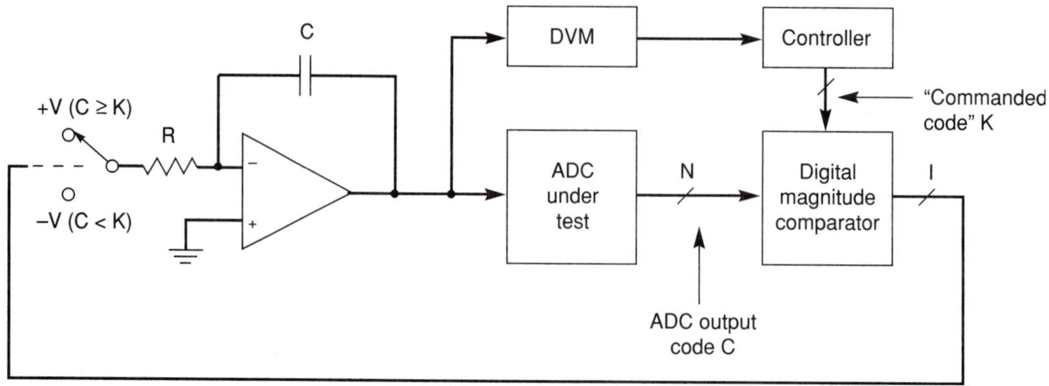

Figure 6.21 Tracking loop used for dc testing of ADC linearity. The integrator forces the transition voltage $T(k)$ at the ADC input.

voltmeter resolution and accuracy must be better than that of the ADC to be measured, but this is generally easy to ensure. The integrator time constant should be chosen so that the peak-to-peak ramp expected (given the ADC conversion time) is small relative to one LSB. If this is not easy to arrange, the DVM should at least average over many periods of the A/D conversion.

The tracking loop technique works well with well-behaved converters that are monotonic. For nonmonotonic converters, the integrator may get "stuck" at a particular transition when another is being requested. This will show as a large error (>1 LSB) until the nonmonotonic region is passed.

Figure 6.22 is a plot of a typical well-behaved A/D converter measured with a tracking loop. What is plotted is the difference between the actual $T(k)$ and the ideal $T(k)$ (in units of LSBs) vs. the code k. Before plotting, the gain and offset errors were removed. In other words, this is a plot of the integral linearity of the converter. The differential nonlinearity can be estimated directly from this plot. Since DNL is the first difference of INL, the value of the DNL for every k is the size of the "jump" in INL from one point to the next. This converter shows DNL errors of about ±½ LSB. DNL could of course be exactly plotted by a precise calculation of these differences or of the errors in the bin widths $W(k)$.

In spectrum analyzer applications, distortion of the converter is a key measure of performance. If a choice must be made, converters with low DNL errors are generally preferred to those with small INL errors, since they give much lower distortion for small input signals. A step in the INL error plot (i.e., a DNL error) will produce a large distortion relative to the signal amplitude when small inputs are applied which happen to span the step in the error curve. On the other hand, a smooth bow-shaped INL error will produce distortion for full-scale inputs, but its magnitude will drop rapidly relative to the signal when the signal amplitude is reduced. This can be accomplished with the input attenuator of the spectrum analyzer.

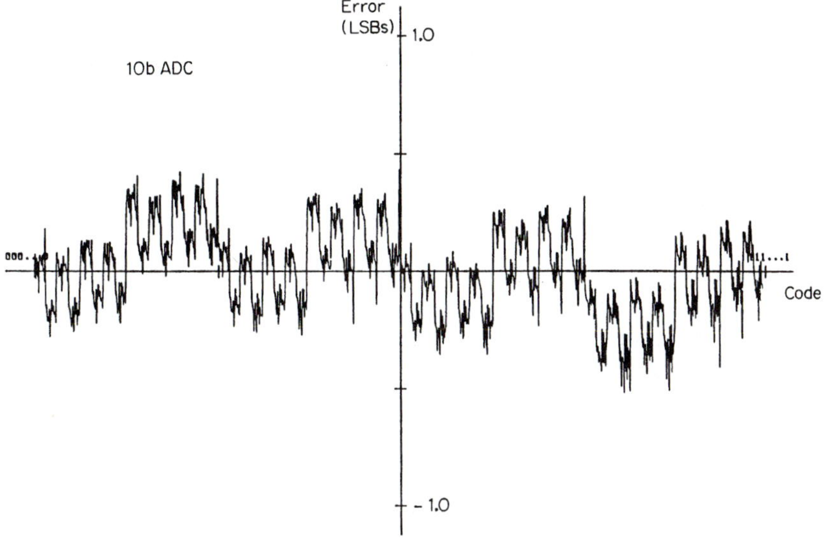

Figure 6.22 Typical integral linearity plot for a 10-bit ADC, generated using the tracking loop of Fig. 6.21.

The histogram is an alternative technique for assessing DNL. In this test a low-frequency, linear, triangle-shaped waveform is applied to the converter. The results of the conversions are stored in memory, and the number of occurrences of each code are counted. With a large enough sample, a good measure of all the $W(k)$ can be obtained by calculating the percentage of the conversions that resulted in each code. In an ideal converter, all the percentages would be the same. Deviations from the nominal percentage can be converted to bin width (DNL) errors.

Sine waves can also be used as the source in histogram testing. In this case a nonuniform distribution of code occurrences will occur, with a heavy accumulation near the peaks of the sine wave. But since the sine-wave shape is well known, this effect can be removed mathematically. A low-distortion sine wave should be used.

6.8 Dynamic ADC Errors and Testing Techniques

Dynamic ADC errors are additional errors which occur when high-frequency signals are applied to the analog input of the converter. The most common dynamic errors are distortion, aperture jitter, and step response anomalies. These errors (and the associated testing techniques) are described in this section. Spurious components, noise, and metastability errors can occur for either static or dynamic analog inputs. They are described here (instead of in Sec. 6.7) since dynamic test techniques are usually the best way to measure them.

6.8.1 Distortion and spurious components

ADC distortion (which produces harmonics of the input signal) are a particularly important measure of performance for spectrum analyzers, since they are often used to find distortion in the signal under test. Spurious components, defined here as clearly identifiable spectral components that are not harmonics of the input signal, are also important for spectrum analyzer applications.

Distortion can be caused by integral and differential nonlinearities (see Sec. 6.7.2) in the converter's transfer curve. These produce distortion with dc inputs, and also for all higher input frequencies. Other distortion, referred to here as dynamic distortion, can occur for high input signal frequencies. Dynamic distortion can be due to limitations in the sample and hold in front of the ADC, or in the ADC itself if no sample and hold is used. A common source is voltage variable capacitance in the converter active circuits. At high frequencies, this capacitance produces distortion when driven by a source with finite output impedance.

Spurious components are spectral lines that are not harmonics of the input signal frequency. They can appear as subharmonics of the clock frequency or as intermodulation products of the input frequency and clock subharmonics, or they can be caused by interference sources nearby in the system, such as digital system clocks or power-line noise sources. ADC distortion is generally specified in negative dB relative to the amplitude of the input signal. Total harmonic distortion (the rms sum of all harmonics) and "largest harmonic" are measures commonly used. Spurious components are usually specified in negative dB relative to ADC full scale.

Distortion can be measured using the converter test arrangement of Fig. 6.23. This block diagram is actually a generic setup for sine-wave testing

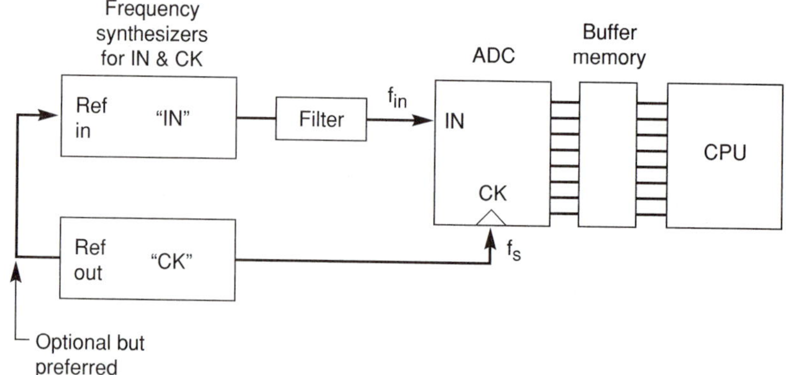

Figure 6.23 Test arrangement for sine-wave testing of ADCs. Frequency synthesizers (preferably with locked frequency references) are used to drive the ADC input and clock. Output data are captured in memory and analyzed by the processor.

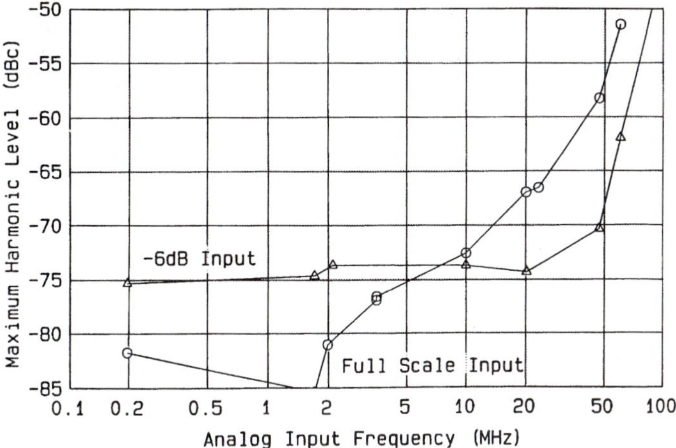

Figure 6.24 Plot of distortion (largest harmonic) at two input levels vs. input frequency for a 20 megasamples/s, 12-bit ADC.

of A/D converters, and will be referred to in many of the discussions which follow. Sine-wave testing requires two sine-wave signal sources (preferably synthesizers with the same frequency reference), one for the clock and one for the input of the ADC. In general, the synthesizer driving the input is filtered to reduce its harmonic distortion to a negligible level. The conversion results are stored in a buffer memory, then transferred to a CPU for analysis.

Distortion can be easily measured in this arrangement by using the fast Fourier transform (FFT) for analysis. Clock and input frequencies appropriate to the application are applied to the ADC, and the data are gathered and transformed by the CPU. The output can be displayed as an amplitude spectrum. Assuming the signal source is suitably free of distortion, spurious components, and noise, the resulting spectrum should represent the performance of the A/D converter.

Distortion will generally be a function of signal amplitude as well as frequency. For this reason, distortion is best displayed graphically as a family of curves. A typical distortion result is shown in Fig. 6.24 for a 20 megasamples/s, 12-bit ADC.[19] The plot is of largest harmonic vs. input frequency for two different input amplitudes, full scale and 6 dB below full scale. The results were obtained with FFT analysis. Two points should be noted. For low input frequencies, the distortion is higher with the –6-dB input. This indicates that differential nonlinearity errors dominate in this region. (If integral nonlinearity errors were dominant, the distortion would have been lower.) At high input frequencies, however, the opposite is true. The increasing distortion is due to dynamic effects which have a smooth nonlinearity characteristic. The distortion is thus lower for smaller input signals. This characteristic is common to many high-speed A/D converters. Spurious components can also be measured using the FFT technique. The largest nonharmonically related component can be determined from the spectrum. Its value relative to the ADC full scale represents the spurious performance of the converter.

6.8.2 Noise

As defined here, noise refers to what is left in a spectrum when the fundamental and all harmonics of the input are removed. This includes random components and also the interference sources referred to as spurious components above. In spectrum analyzer applications, spurious components may be specified separately. In other applications like waveform display, a single more inclusive noise measure (including spurious signals) is desired.

Noise is usually described in the form of signal-to-noise ratio (SNR). SNR is typically specified for a full-scale input to the ADC. Signal-to-noise ratio is defined as

$$\text{SNR} = \frac{\text{rms value of signal}}{\text{rms value of noise}} \tag{6.11}$$

SNR may be calculated from a sine-wave test using the FFT algorithm. The rms value of the fundamental is noted; then the fundamental and all its harmonics are removed from the FFT output data. The rms sum of all remaining components is computed, and the ratio of signal to noise is computed to give SNR.

6.8.3 Aperture jitter

Signal-to-noise ratio can be a function of input signal frequency. This is especially true if there is significant time jitter in clock driver or sampling circuits within the ADC. This problem is often referred to as aperture jitter. Aperture jitter is inconsequential for low-frequency inputs, but it can be transformed into significant voltage noise for rapidly changing inputs. This problem is most severe for very high frequency ADCs. To avoid introducing clock jitter external to the ADC, low-phase noise sources should be used in sine-wave-based testing.

The single-source test setup in Fig. 6.25 can be used for aperture jitter tests. Use of a single source minimizes the effect of the jitter in that source, since it is common to both the clock and input signals. The other effect of using a single source is that the ADC takes one sample per period of the input signal.

The delay is adjusted first so that the ADC samples at the peak of the sine wave (slew rate zero), and a noise measurement is made with the FFT approach. Then the delay is adjusted so that the ADC samples at the zero crossing of the sine wave (slew rate maximum). Under this condition, ADC sampling jitter will be transformed into voltage noise by the slew rate of the

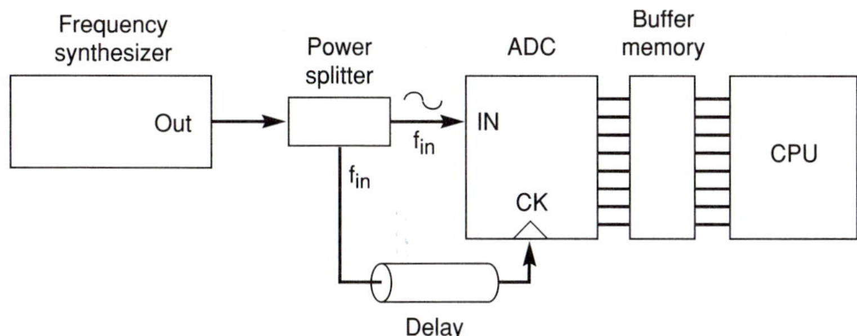

Figure 6.25 Test setup for measuring aperture jitter. Use of one source for both clock and input reduces effect of jitter in the synthesizer.

input. If the noise is larger in this second test, there is significant aperture jitter in the system. Assuming an rms summing of noise sources, and knowing the slew rate of the input sine wave, the aperture jitter can be calculated.

6.8.4 Interleaved ADC testing

A time-interleaved ADC system is one where multiple A/D converters are clocked in sequence to give a higher sample rate than obtainable from a single ADC (see Sec. 6.5.4). In an interleaved system, aperture "jitter" errors can be systematic. That is, if there are timing errors in delivering clocks to the individual ADCs, a systematic error signature will occur. These timing errors cause the composite ADC clock to be nonuniform in period (i.e., to contain subharmonics). In the frequency domain, the errors due to these timing problems will typically appear as a subharmonic of the clock, or as sidebands of the input around subharmonics of the clock.

In the time domain, these errors can be assessed readily using the same test setup as used for aperture testing (Fig. 6.25). When sampling the input once per period near its zero crossing, an ideal ADC would produce the same code every time. Timing mismatches will produce a fixed pattern error in a waveform display, repeating with a period equal to the number of ADCs. The amplitude of this pattern relative to full scale gives a good measure of the severity of the problems due to interleaving.

6.8.5 Signal-to-noise-and-distortion ratio

A very common and all-inclusive measure of ADC performance is the signal-to-noise-and-distortion ratio (SNDR). This is a fairly relevant measure for digitizing oscilloscopes since it includes all the undesired error sources in one number. For spectrum analyzers, noise is usually not as important, so SNDR is less relevant.

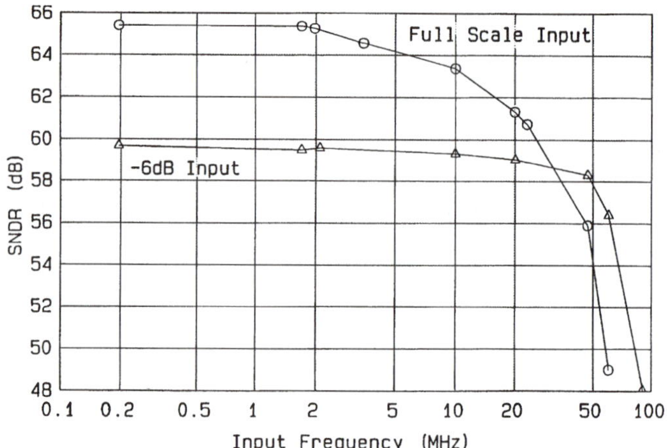

Figure 6.26 Signal-to-noise-and-distortion ratio (SNDR) vs. input frequency for a 20 megasamples/s, 12-bit ADC.

As the name implies, SNDR is calculated by taking the ratio of signal rms value to the rms sum of all distortion and noise contributions:

$$\text{SNDR} = \frac{\text{signal rms value}}{\text{noise + distortion rms value}} \tag{6.12}$$

This is easily done from the FFT results in a sine-wave test. The numerator is the amplitude of the fundamental, and the denominator the sum of everything else.

SNDR varies with both ADC input amplitude and frequency. For this reason, it is best displayed as a family of curves. Figure 6.26 plots SNDR for a 20-MHz, 12-bit ADC[19] for full-scale and −6-dB inputs versus input frequency. For a full-scale input, we see a 65-dB SNDR at low input frequency. The distortion of this converter (plotted earlier in Fig. 6.24) was better than −80 dB under the same conditions. From this it is clear that the SNDR is noise-dominated in this region. At high frequencies the SNDR falls owing to the increasing distortion seen in Fig. 6.24. Not surprisingly, the SNDR is higher for the −6-dB input in this regime because it results in less distortion.

6.8.6 Effective bits

Closely related to the signal-to-noise-and-distortion ratio is the measure known as effective bits. Like SNDR, effective bits attempts to capture the distortion and noise of the converter in a single number. Effective bits represents the resolution of an ideal (error-free) ADC with quantization noise equal to the total errors of the converter under test.

Effective bits is measured in a sine-wave test. The block diagram of Fig. 6.23 is again applicable. The output data are collected in memory, and a sine-wave curve fit routine is carried out by the CPU. The curve fit finds the offset, amplitude, and phase of an ideal sine wave (assuming frequency is known) which best fits the captured data in a least-mean-squared error sense. The difference between the sine wave and the data is then taken, leaving an error signal which includes the quantization noise and all other ADC imperfections. (This subtraction is equivalent to removing the fundamental from the FFT output in calculating SNDR.) The rms value of this error signal is then calculated.

Effective bits E is then defined as follows:

$$E = n - \log\,(\text{base } 2)\,\frac{\text{Actual rms error}}{\text{ideal rms error}} \tag{6.13}$$

where n is the nominal converter resolution, actual rms error is the residue after subtraction of the sine wave, and ideal rms error is the nominal quantization noise. This can be shown to be $0.29Q$, where Q = one least significant bit of the converter.

A converter with no errors other than quantization noise would have an actual rms error of $0.29Q$, and the log term would be zero. In this case, effective bits would be equal to the nominal converter resolution. If the actual rms error was $0.58Q$, the log would have a value of one, and effective bits would be one less than the converter resolution.

For full-scale inputs, effective bits and SNDR are equivalent measures of performance. In fact, they can be related by

$$\text{SNDR} = 6.02\,E + 1.76\,\text{dB} \tag{6.14}$$

where SNDR is expressed in dB and E in bits.

For less than full-scale inputs, SNDR and effective bits are not directly related. This is so because SNDR includes a signal amplitude term in the numerator, but signal amplitude is not used in the effective bits calculation.

6.8.7 Step response

Although SNR, SNDR, effective bits, etc., are useful measures of ADC performance, they do not provide sufficient information to predict the step response of an ADC, which is largely a function of the frequency and phase response of the converter. Lack of gain flatness in the low-frequency regime (which can sometimes be caused by thermal effects) can lead to a sluggish settling to a step input. These effects can last for microseconds or even milliseconds. In general, step response is of most interest in digitizing oscilloscope applications of ADCs.

To characterize step response, it is simpler to measure it directly than to infer it from many sine-wave measurements at different frequencies. The simplest approach is to use a pulse generator to drive the converter, take a long data record which spans the expected thermal time constants, and exam-

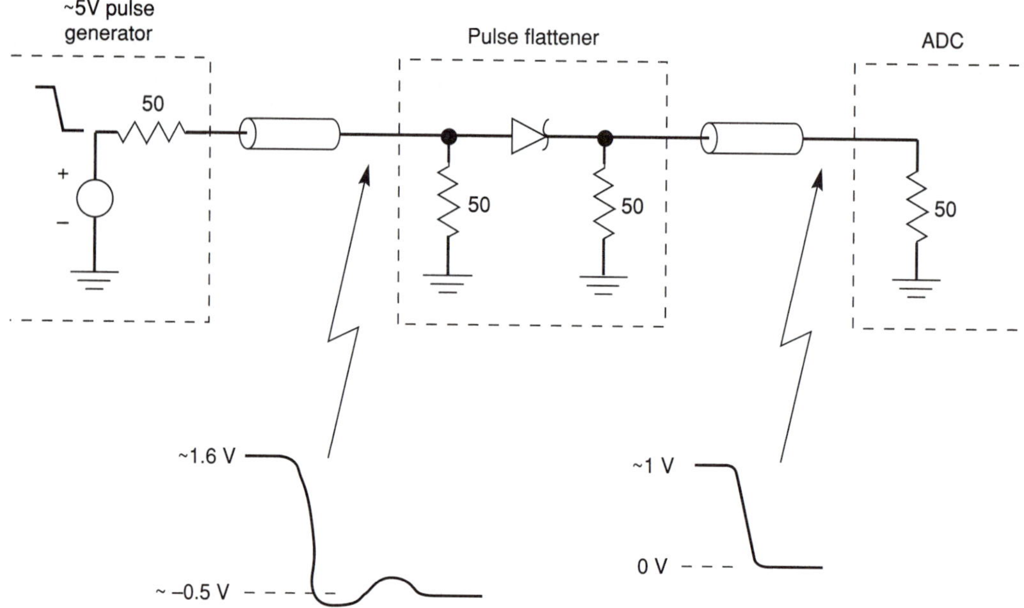

Figure 6.27 Pulse flattener circuit for ADC step response testing. The Schottky diode is used to produce a waveform that settles very rapidly to 0 V.

ine the resulting codes vs. time. For this to work, the pulse generator must have a flat settling behavior.

If a flat pulse source is not available, one can be made using the pulse flattener circuit of Fig. 6.27. The Schottky diode conducts when the pulse generator is high, establishing the base line of the pulse. When the generator output goes low, the Schottky diode turns off, and the 50-Ω resistor quickly pulls the input line to ground. Clearly, for this technique to work, ground must be within the input range of the ADC.

6.8.8 Metastability errors

Metastability errors can occur in ADCs when a comparator sustains a "metastable state." A metastable state is one where the output of the comparator is neither a logic high nor a logic low but resides somewhere in between. This can occur when a comparator input signal is very close to its threshold, and insufficient time is available for the comparator to regenerate to one logic state or the other. Although metastable errors are described here in the section on dynamic errors, they happen just as readily with dc input signals.

Metastable states can cause very large errors in the ADC output, although they usually occur quite infrequently. Large errors can result when logic circuits driven by the comparator interpret the bad level differently. Usually these logic circuits are part of an encoder. Sometimes half full-scale errors

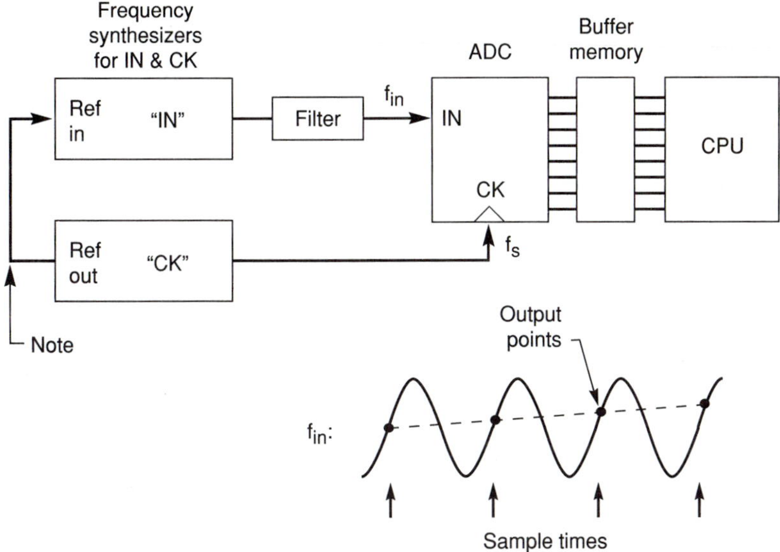

Figure 6.28 Beat frequency test arrangement. The clock and input sources are operated at slightly different frequencies, producing a gradually changing sampling phase and a very high effective sampling rate.

can result. Metastable states are more likely to occur in very high speed converters where less time is available for regeneration.

Testing for metastable states may require gathering large amounts of data. One approach is to apply a very low frequency sine wave as input, so that the output codes change on average very slowly (less than one LSB per conversion). Any change greater than one LSB must be due to noise or metastability. If the random noise level is known, probabilistic bounds can be placed on how large an error can occur owing to noise. Errors beyond these bounds may well have been induced by metastable states.

6.8.9 Beat frequency testing

In prior sections numerous measures of ADC performance have been defined including signal-to-noise ratio, effective bits, and total harmonic distortion. These are good quantitative measures of ADC performance, but they don't necessarily give good qualitative insight into what is causing the problems observed or how to fix them. Beat frequency testing can sometimes help in making those insights.

A block diagram for beat frequency testing is shown in Fig. 6.28. The test setup is identical to that for sine-wave-based testing shown in Fig. 6.23. The difference is that locking of the two synthesizers' frequency references is now quite important. The beat frequency test is carried out by setting the input frequency to a value slightly offset from the clock frequency f_s, say higher by a frequency df. This means the ADC will take roughly one sample per period of the input. But since the input frequency is a little higher than the clock, the phase of the sample taken by the ADC is advanced slightly each period. The result is that the ADC output codes reconstruct a low-frequency sine wave which will have an apparent frequency of df (the "beat frequency"). This sine wave with its imperfections can be displayed to aid in analyzing ADC behavior.

Another way of viewing the process is that the input frequency is effectively being sampled at a very high rate, equal to f_s/df. For low df, this can be made very high. The result is that the ADC will trace out samples of a high-frequency sine wave that is nevertheless heavily oversampled, with perhaps several samples in each code bin. This can reveal many fine scale details of ADC behavior, details which would not be seen in normal operation with perhaps four samples per period of high-frequency inputs.

If the input frequency cannot be operated at the sample rate (and it is desired to keep the sample rate at maximum), the input frequency can be set to (say) $(f_s/4) + df$ and the beat frequency principle can still be employed. In this case, only every fourth output sample is displayed. This once again reconstructs a single oversampled sine wave. Likewise, if it is desired to operate the input well above f_s, it can be set to any integer multiple of f_s, then offset by df, and the beat frequency technique will still be effective.

References

1. J. C. Candy, "A Use of Limit Cycle Oscillations to Obtain Robust Analog-to-Digital Converters," *IEEE Transactions on Communications*, vol. COM-22, pp. 298–305, March 1974.
2. R. J. van de Plassche, "A Sigma-Delta Modulator as an A/D Converter," *IEEE Transactions on Circuits and Systems*, vol. CAS-25, no. 7, pp. 510–514, July 1978.
3. W. L. Lee and C. G. Sodini, "A Topology for Higher Order Interpolative Coders," *Proceedings 1987 International Symposium on Circuits and Systems*, May 1987, pp. 459–462.
4. Wayne C. Goeke, "An 8½ Digit Integrating Analog-to-Digital Converter with 16-Bit, 100,000-Sample-Per-Second Performance," *Hewlett-Packard Journal*, vol. 40, no. 2, pp. 8–15, April 1989.
5. Ken Poulton, John J. Corcoran, and Thomas Hornak, "A 1-GHz 6-bit ADC System," *IEEE Journal of Solid-State Circuits*, vol. SC-22, no. 6, December 1987.
6. T. Wakimoto, Y. Akazawa, and S. Konaka, "Si Bipolar 2GS/s 6b Flash A/D Conversion LSI," *1988 ISSCC Digest of Technical Papers*, February 1987, pp. 232–233.
7. Ken Rush and Pat Byrne, "A 4GHz 8b Data Acquisition System," *1991 ISSCC Digest of Technical Papers*, February 1991, pp. 176–177.
8. Albert Gookin, "A Fast-Reading High-Resolution Voltmeter That Calibrates Itself Automatically," *Hewlett-Packard Journal*, vol. 28, no. 6, February 1977.
9. John J. Corcoran, Knud L. Knudsen, Donald R. Hiller, and Paul W. Clark, "A 400MHz 6b ADC," *1984 ISSCC Digest of Technical Papers*, February 1984, pp. 294–295.
10. T. W. Henry and M. P. Morgenthaler, "Direct Flash Analog-to-Digital Converter and Method," U.S. Patent 4,386,339.
11. Adrian P. Brokaw, "Parallel Analog-to-Digital Converter," U.S. Patent 4,270,118.
12. R. E. J. van de Grift and R. J. van de Plassche, "A Monolithic 8b Video A/D Converter," *IEEE Journal of Solid-State Circuits*, vol. SC-19, no. 3, pp. 374–378, June 1984.
13. Rudy van de Plassche and Peter Baltus, "An 8b 100MHz Folding ADC," *1988 ISSCC Digest of Technical Papers*, February 1988, pp. 222–223.
14. Rob E. J. van de Grift and Martien van der Veen, "An 8b 50MHz Video ADC with Folding

and Interpolation Techniques," *1987 ISSCC Digest of Technical Papers*, February 1987, pp. 94–95.

15. Johan van Valburg and Rudy van de Plassche, "An 8b 650MHz Folding ADC," *1992 ISSCC Digest of Technical Papers*, February 1992, pp. 30–31.
16. Keiichi Kusumoto et al., "A 10b 20MHz 30mW Pipelined Interpolating ADC," *1993 ISSCC Digest of Technical Papers*, February 1993, pp. 62–63.
17. Ken Poulton et al., "A 2 GS/s HBT Sample and Hold," *Proceedings of the 1988 GaAs IC Symposium*, November 1988, pp. 199–202.
18. Robert A. Blauschild, "An 8b 50ns Monolithic A/D Converter with Internal S/H," *1983 ISSCC Digest of Technical Papers*, February 1983, pp. 178–179.
19. Robert Jewett, John Corcoran, and Gunter Steinbach, "A 12b 20MS/s Ripple-through ADC," *1992 ISSCC Digest of Technical Papers*, February 1992, pp. 34–35.
20. Stephen H. Lewis and Paul R. Gray, "A Pipelined 5MHz 9b ADC," *1987 ISSCC Digest of Technical Papers*, February 1987, pp. 210–211.
21. Bang-Sup Song and Michael F. Tompsett, "A 12b 1MHz Capacitor Error Averaging Pipelined A/D Converter," *1988 ISSCC Digest of Technical Papers*, February 1988, pp. 226–227.
22. David Robertson, Peter Real, and Christopher Mangelsdorf, "A Wideband 10-bit, 20MSPS Pipelined ADC Using Current-Mode Signals," *1990 ISSCC Digest of Technical Papers*, February 1990, pp. 160–161.
23. "IEEE Trial Use Standard for Digitizing Waveform Recorders" (IEEE Standard 1057), Waveform Measurement and Analysis Committee, IEEE Instrumentation and Measurement Society, July 1989.

Chapter

7

Signal Sources

Charles Kingsford-Smith
Agilent Technologies
Lake Stevens, Washington

7.1 Introduction

This chapter deals with *signals* and, in particular, the production or generation of signals, rather than the analysis of them.

What is a signal and how is it characterized? The simplest useful definition is that a signal is an electrical voltage (or current) that varies with time. To characterize a signal, an intuitive yet accurate concept is to define the signal's *waveform.* A waveform is easy to visualize by imagining the picture a pen, moving up and down in proportion to the signal voltage, would draw on a strip of paper being steadily pulled at right angles to the pen's movement. Figure 7.1 shows a typical periodic waveform and its dimensions.

A signal source is an electronic instrument which generates a signal according to the user's commands respecting its waveform. Signal sources serve the frequent need in engineering and scientific work for energizing a circuit or system with a signal whose characteristics are known.

7.2 Kinds of Signal Waveforms

Most signals fall into one of two broad categories: periodic and nonperiodic. A periodic signal has a waveshape which is repetitive: the pen, after drawing one *period* of the signal waveform, is in the same vertical position where it began, and then it repeats exactly the same drawing. A sine wave (see below) is the best-known periodic signal. By contrast, a nonperiodic signal has a nonrepetitive waveform. The best-known nonperiodic signal is random noise. Signal source instruments generate one or the other, and sometimes both. This chapter is concerned with periodic signals and provides an overview of

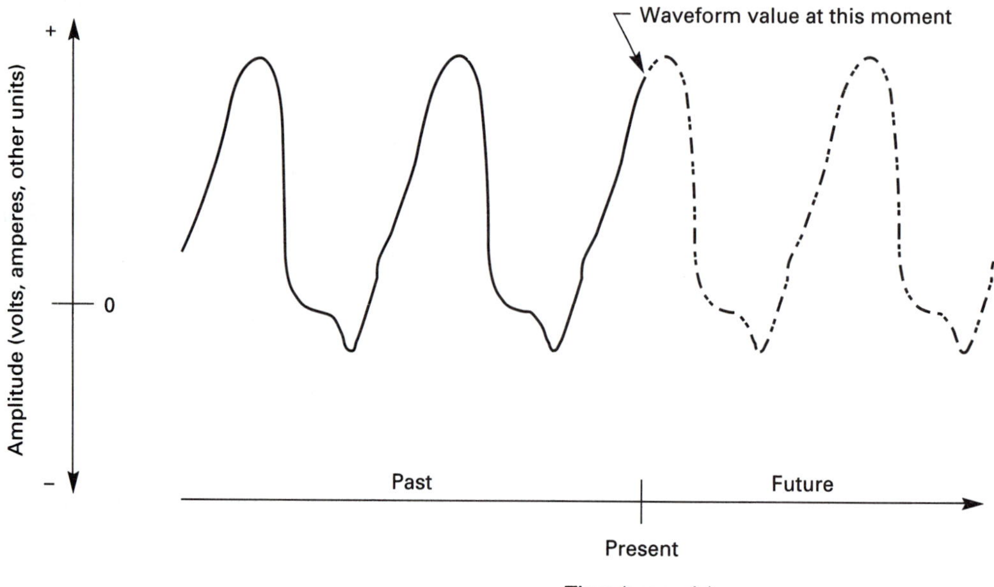

Figure 7.1 Waveform of an active typical periodic signal.

ways to generate them. More specific instrument techniques are covered in Chap. 16, along with how to understand and interpret specifications.

7.2.1 Sine waves, the basic periodic signal waveform

The familiar sinusoid, illustrated in Fig. 7.2*a,* is the workhorse signal of electricity. The simple mathematical representation of a sine wave can be examined to determine the properties which characterize it:

$$s(t) = A\sin(2\pi ft) \tag{7.1}$$

where s represents the signal, a function of time
t = time, seconds
A = peak amplitude of the signal, V or A
f = signal frequency, cycles/second (Hz)

From this expression and Fig. 7.2*a* the important characteristics (or parameters) of a sine wave may be defined.

Phase. This is the argument $2\pi ft$ of the sine function. It is linearly increasing in time and is not available as a signal that can be directly viewed. For mathematical reasons, the phase is measured in radians (2π radians = 360°). However, two sine waves are compared in phase by noting their

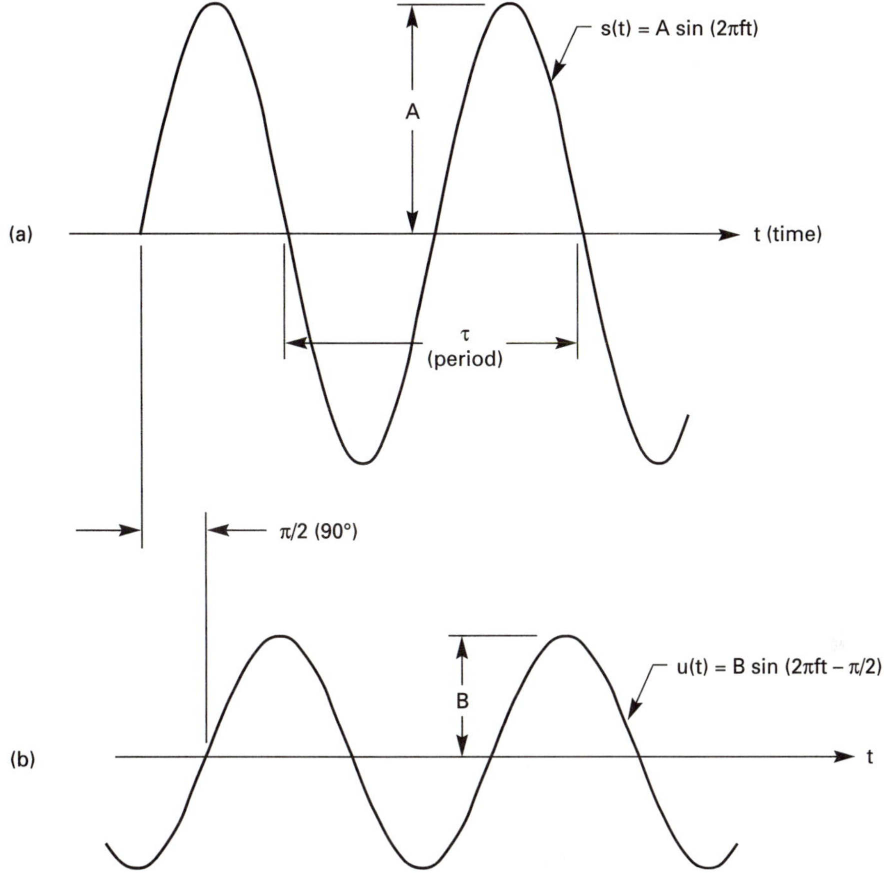

Figure 7.2 Sine-wave basics. (a) A typical sine wave; (b) another sine wave, same period as (a) but different amplitude and displaced in phase.

phase *difference,* seen as a time shift between the waveforms (Fig. 7.2*b*). The waveform $u(t)$ lags $s(t)$ by 90° ($\pi/2$ radians) and, in addition, has a different amplitude.

Period. The time τ between repetitions of the waveform, or the time of one waveform *cycle.* Since the sine wave repeats every 360°, the period is just the time needed for the phase to increase by 2π radians: $2\pi f\tau = 2\pi$; hence $\tau = 1/f$.

Frequency. The number of cycles per second, or the reciprocal of τ: $f = 1/\tau$. The term "hertz" (abbreviated Hz) represents cycles per second.

Amplitude. The coefficient A, describing the maximum excursion(s) of the instantaneous value of the sine wave from zero, since the peak values of the sine function are ± 1.

A principal reason for calling sine waves *basic* is that other waveforms, both periodic and nonperiodic, are composed of combinations of sine waves of different frequencies, amplitudes, and phases.* When a waveform is periodic, an important relation holds: The waveform is made up of sine-wave components whose frequencies are integer multiples—called *harmonics*—of a *fundamental* frequency, which is the reciprocal of the signal period. For instance, a symmetrical square wave (a member of the pulse waveform family introduced below) with a period of 0.001 s is composed of sine waves at frequencies of 1000 Hz (the fundamental frequency), 3000 Hz, 5000 Hz, etc.; all the harmonics are odd multiples of the 1000 Hz fundamental. This is true only if the square wave is symmetrical; otherwise, even-multiple harmonics appear in the composition.

It is insightful to illustrate that complex periodic signals are composed of various sine waves which are harmonically related. Figure 7.3 shows the waveforms which result when more and more of the sine-wave components of a symmetrical square wave are combined. In Fig. 7.3*a,* only the fundamental and third harmonic are present, yet the non-sine-wave shape already is a crude approximation to a symmetrical square wave. In Fig. 7.3*b,* the fifth and seventh harmonics are added, and in Fig. 7.3*c,* all odd harmonics through the thirteenth are present; the resultant waveform is clearly approaching the square-wave shape.

7.2.2 Complex periodic signal waveforms

Waveforms other than sine waves are useful, too. The most common of these are illustrated in Fig. 7.4.

Pulse waveforms. A conspicuous feature of a pulse waveform (Fig. 7.4*a*) is that the maximum levels (elements 2 and 4 of the waveform) are constant-amplitude, or "flat." A "rising edge" (1) joins a negative level to the next positive level, and a "falling edge" (3) does the opposite.

Rise time, fall time. The time duration of an edge is called "rise time" (T_1) and "fall time" (T_3), respectively. One period τ of the waveform consists of the sum of the edge times and level times. The frequency of the waveform is $1/\tau$. The idealized pulse waveform has zero rise and fall times, but this is impossible to achieve with a physical circuit. Why this is so may be deduced by examining Fig. 7.3. The rise and fall times of the approximations become shorter as more harmonics are added. But it takes an infinite number of harmonics—and infinite frequency—to realize zero rise and fall times. In addition, there

* This fact is one result of Fourier analysis theory. For a good introduction or refresher on Fourier analysis for both continuous and sampled signals, a very readable text is "Signals and Systems," by Oppenheim and Willsky, Prentice-Hall, 1983.

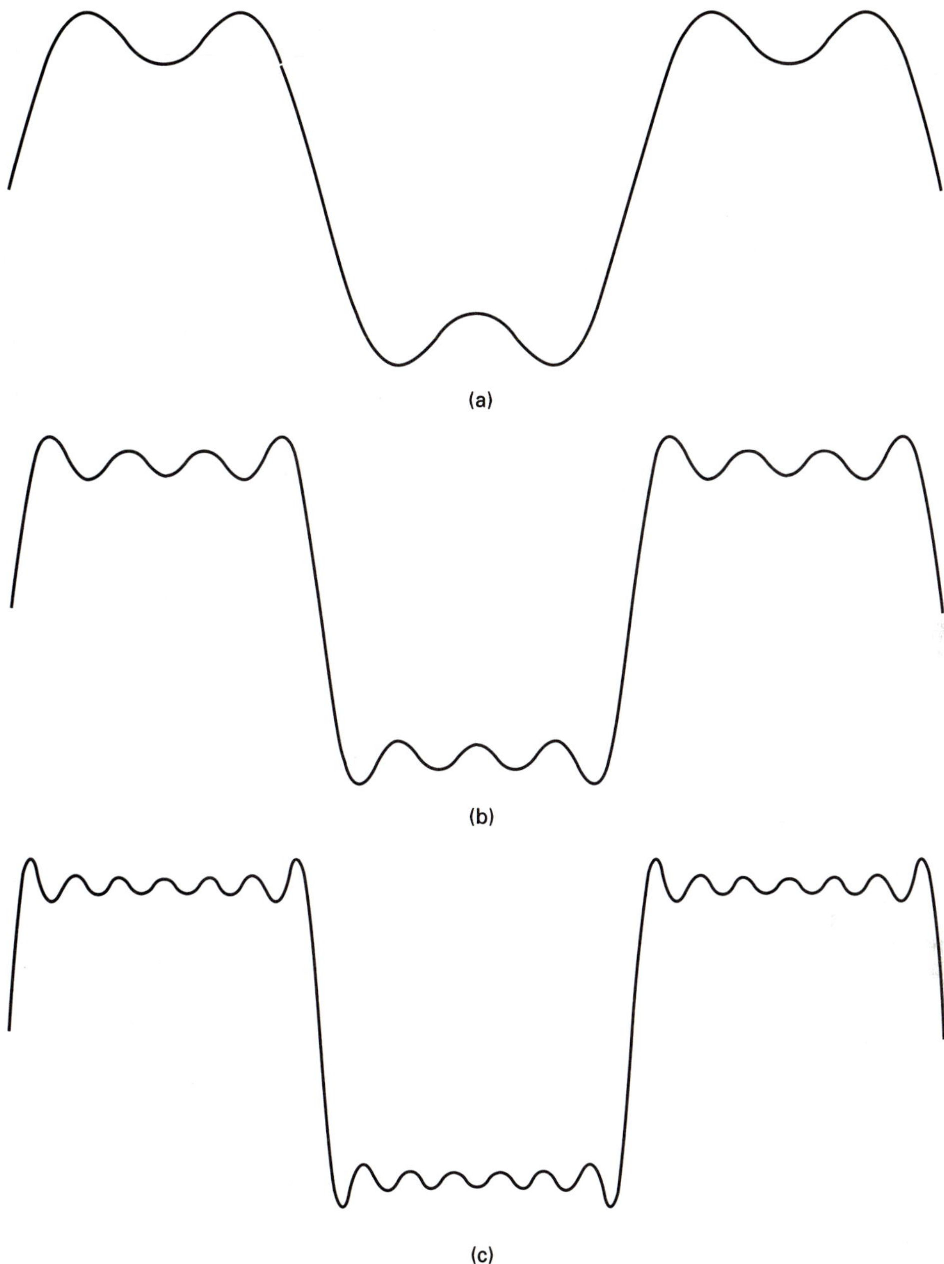

Figure 7.3 Construction of a square wave from its sine-wave components. (a) Components 1 and 3; (b) components 1, 3, 5, and 7; (c) components 1, 3, 5, 7, 9, 11, and 13.

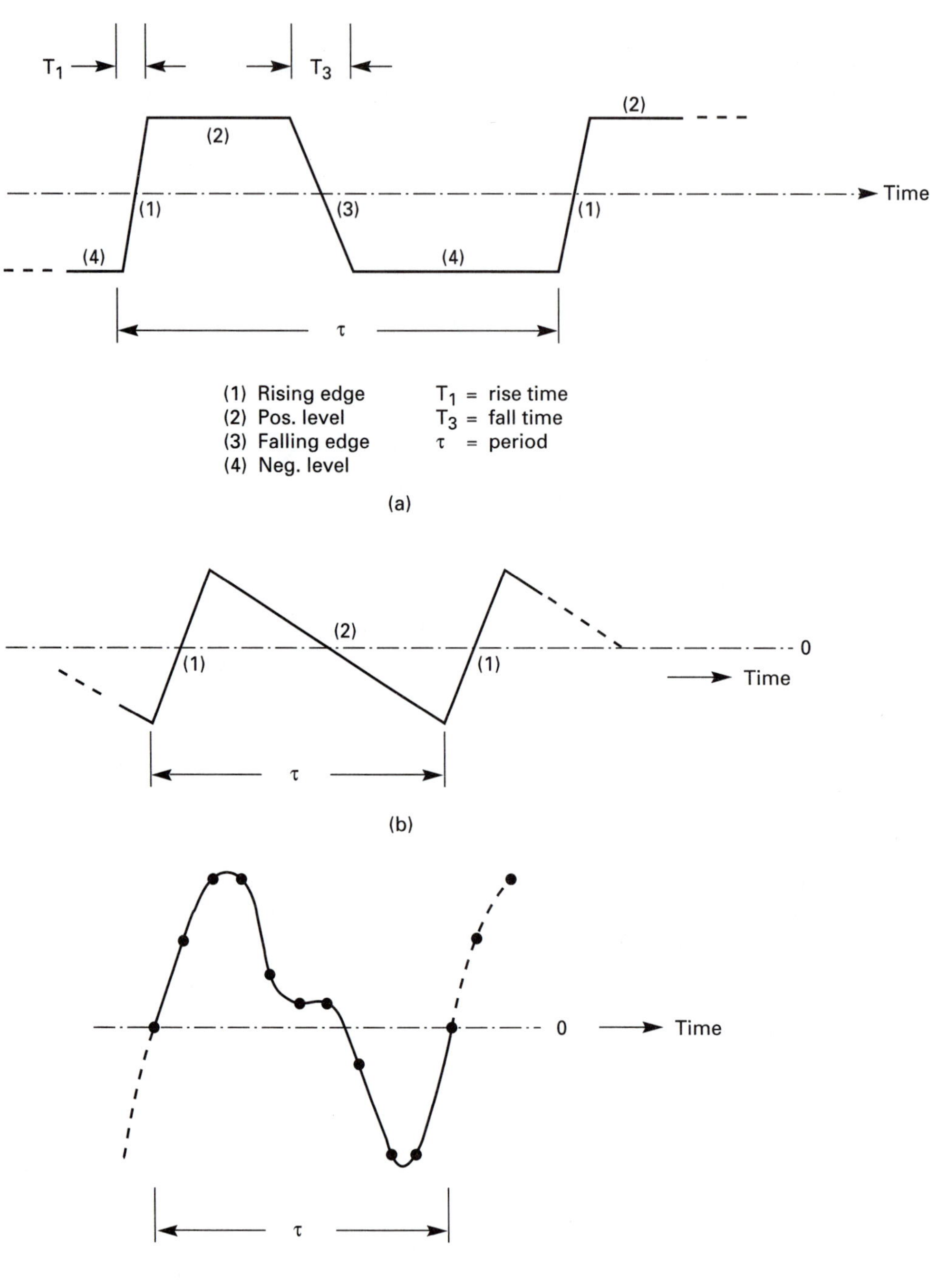

Figure 7.4 Nonsinusoidal, periodic signal waveforms: (a) pulse; (b) triangle; (c) arbitrary.

is often a sound engineering reason for making these times longer than could be achieved with available circuitry: the higher-frequency sine-wave components which are needed for short rise and fall times are often a source of interference energy, as they can easily "leak" into nearby apparatus. Hence, it is prudent to limit rise times to just what is required in the particular application.

Symmetry. Often called "duty cycle," symmetry is another important parameter of a pulse waveform. This is defined as the ratio of the positive portion of the period to the entire period. For the waveform illustrated in Fig. 7.4*a*, the symmetry is $(\frac{1}{2}T_1 + T_2 + \frac{1}{2}T_3)/\tau$. A pulse waveform with 50 percent symmetry, and equal rise and fall times, is an important special case called a "square wave"; it is composed of the fundamental frequency sine wave and only odd harmonics.

Triangle waveforms. Illustrated in Fig. 7.4*b*, ideal triangle waves consist of linear positive-slope (1) and negative-slope (2) segments connected together. When the segment times are equal, the waveforms is called symmetrical. Like square waves, symmetrical triangle waves are composed of the fundamental frequency sine wave and only odd harmonics.

An unsymmetrical triangle wave, as illustrated, is often called a "sawtooth" wave. It is commonly used as the horizontal drive waveform for time-domain oscilloscopes. Segment 2 represents the active trace where signals are displayed, and segment 1 is the beam retrace. In this and similar applications, what is most important is the *linearity* of the triangle wave, meaning how closely the segments of the waveform approximate exact straight lines.

Arbitrary waveforms. The word "arbitrary" is not a catchall term meaning all remaining types of waveforms not yet discussed! Rather, it is a consequence of the widespread use of digital signal generation techniques in instrumentation. The idea is to generate a periodic waveform for which the *user defines* the shape of one period. This definition could be a mathematical expression, but it is much more common to supply the definition in the form of a set of sample points, as illustrated by the dots on the waveform in Fig. 7.4(*c*). The user can define these points with a graphical editing capability, such as a display screen and a mouse, or a set of sample values can be downloaded from a linked computer. The more sample points supplied, the more complex the waveform that can be defined. The repetition rate (that is, frequency) and the amplitude are also under the user's control. Once a set of sample points is loaded into the instrument's memory, electronic circuits generate a waveform passing smoothly and repetitively through the set.

An interesting example of such user-defined waveforms is the synthesis of various electro-cardiogram waveforms to use for testing patient monitors and similar medical equipment.

7.3 How Periodic Signals Are Generated

Periodic signal generation does not happen without oscillators, and this section begins by introducing basic oscillator principles. Some signal generators directly use the waveform produced by an oscillator. However, many signal generators use signal processing circuitry to generate their output. These processing circuits are synchronized by a fixed-frequency, precision oscillator. Such generators are *synthesizers,* and their principles of operation are introduced here also.

7.3.1 Oscillators

The fundamental job of electronic oscillators is to convert dc energy into periodic signals. Any oscillator circuit fits into one of these broad categories:

- AC amplifier with filtered feedback
- Threshold decision circuit

Feedback oscillators. The feedback technique is historically the original, and still the most common form of oscillator circuit. Figure 7.5 shows the bare essentials needed for the feedback oscillator. The output from the amplifier is applied to a frequency-sensitive filter network. The output of the network is then connected to the input of the amplifier. *Under certain conditions,* the amplifier output signal, passing through the filter network, emerges as a signal, which, if supplied to the amplifier input, would produce the output signal. Since, because of the feedback connection, this *is* the signal supplied to the input, it means that the circuit is capable of sustaining that particular output signal indefinitely: it is an *oscillator.* The circuit combination of the amplifier and the filter is called a *feedback loop.* To understand how the combination can oscillate, mentally break open the loop at the input to the amplifier; this is called the *open-loop* condition. The open loop begins at the amplifier input and ends at the filter output. Here are the particular criteria that

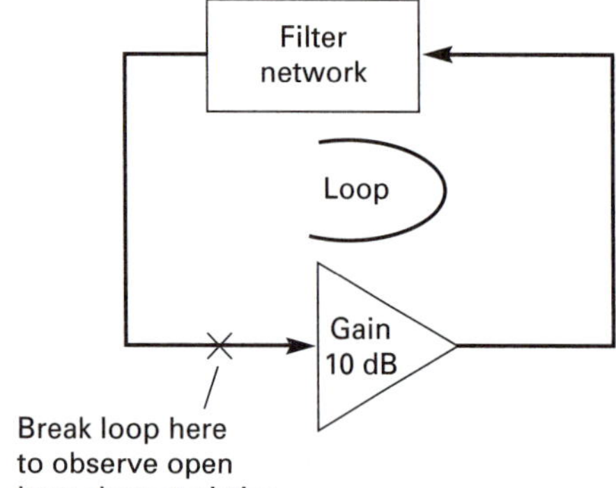

Figure 7.5 Oscillator, using an amplifier and a filter to form a feedback loop.

the open loop must satisfy in order that the closed loop will generate a sustained signal at some frequency f_0:

1. The power gain through the open loop (amplifier power gain times filter power loss) must be unity at f_0.
2. The total open-loop phase shift at f_0 must be zero (or 360, 720, etc.) degrees.

Both criteria are formal statements of what was said previously: the loop must produce just the signal at the input of the amplifier to maintain the amplifier output. Criterion 1 specifies the amplitude and criterion 2 the phase of the requisite signal at the input.

A feedback oscillator is usually designed so that the amplifier characteristics don't change rapidly with frequency. The open-loop characteristics—power gain and phase shift—are dominated by those of the filter, and they determine where the criteria are met. Thus the frequency of oscillation can be "tuned" by varying one or more components of the filter. Figure 7.6 shows a loop formed from a constant-gain amplifier and a transformer-coupled resonant filter. The 10-dB gain of the amplifier is matched by the 10-dB loss of the filter at the resonant frequency (and only there; the open loop has a net loss everywhere else). Likewise the filter phase shift is zero at resonance, and so the combination will oscillate at the resonant frequency of the filter when the loop is closed. Changing either the inductance or the capacitance of the filter will shift its resonant frequency. And this is where the closed loop will oscillate, provided the criteria are still met.

It is impractical to meet the first criterion exactly with just the ideal elements shown. If the loop gain is even very slightly less than (or greater than) unity, the amplitude of the oscillations will decrease (or grow) with time. In practice, the open-loop gain is set somewhat greater than unity to ensure that oscillations will start. Then some nonlinear mechanism lowers the gain as the amplitude of the oscillations reaches a desired level. The mechanism commonly used is saturation in the amplifier. Figure 7.7 is a plot of the input-output characteristic of an amplifier, showing saturation. Up to a certain level of input signal, either positive or negative, the amplifier has a constant gain, represented by the slope of its characteristic. Beyond that level, the gain drops to zero more or less abruptly, depending on the amplifier. The amplifier operates partially into the saturation region, such that the *average* power gain over a cycle is unity. Clearly this means that waveform distortion will be introduced into the output: The waveform tops will be flattened. However, some of this distortion may be removed from the external output signal by the feedback filter.

The second criterion is especially important in understanding the role that filter quality factor Q plays in determining the frequency stability of the oscillator. Q is a measure of the energy stored in a resonant circuit to the energy being dissipated. It's exactly analogous to stored energy vs. friction loss in a flywheel. For the filter, the rate of change of its phase shift at resonance (see

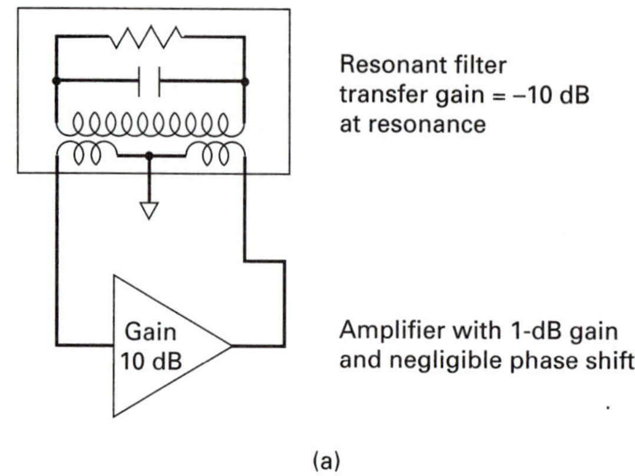

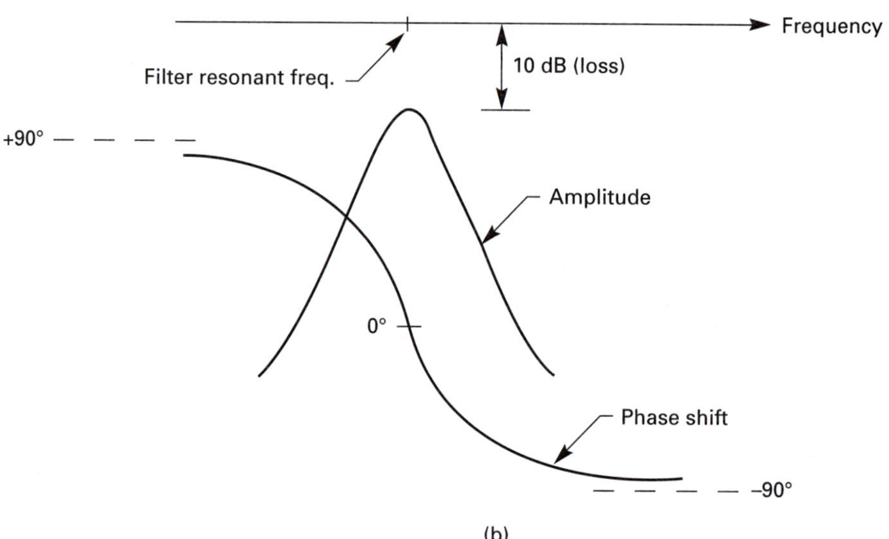

Figure 7.6 Details of a filtered-feedback oscillator: (a) Feedback circuit: inductively coupled resonator; (b) amplitude and phase shift transfer characteristics of resonator.

Fig. 7.6*b*) is directly proportional to Q. During operation, small phase shifts can occur in the loop; for instance, the transit time in the amplifier may change with temperature, or amplifier random noise may add vectorially to the loop signal and shift its phase. To continue to meet the second criterion, the oscillator's instantaneous frequency will change in order to produce a compensatory phase shift which keeps the total loop phase constant. Because the phase slope of the filter is proportional to its Q, a high-Q filter requires less frequency shift (which is unwanted FM) to compensate a given phase disturbance in the oscillator, and the oscillator is therefore more stable.

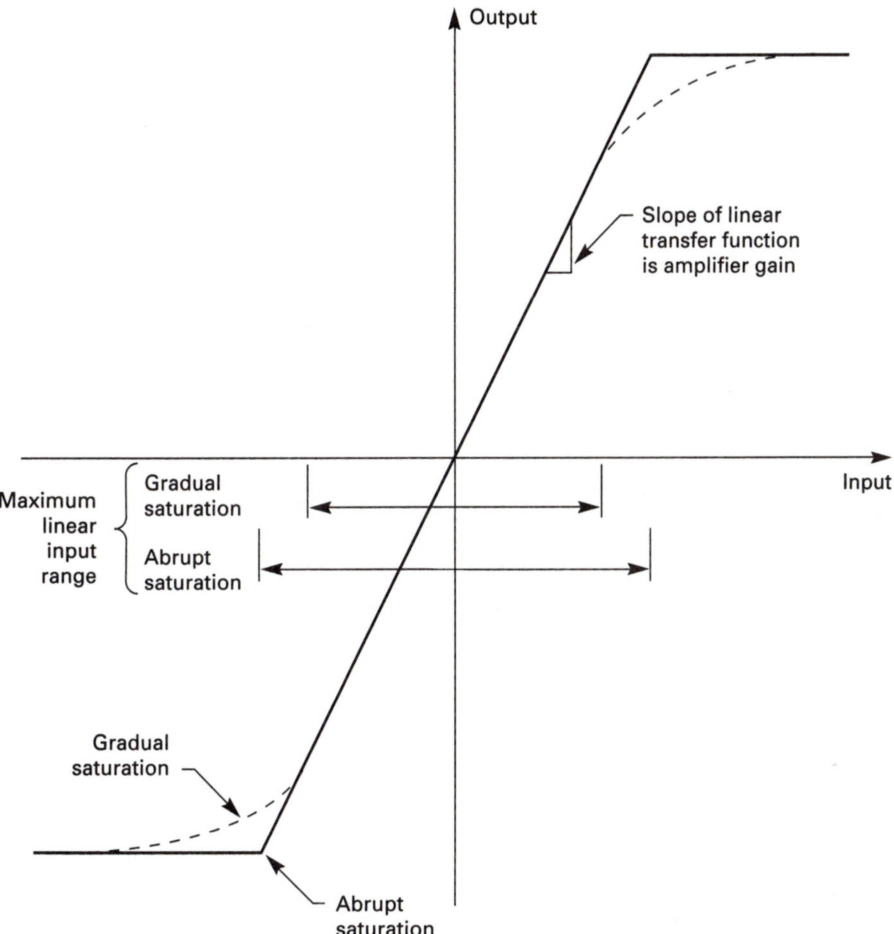

Figure 7.7 Typical amplifier transfer (input-to-output) characteristics, showing both gradual and abrupt saturation. Any particular amplifier usually has one or the other.

From the discussion above, it also should be clear that a tuned feedback oscillator generates a signal with energy primarily at one frequency, where the oscillation criteria are met. *All* the energy would be at that frequency, except for distortion mechanisms (such as saturation) in the amplifier which generate harmonic signals. Such a signal is a sine wave with modest distortion, typically 20 to 50 dB below the fundamental.

Examples of feedback oscillators are described below. Figure 7.8 shows two practical examples of these oscillators.

Tunable LC oscillator. The oscillator in Fig. 7.8*a* has some interesting features. The input of the Q_1, Q_2 differential amplifier is the base of Q_2, and the output is the collector of Q_1. There is approximately zero phase shift both in the

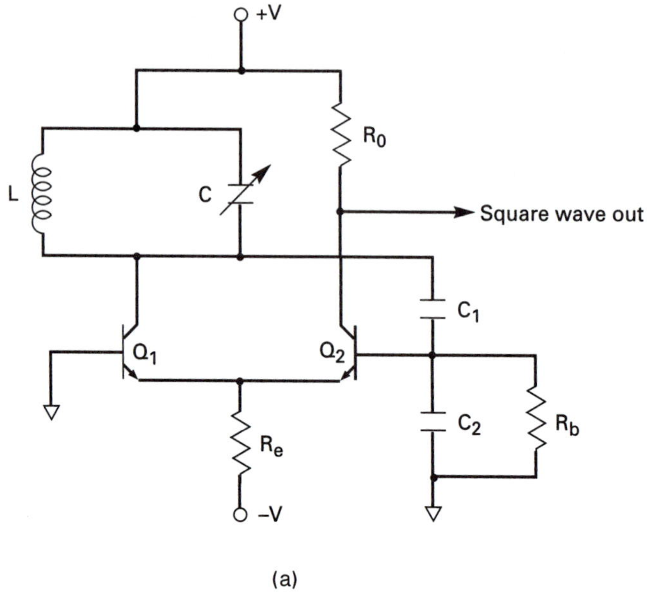

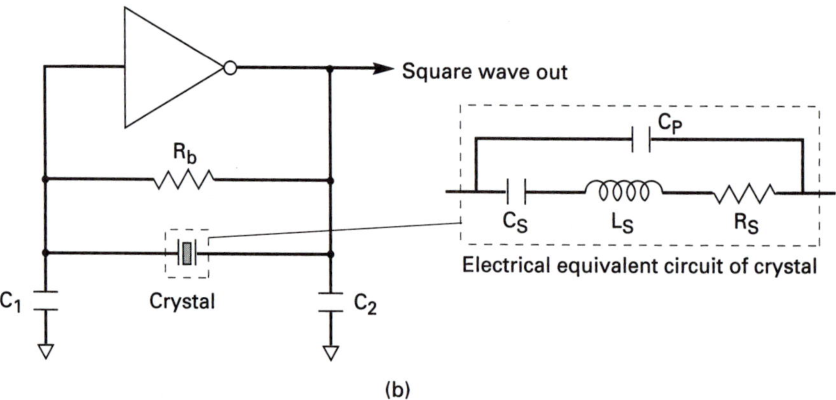

Figure 7.8 Examples of feedback oscillators: (a) Transistor *LC* oscillator. (b) quartz crystal oscillator.

amplifier and in the feedback path through the voltage divider C_1-C_2, so the phase-shift criterion (2), given above in this section, is met. Likewise, there is enough available gain to exceed criterion (1). Therefore, the circuit will oscillate at (or very near) the resonant frequency of the *LC* filter: $1/(2\pi\sqrt{LC})$. The limiting mechanism, needed to stabilize the oscillation amplitude, is found in the well-defined collector current of Q_1: The total emitter current, which is almost constant at approximately $-V/R_e$, is switched between the two transistors, resulting in a square-wave current in each. The fundamental component of this, times the impedance of the *LC* filter (or "tank" circuit), can be controlled so that the collector voltage of Q_1 never saturates. This is also important in reducing resistive loading of the filter, allowing for maximum Q and frequency stability. Another feature is taking the output signal across R_0 in the collector of Q_2. There is very good isolation between this point and the *LC* filter, which minimizes the frequency shift which can occur when a reactive load is placed on an oscillator. Of course, this output signal is a square wave; if this is unsatisfactory, a low-amplitude sine wave can be obtained from the base of Q_2.

Crystal oscillator. Another useful and simple oscillator, using a quartz crystal as the feedback filter, is shown in Fig. 7.8*b*. The amplifier is a digital inverter, preferably CMOS. R_b is needed to bias the inverter into the active region so that oscillations will start. The crystal's electrical equivalent circuit, shown in the dotted box at the right, forms a π network together with C_1 and C_2. The circuit oscillates just slightly higher than the series resonance of the crystal, where the reactance of the crystal is inductive. The phase shift in this network is about 180°, and added to the 180° phase shift of the inverter, the open loop meets the phase criterion for oscillation. The capacitors are made as large as possible while still exceeding the gain criterion. This both decreases the loading on the crystal (thus increasing frequency stability) and limits the voltage swing on the inverter input. Amplitude limiting is, of course, a built-in feature of the digital inverter. Because the output is a logic-level square wave, this and similar circuits are often used for computer clocks.

Threshold decision oscillators. This class is represented in elementary form by Fig. 7.9*a*. The way it generates a periodic waveform is very different from that of the feedback oscillator. A circuit capable of producing a time-varying voltage (or current), such as an *RC* charging circuit, begins operating from some initial state. This circuit is not intrinsically oscillatory. As it changes, its instantaneous state is monitored by a detector, which is looking for a certain threshold condition, such as a voltage level. When the detector decides that the threshold is reached, it acts to reset the circuit to its initial state. The detector also resets, and another cycle starts. Sometimes there are two detectors, and the time varying circuit moves back and forth between two states. There is an example of this in Chap. 16.

Example of a threshold decision oscillator. Consider the operation of the circuit in Fig. 7.9*b*. When power is initially applied, the switch is open and capacitor *C* begins to charge through the resistor *R,* and its voltage rises in the familiar exponential manner (Fig. 7.9*c*). This rising voltage is monitored by the *comparator,* which is empowered to take action when the capacitor voltage becomes equal to a reference voltage, or threshold. When this happens, the comparator momentarily closes the switch, discharging *C* almost instantaneously; *C* then begins to charge again. These actions define a cycle of the oscillator,

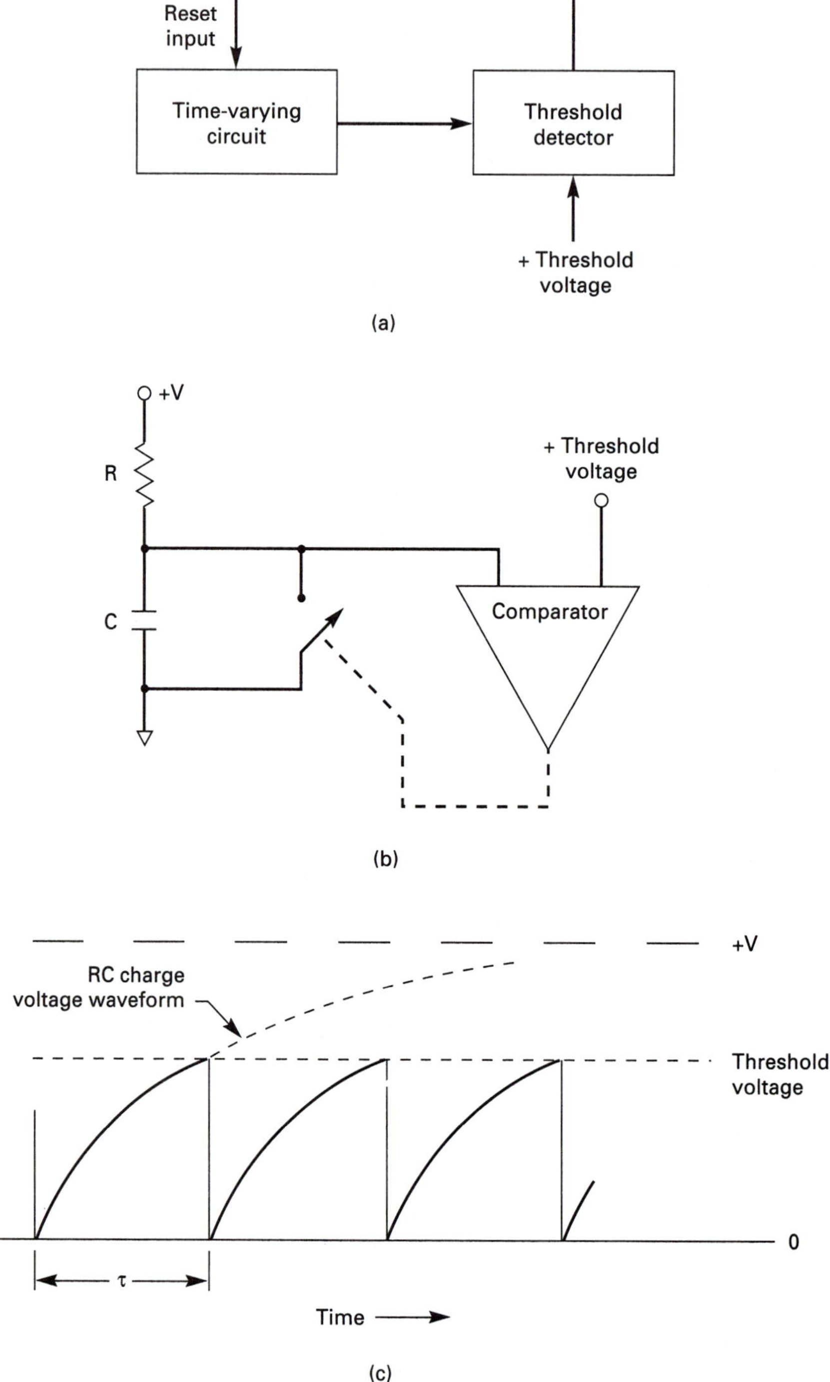

Figure 7.9 Threshold-decision oscillators: (a) Basic circuit functions; (b) simple example; (c) wave-form across C in circuit b.

and they are repeated periodically at a frequency which is determined by the values of R and C and the ratio of $+V$ to the threshold voltage. Quite clearly, the waveform is not a sine wave but is formed of repeated segments of the exponential charging characteristic of the RC circuit.

A threshold decision oscillator is often used when nonsinusoidal waveforms are needed (or can be tolerated) from very low frequencies (millihertz) to a few megahertz. Its frequency is less stable than that of a good feedback oscillator. But, with careful design, the frequency change can be held to less than 1 percent over a wide range of temperature and power-supply variation.

7.3.2 Synthesizers

Although there are two classes of signal generators using the term synthesizer (see below), the technology they share is the use of a fixed-frequency oscillator to synchronize various signal processing circuits which produce the output signal. The oscillator is variously called the "reference" or "clock," the latter term being borrowed from computers. Its frequency accuracy and stability directly affect the generator output quality.

Frequency synthesizers. The emphasis in this class of signal generator is frequency versatility: a very large choice of output frequencies, each "locked" to the reference oscillator. The output frequency of a synthesizer may be expressed as a rational number times the reference frequency:

$$f_{\text{out}} = \frac{m}{n} \times f_{\text{ref}} \tag{7.2}$$

where m, n = integers
f_{out} = synthesizer output frequency
f_{ref} = reference oscillator frequency

In practice, the user types the output frequency on a keyboard or sets it on some switches, or it is downloaded from a computer. For instance, if the reference frequency is 1 MHz and $n = 10^6$, then the user, by entering the integer m, may choose any output frequency within the instrument range to a resolution of 1 Hz.

Synthesizer output waveforms are typically sine waves, with square waves also being popular at lower frequencies. Signal processing techniques for generating the output are described in Chap. 16.

Arbitrary waveform synthesizers. In this technique, the complete period of some desired waveshape is defined as a sequence of numbers representing sample values of the waveform, uniformly spaced in time. These numbers are stored in read-write memory and then, paced by the reference, repetitively read out in order. The sequence of numbers must somehow be converted into a sequence of voltage levels. The device that does this is called, not surprisingly, a digital-to-analog converter (DAC). This device works by causing its digital inputs to switch weighted currents into a common output node. For instance, in a 00–99 decimal DAC, the tens digit might switch increments of 10 mA, and the units digit would then switch increments of 1 mA. Thus a digital input of 68 would cause an output current of $6 \times 10 + 8 \times 1 = 68$ mA. The DAC current output is converted to a voltage, filtered, amplified, and made available as the generator output. Because this is a sampled data technique, there is a limit on the complexity of the waveform. That is, the various curves of the waveform must all be representable with the number of samples available. There is likewise a limit on the waveform frequency, depending on the speed of the digital hardware used in implementing the technique.

A special case of this technique occurs when the only desired waveform is a sine wave whose waveshape samples are permanently stored in read-only memory. This case will be discussed along with other frequency synthesizer techniques.

7.4 Signal Quality Problems

Signal sources, like other electronic devices, suffer impairments due to their imperfectable circuits. Most signal quality problems are the results of noise, distortion, and the effects of limited bandwidth in the circuits which process the signals.

7.4.1 Classes of signal impairments

Noise. This catchall term includes various kinds of extraneous energy which accompany the signal. The energy can be added to the signal, just as audio channels are added together, or it can affect the signal by modulating it. Additive noise includes thermal noise and active device (e.g., transistor) noise, as well as discrete signals like power-supply hum. Synthesizers, in particular, are troubled by discrete, nonharmonic spurious signals referred to as "spurs" by designers. The noise most difficult to control is that which modulates the signal. It usually predominates as phase modulation and is referred to as "phase noise" in the literature and in data sheets. It causes a broadening of the signal spectrum and can be problematic when the signal source is used in transmitter and receiver applications.

Distortion. Owing to small amounts of curvature in transfer functions—the characteristics relating input to output—amplifiers and other signal processing circuits slightly distort the waveshape of the signal passing through them. For sine-wave signals, this means the loss of a pure sinusoid shape, and in turn, harmonics of the signal appear with it. For triangle waveforms, there is degradation in linearity. However, pulse signal sources sometimes purposely use nonlinear (saturating) amplifiers to improve the rise time and flatness specifications of the source.

Bandwidth restrictions. No physical circuit has the infinite bandwidth which is usually assumed in an elementary analysis. Real circuits—such as signal source output amplifiers—have finite passbands. And, within the passband of a real circuit, both the gain and the signal time delay change with frequency. When a complex signal is passing through such a circuit, both the relative amplitudes and relative time positions of the signal components are changed. This causes changes in the shape of the signal waveform. A frequent example of such a change is the appearance of damped oscillations ("ringing") just after the rising and falling edges of a square wave.

7.4.2 Manufacturer's specifications

Manufacturers of signal sources evaluate their products, including the imperfections, and they furnish the customer a set of limits in the form of specifications. Some guidance in interpreting specifications for various signal sources is provided in Chap. 16.

Chapter

8

Microwave Signal Sources

William Heinz
Agilent Technologies
Santa Clara, California

8.1 Introduction

Frequencies usually designated as being in the microwave range cover 1 to 30 GHz. The lower boundary corresponds approximately to the frequency above which lumped-element modeling is no longer adequate for most designs. The range above is commonly referred to as the "millimeter range" because wavelengths are less than 1 cm, and it extends up to frequencies where the small wavelengths compared with practically achievable phyical dimensions require quasioptical techniques to be used for transmission and for component design. The emphasis of the following discussion will be on factors that affect the design and operation of signal sources in the microwave frequency range, though many of the characteristics to be discussed do apply to the neighboring ranges as well. Methods for the generation of signals at lower frequencies employing synthesis techniques are also described, since up-conversion can be performed readily to translate them up into the microwave and millimeter ranges.

Application for such sources include use in microwave signal generators (see Chap. 18), as local oscillators in receivers and down-convertors, and as exciters for transmitters used in radar, communications, or telemetry. The tradeoffs between tuning range, spectral purity, power output, etc. are determined by the application.

Previous generations of microwave sources were designed around tubes such as the Klystron and the backward-wave oscillator. These designs were bulky, required unwieldy voltages and currents, and were subject to drift with environmental variations. More recently, compact solid-state oscillators employing field-effect transistors (FET) or bipolar transistors and tuned by electrically or magnetically variable resonators have been used with additional benefits in ruggedness, reliability, and stability. Frequency synthesis techniques are now used to provide accurate, programmable sources with excellent frequency stability, and low phase noise.

8.2 Solid-State Sources of Microwave Signals

The most common types of solid-state oscillators used in microwave sources will be described in the following subsections.

8.2.1 Transistor oscillators

The generic circuit in Fig. 8.1 illustrates the fundamental operation of an oscillator. The active element with gain A amplifies noise present at the input and sends it through the resonator, which serves as a frequency-selective filter. Under the right conditions, the selected frequency component, when fed back to the input, reinforces the original signal (positive feedback), which is again amplified, etc., causing the signal at the output to grow until it reaches a level determined by the saturation level of the amplifier. When steady state is finally reached, the gain of the amplifier reaches a value that is lower than the initial small signal value that initiated the process, and the loop gain magnitude $\alpha\beta A = 1$. The frequency of oscillation is determined from the requirement that the total phase shift around the loop must be equal to $n \times 360°$.

The Colpitts oscillator (Fig. 8.2) and circuits derived from it that operate on the same basic principle are the most commonly used configurations in microwave transistor oscillator design. The inductor L and the capacitors C, C_1, and C_2 form a parallel resonant circuit. The output voltage is fed back

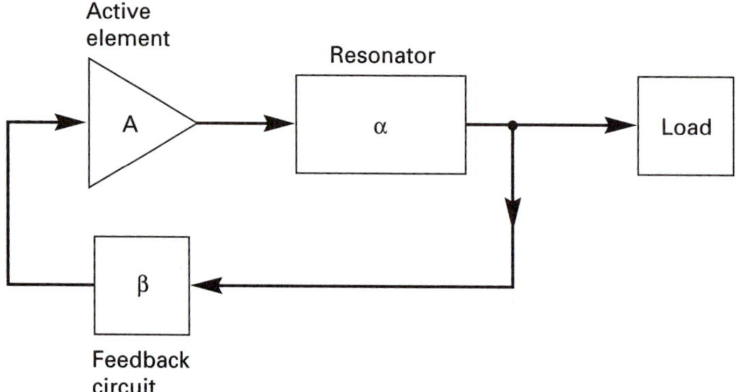

Figure 8.1 Generic oscillator block diagram, where A = forward gain of active element, α = transmission coefficient of resonator, and β = transmission coefficient of feedback circuit.

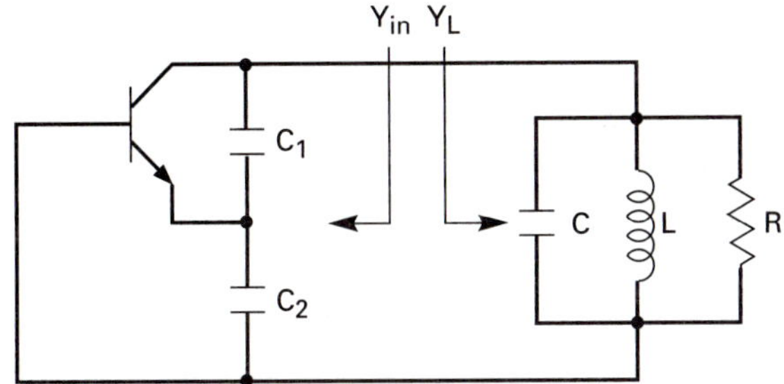

Figure 8.2 Colpitts oscillator circuit.

to the input in the proper phase to sustain oscillation via the voltage divider formed by C_1 and C_2, parts of which may be internal to the transistor itself.

Bipolar silicon (Si) transistors are typically used up to 10 or 12 GHz, and gallium arsenide (GaAs) FETs are usually selected for coverage above this range, though bipolar Si devices have been used successfully to 20 GHz. Bipolar Si devices generally have been favored for lower phase noise, but advances in GaAs FET design have narrowed the gap, and their superior frequency coverage has made them the primary choice for many designs.

It is possible to view the circuit in Fig. 8.2 in a different way that can add insight into the design and operation of transistor oscillators. Since power is being delivered to the resonator by the amplifier, the admittance Y_{in} looking to the left must have a negative real part at the frequency of oscillation. The reactive (imaginary) part of the admittance Y_{in} is tuned out at this frequency by the tank circuit on the right so that $Y_{\text{in}} + Y_L = 0$. The circuitry to the left can be viewed as a one-port circuit with a negative conductance (or resistance) connected to the resonator on the right.

The circuit in Fig. 8.3 (shown for a FET, but similar for a bipolar transistor) can be shown to provide a negative resistance at frequencies above the resonance frequency of L and C_{GS} (i.e., Z_{in} is inductive), so the frequency of oscilla-

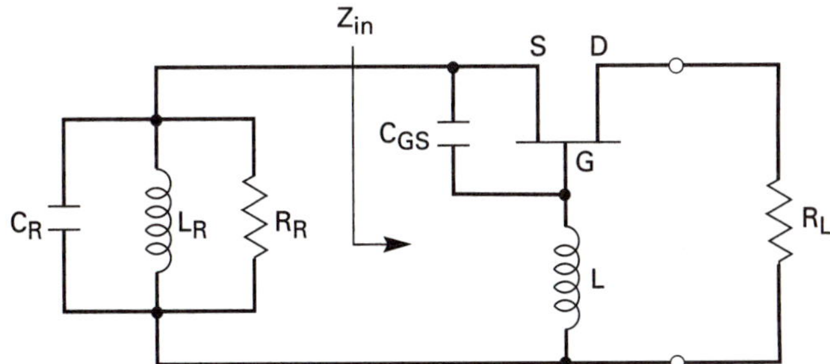

Figure 8.3 Negative-resistance oscillator. C_R, L_R, and R_R represent a resonator. The inductance L is required to achieve a negative resistance in Z_{in}.

tion will be where the resonator looks capacitive (slightly above the resonator's center frequency). This negative-resistance circuit, which can be shown to be a variation of the basic Colpitts circuit in Fig. 8.2 (with the bottom of the resonator connected back to the drain through R_L), is a commonly used building block for microwave oscillators.

8.2.2 Electrically tuned oscillators

The use of an electrically tuned capacitor as C_R in Fig. 8.3 would allow the frequency of oscillation to be varied and to be phase locked to a stable reference (see below). A common technique for obtaining a voltage-variable capacitor is to use a variable-capacitance diode, or "varactor" (Fig. 8.4). This device consists of a reverse-biased junction diode with a structure optimized to provide a large range of depletion-layer thickness variation with voltage as well as low losses (resistance) for high Q. The shape of the tuning curve may be varied by changing the doping profile of the junction. Capacitance variations of over 10:1 are obtainable (providing a theoretical frequency variation of over 3:1 if the varactor provides the bulk of the capacitance C_R in Fig. 8.3), but it is usually necessary to trade off tuning range for oscillator Q to obtain desired frequency stability and phase noise in the microwave frequency range. This can be accomplished by decoupling the varactor from the high-Q resonant circuit by connecting it in series or in parallel with fixed capacitors.

Since the capacitance of the diode is a function of the voltage across it, the rf voltage generated can drive the capacitance. Thus the potential exists for nonlinear mechanisms including generation of high levels of harmonics, AM to FM conversion, and parametric effects. A commonly used method for reducing these effects is to connect two varactors in series in a back-to-back configuration so that rf voltage swings are in equal but opposite directions across them, thereby canceling the odd-ordered components of distortion. This configuration also halves the rf voltage across each diode.

Major advantages of varactor oscillators are the potential for obtaining high tuning speed and the fact that a reverse-biased varactor diode does not dissipate dc power (as does a magnetically biased oscillator, as described be-

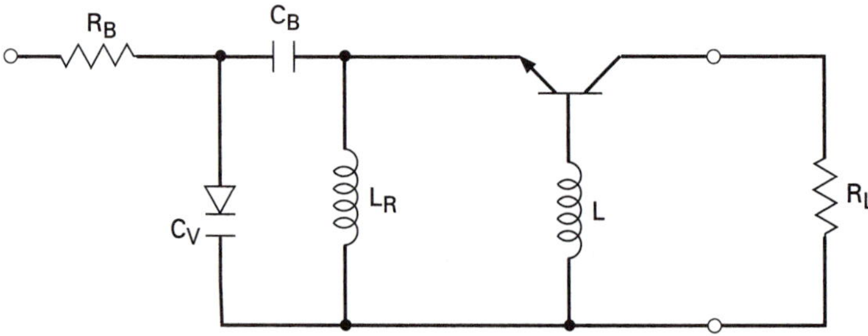

Figure 8.4 Varactor-tuned oscillator.

low). Typical tuning rates in the microsecond range are realized without great difficulty.

8.2.3 YIG-tuned oscillators

High Q resonant circuits suitable for tuning oscillators over very broad frequency ranges can be realized with polished single-crystal spheres of yttrium-iron-garnet (YIG). When placed in a dc magnetic field, ferrimagnetic resonance is attained at a frequency that is a linear function of the field (2.8 MHz/Oe). The microwave signal is usually coupled into the sphere (typically about 0.5 mm in diameter) via a loop, as shown in Fig. 8.5. The equivalent circuit presented to the transistor is a shunt resonant tank that can be tuned linearly over several octaves in the microwave range. Various rare earth "dopings" of the YIG material have been added to extend performance to lower frequency ranges in terms of spurious resonances (other modes) and non-linearities at high power, but most ultrawideband oscillators have been built above 2 GHz. Frequencies as high as 40 GHz have been achieved using pure YIG, and other materials (such as hexagonal ferrites) have been used to extend frequencies well into the millimeter range.

A typical microwave YIG-tuned oscillator is usually packaged within a cylindrical magnetic steel structure having a diameter and axial length of a few centimeters. Because of the small YIG sphere size, the air gap in the magnet structure also can be very small (on the order of 1 mm). The resulting electromagnet thus has a very high inductance (typically about 1000 mH) so that typical speeds for slewing across several gigahertz of frequency range and stabilizing on a new frequency are on the order of 10 ms. To provide the capability for frequency modulation of the oscillator and to enable phase locking to sufficiently high bandwidths to optimize phase noise, a small

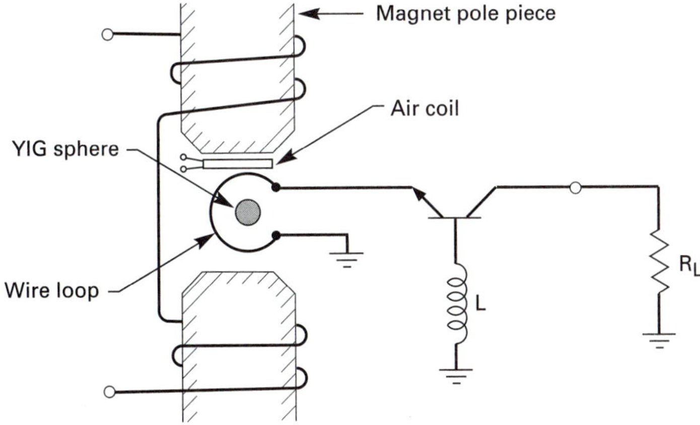

Figure 8.5 YIG-tuned oscillator.

coil is usually located in the air gap around the YIG sphere. Frequency deviations of ±10 MHz with rates up to 10 MHz can be achieved with typical units.

8.2.4 Frequency multiplication

Another approach to signal generation involves the use of frequency multiplication to extend lower-frequency sources up into the microwave range. By driving a nonlinear device at sufficient power levels, harmonics of the fundamental are generated that can be selectively filtered to provide a lower-cost, less complex alternative to a microwave oscillator. The nonlinear device can be a diode driven through its nonlinear *i* vs. *v* characteristic, or it can be a varactor diode with a nonlinear capacitance vs. voltage. Another type of device consists of a *pin* structure (*p*-type and *n*-type semiconductor materials separated by an intrinsic layer) in which charge is stored during forward conduction as minority carriers. Upon application of the drive signal in the reverse direction, conductivity remains high until all the charge is suddenly depleted, at which point the current drops to zero in a very short interval. When this current is made to flow through a small drive inductance, a voltage impulse is generated once each drive cycle, which is very rich in harmonics. Such step-recovery diodes are efficient as higher-order multipliers.

One particularly versatile multiplier configuration is shown in Fig. 8.6. The circuit consists of a step-recovery diode in series with a YIG resonator which serves as a bandpass filter tuned to the resonance frequency of the sphere. The diode is driven by a YIG oscillator covering the range 2 to 6.5 GHz. By forward-biasing the diode, this frequency range can be transmitted directly through the filter to the output. To provide frequency coverage from 6.5 to 13 GHz, the diode is reverse-biased and driven as a multiplier from 3.25 to 6.5 GHz, where the filter selects the second harmonic. Above 13 GHz, the third harmonic is selected and so on through the fourth harmonic. Thus 2- to 26-GHz signals can be obtained at the single output port.

While this YIG-tuned multiplier (YTM) can provide very broadband signals relatively economically, a limitation is in the passing through of undesired

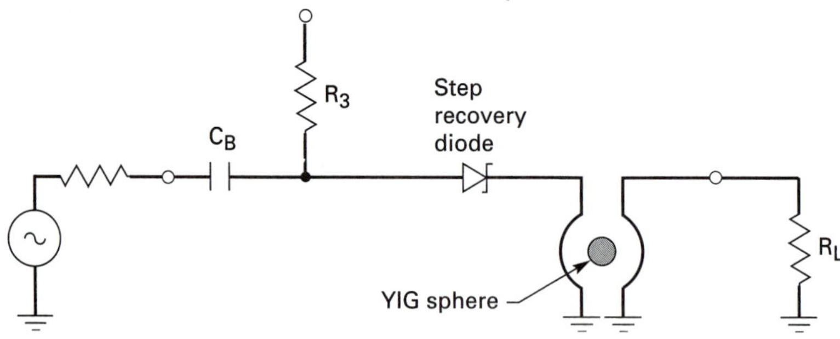

Figure 8.6 YIG-tuned multiplier.

"subharmonics," i.e., the fundamental and lower harmonics are typically attenuated by only about 20 to 25 dB for a single-resonator YIG filter. The availability of very broadband YIG oscillators has eliminated many of the major advantages of the YTM.

8.2.5 Extension to low frequencies

Since broadband YIG oscillators are usually limited to operating frequencies above 2 GHz, for those applications in which frequencies below 2 GHz are needed, it may make sense to extend frequency coverage without adding a separate broadband-tuned oscillator. Two methods that are of interest are heterodyne systems and the use of frequency dividers.

Figure 8.7 shows a heterodyne frequency extension system in which a 2- to 8-GHz YIG-tuned oscillator is extended down to 10 MHz by means of a mixer and a fixed local oscillator at 5.4 GHz. To cover the 10-MHz to 2-GHz range, the YIG-tuned oscillator is tuned from 5.41 to 7.4 GHz. Because it is desirable to operate the mixer in its linear range to minimize spurious signals and possibly to preserve AM, the signal level out of the mixer is generally not high enough, requiring a broadband amplifier to boost the output power. The additional cost and the addition of noise are the unwelcome price to be paid of this approach. Advantages include the otherwise cost efficiency, the ability to get broadband uninterrupted sweeps from 10 MHz to 2 GHz, and the preservation of any AM and/or FM that may be generated ahead of the mixer.

An alternate approach to that in Fig. 8.7 is shown in Fig. 8.8, in which frequency dividers are used. Each octave below 2 GHz requires an additional binary divider, and since the output of these dividers is a square wave, extensive filtering is needed if low harmonics are desired at the output. Each octave from a divider needs to be split into two low-pass filters, the outputs of

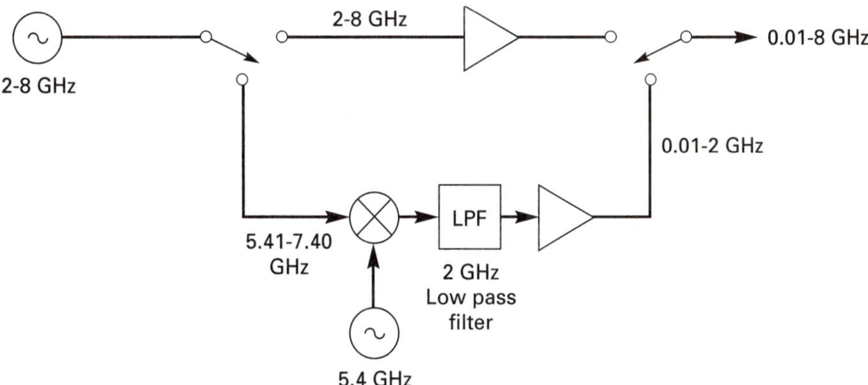

Figure 8.7 Heterodyne frequency extension system. The 2- to 8-GHz frequency range of the oscillator is extended down to .01 GHz.

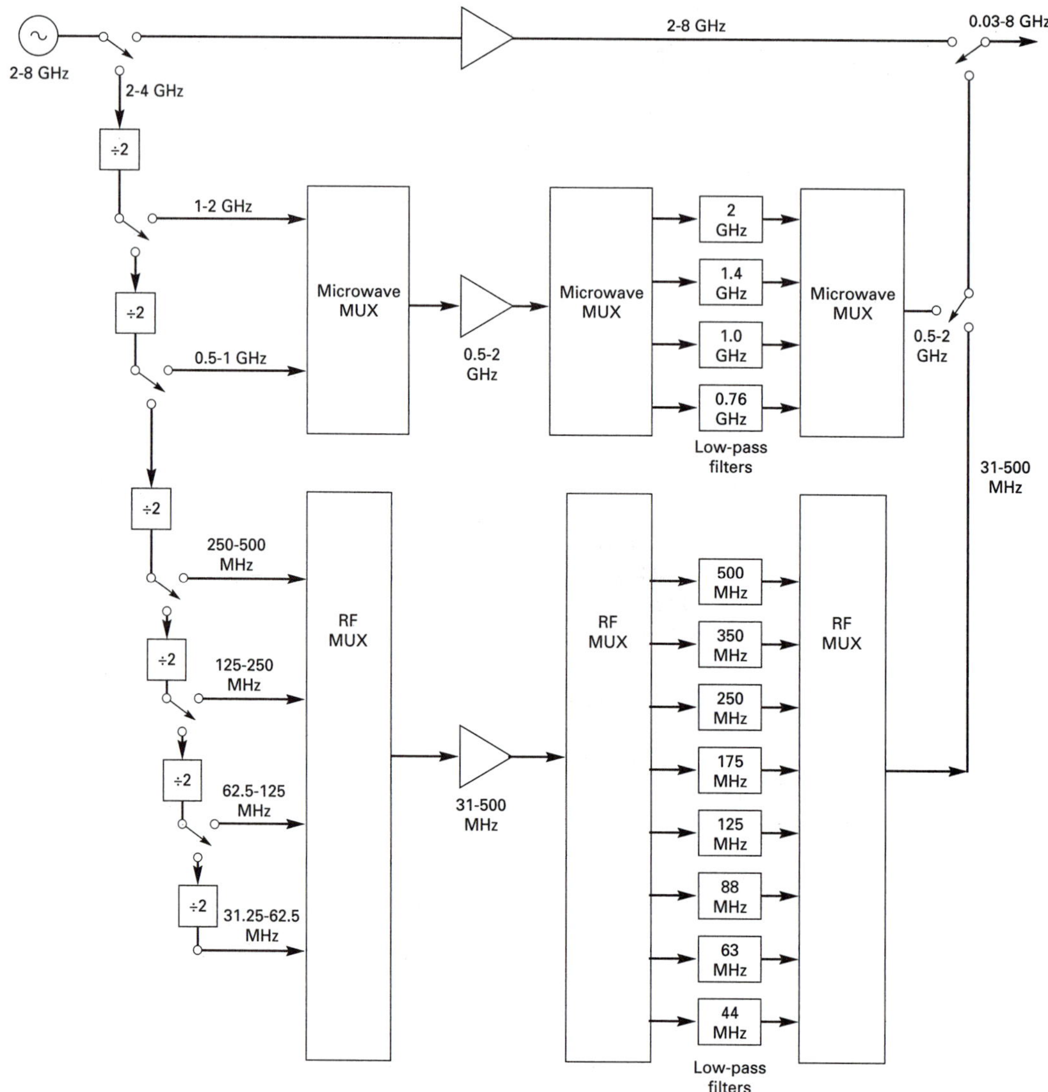

Figure 8.8 Frequency extension using frequency division. The dividers are all divide-by-2. Microwave multiplexers (microwave MUX) and low-pass filters provide 0.5- to 2-GHz and rf MUX's and filters using lumped-element techniques fill in below 500 MHz.

which are then selected and switched over to the output. The divider frequency extension architecture has the advantage of delivering clean, low-phase-noise signals (phase noise and spurs are reduced 6 dB per octave of division) at low cost. Disadvantages are that AM is not preserved and FM deviation is halved through each divider.

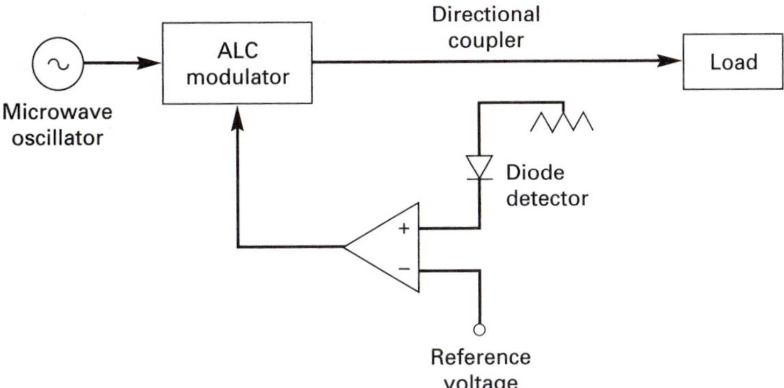

Figure 8.9 Automatic level control of a source.

8.3 Control and Modulation of Signal Sources

The various sources just described can provide signals over broad ranges of the microwave spectrum, but variations in power and frequency with time, load conditions, or environmental changes can be appreciable. Most real applications require the addition of components and feedback circuitry to provide control and stabilization of signal level and frequency. They also may require the addition of frequency, amplitude, and/or pulse modulation.

8.3.1 Leveling and amplitude control

Figure 8.9 shows a commonly used technique for achieving a controllable amplitude that can be kept constant over frequency and under a variety of load conditions. A portion of the output signal incident on the load is diverted over to the diode detector by means of the directional coupler (see Chap. 32). The detector remains relatively insensitive to the signal reflected by the load, depending on the directivity of the coupler. The detected voltage is connected to an input of the differential amplifier that drives the modulator, thus forming a feedback loop. The reference voltage provided to the other input of the amplifier determines the signal level at the output and can be used to vary it. Dc voltages representing corrections for frequency, temperature, etc., can be applied at this point, as can AM signals consistent with loop gain and bandwidth. Loop gain and bandwidth are key design parameters determined also by required switching speed (i.e., amplitude recovery time), the total variation in power from the source that needs to be corrected, sweep rates for a frequency-swept source, the AM noise spectrum of the source, etc., and must be well controlled to ensure loop stability.

Since the directional coupler remains relatively insensitive to the signal reflected by the load back toward the source (depending on the directivity of the coupler), the incident power remains constant. Furthermore, the power re-reflected back from the modulator is detected and corrected for so that the

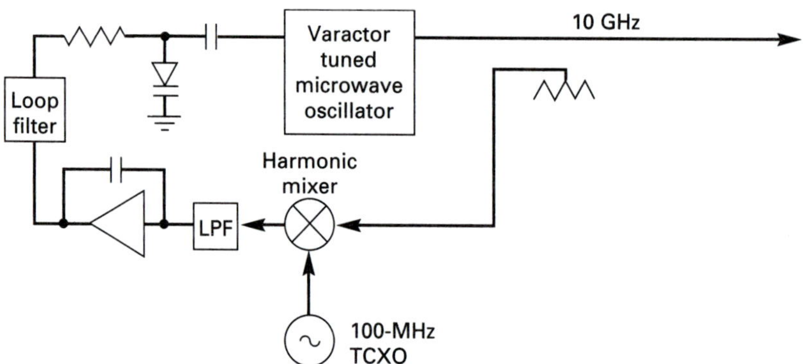

Figure 8.10 Phase-locked loop using harmonic mixer.

effective source match looks perfect (in practice, the finite directivity of the coupler combined with other imperfections in the output path will limit performance).

8.3.2 Frequency control

Phase-locked loops (PLLs) are commonly used to provide frequency stability and to optimize phase noise of microwave sources. By phase locking to a stable reference source, usually a temperature-controlled crystal oscillator (TCXO) at a lower frequency, the long-term stability of the latter can be transferred to the microwave oscillator. Figure 8.10 illustrates how this can be done. A portion of the 10-GHz output signal is connected to a harmonic mixer or sampler which is driven by the 100-MHz TCXO. The 100th harmonic of this signal mixes with the output signal to produce a dc voltage at the mixer IF port that is proportional to the phase difference between them. This voltage is low-pass filtered and fed to the integrating amplifier, which in turn drives the varactor tuning diode to close the loop.

Two observations can be made regarding this approach: First, only output frequencies that are exact multiples of the TCXO frequency can be provided by the source; i.e., the frequency resolution is equal to the reference frequency. Second, the phase noise of the source will be equal to the phase noise of the reference source multiplied by the square of the ratio of the output frequency to the reference frequency (20 log 100 = 40 dB) within the loop bandwidth.

Thus there is a tradeoff between reaching a small step size (reference frequency) and minimizing phase noise (demanding a high reference frequency). There are several ways that this limitation can be overcome. These usually involve multiple loop architectures in which fine frequency resolution is achieved from another voltage-controlled oscillator (VCO) at IF, as is shown

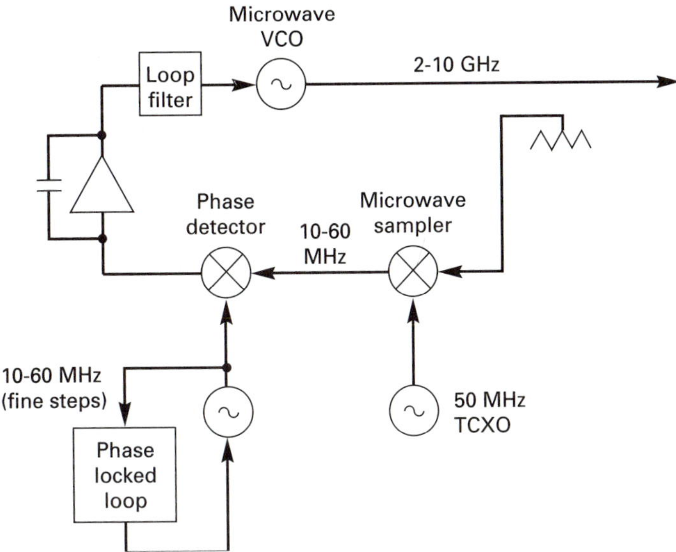

Figure 8.11 Multiple loop architecture to get fine frequency control.

in Fig. 8.11. Since the output frequency is translated down to the intermediate frequency (IF), the frequency resolution is preserved. The total frequency range of the input frequency VCO must be large enough to fill in the range between the appropriate harmonics of the sampler drive. Since it is usually desirable to limit the input frequency range to values well under the sampler drive frequency, the latter also can be varied, with only a relatively small number of discrete frequencies required here.

Because of the second observation above, the phase noise of the sampler driver source and the phase lock loop bandwidth of the microwave VCO become major design considerations for optimizing output phase noise.

8.4 Frequency Synthesis

The preceding subsection on frequency control illustrated the concept, with examples of how a microwave source can be stabilized, producing a finite number of output frequencies phase locked to a reference. The methods by which frequencies can be generated using addition, subtraction, multiplication, and division of frequencies derived from a single reference standard are called "frequency synthesis techniques." The accuracy of each of the frequencies generated becomes equal to the accuracy of the reference, each expressed as a percent.

Three classifications are commonly referred to: indirect synthesis, direct synthesis, and direct digital synthesis (DDS). The basic concepts of these techniques will be described briefly below using representative examples.

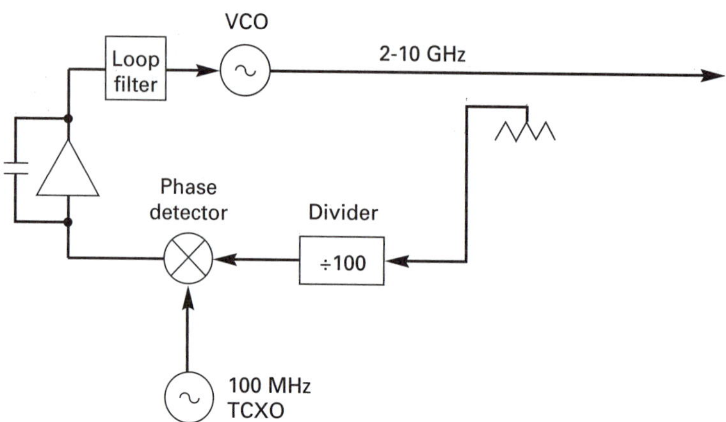

Figure 8.12 Phase-locked loop using divider.

8.4.1 Indirect synthesis

The term "indirect synthesis" is usually applied to methods in which a sample of the output frequency is compared with a frequency derived from the reference and fed back to form a phase-locked loop, such as in Figs. 8.10 and 8.11. The output frequency sample can be translated in frequency and/or divided or multiplied for comparison (usually in a phase detector) with a convenient reference frequency derived (using similar techniques) from the reference standard. The synthesizer can comprise several individual phase-locked loops or synthesizers.

In the example in Fig. 8.12, the output frequency is divided down to the reference frequency and applied to a phase detector. The effect is similar to the circuit in Fig. 8.10 in terms of step size and noise at the output (neglecting any possible excess noise contributions from the divider), but the phase detection is now accomplished at a lower frequency.

Figure 8.13 shows a method for generating small step sizes without the noise degradation inherent in schemes that incorporate large divide ratios. The first divider is a "dual-modulus divider," meaning that the divide ratio can be changed dynamically between two adjacent integers p and $p + 1$ (e.g., 10 and 11) via a control line. Thus the divide ratio can be $p + 1$ for M cycles and p for $N - M$ cycles, and thereafter the process repeats every N cycles. The result is that the output frequency can vary in fractional, constantly varying multiples of the reference frequency and is therefore known as a "fractional-n technique." The dual-modulus divider starts out at $p + 1$ and continues with this value until M pulses have been counted in the frequency-control unit [i.e., after M $(p + 1)$ cycles from the VCO]. The control unit then changes the divide number to p. After $(N - M)p$ more cycles from the VCO, the process repeats. The result is a fractional divide number between p and $p + 1$ (equal to $p + M/N$). While this method solves the problem of forcing

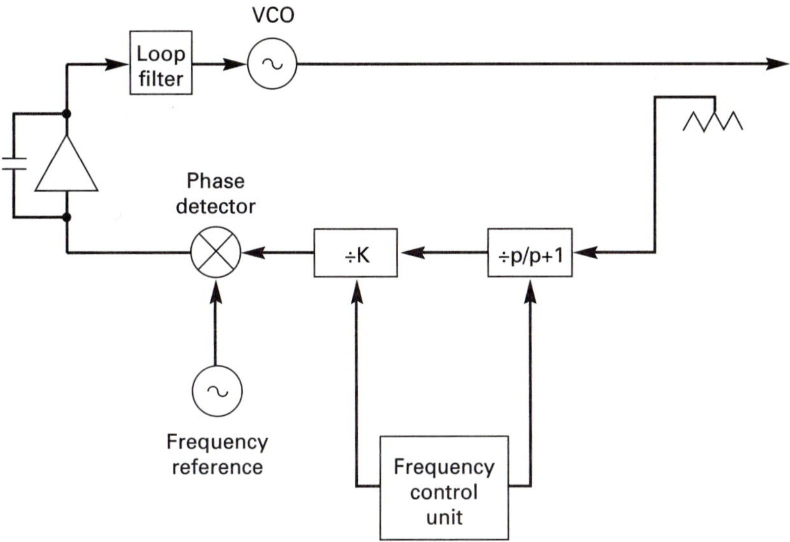

Figure 8.13 Phase-locked loop with fractional *n*.

the reference frequency to a low enough value to provide the required step size without excessive phase noise degradation, spurious signals are introduced about the primary output signal at offset frequencies equal to multiples of the frequency resolution. These can be controlled through appropriate choices in loop bandwidth and through phase-error correction schemes.

8.4.2 Direct synthesis

The set of techniques commonly referred to as "direct synthesis" involves the simultaneous generation of multiple frequencies from a common reference which are then selected and assembled in various combinations to produce each desired frequency at the output. Figures 8.14 and 8.15 are "brute force" examples of direct synthesizers utilizing mixers, multipliers, filters, and switches to generate signals in the range 1 to 9 MHz and 1 to 99 MHz in increments of 1 MHz. A more practical mix and divide technique is described in Chap. 18. These techniques can be extended to provide broad coverage at microwave frequencies with architectures incorporating direct synthesis up-conversion.

Several advantages and disadvantages of the direct synthesis approach become clear upon examination of this example. There is an inherent capability to achieve high speeds for switching frequency due to the access to the frequencies generated concurrently without having to wait for loops to lock. It is primarily the speed of the switches and the time needed to generate the proper commands to drive them that will determine frequency-switching time. Another advantage of this kind of approach is that there will be "phase

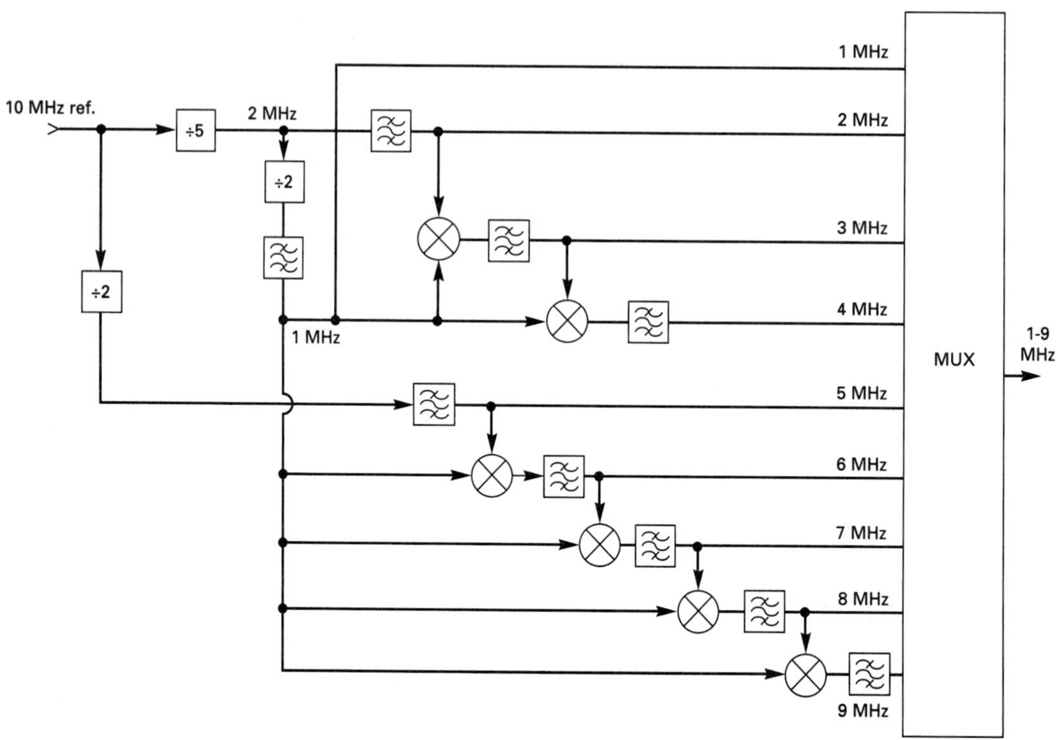

Figure 8.14 Direct synthesizer in the 1- to 9-MHz range.

memory," i.e., if the synthesizer is switched from one frequency to another and then back to the original, the phase will remain identical to what it would have been without switching. It should be noted that if the switching of inputs to dividers is involved, this result cannot be guaranteed.

From the system design point of view, the multiplicity of mixers means that many spurious signals (both harmonics and nonharmonics) will be generated and need to be accounted for in the design. Frequencies need to be chosen carefully, and filtering needs to be provided properly to reduce the number and levels of spurious signals at the output. In the synthesizer in Fig. 8.15, the "local oscillator" frequencies from the lower MUX will be present at the output along with image frequencies. Output filtering requirements can be eased by using the frequencies from 5 to 9 MHz and choosing the appropriate sideband (e.g., 71 MHz is realized by mixing 80 and 9 MHz). Careful isolation of critical components needs to be provided. In general, direct synthesizers tend to be more bulky than indirect synthesizers because of the higher number of components involved, the need for more filtering, and isolation requirements.

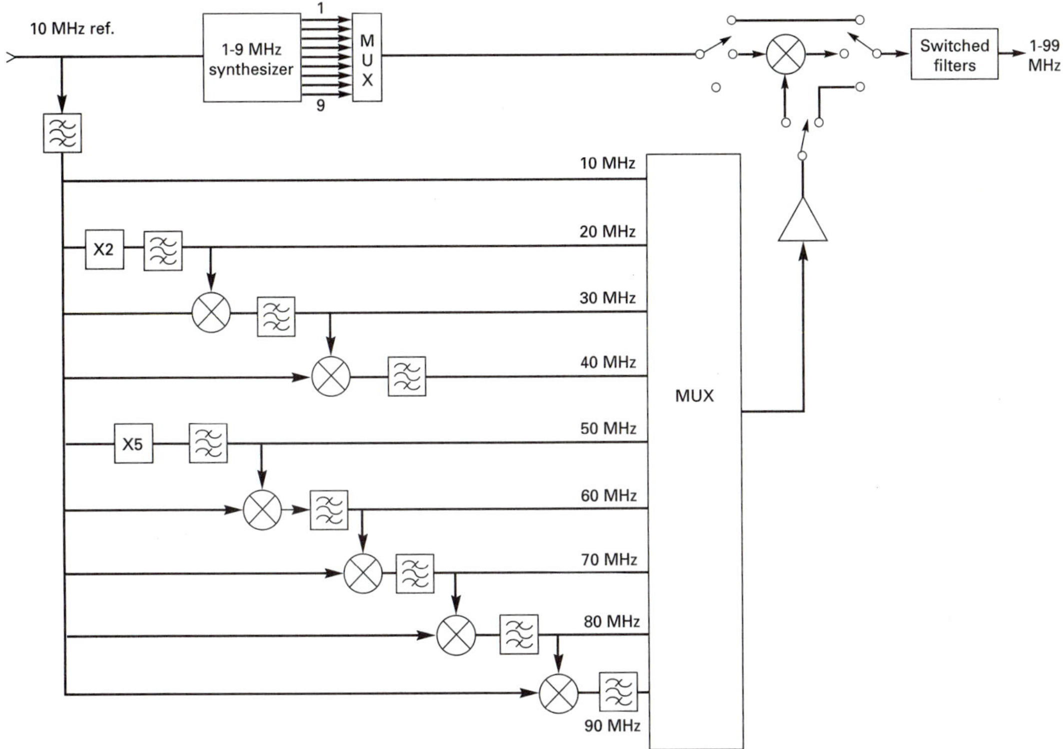

Figure 8.15 Direct synthesizer in the 1- to 99-MHz range.

8.4.3 Direct digital synthesis (DDS)

This method overcomes several shortcomings referred to in the techniques described previously. There are applications where phase-locked loops can be replaced quite effectively, and when they are used in combination with the methods described above, the realization of versatile, more compact high-performance sources has become a reality, with the added capability of high-quality phase and frequency modulation.

The DDS (also referred to as a "numerically controlled oscillator," or NCO) block diagram is shown in Fig. 8.16. An N-bit digital accumulator is used to add an increment of phase to the contents on each clock cycle. An M-bit lookup ROM provides the sine of the accumulated phase. These digital data then drive a digital-to-analog converter (DAC) to generate a series of steps approximating a sine wave. After low-pass filtering, higher-order harmonics, aliasing signals, and other undesired spurious outputs are attenuated, and a relatively clean sine wave emerges. The Nyquist criterion requires a sampling rate (clock frequency) greater than twice the maximum output frequency. Practical design concerns limit the output frequency to about 75 percent of the Nyquist frequency.

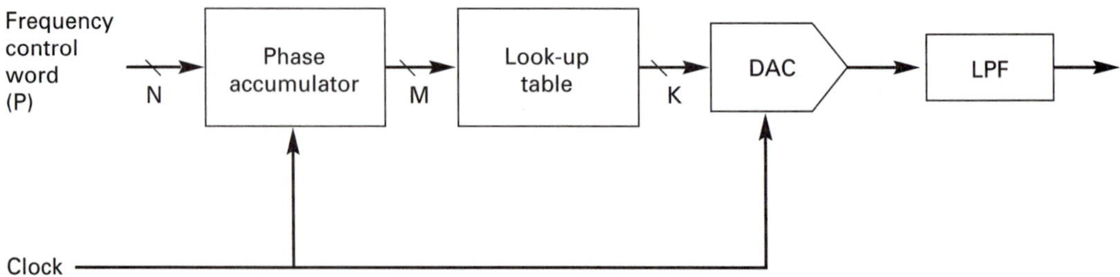

Figure 8.16 DDS block diagram.

Since the accumulator has a modulus of 360°, the process repeats and generates a continuously varying since wave. Figure 8.17 illustrates the process by means of a "phase circle" in which the total 360° of phase is divided into 2^N equal increments for addition in the accumulator. This represents the lowest frequency of operation as well as the minimum frequency increment (step size). Higher frequencies are generated by effectively multiplying (programming the step size of) the minimum increments by the integer *P*, contained in the frequency control word.

Thus the frequency resolution F_{res} that can be obtained is

$$F_{\text{res}} = \frac{F_{\text{clk}}}{2^N} \tag{8.1}$$

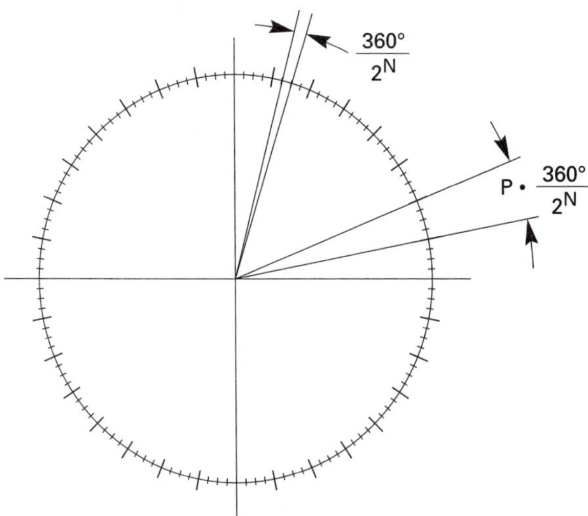

Figure 8.17 DDS phase circle. The phase increment $360°/2^N$ corresponds to the lowest frequency. To generate higher frequencies, this increment is multiplied by the integer *P*, contained in the frequency control word.

where F_{clk} is the clock frequency, and N is the number of bits in the accumulator. For example, for a 50-MHz clock frequency and a 24-bit accumulator, a step size of about 3 Hz is available to output frequencies beyond 18 MHz. It should be pointed out that if it is desirable to generate frequencies with decimal frequency resolutions, a clock frequency equal to a power of 2 (binary) is required.

In addition to fine frequency resolution, DDS is capable of short frequency switching intervals with continuous phase, since this can be accomplished in principle merely by changing the size of the phase increment being added in the accumulator. Pipeline delays in the digital circuitry are the primary limit to speed. Frequency modulation can be done by varying the frequency control word, and phase modulation can be performed by varying the digital phase word provided to the lookup ROM.

The primary factor governing DDS use at high frequencies is spurious signal performance. Spurious performance is determined by several factors, including the DAC switching transients, DAC nonlinearities, and imperfect synchronization of latches and coupling effects along the digital path. Continuing development and improvement in CMOS and gallium arsenide NCOs and DACs are pushing replacement of PLLs and direct synthesizers by DDS to higher and higher frequencies.

Chapter

9

Digital Signal Processing

John Guilford

Agilent Technologies
Lake Stevens, Washington

9.1 Introduction

Digital signal processing (*DSP*) consists of modifying or processing signals in the digital domain. Because of the advances made in the speed and density of IC technology, more and more functions that were once performed in the analog domain have switched over to be processed digitally, such as filtering and frequency selection. Furthermore, digital signal processing has allowed new kinds of operations that weren't possible in the analog domain, such as the Fourier transform. With the high performance and low cost of integrated circuits and microprocessors, digital signal processing has become pervasive. It is built into almost all instruments, as well as such things as cellular telephones, compact disc players, and many automobiles.

This chapter covers what is a signal, ways to characterize signals, ways to characterize signal processing, and the advantages and disadvantages of digital signal processing, compared to analog signal processing. Before many signals can be processed digitally, they must first be converted from analog signals to digital signals in a process called *digitizing* or *analog-to-digital conversion.* Some of the common tasks performed in digital signal processing include filtering, sample-rate changing, frequency translation, and converting from the time domain to the frequency domain via the Fast Fourier Transform. Depending on the cost and performance constraints of a particular design, the signal processing task can be implemented in various forms of hardware, ranging from custom integrated circuits to field-programmable gate arrays to off-the-shelf chips, or it can be implemented in software on equipment ranging from general-purpose computers to special-purpose DSP processors.

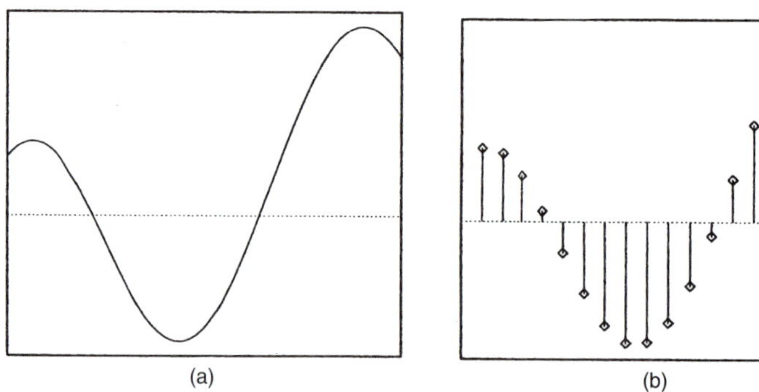

Figure 9.1 A continuous time signal (a). A discrete time signal (b).

9.2 Signal Characterization

Before getting into signal processing itself, first consider what a signal is. Most generally, a signal is something that varies and contains information. Examples of signals include things like the changing air pressure of a sound wave or the changing elevation of terrain as the location changes.

9.2.1 Continuous and discrete time signals

There are various ways of characterizing signals. One way to do this is by the domain or independent variable of the signal. The domain of a signal can be one-dimensional, such as time, or it can be multidimensional, such as the spatial dimensions of an image. For the most part, this chapter will consider signals that vary with time as the independent variable. Signals can vary in a continual manner, where time can take on any value. These are called *continuous time signals* (Fig. 9.1A). Other signals only have values at certain particular (usually periodic) values of time. These are called *discrete time signals* (Fig. 9.1B).

9.2.2 Analog and digital signals

Likewise, the signal's range of values can be characterized. Signals can be one-dimensional as well as multidimensional. A signal can take on continuous values. Such a signal is often called an *analog signal.* An example of this might be the deflection of the meter movement of an analog voltmeter. Alternately, a signal can be restricted to only taking on one of a set of discrete values. This type of signal is often called *digital.* The corresponding example for this would be the voltage displayed on a digital voltmeter. A 3½-digit voltmeter can only show one of about 4000 discrete values (Fig. 9.2).

Although digital signals tend to be discrete time and analog signals tend to be continuous time, analog signals can be continuous or discrete time, and the same is true of digital signals. An example of a discrete time analog signal

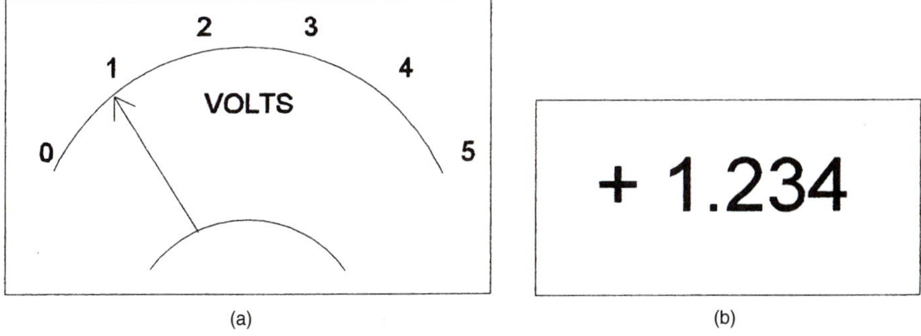

Figure 9.2 Analog signal with continuous values (a). A digital signal with discrete values (b).

is the charge value stored in the CCD array of an imaging chip in a digital camera. Each charge value can be any of a continuum of values while each pixel is discrete. An example of a continuous time digital signal is Morse code, where there are only two different states, tone or no tone, but the timing of the start and ending of the tones is continuous. Furthermore, the same information can be represented by two different signals. For example, the exact same music can be represented by the discrete-time digital signal stored on a compact disc or by the continuous-time analog pressure variations emanating from a speaker.

9.2.3 Physical and abstract

Signals can be something physical, such as the height of an ocean wave, pressure variation in a sound wave, or the varying voltage in a wire, or they can be entirely abstract, merely being a sequence of numbers within a computer.

9.3 Signal Representations

A signal can be represented in many ways. Probably the most familiar way to represent a signal is by showing the value of the signal as a function of time. This is known as the *time-domain representation* of a signal, and it is what an oscilloscope shows. Sometimes other representations of a signal can be more useful. Probably the next most common way to represent signals is by showing the value of a signal as a function of frequency. This *frequency-domain representation* of a signal is what a spectrum analyzer shows (Fig. 9.3). Some aspects of a signal that might be very hard to discern in one representation of a signal might be very obvious in another. For example, a slightly distorted sinusoidal wave might be very difficult to distinguish from a perfect sinusoid in the time domain, whereas any distortion is immediately obvious

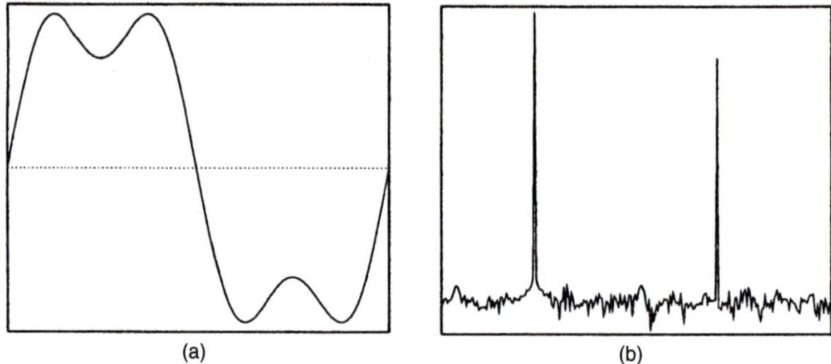

Figure 9.3 A time-domain representation of a signal (a). A frequency-domain representation of the same signal (b).

in the frequency domain. Figure 9.4A shows a perfect sine wave overlaid with a sine wave with 1% distortion. The two traces are indistinguishable when viewed in the time domain. Figures 9.4B and 9.4C show these signals in the frequency domain where the distortion is easy to see.

Other domains can be used to represent signals, but this chapter will only

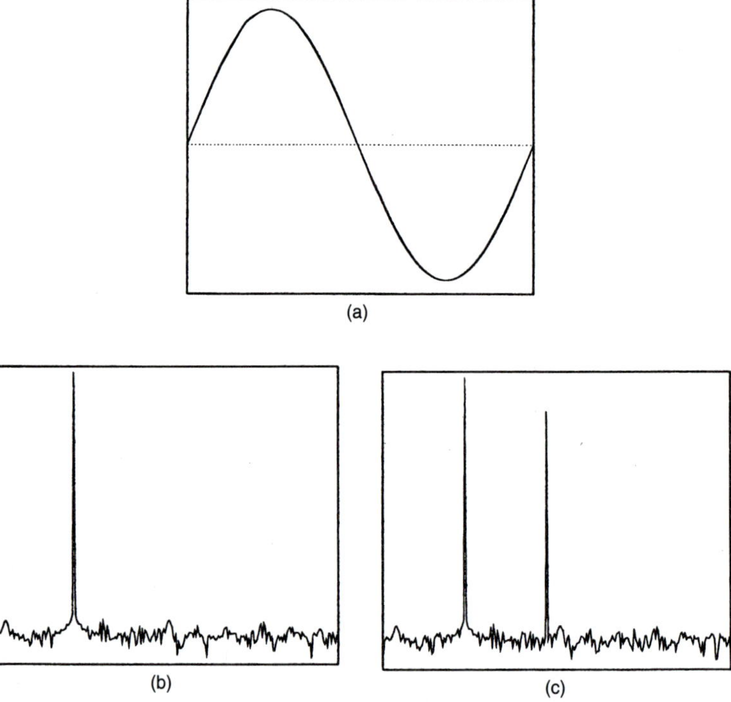

Figure 9.4 A sine wave with no distortion overlaid with a sine wave with 1% distortion (a). A frequency-domain representation of a perfect sine wave (b). A frequency-domain representation of a distorted sine wave (c).

be concerned with the time domain and the frequency domain. See the bibliography for more information on other domains.

9.4 Signal Processing

Signal processing, in the basic sense, means changing, transforming, or analyzing a signal for any of a variety of purposes. Some of the reasons to process a signal include reducing noise, extracting some information from the signal, selecting certain parts of a signal, and accentuating certain parts of a signal. The goal of signal processing is to make the signal more appropriate for some particular application, such as modulating an audio signal onto a carrier so that it can be more efficiently broadcast as radio.

9.4.1 Reversible and irreversible

The different types of signal processing systems can be characterized in several different ways. One can speak of *reversible systems* and *irreversible systems*. In a reversible system, given the output of the system, the input can be uniquely reconstructed. As an example, the Fourier transform of a signal is reversible because, given the Fourier transform, the inverse-Fourier transform can be used to determine the original signal. A system that takes the absolute value of the input, on the other hand, is irreversible. Given the output of that system, the input can't be determined because both positive and negative values map to the same output.

9.4.2 Linear and nonlinear

Another major way to differentiate signal-processing systems is whether they are linear or non-linear.

Linear. A linear system has the property of superposition; if the input to the system consists of a weighted sum of several signals, then the output is the weighted sum of the responses of the system to the various input signals. In particular, a linear system obeys two rules: if the input to the system is scaled by a constant, then the output is also scaled by the same constant; if the input to the system is the sum of two signals, then the response of the system is the sum of the responses of the system applied to each of the two inputs. Mathematically, if $y_1[n]$ is the response of the system to the input ($x_1[n]$), and $y_2[n]$ is the response of the system to the input $x_2[n]$, and a is any constant, then the system is linear if:

1. The response of the system to

$$x_1[n] + x_2[n] \text{ is } y_1[n] + y_2[n] \tag{9.1}$$

2. The response of the system to

$$ax_1[n] \text{ is } ay_1[n] \tag{9.2}$$

where: $x_1[n]$ is an arbitrary input to the system
$x_2[n]$ is another arbitrary input to the system
$y_1[n]$ is the response of the system to $x_1[n]$
$y_2[n]$ is the response of the system to $x_2[n]$
a is any constant

Note: this applies to continuous time systems, too. For example, the response of a linear continuous time system to the input $a \times (t)$ is $ay(t)$. Because this chapter is about digital signal processing, most of the examples are given in terms of discrete time signals, but, in general, the results hold true for continuous time signals and systems, too.

Nonlinear. Multiplying a signal by a constant is a linear system. Squaring a signal is nonlinear because doubling the input quadruples the output (instead of doubling it, as required by the definition of linearity).

Time invariance or shift invariance. Linear systems often have a property called *time invariance* (sometimes called *shift invariance*). *Time invariance* means that if the input is delayed by some amount of time, the output is the same, except that it is delayed by the same amount: if $y[n]$ is the response of a system to the input $x[n]$, then a system is time invariant if the response of the system to:

$$x[n+N] \text{ is } y[n+N] \tag{9.3}$$

where: $x[n]$ is an arbitrary input to the system
$y[n]$ is the response of the system to $x[n]$
$x[n+N]$ is the input shifted by N samples
$y[n+N]$ is the output shifted by N samples
N is an arbitrary amount of shift

A system that is not time invariant is called *time varying*. Multiplying a signal by a constant is time invariant. Multiplying a signal by $\sin[\pi n/2N]$, as in amplitude modulation of a carrier, is not time invariant. To see this, consider the input $x[n]$ to be a unit pulse, starting at $t = 0$, and lasting until $n = 2N$. The response of the system would be the positive half cycle of the sine wave. If the input is delayed by N, then the output of the system would be the second and third quarter cycles of the sine wave, which is not the same as a delayed version of the positive half cycle, as shown in Fig. 9.5.

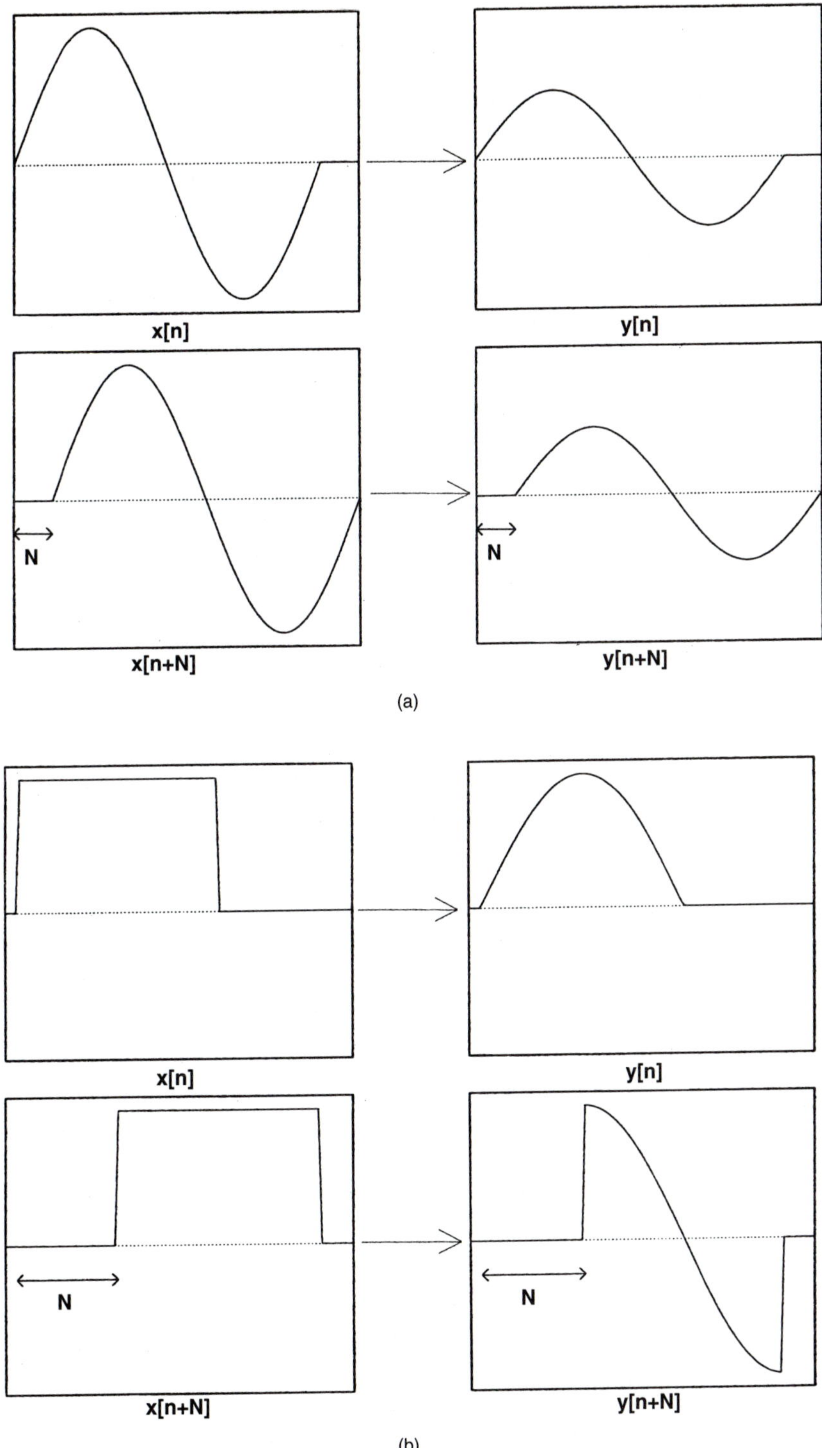

Figure 9.5 An example of a time-invariant system (a). An example of a time-varying system (b).

Linear time-invariant systems and filters. *Linear time-invariant* (*LTI*) systems form an important class of signal-processing systems. It can be shown that an LTI system cannot create new frequency components in a signal. The output of such a system only contains signals with frequency components that were in the input. Generally, these systems are called *filters*. A filter can change the amplitude and/or phase of a particular frequency component, but it cannot create a new frequency that was not in its input. This is important because it allows filters to be analyzed and described by their frequency response.

Causal and anticipatory. A system is said to be *causal* if its response at time t only depends on its input up to time t, or, in other words, its current output doesn't depend on the future input. A system whose output depends on future inputs is said to be *anticipatory*. Mathematically, a system is causal if its response $y[n]$ is only a function of $x[i]$ for $i \leq n$. As a result, if the input to a causal system is zero up to time N, that is to say $x[n] = 0$ for $n \leq N$, then the response of the system will also be zero up to time N, $y[n] = 0$ for $n \leq N$. Furthermore, if two inputs to a causal system are identical up to time N, then the responses of the system will also be identical up until that time. A system that averages the previous three inputs, $y[n] = (x[n] + x[n-1] + x[n-2])/3$ is causal. A system that averages the previous, current, and next input, $y[n] = (x[n+1] + x[n] + x[n-1])/3$ is not causal because the output, $y[n]$, depends on a future input, $x[n+1]$. The real world is causal. Physical systems can't respond to future inputs. However, noncausal systems can exist and can sometimes be very useful. One typical application that uses noncausal systems is image processing, where the domain of the system is spatial, rather than temporal. Clearly, when processing an image, the entire image is accessible and there is no need to restrict oneself to only causal filters. Even when the domain of the signals is time, noncausal filters can be used if post-processing the data. If an entire record of the input is first recorded, the data can be processed later, then the signal processor has access to "future" data and can implement noncausal filters.

Stability. A final important property of signal-processing systems is stability. A system is said to be stable if any bounded input to the system results in a bounded output. A signal is bounded if it doesn't grow without limit. An example of a stable system is one that sums the previous three inputs. If the values of these inputs is bounded $-B \leq x[n] \leq B$, then the output is bounded $-3B \leq y[n] \leq 3B$. An ideal integrator (a system that sums all its previous inputs), however, is not stable because the constant input $x[n] = 1$ will result in an arbitrarily large output eventually. In general, most useful signal-processing systems need to be stable.

9.5 Digital Signal Processing

This chapter deals primarily with the subset of signal processing that here will be called *digital signal processing*. This means discrete time signal processing of quantized or discrete values. This isn't as restrictive as it might at first sound. In many applications, digital signal processing can be used to emulate or simulate analog signal processing algorithms. In addition, digital signal processing can perform many tasks that would be very hard or impossible to do in analog processing.

9.5.1 Advantages

One might wonder why anyone would use digital processing to simulate something that could be done directly in the analog domain. There are several advantages that can be gained by moving the signal processing from the analog domain to the digital domain.

Repeatability. One of the biggest advantages that digital processing has over analog is its repeatability. A digital filter, given the same input, will always produce the same output. This is not necessarily true with analog filters. The components used in analog filters are never perfect or ideal. They all have tolerances and some variability in their true value. In a batch of 100-ohm resistors, all the resistors don't have values of exactly 100 ohms. Most will have value within 1% of 100 ohms (if they are 1% resistors, for example). Because of this, precision filters used in instruments are designed with a number of adjustable components or "tweaks." These are used to adjust and calibrate the filter to meet the instrument's specifications, despite the variability of the analog components used in the filter. Digital filters are "tweakless;" once a filter is properly designed, every copy of it in every instrument will behave identically.

Drift. Another imperfection in analog components that is related to component variation is component stability and component drift. Not only do the values of analog components vary within a batch, but they also tend to drift over time with such factors as changing temperatures or merely aging. Not only does this add complexity to the design of analog filters, but it also requires the periodic recalibration of instruments to compensate for the drift. Digital signal processing systems, however, are drift free.

Cost. Digital signal processing can also have a cost advantage over analog signal processing. The amount of circuitry that can be placed on an integrated circuit has increased dramatically and the cost for a given amount of logic has plummeted. This has tremendously reduced the cost of implementing digital processing hardware. Advances in computer technology have led to the development of microprocessors that have been optimized for digital signal processing algorithms. This has allowed developers to implement DSP systems using off-the-shelf hardware and only writing software. This expanded the range of applications suitable for using DSP to include prototypes and low-volume products that otherwise wouldn't be able to afford the cost of building dedicated, custom integrated circuits.

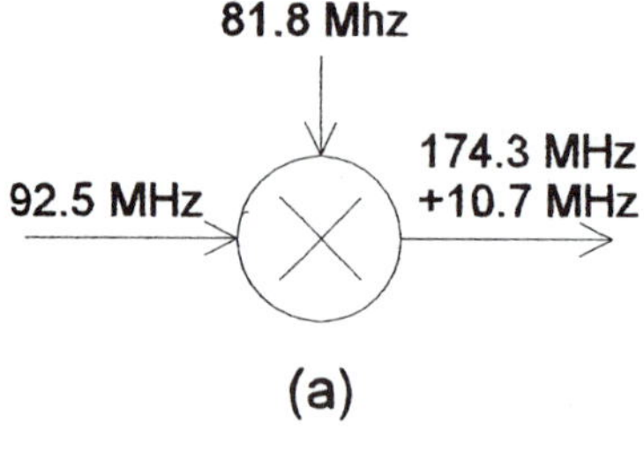

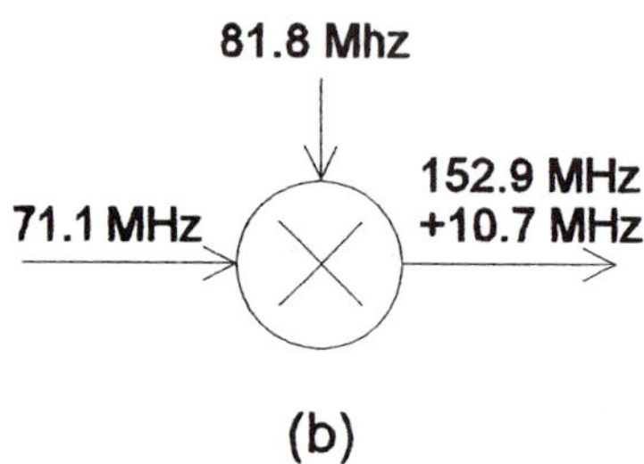

Figure 9.6 A signal at 92.5 MHz mixed with 81.8 MHz produces the desired signal at 10.7 MHz, plus an extraneous signal at 174.3 MHz. (a) A signal at the image frequency of 71.1 MHz also produces an output signal at 10.7 MHz (b).

Precision. Digital signal processing can exceed analog processing in terms of fundamental precision. In analog processing, increasing the precision of the system requires increasingly precise (and expensive) parts, more tweaks, and more-frequent calibrations. There are limits to precision, beyond which it is impractical to exceed. In digital signal processing, higher precision costs more because of the extra hardware needed for higher precision math, but there are no fundamental limits to precision.

As an example, consider mixing a signal with a local oscillator to shift its frequency. This is a common task in many instruments. When a signal is multiplied by a local oscillator, both the sum and difference frequencies are generated. This can cause images of the desired signal to appear at other frequencies and it can cause other, undesired input signals to appear at the same location as the desired signal. These unwanted signals must be filtered out before and after the actual mixing. For example, in an FM (frequency modulation) radio, a station at 92.5 MHz can be mixed down to an IF (intermediate frequency) of 10.7 MHz by multiplying it by a local oscillator at 81.8 MHz. This mixing will also place that station at 174.3 MHz, which is the sum of the local oscillator frequency and the input signal. Furthermore, any input signal at 71.1 MHz would also be mixed to 10.7 MHz (Fig. 9.6). Because of this image problem, many instruments use multiple conversions, each with its own local oscillator and with filtering between each conversion, to translate the desired signal to its final location. One way to avoid these image problems is to use a quadrature mix. In a quadrature mix, the local oscillator is actually two signals equal in frequency, but with exactly a 90-degree

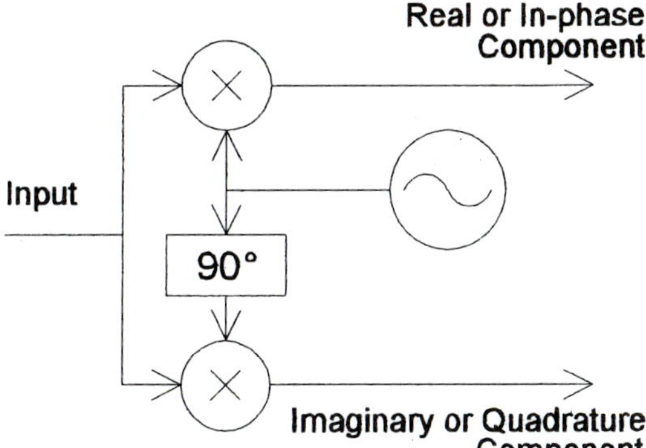

Figure 9.7 Quadrature mixing.

phase difference (Fig. 9.7). A quadrature local oscillator can be thought of as the complex exponential e^{jwt}. Another way to think of the result of a quadrature mix is as the sum of two signals in which the desired signal is in phase and the image signals are out of phase. The two out-of-phase signals cancel, leaving only the desired signal. Getting perfect cancellation of the image signals depends on two things: having identical amplitudes in the two quadrature local oscillators, and having exactly 90 degrees of phase difference between the two local oscillators. As one deviates away from this condition, residual image signals appear. To get 80 dB of image cancellation, the phase difference between the two local oscillator signals must be within 0.01 degrees of 90. This is not feasible to do using analog components. This is easy to accomplish in digital signal processing and is routinely done in many instruments.

9.5.2 Disadvantages

Digital signal processing also has some disadvantages, as compared to analog processing.

Bandwidth. Probably the largest disadvantage is bandwidth. To process higher-bandwidth signals, higher sample rates and faster math is required. Analog-to-digital converter technology sets the limits on the widest bandwidth signals that can be processed with DSP. As technology progresses, the bandwidth of signals that can reasonably be processed in the digital domain has risen. In the 1980s, a high-performance analyzer might have a sample rate of 250 ksamples/s and a maximum bandwidth of 100 kHz. By the end of the 1990s, sample rates had risen to 100 Msamples/s, which is capable of processing signals with bandwidths up to 40 MHz. During this time, however, analog signal processing has been used to process signals up into the many GHz.

Cost. Cost can also be a disadvantage to DSP. Although VLSI technology has lowered the cost of doing math, the process of converting analog signals into the digital domain and then back into the analog domain still remains. In many applications, an analog solution works just fine and there is no reason to resort to DSP.

9.6 Digitizing Process

Inherent in many DSP applications is the process of digitizing, that is, the process of converting an analog signal to the digital domain. For a more-detailed look at the analog-to-digital conversion process, see Chapter 6 (Analog to Digital Converters). Here, the emphasis is on how the digitizing process modifies signals.

9.6.1 Sampling and quantizing

The digitizing process consists of two distinct processes. The first is sampling and the second is the quantizing, each of which has its effect on the signal being digitized.

Sampling. Sampling is the process of examining the analog signal only at certain, usually periodic, discrete times. Typically, the signal is sampled periodically at a rate known as the *sample rate* (f_s), as shown in Fig. 9.8. This function is typically performed by the track-and-hold or sample-and-hold circuit in front of the analog-to-digital converter. After the analog signal is sampled, it is still an analog voltage. It must still be converted to a digital value in the analog-to-digital converter. This process is known as *quantizing* (Fig.

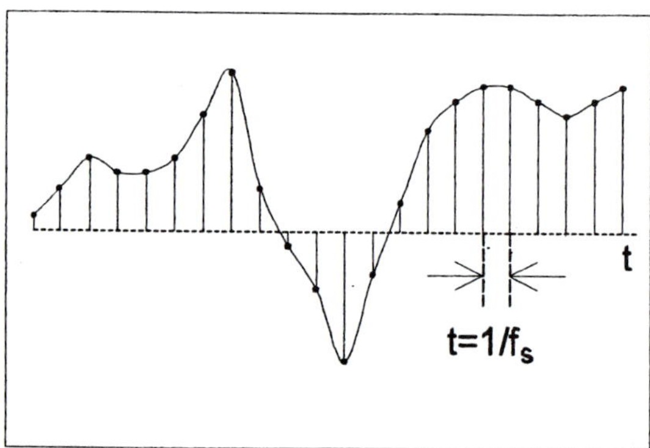

Figure 9.8 Sampling an analog signal at a rate of $f - s$.

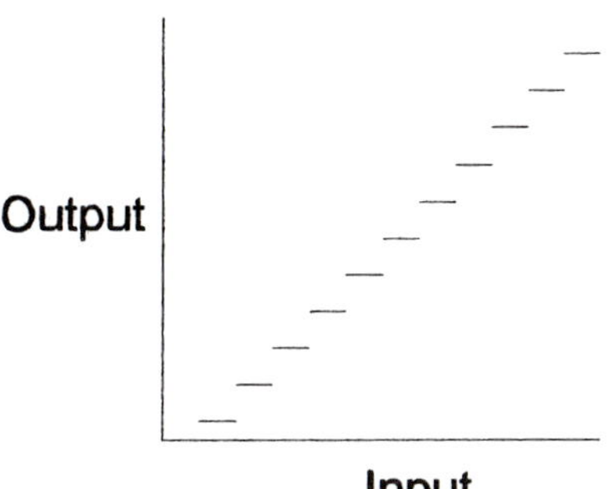

Figure 9.9 Quantizer input/output transfer function.

9.9). Because a digital value can only take on a finite number of discrete values and the analog signal can take any of a continuum of values, the digitized value can't exactly describe the value of the analog signal. Ideally, the analog-to-digital converter finds the closest digital value to the value of the sampled analog signal.

Aliasing. The sampling process, which converts the signal from a continuous time domain to a discrete time domain, can potentially have quite an effect on the signal. A discrete time signal can only describe signals of a limited bandwidth, in particular signals between $\pm f_s/2$, where f_s represents the sampling frequency. Signals outside of this range are folded back, or aliased, into this range by adding or subtracting multiples of f_s. For example, analog signals of frequencies $5f_s/4$ and $f_s/4$ can alias into the same digital samples, as shown in Fig. 9.10. The sampled values cannot indicate which of those analog signals produced that set of samples. Indeed, these two signals, along with all the other signals that generate the same set of samples, are known as *aliases* of each other. Any signal with a bandwidth less than $f_s/2$, which is known as the *Nyquist frequency,* can be sampled accurately without losing information. Note: although it is common that the signal being digitized is often centered about dc, that isn't necessary. The signal can be centered about a frequency higher than the sample rate, as long as the signal's bandwidth is

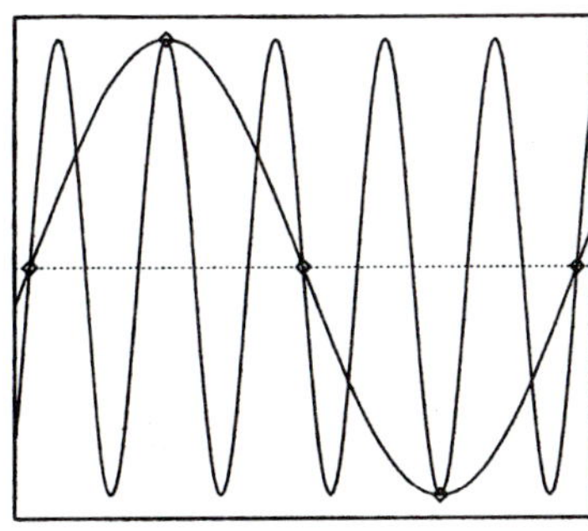

Figure 9.10 An example of two analog signals that alias into the same digital signal.

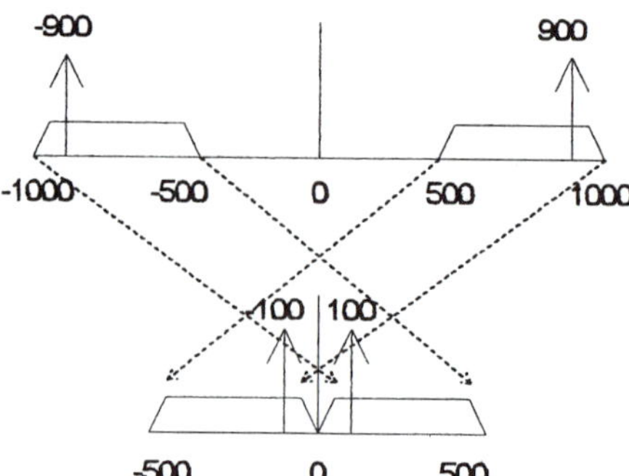

Figure 9.11 The aliasing of a 500-kHz-wide signal centered at 750 kHz when sampled at a rate of 1 Msample/sec.

less than the Nyquist frequency. For example, a spectrum analyzer can down convert its input down to a center frequency of 750 kHz with a bandwidth of 500 kHz. Thus, the signal prior to digitizing has content from 500 kHz to 1000 kHz. This is then sampled at a rate of 1 MHz. Because this signal's bandwidth is half of the sample rate, this is okay. The signal is then aliased into the range of ±500 kHz so that an input signal at a frequency of ±900 kHz would show up in the digitized signal at a frequency of ±100 kHz, as shown in Fig. 9.11. If the bandwidth of the analog signal is greater than the Nyquist frequency, then two or more frequencies in the input signal will alias into the same frequency in the digitized signal, in which case it becomes impossible to uniquely determine the original analog signal. As an example, if the sample rate is 1 MHz, then analog signals at frequencies of 0.1 MHz, 0.6 MHz, and 1.1 MHz all alias to the same value, 0.1 MHz. By observing the digitized signal, it would be impossible to determine whether the input was at 0.1 MHz, 0.6 MHz, or 1.1 MHz. For this reason, an anti-aliasing filter is usually placed before the sampler. This (analog) filter's job is to bandlimit the input signal to less than the Nyquist frequency.

Quantizing. After the signal is sampled, it must be quantized to one of a set of values that are capable of being represented by the analog-to-digital converter. This process can be considered an additive noise process, where the value of noise added to each sample is the difference between the actual value of the sampled signal and the quantized value (Fig. 9.12). The characteristics of this additive noise source can then be examined. If the A/D converter is ideal, then the noise would be zero-mean noise uniformly distributed between ±½ of the least-significant bit of the quantizer output. If the converter is less than ideal, then the noise would be higher. Clearly, a converter with more precision or number of bits would have more quantizing levels with the quantizing levels closer together and would add less noise.

For signals that change a large number of quantizing levels between samples, the added noise is approximately uncorrelated with the input signal. In this case the noise shows up as a broadband, white noise. However, when the

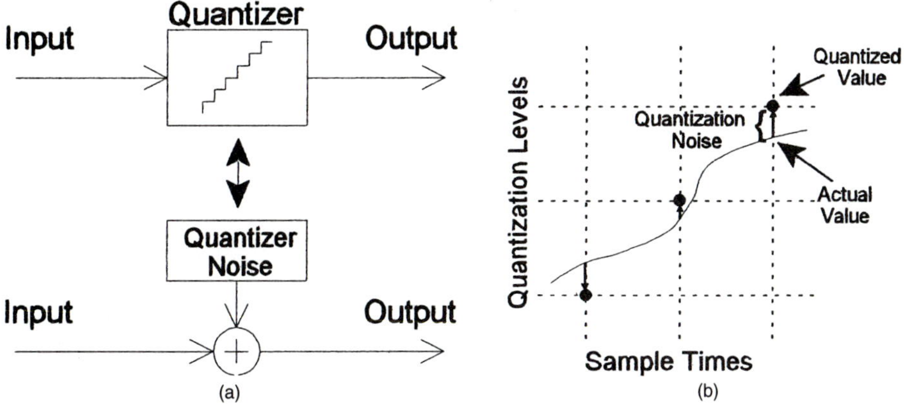

Figure 9.12 A quantizer can be considered to be an additive noise source (a). Quantizer noise is the difference between the signal's actual value and the quantized value (b).

signal only traverses a few different quantizing levels, the noise can become highly correlated with the input signal. In this case, the noise looks more like distortion than broadband white noise (Fig. 9.13). As an extreme example, consider a low-level sinusoidal input that only traverses between two different quantizing levels. The quantized signal would look like a square wave, rather than a sinusoid! The quantizing errors, instead of being spread out across the whole spectrum, are concentrated at multiples of the sinusoid's frequency.

9.6.2 Distortion

Distortion-like noise is more of a problem than wide-band noise because it can look like spurious signals. For this reason, dither is often used in the digitizing process of instruments. Dither is a (typically small) random signal added to the input signal prior to the digitizing process. This same random signal is subtracted from the digitized values after the digitizing process. This added dither has the effect of converting correlated noise into uncorrelated noise. Consider ±½ mV of dither added to a 0- to 1-mV sinusoid that is being quantized by a digitizer with 1-mV quantizing levels. If no dither was added, then when the input sinusoid was below ½ mV, the quantizer would read 0, and when the sinusoid was above ½ mV, the quantizer would read 1, producing a square wave (Fig. 9.14A). With the dither added, the output of the quantizer will be a function of both the value of the input sinusoid and the particular value of the dither signal. For a given input signal value, the output of the digitizer can be different depending on the value of the dither. The digitizer's output becomes a random variable with some distribution. If the sinusoid's value is close to 0, then the digitizer's output is more likely to be 0 than 1. If the sinusoid is at 0.5 mV, then the digitizer's output is equally

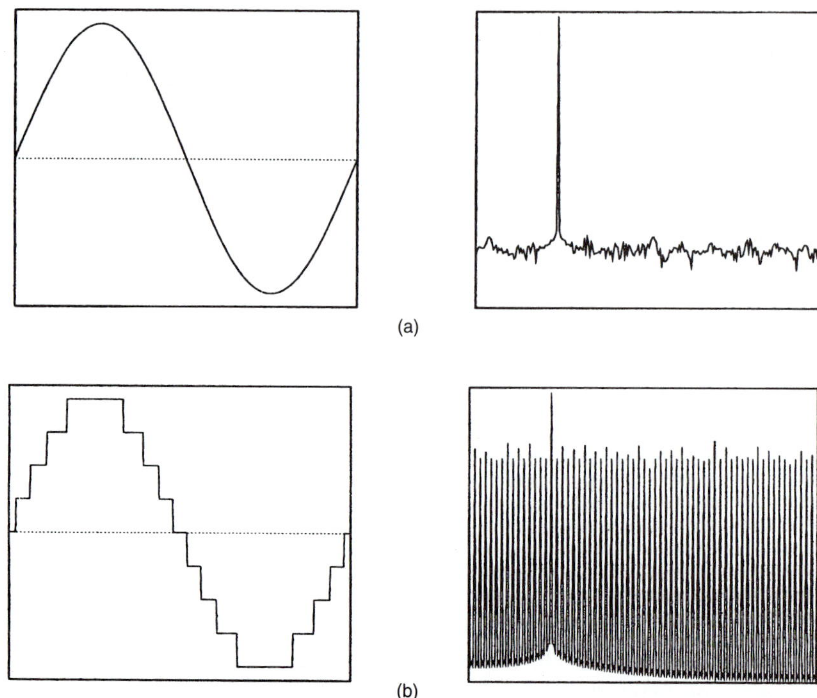

Figure 9.13 A signal that traverses many quantizer levels results in broadband (approximately white) quantizer noise (a). A signal that only traverses few quantizer levels results in distortion-like noise (b).

likely to be 0 or 1. If the sinusoid's value is close to 1, then the output of the quantizer is more likely to be 1. The output of the digitizer might look like Fig. 9.14B, where the output is more often 0 when the sinusoid is low and more often 1 when the sinusoid is high. Looking at the spectrum of this output, the distortion seen in Fig. 9.14A has been converted into broadband noise in Fig. 9.14B. Notice that the noise generated by the digitizing process hasn't been removed; it is still there. It has just been converted from correlated noise to uncorrelated noise.

9.6.3 Quantization noise

For many types of digitizers, the quantization noise that is added to the signal is approximately white noise, which means that each noise sample is uncorrelated with any other noise sample. This means that knowing the noise in one sample shows nothing about the noise in any other sample. One characteristic of white noise is that it is broad band. Furthermore, white noise has a constant power density across the entire spectrum of the signal. If the signal is filtered, the amount of noise in the result will be proportional to the

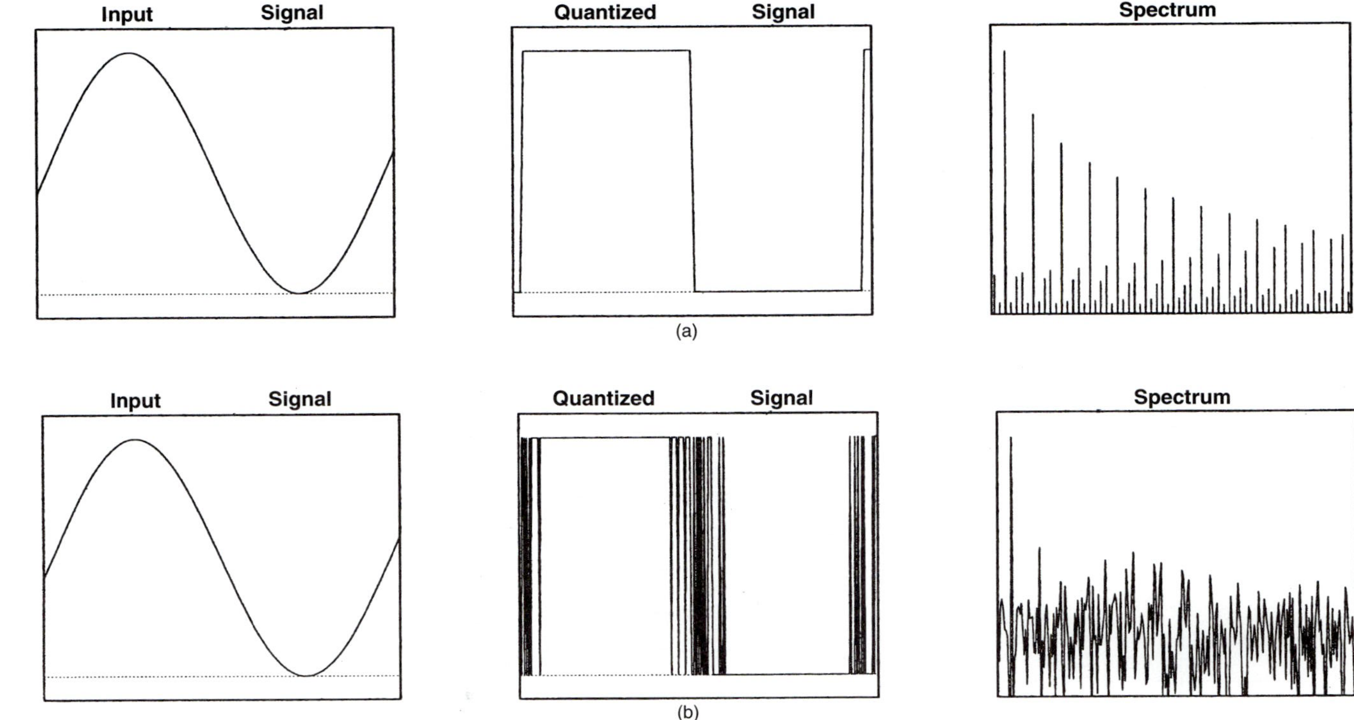

Figure 9.14 (a) A low-level sine wave digitized without dither results in a square wave with distortion-like quantizer noise. When it is digitized with dither, the distortion-like quantizer noise is converted into broadband noise (b).

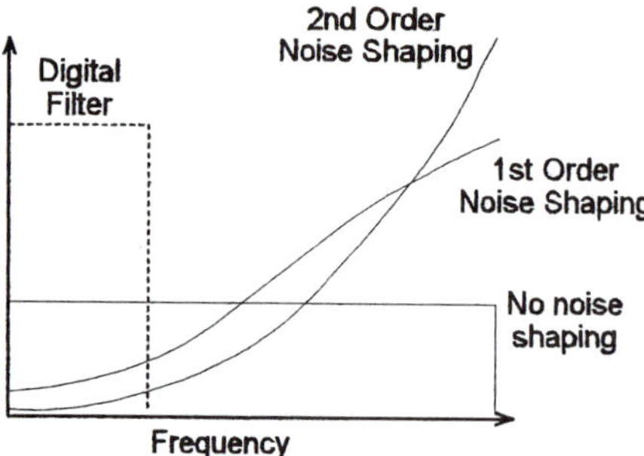

Figure 9.15 A comparison of quantizer noise with no noise shaping, first-order noise shaping, and second-order noise shaping.

bandwidth of the filtered signal. Halving the bandwidth halves the noise power. Because the noise power is proportional to the square of the noise amplitude, this reduces the noise amplitude by the square root of two (a half bit). Every two octaves of bandwidth reduction reduces the noise amplitude by a bit. This effect is familiar to users of spectrum analyzers, who notice that the noise floor drops as their resolution bandwidth is decreased. In this way, a 20-Msamples/s digitizer, which is noise limited to 16 bits, can digitize a lower frequency signal with more precision than 16 bits. If the output of such a digitizer is low-pass filtered by a factor of $2^8 = 256$, to a bandwidth of 78 ksamples/s, then each factor of 2 contributes ½ bit of noise. The resulting signal would have a noise level 4 bits less, a noise-limited performance of 20 bits.

9.6.2 Noise-shaping networks

Not all digitizers produce white quantization noise. Digitizers can be designed with noise-shaping networks so that the quantization noise is primarily in the higher frequencies. This is the basis behind a class of analog-to-digital converters known as *delta-sigma converters.* These converters have the same amount of noise as traditional converters; however, instead of the noise spectrum being flat, the noise spectrum is shaped to push a majority of the noise into the higher frequencies. If the result from this converter is low-pass filtered, the noise performance can be quite a bit better than a half-bit of noise performance per octave.

In a typical delta-sigma converter, the actual quantizer is a one-bit converter running at a sample rate of 2.8 Msamples/s. The noise is shaped (Fig. 9.15) such that when the output from the one bit converter is low-pass filtered to a sample rate of 44.1 ksamples/s, the noise has been reduced by 15 bits, resulting in a performance equivalent to a 16-bit converter running at 44.1 ksamples/s. To see how this might work, consider a traditional one-bit converter and a delta-sigma converter both digitizing a sine wave. Figure 9.16A shows the output of the traditional converter along with its spectrum. The output looks like a square wave and the noise spectrum has components near

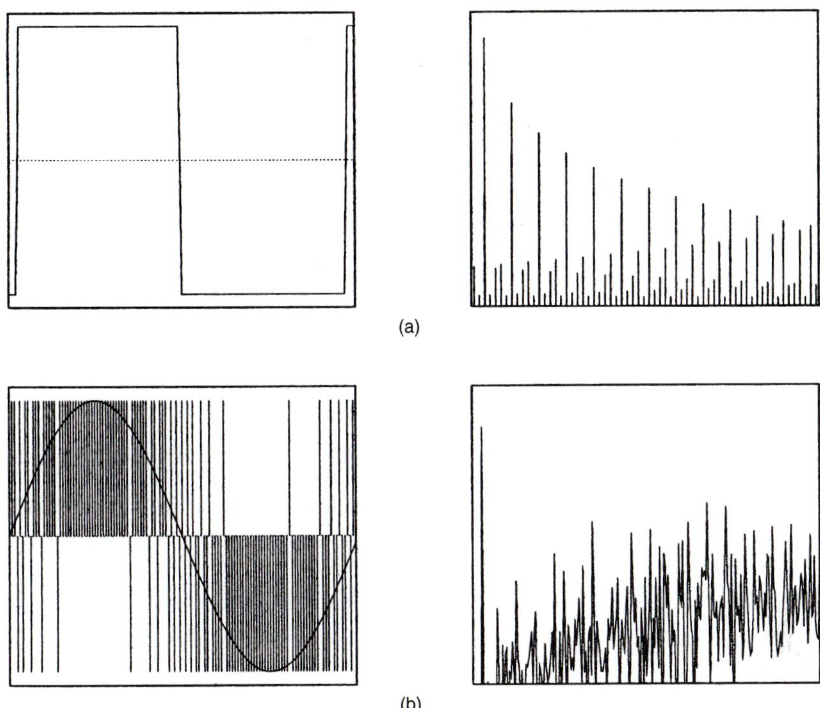

Figure 9.16 Digitizing a sine wave with a one-bit quantizer with no noise shaping (a). Digitizing a sine wave with a one-bit quantizer with first-level noise shaping (b).

the desired signal. The output of a delta-sigma converter would look more like Fig. 9.16B.

A couple of things are apparent. The delta-sigma converter looks noisier, but that noise is mostly high-frequency noise. The spectrum shows that although the high-frequency noise is higher than in the traditional converter, the low-frequency noise is much less. It is obvious that the low-pass filtered result from the delta-sigma converter is a better representation of the sine wave than the result from the traditional converter. The delta-sigma converters make use of the relatively low cost of implementing digital filters using large-scale integrated circuits. A major application for delta-sigma converters are in digital audio, which use a sample rate of 44.1 ksamples/s with bandwidths near 20 kHz. To do this with a traditional ADC would require an analog anti-aliasing filter with a pass-band of 0 to 20 kHz, and a stop band starting at 22.05 kHz. With digital audio requiring a stop-band rejection of 90 dB or more, this is a very difficult (i.e., expensive) analog filter to build. The delta-sigma converter moves the initial sampler up to 2.8 Msample/sec, where a low-order anti-alias filter suffices. The final filtering (with a pass-band to 20 kHz and a stop-band starting at 22.05 kHz) is implemented with a digital filter that is cheaper than the equivalent analog filter.

9.7 Linear Filters

One of the largest class of tasks in signal processing is linear filtering: for example, low-pass, band-pass, and high-pass filters.

9.7.1 Characterizing a linear filter

There are two principal ways to characterize a linear filter: the frequency response of the filter in the frequency domain and the impulse or step response of the filter in the time domain. Either the frequency response or the impulse response uniquely characterizes a linear filter. Depending on the application, one way might be more convenient than the other. A oscilloscope designer might care more about the impulse response of a filter, but a spectrum analyzer designer might care more about the frequency response. The principle of superposition that linear filters possess allows the response of a filter to be analyzed by decomposing the input into a sum of simpler signals and summing the response of the filter to these simple signals. In the case of the frequency response, each input signal is decomposed into a sum of sinusoids with differing amplitudes and phases. Because the response of a linear filter to a sinusoid is another sinusoid with the same frequency, a complete characterization of a filter consists in how the amplitude and phase change between the input and output at each frequency. This characterization is the frequency response of the filter.

In the case of time-domain characterization, the input signal is broken up into a series of discrete time impulses (signals that are non-zero at only one time, and zero everywhere else) and the characterization of the filter consists of the response of the filter to a unit impulse input. This response is known as the *impulse response* of the filter. The impulse response and the frequency response are related via the Fourier transform. The frequency response is the Fourier transform of the impulse response.

9.7.2 Categories of digital filters

There are two broad categories of digital filters. Filters whose impulse responses are non-zero for only a finite number of samples are called *finite impulse response* (*FIR*) *filters.* Other filters, known as *infinite impulse response* (*IIR*) *filters,* have impulse responses that are, in principal, infinite in extent. The fundamental difference between IIR and FIR filters is that IIR filters use feedback within their implementation, whereas FIR filters do not. This difference, feedback or no feedback, profoundly changes some of the characteristics of these filters.

Although the basic building blocks of analog filters are capacitors, inductors, or integrators, the basic building blocks of digital filters are scaling blocks (blocks that multiply a signal by a constant), summing blocks, and unit-delay blocks (blocks that delay a signal by one sample time). These blocks can be combined to create any realizable filter. Analog filters are often

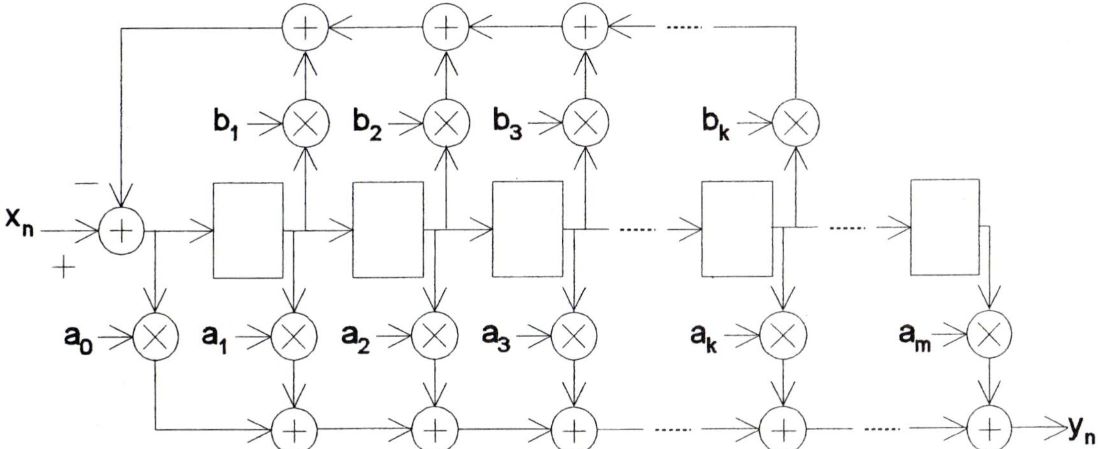

Figure 9.17 A canonical form of an IIR filter.

analyzed with Laplace transforms and poles and zeros in the s-plane; digital filters are analyzed by a similar, but different, transform, the z-transform, which has poles and zeros in the z-plane. The s-plane and the z-plane share many characteristics, although the details vary. For example, in the s-plane, all stable filters have poles in the left half plane. The corresponding criteria in the z-plane is that all stable digital filters have poles within the unit circle. The details of the z-transform are beyond the scope of this chapter. Interested readers can check some of the text books listed at the end of this chapter.

9.7.3 Infinite impulse response (IIR) filters

The class of IIR filters more closely resemble analog filters because both IIR filters and analog filters, in principle, have impulse responses that last forever; in practice, the impulse response will get lost in noise or rounding errors within a finite time. Because these filters use feedback, they are sometimes called *recursive filters.* IIR filters are often designed using the same or similar techniques to the design of analog filters. In many ways, an IIR filter is a discrete time simulation of a continuous time filter.

There are many ways to convert a continuous time filter, such as a Butterworth low-pass filter, to a digital IIR filter. One method, called *impulse invariance,* seeks to make the impulse response of the digital filter equal to equally spaced samples of the impulse response of the continuous time filter. A second technique is to convert the differential equation that described the continuous time filter into a discrete time difference equation that can be implemented in an IIR filter. A third technique converts the Laplace transform to the z-transform by means of a bilinear transform. The canonical form of an IIR filter is shown in Fig. 9.17, where the boxes represent unit delays and the a_i and b_i are filter coefficients.

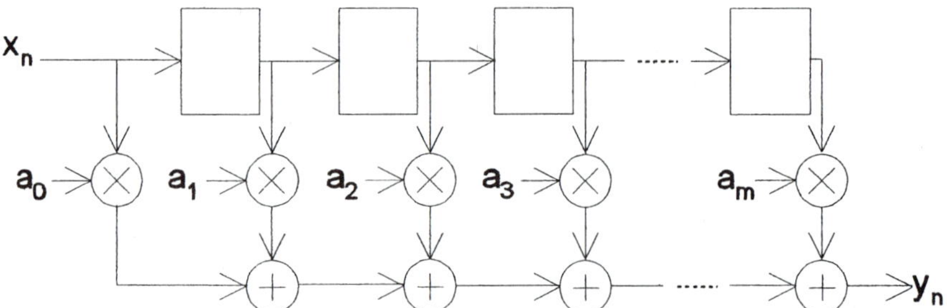

Figure 9.18 A canonical form of an FIR filter.

9.7.4 Finite impulse response (FIR) filters

In contrast to IIR filters, FIR filters do not utilize any feedback. FIR filters could be considered as special cases of IIR filters, namely IIR filters where all the feedback terms are zero. However, because of some special characteristics of FIR filters, it is useful to consider them as a different class. The canonical form of a FIR filter is the same as that for an IIR filter with all the feedback terms set to zero, as shown in Fig. 9.18. From this figure, it is apparent that a FIR filter can be considered to be a delay line, where each delayed value is scaled by some coefficient and then summed. For this reason, FIR filters are sometimes called *tapped delay lines,* and the filter coefficients are sometimes called the *tap weights.* Although IIR filters are often modeled after analog filters, FIR filters are almost always computer generated, often to minimize the pass-band ripple and stop-band rejection. Known efficient algorithms, such as the McClellan-Parks algorithm, can find the optimum FIR filter given a set of constraints. For many applications, FIR filters are designed to be linear-phase filters. This means that the filter's phase response is linear with frequency. A linear phase response corresponds to a pure time delay. Thus, a linear phase filter can be considered to be a filter with zero phase response combined with a time delay. The linear phase response equates to a constant group delay. This constant group delay contrasts with IIR filters, whose group delays tend to vary quite a bit, especially near their band edges. A linear phase response can be generated by having the impulse response symmetric

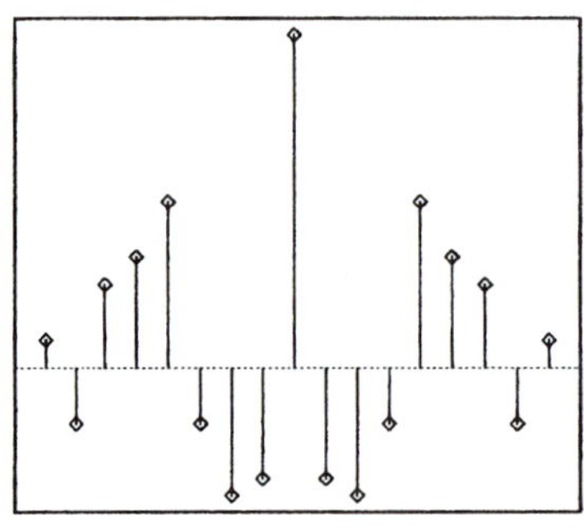

Figure 9.19 A linear-phase impulse response is symmetrical about its midpoint.

about its midpoint. That is to say, the second half of the impulse response is a time-reversed version of the first half, as shown in Fig. 9.19. Symmetric coefficients can be used to simplify the implementation of linear-phase FIR filters by halving the number of multiplications needed, as shown in Fig. 9.20.

9.7.5 Comparison of FIR and IIR filters

Both FIR and IIR filters have advantages and disadvantages and the choice of which type of filter depends on the particular application. IIR filters tend to be more efficient than FIR filters in the sense that typically a lower-order IIR filter can be designed to have the same amplitude performance as a corresponding FIR filter. The disadvantages of IIR filters include their non-linear phase, as well as the possible existence of things called *limit cycles.* Limit cycles can exist in IIR filters because of the effects of finite precision math. In a FIR filter, a constant input eventually results in a constant output. In an IIR filter, a constant input can result in a nonconstant, repeating output. Consider the filter shown in Fig. 9.21, where the results of the multiplications are rounded to the nearest result. The equation describing this filter is:

$$y[n] = x[n] - \frac{3}{4}[n-2] \tag{9.4}$$

where: $x[n]$ is the input at time n
$y[n]$ is the output at time n
$y[n-2]$ is the output two samples before time n

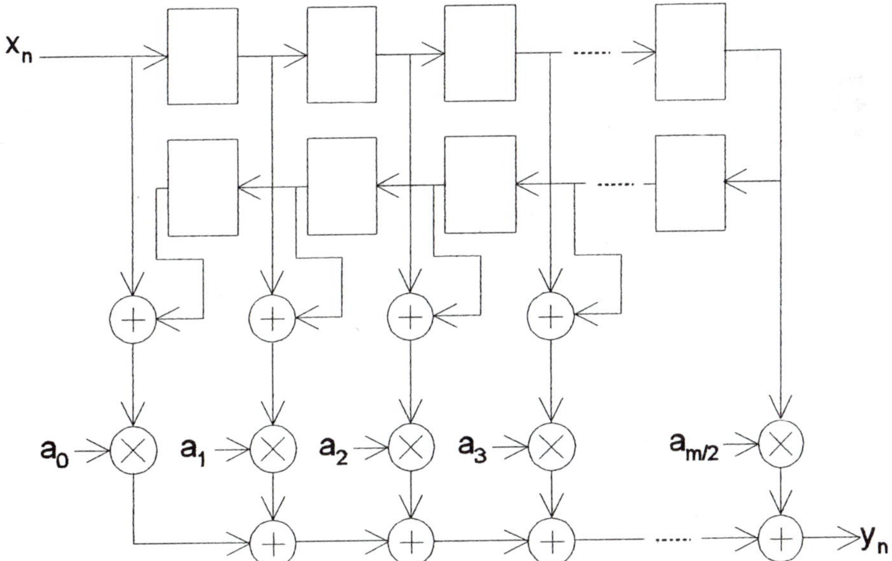

Figure 9.20 A reduced multiplier form of a linear-phase FIR filter.

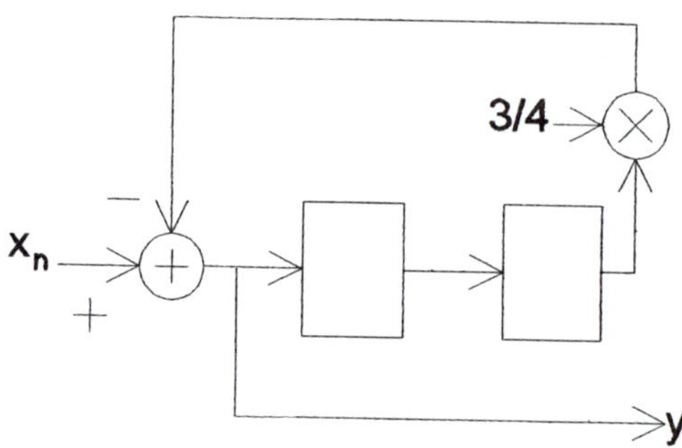

Figure 9.21 An IIR filter with a limit cycle.

For a large input, the output is a decaying sinusoid. If $x[0] = 256$ (with all other input samples zero), then the output would be the sequence 256, 0, −192, 0, 144, 0, −108, 0, −81. . . . However, the response does not decay all the way down to zero. Eventually, it reaches the limit cycle 1, 0, −1, 0, 1, 0, −1, 0. . . . This results from the multiplication of ±1 by ¾ being rounded back up to ±1. In the absence of further inputs to break up the limit cycle, this oscillation will continue forever. Not all IIR filters have limit cycles, and, in many applications, enough noise is in the input data signal to break up limit cycles so that even though a filter might have potential limit cycles, they might not be observed in practice.

FIR filters, although inherently immune to limit cycles and capable of possessing linear phase responses, tend to require a higher-order filter to implement a desired response than an IIR filter.

9.7.6 Coefficient quantizing

Digital filters are implemented with fixed coefficient multipliers. Like the numbers representing the signal being filtered, the coefficients are quantized to a set of distinct values. The precision of the coefficients doesn't have to be the same as the precision of the signal. For example, an audio system might have 16-bit data but only 8-bit filter coefficients. When a filter is designed, typically the ideal coefficient isn't exactly equal to one of the quantized values. Instead, the ideal coefficient is moved slightly to the nearest quantized coefficient. These perturbations slightly change the filter's frequency response. These changes must be taken into account by the filter designer. As an example, consider one second-order section of an IIR filter. Because of the coefficient quantizing, the poles and zeros of the filter section can't be arbitrarily placed in the z-domain. The poles and zeros can be located in only a finite number of locations. The distribution of possible locations isn't necessarily uniform. Depending on the particular topology of the second-order section, the distribution of pole locations might be denser in certain parts of the z-plane than in others. This can affect the choice of which filter topology to use because the denser the possible pole locations, the less an ideal pole location must move to fall on one of the allowable positions.

9.8 Changing Sample Rate

One common task performed in the DSP section of an instrument is changing the sample rate. There are several reasons why the sample rate could be changed. Often as the signal is processed, it is filtered to reduce its bandwidth. As the bandwidth is reduced, the sample rate can be reduced, lowering the computational load of further signal processing blocks. Sometimes a particular sample rate is desired for user convenience. For example, the frequency resolution of a Fast Fourier Transform is directly related to the sample rate of the signal. To get an even resolution (for example, exactly 10 Hz/bin instead of something like 9.875 Hz/bin), the signal might have to have its sample rate changed. As a third example, consider an arbitrary waveform generator whose digital to analog converter runs at a rate of 1 Msample/sec. To play data from an audio compact disc, the disc's sample rate of 44.1 ksamples/s, must be changed to the converter's rate of 1 Msample/sec. The process of changing a signal's sample rate is sometimes known as *resampling the signal.*

Consider reducing a signal's sample rate such that the original sample rate is an integral multiple, N, of the new sample rate. Merely taking every Nth sample would not suffice. The process of taking every Nth sample, called *decimation by a factor of* N, is the same as the sampler process covered earlier (see 9.6.1). Unless the input signal is band limited to less than half of the new sample rate, decimation results in aliasing so that multiple frequencies in the original signal get folded into the same output frequency. To prevent aliasing, the original signal must be filtered to reduce its bandwidth before decimation. This is no different than the anti-alias filter and sampler before an analog-to-digital converter. Typically, the signal is low-pass filtered before decimation, although this isn't necessary, as long as the bandwidth of the signal is reduced sufficiently.

To raise a signal's sample rate by an integral factor, M, the reverse is done. First, the signal is interpolated by adding $M - 1$ zeros between each sample point. However, the spectrum of the interpolated signal has M copies of the original signal's spectrum. All but one copy of this spectrum must be filtered out with a filter after the interpolator. This filter is similar to the reconstruction filter that comes after a digital-to-analog filter.

To change a signal's sample rate by a rational factor M/N, first interpolate the signal by a factor of M, filter, and then decimate by a factor of N. Note that the reconstruction filter and the anti-alias filter can be combined into the same filter; there is no need for two filters. It may seem that the computational load on this filter could be excessive if N and M are both large numbers because the sample rate that the filter must process is M times the original sample rate. However, all of the zero samples added in the interpolation process can ease the task. Furthermore, all of the samples that will be eliminated in the decimation process need not be calculated. These two things can make the computational load reasonable—especially when the filter used is a FIR type.

Sample rates can also be changed by arbitrary, not necessarily rational, or even time-varying rates, although the details of this is beyond the scope of this book.

9.9 Frequency Translation

Changing the frequency of a signal is a common operation in instruments. This might be moving a signal of interest down to an intermediate frequency or even to baseband, or it might be up converting a signal to a higher frequency. As in the analog domain, frequency translation (mixing), is accomplished by multiplying the signal by the output of a local oscillator using a digital multiplier. An example of this was included in section 9.5.1.

9.10 Other Nonlinear Processing

In addition to linear filters there are many nonlinear signal-processing systems. This chapter just touches upon some examples. The general topic of nonlinear signal-processing systems is beyond the scope of this chapter.

9.10.1 Frequency translation

Frequency translation, covered earlier, is a common, nonlinear process in instrumentation. Other nonlinear processes found in analog instruments find their counterpart in digital form. For example, the envelope detector in a swept superheterodyne spectrum analyzer can be implemented in the digital domain by taking the absolute value of the input signal followed by a lowpass filter. Similarly, the logarithmic amplifier (log amp) of such an analyzer can easily be implemented in hardware that calculates the log function. Unlike analog log amps, which only approximately follow a true logarithmic function, a digital log can be made arbitrarily precise (at the expense of more hardware).

9.10.2 Compressors and expanders

Another large class of nonlinear processes are compressors and expanders. A compressor takes a signal and tries to represent it in fewer bits. An expander does the opposite. It takes a signal that has been compressed and reverts it back to its original form. A log amp is one example of a compressor that has a counterpart in the analog domain. By compressing the dynamic range of a signal, the output of a log amp can be expressed using fewer bits than the input. To get 120 dB of dynamic range using a linear scale would require 20 bits. A "logger" can convert this to a dB scale that would only require 7 bits.

The two classes of compressors are: lossless and lossy compressors. A lossless compressor does not lose any information contained in the signal. It removes redundancy in the signal, but doesn't remove any of the signal itself. The output of a lossless compressor can be expanded back into an exact copy of the same signal before it was compressed. A lossy compressor, on the other hand, does remove, in a very controlled manner, some of the information in the input signal. A good lossy compressor only removes unimportant information such that the output can be expanded back to a close approximation of the original signal. Although the expanded signal won't be an exact copy of the input, it can be good enough. The criteria for "close approximation" and "good enough" is dependent on the ultimate use of the signal. Many digital communication systems use sophisticated compression algorithms to reduce the data rate (and hence bandwidth) required to transmit the signal. In this context, a lossy compressor is acceptable with the fidelity criterion being that the expanded signal must sound like the original signal to a listener.

Computers use compressors to reduce the number of bits needed to store information and provide a good example of the difference in lossy versus lossless compression. Three common formats for storing digital images are TIF, GIF, and JPG. The TIF format is uncompressed and generates the largest files. The GIF format uses a lossless compressor. It generates a smaller file than TIF and can still be expanded to make an exact copy of the original image. The JPG format is a lossy compressor. It can generate the smallest file. It can be expanded into an image that looks like the original image while not being an exact copy of the original. A parameter controls the amount of compression that JPG will do. That allows a user to trade-off the amount of compression against the fidelity of the reconstructed image.

9.11 Fourier Transform

The Fourier transform can be used to convert a signal from the time domain into the frequency domain. The *Fast Fourier Transform (FFT)* is a particular implementation of a discrete-time Fourier transform that is computationally very efficient. Calculating the Fourier transform via the FFT requires drastically fewer operations than calculating it in its direct form. A 1000-point Fourier transform calculated in its direct form requires on the order of one million operations. The same transform, when computed via the FFT, requires only on the order of 10,000 operations. In this case, the FFT implementation of the Fourier transform is 100 times more efficient. Note that calculating the Fourier transform by either means generates the same results. They both calculate the Fourier transform. They just vary in the steps used to get from the input to the output. Because of its efficiency, the FFT is by far the most common way to calculate the Fourier transform, which is why the term "FFT analyzer" is sometimes used instead of "Fourier transform analyzer."

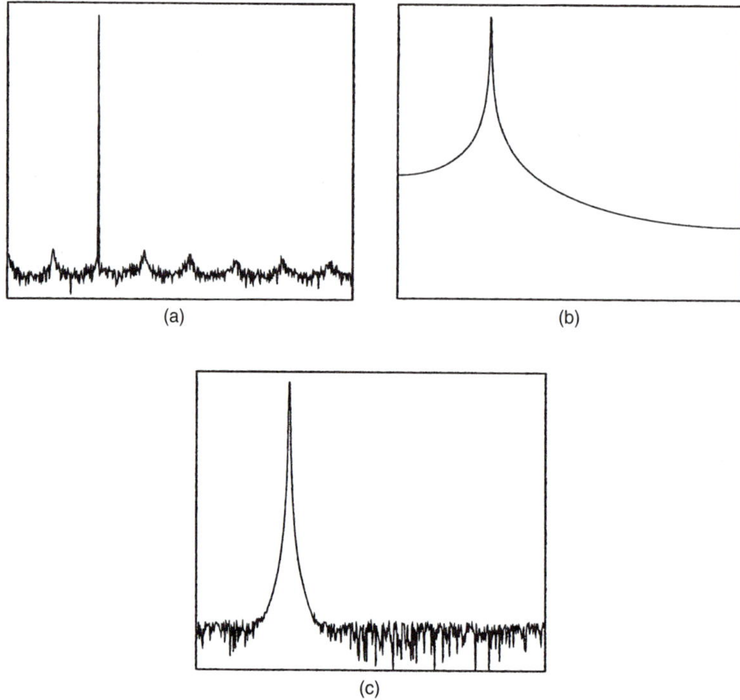

Figure 9.22 An FFT of a signal consisting of 137 periods of a sine wave (a). An FFT of a signal consisting of 137½ periods of a sine wave with uniform (no) windowing (b). An FFT of a signal consisting of 137½ periods of a sine wave windowed with a Hanning window (c).

An N-point FFT calculates N complex output points (points consisting of both real and imaginary components) based on N input points. If the input signal is complex, then all N output points are independent. If the input signal is real, then half of the output points are redundant and are typically discarded resulting in $N/2$ independent points. By only considering N input points, the FFT treats the input signal as being zero everywhere outside of those N points. This is known as *windowing*. It is as if the input signal was multiplied by a window function that was zero everywhere except for those N points. By taking larger-sized FFTs, more of the input signal can be considered, but the FFT can only consider a finite portion of input. Because of this windowing, an infinitely long sinusoid looks like a tone burst with a definite start and stop. The FFT calculates samples of the continuous time Fourier transform of this windowed input. The sharp transitions at the edges of the window cause a smearing or leakage of the frequency spectrum, unless the input signal is periodic with a period the same as the length of the FFT. Figure 9.22A shows the log magnitude of the FFT of a sinusoid that happens

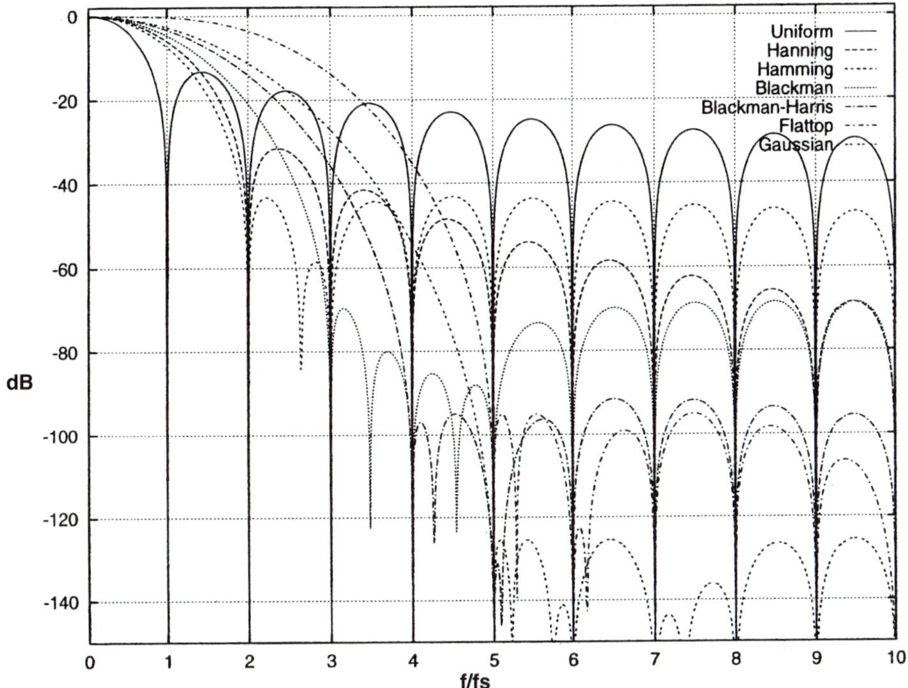

Figure 9.23 A spectrum of seven popular window functions.

to be periodic with exactly 137 periods in the length of the FFT. As expected, only one point is nonzero, corresponding to the input sinusoid. Figure 9.22B shows the FFT for a slightly different signal, a sinusoid with 137½ periods in the length of the FFT. This time, the output is smeared all over the spectrum. To combat this problem, the input to an FFT is often multiplied by a window function more appropriate than the default uniform window function. Mathematically, the spectrum of the window function is convolved with the spectrum of the input signal. Thus, the spectrum of the window function is indicative of the amount of spectral smearing that will result from the FFT.

Figure 9.23 shows the spectrum of seven popular window functions, shown in Fig. 9.24, with their amplitude plotted with a scale of dB to show the relative heights of their side lobes. Here, the high side lobes of the uniform window can be seen. The other window functions trade off lower side lobes for a broader central peak. For example, the central peak of a Hanning window is twice as wide as a uniform window, but its first side lobe is at −32 dB, as opposed to the uniform window's side lobe at −13 dB. Table 9.1 lists some properties of various windows. Generally, the lower the side lobes, the wider the central peak. The choice of which window to use depends on what charac-

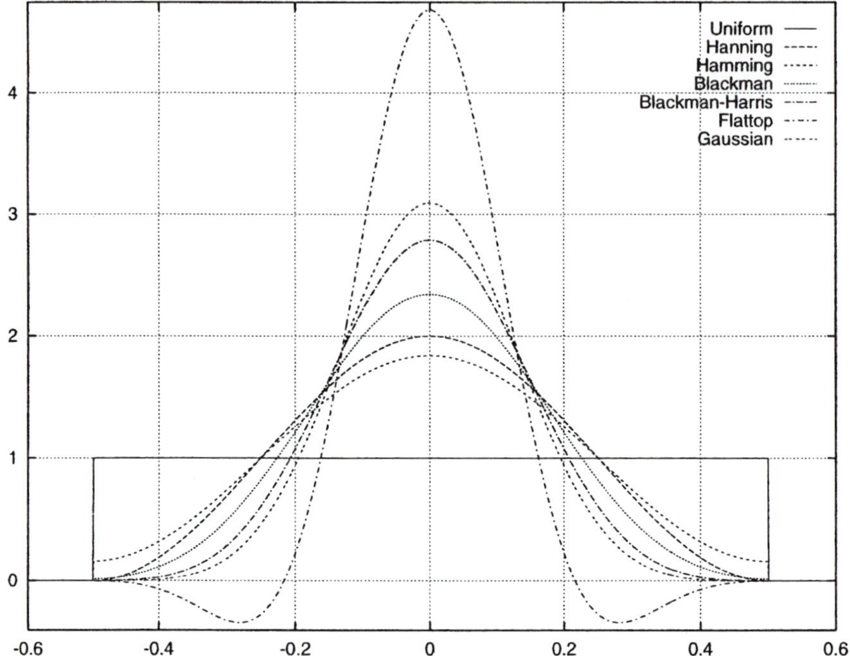

Figure 9.24 The shape of seven popular window functions.

teristics of the signal the user is trying to discern. To discern signals of widely different amplitudes, choose a window with low side lobes, such as the Gaussion. To discern signals near in frequency, choose a window with a narrow central peak, such as the Hanning. To accurately measure the amplitude of a sinusoid, pick a window whose central peak is relatively constant near its peak, such as the flattop. For signals that are naturally zero outside the window, such as the acoustic signal from a gun shot, no additional windowing is needed (which is equivalent to using the uniform window).

TABLE 9.1 Window Function Properties

Window name	Sidelobe level	Amplitude flatness (+/− ½T)	First null	Noise equivalent bandwidth	Width (3 dB)	Width (10 dB)
Uniform	−13 dB	−3.92 dB	1	1	0.9	1.5
Hanning	−32 dB	−1.42 dB	1	1.5	1.4	2.5
Hamming	−42 dB	−1.78 dB	2	1.4	1.3	2.7
Blackman	−68 dB	−1.15 dB	3	1.7	1.6	2.9
Blackman-Harris	−92 dB	−0.83 dB	4	2.0	1.9	3.4
Flattop	−95 dB	−0.01 dB	5	3.8	3.8	5.4
Gaussian	−125 dB	−0.68 dB	5	2.2	2.1	3.8

9.12 Digital Signal Processing Hardware

Determining the DSP algorithm is only part of the problem that faces a signal-processing designer. The algorithm must also be implemented, in hardware or software, before it can actually process signals. The numerous tradeoffs must be determined when deciding how to implement the algorithm.

9.12.1 Signal representation

When implementing a DSP algorithm, one of the fundamental choices is how to represent the signal. Although, conceptually, you might think of a signal as a sequence of values, in hardware, the signal values are represented by the 1s and 0s of the digital hardware. Various patterns of these 1s and 0s represent the different values that a signal can take. The two most common representations used are fixed point and floating point.

Fixed-point numbers. A *fixed-point number* is a binary number with an implied (fixed) decimal point. They are called *fixed point* instead of *integers* because they can represent fractional values. This implied decimal point determines how to align two fixed-point numbers when they are added. When two fixed-point numbers are multiplied, the implied decimal points determine the scaling of the result. For example, consider the product of two 8-bit numbers: 00001100 and 00001010, which are decimal 12 and 10. If the implied decimal point is on the far right, such as 00001100. and 00001010., then the numbers do represent 12 and 10, and the product is 120 represented by 01111000. However, if their implied decimal point is two positions to the left, then 000011.00 represents 3.0 and 00010.10 represents 2.5. The product is now 7.5, which is represented by 000111.10.

Floating-point numbers. *Floating-point numbers* are the binary equivalent of scientific notation. A floating-point number consists of a fractional part, representing some value between ½ and 1, and an exponent part consisting of the power of two that the fractional part must be multiplied by to get the real value. Consider the floating-point representation of the number 13. Because $13 = 2^4 \times 0.8125$, the exponential part would be 4 and the fractional part would be 0.8125 or 1101.

Advantages and disadvantages. The advantage of floating-point numbers is that they can represent values with larger dynamic ranges than fixed-point numbers. For example, a floating-point number with nine bits of exponent can represent values from 10^{-77} to 10^{77}. A 512-bit fixed-point number would be necessary to represent all these values. The disadvantage of floating point numbers is that more hardware is required to implement floating-point math compared to fixed-point math. This means that floating-point math can be more expensive than fixed point.

Other representations. There are other signal representations besides fixed point and floating point. For example, in the telecommunications industry there are two formats known as "mu-law" and "A-law" encoding. They are similar to a cross between fixed point and floating point. Like fixed-point numbers, there is no explicit exponent. Unlike fixed-point numbers, the spacing between adjacent values isn't a constant. The spacing increases with increasing signal. These encodings allow a larger dynamic range to be encoded into 8 bits. However, doing math on either of these formats is very difficult and they are converted to fixed- or floating-point numbers before being operated on by any signal processor.

9.12.2 Overflow and rounding

Two issues that confront implementors of signal-processing systems are numerical overflow and rounding.

Overflow. Because the numbers used inside signal-processing systems can only represent a certain range of values, care must be taken to ensure that the results of operations fall within these ranges. If the result of some operation exceeds the range of possible output values an overflow is said to occur. When an overflow occurs, the hardware can either signal an error and stop processing, or it can assign some other (incorrect) value as the result of the operation. For example, if a system is using 8-bit numbers representing the values $-128 \ldots 127$ tried to add $100 + 30$, it would get 130, which isn't in the range $-128 \ldots 127$. If the math is implemented with what is known as *saturating logic,* when the output tries to exceed the maximum representable value, then the output saturates at that maximum value. In this case, the adder would output 127, which is the largest it can. Different hardware can output some other value. If the hardware performed a normal binary add and ignored overflows, then adder would add binary 01100100 (decimal 100) and 00011110 (decimal 30) to get 10000010, which represents -126. Clearly, this is undesirable behavior.

Signal-processing systems can be designed to be "safe scaled." In such a system, all the operations are scaled so that no valid input signal can cause an overflow to occur. This can be performed through a combination of increasing the number of bits used to represent the results of operations and scaling the results. To continue the example of adding two 8-bit numbers, if the output of the adder had 9 bits instead of 8, then no overflow could occur. Another way to avoid the overflow is if the result of the addition was scaled by ½. In that case, the result could never exceed the output range.

Rounding. The results of math operations require more precision than their inputs. Adding two N-bit numbers results in $N + 1$ bits of result. Multiplying two N-bit numbers results in $2N$ bits of result. To keep the required precision of numbers from growing without bound, the results of a math operation must frequently be rounded to a lower precision. There are several different ways of rounding numbers. You could round down to the next lower number, round up to the next larger number, round toward zero, round to the nearest number, or randomly round up or down. The act of rounding is equivalent to injecting a small amount of noise equal to the difference between the real result and the rounded result. Depending on the particular rounding scheme used, different amounts of noise are added. One measure of a rounding scheme is the noise power injected. Another measure is the bias or dc value added. Some rounding schemes are unbiased. This means that, on average, they add just as much positive noise as negative noise. A biased rounding scheme might preferentially add more positive noise than negative. Depending on the use of the data, whether a rounding scheme is biased might not be significant. Another measure of a rounding scheme is the amount of hardware that it takes to implement it. For example, rounding down takes no additional hardware because rounding down is merely truncation. Rounding to nearest can potentially change all the bits of the result, requiring significant hardware. One easy to implement unbiased rounding scheme is called *OR'd rounding.* In this rounding scheme, the least-significant bit of the result is found by logically ORing together all the bits that are being discarded. Table 9.2 shows the results of rounding the numbers from 0 to 8 to multiples of 4 for various rounding schemes and notes the mean and variance for the added noise.

9.12.3 Hardware implementations

A digital signal-processing system can be implemented in fundamentally two different ways: in hardware or in software. A hardware implementation consists of dedicated hardware designed specifically to do some signal-processing task. A software implementation consists of a more general-purpose piece of programmable hardware and the software necessary to implement the signal-processing system.

Dedicated hardware. Dedicated hardware is hardware that is specifically designed to perform one task. Because it is special purpose, it can be made exceedingly fast or it can be made using the absolute minimum of hardware. Dedicated hardware can be especially appropriate in high-volume applications, where the design costs of a custom integrated circuit can be amortized over many units. Dedicated hardware is, in one sense, the most flexible way to implement signal-processing algorithms because while a chip is being designed, virtually any function can be put into it. In another sense, dedicated hardware is very inflexible because once a chip is designed and built, it is very difficult or impossible to change its function.

Designing dedicated hardware presents the designer with many choices. These choices include the decision of whether to implement the design as a parallel implementation or a serial implementation. In a parallel implemen-

TABLE 9.2 Comparison of Various Rounding Schemes

	Rounded values			
Input	Round up	Round down	Round to nearest	OR'd rounding
0 (0000)	0 (0000)	0 (0000)	0 (0000)	0 (0000)
1 (0001)	4 (0100)	0 (0000)	0 (0000)	4 (0100)
2 (0010)	4 (0100)	0 (0000)	4 (0100)	4 (0100)
3 (0011)	4 (0100)	0 (0000)	4 (0100)	4 (0100)
4 (0100)	4 (0100)	4 (0100)	4 (0100)	4 (0100)
5 (0101)	8 (1000)	4 (0100)	4 (0100)	4 (0100)
6 (0110)	8 (1000)	4 (0100)	8 (1000)	4 (0100)
7 (0111)	8 (1000)	4 (0100)	8 (1000)	4 (0100)
8 (1000)	8 (1000)	8 (1000)	8 (1000)	8 (1000)
	Errors or added noise			
Input	Round up	Round down	Round to nearest	OR'd rounding
0 (0000)	0	0	0	0
1 (0001)	+3	−1	−1	+3
2 (0010)	+2	−2	+2	+2
3 (0011)	+1	−3	+1	+1
4 (0100)	0	0	0	0
5 (0101)	+3	−1	−1	−1
6 (0110)	+2	−2	+2	−2
7 (0111)	+1	−3	+1	−3
8 (1000)	0	0	0	0
Average Value	+3/2	−3/2	+1/2	0
RMS Value	1.87	1.87	1.22	1.87

tation, each element shown in the signal flow diagram of the algorithm is implemented in a distinct piece of hardware. For example, a 32-tap FIR filter would require 31 registers, 32 multipliers, and 31 adders, as shown in Fig. 9.25A. This implementation could process one output sample per clock cycle. In a serial implementation, a piece of hardware is re-used to accomplish many tasks over a number of clock cycles. To continue the example, the same 32-tap FIR filter could be implemented using only a single multiplier and a single adder/accumulator, as shown in Fig. 9.25B. In this case, it would take 32 clock cycles to calculate one output point. The serial hardware is 32 times slower than the parallel hardware, but is uses only 1/32 of the adders and multipliers. Parallel implementations allow very high performance at the cost of much hardware. Serial implementations use less hardware at the cost of lower performance. The mix between a full parallel implementation and a full serial one is determined by the performance requirements of the application compared to the performance available in the hardware technology.

Implementing dedicated hardware. A wide range of technologies can be used to implement dedicated hardware, including full custom integrated circuits, semi-custom integrated circuits (including sea-of-gates and gate-array de-

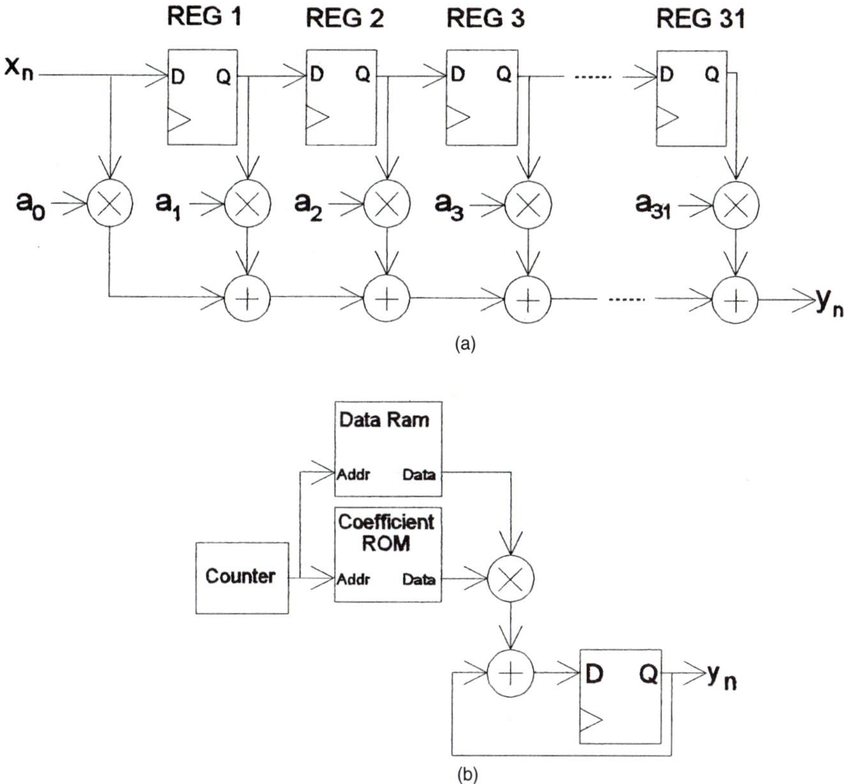

Figure 9.25 A 32-tap FIR filter implemented with parallel hardware (a). A 32-tap FIR filter implemented with serial hardware (b).

signs), and field-programmable gate arrays (FPGAs). A full custom design has all custom masks in the IC fabrication process and allows essentially any circuit to be designed. In sea-of-gates or gate-array designs, small groups of transistors or gates are laid out in an array and only the higher-level interconnect is customized. A FPGA has all the logic gates already laid out in the chip. The connections between these gates can be programmed in the field—either by blowing fuses or programming static switches. A full custom design has the highest performance and the lowest per part cost, but it has a very high development cost. A FPGA has lower performance and a higher part cost, but it has a very low development cost. Semi-custom logic lies in between the two.

Some applications occur frequently enough that companies have developed standard products addressing these applications. This approach combines the high performance of custom logic with the economies of scale of commercially available chips. Such devices include digital filters, numerically controlled oscillators, mixers, and even chips that can do a Fourier transform.

9.13 Software Implementations

Instead of using dedicated hardware, digital signal processing algorithms can be implemented as software programs. Depending on the performance requirements of the application, the software might run on anything from an 8-bit microcontroller to a 32-bit microprocessor, such as a Pentium to special-purpose digital signal processors that are optimized for signal-processing tasks. One such optimization is in multiplication. Signal-processing algorithms tend to have a higher percentage of multiplications than general-purpose software. Thus, processors designed for use in signal-processing applications tend to have fast multipliers—often multiplying in a single clock cycle. Another example of an optimization is the MMX extensions that Intel created for the Pentium and Pentium II processors. These extensions allow multiple, parallel operations to occur simultaneously on several pieces of data, thus speeding up throughput.

For nonreal-time signal processing, general-purpose computers running standard operating systems, such as Unix or Windows provide one of the most convenient environments for signal processing with their high-level language-development tools, extensive signal-processing libraries, and built-in graphical user interfaces.

9.14 Design of Signal-Processing Algorithms

Traditionally, signal-processing systems were developed by highly skilled people who had invested much time in learning the theory behind signal processing. Although hand crafting the hardware or software implementation of a signal processing task can result in a highly optimized result, it can also be very time consuming. For unique applications or very sophisticated algorithms, it can still be the best choice. For more-common signal-processing tasks, other design methods are available.

Numerous design packages are available that ease or even automate the generation of the detailed hardware or software design. These range from libraries that a software writer can call to do common tasks, such as filter or FFT, to complete graphical environments for the design and analysis of signal-processing algorithms. It is possible to draw the block diagram of a system and let the software take care of simulating and implementing it. For designers who only want help with a specific part of the design, some software packages will design a digital filter given the desired specifications.

Many of the gate-array design tools have module generators available. For example, one vendor's tools will automatically make a module for a multiplier, FIR filter, or even an FFT transform. These modules can then be placed in a design as easily as an adder or a register.

Using dedicated DSP chips in a design is another way to implement a signal-processing design without having to delve into the details of how that particular function is done. Commercially available chips can perform such tasks as filtering, operating as a local oscillator, mixing, convolving, or resampling. Complete DSP systems can be built using off-the-shelf chips.

9.15 Bibliography

Bracewell, Ronald N., *The Fourier Transform and Its Applications*, McGraw-Hill, New York, 1986.

Brigham, E. Oran, *The Fast Fourier Transform and Its Applications*, Prentice-Hall, Englewood Cliffs, NJ, 1988.

Crochiere, Ronald E., *Multirate Digital Signal Processing*, Prentice-Hall, Englewood Cliffs, NJ, 1983.

Frerking, Marvin E., *Digital Signal Processing In Communication Systems*, Van Nostrand Reinhold, New York, 1994.

Jones, N. B. and Watson, J. D. McK., *Digital Signal Processing: Principles, Devices, and Applications*, Peter Peregrinus Ltd., London, 1990.

Oppenheim, Alan V. and Schafter, Ronald W., *Digital Signal Processing*, Prentice-Hall, Englewood Cliffs, NJ, 1975.

Oppenheim, Alan V. and Willsky, Alan S., *Signals and Systems*, Prentice-Hall, Englewood Cliffs, NJ, 1983.

Stanley, William D., Dougherty, Gary R., and Dougherty, Ray, *Digital Signal Processing*, Prentice-Hall, Englewood Cliffs, NJ, 1984.

Chapter

10

Embedded Computers in Electronic Instruments

Tim Mikkelsen
Agilent Technologies
Loveland, Colorado

10.1 Introduction

All but the simplest electronic instruments have some form of embedded computer system. Given this, it is important in a handbook on electronic instruments to provide some foundation on embedded computers. The goals of this chapter[1] are to describe:

- What embedded computers are.
- How embedded computers work.
- The applications for embedded computers in instruments.

The embedded computer is exactly what the name implies: a computer put into (i.e., embedded) in a device. The goal of the device is not to be a general-purpose computer, but to provide some other function. In the focus of this chapter and book, the devices written about are instruments. Embedded computers are almost always built from microprocessors or microcontrollers. Microprocessors are the physical hardware integrated circuits (ICs) that form the central processing unit (CPU) of the computer. In the beginning, microprocessors were miniature, simplified versions of larger computer systems.

[1] This chapter includes material developed from Joe Mueller's chapter "Microprocessors in Electronic Instruments" from the second edition of this handbook.

Because they were smaller versions of their larger relatives, they were dubbed *microprocessors*. *Microcontrollers* are a single IC with the CPU and the additional circuitry and memory to make an entire embedded computer.

In the context of this chapter, *embedded computer* refers to the full, embedded computer system used within an instrument. *Microprocessor* and *microcontroller* refer to the IC hardware (i.e., the chip).

10.2 Embedded Computers

Originally, no embedded computers were in electronic instruments. The instruments consisted of the raw analog and (eventually) digital electronics. As computers and digital electronics advanced, instruments began to add limited connections to external computers. This dramatically extended what a user could do with the instrument (involving calculation, automation, ease of use, and integration into systems of instruments). With the advent of microprocessors, there was a transition to some computing along with the raw measurement inside the instrument—embedded computing.

There is a shift to more and more computing embedded in the instrument because of reductions of computing cost and size, increasing computing power, and increasing number of uses for computing in the instrument domain. Systems that were previously a computer or PC and an instrument are now just an instrument. (In fact, what used to require five bays of six-foot-high equipment racks, including a minicomputer, is now contained in a 8″×18″×20″ instrument.) So, more and more computing is moving into the instrument.

This transition is happening because of the demand for functionality, performance, and flexibility in instruments and also because of the low cost of microprocessors. In fact, the cost of microprocessors is sufficiently low and their value is sufficiently high that most instruments have more than one embedded computer. A certain amount of the inverse is also happening; instrumentation, in the form of plug-in boards, is being built into personal computers. In many of these plug-in boards are embedded computers.

10.2.1 Embedded computer model

The instrument and its embedded computer normally interact with four areas of the world: the measurement, the user, peripherals, and external computers. The instrument needs to take in measurement input and/or send out source output. A *source* is defined as an instrument that generates or synthesizes signal output. An *analyzer* is defined as an instrument that analyzes or measures input signals. These signals can consist of analog and/or digital signals (throughout this chapter measurement means both input analysis and output synthesis/generation of signals). The *front end* of the instrument is the portion of the instrument that conditions, shapes, or modifies the signal to make it suitable for acquisition by the analog-to-digital converter. The in-

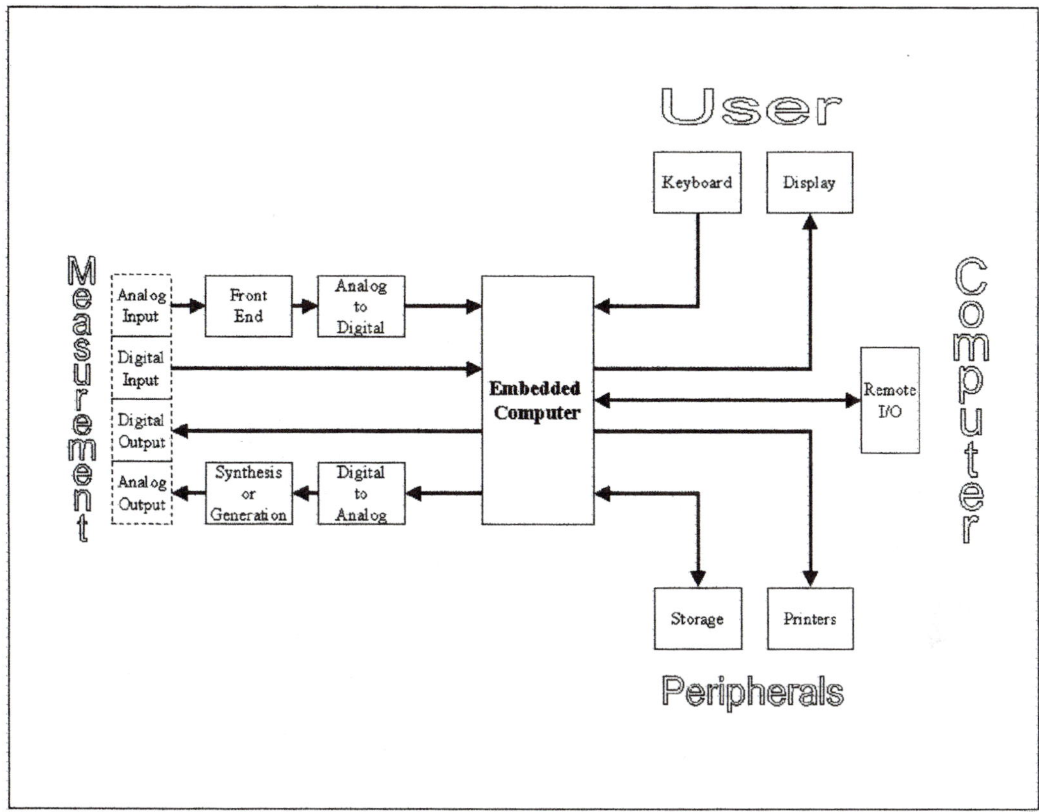

Figure 10.1 An embedded computer generalized block diagram.

strument normally interacts with the user of the measurement. The instrument also generally interacts with an external computer, which is connected for control or data-connectivity purposes. Finally, in some cases, the instrument is connected to local peripherals, primarily for printing and storage. Figure 10.1 shows a generalized block diagram for the embedded computers and these aspects.

10.2.2 Embedded computer uses

The embedded computer has taken a central role in the operation and function of an instrument. Embedded computers have a wide range of specific uses (related to the measurement, user, external computer, and local peripheral aspects) within the instrument:

- External computer interfaces
- User interfaces

- User configuration and customization
- User-defined automation
- Measurement or source calculation
- Measurement or source control
- Measurement or source calibration and self tests
- Measurement or source error and failure notification
- Local peripheral control
- Coordination of the various tasks

Table 10.1 describes these uses.

10.2.3 Benefits of embedded computers in instruments

In addition to the direct uses of embedded computers, it is instructive to think about the value of an embedded computer inside an instrument. The benefits occur throughout the full life cycle of an instrument, from development through maintenance. These benefits are described in Table 10.2.

10.3 Embedded Computer System Hardware

10.3.1 Microprocessors as the heart of the embedded computer

The embedded computer in an instrument requires both hardware and software. The microprocessor is just one of the hardware components. A full embedded computer also requires support circuitry, memory, and peripherals (including the instrument hardware). The microprocessor provides the central processing unit (CPU) of the embedded computer. Some microprocessors are very complex, but others are fairly rudimentary. There is variation in the amount of integration of functionality onto microprocessors; this can include memory, I/O, and support circuitry.

10.3.2 How microprocessors work

A microprocessor is basically a state machine that goes through various state changes, determined by its program and its external inputs. This program is a list of machine-language instructions that are stored in memory. The microprocessor accesses the program by generating a memory storage location or address on the *address bus*. The memory then returns the appropriate instruction (or data) from that location over the *data bus*. The machine-language instructions are numbers that are returned to the microprocessor.

10.3.3 Program and data store

The address bus and the data bus are physical connections between the microprocessor and memory. The number of connections on each bus varies and

TABLE 10.1 Uses of Embedded Computers

Use	Description
External computer interfaces	The embedded computer is typically involved in the control and transfer of data through external interfaces. This allows for the connection of the instrument to external PCs, networks, and peripherals. Examples, which are described later in this chapter, include IEEE 488 (also known as *GPIB* or *HP-IB*), RS-232 (serial), Centronics (parallel), Universal Serial Bus (USB), IEEE 1394 (FireWire), etc.
User interfaces	The embedded computer is also typically involved in the display to and input from the user. Examples include keyboards, switches, rotary pulse generators (RPGs, i.e., knobs), LEDs (single or alpha-numeric displays), LCDs, CRTs, touch screens, etc.
User configuration and customization	Many instruments often have a large amount of configuration information because of their advanced capabilities. The embedded computer enables saving and recalling of the instrument state. Also, the embedded computer sometimes is used for user customization of the instrument. This can range from simple configuration modifications through complete instrument programmability.
User-defined automation	With very powerful embedded computers available in instruments, it is often unnecessary to connect the instrument to an external computer for more-advanced tasks. Examples include go/no-go (also known as *pass/fail*) testing and data logging.
Measurement or source calculation	The embedded computer almost always performs calculations, ranging from very simple to very complex, that convert the raw measurement data to the target instrument information for measurement or vice versa for source instruments. For example, such electrical transducers as thermocouples don't produce results that are in the terms that the users want. Embedded computers do the calculations to convert the measured voltage to the desired temperature reading.
Measurement or source control	The embedded computer generally controls the actual measurement process. This can include control of a range of functions, such as analog-to-digital conversion, switching, filtering, detection, shaping, etc. Note that this is not the analog or digital electronics, but the control of these components that do the measuring or synthesizing.
Measurement or source calibration and self tests	Many instruments are very complex. The embedded computer is almost always used to do at least a small amount of self-testing. Most instruments use embedded computers for more extensive calibration tests.
Measurement or source error and failure notification	As with calibration, many instruments are very complex—not only in the basic hardware and system, but also in the type of measurements. Instruments often do sampling and statistical analysis. Acceptable measurement data can be refined and verified with a microprocessor in the system. Notification of marginal, out-of-bounds, or failure conditions can be given.
Local peripheral control	Many instruments have built-in storage devices. These are often floppy disk drives. There are sometimes built-in printers, but more often there is a connection to an external, but local, printer. Less common are local measurement-related peripherals, such as switches or attenuators, that are controlled by the embedded computer.
Coordination of the various tasks	The previous uses interact in various ways. A key use of embedded computers is to organize, synchronize, and control the various aspects of the instrument.

TABLE 10.2 The Benefits of Embedded Computers Through Their Lifecycles

Lifecycle phase	Benefit of embedded computers
Development	One of the biggest advantages of embedding computers inside an instrument is that they allow several aspects of the hardware design to be simplified. In many instruments, the embedded computer participates in acquisition of the measurement data by servicing the measurement hardware. Embedded computers also simplify the digital design by providing mathematical and logical manipulations, which would otherwise be done in hardware. They also provide calibration both through numerical manipulation of data and by controlling calibration hardware. This is the classic transition of function from hardware to software.
Manufacturing	The embedded computer allows for lower manufacturing costs through effective automated testing of the instrument. Embedded computers also are a benefit because they allow for easier and lower-cost defect fixes and upgrades (with a ROM or program change).
Installation	When used as a stand-alone instrument, embedded computers can make the set-up much easier by providing on-line help or set-up menus. This also includes automatic or user-assisted calibration. Although many are stand-alone instruments, a large number are part of a larger system. The embedded computers often make it easier to connect an instrument to a computer system by providing multiple interfaces and simplified or automatic set up of interface characteristics.
Use	Given the complexity of many instruments, it becomes more difficult to coherently present functionality to the user. The embedded computers allow for user interfaces that are easier to learn and use. This also applies to local language, data, and numerical format customization of the instrument.
Maintenance	Given the complexity of function, it is often difficult to know if the instrument is operating properly. The embedded computer allows for a system that will check itself and its measurement hardware at various times (at power on, periodically, before each measurement) and describe not only the presence of a failure, but also repair suggestions.

has grown over time. In a high-level architectural sense, the two general classes of memory are program store and data store. The *program store* is, as the name implies, the memory where the program is stored. Similarly, the *data store* is where data values used by the program are stored and read. The general structure of the microprocessor connected to program and data store can be seen in Fig. 10.2. In most cases, the *input/output (I/O) devices* are connected via this same address and data bus mechanism.

The data store is very similar to the program store. It has two primary functions: It is used to store values that are used in calculations and it also provides the microprocessor with a subroutine stack. The read/write line shown in Fig. 10.2 is the signal used by the data store memory to indicate a memory read or a memory write operation. Depending on the microprocessor system, other control lines (beyond the scope of this chapter) are used to control the memory devices and the address and data bus. The subroutine stack

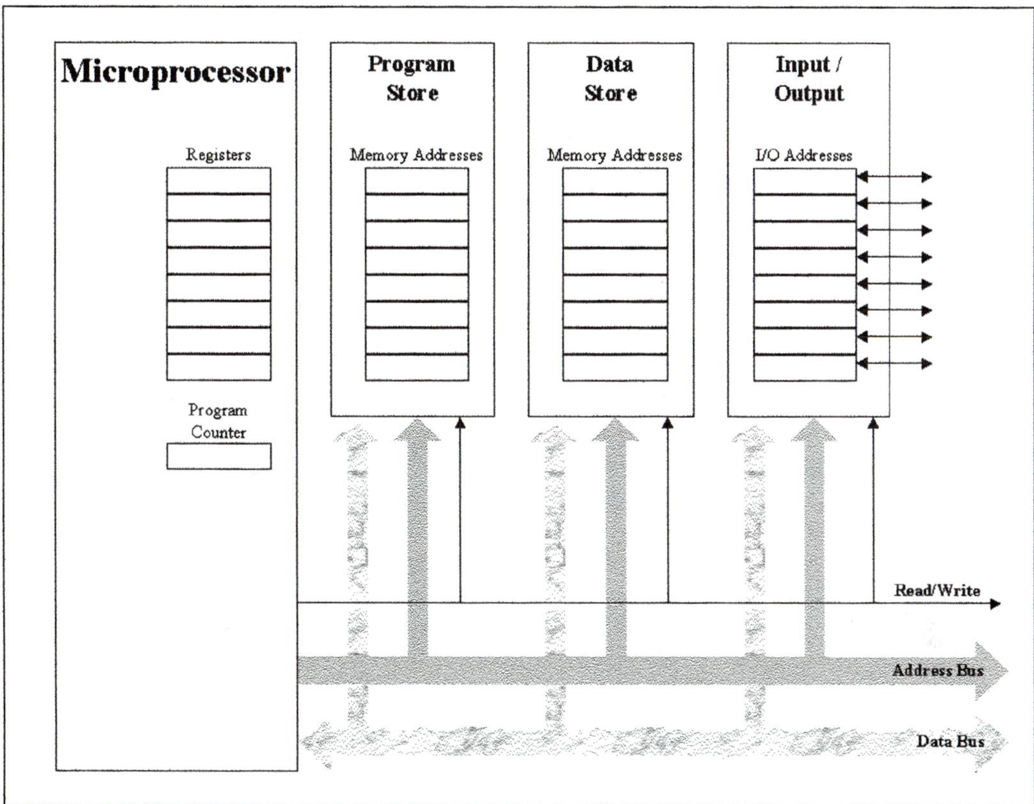

Figure 10.2 A microprocessor basic block diagram.

is an integral part of the computer that provides for subroutine call-and-return capability. The return address (the next address for the instruction pointer after the call instruction) is pushed onto the stack as the call instruction is executed so that when the return instruction is executed, it gets the appropriate next instruction address. Calls generally push the instruction pointer onto the stack and it is the job of the developer or the high-level language compiler to deal with saving registers. Pushing information onto the stack is one of the common ways to pass parameters to the subroutine. For interrupts, the information pushed on the stack often includes not only the return address information, but also various microprocessor registers.

Although the data store needs to be writable, there does not need to be a physical difference between program store and data store memory. Therefore, most microprocessors do not differentiate between program and data. Microprocessors like this are referred to as *Von Neumann machines* and have program and data in the same memory space. Also, quite a few microprocessors have separate program and data space (inside or outside of the microproces-

TABLE 10.3 Types of Microprocessor Operations

Type of operation	Examples
Data	Load, store, clear
Stack	Push, pop
Integer math	Add, subtract, multiply, divide
Boolean logic	And, or, not, nor, xor, shift, rotate
Branching	Comparisons, conditional branch, unconditional branch
Subroutine	Call, return
Floating point math	Add, subtract, multiply, divide
Interrupts	Enable, disable, generate interrupt

sor), called *Harvard machines*. This differentiated memory is useful because it allows for simpler and/or faster CPU design.

10.3.4 Machine instructions

The machine-language instructions indicate to the microprocessor what sort of actions it should take. Table 10.3 shows examples of the types of operations that a microprocessor can support.

10.3.5 Integer and floating-point instructions

All microprocessors have integer mathematical operations, but many do not have floating-point operations built in. For microprocessors with no floating-point facilities, the developer is required to use software routines, which are very costly in time, to handle floating-point operations. Some microprocessors have predefined floating-point operations, but without built-in floating-point hardware. In these cases, when the floating-point instructions are encountered, the microprocessor will generate internal signals that cause an optional floating-point coprocessor to perform the operation. If no coprocessor is present, a CPU exception or trap is generated, which causes a software emulation of the floating-point operation to occur. The software emulation is slower than hardware floating point, but it provides the advantage that the software can be identical across a range of microprocessor hardware.

10.3.6 Internal registers

Microprocessors also have various internal registers. Common registers store the instruction address (also called the *instruction pointer* or program counter register), data addresses, and data. There is a wide variety in the number of general address and data registers. More-advanced microprocessors have both integer and floating-point registers. Microprocessors also have a variety of status registers available to the programmer.

10.3.7 Interrupts

Another crucial part of microprocessor architecture is the interrupt. The concept is that a microprocessor is executing a program and it needs to respond to an important event. An *interrupt* is a hardware signal sent to the microprocessor that an important event has occurred. The interrupt forces the microprocessor at the end of the current instruction to do a subroutine call to an interrupt service routine. The microprocessor performs the instructions to handle the event and then returns to the previous program. Although this is a crucial part of computers used in general applications, it is especially important in instrument applications. Interrupts can be the trigger of a measurement, data-transfer completion, error conditions, user input, etc. Just as instruction sets vary, the interrupt system design of a microprocessor can vary greatly from a single-interrupt approach to multiple interrupts with various priorities or levels.

10.3.8 Cache

Many current and high-performance microprocessors have a feature called *cache*. This is a portion of high-speed memory that the microprocessor uses to store copies of frequently used memory locations. This is helpful because the main semiconductor memory of computer systems is often slower (often around 100 nanoseconds access time) than the microprocessor can access and use memory. To help with this imbalance, cache memory stores or prestores memory locations for these frequently used areas, which could be either instructions (in *instruction cache*) or data (in *data cache*). Cache tends to be in the 10-nanosecond access time range. The cache memory can be located on the microprocessor and is called *primary cache memory* (usually less than 64 Kb of memory). The cache memory can also be located outside of the microprocessor. This external cache is called *secondary cache memory* (usually in the 256-Kb to 512-Kb range). Cache has a huge performance benefit—especially in loops and common routines that fit in cache. It also can represent a challenge for the instrument designer because cache causes the microprocessor to operate nondeterministically with respect to performance—it can run faster or slower, depending on the data stream and inputs to the computer system.

Some microprocessors have internal bus widths that are different than what is brought out to the external pins. This is done to allow for a faster processor at a lower cost. In these cases, the external (to the microprocessor) data paths will be narrower (e.g., 16 data bits wide) and the internal microprocessor data paths will be full width (e.g., 32 data bits wide). When the microprocessor gets data into or out of a full-width register it would perform two sequential read or write operations to get the full information. This is especially effective when coupled with caching. This allows for segments of code or loop structures to operate very efficiently (on the full-width data)—

even though the external implementation is less-expensive partial-width hardware.

10.3.9 RISC versus CISC

Another innovation in computing that has been making its way into instruments is microprocessors based on *Reduced Instruction Set Computers* (*RISC*). This is based on research that shows that a computer system can be designed to operate more efficiently if all of its instructions are very simple and they execute in a single clock cycle. This is different from the classic *Complex Instruction Set Computer* (*CISC*) model. Such chips as the Intel x86 and Pentium families and the Motorola 68000 family are CISC microprocessors. Such chips as the Motorola PowerPC and joint HP and Intel IA-64 microprocessors are RISC systems. One of the challenges for instrument designers is that RISC systems usually have a very convoluted instruction set. Most development for RISC systems require advanced compiler technologies to achieve high performance and allow developers to easily use them. Another characteristic of RISC systems is that their application code tends to be larger than CISC systems because the instruction set is simpler.

10.4 Elements of an Embedded Computer

10.4.1 Support circuitry

Although requirements vary, most microprocessors require a certain amount of support circuitry. This includes the generation of a system clock, initialization hardware, and bus management. In a conventional design, this often requires 2 or 3 external integrated circuits (ICs) and 5 to 10 discrete components. The detail of the design at this level depends heavily on the microprocessor used. In complex or high-volume designs, an *Application-Specific Integrated Circuit* (*ASIC*) can be used to provide much of this circuitry.

10.4.2 Memory

The microprocessor requires memory both for program and data store. Embedded computer systems usually have both ROM and RAM. *Read-Only Memory* (*ROM*) is memory whose contents do not change—even if power is no longer applied to the memory. *Random Access Memory* (*RAM*) is a historical, but inadequate, term that really refers to *read/write memory*, memory whose contents can be changed.

RAM memory is volatile; it will lose its contents when power is no longer applied. RAM is normally implemented as either static or dynamic devices. *Static memory* is a type of electrical circuit[2] that will retain its data, with or

[2] Static memory is normally built with latches or flip-flops.

without access, as long as power is supplied. *Dynamic memory* is built out of a special type of circuit[3] that requires periodic memory access (every few milliseconds) to refresh and maintain the memory state. This is handled by memory controllers and requires no special attention by the developer. The advantage of the dynamic memory RAM is that it consumes much less power and space.

ROM is used for program storage because the program does not usually change after power is supplied to the instrument. A variety of technologies are used for ROM in embedded applications:

- *Mask ROM*—Custom programmed at the time of manufacture, unchangeable.
- *Fusible-link Programmable ROM* (*PROM*)—Custom programmed before use, unchangeable after programming.
- *Erasable PROM* (*EPROM*)—Custom programmed before use, can be reprogrammed after erasing with ultraviolet light.
- *Electronically Erasable PROM* (*EEPROM*)—Custom programmed before use, can be reprogrammed after erasing. It's like an EPROM, but the erasing is performed electrically.

10.4.3 Nonvolatile memory

Some instruments are designed with special *nonvolatile RAM*, memory that maintains its contents after power has been removed. This is necessary for storing such information as calibration and configuration data. This can be implemented with regular RAM memory that has a battery backup. It can also be provided by special nonvolatile memory components, most commonly, flash memory devices. Flash memory is a special type of EEPROM that uses block transfers (instead of individual bytes) and has a fairly slow (in computer terms) write time. So, it is not useful as a general read/write memory device, but is perfect for nonvolatile memory purposes. Also, a limited number of writes are allowed (on the order of 10,000).

All embedded systems have either a ROM/RAM or a flash/RAM memory set so that the system will be able to operate the next time the power is turned on.

10.4.4 Peripheral components

Microprocessors normally have several peripheral components. These *Very Large Scale Integration* (*VLSI*) components provide some major functionality.

[3] Dynamic memory is normally built from a stored-charge circuit that uses a switched capacitor for the storage element.

These peripheral components tend to have a small block of registers or memory that control their hardware functions.

For example, a very common peripheral component is a Universal Asynchronous Receiver/Transmitter (UART). It provides a serial interface between the instrument and an external device or computer. UARTs have registers for configuration information (like data rate, number of data bits, parity) and for actual data transfers. When the microprocessor writes a character to the data-out register, the UART transmits the character, serially. Similarly, when the microprocessor reads the data in register, the UART transfers the current character that has been received. (This does require that the microprocessor checks or knows through interrupts or status registers that valid data is in the UART.)

10.4.5 Timers

Another very common hardware feature is the timer. These devices are used to generate periodic interrupts to the microprocessor. These can be used for triggering periodic operations. They are also used as *watchdog timers*. A watchdog timer helps the embedded computer recover after a non-fatal software failure. These can happen because of static electricity, radiation, random hardware faults or programming faults. The watchdog timer is set up to interrupt and reset the microprocessor after some moderately long period of time (for example, one second). As long as everything is working properly, the microprocessor will reset the watchdog timer—before it generates an interrupt. If, however, the microprocessor hangs up or freezes, the watchdog timer will generate a reset that returns the system to normal operation.

10.4.6 Instrument hardware

Given that the point of these microprocessors is instrumentation (measurement, analysis, synthesis, switches, etc.), the microprocessor needs to have access to the actual hardware of the instrument. This instrument hardware is normally accessed by the microprocessor like other peripheral components (i.e., as registers or memory locations).

Microprocessors frequently interface with the instruments' analog circuits using *analog-to-digital converters* (*ADCs*) and *digital-to-analog converters* (*DACs*). In an analog instrument, the ADC bridges the gap between the analog domain and the digital domain. In many cases, substantial processing is performed after the input has been digitized. Increases in the capabilities of ADCs allow the analog input to be digitized closer to the front end of the instrument, allowing a greater portion of the measurement functions to occur in the embedded computer system. This has the advantages of providing greater flexibility and eliminating errors introduced by analog components. Just as ADCs are crucial to analog measuring instruments, DACs play an important role in the design of source instruments (such as signal generators). They are also very powerful when used together. For example, instruments can have automatic calibration procedures where the embedded computer adjusts an analog circuit with a DAC and measures the analog response with an ADC.

10.5 Physical Form of the Embedded Computer

Embedded computers in instruments take one of three different forms: a separate circuit board, a portion of a circuit board, or a single chip. In the case of a separate circuit board, the embedded computer is a board-level computer that is a circuit board separate from the rest of the measurement function. An embedded computer that is a portion of a circuit board contains a microprocessor, its associated support circuitry and some portion of the measurement functions on the same circuit board. A single-chip embedded computer can be a microcontroller, digital signal processor or microprocessor core with almost all of the support circuitry built into the chip. Table 10.4 describes these physical form choices in some additional detail:

A *digital signal processor* (*DSP*) is a special type of microcontroller that includes special instructions for digital signal processing, allowing it to perform certain types of mathematical operations very efficiently. These math operations are primarily multiply and accumulate (MAC) functions, which are used in filter algorithms. Like the microcontroller, the space, cost, and power are reduced. Almost all of the pins can be used to interface to the instrument system.

Microprocessor cores are custom microprocessor IC segments or elements that are used inside of custom-designed ICs. In this case, the instrument designer has a portion of an ASIC that is the CPU core. The designer can put much of the rest of the system, including some analog electronics, on the ASIC, creating a custom microcontroller. This approach allows for minimum size and power. In very high volumes, the cost can be very low. However, these chips are very tuned to specific applications. They are also generally difficult to develop.

10.6 Architecture of the Embedded Computer Instrument

Just as an embedded computer can take a variety of physical forms, there are also several ways to architect an embedded computer instrument. The architecture of the embedded computer can impact the instrument's cost, performance, functionality, ease of development, style of use, and expandability. The range of choices include:

- Peripheral-style instruments (externally attached to a PC)
- PC plug-in instruments (circuit boards inside a PC)
- Single-processor instruments

- Multiple-processor instruments
- Embedded PC-based instruments (where the embedded computer is a PC)
- Embedded workstation-based instruments

Table 10.5 describes these architectural choices in some additional detail.

TABLE 10.4 Classes of Embedded Computers

Embedded computer class	Description	Advantages and Disadvantages
Board-level computer	*A board-level computer* is an embedded computer that is built on a circuit board (or sometimes multiple boards) that is separate from the rest of the measurement function. These usually include CPU, memory, and user I/O. The designer needs to add disk, display, power, and the measurement hardware. These can be a specially designed board, an integrated PC motherboard, or even a RISC workstation motherboard.	*Advantages:* The computer is all there. The system is standard (either PC or workstation) and already designed. The system, because it is based on existing computers, has development tools, which aid software development. *Disadvantages:* Because it is based on a more general-purpose computer board, it usually is larger in size and power. These boards often require a hard disk drive. There is also the potential for radio-frequency interference (RFI).
Standard microprocessor	This is a normal, off the shelf, microprocessor (like a Motorola PowerPC or an Intel Pentium). The computer system around the microprocessor must be designed and integrated with the instrument functions.	*Advantages:* Because it is being designed specifically for the instrument, it can be tightly integrated into the instrument application. This can allow for less space, cost, and power. *Disadvantages:* The embedded computer must be both designed and manufactured.
Single chip microcontroller	A microcontroller is a single IC that contains almost the entire embedded computer. It typically includes enough ROM and RAM for the application along with several other peripherals and I/O lines that can be used for digital input or output. Microcontrollers frequently include ADCs (analog-to-digital converters), DACs (digital-to-analog converters), and interface support, such as serial ports.	*Advantages:* Because it is all there, there is minimal space, cost, and power. Almost all of the pins can be used to interface to the instrument system because no pins are necessary for an address bus. Sometimes these lines are included, but are often shared with other I/O uses. *Disadvantages:* By their nature, they tend to be very limited in what applications they can be used in (primarily by the limited ROM and RAM). They are more difficult to develop applications for because they don't bring out address and data bus information, which facilitate development tools. This means that they don't have the support that makes it easy to develop firmware with the development systems.

Some instruments are based on embedded workstations. This is very much like using an embedded PC, but this type of instrument is based on a high-performance workstation. Normally, these are built around a RISC processor and UNIX. The key advantage is very high performance for computationally intensive applications (usually floating-point mathematical operations). Like the embedded PC, standard high-level components can be used for the hardware and software. There are also established and highly productive software-development tools. Like the embedded PC, the key challenge is dealing with larger and constrained physical form factor. Embedded workstations also tend to be more expensive.

10.7 Embedded Computer System Software

As stated earlier, the embedded computer in an instrument requires both hardware and software components. Embedded computer system software includes:

Operating system—The software environment that the (instrument) applications run within.

Instrument application—The software program that performs the instrument functions on the hardware.

Support and utility software—Additional software the user of the instrument requires to configure, operate, or maintain the instrument (such as reloading or updating system software, saving and restoring configurations, etc.).

10.7.1 How embedded computers are programmed

As mentioned previously, the instructions for the computer are located in program store memory. *Program development* is fundamentally the process that the software developers use to generate the machine-language instructions that perform the intended instrument functions.

Some development for embedded computers for instruments has been done in assembly language. This is a convenient representation of the machine language of the embedded computer. It still is at the level of the microprocessor, but it is symbolic, rather than ones and zeros. Table 10.6 shows a simple program fragment in assembly language and the resulting machine code (on the right). This program fragment is intended to show reading some character data and processing the character data when a space is encountered.

The resulting machine code for embedded computers is often put into ROM. Because this programming is viewed as being less changeable than general software application development, it is referred to as *firmware* (*firmware* being less volatile than *software*).

TABLE 10.5 Architecture Choices for Embedded Computers

Architecture	Description	Advantages and disadvantages
Peripheral-style instruments	This is a very simple instrument that is designed to always be used with a PC. They usually have a low-end microcontroller and little or no human interface because they are designed to connect to a PC. They generally have very simple system software.	*Advantages:* Because they are "faceless," they have much lower cost, power, and size. There is much simpler design because of the simplicity of HW (mostly the measurement part) and simplicity of the embedded SW (because there is very little). There is also a benefit because the software is up to date because the computer is external. The data is already in the PC, so exporting it to application programs is easier. *Disadvantages:* They require a PC (or laptop). They might not be as portable as a stand-alone instrument. External system and application changes and versions require ongoing updates. The interface to the PC needs to be chosen carefully to prevent obsolescence.
PC plug-in instruments	These are plug-in measurement boards that go into a PC. They are similar in many respects and are a variant of peripheral-style instruments. The difference is that the interface is the internal PC bus as opposed to an external I/O interface. Plug-in board instruments still have a microprocessor and simple system software.	*Advantages:* Similar to the peripheral-style instrument because they are "faceless," they have much lower cost, power, and size. There is much simpler design because of the simplicity of HW (mostly the measurement part) and simplicity of the embedded SW (because there is very little). *Disadvantages:* They have to be in a PC. This also presents some challenges from a support point of view, with respect to different PC configurations and reliability. There are ongoing challenges because of changes in PC buses. There are also changes in PC form factors. There have to be ongoing updates to deal with external system and application changes and versions. Another issue is in dealing with radio-frequency interference from the PC.
Single processor instruments	The normal microprocessor-based instrument. They have a local human interface and most have some form of remote computer interface.	*Advantages:* They can operate stand-alone and don't require a PC. This helps with portability and with some of the challenges of PC connectivity, operating systems, and applications. *Disadvantages:* They are costlier because of the user interface, extra power, size, etc. There has to be some instrument connectivity mechanism (i.e., software and drivers) to import instrument data into the PC.

TABLE 10.5 Architecture Choices for Embedded Computers (*Continued*)

Architecture	Description	Advantages and disadvantages
Multiple-processor instruments	High-end instruments often include multiple processors for parallel measurement streams or tasks (such as measurement, communication, front panel, and computation). Usually more than one type of operating system is in multiple-processor instruments—typically with at least one real-time operating system (which is described later in this chapter).	*Advantages:* The key advantage is performance—there is more processor power dedicated to tasks. The system can be optimized across tasks—for example using simple microcontrollers for I/O intensive operations, normal processor for RAM/ROM intensive operations. It also has some electrical advantages because the processors can be located closer to measurement hardware. *Disadvantages:* Multiple-processor instruments are more expensive and larger. There has to be some instrument connectivity mechanism (i.e., software and drivers) to import instrument data into the PC.
Embedded PC-based instruments	A full PC is inside the instrument as the embedded computer. This can be a normal PC motherboard. The embedded PC operating system is usually Windows 95 or NT. Sometimes DOS is used and Windows CE is an emerging technology.	*Advantages:* Standard high-level components can be used for the embedded computer hardware and software. There is no hardware computer development and the hardware tends to be lower in cost. It has more computing power because of ongoing improvements in PC performance. Established and highly productive software-development tools are available. *Disadvantages:* A key challenge is dealing with larger and constrained physical form factor. The designer also has to deal with software form factor (software footprint) because a full OS tends to have many things that are not needed in the instrument. The designer has to deal with obsolescence of the motherboard and operating system changes and/or obsolescence. Some instruments use special form-factor variants of PCs, but even though this helps with the form-factor, it is no longer a standard motherboard.

10.7.2 Assembly-language development

The assembly-language source code is run through an assembler. The *assembler* is a program that translates the assembly-language input into the machine language. The assembler is normally run on a *development computer*, normally a PC that is used to create the software for the embedded computer. Note that the development computer is not usually the *target computer*, which is the embedded computer in the instrument application. The machine code

TABLE 10.6 An Assembly-Language Program Example

	Assembly language		Machine code
MOVE	(3F),A	; read from port at address 3F	54 3F
AND	#7F,A	; mask off the upper bit	21 7F
COMP	#20,A	; if the input is equal to 20H (a space)	56 20
BEQ	ParseIt	; then go to the parse routine	85 24 35

produced by the development computer is for the target computer microprocessor (the processors of the development and target computers are usually different). This machine code is then transferred to the board containing the microprocessor. For smaller, simpler, systems this is often done by transferring a memory image to a *PROM programmer*, a device that writes the image into a PROM. This PROM is then physically inserted onto the board. This process is shown in Fig. 10.3.

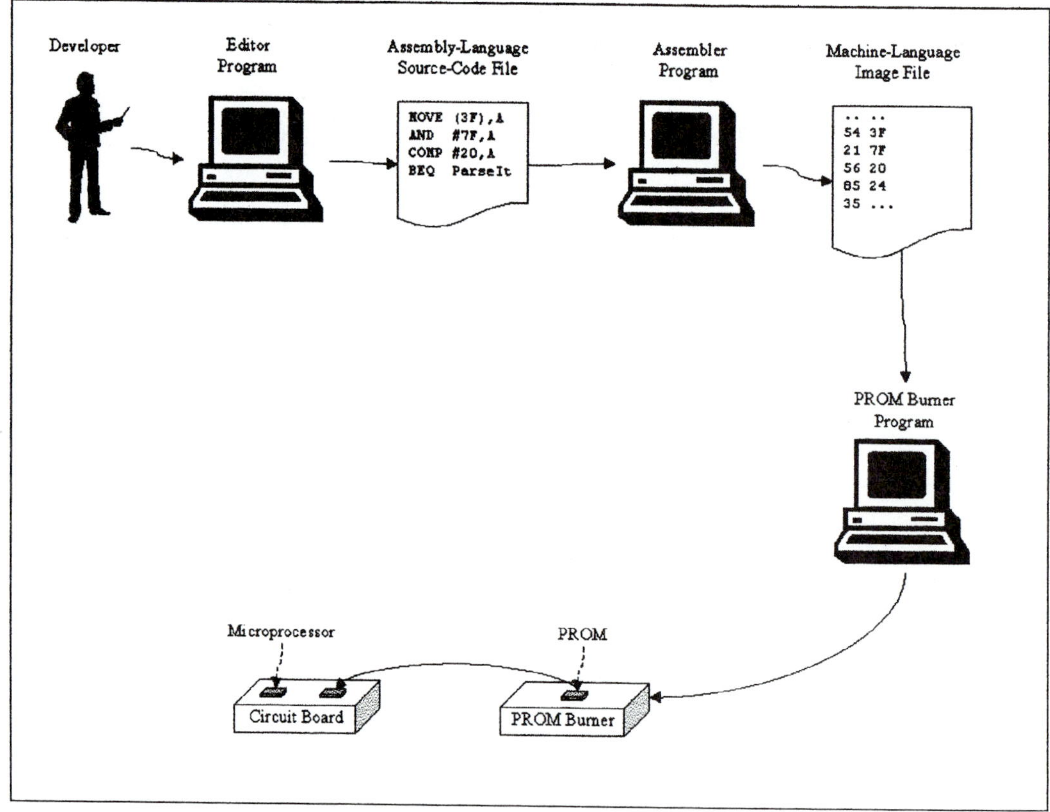

Figure 10.3 The process of converting assembly language to machine code.

TABLE 10.7 Common Cross-Development Tools

Tool	Description	Comments
Development boards	Hardware development boards and systems are connected to the microprocessor. They bring out the microprocessor signals. This allows for the developer to see the address lines and the flow of execution. They normally allow for single stepping, limited breakpoints, interrupts, and processor resets.	Development boards tend to be very inexpensive, but have limited debugging capabilities. They also are not as effective with some classes of microcontrollers, particularly those with limited access to the microprocessor signals.
Monitors	The development prototype of the instrument is built with RAM for program storage and a computer interface—dedicated for development use. The program loaded into the RAM contains some code called a *monitor*. This monitor provides a basic external interface to the instrument over the dedicated interface. The monitor programs generally provide for memory access, starting execution and limited breakpoints.	Monitor-based development tends to be inexpensive. The monitors tend to have limited capabilities. They also depend on the microprocessor running properly. In case of a serious hardware or software problem, the monitor often is not able to work properly.
Logic analyzers	These digital logic analyzers are built or configured for use with microprocessors. Normally, they are general-purpose logic analyzers with microprocessor pods. The developer attaches the clip onto the microprocessor on the prototype or production PC board. The analyzer will convert address and data-line signals into the appropriate code. They generally show some of the other signals as well.	Logic analyzer development is very common. It is a middle ground in terms of cost and functionality between emulators and monitors or development boards.
Emulators	An emulator is a device connected to the instrument being developed in place of the microprocessor. It behaves just like the microprocessor, except that the developer has full control over execution: memory access, register access, breakpoints, single stepping, reset, etc.	Emulators are very powerful, but are rather expensive. They also tend not to be immediately available for the latest microprocessors.

This approach of developing code on a development system and placing it into the embedded computer is called *cross development*. As mentioned earlier, embedded computer firmware and software are fairly complex. Because of this complexity, errors are introduced during development. This means that tools must be available to debug the program code. Some of the common tools are described in Table 10.7.

Emulator development is very powerful, but it requires an expensive system for each development station. Monitors or development boards are less expensive, but not as powerful. Most major instrument development is done with a combination of these techniques.

As embedded computers have become more powerful and complex, there has been a shift from ROM-based firmware to systems with modifiable program storage memory. This modifiable program storage ranges from flash memory to integrated hard disk drives. In these systems, the program is loaded into RAM at instrument power up. This has many advantages. It allows the instrument to be modified. It allows for shorter development because there is no step producing the mask-programmable ROMs. It also allows the instrument software to be corrected (or patched) after release. To achieve these characteristics, it is necessary for the embedded computer to include a computer interface and a protocol for downloading new software. And this mechanism needs to be done in a way that the user does not accidentally destroy the instrument software (making for very expensive paperweights).

10.7.3 High-level language development

Another change, as embedded computers have gotten more powerful, is that most instruments are programmed in high-level languages. A *high-level language* is one that is at a higher level of abstraction than assembly or machine language, where the high-level language constructs and capabilities are more complex and conceptual. In a high-level language, the developer writes code to add (e.g., $A = B + C$). In assembly language, the developer has to load a number from a memory location, add a number from a different memory location, and then store the resulting value in a third memory location. This high-level language enables developers to have higher productivity and functionality per line of code written because they are dealing with the application at a level closer to the way they think about it.

Several high-level languages have been and are currently being used for embedded computing. The most popular languages include C, C++, Java, and Pascal. Table 10.8 compares these languages and assembly language. The example program fragment is based on the same example shown previously with assembly language and is intended to show reading character data and processing the character data when a space is encountered.

Some instruments have been built with special-purpose programming languages (or standard languages that were modified or extended). There are development speed benefits to having a language tuned to a specific application, but they come at the expense of training, tool maintenance, and skilled developer availability.

10.7.4 High-level language compilers

The high-level language source code is normally run through a compiler. The *compiler*, like the assembler, is a program that runs on a development computer. It translates the high-level language input into object code files, files that contain some symbolic information and machine-language instructions (or byte codes, in the case of Java). Normally, a program called a *linker* links together the various object-code language and system libraries used during

TABLE 10.8 Common Embedded Computer Languages

Language	Example	Comments

C and C++	`if ((PortA & inputMask) == Space )`
	`{ ParseInput();`
	`}`
		These are the current popular languages for embedded programming because of space and run-time efficiency. C++ is an object-oriented extension to C. They do not inherently encourage good programming practices.

Assembly	`MOVE (3F),A ; read from port 3F`
	`AND #7F,A ; mask upper bit`
	`COMP #20,A ; if equal to space`
	`BEQ ParseIt ; then go parse it`
		Assembly is not normally used in most instruments (because of the complexity of the software). It is not portable, but allows for very high performance and small size because it is symbolic machine code. Often, developers will use a small portion of assembly language for crucial sections of code, with the rest written in a high-level language.

Java	`Port a = new Port("3F");` `if (a.read == ' ') ParseInput();`	Java is an object-oriented language that compiles to a byte code that is run on a virtual machine. Because of this, it has small executables and is very portable. It does run somewhat slower than classic compiled languages and requires a Java Virtual Machine (JVM), which is often ½ megabyte in memory.

Pascal	`CONST port = #3F; CONST space = 32;`
	`BEGIN`
	`IF ( rdport(port) & mask) = space`
	`THEN CALL parseit;`
	`END;`
		Pascal is an older, block-structured language that has been used in some instruments.

the development process and the object-code files that the developer has produced. This linked object code is then transferred to the microprocessor board. For moderate to large systems, this is often done by downloading the memory image to the computer via a communication interface (such as RS–232 or LAN) or via removable media (such as a floppy disk). This process is shown in Fig. 10.4.

10.7.5 Operating system software

The *Operating System* (*OS*) provides the software environment for the programs in the computer (embedded or otherwise). Although not all uses of embedded computers involve a true operating system, those used in most instruments do. The operating system is fundamentally the overarching program that enables the computer to do its job of executing the application program(s). Operating system services include:

- System initialization
- Resource management—Memory, input, and output devices.
- Task management—Creation, switching, priorities, and communications.

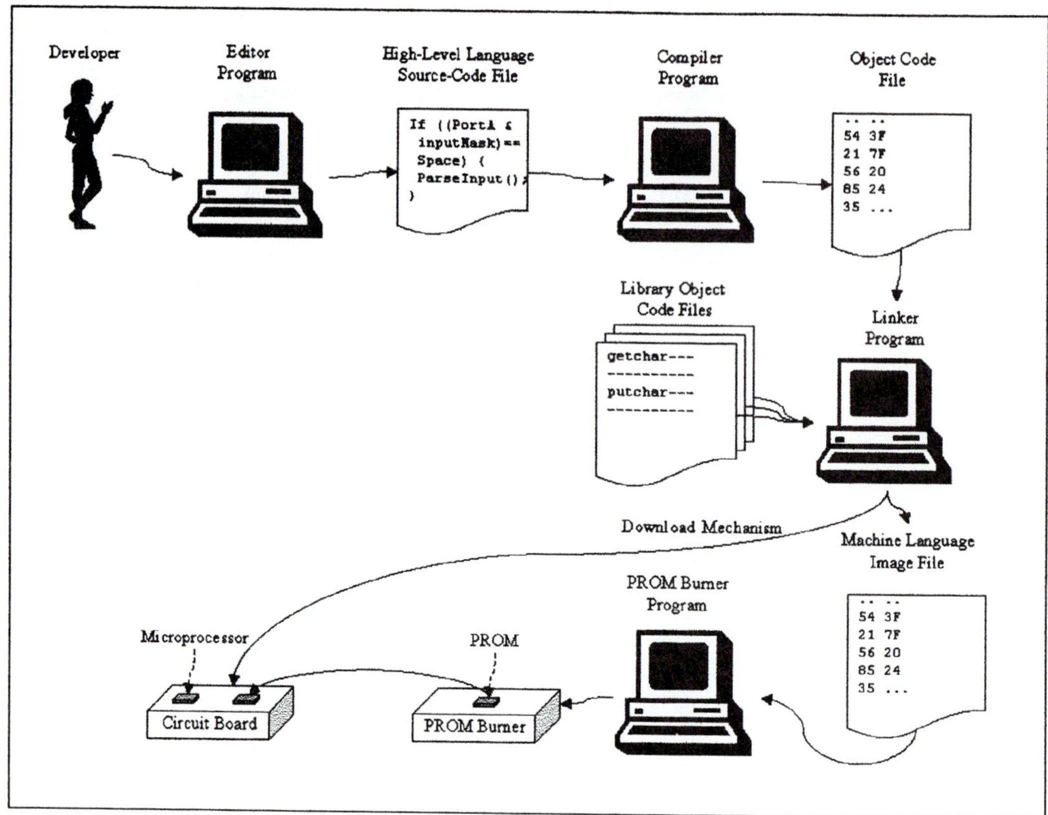

Figure 10.4 The process of converting high-level language to machine code.

There are different types of operating systems and application environments described in Table 10.9.

A big part of the value and benefit of operating systems is that they support many things going on at the same time. The many different things are called *processes, tasks,* or *threads*. There is no general agreement between operating-system vendors about what each of these terms means. In general, the terms *process* and *task* both refer to a program and its associated state information during the execution within an operating system. A thread is a part of a program that can be run independently of the rest of the program. (Programs in an operating system can be single-threaded or multiple-threaded.) Threads tend to be light-weight and processes tend to be more complete and costly in terms of the amount of operating system overhead.

The basic concept behind all of these processes, tasks, and threads is that many things occur simultaneously. Rather than have multiple processors or very complex code, the operating system allows for multiple processes to request execution, get slices of computer processing time, and request and free

TABLE 10.9 Embedded Computer Operating Systems and Application Environments

OS/environment	Description	Examples
Bootstrap loader	This is not truly an operating system, but a simple program that loads another program to run. Most often, this is called a *boot ROM* and is always in some form of nonvolatile memory (ROM or Flash). They tend to be relatively small.	Generally homegrown.
Simple executive	This is a very simple type of operating system. They tend to be simple software loops that support limited multi-tasking and scheduling. They are normally used in very simple cases.	Generally homegrown.
Real-time OS (RTOSs)	This is a special type of operating system with specific extensions designed to support high performance and real-time characteristics (described in section 10.7.6). This OS is very commonly used in instruments. RTOSs tend to be modular and allow for relatively easy scaling.	pSOS+, VxWorks, Lynx, . . .
Device OS	This is a commercial operating system designed for appliances or devices, but not hard-core real-time characteristics. They tend to be moderate in size, are ROMable, and do not require a hard drive.	Windows CE, JavaOS, . . .
Desktop OS	This is a full commercial operating system designed for PCs. They tend to be larger in size and normally require some sort of hard drive (and are not normally ROMable). They do not have hard real-time capabilities. Real time is sometimes achieved with add-on processors or underlying software extensions.	MSDOS, DRDOS, Windows 95/98/xx, Windows NT, MacOS
Server or Multi-user OS	These higher-end operating systems are designed for multiple users. These are occasionally used in instruments, primarily in multiple-processor designs.	Unix and its possesive variants (HP-UX, Linux, SunOS, . . .), Windows NT

system resources (such as memory and hardware devices). The management of system resources is important because it manages competing access to resources. In a normal single-processor embedded system, it is important to remember that the tasks only execute one at a time (because there is a single processor). They appear to be simultaneous because of the speed of execution and the slices of computer execution given to these various processes, tasks, or threads. Typical processes, tasks, or threads are shown in Table 10.10.

10.7.6 Real time

A key aspect that an embedded computer designer/developer has to address is the required level of real-time performance for the embedded computer. In

TABLE 10.10 Typical Embedded Computer Tasks

Measurement	■ Calibration and correction ■ Hardware setup
Calculation	■ Measurement algorithms ■ Numerical manipulation
User interface	■ Keyboard ■ Knobs and switches ■ Display (LED, LCD, CRT, video)
External interface	■ Local peripherals ■ Computer interface (RS-232, IEEE 488, parallel, . . .) ■ Network interface (LAN, . . .)

the context of operating systems, the term *real time* is used to specify the ability of an operating system to respond to an external event within a specified time (*event latency*). A *real-time operating system* (*RTOS*) is an operating system that exhibits this specified and/or guaranteed response time. This real-time performance is often crucial in an instrument where the embedded computer is an integral part of the measurement process and it has to deal with the flow of data in an uninterrupted fashion. Most general-purpose operating systems are not real time.

Because PC operating systems need, in general, to be responsive to the user and system events, terminology about real time has become somewhat imprecise. This has brought about some additional concepts and terms. An operating system has a time window bounded by the best and worst response times. *Hard real time* is when the event response always occurs within the time window, independent of other operating system activity. Note that hard real time is not necessarily fast. It could be microseconds, milliseconds, or days—it is just characterized and always met. *Soft real time* is when the event response is, on average, normally in the time window. So, the system designer needs to determine if there is a need for the real-time performance and whether it is hard real time or soft real time. Often, an instrument designer will use separate processors (especially DSPs) to encapsulate the real-time needs and, therefore, simplify the overall system design.

10.8 User Interfaces

Originally, instruments used only *direct controls*, in which the controls are connected directly to the analog and digital circuits. As embedded computers became common, instruments used *menu* or *keypad-driven systems*, in which the user input was read by the computer, which then modified the circuit operation. Today, things have progressed to the use of *Graphical User Interfaces* (*GUIs*). Although some instruments are intended for automated use or are *faceless* (have no user interface), most need some way for the user to

TABLE 10.11 Typical Instrument Embedded Computer Input Devices

Input device	Description
Buttons/switches	A very common input device is a simple switch connected to I/O ports on the microprocessor. Sometimes the switch will have some preconditioning to remove the "bounce" associated with key closure (which could cause spurious signals).
RPGs	*Rotary Pulse Generators* (*RPG*s) are essentially digital knobs. They generate signals that are proportional to the rate, amount, and direction of rotation. These are frequently used in a soft-control application in place of potentiometers. Some instruments have been designed with a single multi-purpose RPG, whose function is set by the current instrument mode.
Keyboard	Most instruments do not have full alphanumeric keyboards, in part because of size and that they are not needed. They are used in some applications, such as telecommunications. However, numeric keypads are very common. In most cases, the keyboard has instrument application specific keys (e.g., ATTENUATION). Generally, a special keyboard controller chip takes care of scanning and processing key presses. Usually, the key codes are mapped into a normal ASCII key code for use by the microprocessor.
Mouse/pointing device	Many instruments have some form of pointing device. A mouse is useful, even necessary with a GUI, but it can be a problem, given the physical environment where some instruments are located. Often, mice have space or reliability problems. In these cases, trackballs, touch screens, and touch pads, are effective replacements.

interact with the measurement or instrument. Tables 10.11 and 10.12 show examples of the instrument input and output devices.

All of these user interface devices can be mixed with direct control devices (like meters, switches, potentiometers/knobs, etc.). There are a variety of design challenges in developing effective user interfaces in instruments. However, a full description of instrument user interfaces is beyond the scope of this chapter (see Chapters 12 and 44).

10.9 External Interfaces

Most instruments include external interfaces to a *peripheral*, another instrument device or to an external computer. An interface to a peripheral allows the instrument to use the external peripheral, normally printing or plotting the measurement data. An interface to another device allows the instrument to communicate with or control another measurement device. The computer interface provides a communication channel between the embedded computer and the external computer. This allows the user to:

- Log (capture and store) measurement results.
- Create complex automatic tests.

- Combine instruments into systems.
- Coordinate stimulus and response between instruments.

The external computer accomplishes these tasks by transferring data, control, set-up, and/or timing information to the embedded computer. At the core of each of these interfaces is a mechanism to send and receive a stream of data bytes.

10.9.1 Hardware interface characteristics

Each interface that has been used and developed has a variety of interesting characteristics, quirks, and tradeoffs. However, external interfaces have some common characteristics to understand and consider:

- *Parallel or serial*—How is information sent (a bit at a time or a byte at a time)?
- *Point to point or bus/network*—How many devices are connected via the external interface?
- *Synchronous or asynchronous*—How is the data clocked between the devices?
- *Speed*—What is the data rate?

Table 10.13 describes these key characteristics.

TABLE 10.12 Typical Instrument Embedded Computer Output Devices

Output device	Description
LEDs	*Light Emitting Diodes* (*LEDs*) are easily connected to the microprocessor's I/O ports.
Multi-line displays	Multiple-line displays are very common in simpler instruments with moderate user interface needs. They are often implemented with LEDs, *Liquid Crystal Displays* (*LCDs*), and sometimes with plasma display technology. Some are 7-segment displays (such as calculators), allowing for numbers and a few characters. Some are 16-segment devices that display alphanumeric characters. Dot-array style displays from simple 5×7 to full addressable pixels (*picture elements*) that allow for simple graphics. Except for some of the simple 7-segment displays, they generally have built-in control circuits so that they can be directly connected to one of the microprocessor's I/O ports. This controller is very helpful, from a software point of view, because it takes over the task of scanning the display.
Video displays	Some instruments use *Cathode Ray Tubes* (*CRTs*) to display information. Most current displays are computer-style displays. These can be LCD or CRT technology. LCDs have advantages in weight, power, and size and have improved in visual quality.
Speaker	Audio output is occasionally used in instruments, mostly for error or feedback purposes. The audio output is difficult to hear in some instrument environments.

TABLE 10.13 External Interface Characteristics

Characteristic	Description
Parallel or serial	Probably the most fundamental characteristic of hardware interfaces is whether they send the data stream one bit at a time (*serial*) or all together (*parallel*). Most interfaces are serial. The advantage of serial is it limits the number of wires down to a minimum of two lines (data and ground). However, even with a serial interface, additional lines are often used (transmitted data, received data, ground, power, request to send, clear to send, etc.). Parallel interfaces are normally 8 bit or 16 bit. Some older instruments had custom *Binary Coded Decimal* (BCD) interfaces, which usually had six sets of 4-bit BCD data lines. Parallel interfaces use additional lines for handshaking—explicit indications of data ready from the sender and ready for data from the receiver.
Point to point or bus/network	The other key characteristic of hardware interfaces is whether the interface is point to point (between the external computer and device/instrument) or some multiple-device interface. The point-to-point interfaces can be very simple. The multiple-device interfaces are sometimes implemented as a bus or a network. These bus or network interfaces require some form of device addressing. Many also need some form of conflict resolution to determine who currently has the right to send data. Network style interfaces are usually packet oriented, they send addressed blocks of data.
Synchronous or asynchronous	*Synchronous* refers to interfaces that use a clock (either a separate line or one that can be reconstituted from the signal) to determine when the data is to be read. *Asynchronous* refers to interfaces that do not have a separate clock signal, so they depend on the data coming at a relatively consistent rate. Synchronous interfaces are generally faster than asynchronous interfaces.
Speed	The raw speed of the interface is a key characteristic. This is normally specified in *baud* (signaling elements per second) or *bps* (bits per second) for serial interfaces and Mbits (millions of bits per second) for network interfaces. Asynchronous serial interfaces tend to be in the 110-bps to 56-Kbps range. Network interfaces tend to be in the 1-Mbit to 100-Mbit range.

10.9.2 Hardware interface standards

For the common hardware interfaces used with instruments, see Table 10.14.

10.9.3 Software protocol standards

The interfaces in Table 10.14 provide the hardware interface between devices and computers. The physical layer is necessary, but not sufficient. To actually exchange information, the devices (and/or computers) require defined ways

TABLE 10.14 Common Instrument External Interfaces

Interface	Description	Serial/parallel	Point/network
IEEE 488.1, *Hewlett-Packard Instrument Bus* (HP-IB), *General-Purpose Instrument Bus* (GPIB)	The IEEE 488 interface has been one of the primary instrument interfaces. It allows (electrically) up to 15 devices up to 2 meters apart (each). It can achieve Mb/s transfer rates.	Parallel ■ Eight data bits ■ Three-wire handshake (DAV, NRFD, and NDAC) ■ Five additional lines (ATN, IFC, EOI, SRQ, and REN)	Bus ■ Up to 2 meters apart ■ 31 addresses ■ Talker ■ Listener(s) ■ Controller
RS-232C, EIA-232-D, RS-422A, RS-449, RS-530, CCITT V.24, IEEE 1174, MIL-188C	*RS-232C* (Recommended Standard 232C) is the venerable serial interface from the Electronic Industries Association (EIA). Almost all computers have some form of RS-232 interface. An ongoing challenge with RS-232 is configuration, which is not defined. This configuration normally includes data rate, number of stop bits, connector size (9 pin, 15 pin), connector style (male or female), hardware handshake settings, which pin has the transmit data and which pin has the receive data (pin 2 and 3 problems). The other interfaces on the left are variations of the RS-232 standard with variations for data rates, configuration and connectors. Data rates for RS-232 are up to 56 Kb/s. The variants can get above 10 Mb/s.	Serial ■ 2 data lines: transmitted data: TD received data: RD ■ 6 optional additional lines (RTS, CTS, DTR, DSR, DCD, and RI)	Point to point ■ Up to ~75 feet apart
Centronics, EPP, ECP, IEEE 1284	The Centronics interface is a standard output (one-direction) parallel interface used for connecting computers and instruments to printers or other peripherals. It has also been used as an instrument interface (connecting the computer to the instrument). The original interface has been extended to the ECP (Extended Capabilities Port) and EPP (Enhanced Parallel Port). IEEE Std. 1284-1994 (Standard Signaling Method for a Bi-directional Parallel Peripheral Interface for Personal Computers) defines a superset that supports Centronics, ECP, EPP and two other parallel interface modes (nibble and bidirectional). The EPP and ECP extensions are both bidirectional interfaces that are about 10 times faster. In general, data rates tend to be on the order of 50–100 Kb/s, but it is possible to achieve up to 2 Mb/s.	Parallel ■ Eight data bits ■ Three-wire handshake (Strobe, Busy, and Acknowledge) ■ Six additional lines (Paper out, Error, Auto Feed, Select In, Select Out, and Initialize)	Point to point ■ Up to 10 meters apart

TABLE 10.14 Common Instrument External Interfaces (*Continued*)

Interface	Description	Serial/parallel	Point/network
Universal Serial Bus	USB (Universal Serial Bus) is an interface between a PC and devices (targeted at consumer peripherals). The USB peripheral bus is a multi-company industry standard. USB supports a data speed of 12 megabits per second.	Serial ■ One differential data line pair: data− & data+ ■ Supplied voltage	Network ■ Up to ~3 to 4 meters apart ■ 127 addresses (not including bridges or extenders) ■ Peer to peer
FireWire (IEEE 1394)	FireWire (Apple's version) or IEEE 1394 High-Performance Serial Bus is another interface between a PC and devices. It provides a single plug-and-socket connection on which as many as 63 devices can be attached with data transfer speeds up to 400 Mb/s (and eventually more). It is intended for high-performance devices.	Serial ■ Two differential data line pairs: TPA, TPA# TPB, TPB# ■ Supplied voltage	Network ■ Up to 4.5 meters apart ■ 63 addresses (not including bridges or extenders) ■ Peer to peer
Ethernet or LAN (IEEE 802.3)	Ethernet is a LAN protocol and hardware specification developed in 1976 for communication between computers. It supports transfer rates of 10 Mb/s. Although originally on medium to large computers, it is available on PCs and is also implemented in several instruments. 100Base-T is one of several variations of Ethernet that supports 100 Mb/s.	Serial	Network ■ Up to ~100 to 500+ meters apart ■ 2^{48} addresses

to communicate, called *protocols*. If a designer is defining and building both devices that communicate, it is possible to define special protocols (simple or complex). However, most devices need to communicate with standard peripherals and computers that have predefined protocols that need to be supported. The protocols can be very complex and layered (one protocol built on top of another). This is especially true of networked or bus devices. Some of the common protocols used in instruments are shown in Table 10.15. A complete description of interface protocols is beyond the scope of this chapter.

10.10 Numerical Issues

The hardware used to construct microprocessors is only capable of manipulating ones and zeros. From these, the microprocessors build up the ability to represent integers and real numbers (and characters for that matter). Microprocessors represent all information in a binary form.

TABLE 10.15 Common Instrument Software Protocols

Software protocol	Description
IEEE 488.2	The *IEEE 488.2, Codes, Formats, Protocols and Common Commands for Use with IEEE 488.1* is a specification that defines: 39 common commands and queries for instruments, the syntax for new commands and queries, and a set of protocols for how a computer and the instrument interact in various situations. Although this is a companion to the IEEE 488.1 interface, it is independent of the actual interface, but it does depend on certain interface characteristics.
SCPI	The *Standard Commands for Programmable Instruments* (*SCPI*) is a specification of common syntax and commands so that similar instruments from different vendors can be sent the same commands for common operations. It also specifies how to add new commands that are not currently covered in the standard.
TCP/IP	The *Transmission Control Protocol/Internet Protocol* (*TCP/IP*) specification is the underlying protocol used to connect devices over network hardware interfaces. The network hardware interface can be a LAN or a WAN.
FTP	The *File Transfer Protocol* (*FTP*) specification is a protocol used to request and transfer files between devices over network hardware interfaces.
HTTP	The *HyperText Transfer Protocol* (*HTTP*) specification is a protocol used to request and transfer Web (HyperText Markup Language, HTML) pages over network hardware interfaces.
VXI-11	The VXI-11 plug-and-play specification is a protocol for communicating with instruments that use the VXI-bus (an instrument adaptation of the VME bus), GPIB/HP-IB, or a network hardware interface.

10.10.1 Integers

Integers are directly represented as patterns of ones and zeros. Each bit corresponds to a subsequent power of two, from right to left. So, "0101" is equal to $2^2 + 2^0$ or 5. Most operations on integer data in microprocessors use unsigned binary or two's complement representation. The unsigned binary has the left-most bit representing a power of two; in a 16-bit number, a 1 in the left-most bit has the decimal value of 32768. In unsigned 16-bit binary, "1000 0000 0000 0001" is equal to $2^{15} + 2^0$ or 32769. The two's complement binary has the left-most bit representing a negative number. Again, in a 16-bit number, the left-most bit has the value of −32768, which is then added to the rest of the number.[4] So, in two's complement 16-bit binary, "1000 0000 0000 0001" is equal to $-2^{15} + 2^0$ or −32768 + 1, which is −32767. The range of numbers in 16-bit binary two's complement and unsigned representation is shown in Figure 10.5.

[4] This can also be viewed that a 1 in the left-most bit of a two's complement number indicates a negative number where the rest of the bits need to be flipped and 1 added to the result.

Number	Two's complement
32767	0111 1111 1111 1111
32766	0111 1111 1111 1110
32765	0111 1111 1111 1101
...	...
2	0000 0000 0000 0010
1	0000 0000 0000 0001
0	0000 0000 0000 0000
-1	1111 1111 1111 1111
-2	1111 1111 1111 1110
...	...
-32766	1000 0000 0000 0010
-32767	1000 0000 0000 0001
-32768	1000 0000 0000 0000

Number	Unsigned
65535	1111 1111 1111 1111
65534	1111 1111 1111 1110
65533	1111 1111 1111 1101
...	...
32769	1000 0000 0000 0001
32768	1000 0000 0000 0000
32767	0111 1111 1111 1111
32766	0111 1111 1111 1110
32765	0111 1111 1111 1101
...	...
2	0000 0000 0000 0010
1	0000 0000 0000 0001
0	0000 0000 0000 0000

Figure 10.5 A 16-bit two's complement and unsigned binary representation.

10.10.2 Floating-point numbers

Given the nature of the vast majority of instruments, representing and operating on real numbers is necessary. Real numbers can be fixed point (a fixed number of digits on either side of the decimal point) or floating point (with a mantissa of some number of digits and an exponent).

In the design of microprocessor arithmetic and math operations, numbers have a defined number of bits or digits. The value of floating-point numbers is that they can represent very large and very small numbers (near zero) in this limited number of bits or digits. An example of this is the scientific notation format of the number 6.022 1023. The disadvantage of floating point numbers is that, in normal implementation, they have limited precision. In the example, the number is only specified to within 1020, with an implied accuracy of ±5 1019.

Although there are several techniques for representing floating-point numbers, the most common is the format defined in IEEE 754 floating-point standard. The two formats in common use from this standard are a 32-bit and a 64-bit format. In each, a number is made up of a sign, a mantissa, and an exponent. A fair amount of complexity is in the IEEE floating-point standard. The following information touches on the key aspects. The aspects of IEEE floating-point representation are beyond the scope of this chapter. In particular, a variety of values represented indicate infinity, underflow, and not a

Number	bit 31 Sign	bits 30-23 Exponent	bits 22-0 Mantissa
+ Not a Number (+NaN)	0	1111 1111	Non-zero
+ infinity	0	1111 1111	0000 0000 0000 0000 0000 000
...	0	...	...
+ 3.2767×10^4 (+32767)	0	1000 1101	1111 1111 1111 1100 0000 000
...	0	...	...
+ 0.125×10^0 (+1/8)	0	0111 1100	0000 0000 0000 0000 0000 000
...	0	...	...
+ 0	0	0000 0000	0000 0000 0000 0000 0000 000
- 0	1	0000 0000	0000 0000 0000 0000 0000 000
...	1	...	...
- 0.125×10^0 (-1/8)	1	0111 1100	0000 0000 0000 0000 0000 000
...	1	...	...
- 3.2767×10^4 (-32767)	1	1000 1101	1111 1111 1111 1100 0000 000
...	1	...	...
- infinity	1	1111 1111	0000 0000 0000 0000 0000 000
- Not a Number (-NaN)	1	1111 1111	Non-zero

Figure 10.6 A 32-bit IEEE 754 floating-point binary representation.

number. For more-detailed information on these topics (and more), refer to the IEEE standard.

For the 32-bit format (see Fig. 10.6), the exponent is 8 bits, the sign is one bit, and the mantissa is 23 bits. If the exponent is nonzero, the mantissa is normalized (i.e., it is shifted to remove all the zeros so that the left-most digit is a binary 1 with an implied binary radix point following the left-most digit). In this case, the left-most digit in the mantissa is implied, so the mantissa has the effect of a 24-bit value, but because the left-most digit is always a 1, it can be left off. If the exponent is zero, the unnormalized mantissa is 23 bits.

The exponent for 32767 is slightly different than expected because the exponent has a bias of 127 (127 is added to the binary exponent). The 64-bit format has an 11-bit exponent (with a bias of 1023) and a 52-bit mantissa.

number	Fixed point scaling
7.9375	0111 1111
7.8750	0111 1110
...	...
0.0625	0000 0001
0.0000	0000 0000
-0.0625	1111 1111
...	...
-7.9375	1000 0001
-8.0000	1000 0000

Figure 10.7 An 8-bit fixed-point scaling binary representation example.

number	Arbitrary scaling
error	1111 1111
...	...
error	1111 1011
120.0	1111 1010
119.5	1111 1001
...	...
0.5	0000 1011
0.0	0000 1010
-0.5	0000 1001
...	...
-4.5	0000 0001
-5.0	0000 0000

Figure 10.8 An 8-bit fixed-point arbitrary-scaling binary-representation example.

10.10.3 Scaling and fixed-point representations

Microprocessors deal with floating-point numbers—either through built in floating-point hardware, a floating-point coprocessor, or through software. When software is used, there are significant performance penalties in speed and accuracy; it might take thousands of instructions to perform a floating-point operation. When the microprocessor has floating-point capability (directly or through a coprocessor), there is added cost. So, in most instrumentation applications, integers are used whenever possible. Often, instrument developers will use integer scaling (an implied offset or multiplier) to allow for some of the benefits of floating point with integer performance.

Fixed-point number representation is one example of this technique (see Fig. 10.7). In this case, an implied radix point is at a programmer-defined location in the binary number. For example, a pressure sensor needs to represent pressure variations in the range of −4 to +6 Pascals in steps of 1⁄16th Pascal. A good fixed-point representation of this is to take a two's complement signed 8-bit binary number, and put the implied radix point in the middle. This gives a range of −8 to 7.9375 with a resolution of 1⁄16th.

It is also possible to have arbitrary scaling (see Fig. 10.8). A specific example of this is a temperature measurement application. The thermocouple used can measure −5 degrees Celsius to 120 degrees Celsius, a range of 125 degrees. The thermocouple has an inherent accuracy of ½ degree Celsius. The analog hardware in the instrument measures the voltage from the thermocouple and the embedded computer in the instrument converts it into a temperature reading. It would be possible to store the temperature as a 64-bit floating-point number (taking additional computing power and additional storage space). It is also possible to store the temperature as a single integer: 125 degrees × 2 for the degree accuracy means that 250 distinct values need to be represented. The integer can be an unsigned 8 bit number (which allows for 256 values).

MSB			LSB
0000 0000	0000 0000	0000 1000	0000 0010
2^{31} 2^{24}	2^{23} 2^{16}	2^{15} 2^{8}	2^{7} 2^{0}

Figure 10.9 A 32-bit integer for big- and little-endian examples.

10.10.4 Big-endian and little-endian

Big-endian and *little-endian*[5] refer to the byte order of multi-byte values in computer memory, storage and communication (so this applies to computers and to protocols). The basis for the differences is determined by where the *least-significant bit* (*LSB*) and the *most-significant bit* (*MSB*) are in the address scheme. If the LSB is located in the highest address or number byte, the computer or protocol is said to be *big-endian*. If the LSB is located in the lowest address or number byte, the computer or protocol is said to be *little-endian*.

To illustrate this, look at a segment of computer memory (Fig. 10.9) that contains the 32-bit integer value, 2050. This value, put in a big-endian computer's memory would look like that in Fig. 10.10. This value, put in a little-endian computer's memory would look like Fig. 10.11.

address	data	
0	0000 0000	MSB
1	0000 0000	
2	0000 1000	
3	0000 0010	LSB

Figure 10.10 A 32-bit big-endian memory layout.

Although very arbitrary, this is a serious problem. Within a computer, this is typically not an issue. However, as soon as the computer is transmitting or another external computer is accessing binary data, it is an issue because a different type of computer might be reading the data. Mainframes tend to be big-endian. PCs and the Intel 80 × 86 and Pentium families are little-endian. Motorola 68000 microprocessors, a popular embedded microprocessor family, are big-endian. The PowerPC is unusual because it is *bi-endian*; it supports both big-endian and little-endian styles of operation. Operating systems provide for conversions between the two orientations and the various network and communication protocols have defined ordering. Instrument users are not

[5] The terms *big-endian* and *little-endian* are drawn from the Lilliputians in *Gulliver's Travels,* who argued over which end of soft-boiled eggs should be opened.

address	data	
0	0000 0010	LSB
1	0000 1000	
2	0000 0000	
3	0000 0000	MSB

Figure 10.11 A 32-bit little-endian memory layout.

normally aware of this issue because developers deal with the protocols and conversions at the application software and operating system level.

10.11 Instrumentation Calibration and Correction Using Embedded Computers

Instruments normally operate in an analog world. This analog world has characteristics (such as noise and nonlinearity of components) that introduce inaccuracies in the instrument. Instruments generally deal with these incorrect values and try to correct them by using software in the embedded computer to provide calibration (adjusting for errors inside the instrument) and correction (adjusting for errors outside of the instrument).

10.11.1 Calibration

Calibration in the embedded computer adjusts the system to compensate for potential errors within the instrument and the instrument's probes. Embedded computer-based calibration makes hardware design much easier. In the simple case, the embedded computer can apply a calibration to correct for errors in hardware. Hardware is simpler because a linear circuit (with a reproducible result) can be corrected to the appropriate results. The calibration software in an instrument can deal with both the known inaccuracies in a family of instruments and also for the characteristics of a single instrument.

A good example of calibration in instruments is the use of Digital to Analog Converters (DACs) to calibrate analog hardware in an instrument. A DAC takes a digital value from the embedded computer and converts it to an analog voltage. This voltage is then fed into the analog circuitry of the instrument to compensate for some aspect of the instrument hardware. In a conventional oscilloscope, a probe has an adjustment to match the probe to the input impedance of the oscilloscope. A user connects the probe to a reference square-wave source and then adjusts the probe until the trace appearing on the screen is a square wave. In oscilloscopes with embedded computers, this compensation is done automatically by replacing the adjustment on the probe with a DAC and having the embedded computer adjust the DAC until it has achieved a square wave. In many cases, such an automatic adjustment is more accurate. (To fully automate this, the embedded computer also needs to switch the input between the reference input and the user's input.)

Another common example of calibration is *auto zeroing*. Many measurements can be improved by eliminating offsets. However, the offsets can be tremendously variable, depending on many factors, such as temperature, humidity, etc. To achieve this, such measuring devices as a voltmeter will alternate between measuring the user's input and a short. The embedded computer can then subtract the offsets found in the zero measurement from the actual measurement, achieving a much more accurate result.

10.11.2 Linear calibration

Linear calibrations are one of the most basic and common algorithms used by embedded computers in instrumentation. Typically, the instrument will automatically measure a calibration standard and use the result of this measurement to calibrate further measurements. The calibration standard can be internal to the instrument, in which case, the operation can be completely transparent to the user. In some cases, the user might be prompted to apply appropriate calibration standards as a part of the calibration procedure.

A linear calibration normally requires the user to apply two known values to the input. The embedded computer makes a measurement with each input value and then calculates the coefficients to provide the calibrated output. Typically, one of the two known values will be at zero and the other at full scale. The linear calibration will be based on a formula like the following equation (Equation 10.1):

$$V_{\text{calibrated}} = \frac{V_{\text{raw}} - V_{\text{short}}}{m} \tag{10.1}$$

where: m = the calibration coefficient empirically derived from the device.

This simple formula is relatively quick to perform. However, sometimes, there are nonlinear errors in the measurement system. For example, a parabola in the response curve might indicate a second-order error. Even though a linear calibration will result in a better reading in this case, a linear calibration cannot correct a non-linear error. Many instruments address this problem with non-linear calibrations. This can be done using a higher-order calibration formula like the following equation (Equation 10.2):

$$V_{\text{calibrated}} = a \times V_{\text{raw}}^2 + b \times V_{\text{raw}} + c \tag{10.2}$$

An alternative to a polynomial is to use a piece-wise linear calibration. A *piece-wise linear correction* applies a linear correction, but the constants are varied based on the input. This allows a different correction to be applied in different regions of input. Piece-wise corrections have the advantage of not requiring as much calculation as the polynomial. Regardless of the technique used, high-order corrections require more calibration points and, therefore, a more-involved calibration procedure for the user.

10.11.3 Correction

Correction in the embedded computer adjusts values to correct for an influence external to the instrument (e.g., in the user's test setup). For example, network analyzers will typically compensate for the effects of the connection to the *device under test* (*DUT*) displaying only the characteristics of the device under test. Often, the correction is performed by having the instrument make a measurement on an empty fixture and then compensating for any effects this might have on the measured result. Correction also can be used to compensate for the effects of a transducer. Typically, the sensor in a radio-frequency power meter is corrected and when transducers are changed, a new correction table is loaded.

10.12 Using Instruments That Contain Embedded Computers

In the process of selecting or using an instrument with an embedded computer, a variety of common characteristics and challenges arise. This section covers some of the common aspects to consider:

- *Instrument customization*—What level of instrument modification or customization is needed?
- *User access to the embedded computer*—How much user access to the embedded computer as a general-purpose computer is needed?
- *Environmental considerations*—What is the instrument's physical environment?
- *Longevity of instruments*—How long will the instrument be in service?

10.12.1 Instrument customization

People who buy and use instruments sometimes want to modify their instruments. This is done for a variety of reasons. One of the most common reasons is extension: allowing the user to extend the instrument so that it can perform new instrument or measurement operations. Another very common reason is ease of use: customizing the instrument so that the user doesn't have to remember a difficult configuration and/or operation sequence. A less-common, but still important, customization is "limitation": this modifies the instrument by preventing access to some instrument functionality by an unauthorized user. Often, this type of customization is needed for safety or security reasons, usually in manufacturing or factory-floor situations.

Several common approaches are used in instruments to accommodate customization:

- Instrument configuration is the built-in mechanism to set the options supported by the instrument. These can include modes of operation, user language, external interfacing options, etc. User defined keys, toolbars, and

TABLE 10.16 Common User-Accessible Embedded Programming Languages

Language	Example	Comments

Basic	`10 Full = 127`
	`20 FOR I=1 TO 10`
	`30 MyData = Read(Port)`
	`40 IF MyData > Full GOTO MyError`
	`50 NEXT I`
		Basic has been one of the most common languages included inside instruments. It is easy to learn and use. It is almost always interpreted, which generally means there are performance limitations. Unfortunately, the many dialects of Basic have minor variations and extensions to the language.

Java	`int full = 127;`
	`for(int i = 1; i <= 10; i ++) {`
	`  int myData = read(port);`
	`  if (myData > full) throw(myError);`
	`}`
		Although Java can be used in the instrument software development, it can also be used as a user-accessible language. It is interpreted, so it also has performance limitations (but usually not as severe as Basic). To be useful, extensions to the language through class libraries are required. There are difficulties in that the extensions require access to the underlying instrument architecture which expose the inner workings (which are often proprietary).

C/C++	`int i, myData, full;`
	`full = 127;`
	`for(i = 1; i <= 10; i ++) {`
	`  myData = read(port);`
	`  if (myData > full) myError();`
	`}`
		Although C and C++ are very common instrument implementation languages, they can also be used in an instrument as a user-accessible language. They are very good at producing customizations that are small and fast. They are not very appropriate for end users because of the difficulty of programming. They are mostly useful for manufacturers and instrument system integrators. There are also difficulties in that the extensions need access to the underlying instrument architecture, which expose the inner workings (which are often proprietary).

menus are instrument user-interface options that allow a user to define key sequences or new menu items customized to suit their needs.

- Command language extensions are instrument external interface commands that the user can define. These are normally simple command sequences: macros or batch files. A *macro* is defined as a symbol or command that represents a series of commands, actions, or keystrokes.
- Embedded programming languages are mechanisms built into the instrument that allow the instrument to receive and run new software (either complete applications or extensions to the current applications). In some cases, these programs are entered from the front panel. In most cases, they are provided by the manufacturer on a storage media (3.5-inch floppy, PCMCIA memory card, etc.) or downloaded over an external interface (IEEE 488, RS-232, or LAN).

This modification set is also important to instrument manufacturers. It allows the creation of instrument personalities: modifying or extending instruments for custom applications or markets. Personalities are very effective for manufacturers because it allows the development of what is essentially a new product without going through the process and expense of a full release cycle. One aspect to keep in mind about all of these customization technologies is that they do inherently change the instrument so that it is no longer a "standard" unit. This might cause problems or have implications for service and also for use because the modified instrument might not work or act like the original, unmodified, instrument.

The user-accessible embedded programming language mechanism has been popular in instruments. Typical accessible languages are Basic, Java, and C or C++. Table 10.16 describes these languages. It also shows a simple example program in each language that reads some data from a port 10 times and checks for an error condition.

In addition to these standard languages, some instrument designers have created special-purpose languages to extend or customize the instrument. Similar to custom embedded development languages, there can be ease-of-use benefits to having a custom language. However, these custom languages come at the expense of user training, user documentation, and internal instrument language development and maintenance costs.

10.12.2 User access to an embedded computer

As described earlier, many instruments now use an embedded PC or workstation as an integral part of the system. However, the question arises: "Is the PC or workstation visible to the user?" This is also related to the ability to customize or extend the system.

Manufacturers get benefits from an embedded PC because there is less software to write and it is easy to extend the system (both hardware and software). The manufacturers get these benefits—even if the PC is not visible to the end user. If the PC is visible, users often like the embedded PC because it is easy to extend the system, the extensions (hardware and software) are less expensive, and they don't require a separate PC.

However, there are problems in having a visible embedded PC. For the manufacturer, making it visible exposes the internal architecture. This can be a problem because competitors can more easily examine their technologies. Also, users can modify and customize the system. This can translate into the user overwriting all or part of the system and application software. This is a serious problem for the user, but is also a support problem for the manufacturer. Many instrument manufacturers that have faced this choice have chosen to keep the system closed (and not visible to the user) because of the severity of the support implications.

The user or purchaser of an instrument has a choice between an instrument that contains a visible embedded PC and an instrument that is "just"

TABLE 10.17 Environmental Considerations for Embedded Computers

Environmental characteristic	Description
Temperature and humidity	The embedded computer and its internal peripherals have to withstand the temperature and humidity of the instrument environment. Often, a fan is required, given the current power used by embedded computers. Fans themselves can cause issues with failure, dust, and blockage, which lead to overheating problems.
Shock and vibration	Many instruments need some form of mass-storage device. Most of these devices are very sensitive to shock. It is often necessary to use special mounting or to use an alternative device (such as flash memory).
Operator interface	If the system is installed in a challenging environment, the human interface has to operate with those aspects. For example, the system could be located near high amounts of dust, dirt, industrial chemicals, industrial gases, coffee, soft drinks, untrained operators, etc. The human-interface components that need to survive these threats include the keyboard, pointing devices, display, switches, knobs, and removable media (such as floppy disks).
Power quality	Often, the quality of power at the measurement site is suspect or poor. Most instrument power supplies are very carefully designed because of the analog aspects of the system. If the embedded computer has a separate power supply, it needs to be properly regulated, taking the power quality into account. This is especially true of embedded PCs or workstations.
Power fail	A separate issue from power quality is power failure. For simpler instruments, this tends to be less of a problem. However, for more and more instruments, the embedded computer makes power failures a real challenge. These more complex systems have a variety of issues: configuration memory in the middle of modification, buffers that need to be written to disk, etc.
Radio-frequency interference	RFI is a problem both from and to the embedded computer. Most embedded computers are operating at clock frequencies of tens and hundreds of MHz, with wide buses (transmission lines). This causes high levels of generated RFI within, and potentially outside of the box. The designer needs to ensure that the embedded computer does not affect the instrument function or the function of other instruments nearby. Similarly, the computer is susceptible to RFI generated by the rest of the instrument and by devices outside of the instrument.

an instrument (independent of whether it contains an embedded PC). It is worth considering how desirable access to the embedded PC is to the actual user of the instrument. The specific tasks that the user needs to perform using the embedded PC should be considered carefully. Generally, the instrument with a visible embedded PC is somewhat more expensive.

TABLE 10.18 The Implications of Long-Term Instrument Service

Microprocessor	The microprocessor chip used in the instrument needs to be available for the manufacture of the instrument through the life of the product. Also, chips must be available for maintenance.
Language	If the languages used for developing or customizing the instrument are in the mainstream, they will probably be around later in the instrument's life. Custom and nonmainstream languages will become support challenges for the manufacturer and the user.
Firmware/software	Software (applications, operating systems, and utility software) will have defects and changes. Instruments will usually get different software over the life of a production run. Customer units will often be updated with these to enable general improvements or to fix major defects. This can, over time, become a serious problem for a variety of reasons: ■ Is enough memory available for the upgrade? ■ Is enough disk or flash space available for the upgrade? ■ What software version is the instrument being upgraded from? ■ What is the hardware configuration of the instrument? ■ Where are the upgrades for an old instrument archived?
Interfaces	Over time, computer and instrumentation interfaces change. For example, BCD interfaces were previously not uncommon for instruments. Some good questions to ask are: ■ What performance is needed over the instrument's life? ■ What will be common on PCs over time? ■ Will the interface choice (hardware and drivers) be available on PCs over time?
Removable media	As with interfaces, computer media changes, as well. For example, 5.25″ floppy disk media was a popular choice. It is worth considering what is media choice over the life of the instrument, taking into account size, cost, capacity, writability, reliability, and long-term viability of the media format.
People	It is very important to adequately document the instrument design, support, manufacture, customization, and use. Over the course of decades, people move on and knowledge can be forgotten or lost (or even destroyed).

10.12.3 Environmental considerations

The embedded computer is (by definition and design) inside the instrument. Therefore, the computer has to survive the same environment as the instrument. Although many instruments are in "friendly" environments, this is certainly not always true. Some of the specific factors that should be considered for the embedded computer by the instrument user and designer are described in Table 10.17.

10.12.4 Instrument longevity

Instruments tend to be manufactured and used for very long periods of time. Unlike the consumer or personal computer markets, it is not unusual for a popular instrument to be manufactured for 15 years and in service for 20 years or longer. The implications on the embedded computer in the instrument for the users and manufacturers are shown in Table 10.18.

Chapter

11

Power Supplies

James S. Gallo
Agilent Technologies
Rockaway, New Jersey

11.1 Function and Types of Power Supplies and Electronic Loads

A regulated power source provides electrical energy which is precisely controlled. The direct-current (dc) source converts electrical energy from the commercial electrical power distribution system from alternating voltage to a tightly controlled source of constant voltage or constant current. The alternating-current (ac) source converts an unregulated source of electrical energy to a regulated source of alternating current. The ac power source usually has the means to vary both the amplitude and frequency of its output. In addition, many ac sources available also can simulate disturbances in the power line waveform and make measurements of line current, power, and power factor.

The "electronic load" is an instrument capable of absorbing direct current in a controlled fashion. It can function as a variable current sink, variable power resistor, or shunt voltage regulator; i.e., it will maintain a fixed voltage as it absorbs a variable current.

The instruments referred to above and in this chapter all use solid-state semiconductor devices to regulate or control sourcing or absorption of energy. Typically, they are available commercially with power ratings from tens of watts to tens of kilowatts, with maximum voltages from a few volts to tens of kilovolts and with maximum currents from milliamperes to several thousand amperes. These power instruments are more than sources of power. They are instruments that can be used on laboratory benches or automated test systems as sources of precisely controlled power. They also can make precision measurements of current, power, and voltage and can communicate these as well as instrument status to operators and computers. The following sections describe these capabilities.

11.2 The Direct-Current Power Supply

Nearly all electronic products manufactured today require a source of direct current for operation and usually have internal fixed output voltage sources to provide bias and reference voltages for the internal circuit elements. The fixed-voltage source is a subset of the typical power supply electronic instrument that provides a dc voltage that can be set to a desired value within a specified range. Typical applications for dc power supplies are for biasing electronic circuits under development in a design laboratory, powering lasers or electromagnets, testing motors, testing electronic or electromechanical devices (e.g., disk drives) in a production environment, charging batteries, dc metal plating chemical processes, and energizing special gas discharge lamps.

11.2.1 Direct-current voltage sources

The simplest dc source consists of a transformer and set of rectifiers and filters to convert a source of alternating current to direct current. The circuit and output voltage waveform is shown in Fig. 11.1. This type of output voltage is not adequate for many applications because of the large ripple voltage and because the output value will vary with changes to the input ac voltage

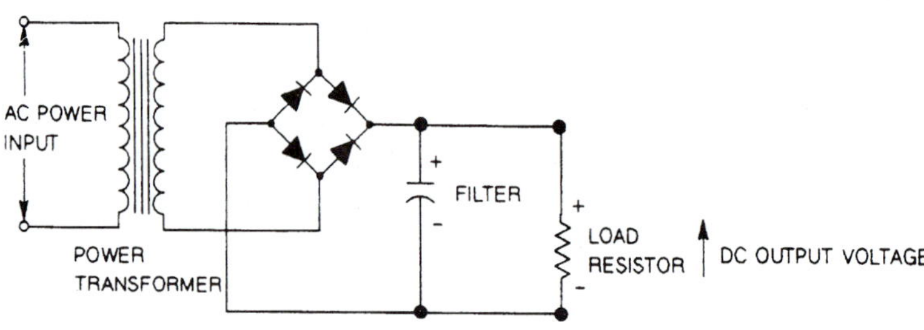

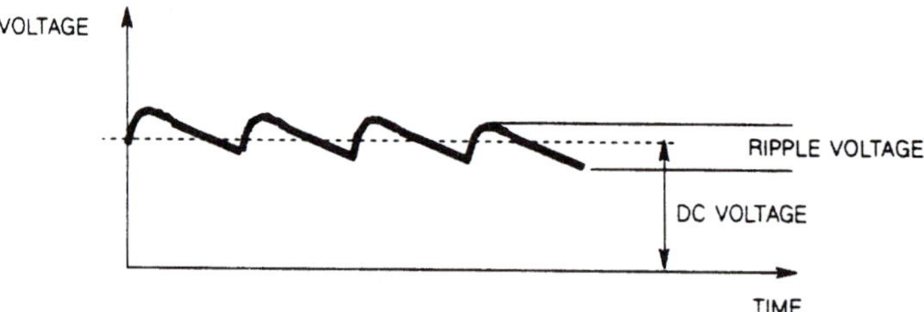

Figure 11.1 Simple dc source.

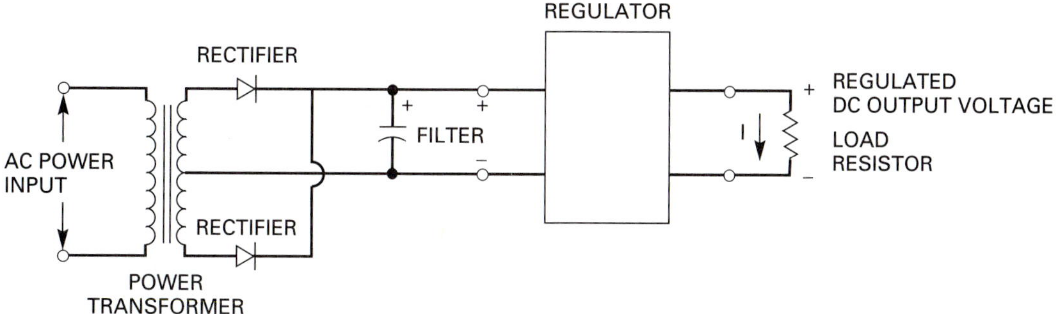

Figure 11.2 Regulated dc source.

magnitude, load resistance, and ambient temperature. A regulator is placed between the load being powered and the unregulated source to remove the ripple voltage shown in Fig. 11.1 and to control the magnitude of the output voltage. Figure 11.2 illustrates the conversion from the unregulated source to the regulated source. The regulator in Fig. 11.2 is shown as a simple "black box" or four-terminal network whose function is to provide a ripple-free and precisely controlled output voltage. Details on the types of regulators and their operation are provided later in this chapter. At this point we shall focus on the function of the regulator and the resulting output voltage and current characteristics.

The ideal constant-voltage source will maintain a constant output voltage regardless of load current, ac input voltage, or temperature; however, every practical instrument has current and output power limitations. Therefore, well-designed instruments have regulators that limit the output current and voltage in some manner.

11.2.2 Constant-voltage/constant-current or current-limiting sources

Figure 11.3 shows the voltage/current characteristics of different types of regulators. Figure 11.3*a* shows the voltage/current characteristics of an ideal current source. The current is maintained at its set point I_{out} by the regulator regardless of load impedance. Whether there is a short circuit or a finite load, the regulator will adjust output voltage to maintain I_{out} constant. For most current sources, the set-point current I_{out} may be adjusted to any desired value up to the maximum value I_{max}. Current sources have various applications. One example is in the testing of zener diodes. The diode is connected to the current source, and the source is varied while measuring the voltage across the zener. Another application is in testing of the current gain of a power transistor. Figure 11.4 shows how a power transistor's characteristics are determined by using a constant-voltage source E_S and constant-current source I_S and measuring I_C versus I_b for various values of E_S.

Figure 11.3*b* shows the voltage/current characteristics of the ideal con-

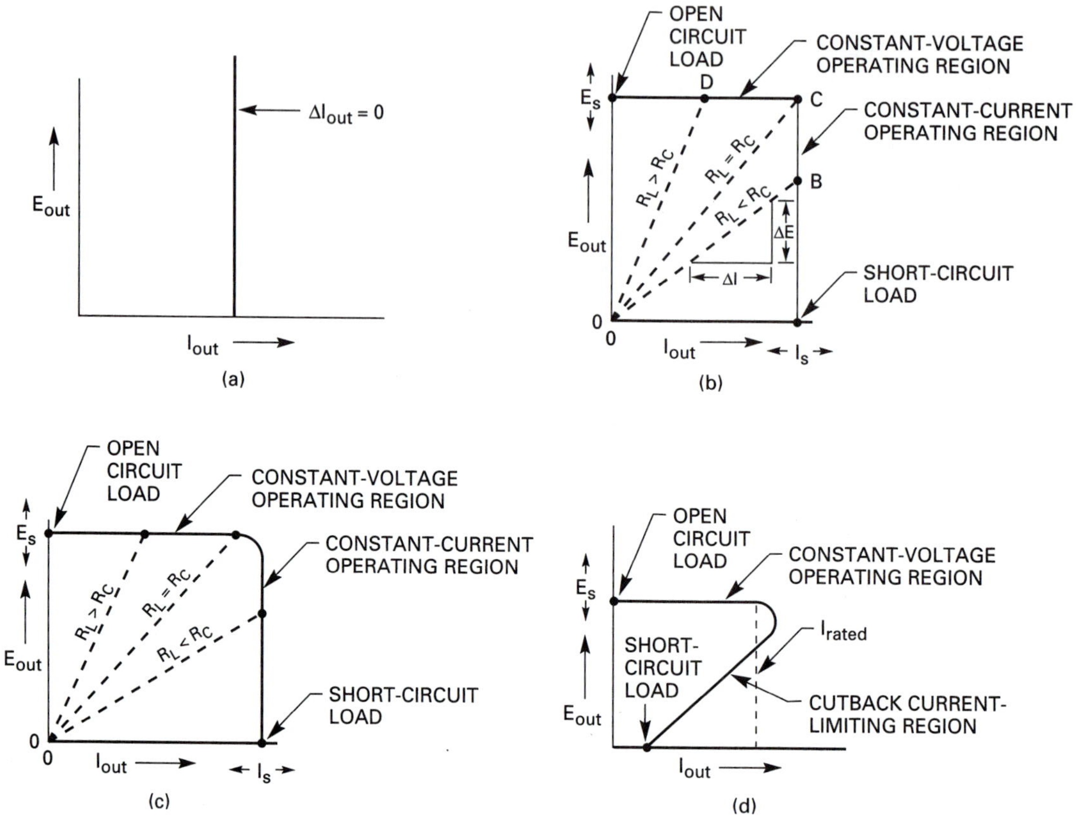

Figure 11.3 (a) Ideal constant-current power supply output characteristics; (b) ideal constant-voltage/constant-current source characteristics; (c) constant-voltage/current-limiting source characteristics; (d) constant-voltage/current-cutback source characteristics.

stant-voltage/constant-current source. In this type of regulator, the output voltage E_S and output current I_S are selected by the operator controls. The regulator will operate in the constant-voltage mode for load resistances greater than R_C (such as point D) and constant-current mode for load resistances less than R_C (such as point B). R_C is the "critical resistance," defined

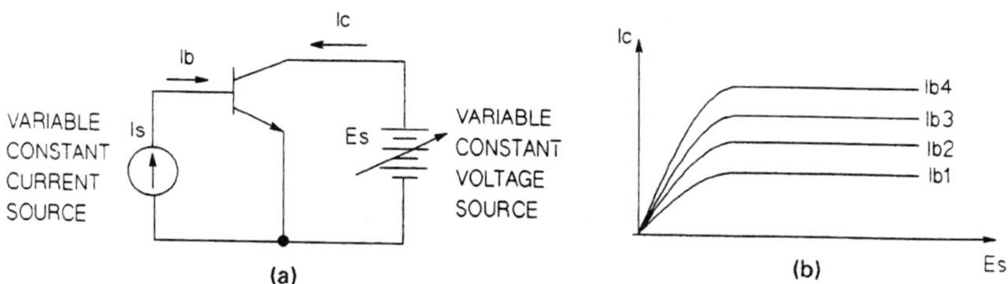

Figure 11.4 Using constant-current (a) and constant-voltage (b) sources to determine transistor current gain.

as $R_C = E_S/I_S$. R_C is not fixed but is dependent on the voltage and current settings of the power supply. Figure 11.3*c* shows the voltage/current characteristics of the constant-voltage/current-limiting power supply. This characteristic is used to protect the power supply from overload. The current-limit set point is usually a fixed value and cannot be changed by the operator. Because a current-limiting supply uses less sophisticated circuitry in the current-regulating loop, regulation is of poorer quality than in true constant-current operation. This is reflected by the finite slope of the current-limiting curve in this figure. Figure 11.3*d* shows the constant-voltage/cutback-current-limiting output characteristic. This characteristic is designed to protect the load being powered and the power supply. It is usually used with fixed-output power supplies. The crossover current I_{rated} is fixed by the maximum current that can be supplied by the power supply. The short-circuit current is typically 10 to 20 percent of I_{rated} and is the maximum current that will flow with a short across the power supply output terminals. This characteristic is useful when powering a relatively fixed distributed load such as many circuits on a printed circuit board assembly. In the event that a short occurs in one location on the board, all the current flows through the short. This will cause local heating that may damage the board beyond repair. By limiting the current to a small percentage of rated current, the load is protected during short-circuit conditions.

11.3 The Electronic Load

The "electronic load" is an instrument that functions as a programmable electrical energy absorption device. The primary application is for testing dc power supplies; however, it is also used in applications such as battery testing during manufacturing or research and development, solid-state semiconductor power component testing, dc motor testing, dc generator testing, and testing of solid-state motor controls. The electronic load typically has a high output impedance that allows the output voltage and current to change rapidly. Since the electronic load absorbs energy, it is often referred to as a "current sink." Typically, electronic load ratings vary from tens of watts to kilowatts, with current ratings from amperes to hundreds of amperes and voltage ratings from several volts to nearly 1 kV.

11.3.1 Modes of operation

The electronic load has three modes of operation: constant current, constant voltage, and constant resistance. In all three, the device is capable of absorbing significant power. A simplified circuit for an electronic load is shown in Fig. 11.5. In this circuit, the power field effect transistor T_1 functions as the energy absorption element, V_{ref} is the reference voltage source, and A is an operational amplifier. The battery E_L represents the external source under test, and R_L represents the source internal resistance.

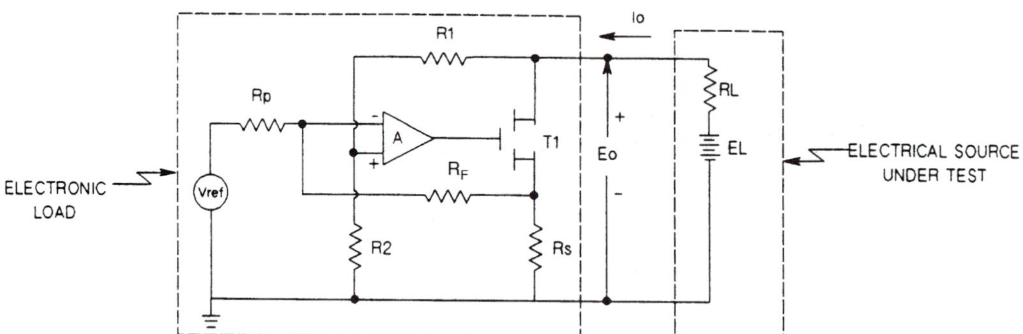

Figure 11.5 Simplified electronic load circuit.

Constant-current mode. In this mode, the electronic load acts as a controlled current sink. Referring to Fig. 11.5, R_1 is infinite (open circuit) in this mode of operation. Assuming 0 V across the operational amplifier A input terminals and the R_P and $R_F >> R_S$ then

$$\frac{I_o R_S}{R_F} = \frac{-V_{\text{ref}}}{R_P} \tag{11.1}$$

$$I_o = \frac{-V_{\text{ref}}}{R_S} \frac{R_F}{R_P} \tag{11.2}$$

Therefore, the magnitude of the load current I_o may be controlled by varying V_{ref} or R_F. Typically, R_P is a fixed resistor and R_S (which is a precision, very low value, high-power resistor with a very low temperature coefficient of resistance) is used to measure the output current.

Constant-resistance mode. In this mode of operation, R_1 is included to provide voltage feedback, and the reference voltage V_{ref} is replaced by a short circuit. The relationship of interest when making the same assumptions as above for R_P, R_F, and R_S is

$$E_o \cdot \frac{R_2}{R_1 + R_2} = I_o R_S \cdot \frac{R_P}{R_F + R_P} \tag{11.3}$$

or

$$\frac{E_o}{I_o} = R_o = R_S \cdot \frac{R_P}{R_2} \cdot \frac{R_1 + R_2}{R_F + R_P} \tag{11.4}$$

$$R_o = \frac{R_S R_P}{R_2} \tag{11.5}$$

where $R_1 + R_2 = R_F + R_P$

This means that the equivalent output resistance R_o is a linear function of R_P. Thus a programmable power resistor is obtained.

Constant-voltage mode. In this mode the electronic load acts as a shunt regulator; i.e., it will continue to increase the current I_o until the terminal voltage E_o drops to the desired value. In this mode, the circuit is reconfigured to connect R_P to the noninverting input of amplifier A. R_F and R_2 are also removed from the circuit, and the inverting input of amplifier A is grounded. The output voltage then becomes

$$E_o = -\frac{R_1}{R_P} \cdot V_{\text{ref}} \tag{11.6}$$

The output voltage may be linearly controlled by varying R_1 or V_{ref}.

11.3.2 Electronic load ratings

Since the electronic load absorbs power, its rating is limited by the maximum absorbed power as well as the maximum current and maximum voltage. In addition, most electronic loads are derated in current when the terminal voltage across the load falls below 3 V. A typical maximum operating characteristic for a 300-W, 60-V, 60-A electric load is shown in Fig. 11.6. The maximum voltage and current are shown as limited to 300 W. Provided the locus of voltage-current operating points lies within or on this boundary, the elec-

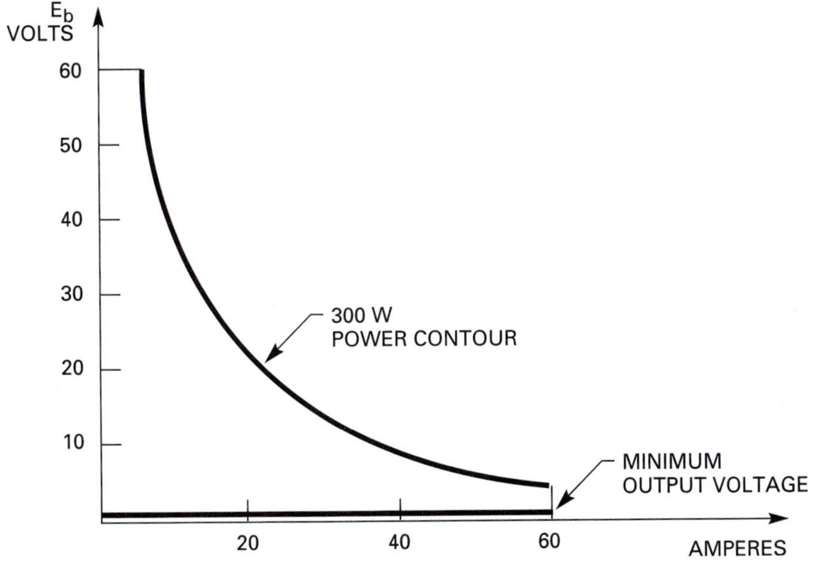

Figure 11.6 Electronic load output characteristics.

tronic load will operate as defined by the preceding equations for the various modes: constant voltage, constant current, or constant resistance.

11.4 The Alternating-Current Power Source

Modern ac power sources are complex instruments that produce and measure ac power. These instruments are used to evaluate how electronic or electromechanical products such as power supplies (in televisions or computers), ac motors, appliances (such as air-conditioners or refrigerators), or complete systems (such as aircraft avionic systems) behave under different conditions of ac voltage magnitude or waveform.

These instruments are rated by voltampere capability and for the preceding applications range in maximum VA from several hundred to several tens of thousands of voltamperes. Output voltages range up to 300 V for both single- and three-phase products. Frequencies of output power cover the range of 45 to 1000 Hz. The basic block diagram and circuits in this instrument are covered in more detail in Sec. 11.5.

11.4.1 Key features and modes of operation

The modes of operation and key features available in these instruments are related to the types of disturbances typically found on ac power lines. During the design and development of any product using ac power, the designer must consider the types of disturbances and the effect they have on the product, the amount of power used, and the efficiency of the product. In addition, the effect that the product has on the power distribution system is important. The features and modes allow the designer to evaluate these effects.

Types of power line disturbance features and modes

Steady-State Source Mode

Voltage amplitude variation

Voltage frequency variation

Voltage waveform variation

Voltage phase variation

Transient Source Mode

Transient voltage surges. Voltage rises above the steady-state value for a specified time.

Voltage cycle dropout. Any number of whole half cycles and/or any portion of a half cycle of voltage is eliminated.

Turn-on phase control. This allows the user to apply the initial turn-on voltage to the product under test at any desired phase angle of the voltage waveform.

Measurement capabilities. Typically, ac source instruments have the capability to measure the following:

Real power	Peak repetitive current
Reactive power	RMS current
Power factor	RMS voltage
Voltamperes	Peak voltage
Peak transient current	Harmonic analysis of input current

The last measurement is important because some products draw current in pulses. This produces higher-frequency current components, which produce voltage waveform disturbances on the power line that affect other equipment. The harmonic current also adversely affects the power distribution system wiring and other components due to increased heating. Both users of equipment and manufacturers are becoming more sensitive to this and are designing and purchasing equipment that does not produce harmonic power line currents.

11.5 General Architecture of the Power-Conversion Instrument

The basic functions in all power-conversion instruments are shown in Fig. 11.7. There are three main functions in this type of instrument: (1) to convert electrical power or energy from one form to another in a controlled fashion, (2) to measure the physical elements being controlled, and (3) to provide a means for an operator, computer, or external circuit to control the power input or output. In the figure, the flow of power is designated on lines with the letter *P,* while control and monitoring signals are designated by lines with letters *C* and *M,* respectively. The following subsections describe the function of each block and some of the types of circuits used to perform the function. In general, the ac input interfaces with the commercial source of power, the ac power distribution system, and converts this to an unregulated dc voltage source. The power-conversion circuits provide for the transformation of this energy to an alternate form or magnitude, e.g., low-voltage isolated dc power in the case of a dc power supply, 400-Hz, 120-V ac power in the case of an ac power source, or the absorption of energy in the case of the electronic load instrument. Output filters smooth the waveform by removing undesirable harmonics. Other circuit blocks shown provide the means of controlling the output voltage, current, or power by accepting some type of control input and giving back information to a computer, human operator, or external instrument for the control and measurement of the electrical output or input quantities.

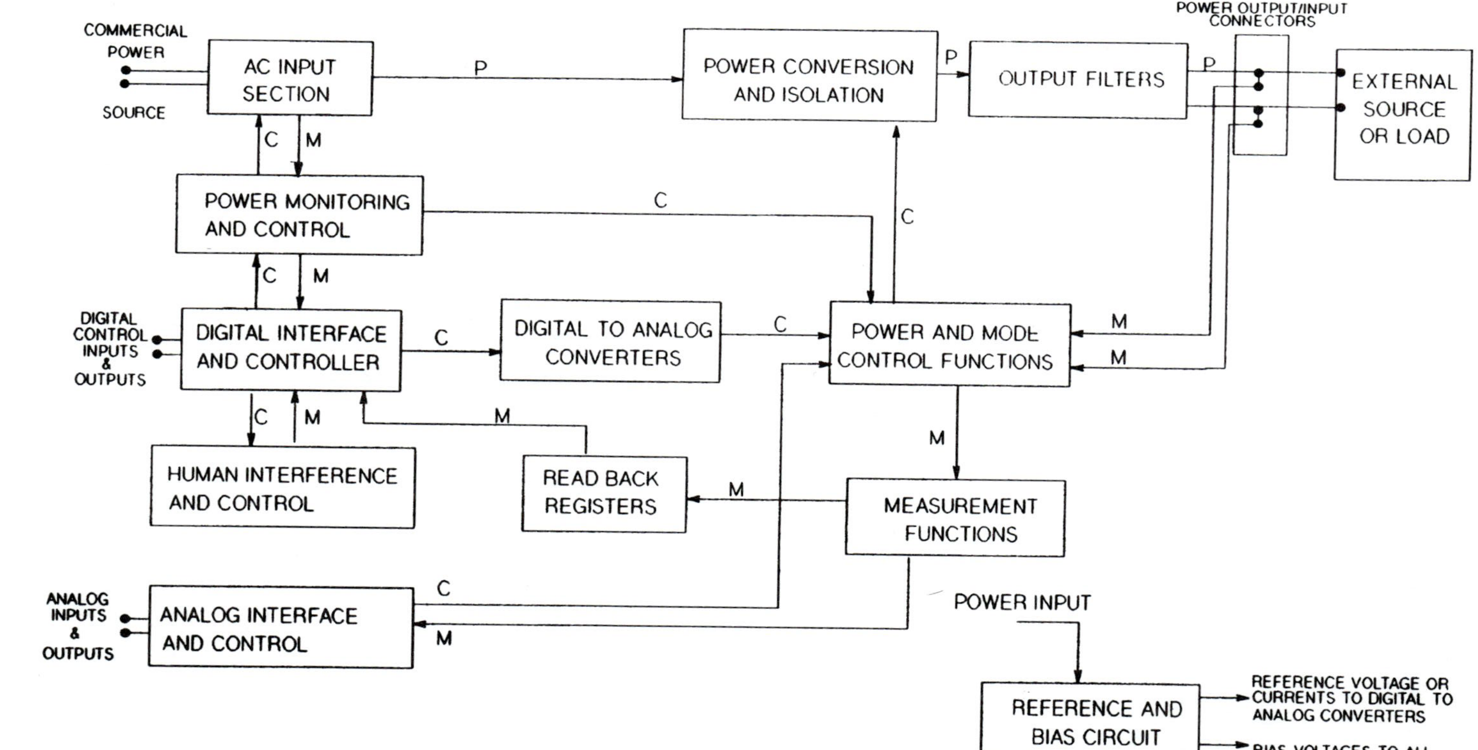

Figure 11.7 General block diagram of a power-conversion instrument (P = power flow; C = control signals; M = monitoring signals).

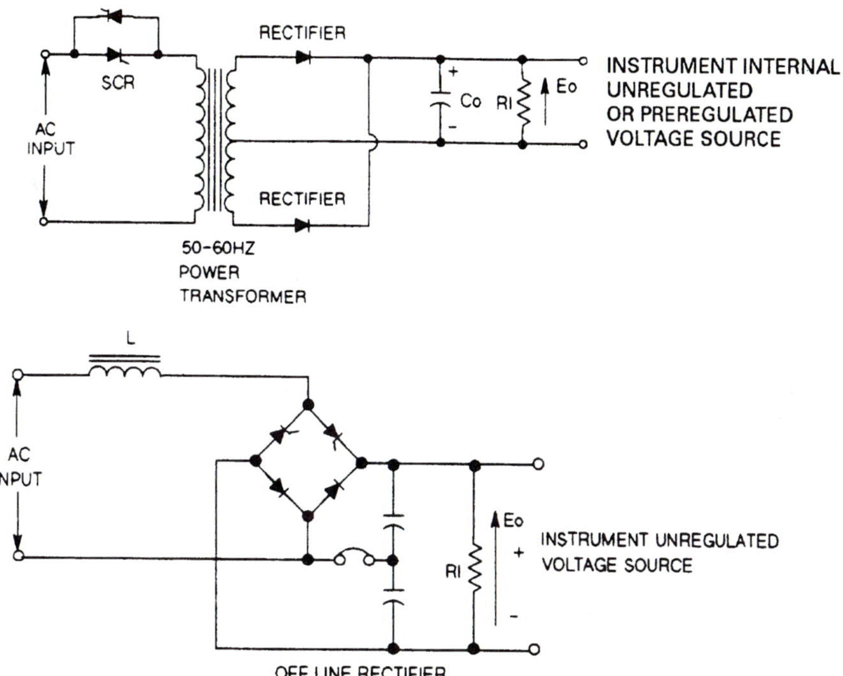

Figure 11.8 Alternative ac input sections for power conversion.

11.5.1 Alternating-current input

This circuit block accepts the input ac commercial power and converts it to a source of dc voltage that becomes the instrument's internal energy source. The manner in which this is done depends on the type of power converter in the instrument. There are basically two methods in common use: the 50- or 60-Hz power transformer and the off-line rectifier. Both are shown in Fig. 11.8.

The power transformer provides isolation of the unregulated voltage source from the power line and transformation of the input ac line voltage to an appropriate voltage for rectification. In some designs, the two silicon control rectifiers (SCRs) are inserted in series with one line as shown. Additional control and SCR firing circuits are added to selectively turn on the SCRs at an appropriate phase in the line-voltage waveform. This has the advantage of providing a "pre-regulated" voltage E_o which will not vary with variations in load R_L or amplitude of the ac input voltage.

The off-line rectifier shown in Fig. 11.8 rectifies the ac input voltage directly. This circuit functions as a full-wave bridge when the input voltage is greater than 200 V ac. When the jumper shown is installed, it functions as a voltage doubler for 100- to 120-V ac input. In this manner, the voltage E_o is nearly the same for either the higher or lower ac input. The two SCRs in the

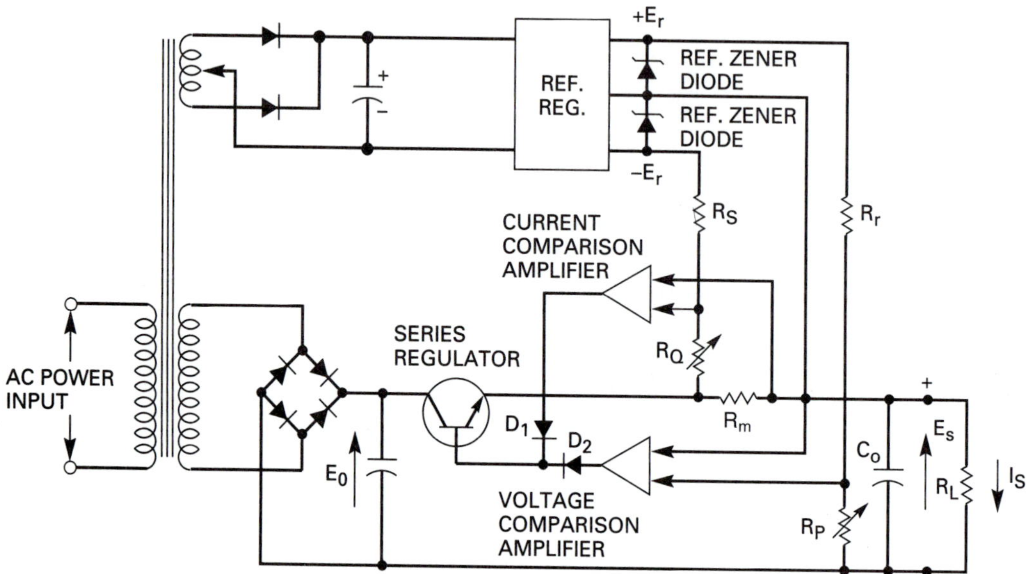

Figure 11.9 Series-regulated constant-voltage/constant-current dc power supply.

bridge are again used if the voltage E_o is to be "preregulated" so as to make its value independent of the variations in magnitude of the ac line input and load. This does require additional circuitry to control the SCR firing time. If it is not required to "preregulate" E_o, then the SCRs are replaced by diodes. Notice that this circuit does not provide isolation from the ac power line. Isolation of the output is provided in the power-conversion circuit.

11.5.2 Power conversion, control, and filtering

Referring to Fig. 11.7, the power conversion and isolation block accepts the unregulated or preregulated voltage E_o (Fig. 11.8) from the ac input section and provides the regulated output voltage and current. The power and mode control functions block provides the control signals to the power-conversion circuits for control of the output voltage and current. The output filters block smooths the voltage and current to remove ripple and provide energy storage. The following paragraphs describe some of the common power conversion circuits, control circuits, and filter circuits.

Series regulator power conversion and regulation for the dc source. Figure 11.9 shows a series-regulated constant-voltage/constant-current power supply. The series regulator, functioning as the power-conversion element in Fig. 11.7, takes the voltage E_o from the ac input section and converts it to the output voltage E_S. In this circuit, the series regulator conducts a current I_S and has a voltage across it equal to $E_o - E_S$. E_S is controlled by varying the current in the series regulator transistor. The transistor has voltage across it and current through it simultaneously and is therefore being operated as a class A amplifier. The constant-voltage and constant-current comparison amplifiers provide the power and mode control functions in Fig. 11.7. The capacitor C_o is the output filter in this circuit. The voltage references $+E_r$ and $-E_r$ are shown in Fig. 11.9 as being derived from zener diodes; however, in the general-purpose power-conversion instrument shown in Fig. 11.7, these references are provided by the digital-to-analog converters. Referring to Fig. 11.9, when R_L is greater than R_C (see Sec. 11.2 for a definition of R_C), the voltage-comparison amplifier is active and the current amplifier is effectively removed from the circuit by diode D_1 being reversed-biased. In this case, there is 0 V across the voltage-comparison input amplifier terminals, and therefore,

$$\frac{E_r}{R_r} = \frac{E_S}{R_P} \tag{11.7}$$

or

$$E_S = \frac{E_r R_P}{R_r} \tag{11.8}$$

E_S, the output voltage, is controlled by varying E_r or R_P. When $R_L < R_C$ the voltage-comparison amplifier is effectively removed from the circuit by diode D_2 being reversed-biased. In this case, the current-comparison amplifier becomes active, and there is 0 V across its input terminals. Therefore,

$$\frac{E_r}{R_S} = \frac{I_S R_m}{R_q} \tag{11.9}$$

or

$$I_S = \frac{E_r}{R_S} \cdot \frac{R_q}{R_m} \tag{11.10}$$

I_S, the output current limit set point, is controlled by varying E_r or R_q. The output capacitor C_o serves as an output filter or energy storage device and provides a low output impedance at high frequencies. This is desirable for a constant-voltage source. C_o also helps to stabilize the output voltage and keep the circuit from oscillating with various load impedances in place of R_L. Series regulators are used in applications typically requiring 500 W or less in output power because of size, weight, and power-conversion efficiency considerations.

Switching regulator power conversion for the dc source. Figure 11.10 shows one of many types of switching regulated power supplies and the associated circuit waveforms. One may understand how regulation is achieved by referring to waveform E in Fig. 11.10. This shows the waveform of the voltage across the primary of the transformer. The voltage is rectified on the secondary and

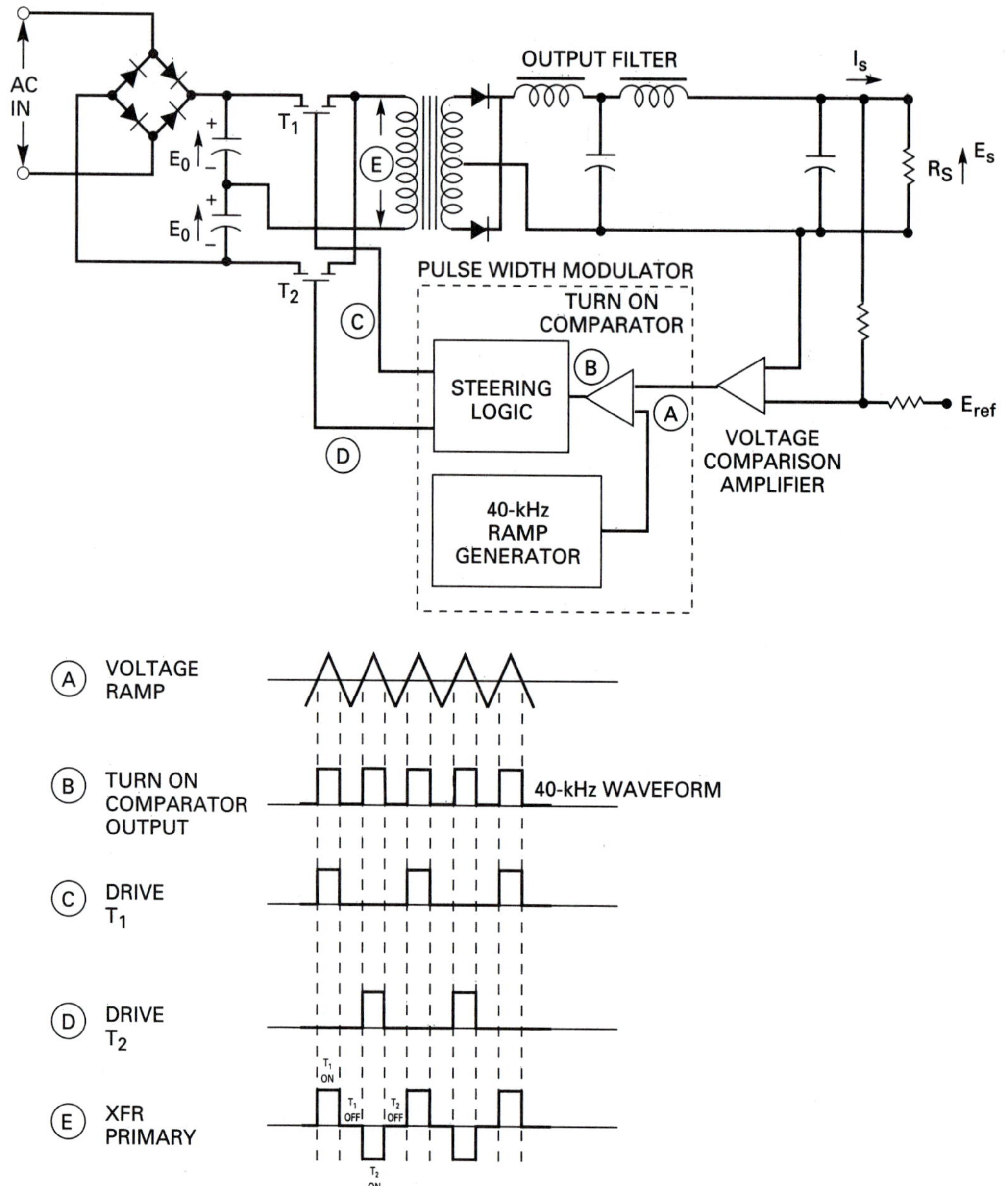

Figure 11.10 Switching-regulated dc power supply.

smoothed by the output filter. The output voltage is the average value of the rectified voltage waveform on the primary modified by the transformer turns ratio. By varying the pulse width (T_1 and T_2 on time), the average value of the output voltage is changed, and a precisely regulated output voltage is achieved.

The power field effect transistors T_1 and T_2 are connected as a half-bridge inverter and are switched on and off at a high frequency rate, as shown in waveforms C and D. By operating in a switched mode (i.e., on or off), they are much more efficient than the series regulator circuit in the preceding subsection. When conducting current, they have nearly 0 V from drain to source, and when they have full drain to source voltage, they are not conducting. They therefore dissipate relatively little power, since power in the regulator transistors is the average value of voltage across them times current through them. The waveforms shown in Fig. 11.10 are for 20-kHz operating frequency; however, while this is typically constant for an individual design, switching frequencies vary from 20 to 200 kHz, with some designs operating as high as 4 MHz.

Control of the output voltage E_S is obtained by comparing it with the dc reference voltage E_{ref} by the voltage-comparison amplifier (Fig. 11.10). The output of this amplifier is compared with the 40-kHz ramp voltage (waveform A) to produce the pulse width modulated voltage (waveform B) to the steering logic which steers alternate pulses to transistors T_1 and T_2. The pulse width increases in proportion to the average output voltage, which is controlled by varying the magnitude of E_{ref}.

The high-frequency switching results in smaller magnetic components (transformers and filter inductors) and smaller capacitors. Because of the smaller size components and high efficiency (typically 85 percent compared with typically 40 percent in a series regulator), this type of regulator has higher power density than the series regulator. Switching regulators are used when size and efficiency are prime considerations.

Switching regulators typically have poorer dynamic characteristics (transient response) than the linear class A amplifier used in the series regulator and have higher noise voltages on the output. The switching regulator may be configured as a constant-voltage/constant-current source by sampling the output current and comparing it with a reference voltage, as is done in the series regulator. Both voltage-comparison and current-comparison amplifiers are typically included in most switching regulators used in power supply instrument applications.

Switching regulator power conversion for the ac source. An ac 45- to 1000-Hz power source may be obtained from Fig. 11.7 by changing the configuration of the circuit in Fig. 11.10. The diodes are removed from the output filter, the filter is connected across the entire transformer secondary, and the dc reference voltage E_{ref} is replaced by an ac reference voltage with a frequency equal to the desired output frequency and with an amplitude proportional to the

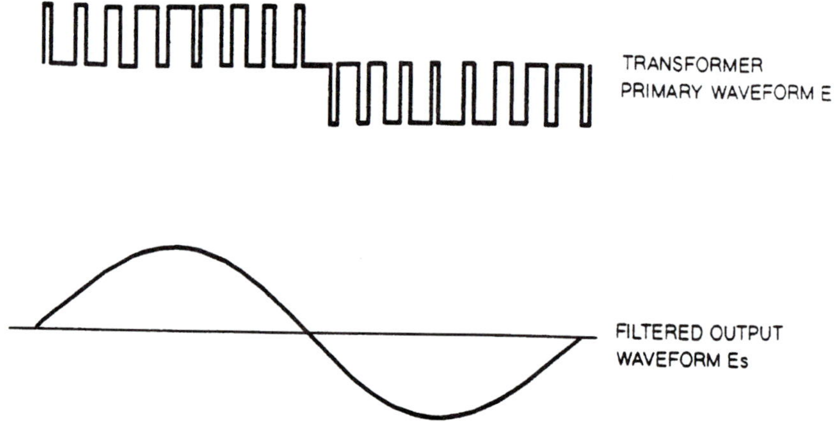

Figure 11.11 Switching-regulated ac power supply waveforms.

desired output amplitude. The steering logic is modified so that T_1 and T_2 conduct high-frequency current pulses on alternate half cycles of the output waveform. Figure 11.11 shows the waveform across the primary of the power transformer and the filtered output waveform. It can be seen from the figure that the filtered output waveform is the average of the pulse-width-modulated voltage across the transformer primary. Current limiting is obtained by adding current-sampling circuits, a programmable current reference, and a current-comparison amplifier in a manner similar to the control circuits used in the series regulator.

Silicon control rectifiers in regulators for the dc source. For some high-power applications, typically greater than 2 kW, the silicon control rectifier (SCR) is used in power-conversion regulators. SCRs are rated at much higher power than power transistors; therefore, for very high power, they are more economical because fewer are required. They cannot operate at the high frequencies that power transistors operate at, so most designs operate at ac power line frequencies (50 to 60 Hz). This results in a product that is heavier than a switching power supply and larger and slower to respond to changing loads. Figure 11.12 shows a simplified schematic of this type of regulator. Regulation is maintained by firing (turning on) the SCR at various phase angles in the input ac line-voltage waveform. This is accomplished by either the voltage- or current-comparison amplifier (depending on whether in constant-voltage or current-limit mode) providing a signal to the SCR control unit to fire the SCR. This is done earlier in the ac line-voltage waveform for higher output voltages and later in the waveform for lower output voltages or currents. For this circuit, filtering of the voltage ripple is provided by the output capacitor C_o in conjunction with the leakage inductance which is internally built into the input transformer. This type of regulator usually has high output

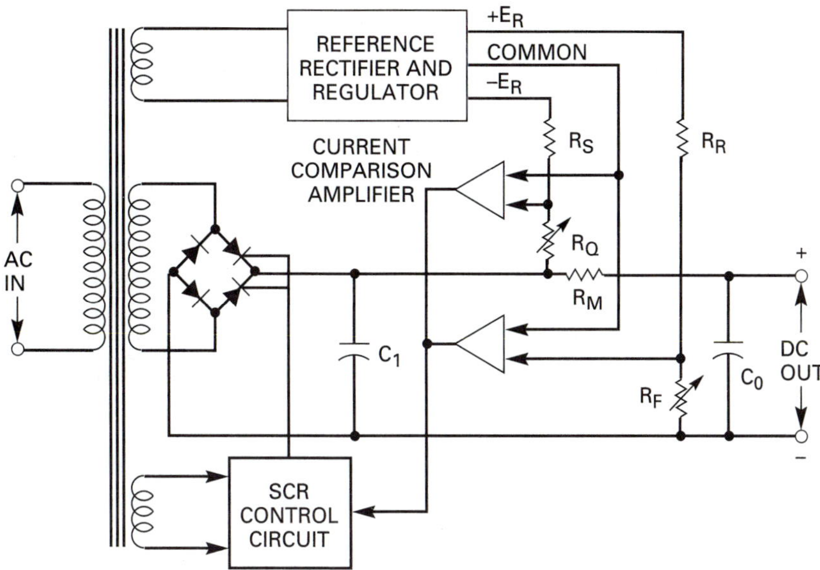

Figure 11.12 Silicon control rectifier regulated dc power supply.

ripple voltage and poor transient response, since corrections to the output voltage cannot occur until at least the next half cycle of line voltage. These regulators are typically used in applications such as electroplating, where these characteristics are acceptable. In very high power applications, e.g., 20 kW, the SCRs are placed in the primary of the power transformer because three-phase input power is required.

11.5.3 Measurement, readback, and power monitoring

Refer to Fig. 11.7 for a discussion of these functions. Most power-conversion instruments have the ability to measure the output parameters being provided to the load—dc voltage or current, ac voltage or current, power, power factor, voltamperes, or input resistance in the case of the electronic load instrument. These measurements are usually made with analog-to-digital converters. Multipliers and other computational functions are incorporated for the more complex measurements, such as power or harmonic content of the output current. These measurements are important in many applications, e.g., automatic production testing of motors, where measurement of the voltage or current is indicative of whether the unit under test is operating properly. These measurements are converted to a digital form and stored in the readback registers, where their value can be read by the digital controller or human interface and controller functions in Fig. 11.7.

The power monitoring and control functional block in Fig. 11.7 has several purposes. First, it monitors the input power to ensure that all voltages, references and bias, are at appropriate levels before passing a signal to the power and mode control functions to allow the start of the power converter. This is required so that the output voltage comes up smoothly during turn-on or removal of input ac power. This ensures that there is no turn-on or turn-off overshoot or undershoot transients (during application or removal of power) which could damage the load or device under test. Second, the power monitor and control block also monitors the status of various conditions within the power-conversion instrument. For example, it monitors the mode the instrument is operating in, constant voltage or constant current for a dc power supply or electronic load and voltage or current limit for an ac source. It is also usually capable of monitoring the status of the ac power input to determine if there has been a momentary interruption which caused the instrument to drop out of regulation. Certain fault conditions within the regulator such as overtemperature, overvoltage, or out of regulation are also typically monitored. These conditions are reported to the digital interface and controller and are presented to the human operator or external computer controlling the instrument. These conditions are important to monitor, especially during automatic testing applications, so that the operator or computer in control can take appropriate action to remedy the fault. In automatic test applications, a service request may be generated for any of these fault conditions. Reading the appropriate status registers in the instrument allows the computer to determine what fault has occurred so that the appropriate action may be taken.

11.5.4 Interface and control functions

Referring to Fig. 11.7, it can be seen that there are three different methods of monitoring and controlling the output voltage, current, resistance, etc. and selecting the desired operating mode or instrument function. One method is through the digital input/output (I/O) connectors. Typically, the digital I/O interface is IEEE 488. A computer with an I/O port connected to this interface supplies the necessary digital commands to program the output, select the mode of operation, and monitor the status of the power-conversion instrument. A second method is via the analog input port. In this case, an input voltage, typically in the range of 0 to 5 V, is used to control either voltage output or current limit or both. The output voltage or current is linearly proportional to the input control voltage. Also available at this port are two output voltages that are proportional to the actual voltage and current output of the power supply or electronic load. A third method of control is by the operator through the human interface. The digital control and human interface are discussed in more detail below.

Human interface and control. Figure 11.13 is a drawing of a typical power supply front panel. This consists of a display that shows the output voltage and current magnitude or the function being programmed when executing a com-

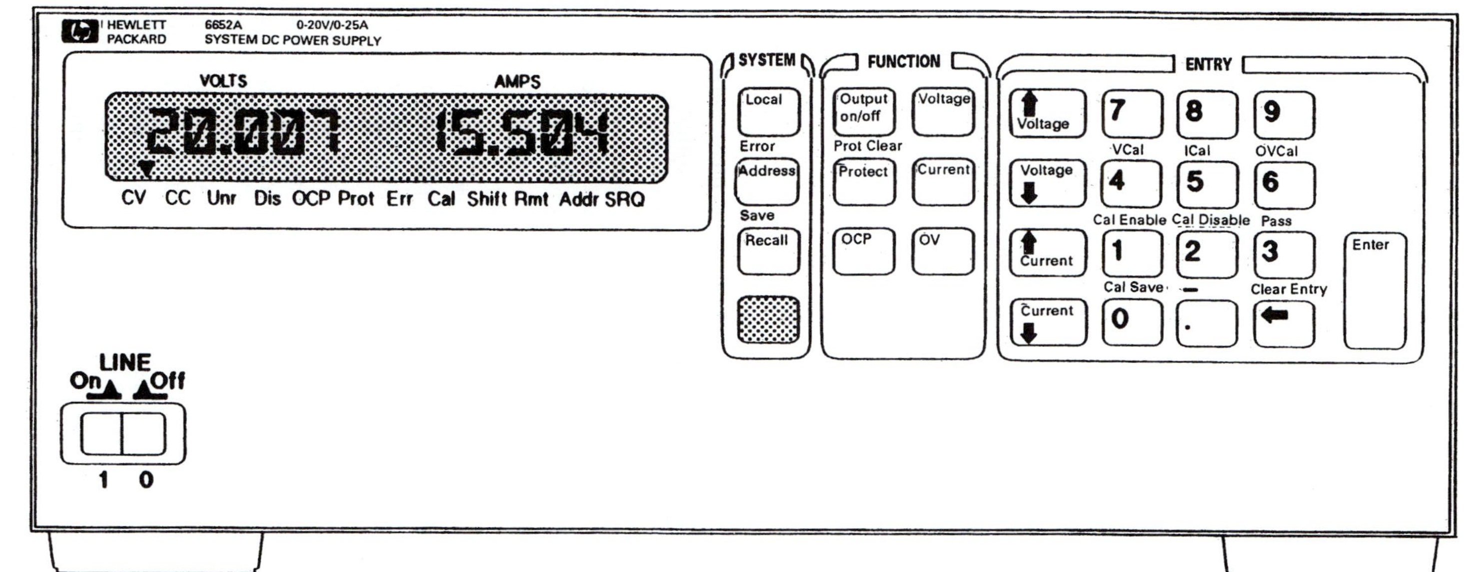

Figure 11.13 Typical dc power supply front panel.

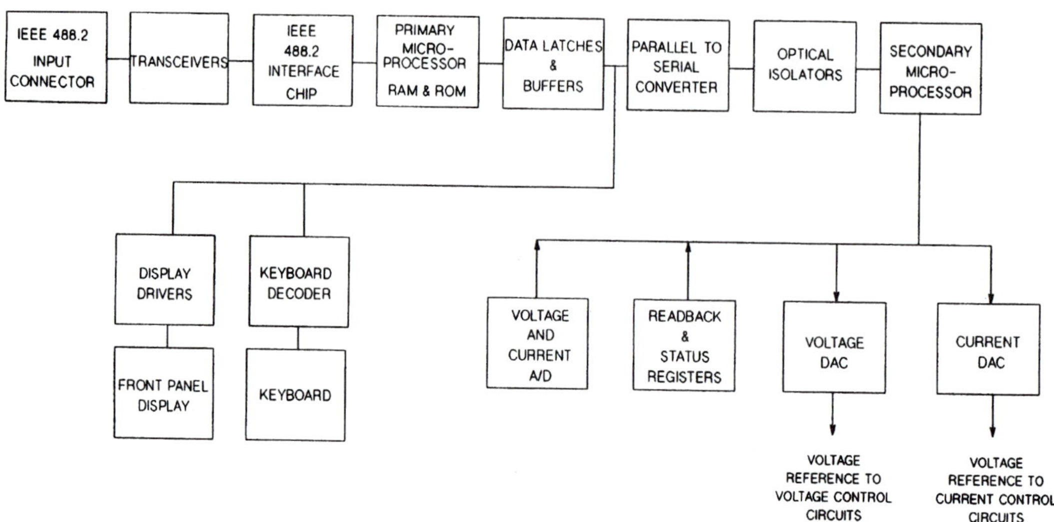

Figure 11.14 Digital interface and control for power-conversion instruments.

mand from the front panel. A keyboard with system keys, function keys, and numeric entry keys is usually available. The system keys are used to select local (front panel) or remote (digital interface) control, display or change the IEEE 488 address, and save or recall instrument states in nonvolatile memory. The function keys are used to turn the output on or off, program voltage or current magnitude, and select protection features such as overcurrent or overvoltage protection. The entry keys are used to raise or lower the output voltage or current through "up-down" function keys or to enter a precise numeric input for the desired value.

Digital interface and control. Figure 11.14 is a block diagram which is representative of the type of digital interface in power-conversion instruments. The digital input command is received over the IEEE 488 interface by the transceiver circuits. This is processed by the IEEE 488 interface chip and presented to the primary microprocessor. This processor interprets the command and stores it in the appropriate data buffers and latches. The data are converted from parallel digital words to serial form and transmitted via optical isolators to the secondary microprocessor. The secondary microprocessor reconstructs the command and presents it to the appropriate converter, voltage or current, which converts the digital word to an analog reference voltage for controlling the power supply output voltage or current. This reference voltage is referred to one of the output terminals of the power supply. These output terminals are floated with respect to earth ground and may be several hundreds of volts from ground. The primary microprocessor is tied to or referenced to earth ground via the IEEE 488 interface. The output of the power supply, all the control circuits, and the secondary microprocessor are therefore electrically isolated from the primary circuits through the optical isolators. All information regarding the power supply output voltage and current values and operational status are read back from the analog-to-digital converter and status registers through the secondary microprocessor to the primary microprocessor. The primary microprocessor also controls the display and keyboard. It presents the voltage, current, and status information to the display and accepts commands from the keyboard when in local mode of operation.

11.6 Power Supply Instruments for Specific Applications

Although the specific applications for power supplies are numerous, some of the recent applications arising from the development of digital cellular phones, portable computers, and the new battery technologies that power these products are worth mentioning. These applications require power-supply characteristics or features that have not existed prior to the introduction of these products.

These digital, portable products operate from new battery technologies, such as lithium ion or nickel metal hydride. Typically, these products operate from constant voltage sources, batteries; however, because of their digital nature, they draw current in short pulses. During automated production testing, the dc power supply is used to replace the battery and the battery charger that are part of these products. Testing these products requires that the dc voltage at the test fixture is precisely controlled and maintained constant—even in the presence of the high-current pulse loads that are drawn. The ability of a power supply to control a voltage at a remotely located load is performed by remote sensing. That is, a pair of sense wires is used in addition to the current carrying or load wires. These sense wires are connected from the power-supply voltage-sense amplifier to the point at which the voltage is to be precisely controlled. Most dc power supplies have this remote-sense ability and can control the dc voltage in the presence of a steady or dc current. Because these products draw current in short pulses with fast rise times; the power supply must now have the ability to maintain the voltage constant in the presence of these current pulses. This requires a power supply with extremely wide bandwidth, several hundred kHz, in order for the voltage at the remote location to remain constant. Most dc power supplies do not have this remote-sense bandwidth.

The new battery technologies developed for these portable digital products require precise control of the battery charge rate and time. The charger-control characteristics are usually built into the products. However, during production testing, calibration of the charger characteristics requires that the power supply is able to sink or absorb current, rather than source it. Therefore, for these applications, the dc power supply must be able to serve both as a source and sink of current.

Furthermore, it is usually required that the current drawn during different portions of the test be accurately measured. Because the current is not dc, but of a pulse nature with a high peak-to-average ratio, new techniques are being used to measure the characteristics of the current waveform. The advent of low-cost, high-speed ADCs and DSP (digital signal processors) now allow these measurements to be made by the power-supply instrument.

In summary, the introduction of portable digital communication and computer products has resulted in some unique requirements for dc power supply characteristics; namely, high-bandwidth remote-sense, current-sink, and source capability, and measurement of pulse-current waveform properties.

11.7 Selecting Your Power-Conversion Instrument

There are many considerations in selecting your power-conversion instrument. Following is a list of important common considerations:

- Voltage, current, and power required for the application
- Systems or bench application—That is, will the product be used in an automatic test system that requires the instrument to have an interface to a computer or controller, or will it be used strictly in a manual mode by an operator?
- Key performance specifications and features of importance to the application
- User friendliness of the human interface, i.e., ease of use
- Ease of programming or control by computer—Does it use a standard programming or command language?
- The size, weight, height, power, cooling requirements, and power-conversion efficiency
- Ease of calibration or service
- Safety and regulatory requirements, e.g., international EMC (electromagnetic compatibility) requirements and designed to recognized safety agency standards
- Protection features required for the load under test, e.g., overcurrent, overvoltage, or overtemperature protection
- Key measurement and monitoring requirements, e.g., visual display of voltage, current, and status (overvoltage, etc.) and ability to report measurement and status over the digital interface

The following subsections address these considerations.

11.7.1 Power-conversion instrument specifications

The first key consideration in selecting a power-conversion instrument is the rating and performance specifications. “Ratings” deal with maximum voltage, current, power, and resistance (for the electronic load) capabilities of the product. “Specifications” deal with the performance capabilities or qualities of the instruments and are usually separated into guaranteed performance specifications (over a broad ambient temperature range) and operating characteristics which are typical (not guaranteed 100 percent) performance or qualities of the instrument.

Definition of key specifications

Load regulation. The change in output voltage for a maximum change in output current while in constant-voltage mode or the change in output current while in constant-voltage mode or the change in output current for a maximum change in output voltage while in constant-current mode.

Line regulation. The change in output voltage in constant-voltage mode or change in output current in constant-current mode for a change in input ac line voltage from minimum specified input to maximum specified inputs.

Transient response. Time to recover within a specified percentage of output voltage following a specified magnitude step change in output current while in constant-voltage mode or time to recover within a specified percentage of output current for a specified magnitude step change in output voltage while in constant-current mode.

Ripple and noise. Magnitude of the variation in the output voltage or current from the steady-state output.

11.7.2 Key features

“Key features” are capabilities that the power-conversion instruments have which are important for specific applications. The following paragraphs provide examples of key features.

Remote sensing. When the load is connected remotely from the power source or electronic load, the load regulation will significantly degrade due to the impedance of the leads connecting the power supply output terminals to the load. By providing a separate pair of voltage sense terminals and connecting them to the load, the voltage is regulated at the load instead of at the output terminals of the power supply. This requires that the load be connected using four leads, two to carry the current and two to sense the output voltage at

TABLE 11.1 Stranded Copper Wire Ampere Capacity Resistance

Wire size			
AWG	Area, mm²	Ampere capacity	Resistance per unit length, Ω/ft
18	0.82	15.4	0.0064
16	1.31	19.4	0.004
14	2.08	31.2	0.0025
12	3.31	40	0.0016
10	5.26	55	0.001
8	8.37	75	0.0006
6	13.3	100	0.0004
4	21.1	135	0.00025

NOTES:

1. AWG wire ratings derived from MIL-W-5088B.
2. Ampere capacity of aluminum wire is approximately 84 percent that of copper wire.
3. With bundled wires, use the following percentages of the rated current:

   2 conductors, 94 percent
   3 conductors, 89 percent
   4 conductors, 83 percent
   5 conductors, 76 percent
4. Maximum ambient temperature 50°C; maximum conductor temperature 105°C.

the load. Table 11.1 shows the resistance of different-sized copper conductors and the current ratings based on maximum wire temperatures. The load-carrying conductors should be chosen to limit the voltage drop and temperature rise in the leads. Voltage drop in the load leads is limited by the design of the power supply. Typically, 1 to 2 V is maximum, but wide variations appear in practice. Also remember that voltage drop in the leads subtracts from the voltage available at the load and also slightly degrades the load regulation of the unit.

Overvoltage protection (OVP). Separate manual and programmable controls limit the output voltage to a specified value. If the power supply output voltage exceeds this specified value either because of a program command, a manual control setting change, or a fault in the instrument, the power supply will automatically shut down until the fault is removed and the power is recycled or the OVP sense circuit is reset. This feature is useful to protect sensitive loads that may be damaged by excessive voltage. The circuitry to implement this feature varies depending on the type of regulator in the power-conversion section of the instrument.

Overcurrent protection (OCP). Most power supplies have dual-mode capability, i.e., they have both constant-voltage and constant-current capability. When the power supply load current exceeds the programmed current limit set point, the power supply will enter the constant-current state and begin to reduce the output voltage as the load increases. There are some applications in which it is desirable to reduce the current to zero if the programmed current limit value is reached. This may be the case, for example, when testing a printed-circuit assembly. A local short or low impedance will draw all available current, which previously was uniformly distributed, through the local short. Self-heating may be sufficient to cause the printed circuit to be damaged beyond repair. By enabling the overcurrent protect feature, the power supply enters a high output impedance state if the load current exceeds the programmed value. The power supply will remain in this state until the fault is removed and the OCP reset command is given.

Parallel and series operation. In some applications, the maximum current or maximum voltage required may exceed what is available in a single power supply. Many power supplies have control circuits accessible to the user that allow the user to connect power supplies in parallel and have them equally share the load or to connect them in series and have them equally share the voltage. These configurations are called "autoparallel" and "autoseries," respectively. Some supplies are designed to operate in parallel without the autoparallel capability. In this case, they do not equally share the load. Always check the operating manual to determine which of these features is available prior to connecting power supplies in parallel or series.

Reverse-voltage, reverse-current capability. This is a protection feature designed into many power supplies that allows reverse current or voltage from an active source or another power supply to flow through the supply without damaging the supply. This can occur, for example, when two supplies are connected in series with a load, and the ac power is removed from one supply. If the power supply has reverse-current protection, it will pass through the current from the active supply to the load without being damaged.

EMC (electromagnetic compatibility). All power-conversion instruments are sources of high-frequency noise. Noise energy may be conducted on the output leads or on the power line input cord. It also may be radiated energy. This noise may interfere with precision measurements. Therefore, in addition to the ripple voltage specification, many power-conversion products are designed and specified to meet certain national and international EMC standards. The standards cover both conducted and radiated interfence and set specific maximums for both over a broad frequency range. Also covered is the susceptibility of the instrument to radiated and conducted interference; i.e., the product must meet specifications when subject to electromagnetic fields or high-frequency noise on the ac power line. Many countries have enacted laws that require power-conversion instruments to meet the applicable standards.

Safety features. Safety is an important consideration in the selection and use of a power-conversion instrument. The instrument should be designed to a recognized national or international standard. The power output terminals are physically and electrically isolated from the ground and therefore may be floated from ground up to a maximum voltage fixed by instrument mechanical design. If there is a need to float the output terminals, make sure the product is designed to do this safely.

System features

Programming command set. All instruments that can be controlled through a digital interface have a programming language or command set. Most instruments are controlled by way of IEEE 488, *Standard Digital Interface for Programmable Instruments.* Many power-conversion instruments use a standard programming language called SCPI (Standard Commands for Programmable Instruments). A standard language implies a standard command for any instrument that implements a particular function. Once the standard is understood, programming a variety of instruments is much easier. The commands are broken down into common commands which perform common interface functions for all programmable instruments and subsystem commands which are specific to power-conversion instrument functions. Examples of specific subsystem commands for setting voltage and current are

:SOURCE:CURRENT 5.0; VOLTAGE 25.0

This will set current limit to 5.0 A and voltage to 25.0 V.

:MEAS:CURR?; VOLT?

This will measure the actual current and voltage of the power supply. The programming guide provided with the power-conversion instrument provides all the subsystem commands.

Waveform variations. Waveform variation is particularly important for ac sources and electronic loads. These variations should be controllable from the human or digital interface. Typical waveform variations for electronic loads consist of the ability to program current or voltage pulses with selectable frequency, duty cycle, and rise and fall times. These capabilities are important because they give the electronic load the ability to generate fast transients, useful for electronic component testing and power supply testing.

For ac sources, the waveform variations include surge voltage magnitude and duration, cycle dropout (the elimination of a portion of or entire line-voltage cycles), and turn-on phase angle (the ability to apply voltage at a selected phase angle of the ac line). These features provide the ac source with the ability to simulate actual fault or turn-on conditions that occur on the power line.

Chapter

12*

Instrument Hardware User Interfaces

Jan Ryles
Agilent Technologies
Loveland, Colorado

12.1 Introduction

Considering the design of the instrument user interface† is very important when selecting an instrument for a specific measurement task. The instrument user interface may be comprised of both hardware components and graphical user interface elements. Chapter 12 focuses on traditional hardware user interface components, including output and input devices. Chapter 44 focuses on the newer graphical user interfaces, which are becoming more commonplace in some instrument configurations.

12.2 Hardware-User Interface Components

The quality of the hardware determines how the user will both learn to use and use the instrument system. For example, physical size, display resolution, processor speed, and ruggedness determine under what conditions users will be able to use the instrument and what measurements they will be able to make with it. For example, a telephone repair person will not carry a rack of heavy instruments up a telephone pole to do protocol testing on the lines. In addition, the available control knobs and built-in physical displays on the

* Adapted from Coombs, Electronic Instrument Handbook, 2nd Ed. McGraw-Hill, New York, 1996, Chapter 12, originally authored by Janice Bradford.

† An "instrument-user interface" is any part of an instrument that the user comes into contact with, either physically, perceptually, or conceptually.

instrument determine both the ways the user will interact with it to perform measurements and the format of the data returned to the user, thereby dictating what measurements the user can and cannot make and what information the user can and cannot get from the instrument.

12.2.1 Configuration of instruments

Beginning at the global level and considering the number and configuration of physical hardware components in an instrument, we find four common instrument hardware configurations (Fig. 12.1).

Portable hand-held instruments. Portable hand-held instruments are meant to be carried by the user to a location away from the usual laboratory environment for the purpose of making measurements on the device under test (DUT) in its environment. Portable instruments are characterized by their small size and weight, a limited measurement functionality but one that has been carefully designed to meet the most important measurement needs of the user, a design with low power requirements so that they can be battery operated, and ruggedness to withstand the inevitable bumps and knocks that will occur because they are transported to the measurement site.

Stand-alone instruments. The second category of common instrument configurations is the traditional stand-alone instrument most commonly found on the laboratory bench, the oscilloscope being the canonical example. These instruments have a traditional rectangular box frame with all the hardware-user interface controls and displays located on the front of the instrument, called the "front panel." Just as with the hand-held portable instruments, instruments in the stand-alone category can be used independently of other instruments to get useful measurement results.

Rack-and-stack and cardcage instruments. By itself, the stand-alone instrument performs useful measurements, but it can be combined with other instruments in a configuration known as a "rack-and-stack instrument," the third category. In a rack-and-stack instrument, the stand-alone measurement components are integrated into a sophisticated measurement system capable of making all measurements required for a specific measurement application by means of a metal frame, connecting cables, and a controller unit. The components are selected based on the functional measurement requirements of the system as a whole. Control of the individual units to make them appear as a single, integrated measurement system is usually done by means of a computer running software designed specifically for the measurement application of the user. An example of a rack-and-stack instrument would be a network analyzer composed of multiple synthesizer sweepers, test sets, and analyzers with a computer that acts as a controller and user interface, all packaged together into a single test system.

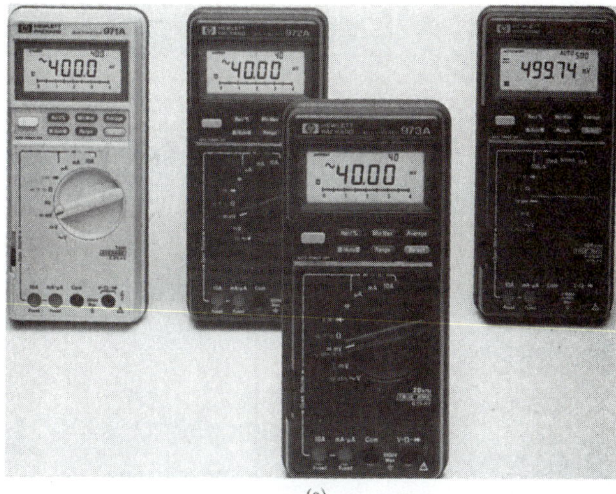

(a)

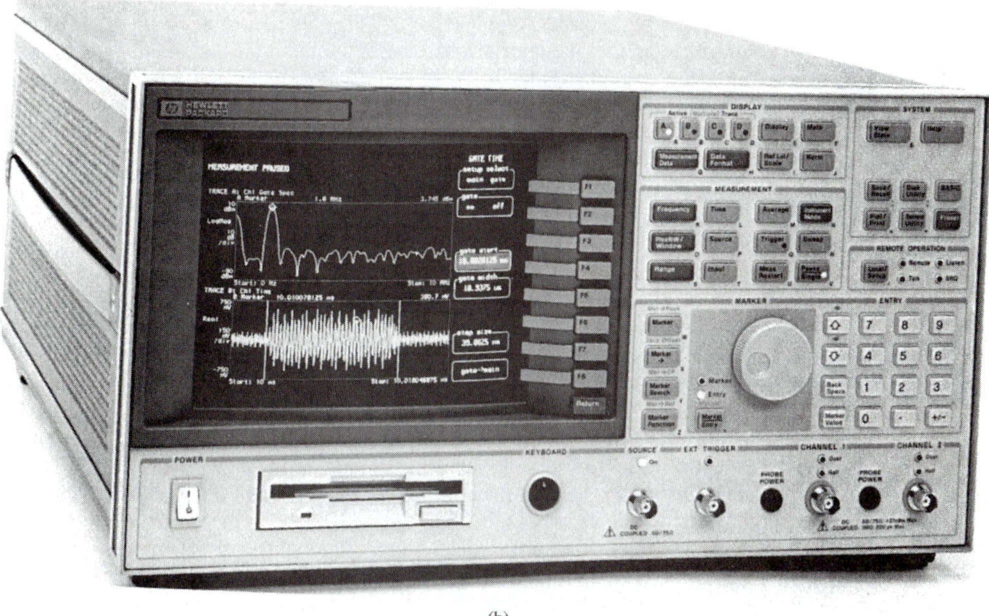

(b)

Figure 12.1 Examples of instrument configurations. (*a*) Portable hand-held instrument. (*b*) Stand-alone bench instrument. (*c*) Rack-and-stack instrument with a computer controller. The controller software has a text-based command language interface, and the user issues commands through the keyboard. (*d*) Cardcage instrument with a computer controller. The controller software has a graphic user interface, and the user interacts with it through the keyboard and mouse. (*e*) Distributed measurement system.

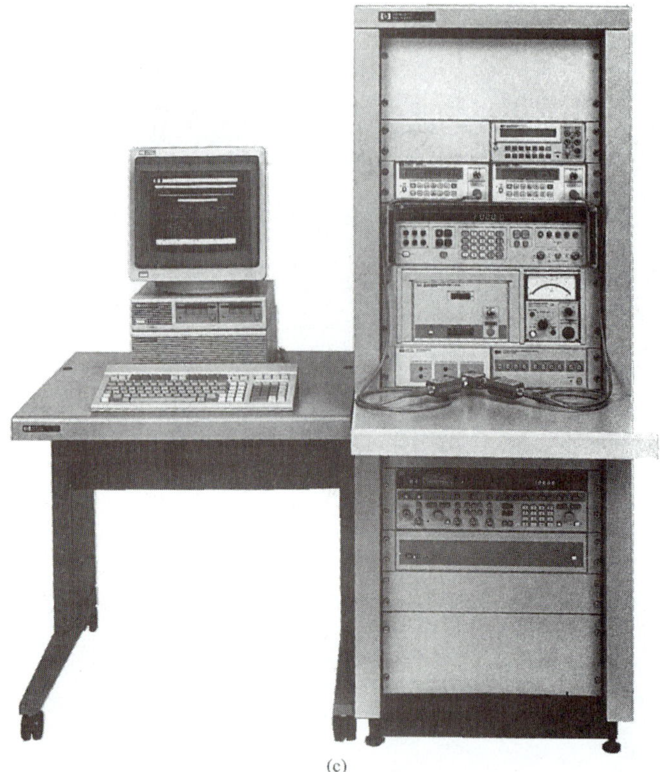

(c)

(d)

Figure 12.1 *(Continued)*

(e)

Figure 12.1 *(Continued)*

"Cardcage instruments" are a variation on the rack-and-stack instrument with two main differences. Since rack-and-stack instruments are composed of stand-alone instruments, each will still have the front panel-user interface. The individual units in a cardcage instrument are designed for the purpose of being put together to form a more sophisticated measurement system, so they do not have a front panel-user interface. The controls and displays of their user interface are implemented in software on the computer controller unit, providing a single physical location for that portion of the user interface. The second difference is that instead of using cables to connect the different measurement components, the units of a cardcage instrument plug directly into a backplane board designed to handle communication between the units and between the units and the computer controller, thus providing more efficient communication.

Distributed measurement systems. The fourth category is "distributed measurement systems," which have multiple measurement units, just as the rack-and-stack and cardcage instruments do, but differ in that their components are physically distributed over a large testing area. An example is a manufacturing line with multiple test stations monitoring process control. Again, a computer acts as the controller unit.

A noticeable difference in the configurations of these four categories of in-

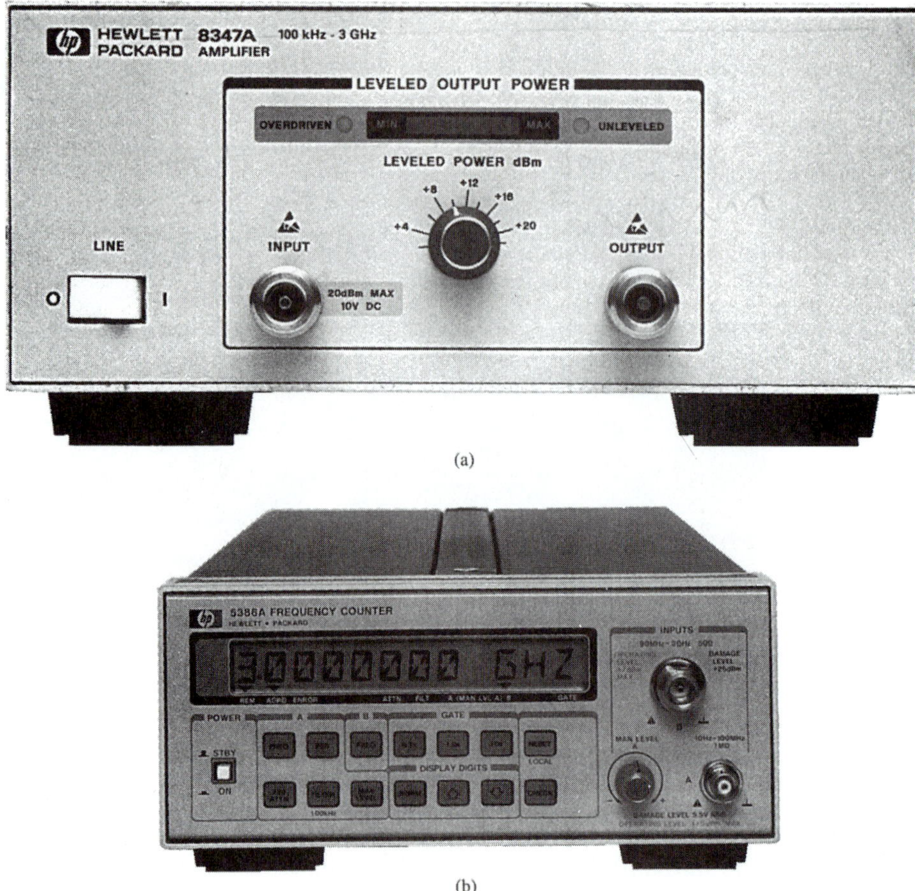

(a)

(b)

Figure 12.2 Three display types and technologies commonly found in instruments. (*a*) Indicator lights. (*b*) Small alphanumeric display implemented with liquid-crystal display (LCD) technology. (*c*) Small alphanumeric displays implemented with light-emitting diode (LED) display technology. (*d*) Graphics and text display implemented with cathode-ray tube (CRT) display technology.

struments is that the user interface hardware components of instruments may be physically grouped in a small area or distributed across a wide area. In the case of large, distributed measurement systems, there may be multiple, concurrent users, requiring the interface to handle coordination and communication between them.

A second major difference in the user interfaces of the different instrument configurations is that some instruments include a computer as a component. In this case, instrument controls and displays are usually accomplished on the computer display through software that runs on the computer in either a text-based command language or graphic user interface instead of the physi-

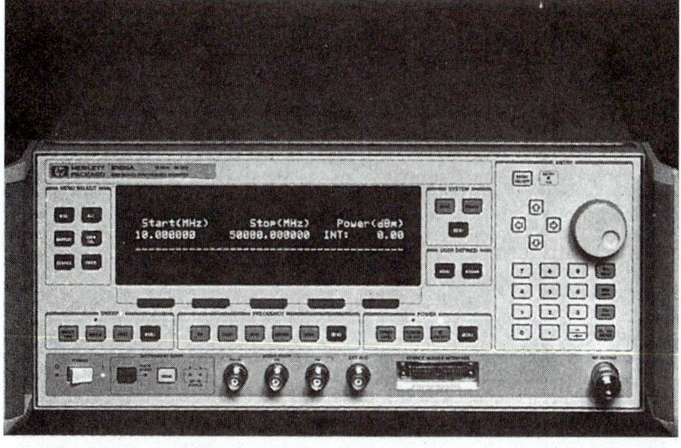

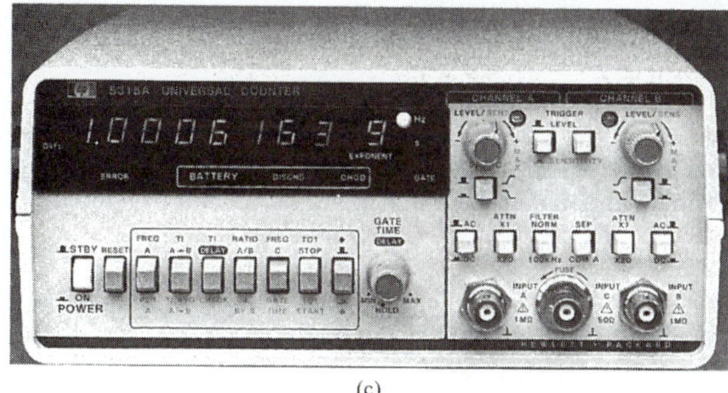

(c)

Figure 12.2 *(Continued)*

cal knobs, buttons, and meters traditionally found on the front panel of an instrument.

12.2.2 Hardware-user interface components: output devices

All instrument front panel-user interface components can be classified as either input devices, such as dials, keys, and switches, used for the purpose of allowing the user to input commands to the instrument, or output devices, such as lights, displays, and meters, used for the purpose of displaying information to the user. Just as there are differences in the global configuration of instruments, each physical interface component (i.e., the instrument controls and information displays) can be designed with many variations appropriate to different uses.

Display devices. Display devices on instruments come in a wide range and variety depending on the use of the display, but they can be grouped into three basic categories (Fig. 12.2). The simple, single-purpose "indicator light"

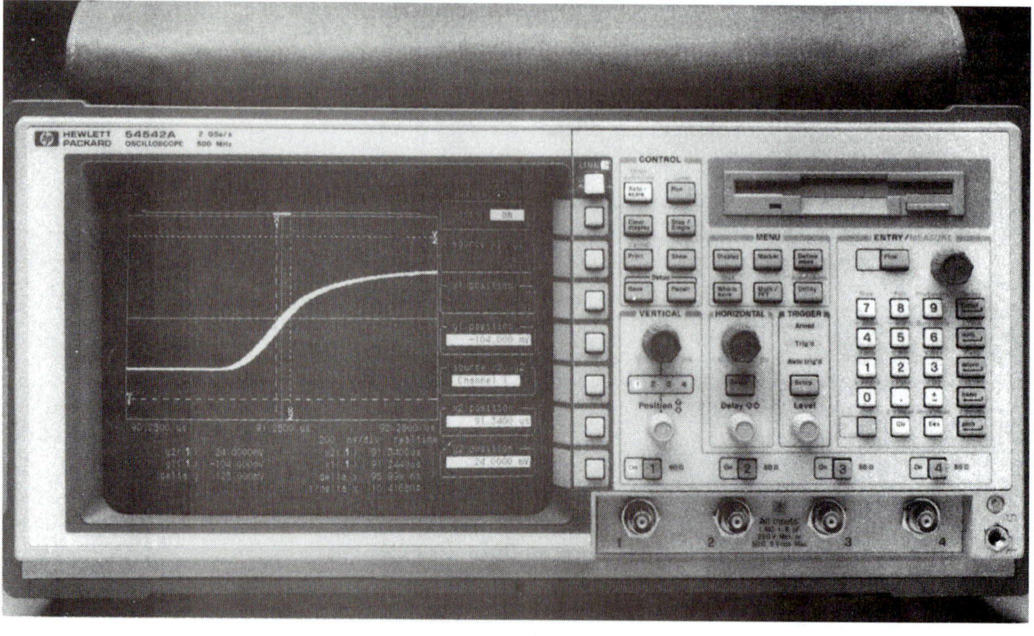

(d)

Figure 12.2 *(Continued)*

is used for conveying information that has a binary or threshold value, such as warnings, alerts, and go/no-go messages. "Small alphanumeric" displays are quite common on instrument front panels. They are useful for displaying information that has relatively short messages composed of text only. The third category of display handles both "graphics and text" and is found on instruments whose measurement results are easier to interpret if the information is in a graphic form, such as the oscilloscope. Computer displays also will be in this third category, since they use graphic representations to signify instrument controls, as well as displaying graphic results.

No matter what kind of display is used, its purpose is to convey information in a timely and nonpermanent manner with a high enough quality of presentation that the user can extract the information efficiently and accurately. In instrumentation, a display allows the user to observe measurement data in a more immediately interactive fashion and actively participate in the measurement itself.

A display must have a high enough image quality for its purpose to allow the user to extract the information presented there efficiently and accurately. Factors affecting the image quality of a display include the following: amount of glare, resolution (i.e., the degree of jaggedness of the strokes), design of the characters, stability of the screen image (i.e., amount of flicker and jitter),

TABLE 12.1 Comparison of Common Display Technologies Found in Instruments

			CRT	
	LED	LCD	Raster	Vector
Common use	Indicator light; small alphanumeric display	Small alphanumeric display, graphics and text	Graphics and text	Graphics and limited text
Color capabilities	Single color	Black and white, gray scale, or color	Black and white, gray scale, or color	Black and white
Resolution	Poor	Good	Good to excellent	Good to excellent
Contrast	Excellent	Fair	Good to excellent	Good to excellent
Viewing angle	Good	Moderate	Excellent	Excellent
Speed	Very fast	Comparatively slow	Fast	Fast
Brightness	Brightest	Variable brightness	Bright	Bright
Power consumption	Highest	Low	High	High
Weight	Low	Low	Heavy	Heavy
Size	Small, light	Flat panel	Bulky tube form	Bulky tube form
Ruggedness	Good	Good	Fair to poor	Fair to poor

contrast in the image, color selection, and image refresh rate. The optimal choice of values for these parameters depends on the particular conditions of use of the display and the type of information presented on it. The distance of the user from the display, ambient lighting conditions of use, user task requirements, and the content of the screen all affect the best choice of values for these parameters in order to get the desired image quality.

Designers have developed a number of display technologies in the effort to find the perfect display that has all the desirable features of high image quality, low cost, low power consumption, light weight, ruggedness, and rapid refresh rate. Current display technology is largely a compromise between what is desired and what can be achieved. The technologies used most commonly in instrumentation are cathode-ray tubes (CRTs), light-emitting diodes (LEDs), and liquid-crystal displays (LCDs) (see Fig. 12.2*b–d*). The different display technologies are discussed below and compared in Table 12.1.

All display technology works by illuminating portions of the display screen while leaving other portions unilluminated. The technologies differ in their method of illumination. The basic unit of a display is the smallest area that can be illuminated independently, called a "pixel"* (Fig. 12.3). The shape and number of pixels in a display are factors in the resolution that the display can achieve. The number of available pixels determines the maximum information content of the display.

Light-emitting diode (LED) display technology. LEDs illuminate pixels by converting electrical energy into electromagnetic radiation ranging from green to near-infrared (about 550 to over 1300 nm). Fig. 12.2*c* shows two instrument

* The word "pixel" is derived from "picture element."

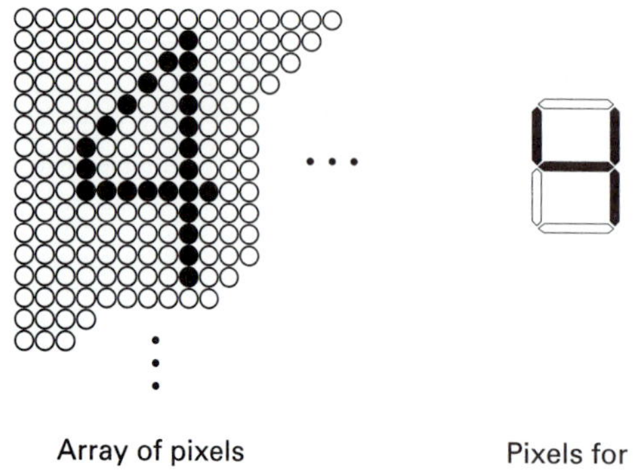

Figure 12.3 Pixels come in different shapes and sizes. A pixel is the smallest unit of a display screen that can be illuminated independently. The shape and number of pixels relative to the area of the screen are a factor in screen resolution. The number of available pixels determines the maximum information content of the display.

LED displays with different pixel shapes and sizes for different resolution capabilities.

LEDs in instruments are used most commonly as indicator lights and small alphanumeric displays. Relative to CRTs, LEDs are smaller, more rugged, have a longer life, and operate at a lower temperature (see Table 12.1).

Liquid-crystal display (LCD) technology. Where LED technology is emissive (i.e., it emits light) LCD technology channels light from an outside source, such as a fluorescent lamp behind the display, through a polarizer and then through liquid crystals aligned by an electric field. The aligned crystals twist the light and pass it through the surface of the display. Those crystals which have not been aligned, do not pass the light through. This creates the pattern of on/off pixels that produces the image the user sees.

The advantages of LCD technology are low power consumption, physical thinness of the display, light weight, ruggedness, and good performance in bright ambient light conditions (LCDs outperform both LED and CRT technology for readability in bright light). LCDs are used for both small alphanumeric displays and larger graphics and text displays.

Cathode-ray tube (CRT) display technology. CRT technology was the first to be developed of the three discussed here. While it has some drawbacks, such as size, weight, and power consumption, in some applications it is still superior to the other technologies. In instruments, CRT display technology was first used in oscilloscopes, and it continues to be the best choice of display technologies for displaying information rapidly with a high graphic waveform content requiring high visibility.

In a CRT, an electron beam is guided over a phosphor screen inside a vacuum tube; when the electron beam is turned on, it illuminates the luminescent material it touches. The luminescence persists only for a short period of time, so it must be refreshed continuously by passing the electron beam over the screen again and again.

There are two methods of guiding the beam over the display, "raster scan"

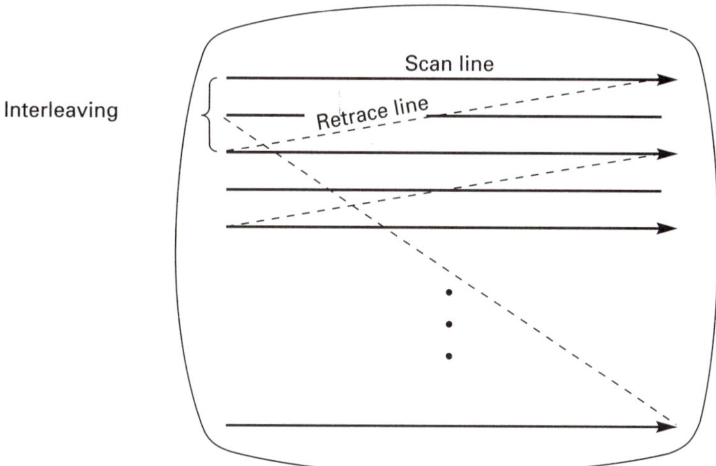

Figure 12.4 CRT raster scan. The lines show the path of the electron beam on the phosphor screen. Half the scan lines are written in one pass of the screen, from top to bottom; the other half of the lines are written in the next pass.

and "vector scan." The raster-scan method (Fig. 12.4) is used in computer CRT displays and other commonly found CRT displays such as television screens. The electron beam is swept horizontally from side to side over the back surface of the screen, tracing out the image by turning on the necessary pixels as it comes into contact with them. As the beam sweeps across the raster scan line, it is turned off and on according to whether that pixel should be illuminated or not. Each horizontal sweep is stepped lower than the preceding one until the bottom of the screen is reached, and the beam is returned to the upper corner of the screen. Typically, there are 480 raster lines, and the vertical steps down the screen are done in increments of 2 lines so that on one scan of the screen, top to bottom, 240 raster lines are covered, and in the next scan the other 240 lines are covered. On high-speed graphic displays, this vertical interlacing is omitted in order to control flicker on the screen. Color images are accomplished by the use of three electron beams swept together, one to excite each of the three color phosphors on the back surface of the screen. In this case, a single pixel has red, green, and blue phosphor components.

In the vector-scan method, the electron beam is guided across the back of the screen in a pattern that matches the pattern of the lines of the image it is displaying. It draws the image by the lines and strokes contained in it. Figure 12.5 compares the two methods and shows how a particular image is accomplished by both methods. The vector-scan method is preferred for applications requiring high writing speed such as oscilloscopes and spectrum analyzers.

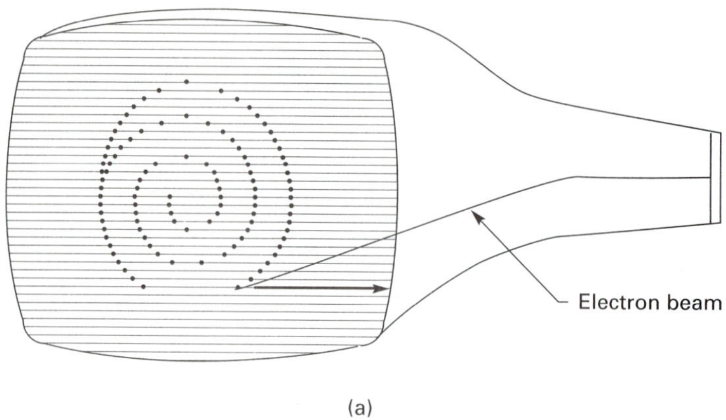

(a)

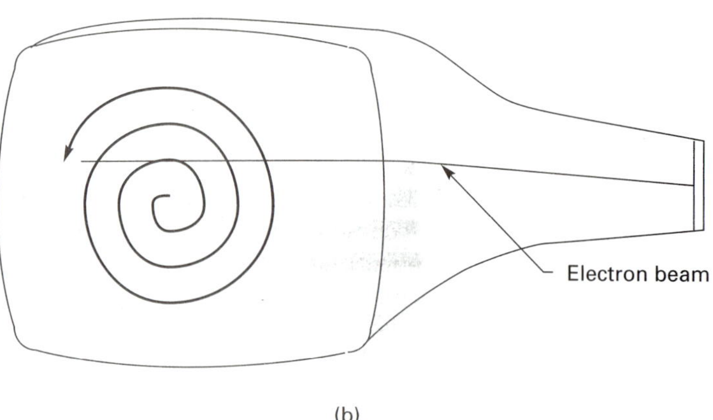

(b)

Figure 12.5 Comparison of (*a*) raster-scan and (*b*) vector-scan CRT display methods. The same image is shown being displayed on a raster-scan CRT and on a vector-scan CRT. The raster-scan method creates the image from top to bottom as the electron beam turns on those pixels which map to the image, one raster line at a time. The vector-scan method creates the image line by line as the electron beam traces the lines of the image.

A drawback of the vector-scan method is that it is very difficult to use the method with color CRT screens because of difficulty in aligning the three electron beams required to excite the three different colored phosphors on the screen, so vector-scan CRTs are limited to black and white displays.

There are several types of CRT hardware displays in the PC (personal computer) world that have evolved historically (Table 12.2). These are, in chronologic order of development, as follows: the MDA (monochrome display adapter), CGA (color graphics adapter), EGA (enhanced graphics adapter), and the VGA (video graphics array). The different hardware implementations

TABLE 12.2 Comparison of Different PC Hardware Display Capabilities

	MDA	CGA	EGA	VGA	Super-VGA
Alphanumeric text	Yes, 80 columns by 25 rows	Yes	Yes	Yes	Yes
Graphics	No	Yes	Yes	Yes	Yes
Resolution	720 × 350	640 × 200	640 × 350	720 × 400 text mode, 640 × 480 graphics mode	1024 × 768
Colors	Monochrome	16	16 of a palette of 64	16 of a palette of 262,144	256 of a palette of 262,144

differ in their resolution and color capabilities, with the MDA capable of displaying only monochrome text, the CGA capable of displaying graphics and text in 16 colors but at a low resolution, and the EGA capable of displaying 16 colors but at a higher resolution than CGA. VGA and super-VGA use an analog display signal instead of the digital signal that MDA, CGA, and EGA use, requiring a monitor designed specifically for use with VGA. VGA and super-VGA are capable of high resolution and use of 256 simultaneous colors out of a palette of 262,144.*

Meters. The advent of the modern computer has brought about a movement from analog design and representation of information toward digital design and representation of information. However, meters such as permanent-magnet moving-coil (PMMC) meters, which operate on and display analog values, are still often found on instrument front panels.

Meters are a type of single-number output device whose movement and data indication depend on a direct current passing through the terminals of the meter (Fig. 12.6). The deflection of the pointer is linearly proportional to the average value of the current. The deflection of the pointer and the scale may be calibrated to show any desired quantity, such as current, voltage, resistance, power, temperature, displacement, or velocity. The instrument produces a current whose value is a function of the quantity being measured; the PMMC movement responds to this direct current.

Audio. The bandwidth of the user interface can be increased by using other human sensory input and output channels. The use of audio cues, addressing the human sense of hearing, for warning or confirmation, is a good way to relieve the human visual channel, which is already overloaded in most user interfaces. Also, audio cues can be useful where visual cues might fail if they come at a time when the user has his or her visual attention focused else-

* For a detailed discussion of CRT technology and construction, see Chap. 14.

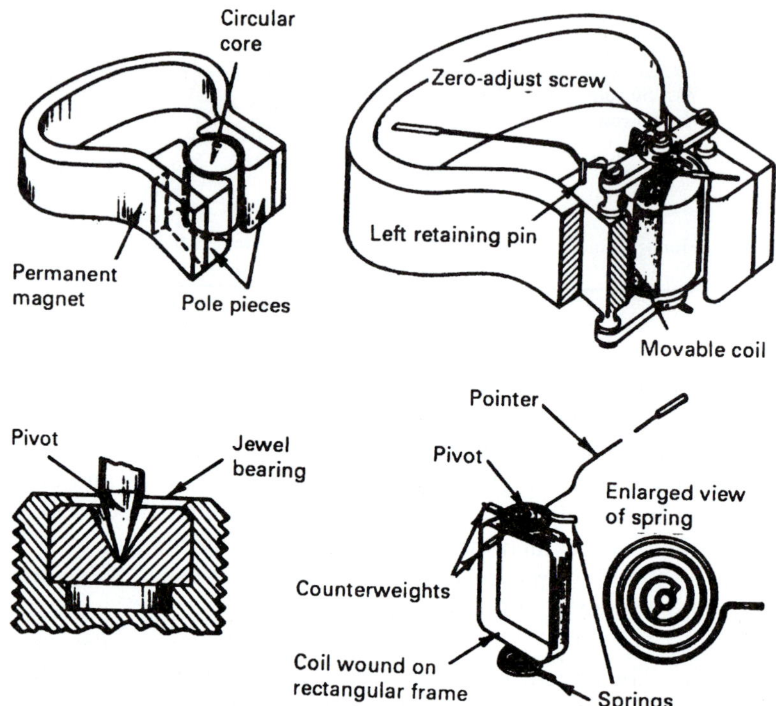

Figure 12.6 The permanent-magnet moving-coil (PMMC) meter movement has a coil of wire suspended in the field of a permanent magnet. When a current is passed through the coil, a torque is produced which causes the coil to rotate; the rotation is opposed by the restraining spring. The torque is proportional to the current in the coil; the deflection of the pointer–rotation of the coil–is proportional to the average value of the current, and a value for the average current may be read against a linear scale. If the shape of the magnet poles is altered, the deflection will no longer be proportional to the average current in the coil. This is used to expand a selected portion of the scale or to cause the deflection versus current relation to approximate some relationship other than a linear one.

where. Finally, audio cues contain spatial information that may be meaningful if the interface designers have taken advantage of it.

12.2.3 Hardware-user interface components: input devices

Input devices allow the user to issue commands directing the instrument to make measurements, set values for the measurement parameters, and respond to feedback on the instrument's output displays, among other things. The front panel of an instrument interface may provide three common types of physical devices for the user to control the instrument's measurement functions and information output displays. The three categories are keys, switches, and dials (Fig. 12.7).

The category of "keys" includes alphanumeric keys found on keyboards and

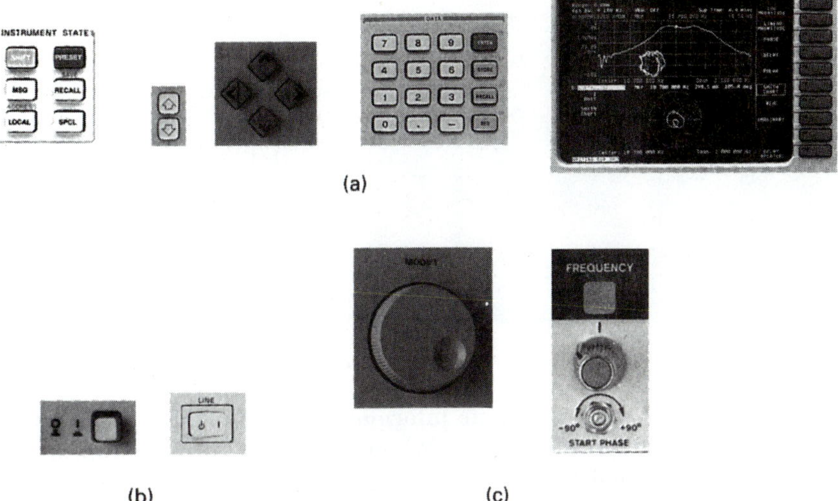

Figure 12.7 Keys, switches, and dials are three categories of hardware user input devices commonly found on instrument front panels. (*a*) Keys become active when they are pressed. Some keys remain active as long as they are pressed, such as the arrow keys shown above. User interactions with the instrument following a key press may override the key press action. Fkeys, shown above lining the CRT display, are a special class of keys. Fkeys, also called "softkeys," are programmable so that their meaning changes according to the label displayed next to them on the CRT. (*b*) Switches toggle the instrument between well-defined states, such as ON and OFF. The position of the switch gives visual feedback to the user as to the state the machine is in. In order to change the state of the machine, the user will have to change the position of the switch; other types of user interactions will not override a switch action. (*c*) Dials allow the user to control a parameter value in an analog, or qualitative, fashion. They usually have a physical range through which the user can turn the dial, giving good feedback as to the endpoints of the range of values allowed for the parameter. Dials also may be labeled with the range of values. As the dial is turned and the value of the parameter is affected, visual feedback about the value is often given to the user on the display.

keypads; keys used to set up the instrument for a measurement, such as the key labeled "Octave Analysis," which requests that the FFT (fast Fourier transform) be performed in a specific way; and programmable function keys,* whose meanings are derived from the user's previous actions.

"Switches" are used to toggle the instrument between a small number of well-defined states; an example is the on/off switch. "Dials" are used to control an analog parameter of the instrument such as voltage or frequency.

12.2.4 Design of the hardware-user interface components

Choice of controls. In order to be useful, display controls must be appropriate to the type of information they can present and the type of information de-

* Also called "fkeys."

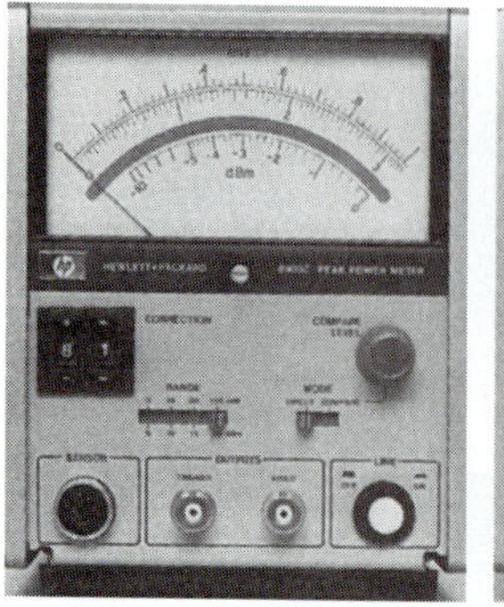

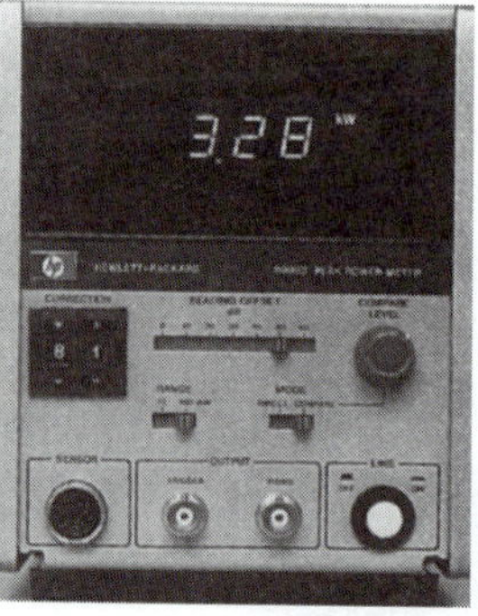

Figure 12.8 Analog and digital displays. Analog and digital displays contain different types of information, and each is the best choice under differing conditions of use. The analog display contains qualitative information of relative position and direction, while the digital display is a precise, numeric representation of the value being measured.

sired. For example, some information can be displayed in either an analog or digital format (Fig. 12.8). When choosing the most appropriate display, the designer must be aware that digital and analog displays have different information contents. An analog display contains qualitative and directional information about the location of the needle relative to the endpoint value of the range of possible values. Precise values are difficult to obtain from an analog display, however. A digital display is best for conveying precise information. In choosing between an analog or digital display, the designer also must understand the characteristics of the measurement the user will be performing. For example, if the user will be attempting to track a rapidly changing value, perhaps due to noise in the measurement, the user will be unable to read it on a digital display, so an analog dial is better in this case. Some instrument designs contain both analog and digital displays of the same information source, allowing the user to choose the appropriate display for the need at the moment.

Front panel design. Good design of the instrument front panel includes choosing the particular displays, keys, switches, and dials according to their appropriate use and the purpose the user will want to put them to. Selecting an LED display for graphic information and using a switch to set frequency range are inappropriate uses of the interface components.

For ease of use, switches, keys, and dials will be grouped according to the user tasks, either by locating them physically together or making all controls for one measurement look or feel distinct from other controls but similar to each other. For measurement tasks where the user's eyes will be located elsewhere, the controls should be distinguishable from each other without visual cues. Controls relevant to a user's particular task should be readily accessible.

Use of a control should give feedback to users that lets them know the consequence of their action. This feedback should occur immediately and be obvious to the users.

In general, good ergonomic design of the front panel interface can fail if there are too many input keys, switches, and dials, and/or if they are organized poorly. Some strategies in front panel designs attempt to overcome the "real estate" problem. Sometimes a key is given multiple meanings by having the user press another "shift key" before pressing the key in question. Another strategy is to use command menus displayed on the instrument's CRT display and accessed by function keys lined up vertically by the edge of the screen next to the menu selections. Implementing the front panel in software on a computer controller overcomes some of these limitations.

Modal interfaces. A user interface is "modal" if a user action, such as pushing a particular control key, will have different results depending on the other keys the user pushed immediately before. For example, an ATM has a modal interface. Pressing the "Primary Savings Account" key can mean either transfer money into this account or withdraw money from this account based on the user commands just prior to this key press.

Most instrument front panel interfaces are modal because of the limited space for all the control keys needed to make their use unique. Modes should be avoided where possible, but when not, the design of the interface should make it very clear to users which mode they are actually in and what the result of their interaction with the instrument controls will be. For example, the ATM provides good, textual feedback to users on the display describing unambiguously which mode they are in.

Ruggedness and form factor. Instruments must be designed to perform in their measurement environment. Contact with water, humidity, or dust and other environmental conditions must be protected against if they are commonly found in the environment of use. The instrument must be rugged enough to withstand the falls and knocks likely to occur while being used or set up to be used. And finally, the ergonomics of use dictate the size, power source requirements, etc. that the instrument should be designed to meet.

Chapter

13

Voltage, Current, and Resistance Measuring Instruments

Scott Stever
Agilent Technologies
Loveland, Colorado

13.1 Introduction

Voltage (both ac and dc), ac and dc current, and resistance are common quantities measured by electronic instruments. Meters are the easiest to use instrument for performing these measurements. In the simplest case, each measurement type is performed by an individual instrument—a voltmeter measures voltage, an ammeter measures current, and an ohmmeter measures resistance. These instruments have many elements in common. A multimeter combines these instruments—and sometimes others—together into a single general-purpose multifunction instrument.

13.1.1 Categories of meters

There are two primary types of meters—general purpose and specialty. General-purpose meters measure several types of electrical parameters such as voltage, resistance, and current. A "digital multimeter" (DMM) is an example of a common variety of general-purpose meter. Specialty meters are generally optimized for measuring a single parameter very well, emphasizing either measurement accuracy, bandwidth, or sensitivity. Each type of instrument is tuned for a different group of users and measurement applications. Table

TABLE 13.1 Types of Meters and Features Commonly Compared when Choosing a Meter

Type of meter	Multi-function	Measuring range	Frequency range	Speed, max readings/second	Best accuracy	Digits	Price range
			General Purpose				
Handheld DMM	Y	10 μV–1000 V 1 nA–10 A 10 mΩ–50 MΩ	20 Hz–20 kHz	2	0.1%	3½–4½	\$30–\$400
Bench DMM	Y	10 μV–1000 V 1 nA–10 A 10 mΩ–50 MΩ	20 Hz–100 kHz	10	0.01%	3½–4½	\$250–\$500
System DMM	Y	10 nV–1000 V 1 pA–1 A 10 μΩ–1 GΩ	1 Hz–10 MHz	50–100,000	0.0001%	4½–8½	\$600–\$8000
			Specialty				
ac Voltmeter	N	100 μV–300 V	20 Hz–20 MHz	1–10	0.1%	3½–4½	\$2000–\$4000
Nanovoltmeter	N	1 nV–100 V		1–100	0.005%	3½–7½	\$3000–\$6000
Picoammeter	N	10 fA–10 mA		1–100	0.05%	3½–5½	\$2000–\$4000
Electrometer	pA, high Ω	1Ω–1000 MΩ					
Microohmmeter	N			1–100	0.05%	3½–4½	\$1000–\$3000
High resistance	N	>10 TΩ		1–10	0.05%	3½–4½	\$2000–\$4000

13.1 presents a comparison of various types of meters and selected measuring capabilities.

The general-purpose multimeter is a flexible, cost-effective solution for most common measurements. DMMs can achieve performance rivaling the range and sensitivity of a specialty meter while delivering superior flexibility and value, as shown in Table 13.1. It is important to remember that the presence of many display digits on a meter does not automatically mean that the meter has high accuracy. Meters often display significantly more digits of resolution than their accuracy specifications support. This can be very misleading to the uninformed user.

13.2 General Instrument Block Diagram

The function of a meter is to convert an analog signal into a human- or machine-readable equivalent. Analog signals might be quantities such as a dc voltage, an ac voltage, a resistance, or an ac or dc current. Figure 13.1 illustrates the typical signal-conditioning process used by meters.

13.2.1 Signal conditioning: ranging and amplification

The input signal must first pass through some type of signal conditioner—typically comprising switching, ranging, and amplification circuits, as shown in Fig. 13.1. If the input signal to be measured is a dc voltage, the signal conditioner may be composed of an attenuator for the higher voltage ranges and a dc amplifier for the lower ranges. If the signal is an ac voltage, a converter is used to change the ac signal to its equivalent dc value. Resistance measurements are performed by supplying a known dc current to an un-

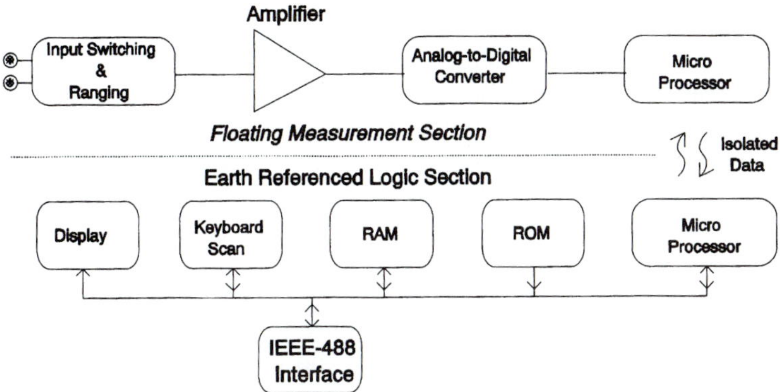

Figure 13.1 Generalized block diagram of most modern meters. The signal conditioning circuits (comprised of the input switching and ranging and amplifier sections), and ADC sections vary for different types of meters.

known resistance—converting the unknown resistance value to an easily measurable dc voltage. In nearly all cases, the input signal switching and ranging circuits along with the amplifier circuits convert the unknown quantity to a dc voltage which is within the measuring range of the analog-to-digital converter (ADC).

13.2.2 Analog-to-digital conversion

The job of the ADC is to take a prescaled dc voltage and convert it to digits. For example, the ADC for a 6½ digit resolution (21-bit) instrument is capable of producing over 2.4 million unique reading values. You can think of this as a bar chart with 2.4 million vertical bars—each bar increasing in size from the previous bar by an identical amount. Converting the essentially infinite resolution of the analog input signal to a single bar in our chart is the sole function of the ADC. The continuum of analog input values is partitioned—quantized—into 2.4 million discrete values in our example.

The ADC used in a meter governs some of its most basic characteristics. These include its measurement resolution, its speed, and in some cases its ability to reject spurious noise. The many methods used for analog-to-digital conversion can be divided into two groups—integrating and nonintegrating. "Integrating" techniques measure the average input value over a relatively long interval, while "nonintegrating" techniques sample the instantaneous value of the input—plus noise—during a very short interval. ADCs are designed strictly for dc voltage inputs. They are single-range devices—some take a 3-V full-scale input, while others take a 12-V full-scale input. For this reason, the input switching and ranging circuits must attenuate higher voltages and amplify lower voltages to give the meter a selection of ranges.

13.2.3 Managing the flow of information

The first three blocks in Fig. 13.1 combine to produce a meter's overall analog performance characteristics—measuring functions, ranges, sensitivity, reading resolution, and reading speed. The two microprocessor blocks manage the flow of information within the instrument—ensuring that the various subsystems are properly configured and that internal computations are performed in a systematic and repeatable manner. Convenience features such as automatic range selection are managed by these microprocessors. Electrical isolation is also provided between the earth-referenced outside world and sensitive measuring circuits. The earth-referenced microprocessor also acts as a communications interpreter—managing the outward flow of data to the display or IEEE-488 computer interface while accepting keyboard and programming instructions. These microprocessors govern the instrument's overall measurement functionality, responsiveness, and friendly, intuitive human interface characteristics which are uniquely tuned for each type of measuring instrument.

13.3 DC Voltage Measurement Techniques

Signal conditioning and analog-to-digital conversion sections have the greatest influence on the characteristics of a dc meter. The ADC measures over only one range of dc voltage, and it usually has a relatively low input resistance. To make a useful dc meter, a "front end" is required to condition the input before the analog-to-digital conversion. Signal conditioning increases the input resistance, amplifies small signals, and attenuates large signals to produce a selection of measuring ranges.

13.3.1 Signal conditioning for dc measurements

Input signal conditioning for dc voltage measurements includes both amplification and attenuation. Figure 13.2 shows the typical configuration for the dc input switching and ranging section of a meter. The input signal is applied directly to the amplifier input through switches K_1 and S_1 for lower voltage inputs—generally those less than 12 V dc. For higher voltages, the input signal is connected through relay K_2 to a precision 100:1 divider network formed by resistors R_4 and R_5. The low voltage output of the divider is switched to the amplifier input through switch S_2.

The gain of amplifier A_1 is set to scale the input voltage to the full scale range of the ADC, generally 0 ± 12 V dc. If the nominal full-scale input to the ADC is 10 V dc, the dc input attenuator and amplifier would be configured to amplify the 100-mV range by $\times 100$ and to amplify the 1-V range by $\times 10$. The input amplifier would be configured for unity gain, $\times 1$, for the 10-V measuring range. For the upper ranges, the input voltage is first divided by 100, and then gain is applied to scale the input back to 10 V for the ADC—inside

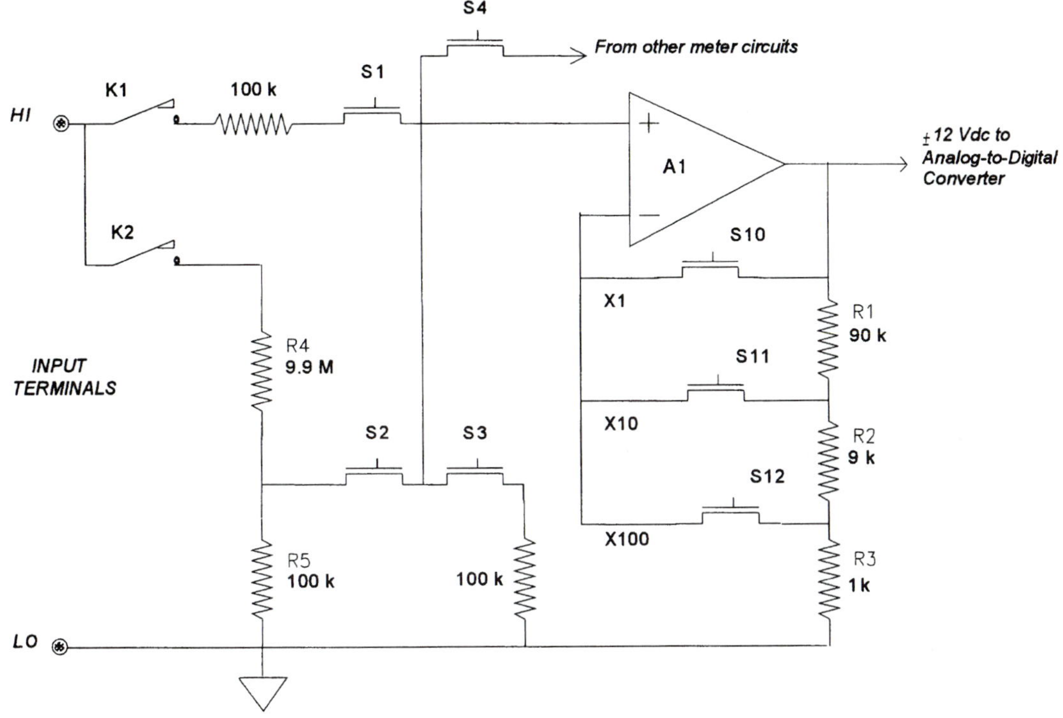

Figure 13.2 Simplified schematic of the input switching, measuring range selection, and amplifier for a dc voltage meter.

the meter, 100 V dc is reduced to 1 V dc for the amplifier, and 1000 V dc is divided to become 10 V dc.

For the lower voltage measuring ranges, the meter's input resistance is essentially that of amplifier A_1. The input amplifier usually employs a low-bias current—typically less than 50 pA—FET input stage yielding an input resistance greater than 10 GΩ. The meter's input resistance is determined by the total resistance of the 100 : 1 divider for the upper voltage ranges. Most meters have a 10-MΩ input resistance for these ranges.

13.3.2 Amplifier offset elimination: autozero

The main performance limitation of the dc signal-conditioning section is usually its offset voltage. This affects the meter's ability to read zero volts with a short applied. Most meters employ some method for automatically zeroing out amplifier offsets. Switch S_3 in Fig. 13.2 is used to periodically "short" the amplifier input to ground to measure the amplifier offset voltage. The measured offset is stored and then subtracted from the input signal measurement to remove amplifier offset errors. Switches S_1 and S_2 are simultaneously opened during the offset measurement to avoid shorting the meter's input terminals together.

In a multifunction instrument, all measurements are eventually converted into a dc voltage which is measured by an ADC. Other dc signals are often routed to the ADC through a dc voltage measuring front end—switch S_4 in Fig. 13.2 could be used to measure the dc output of an ac voltage function or a dc current measuring section.

13.4 AC Voltage Measurement Techniques

The main purpose of an ac front end is to change an incoming ac voltage into a dc voltage which can be measured by the meter's ADC. The type of ac voltage to dc voltage converter employed in a meter is very critical. There are vast differences in behavior between rms, average-responding, and peak-responding converters—these differences are discussed in detail later in this section. Always be sure you understand the type of ac converter your meter employs and what its capabilities and limitations are.

13.4.1 Signal conditioning for ac measurements

The input signal conditioning for ac voltage measurements includes both attenuation and amplification just like the dc voltage front end already discussed. Figure 13.3 shows typical input switching and ranging circuits for an

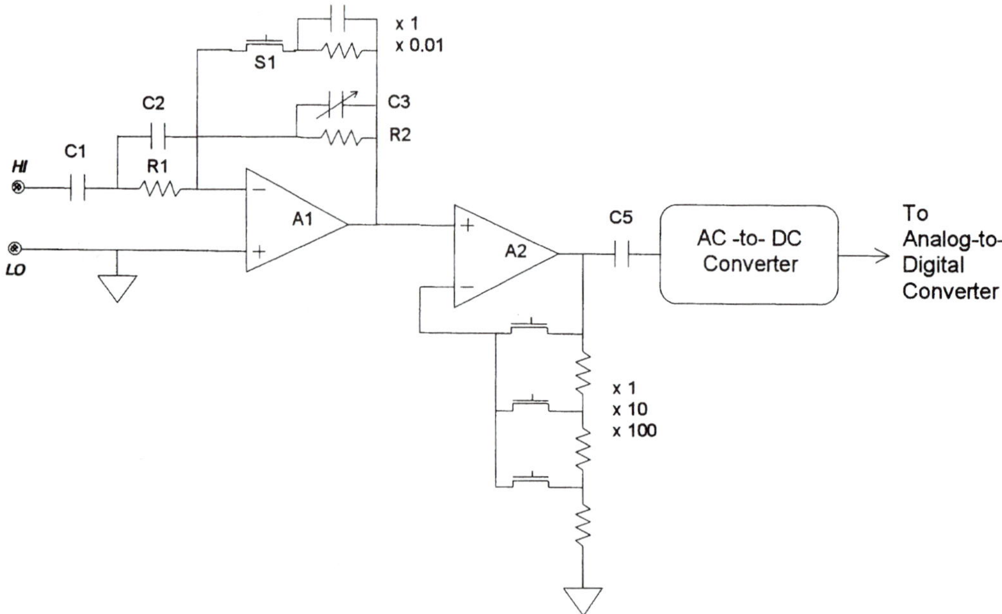

Figure 13.3 Simplified schematic of the input switching and ranging sections of a typical ac voltage measurement section.

ac voltage instrument. Input coupling capacitor C_1 blocks the dc portion of the input signal so that only the ac component is measured by the meter. Ranging is accomplished by combining signal attenuation from first-stage amplifier A_1 and gain from second-stage amplifier A_2.

The first stage implements a high input impedance—typically 1 MΩ, switchable compensated attenuator. The value of capacitor C_3 is adjusted so that the R_2C_3 time constant precisely matches the R_1C_2 time constant—yielding a compensated attenuator whose division ratio does not vary with frequency. Switch S_1 is used to select greater attenuation for the higher input voltage ranges. The second stage provides variable gain, wide bandwidth signal amplification to scale the input to the ac converter to the full-scale level. The output of the second stage is connected to the ac converter circuit. Any residual dc offset from the attenuator and amplifier stages is blocked by capacitor C_5.

An ac voltage front end similar to the one discussed above is also used in ac current measuring instruments. Shunt resistors are used to convert the ac current into a measurable ac voltage. Current shunts are switched to provide selectable ac current ranges.

Amplifier bandwidth and ac converter limitations provide the main differences between various ac front ends. As mentioned earlier, the type of ac to dc converter circuit will have a profound effect on overall measurement accuracy and repeatability. True rms converters are superior to both average-responding and peak-responding ac converters in almost every application.

13.4.2 AC signal characteristics

The most common ac voltage or current signal is the sine wave. In fact, all periodic wave shapes are composed of sine waves of varying frequency, amplitude, and phase added together. The individual sine waves are harmonically related to each other—that is to say, the sine wave frequencies are integer multiples of the lowest, or fundamental, frequency of the waveform. Moreover, ac waveforms exhibit many interesting characteristics. Unlike dc signals, the amplitude of ac waveforms varies with time, as shown in Fig. 13.4.

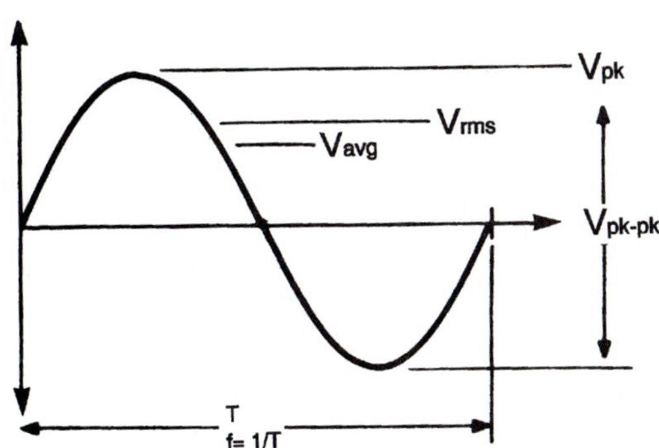

Figure 13.4 The voltage of a sine wave can be uniquely described by any of the parameters indicated—the peak-to-peak value, peak value, rms value, or average value and its period T or frequency I/T.

TABLE 13.2 Table of Wave Shapes, rms Values, and Measurement Error for an Average Responding ac Voltmeter

Waveform shape	Crest Factor (C.F.)	AC RMS	AC + DC RMS	Average responding error
	1.414	$\frac{V}{1.414}$	$\frac{V}{1.414}$	Calibrated for 0 error
	1.732	$\frac{V}{1.732}$	$\frac{V}{1.732}$	−3.9%
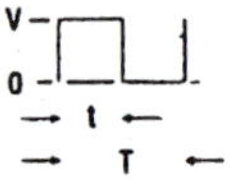	$\sqrt{\frac{T}{t}}$	$\frac{V}{C.F.} \times \sqrt{1-\left(\frac{1}{C.F.}\right)^2}$	$\frac{V}{C.F.}$	−46% for C.F. = 4

The magnitude of a sine wave can be described by its rms value—effective heating value—average value, or peak value. Each describes the amplitude of a sine wave. Table 13.2 shows several common waveforms along with a tabulation of their respective peak, ac rms, and ac + dc rms values. Approximate errors are also shown for measurements performed with an averaging-type ac meter along with the crest factor ratio for the various wave shapes. "Crest factor" (CF) defines the peak amplitude to rms amplitude ratio: CF = V_{peak}/V_{rms}.

13.4.3 Rms value

The "rms value"—or "root-mean-square value"—is the only amplitude characteristic of a wave-form which does not depend on shape. Therefore, the rms value is the most useful means to quantify signal amplitude in ac measurements. The rms value measures the ability of an ac signal to deliver power to a resistive load—thus measuring the equivalent heating value of the signal. This means that the rms value of an ac waveform is equal to the dc value which produces the same amount of heat as the ac waveform when connected to the same resistive load.

For a dc voltage V, this heat is directly proportional to the amount of power P dissipated in the resistance R: $P = V^2/R$. Note that the power is proportional to the square of the dc voltage. For an ac voltage, the heat in a resistive load is proportional to the average of the instantaneous power dissipated in the resistance:

$$P_{ave} = V_{rms}^2/R = 1/T \int_0^T V_i^2/R\,dt \tag{13.1}$$

The definition of rms voltage is derived from Eq. 13.1 by solving for V_{rms}:

$$V_{rms} = \left(\frac{1}{T}\int_0^T TV_i^2\,dt\right)^{1/2} \tag{13.2}$$

where V_i is the input waveform, and T is the period of the waveform.

This means that the rms value of an ac waveform is equal to the square root of the sum of the squares of the instantaneous values averaged over the period of the waveform. This, of course, has meaning only for periodic signals. Said another way, the rms value of a periodic waveform is obtained by measuring the dc voltage at each point along one complete cycle, squaring the measured value at each point, finding the average value of the squared terms, and finally, taking the square root of the average value. This technique leads to the rms value of the waveform regardless of the wave shape.

The most straightforward way to measure the rms value of a waveform is to measure the heat it generates in a resistive load and compare it with the heat generated by a known dc voltage in an equivalent load. Devices that perform this measurement are called "thermal rms detectors." Many varieties have been developed, some as straightforward as small resistors manufactured with a thermocouple or thermistor attached. Other detectors are as sophisticated as silicon integrated circuits fabricated using micromachining techniques. These sensors provide matched detectors comprised of a precision thin-film resistor and a solid-state temperature sensor. Thermal rms detectors are capable of measuring rms values for signal frequencies in excess of a few hundred megahertz.

Equation 13.2 also can be solved using analog computing techniques. Nonlinear analog feedback circuits are combined to solve the rms equation. Several analog integrated circuit suppliers market parts designed to compute the ac + dc rms value. Their performance is generally usable for sine wave frequencies in excess of a few hundred kilohertz.

The rms value also can be computed precisely using data sampled from the waveform. Samples must be acquired at a rate greater than twice the highest harmonic frequency of the signal. The samples are squared, the squared values are summed over some averaging interval *T,* and then a square root is performed on the sum of squared values—thus yielding the rms value of the sampled signal. These mathematical operations can be performed either directly in a digital computer or in digital signal processing (DSP) hardware. Many instruments, including system DMMs and oscilloscopes, use this sampling technique. The signal-frequency range of this technique is theoretically limited solely by available sampler and ADC rates.

13.4.4 Average value

The average value of an ac waveform is simply the average of the instantaneous values measured over one complete cycle: $V_{\text{ave}} = 1/T \int_0^T V_i \, dt$. For sine waves, the average amplitude is zero because the waveform has equal positive and negative half cycles. Since the quantity of interest is the heating value of the signal, the average value of a sine wave is taken to mean the average of the full-wave rectified waveform. For sine wave shapes, the rms value can be calculated from the average by $V_{\text{rms}} = (\pi/2\sqrt{2}) \times V_{\text{ave}} = 1.11 \times V_{\text{ave}}$.

This relationship does not hold true for other wave shapes, as indicated in

Table 13.2. Average responding ac meters will generally indicate the rms value by multiplying the measured average value by 1.11. Of course, this correction yields accurate rms values only for sine wave signals. Therefore, average-responding meters are not well suited for precision measurements when nonsinusoidal wave shapes might be applied. Indeed, even small amounts of odd harmonic distortion of a sine wave input can cause large errors in the readings of an average-responding ac converter.

13.4.5 Peak value

The peak amplitude of a complex ac waveform does not necessarily correlate with the rms heating value of the signal. However, if the wave shape is known, the rms value can be approximated.

Like average-responding meters, peak-responding meters usually display the rms value by multiplying the peak reading by 0.707—the inverse of 1.414, the sine wave crest factor value. This correction is accurate for pure sine wave signals only. Peak-responding meters will exhibit several percentage points of error when measuring sine waves with only a few percentage points of second or third harmonic distortion. Again, peak-responding meters are not well suited for precision measurements when nonsinusoidal wave shapes might be applied.

13.4.6 Signal-coupling effects on measured amplitude

Many signals can be thought of as a combination of a dc component and an ac component. Some ac meters measure only the ac signal, while others measure both the ac and dc signal components. It is very important to know which a meter measures. Average-responding and peak-responding meters are usually ac-coupled—that is, they measure only the ac signal component, while dc is rejected. On the other hand, rms responding meters can be either ac-coupled or dc-coupled. A dc-coupled rms responding meter—sometimes called "ac + dc coupled rms"—measures the true "heating value" of the *entire* signal. An ac-coupled rms responding meter measures the "heating value" of *only* the ac components of a waveform. The ac rms and ac + dc rms values are equal for many common wave shapes such as sine waves, triangle waves, and square waves. Other waveforms, such as pulse trains, contain dc voltages which are rejected by ac-coupled rms meters.

An ac-coupled rms measurement is desirable in situations where you are measuring small ac signals in the presence of large dc offsets. For example, this situation is common when measuring ac ripple present on dc power supplies. There are situations, however, where you might want to know the ac + dc rms value. You can determine this value by combining results from an ac-coupled rms measurement and a dc-only measurement, as shown in Eq. 13.3. The dc measurement should be performed using a meter incorporating an integrating ADC capable of integration times of at least 100 ms for best rejection of the ac components.

$$\mathrm{rms}_{\mathrm{ac+dc}} = \sqrt{\mathrm{ac}^2 + \mathrm{dc}^2} \tag{13.3}$$

13.5 Current Measurement Techniques

An ammeter senses the current flowing through its input connections—approximating a short circuit between its input terminals. A conventional ammeter must be connected in series with the circuit or device being measured such that current flows through both the meter and the test circuit. There are two basic techniques for making current measurements—in-circuit methods and magnetic field sensing methods.

13.5.1 In-circuit methods

In-circuit current sensing meters use either a current shunt or virtual ground amplifier technique similar to those shown in Fig. 13.5*a* and *b*. Shunt-type

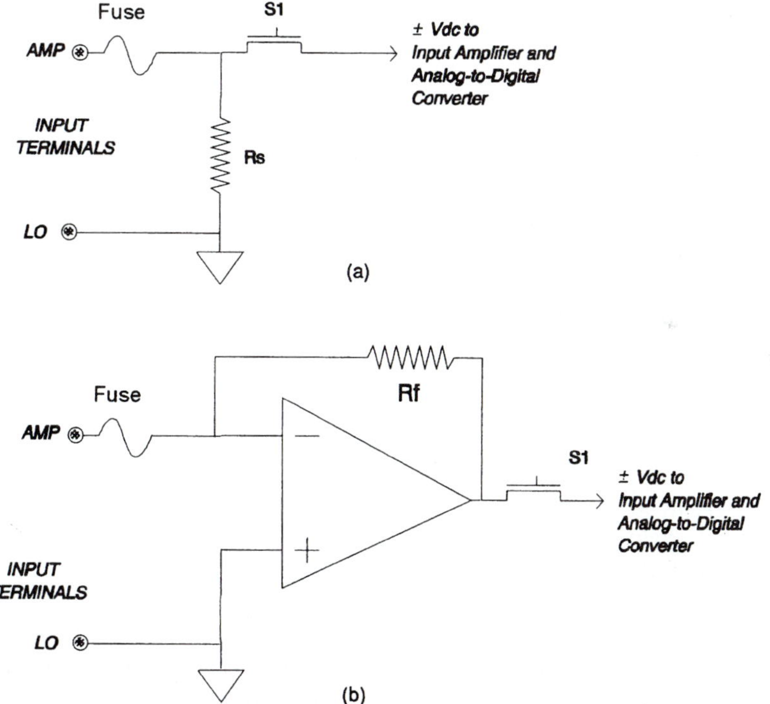

Figure 13.5 Two common methods for in-circuit current measurements. (*a*) Shunt resistor R_s is connected across the input terminals, developing a voltage proportional to the input current. (*b*) The input current is forced to flow through R_f while the meter burden voltage is limited to the drop across the fuse and the amplifier offset voltage.

meters are very simple—a resistor, R_s in Fig. 13.5*a*, is connected across the input terminals such that a voltage drop proportional to the input current is generated. The value of R_s is kept as low as possible to minimize the instrument's "burden voltage," or *IR* drop. This voltage drop is sensed by an internal voltmeter and scaled to the proper current value to complete the measurement.

Virtual ground-type meters are generally better suited for measuring smaller current values—usually 100 mA to below 1 pA. These meters rely on low-noise, low-bias-current operational amplifiers to convert the input current to a measurable voltage, as illustrated in Fig. 13.5*b*. Negligible input current flows into the negative input terminal of the amplifier—therefore, the input current is forced to flow through the amplifier's feedback resistor R_f, causing the amplifier output voltage to vary by IR_f. The meter burden voltage—the voltage drop from input to LO—is maintained near zero volts by the high-gain amplifier forming a "virtual ground." Since the amplifier must source or sink the input current, this technique is generally limited to lower current measurements.

13.5.2 Magnetic field sensing methods

Current measurements utilizing magnetic field sensing techniques are extremely convenient. Measurements can be performed without interrupting the circuit or producing significant loading errors. Since there is no direct contact with the circuit being measured, complete dc isolation is also ensured. These meters utilize a transducer—usually a current transformer or solid-state Hall-effect sensor—to convert the magnetic field surrounding a current-carrying conductor into a proportional ac or dc signal. Sensitivity can be very good, since simply placing several loops of the current-carrying conductor through the probe aperture will increase the measured signal level by the same factor as the number of turns.

13.5.3 AC current

Ac current measurements are very similar to dc current measurements. However, ac measurements employ shunt-type current-to-voltage converters almost exclusively. The output of the current-to-voltage sensor is measured by an ac voltmeter. Signal and ac converter issues discussed in Sec. 13.4 are relevant to ac current measurements as well.

The input terminals of in-circuit ac current meters are always direct coupled—ac + dc coupled—to the shunt so that the meter maintains dc continuity in the test circuit. The meter's internal ac voltmeter section can be either ac coupled or ac + dc coupled to the current-to-voltage converter. Performing ac current measurements demands additional care. "Burden voltage"—loading—varies with frequency and meter and input lead inductance, often causing unexpected behavior in the test circuit.

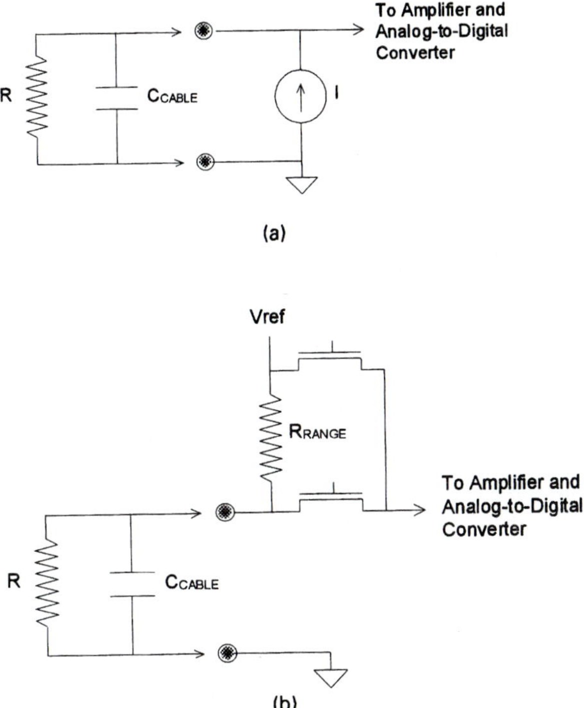

Figure 13.6 Two common types of ohms converter circuits used in meters. (*a*) The current-source ohms converter employs a constant current source, forcing current I through unknown resistance R, developing a voltage to be measured by a dc voltage front end. (*b*) The voltage-ratio-type ohms converter calculates the unknown resistor value R from dc voltage measurements in a voltage divider circuit.

13.6 Resistance Measurement Techniques

An ohmmeter measures the dc resistance of a device or circuit connected to its input. As mentioned earlier, resistance measurements are performed by supplying a known dc current to an unknown resistance—converting the unknown resistance value to an easily measured dc voltage. Most meters use an ohms converter technique similar to the current source or voltage ratio types shown in Fig. 13.6.

13.6.1 Signal conditioning for resistance measurements

The current-source method shown in Fig. 13.6*a* employs a known current-source value I which flows through the unknown resistor when it is connected to the meter's input. This produces a dc voltage proportional to the unknown resistor value: by Ohm's law, $E = IR$. Thus dc voltmeter input-ranging and

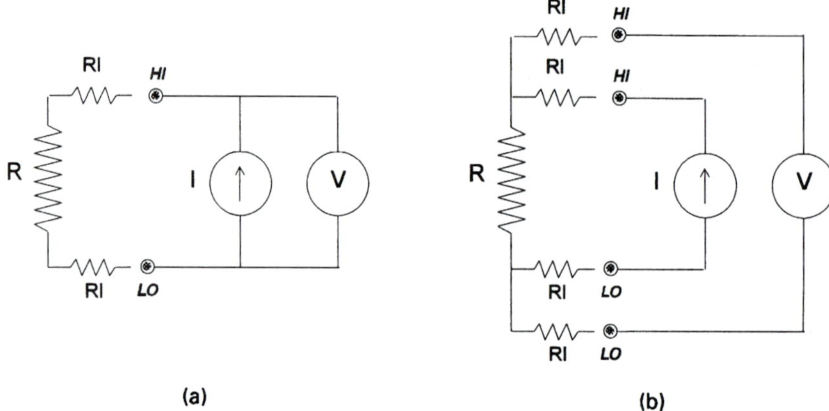

Figure 13.7 (*a*) Simplified schematic for two-wire sensing. Lead resistances R_l are inseparable from the unknown resistance measurement. (*b*) Simplified schematic for four-wire sensing. This measurement is relatively insensitive to lead resistances R_l in the high-impedance input of the dc voltmeter. Voltage drops in the current source leads do not affect the voltmeter measurement; however, it can affect the accuracy of the current source itself.

signal-conditioning circuits are used to measure the voltage developed across the resistor. The result is scaled to read directly in ohms.

Figure 13.6*b* shows the voltage ratio-type ohms converter technique. This method uses a known voltage source V_{ref} and a known "range" resistor R_{range} to compute the unknown resistor value. The range resistor and the unknown resistor form a simple voltge divider circuit. The meter measures the dc voltage developed across the unknown resistor. This voltage, along with the values of the internal voltage source and range resistor, is used to calculate the unknown resistor value.

In practice, meters have a variety of resistance measuring ranges. To achieve this, the ohms test current—or range resistor—is varied to scale the resulting dc voltage to a convenient internal level, usually between 0.1 and 10 V dc.

13.6.2 Two-wire sensing

The ohms converters discussed above utilize two-wire sensing. When the same meter terminals are used to measure the voltage dropped across the unknown resistor as are used to supply the test current, a meter is said use a two-wire ohms technique. With two-wire sensing, the connection lead resistances R_l shown in Fig. 13.7*a* are indistinguishable from the unknown resistor value—causing potentially large measurement errors for lower-value resistance measurements. The two-wire technique is widely used by all types of ohmmeters due to its simplicity. Often, meters provide a "relative" or "null" math function to allow lead resistances to be measured and subtracted from subsequent resistance measurements. This works well unless the lead resistances vary due to temperature or connection integrity. The four-wire ohms technique—or "Kelvin sensing"—is designed to eliminate these lead resistance errors.

13.6.3 Four-wire sensing

The four-wire sensed-resistance technique provides the most accurate way to measure small resistances. Lead resistances and contact resistances are automatically reduced using this technique. A four-wire converter senses the voltage dropped just across the unknown resistor. The voltages dropped across the lead resistances are excluded from measurement, as shown in Fig. 13.7*b*. The four-wire converter works by using two separate pairs of connections to the unknown resistor. One connection pair—often referred to as the "source leads"—supplies the test current which flows through the unknown resistor, similar to the two-wire measurement case. Voltage drops are still developed across the source lead resistances. A second connection pair—referred to as the "sense leads"—connects directly across the unknown resistor. These leads connect to the input of a dc voltmeter. The dc voltmeter section is designed to exhibit an extremely large input resistance so that virtually no current flows in the sense input leads. This allows the meter to "sense" the voltage dropped just across the unknown resistor. This scheme removes from the measurement voltage drops in both the source leads and the sense leads. Generally, lead resistances are limited by the meter manufacturer. There are two main reasons for this. First, the total voltage drop in the source leads will be limited by the design of the meter—usually limited to a fraction of the meter measuring range being used. Second, the sense lead resistances will introduce additional measurement noise if they are allowed to become too large. Sense leads less than 1 kΩ usually will contribute negligible additional error.

The four-wire technique is widely used in systems where lead resistances can become quite large and variable. It is often used in automated test applications where cable lengths can be quite long and numerous connections or switches may exist between the meter and the device under test. In a multichannel system, the four-wire method has the obvious disadvantage of requiring twice as many switches and twice as many wires as the two-wire technique. The four-wire method is used almost exclusively for measuring lower resistor values in any application—especially for values less than 10 Ω.

13.6.4 Offset compensation

Many components utilize materials which produce small dc voltages due to dissimilar metal thermocouples or electrochemical batteries. Unexpected dc voltages will add error to resistance measurements. The offset-compensated resistance technique is designed to allow resistance measurements of components in the presence of small dc voltages.

Offset compensation makes two measurements on the circuit connected to the meter input terminals. The first measurement is a conventional resistance measurement. The second is the same except the test current source is turned off—this is essentially a normal dc voltage measurement. The second dc voltage measurement is subtracted from the first voltage measurement prior to scaling the result for the resistance reading—thereby giving a more accurate resistance measurement.

13.7 Sources of Measurement Error

Meters are capable of making highly accurate measurements. In order to achieve the greatest accuracy, you must take the necessary steps to eliminate extraneous error sources. This section describes common problems encountered and offers suggestions to help you minimize or eliminate these sources of measurement error.

13.7.1 Thermal EMF errors

Thermoelectric voltages are the most common source of error in low-level dc voltage measurements. Thermoelectric voltages are generated when you make circuit connections using dissimilar metals at different temperatures. Each metal-to-metal junction forms a thermocouple, which generates a voltage proportional to the junction temperature. The net voltage generated by the dissimilar metals is proportional to the temperature difference at the two metal-to-metal junctions. You should take the necessary precautions to minimize thermocouple voltages and temperature variations in low-level voltage measurements. The best connections are formed using copper-to-copper crimped connections. Table 13.3 shows thermoelectric voltages for connections between copper and various common dissimilar metals.

TABLE 13.3 Thermoelectric Potentials

Junction materials: Copper to	Thermoelectric voltages, $\sim\mu$V/°C
Copper	<0.2
Cadmium-tin solder	0.2
Silver	0.3
Gold	0.5
Brass	3
Beryllium copper	3
Tin-lead solder	3
Aluminum	5
Kovar or alloy 42	40
Silicon	500
Copper-oxide	1000

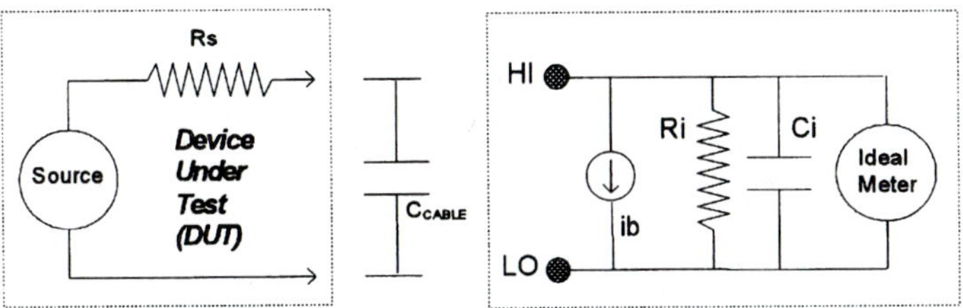

Figure 13.8 Simplified schematic of a meter's input impedance and the device under test (DUT). A voltage divider is formed between source resistance R_s and the meter input impedance R_iC_i which introduces additional measurement error.

13.7.2 Loading errors

Measurement loading errors occur when the impedance of the device under test (DUT) is an appreciable percentage of the meter's own input impedance for the selected measuring function and range. The meter input impedance R_iC_i forms a voltage divider with the source impedance R_s of the DUT, as shown in Fig. 13.8. Measurement errors can be quite large and unexpected. These errors are largely determined by the DUT and can be difficult to detect because most calibration sources have near-zero output impedance R_s.

The meter input resistance R_i generally varies with the selected measuring range for the dc voltage function, typically from 10 MΩ to > 10 GΩ. Handheld meters often exhibit a fixed 10-MΩ input impedance for these ranges. The input capacitance (C_i + C_{cable}) will charge up to V_{source} when R_i is large. The meter will actually hold the last input voltage after it is removed until R_i discharges the capacitance or until the meter's input bias current i_b charges the input capacitance to a new value. This can cause errors in multiplexed systems when the system is charged to the voltage on the previous channel and then the next channel is open circuited or measuring a device with high output impedance. Note that the bias current i_b will charge C_i even without an input appliled. This effect is sometimes mistaken for a noisy or broken meter. High-performance meters such as system DMMs, picoammeters, high-resistance meters, and electrometers are designed for R_i > 10 GΩ.

The input impedance of a meter's ac voltage function will usually be different from the dc function, typically a constant 1-MΩ R_i in parallel with approximately 100-pF C_i. Meter input capacitance is the dominant impedance factor for most ac measurements. The wiring that you use to connect signals to the meter will add additional capacitance C_{cable} and ac loading effects. Table 13.4 shows how a typical meter's ac input impedance can vary with input frequency.

13.7.3 Settling time errors

Settling time errors are most common when making resistance measurements in excess of 100 kΩ. Meter input capacitance, cabling capacitance, and

TABLE 13.4 Equivalent Input Impedance of an ac Meter Varies with Frequency Causing Additional Loading Errors

Frequency	Input impedance	Frequency	Input impedance
100 Hz	1 MΩ	10 kHz	160 kΩ
1 kHz	850 kΩ	100 kHz	16 kΩ

other stray capacitances quickly add to unexpectedly high values. Settling due to *RC* time constant effects can be quite long. Some precision resistors and multifunction calibrators use large capacitors (1000 pF to 0.1 μF) with higher resistance values to filter out noise currents injected by their internal circuitry. Nonideal capacitances in cables and other devices may exhibit significantly longer settling times than expected due to dielectric absorption—"soak"—effects. Settling times often exhibit a linearly decreasing error with time instead of the expected exponentially decreasing error. For best accuracy, allow an appropriate settling time following initial connection and following a change of range or measuring function.

Thermal settling errors are also common. They can be produced by either thermal offset voltages or power coefficient effects. Small thermal offsets are usually generated following changes in connections or changes in instrument configuration. Additionally, measurements performed immediately following a measurement that produced a significant power coefficient error in the input signal-conditioning circuits may exhibit slight gain and offset errors which will vary with time. For best accuracy, always allow as much time as is practical in your application for instrument and circuit settling effects to stabilize before performing your measurement.

13.7.4 Power coefficient errors

Nonlinearities caused by the signal-conditioning circuits produce smoothly varying errors or integral nonlinearities. Power coefficient errors are a common source of nonlinearities in signal-conditioning circuits. These errors occur in high-voltage dividers or amplifier feedback dividers when large differences in power dissipation exist between two or more resistors. One resistor changes value due to a slight increase in its operating temperature from self-heating, while the other resistor value remains constant—thus changing the divider ratio in proportion to the square of the input voltage. Power coefficient nonlinearities are generally the major error source in voltage measurements above several hundred volts. This effect often causes insidious, time-dependent gain or offset errors in other measurements performed following a higher-voltage measurement. This is especially true in instruments which exhibit larger power coefficient errors on at least one measuring range or function.

13.7.5 Input bias current errors

All meters exhibit some finite input "leakage" or bias current as shown by i_b in Fig. 13.8. This current is caused by the signal-conditioning, ranging, and input protection circuits inside the meter. The magnitude of this current will range from a few picoamps (1 pA = 1×10^{-12} A) to as much as a few nanoamps (1 nA = 1×10^{-9} A). This current will usually exhibit a strong temperature sensitivity when the instrument is operated above about 30°C, often doubling in size for every 8 to 10°C change in operating environment.

Instrument bias currents will generate small voltage offsets which are dependent on the source resistance R_s of the DUT. This effect becomes particularly evident for $R_s > 100$ kΩ or when the meter's operating environment is significantly warmer than 30°C.

It is common for the meter's input capacitance to "charge up" due to bias current when the input is open-circuited. If the input resistance of the meter is quite large, $R_i > 10$ GΩ, the meter's open circuit voltage will slowly ramp up—or down—as charge ($i_b \times$ time) is accumulated on the input capacitance C_i. The voltage will continue to change until an equilibrium condition is reached. This effect is sometimes mistaken for a broken or noisy instrument when in fact it is a natural indication of a high-quality high-input-resistance instrument. An autoranging meter with open input terminals will occasionally change ranges while the input voltage is ramping, alternately discharging C_i with the meter's internal 10-MΩ high-voltage divider and then charging C_i on the meter's very high R_i low-voltage ranges.

13.7.6 Normal mode noise rejection (NMR): Rejecting power-line noise

Noise is a serious problem for any measurement. It can be an especially severe problem for instruments with high resolution or high sensitivity. Noise may be defined by its origin relative to the signal input connections. "Normal mode" noise enters with the signal and is superimposed on it. "Common mode" noise is common to both the high and low signal inputs. Normal mode noise usually originates from power-line pickup, magnetic coupling, or noise from the device being measured. Noise can be sinusoidal, spikes, white noise, or any unwanted signals.

Filtering is used to reduce normal mode noise. Filtering can be accomplished in a variety of ways—by passive RC filters, averaging of sampled data, or utilizing the intrinsic filtering characteristics of an integrating ADC. Integration makes a continuous measurement over a fixed time interval during which amplitude variations are "averaged out." If the integration time includes an integral number of periodic noise cycles, the noise will be completely averaged out. A meter with a 1/60 s integration period would average out one complete 60-Hz noise cycle, two complete 120-Hz cycles, four complete 240-Hz cycles, etc., as illustrated in Fig. 13.9. Normal mode noise rejection (NMR) is specified at specific frequencies—the noise frequencies of greatest interest are the 50- and 60-Hz power-line frequencies. Power-line frequencies

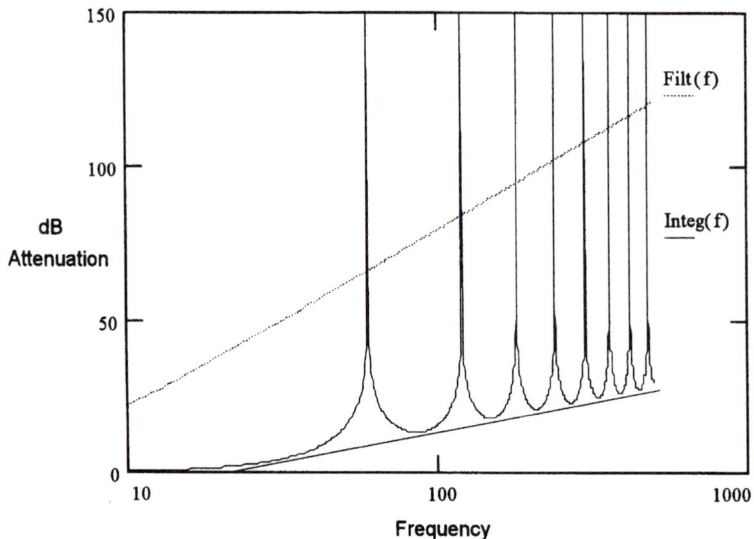

Figure 13.9 Noise rejection for passive filtering and for an integrating analog-to-digital converter with an integration time of 1/60 s.

are subject to short-term variations, typically less than ±0.1 percent—although this variation can be greater in some countries or extreme situations. Little or no noise rejection will occur if the analog-to-digital integration period is less than one period of the noise signal, as shown in Fig. 13.9.

Integration is best for rejecting line-related and periodic noise sources—filtering is best for rejecting broadband noise sources, as illustrated in Fig. 13.9. An integrating ADC exhibits extremely high noise rejection near each integer multiple of $1/T_{\text{integrate}}$, the integration time of the ADC, as shown in Fig. 13.9. Note that the asymptotic noise rejection (shown by the solid line) of an integrating ADC exhibits a single-pole (6 dB/octave) noise rejection starting at $\approx 1/(2 \times T_{\text{integrate}})$.

For signals containing line-frequency-related noise, an integrating ADC provides the shortest measurement settling time—fastest measurement rate—and greatest line-frequency noise rejection of any technique. The three-pole 6-Hz passive filter response shown in Fig. 13.9 (dotted line) achieves noise rejection comparable with an integrating ADC at frequencies near 60 Hz. Meters utilizing passive filtering sufficient to achieve line-frequency noise rejection similar to integration will require at least 10 times the measurement settling time after a full-scale input signal is applied—slowing fully settled measurement speeds to at least one-tenth that of a meter incorporating an integrating ADC.

Normal mode noise is specified in terms of a decibel voltage ratio. Both input noise and measured error must be characterized the same way—peak, peak-to-peak, or rms. For example, a 10-V peak ac noise signal is applied to the input of an integrating ADC. If the meter reading is observed to vary 100 μV peak (error), then the normal mode noise rejection is calculated to be 20 log 10 V/100 μV = 100 dB.

13.7.7 Common mode rejection (CMR)

Common mode signals appear in both the HI and LO input leads of a meter. These signals can be either ac or dc voltages, but common mode signals are often ac line-related. Noise signals frequently originate from grounding differences between the meter and the signal being measured. Common mode voltages can vary from a few millivolts to several hundred volts.

Ideally, a multimeter is completely isolated from earth-referenced circuits. However, there is generally a finite impedance between the meter's input LO terminal and earth ground, as shown in Fig. 13.10*a*. Common mode voltages V_{cm1} and V_{cm2} will cause small currents to flow which will develop an error voltage across R_b that can be difficult to identify. Once the common mode error voltage is developed across R_b, it becomes a normal mode noise voltage—that is, it appears across the meter input HI and LO terminals just like any other input signal. If the resulting normal mode signal is ac, it can be rejected by filtering in the meter's input circuits or by the filtering characteristic of an integrating ADC as described in the preceding section.

A voltage divider is formed between the meter's isolation impedance to earth R_l in parallel with C_l and the input LO terminal lead resistance. For comparison purposes, R_b is generally assumed to be 1 kΩ by most instrument manufacturers. The values of R_l and C_l are determined primarily by the instrument's mechanical construction, electrical protection components, and power supply design.

Common mode rejection is specified in terms of the decibel voltage ratio 20 log VR_b/V_{cm}. The 1-kΩ lead resistance imbalance R_b is critical to the common mode rejection specification. If its size is decreased to 100 Ω, the common mode rejection will appear to improve by 20 dB.

13.7.8 DC CMR and AC CMR

Dc common mode rejection is a measure of the meter's measuring circuit's dc isolation resistance from earth. Since the isolation resistance R_l is much, much greater than the LO lead resistance R_b, the dc CMR is approximately equal to 20 log R_l/R_b.

Similarly, ac common mode rejection measures the ac isolation impedance from the measuring circuit LO input terminal to earth—essentially a measure of the capacitance C_l from the LO terminal of the meter to earth. The ac CMR is approximately equal to 20 log $\frac{1}{2\pi f C_l}R_b$, where f is the frequency of the common mode input signal.

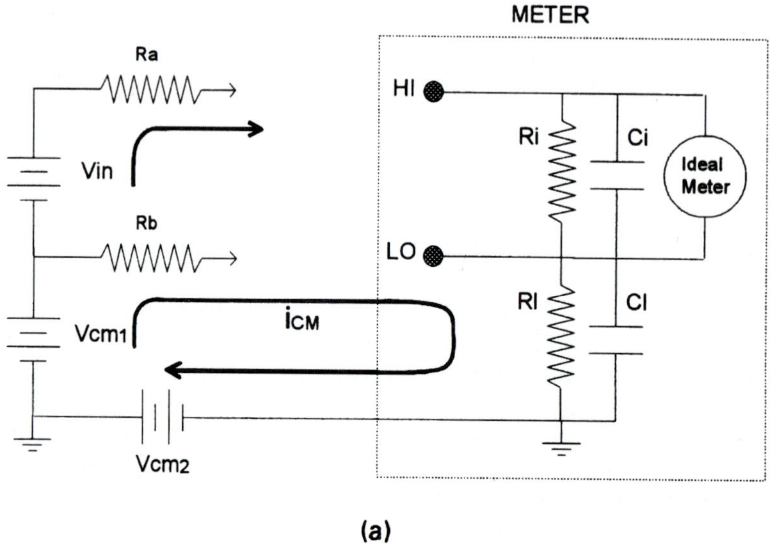

(a)

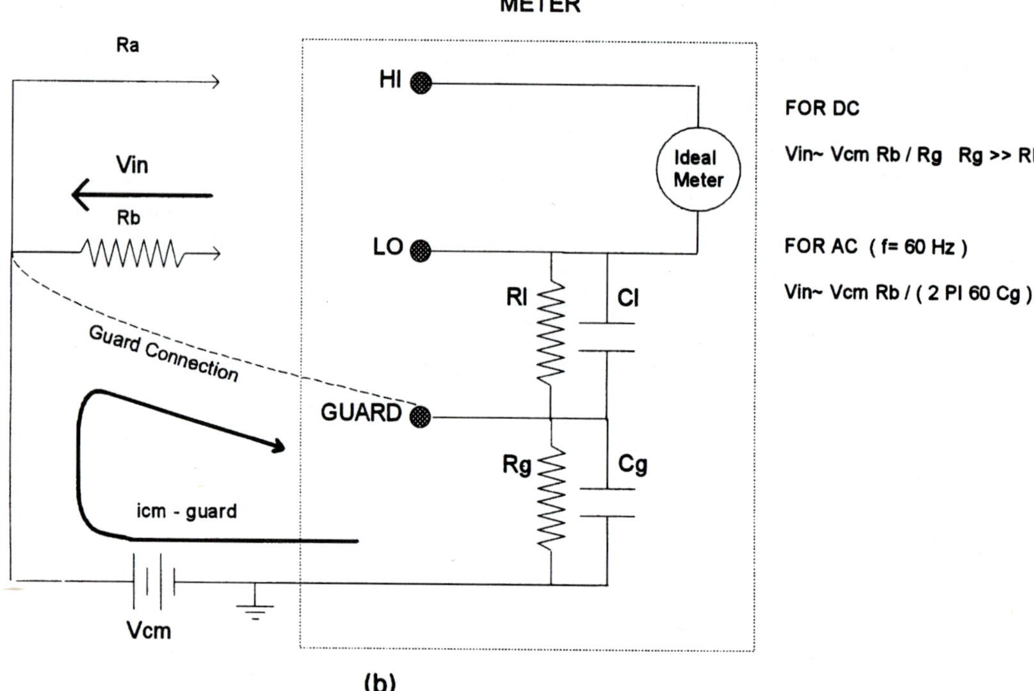

(b)

Figure 13.10 (*a*) A floating instrument with common mode voltage originating between grounds (V_{cm2}) or as an input V_{in} floating off of ground by V_{cm1}. HI and LO lead resistances R_a and R_b are also shown. (*b*) Indicates how a common mode current i_{cm} is converted into a normal mode input signal $i_{cm}*R_b$. Guard impedances R_gC_g and typical guard connections are also shown.

13.7.9 AC reversal errors

Common mode errors are compounded for ac measurements. A frequent situation where *unnecessary* common mode voltages are created is when the output of a voltage source is connected "backwards" to a multimeter—connecting the source's output directly across the relatively large LO to earth capacitance C_l of the meter. Ideally, a multimeter reads the same regardless of how the source is connected. However, both meter and source effects can degrade this ideal situation. The isolation capacitance C_l will load the source differently depending on how the input is applied. The magnitude of the resulting error depends on how the source responds to this load. A well-constructed multimeter will provide internal shielding for its measuring circuits to minimize sensitivity to ac reversal errors. These errors will be greatest for higher-voltage and higher-frequency inputs.

13.7.10 Effective common mode rejection (ECMR)

Effective common mode rejection is often confused with common mode rejection. As previously discussed, a common mode signal will be converted into a normal mode signal by the resistance in the L_0 input lead of a meter. The size of the normal mode signal depends on the common mode rejection of the meter, as shown in Fig. 13.10*b*. If the resulting normal mode noise voltage is an ac signal, filtering in the meter's HI to LO input will produce additional noise rejection. For power-line-related common mode signals, then, ECMR represents the combined rejection from ac CMR and from the NMR rejection of the meter's input filtering or ADC. Meters typically exhibit near 70 dB of ac CMR at 60 Hz. In addition, an integrating ADC will produce near 60 dB of rejection (NMR) for 60-Hz signals. Therefore, most meters employing an integrating ADCs will produce greater than 130 dB (70 dB + 60 dB) of "effective" rejection of power-line-related common mode input signals.

13.7.11 Guarding

In some measurement situations, the floating input of a meter may not yield enough common mode noise rejection. This is especially true for bridge measurements, where the output level may be a few millivolts and the source resistance seen by both meter input terminals may be high. "Guarding" is a passive technique employed by some meters for achieving additional common mode noise rejection. Guarding uses an additional internal shield enclosing the measuring circuits. The guard shield effectively splits the impedance between the LO input terminal and ground, as shown in Fig. 13.10*b*—adding isolation impedances R_gC_g between the shield and ground and effectively increasing the isolation impedance R_lC_l between the input LO terminal and the guard shield. An additional input terminal—Guard—is provided, connecting directly to the internal guard shield at the junction of R_gC_g and R_lC_l. The guard terminal is used to provide an alternate connection directly to the low side of the input signal. Since the guard terminal and input LO terminal are at virtually the same potential, little current actually flows in impedance $R_l C_l$. The guard connection essentially shunts the common mode current—at its source—from flowing through R_b, thus significantly reducing measurement error. An improperly connected guard terminal can actually increase common mode noise error. The guard terminal should always be connected such that it and the input LO terminal are at the same potential or as close to the same potential as possible—reducing the common mode current which flows through any resistance that is across the input terminals of the meter.

13.7.12 Noise caused by ground loops

When measuring voltages in circuits where the multimeter and the DUT are both referenced to a common ground, a "ground loop" is formed as shown in Figure 13.10*a*. A voltage difference between the two ground reference points V_{cm2} causes current to flow through the measurement leads. These noise and offset voltage errors, often power-line-related, will add to the desired input voltage.

The best way to eliminate ground loops is to maintain the multimeter's isolation from earth—whenever possible, do not connect the meter's LO terminal to ground. Isolating the meter's LO terminal places isolation impedances R_l and C_l in series with V_{cm2} to limit the error current which will flow in R_b. If the meter LO terminal must be earth-referenced, be sure to connect it and the DUT to the same common ground point. This will reduce or eliminate any voltage difference—V_{cm2}—between the devices. Also make sure that the meter and DUT are connected to the same electrical outlet whenever possible to minimize ground voltage differences.

13.7.13 Noise caused by magnetic loops

If you are making measurements near magnetic fields, you should take the necessary precautions to avoid inducing voltages in the measurement connections. Moving leads or time-varying magnetic fields will generate a spurious voltage in series with the input signal. Tens of nanovolts can be generated by poorly dressed, unshielded input leads moving in the earth's weak magnetic field. Equation 13.4 describes the relationship between the magnitude of the induced error voltage, the circuit loop area, and the magnetic field strength:

$$E = A\,\delta B/\delta t + B\,\delta A/\delta t \tag{13.4}$$

where E is the magnitude of the induced error voltage, and A is the area enclosed by the input leads through which the magnetic field B passes. Both B and A may vary with time.

Make sure your test leads are tied down securely when operating near mag-

TABLE 13.5 Typical Near-Field Magnetic Field Attenuation for Common Shielding Materials of Thickness between 10 and 50 Mils

Shield material	Magnetic field attenuation (typical)		
	1 kHz	10 kHz	100 kHz
Mumetal	15–20 dB	20–25 dB	25–30 dB
Steel	10–20 dB	30–35 dB	35–40 dB
Copper	3–8 dB	10–25 dB	30–35 dB
Aluminum	2–6 dB	10–20 dB	25–30 dB

netic fields. Use twisted-pair connections to the multimeter to reduce the noise pickup loop area, or dress the test leads as close together as possible. Loose or vibrating test leads also will induce small error voltages in series with the measurement. You should be especially careful when working near conductors carrying large currents. Whenever possible, use magnetic shielding materials such as those shown in Table 13.5 or physical separation to reduce problem magnetic field sources.

13.7.14 Crest factor errors: Nonsinusoidal inputs

A common misconception is that "since an ac meter is true rms, its sine wave accuracy specifications apply to all waveforms." Actually, the shape of the input signal can dramatically affect measurement accuracy. As noted earlier, crest factor is a common way to describe signal wave shapes. All ac meters have limitations on their ability to accurately measure ac signals that exhibit high crest factor ratios. This is due primarily to two factors. First, signal conditioning circuits may become saturated, distorting the input signal before it reaches the ac converter. Second, high crest factor signals contain significant energy in harmonics well above the fundamental frequency of the signal. These harmonics may be at frequencies beyond the bandwidth of the signal conditioning or ac converter circuits. The energy in these harmonics will not be measured accurately—if at all—and significant reading errors will result. This is particularly true when the fundamental frequency of a high crest factor signal—CF > 2—is within 1/100 of the meter's sine wave -3-dB bandwidth. For example, a true rms ac meter with a 100-kHz, -3-dB bandwidth specification will indicate increasing error for pulse train inputs above 1 kHz.

You can estimate the measurement error of an rms responding meter due to high crest factor signals as shown in Eq. 13.5.

$$\text{Error}_{\text{total}}(\text{approximately}) = \text{error}_{\text{sine}} + \text{error}_{\text{crest factor}} + \text{error}_{\text{bandwidth}} \qquad (13.5)$$

where $\text{error}_{\text{sine}}$ = meter's sine wave accuracy specification

$\text{error}_{\text{crest factor}}$ = meter's additional crest factor error specification; typical values might be

CF = 1–2 0.1%

$$\begin{aligned} \text{CF} &= 2\text{–}3 \quad 0.2\% \\ \text{CF} &= 3\text{–}4 \quad 0.4\% \\ \text{CF} &= 4\text{–}5 \quad 0.8\% \end{aligned}$$

$$\text{error}_{\text{bandwidth}} = \text{estimated bandwidth error} = \frac{F \times (\text{CF})^2}{4\pi \times BW}$$

where F = input signal fundamental frequency
CF = input signal crest factor
BW = meter's − 3-dB bandwidth specification (or measured)
π = 3.14159

13.7.15 Low-level ac voltage measurement errors

When measuring ac voltages less than 100 mV, be aware that these measurements are especially susceptible to errors introduced by extraneous noise sources. An exposed test lead will act as an antenna, and a properly functioning meter will measure the signals received. The entire measurement path, including the power line, acts as a loop antenna. Circulating currents in the loop will create an error voltage across any impedance in series with the meter's input. For this reason, you should apply low-level ac voltages through shielded cables. The shield should be connected to the meter's input LO terminal to minimize noise pickup.

Make sure that the meter and the ac source are connected to the same electrical outlet whenever possible. You also should minimize the area of any ground loops that cannot be avoided. A high-impedance source is more susceptible to noise pickup than a low-impedance source. You can reduce the high-frequency impedance of a source by placing a capacitor in parallel with the meter's input terminals—you will have to experiment to determine the correct value for your application.

Most extraneous noise is not correlated with the input signal. You can calculate the expected reading for rms measuring instruments as

$$V_{\text{measured}} = \sqrt{V_{\text{in}}^2 + V_{\text{noise}}^2}$$

Correlated noise, while rare, is especially detrimental. Correlated noise will always add directly to the input signal. Measuring a low-level signal with the same frequency as the local power line is a common situation that is prone to this error.

13.7.16 Self-heating effects in resistance measurements

When measuring resistors designed for temperature sensing—or other resistors with large temperature coefficients—be aware that the ohmmeter will dissipate some power in the DUT, causing self-heating and a change in its measured value. If self-heating is a problem, you should select the meter's

TABLE 13.6 Typical Ohmmeter Power Dissipation in the Device Under Test for Full-Scale Resistance Values

Meter range	Test current	Power dissipation at full scale
100 Ω	1 mA	100 μW
1 kΩ	1 mA	1 mW
10 kΩ	100 μA	100 μW
100 kΩ	10 μA	10 μW
1 MΩ	5 μA	25 μW
10 MΩ	500 nA	2.5 μW

next higher range to reduce the measurement current and hence reduce the power dissipation error to an acceptable level. Table 13.6 shows resistor power dissipation values for a typical meter.

13.7.17 High-resistance measurement errors

Significant errors can occur when measuring large resistances due to cabling insulation resistance and surface cleanliness. You should always take the necessary precautions to maintain a "clean" high-resistance system. Test leads and fixtures are susceptible to leakage due to moisture absorption in insulating materials and "dirty" surface films. Handling should be minimized because oils and salts from the skin can degrade insulator performance. Contaminants in the air can be deposited on an insulator's surface, reducing its resistance.

Nylon and PVC are relatively poor insulators when compared with PTFE (Teflon) insulators, as shown in Table 13.7. Leakage from nylon and PVC insulators can easily contribute greater than 0.1 percent error when measuring a 1-MΩ resistance in humid conditions.

Conventional ohmmeters can be very sensitive to external noise when small test currents are used. Electrostatic coupling through minute stray capacitances can contribute noise currents when either the voltage source changes

TABLE 13.7 Impedance Characteristics of Various Common Insulating Materials

Insulating material	Resistance range	Moisture absorbing
Teflon (PTFE)	1 TΩ–1 PΩ	N
Polystyrene	100 GΩ–1 PΩ	N
Ceramic	1 GΩ–1 PΩ	N
Nylon	1 GΩ–10 TΩ	Y
PVC	10 GΩ–10 TΩ	Y
Glass epoxy (FR-4, G-10)	1 GΩ–10 TΩ	Y
Phenolic, paper	10 MΩ–10 GΩ	Y

$P = 10^{15}$.

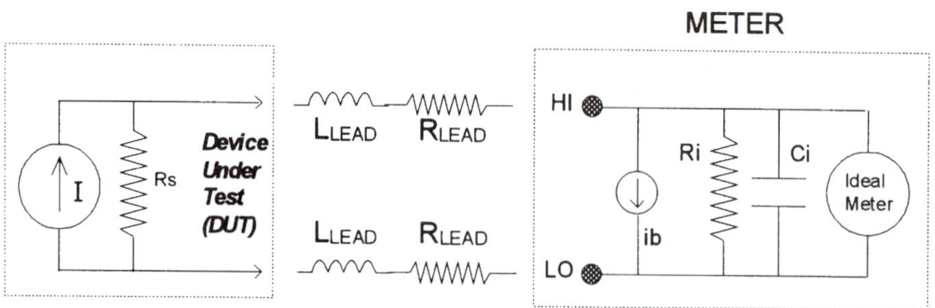

Figure 13.11 Current meter, DUT, and interconnect wiring equivalent circuits illustrating loading error due to shunt resistance R_i.

or the stray capacitance changes. Vibrating or moving leads contribute to changing stray capacitances, while moving, statically charged bodies—like humans—can easily generate noise signals varying by hundreds of volts. To minimize this effect, meters designed specifically for measuring high resistance usually employ a different measurement technique. A large, variable test voltage source creates a current in the high resistance. The output current is monitored, and the resistance value is calculated from these quantities. Since all external connections are low resistance—the output resistance of the voltage source—external noise pickup is minimized.

13.7.18 Burden voltage: Loading error in current measurements

Whenever you connect a multimeter in series with a test circuit to measure current, a measurement error is introduced. The error is caused by the multimeter's series "burden voltage" ($I \times R_i$). A voltage is developed across the wiring resistance and current sensing circuit—often a shunt resistor—as shown in Fig. 13.11.

Burden voltage effects—loading—cause errors in both ac and dc current measurements. However, the burden voltage for alternating current is larger due to the series inductance of the meter and the input connections, as shown in Fig. 13.11. The ac burden voltage increases as the input frequency increases. Some DUT circuits may oscillate when performing current measurements due to this series inductance—use an oscilloscope to check for oscillation if erratic or unusual measurements are observed.

13.7.19 Low-current errors

Low-current measurements are sensitive to many subtle sources of error. All the leakage errors discussed earlier also apply to low-current measurements. Electrostatic shielding can significantly reduce many common noise sources. Strapping connecting leads to fixed surfaces can dramatically reduce electrostatic noise due to movement and vibration.

Additional low-current errors can be generated by piezoelectric, triboelectric, and electrochemical effects in your cabling and test fixture. Many materials will generate a piezoelectric current when mechanical stress is applied. Similarly, triboelectric currents are generated when conductors and insulators rub together, tearing loose free electrons. Cables subjected to vibration, bending, or even thermal stresses commonly produce small triboelectric currents. Ionic contamination can produce weak electrochemical "batteries" capable of sourcing tens or even hundreds of picoamps when exposed to the right conditions. High humidity and high temperature accelerate the movement and spread of free ions over large distances. Likewise, free ions will be attracted by large, positive bias voltages in your test setup. Ion contamination can result in permanent "growth" of a relatively low-resistance conductive path between conductors in your circuit when the right—or maybe wrong—conditions exist. It can be virtually impossible to remove an electrochemical battery or conductive path once it forms. Aggressive and repeated scrubbing with pure alcohol and rinsing with de-ionized water can be successful in these situations.

13.8 Interpreting Specifications

Understanding manufacturers' specifications can be confusing and frustrating and is often an error-prone process itself. Most meter specifications begin with a similar format—(±% reading ±% range). For each measurement situation, however, many additional factors will affect the ultimate precision of your measurements. This section explains basic specification terminology used by manufacturers and suggests several error sources that should be considered when interpreting and comparing meter specifications.

13.8.1 Number of digits and overrange

The "number of digits" specification is the most fundamental and sometimes the most confusing characteristic of a meter. The number of digits is equal to the *maximum* number of 9s the meter can measure or display. This indicates the number of "full digits." Most meters have the ability to overrange and add a partial or "half" digit.

For example, some multimeters can measure 9.99999 V dc on a 10-V measuring range. This represents six full digits of resolution. Usually, the meter also will overrange on the 10-V range and measure inputs up to a maximum of 12.00000 V dc. This corresponds to a 6½-digit measurement with 20 percent overrange capability.

The amount of overrange varies widely, with 20 percent representing the minimum overrange capability. Some hand-held multimeters provide as much as 500 percent overrange.

13.8.2 Sensitivity

"Sensitivity" is the minimum level that the meter can detect for a given measurement. Sensitivity defines the ability of the meter to respond to small changes in the input level. For example, suppose you are monitoring a 1-mV dc signal and you want to adjust the level to within ±1 μV. To be able to respond to an adjustment this small, this measurement would require a meter with a sensitivity of at least 1 μV. You could use a 6½-digit multimeter *if* it has a 1-V dc or smaller measuring range. You could also use a 4½-digit multimeter if it has a 10-mV dc range.

The smallest value that can be measured is usually not the same as the sensitivity for ac voltage and ac current measurements—especially for rms meters. For example, the rms ac voltage measuring function of a multimeter is often specified *only* for measurements from 1 percent of range to 120 percent of range—inputs greater than 1 mV ac on the 100-mV ac measuring range, for example—due to the broadband rms noise floor of the signal-conditioning circuits and the rms detector.

Sensitivity and resolution are often confused. As described below, "resolution" is essentially the smallest displayed digit in a measurement. Sensitivity represents the smallest actual input value which can be measured or discerned. Often there is little difference between sensitivity and resolution. A well-designed dc meter's sensitivity should be limited only by internal random noise. Dc voltage changes much below the peak noise level are indistinguishable from normal reading variations—and therefore not discernible. Generally, ac meters exhibit a significant difference between their sensitivity and their resolution. Ac converters often require signal levels greater than 1/400 of their full-scale range in order to respond at all to an input signal. However, once sufficient signal level is applied, their measurement resolution is generally limited by the system amplifiers and ADC much like a dc meter. In this case, the minimum "measurable" input to the ac meter would be 1/400 of the smallest full-scale range. For example, a typical 5½-digit ac meter might provide a smallest full-scale range of 100 mV, a sensitivity of 0.25 mV on that range, and overall reading resolution to 1 μV—so long as the input exceeds 0.25 mV.

13.8.3 Resolution

"Resolution" is the numeric ratio of the maximum displayed value divided by the minimum displayed value on a selected range. Resolution is often expressed in percentages, parts per million (ppm), counts, or bits. For example, a 6½-digit multimeter with 20 percent overrange capability can display a measurement with up to ±1,200,000 counts of resolution. This corresponds to about 0.0001 percent, or 1 ppm, of full scale—about 21 bits of resolution, including the sign bit. All these methods for describing the meter's resolution are equivalent.

13.8.4 Accuracy

"Accuracy" is a measure of the "exactness" to which the multimeter's measurement uncertainty can be determined *relative to the calibration reference*

TABLE 13.8 Probability of Nonconformance, or Failure, of an Individual Specification for Various Common Statistical Criteria

Specification criteria	Probability of each failure
Mean $\pm$ 2σ	4.5%
Mean $\pm$ 3σ	0.3%
Mean $\pm$ 4σ	0.006%

used. Absolute accuracy includes the multimeter's relative accuracy specification plus the known error of the calibration reference relative to national standards. To be meaningful, accuracy specifications must include the conditions under which they are valid. These conditions should include temperature, humidity, and time.

There is no standard convention among manufacturers for the confidence limits at which specifications are set, although most reputable manufacturers choose at least a 95 percent confidence or 2σ level. Table 13.8 shows the probability of nonconformance for each specification point under the assumptions listed.

When comparing similar instruments with similar specifications, *actual* variations in performance from reading to reading and instrument to instrument will be lower for instruments whose specifications are set to higher confidence levels—greater σ values. This means that you will achieve greater "actual" measurement precision than indicated by the instrument specification when higher confidence limits are used by the manufacturer—the primary reason why some manufacturers' instruments seem to perform better than expected. This is also the reason that new equipment purchases should *never* be based solely on manufacturers' published specification numbers, especially when comparing instruments from different manufacturers.

For example, suppose that an instrument manufacturer characterizes its product's performance—probably with a small sample of units—and then publishes the bounds of what it observes as the instrument's accuracy specification. A measurement using this meter typically will exhibit an error of about 70 percent of its published specification.

A second manufacturer, *building an identical product,* characterizes a large quantity of its units. This manufacturer performs a statistical analysis of the data and uses this information to develop a mathematical model—incorporating known errors and the statistically characterized variation. The second manufacturer's calculated accuracy specifications are set assuming a 4σ confidence limit for each specification. As a result, assume that the second manufacturer's product specifies accuracy numbers that are four times that of the first manufacturer. A measurement using this meter will typically measure the same as the first meter—but it will be about 18 percent of the unit's published specification.

Which meter is more accurate? Neither—they both have similar "actual"

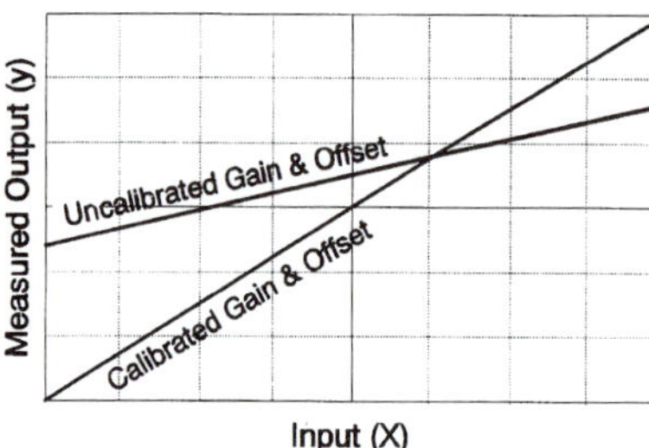

Figure 13.12 The process of calibration determines and corrects for the linear errors through the gain coefficient m and the offset coefficient b in the equation $y = mX + b$.

accuracy for the measurement. It appears that the first product is more accurate than the second by a factor of 4 by simply comparing published specifications for these instruments. However, remember that for this example these were identical instruments—only the philosophy and methodology used for setting specifications varied.

Normally, you will be comparing specifications for products that *are not* identical. So how do you compare specifications? Unless you know the assumptions used by the manufacturer, you really can't compare. Many instrument manufacturers will tell you about their methods if you ask—many cannot. If you can't find answers to this question, performing a careful evaluation and comparison of actual measurement accuracy and repeatability provides the best indicator of instrument accuracy.

13.8.5 Calibration

Measurement errors in a meter are specified in two categories—gain error and offset error. "Calibration" is a process in which each individual measuring function range gain and offset values are determined, manually or automatically, to yield a minimum difference relative to an applied input, as shown in Fig. 13.12. Calibration is a two-step process—adjustment and then verification. After adjusting each gain and offset, verification checks are usually performed to ensure that accuracy specifications are achieved.

13.8.6 Gain error or reading error

"Gain" represents the slope m of the V_{measure} versus V_{input} line in Fig. 13.12. Gain errors in a meter result from changes in amplifier gains, divider ratios, or internal reference voltages. Each gain term exhibits a temperature coefficient and some finite aging rate and can potentially change value following exposure to high-humidity environments or severe mechanical shock or vibration.

Gain errors are expressed in percentages or parts per million (ppm) of the input or reading. Gain error is a constant percentage of the applied input for each range. For example, a meter has a 0.001 percent of reading gain error for its 12-V dc range. If you apply 10 V dc, the gain error will be 0.001 percent of 10 V, or 100 μV. If you apply 1 V on this range, the gain error will be 0.001 percent of 1 V, or 10 μV.

13.8.7 Offset error or range error

"Offset errors" result from amplifier offset voltages and noise, leakage currents ($l \times R$), and thermocouple effects generated by dissimilar metals used in component construction or inter-connection. Offset errors are expressed in percentages or parts per million (ppm) of the range being considered. They are also specified in counts, where one count is the smallest displayed quantity for the range being considered (see "Number of Digits and Overrange," above). Offset error is a fixed error quantity for the specific range. For example, a meter has a 0.0005 percent of range offset error for its 12-V dc range. If you apply 10 V dc, the offset error will be 0.0005 percent of 12 V, or 60 μV. If you apply 1 V on this range, the offset error will still be 0.0005 percent of 12 V, or 60 μV. Notice that as the input is reduced on a fixed range, offset error introduces a greater and greater error relative to the input signal.

13.8.8 Linearity

"Linearity error" is a measure of an instrument's ability to linearly respond to changes in input signal level. Nonlinearities—"linearity error"—can be produced either by an instrument's ADC or by the signal-conditioning circuits of an instrument. Linearity errors are usually included in an instrument's overall gain accuracy and offset accuracy specification. Sometimes these errors are also specified separately to clarify how the instrument will perform in specific measurement applications.

As mentioned in Sec. 13.2, you can think of the possible outputs of an ADC as a vertical bar chart with each bar increasing in height from the previous bar by an identical amount. Converting the essentially infinite resolution of an analog input signal to a single bar in the chart is the function of the ADC. The continuum of analog input values is partitioned—quantized—into discrete values. In an ideal ADC, each successive bar, or quantized value, increases linearly from its predecessor. If the ADC linearity is perfect, each step from one bar to the next will be a constant. The relationship between steps or codes describes the nonlinearity error of an ADC.

A gradually increasing nonlinearity error that becomes largest between zero and half scale and then reduces toward full scale is called an "integral linearity error." Integral linearity errors will appear like gain errors which vary throughout the measuring range and are expressed as a percentage of the reading representing the maximum error over the full range.

Linearity errors which cause abrupt reading changes for small changes in the input are called "differential nonlinearities." These errors usually can occur anywhere throughout the full span of the ADC and usually appear like an offset or additional noise in your reading. Differential linearity errors are specified by a percentage of range error like offset voltages. If the differential linearity error is large enough, it is possible for one or more quantized values to be missed and therefore never produced by the ADC.

The linearity error of an instrument's ADC should be small enough to barely contribute to the instrument's overall measurement error. Linearity error, especially differential nonlinearity, can severely limit an instrument's usefulness in many applications. Some manufacturers do not specify ADC linearity error for their instruments—you may want to evaluate and compare various instruments' linearity performance when selecting a meter for your application.

13.8.9 Long-term stability

"Stability" is a measure of an instrument's ability to remain within its rated accuracy for some specified time period. Stability may be specified in two parts, long-term stability and short-term stability or transfer accuracy. All stability specifications contain a set of operational conditions under which the specification will apply. Short-term stability specifications will generally have more bounding conditions than will long-term stability specifications. Meter long-term stability specifications are usually given for 90-day and 1-year intervals. These specifications are dominated by internal voltage reference or resistance reference standards and by environmental effects such as temperature and humidity. Short-term stability specifications are generally limited to a few tens of minutes and are intended for measuring two nearly equal values—"transferring" known accuracy from one device to another through the meter. Transfer accuracy specifications are dominated by short-term temperature variations, meter noise and linearity error, and general metrology technique.

13.8.10 Temperature coefficients

The temperature of an instrument's operating environment will affect measurement accuracy. Both gain and offset errors are affected by temperature. For this reason, instrument specifications are given over a defined temperature range, usually from 18 to 28°C. Measurement accuracy is degraded when the instrument is operated outside this temperature range. The temperature coefficient specification is used to describe the degradation in instrument accuracy that occurs as the instrument's operating environment exceeds the specified temperature range. These temperature coefficient errors *add* to the instrument's accuracy specifications unless otherwise noted by the manufacturer.

13.9 Considerations When Selecting a Meter

As with any instrument, there are a wide variety of products to choose from when selecting a voltage, current, or resistance measuring instrument. It is critical that you carefully define your application needs before attempting to choose an instrument. You should consider these questions carefully:

1. Who will be the primary user of this instrument, an engineer, a technician or skilled worker, or a nontechnical user?
2. Are several people going to share this instrument, or will it be used by a single individual?
3. Am I buying this instrument for a specific project need or for general-purpose use within my department?
4. Do I expect this instrument to provide enough flexibility or added capability to satisfy future requirements or to simply solve a measurement need today?
5. Do I have a specialized or otherwise difficult measurement problem or a more common general-purpose need?
6. Will this instrument be part of an automated test system, or will it be used manually? Will this still be true 2 or 3 years from now?

Almost all decisions will begin by considering your basic measurement requirements—type of measurement, ranges needed, and accuracy requirement. Often, selection criteria stop here, well before many important issues suggested by the preceding questions have been answered appropriately.

13.9.1 Benchtop applications

In manual—benchtop—applications, careful consideration of the instrument's size, form factor, and human interface is needed. Instrument displays with poor viewing angles or small digits and measurement unit labels can cause confusion for the user. Simple-to-use, clearly marked controls help the user to quickly gather measurement data. Instruments which require the user to remember the right "trick" to access a function or mode can cause wasted time and frustration—especially when an instrument is shared by several users. Features such as built-in math operations can simplify repetitive measurements and reduce the chance of accidental errors. Don't forget to consider such mundane aspects as size, weight, and availability of a handle for easy movement whenever the instrument will be shared by several users. Safety features such as maximum input ratings, isolation voltages, safety terminals, and active user warnings can be critical in applications involving power-line voltages or greater—many times these features aren't even considered before the purchase decision has been made.

13.9.2 Automated test applications

Additional features and capabilities should be evaluated when you are selecting a meter for a system or computer-controlled application. The type of computer interface is critical. The IEEE-488 interface—also known as HP-IB or GP-IB—is the most common computer interface for instruments. The RS-232 interface is also available on some instruments. The instrument programming language can be important for some applications. The SCPI (Standard Commands for Programmable Instruments) language is an industry standard used by many manufacturers today. Its use can make programming easier while also protecting your software development investment through multivendor support. "Throughput"—the speed at which an instrument can execute various command sequences—is often crucial in automated test applications. Instrument throughput performance varies widely from instrument category to category and from manufacturer to manufacturer. Not all instrument manufacturers specify their throughput performance. You may have to ask for throughput data or perform benchmarks yourself if adequate specifications are not available. Additional system features that you may want to consider include: Does the meter have its own internal reading buffer, and how big is it? What type of measurement time base and triggering capability does the instrument have? How does the instrument synchronize with other instruments or switches in the system?

You can save time and money and improve your satisfaction with the instrument you choose by taking time to consider these questions. They are intended to help you decide what features are important for you now and in the future—and to guide your thoughts when comparing specifications and information from various competitive products.

Chapter

14

Oscilloscopes

Alan J. De Vilbiss
Agilent Technologies
Colorado Springs, Colorado

14.1 Introduction

The word "oscilloscope" has evolved to describe any of a variety of electronic instruments used to observe, measure, or record transient physical phenomena and present the results in graphic form. Perhaps the popularity and usefulness of the oscilloscope spring from its exploitation of the relationship between vision and understanding. In any event, several generations of technical workers have found it to be an important tool in a wide variety of settings.

14.1.1 Basic functions

The prototypical oscilloscope produces a two-dimensional graph with the voltage presented at an input terminal plotted on the vertical axis and time plotted on the horizontal axis (Fig. 14.1). Usually the graph appears as an illuminated trace on the screen of a cathode-ray tube (CRT) and is used to construct a useful model or representation of how the instantaneous magnitude of some quantity varies during a particular time interval. The "quantity" measured is often a changing voltage in an electronic circuit. However, it could be something else, such as electric current, acceleration, light intensity, or any of many other possibilities, which has been changed into a voltage by a suitable transducer. The "time interval" over which the phenomenon is graphed may vary over many orders of magnitude, allowing measurements of events which proceed too quickly to be observed directly with the human senses. Instruments of current manufacture measure events occurring over intervals as short as tens of picoseconds (10^{-12} s) or up to tens of seconds.

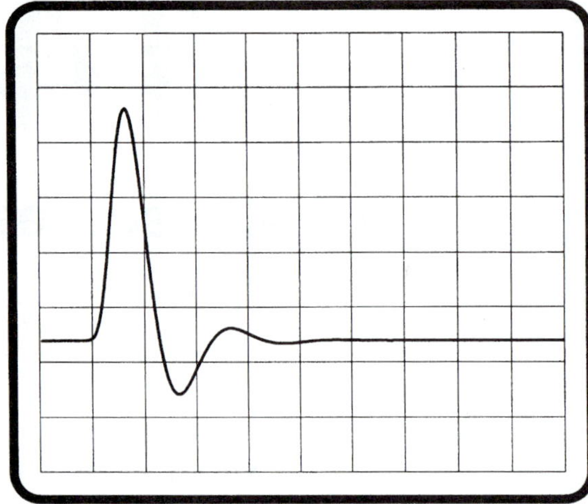

Figure 14.1 Voltage is plotted on the vertical axis and time horizontally on the classic oscilloscope display.

The measured quantities can be uniformly repeating or essentially nonrecurring. The most useful oscilloscopes have multiple input channels so that simultaneous observation of multiple phenomena is possible, allowing measurement of the time relationships of related events. For example, the time delay from clock to output of a D type logic flipflop can be measured (Fig. 14.2). This type of flipflop copies the logic state on the input D to the output Q when the clock has a state change from low to high. The lower trace shows the flipflop clock, while the upper trace shows the resulting state change propagating to the Q output. The difference in the horizontal position of the two positive transitions indicates the time elapsed between the two events.

With the aid of an oscilloscope, rapidly time-varying quantities are captured as a static image and can be studied at a pace suitable for a human

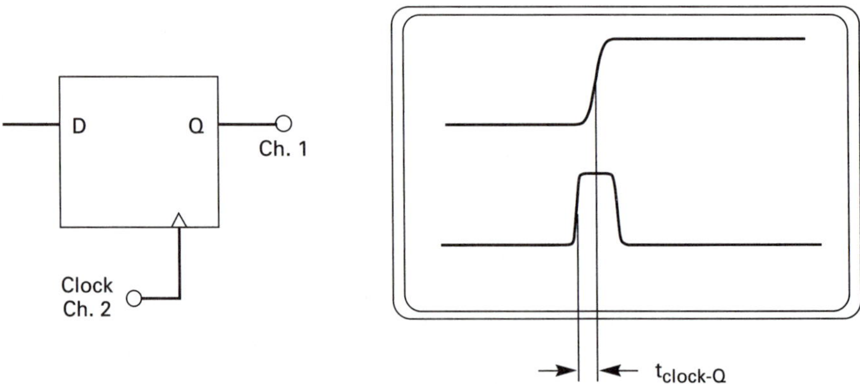

Figure 14.2 A time-interval measurement using a two-channel oscilloscope.

observer. The image can be preserved as a photograph or on paper with a plotter or printer. The record can be transferred into the memory of a digital computer to be preserved or analyzed.

14.1.2 Applications

The oscilloscope has for many years been used for a wide variety of measurements by engineers, scientists, and technicians. Many would describe it as among the most versatile and useful of the general-purpose electronic measuring tools, limited only by the imagination of those who use it.

An oscilloscope is ordinarily used to characterize the way a voltage varies with the passage of time. Therefore, any time-varying quantity which can be converted into a voltage also can be measured. Devices which change energy from one form to another are called "transducers" and are used for many purposes. For example, a loudspeaker changes electrical energy into sound waves, and a microphone does the opposite conversion. The familiar glass tube thermometer uses heat energy to expand a column of mercury, giving a visual indication of the temperature of its surroundings.

The transducers useful in oscilloscope applications are those which convert energy into an electric voltage, and many such devices have been designed for or adapted to this purpose. The microphone, although originally designed to allow the transmission of sound by telephone and radio, is now used in conjunction with an oscilloscope to explore in detail the duration and intensities of sound waves. An electric current probe, although just a special form of a transformer, is an example of a transducer designed specifically for use with an oscilloscope. The current to be sensed is linked though a magnetic core, and a voltage proportional to the input current is developed across the multiturn secondary winding and transmitted to the oscilloscope input for display. The details of current probe operation are explored more completely in the general discussion of probes.

Oscilloscopes are probably most often used for direct measurement of the transient voltage signals occurring within equipment that directly depends on electricity for its operation, such as computers, automation controls, telephones, radios, television, power supplies, and many others. In equipment that functions by the use of rapidly changing electrical signals, pulses, or wave trains, the oscilloscope is useful for measuring the parameters that may be important in determining its correct operation: signal timing relationships, duration, sequence, rise and fall times, propagation delays, amplitudes, etc.

14.2 General Oscilloscope Concepts

General-purpose oscilloscopes are classified as analog oscilloscopes or digital oscilloscopes. Each type has special applications in which it is superior, but many measurements could be performed satisfactorily with either.

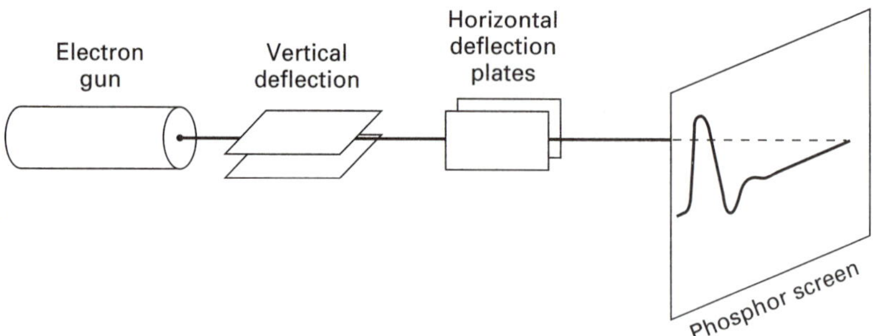

Figure 14.3 Analog oscilloscope cathode-ray tube.

14.2.1 Analog and digital oscilloscope basics

The classic oscilloscope is the analog form, characterized by the use of a CRT as a direct display device. A beam of electrons (the "cathode rays") is formed, accelerated, and focused in an electron gun and strikes a phosphor screen, causing visible light to be emitted from the point of impact (Fig. 14.3). The voltage transients to be displayed are amplified and applied directly to vertical deflection plates inside the CRT, resulting in an angular deflection of the electron beam in the vertical direction. This amplifier system is conventionally referred to as the "vertical amplifier." The linear vertical deflection of the point at which the electron beam strikes the screen is thus proportional to the instantaneous amplitude of the volgate transient. Another voltage transient, generated inside the oscilloscope and increasing at a uniform rate, is applied directly to the horizontal deflection plates of the CRT, resulting in a simultaneous, uniform, left-to-right horizontal motion of the point at which the electron beam strikes the phosphor screen. The electronic module that generates the signals that sweep the beam horizontally and control the rate and synchronization of those signals is called the "time base." Thus the point on the phosphor screen illuminated by the electron beam moves in response to those voltages, and the glowing phosphor traces out the desired graph of voltage versus time.

The digital oscilloscope has been made practical and useful by recent advances in the state of the art of the digitizing devices called "analog-to-digital converters" (ADC). For the purposes of this discussion, an ADC is a device which at suitable regular intervals measures ("samples") the instantaneous value of the voltage at the oscilloscope input and converts it into a digital value (a number) representing that instantaneous value (Fig. 14.4). The oscilloscope function of recording a voltage transient is achieved by storing in a digital memory a series of samples taken by the ADC. At a later time, the series of numbers can be retrieved from memory, and the desired graph of volts versus time can be constructed. The graphing or display process, since it is distinct from the recording process, can be performed in several different ways. The display device can be a CRT using direct beam deflection methods

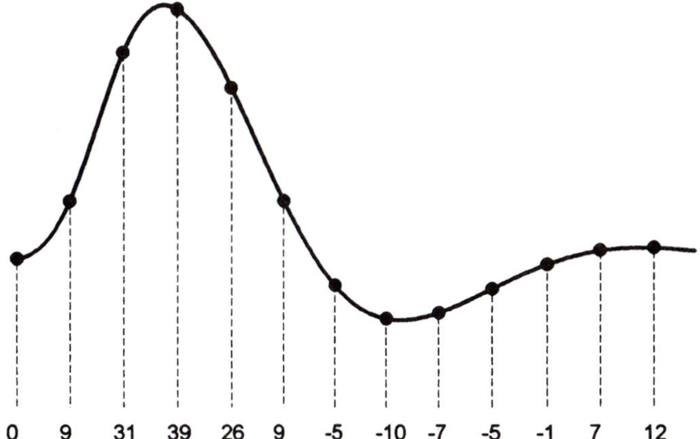

Figure 14.4 Sampling in a digital oscilloscope.

(in effect using an analog oscilloscope as the display device in a digital oscilloscope). Alternatively, a "raster-scan display," similar to that used in a conventional television receiver or a computer monitor, can be used. Or the samples could be plotted on paper using a printer with graphics capability or a plotter.

The digital oscilloscope is usually configured to resemble the traditional analog instrument in the arrangement and labeling of its controls, the features included in the vertical amplifier, and the labeling and presentation of the display. In addition, the system that controls the sample rate and timing of the data-acquisition cycle is configured to emulate the functions of the time base in the analog instrument. This preservation of the measurement model developed around the analog oscilloscope allows one familiar with its use in that context to quickly become proficient with the digital version. Even though there are fundamental and important differences in the two measurement technologies, many common elements and shared requirements exist. In the following discussion, the basic concepts that apply to both types are treated.

14.2.2 Control panel

The control panel of the oscilloscope contains three important groups: (1) the display screen and its associated controls, such as focus and intensity adjustments, (2) the vertical amplifier input signal connectors, sensitivity adjustments, and other input signal conditioning controls, and (3) the horizontal or time-base controls, which set the speed and timing of the signal capture. These controls are provided so that the operator can adjust the oscilloscope to frame a precise window in voltage and time to capture and display the desired voltage transient.

The microcomputer has become an important part of the internal control system of electronic instruments, and the influence of embedded control technology is evident in the arrangement and operation of the user-control interface in the modern oscilloscope. Once there were only mechanically actuated switches and variable resistors on the front panel. Now oscilloscopes are available in which the control repertoire has been augmented with data-entry keypads, knobs (with no position markings) that can be used to control several different functions, and menus which appear on the display to access seldom-used or complex features (Fig. 14.5).

14.2.3 Display

The oscilloscope display is traditionally subdivided by a grid of horizontal and vertical lines, called a "graticule," to aid the operator in making measurements, voltage or time, on the signals displayed on the screen. There are 10 major divisions horizontally, the time axis. In the vertical, or voltage, direction, there are between 4 and 10 divisions. The graticule is sometimes scribed on a sheet of transparent material which is then placed over the phosphor screen, and the operator views the signal trace through the grid. This method has the disadvantage of introducing errors due to parallax, because the plane containing the graticule is not coincident with the plane containing the signal trace, and the user's eye position while making the measurement becomes an important factor. Alternatively, the graticule can be etched on the inside surface of the glass face plate of the CRT during its manufacture. Then the phosphor material is placed in contact with the plane of the graticule, minimizing distortion from parallax in the resulting oscilloscope display. A third method is to use the electron beam of the CRT to "draw" the graticule by illuminating lines in the phosphor screen.

14.2.4 Modular construction

Oscilloscopes are available in different configurations, depending on cost and intended application. Many models are designed to perform a specific set of measurements optimally, such as field service, telecommunications equipment testing, or research and development (R&D) laboratory measurements. By the nature of the oscilloscope, even the more specialized instruments retain a general-purpose flavor. Even so, significant design compromises to suit a particular measurement arena are in order. For example, a field service oscilloscope needs to be rugged, small, and light, and these requirements mean that control panel space and display size are limited. An oscilloscope that will be used in a laboratory can be larger and heavier but is required to be more versatile and have more measurement power.

Modular construction is used to enhance the versatility of a particular oscilloscope model. A common implementation is to construct the vertical amplifier as a removable module, called a "plug-in." The remainder of the instrument, called a "mainframe," has an opening into which the plug-in is inserted. The advantage of this arrangement is that several different types of vertical amplifiers can be designed to operate in the same mainframe at dif-

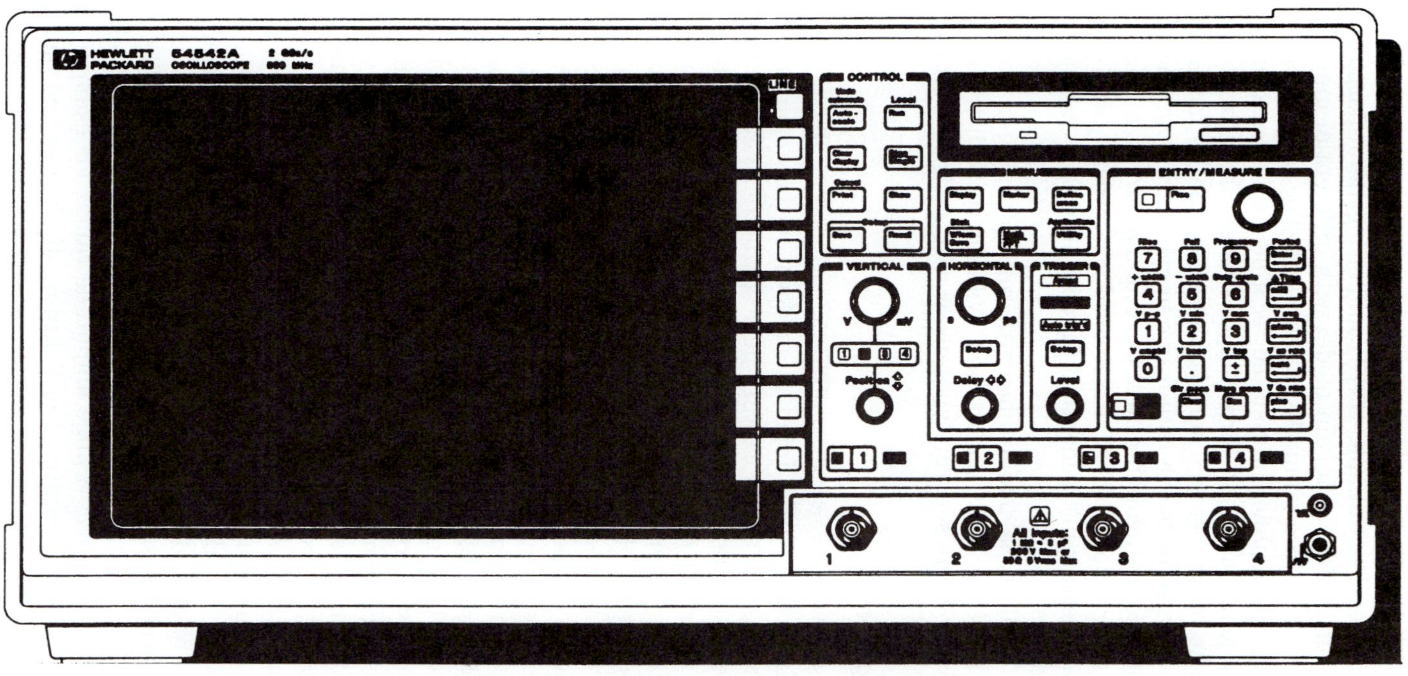

Figure 14.5 The buttons and knobs on a digital oscilloscope send signals to the embedded controller (computer) for action.

ferent times. For example, there could be a general-purpose two-channel unit, a low-bandwidth, high-gain model with differential inputs, and a four-channel plug-in with a simplified feature set. The oscilloscope can be used with the vertical amplifier best suited for a particular measurement, and the cost for one mainframe and three plug-ins is less than the cost of three different, complete instruments. There is also the potential for inexpensive extension of the service life of the mainframe by purchasing an upgraded vertical plug-in that becomes available after the original purchase is made. However, modular construction results in increased size, weight, and cost when compared with an instrument of identical functionality that does not have these modular features.

An oscilloscope mainframe with multiple plug-in capability offers even more flexibility. Instruments are available with both the vertical amplifier and time base in separate plug-ins or with multiple vertical amplifier slots.

14.3 Vertical Amplifier

The oscilloscope vertical amplifier is designed for linear scaling of the voltage signals received at its input up or down to the amplitude range required by the oscilloscope signal capture device, whether CRT or ADC. The user then is able to set the "window" in input voltage to that which the signal requires. It is important that the vertical amplifier conform to an operational model that is easily understood by the user. The trace viewed on the display is a representation of the input signal *after* the application of the vertical amplifier transfer function. In visualizing what the signal "really looks like," the user must compensate for any significant changes in the signal characteristics imposed by vertical amplifier. If this process is needed, it is usually done "in one's head" as the operator views the waveforms and should be kept as simple as possible. Indeed, it is the goal that for most measurements *no* compensation by the user for vertical amplifier response will be necessary; the waveform displayed on the screen can be accepted as a faithful representation of the user's signal.

In striving to achieve a simple and standard operational model for the vertical amplifier, several desirable characteristics have been recognized as important: (1) the amplifier gain should not vary appreciably with the frequency of the input signal within the amplifier pass band, (2) the pass band should include dc signals, (3) the vertical amplifier cutoff frequency should be guaranteed to exceed some stated value, and (4) these characteristics should be substantially the same at each and every gain setting (Fig. 14.6).

The signal voltages which may be analyzed appropriately with an oscilloscope can vary over a huge amplitude range, for example, megavolt (10^6 V) transients are encountered in lightning discharges and pulsed power sources for high-energy lasers, while the signals generated by nerve activity in biologic studies may be of only a few microvolts (10^{-6} V) peak-to-peak variation. In these extreme cases, it is reasonable to use attenuators or amplifiers exter-

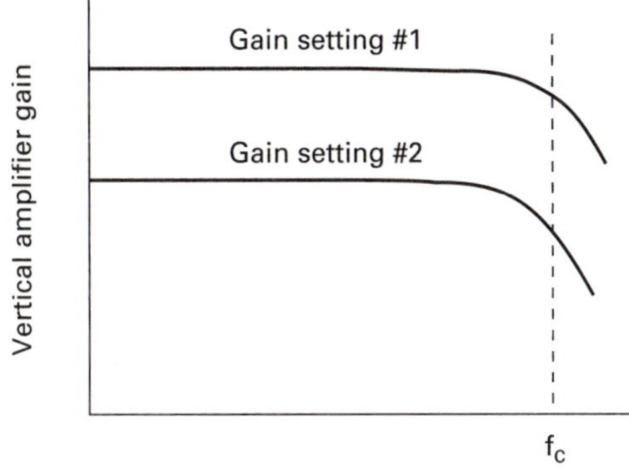

Figure 14.6 Vertical amplifier gain versus frequency for two different gain settings.

nal to the oscilloscope, sometimes of specialized design, to transform the signal amplitude range to one that is within the display capability of the general-purpose oscilloscope. This course is indicated in consideration of operator safety, noise contamination of the signals, as well as optimization of the total cost of the equipment required to perform a specific measurement. Nevertheless, it has been found useful and desirable to have the vertical amplifier be capable of a wide range of adjustments, minimizing the need for external components in measurement of the signals most commonly encountered.

14.3.1 Deflection factor

The most important and frequently used vertical amplifier control determines the change in voltage at the vertical channel input required to move the display trace one major division of the graticule in the vertical direction on the display screen. This control is labeled "Volts/Division" or "Sensitivity" and is said, by convention, to control the "deflection factor" of the vertical channel. It is desirable for this control to have a wide range of possible settings to conveniently accommodate the measurement of many different signal amplitudes. However, a wide range of settings increases the expense of manufacture of the vertical amplifier (and the price of the oscilloscope) and can compromise other amplifier characteristics, such as noise and bandwidth. A general-purpose instrument might provide deflection factor settings from 1 mV per division up to 5 V per division. Such a large variation, a ratio change in deflection factor of 5000 times, is implemented by using attenuators and different gain factor amplifiers that are selectively switched in or out of the vertical amplifier paths in various combinations.

Also of interest is the number and spacing of the intermediate settings on the "Volts/Division" control. The traditional configuration provides steps in a 1, 2, 5 sequence (e.g., 1 mV/div, 2 mV/div, 5 mV/div, 10 mV/div, 20 mV/div,

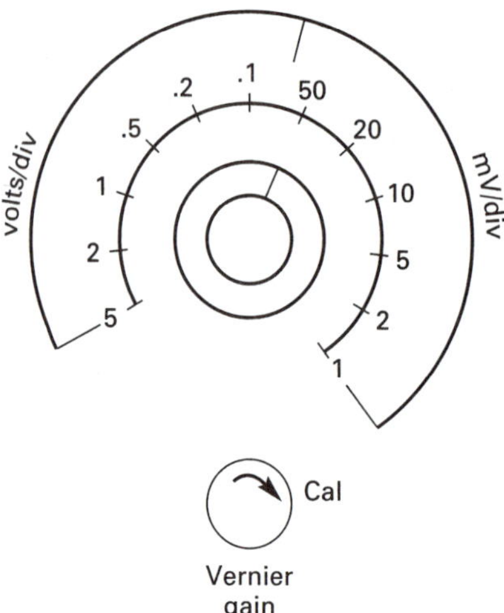

Figure 14.7 Vertical deflection factor control.

. . . , 2 V/div, 5 V/div) (Fig. 14.7). With this approach, a secondary "Volts/Division" control is provided, called the "vernier gain" adjustment. This control has a "calibrated," detented position (full clockwise rotation). Rotating this control counterclockwise reduces the vertical amplifier gain (increases the deflection factor) smoothly and continuously until, at full counterclockwise rotation, the vertical amplifier gain has been reduced by a factor of at least 2.5 times from that in the "cal" position.

Other oscilloscope models use a different strategy for controlling the deflection factor of the vertical amplifier. The control knob has no markings to indicate the deflection factor setting, and the current setting for the deflection factor appears as a number in a digital register, usually on the main display screen of the oscilloscope. Rotating the control clockwise causes the deflection factor to decrease; rotating it counterclockwise causes it to increase, and the display register changes to reflect the current setting. A desired deflection factor setting also can usually be entered directly using a numeric keypad on the instrument control panel. This configuration is capable of many discrete, calibrated settings. For example, between 1 mV per division and 100 mV per division the deflection factor might be adjustable in 1 mV per division increments.

It is intuitive that the greatest measurement accuracy can be achieved if the signal to be measured is displayed with as large a vertical deflection on the display screen as possible. This view is supported by reading the oscilloscope manufacturer's specifications regarding vertical accuracy. The guaranteed accuracy invariably is specified as a fraction or percentage of the full-scale vertical deflection (the total number of divisions multiplied by the volts per division). Thus accuracy, expressed as a percentage of the *measured signal amplitude,* is best when the display deflection approaches full screen and is degraded when less of the vertical deflection space is used.

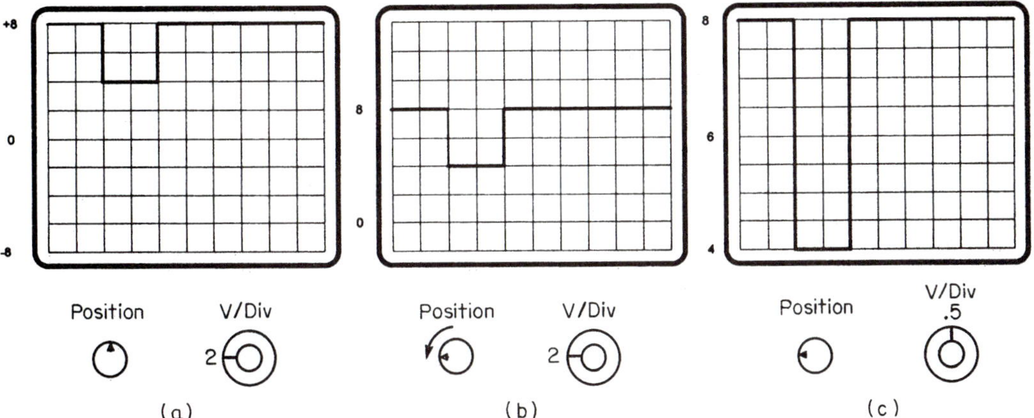

Figure 14.8 Using the "Position" and "Volts/Division" controls to adjust the display.

14.3.2 Position and dc offset

Also provided as part of the vertical amplifier is a means of moving the vertical space that a signal occupies up or down on the display without changing the vertical size of the displayed signal. This control function allows the user (1) to position the signal so that a specific part of the waveform can be brought into coincidence with a particular horizontal graticule line to aid in performing a voltage or time measurement, (2) to independently position two different signals being viewed simultaneously to remove ambiguity or to assist in making relative time measurements, or (3) to overcome the effect of a large dc component in the signal being measured.

As an example of the use of the vertical position control in conjunction with the deflection factor control, consider the measurement of a pulse of voltage switching between +4 and +8 V. In Fig. 14.8*a*, the position control is centered (the central horizontal graticule mark corresponds to zero volts), the deflection factor control is set to 2 V per division, and the display shows a pulse amplitude of two divisions. Changing the deflection factor to 1 V per division to increase the size of the displayed pulse would cause the pulse to disappear from the screen (the center graticule would still be zero volts, the top graticule would be +4 V). Instead, move the pulse to the center of the display with the position control as shown in Fig. 14.8*b*. Now change to 0.5 V per division, and pulse amplitude expands to a full eight divisions (Fig. 14.8*c*).

A control that moves the voltage window vertically can be labeled either "Position" or "Offset." A position control always moves the displayed waveform the same number of divisions per unit of rotation and is unaffected by the volts per division setting. The offset control moves the trace a certain number volts per unit of rotation, so the sensitivity of trace motion varies as the deflection factor is changed. The "offset voltage" is often displayed on a calibrated knob or in a display register.

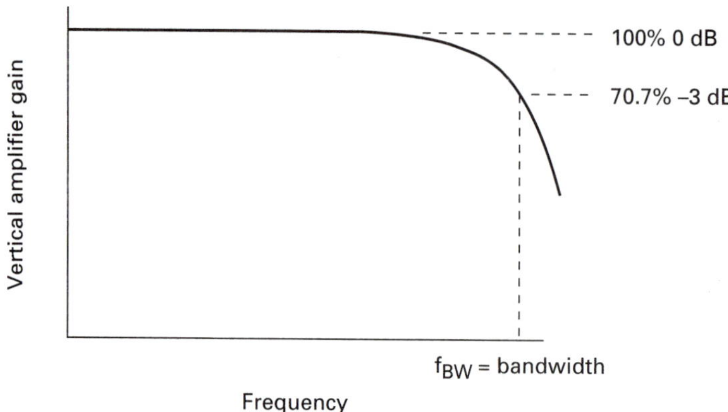

Figure 14.9 Vertical amplifier response versus frequency.

14.3.3 Bandwidth and rise time

Figure. 14.9 shows the frequency response of a vertical amplifier at a particular gain setting, exhibiting a constant-amplitude response to the input signal regardless of the frequency of that signal if it is within the pass band of the amplifier. Above the cutoff frequency f_{BW}, the amplifier response is decreased. For every amplifier, there is a limit to the frequencies which can be amplified, imposed by limits of cost, weight, and power consumption of the oscilloscope in which it is contained. The frequency-response capabilities of various oscilloscope models vary over a wide range depending on the application for which they were intended. One fact is certain, however: there is an ultimate limit imposed by the laws of physics and economics. This limit is traditionally referred to as the "bandwidth" of the oscilloscope, the frequency at which the response drops to 70.7 percent (a factor of $1/\sqrt{2}$,) of the low-frequency value, or down 3 dB from the low-frequency reference level of zero decibels.

Bandwidth is one of the more important oscilloscope performance specifications, because manufacturers and users can and do treat it as the single quantity that communicates the ability of the instrument to deal with fast transients. However, oscilloscopes are seldom used to observe pure sine waves; the natural and frequent use is to observe more generalized voltage transients which vary in a less regular way. This type of measurement is referred to as "time domain," in contrast with the spectrum analyzer, which measures in the "frequency domain."

Thus it is useful to characterize the oscilloscope response to a well-defined time-domain signal, the ideal voltage step function (Fig. 14.10). The voltage is initially V_1 and instantaneously switches to V_2. The ideal response of the oscilloscope to this stimulus is depicted in Fig. 14.11. The oscilloscope has a finite bandwidth, so a certain amount of time is required for the trace to respond to the instantaneous step, and this interval is called the "rise time." The usual definition of rise time is the time taken for the trace to move be-

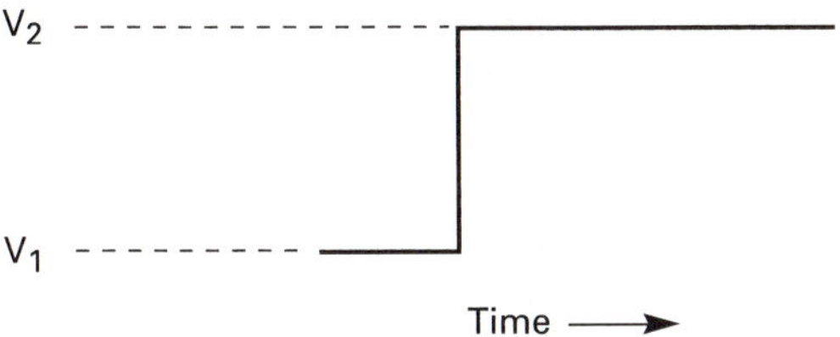

Figure 14.10 The ideal voltage step function.

tween the points corresponding to 10 and 90 percent of the total transition. Other definitions are possible, but in this discussion the 10 to 90 percent definition will be used.

At this point it is important to note that the response of the vertical amplifier to the ideal voltage step could in theory follow any of an infinite number of possible paths from the initial to the final value, even for one rise-time value. Specifically, the shape of the response before crossing through the 10 percent point and after crossing the 90 percent point is not constrained in principle by the rise-time definition. Even so, the ideal oscilloscope step response is a smooth, symmetric, and unidirectional transition from one steady-state level to another, as in Fig. 14.11. By "unidirectional" it is meant that the response to a voltage step input will move exclusively in one direction throughout the transition.

Examples of other possible, but less desirable, oscilloscope step responses

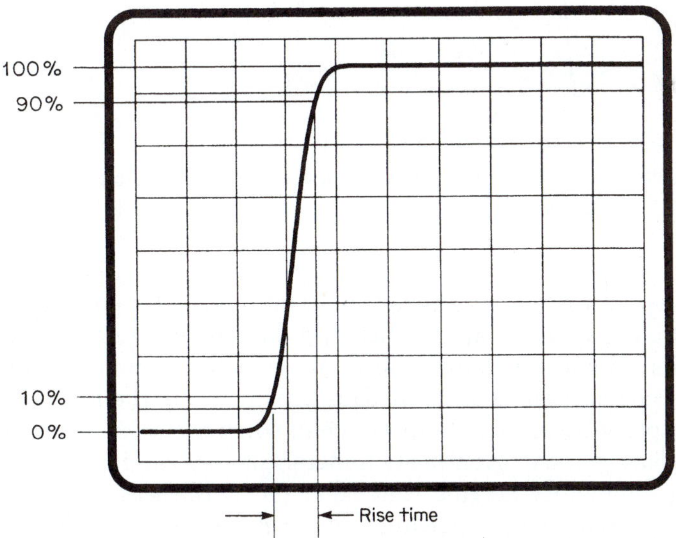

Figure 14.11 The ideal oscilloscope response to the voltage step function.

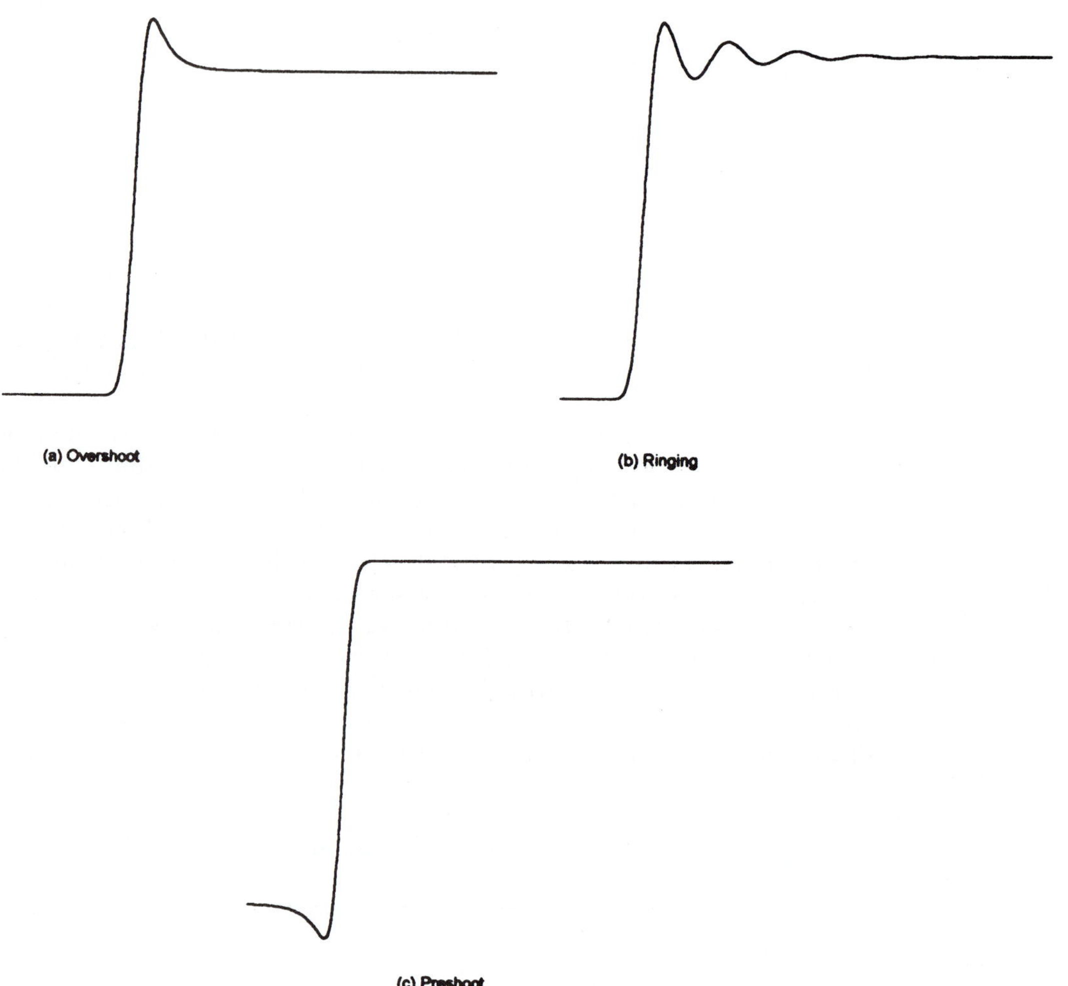

Figure 14.12 Undesirable vertical amplifier step responses. (*a*) Overshoot; (*b*) ringing; (*c*) preshoot.

are shown in Fig. 14.12. Deviations from the ideal response can be caused by parasitic coupling or an insufficiently damped resonance within the vertical amplifier. The pulse shapes shown are common occurrences in the signals an oscilloscope is used to observe. This prevalence is the reason the oscilloscope step response should be unidirectional: when complex shapes are observed on the display screen (overshoot, undershoot, or, ringing), the user should be able to confidently ascribe that behavior to the signal being measured and not to the oscilloscope vertical amplifier response. Then the only accommodation that needs to be made for the limited transient response capability of the vertical amplifier is to note that any rise time observed on the display screen

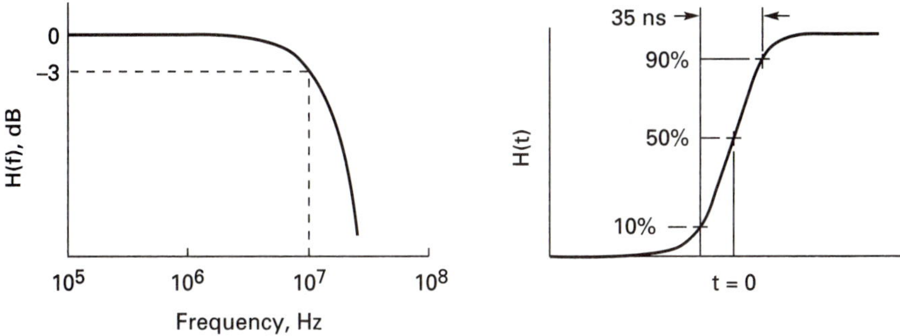

Figure 14.13 Gaussian response for a 10-MHz bandwidth vertical amplifier.

that is close to the known rise time of the vertical amplifier may not represent the actual rise time of the signal being observed.

After examination of the vertical amplifier response in both time-domain (transient response) and frequency-domain characterizations, it is intuitive that the two are related. In fact, it is a well-known result from filter theory that the complete pulse response can be calculated if the frequency response is completely known by using the Fourier transform, and vice versa with the inverse Fourier transform. The mathematical relationship between pulse response and frequency response makes it possible to obtain several relationships that are useful in oscilloscope applications.

Figure 14.11 shows one possible version of the ideal vertical amplifier pulse response, having a "smooth, symmetric, and unidirectional transition" between the two voltage levels. While there are any number of different pulse shapes that satisfy this description, the most desirable pulse shape results when the vertical amplifier response is "gaussian." This is a frequency response given by the equation

$$H(f) = \exp\left[-0.3466\left(\frac{f}{f_{\mathrm{BW}}}\right)^2\right] \tag{14.1}$$

where $H(f)$ = amplifier response
f = input signal frequency
f_{BW} = cutoff frequency where response is down 3 dB

Equation 14.1 and its corresponding pulse response for f_{BW} = 10 MHz are plotted in Fig. 14.13.

The filter defined by Eq. 14.1 has a number of useful and interesting properties. The pulse response is completely free of overshoot or ringing throughout the step response. Moreover, it can be shown that this is the fastest possible cutoff rate in the frequency domain for any filter that is free of overshoot and ringing. If the cutoff rate is increased over any nonzero band of frequencies, the resulting filter will have ringing (nonunidirectional behavior) somewhere in its step response. Also, the cutoff frequency of two cascaded gaussian filters with different cutoff frequencies can easily be calculated, because the resulting response is gaussian. The cutoff frequency of the equivalent single gaussian filter is given by

$$f_3 = \frac{1}{\sqrt{(1/f_1)^2 + (1/f_2)^2}} \tag{14.2}$$

where f_1 and f_2 = cutoff frequencies of the cascaded filters
f_3 = cutoff frequency of the equivalent filter

It was stated previously that the step response of the gaussian filter can be derived from the frequency-domain description of Eq. 14.1 with use of the Fourier transform. It also can be shown that the filter rise time is inversely related to the cutoff frequency by the following equation:

$$\tau_r = \frac{0.35}{f_{\mathrm{BW}}} \tag{14.3}$$

where τ_r = rise time, in seconds
f_{BW} = cutoff frequency, in hertz

Then, using Eqs. 14.2 and 14.3, it can be shown that the composite rise time of the two cascaded gaussian filters can be expressed in terms of the rise times of the individual filters by

$$\tau_3 = \sqrt{(\tau_1)^2 + (\tau_2)^2} \tag{14.4}$$

where τ_1 and τ_2 = rise times of the cascaded filters
τ_3 = composite rise time

These results are useful directly in understanding the operation of a vertical amplifier that has been adjusted to a gaussian response: (1) the shape of the step response is known, (2) the rate at which the frequency response decreases above the cutoff frequency is known, and (3) the rise time can be calculated if the cutoff frequency is known, and vice versa. They are also useful in estimating the effect of vertical amplifier rise time on the measurement of the rise time of a user's signal. Let the response $H_2(f)$ represent a gaussian vertical amplifier, and let $H_1(f)$ represent the transfer function of the external circuit responding to a voltage step. The rise time τ_3 is measured on the display screen. The vertical amplifier rise time τ_2 is known, so the actual rise time of the input signal τ_1 can be calcualted using a rearrangement of Eq. 14.4:

$$\tau_1 = \sqrt{(\tau_3)^2 - (\tau_2)^2} \tag{14.5}$$

and it will be the correct value *if* both the vertical amplifier *and* the user's circuit have a gaussian response. Equations 14.3, 14.4, and 14.5 are widely known to oscilloscope users, and they have a certain utility in estimating the rise time of a signal that approaches the rise time of the oscilloscope in use at the moment. That these equations were derived for gaussian systems is less than widely appreciated. There is the possibility of a significant error if the signal or the vertical amplifier is not gaussian. Use these equations with caution, particularly if ringing is observed on the display screen or if the calculated correction is more than a few percent of the measurement. If an accurate measurement of rise time is needed, use an oscilloscope with a rise time much shorter than that of the input signal. For 5 percent accuracy, a ratio of 3 or 4 : 1 is sufficient.

14.3.4 Signal conditioning

The general-purpose vertical amplifier usually includes three frequency-sensitive filters that can be selected from the oscilloscope control panel. They are provided to make the oscilloscope more adaptable to common measurement situations that would otherwise require the use of external filters. Each input channel on a multichannel oscilloscope has a complete and independent set of these controls.

Ac coupling. The first control in this group is labeled "ac coupling." When activated, this function disables the normal dc coupling of the vertical amplifier and prevents any dc component of the input signal from affecting the display. The frequency response with ac coupling selected is shown in Fig. 14.14. Frequencies below cutoff, normally between 1 and 10 Hz, are rejected or filtered from the input signal before amplification. A 1-kHz square wave has the same shape in both ac and dc coupling modes (Fig. 14.15*a*), but the position on the display screen is different in the two cases if the square wave

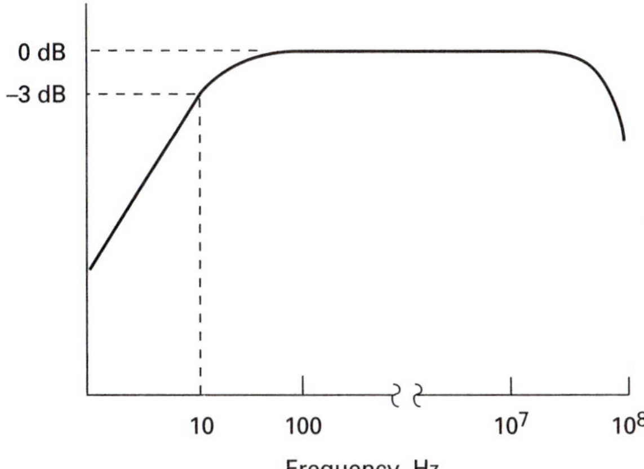

Figure 14.14 Vertical amplifier frequency response with ac coupling activated.

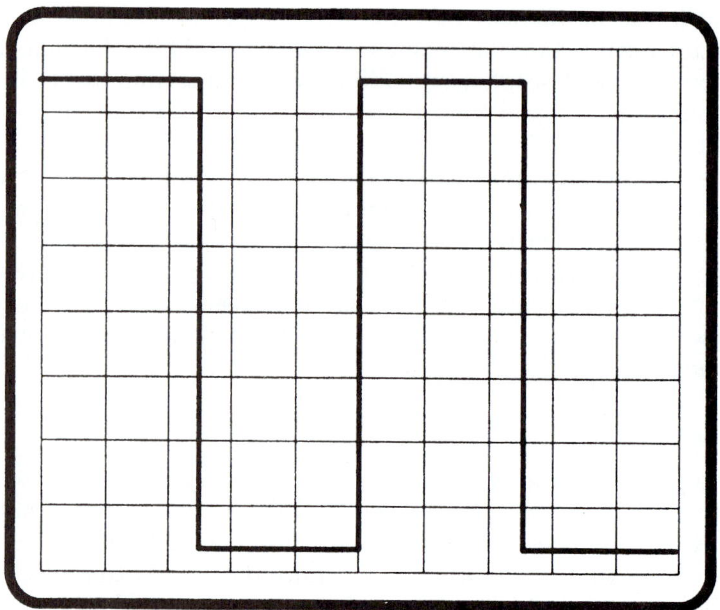

(a) One KHz square wave

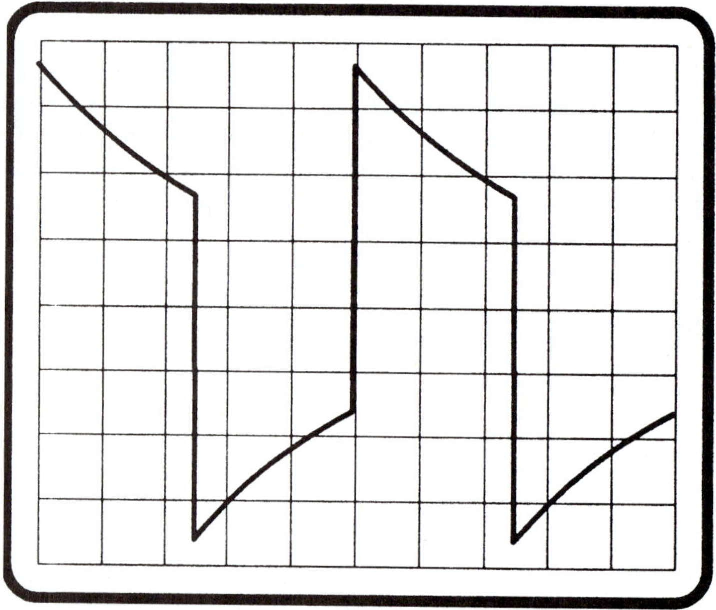

(b) 40 Hz square wave

Figure 14.15 Two square waves in ac-coupled mode. (*a*) 1-kHz square wave; (*b*) 40-Hz square wave.

has a dc component. The display resulting from a 40-Hz square wave is shown in Fig. 14.15*b*.

The ac coupling feature is used when a dc component in the input signal might interfere with a measurement. As an example, the ac coupling mode could be used to observe the noise, power line frequency ripple, or switching transients appearing on a 10-V power supply output terminal. With ac coupling selected, the 1 mV per division deflection factor setting could be used if the transients were very small. In dc coupling mode, the 10-V component would cause the signal to be deflected off the display screen on the most sensitive deflection factor settings beyond the adjustment range of the position or offset controls; increasing the deflection factor to 1 or 2 V per division would allow the signal to appear on the display screen, but the user would not be able to see small transients.

Low-frequency reject. A second control, labeled "Low-Frequency Reject" or "LF Reject," is similar to ac coupling, except that the rejection band extends to a higher frequency, usually between 1 and 50 kHz. Of course, LF reject blocks any dc component in the input signal, but its main use is to remove power-line frequency signals (hum) from the input signal.

Bandwidth limit. The third control in this group is labeled "Bandwidth Limit" or "BW Limit" (Fig. 14.16). This is a low-pass filter that reduces the bandwidth of the vertical amplifier, usually by a factor of between 10 to 30. It is used to remove, or at least attenuate, unwanted high-frequency signals or noise contained in the input signal. The carrier signal from a nearby radio broadcast transmitter and a local oscillator are possible sources of such unwanted signals. The filter that is switched into the vertical amplifier path by this control is usually not very complex, and the pulse response with BW limit selected is not a particularly good approximation to the gaussian goal.

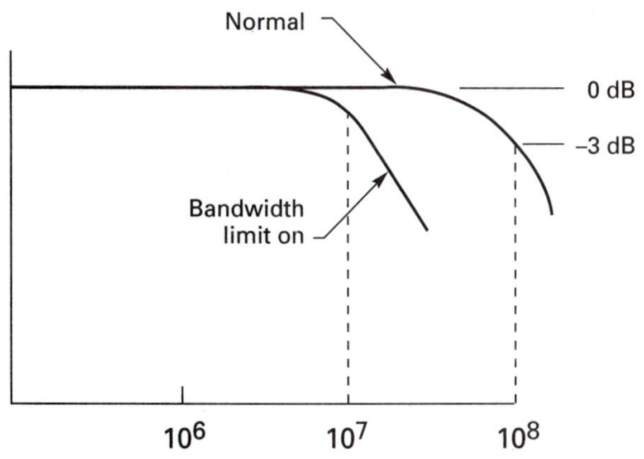

Figure 14.16 Vertical amplifier response with and without bandwidth limit.

14.3.5 Input impedance

The final topic in the discussion of the standardized vertical amplifier model is the interface to the user's signal at the oscilloscope input connectors. In order to observe the user's signal, some electrical energy must be diverted to the oscilloscope from the circuitry being measured, and there is the possibility that this act of measurement could significantly affect the waveform or even alter the operation of the user's circuit. Several features relating to the design and parametric specification of this input have evolved to address these concerns. A large value, compatible with the oscilloscope bandwidth and intended application, is chosen for the impedance at the vertical amplifier input connector to minimize loading. The nominal value and tolerance of the input impedance are stated in the oscilloscope specifications so that the user can calculate loading effects. The circuitry in the vertical amplifier is designed to maintain the input impedance at that specified value, regardless of changes to the oscilloscope gain, attenuator, offset, or signal conditioning control settings. The input impedance also should remain constant if the input signal is very large, even many times that required for full-vertical-scale deflection. Any leakage current or offset voltage appearing at an input connector that originates in the vertical amplifier circuitry also must be limited to a negligible amount.

The need for a well-defined interface is clear when the oscilloscope user connects a circuit directly to the vertical amplifier input, but these characteristics are also important when passive probes, amplifiers, or other transducers are used to make the connection between oscilloscope and circuit. An accurately known terminating or load impedance is important to the operation of many of these devices. This topic is explored further in the discussion of probes later in this chapter.

In theory, many different values could be used for the vertical amplifier input resistance, as long as the previously stated conditions are met. Actually, most oscilloscopes are designed to conform to one of two distinct models.

High-impedance input. In the "high impedance" input model, the oscilloscope input has a resistance of 1 MΩ (10^6 Ω) shunted with a small capacitance, between approximately 7 and 30 pF depending on model and application (Fig. 14.17*a*). The advantage of this configuration is that only very small dc or low-frequency currents are drawn from the user's circuit. The disadvantage is that the input capacitance causes the high-frequency loading to be relatively much higher. The maximum bandwidth of instruments built this way is limited to approximately 500 MHz, mainly because of technological limitations in the switchable lumped-element attenuators compatible with this approach.

50-Ω Input. The "50-Ω input" (Fig. 14.17*b*) provides a constant 50-Ω resistance to ground and is designed to accurately terminate a transmission line (coaxial cable) having a characteristic impedance of 50 Ω. Switchable attenuators and other circuitry built to this model are designed as distributed ele-

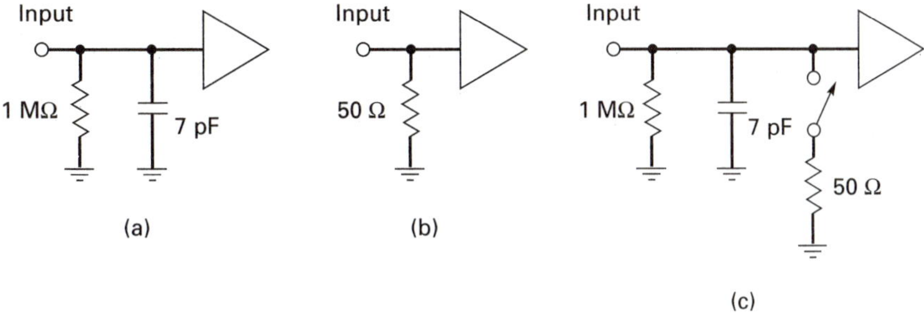

Figure 14.17 Vertical amplifier input impedance models. (*a*) High-impedance input; (*b*) 50-Ω input; (*c*) combination input.

ments and are capable of operation at much higher frequencies than the high-impedance types. The constant 50-Ω resistance is a serious disadvantage in general-purpose applications because of the relatively large currents drained from the user's circuits. Thus oscilloscopes with 50-Ω inputs are optimized for very high frequency measurements, from 1 GHz and above.

A selectable 50-Ω setting is a popular feature for oscilloscopes having high-impedance inputs (Fig. 14.17*c*). An internal 50-Ω resistor can be optionally connected in parallel with the normal 1-MΩ input resistance, controlled from the oscilloscope-user interface. With the 50-Ω resistor disconnected, operation is identical to that of a high-impedance input. With the resistor connected, the input can be used as a transmission-line termination while making measurements with some of the advantages of the dedicated 50-Ω input systems. However, the capacitive component remains, degrading the quality of the termination, and the system bandwidth is subject to the same limits as the unterminated high-impedance input.

14.4 Horizontal or Time Base and Trigger

The time-base section of the oscilloscope is used to control the horizontal or time axis of the display, so this group of controls is labeled either "Time Base" or "Horizontal." The settings of these controls are used to set the "window" in time, determining precisely when, for how long, and how often the user's signal appears on the display.

The variation with time of the voltage appearing at the vertical amplifier input can follow any of an infinite number of possibilities. It could, for example, be a constant-frequency sawtooth wave (Fig. 14.18*a*), a series of identical pulses occurring at irregular intervals (Fig. 14.18*b*), or a transient that occurs only once (Fig. 14.18*c*). The oscilloscope user might want to observe many cycles of the sawtooth of Fig. 14.18*a* to estimate the period or to observe the pulses in Fig. 14.18*b* to measure the overshoot of the trailing edge of the pulse. The time-base controls are designed to permit precise control over the timing aspects of signal capture. It is necessary to determine when signal

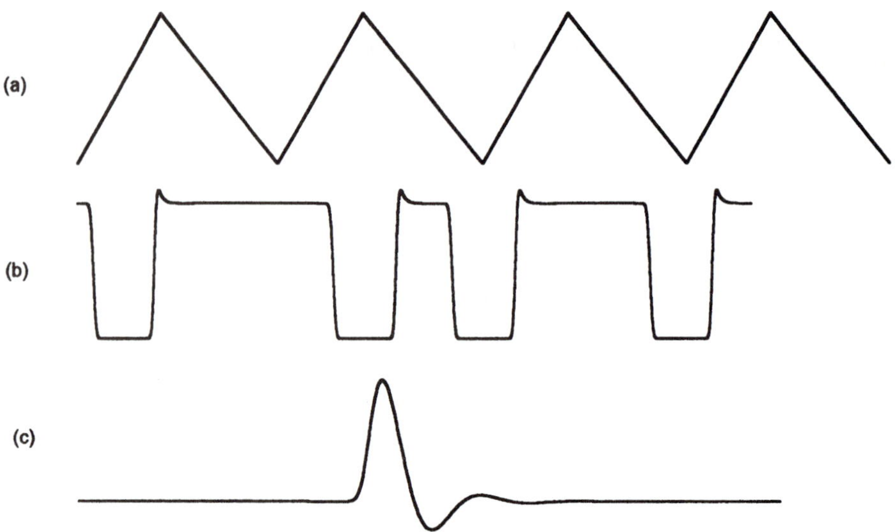

Figure 14.18 Example signal waveforms.

capture begins, the time interval over which the signal is recorded, and how long to wait after each signal capture before repeating the acquisition cycle. Figure 14.19 illustrates the signal-capture sequence on the pulse train from Fig. 14.18*b*. The time intervals during which signal capture occurs are indicated by the upper row of shaded bands. Immediately after each signal capture ends, a fixed-duration "holdoff" period begins while the acquisition circuits are being reset, shown as the lower row of shaded bands. A subsequent signal capture cannot begin until the end of the holdoff period. After the holdoff period, there is a wait for a trigger pulse signal to begin the next capture sequence. The resulting oscilloscope display, a composite of the multiple signal captures, is shown at the bottom of Fig. 14.19.

14.4.1 Triggering

The process of determining the exact instant in time to begin signal capture is called "triggering" the oscilloscope. The most straightforward triggering method, called "edge triggering," is discussed in the next section. Extension of the triggering concept to combinations of multiple signal sources is presented in Sec. 14.4.7.

Edge triggering. In the edge-triggering method, circuitry inside the time base identifies the instant in time that the input signal passes through a threshold voltage, called the "trigger level," in a particular direction (positive or negative slope) and responds by generating a precisely shaped pulse that causes the start of the signal acquisition cycle. Refer to Fig. 14.20*a*, in which a trig-

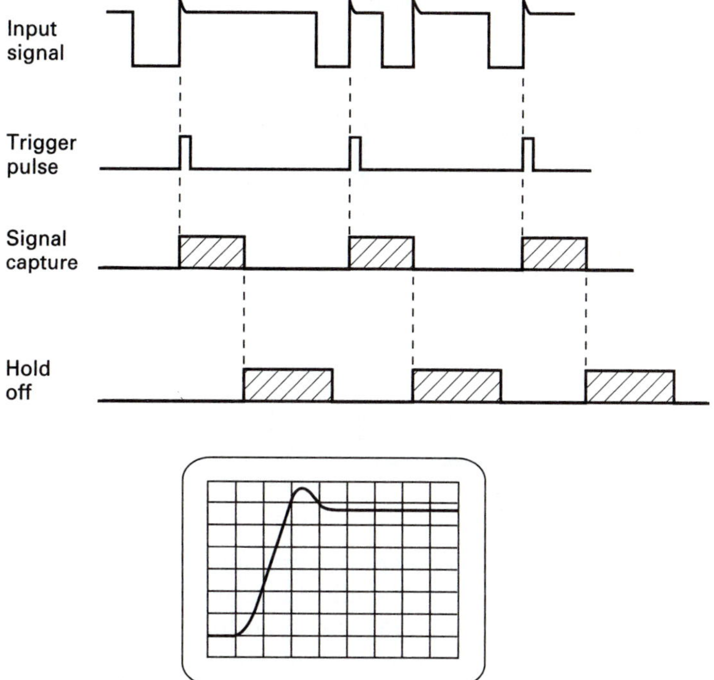

Figure 14.19 Signal capture on an irregular pulse train.

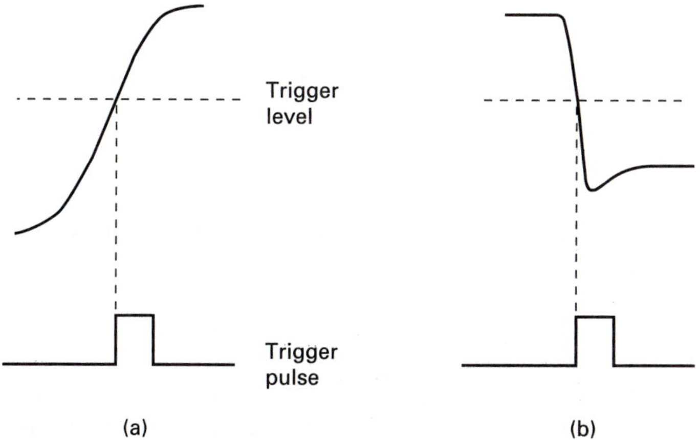

Figure 14.20 Edge triggering. (*a*) Positive-slope trigger; (*b*) negative-slope trigger.

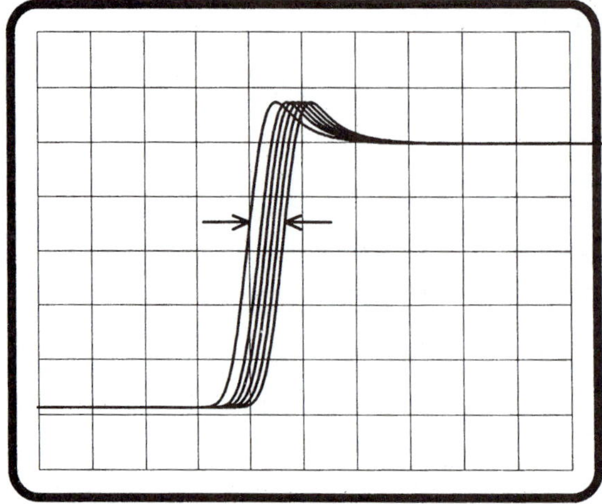

Figure 14.21 Jitter in the display caused by variation in the time delay from trigger to signal capture.

ger pulse follows a positive-slope trigger, and Fig. 14.20*b*, with a negative-slope trigger. The trigger level is adjustable and the trigger slope is selectable from front panel controls, giving the operator the ability to trigger on any part of a positive or negative edge anywhere in the signal.

It is important that the time elapsed from trigger level threshold crossing to the start of signal acquisition be exactly the same on each capture. Instability in this interval will show up as jitter on the oscilloscope display (Fig. 14.21). Of course, jitter cannot be eliminated entirely from the trigger and time-base circuits, and the residual value is stated in the specification sheet for each oscilloscope model. This is an important specification because the oscilloscope is sometimes used to measure jitter in the user's signal.

14.4.2 Metastability

The sequence from trigger to signal capture to holdoff period to next trigger is illustrated in Fig. 14.19. Implicit in this process is the idea that the trigger circuit is inhibited from generating a trigger pulse during the signal capture and holdoff periods and enabled or "armed" at the end of the holdoff periods. However, the time until the next threshold crossing which could generate the next trigger is determined by the input signal and is not ordinarily under the control of the time base. Display jitter can result if a trigger threshold crossing sometimes occurs just at the instant in time that the trigger circuit is being armed. If the threshold crossing occurs just before the end of the holdoff period, a trigger pulse is not generated; if just after, a trigger pulse is generated. Thus there *must* be a critical instant just as the trigger circuit is being armed when a trigger threshold crossing will result in indecision as to whether a trigger pulse should be generated. The indecision can result in extra delay or a modified shape for the trigger pulse. The two simultaneous events are called a "collision," and the uncertain result is called "metastability."

14.4.3 Holdoff

The time base has a control labeled "Holdoff" that allows the operator to modify the length of the holdoff period. If display jitter is observed and trigger generator metastability is suspected, then adjusting the holdoff period will modify the collision statistics and eliminate, or at least change, the jitter pattern.

The holdoff control is occasionally used to obtain a stable display when there are multiple recurrence patterns in a complex input signal, as in Fig. 14.19. A waveform may appear to have a stable trigger point, but a second pulse may be seen occasionally in the display, depending on whether the trigger is armed in the short interval between pulses or during the long interval.

Thus the appearance of the display may depend on the holdoff control setting and chance. These anomalous results can be called "aliasing" and happen because the oscilloscope, both analog and digital forms, is a sampled data system; data are recorded during intervals interspersed with intervals during which no data are recorded. Viewing the input signal with a much longer capture interval and varying the holdoff period would clarify this situation, unless there were some even longer-term periodicity in the pulse train. Aliasing should be suspected whenever the display changes in an unexpected way, as happened in the examples with different settings of the holdoff control. Aliasing occurs in a number of situations in oscilloscope measurements and is also addressed later in this chapter during the discussion of multichannel analog and real-time digital oscilloscopes.

14.4.4 Hysterisis

An oscilloscope trigger system compares the input signal with a voltage set by the trigger level control using a comparator circuit in the generation of the trigger pulse. The type of comparator that performs best as a trigger circuit has a small amount of "hysteresis," a characteristic that causes the input voltage at which the comparator output switches to depend on the output state (Fig. 14.22). When the comparator output is in the low state, the output will change state when the input signal passes through the upper trigger level. When in the high state, the output switches as the lower threshold is traversed. The two levels are usually separated by a voltage corresponding to between approximately 10 and 30 percent of the vertical amplifier volts per division setting. The amount of hysterisis is usually preset by the oscilloscope manufacturer, although it is sometimes adjustable by a front panel control.

This use of hysterisis has several important consequences. First, hysterisis requires that the comparator have positive feedback, which makes the output state change more definite and less influenced by the rate at which the input signal crosses the trigger level. Also, the separation of the two thresholds

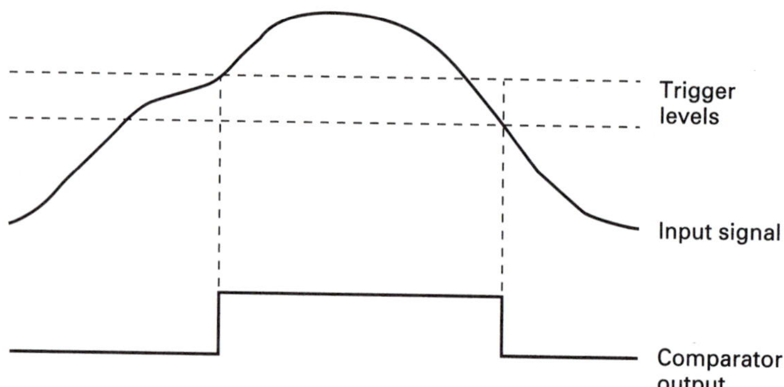

Figure 14.22 Hysterisis in the trigger level comparator.

determines the minimum peak-to-peak amplitude input signal required to cause the oscilloscope to trigger. The input signal must cross both thresholds for a trigger to be generated. Finally, the threshold separation gives the trigger circuit the ability to prevent a small amount of noise on the input signal from generating unwanted triggers. To understand why this is needed, refer to Fig. 14.23. With positive trigger slope selected, a trigger pulse is expected as the input signal passes through the trigger level going positive. However, as the input signal crosses the threshold in the negative direction, a small noise burst will generate a trigger pulse here also if the hysterisis is set too low. The display of a slightly noisy low-frequency sine wave on an oscilloscope with no trigger hysterisis would look like Fig. 14.23, and changing the trigger level only causes the intersection of the two sine waves to move up or down. Hysterisis is used for accurate setting of the minimum-amplitude trigger signal so that most low-amplitude noise does not cause unwanted false triggers.

14.4.5 Signal conditioning

The trigger system ordinarily has available control settings to selectively engage filters that exclude some part of the input signal that may interfere with achieving the desired trigger point. These are similar to the filters available in the vertical amplifier, although different cutoff frequencies might be used. “HF reject” is a low-pass filter that reduces high-frequency noise or carrier signals. “LF reject” blocks dc signals and power-line frequency hum.

14.4.6 Trigger source

Multichannel oscilloscopes have switch settings that permit the signal appearing on any of the vertical amplifier inputs to also be used as the trigger signal. In addition, there is usually a separate input signal connector located near the time-base controls labeled “External Trigger,” which can be selected as the trigger source. “Line Trigger” is also usually included as an optional

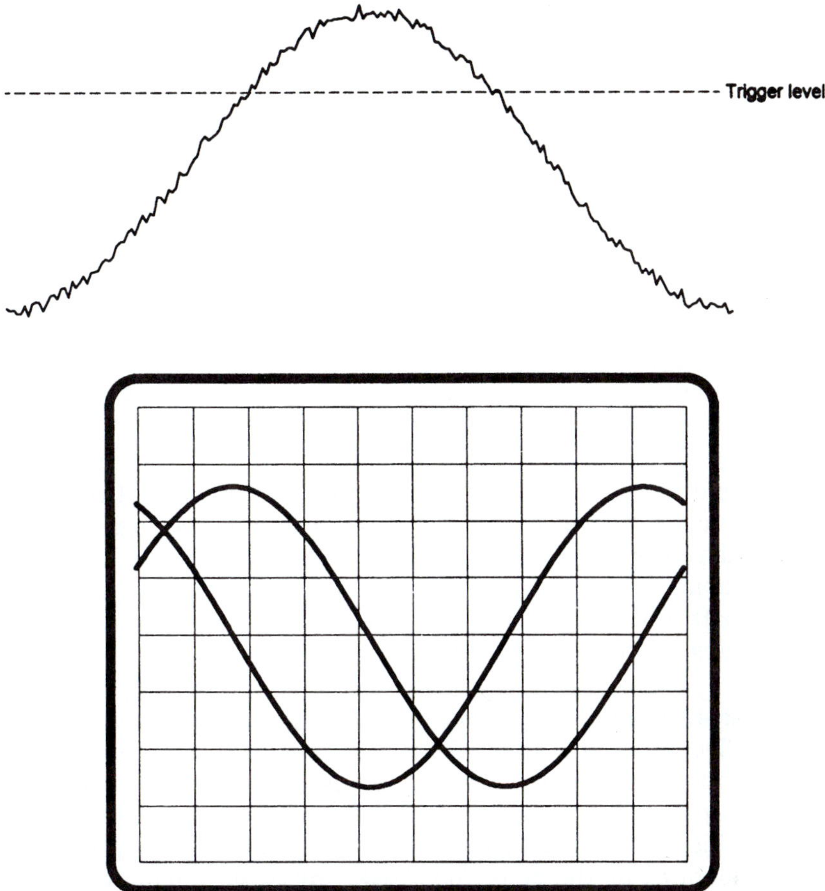

Figure 14.23 Triggering on a slightly noisy sine wave using a trigger circuit with no hysterisis.

source to allow triggering on the power-line frequency. This triggering mode is helpful in identifying power-line frequency-related components (hum) in the signals viewed on the oscilloscope.

14.4.7 Logic and glitch trigger

Edge triggering has been a standard feature on oscilloscopes for many years and is adequate for capturing most signals. However, the complex multichannel pulse trains encountered in digital systems present more of a challenge, and the oscilloscope trigger system has been extended to allow combinations of all the trigger sources to be used simultaneously in the identification of the desired trigger point. A trigger system capable of using multiple sources

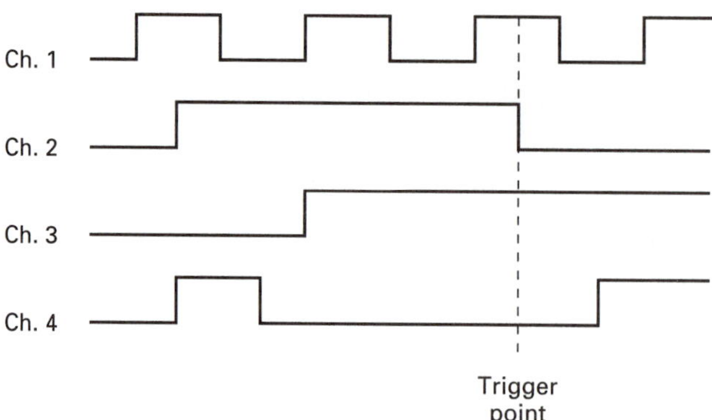

Figure 14.24 Trigger point when pattern trigger is set to "entering HLHL."

simultaneously is called a "logic trigger." Each vertical input channel and the external trigger signal are treated as digital logic signals. Each has an independent trigger level control, which defines the logic threshold for that channel. The trigger circuit response to each signal is determined by its state, whether the signal is above the threshold (H) or below (L), or by a state change, L to H (↑) or H to L(↓).

Pattern trigger. A trigger circuit configuration that considers only states is called a "pattern trigger." The user defines the trigger condition as a set of states for the inputs, and a trigger is generated when the pattern appears (or disappears). As an example, inputs to a four-channel oscilloscope are shown in Fig. 14.24. If the pattern trigger is set to "entering HLHL," the trigger is generated at the point indicated. Any of the inputs can be ignored for the purpose of triggering by setting its trigger condition to "don't care" (X). This triggering method is useful in isolating simple logic patterns in digital systems, such as decoder inputs or I/O addresses.

Time qualifier. For more power, a "time qualifier" is added to the pattern trigger. In addition to the pattern specification, the user sets a constraint on how long the pattern must be present before the trigger is generated. The user can specify that, for triggering to occur, the pattern must be present for (*a*) longer than a stated time, (*b*) shorter than a stated time, or (*c*) both longer than one stated time and shorter than another stated time. To see how this works, see Fig. 14.25. For case (*a*), the pattern is "XXHX present > 40 ns," trigger occurs at the point indicated, and a long pulse can be captured from a train of narrower pulses. In case (*b*), the polarity of the pulse train is inverted, so the pattern "XXLX present < 20 ns" generates a trigger on the narrow pulse in a train of 30-ns pulses. This case is often referred to as "glitch trigger," but any of these modes designed to pick out unusual events from a

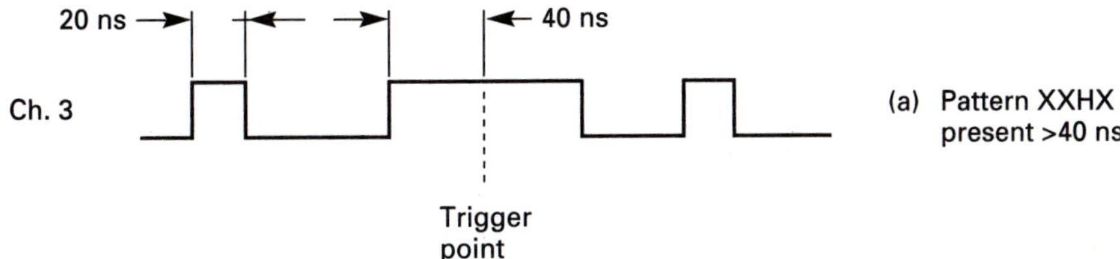

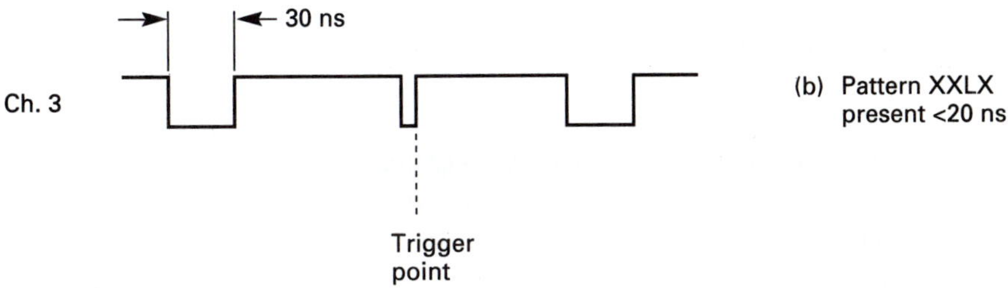

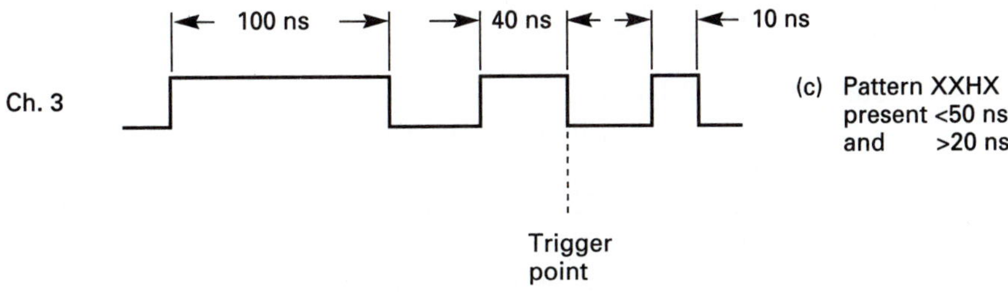

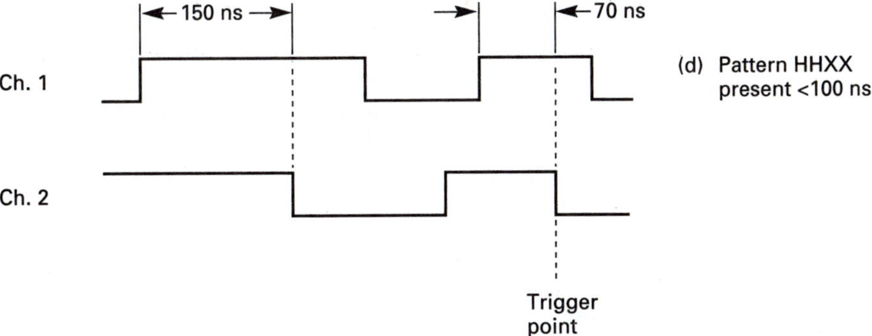

Figure 14.25 Time-qualified or "glitch" triggering.

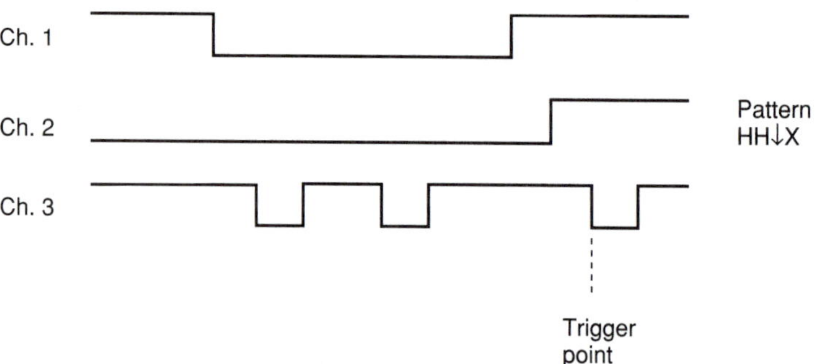

Figure 14.26 State trigger.

pulse stream could properly be called a "glitch trigger." Finally, case (*c*) shows a trigger point on a 40-ns pulse surrounded by both shorter and longer pulses. The previous examples illustrate the time qualification feature on a simple pattern. The states of the other channels could be used as qualifiers (e.g., find an abnormally narrow clock pulse on channel 3 when channels 1 and 2 are both in state H), but the time qualifier need not be determined entirely by a pulse on a single channel. In Fig. 14.25*d*, the pattern "HHXX present $<$ 100 ns" picks out a trigger point determined by activity on two channels.

State trigger. Edge trigger is combined with pattern dependence to implement a triggering mode called "state trigger." One channel is designated as the clock signal, and the other channels can have level requirements for triggering. The pattern HL↓H generates a trigger on the negative edge (clock) of channel 3 only when the other channels are in the specified state. This mode is useful in observing the operation of clocked digital systems. The example in Fig. 14.26 could be tracking the operation of loading a register; channel 3 is the clock and channels 1 and 2 are connected to enable inputs for the register.

Logic analyzer for triggering. The concepts used in the logic trigger are also used in logic analyzers. Thus it is a natural step to use a logic analyzer to trigger an oscilloscope when the complexity of the target digital system exceeds the capability of the oscilloscope logic trigger. Logic analyzers are able to process many channels simultaneously, have very complex and versatile triggering specifications, and output a pulse when the trigger condition is met. Just connect the logic analyzer trigger output to the oscilloscope external trigger input, and use the two instruments together.

14.4.8 Sweep speed and delay

Calibration of the horizontal axis of the display is necessary for accurate time measurements, and the operation of several controls must be considered. The "sweep speed" control is labeled "Time/Division," and the setting changes in a 1, 2, 5 sequence (e.g., 1 μs/div, 2 μs/div, 5 μs/div, 10 μs/div) to control the duration of the oscilloscope window in time. The slowest sweep speed setting is 1 s per division or longer, and the fastest is equal to or somewhat greater than the rise time of the vertical channel. For example, the fastest sweep speed on a 100-MHz bandwidth oscilloscope (3.5-ns rise time) is 5 ns per division.

In all previous examples, signal capture started as soon as the trigger pulse was generated. However, it is sometimes desirable to wait a predetermined time after trigger before signal capture begins. The "delay" control allows the operator to set that time interval, with a readout in divisions (corresponding to the current sweep speed setting in time per division) or directly in units of time. This signal-capture method is useful in setting up a measurement so that the desired part of the waveform is in the display window and is displayed at the optimal sweep speed. The delay control also can be used as a calibrated horizontal position control so that the waveform can be scrolled through the display window to view a large part of the signal at a high sweep speed.

14.4.9 Sweep control

Three different modes of operation are provided to control the way the oscilloscope repeats the signal-acquisition process: (1) single, (2) normal, and (3) autotrigger.

Single. In the single sweep mode, the trigger circuit is enabled by depressing an "arm" button. Then, when the trigger point is recognized, the signal is acquired and displayed once, and then signal acquisition is stopped until the "arm" button is depressed again.

Normal. In normal mode, the trigger is automatically armed after each signal capture and holdoff period; therefore, as long as trigger events happen regularly, the display is continually receiving new traces that represent the current input signal status. If the triggers stop for any reason, then the signal capture and display updates stop. Until the trigger condition is met, no measurement is possible, not even a gross estimate of the dc voltage present.

Autotrigger. The autotrigger mode is the same as normal mode, except that a trigger will be forced whenever a preset time interval from the last trigger is exceeded. The trigger timeout in autotrigger mode is set to approximately 25 ms. The signal-capture cycle following a forced trigger event will not be synchronized according to the current trigger specification, and the display will have a mixture of triggered and untriggered waveforms. If, however, trigger points are occurring regularly (and more frequently than a 40-Hz rate), the display will look the same as normal mode. The advantage of autotrigger mode is that the display will never be totally blank, even if the input signal is disconnected. Viewing an unsynchronized waveform on the display helps the operator determine what action is necessary to trigger the oscilloscope. That the signal is absent or has an unexpected range or amplitude is immediately obvious, and the appropriate action can be taken to establish a triggered condition. For this reason, autotrigger mode is the best choice to use in setting up a new measurement until a good trigger point is identified. Then, if the triggers are occurring at a slow enough rate to cause an occasional untriggered signal capture, normal mode can be selected to eliminate the unwanted triggers.

14.5 The Analog Oscilloscope

The preceding discussion covered the operation of various parts of the oscilloscope that are essentially the same in the analog and digital models. In this section the unique aspects of the analog oscilloscope are described.

14.5.1 The cathode-ray tube

The cathode-ray tube (CRT) and the analog oscilloscope have been linked in a process of mutual invention, development, and refinement for many years, leading to the instruments in use today. In the CRT, a beam of electrons is formed, focused, directed, accelerated, and strikes a phosphor screen causing light to be emitted at the point of impact. The physical structures that accomplish this are assembled inside a glass enclosure which holds them in the desired positions. The space inside the CRT is maintained at a high vacuum to minimize collisions between gas molecules and the electron beam.

Operation of a CRT critically depends on controlling the velocity and direction of moving electrons, and in the analog oscilloscope this is done exclusively with the use of electric fields. The force exerted on an electron by an electric field is given by the equation

$$\mathbf{F} = q_e \mathbf{E} \tag{14.6}$$

where $\mathbf{F}$ = force on the electron
q_e = charge on the electron
$\mathbf{E}$ = electric field strength

Most of the electric fields in a CRT have cylindrical symmetry about the central axis. Such a field in which the radial components are directed toward the axis and increase in strength at greater distances from the axis can be arranged so that electrons initially near and moving parallel to the axis through the field will all be directed toward a single point on the axis. This kind of electric field region is called an “electrostatic lens” and is said to focus an electron beam, but the analogy to light optics, although useful, is inexact.

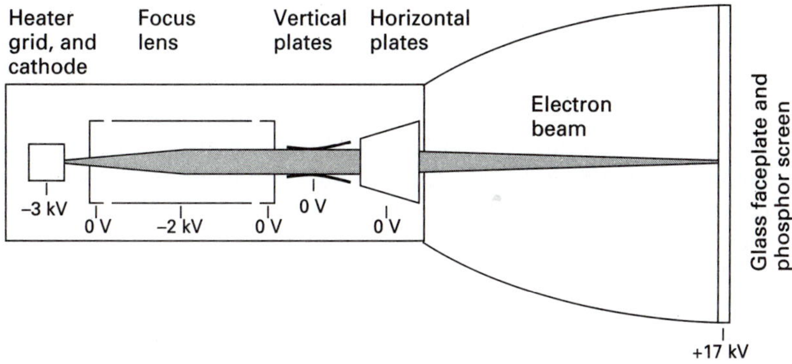

Figure 14.27 Schematic diagram of a CRT.

In Fig. 14.27, a CRT is shown in schematic form, with typical operating voltages indicated for the various key component parts. Everything in the diagram is understood to be a surface of rotation about the central axis of the CRT except the vertical and horizontal deflection plates.

The triode. The surface temperature of the cathode is raised to approximately 1000 K by an electrically powered heating element so that electrons can acquire enough thermal energy to escape from the hot cathode surface, a process called "thermionic emission" (Fig. 14.28). The grid and cathode operate at a voltage of approximately −3000 V, and the first anode (actually one of the electrodes in the focus lens) is at zero volts, so an electric field is established in the space between them, causing the electrons to be accelerated toward the first anode. The dotted lines indicate equipotential lines in this region. The grid, cathode, and first anode are called the "triode section" in the CRT.

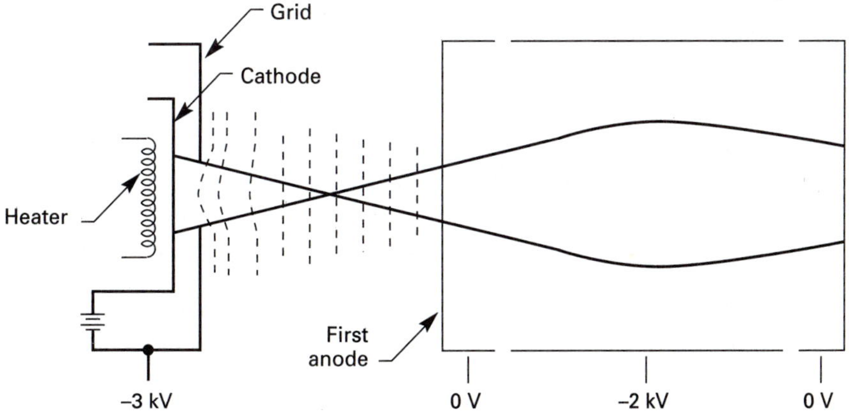

Figure 14.28 The CRT triode and focus lens.

The grid has a central circular hole through which the electrons emerge. The equipotential lines curve toward the cathode near the grid aperture, forming a converging electrostatic lens. The initially slow moving electrons are swept away from the cathode surface and tend to follow the electric field lines, which are always normal to the equipotential lines. Thus the electron beam is formed and then focused to a small-diameter spot called the "first crossover." Ultimately, it is an image of this first crossover spot that is projected onto the phosphor screen.

The grid is always biased at a negative voltage relative to the cathode to prevent current flow to the grid electrode. The magnitude of the voltage applied to the grid relative to the cathode regulates the magnitude of the beam current by controlling the intensity and direction of the electric field at the cathode surface. The beam current is maximized when the grid and cathode are at the same voltage; the beam current is completely cut off when the grid is 60 to 100 V more negative than the cathode, depending on the geometric details of the triode section. The grid-to-cathode voltage is used to control beam current and display intensity. It is also used to completely turn off the beam ("blanking") between acquisitions while waiting for a trigger to occur.

Beam focusing. The diverging electron beam enters the focus lens assembly through a central circular hole in the first anode. The middle element of the focus lens is a cylinder biased at a negative voltage relative to the electrodes at each end, forming a converging electrostatic lens. The magnitude of the negative voltage applied to the focus lens cylinder determines the strength of the lens and its focal length, so it is this voltage that the oscilloscope focus adjustment controls. The operator adjusts the focus to cause the beam leaving the focus lens to converge at just the rate required to converge to a small spot at the point of impact on the phosphor screen.

Of course, a small-diameter spot is desirable for good resolution of the display image, and almost every design parameter of the CRT has some influence on the minimum spot diameter which can be achieved. However, there are two physical processes which preclude a perfectly focused spot, even with an ideal electrostatic lens. The first is the result of the process of thermionic emission of electrons from the cathode surface. An electron "boiled" off the surface leaves with an initial velocity vector that has a random direction and magnitude (corresponding to an energy of up to a few electronvolts), and this randomness is an important factor limiting the minimum spot diameter (Fig. 14.29). An electron that happens to leave the cathode surface with a negligible initial velocity will pass through the CRT axis precisely at first crossover, regardless of the position on the cathode surface from which it was emitted. However, an electron emitted from the same point with a significant radial velocity component will be off the central axis at first crossover, with an error magnitude and direction that are proportional to the magnitude and direction of the initial radial velocity. The minimum spot diameter on the phosphor

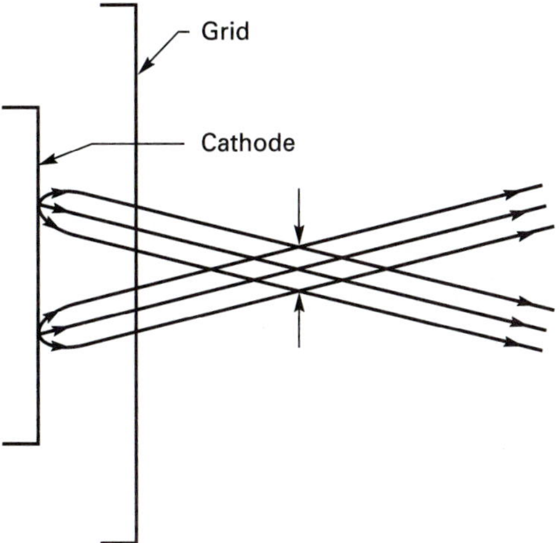

Figure 14.29 Random initial velocity vector of thermionically emitted electrons limits the minimum diameter at first crossover.

screen is proportionately limited by the minimum diameter at first crossover, since the first is the focused image of the second.

The other effect that limits spot diameter is called "space charge defocusing" and is an important effect at large CRT beam currents. All the electrons in the beam repel each other because of their identical negative charge. The defocusing occurs because the electrons traveling in the outer parts of the beam are subject to a net outward radial force that varies as a function of the beam charge density. The unbalanced radial force causes a net outward motion during the trip from cathode to screen, and this defocusing effect can only be partially compensated by an increase in the strength of the main focus lens.

Phosphor screen. Ultimately, the purpose of the focused electron beam in the CRT is to generate a lighted spot on the display screen. When a moving electron strikes an atom, one or more of its electrons may be raised to a higher energy state. As an excited electron reverts to its initial energy state, it gives up the extra energy as a quantum of electromagnetic radiation which, depending on the atom and the excitation energy, may or may not be visible light. The phosphors used in CRTs are a group of materials which are chosen for their ability to convert energy from electron beams of a convenient energy level into visible light. If the luminescence stops immediately when the electron beam current is interrupted, the material is called "fluorescent." If the light emission continues after the excitation is removed, the material is said to be "phosphorescent." The length of time required for phosphorescence to decay is referred to as the "persistence" of the phosphor. Many different phosphor formulations are commonly used, and several factors are considered in the choice of a phosphor for a particular application: the dominant color of the light, the persistence, and the overall efficiency of energy conversion.

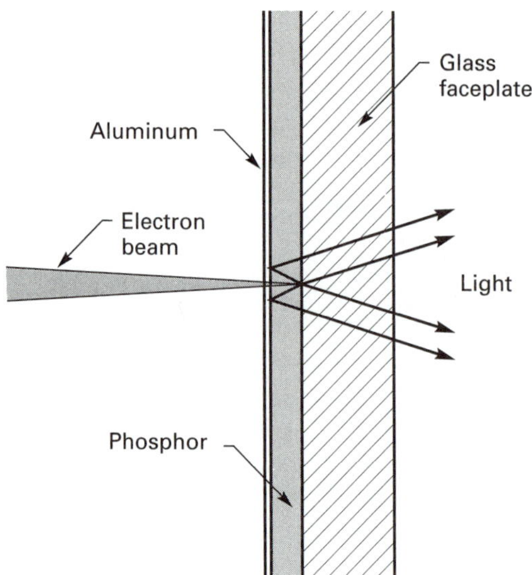

Figure 14.30 Part of the electron beam energy is converted into light in the phosphor layer.

The phosphor is deposited in a thin layer on the inner surface of the glass face plate of the CRT and is covered with a very thin coating of aluminum (Fig. 14.30). The aluminum layer serves several purposes. It makes the entire area that the electron beam strikes uniformly conductive, preventing accumulation of electric charge. The metallic layer is also an efficient reflector of light and prevents any extraneous radiation inside the oscilloscope from exciting the phosphor or itself escaping through the viewing screen, preserving a uniformly dark background for viewing or photographing the waveform trace. And all the light resulting from the action of the electron beam on the phosphor is radiated outward, almost doubling the trace luminous intensity. A small part of the electron beam energy is lost in penetrating the aluminum layer, but the benefits of the aluminum more than compensate.

Each electron that strikes the phosphor screen has a kinetic energy proportional to the voltage difference between the cathode and the phosphor screen, called the "total acceleration potential." A high final velocity is desirable because more energy is then available to stimulate the phosphor, and the trace is brighter at a given beam current. An upper limit on acceleration voltage is imposed by the inherent difficulty of generating and containing high voltages and absorbing the soft × radiation resulting from stopping energetic electrons. The total acceleration potential in a high-frequency analog oscilloscope is in the range of 12 to 20 kV.

Post acceleration. Acceleration occurs in the triode section, but the velocity is chosen to optimize the first crossover spot size and deflection system operation. Increasing the acceleration potential in the triode generally leads to a

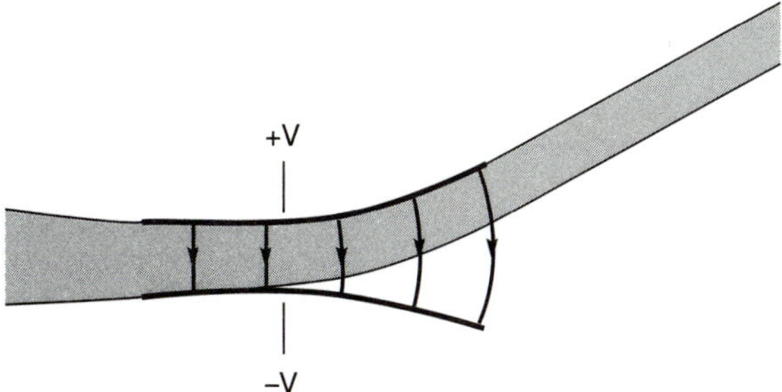

Figure 14.31 The electric field between the deflection plates bends the electron beam.

smaller spot size on the display screen, while making beam deflection more difficult. This design compromise results in a triode acceleration that is close to 3000 V. The extra energy required for satisfactory trace brightness is accomplished by elevating the voltage of the phosphor screen, accelerating the electron beam after it has been deflected, a procedure called "post (deflection) acceleration."

Deflection plates. The trajectory of a moving electron can be changed with either an electric or a magnetic field, but electrostatic deflection is used in an oscilloscope CRT because an electric field can be changed more quickly and easily than can a magnetic field. The electrodes used to establish an electric field transverse to the electron beam are called "deflection plates," and each is shaped so that the surface adjacent to the electron beam is curved in only one direction (Fig. 14.31). Electrons in the beam leave the main focus lens at a velocity of 3.2×10^7 m/s, corresponding to an energy of 3000 eV, and coast through the region where the deflection plates are located because the average voltage through this part of the CRT is held constant. If the potential changed in this region, the radially asymmetric deflection plates would form an astigmatic electrostatic lens and distort the shape of the spot on the display screen. The amplifier that drives the deflection plates has two outputs, one for each deflection plate, an arrangement that is called "differential outputs." Whenever the voltage on one output changes, the other changes by the same magnitude in the opposite direction, and the *average* voltage on the deflection plates remains constant for all deflection angles.

Establishing a voltage difference between the two deflection plates forms a transverse electric field in the space between them, and electrons passing between them are subjected to a force toward the relatively positive plate that is proportional to the intensity of the field. The integrated effect of the force over the duration of the passage causes the direction of the beam, and the ultimate landing spot on the display screen, to change. The plates are located close together to maximize the field strength for a given deflection voltage but must not obstruct the electron beam. The deflection plates are shaped so that the converging electron beam barely grazes the plate at the deflection angle that places the spot at the edge of the display screen.

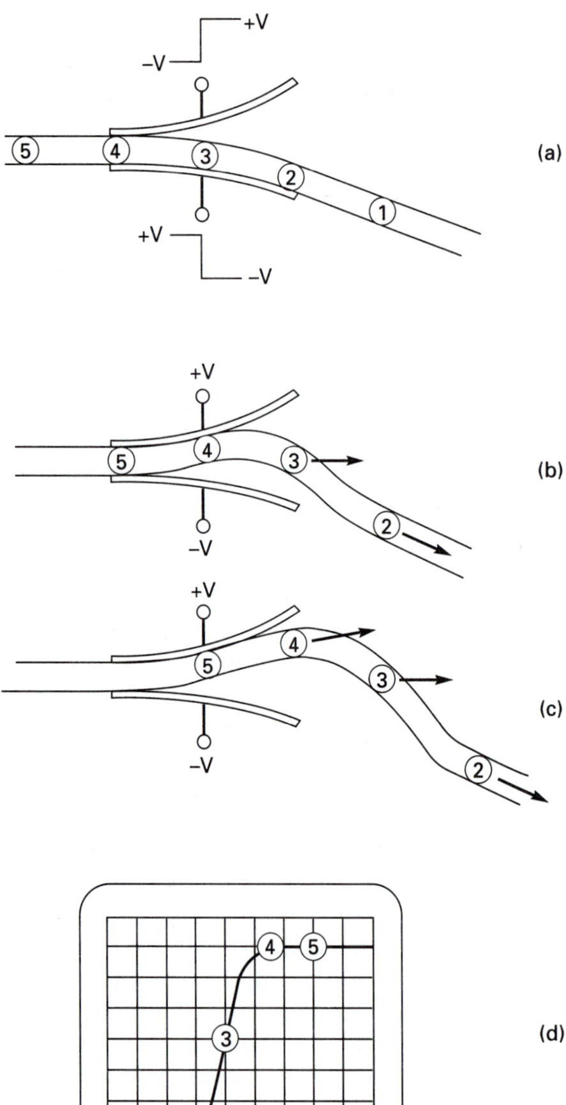

Figure 14.32 The rise time of a CRT with solid deflection plates can be no less than the time needed for a single electron to pass between the plates.

Good sensitivity is a desirable property for a CRT used in a high-frequency oscilloscope because deflection from one screen edge to the other then requires a smaller voltage swing and less amplifier gain. Lengthening the deflection plates or locating them farther from the display screen lowers the deflection factor (volts per division) at the deflection plates, at the expense of overall length of the CRT. It is for this reason that the vertical deflection plates are located adjacent to the focus lens assembly and the horizontal plates closer to the display screen (see Fig. 14.27). The movements of the spot are much more rapid in the vertical direction than in the horizontal direction, so the lowest deflection factor is best used in the vertical path.

Distributed deflection plates. A certain amount of time is required for one electron to pass between the vertical deflection plates, and this transit time places a limit on the rise time of a CRT with conventional deflection plates. Figure 14.32 illustrates the effect. In Fig. 14.32*a*, the deflection plate voltages have been steady for some time, and each electron (numbered for convenience) is following the same trajectory at regular intervals on the path to the screen, when the voltages switch instantaneously. A short time later (Fig. 14.32*b*), electron 2, which was leaving the plates when the voltage changed, continues to the screen along its original path. Electron 3, which was midway when the plate voltage switched, emerges with no net angular deflection, since it experienced a zero average deflection field. Electron 4 and all later electrons see only the unchanging positive field and follow the path to the top of the screen (Fig. 14.32*c*). The points of impact on the display screen are indicated in Fig. 14.32*d*, and it can be seen that the time required for the beam to the completely deflected *is* the deflection plate transit time.

Eight centimeters is a typical vertical deflection plate length for a CRT used in a high-frequency oscilloscope, giving a transit time of 2.5 ns at a beam velocity of 3.2×10^7 m/s. Solid deflection plates in such a CRT would lead to an upper frequency limit of approximately 150 MHz from this effect alone.

Shortening the deflection plates or increasing the accelerating potential and beam velocity will shorten the transit time, but at the expense of an increased deflection factor. A better solution is to separate the deflection plates into segments and propagate the deflecting voltage along the structure at a rate equal to the beam velocity. This arrangement is called a “distributed deflection system.” A number of different implementations of this idea have been used, but the following is a description of distributed deflection plates built from two narrow strips of metal, each wound into a flattened helix (Fig. 14.33). They are shaped and located to provide space between for the electron beam, and the vertical amplifier output signal is applied to the ends closest to the focus lens, where the electron beam enters. An electrical equivalent circuit for the structure (Fig. 14.34) is in the familiar form of a lumped parameter differential transmission line. The line has a characteristic resistance R_0 and is terminated with a resistance of that value to minimize reflections of the deflection voltage signals.

Figure 14.35 illustrates the operation. In Fig. 14.35*a*, the upper plate is at $-V$ volts, the lower plate is at $+V$ volts, and the electrons all follow the path shown. Now when the voltage step arrives, only the segments at the entrance and switch to the new values, $+V$ on the upper and $-V$ on the lower, and the voltage step propagates along the plates at the same velocity as the beam. In

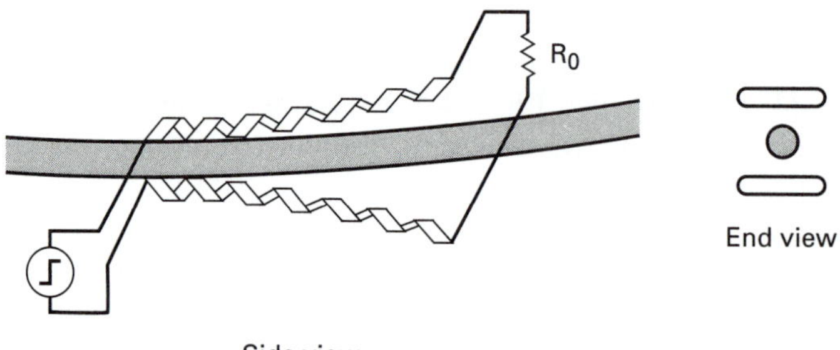

Figure 14.33 Distributed deflection plates.

Fig. 14.35*b*, half of each plate has reached its new voltage, but the trajectory of electron 3 has not been influenced by the wave front; electron 4 is the first to experience the new electric field state. Figure 14.35*c* shows the situation when the propagating signal reaches the exit end, and the new equilibrium between the plates is reached. The beam has been almost severed in the distributed structure, and the apparent rise time in Fig. 14.35*d* is much shorter than the transit time through the entire vertical deflection structure. Actually, the apparent rise time is slightly greater than the transit time through a single plate segment, an n-fold reduction compared with solid plates, where n is the number of segments in the distributed deflection system.

14.5.2 Analog oscilloscope block diagram

A complete block diagram for a basic two-channel analog oscilloscope is shown in Fig. 14.36. Most of the components and subsystems were discussed earlier. In this section the remaining parts are introduced, and the interdependent operation of the subsystems is treated.

Trigger. In Fig. 14.37, the signals occurring at the indicated nodes in Fig. 14.36 are shown for a single acquisition cycle of a signal connected to the channel one input. The trigger source is set to the "internal, channel one" position, and a positive slope edge trigger is selected, so a trigger pulse is generated as the input signal crosses the indicated trigger level. In response to the trigger pulse, the ramp signal starts the motion of the spot position on the display from left to right, and the unblanking gate signal reduces the

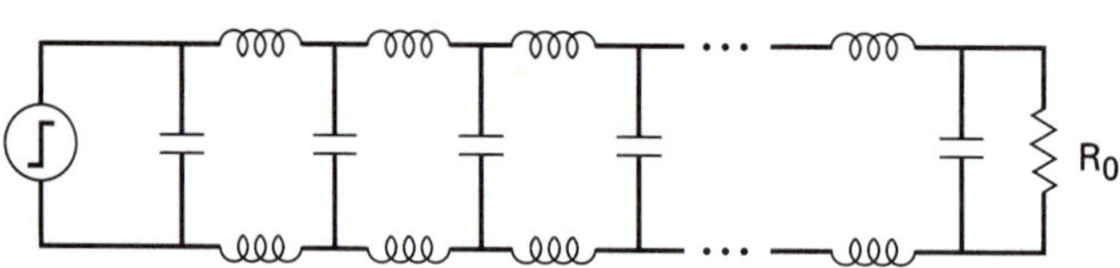

Figure 14.34 Electrical equivalent circuit of distributed deflection plates.

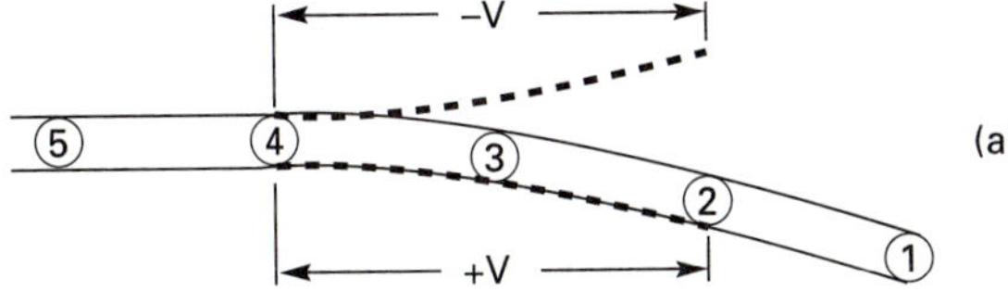

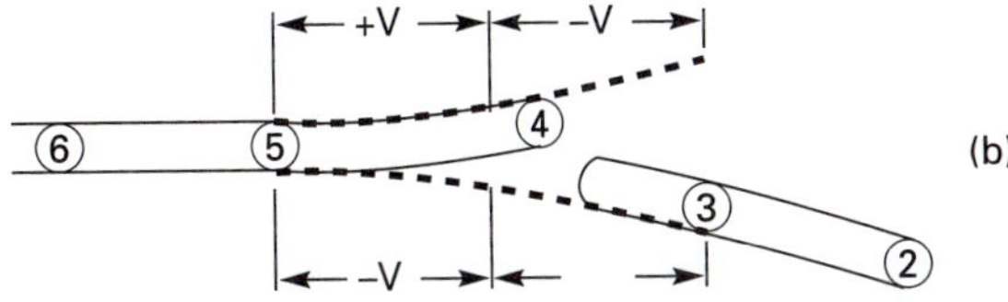

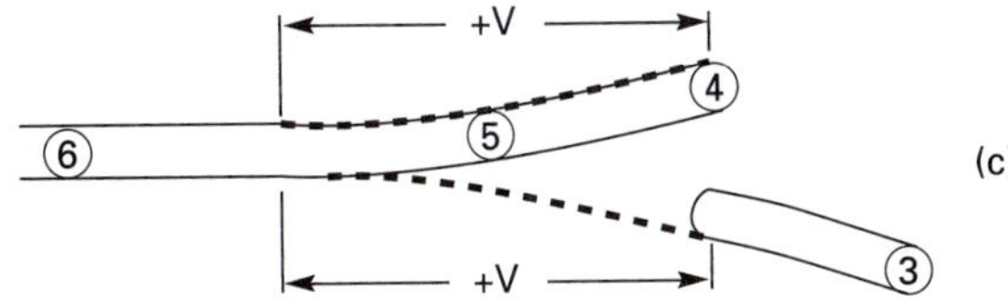

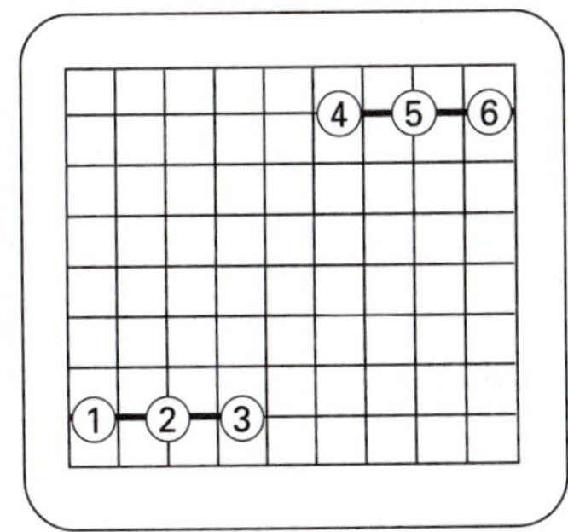

Figure 14.35 Electron beam motion in a distributed deflection system.

CRT grid-to-cathode negative bias, causing the spot to become visible on the display screen. The ramp increases at a precisely controlled rate, causing the spot to progress across the display screen at the horizontal rate determined by the current time per division setting. When the spot has moved 10 divisions horizontally, the unblanking gate switches negative, the trigger holdoff period begins, and the ramp retraces the spot to the starting point. At the end of trigger holdoff, the system is ready to recognize the next trigger and begin the next signal acquisition.

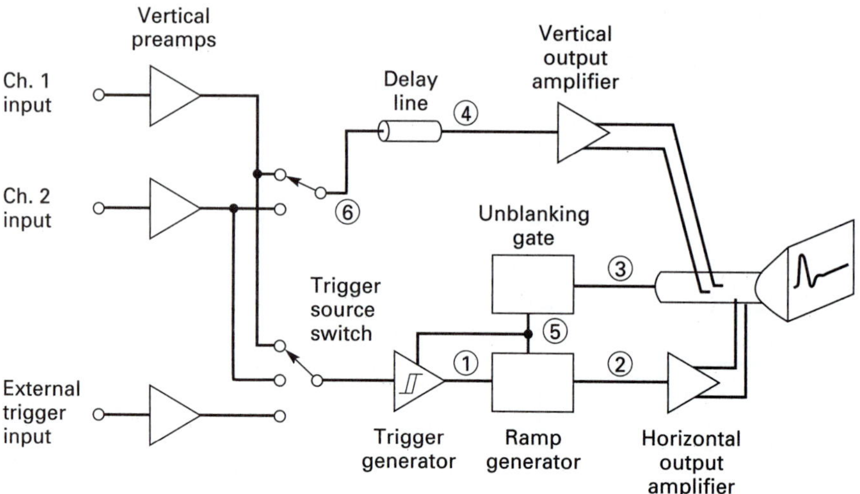

Figure 14.36 Analog oscilloscope block diagram.

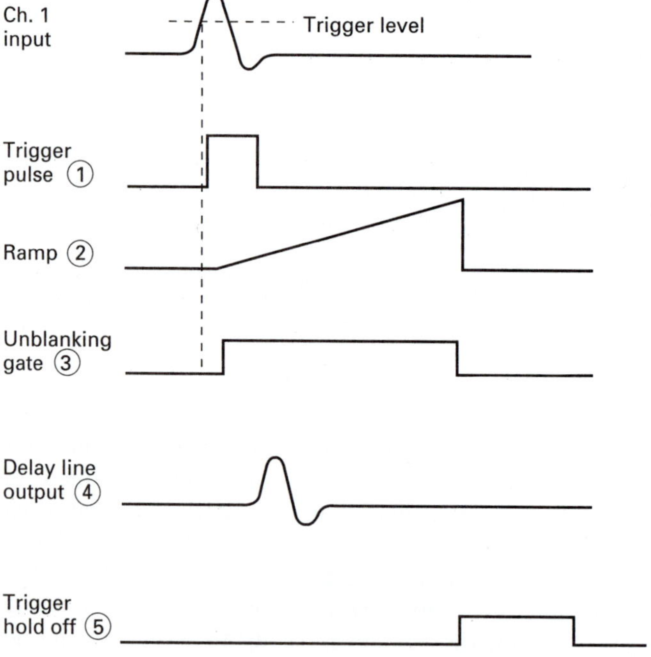

Figure 14.37 Waveforms from a single acquisition cycle.

Delay line. Clearly, some time is necessary after initiation of the trigger pulse for the sweep to attain a linear rate and for the display spot to reach full brightness. Since it is desirable to be able to view the voltage transient that caused the trigger, a means of storing the input signal during the startup delay is needed. This is accomplished by placing a delay line in the vertical path. Many oscilloscope models are equipped with an internal device for this purpose, providing enough delay time to display the trigger transient at a position one or two divisions after the display trace starting point on the fastest time per division setting. The triggering signal at the output of the delay line is visible on the display screen while the unblanking gate signal is at its most positive level (Fig. 14.37).

A total delay time of between 25 and 200 ns is required depending on the oscilloscope model. High-bandwidth units are equipped with correspondingly fast trigger, sweep, and unblanking gate circuits and therefore can use a relatively short delay line. In some applications, the delay line can be a length of coaxial cable, but delays in the longer part of the stated range of values are implemented using lumped-element delay structures. However the delay is implemented, it is critically important that the transfer function of the entire vertical path does not vary with frequency (see Fig. 14.9 and the accompanying discussion). The transmission efficiency of delay lines characteristically decreases with increasing signal frequency, and this effect must be carefully compensated elsewhere in the vertical amplifier.

Dual-trace operation. Oscilloscopes have two or more channels because the most important measurements compare time and amplitude relationships on multiple signals within the same circuit. However, a conventional analog oscilloscope CRT has only one write beam and thus is inherently capable of displaying only one signal. Thus the channel switch (see Fig. 14.36) is used to time share, or "multiplex," the single display channel among the multiple inputs. The electronically controlled channel switch can be set manually first to channel one and then later switched by the user to channel two. However, a better emulation of simultaneous capture is attained by configuring the oscilloscope to automatically and rapidly switch between the channels. Two different switching modes are implemented, called "alternate" and "chop," but it is important to realize that both methods display a single signal that is a composite of the multiple input signals.

In "alternate mode," the channel switch changes position at the end of each sweep during retrace while the write beam is blanked (Fig. 14.38). This method works best at relatively fast sweep speeds and signal repetition rates. At slow sweep speeds the alternating action becomes apparent, and the illusion of simultaneity is lost.

In "chop mode," the channel switch is switched rapidly between positions (Fig. 14.39) at a rate which is not synchronized with the input signals or the sweep. During each channel switch setting change, the unblanking gate pulses negative for a short interval, dimming the CRT display spot so that

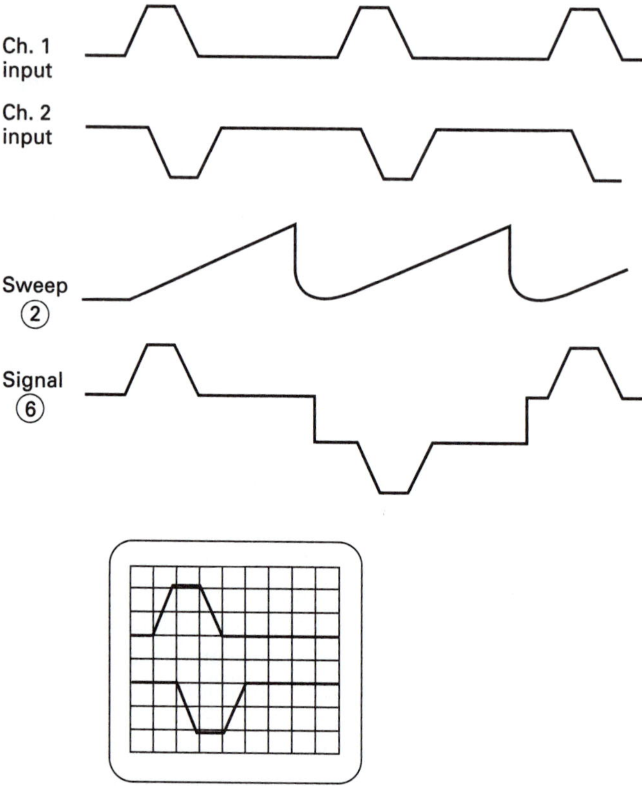

Figure 14.38 Alternate mode operation.

the transition will not be visible on the display screen, and both signals are drawn as dashed traces on every sweep. If the chop rate is fast enough to make at least a few hundred transitions on each sweep, then the individual dashes are not noticeable, even on a single sweep acquisition, and simultaneous capture is effectively realized. Increasing the chop rate satisfies this condition on faster sweep speed settings and is desirable for this reason. However, the vertical output amplifier must accurately amplify the chopped signal (6 in Fig. 14.36 and Fig. 14.39), which is a more difficult task at higher chop rates. As a compromise between these two conflicting requirements, the chop frequency is set between 1/1000 and 1/100 of the upper cutoff frequency (bandwidth) of the vertical amplifier in each oscilloscope model.

In addition to time-multiplexed operation, a mode setting is provided which displays the sum of two input signals as a single trace. One or both channels is capable of inverted (reversed polarity) operation, so the difference between two input signals can be captured as a single trace.

Delayed sweep. Some analog oscilloscope models include a second ramp generator, called a "delayed sweep," and a second trigger generator, called a "de-

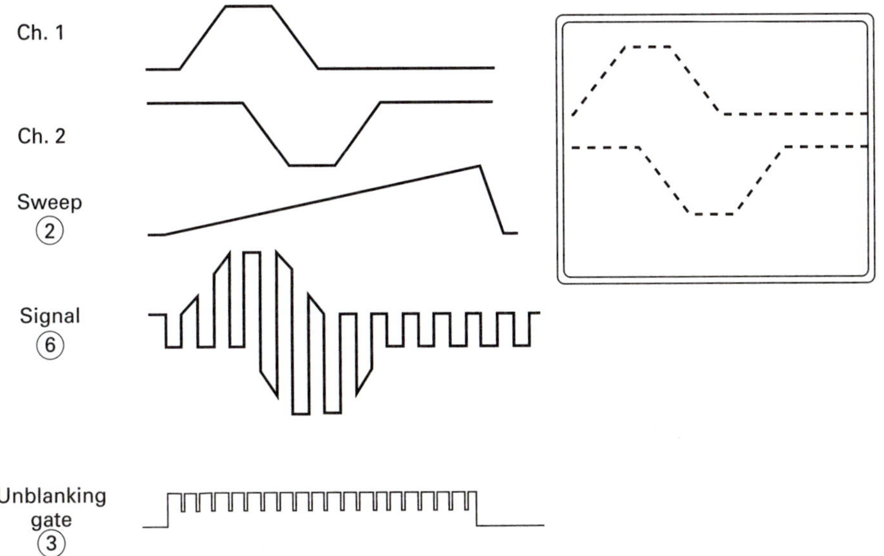

Figure 14.39 Chop mode operation.

layed trigger" (Fig. 14.40), providing an additional method for controlling the placement of the "window" in time relative to the main trigger event. The horizontal amplifier can still be connected to the output of the main ramp generator, in which case the operation is identical to that of the standard oscilloscope configuration described earlier. A comparator circuit is added whose output (4) switches when the main ramp signal exceeds a voltage

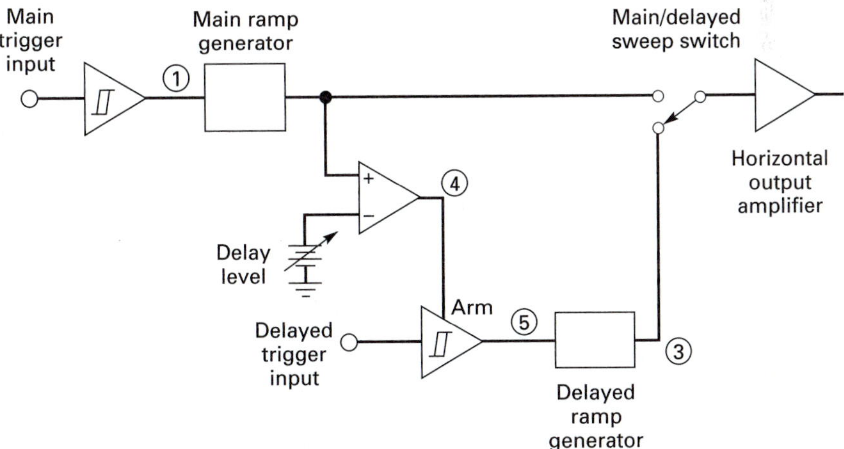

Figure 14.40 Main and delayed sweep generator block diagram.

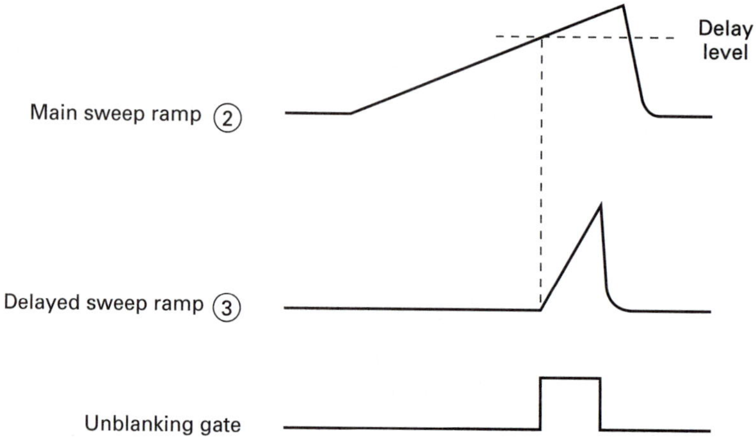

Figure 14.41 Delayed sweep starting when the main sweep ramp exceeds the delay level.

called the "delay level" (Fig. 14.41). This dc voltage can be adjusted by the oscilloscope operator using a calibrated front panel control labeled "Delay." The main ramp is initiated by the main trigger pulse and increases at a precisely controlled rate (volts per second) determined by the sweep speed setting. Therefore, the time elapsed between the trigger pulse and comparator output state change is the reading on the delay control (in divisions) multiplied by the sweep speed setting (in seconds per division).

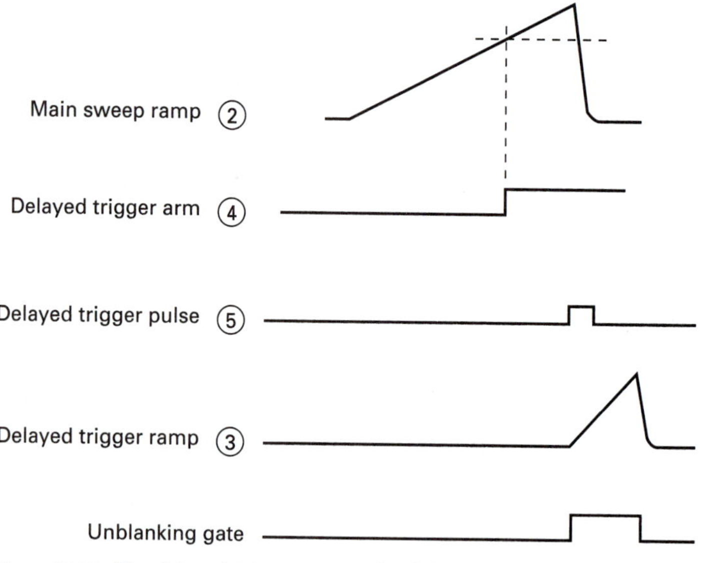

Figure 14.42 The delayed trigger starts the delayed sweep.

The delayed sweep ramp is initiated after the delay period in one of two ways. The “normal” method immediately starts the delayed sweep when the delay comparator switches (see Fig. 14.41). The “delayed trigger” mode uses the delay comparator output to arm the delayed trigger circuit. Then when the delayed trigger condition is met, the delayed sweep starts (Fig. 14.42). The delayed sweep rate always has a shorter time per division setting than the main sweep.

14.6 The Digital Oscilloscope

The preceding discussion covered the analog oscilloscope and the operation of various parts of the oscilloscope that are essentially the same in the analog and digital models. In this section the unique aspects of the digital oscilloscope are described.

14.6.1 Block diagram

A block diagram of a two-channel digital oscilloscope is shown in Fig. 14.43. Signal acquisition is by means of an analog-to-digital coverter (ADC) or digitizer, which at uniformly spaced time intervals measures (samples) the instantaneous amplitude of the signal appearing at its input and converts it to a digital value (a number), which in turn is stored in a digital memory. When the trigger condition is satisfied, the sampling process is interrupted, the stored samples are read from the acquisition memory, and the volts versus time waveform is constructed and graphed on the display screen. The time interval between samples t_s is called the “sample time,” and its reciprocal is called the “sample frequency” f_s. The signal that regulates the sampling process in the ADCs is called the “sample clock,” and it is generated and controlled by the time-base circuit. A crystal oscillator is used as a reference for

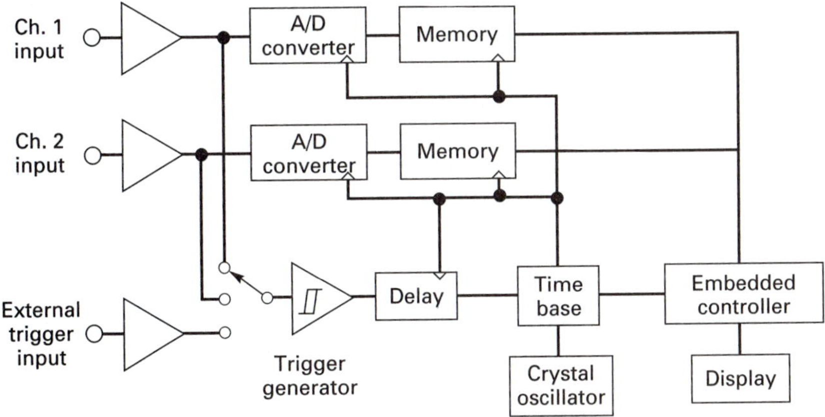

Figure 14.43 Digital oscilloscope block diagram.

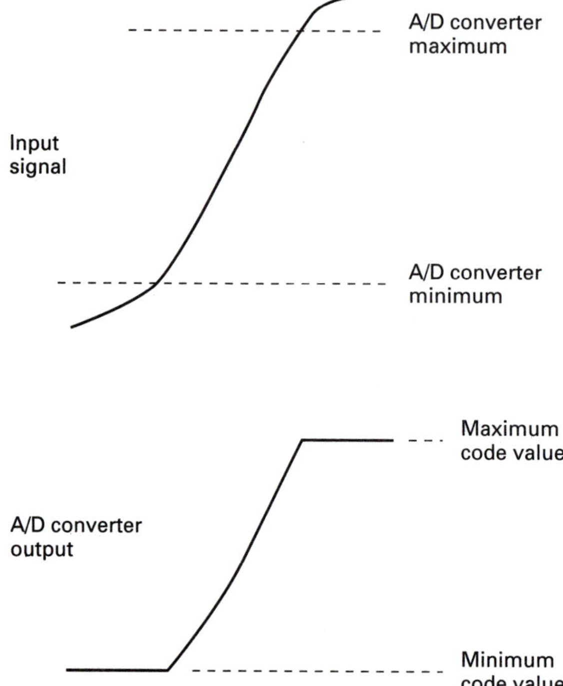

Figure 14.44 An ADC quantizes a continuous signal into discrete values having a specific range and spacing.

the time base to ensure the accuracy of the sample interval and ultimately of the time measurements made using the digital oscilloscope.

Quantization. Converting a continuous waveform into a series of discrete values is called "quantization," and several practical limitations apply to this process as it is used in the digital oscilloscope. The signal is resolved into discrete levels only if it is within a specific range of voltages (Fig. 14.44). If the input signal is outside this range when a sample is taken, either the maximum code or the minimum code will be output from the ADC. The limited "window" in voltage is similar to that encountered in the analog oscilloscope CRT; a sufficiently large vertical deflection signal causes the trace to disappear off the edge of the display screen. As in the analog instrument, the vertical position and deflection factor controls are used to offset and scale the input waveform so that the desired range of voltages on the waveform will fall within the digitizer voltage window.

Resolution. Voltage resolution is determined by the total number of individual codes that can be produced. A larger number permits a smoother and more accurate reproduction of the input waveform but increases both the cost of the oscilloscope and the difficulty in achieving a high sample frequency. ADCs are usually designed to produce a total code count that is an integer

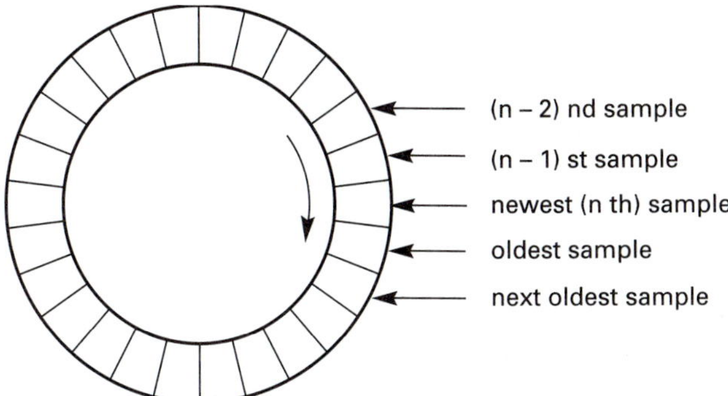

Figure 14.45 Acquisition memory is arranged in a circular array so that the next sample always overwrites the oldest sample in memory.

power of 2, and a unit capable of 2^n levels of resolution is called an "n-bit digitizer." Digital oscilloscopes are available in resolutions from 6 up to 12 bits, with the resolution varying generally in an inverse relationship to sample rate. Eight bits is the most frequently used resolution. The best possible voltage resolution, expressed as a fraction of the full-scale deflection, is 2^{-n}, e.g., 0.4 percent, for an 8-bit ADC.

Acquisition memory. Each sample code produced must be stored immediately in the acquisition memory, and so it must be capable of accepting data from the digitizer continuously at the sample frequency. For example, the memory in each channel of an 8-bit, 2-gigasample/s oscilloscope must store data at the rate of 2×10^9 bytes/s. Memory that operates with such a fast write cycle rate is difficult to design and expensive to produce and so is a limited resource. The memory is arranged in a serially addressed, conceptually circular array (Fig. 14.45). Each storage location is written to in order, progressing around the array until every cell has been filled. Then each subsequent sample overwrites what has just become the oldest sample contained anywhere in the memory, the progression continues, and a memory with n_m storage locations thereafter always contains the most recent n_m samples. The n_m samples captured in memory represent a total time of waveform capture of n_m multiplied by the sample time interval. For example, a digitizer operating at 2 gigasample/s with a 32,000-sample acquisition memory captures a 16-μs segment of the input signal.

If sampling and writing to acquisition memory stop immediately when the trigger pulse occurs (Delay = 0 in Fig. 14.43), then the captured signal entirely precedes the trigger point (Fig. 14.46*a*). Setting delay greater than zero allows sampling to continue a predetermined number of samples after the trigger point. Thus a signal acquisition could capture part of the record before and part after the trigger (Fig. 14.46*b*), or capture information after the trigger (Fig. 14.46*c*).

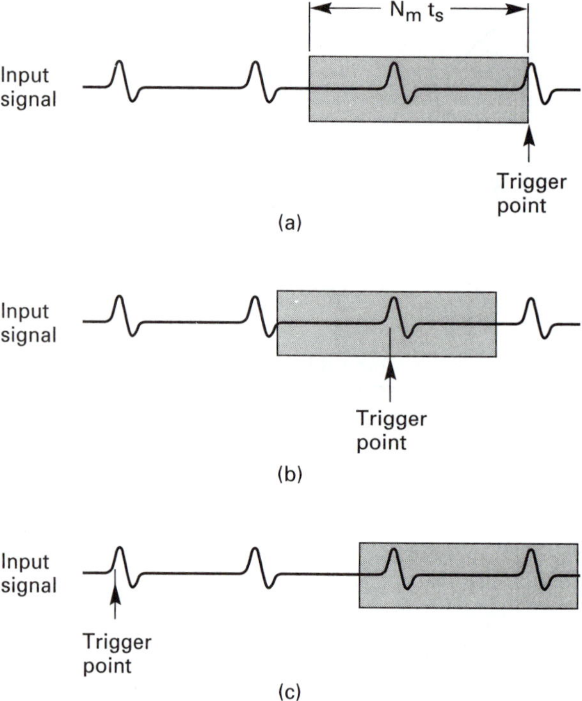

Figure 14.46 The trigger point can be placed anywhere within or preceding the captured record. (*a*) Delay = 0; (*b*) delay = $n_m/2$; (*c*) delay > n_m.

14.6.2 Acquisition methods

The preceding discussion applies generally to all digital oscilloscopes models, but the digital signal capture is performed in three distinct modes depending on the intended application. The three acquisition methods are called "real-time sampling," "sequential repetitive sampling," and "random repetitive sampling."

Real-time sampling. Real-time sampling is conceptually the most straightforward application of digital signal capture technology. In this method, a complete record of n_m samples is simultaneously captured on each and every channel in response to a single trigger event. Each waveform plotted on the display is derived entirely from the samples recorded in a single acquisition cycle and might represent the capture of a single nonrepeating transient.

Three important attributes of real-time sampling are apparent: (1) the complete waveform image is stored in memory, and display or analysis can take place at a later time, (2) it is straightforward to capture signals that happen before the trigger event, and (3) a truly simultaneous capture of multiple signals is automatic. The ability to view a signal that occurred before the trigger is called "negative time" and is especially valuable in fault analysis. The oscilloscope is set to trigger on the abnormal state, and the user can look back in time to see potentially unique waveforms on all channels that led to the fault. Real-time sampling can be used in a continuously repeating mode, but each waveform displayed is from a single capture and display cycle.

Both a fast maximum sample rate capability and a large memory depth are beneficial in real-time sampling. When a long-time-interval capture is needed, then a larger memory allows a higher sample rate to be selected. It is intuitive that the minimum rise-time transient that can be captured is related to the sample time actually used, since signal fluctuations occurring entirely between samples will not be captured in the sample record. The mathematical relationship between sample frequency and the permissible signal-frequency spectrum has been rigorously developed in sampled data system theory, and some of the results are presented in the following discussion.

Sampling theorem. Given a continuous function $h(t)$ that is band limited to a frequency $f_c = f_s/2$ (that is, $H(f) = 0$ for all $f \geq f_c$), and given a series of values h_i representing samples of $h(t)$ taken at evenly spaced time intervals $t_s = 1/f_s$ between samples, then $h(t)$ is specified completely by the samples h_i and in fact is given by the formula

$$h(t) = \sum_{i=-\infty}^{+\infty} h_i \frac{\sin\left[\pi\left(\frac{t}{t_s} - i\right)\right]}{\pi\left(\frac{t}{t_s} - i\right)} \tag{14.7}$$

where t = time for which a reconstructed value is desired
$h(t)$ = reconstructed value
t_s = sample interval
i = an integer ranging from $-\infty$ to $+\infty$
h_i = samples

In other words, if a signal is sampled at greater than twice the frequency of the highest-frequency component in the signal, then the original signal can be reconstructed *exactly* from the samples. Half the sample frequency f_c is known as the "Nyquist limit" or the "Nyquist critical frequency."

Upon first reading this remarkable theorem, it is easy to jump to the conclusion that it is a reasonable undertaking to, for example, attempt to make 1-GHz bandwidth real-time measurements with a 2-gigasample/s digitizer. In practice, operating very close to the Nyquist limit while maintaining good accuracy is exceedingly difficult. First, consider the reconstruction problem, which is calculating the values of $h(t)$ between samples. Equation 14.7 describes what must ideally be done: For each value of t, *every* sample h_i in the infinitely long record is multiplied by a weighting factor and summed, yielding *one* interpolated point on $h(t)$. Of course, the weighting factors for samples far removed in time from the current t are small and can eventually be ne-

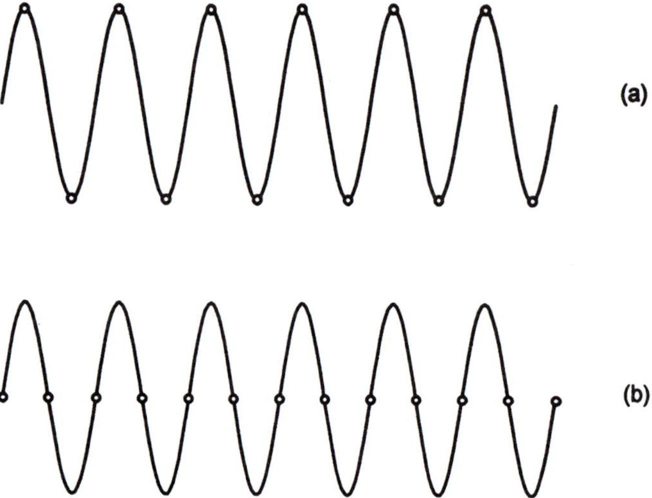

Figure 14.47 Two different sampling patterns resulting from sampling a sine wave near the Nyquist critical frequency.

glected, reducing the summation to a finite number of terms if a reasonable estimate for the interpolated value is acceptable. However, even settling for 8-bit accuracy (0.4 percent), on the order of 512 multiplications and sums are needed *per interpolated point.* Using Eq. 14.7 directly is referred to as "sin(x)/x reconstruction," but the computation burden is so large that it is never used in a real-time digital oscilloscope. Reconstructed waveforms would be plotted very slowly.

Practical, and reasonably fast, reconstruction algorithms use a weighted sum of a smaller number of "nearby" samples. For real-time digital oscilloscope applications, a typical number is eight total samples, four earlier and four later, which works well on slow moving parts of the signal but encounters difficulty dealing with rapid transitions. To understand this, consider the problem of reconstructing a pure sine wave having a frequency just under f_c from a sampled data record (Fig. 14.47). There are two samples per cycle, and the situation in Fig. 14.47*a*, in which the samples by chance fall on the peaks of the sine wave, looks easy to handle. However, in Fig. 14.47*b*, the two samples per cycle happen to fall on the sine wave zero crossings, and all have the same value. Interpolation using the eight closest samples clearly cannot reconstruct a sine wave of the correct amplitude, but the ideal method [sin(x)/x] can because it uses many samples in both directions. Remember that the sampling theorem requires samples for i from $-\infty$ to $+\infty$, and all of them are not at zero crossings because the signal frequency is slightly less than f_c. This example illustrates the point that accurate reconstruction of signals near the Nyquist limit *must* be very expensive computationally.

Aliasing. Next, consider the effect of sampling a signal that is not band limited to f_c. Any signal component having a frequency higher than f_c is aliased,

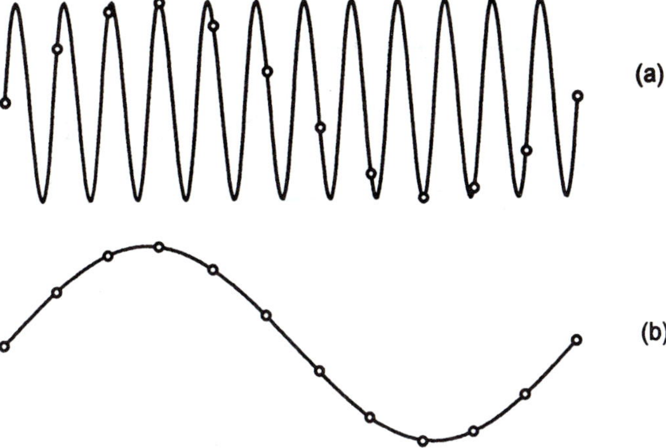

Figure 14.48 Aliasing: sampling a signal that is above $f_s/2$ produces a sample pattern that will be interpreted as a signal that is below $f_s/2$.

or falsely translated, to another frequency somewhere between dc and f_c by the act of sampling. Ideal reconstruction must produce a band-limited signal from *every* set of samples, so in the process of digitizing a signal that is entirely above f_c, any response in the sample record must be aliased to entirely below f_c after reconstruction. In Fig. 14.48, a sine wave having a frequency just below f_s and sampled at f_s produces the same sample record as a low-frequency sine wave also sampled at f_s. In general, a signal of frequency $f_c + \Delta$ will be aliased to $f_c - \Delta$ for $\Delta < f_c$.

It is natural to depend on the vertical amplifier response to limit the amount of signal above f_c that reaches the ADC (Fig. 14.49), thus the 3-dB bandwidth of the vertical amplifier should be less than f_c at the fastest sample

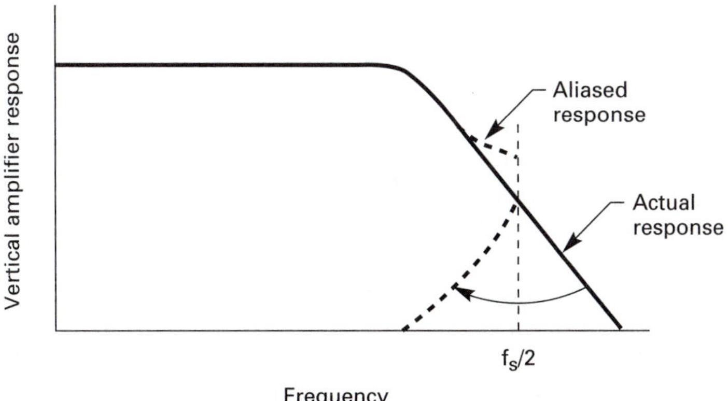

Figure 14.49 Vertical amplifier response above the Nyquist critical frequency is aliased down in frequency.

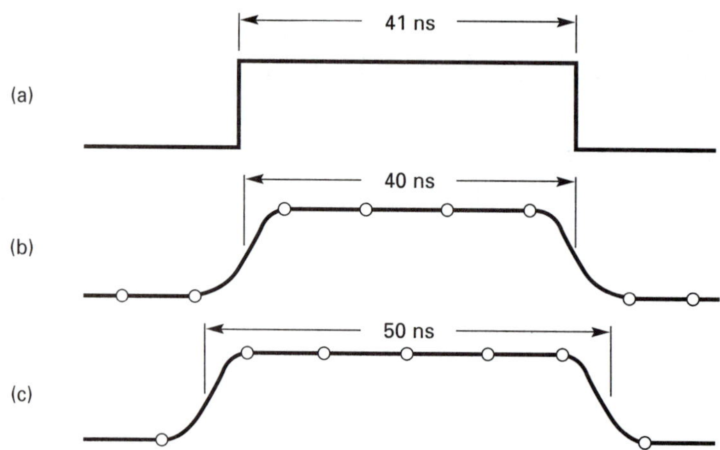

Figure 14.50 Sampling and reconstruction of a pulse that is not band limited. (*a*) Input signal; (*b*) reconstructed waveform, case 1; (*c*) reconstructed waveform, case 2.

rate. A difficult compromise is in involved here. At a specific sample rate, more bandwidth lets through more of the user's signal at the expense of increased aliasing. Increasing the cutoff rate reduces aliasing but increases overshoot and ringing on the vertical amplifier pulse response. A popular compromise, though by no means universal, is to set the 3-dB bandwidth to $f_s/4$ with about 5 percent overshoot.

Time-domain aliasing. The sampling theorem and the effects of aliasing are traditionally developed and perhaps are most easily understood using frequency-domain terminology, but the user of a real-time digital oscilloscope also needs to understand how aliasing affects time measurements and time-domain displays. Figure 14.50 is an example of a heavily aliased pulse-width measurement. A pulse of duration 41 ns having negligible rise and fall times, compared with 10 ns, is sampled at f_s = 100 megasamples/s. The frequency spectrum associated with the fast edges far exceed f_c = 50 MHz. Only two sample patterns are possible: four samples somewhere on the pulse top (Fig. 14.50*b*) or five (Fig. 14.50*c*), giving pulse-width measurements of 40 and 50 ns as the only possible answers. Note that the *shapes* of the reconstructed pulse edges are all the same, but that horizontal placement of each edge on the display is uncertain over a one-sample time interval, and that a pulse-width measurement is always an integer number of sample times. The average pulse-width measurement would converge to 41 ns after many acquisition cycles. However, if only one measurement of an unknown pulse width is made, the error can be as large as one sample time. If, for example, the single capture produced a pulse-width measurement of 40 ns, then it is known only that the pulse width is greater than 30 ns and less than 50 ns.

In the next example, the input pulse in Fig. 14.50 has been filtered to a

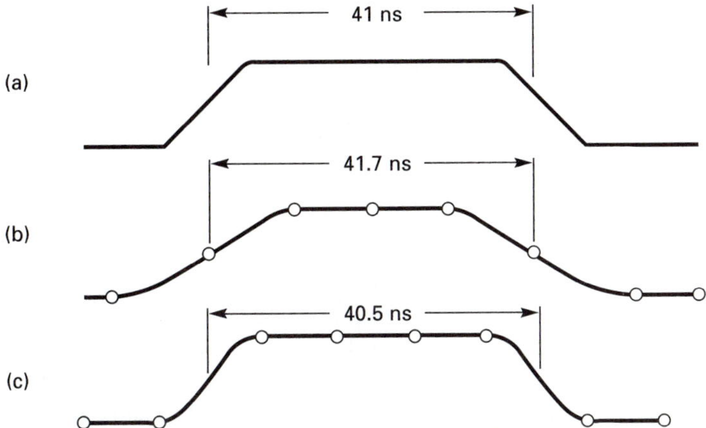

Figure 14.51 Sampling and reconstruction of a band-limited pulse. (*a*) Input signal; (*b*) reconstructed waveform case 1; (*c*) reconstructed waveform, case 2.

gaussian response, a bandwidth of 25 MHz, and rise and fall times of 14 ns (Fig. 14.51*a*). Gaussian band limiting to $f_s/4$ reduces, but by no means eliminates, the pulse spectrum above f_c. Now two different sample patterns (Fig. 14.51*b* and *c*) yield reconstructed pulses which have slightly different shapes, but both estimates of pulse width shown are much closer to the correct value. The partial band limiting slows edge speeds sufficiently to ensure that at least one sample is taken somewhere on each transition, so time placement of the 50 percent points is more accurate. A real-time oscilloscope using the paradigm of $f_s/4$ gaussian band limiting and a practical reconstruction algorithm is capable of meeting a worst-case single-shot time-interval accuracy specification of $\pm 0.15t_s$.

To complete the time-domain illustration, the ideal case is shown in Fig. 14.52. The input pulse of Fig. 14.50*a* is now ideally band limited to $f_c = 50$

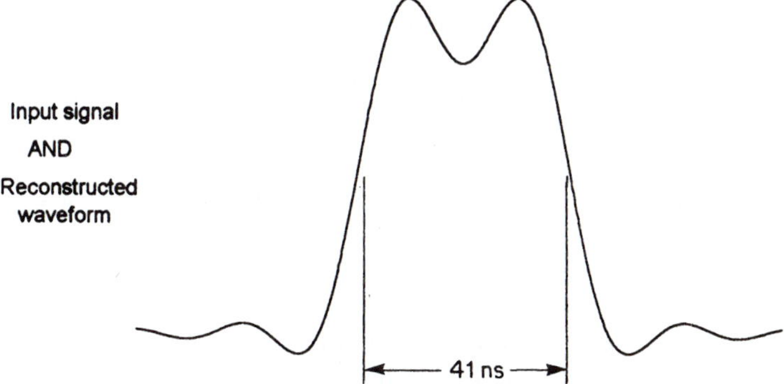

Figure 14.52 Sampling and reconstruction of an ideally band-limited pulse.

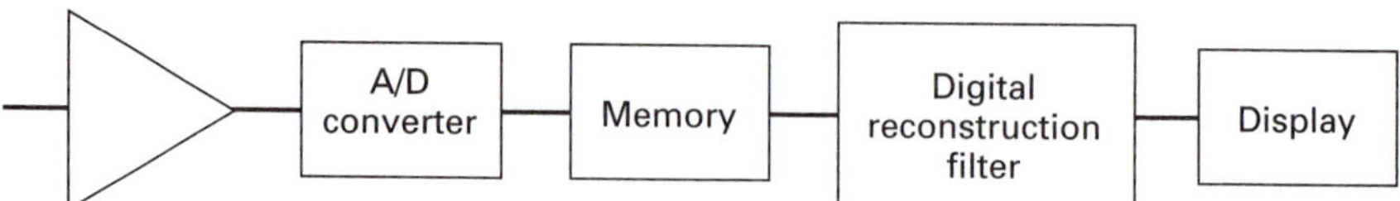

Figure 14.53 Reconstruction of the input signal from the sampled data record is done in a digital filter.

MHz. A separate plot of the reconstructed waveform is not needed since, regardless of where the samples happen to land, ideal reconstruction using Eq. 14.7 always precisely reproduces the band-limited input waveform. This ideal method is not implemented because, in addition to the great difficulty of closely approximating ideal band limiting and reconstruction, there is the additional impediment of convincing the user that the waveform in Fig. 14.52 is a faithful and useful representation of a square pulse.

Waveform reconstruction. Several different methods of estimating the original $h(t)$ from a sampled data record h_i are available, but the repeated use of weighted sums of the "nearby" samples is the procedure most often utilized in real-time digital oscilloscopes. The actual computation is executed as a subprogram in the system-embedded controller (see Fig. 14.43) or in a separate digital signal processor which has hardware and programming capabilities optimized for this task.

In this context, reconstruction consists of obtaining an interpolated value at each of a number of uniform subintervals of the time between two adjacent samples. The number of interpolated values needed per sample interval is determined by the sample rate, the sweep speed, and the resolution of the display screen. Usually a maximum expansion ratio of between 20 and 50 is necessary. The weighting coefficients are stored as constants in a table in memory, and the total number of coefficients required is the product of the maximum expansion ratio and the number of samples summed in the calculation of each interpolated point.

This reconstruction method can be described and analyzed as a digital filter (Fig. 14.53). The filter characteristics, such as bandwidth, flatness, and cutoff rate, are determined by the values of the coefficients. The filter can be described by its frequency response or in the time domain by its impulse response. The impulse response of a digital filter is the waveform it produces from a sampled data record h_i, for which $h_0 = 1$ and $h_i = 0$ for all $i \neq 0$. The impulse and frequency responses of three filters is shown in Fig. 14.54. The ideal case is defined by Eq. 14.7, and only the central part of the infinite impulse response can be shown (Fig. 14.54*a*). Figure 14.54*b* shows the responses of a filter suitable for general-purpose reconstruction in a real-time digital oscilloscope, in which each point is reconstructed from the eight nearest samples. Linear reconstruction, or "connect the dots," is shown in Fig. 14.54*c*. This method is used because it is easily understood and computes

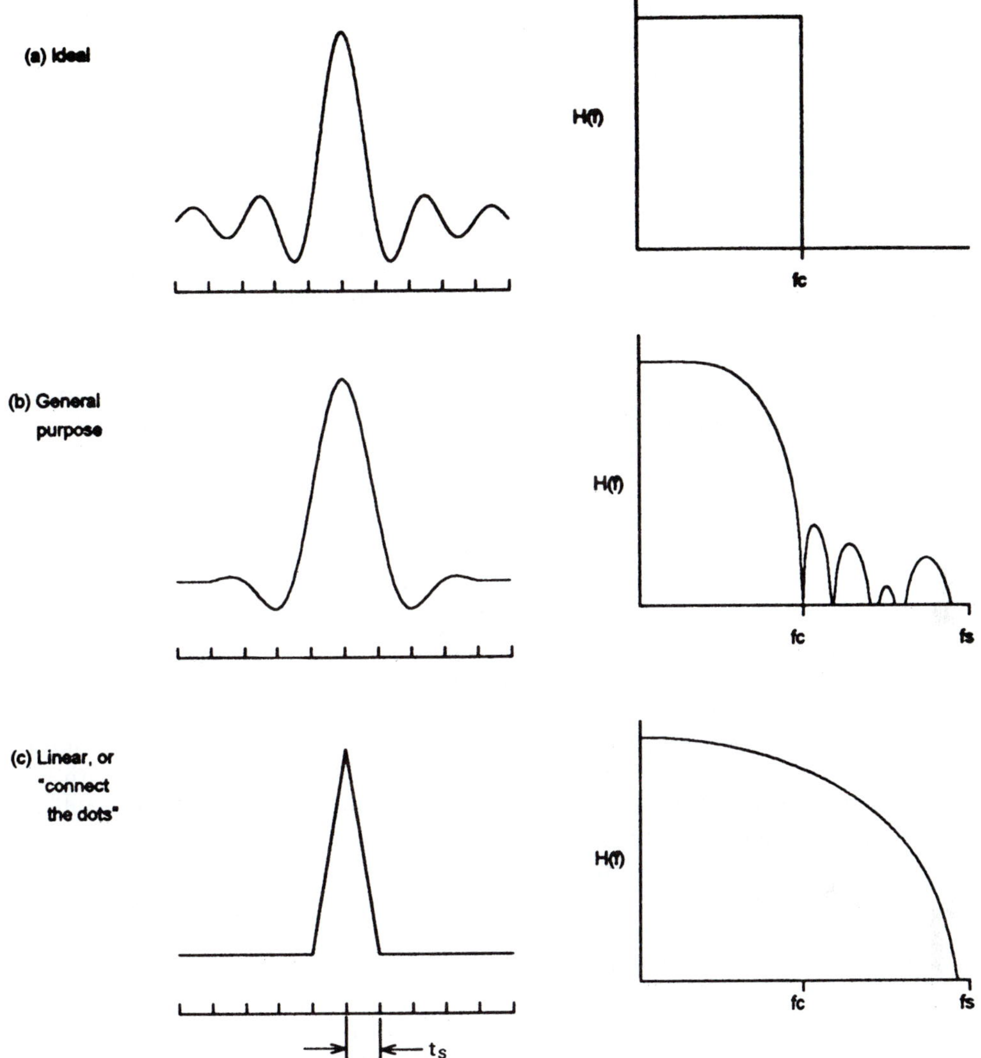

Figure 14.54 Impulse and frequency responses of three digital reconstruction filters. (*a*) Ideal; (*b*) general purpose; (*c*) linear or "connect the dots."

relatively quickly, since it connects adjacent samples with a straight-line segment, but reconstruction accuracy is limited because of the poor attenuation above f_c.

Alias detection. The unique measurement advantages provided by the real-time digital oscilloscope are somewhat counterbalanced by the possibility of unexpected aliasing effects. To prevent measurement errors, the user needs to understand thoroughly how a digital oscilloscope functions and to watch carefully for indications of aliasing in the waveforms.

At the fastest sample rate settings, the band limiting of the vertical amplifier in combination with natural rise-time limits in the signals measured is usually sufficient to prevent large errors, so difficutly is most frequently encountered when measurements are made over long time intervals (slow sweep speed and low sample rate) on circuits which have rapidly changing signals. Sampling a signal transition that has a rise time that is short compared with the current sample time (see Fig. 14.50) limits time measurement accuracy. If reconstruction is in use, such an undersampled edge will exhibit the characteristic step response of the reconstruction filter. For this reason, the operator of a real-time digital oscilloscope should become familiar with how the instrument displays a very fast edge, e.g., from a pulse generator, at slow sample rates to aid in recognizing this situation during measurement of signals. In general, if measurement of a pulse edge on the oscilloscope display screen indicates a rise time of less than one sample time, then assume that time interval measurements involving that pulse edge are uncertain to ±one sample time.

Aliasing of the type depicted in Fig. 14.47 also can be a problem at low sample rates. If the signal is oscillating at or near an integer multiple of the sample rate, then each successive sample is taken at nearly the same point on the periodic signal, and the display will erroneously show a dc or slow-moving signal. This phenomenon could show up on signals other than continuous sine waves, e.g., while monitoring a signal that has a short burst of high-speed clock pulses. Aliasing is also the problem when pulses with a duration shorter than the sample time are sporadically missing from the display or show at reduced amplitude.

Increasing the sample frequency, if possible, is the best remedy to reduce aliasing effects. In capturing a specific length time interval, a larger memory allows a higher sample rate to be selected, so deep memory is an advantage in avoiding aliased displays.

Peak detection. An antialiasing feature called “peak detection” is available on certain real-time digital oscilloscope models. This is a special acquisition mode in which the ADC samples at a fast rate that is independent of the sweep speed setting. Instead of immediately storing the sample values in acquisition memory, each sample value is processed by magnitude comparison circuitry. Two intermediate storage registers are provided; one is to contain the largest sample value processed, and the other, the smallest. Each sample taken by the ADC is compared with the current maximum value and, if it is larger, replaces it. A similar procedure saves the smallest sample value. At a rate called the “effective sample frequency,” both register contents are transferred to the acquisition memory, and the registers are reset. The effective memory depth is reduced by half, since two sample values, a maximum and minimum, are stored for each effective sample taken.

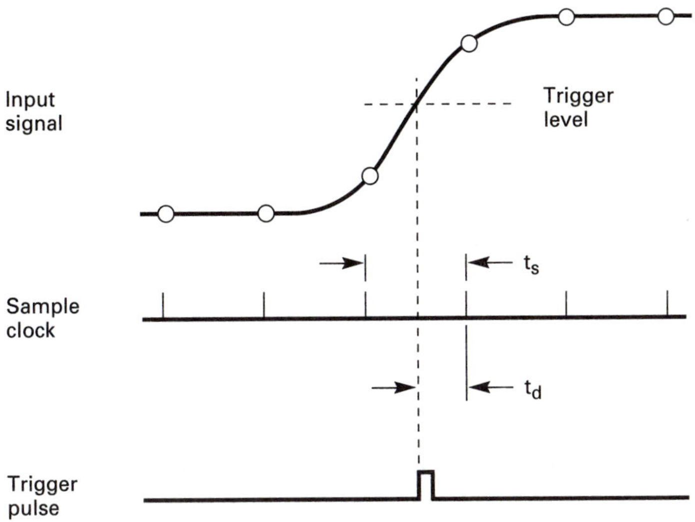

Figure 14.55 The trigger interpolator measures the time from trigger to the next sample clock pulse.

Of course, the effective sample rate is determined by the sweep speed setting and the effective memory depth. When the waveform is plotted on the display screen, the maximum and minimum values are placed at the same horizontal position and connected with a vertical line. Thus a single pulse narrow enough to be missed between samples in the normal acquisition mode will be captured and displayed as a vertical line having a length proportional to the pulse peak-to-peak amplitude. Acquisition in peak detect mode generates a display that emulates that produced by an analog oscilloscope having the same bandwidth and sweep speed setting.

Trigger interpolation. In a real-time digital oscilloscope, sampling is synchronized to a crystal oscillator, and samples are captured continuously while waiting for the trigger condition to be met. The time of the trigger event is determined exclusively by activity in the user's circuit and is not synchronized to the sample clock. Therefore, the time interval from the trigger pulse until the next sample is taken, t_d in Fig. 14.55, can have any value from zero to t_s. A circuit called a "trigger interpolator," located in the time base (see Fig. 14.43), accurately measures this time interval on each acquisition cycle. Sampling stops on the next sample pulse after the trigger pulse if delay $= 0$, or if delay > 0, a number of extra samples equal to the delay setting are taken before stopping. In any case, the time interval t_d is needed so that the trigger point can be plotted at the same horizontal position on the display screen on successive data capture cycles. The time interval t_d is measured to an accuracy of a few percent of the minimum time per division setting for a given oscilloscope model.

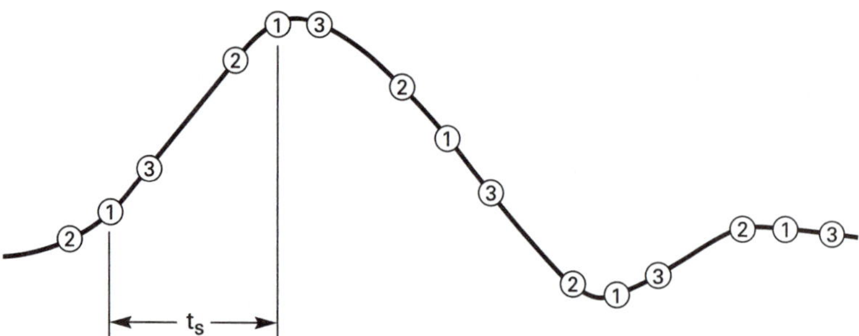

Figure 14.56 Random repetitive sampling builds up the waveform from multiple data capture cycles. All dots from the same acquisition bear the same number.

Random repetitive sampling. The only major disadvantage of real-time sampling is that bandwidth is limited to approximately $f_s/4$, so increasing bandwidth means increasing the sample rate, and very fast sample rate digitizers and memory are expensive. Ultimately, at any given state of the art there is a sample rate and bandwidth that cannot be obtained at any price. Fortunately, two alternative sampling techniques are available in which bandwidth is not limited by sample rate. They can be used if the signal to be observed is repetitive and if a stable trigger is available.

In the method of this type that is called "random repetitive sampling," repeated real-time data-acquisition cycles are performed, but each sample value is plotted individually on the display as a dot. No interpolation between samples is done. Each acquisition cycle produces a random time interval t_d between trigger point and sample clock (see Fig. 14.55), and samples from that capture are plotted on the display at intervals t_s with an offset from the trigger point of t_d. Each successive acquisition is plotted at its measured random offset, progressively filling in the picture of the waveform (Fig. 14.56). This method is also known as "equivalent time sampling." As the waveform fills in, the gaps (time) between dots (samples) become smaller and smaller, and the apparent or effective sample rate rises. The effective sample rate is ultimately limited by the accuracy of the trigger interpolator measurement of t_d.

Random repetitive sampling retains the storage and negative time attributes of real-time sampling, but the ability to capture a nonrecurring transient is lost. Certain oscilloscope models can be set to sample in either the real-time mode or the random repetitive mode.

Sequential repetitive sampling. In random repetitive sampling, the average number of dots plotted on the display screen during each capture and plot cycle is the product of the sample rate in samples per second and the width of the display window in seconds. Measurements made with a fast sample rate digitizer and moderate sweep speed setting will completely fill in the waveform in a few acquisitions. However, fast sweep speeds and a slow sample rate ADC get painfully slow. For example, using a 20 megasample/s sam-

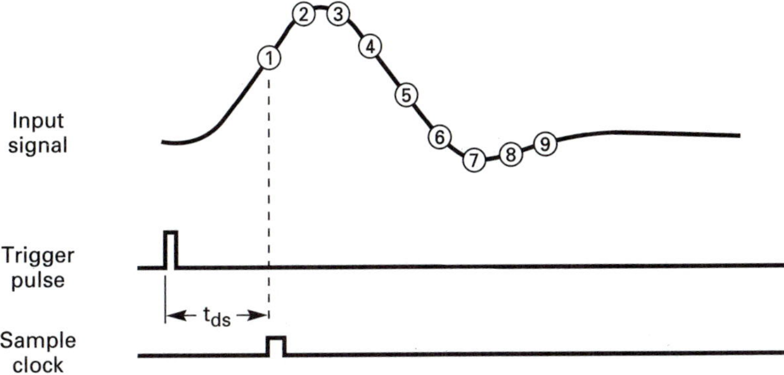

Figure 14.57 Sequential repetitive sampling captures one sample per trigger point and increases the delay t_{ds} after each trigger.

pler on a sweep speed setting of 1 ns per division captures only 0.2 samples per acquisition on average. In addition, a microwave-bandwidth ADC is feasible only at relatively low sample rates. An oscilloscope operating in this range bandwidths of 20 to 50 GHz, needs sweep speed settings as fast as 10 ps per division and consequently cannot satisfactorily use random repetitive sampling.

Sequential repetitive sampling is used in this situation to improve update rate by capturing only samples that will map onto the display screen. This is accomplished by capturing one sample value per trigger event at a carefully controlled time delay t_{ds} after the trigger pulse (Fig. 14.57). After each point is captured, the delay is increased by a small amount t_{se}, and the single sample acquisition cycle is repeated until the entire waveform has been plotted. The delay increment between samples, t_{se} is the effective sample time.

The delay t_{ds} from trigger pulse to sample clock must have a minimum (and positive) value, so this sampling method cannot capture the trigger event or any pretrigger information. The lack of negative time capability is a serious handicap; hence this sampling method is used only in microwave-bandwidth digital oscilloscopes.

14.6.3 Automatic measurements

It is difficult to overestimate the impact that microcomputer technology has had on improving the functionality of all types of electronic instrumentation, but it has been a particularly useful addition to the digital oscilloscope. Embedded controllers are beneficial and widely used in analog oscilloscopes, e.g., in implementing a user interface that is powerful yet easily understood. However, in the digital configuration, the waveform data are accessible to the computer, and this opens up a large array of useful features. In the discussion that follows, examples of some of the ways that this capability has been exploited will be presented. Be aware, however, that the extent and manner of implementation of these ideas vary widely in different digital oscilloscope models, even from the same manufacturer. A careful review of the automatic measurement feature set should be a part of the evaluation of a particular instrument.

Calibration. The ability to take measurements and analyze the results automatically is used to streamline the process of calibration of the digital oscilloscope. Previously, calibration was done by a trained technician who viewed the response to various standard signals and adjusted internal controls to bring the instrument responses into conformance with the manufacturer's specifications, but this is a costly and time-consuming process potentially involving many tens of measurements and adjustments. The automated process in a digital oscilloscope uses the system-embedded controller to measure and analyze the response to standard signals, and the results are stored in calibration tables in nonvolatile memory. (Nonvolatile memory retains its stored information when instrument power is removed.) For example, the actual gain of the vertical amplifier at every gain setting can be measured and retained. Then, when the volts per division control is changed, the correct new internal gain setting for that particular vertical amplifier channel can be selected. This design philosophy greatly reduces the need for manual internal adjustments, and some oscilloscope models require no internal adjustments for routine calibration. Certain models have the reference signal sources built in, and a complete calibration can be done on the user's bench by following instructions on the calibration menu, significantly lowering the oscilloscope operating cost.

Auto scale. Automatic measurement is also used to implement "auto scale." Invoked by pressing a button on the control panel, auto scale causes the oscilloscope to scan all channels for active signals. If any are found, the control settings are automatically adjusted to display a few cycles in the display window. This helps the operator quickly get a picture on screen and is especially useful when the signal has an unknown amplitude, dc offset, or frequency.

Measurement. Measurement of signal parameters, such as amplitude or period, is easily done directly from the display screen, but in addition, a number of preprogrammed parametric measurements can be selected which are calculated directly from the sampled data record. In some cases, the automatic computation is merely a convenience for the operator. However, the measurement may be difficult to execute or may need to be repeated many times, and using the computing power built into the oscilloscope can be a real time saver. The following is a representative list of built in measurements: rise time, fall time, frequency, period, positive pulse width, negative pulse width, duty cycle, delta time, peak-to-peak voltage, maximum voltage, minimum voltage, average voltage, and rms voltage.

Mathematical operations. Occasionally, it is desirable to do a mathematical operation on the entire sampled data record captured on a single channel, with the result plotted on the display screen. Examples of this type of analysis, which is available on some oscilloscope models, are inversion, integration, differentiation, or fast Fourier transform. Additionally, it may be of interest to combine data captured on two different channels. If the two records are labeled A and B, then the result of the following operations could be plotted on the display screen as a single waveform: $A + B$, $A - B$, $A \times B$, and A versus B.

14.6.4 Data communication

Captured waveforms can be stored in the embedded controller memory and later recalled, as can the oscilloscope control settings needed to perform a particular measurement, called a "setup." But there is also a need to transmit waveforms and setups to and from an external computer or computer peripheral device. Information can be copied to a flexible disk memory in the oscilloscope for storage or transfer, or it can be transmitted over an I/O bus. RS-232 and IEEE-488 (HP-IB or GP-IB) are two bus protocols commonly used for this purpose.

A printer or plotter produces a permanent copy of a waveform when connected to a digital oscilloscope using an I/O bus. Connection to an external computer brings even more power and flexibility. Waveforms and setups can be shared. Sampled data records can be analyzed more completely or in more specialized ways than is done in the general-purpose measurements built into the oscilloscope. In fact, the digital oscilloscope is capable of performing its measurements within the context of a fully automated test environment.

14.7 Comparing Analog and Digital Oscilloscopes

Table 14.1 briefly summarizes the advantages and disadvantages of the oscilloscope architectures discussed earlier. It is intended to aid in understanding

TABLE 14.1 Comparison of Analog and Real-Time Digital Oscilloscopes

Type	Advantages	Disadvantages
Analog	Responsive display Direct access controls Easily understood Low cost	No negative time No multichannel simultaneous capture Dim display (low repetition rate signals) Camera hard copy No waveform I/O
Real-time digital	Multichannel simultaneous capture Negative time Easy single transient capture Stored display image Easy calibration Printer or plotter hard copy Automatic measurements support Digital waveform I/O	Low throughput (waveforms/second) Possibility of aliased displays High cost

their general characteristics and does not apply uniformly to every oscilloscope model. The data sheet and specifications for a particular instrument under consideration should be read carefully.

The random repetitive digital oscilloscope has the same attributes as the real-time scope except (1) lower cost for same bandwidth, (2) less aliasing, and (3) no single transient capture. The sequential repetitive digital oscilloscope has the same attributes as the real-time scope except (1) microwave bandwidth capability, (2) less aliasing, (3) no single transient capture, and (4) no negative time.

14.8 Oscilloscope Probes

A probe is used to connect a signal to an oscilloscope input, and in making this connection, two issues are important. The load on the user's circuit must be acceptable, and the signal must be transmitted to the oscilloscope with a minimum of distortion. Noise pickup is avoided by using only shielded cable for probe construction.

14.8.1 High-impedance passive probes

The simplest probe, in effect just a short length of shielded cable, is known as a "one-to-one probe" (Fig. 14.58). One end is terminated in a connector designed to mate with the shielded coaxial connector at the oscilloscope input. At the other end is an insulated probe body which has a sharpened metal point connected to the cable center conductor and a short ground lead connected to the cable shield. The user's signal is modeled by the voltage source V_s and the source impedance Z_s. The resistive component of the probe loading is just the oscilloscope input resistance of 1 MΩ, ordinarily not a problem. The total capacitive load is the oscilloscope input capacitance, usually between 7 and 30 pF, plus the cable capacitance. A 1-m-long probe cable is approximately 50 pF. This much capacitance will seriously affect the operation of many circuits and attenuate fast transients; hence this probe configuration is seldom used. For example, if Z_s is a 200-Ω resistor and the total load capacitance is 60 pF, then the response is bandwidth limited to 13 MHz.

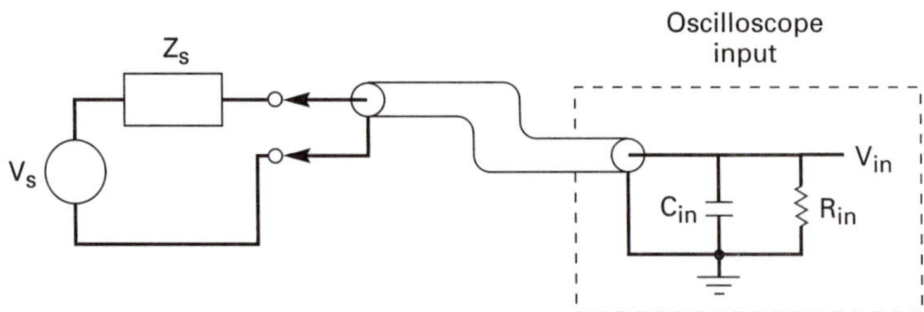

Figure 14.58 Connecting to the user's circuit with a 1:1 probe.

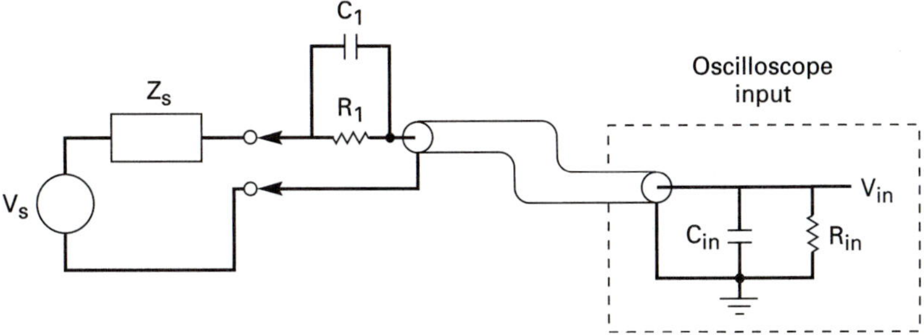

Figure 14.59 Connecting to the user's circuit with a 10:1 passive probe.

The "ten-to-one (10 × or 10:1) probe" reduces the capacitive load on the user's circuit at the price of a 10-fold reduction in signal amplitude (Fig. 14.59). A parallel *RC* circuit, R_1 and C_1, built into the probe tip combines with the probe cable capacitance and the oscilloscope input resistance and capacitance to form a voltage divider. A further idealization of the divider circuit is shown in Fig. 14.60. Making $R_1 = 9R_2$ sets the dc attenuation factor at 10, and choosing $C_1 = C_2/9$ matches it at high frequency. In other words, the voltage divider ratio is constant with frequency when $R_1C_1 = R_2C_2$. With this type of probe, the total load on the user's circuit is 10 times the oscilloscope input resistance in parallel with a capacitor that is one-tenth the sum of the cable capacitance and the oscilloscope input capacitance. Compared with the 1:1 probe, the capacitive loading is greatly reduced, but the vertical amplifier gain must be increased by a factor of 10 to make up for the signal attenuation of the probe. The 10:1 high-impedance passive probe is the most widely used oscilloscope probe type and has been the prevailing configuration for many decades.

One of the capacitors, either C_1 or C_2, on the 10:1 probe is made adjustable. Before using a 10:1 probe, the operator should always check the frequency compensation of the probe, that is, to ensure that $R_1C_1 = R_2C_2$. A square wave

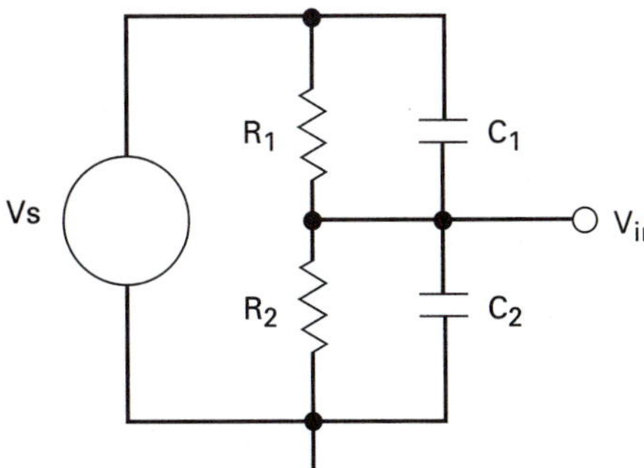

Figure 14.60 The idealized compensated divider equivalent circuit for the 10:1 passive probe.

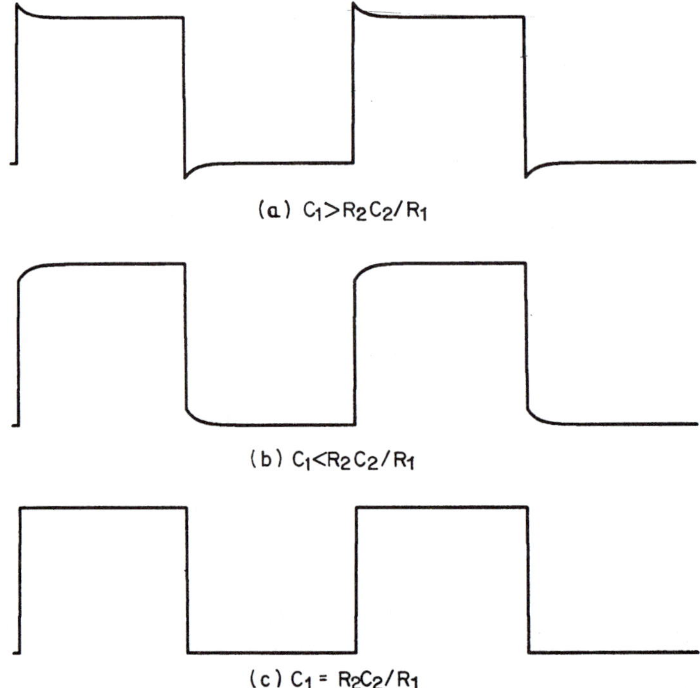

Figure 14.61 Calibrator response on the display screen during high-impedance 10 : 1 probe compensation. (*a*) $C_1 > R_2C_2/R_1$; (*b*) $C_1 < R_2C_2/R_1$; (*c*) $C_1 = R_2C_2/C_1$.

pulse generator of suitable amplitude and frequency for this purpose, called a "calibrator," is usually built into the oscilloscope. The probe is connected to the calibrator output, and the response to the square wave is displayed. The capacitor is adjusted to the setting that produces a square wave response, with neither overshoot nor undershoot (Fig. 14.61). The input capacitance is not necessarily the same on all oscilloscope inputs; there can even be significant variation among different inputs on the same instrument. Thus the probe compensation should always be checked whenever a 10 : 1 probe is first connected to an oscilloscope input because it may have been adjusted previously to match a different input.

In addition, 50 : 1 and 100 : 1 versions of the high impedance passive probe are also available. The primary benefit of these probes is a very small input capacitance and a correspondingly higher measurement bandwidth, but the large attenuation ratio requires a very high gain vertical amplifier to observe ordinary signal amplitudes of a few volts or less.

14.8.2 Resistive divider passive probe

A type of passive probe called a "resistive divider," fabricated using 50-Ω shielded cable, is available for making measurements in low-impedance,

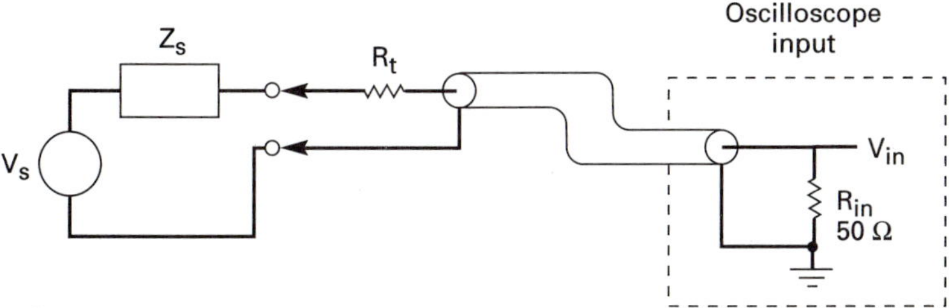

Figure 14.62 Resistive-divider passive probe.

high-frequency circuitry (Fig. 14.62). To use this probe, the oscilloscope input is shunted with a 50-Ω resistor to terminate the 50-Ω transmission line. The probe tip contains a series resistor R_t, that forms a voltage divider with the 50-Ω load from the terminated probe cable (Fig. 14.63). The most popular divider ratio is 10:1, for which $R_t = 450\ \Omega$, but other resistor values and divider ratios are also commonly used. The tip resistor can be made in such a way that parasitic capacitances are small enough to be neglected, effectively eliminating the bandwidth-limiting effect from capacitive loading of the user's circuit. In fact, probes of this type are ordinarily bandwidth-limited by high-frequency transmission losses in the 50-Ω cable or from parasitic inductance in the ground lead connection at the probe tip. A well-designed passive divider probe can have a bandwidth of several gigahertz.

Of course, the disadvantage of this probing technique is that the effect of the pure resistance that is so advantageous at high frequencies extends all the way to dc. Probing a point in the user's circuit with the 10:1 version connects a 500-Ω resistor from that point to ground. While this is an acceptable load for some circuits, for many it is not. A larger divider ratio increases the resistor size proportionately at the price of reduced signal amplitude at the oscilloscope input. Or the dc loading can be eliminated entirely by placing

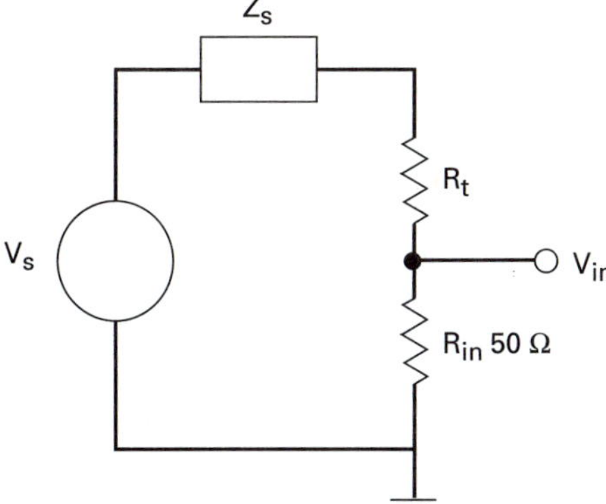

Figure 14.63 Idealized equivalent circuit for the resistive-divider probe.

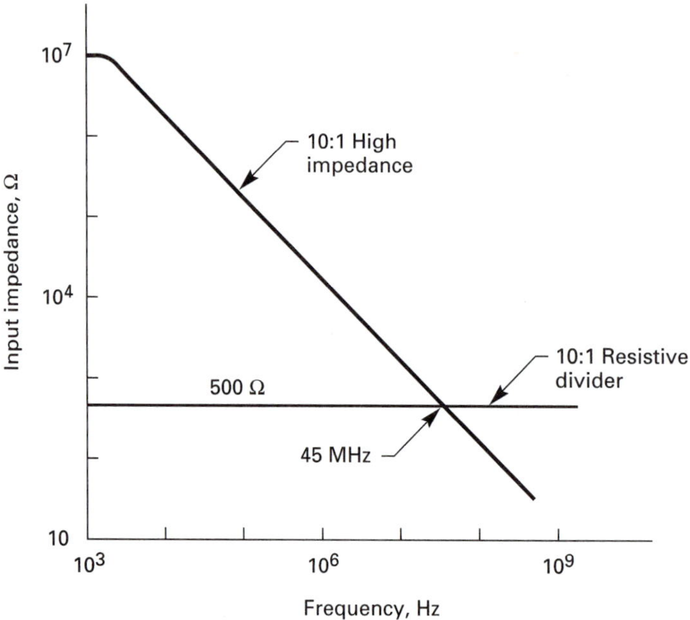

Figure 14.64 Input impedance at the probe tip for high-impedance and resistive-divider 10 : 1 probes.

a coaxial blocking capacitor between the probe cable and the oscilloscope input, but then the dc and some low-frequency ac signal information is lost.

Figure 14.64 is a graph that compares the loading characteristics of 10 : 1 high-impedance and 10 : 1 resistive-divider passive probes. The high-impedance probe has an input capacitance of 7 pF and has a *lower* input impedance than the resistive-divider probe at frequencies above approximately 45 MHz.

14.8.3 Active probes

Both types of 10 : 1 passive probes discussed above seriously load the user's circuit when high-frequency measurements are made. An active probe offers a lower input capacitance and a high input resistance combined in the same unit. It has a physically small amplifier built into the probe body very near the probe tip so that the probe input capacitance can be kept small, usually less than 2 pF. The amplifier output drives a 50-Ω transmission line which is terminated at the oscilloscope input with a 50-Ω resistor. Active probe models with different divider ratios are available, ranging from 1 : 1 to 10 : 1. The amplifier input is connected directly to the probe tip on a 1 : 1 model, while units having the largest attenuation incorporate a compensated divider built into the probe tip ahead of the amplifier. The divider uses the same $R_1C_1 = R_2C_2$ principle used in 10 : 1 high-impedance passive probes, so a model with larger attenuation will have a correspondingly smaller input capacitance. Some active probe models provide a removable 10 : 1 compensated divider.

Compared with passive probes, active probes are generally larger, heavier, more expensive, and less rugged. Because the active probe uses an amplifier, the signal dynamic range is limited, and peaks above a certain amplitude, usually a few volts, will be clipped.

14.8.4 Differential probes

A "differential" probe is an active probe which has two inputs, one positive and one negative, and a separate ground lead, and it drives a single terminated 50-Ω cable to transmit its output to one oscilloscope channel. The output signal is proportional to the difference between the voltages appearing at the two inputs. Both inputs can have active signals simultaneously, but they must be within a few volts from ground to stay within the dynamic signal range of the probe. The average of the two input voltages is called the "common mode signal." A differential probe is designed to reject (not respond to) the common mode signal, but inevitably there will be a small error response. Common mode rejection capability is easily measured by connecting both inputs to the same signal simultaneously and observing the probe response. The rejection is best at dc and low frequencies and deteriorates with higher-frequency signals. Some active probe models provide a removable 10 : 1 balanced two-input compensated divider. The differential probe has disadvantages similar to those listed for the single-input active probe.

14.8.5 Current probes

A transducer that generates a voltage output proportional to a current in the user's circuit is called a "current probe." A resistor and a 1 : 1 voltage probe do that (Fig. 14.65) and they are a useful method in certain situations. However, inserting a resistor into the user's circuit has some disadvantages. Generating a large enough voltage drop to register adequately on the oscilloscope could adversely affect the circuit operation, and so might connecting the probe ground lead to the circuit at the point current is to be monitored. Using a differential probe would permit measuring current into or out of nodes which cannot be grounded.

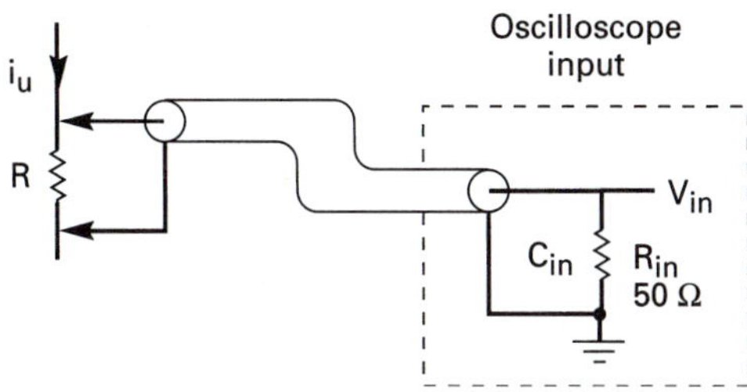

Figure 14.65 Adding a resistor in series with a branch in the user's circuit to measure current.

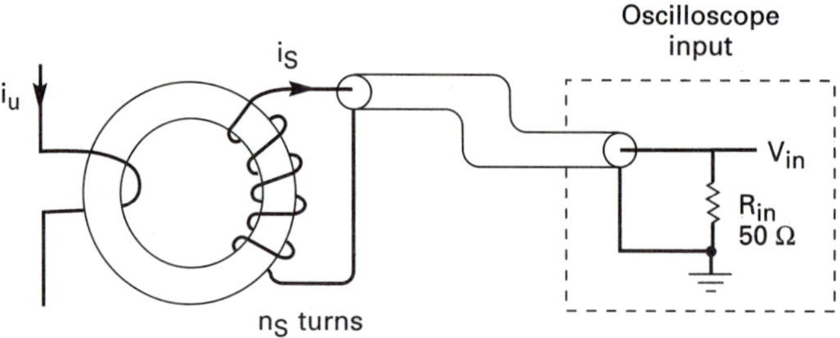

Figure 14.66 The transformer-based current probe.

Current probes are available that use a transformer to, in effect, insert a small resistor into the user's circuit (Fig. 14.66). Located in the probe tip, a transformer having a secondary winding of n_s turns drives a 50-Ω probe cable which is terminated at the oscilloscope input with a 50-Ω resistor. The current to be measured in the user's circuit is conducted through a single-turn primary winding. Since it depends on a transformer, this method does not measure dc currents; i.e., it is ac-coupled. Using the equations for an ideal transformer, it is straightforward to show that (1) $i_s = i_u/n_s$, (2) the sensitivity of the probe is R_{in}/n_s V/A, and (3) the apparent resistance in the primary winding is R_{in}/n_s^2. When the transformer current probe is used to sense current in a wire in the user's circuit, a resistance R_{in}/n_s^2 appears to be added to that wire so that the loading effect of the current measurement can be determined. Unfortunately, probe manufacturers do not always state the number of turns on the transformer secondary winding explicitly, but the termination resistance and sensitivity are given, and the number of secondary turns can be easily calculated.

Current probes are also available that use "Hall effect" or a "Hall generator" to directly sense magnetic flux intensity in the transformer core to generate a voltage signal that is amplified and fed to an oscilloscope input. This method allows measurement of dc currents but is inherently limited to measuring relatively low frequency signals.

Hybrid current probes combine a transformer and a Hall generator into an integrated unit and combine their outputs to provide the best features of both types. Measurements bandwidths from dc to approximately 50 MHz are available.

In one type of transformer current probe, a short length of wire is fed through a hole in the probe body, passing through the transformer core, and the wire is then inserted into the branch of the user's circuit to be measured. Other models arrange the transformer core material into two movable segments so that the transformer can be placed around an existing wire without first disconnecting it. The transformer turns ratio, and probe sensitivity, can be changed by looping two or more turns of the current-carrying wire through the transformer. Addition or substraction of currents in different branches of the user's circuit can be performed by linking the currents through the current probe simultaneously, but there will be some cross-coupling between the branches measured.

14.9 How to Buy an Oscilloscope

In deciding whether to buy an oscilloscope or which model to buy, the first step is to carefully define the measurement requirements it is to satisfy. Ask and answer questions like these:

What quantities need to be measured and to what accuracy?

Where will it be used (laboratory, field service, etc.)?

How long will measurement requirement exist?

Will the measurement requirement become more difficult?

What is the technical skill level of the operator(s)?

What is the equipment purchase budget?

What are the probing requirements?

As these questions are addressed, other significant considerations or questions may be identified. Then a list of specifications that the oscilloscope must meet can be generated, perhaps with an additional list of desirable features.

The next step is to identify and evaluate the oscilloscope models that could potentially satisfy the measurement requirements. Contact the various oscilloscope manufacturers' sales representatives for information. The manufacturers' catalogs list the current models and usually include a chart or table comparing the various types of oscilloscopes. A technical data sheet and a sales brochure are available for each oscilloscope model, and they do a good job of presenting key feature and capabilities. Field sales engineers know the capabilities of their products, will assist in the selection process, and will demonstrate the operation of their products, will assist in the selection process, and will demonstrate the operation of their products. If there is a specific key measurement task, get a demonstration of the competing models each making that measurement. Each oscilloscope manufacturer has a library of application notes which describe in detail how to perform specific measurements. These notes cover a broad range of technical topics, some quite specialized, on using the manufacturer's products to solve actual measurement problems.

As an alternative to buying an new oscilloscope, consider buying a used instrument or renting. Rental agencies stock the most popular current models and can be a good source to satisfy a short-term measurement requirement or to do an evaluation lasting several months. Rental agencies also sell used instruments to keep their inventory current, and there are companies that specialize in buying and selling used electronic instruments.

Chapter

15

Power Measurements

Ronald E. Pratt
Agilent Technologies
Santa Clara, California

15.1 Introduction

A power meter may be used to measure the electrical power that a source is capable of delivering to a specified load impedance, or it can be used to measure the electrical power that a source furnishes to an arbitrary load impedance. Many factors influence the selection of a power measurement technique; these include frequency range, power level, the spectral content of the signal, and the required accuracy. No single instrument satisfies all these requirements, and a wide variety of power meters and measurement techniques are available which satisfy the specific objectives of a given set of measurement requirements.

15.2 Basic Power Definitions

Power is defined to be the rate at which energy is transferred between systems or elements within a system. A given element may either supply energy (power is furnished by the element which is acting as a source) or energy may be absorbed by the element (power is delivered to an element which acts as a load). Two broad categories of power measurements emerge from this discussion:

1. Meters or measurement techniques which determine the power *transmitted* through the meter or measurement system from the source to the load (transmission measurements)
2. Meters or measurement techniques which *absorb* the measured power (absorption measurements)

The basic definition of power is described by the following relations:

$$\text{Power} = d(\text{energy})/dt \quad \text{and} \quad \text{Energy} = \int (\text{power})\, dt \tag{15.1}$$

Monitoring the rate of change of energy is the basis of techniques to determine transmitted power and the conversion of electrical to thermal energy, and its attendant temperature rise forms the basis for many techniques of measuring absorbed power. Other relationships describe the basis for alternative methods for electrical power measurements.

The instantaneous power may be found from the relationship:

$$\text{Power} = \text{voltage} \times \text{current} \tag{15.2}$$

where the values of the power, voltage, and current are the instantaneous values at any given time. If the voltage and current do not vary with time (dc) the instantaneous power is constant and becomes the quantity that is measured. Power meters used with ac signals provide a measurement of the *average* power, which is the net rate of change in energy occurring during the time of one period of the signal. Average power is related to instantaneous power by

$$P = \frac{1}{T}\int_0^T v(t) \times i(t)\, dt \tag{15.3}$$

where P = average power
$v(t)$ = instantaneous voltage
$i(t)$ = instantaneous current
T = period of the signal

While very fast power sensing elements exist which are able to profile a pulse of rf power having rise times of a few nanoseconds, they are still not responding to instantaneous power and their response is calibrated to indicate the average power that exists during time intervals of a few nanoseconds.

Other types of power meters respond to the long-term average of the power of the signal. If the voltage and current are represented by continuous-wave (CW) sinusoidal waveforms the expression for average power becomes

$$P = VI \cos\theta \tag{15.4}$$

where P = average power, W
V = rms (root-mean-square) value of the voltage
I = rms value of the current
θ = phase of the voltage relative to the current

In practice power sensing elements which are designed to absorb power

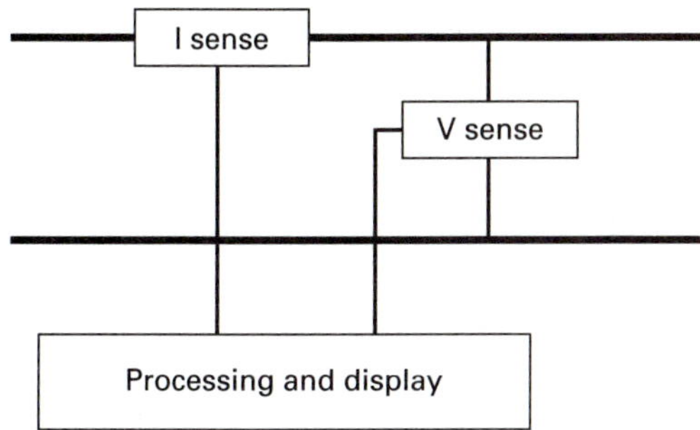

Figure 15.1 Basic principle of a low-frequency power meter.

present a resistive load so the voltage and current are in phase ($\theta = 0$), and since $V = IR$, the following equations apply:

$$P = \frac{V^2}{R} \tag{15.5}$$

or

$$P = I^2R \tag{15.6}$$

In some designs the power sensing element is operated as a *square-law* detector. These detectors have a response mechanism which yields an output proportional to the square of the applied voltage or current, thereby obeying the power relationships expressed by Eqs. 15.5 and 15.6. Another way of indicating the use of square-law mechanisms is by specifying that a power sensor has *true rms* response. If a signal containing many frequency components is to be measured, these sensors will correctly respond to the total power expressed by

$$P_{\text{total}} = \frac{V_1^2 + V_2^2 + V_3^2 + \cdots + V_n^2}{R} \tag{15.7}$$

which indicates that the total power is determined by the sum of the power content of each sinusoidal signal represented by V_1 through V_n. These may be in the form of modulation sidebands, harmonics, or a multitude of frequencies. Examples of square-law sensors include those which employ thermal principles or use diodes which are operated at power levels less than 10 μW.

15.3 Transmission-Type Power Measurements

A transmission-type power meter is designed to be connected between a source and a load. For low-frequency applications the meter may contain current and voltage sensing elements as indicated in Fig. 15.1, and the output

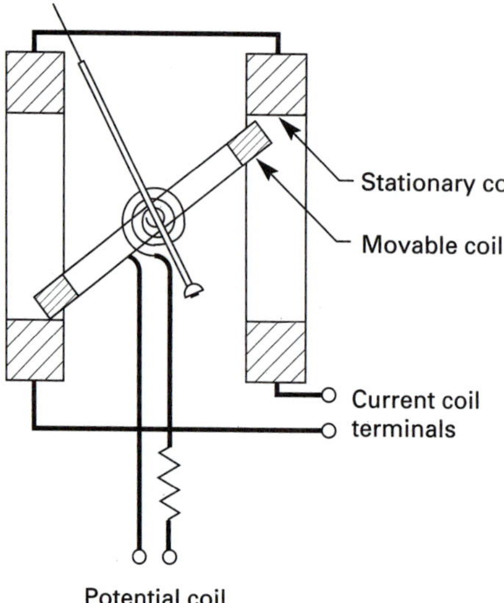

Figure 15.2 Drawing of the electrodynamometer power meter.

of these sensing elements is processed by the meter to produce the power measurement. The electrodynamometer meter shown in Fig. 15.2 is an electromechanical realization of the principle illustrated in Fig. 15.1. The current sensing element is a fixed or stationary coil, and the voltage sensing element is a movable or potential coil which is placed in the magnetic field of the fixed coil. The torque on the movable coil is opposed by a spring and is proportional to the average value of the instantaneous product of the currents in the two coils. The meter is connected such that the load current passes through the current coil terminals and the potential coil terminals are connected across the load. Since the current through the movable coil is proportional to the load voltage, the amount it rotates is a function of the average value of the instantaneous voltage-current product, and the attached pointer will indicate the average load power. Other low-frequency power meter designs rely on digitizing the responses from the voltage and current sensors and they must include the phase information. Features may include the capability of displaying volt-amperes (VA) and volt-amperes reactive (VAR) as well as the power in watts. Figure 15.3 illustrates the relationship among power, VA, VAR, and the phase angle. The requirement of having knowledge of the phase relationship between the voltage and current limits the frequency range of this approach to power measurements of a few kilohertz. Power levels ranging from a few watts to several kilowatts may be measured with these types of meters.

Transmission-type power meters which operate at rf frequencies employ a coupling element which responds to the direction of *power flow* between the source and load. As Fig. 15.4 illustrates, these meters may be connected to

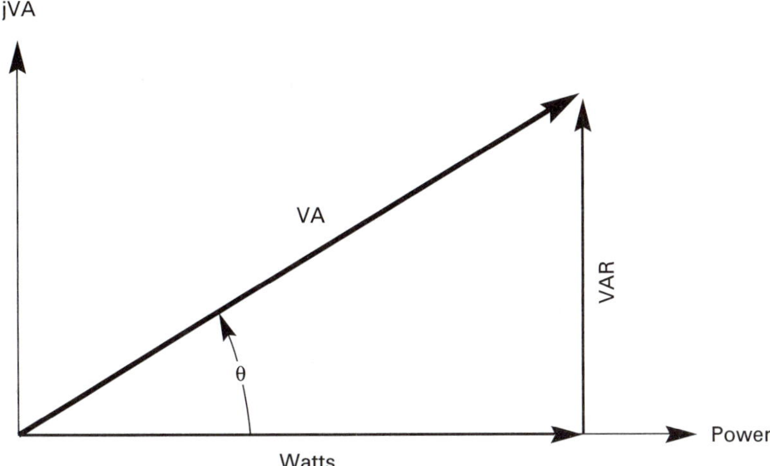

Figure 15.3 Relationship of power in watts to volt-amperes (VA), volt-amperes reactive (VAR), and the phase angle θ.

indicate the power flowing from the source to the load (the *incident* power) and the connections to the meter may be reversed so it will measure the power flowing back to the source (the *reflected* power). Since the coupling element responds to power flowing in one direction, it is not necessary to have knowledge of the phase of the coupled signal and it may be measured by approaches similar to those employed by absorption-type meters. At microwave frequencies it is common practice to use a *directional coupler* in conjunction with an absorption power meter to measure the power transmitted by the source. The power range of these approaches varies from a few milliwatts to many kilowatts, limited mainly by the coupling factor.

Two basic approaches to transmission-type power measurements have been described. Low-frequency power meters sense the total voltage and total current so this type of meter will directly measure the power transmitted to the load. In contrast, high-frequency techniques measure power flow and the user

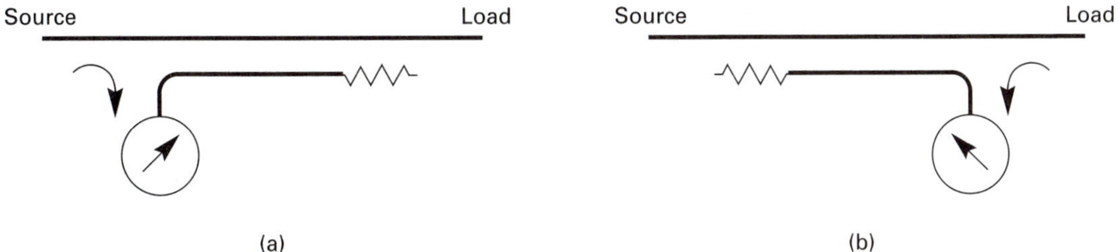

Figure 15.4 Transmission-type rf power meters respond to the direction and magnitude of power flow. (*a*) Measure power flow from source to load; (*b*) measure power flow from load to source.

must determine the power delivered to the load from the following relationship:

$$P_{\text{load}} = P_{\text{incident}} - P_{\text{reflected}} \tag{15.8}$$

The effect of the coupling element must also be considered. Transmission-type rf power meters often have the coupling element built into the meter so its effect is accounted for when the meter is calibrated. In microwave systems the effect of the coupling factor of the external couplers must also be considered in order to determine the power absorbed by the load:

$$P_{\text{load}} = \frac{P_{\text{incident}} - P_{\text{reflected}}}{C} \tag{15.9}$$

where C is the *coupling factor,* which is the ratio of the sampled power to the mainline power and is often expressed in units of dB loss.

For example, if incident and reflected power readings of 100 and 60 mW, respectively, are obtained the power delivered to the load would be 40 mW if the effect of the coupler was included in the calibration of the meter. If the readings were obtained using an external coupler having a coupling factor of 30 dB ($C = 0.001$) the actual load power would be 40 W.

The coupling factor is determined by the design of the coupling element, and its value is stated by the manufacturer either as a single value or as data presented across the frequency range of operation. The range of coupling factors which are available varies from 3 dB to more than 60 dB.

15.4 Absorption-Type Power Measurements

Absorption-type power meters are generally comprised of two components.

1. A *power sensor* which terminates the transmission line, dissipates the power, and contains a means of producing a dc or low-frequency signal which is proportional to the applied power.
2. A *power meter* which contains amplifiers and processing circuits which produce a readout of the power level. It is quite common to find a family of power sensors which are differentiated by frequency range or power-handling capability which all work with the same meter. The complexity of the power meter varies from simple manually controlled analog instruments to multichannel meters employing microprocessor control.

The frequency range of a power sensor varies from less than an octave ($<2:1$ range) for waveguide sensors to multiple decades ($10:1$ range) with coaxial sensors (50 MHz to 40 GHz). Absorptive and mismatch losses within the sensor will produce some variation in their sensitivity which may be corrected for by use of the *calibration factor* of the sensor. The definition of calibration factor K_b is

$$K_b = \eta(1 - \rho^2) \times 100 \tag{15.10}$$

where K_b = calibration factor as a percentage
η = effective efficiency (absorptive loss)
$1 - \rho^2$ = mismatch loss
ρ = magnitude of the sensor reflection coefficient

The power meter has provisions for entering the value of the calibration factor and results in either a gain change or a mathematical correction made to the power meter reading which compensates for the value of the calibration factor.

15.5 Thermistor Sensors and Meters

A thermistor is a resistor fabricated from a compound of metallic oxides and exhibits a large resistance change as a function of temperature. If a thermistor is used to form the termination in a power sensor its resistance becomes a function of the temperature rise produced by the applied power. The data presented in Fig. 15.5 illustrate typical power resistance characteristics of a thermistor over a wide range of temperatures.

The basic concept of the thermistor power meter is illustrated in Fig. 15.6. The circuit in Fig. 15.6*a* shows how two thermistors can be arranged to be connected in parallel for signals appearing at the rf input and be connected in series to obtain the connection to the meter. The meter connection is taken across an rf bypass capacitor to avoid rf leakage beyond the thermistors. The power meter employs a circuit known as a self-balancing bridge (Fig. 15.6*b*) which provides a dc bias power which maintains the thermistor resistance R_T constant at a value of R. If the rf power in the thermistor increases, the

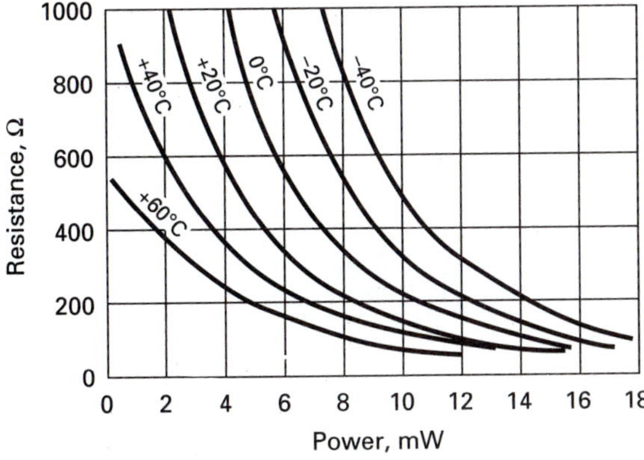

Figure 15.5 Characteristic curves of a typical thermistor used for sensing power.

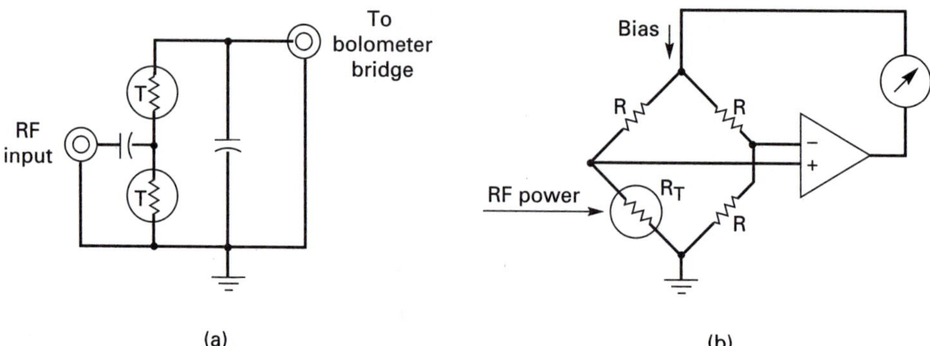

Figure 15.6 Basic concepts of the thermistor power meter. (*a*) Thermistor power sensor; (*b*) self-balancing bridge.

bridge reduces the bias power by a like amount. A decrease in the rf power causes the bridge to increase the bias power to maintain a constant resistance in the thermistors. Additional circuitry in the meter processes this change in dc power to obtain the power readout.

The resistance of the thermistor is a function of the ambient temperature as well as the rf and dc power, so any change in temperature would produce a change in the power reading. Modern thermistor power sensors minimize this problem by employing a second set of thermistors which are thermally linked to the rf sensing thermistors but electrically isolated. The circuit of a temperature-compensated thermistor sensor is shown in Fig. 15.7. This type of sensor requires a specialized meter containing two self-balancing bridges and circuitry which obtains a power reading based on the bias signals applied to the detection and the compensation thermistors. The details of such a meter are shown in Fig. 15.8.

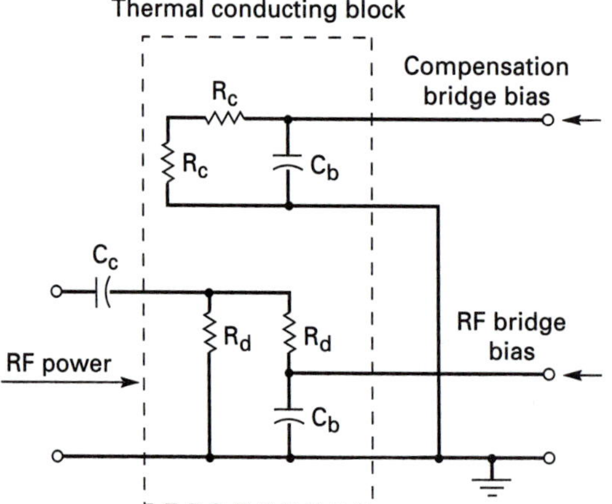

Figure 15.7 Temperature-compensated thermistor sensor.

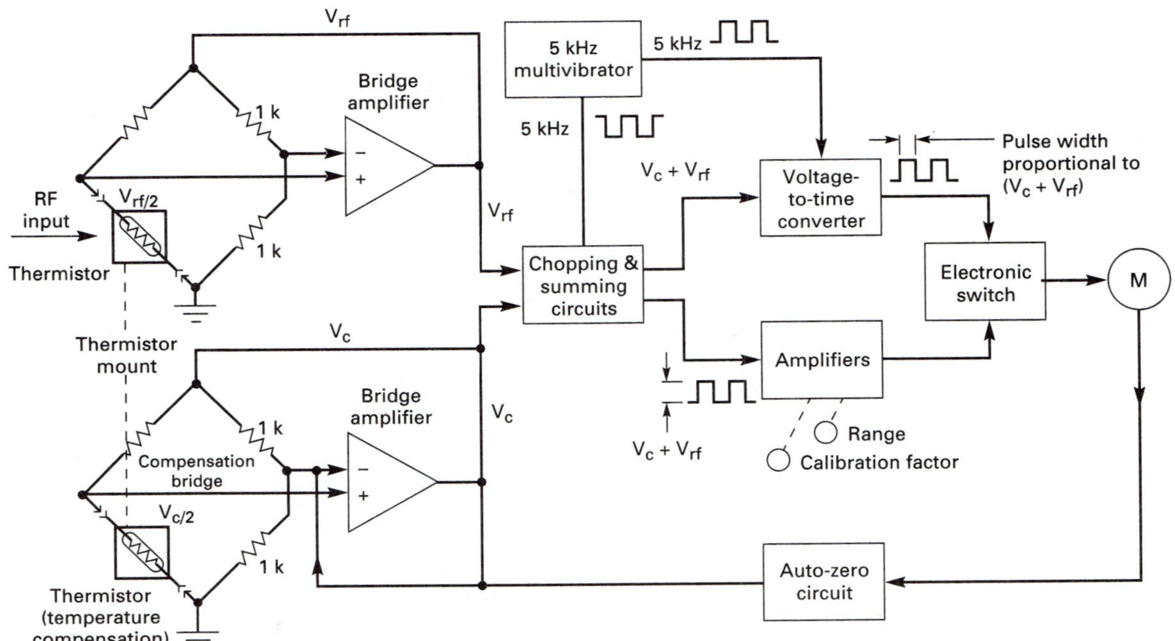

Figure 15.8 Power meter used with a temperature-compensated thermistor sensor.

By comparing the bias power applied to the rf bridge relative to the power applied to the compensation bridge, the rf power is given by

$$P_{rf} = \frac{V_c^2 - V_{rf}^2}{4R} \tag{15.11}$$

where P_{rf} = rf power
V_c = voltage applied to compensating bridge
V_{rf} = voltage applied to rf bridge
R = resistance of thermistor sensor at balance

The power meter contains an electronic switch which is closed for a period of time proportional to the sum of V_c and V_{rf}. This allows a current to flow in the meter movement M which is proportional to the difference between V_c and V_{rf} and the average value of the current through the meter current satisfies the relationship in Eq. 15.11. With no rf power applied, $V_c = V_{rf}$ and this condition is obtained when the user activates the auto-zero circuit. A modern thermistor power meter provides measurement capability over an input power range from 10 mW to 1 μW (40 dB), and sensors are available which operate in various bands across a frequency range of 100 kHz to 100 GHz.

At one time thermistor sensors were widely used for general-purpose power measurements but they have been replaced by other power sensing methods which offer better performance. The calibration of power meters and sensors is a major application for the thermistor sensor which is not addressed by most other methods of measuring power.

15.6 Thermocouple Power Meters

It is very desirable that a power sensor offer wide dynamic range, low drift and SWR, and accept a wide range of frequencies in a single unit. Power sensors which employ thermocouples are available which satisfy these needs.

The connection of two dissimilar conductors produces a pair of thermocouple junctions and a voltage will be produced by any temperature gradient which exists across those junctions. The thermocouple structure of a power sensor is designed to include a resistor which dissipates the majority of the applied power. The temperature of the resistor increases and produces a gradient across a nearby thermoelectric junction which produces a voltage proportional to power. Two sets of these structures can be physically oriented such that the temperature rise produced from the power dissipated by the resistors will cause the two thermocouple structures to produce an additive thermoelectric voltage while a gradient produced by a change in ambient temperature causes thermoelectric voltages which cancel, thereby minimizing drift in the zero reading. The ohmic value of the resistors is designed to provide a well-matched termination for the transmission line.

Thermocouple elements used in power sensors may be comprised of a com-

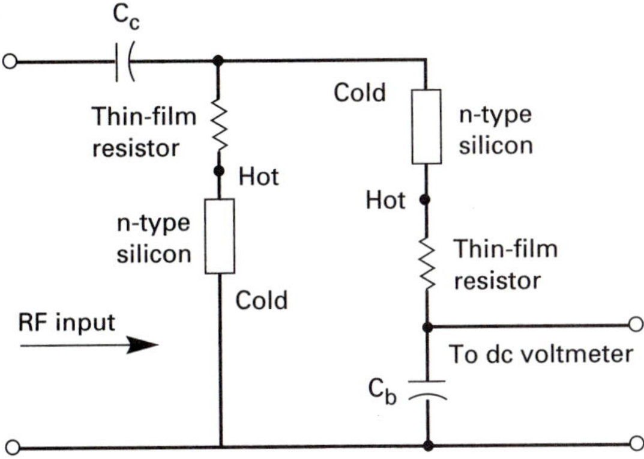

Figure 15.9 Thermocouple power sensor.

bination of gold, n-type silicon, and tantalum nitride resistive material, and thin-film construction provides the small size and precise geometry required for operation at frequencies in excess of 40 GHz. A schematic diagram of the thermocouple sensor employing these techniques is illustrated by Fig. 15.9.

The sensitivity of a thermocouple may be stated in terms of the magnitude of its dc output voltage relative to the amount of rf power dissipated by the sensor. A typical sensitivity is about 160 μV/mW, and power levels as low as 1.0 μW may be measured with this sensor. The dc voltage that must be measured may be as low as 0.16 μV, so a large amount of gain must be provided by the amplifiers inside the power meter. It is essential that these amplifiers do not add any additional dc offsets which add to or subtract from the fraction of a microvolt being measured.

The chopper input amplifier and synchronous detector shown in Fig. 15.10 satisfy this requirement. The chopper is operated by a square-wave drive sig-

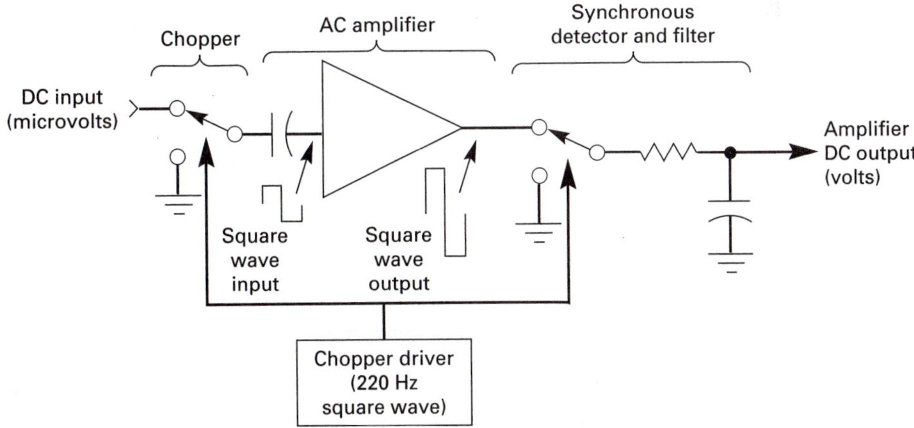

Figure 15.10 Chopper input amplifier and synchronous detector.

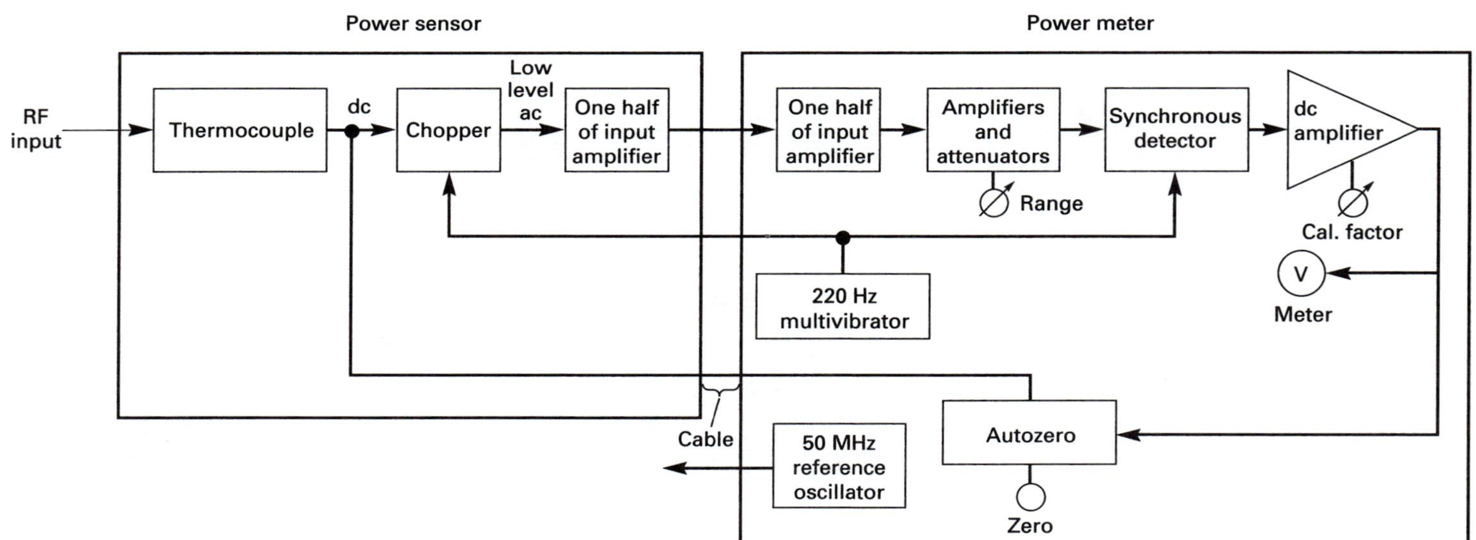

Figure 15.11 Block diagram of a thermocouple power sensor and the supporting power meter.

nal, and it simply switches the input capacitor of an ac coupled amplifier to either the output of the sensor or ground. The input capacitor is charged by the dc input voltage and is discharged by the ground connection so the input signal to the amplifier becomes a square wave whose amplitude is proportional to the output of the sensor. The ac coupled amplifier has sufficient gain to produce an output square wave of several volts and it contains no offset voltages. The synchronous detector is another switch which is operated by the same signal as the chopper, and it connects the amplifier output to an *RC* (resistor-capacitor) filter or it grounds the filter input. Since the output switching is synchronized to the input chopper, the filter capacitor is charged by the same half cycle of the square wave that was produced by the input dc voltage. The output of the filter is a dc voltage which may be easily processed and displayed.

The block diagram in Fig. 15.11 illustrates some of the details about the implementation of a power meter and shows how the meter circuitry can be partitioned. The chopper and a portion of the first amplifier are contained in the sensor so a relatively high level signal is sent to the meter where it is amplified, converted back to direct current by the synchronous detector, and displayed by a meter. Similar circuitry will be found in digital or microprocessor-based power meters which are used with thermocouple sensors. A modern thermocouple power meter provides power measurement capability over an input power range from 100 mW to 1 μW (50 dB).

Most thermocouple power meters provide a precision reference source having a calibrated output power which is to be used while adjusting the gain of the system to compensate for unit-to-unit variation in the sensitivity of the thermocouples. The user should perform this adjustment whenever a different sensor is connected to the meter, and this can be as simple as connecting the sensor to the reference source and pressing a button.

15.7 Diode Power Sensors

With the use of a semiconductor diode as the sensing element it is possible to measure very low power levels. The simplest form of a diode sensor is shown in Fig. 15.12 and can be seen to contain a dc blocking capacitor, a terminating

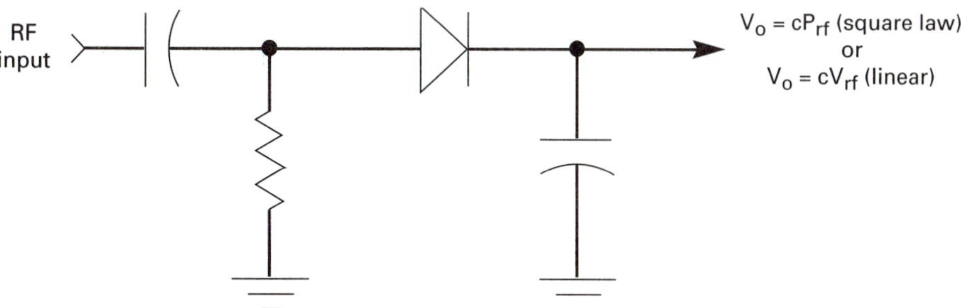

Figure 15.12 Detector circuit for a diode power sensor.

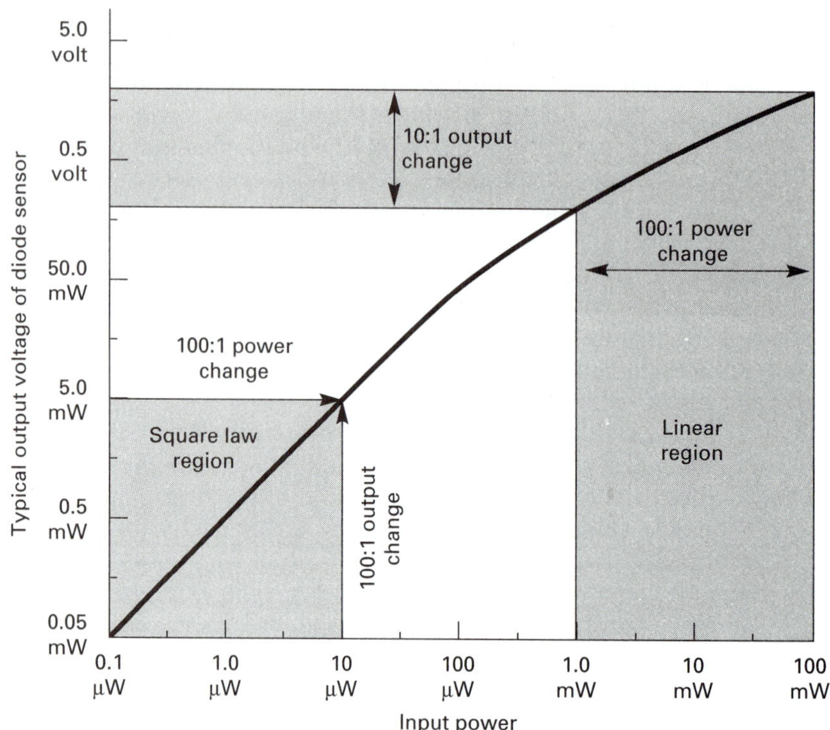

Figure 15.13 Output voltage of a diode power sensor as a function of input power.

resistor, the diode, and an rf bypass capacitor. The current flow through the diode is a nonlinear function of the applied voltage appearing across the load resistor. Some diodes will conduct significant current (microamperes) with very low applied voltage (millivolts) but the nonlinear relationship still exists and causes the rectified output to follow the square of the applied voltage (i.e., square-law response), thereby obeying the power relationship in Eq. 15.5. The data presented in Fig. 15.13 illustrate that for operation within the square-law region the output of the sensing diode directly follows the changes in input power. Since the detection mechanism obeys the power relationships, the square-law diode sensor will indicate the correct value of the total power of a complex waveform as expressed by Eq. 15.7.

To assure that the diode is responding to the signal power, some power sensor designs restrict the measurement range to be within the square-law region. These sensors are capable of measuring power levels as low as 0.10 nW (−70 dBm), and they will produce accurate power measurements independent of the waveform of the applied signal. The usable dynamic range of square-law operation is about 50 dB, so square-law diode sensors may use the same meters as employed with thermocouple sensors.

Power meters which extend the operation of a diode sensor to higher power levels (10 to 100 mW) can provide measurement capability having a very wide dynamic range (70 dB or more) but the readings obtained on the ranges above 10 μW may only be valid for a CW (continuous-wave) sinusoidal signal. At high power levels the diode is operating as a linear detector which responds to the peak value of the applied voltage. Figure 15.13 shows that the output of the diode is indicating the 10-to-1 change in voltage necessary to produce a 100-to-1 power change. In this operating range, the output of the sensor diode must be squared before it becomes an indication of the power.

The meters which are used with diode sensors operating in the linear range contain a means of squaring the output voltage of the diode to make the reading correspond to the power of a CW sinusoidal signal. A meter designed to measure the average power of CW signals will not accurately measure the power of a signal with any form of amplitude modulation. The solution to this problem is to reduce the amplitude of the signal until the diode is operating in the square-law region where the diode responds to the total power.

Harmonics of the carrier frequency may introduce significant measurement error if the sensor is operating in the linear range. For example, if the harmonic is 20 dB below (10 percent harmonic voltage) the fundamental—it is contributing 1 percent of the total signal power. A sensor with square-law response would indicate the correct value of total power. The voltage of the harmonic could add to or subtract from the peak voltage of the fundamental so a linear detector might have an output which varies between 0.9 and 1.1

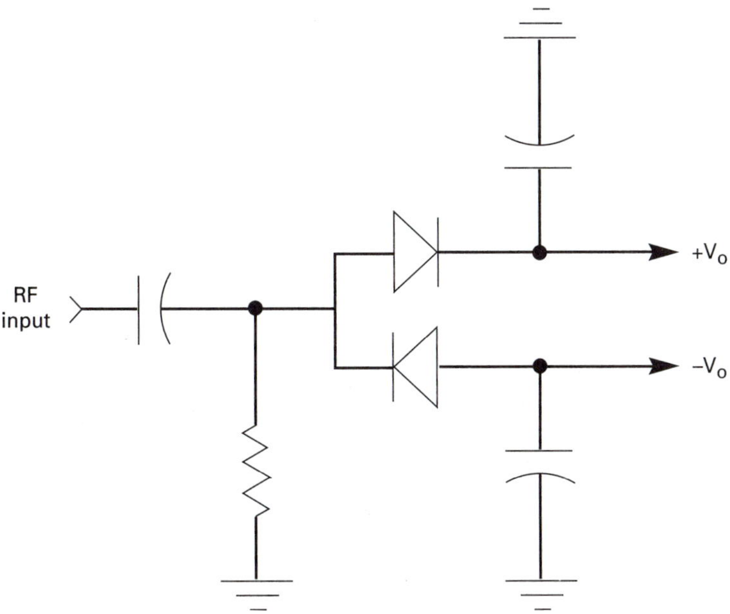

Figure 15.14 A full-wave detector for a diode power sensor.

times the voltage of an undistorted signal. Since the output of the sensor is then squared, the indicated power could be 20 percent higher or lower than the true value. The actual peak voltage depends on the phase relationship of the fundamental and its harmonic so there is no way to correct for this error. The full-wave detector illustrated by Fig. 15.14 senses the peak-to-peak voltage and adds appreciable error only if the signal contains odd harmonics.

15.8 Peak Power Measurements

Specialized power meters are available for measuring pulse-modulated signals and they often include a display similar to an oscilloscope to present the measured results as a function of time. The sensing element is generally a diode which is designed to have a fast output response time (10 ns). The output of this type of sensor closely follows the envelope of a modulated signal and the meters which are used with these sensors combine the attributes of a CW power meter with those of an oscilloscope. These sensors will be sensitive to the effects of harmonics which were previously discussed.

15.9 Effect of Multiple Reflections

The techniques used for rf and microwave power measurements generally rely on the ability of the power sensor to provide a matched (nonreflecting) termination for the transmission line. Ideally the source being measured is also matched. If either the source or the load is matched, the accuracy of a measurement can be obtained by considering the specifications of the sensor and power meter as described in Sec. 15.10. In general, neither the source nor the sensor provides a perfect match and these two reflections will interact to produce a measurement uncertainty which cannot be specified because the manufacturer has no knowledge of the source mismatch present during the measurement. If the user of the power meter is performing a detailed analysis of the accuracy of a given measurement, consideration must be given to the interaction of the source and sensor reflections.

The power reading displayed by the meter will be the true value P_0 if either the source or the sensor has a perfect match. If neither source nor sensor is matched, the reflections will interact and produce a power reading described by

$$P_{\text{meas}} = \frac{P_0}{|1 - \Gamma_g \Gamma_l|^2} \tag{15.12}$$

where P_{meas} = power displayed by meter
P_0 = power with a perfect match
Γ_g = complex reflection coefficient of source
Γ_l = complex reflection coefficient of load

The reflections produced by the source and sensor are often stated in terms

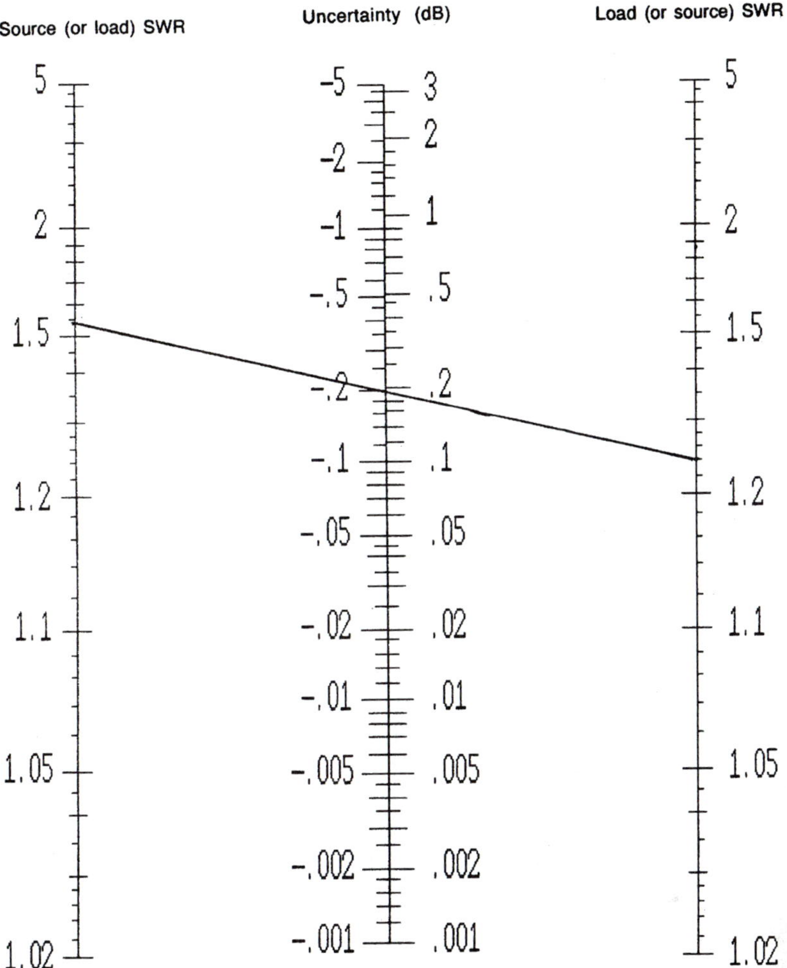

Figure 15.15 Multiple mismatch uncertainty limits.

of SWR (standing-wave ratio), and without knowledge of the phase of these reflections, Eq. 15.12 cannot be applied. A worst-case estimate of the effect of the interacting reflections can be made and the results should be added in as part of the overall accuracy which can be obtained from the power meter specifications.

Figure 15.15 may be used to perform this estimate and illustrates the uncertainty in dB resulting from the interaction of two reflections expressed as standing-wave ratios. In the example shown, a line drawn from an SWR of 1.54 (plotted on the left hand SWR scale) to a 1.24 SWR (found on the right hand SWR scale) intersects the uncertainty scale at approximately ±0.2 dB. In terms of percentage, the desired power P_0 may be 4.7 percent higher than the display value or 4.5 percent lower. This uncertainty range must be added to the other specifications for the power sensor and meter in order to determine the overall accuracy of the measurement.

15.10 Specifications

A set of specifications for power meters includes information which relates to the characteristics of the sensor as well as the meter, and they may be specified separately.

Specifications relating to the power sensor include:

1. *Frequency range.* The input signal must be within the specified frequency range of the sensor. It is not good practice to assume that the sensor will continue to provide accurate measurements of signals outside of its specified frequency range. Sensors can cover very wide frequency ranges but they are more costly and may use connectors which are more subject to wear and breakage than those used with sensors operating at lower frequencies.
2. *Maximum applied power.* This specification should be carefully observed when the sensor is being used. The sensor will be damaged or destroyed by excessive power.
3. *SWR (standing-wave ratio).* One should always seek the lowest possible SWR. This will minimize the uncertainty in the measurement which is caused by the interaction of the sensor's reflection with other reflections in the measurement system.
4. *Calibration accuracy.* The accuracy of the calibration data is a direct contribution to the overall measurement accuracy. Sometimes it is necessary to request (and pay for!) a more accurate calibration if the sensor is to be used for a critical measurement.
5. *Response time (peak power sensors).* There is a tradeoff between the response time of a peak power sensor and the lowest frequency of operation. The lower the operating frequency, the more slowly it will respond.
6. *Coupling factor (in-line power measurements only).* The value of the coupling factor of an external coupler must be known in order to correct the power reading for its presence (Eq. 15.9). The accuracy of this value contributes directly to the overall measurement accuracy.
7. *Directivity (in-line power measurements only).* The directivity specification indicates the degree to which the directional coupler separates the incident wave from the reflected wave and is generally specified in terms of dB loss. The directivity is another specification which needs to be considered when determining the overall accuracy of the measurement. To minimize errors produced by imperfect separation, one should always attempt to use couplers having high directivity.

8. *Loss (in-line power measurements only).* The power absorbed by the coupler will deduct from the power delivered to the load. If the amount of loss is known, the reading of the power meter may be corrected in a manner similar to that used for the coupling factor.

Meter-related specifications include:

1. *Instrumentation accuracy.* This specification relates to the stability of the gain and accuracy of the range changes in the meter amplifiers and is only one of several factors which determine the overall accuracy of the measurement.
2. *Zero drift and noise.* Both of these specifications influence the overall accuracy of the measurement and apply mainly when the meter is operated on its most sensitive range. Using a more sensitive power sensor may help minimize the contribution of these specifications.
3. *Dynamic range.* This specification defines the maximum to minimum range of power levels which can be measured. The maximum level is sensor dependent, and the minimum level is governed by the effects of drift and noise.
4. *Response time.* This specification indicates the processing time required by the meter to update the displayed power. Typically the response time increases on the lowest power range of the meter, and this might be improved by use of a more sensitive power sensor.
5. *Rise time (peak power meters).* The rise time of a peak power meter is the fastest change in power which can be measured. The rise time of the peak power meter sets a lower limit on the overall rise time of the measurement system.

The wide variety of power meters and sensors available today make it impractical to quote typical values for each specification. An examination of the specified quantities will only give an indication of the accuracy of a given power measurement. Other effects such as mismatch interaction and losses external to the power meter can be dominant factors which influence the true accuracy. Other common problems which influence the accuracy are sensor damage caused by connector wear, excessive power, or mechanical shock. It is good practice to periodically perform a calibration on the power meter and its sensor to verify their performance relative to the specifications.

15.11 Calibration

Power meters are calibrated by comparison of the reading obtained from the unit under test with that obtained from a more accurate instrument. Low-frequency power meters may be calibrated by using a precision load and an accurate voltmeter or ammeter. High-frequency power meters are generally compared against a transfer standard which is a power sensor whose calibration is further up the traceability chain to national standards. Facilities like the U.S. National Institute of Science and Technology use specialized instrumentation to calibrate the reference standards used by industry. The calibration of a reference standard is then carried over to the transfer standards used for day-to-day calibrations.

The thermistor sensor is often used as a reference or transfer standard. Since the thermistor relies on rf to dc substitution, the error associated with the meter circuits can be minimized by direct measurement of the dc substitution power. Transfer standards are also used to calibrate the output of the power reference contained in thermocouple and diode power meters. It is good practice to also perform a check of the SWR during the calibration process.

Chapter

16

Oscillators, Function Generators, Frequency and Waveform Synthesizers

Charles Kingsford-Smith
Agilent Technologies
Lake Stevens, Washington

16.1 Introduction

This chapter covers the more commonly used types of signal source instruments introduced in Chap. 7. It surveys operating principles of these types and provides some guidance in understanding the specifications provided by manufacturers of these instruments.

16.2 Sine-Wave Oscillators

As used here, the term *sine-wave oscillators* refers to oscillator circuits which naturally produce sinusoidal waveforms. These circuits consist of an ac amplifier with a positive feedback path from output to input. In the feedback path is a filter network: Its input-output (transfer) gain is very low except close to the desired frequency of oscillation. Hence the ability of the circuit to operate as an oscillator (that is, to operate regeneratively by producing its own amplifier input) is restricted to a narrow band around this frequency.

16.2.1 Radio-frequency (RF) signal generators

This important class of instrument began in the late 1920s when the need was recognized for producing radio receiver test signals with accurate frequencies (1 percent) and amplitudes (1 or 2 dB) over wide ranges. The basic block diagram of these instruments remained nearly unchanged until about 1970, when frequency synthesizer circuits began to usurp "free-running" oscillators in the waveform generator section of the instruments.

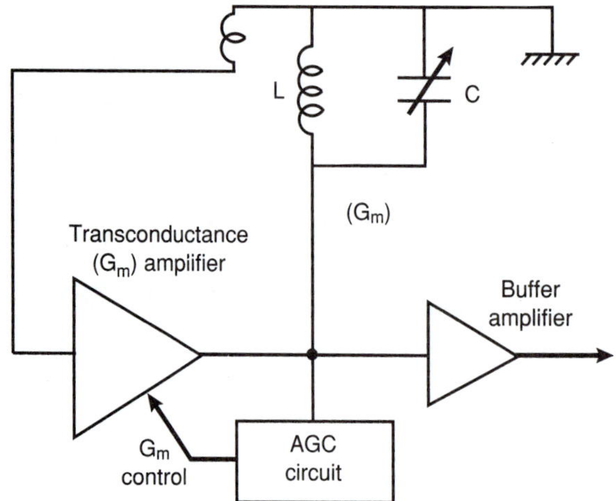

Figure 16.1 Radio-frequency sine-wave oscillator.

Figure 16.1 shows a simplified form of a typical rf sine-wave oscillator used in a signal generator. An amplifier producing an output current proportional to its input voltage (transconductance amplifier) drives a tunable filter circuit consisting of a parallel *LC* resonant circuit with a magnetically coupled output tap. The tap is connected to the amplifier input with the proper polarity to provide positive feedback at the resonant frequency of the filter. To assure oscillation, it is necessary to design the circuit so that the gain around the loop—the amplifier gain times the filter transfer function—remains greater than unity over the desired frequency tuning range. But since the output signal amplitude stays constant only when the loop gain is exactly unity, the design must also include level-sensitive control of the loop gain. Such a level-responding circuit is *nonlinear,* because it does nothing until the amplifier output signal reaches some predetermined amplitude, rather than responding proportionately, like a linear circuit. In the oscillator shown here, the automatic gain control (AGC) circuit provides the level control function. The AGC monitors the sine-wave amplitude at the amplifier output and acts to smoothly decrease its gain when the desired amplitude is reached.

This oscillator circuit produces a low level of harmonics in its output. This is because the sine-wave amplitude can be kept within the linear operating range of the amplifier by the AGC circuit. Many other oscillator circuits depend on amplifier saturation as the gain-limiting mechanism, and this generates a good deal of harmonic energy, which degrades the signal purity.

16.2.2 Audio oscillators

At audio frequencies there is more emphasis on purity of waveform than at rf, so it is natural to use an oscillator circuit which generates a sine wave as its characteristic waveform: an amplifier with filtered feedback. For audio

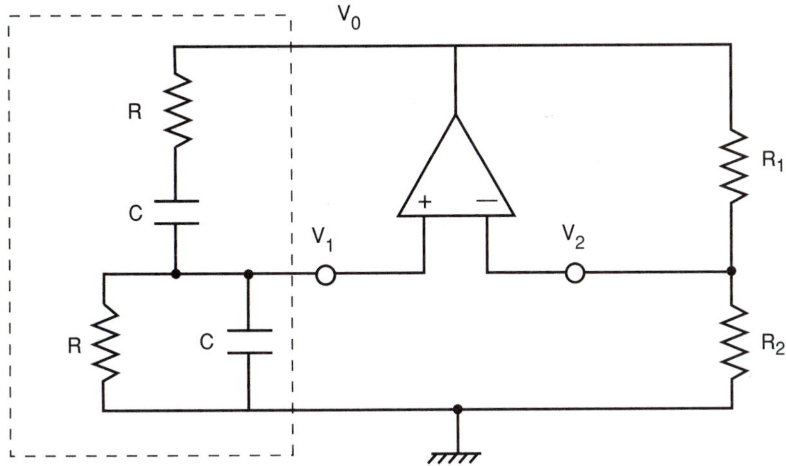

Figure 16.2 Wien bridge oscillator.

frequencies, however, the elements of an *LC* resonant circuit for the filter become expensive and large, with iron or ferrite core construction necessary to get the large values of inductance required. Another handicap is that ferrocore inductors have some nonlinear characteristics which increase harmonic output when they are used in a resonant filter circuit.

Using resistors and capacitors, it is possible to get a voltage transfer function which resembles that of a resonant circuit in both its phase and amplitude response. The four-element *RC* network in the dashed portion of Fig. 16.2 is characterized by an input-output voltage ratio V_1/V_0 (called a transfer function) which varies with frequency. This transfer function has an amplitude peak of ⅓ V_0 at the "resonant" frequency $1/(2\pi RC)$ and a phase shift passing through zero at this frequency. The value of Q, a measure of the bandwidth of the transfer function, is only ⅓, normally a poor value for use in a feedback oscillator. (Low resonant circuit Q means frequency drift and noise in the output signal.) However, if R_1 and R_2 are adjusted so that their transfer function V_2/V_0 is about the same as V_1/V_0 at resonance (that is, ⅓), then the combined transfer function $(V_1 - V_2)/V_0$ exhibits a much sharper (higher Q) character. This topology, with resistors R_1 and R_2 added to the four-element network, is now a bridge circuit, and it was called the "Wien bridge" by William R. Hewlett. If a high-gain amplifier is connected as shown, oscillations can be generated at the resonant frequency, provided the bridge is just slightly unbalanced to provide net positive feedback for the amplifier.

The frequency of the Wien bridge oscillator is tuned by using ganged variable capacitors for the two *C*'s. Since a 10 : 1 capacitance ratio is readily obtained in a mechanical capacitor, this provides continuous tuning over a decade of frequency. This is, in itself, a decided advantage over *LC* resonant circuit tuning, which exhibits a square-root variation of resonance vs. capacitance tuning. Decade band switching is provided by switching the R's in pairs.

As always with feedback oscillators, once oscillation starts, it is necessary to stabilize the loop gain at unity to provide a constant sine-wave amplitude. This problem was solved by replacing R_2 with a nonlinear resistor, such as a small Christmas-tree light bulb. Since tungsten has a positive temperature coefficient, an increasing sine-wave amplitude causes the bulb resistance to increase, thus increasing the negative feedback in the amplifier. Because the filament thermal time constant is a second or more, its resistance does not change over the much shorter time of one cycle of the sine-wave signal, and so the filament resistance contributes negligible distortion to the output signal.

16.2.3 Performance and specifications

Since there are no ideal signal sources producing perfectly stable, noiseless waveforms, it is necessary to describe the actual performance of instruments so that the user knows what (and what not) to expect. Several parameters, because of their utility, have become standardized for describing the quality of waveforms produced by signal sources in quantitative terms. These parameters are discussed in this and similar sections in this chapter.

Frequency accuracy. The frequency of rf and audio oscillators is usually controlled by mechanically varied capacitors. The relationship between frequency and capacitance is recorded as a dial calibration and can be read and set to better than 1 percent. However, other factors usually decrease this accuracy: Manufacturing tolerances, component aging, and ambient temperature are the main ones.

Amplitude accuracy. The amplitude of the output waveform is usually expressed in volts or dBm (decibels referred to 1 mW of power). Specifications for amplitude vary widely; as might be expected, more expensive instruments usually come with more precise numbers. Rf signal generators, which often cost more than audio sources, also usually have better specifications. Some signal generators have a meter for reading the output amplitude, and a vernier control to set it to a standard level. Other generators have a digital readout and an automatic circuit for keeping the amplitude constant. There is usually a switchable attenuator to reduce the output level in 10-dB (or smaller) steps. The accuracy of the output level degrades over a wide frequency range (this is called "flatness" on data sheets) and with larger values of attenuation. Both effects are mainly due to parasitic impedances in the attenuator. Output levels are specified with a standard resistive load placed at the generator output. This is usually 50 Ω for rf sources but may be 75 Ω if the source is to be used in video applications.

Typically, audio oscillators also have a step attenuator in the output, but levels between the steps are set with an uncalibrated potentiometer. Both flatness with frequency and attenuator accuracy are usually specified. The standard load is 600 Ω.

Frequency stability. The ideal sine wave is noiseless and absolutely constant in both frequency and amplitude. Nonideal oscillators, of course, depart from this, and *stability* is a measure of the degree to which they approximate the ideal. Stability is further classified as *long-term* and *short-term,* depending on whether significant frequency change occurs on a time scale of minutes or fractions of a second. An example of long-term frequency stability is "warmup drift," a change in frequency usually caused by dimensional change in inductors and capacitors as they are heated over minutes or hours by circuit power dissipation. Such stability is specified in terms such as "less than 10 kHz change 30 minutes after power-on." On the other hand, short-term stability is affected by physical factors which have higher frequency content, such as random noise, power supply ripple, and microphonics. These lead to more rapid frequency variations of the output signal. The effects of short-term factors are combined on an rms basis and then expressed as frequency modulation (FM) of the oscillator. A typical example of such a specification is "hum and noise less than 10 Hz rms deviation."

Harmonic distortion. For a signal trying to be a perfect sine wave, one measure of how closely it approximates the ideal is its harmonic distortion. A "harmonic" is a sine wave having a frequency which is an integer multiple of the basic signal frequency. If a sine wave is accompanied by some of its harmonics, the resultant waveform is no longer sinusoidal; that is, it is *distorted.* As the number and the amplitude of harmonic terms increase, so does the distortion. The amount of harmonic distortion is usually specified in one of two ways. Less common is to give an upper bound for any harmonic term; e.g., "all harmonic terms less than −80 dBc" means every harmonic is no larger than 80 dB below the "carrier" or output signal. A weakness in this specification is that there can be a large number of harmonic terms and hence a lot of total harmonic energy. More commonly, the manufacturer combines the energy of all harmonics and expresses the rms value as a percentage of the fundamental, such as "harmonic distortion less than 0.5 percent." In the latter case, sometimes residual noise in the output is added to the harmonic energy to give a noise-plus-distortion specification.

16.3 Function Generators

This term is used to describe a class of oscillator-based signal sources in which the emphasis is on *versatility.* Primarily, this means providing a choice of output waveforms. It also includes continuous tuning over wide bands with max-min frequency ratios of 10 or more, sub-Hz to MHz frequencies, flat output amplitude, and sometimes modulation capabilities: frequency sweeping, frequency modulation (FM), and amplitude modulation (AM). In their frequency accuracy and stability, function generators are inferior to sine-wave oscillators, but their performance is quite adequate for many applications.

Some frequency synthesizers (Sec. 16.4) are also called "precision" function generators by their manufacturers. What this means is that circuitry has been added to the synthesizer to produce other waveforms in addition to sine waves.

16.3.1 Threshold-decision oscillators

To realize the versatility described above—especially the wide frequency tuning range—function generators nearly always use a basically different oscillator mechanism than filtered feedback around an amplifier. This is the threshold-decision oscillator. Section 7.3.1 has additional discussion on this oscillator.

The threshold-decision oscillator requires three basic elements:

1. A circuit whose state (voltage, current, etc.) changes with time
2. A way to reset this circuit to an initial state
3. A method for deciding when to perform the reset

Figure 16.3*a* shows a simplified form of such an oscillator. An *RC* circuit charges from a positive supply. The time-changing state of this circuit is the voltage across the capacitor. A comparator monitors this voltage and, when this reaches a reference level (the decision criterion), momentarily closes the switch. The switch discharges the capacitor, restoring the *RC* circuit to its initial state, and the cycle restarts. Figure 16.3*b* shows a couple of cycles of the oscillator waveform.

The typical function generator oscillator circuit, shown simplified in Fig. 16.3*c,* is a little more complex than the circuit of Fig. 16.3*a* in order to gain some versatility. Two current sources $i+$ and $i-$ are available to charge and discharge capacitor *C*. A switch, under control of a bistable flip-flop, determines which current source is connected. The voltage across *C* is monitored by two comparators. Assume that $i+$ is initially connected to *C*. This causes its voltage to rise linearly. This property is important, since many applications require a linear ramp waveform. When the voltage reaches the high reference, the upper comparator responds, resetting the flip-flop. This actuates the switch, connecting $i-$ to *C,* which then begins to discharge linearly. When the voltage of *C* reaches the low reference, the lower comparator responds, resetting the flip-flop and beginning a new cycle.

One cycle of oscillation consists of two linear voltage segments across *C*. If the current sources are the same magnitude, the slopes of these segments will be of equal magnitude but opposite sign. The voltage across *C* will therefore be a symmetrical triangle waveform. The frequency of oscillation can be

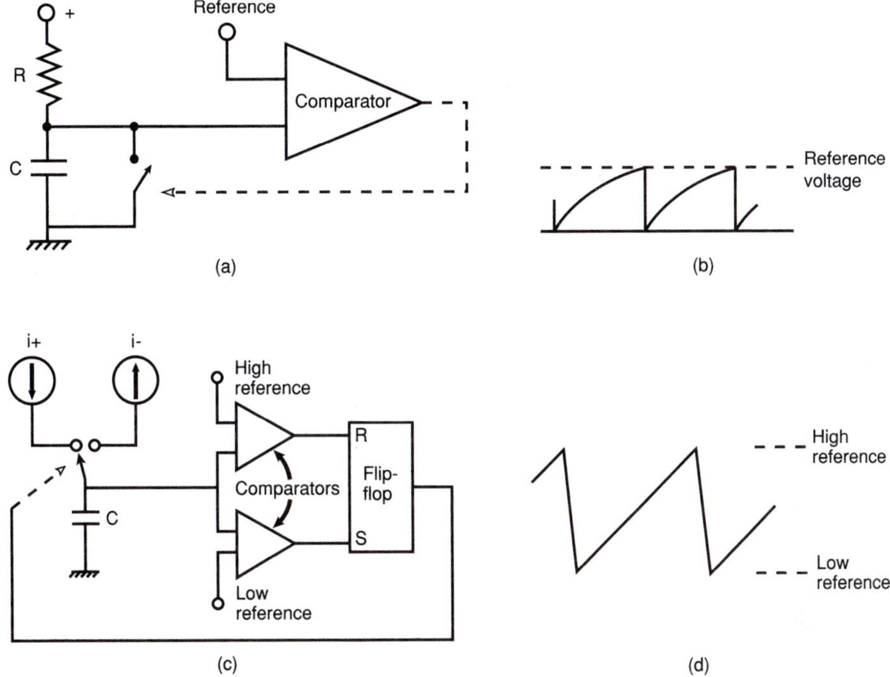

Figure 16.3 Threshold-decision oscillator. (a) Simple circuit; (b) waveform across C in simple circuit; (c) function generator version; (d) asymmetrical triangle produced in function generator.

computed as the inverse of the period, and the period is the sum of the durations of the positive and negative segments. For the symmetrical case, this is

$$f_{\text{osc}} = \frac{1}{2T} = \frac{i}{2C(V_h - V_l)} \tag{16.1}$$

where f_{osc} = frequency of oscillation
T = time duration of either segment
i = charge (and discharge) current
C = capacitor, farads
V_h = high reference voltage
V_l = low reference voltage

The triangle waveform is one of the basic waveforms generated by this oscillator. The other is the square wave at the output of the flip-flop. Narrow pulses are also available from the comparators, but these are seldom used.

Varying the frequency of this oscillator is simple. Since, from Eq. 16.1, the frequency of oscillation is proportional to the charging current, both current sources are designed so that their magnitudes vary linearly with a common control voltage. This makes the oscillator frequency directly proportional to this voltage. In practice, continuous frequency control is obtained by a potentiometer which produces the control voltage, while frequency bands are switched by changing the value of C in decade steps.

The linearity of the triangle waveform segments is directly related to the quality of the current sources: how constant their current remains as the voltage across C varies.

Varying the duty factor of the output waveforms is just a simple matter of varying the ratio of $i+$ to $i-$. Duty factors as asymmetrical as 100:1 are readily achievable. A little design effort is required to maintain a constant waveform frequency while varying the duty factor: A nonlinear relation is needed between the increase in one current source and the decrease in the other. Figure 16.3*d* shows the waveform across the capacitor when the magnitude of $i-$ is 10 times that of $i+$.

16.3.2 Producing sine waves with a function generator

From this description of a function generator oscillator, it is evident that triangle waves and square waves are its natural or intrinsic waveforms, while sine waves are not. Since a versatile source should include the most basic signal—a sine wave—in its repertoire, it is necessary to create sine waves from the waveforms which *are* present.

One possible method would be to select the fundamental from the triangle waveform, using a low-pass or bandpass filter. However, this is impractical, because such a filter would need to be *tunable* to track the wide tuning range of the oscillator. Instead, what is usually done is to apply a triangle wave to the input of a nonlinear device whose *transfer function* approximates a sinusoidal shape. A big advantage of this approach is that such a device is frequency-independent, thus doing away with the need for filtering. Distortion in the output waveform is determined by the accuracy of the approximation to a true sinusoidal transfer function.

A practical method of implementing such a transfer function is shown in Fig. 16.4*a*. This circuit effects a piecewise-linear approximation to a sinusoidal transfer function. As the voltage of the incoming triangle waveform departs from zero toward its positive peak, it turns on successive biased diodes, beginning with D_1. As each diode conducts, it connects its associated resistor across the bottom leg of the voltage divider formed by these resistors and R_{in}. The value of the voltage division is the transfer function. Consequently, as more diodes conduct and connect their resistors in parallel, the resistance of the bottom leg of the divider decreases, and so does the value of the transfer function. The same action occurs on negative excursions, with D_1' being the first diode to conduct. The result of these actions is a transfer function composed of a series of straight-line segments. Bias voltages controlling diode conduction points are produced by the bottom row of resistors. The R_i and the bias voltages are chosen to cause the straight-line segments to approximate

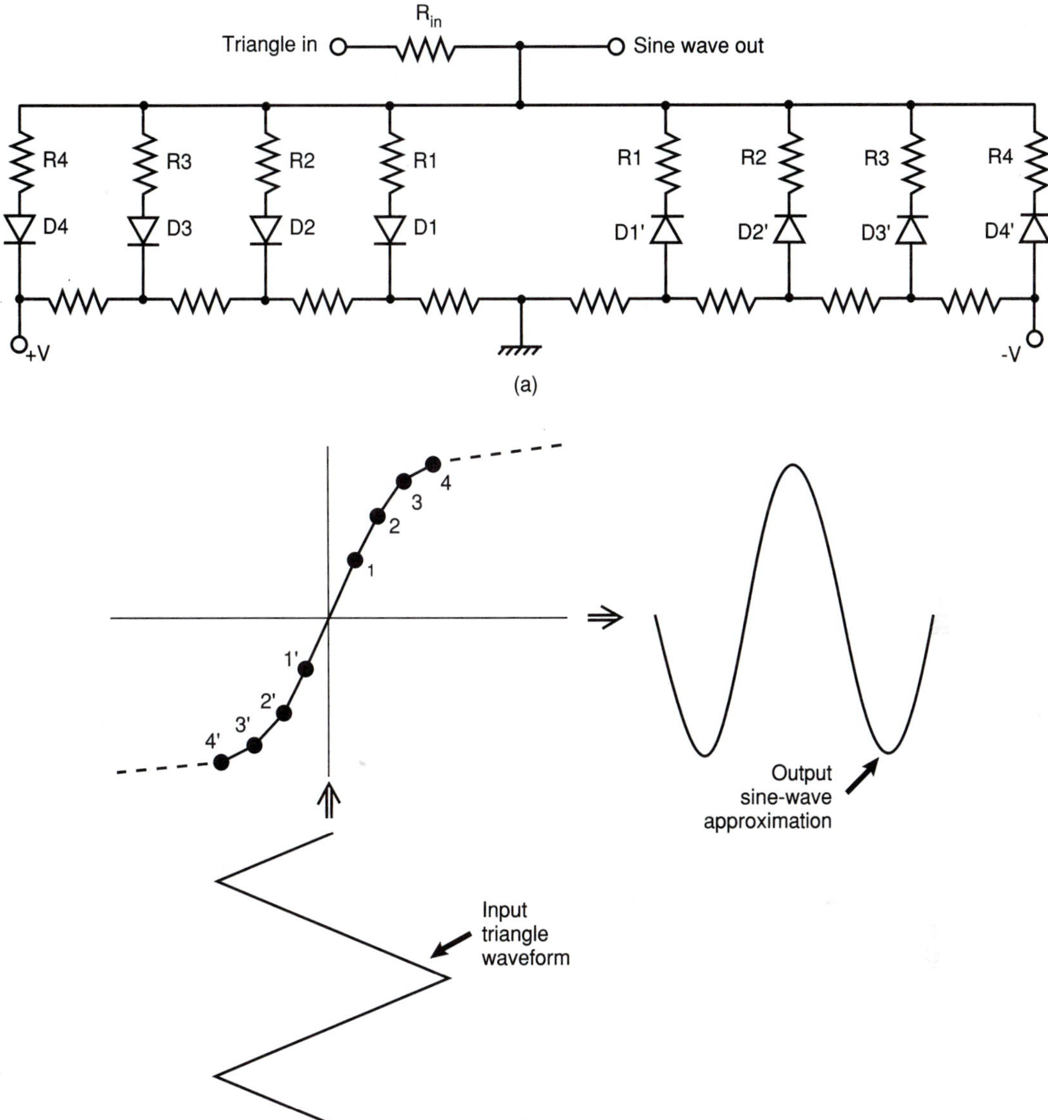

Figure 16.4 Sine waves from triangle waves. (a) Converter circuit; (b) converter piecewise-linear transfer function.

a sinusoidal waveform, as seen in Fig. 16.4*b,* where the dots indicate the breakpoints, or divisions between segments. Only four breakpoints per half waveform are shown here for simplicity; usually six are used. With the proper placement of six breakpoints, distortion of the output sine approximation can be as low as 0.25 percent.

16.3.3 Modulation

Most function generators provide a terminal where the user can apply a voltage to vary the frequency of the generator. The user's voltage is added to the frequency control voltage coming from the front panel potentiometer. This is a convenient way to produce frequency modulation (FM). The quality of the resultant FM will be directly related to how linearly the oscillator frequency responds to the control voltage, since nonlinearity will produce distortion in the modulation.

A special case of FM is frequency sweeping. The ability of function generators to sweep frequency makes possible some interesting applications. With an oscilloscope and a function generator, one can make a rudimentary network analyzer (Chap. 28), an instrument used to display the frequency transfer functions of filters and other circuits. One connects the frequency control terminal to the horizontal scanning waveform of an (analog) oscilloscope. This is an asymmetrical triangle wave, much like Fig. 16.3*d*. By doing this, the scanning waveform causes the frequency of the function generator to track with the horizontal position of the scope trace. In turn, this maps a linear frequency scale onto the *X* axis of the scope. By then connecting a test circuit between the function generator signal and the vertical scope input, one may view (repetitively) the amplitude vs. frequency characteristic of the test circuit.

Amplitude modulation (AM) is a little harder to produce in these oscillators than FM. Simultaneously increasing the switching thresholds *and* the charging currents will vary the amplitude of the triangle waveform, but there is likely to be a lot of simultaneous FM unless this is precisely done. A better method is to apply the oscillator output signal to an amplifier whose gain may be varied with a control signal. Because of this additional complexity, AM is usually included only in the more expensive instruments.

16.3.4 Specifications

Frequency accuracy of function generators is usually defined in terms of frequency dial accuracy, and 5 to 10 percent of full scale would be representative of many sources. Residual FM is usually not specified. It is higher than that of a feedback oscillator, but this fact is not too important for the usual applications of function generators. On the other hand, amplitude flatness is quite good. Over the full range of the instrument, it is usually better than 1 dB. However, absolute amplitude accuracy is usually not given, and only an uncalibrated control regulates the output level. For sine waves, distortion is specified, usually as an rms percentage of the amplitude of the fundamental.

Rise and fall times of pulse waveforms are nearly always given. Linearity of triangle waveforms—that is, the deviation of the segment voltages from a straight-line shape—is less frequently specified. However, deviations of less than 1 percent are common, especially in the lower-frequency ranges of an instrument.

16.4 Frequency Synthesizers

Starting in the early 1960s, frequency synthesis techniques have grown steadily in instrument applications and have become the dominant technology in signal sources. Synthesis refers to the use of a fixed-frequency oscillator called the *reference oscillator* or, sometimes, *the clock.* The latter term is borrowed from computers and refers to the role of this oscillator in pacing or synchronizing all the other circuits in the instrument. This is usually a precision crystal oscillator with an output at some cardinal frequency, such as 10 MHz. Various signal processing circuits then operate on the reference signal (or are paced by it) to produce a large choice of output frequencies. Every possible output frequency is derived from the reference oscillator frequency by multiplying its frequency by the fraction m/n, where m and n are integers. The integer m, and sometimes n, are supplied by the user via the front panel keyboard or an equivalent input. Suppose the reference frequency is 10 MHz and n is 10,000. Then, by varying m, the user can generate a range of output frequencies spaced 1 kHz apart.

16.4.1 Direct synthesis

By assembling a circuit assortment of frequency dividers and multipliers, mixers, and bandpass filters, an output m/n times the reference can be generated. There are many possible ways to do this, and the configuration actually used is chosen primarily to avoid strong *spurious signals,* which are low-level, non-harmonically related sinusoids. Figure 16.5 shows one way to produce a 13-MHz output from a 10-MHz reference. The inputs to the mixer are 10 and 3 MHz, the mixer produces sum and difference frequency outputs, and the bandpass filter on the output selects the 13-MHz sum. Notice that another bandpass filter could have been used to select the 7-MHz difference, if that were wanted.

The principal advantage of the direct synthesis technique is the speed with which the output frequency may be changed. Against this are a variety of disadvantages which, together with better alternative techniques (see below), have caused direct synthesis to fall out of favor. It is very hardware-intensive, and therefore expensive. When switching frequencies, phase continuity is lost, unless certain constraints are placed on when to switch, such as only at exact millisecond intervals in the $m/10{,}000$ example given above. It is also very prone to *spurious* signals in the output.

16.4.2 Indirect synthesis

This name derives from the use of an oscillator other than the reference to generate the output. However, by placing the oscillator in a phase locked loop, its frequency is controlled so that its output is the desired m/n times the reference frequency.

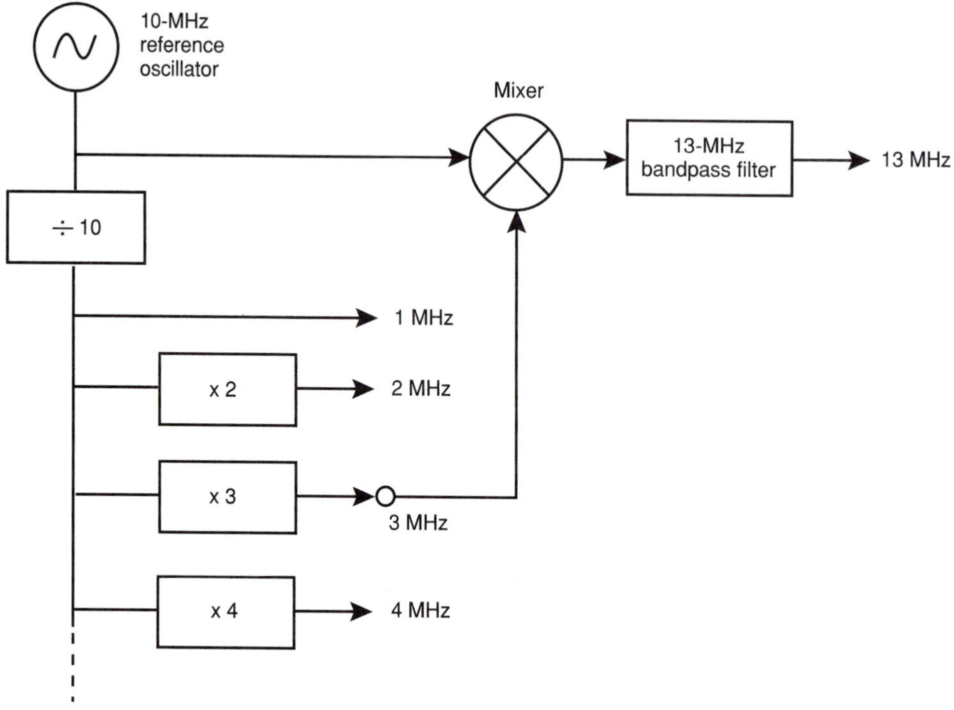

Figure 16.5 Direct frequency synthesis.

Phase locked loop synthesis. This most popular technique came into wide use with the availability of programmable divider ICs. These are counters whose count modulus (the number they reach before starting over) is externally programmable. With such a divider, a phase-locked loop (PLL) becomes a variable-modulus frequency multiplier. In addition, other PLLs act as bandpass filters, selecting the sum or difference frequencies in a mixer output. Figure 16.6 shows how one could synthesize the same output as the circuit of Fig. 16.5 by using PLLs. The reference is divided to 1 MHz, which is applied to the loop phase detector. With the variable modulus divider programmed to 13, the loop will stabilize when the voltage controlled oscillator (VCO) outputs exactly 13 MHz. Programming the divider to another number will cause the loop to lock to that number times 1 MHz. This observation reveals that the basic *frequency resolution*—the spacing between available output frequencies—is equal to the loop reference frequency, or 1 MHz in this case.

This fact opens up an interesting possibility: If we want a synthesizer frequency resolution of, say, 1 Hz, why not divide the reference oscillator to 1 Hz and use that frequency as the loop reference? Of course, it would also be necessary to use a very large modulus of 13,000,000 in the loop divider in the above example, but this could be an LSI part of modest cost.

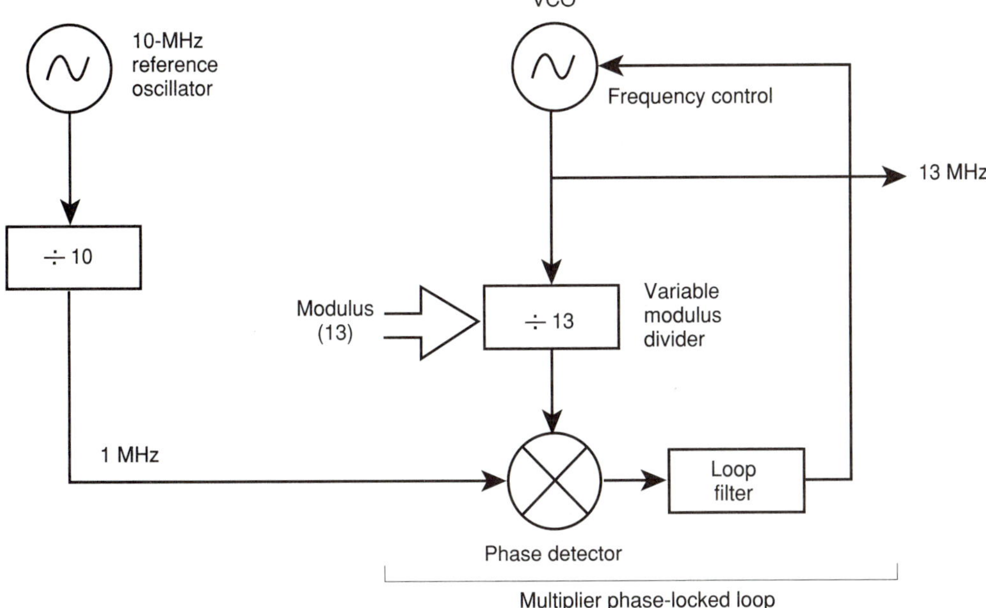

Figure 16.6 Indirect frequency synthesis.

Two important facts foreclose this possibility. One is that the loop frequency switching speed is of the order of 10 times the period of the loop reference, or 10 s in this case! The other is that frequency multiplier loops also multiply noise at the phase detector, which appears as noise sidebands on the oscillator output. This noise would be raised 142 dB in this example!

In practice, the maximum modulus used is a few thousand. When fine resolution is needed, sequences of multiplication, division, and addition are used, involving a number of loops. To introduce the idea of a multiple-loop synthesizer, Fig. 16.7 shows one scheme for synthesizing 13.1 MHz. The multiplier PLL is the same as the one in Fig. 16.6, only its modulus is 31, thereby producing a 31-MHz output. This is divided by 10, and the resulting 3.1 MHz is added to 10 MHz by the summing PLL at the right. Techniques like this can be extended to reach any desired resolution. Since multiplication numbers are low and the loop reference frequency is high, the output will have low noise sidebands and be capable of fast switching. All this occurs at a cost of greater circuit complexity.

Fractional-*N* synthesis. Various attempts have been made to achieve fine frequency resolution in one PLL while avoiding the penalties of poor noise and switching speed. A commercially successful technique which achieves this is known as "fractional-N" synthesis, after the fact that the PLL locks to a noninteger multiple of the loop reference. This multiple is written as $N \cdot F$, where

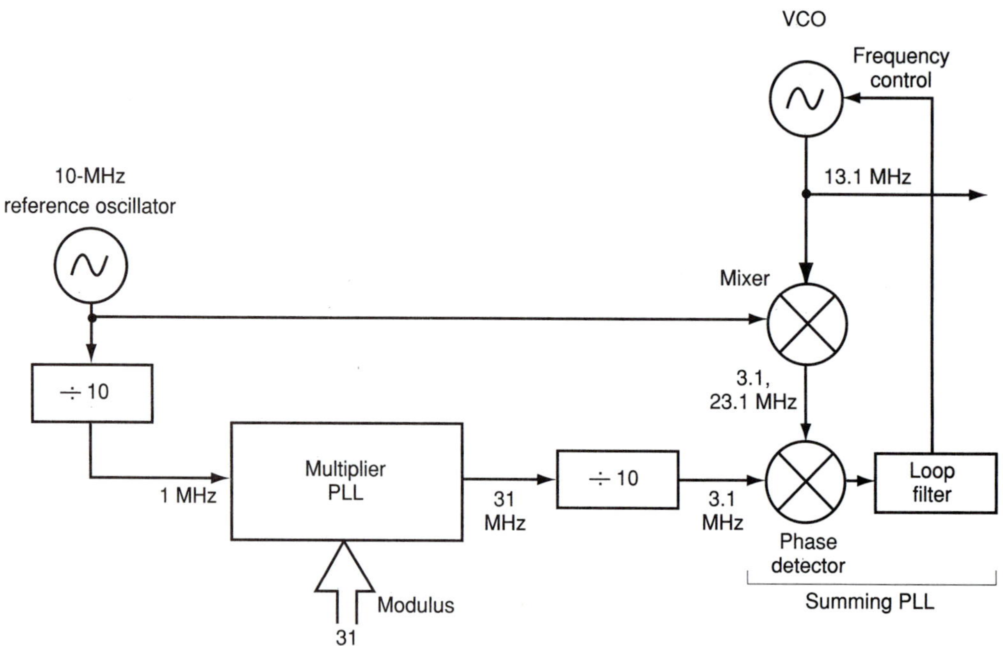

Figure 16.7 Multiple loop indirect synthesis.

N is an integer and F is a rational number less than 1. Using loop reference frequencies of 100 kHz and higher, fractional-N synthesizers achieve resolutions in the microhertz region and switching speeds of a millisecond or less.

To introduce this technique by using a simple choice of numbers, assume that 1.1 is the desired multiplier: that is, the PLL should lock at 1.1 MHz for a loop reference of 1 MHz. Figure 16.8*a* shows a block diagram of a fractional-N loop configured for this case. To explain how it works, it is useful to assume that the loop is *already* locked to 1.1 MHz, and then to examine its operation to see the conditions which make this true.

First, consider the phase detector. The divider is set to 1 (the value of N), so its output pulse rate is the same as its input, the VCO. Since there are 11 cycles of the 1.1-MHz VCO for every 10 of the 1.0-MHz reference, then, for each cycle of the reference, the VCO phase advances 0.1 cycle, or 36°, relative to the reference. The phase detector samples this phase difference every reference cycle, and so its output voltage increases positively, let's say, by a step corresponding to 36° every microsecond. Ordinarily, this condition would quickly drive most phase detectors out of range, since their range is, at most, 360°.

At the bottom of Fig. 16.8*a* is an accumulator. This is a decimal register clocked by the reference, to whose contents is added an amount F each cycle. The accumulator is scaled so that 0.999. . . is the maximum value it can

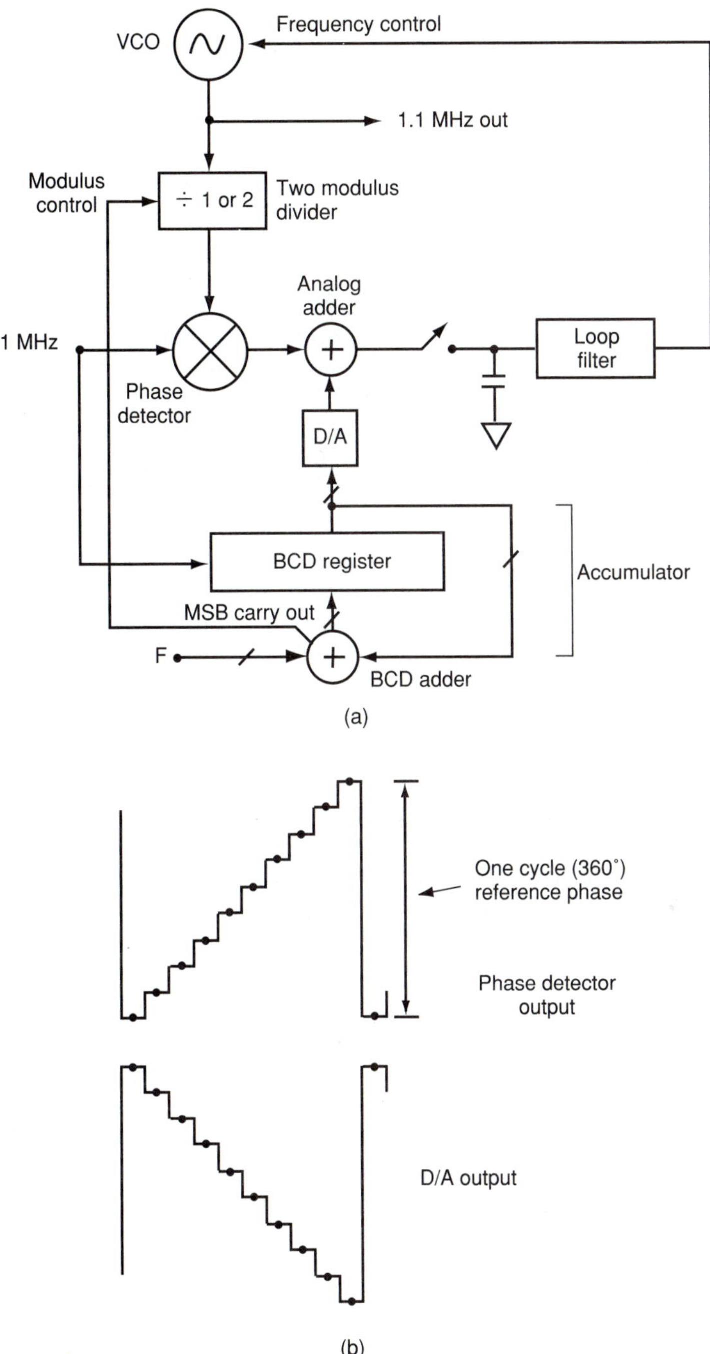

Figure 16.8 Fractional-N frequency synthesis. (a) Fractional-N circuit; (b) waveforms in fractional-N circuit.

hold. In the example, F has the value 0.1, so, beginning with zero, the accumulator will overflow on the tenth clock pulse. However, 10 clock pulses also correspond to the time for the phase detector output to change by a voltage equal to 360°.

Along with the accumulator overflow there is a carry output from the adder. This carry is connected to the divider and momentarily changes the divider modulus from 1 to 2—in effect skipping one cycle from the VCO. For the phase detector, the loss of one divider pulse is equivalent to abruptly retarding the VCO phase *by 360°*. So, instead of advancing another 36°, the phase detector output falls negatively an amount corresponding to 324° (0.9 cycle) at the same time that the accumulator contents change from 0.9 to 0.0.

Although the periodic pulse removal keeps the phase detector output within its linear range so that the loop can lock, the sawtooth waveform, shown in Fig. 16.8*b,* in the phase detector output is very undesirable, as it produces phase modulation sidebands on the oscillator. But, when this loop is locked to the offset frequency of 1.1 MHz, the *accumulator contents predict the phase detector sawtooth* which results from the frequency difference between the two phase detector inputs. By properly scaling and polarizing the output from a D/A converter connected to the accumulator, the D/A output waveform may be used to cancel the phase detector sawtooth. (See Chap. 7 for an explanation of D/A converters.) The two waveforms (shown in Fig. 16.8*b*) are added, sampled at times indicated by the dots, and filtered to provide the oscillator control voltage.

In the general case of a fractional-N synthesis loop, the integer part of the desired multiplier is supplied to the digital divider and the fractional part to the accumulator. The sudden removal of 360° of oscillator phase (as seen by the phase detector) is accomplished either by changing the modulus momentarily to $N + 1$ or by "swallowing" one cycle from the oscillator input to the phase detector. The accumulator may be made arbitrarily long, allowing very small frequency offsets (e.g., 1 μHz) to be used.

Limitations of indirect synthesis. The most important concern for a user of an indirect synthesizer is usually its behavior when a frequency change is commanded, that is, its frequency switching speed. Because PLLs are feedback control circuits, there is a finite response time when an operating change is imposed. For some applications, this is of little importance. For others, such as frequency hopping, the time required to change frequencies may be intolerably slow. It is highly advisable to compare carefully the synthesizer specifications with the needs of the proposed application.

16.4.3 Sampled sine-wave synthesis

Many technical workers are familiar with sampling theory, in particular the central precept of this theory: A waveform may be exactly reconstructed from a sequence of uniformly spaced samples of its values, provided the sampling rate (i.e., frequency) is at least twice that of the highest-frequency component in the waveform. Applying this to frequency synthesis means generating samples of a sine wave and interpolating among the sample values to obtain a smooth waveform, in accordance with sampling theory. Although the sine wave values could be computed as the signal is generated, in practice they are stored in a lookup table. Values of the sine-wave phase are used as the argument for the table lookup. The phase values are determined by an accumulator, similar to that in the fractional-N circuit, in which the phase increment used is a number proportional to the desired frequency. The numeric output from the lookup table must be converted to an electrical signal and smoothed for output, as the next section shows.

Advantages and limitations. There are great advantages to this approach. The most apparent is instantaneous frequency switching. This is possible because the size of the angle increment between table lookups may be changed instantaneously. For instance, a tabular increment of 36° would generate an output frequency of 0.1 times the sample rate. If the increment is changed to 72°, the output frequency becomes 0.2 times the sample rate. The waveform phase at which this happens may also be chosen arbitrarily. With enough computing ability to determine the lookup argument, various precise yet complex modulation formats may be generated, such as FM, PM, and linear sweeps.

Another advantage is that the frequency of a sine wave synthesized in this manner (point-by-point) is *accurate,* just as with frequency synthesis already described. This is due to two causes: first, the phase increment is usually accurate to 6 or 8 digits, and second, the time intervals at which new phase information is computed and the sine function evaluated are precise, coming from the precision source. Since a sine-wave frequency is defined in terms of the derivative of phase (time rate of change of phase), both of these quantities—time and phase—are accurately controlled in this technique.

The principal disadvantage of this technique is its limited frequency range. A fair amount of digital logic and arithmetic must occur in each sample. In practice, 2.5 samples per output cycle is about the minimum sampling rate. With this requirement, and using TTL logic to implement the circuits, the maximum output frequency is probably less than 10 MHz.

Typical block diagram. A typical, although simplified, block diagram of a sampled sine-wave synthesizer is shown in Fig. 16.9*a*. An accurate reference clock regulates the timing of a digital accumulator: that is, a register whose stored numerical value increases by an externally supplied constant value during each clock cycle. This constant represents the increase in the phase, during the time of a clock period, of the desired synthesized signal. Thus it is labeled "frequency constant" since, in the limit, frequency is the rate of change of the phase of a sine wave. The contents of the accumulator therefore correspond to the sample values of the phase of the desired signal, modulo 2π radians

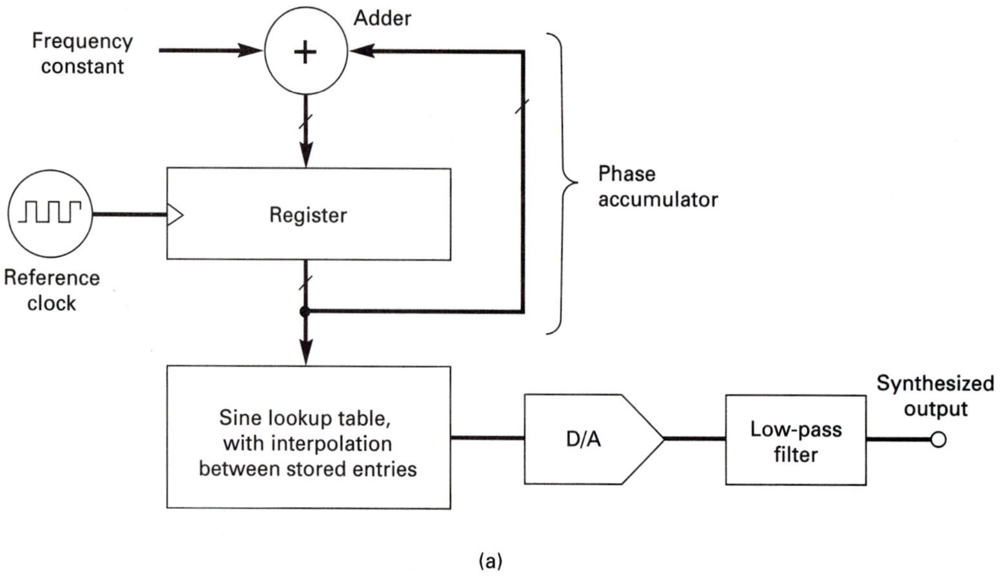

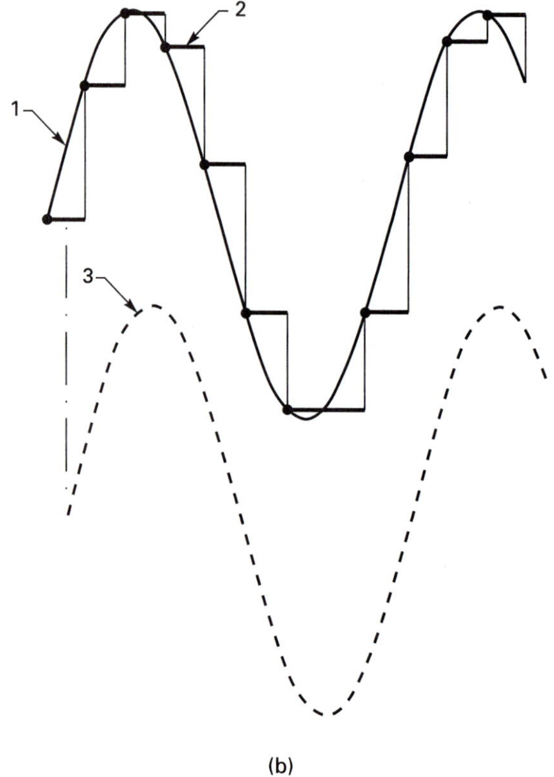

Figure 16.9 Sampled sine-wave frequency synthesis. (a) Block diagram of technique; (b) sampled sine waveforms: (1) ideal signal to be synthesized, (2) D/A output, and (3) filtered D/A output.

(the remainder after dividing the phase by 2π). These contents, representing phase, are supplied to the sine lookup table, which produces a number corresponding to the sine of its input. The sine number is then changed into a voltage by the digital-to-analog converter, whose output is a series of steps corresponding to the samples. Some form of output level control, such as an attenuator, is generally provided, although not shown in the figure.

Constructing the sine wave. Typical waveforms which are important in carrying out this process are shown in Fig. 16.9*b*. Waveform (1) represents the ideal sine wave to be synthesized. This waveform doesn't exist within the instrument, but sample values of it are computed and are shown by the dots along the sine wave. Note the uniform time axis spacing of these samples. The sample values are supplied to a D/A converter, whose output is a series of voltage steps, shown as a discontinuous waveform (2). The stepped waveform (2) is smoothed by an interpolator, which is just a low-pass filter whose cutoff frequency is less than half the clock rate. The sine-wave output from this filter is shown as waveform (3). Note that this reconstructed sine wave is delayed, with respect to waveform (1), by one-half the sampling interval; this is characteristic of the sample-and-hold process.

A couple of interesting details about constructing the sine lookup table are not evident in the simplified diagram. One is that the stored values of the sine function need only represent one quadrant (a fourth part of a cycle, or 0 to $\pi/2$ radians), because of sine-wave symmetry. Simple digital logic, including inverting and complementing functions, is used to generate the data for the other three quadrants.

Another detail concerns the resolution of the stored table. It is desirable to represent the frequency by six or eight decimal digits. Yet to store the sine function, even only one-fourth of it, to this degree of resolution would require a huge ROM or other storage device. In practice, the ROM input argument is only two or three digits, which are supplied by the most significant digits in the accumulator. The remaining resolution is computed by relying on the fact that linear interpolation of the sine function is quite accurate for small distances between arguments. The lookup circuit therefore uses the least significant digits from the accumulator and linearly interpolates between table entries. Such interpolation requires determining the slope of the function, but these data are already available: the derivative of the sine is the cosine, so a table lookup $\pi/2$ radians from the current position gives this number!

Limitations of sampled sine-wave synthesis. Being a sampled-data system, this synthesis technique is subject to the inherent limitations of such systems:

Quantization noise. This is caused by finite word-length effects. Usually enough resolution is provided in these synthesizers to maintain the noise floor comparable with that of other techniques.

Aliasing. For output frequencies close to half the sampling rate, the output low-pass filter may not adequately remove the sampling image ("alias"), which will appear as a spurious component.

Spurious components. A more serious problem in the output waveform is due to imperfections in the D/A converter. Signal degradation derives from two kinds of imperfections: level inaccuracies due to bit weight errors, and transient energy generated when the D/A changes levels. The latter problem can be minimized by adding a sample-and-hold circuit to the output. This is a circuit which assumes the output voltage of the D/A at a time when the levels are not changing, and does not "look" at the D/A during the time when they are changing.

16.4.4 Synthesized function generators

Signal sources with this name are usually indirect frequency synthesizers, and so their basic output signal is a sine wave. However, their designers have made an effort to approach the waveform versatility of classic function generators by adding circuitry to produce pulse and triangle waveforms, and sometimes modulation capability. The result is a most useful combination: a function generator with the frequency precision of a synthesizer.

A particularly interesting feature is the ability to generate a very precise ramp, which is a triangle wave with one segment of negligible width. This is achieved by supplying a linear phase detector with two accurate signals with a small frequency offset. For instance, if the two signals are 20.000 and 20.001 MHz, the detector output is a 20-MHz pulse train whose pulse widths are linearly increasing, returning to zero 1000 times per second. With a low-pass filter, a precise 1-kHz ramp is recovered whose linearity is superior to the best analog function generator. Figure 16.10*a* shows the block diagram of this technique, Fig. 16.10*b* shows the phase detector output before entering the low-pass filter, and Fig. 16.10*c* shows the linear ramp at the output of the low-pass filter.

16.4.5 Synthesizer specification

Since frequency synthesizers are complex instruments, a number of categories are required to characterize their performance and signal quality. Following are the most important of these.

Frequency range and resolution. Although manufacturers naturally herald the total frequency range of their products, it is worthwhile to examine the specifications carefully to see whether this range is covered in a single band or a series of contiguous bands. In the latter case, adjacent bands may have different characteristics such as noise. Likewise, when a frequency transition crosses the boundary between two of these bands, the output transient may be quite a bit larger than a normal transition which occurs for a transition

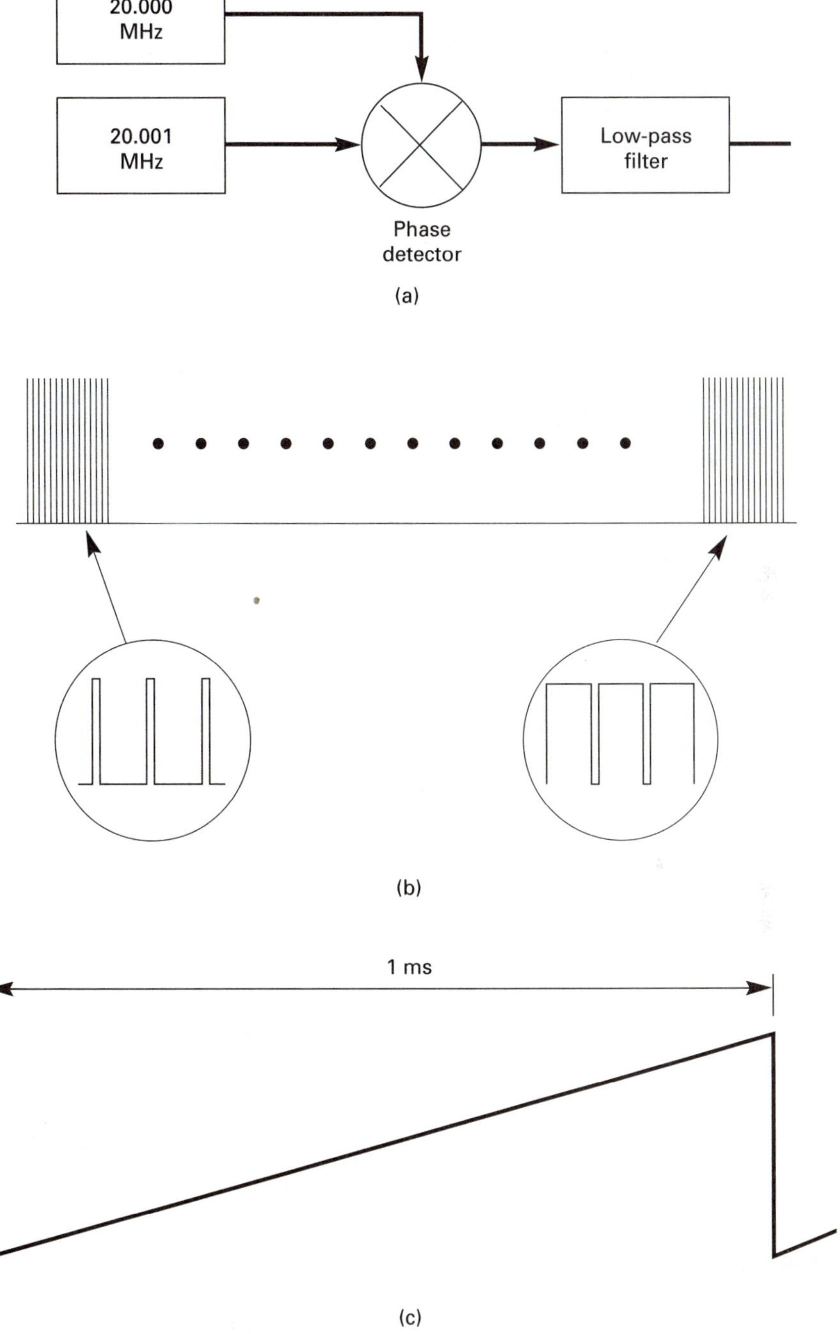

Figure 16.10 Precision ramp generator. (a) Block diagram of technique; (b) unfiltered phase detector output; (c) filtered output.

within a band. Resolution is most commonly a constant number, such as 0.1 Hz, over the entire frequency range. However, some commercial instruments feature a much smaller resolution, such as 1 μHz, over their lowest frequency ranges.

Frequency switching speed. This specification is a measure of the time required for the source to stabilize at a new frequency after the change command is issued. Since the majority of synthesized sources use phase-locked loops, the specification really reflects the loop transient response characteristics. Because these transients die off asymptotically, specifications often use nearness-of-approach language, such as "to within 100 Hz of the final frequency." Such numbers are usually in the order of tens of milliseconds, although some sources designed for fast switching claim switching times in hundreds of microseconds. As mentioned above, digital sine-wave synthesizers can change frequency nearly instantaneously, usually in just a few reference clock cycles.

Signal purity. This specification is most easily understood as applying to the frequency domain. It describes how well the output signal approximates the ideal single spectral line. There are two principal contributors which degrade this ideal: *phase noise* and *spurious signals* ("spurs").

Phase noise. This term is somewhat peculiar to synthesizers. It refers to the sidebands which result from phase modulation of the carrier by noise. AM sidebands from noise modulation exist also, but these are usually insignificant, because of the amplitude-limiting mechanism of oscillators. Phase noise is specified most simply as the total sideband power, in dB with respect to the carrier (dBc). In higher-performance instruments, the noise power is specified in the frequency domain, using frequency offset from the carrier as the abscissa. The specification is in the form of a loglog plot of noise power in dBc per Hz BW, or its verbal equivalent.

Spurious signals. These are non-harmonically related tones which exist in the output along with the desired sine wave. There are several sources of spurs, including unintentional coupling among circuits in the instrument and distortion products in signal mixers. An undesirable property of spurs is that they generally move in frequency at various rates as the signal frequency is changed, often crossing over the output. Such "crossover spurs" can't be removed by filtering. It is customary to specify spurious signals by declaring an upper bound for their amplitudes in dBc.

16.5 Arbitrary Waveform Synthesizers

By generalizing the sampled sine-wave synthesis technique to allow other waveforms to be stored in memory, a versatile signal source instrument can be achieved: the arbitrary waveform synthesizer. In this instrument, the user determines the shape of one period of the waveform and its repetition rate. It is easy to imagine how useful this ability would be in many applications. A good example is the testing of patient cardiac monitors by simulating various ECG waveforms to which the monitors must respond in a prescribed manner.

16.5.1 Principles of operation

The operation of the arbitrary waveform synthesizer is similar to that of the stored sine-wave synthesizer, in that stored digital sample values of the waveform are converted to sequential voltage values of a signal by a digital-to-analog (D/A) converter. However, there are a few distinctions between the two instruments.

Figure 16.11*a* shows a basic block diagram of an arbitrary waveform synthesizer. It is very much like that of the sampled sine-wave synthesizer, but there are a few differences, both in the hardware and in operation. The heart of the instrument is the random-access memory (RAM). This memory has a number of locations—for instance, 256—in which the user can store sequential amplitude values of a particular waveform to be generated. Only one period of the waveform is stored, as the example in Fig. 16.11*b*. This particular waveform is a combination of a sine wave plus its third harmonic, typical of a power transformer's magnetizing current. The sample values are assumed to be evenly spaced in time.

Provision must be made for the user to load the sample values into the RAM. In the simplest case, the user can enter the successive values from a keyboard. A more useful method, less prone to numerical mistakes, is for the user to draw the desired waveform on paper and then to trace over the waveform with a digitizer stylus which inputs the numerical values directly. Some more expensive instruments provide a graphical editor by which the user can construct and smooth a waveform.

Once the sample values are loaded into the RAM, they are ready to be stepped through at a rate which will produce the desired repetition rate (frequency) of the waveform. This is controlled by the size of the "frequency word" input to the phase accumulator, exactly like the sine-wave synthesizer. However, one difference is immediately apparent in the operation of interpolating between adjacent sample entries. For arbitrary functions, the slope of each line segment between entry values, which is needed for linear interpolation, must be computed rather than looked up as the cosine of that value. This point-by-point process can be made more efficient by enlarging the RAM, and computing and storing the slope values as the points are entered. Thus the slopes need be computed only once, after which they are looked up as needed.

As the interpolated RAM contents are output in sequence, the sample numbers are supplied to the D/A converter, which produces an output voltage proportional to each number. Hence the D/A output is not a smooth waveform but rather a stepped approximation, similar to Fig. 16.9*b*, but with many

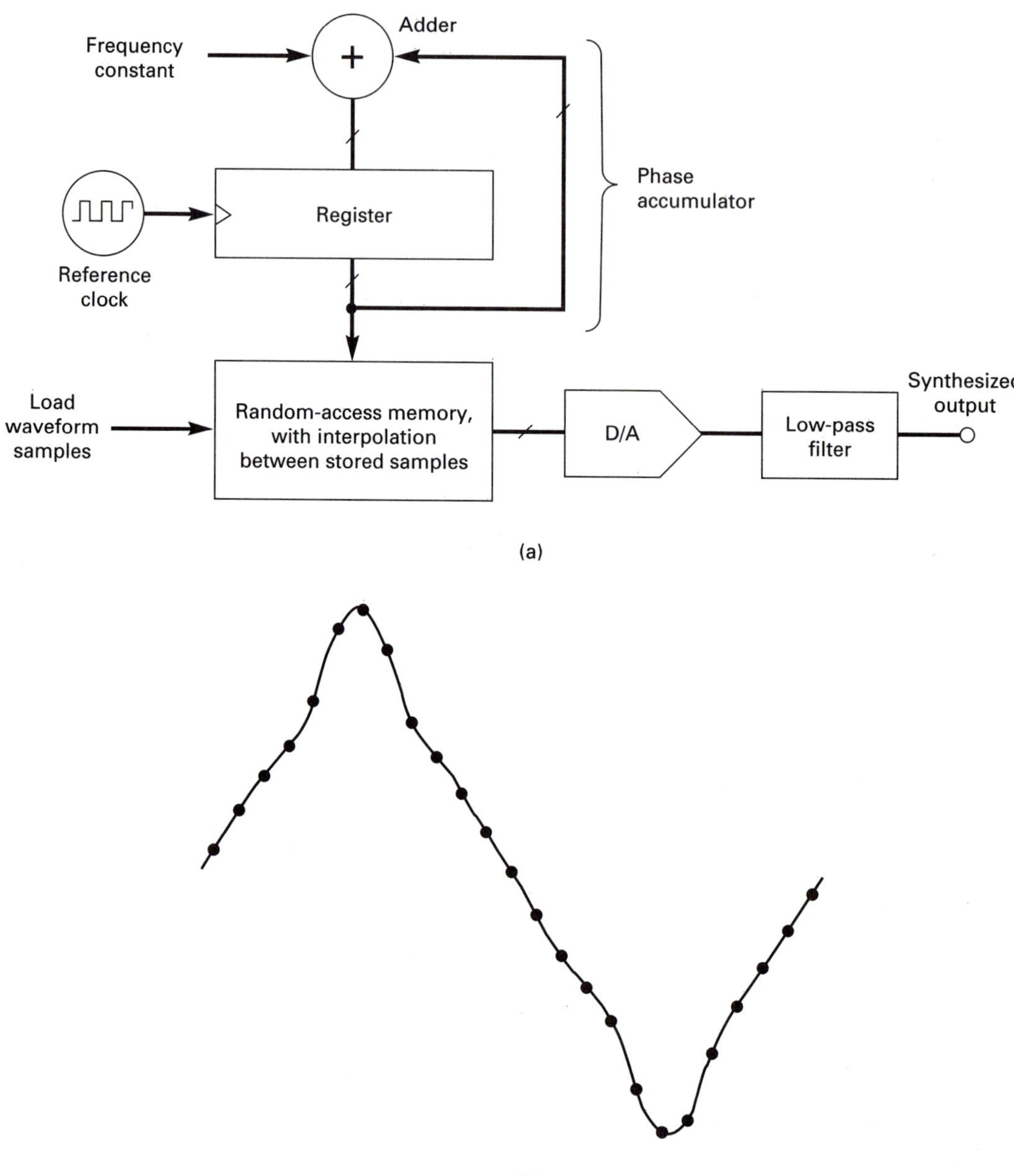

Figure 16.11 Arbitrary waveform synthesis. (a) Block diagram of technique; (b) typical arbitrary waveform, showing a few stored values.

more and finer steps. The steps, which contain high-frequency energy resulting from the sampling process, are removed by the low-pass filter, and the output from the filter is the desired smooth waveform.

It is apparent that the complexity of the waveform to be synthesized by this process is limited by the number of samples. With 256 samples per period as assumed here, fairly complex waveforms can be generated, but even so it is obvious that any fine waveform structure which might exist between samples will not be "seen" by the sampling process. In fact, the situation is actually more restrictive than this. As a working rule, a minimum of about three to four samples per cycle of the highest frequency in the waveform should be used. This rule of thumb derives from the need to filter the sampling image ("alias") from the output. The desired signal and the image approach each other as the signal approaches one-half the sampling frequency (that is, 2 samples per cycle). Therefore, if there are, say, 256 samples and the sampling (clocking) rate is 1 MHz, the highest frequency in the waveform should be about one-third to one-fourth of the sampling rate, or about 300 kHz.

Another way to state this limitation, which is basic and comes from the sampling theorem, is that the waveform to be synthesized must be band-limited. In theory, this limitation would eliminate synthesizing common waveforms, such as square waves, which are not band-limited. However, if the fundamental frequency of the waveform is low, enough harmonics can be included to generate a good approximation, usable for most purposes. A 1-kHz square wave, in the example above, would be represented by about 150 different frequency components and would have a rise time of about 1.5 μs.

16.5.2 Arbitrary waveform generator specifications

Because an ARB is a generalized form of the sampled sine-wave frequency synthesizer, the same limitations and types of specifications apply. In addition, it is important to know (1) the clock rate and (2) the number of stored samples. The ratio of these will determine the highest *fundamental* frequency of the synthesized waveform. For the numbers used in the discussion above, this would be 1 MHz/256, or about 4 kHz. It is possible to overcome this rather low frequency limit if provision is made for the instrument to use *fewer* samples per cycle. This, however, means synthesizing simpler (less complex and harmonic-rich) waveforms, since the upper frequency limit of about 300 kHz *for any component* cannot be exceeded.

Chapter

17

Pulse Generators

Andreas Pfaff
Agilent Technologies
Böblingen, Germany

17.1 Introduction

The first part of this chapter describes what a pulse generator basically is and explains a couple of terms that are often used in conjunction with these instruments or their applications.

17.1.1 Basic description

A pulse generator is an instrument that can provide a voltage or current output whose waveform may be described as a continuous pulse stream. A pulse stream is a signal that departs from an initial level to some other single level for a finite duration, then returns to the original level for a finite duration. This type of waveform is often referred to as a rectangular pulse or pulse stream.

A pulse generator may be used to stimulate a device, such as an integrated circuit (IC), a multichip module (MCM), or a passive component like a connector or a cable. In some cases, it can even stimulate a complete system, such as the clock distribution network of a personal computer (PC) or a workstation. The reactions of these devices, when properly measured, can be used to describe or specify many of their characteristics.

In order to achieve meaningful results, the suitability of the measuring device, along with the method of measurement, is as important as the suitability of the pulse generator. In most applications, the measuring device will be either a real-time or a sampling oscilloscope (see Chap 14), because oscilloscopes can conveniently display a wide range of voltage levels on one while representing time on the other axis. They also can easily perform automatic timing measurements as well as voltage measurements.

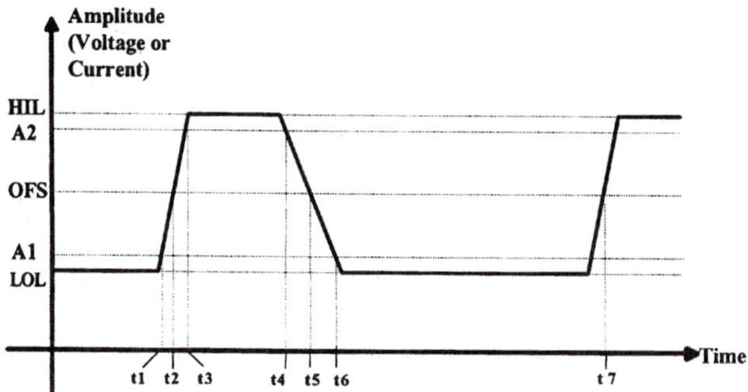

Figure 17.1 Ideal pulse nomenclature. HIL = high level (voltage or current); LOL = low level (voltage or current); AMP = HIL − LOL = amplitude (voltage or current); OFS = (HIL + LOL)/2 = offset (voltage or current); A1 = LOL + X∗AMP, A2 = HIL − X∗AMP, where X is defined as 0.1 (normally) or as 0.2 (for ECL devices). PER = t7 − t2 = pulse period; FREQ = 1/PER = pulse frequency; PWID = t5 − t2 = positive pulse width; NWID = t7 − t5 = negative pulse width; DCYC = PWID/PER = duty cycle; TRISE = t3 − t1 = rise time; TFALL = t6 − t4 = fall time. t1,t7 = rising edge crosses A1; t2 = rising edge crosses OFS; t3 = rising edge crosses A2; t4 = falling edge crosses A2; t5 = falling edge crosses OFS; t6 = falling edge crosses A1. Remarks: ECL stands for emitter-coupled logic. Rise and fall time are sometimes also referred to as transition times or edge rates.

17.1.2 Pulse nomenclature

The common terms that describe an ideal pulse stream are shown in Fig. 17.1. However, in order to describe *real* pulses, some additional terms are needed. These terms are explained in Fig. 17.2.

If a pulse generator offers more than just one channel, the outputs do not

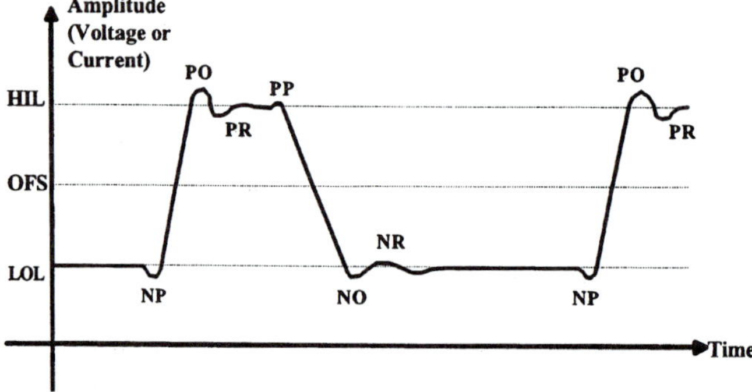

Figure 17.2 Real pulse nomenclature. NP = negative preshoot; PO = positive overshoot; PR = positive ringing; PP = positive preshoot; NO = negative overshoot; NR = negative ringing. Remark: Preshoot, overshoot, and ringing are sometimes also referred to as pulse distortions.

necessarily have to be time-synchronous. The majority of these instruments have the capability of delaying the channels with respect to each other. This delay may be entered as an absolute value (e.g., 1 ns) or as a phase shift (e.g., 90° = 25 percent of the period). Sometimes, especially in one-channel instruments, these values are referenced to the *trigger output* (see Sec. 17.2). The maximum delay or phase range is limited and in most cases dependent on the actual pulse period.

All pulse generators offer variable levels (HIL, LOL, AMP, OFS), variable pulse period (PER, FREQ), and variable pulse width (PWID, NWID, DCYC). Some instruments also offer variable transition times (TRISE, TFALL). The minimum and maximum values and the resolution with which these parameters can be varied are described in the data sheet of a pulse generator.

17.2 Pulse Generator Basics

This section describes how a pulse generator works, how it looks, and how it can be operated.

17.2.1 Basic block diagram

Figure 17.3 shows the basic block diagram of a typical two-channel pulse generator. Normally, the internal period generation circuit consists of a voltage-controlled oscillator (VCO) and a programmable divider, which can divide the frequency generated by the VCO by powers of 2 (1, 2, 4, 8, . . .). This is necessary because most VCOs can only be tuned within a range of f and $f/2$

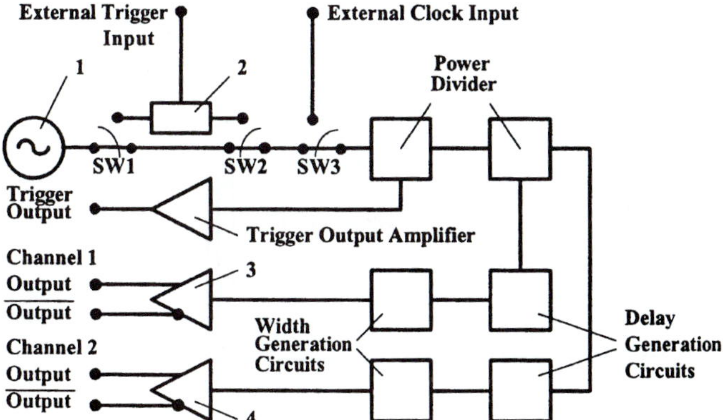

Figure 17.3 Basic block diagram of a typical two-channel pulse generator. 1: Internal clock generation circuit; 2: start/stop generation circuit; 3: output amplifier channel 1; 4: output amplifier channel 2. Switch SW1: selects continuous pulse mode or start/stop mode; Switch SW2: selects continuous pulse mode or start/stop mode; Switch SW3: selects internal or external clock generation. Output: normal output; $\overline{\text{Output}}$: inverted or complement output. Remark: SW1 and SW2 are always switched simultaneously.

or $f/4$, where f is the maximum frequency (or $1/f$ is the minimum period) of the VCO.

The output amplifiers of channel 1 and channel 2 can be either single-ended or differential. Differential amplifiers provide a normal and an inverted output (or complement output, see Fig. 17.3); single-ended amplifiers (like the trigger amplifier in Fig. 17.3) only provide a normal output. In most cases, the output amplifiers are current sources with a 50-Ω reverse termination resistor. Thus not only driving a 50-Ω-load but also driving a short or an open circuit results in a pulse with acceptable performance, because the reflection at the DUT (device under test) is being absorbed by this resistor, if the load is connected to the pulse generator with 50-Ω cables.

By means of SW3 (controlled by the microprocessor of the pulse generator), the internal period generation can be disabled, so that the instrument can be driven by an external source. With SW1 and SW2 (also controlled by the microprocessor), a start/stop generation circuit is switched into the signal path. With this circuit, the pulse stream can be started (*triggered mode*) or started and stopped (*gated mode*) with an external trigger signal. Normally, a positive edge at the corresponding input starts the pulse stream and a negative edge stops it.

Thus pulse generators with this feature can also produce nonrepetitive signals. For instance, if the internal period of a pulse generator that operates in the gated mode is set to 10 ns and a positive pulse with a duration of 100 ns is stimulating the external trigger input, the instrument puts out a single *burst* of 10 pulses. In many cases, a pulse generator does not have separate connectors for the external clock and the external trigger input, so that these two modes are mutually exclusive.

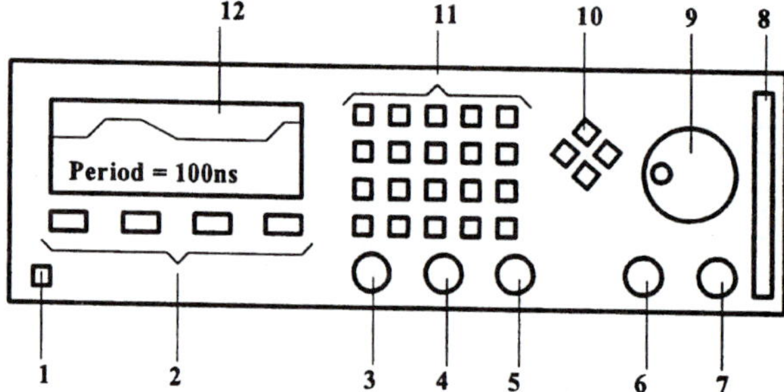

Figure 17.4 Typical front panel of a two-channel pulse generator. 1: Line switch; 2: soft keys; 3: external input (clock and trigger); 4: strobe output (see Sec. 17.3); 5: trigger output; 6: output channel 1; 7: output channel 2; 8: mass storage device (memory card); 9: knob (used to verify the selected parameter); 10: cursor keys (used to select a parameter); 11: data entry keys (used to select or vary a parameter); 12: (graphic) display.

17.2.2 Front and rear panel

Figure 17.4 shows the front panel of a typical pulse generator. Note that not all instruments provide a mass storage device and that only a few have a strobe output. Pulse generators with such an output also have data capabilities, as described below. The vast majority of pulse generators not only can be operated from the front panel but are also programmable. The computer interface (typically GPIB = general-purpose interface bus) is always located at the rear panel. Many pulse generators can be programmed according to the SCPI norm (standard commands for programmable instruments). Also, a lot of pulse generators are available with a "rear panel option," which means that all inputs and outputs (see Fig. 17.4, 3 to 7) are then not located at the front but at the rear panel. This plus the availability of a "rack mount kit" allows it to easily integrate the instrument into an automatic test system.

17.3 Special Pulse Generators

This section describes pulse generators with special capabilities and distinguishes them from instruments called data generators.

17.3.1 Pulse generators with data capabilities

Programmable data. Some pulse generators can generate not only repetitive pulse but also serial data streams. These instruments have a memory with a certain depth (typically 1 . . . 4 kbit) and an address counter in the signal path of each channel, as shown in Fig. 17.5. Thus single-shot or repetitive

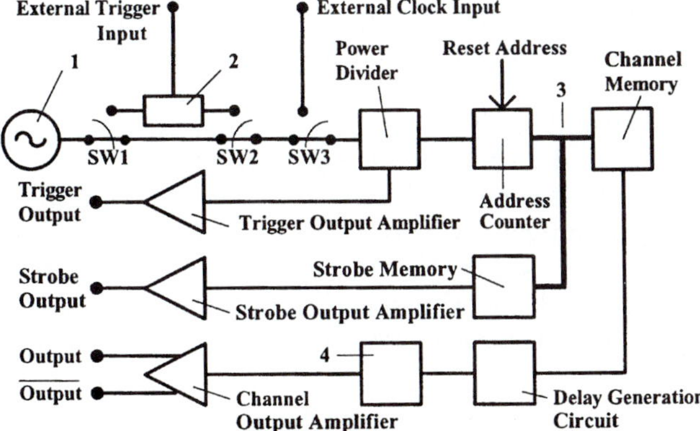

Figure 17.5 Block diagram of a one-channel pulse generator with data capabilities. 1: Internal clock generation circuit; 2: start/stop generation circuit; 3: memory address bus (n lines); 4: width generation circuit (NRZ, RZ with variable duty cycle). Data stream length (selected by the user): m bit ($m \leq 2^n$); memory depth: 2^n; reset address: 2^{n-m}. Remark: When the address counter value reaches 2^{n-1}, the counter resets itself to the reset address (calculated and programmed by the microprocessor).

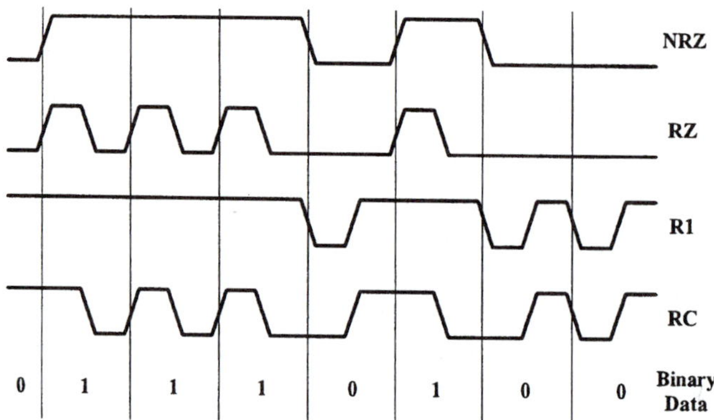

Figure 17.6 Different data formats. NRZ: Non return to zero (zero = logic "0" = LOL); RZ: return to zero (zero = Logic "0" = LOL); R1: return to "1" (Logic "1" = HIL); RC: return to complement ("0" → "1" and "1" → "0").

data streams of a programmable length can be generated (the maximum length of such a data stream is limited by the memory depth). The strobe output provided by these instruments can be used to generate a trigger signal that is synchronous to one specific bit of this data stream, which is useful when this bit (or the corresponding reaction of the DUT) has to be observed with an oscilloscope.

There are a couple of different *data formats,* as shown in Fig. 17.6. The most popular one is NRZ, because the bandwidth of a data stream with that format and a data rate of, for instance, 100 Mbit/s, is only 50 MHz (100 MHz if the format is RZ, R1, or RC). RC is used if the average duty cycle (see Fig. 17.1) of the data stream has to be 50 percent, no matter what the data that are being transmitted look like. For instance, if the data link that has to be tested cannot transmit a dc signal, an offset (or threshold) of 0 V (or 0 A, respectively) and the data format *RC* has to be chosen.

However, most pulse generators that have data capabilities offer only NRZ and RZ data format; only a few also offer R1 and RC. Sometimes, when selecting RZ, users can even select the duty cycle of the signal. If they choose 25 percent, for instance, a logic "1" will return to "0" after 25 percent of the pulse period (in Fig. 17.6, a duty cycle of 50 percent was chosen).

Random data. Some pulse generators with data capabilities can generate random data streams, which are also referred to as pseudo random binary sequences (or pseudo random bit sequences, PRBS). These sequences are often generated according to an international standard.

Difference between data/word generators and pulse generators. In contrast to pulse generators, most data generators are modular systems that can provide up to 50 channels and more. Data generators usually have a much higher memory depth and more functionality in terms of sequencing (looping, branching). Some data generators even can provide asynchronous data streams (generated with several independent clock sources). This means that a pulse generator with data capabilities can *not* replace a data generator, especially if more than one or two channels are needed. However, having some data capabilities in a pulse generator can be very beneficial, as will be shown in Sec. 17.4.

17.3.2 Pulse generators with distortion capabilities

Normally, a pulse generator is designed in a way that the pulse distortions described in Fig. 17.2 are as small as possible. However, the pulses that stimulate a DUT in its *real* environment are not always as clean. This means that a device that works fine as long as it is stimulated with the almost ideal pulses produced by a pulse generator does not necessarily work when stimulated by the real devices in its regular environment. Pulse generators with distortion capabilities can intentionally distort its output pulses by adding noise spikes or ringing, increasing the overshoot and so on. Thus a DUT can be tested with "real life" pulses, so that its sensitivity against noise, etc., can be determined.

17.4 Applications

This section describes some applications of basic or special pulse generators, explaining the measurements as well as discussing the results.

17.4.1 Characterizing the timing parameters of a state device

A digital circuit that requires a clock input is called a state device. Flip-flops, shift registers, and memories, for instance, are state devices. Their timing behavior can be described with parameters like *setup time, hold time,* and *propagation delay.* The least complex state device is a flip-flop. Besides the clock input (C), it requires a data input (D) and has a data output (Q). The setup time is the time that the data signal has to arrive at the flip-flop *prior* to the clock edge in order to be recognized as valid data. The hold time is the time that the data signal has to be stable *after* the clock edge arrived. The propagation delay is the time that it takes for the data signal to "travel" from D to Q, if the delay between the data and the clock input equals the setup time. Figure 17.7 illustrates how the setup time of a state device can be measured with a two-channel pulse generator and a digital sampling oscilloscope. The pulse generator does not have to have data or distortion capabilities, and variable transition times are not needed either. However, its maximum pulse frequency, level capabilities, and typical transition times must be suitable for the DUT that has to be characterized (see Sec. 17.5).

The bandwidth of the oscilloscope that is needed for this test is *not* deter-

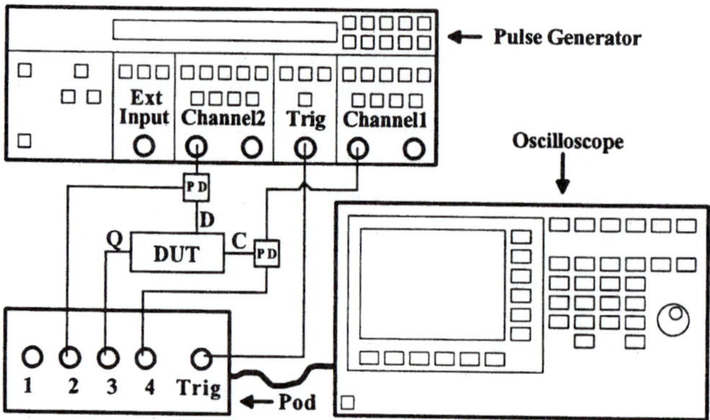

Figure 17.7 Test setup for the timing characterization of a state device. PD = power divider. Remark: Before measuring any timing parameters, the test setup has to be deskewed, owing to different cable lengths. In this case, the delay setting that results in a zero delay between the clock and the data signal at the input of the buffer has to be found, using an oscilloscope probe. Some pulse generators have a separate deskew feature that allows them to do that without affecting the delay value.

mined by the maximum pulse frequency with which the device has to be tested but by the typical transition times of the pulse generator:

$$\text{Bandwidth of oscilloscope} \geq \frac{10}{3 \times \text{transition time}}$$

So, if the typical transition time is 1 ns, the oscilloscope must have a bandwidth of at least 3 to 4 GHz.

Figure 17.8 shows how the waveforms on the screen of the oscilloscope can look. The setup time of the flip-flop is the minimum delay between the data and the clock signal that causes the output to behave as shown in case 1. Note that the premetastable behavior shown in case 2 has to be avoided (although it does not cause any failures in this device!), because its propagation delay increases as well as its output transition times, which may cause downstream failures: Owing to these effects, the data arrive later at the next devices in the chain (for instance, in clock distribution network), possibly causing metastability.

17.4.2 Measuring crosstalk between two pins of a connector

If a signal that stimulates one pin of a connector also affects the signal at an adjacent pin, this (undesired) phenomenon is called crosstalk. That effect is caused by the parasitic capacitance between the two pins. Crosstalk also occurs between different lines of a cable or between various striplines on a printed circuit board, for instance, in a bus system on the backplane of a computer.

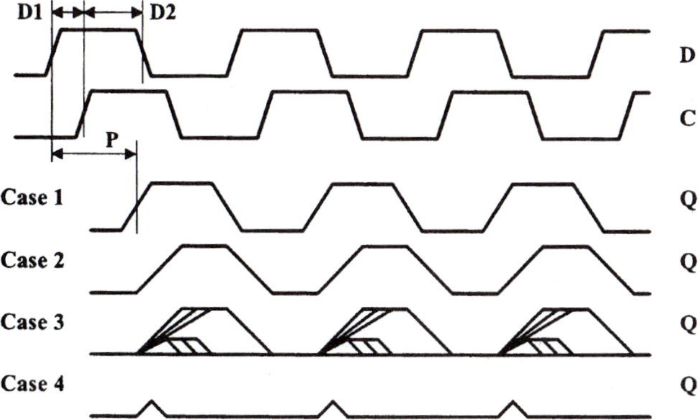

Figure 17.8 Timing characterization of a state device (waveforms). Case 1 (D1 ≫ setup time): stable output (0 percent failures); case 2 (D1 ≈ setup-time): premetastable behavior (still 0 percent failures!); case 3 (D1 ≲ setup time): full blown metastability (about 50 percent failures); case 4 (D1 ≪ setup time): stable output (100 percent failures). Setup time = D1 min for case 1); hold time = D2 min (for case 1); P = propagation delay.

Crosstalk is heavily dependent on the transition times of the signals. The faster the edges are, the worse is the crosstalk value that can be expected. Therefore, a pulse generator that provides variable edges is desirable if a connector has to be characterized in terms of crosstalk. Unfortunately, fast *and* variable edges in a pulse generator are more or less mutually exclusive. So, if the measurements have to be performed up to the typical edge rates of ECL devices (about 250 ps) or even up to faster rates, a pulse generator with ultra-fast but fixed transition times has to be chosen. Its fast edges can then be slowed down with additional passive low-pass filters called TTCs (transition time converters). Data or distortion capabilities do not have to be provided by the pulse generator used for this kind of measurement.

Figure 17.9 shows how a signal that has been distorted by crosstalk typically looks. A crosstalk measurement should be performed in a 50-Ω environment in order to avoid reflections that could be mistaken for crosstalk.

17.4.3 Noise immunity tests

Clock signals as well as data streams can be distorted by effects like crosstalk or ground bounce (a phenomenon caused by parasitic inductances). Although the designers of connectors, cables, and printed circuit boards try to minimize these effects by performing measurements as described in Sec. 17.4.2, they can never be avoided completely. Therefore, devices like buffers, memories, or shift registers must be tested with distorted signals in order to find out how sensitive they are against noise spikes caused by crosstalk or ground bounce. In order to produce this kind of signal (see Fig. 17.10), a pulse generator with data and distortion capabilities is needed. The *positive* noise spike in

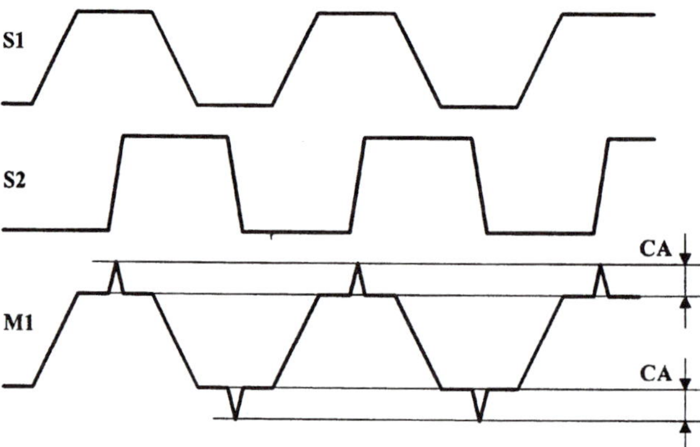

Figure 17.9 Crosstalk measurements. S1: Signal that stimulates the connector at pin 1 (pulse generator channel 1); S2: signal that stimulates the connector at pin 2 (pulse generator channel 2); M1: signal that can be measured at pin 1 of the connector (oscilloscope channel 1); CA: crosstalk amplitude.

the data signal D1 might cause the shift register to capture a logic "1" instead of a logic "0" when the corresponding clock edge arrives. The clock signal C2, which has been distorted by a *negative* noise spike, can cause the device to not capture the corresponding logic "1" of the data signal D2. By varying the amplitude, the pulse delay, and the pulse width of the distorting spike(s), the noise immunity of the DUT can be accurately characterized.

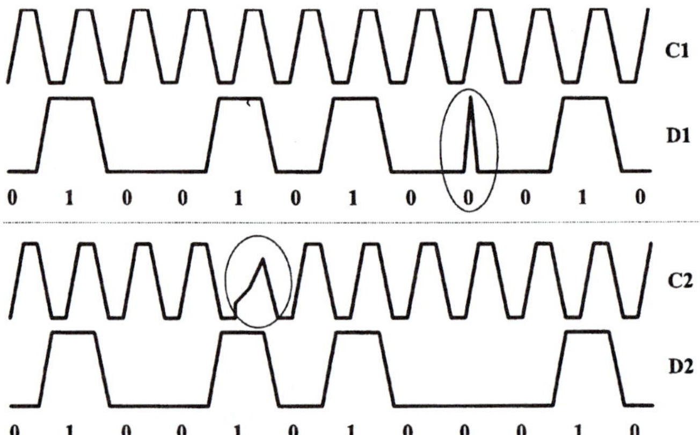

Figure 17.10 Testing the noise immunity of a shift register. C1: Clean clock signal; D1: distorted data signal; C2: distorted clock signal; D2: clean data signal.

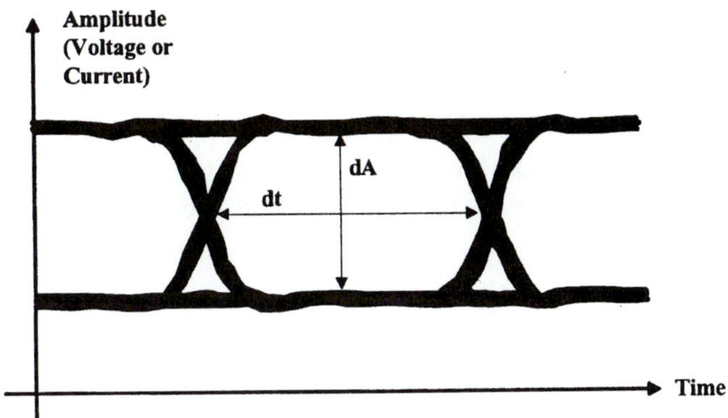

Figure 17.11 Typical eye diagram. *dA:* Vertical eye opening; *dt:* horizontal eye opening.

17.4.4 Testing high-speed drivers

Many devices, especially high-speed amplifiers (or drivers), may work fine when stimulated with a repetitive data pattern (or pulse stream), produced by a normal pulse generator. However, they might still not be applicable for transmitting *random* data streams at the same or at even lower data rates, owing to duty cycle-dependent effects. If a high-speed driver is stimulated with a data pattern that contains many zeros, followed by a single "1," the amplitude of the corresponding "1" at its output may be lower and its pulse width may be smaller than expected. A good way of characterizing these effects is to use a pulse generator with PRBS capabilities. If the oscilloscope that is used to look at the output signal of a driver which is stimulated with such a pattern is triggered with a clock signal whose frequency equals the data rate of the PRBS, the picture that can be observed on its screen is called an *eye diagram* (see Fig. 17.11). When performing such a measurement, it is recommended that the oscilloscope be operated in the *infinite persistence mode,* if possible.

Eye diagram measurements are very powerful, as they can show all duty cycle-dependent effects described above plus reflections due to impedance mismatches and more all at the same time. The bigger the *eye opening* (described by *dA* and *dt*) is, the higher is the quality of the output signal of the driver.

17.5 Important Specifications

This section is written for people who consider buying a pulse generator because they have some measurement problems that need to be solved. It should help them to find the right instrument, considering performance, functionality, and price.

In the past, most people tended to simply buy the instrument with the best performance and functionality available on the market, because they could be sure that it would solve their problems. However, this instrument was also one of the most expensive ones, and price has become an important issue over the last year. This is why today many people look for an instrument that gives them "just enough performance and functionality." This section should help them to identify their needs and thus find the instrument they are looking for.

17.5.1 Performance

Maximum pulse frequency. One of the first things that has to be taken into account is: What is the maximum pulse frequency required for the measurements that have to be performed? If it is 100 MHz, for instance, a 150-MHz instrument may be sufficient to fulfill all needs. However, as a pulse generator approaches its maximum speed, the delay and duty cycle capabilities may be very limited, so that a 300-MHz or even a 500-MHz instrument might be a better fit.

Level capabilities. The required level capabilities are dependent on the technology of the DUTs that have to be characterized. A CMOS device needs a LOL of 0 V and a HIL of 5 V; ECL circuits are tested with −1.8 V and −0.8 V (termination into −2 V) or with 0.2 V and 1.2 V (termination into ground). However, some margin is desirable, especially if stress tests have to be performed. So, for ECL ICs, an instrument that provides an amplitude of at least 2 V, a minimum LOL of −2.5 V and a maximum HIL of 2 V should be chosen.

Caution. Mostly, the level capabilities of a pulse generator are specified into a 50-Ω load. If a DUT with an (almost) infinite impedance has to be tested, the levels at the outputs of the pulse generator are normally twice as high as the programmed values. Sometimes the levels may be "clipped" when working into an impedance higher than 50 Ω, in order to protect the instrument's output amplifiers. In this case, the *real* HIL may be 4 V when 2 V is programmed but only 6.5 V, for instance, when 5 V is programmed, owing to the clipping limit.

Minimum transition time. The minimum transition time (if variable edge rates are provided) or the typical edge speed of a pulse generator with fixed transitions should be about twice as fast as the typical transitions that the DUT has to work with in reality. If they are much faster, crosstalk problems may be observed that do not occur in the DUT's regular environment and are therefore unimportant. If they are much slower, crosstalk problems that *do* occur in reality may not be found when the DUT is characterized.

Duty cycle capabilities. Although the duty cycle is often specified to be variable from 0 to 100 percent, the minimum and maximum value that can *really* be achieved is heavily dependent on the frequency the instrument works at, because the minimum pulse width that can be produced is limited by its bandwidth. For instance, if the minimum transition time of a pulse generator is 1 ns, its bandwidth is about 1/(3ns) or 333 MHz, so the minimum pulse width that this instrument can provide is approximately 1.5 ns. If the programmed frequency is 67 MHz (which equals a pulse period of about 15 ns), the minimum/maximum duty cycle will be about 10 percent/90 percent. If the frequency is 333 MHz (3 ns period), the output signal will always be a square wave (50 percent duty cycle, not variable).

So the duty cycle specification of a pulse generator may be misleading sometimes; the important question is: what is the minimum pulse width provided by the instrument? For example, if a duty cycle 10 percent is needed at a frequency of 100 MHz, the pulse generator must provide a minimum pulse width of 1 ns and thus a minimum transition time of 600 to 700 ps. A pulse generator with a maximum frequency of 300 MHz and a minimum transition time of 1 ns can *not* provide a duty cycle of 10 percent at 100 MHz, even if the duty cycle specification is 0 percent . . . 100 percent.

Delay range. The delay range provided by a pulse generator is often dependent on the programmed frequency. Some instruments have a delay range specification of, for instance, one period minus 10 ns. If this is the specification of a 100-MHz pulse generator, the delay range at a frequency of 1 MHz is 1 μs minus 10 ns (or 990 ns). However, at 100 MHz, the instrument does not provide any delay capabilities, which means that timing measurements as described in Sec. 17.4.1 cannot be performed. It is therefore not advisable to use such a pulse generator at its maximum frequency.

Timing resolution and jitter. The resolution that is required for the timing parameters (period, delay, and pulse width) is dependent on the kind of measurement that has to be performed. For functional tests, a coarse resolution can be sufficient, whereas precise timing measurements may require a very fine resolution. For instance, if the metastable "state" of a state device (see Sec. 17.4.1) has to be characterized, a resolution of 10 ps may not be sufficient, even if the DUT only works at a frequency of 100 MHz, because the delay *window* between the clock and the data input, in which the metastability occurs, can be only 5 ps wide.

Another important consideration is jitter. Jitter describes the *uncertainty* of a timing parameter, as shown in Fig. 17.12. For instance, if a period of 10 ns is programmed, the internal oscillator of a pulse generator may produce one period with a duration of almost exactly 10 ns, then one with a duration of 10.01 ns, followed by one with a duration of 9.99 ns and so on. This phenomenon is basically caused by inadequate power supply regulation (noise). Jitter is often specified as "rms" value (root mean square). The peak-to-peak value that can be expected is roughly six times as high, so if the jitter specification is 10 ps rms, the peak-to-peak jitter should be no worse than 60 ps. If the pulse generator will be used as a high-speed clock source, for example,

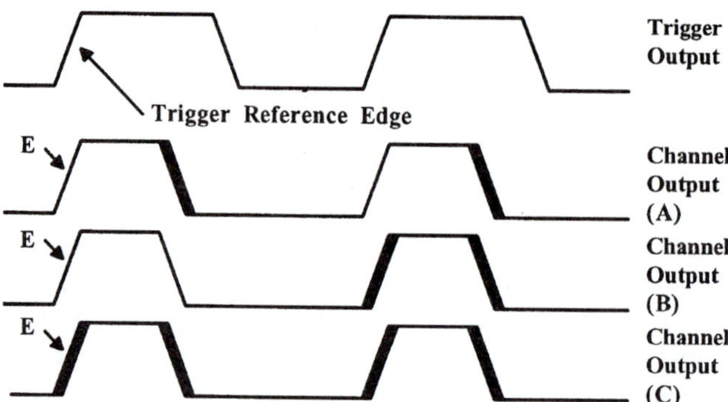

Figure 17.12 Different types of jitter. (*A*): Pulse width jitter; (*B*): pulse period jitter; (*C*): pulse delay jitter. Remark: When a jitter measurement is performed, the first thing that has to be done is to find the edge (*E*) of the channel that "belongs" to the trigger reference edge, which is the one that was generated out of the same edge of the period generator (oscilloscope is triggered with the trigger output of the pulse generator).

low jitter is very essential: When testing parts of a new workstation that should run at a clock speed of 400 MHz, a jitter specification of only 50 ps rms will result in a peak-to-peak jitter of 300 ps, which equals 12 percent of a period! If the jitter specification were 10 times as good, the workstation could be tested at a clock rate that is 10 percent faster, because a much lower noise budget would be required.

17.5.2 Functionality

Channel count. Most pulse generators are one- or two-channel instruments. If more channels are needed occasionally, an instrument that allows master-slave configurations must be chosen.

Variable transition times. If variable transition times are *not* needed, a pulse generator with fixed edge rates should be selected. If the edge speed must be variable, it has to be taken into account that ultra-fast and variable transition times are basically mutually exclusive. This means that a pulse generator with variable edge rates can be chosen, if the fastest transition time that is needed is only 1 ns or even slower. If much faster edge rates are required, an instrument with ultra-fast but fixed transitions plus additional TTCs should be selected.

Trigger capabilities. Some pulse generators can be synchronized to an external clock; some can be started (and stopped) with an external trigger signal. A few instruments even have an *external width mode,* which means that the duty cycle of its output signal will be the same as the duty cycle of the external clock, if this mode is selected.

Data capabilities. Instruments with data capabilities can produce not only repetitive pulse but also serial data streams (programmable or random). If this is not needed occasionally, an instrument that does not provide this feature should be chosen.

Distortion capabilities. Pulse generators with distortion capabilities can intentionally distort its output signals by adding overshoot, noise spikes, and so on. This is necessary if the noise immunity, etc., of a DUT has to be characterized. If only functional tests or timing measurements have to be performed, an instrument that does not provide this feature should be selected.

Chapter

18

Microwave Signal Generators

William Heinz

Agilent Technologies
Santa Clara, California

18.1 Introduction

The need for sources of clean, calibrated electrical signals arose with the development of receivers, which required testing and characterization to determine performance and to make adjustments, e.g., alignment of intermediate frequency (if) blocks, local oscillators (LOs), etc. As microwave receiver technologies and capabilities became more sophisticated, the corresponding signal generator performance demands grew far beyond what could be provided by the unwieldy, manually tuned klystron signal generators of the 1950s.

Modern communications, satellite, radar, navigational, electronic warfare (EW), and surveillance receivers require various degrees of amplitude and frequency precision, frequency range, dynamic range, modulation capabilities, tuning speed, and freedom from undesired spurious signals. Many require programmability under computer control in production test environments.

Accurate component and subsystem testing requires precision swept sources to operate with scalar and vector network analyzers. The modern signal generator combines synthesis and solid-state oscillator technologies (see Chap. 8) with microprocessor control to provide sophisticated frequency and level control, programmability, self-test, and diagnostics, as well as internal and external calibration. This results in high levels of accuracy, useful feature sets, and low cost of ownership.

In this chapter, various types of microwave signal generators are described together with some of their fundamental applications. Implementations are discussed at a block diagram level, and with reference to material in Chap. 8 on microwave sources. Signal generator performance specifications and their relevance to certain applications are described.

18.2 Types of Signal Generators

Several varieties of microwave signal generators, each optimized for certain ranges of applications, are described in this section.

18.2.1 CW signal generators

For certain applications, a continuous-wave (CW) signal generator without modulation capability may be all that is required, providing a cost savings over more sophisticated models. Output power is of importance in typical applications such as where the signal generator serves as an LO driving a mixer in an up or down converter. The signal level needs to be high enough to saturate the mixer to assure good amplitude stability and low noise. If the phase noise of the converted signal is not to be degraded, the phase noise of the signal generator must be sufficiently lower.

Other applications for CW sources include use as exciters in transmitter testing or as sources driving amplifiers and/or modulators. High-level accuracy, low spurious and harmonic signal levels, and good frequency resolution may all be important specifications in these and other applications.

18.2.2 Swept signal generators

Frequency swept signal generators are used for the test and characterization of components and subsystems and for general-purpose applications. They can be used with scalar and vector network analyzers (see Chap. 28) or with power meters or detectors. Sweep can be continuous across a span, or in precise discrete steps. Techniques exist for phase locking throughout a continuous sweep, allowing for high accuracy in frequency. Step sweep techniques, in which the source is phase locked at each discrete frequency throughout the sweep, are more compatible with computer controlled measurement systems.

Throughput demands on systems require fast switching speeds where it is necessary to have small frequency step sizes to avoid missing narrow band perturbations. Real-time displays for network analyzers in which hundreds of frequency points per sweep are refreshed at rates of at least 10 sweeps per second also drive switching time requirements down to millisecond rates and below. Applications such as antenna test, in which many frequency data points need to be measured at each of a large number of spatial locations, demand high measurement speeds and therefore fast signal generator switching speeds as well. Swept signal generators are available with appropriate interfaces for scalar and vector network analyzers and have also been integrated into self-contained network analyzers. See Chap. 28 for more information on network analyzers.

18.2.3 Signal generators with modulation

Microwave signal generators designed to test increasingly sophisticated receivers are being called upon to provide a growing variety of modulations at accurately calibrated signal levels over a wide dynamic range without generating undesired spurious signals and harmonics. Application-oriented signal generators are available which provide combinations of modulation formats. These range from simple amplitude and frequency modulation to those that employ a wide variety of digital modulation formats in which discrete symbols are transmitted as combinations of phase and amplitude of the carrier.

18.2.4 Signal generators with frequency agility and high-performance modulation

For applications where simulations of complex scenarios involving multiple targets or emitters need to be carried to test radar or EW receivers, or to perform certain tests on satellite and communications receivers, higher-performance signal generators featuring very fast frequency switching and/or sophisticated modulation capabilities are used. This additional performance may be obtained by employing DDS techniques at vhf and using direct synthesis for upconverting into the microwave frequency range (see Chap. 8). These sources may also feature a variety of software interfaces that can provide “personalities” for various receiver test applications, allowing entry of parameters in familiar form and allowing the creation of complex scenarios involving lengthy sequences of a number of signals.

18.3 Types of Modulation

Some of the more commonly required modulation formats are described below, along with some common applications. Brief descriptions are given of characteristics of signal generators providing these modulations.

18.3.1 Pulse modulation

Pulse modulation is used to simulate target returns to a radar receiver, to simulate active radar transmitters for testing EW surveillance or threat warning receivers, or to simulate pulse code modulation for certain types of communications or telemetry receivers. Microwave signal generators can have inputs for externally applied pulses from a system under test or from a pulse generator. Input impedances are typically 50 Ω to properly terminate and thereby avoid undesired reflections on the most commonly used cables. Some microwave signal generators have built-in pulse sources which may be free-running at selectable rates or can be triggered externally with selectable pulse widths and delays, the latter being used to simulate a variety of distances to a radar target. Various modes of operation of such internal pulse sources include one in which the internal source is gated by an external source, or doublet mode whereby an external pulse modulates the signal generator, followed by an internally generated pulse of selectable width and de-

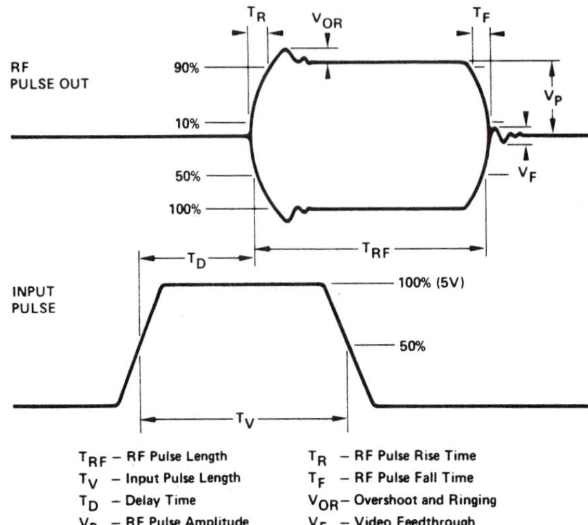

Figure 18.1 Definitions of commonly used pulse-modulation specifications.

lay. Figure 18.1 shows how the most commonly used pulse-modulation specifications are defined.

When broadband EW receivers are being tested, it is important that harmonics of the fundamental microwave frequency are attenuated sufficiently to avoid false signal indications. This requires good internal filtering without degrading the leading or trailing edges of the pulse waveforms and without causing reflections that may produce delayed replicas of the primary pulses.

18.3.2 Amplitude modulation (AM)

In addition to the simulation of microwave signals having AM and for AM to PM (amplitude to phase modulation) conversion measurements, amplitude modulation of a microwave signal generator may be needed for simulation of signals received from remote transmitters in the presence of fading phenomena in the propagation path, or from rotating radar antennas. AM should be externally applicable or internally available over a broad range of modulation frequencies without accompanying undesired (incidental) variations in phase.

Figure 18.2 shows how modulation index is defined for linear AM. Note that the voltage is $(1 + m)V_0$ at the peak and $(1 - m)V_0$ at the trough. Thus there needs to be headroom of 20 log $(1 + m)$ dB above the unmodulated signal, or a maximum of 6 dB as the modulation index m approaches 100 percent. The depth of the trough, on the other hand, is 20 log $(1 - m)$, which would be 20 dB, for example, for $m = 90$ percent. The ability of the signal generator to faithfully produce an accurate AM signal will depend on the dynamic range of the automatic level control (ALC) system, since the AM is usually achieved by summing the modulating signal into the ALC loop.

In certain instances it may be more desirable to have a log AM input to the signal generator, e.g., −10 dB/V applied. Such applications include simula-

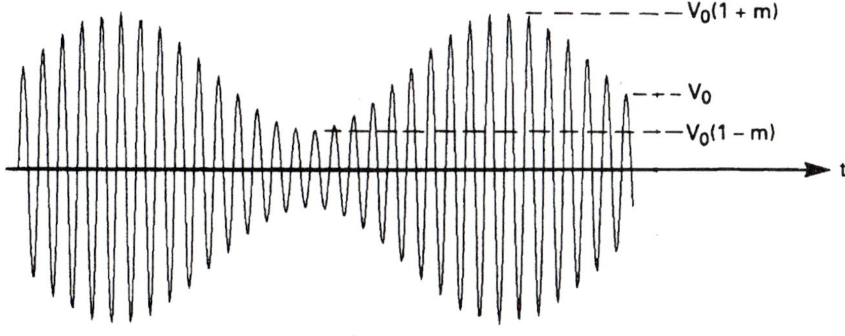

Figure 18.2 AM waveform and equation for voltage versus time, where V_0 = voltage peak level for unmodulated carrier and m = modulation index.

tion of signals with large variations in level such as rotating antenna patterns or scintillation effects in radar target returns.

18.3.3 Frequency modulation (FM)

For applications where signals with FM need to be provided, signal generators with an external input and/or an internal source are available. The modulation index β is defined as

$$\beta = \frac{\Delta f}{f_m} \tag{18.1}$$

where Δf is the peak frequency deviation and f_m is the modulation frequency.

For signal generators employing phase locked loops (PLLs) to stabilize a microwave oscillator there will usually be limitations on the maximum value of β achievable due to limitations in the range of allowable inputs to the phase detectors. This means that there will be a minimum modulation frequency (for a particular deviation) equal to $\Delta f/\beta_{\max}$. Since direct application of FM to a phase locked VCO will be counteracted by the loop, there needs to be a way to avoid this effect (see Sec. 18.4.2 below for how this is done). Some signal generators featuring "dc FM" (FM rates down to dc) may implement this mode of operation by allowing the oscillator to operate unlocked, thereby losing stability and introducing high levels of phase noise. These limitations do not exist for systems employing DDS (see below).

18.3.4 I/Q (vector) modulation

Digital modulation techniques have essentially supplanted analog modulation methods for communications and broadcasting applications. Modulation is said to be digital if the signal is allowed to assume only one of a set of discrete states (or symbols) during a particular interval when it is to be read.

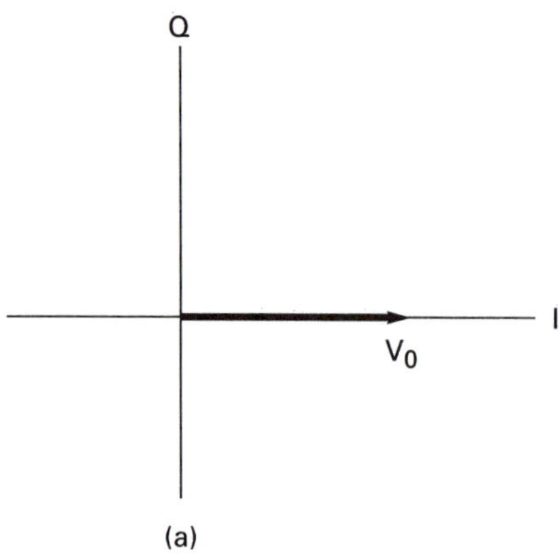

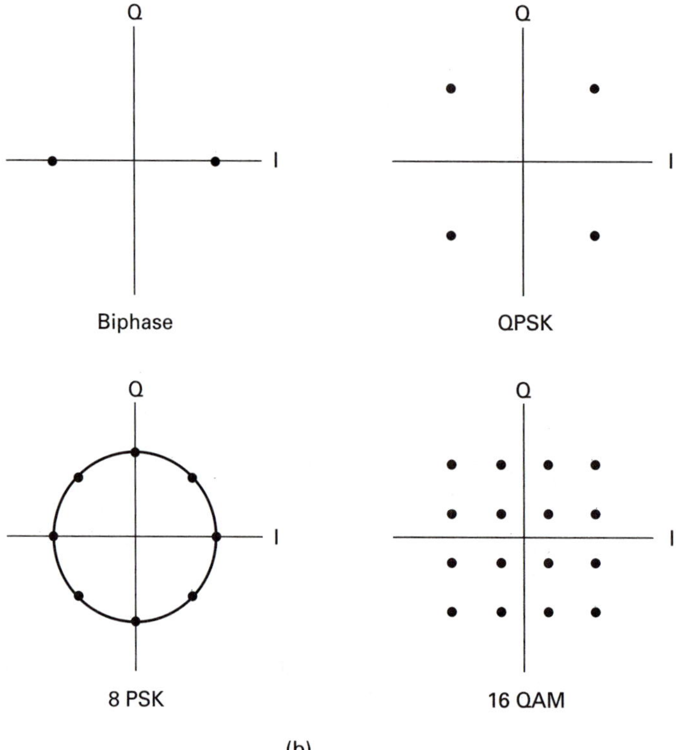

Figure 18.3 (a) Unmodulated carrier displayed as a vector with zero phase. (b) Some examples of digital modulation formats. In biphase modulation, there are two states characterized by a carrier at 0° or 180° of relative phase. QPSK (quadrature phase shift keying) has the four states shown, and 8-PSK (eight-phase shift keying) has the eight states of equal amplitude. The last example shows 16 QAM (quadrature amplitude modulated) where there are three different amplitudes.

Data are transmitted sequentially at a rate of n bits per symbol, requiring 2^n discrete states per symbol. By representing the unmodulated microwave carrier as a vector with unity amplitude and zero phase as shown in Fig. 18.3*a*, we can display various modulation formats on a set of orthogonal axes commonly labeled I (for in-phase) and Q (quadrature phase). A few examples are shown in Fig. 18.3.

Microwave signal generators providing a variety of digital modulation formats such as biphase, QPSK, m-phase, and QAM are available with modulation bandwidths of several hundred MHz. Serial and/or parallel data ports are usually provided for external data sources. The unmodulated carrier is typically available from a separate port for various test applications, and internal pseudorandom bit sequence (PRBS) generators can drive the data ports to provide data streams for various purposes such as bit error rate measurements.

18.4 Microwave Signal Generator Architectures

In this section, some common realizations of the signal generator types mentioned above are discussed at the block diagram level. It is not to be inferred from the examples which follow that all the particular configurations and combinations of features are available as existing products. They are intended as conceptual examples to aid in defining and differentiating fundamental categories of signal generators and the types of features which could be derived from each.

18.4.1 Basic signal generators without modulation

Figure 18.4 shows a simplified generic block diagram of a microwave source consisting of an electrically tunable microwave voltage controlled oscillator (VCO) stabilized by a frequency control circuit typically containing several phase locked loops and a stable reference oscillator (see Chap. 8). This reference oscillator (typically at 10 MHz) determines the frequency stability and accuracy of the signal generator. A temperature controlled crystal oscillator (TCXO) may be included, or a lower-cost oscillator may be sufficient where this level of performance is not required (see Sec. 18.5.1). It is sometimes necessary to drive several instruments (e.g., use with other signal generators or with a spectrum analyzer) from a common 10-MHz reference, and this capability is achievable on most signal generators via an input port on the rear panel.

If the signal generator has sweep capability, there may be provision for application of linear ramp voltages to the VCO, and/or step sweep may be available in firmware, whereby phase lock is achieved at each frequency of hundreds in a typical sweep. Other methods allowing for true synthesis throughout a broadband sweep are also available when a high degree of frequency accuracy is required during a continuous sweep.

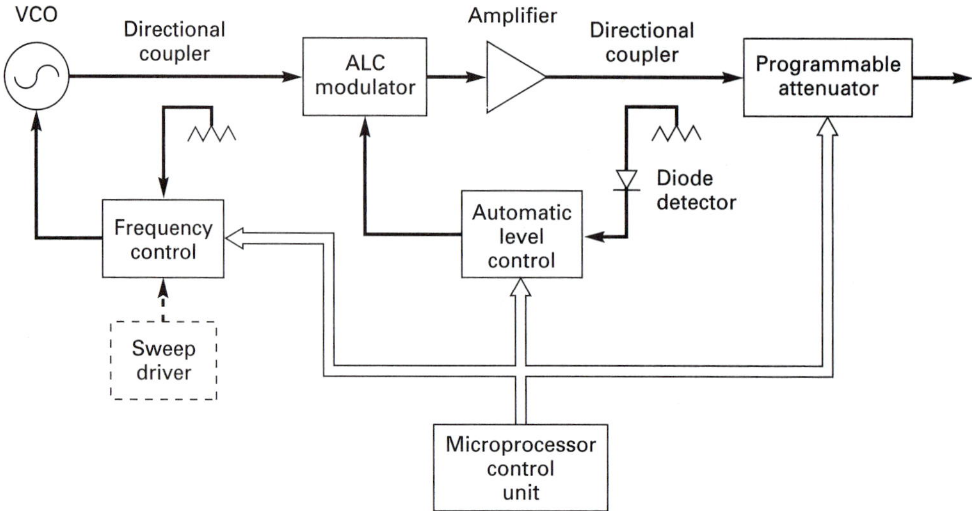

Figure 18.4 Simplified block diagram of basic signal generator with fundamental microwave oscillator.

Output level is stabilized and controlled by an automatic level control loop consisting of a directional coupler and detector to sample the output signal and compare it with a reference voltage to control the output level. The difference signal is used to drive the ALC modulator and thereby close the loop. The amplifier, which is inside the ALC loop, is used to boost the signal power from the oscillator (typically a YIG-tuned fundamental oscillator) and to overcome the losses in the couplers, ALC modulator, and any other elements in the microwave chain used for control, filtering, or modulation.

The programmable attenuator at the output is usually a step attenuator. Thus typical dynamic ranges of +10 to −100 dBm of output power are covered with a combination of coarse 10-dB steps from the attenuator together with fine (e.g., 0.1-dB steps) from a vernier control driven by a DAC which is used to derive the ALC reference voltage. The latter may include correction factors for frequency and temperature stored in a table in the instrument's memory. The data are derived from a factory calibration procedure using a power meter at the output port of the signal generator. A useful feature is to allow the user to generate a stored calibration table with data derived from a remote power sensor located within an external microwave network, thus providing constant power at a port selected by the user. Algorithms resident in the instrument's firmware may be used for interpolating frequency correction data between the cardinal frequency points used in the calibrations.

18.4.2 Addition of am, fm, and pulse modulation

Figure 18.5 shows the same basic configuration of Fig. 18.4 with various modulation capabilities added. Several approaches exist for providing FM to the

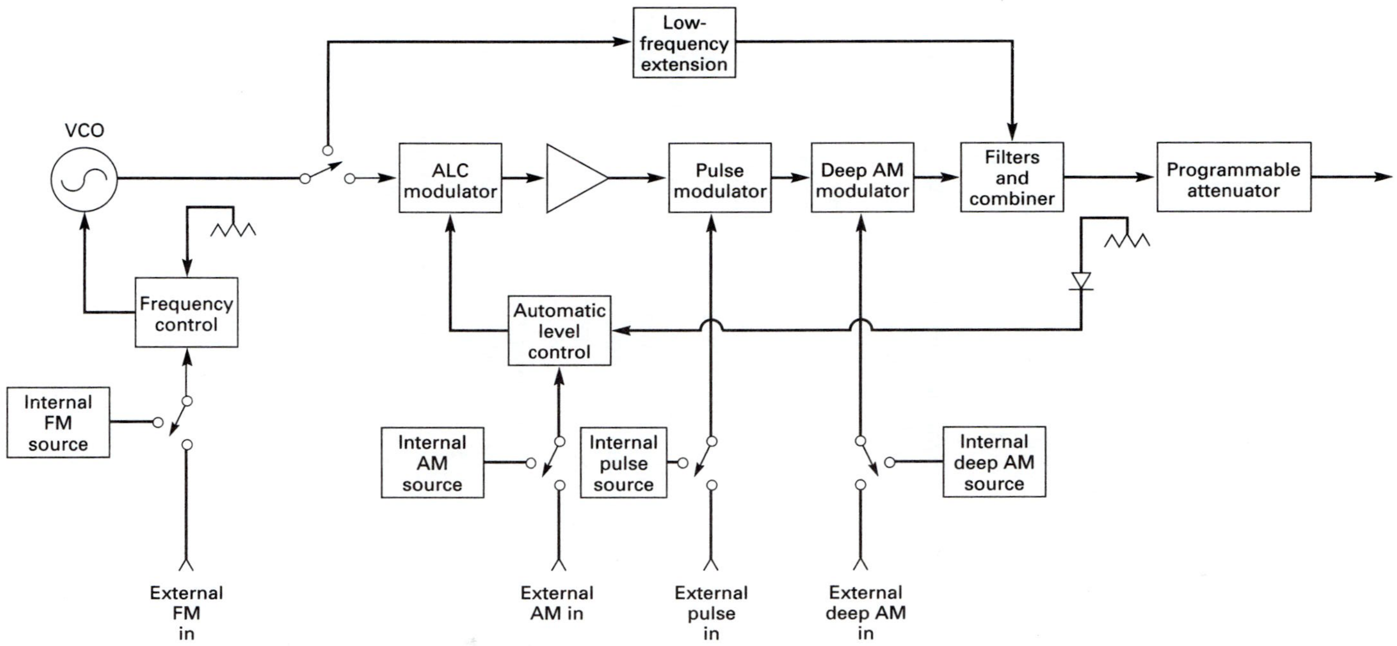

Figure 18.5 Microwave signal generator with modulation.

source. In one technique, the phase lock loop bandwidth in the frequency control circuit is reduced to a sufficiently low value and the FM signal is applied directly to the VCO at frequencies above this value in open loop fashion. The VCO remains phase locked and is thus stabilized, but the lowest-frequency FM rates may be limited to, for example, 1 kHz. Phase noise in this FM mode of operation is degraded from what can be achieved with the optimum PLL bandwidth. Other techniques involve integration of the FM signal at frequencies below the optimum PLL bandwidth and summing this phase modulation component into the PLL, thereby extending rates down to perhaps 10 Hz. Frequencies above the "crossover" frequency for optimum phase noise (typically about 10 kHz for microwave YIG oscillators) are handled as described in the first technique. Some signal generators incorporating accumulator based PLL circuitry may permit dc FM with limited rates and deviations, and some provide a mode of dc FM operation in which the oscillator is unlocked, thereby degrading frequency stability and phase noise considerably.

Linear AM is shown in Fig. 18.5 as being provided via the ALC circuitry by summing it in with the ALC reference voltage. As mentioned previously, the maximum available modulation index will be limited by the ALC dynamic range, which, among other things, will depend on the level of voltage available from the detector relative to noise, at the minimum levels of microwave signal. This in turn will depend on the coupling factor for the directional coupler and the sensitivity of the detector. Because there are many other demands on ALC dynamic range, including the required vernier range and frequency correction factors, a typical maximum AM modulation index for such linear AM systems will be about 90 percent. Typical maximum rates are 100 kHz. Figure 18.5 shows that the user can select an internal source or an external signal to drive the AM, as is the case for the FM. If high index or deep AM is required, linear or logarithmic AM can be applied via an open loop PIN diode or FET (field-effect transistor) modulator. In the example in Fig. 18.5, the ALC loop includes the deep AM modulator as well as the pulse modulator. When these are in operation, the ALC loop is switched into a "hold" mode prior to applying the modulation, in which the loop is opened with the ALC modulator held at the same level that it had while the loop was closed. In this way, both pulse and deep AM modulations may be applied simultaneously, as would be required for typical simulations of remote radar transmitters.

Broadband signal generators will require filtering of harmonics at frequencies within the bands of coverage. Thus switchable low-pass filters may be used as shown in Fig. 18.5, or tunable band stop or bandpass filters may be employed. Pulse fidelity (rise time, overshoot, or flatness) can be degraded if the carrier frequency is too close to the filter skirt so that switch points need to be chosen carefully.

As discussed in Chap. 8, extension to frequencies below the typical YIG oscillator frequency of 2 GHz can be provided using either a heterodyne tech-

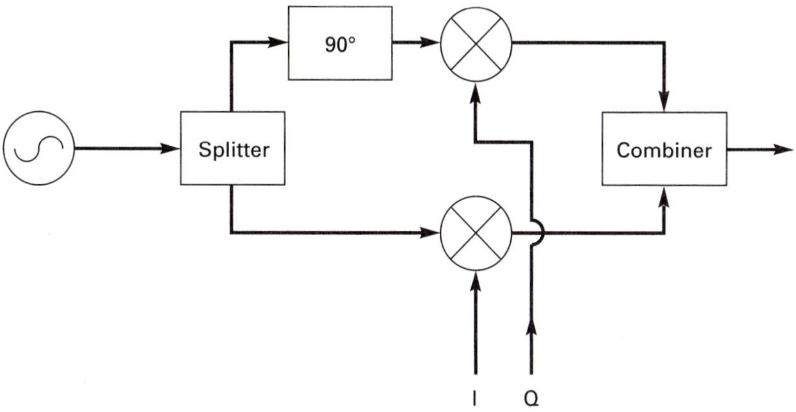

Figure 18.6 *I/Q* modulator block diagram.

nique or dividers. The ALC, AM, and pulse modulation techniques shown in Fig. 18.5 for frequencies above 2 GHz will have to be duplicated for the low band.

18.4.3 Signal generators featuring vector modulation

Figure 18.6 shows a method for achieving vector modulation of a carrier provided by an oscillator. The carrier is split into two paths, with one path (the *Q* channel) having 90° of phase shift relative to the other (the *I* channel). The *I* and *Q* channel modulating signals are provided to the mixers as shown, and the resulting signals are recombined (added) at the output.

This scheme results in a way to generate phase and amplitude modulation at very high rates for a broad range of applications. The accuracy of modulation will depend upon the accuracy of the 90° phase shift, the linearity and tracking of the mixers, the combiner, and the degree of balance in the two channels. While this can be optimized and adjusted at one carrier frequency, it would be more difficult to achieve over a broad range of carrier frequencies that may be required in a vector signal generator. For this reason, the approach shown in Fig. 18.7 can be used.

In this scheme the *I/Q* modulator operates at one frequency, e.g., 8 GHz, and a heterodyne technique is used to provide frequency coverage from 0.01 to 3 GHz by using a local oscillator covering 8.01 to 11 GHz. These frequencies are chosen to minimize mixer products that can result in in-band spurious signals. Instrument firmware can include internal calibration routines to assure amplitude balance and quadrature. Figure 18.7 also shows how digital modulation can be generated by mapping parallel digital lines into sets of digital formats such as QPSK, 16-QAM, etc. Appropriate Nyquist filtering in the *I* and *Q* paths can be connected externally via connectors accessible from the front or rear panel. A serial-to-parallel digital converter may be included for serial data inputs, and an internal PRBS generator may be installed.

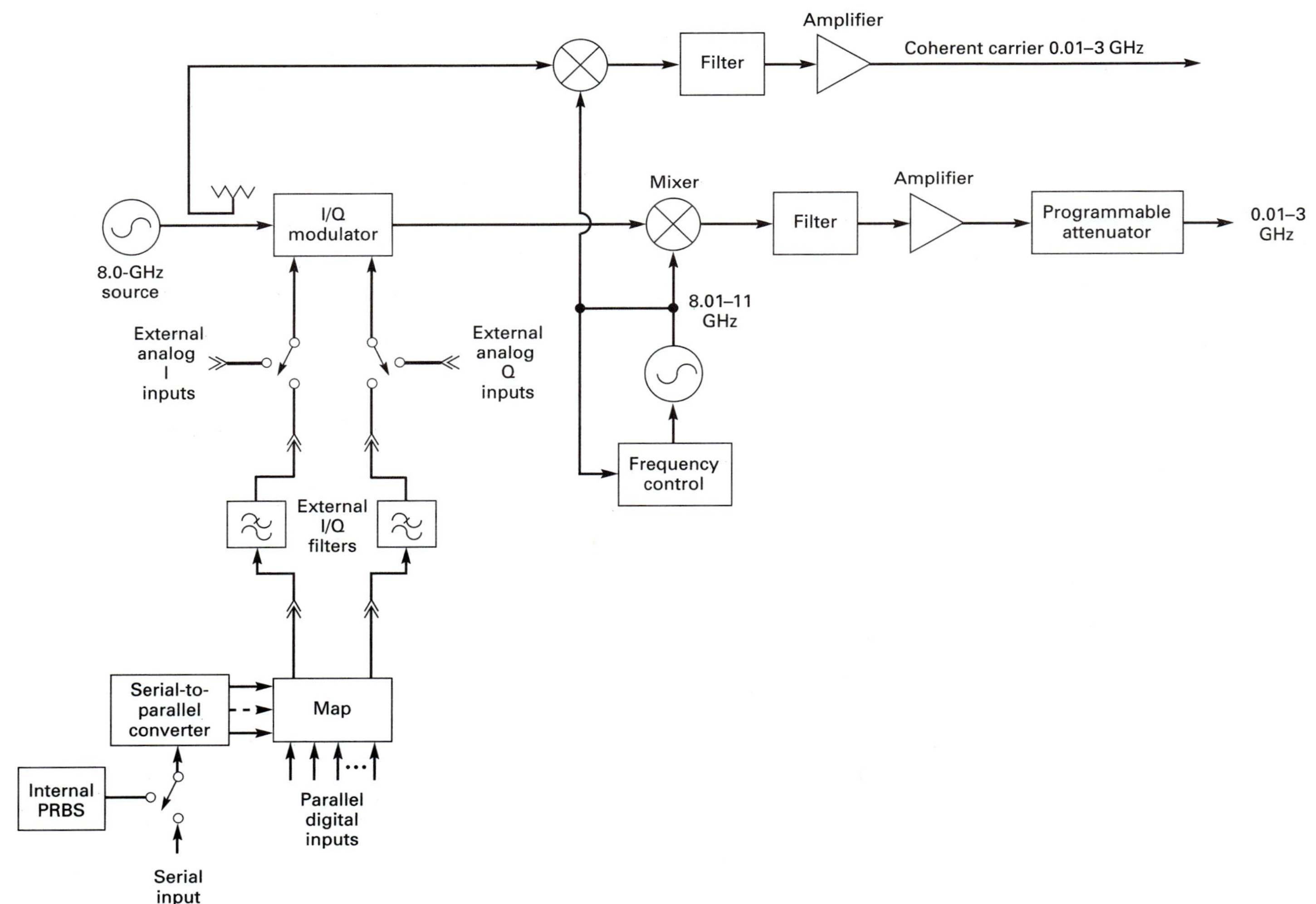

Figure 18.7 Block diagram of vector signal generator covering 0.01 to 3 GHz.

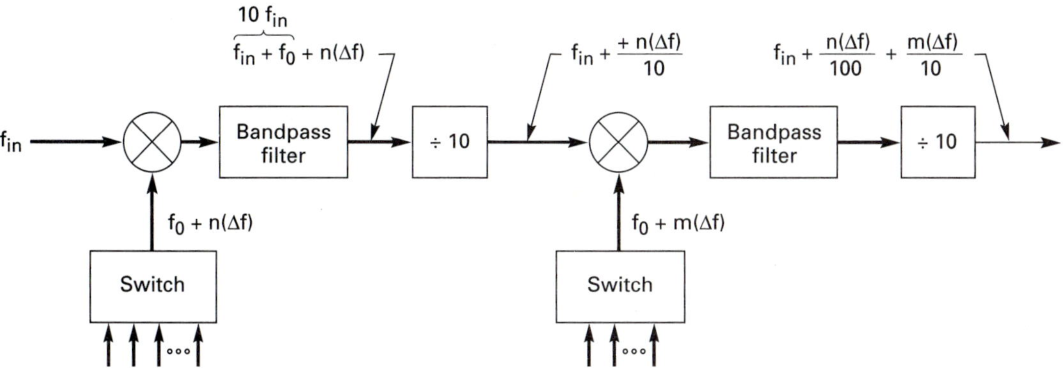

Figure 18.8 Agile (fast-switching) synthesizer using mix and divide technique. Each switch selects one of the 10 frequencies f_1 to f_{10} which satisfy the relationship $f_k = f_0 + k(\Delta f)$.

18.4.4 Signal generators with frequency agility

By employing direct synthesis techniques (see Chap. 8), broadband microwave frequency coverage can be obtained with fast frequency switching capability. Switching speeds in the range 100 ns to 1 μs are available in signal generators with frequency resolutions (step sizes) of 1 Hz.

Figure 18.8 illustrates a scheme utilizing a mix and divide technique to provide decades of fine frequency control in a modular architecture. Each of the identical modules contains a mixer, a switch selecting one of 10 input frequencies, a bandpass filter which selects the upper sideband from the mixer, and a divider (divide by 10). The same 10 frequencies f_k are supplied to each module and satisfy the following relationships:

$$f_k = f_0 + k(\Delta f) \tag{18.2}$$

where k is an integer $1 \le k \le 10$, Δf is the constant interval between each of the input frequencies, and the frequency f_0 is chosen such that

$$f_{\text{in}} + f_0 = 10 f_{\text{in}} \tag{18.3}$$

Thus the output frequency selected by the bandpass filter after the first mixer is

$$f_{\text{in}} + f_0 + n(\Delta f) = 10 f_{\text{in}} + n(\Delta f) \tag{18.4}$$

and the output frequency after the divider becomes $f_{\text{in}} + n(\Delta f)/10$. When this frequency is fed to the second modular stage, its output frequency becomes

$$f_{\text{out}} = f_{\text{in}} + \frac{n(\Delta f)}{100} + \frac{m(\Delta f)}{10} \tag{18.5}$$

thus illustrating how decades of frequency step size can be realized in a modular fashion.

AM and pulse modulation can be added in a manner similar to the techniques employed in Fig. 18.5. Since ALC loop bandwidths are typically limited to several hundred kHz, the time required for the signal level to stabilize may not always match the frequency switching times.

18.4.5 Signal generators with frequency agility and high-performance modulation

To meet demands requiring simulation of multiple signals having complex phase, frequency, pulse, and amplitude variations, systems are available which combine direct synthesis techniques with direct digital synthesis (DDS, see Chap. 8). The DDS portion generates a carrier with broadband modulation in the vhf frequency range, and direct synthesis, agile upconversion is used to translate it up into the microwave frequency range.

Figure 18.9 shows the DDS portion which generates modulated carrier frequencies in the range of approximately 15 to 60 MHz. Complex scenarios of frequencies and modulations are entered via the computer, which loads four random access memories (RAMs) to provide the required data. The phase accumulator and digital-to-analog converter (DAC) are clocked at a binary frequency of 2^{27}, about 134 MHz. Data stored in RAM determine the size of each phase increment added in the accumulator, and thus determine instantaneous frequency. The FM adder adds in any FM modulation called for, and phase modulation is added in just before the sine computer. Pulse modulation is provided by gating the data to the DAC.

Figure 18.10 shows how direct synthesis is used to provide agile frequency coverage from 0.01 to 3 GHz. The DDS signals are first converted up to 1.4 GHz and then on up to cover the 7.7- to 8.2-GHz range using 14 frequencies generated in the rf reference generator which are applied to the direct intermediate synthesizer. This contains switches and mixers that generate the 4.2-MHz steps covering the 6.3- to 6.8-GHz LO range. The microwave reference generator then provides the six frequencies required for the final downconversion.

Figure 18.11 shows the direct synthesizer used to extend frequency coverage from 0.5 to 18 GHz. The output shown in Fig. 18.10 is split, upconverted into two contiguous bands, recombined, amplified, and passed through a step attenuator. The architecture used in this design is determined by considerations based on minimizing spurious frequencies at the output. Switched filters are used to attenuate harmonics and those mixer spurs that are generated.

Systems of the type described above are available with software allowing the user to enter various modulations independently, and to provide user-defined waveforms for this purpose. Scenarios are built up and then played. It is also possible to operate in a mode whereby stored data are sequenced dynamically, i.e., as some external event takes place, and it is possible with

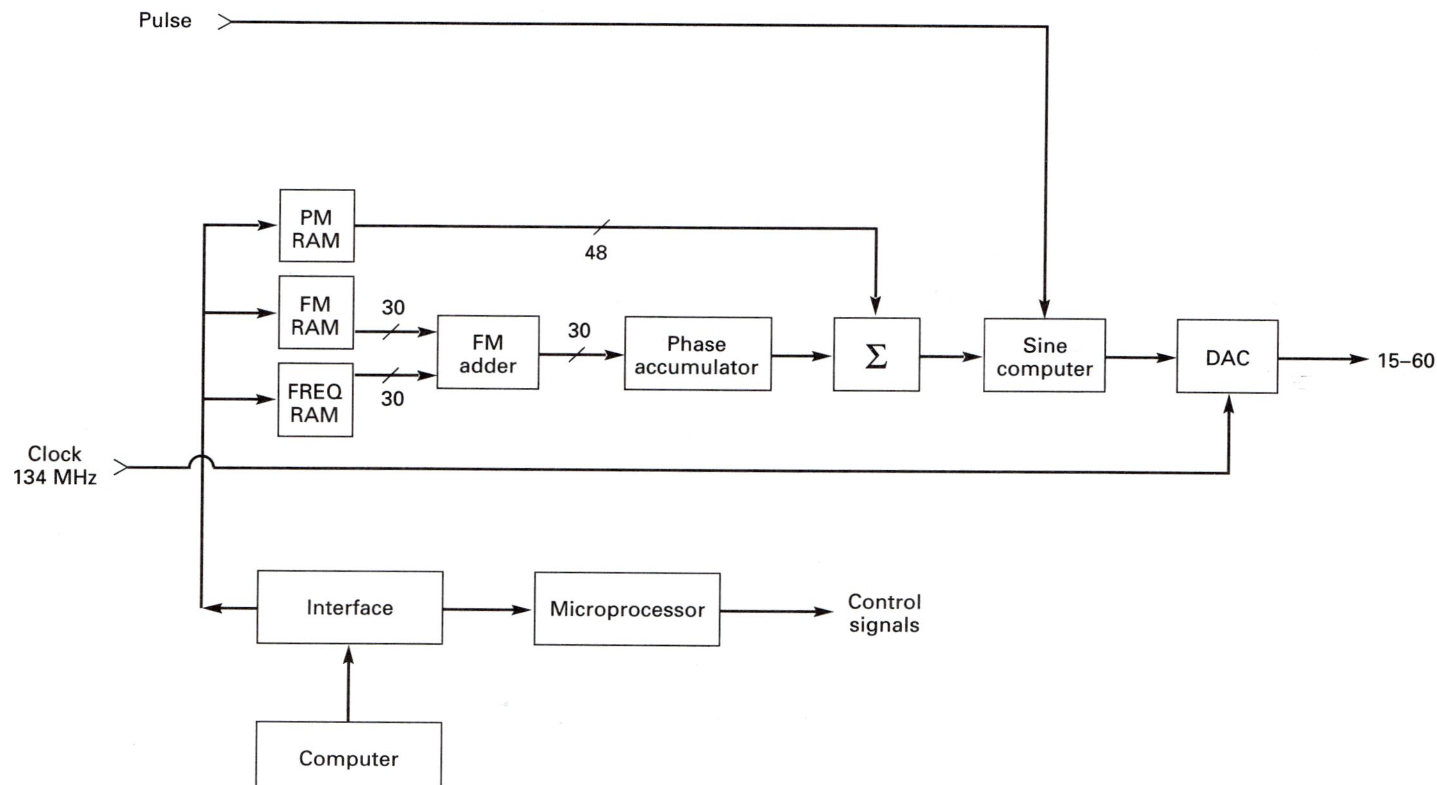

Figure 18.9 Block diagram of DDS carrier and modulation synthesizer.

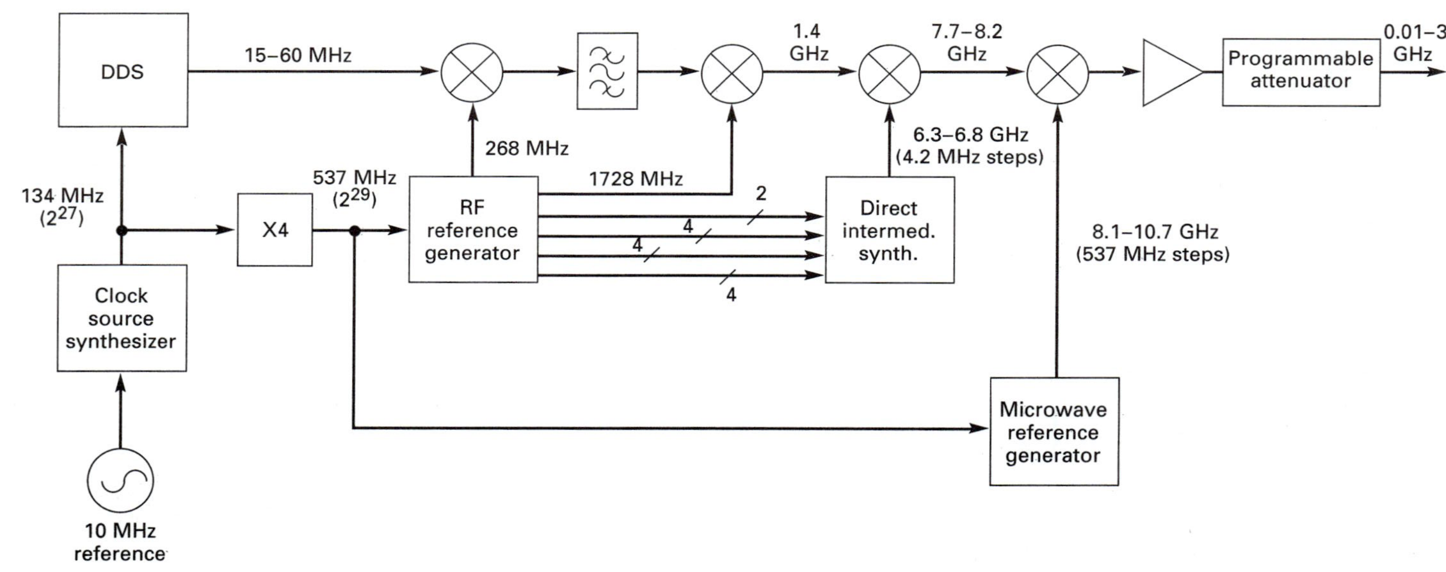

Figure 18.10 Block diagram of agile upconverter of DDS signals.

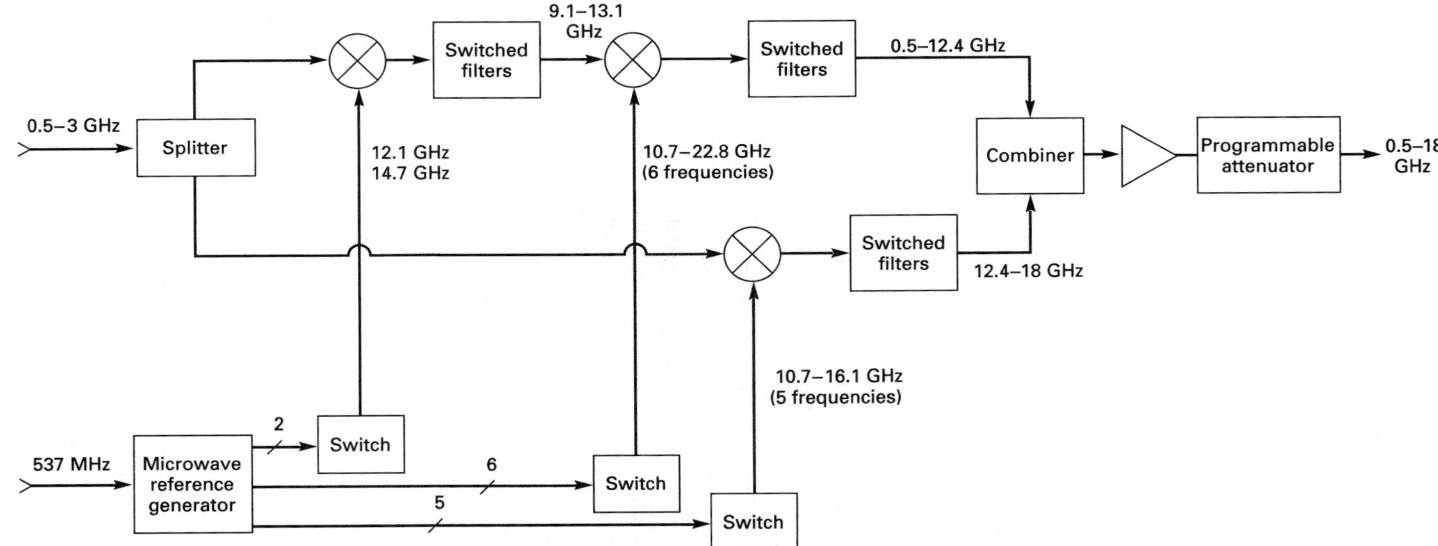

Figure 18.11 Block diagram of microwave agile upconverter.

enhanced computing capability to provide real-time data from an external signal processor.

18.5 Basic Microwave Signal Generator Specifications

The following discussion describes key specifications, their definition, and how they may be important in certain applications. It is not intended to be a complete list of specifications covering all the varieties of signal sources discussed above but is meant to provide the reader with an overview of how to interpret specifications found on a typical data sheet.

Specifications that are guaranteed by the manufacturer should be applicable over a specified temperature range. They should also include a margin for the uncertainties associated with the measurement procedures used.

18.5.1 Frequency

Stability. Frequency stability for the types of signal generators discussed above will depend upon the stability of the time base or reference oscillator, usually a 10-MHz TCXO. The primary factors affecting drift are variations in temperature, line voltage, and the aging rate of the crystal, and these are usually specified individually. These numbers are typically of the order of 10^{-9} each for a 1°C temperature change, for a 10 percent line voltage change, and per day of aging after a 24-h warmup.

Frequency resolution. This is the minimum step size that can be obtained. This number may increase at the higher frequencies of operation if frequency multipliers are used for frequency extension.

Frequency switching speed. This is the time for the new frequency to be stable within a frequency tolerance that should be clearly related to the frequency-resolution specification. The point in the programming sequence that is used to define the start also needs to be clearly delineated.

18.5.2 Microwave output

Maximum output power. This can be specified under various conditions. For generators with active leveling (ALC), maximum leveled power should be specified, and this may be broken up into frequency bands. Generally more power will be available at certain frequencies, and the data sheet may show this as a plot of maximum available power as a function of frequency presented as a typical but nonguaranteed characteristic.

Minimum output power. This is usually expressed as minimum leveled power and will be determined by the range of the step attenuator (if there is one) and the ALC vernier range.

Level accuracy. This needs to be expressed over the corresponding range of frequency and power level. Factory calibrations stored in ROM and effective algorithms for corrections at frequencies in between have led to claims of high accuracy on data sheets. Actual accuracies are limited by the uncertainties of the power measurement. At low power levels (below −90 dBm), these uncertainties are higher and leakage and radiation phenomena need to be carefully guarded against.

Flatness. This is the total peak-to-peak variation in power across a frequency band without relating it to absolute power.

Level switching time. This may be broken into two cases: level changes not requiring the step attenuator to change range and thus being primarily ALC loop bandwidth dependent, and those where the attenuator must be switched, typically adding 20 to 50 ms for an electromechanical device.

18.5.3 Spectral purity

Phase noise. This is usually expressed as single sideband noise in a 1-Hz bandwidth (see Chap. 22) at several offset frequencies from the carrier (usually in the 10-Hz to 1-MHz range). A number of microwave output frequencies may be specified. The data sheet will usually specify phase noise only in the optimum mode of operation, and not in modes where compromises in PLL bandwidth may need to be made to enable FM, for example. Phase noise is an important consideration, e.g., when the instrument is to be used as an LO where IFs fall within close offsets from the carrier, or when "spillover" from adjacent channels cannot be tolerated in multichannel communications system measurements.

Residual fm. This is the total phase noise resulting from an integration over a specified offset frequency range (e.g., 0.05 to 15 kHz) from the carrier, expressed as an equivalent frequency deviation. A synthesized microwave source should have residual FM values of several tens to 100 Hz through 10 GHz.

Harmonics. Second, third, and higher harmonics are usually specified in dBc, i.e., dB below the fundamental. They should be specified at or above some vernier output power level since amplifiers within the signal generator will generally add distortion at higher signal levels. Low harmonics are important when making broadband receiver measurements. Subharmonics may also need to be specified if a frequency multiplier is used in the source or for frequency extension.

Nonharmonic spurious. This can include all spurious signals in a particular offset range about the carrier, including line voltage frequencies and their harmonics, as well as any possible frequencies related to the frequency synthesis process.

18.5.4 Amplitude modulation

Rates. This refers to the range of modulation frequencies that can be applied, usually represented by the −3-dB response values. For logarithmic AM, the bandwidth or slew time may be called out by specifying a rise or fall time in response to the application of a step.

Depth. This is the maximum modulation index in percent that can be achieved for linear AM. For logarithmic AM, this can be expressed in decibels.

Sensitivity. This refers to the external signal level required to generate a particular modulation level, expressed as a voltage for full scale, or in decibels per volt for log AM.

Distortion. This is usually specified in percent as a total harmonic distortion (THD) value.

Incidental fm or pm. This is a measure of the amount of undesired FM or phase modulation created in the process of generating AM. It can be specified in radians of phase variation at a particular AM index.

18.5.5 Frequency modulation

Rates. The range of modulation frequencies usually expressed in terms of the −3-dB response frequencies.

Maximum deviation. The maximum frequency excursion that the microwave source can be driven to without causing excessive distortion. This value may be specified over a frequency range of operation and may decrease with the use of frequency dividers for low-frequency extension.

Modulation index. There will in general be limitations on the maximum modulation index depending upon the method used for generating FM. This will also be reduced with the use of frequency dividers.

Sensitivity. The level of external drive voltage to achieve a particular frequency deviation, commonly expressed in kHz or MHz per volt. There may also be an accuracy specification at a particular rate and deviation.

Distortion. This can be expressed in percent THD at a particular rate and deviation.

Incidental am. This refers to the level of undesired AM created while generating FM. It is expressed as a modulation index in percent.

18.5.6 Pulse modulation

On-off ratio. The ratio of carrier level during the pulse to the level during the rest of the cycle expressed in dB. For radar applications this number should be at least 80 dB.

Level accuracy. The level accuracy during the pulse, usually expressed relative to CW.

Rise time/fall time, overshoot, ringing, delay. See Fig. 18.1.

18.5.7 Vector modulation

Frequency response. The maximum modulation rate that can be applied to the I/Q inputs, usually expressed at the −3-dB point.

Sensitivity. The voltage required to achieve full scale of magnitude for I or Q.

Accuracy. This can be expressed in percent of full scale at a particular output frequency at dc modulation rate.

Crosstalk. The amount of coupling between I and Q channels expressed as a percent at various rates.

Chapter

19

Electronic Counters and Frequency and Time Interval Analyzers

Rex Chappell*

Agilent Technologies
Santa Clara, California

19.1 Introduction

For measurements of frequency, time interval, phase, event counting, and many other related signal parameters, the ideal instrument to use is an electronic counter or its cousin, the frequency and time-interval analyzer. These instruments offer high precision and analysis for research and development applications, high throughput for manufacturing applications, and low cost and portability for service applications.

This chapter describes several classes of counters, including the basic frequency counter, the universal counter, and the microwave and pulsed microwave counter. In addition, frequency and time-interval analyzers are presented, describing this measurement technology and illustrating its application to solving tough measurement problems. Also, several different architectures are covered, showing how each of the types of counters works.

19.2 Instrument Types and Application Overview

Electronic counters come in a variety of forms. This section provides a short explanation of some of the more-common types and outlines several of their applications.

---

* Adapted from Chapter 19, "Electronic Instrument Handbook, 2nd Edition" by Gary D. Saski, McGraw-Hill, 1995.

19.2.1 Frequency counters

The earliest electronic counters were simply that–counters that were used to count such things as atomic events. Before counters were invented, frequency measurement was accomplished with a frequency meter, a tuned device with very low accuracy. Frequency counters were one of the first instruments to digitally measure a signal parameter for a more precise measurement. Today, they are still one of the most precise instruments. Whereas a seven-digit digital voltmeter (DVM) is a relatively precise high-end instrument, a seven-digit frequency counter is considered to have relatively low precision.

Characterizing transmitters and receivers is the most common application for a frequency counter. The transmitter's frequency must be verified and calibrated to comply with government regulations. The frequency counter can measure the output frequency and key internal points, such as the local oscillator, to be sure that the radio meets specifications.

Other applications for a frequency counter can be found in the computer world, where high-performance clocks are used in data communications, microprocessors, and displays. Lower-performance applications include measuring electromechanical events and switching power supply frequencies. Finally, one of the earlier applications for a frequency counter was as a feedback component to control the frequency of a swept signal generator. A few of the digits from a frequency counter's display would be converted into an analog signal, and this would be used to steer the signal generator's frequency until the desired frequency was reached. Today, it is more common to use a synthesized signal generator, in which the frequency counter is a hidden component within the box. An exception is millimeter applications (such as at 94 GHz, used for some radars), where synthesized signal generators are less affordable.

19.2.2 Universal counters

Electronic counters can offer more than just a frequency measurement. When a counter additionally offers a few simple functions, such as period, the reciprocal of frequency, the instrument is sometimes called a *multifunction counter*. When two-channel functions, such as time interval, are provided, the instrument is usually called a *universal counter*. This name reflects this instrument's wide variety of applications. Several measurement functions provided by universal counters are shown in Fig. 19.1.

Time interval measures the elapsed time between a start signal and a stop signal. The start signal is usually fed into one channel (*A*) while the stop signal is fed into a second channel (*B*). The function is often called *time interval A to B*. The resolution of the measurement is usually 100 ns or better, sometimes down to picoseconds. Applications range from the measurement of propagation delays in logic circuits to the measurement of the speed of golf balls.

Variations of time interval that are particularly useful for digital circuit

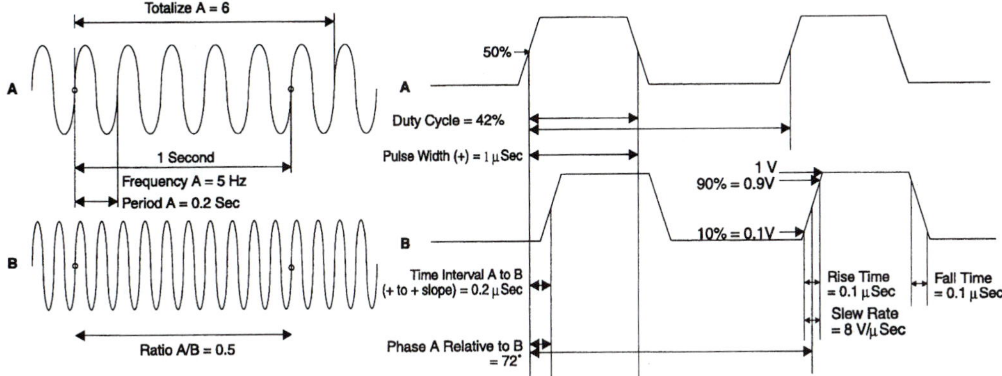

Figure 19.1 Several measurements that a universal counter can make. The A and B refer to the channel where the signal is coming in. Frequency counters often have just one input channel, but universal counters need two for doing timing and signal comparison measurements.

measurements are *pulse width, rise time*, and *fall time*. And, if the trigger points are known for rise and fall time, a calculation of the *slew rate* (volts per second) can be displayed. In all of these measurements the trigger levels are set automatically by the counter.

Time interval average is a function that yields more resolution and can be used to filter out jitter in signals. Another variant is ±*time interval*, which measures the relative timing between two signals—even as they drift back and forth with respect to each other. This is useful in adjusting timing skews to ensure the coincidence between two signals (Fig. 19.2).

Totalize is the simple counting of events. It is useful for counting electronic or physical events, or where a digitized answer is needed for an automated test application. The counter accumulates and displays event counts while the gate is open (see the basic counter architecture in the next section for an explanation of the gate). In some instruments, this function is called *count*.

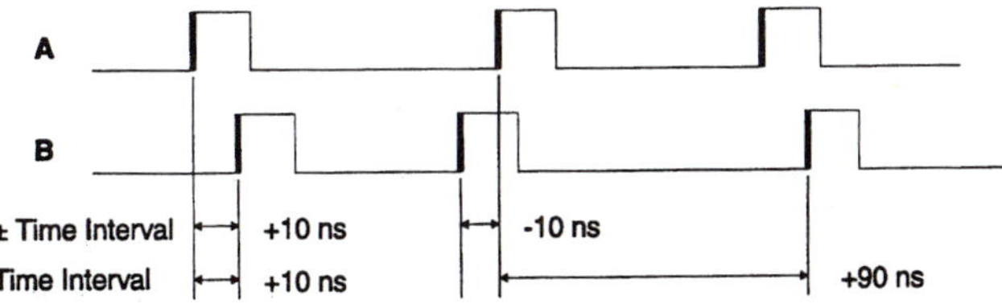

Figure 19.2 The difference between ordinary time interval and ±time interval. With ±time interval, the stop (B) edge can drift back and forth relative to the start (A) edge and the expected answers of +10 and −10 ns. With ordinary time interval, the unexpected answer of +90 shows up when the stop (B) edge drifts left. However, notice that some counters can measure ordinary time interval down to about 0 to −5 ns.

Totalize sometimes has some variations. Normally, to read the count to a computer, the gate must be closed first. This means that some counts can be lost while doing so. Some counters have *totalize on the fly*, which is the ability to read the count into a computer while the counting is still taking place. A few counters also have an ability to do *up-down totalizing*. Events in one channel cause counts to increment, but events in the second channel cause counts to decrement. One application is for comparing the rotation of two independent shafts that need to be compared with each other.

Frequency ratio A/B compares the ratio between two frequencies. It can be used to test the performance of a frequency doubler or a prescaler (frequency divider). *Phase A relative to B* compares the phase delay between two signals with similar frequencies. The answer is usually displayed in degrees. Some instruments display only positive phase, but it is more convenient to allow a display range of $\pm 180°$, or $-180°$ to $+360°$.

19.2.3 CW and pulse microwave counters

CW (continuous wave) microwave counters can be used for frequency measurements in the microwave bands. Applications include calibrating the local oscillator and transmitting frequencies in a microwave communication link. These counters can measure to 20 GHz, 26 GHz, 40 GHz, and even 110 GHz (or higher). They offer relatively good resolution, such as down to 1 Hz, in a short measurement time. This makes them popular for manufacturing test applications. Their low cost and high accuracy, relative to a spectrum analyzer, make them popular for service applications. Some microwave counters also include the ability to measure power. This is a typical measurement parameter in service applications.

The *CW microwave counter* gets the "CW" label because it is made to work on nonpulsed signals. However, some communication signals and many radar signals are pulsed. To measure these signals, the counter must be able to automatically gate the measurement at the appropriate times. Some can do this for moderately wide pulses, such as 500 μs, but many radars have pulse widths of 100 ns or less. To measure these signals, a pulse microwave counter must be used. Not only can these counters measure the carrier frequency, but they can also measure several parameters of the signal's pulse amplitude envelope, such as *pulse repetition frequency* (*PRF*, frequency of pulse bursts), *pulse repetition interval* (*PRI*, the timing of pulses), pulse width, and pulse off time. These are common measurements used in the design, testing, and analysis of radar signals (Fig. 19.3).

19.2.4 Frequency and time-interval analyzers

Modern applications, particularly those driven by digital architectures, often have more-complex measurement needs than those satisfied by a counter. These measurements are often of the same basic parameters of frequency, time, and phase, but with the added element of being variable over time. For example, digital radios often modulate either frequency or phase. In *frequency*

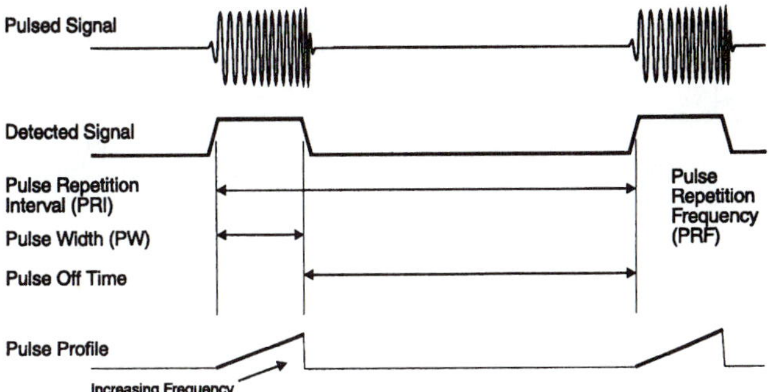

Figure 19.3 Pulse microwave counters can make several pulse measurements on the RF envelope. The amplitude envelope is obtained through the use of a diode detector.

shift keying (*FSK*), one frequency represents a logic "zero" and another frequency represents a "one." It is sometimes necessary to see frequency shifts in a frequency-vs.-time graph. A class of instruments called *frequency and time-interval analyzers* (*FTIA*) can be used for this measurement. An FTIA gives a display that is similar to the output of a frequency discriminator displayed on an oscilloscope, but with much greater precision and accuracy (Fig. 19.4). This type of instrument is also referred to as a *modulation domain analyzer*.

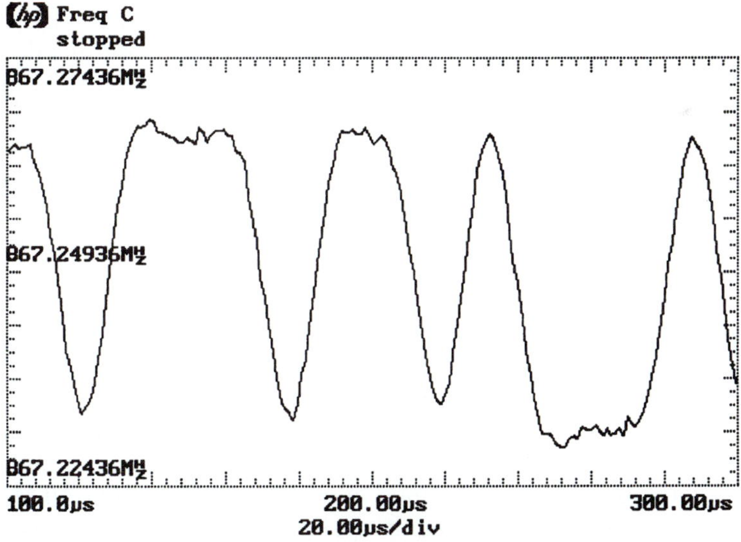

Figure 19.4 The display of a frequency-shift-keyed digital radio signal on a frequency and time-interval analyzer. Although it resembles an oscilloscope display, it is actually showing frequency vs. time, not voltage vs. time. The two frequencies, representing 0 and 1, are about 33 kHz apart. Each bit is about 15 μs long.

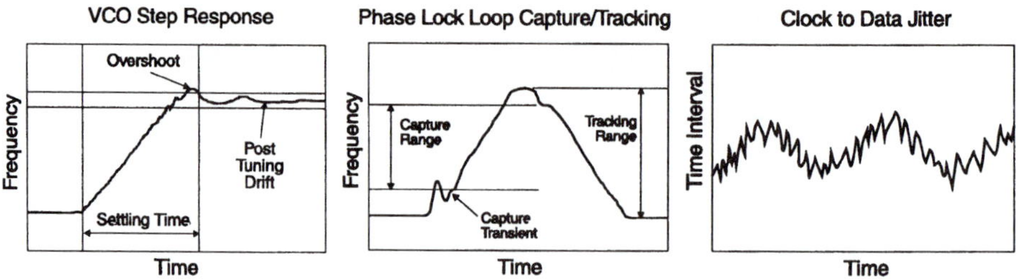

Figure 19.5 VCOs and PLLs are key components in a number of communication circuits, and many of the response parameters can be easily characterized with an FTIA using a frequency vs. time plot. Both of these can be viewed in a phase vs. time plot, too. For example, the response time of a PLL is best characterized this way. The third graph shows a time interval vs. time plot, which is useful for showing clock-to-data jitter. It shows periodic components to the jitter, plus an idea of jitter magnitude. These data can also be passed through an FFT to get the jitter spectrum.

Other applications to which an FTIA can be applied include *voltage-controlled oscillator (VCO)* analysis, *phase lock loop (PLL) analysis*, and the tracking of frequency agile or hopping systems. Some of these are shown in Fig. 19.5.

Phase can be displayed in a similar way. A graph similar to Fig. 19.4 could just as easily be shown for a *phase-shift keyed (PSK)* digital radio system. And, for phase-modulation based (direct sequenced) spread-spectrum radios, the phase distortion and modulation can be analyzed.

Time interval can also be displayed vs. time. One application of this is the analysis of *clock-to-data line jitter* in designs, such as high-speed computer circuits and digital communications. Excessive jitter causes data errors, and a fuzzy trace on an oscilloscope does not give sufficient insight into the source of the jitter or its statistical impact on the system. An analysis of time interval vs. time might show that there is a periodic component to the jitter with a frequency component that can be associated with some unintended part of the circuit.

Because frequency, phase, or time interval is being digitized, this information can also be displayed in the form of a histogram. Fig. 19.6 is a histogram of the signal in Fig. 19.4. This view of the signal can be used to calculate the center frequency by taking the midpoint of the two dominant frequencies, a measurement standard for some types of FSK radios.

When jitter is displayed in histogram form, it can be used for calculating the margin of the system. In order for this to be an effective measurement, however, millions of samples are required. Many frequency and time-interval analyzers can take these samples at rates of 1 Msample/s, 10 Msample/s, and higher. This is orders of magnitude faster than counters. A very specific example of this measurement is *window margin analysis*, used in testing disk drives. This and other measurement examples are shown in Fig. 19.7.

The digitized data can also be processed through an FFT to show the fre-

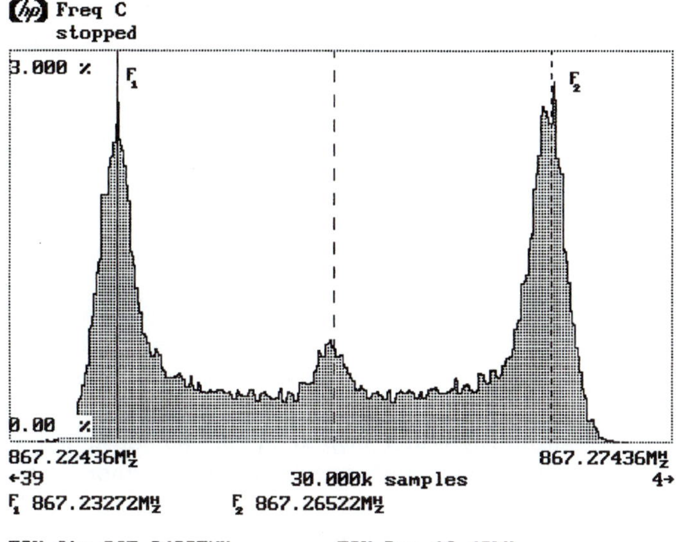

Figure 19.6 A frequency histogram display of the same signal from Figure 19.4 as shown on a frequency and time interval analyzer. Notice how the two frequencies are displayed. The lower frequency (solid marker, F_1) is 867.23272 MHz. The higher frequency (dotted marker, F_2) is 867.26522 MHz. The definition of the FSK center frequency is $(F_1 + F_2)/2$, and this is shown with the dashed marker and is displayed as FSK Ctr 867.24897 MHz. FSK deviation is also displayed.

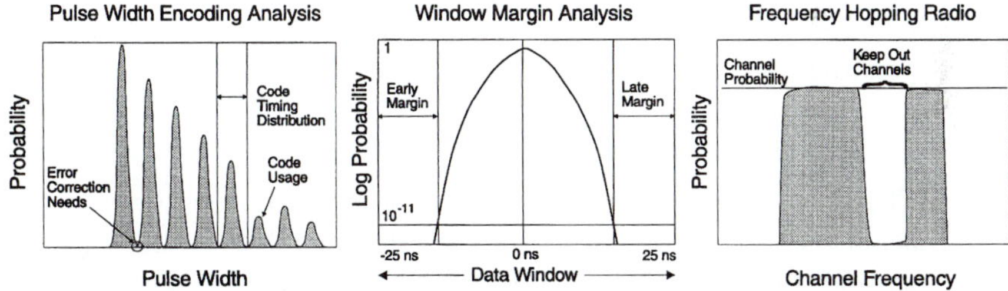

Figure 19.7 This shows three different measurement applications for the histogram mode in an FTIA. The first shows a pulse-width-encoded system (such as in a compact disc player). The code usage and errors are shown in probabilistic terms. The second shows a window margin analysis measurement, used in data-storage applications. This display shows the probability (log plot) of a data edge coming in at any particular time within a specified data window. The bit-error-rate expectations are very demanding (parts in 10^{10} or better), and the amount of early and late margin in the system is important to know. The third measurement shows how often each frequency channel is used in a frequency-hopping radio. The expectation is to have a flat distribution of channel utilization, except where specific "keep out" windows are specified. The wide bandwidth of the FTIA is a good match to frequency-agile systems. This cannot be done on a swept-spectrum analyzer because the signal is not at any particular frequency long enough for it to measure.

quency components of jitter, or the *jitter spectrum*. Other measurements include *jitter magnitude* and *jitter transfer*. This can be thought of as being a spectrum analyzer for digital circuits. Trying to do this same measurement on a spectrum analyzer would produce too many confusing spectral lines from the logic signal itself.

19.2.5 Some variations

A variation of the frequency and time-interval analyzers simply measures time. These are called *time-interval analyzers* (*TIA*), and they tend to be focused on data and clock jitter applications. These are generally a subset of an FTIA, although some may offer special arming and analysis features tuned to particular applications, such as testing the read/write channel timing in hard disk drives.

Another variation is the *scaler* or *photon counter*, designed specifically for scientific work, such as flight mass spectroscopy and fluorescence decay measurements. This has some similar properties to a frequency and time-interval analyzer. It can count (totalize) events in a continuous count vs. time mode. Counts are collected in bins, with each bin being a preset period of time. For example, once a trigger is sent, bin 1 could be during the first 1 ms, and bin 2 would be the next 1 ms, etc. Some instruments can display the results in graphical form and perform analysis on the data.

19.3 Basic Instrument Architectures and Operation

To understand the specifications and operation of counters and analyzers, it is useful to understand a few elements of their architectures. This section covers several different architectures, starting with the older basic counter and ending with the most recent frequency and time-interval analyzer. As will be seen, measurement resolution is one of the specifications most depending upon the instrument's architecture.

19.3.1 The basic counter architecture

The basic counter architecture is still used in very low-cost counters and in some DVMs, oscilloscopes, and spectrum analyzers that include a frequency or time function. This architecture has the advantages of a simple low-cost design and rapid measurement, but has lower resolution and reduced measurement flexibility.

The simplest way to measure frequency is to count the number of periods of a signal that occur during a 1-s gate time. To implement this, the signal is fed into an AND gate that is enabled for 1 s, and the output goes into a counting block. The counter's contents are then displayed directly as frequency with a 1-Hz resolution (thus, basic counters are also called *direct counters*). A gating circuit generates the 1-s gate time by sending a 10-MHz quartz time-base signal through a series of decade dividers.

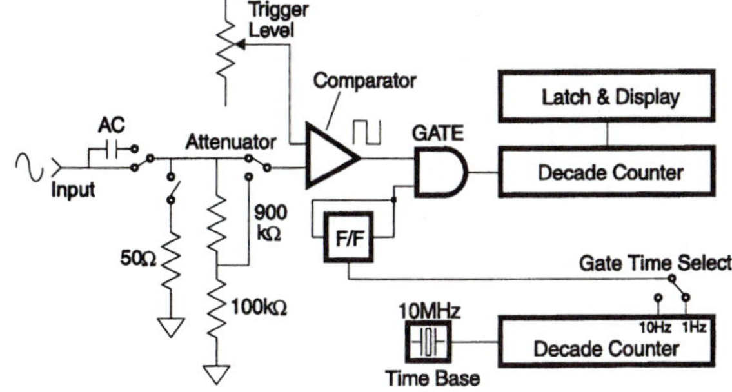

Figure 19.8 Simplified block diagram of a basic frequency counter.

To make it practical to measure real-world signals that do not always come in standard logic levels, the instrument usually has input signal conditioning (front-end amplifiers and comparators). The front end can have added switching to allow for different impedances, voltage ranges, and offsets. All of these elements are shown in Fig. 19.8. In reality, the actual instrument needs some synchronization and control to perform a measurement properly. For example, the display has a set of latches so that it can display the previous measurement while the next measurement is being counted. A very similar block diagram is used for the other measurements. For totalize, simply keep the gate open and display the contents of the counter continuously. For period, the input signal is used to open the gate on the first rising edge and close it on the second rising edge while the 10-MHz signal is fed directly to the counter through the gate. The counter shows the period with 100-ns resolution. Period average is a way for the basic counter to extend this resolution. This is done by opening the gate for 10, 100, 1000, etc., periods of the input signal. A decade of resolution is gained for each decade of averaging, so opening the gate for 100 periods yields a 1-ns resolution. With a basic counter, because of the potentially higher resolution given when measuring low-frequency signals, both period and period average are sometimes used to measure frequency. However, this means manually calculating the reciprocal of the period. For time interval, a second front-end circuit is added, and one channel is used to open the gate while the second channel is used to close the gate. While the gate is open, the 10-MHz time base signal is counted for eventual display, again with 100-ns resolution.

For all of these measurements, if more or less resolution is needed, the time-base portion of the block diagram can be changed. For example, to gain added frequency resolution, keep the gate open longer. A 10-s open gate provides 0.1-Hz resolution. If a shorter measurement time is desired, a 0.1-s gate provides 10-Hz resolution. If higher period or time-interval resolution is needed, multiple measurements can be averaged. However, for time interval, the resolution scales as the square root of the number of measurements averaged (note that this is in contrast to period averaging because the TI average is performed on unrelated samples). A set of 100 time-interval measurements provides a 10-fold increase of resolution to 10 ns. Most counters will automatically truncate the display to reflect the true resolution, but be aware that some will not, thus giving a false impression of instrument precision.

An alternative approach to higher resolution is to use a higher-frequency time base, such as 100 MHz. This provides 10-ns resolution. In a basic counter, 100 MHz is a practical upper limit; to gain additional resolution, a change in architecture is necessary.

19.3.2 Modern counter architectures

The basic counter has several disadvantages. For example, resolution is limited, particularly for low frequencies, where resolution greater than 1 Hz is desired (assuming measurement times greater than 1 s are impractical). The reciprocal counter was invented to address this and other limitations. This was further improved with the advent of the continuous-count counter.

Reciprocal counters. The basic counter was invented well before digital logic was able to economically perform an arithmetic division. Division is needed because a general way of calculating frequency is to divide the number of periods (called *events*) counted by the time it took to count them.

$$\frac{Events}{Time} = Frequency$$

The basic counter accomplishes the division by restricting the denominator (time) to decade values. Dividing by 10, 100, or 1000 is just a matter of positioning a decimal point in the display. After it became practical to perform a full division, the instrument no longer had to be restricted to decade gate times. More importantly, the time base could be synchronized to the input signal and not the other way around. This means that the resolution of the measurement can be tied to the time base and not to the input frequency. This particularly speeds up the measurement of low frequencies. This class of instruments is called *reciprocal counters*.

So, when using a reciprocal counter to measure a 60-Hz signal, the resolution with a 10-MHz time base is seven digits. This is 60.00000 Hz for a 1-s gate time. Note what happens if the frequency is 60.00001 Hz. In a basic counter, the answer is 60 Hz because the event counter would contain a count of 60 events, but the time would be measured as 0.9999983 s, giving an answer of 60.00001 Hz. Further, the measurement gate does not have to be open for exactly 1 s, so a ½ s gate time is also possible. This is important if the signal is pulsed or changing.

Note that, with a reciprocal counter, the number of digits of resolution scales with the gate time. If a 1-s gate gives nine digits of resolution, a 0.1-s gate gives eight digits, etc. Because the resolution does not always work out to ±1 Hz or ±10 Hz, but many times something like ±5 Hz, most counters will not show a digit, unless it is better than ±2 counts of the *least-significant*

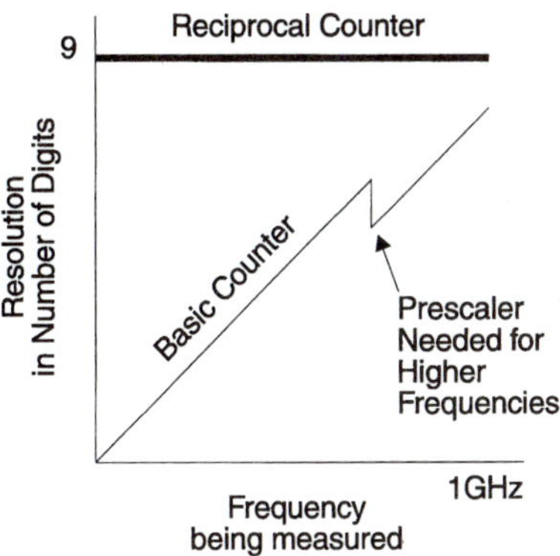

Figure 19.9 Resolution of a reciprocal counter with 1-ns interpolators vs. a basic counter, given a 1-s gate time. The reciprocal counter has the advantage of a constant number of digits of display, no matter what the frequency is.

digit (*LSD*) to avoid giving a false impression of the measurement's resolution.

Because resolution is determined by the ability to measure time, one way to extend resolution is to use a higher-frequency time base, such as 100 MHz. This approach is practical for time bases up to about 500 MHz, giving 2 ns resolution, or nearly 9 digits of resolution, for a 1-s gate time. See Fig. 19.9.

To gain even more resolution, a technique called *interpolation* can be used (Fig. 19.10). Because the input signal's event edges rarely line up with the time-base edges, interpolators are used to measure the time between these edges, shown as A and B. In this example, assume that the basic time base is 10 MHz, so each time base tic represents 100 ns. Now assume interpolators that can resolve time to 1 ns. Given this, the counter can measure the time as 659 ns (±1 ns), instead of the 700 ns (±100 ns) that a counter without interpolators would measure. This provides the equivalent of a 1-GHz time base, or nine digits of frequency resolution with a 1-s gate time. A simplified block diagram of a reciprocal counter with interpolators is shown in Fig. 19.11.

So, why this attention to gaining increased resolution? Aside from some applications, such as synthesizer design (which needs finer resolution), many people have the need to increase measurement throughput or responsiveness.

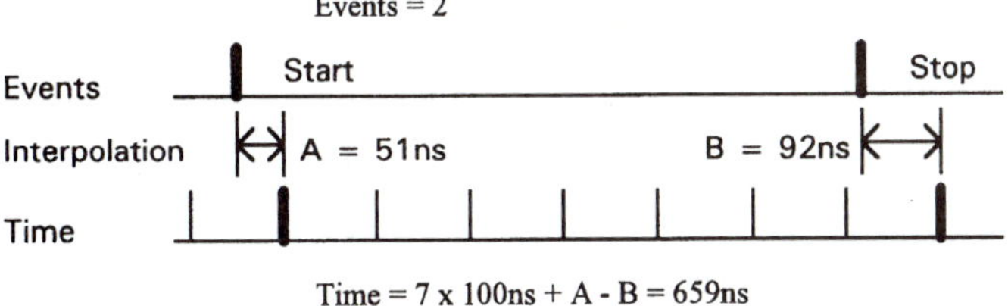

Figure 19.10 Time interpolation, showing how 1-ns time resolution is achieved through interpolation at the start and the stop of the measurement.

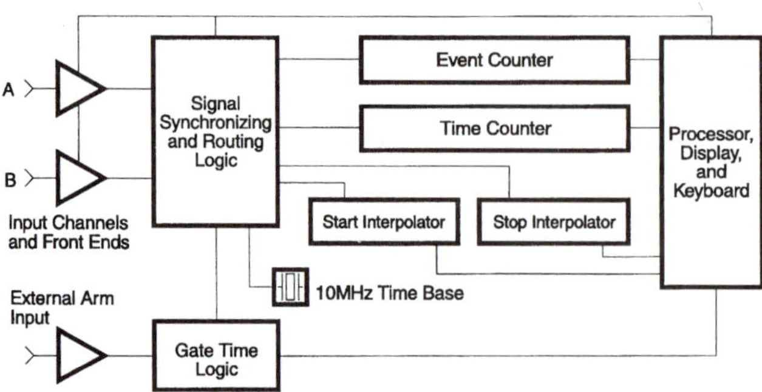

Figure 19.11 A simplified reciprocal counter with interpolators. The processor is central to the control of just about everything. The two input channels (A and B) pass through the front-end signal conditioning and trigger-level circuitry, and go into the signal synchronizing logic. This logic routes the A, B, and time-base signals to the event and time counters. Signal routing is determined by the measurement function. The synchronizer also sends the proper signals to the interpolators to measure. The external arm (covered later) optionally controls the starting and stopping of the measurement.

In an automatic test system in manufacturing or service, a shorter measurement time means higher throughput. With a reciprocal counter that has a 1-ns interpolator, a 10-ms gate time provides seven digits of resolution. Resolution of 1 Hz is given when measuring a 10-MHz signal, more than enough for most purposes. Allowing for the counter's processing time, it might provide around 25 measurements per second. Not only does this speed up testing, but it makes it much easier for a technician to adjust the frequency of the device under test.

The interpolator helps any measurement involving time. Frequency, time interval, pulse width, period, etc., all directly benefit. This higher resolution makes two types of measurement situations possible. First, throughput can be increased. A shorter gate time is available for frequency measurements. Also, for time interval measurements, the need to average to gain more resolution can be avoided. These both mean shorter measurement times and, therefore, more throughput. Second, some measurements need to be made single-shot because the signal is changing or transient. There might be only one chance to get the resolution needed, and the extra resolution gained through the interpolator can be key to making the measurement successfully.

Averaging still has a place in interpolator-based counters. First, even more resolution than that given by the interpolators might be desired. Second, averaging can take out some of the noise in the signal. Third, understanding the *standard deviation, min.* (*minimum*) and *max* (*maximum*) values from a sample of multiple measurements can also be useful. These statistics functions are often provided by the counter.

Continuous count counters. Up to this point, all the counters looked at share one common trait. They open and close a gate, then read the contents of

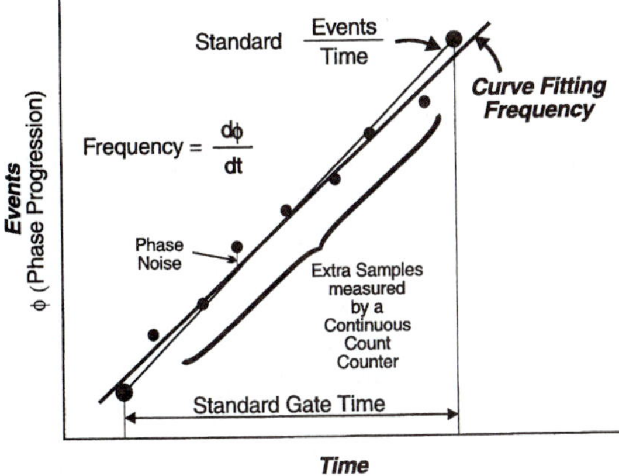

Figure 19.12 This is called a "phase progression plot." The Y axis shows the events counted and is equivalent to phase progression. The frequency is equal to the slope of the lines. A standard reciprocal counter has only two data points (shown with enlarged dots) with which to make a frequency calculation (events per time). A continuous count counter has several more data points, up to several thousand, with which to make a frequency calculation. Thus, more sophisticated curve-fitting algorithms can be used. This more effectively eliminates the errors because of phase noise on the signal.

the counter for eventual display. The time when the gate is closed between measurements is called *dead time*. Counters with dead time miss information in the signal. First, while the gate is closed, the signal can change, which could be important to the measurement. A simple example is with totalize. If closing the gate is necessary to accurately read the counters, events might be missed while the gate is closed. Second, each measurement is a disjoint sample of the signal. It is not known when measurement 1 was made relative to measurement 2. Therefore, the instrument cannot take any advantage of the relationship measurement 1 might have to measurement 2. One example is frequency measurement. If the relationship of each measurement to each other is known, curve-fitting algorithms can be applied to gain additional resolution beyond the level obtained in a standard reciprocal counter. Curve-fitting techniques, in fact, conform more closely to the definition of frequency.

The next level of counter architecture addresses these issues. This is called a *continuous count architecture*. It shares most of the attributes of a reciprocal counter, but adds the ability to read the event, time, and interpolator counters synchronously on the fly, without having to close a gate. So, when totalizing events, the count can now be reliably read at any time, and without missing any events coming into the counter while doing so. This is called *totalize on the fly*. This same concept can be applied to a number of other measurements. Also, curve-fitting algorithms can now be employed to gain additional resolution. In one instrument, this means the difference of 12 digits of resolution vs. 10 digits on a frequency measurement with a 1-s gate time, with no change in interpolators. An application of curve fitting to a frequency measurement is illustrated in Fig. 19.12.

19.3.3 Microwave counter architecture

Standard counter circuitry usually works up to frequencies of 100 to 500 MHz. To measure the higher RF and microwave frequencies, some form of frequency extension is needed. Two major approaches called *prescaling* and *downconversion* are covered here. A third technique, called a *transfer oscillator*, is not widely used and is not included.

Prescalers. In the simplest form, a prescaler consists of a set of fast D flip-flops configured as a divider. This divider would be placed between the front-end amplifier and the gating circuitry of the counter. This arrangement allows the divider to operate at a higher frequency because it is not involved in any of the complex synchronization circuitry. This also means that the divider does not have to operate in a static mode, so a type of self-oscillating prescaler that is capable of going much higher in frequency can be used. In other words, the type of circuitry used in many prescalers is fine for simple frequency division when a steady frequency is fed into it, but too unstable (owing to its tendency to self-oscillate) to be used for pulsed-frequency applications.

Prescalers in counters are available to 1 GHz, 3 GHz, and 5 GHz, for example. They are generally offered as an optional "C channel" (or channel 3) or a third channel that is separate from the standard A and B channels. When used in a basic counter, the prescaler causes a reduction in resolution. This is because the frequency resolution of a basic counter depends on the contents of the event counter; because the prescaler is placed before the gate, the prescaler contents cannot be read. When used in a reciprocal counter, the resolution is not changed. This is because the frequency resolution depends on the time resolution. So, a counter that displays nine digits in 1 s will still display nine digits in 1 s with a prescaler. The least-significant digit, however, might now be 1 Hz instead of 0.1 Hz because the counter is now reading a higher frequency.

Downconversion. Conceptually, the simplest downconverter is a fundamental mixer. The input signal is fed into a mixer along with a *local oscillator* (*LO*). The resulting output (intermediate frequency, IF) is the difference between the input and the LO, which is a lower frequency that the counter can count. The counter can then calculate the input frequency as $LO + IF$. Covering a frequency range of 20 GHz or more, however, makes the LO too expensive for microwave counters.

To overcome this limitation, a harmonic sampler is used. A simplified block diagram is shown in Fig. 19.13. In this approach, a low-frequency LO (for

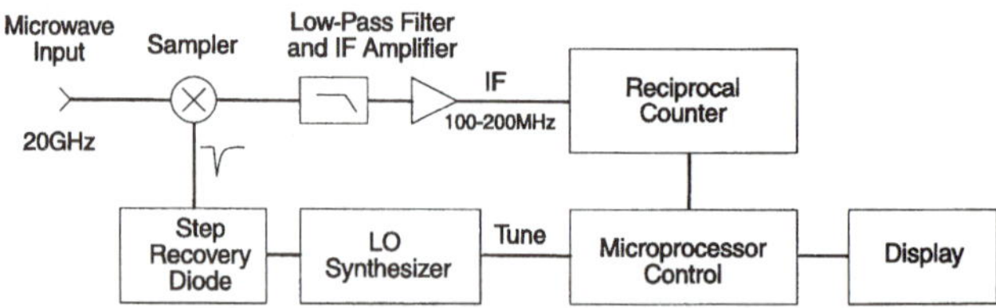

Figure 19.13 Simplified sampler-based microwave counter.

example, 200 MHz) is used to drive a step-recovery diode. This produces a very sharp pulse with usable harmonics up to 20 to 50 GHz. This signal is used to drive a sampler, which essentially samples points of the input signal. The resulting output (IF) is low-pass filtered and counted. Now, however, the counter still does not know which harmonic of the LO was being mixed with the input frequency to provide the IF counted. This harmonic, called N, is the missing ingredient to the calculation: $Input = (N \times LO) + IF$.

The lowest-cost approach to determine N is to shift the LO's frequency slightly (to LO_2) and measure the new IF, which is called IF_2. Now there are two equations and two unknowns. Calculate N with the formula $N = (IF - IF_2)/(LO - LO_2)$, and then calculate the input frequency. Note that the instrument is not always lucky enough to have placed the original LO frequency in just the right spot to have the IF frequency within a tolerable range for all this to work, so some initial scanning by the LO is generally done.

All of this process is hidden from the user, except that it takes some time and certain measurement limitations are imposed. The time required to vary the LO, measure the IFs, and make the calculations is called *acquisition time*, and it is usually several tens of milliseconds. To save this time, if the operator already has a general idea where the signal is, many instruments allow manual setting of the LO by specifying the input's frequency range.

One of the limitations imposed by this measurement approach is FM tolerance. If there is too much FM, the signal would jump outside the tolerable IF range, possibly giving a wrong answer. Microwave counters are designed to handle most forms of communication signals, but frequency-hopping radios and wide-bandwidth radars represent signals that might not be measurable.

Another approach to downconversion is to use a preselector. This is typically a tunable YIG (yttrium iron garnet) filter that is placed in front of the sampler. The YIG's center frequency is generally known, so the harmonic number N can be inferred directly. The YIG also has the advantage of filtering out signals outside of its bandpass range. This allows the measurement of a signal in a multiple-signal environment, which can be an advantage in certain field measurements.

The YIG has its own acquisition-time specifications because it must scan the entire frequency range until it finds the signal, and YIGs also require a bit of time to tune. The process is generally a bit slower than a straight harmonic sampler, so acquisition times are in the 250-ms range. YIGs also have an FM tolerance specification that is determined by the bandpass characteristic of the filter. When a downconverter is used with a reciprocal counter, there is a gain in resolution. This is because the measurement is being made on the IF frequency and not the input frequency. As an example, imagine measuring a 10-GHz signal, and that the counter's IF is at 100 MHz. If the reciprocal counter can resolve nine digits per second, it will be able to measure the IF with 1-Hz resolution, given a 1-s gate time. When the LO frequency is added back to get the result of 10 GHz, the same 1-Hz resolution is kept. This is like getting 11 digits of resolution.

For frequency measurements above 50 GHz (for some instruments, above 26 GHz), the internal sampler might not have sufficient performance. So, for these frequencies, an external mixer that sits in an external pod is used. These mixers are "banded" and will cover only a certain frequency range. The output of these mixers is then fed into the microwave input of the counter.

Connectors. At this point, there is also the issue of connector choice. Up to about 20 GHz, a type-N connector is often used because of its cost and ruggedness. To about 40 GHz, an APC 3.5 type of connector (compatible with SMA connectors) can be used. The connector is actually rated to 26.5 GHz before it starts to mode (distort), but because a counter only "cares" about zero crossings, the distortion is not a factor. To about 50 GHz, a 2.4-mm connector can be used. Waveguides are needed at higher frequencies. External mixers generally use waveguides. The disadvantage of waveguides is that they work over a limited range of frequencies, so choose a counter that uses a coax connector, if flexibility is needed.

19.3.4 Pulse microwave counters

Pulsed microwave counters build upon the previously described architectures by adding gating circuitry to position the gate comfortably within the pulse. The leading and sometimes trailing edges of the pulse usually have some form of distortion, so the instrument positions the gate to avoid the very beginning and end of a pulse. Delaying the opening of a gate is a simple matter, but closing the gate before the pulse ends is trickier.

One method that has been used is based on a delay line. The IF signal is delayed so that the instrument can detect the end of the pulse before the delay line and measure the signal after the delay line. Another method is digitally based. The pulse widths are observed and the gate width is digitally set accordingly. In both cases, the variable gate time that a reciprocal counter affords is necessary. The margins needed at the start and end of the pulses, plus the choice of IF frequency, help to determine how short a pulse can be measured. To handle most radars, this specification should be 100 ns or less.

The resolution from the measurement of one pulse is rarely enough for most applications, so many measurements need to be averaged. This makes the measurement time depend on the duty cycle of the input signal. If pulses come infrequently and are short, the measurement time is longer than if they come more frequently and/or are wide.

19.3.5 Frequency and time interval analyzer architecture

Building upon the architectural concepts of the continuous count counter is the *frequency and time-interval analyzer* (*FTIA*). The FTIA shows much more about a signal than just one parameter. As explained earlier, graphs showing frequency vs. time, or histograms, can be displayed on a CRT. Time and

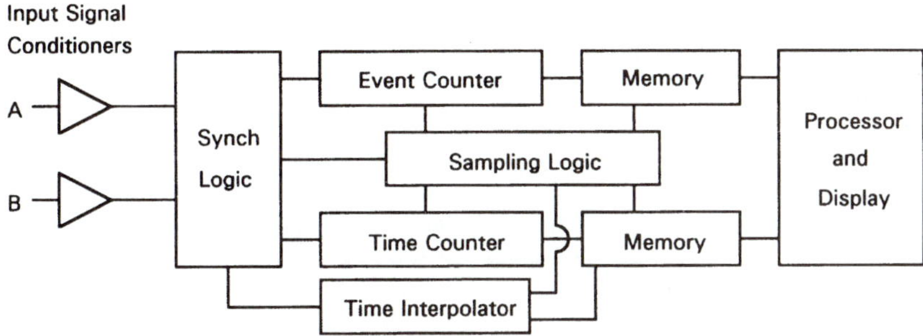

Figure 19.14 The main difference between this and a counter is the sampling logic and memory elements of the hardware, plus the analysis and display of the measurements. When the sampling logic triggers a sample, the contents of the counters and interpolator is read into the next free memory location. The interpolator is then reset for the next sample point. Samples are taken until the specified number is taken, and the memory is then read by the processor for calculations and display. Unlike a counter, the display is often graphics based.

phase can also be displayed these ways. Figure 19.14 shows a simplified block diagram.

To best understand FTIAs, think of them not as a counter, but as a signal digitizer. However, unlike an ADC-based digitizer, which digitizes the voltage waveform, an FTIA only digitizes the timing of the zero crossings of the signal. Another way to look at this is that it digitizes the signal's phase progression. From this information the parameters of frequency, time, and phase can be calculated. Figures 19.15 and 19.16 show how the signal is digitized and how the parameters of frequency and phase are derived.

Note that another common way to measure phase change is with a quadra-

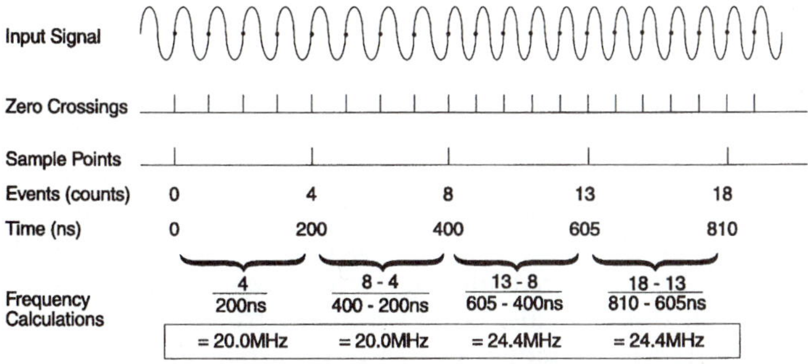

Figure 19.15 Five sample points are shown, which make up four frequency calculations in a continuous-count-based frequency and time-interval analyzer. The signal is shown changing in frequency during the third interval. The "sample points" are specified by the user; in this case, it was probably set to sample at a 5-MHz rate. The exact points in time that the samples are taken vary a bit, depending upon the timing of the input signal relative to the sample points, as can be seen in the last sample.

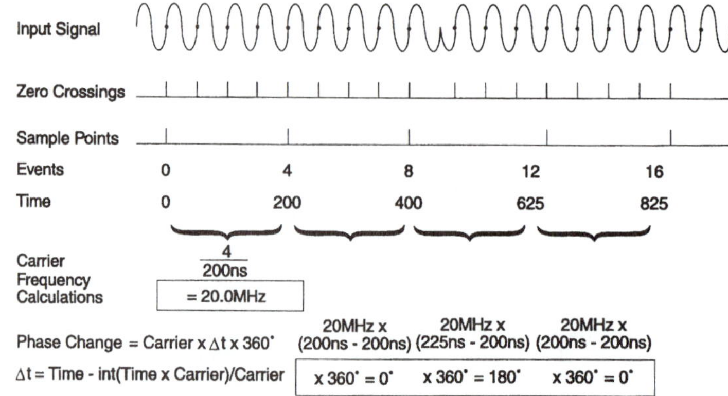

Figure 19.16 Phase modulation in a carrier can be measured by first determining the carrier frequency. In this example, the carrier is 20 MHz, determined by examining the frequency sample measurements. Then, using the relationship $d\phi/dt = F$, calculate phase changes by looking at the changes in timing. The change in timing is Δt, and the measurement is looking for how much the signal has shifted, left or right. This example shows how a 180° phase change was measured and calculated.

ture detector, but that this approach requires a local oscillator that is coherent to the carrier. The FTIA does not require any prior knowledge of the carrier and does not need a coherent LO. In fact, the LO can be a frequency hopper, and the FTIA will calculate phase properly. Important block-diagram parameters that need to be understood in an FTIA include the maximum and minimum sample rates, the measurement resolution at each sample rate, and the depth and width of the memory. These topics are covered in Sec. 19.4.

Enhancements. Histograms of the data, as in Fig. 19.6, show useful information for a variety of applications, but if the operator has to wait a long time to collect a useful set of data, the value of the measurement diminishes. If the instrument collects histogram data through a hardware histogram method, the measurement rate is many times faster than if it is done through software. For measurements requiring resolution of very small probabilities, such as are found in many data communication and data-storage applications, a hardware-based histogram is a much more practical approach.

Frequency extensions are also useful for measuring both RF and baseband digital signals. If a prescaler is added in the front end, the analysis of widely varying signals can be made. For example, some radars are frequency agile over a 2-GHz range. With a 2-GHz prescaler in an FTIA, it is easy to measure and analyze how the frequency changes over this entire range. No other type of instrument can do this. The prescaler can also be used on jitter measurements. It is a common misperception that timing information for jitter measurements is lost if a prescaler is used, but, if properly designed, this is not true.

Constant event sampling is the ability to sample at every *N*th zero crossing. This is most useful for time interval analyzers that do not count the number of events between samples. Normally, if a TIA can sample at maximum of 10 Msample/s, the maximum input frequency that it can handle is 10 MHz because it must assume only one event per sample. With constant event sampling, the TIA places a programmable divider in front of the sampling logic so that only every *N*th event is seen. This is similar to having a prescaler. As an example, it could be set up so that only every tenth pulse was sampled. The result is that the TIA will now be able to measure pulses that come in at up to a 100-MHz rate. This opens up more frequency dynamic range and is useful when TIAs look at high-speed data and clock signals, as well as for frequency measurements.

Continuous time is the ability to mark the timing of each edge relative to a single reference point. This is similar to *split time* in sports. In contrast, *time interval* measures the time between two edges. Continuous time is useful when the timing of every edge on a single signal needs to be known. Examples include looking at serial data streams in disk drives and staggered pulse repetition intervals in radars.

19.3.6 Other architectures

Other approaches exist for measuring frequency and time. In some oscilloscopes, the markers can be used to pick two points on the display and the processor calculates time by knowing the time between these markers. Frequency can be calculated by taking the inverse of this time and assuming the markers represent a full period of the signal. This method, however, is slow and not very precise, so many newer oscilloscopes include some form of basic or reciprocal counter built in.

In spectrum analyzers, the markers can be used to read out the frequency of any signal or spur. Again, the measurement is slow and less precise, although most analyzers have a feature that moves the marker to the largest signal automatically. Some spectrum analyzers also have built-in counters, but some others work with external counters for better precision. In integrated circuit test systems, time is not measured, but is go/no-go tested. A timing window is set up that looks for a logic signal edge; if the edge comes within this window, the test is considered passed. In an instrument called a *phase margin analyzer*, a similar window approach is taken for time measurements, only the window is expanded and contracted to get a picture of the amount of jitter on the signal. Multiple passes are needed to get a complete picture. This type of instrument is used to test the read-write channels in hard disk drives, and it includes several other measurement functions.

19.4 Specifications and Significance

Once there is some appreciation for the architectures of the instrument, the next level of understanding is to look at the detailed specifications. The product data sheets and manuals often provide fairly detailed specifications, so this section covers how to interpret some of these and understand their significance in sample applications. Specifications for counters are covered first, followed by additional specifications found in frequency and time-interval analyzers. For a description of specific measurement functions, see Sec. 19.2.

19.4.1 Universal counter specifications

Many of the specifications found in a universal counter data sheet can be found in other types of counters and analyzers.

Sensitivity and bandwidth. *Sensitivity* refers to how small a signal the instrument can measure. It is usually given in mV rms and peak-to-peak (p-p) values (in RF and microwave counters, it is given in dBm). Because of the element of noise and the broadband nature of a counter's front end, the sensitivity is rarely better than 10 mV rms or 30 mV p-p. For frequency measurements, the sensitivity can be important if measuring a signal off the air with an antenna. Under most situations, however, there is enough signal to measure and sensitivity is not an issue.

The counter's front-end bandwidth is not always the same as the counter's ability to measure frequency. If the bandwidth is lower than the top frequency range, the sensitivity at the higher frequencies will be reduced. For example, a counter that can measure 100 MHz might have a front-end bandwidth of 50 MHz. In this case, the sensitivity might be 25 mV rms over 0 to 50 MHz, and 50 mV rms over 50 to 100 MHz. This is not necessarily a problem if there is enough signal strength at the frequencies being measured.

For timing measurements, however, bandwidth can affect accuracy. An input channel with a low front-end bandwidth can restrict the minimum pulse width of the signal it can measure, and, for very precise measurements, it can distort the signal. One example is rise time. The following formula shows how bandwidth can slow down the edge of a signal and affect the accuracy of a rise-time measurement.

$$\textit{Actual rise time} \cong \sqrt{(\textit{displayed rise time})^2 - (0.35 \times \textit{front-end bandwidth})^2}$$

So, for a 100-MHz bandwidth front end, and a 10-ns displayed rise time, the actual rise time is closer to 9.4 ns.

A more-subtle example is when measurements are being made on digital signals with data present. One logic edge can affect an adjoining edge and distort the timing of the edges. When data are present, the spacing of these edges is unpredictable; therefore, errors can be introduced. Counters with higher bandwidths have less trouble with this. Also, this is not a problem when measuring such regular waveforms as clock signals.

The bandwidth and the counter's counting circuitry also determine the minimum pulse width that can be handled. If pulses are too small, they are missed by the counter. So, if precise timing measurements are being made it is best to get a counter with as high a bandwidth as possible. Because bandwidth is usually not specified, it must be inferred from the sensitivity specifications. A clue can be found by looking at the sensitivity specification across the frequency range of the counter.

Resolution. Counters are commonly used because they can provide reasonably good resolution in a quick, affordable way. This makes resolution an important specification to understand. It is also a specification that can influence throughput.

Least-significant digit. The first factor to understand is called the *least-significant digit* (*LSD*). The LSD is usually very close to the resolution of the measurement, which is often ±1 count of the LSD. This is generally self-evident, but, in some instances, the LSD might not be what it appears in some counters. This is because the LSD can be set somewhat arbitrarily by the instrument manufacturer.

The first case is in averaging. If a counter averages N measurements, it gains resolution only by a factor of the square root of N. So, taking 100 measurements gains only one more digit of resolution, not two. Most counters automatically throw away the irrelevant digit(s), but be aware that some do not.

The second case is with many counters that use interpolators. The timing resolution of interpolators is not always a decade value like 1 ns, but might be 2 ns or 20 ps. In such cases, because decade-value gate times are often used, the counter might show an LSD that is good to ±2 counts or so. The third case is the complement of the second, where the counter's displayed LSD resolution is worse than the theoretical resolution. For example, the counter might be able to display ±500 ps, but chooses to truncate to 1-ns displayed resolution.

Trigger error and noise. Noise can be found in the counter's front end, as well as in the signal being measured. The rms value of these noise factors, coupled with the input signal's slew rate, determines another element of resolution.

$$\textit{Trigger error} = \frac{\sqrt{(\textit{rms noise of input channel})^2 + (\textit{rms noise of signal})^2}}{\textit{slew rate}}$$

Note that trigger error is in units of time. The LSD and trigger error are usually added together to determine the resulting displayed resolution. For a frequency measurement:

$$\begin{gathered}\textit{Displayed frequency resolution} = \\ \pm N \times LSD \pm \sqrt{2} \times \left(\frac{\textit{trigger error}}{\textit{gate time}}\right) \times \textit{frequency}\end{gathered}$$

N is the number of counts that the LSD is allowed to move, owing to the interpolator resolution, and the factors in the error at the opening and closing of the gate. For example, say that the counter has a 2-ns resolution interpolator, input channel noise is 250 mV rms, the input signal has 1-mV rms noise (over the bandwidth of the front end of the counter), and that it is measuring a 1-MHz sine wave with a 1-s gate. The LSD will be 0.001 Hz (assume that the counter allows a ±2 count display), and the trigger error is

$$\textit{Trigger error} = \frac{\sqrt{250\,\mu\mathrm{V}^2 + 1000\,\mu\mathrm{V}^2}}{3.5 \times 100\,\mathrm{mV}/1\,\mu\mathrm{s}} = 3 \times 10^{-9}$$

so:

$$\textit{Displayed frequency resolution} = \pm 0.002\ \mathrm{Hz} \pm 1.4 \times 3 \times 10^{-9} \times 1\ \mathrm{MHz} = \pm 0.006\ \mathrm{Hz}$$

This means that the display's least-significant digits will probably change by about ±6 counts. The actual amount depends upon how conservative the instrument manufacturer was in setting the specifications. For example, the resolution might represent one standard deviation or, perhaps, three standard deviations.

In some products, the noise component of the LSD is separated out and a slightly modified frequency-resolution equation is used. This provides an understanding of the truer resolution of the measurement without having it masked by the counter's choice of LSD.

$$\textit{Frequency resolution} = \pm \frac{\sqrt{(\textit{interpolator noise})^2 + (2 \times \textit{trigger error}^2)}}{\textit{gate time} \times \textit{frequency}}$$

Notice that this is not necessarily the displayed resolution, particularly for a nonaveraged measurement. The display might show less resolution, and the full theoretical resolution might be realized only when averaging several measurements together.

Notice that when measuring digital signals, the trigger error component goes to 0 because the slew rate becomes a large number. Also, some manufacturers do not specify the counter's input noise specification. Noise in the signal can be partially addressed with input signal conditioning by decreasing the sensitivity and/or low-pass filtering the signal. Another way to filter out noise, and gain a bit more resolution, is with averaging. Another form of "noise" is modulation. Because a counter counts "zero crossings," the instrument can be confused if the waveform is not "well behaved." The most common occurrence for this is in digital radios and phase-modulated radars. If the phase or frequency transitions are not phase continuous, there is a greater chance of confusing the counter. Depending upon the situation, the counter would either over- or undercount the signal, producing either a low or a high answer. If the modulation rate is such that very high sampling of the signal is possible by either a continuous count counter or a frequency and time analyzer, the effect of the modulation can be taken out and even analyzed.

Accuracy and time bases. Accuracy is very tightly related to resolution, but it is not identical to resolution. Other factors must be added to the resolution specification to determine the ultimate accuracy of the measurement. One of these factors for most measurements is the time base. Three main issues affect the time base: a 10-MHz quartz crystal-calibration, environment, and stability. Obviously, if the time base was miscalibrated, the entire measurement will be biased. For example, if the time base is off by 1 part in 10^6, a frequency measurement can only give six digits' worth of accuracy—even though nine digits might be shown on the display.

Even if the time base was calibrated properly, it was probably done at room temperature. A typical uncompensated time base can change by as much as 5×10^{-6} over a 0 to 50° range. So, to compensate for this, most counters at least offer a *temperature-compensated quartz oscillator* (*TCXO*), which attempts to electronically compensate for changes in temperature. An even-better solution, of course, more expensive, is an *oven-compensated quartz oscillator* (*OCXO*) that keeps the quartz at a constant elevated temperature all the time. A good OCXO can keep the stability to $<7 \times 10^{-9}$ over a 0 to 50° range. The newest generation of time base is called a *microprocessor-compensated quartz oscillator* (*MCXO*), which falls between the TCXO and OCXO in both price and accuracy.

Once the time base is calibrated, it will still drift over time. This is the aging-rate specification. For a good OCXO the aging rate is $<5 \times 10^{-10}$/day, but a standard quartz time base might age at 1×10^{-8}/day. After 6 months, a typical calibration cycle period, the OCXO would drift by up to 1×10^{-7}, and the standard time base might drift by 2×10^{-6}.

To bring this all together, use the previous example of a 1-MHz frequency measurement and now add the time-base effects to determine accuracy. Assume that the measurement is being taken in the field (thus, temperature is an issue), three months after a calibration, and that the counter uses a good-quality OCXO.

$$\textit{Frequency accuracy} = \textit{resolution} \pm (\textit{time-base error}) \times \textit{frequency}$$

$$\textit{Time-base error} =$$

$$(\textit{temp. effects}) + (\textit{aging rate}) = 7 \times 10^{-9} + 5 \times 10^{-8} = 5.7 \times 10^{-8}$$

$$\textit{Frequency accuracy} = \pm 0.006 \text{ Hz} \pm (5.7 \times 10^{-8} \times 1 \text{ MHz}) = \pm 0.066 \text{ Hz}$$

Notice how the calibration period and aging rate dominated the calculation. If using a system that has access to an even higher-stability time base, such as one based on cesium (a primary standard), rubidium, or a network, such as the *global-positioning system* (*GPS*), this problem is reduced dramatically. See *external time base in*, described in the section "The Rear Panel" later in this chapter. For more about time bases, see Chapter 20.

It is possible that the time-base accuracy is not important. Such is the case for short time-interval measurements. For example, when measuring a 100-ns event to 100-ps resolution, only 1 part in 10^4 time-base accuracy is needed. Just about any kind of time base will deliver this—even without calibration. In fact, for precise time-interval measurements, other factors become more important. An important factor is often channel matching. If the start and stop channels are not of equal electrical length, the timing measurement will be off. The error caused by this problem is often not specified, but it can usually be measured. First, the counter must be able to measure down to 0 ns, and ideally be able to handle $\pm TI$. Then, using a signal splitter, the same signal must be sent to both channels and measurements should be taken with different combinations of slopes and trigger levels. A comparison of these measurements and some simple math give a close indication of the systematic error in the counter because of channel mismatch. Although the mismatch between the channels in a counter is typically small, the signal paths within the overall test setup might have much greater mismatches that should be calibrated out.

Trigger level accuracy is another factor. This is the precision with which the trigger levels can be set. If the trigger level is set in error, the timing of the trigger point can be changed. This is most important for signals having slow rise and fall times.

A hidden, related factor in time-interval accuracy is the hysteresis of the counter's input channel. Hysteresis can effectively offset the point where the counter triggers and, therefore, introduce a timing offset, called *trigger-level timing error*, that is illustrated below. To combat this problem, some counters have hysteresis compensation. This feature automatically adjusts the trigger level up or down to compensate for the counter's hysteresis (Fig. 19.17).

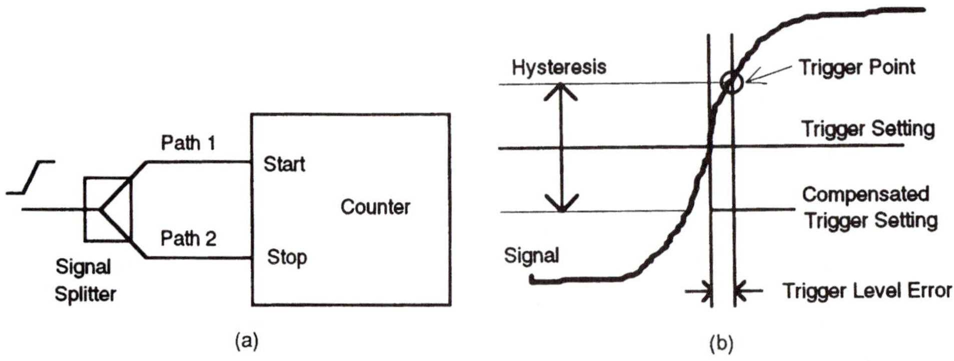

Figure 19.17 A simple setup for determining channel mismatch systemic errors for time-interval measurements (a). Showing how the hysteresis in the counter's input channel causes a delay when the signal triggers the measurement (b). If the delay is the same for both the start and stop channels, this is not a problem. If the counter has hysteresis compensation, the trigger level is adjusted to compensate for the hysteresis to a first-order approximation.

Another hidden factor is *metastability*. Metastability, sometimes called *teeter*, occurs when the input signal comes precisely on top of a time base edge. In simple terms, it is like having a tossed coin land on its edge before falling to one side or the other. The flip-flop might hesitate before flipping to the proper state; this hesitation can sometimes add significant delays. In a well-designed counter, this factor is a near nonexistent problem, although it will always be there. This is almost never specified, but it can be observed when making millions of measurements.

Throughput. There is often a relationship between resolution and throughput. If more resolution is desired and the signal is around long enough, the measurement time can be extended. This slows down throughput. For example, extending the gate time of a frequency measurement by a factor of 10 increases the resolution by a factor of 10, but also slows throughput by about 10. Continuous-count counters are a bit more complicated, but the basic idea still applies.

For timing measurements, one way to extend the resolution is through averaging, which also increases the measurement time by a factor of N^2. This is particularly key if throughput and resolution are essential in a timing measurement; a counter with 1-ns resolution will have 100 times more throughput than one with 10-ns resolution if 1 ns of resolution is required. Also, it has been assumed the signal is around long enough to average. If this is not the case, the measurement might have to rely on the basic single-shot resolution of the counter.

For some measurements, such as totalize, it makes little difference what the counter's architecture is. This is true for any measurement that does not use the counter's time base. Other factors that affect throughput are not related to resolution, but are more related to the speed of the microprocessor and interface system. Two factors to look for are how many measurements per second that the counter can deliver through the interface and how quickly the counter can switch between different functions or setups. If short gate times are being used and/or measurements are being switched back and forth between different functions, these factors are important. If long gate times are being used, these might not be much of a factor. If the counter has the ability to prestore multiple measurement setups, the switching time between functions can be shortened.

Input signal conditioning. The counter's "front end" consists of a set of input signal-conditioning circuitry that is similar in some ways to that of an oscilloscope. In fact, some of the same circuitry might be used in the better counters. However, there are some differences. Here are the most-common controls located on a universal counter or a frequency and time-interval analyzer.

Input impedance. Allows switching between 50 Ω and 1 MΩ. These impedances are not exact. The VSWR of the 50-Ω setting is often 2 or greater. For the 1-MΩ setting, it is important to note the input-capacitance specification because this effectively reduces the input impedance at higher frequencies. This is also important to note if a probe is used. The input capacitance of a counter is often higher than that of many oscilloscopes. Something to be aware of on some counters is what happens when the channels are put into *common* (see the section below). This causes some counters to change the impedance from 1 MΩ to 500 kΩ, and the input capacitance increases.

X10 attenuator. When in the X10 position the signal is attenuated in amplitude by 10. The designation of "times 10" comes from the thinking that this allows input of signals that are 10 times larger. For timing measurements, this widens the effective *dynamic range*, or maximum peak-to-peak signal, measurable by the counter. It also increases the effective hysteresis, if it is necessary to filter out some noise on the signal. The attenuator circuit is often not as well controlled at high frequencies as that of an oscilloscope. The attenuator is not always X10, and some counters offer other settings.

Ac/dc coupling. If the input signal has a large dc offset on it, the ac coupling setting can be used to block the dc component. This works best for frequency and period measurements, and is less useful for time-interval measurements because the latter are often made on logic signals. For counters that only make frequency measurements, the input might always be ac-coupled with the trigger level set at 0 V.

Slope. This is most useful for time-interval measurements. It allows the selection of which edge of the input signal to use. The slope can be set independently for both the start and stop channels. If the counter is designed properly, the slope setting can also be used to select which edge to use for other functions, even frequency and totalize. This is useful if an asymmetrical waveform (duty cycle is not 50 percent) is being measured, or in the case of totalize, if one edge or the other has significance. Note that, in some instruments, the slope is automatically selected for certain measurements, such as rise and fall time.

Separate and common. To make a pulse-width or rise-time measurement it is useful to feed the same signal to both channels. This can be done if the counter has a separate and common switch.

Filter. Some counters offer a switchable low-pass filter. The filter is usually a simple one- or two-pole filter with a 3-dB point in the 100-kHz range. This has been added as a way to filter out noise on low-frequency signals, which is often encountered in power and electromechanical applications.

Trigger level. This sets the point on the signal that the counter triggers on. Other names for this are *sense-event level* (a result of conforming to the SCPI standards) and threshold. Early counters set this with an analog potentiometer, and the only way to know what it was set at was through a voltmeter reading. As such, some counters include a simple voltmeter for these purposes

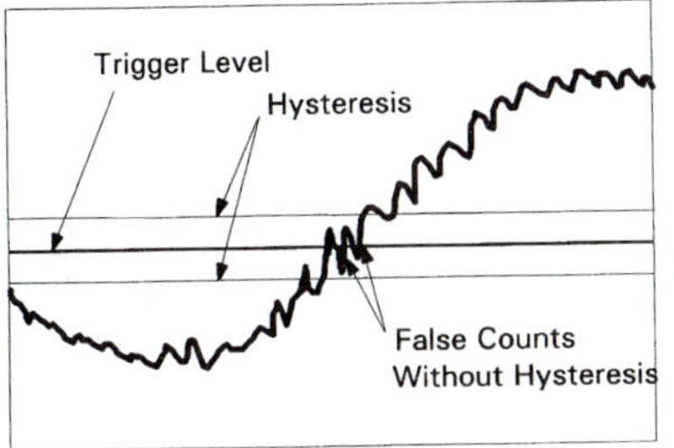

Figure 19.18 Without hysteresis, the noise on this signal would cause multiple false counts at the trigger-level point. With hysteresis, the counter will "see" only one transition.

(note that some of these voltmeters do not have floating inputs). Newer counters, intended for automated applications, use a DAC to set the trigger levels. The setting resolution is usually around 10 mV. Check the range of the trigger level setting to see if it accommodates the application's needs. A ±5-V range is a good range.

Auto trigger. Some counters offer the ability to have the trigger level set automatically. This is done by first scanning the signal to detect the peaks and then setting the level to the 50% point. There are often some limitations to how low a frequency this works on.

Sensitivity adjust. In some counters, the sensitivity of the front end can be set. This offers another way to deal with noise on the signal (Fig. 19.18). On some counters, the sensitivity is controlled by an *automatic gain control* (*AGC*), which varies the gain of the front-end amplifiers to maximize sensitivity while minimizing the effect of noise.

Cross talk. This is not a setting, but a specification that can be given to indicate how much a signal fed into one channel leaks into the other channel. This can be a problem with large signals. If this is not specified, an estimate of what it is can be gotten by feeding a high-frequency signal of large amplitude into one channel and seeing when it causes the other channel to trigger.

Input protection. It is possible to put too much signal into the counter. Counters usually have some form of diode protection to prevent this problem. This is trickier for microwave counters. Limiters are sometimes used to offer some protection. Some frequency counters also offer fuses to prevent damage to the more-expensive circuitry.

Measurement gating and arming. The precise moment that the measurement is made is often not an issue, so the only important issue is how long to set the gate time. For several applications, however, the precise moment that the measurement is made is crucial. Look at a simple application, that of the measurement of a pulsed frequency signal.

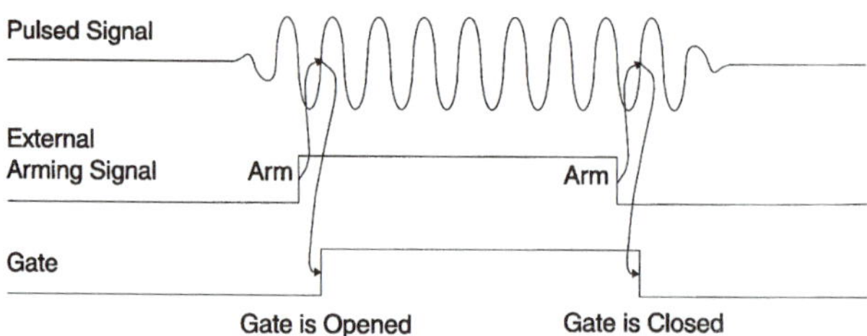

Figure 19.19 This shows the relationships among the input signal, the arming signal, and the opening and closing of the counter's gate.

Figure 19.19 shows a pulsed frequency signal. An ordinary frequency measurement would probably miscount on some of the leading-edge transients, and it might not properly close the gate before the signal went away. To avoid these problems, the gate should be armed. In the same figure, an arming signal is shown. This signal might come from some other point in the device under test (DUT) or from a pulse or timing generator. The arming signal is fed into the counter's external arm input. The counter is then set to gate based on the external arm. This is done in a variety of ways, depending on the counter. In this case, the gate is set to open after a positive edge from the external arm signal, and close after receiving a negative edge. This assumes that the counter has the flexibility to select slopes for external arming. The gate will now wait for a positive edge from the external arm signal. Once this is received, it will wait for the next transition from the input signal. When all of this has happened, the gate will be opened. In a similar fashion, the gate is closed. First, the arming condition must be met; then the input signal must make a transition. Assume that it is a reciprocal counter, so the gate times can be arbitrary. If all this happens successfully, the frequency reading is made.

The external arm settings on more-advanced counters can be set so that any combination of open and closed conditions can be set. For example, some counters allow the opening of the gate to be armed, but let the closing of the gate be determined by the counter's gate time setting and, therefore, not be armed.

Some counters also offer built-in methods for offsetting the gate from the arming signal. The delay can be set by time or events. These features allow easier arming for more-complex measurements. For example, there may be a need to open the gate 10 ms after the arming signal because some kind of header signal needs to be ignored. Situations like this occur in the measurement of digital communication and data-storage signals.

Pulsed microwave counters have automatic arming. This means that measurements can be made on pulsed frequency signals without the need for external arming.

Some counters have a sample-rate control. This determines the pace of the measurement, slowing it down if needed for easier reading by an operator. The counter can usually also be set to *single* or *hold* to freeze a particular measurement.

The rear panel. A number of signal connections are often found on the rear panel of a counter. Some of the most common ones are described here.

External time base in. Often in a test system it is desirable to lock all of the instruments to a common higher-accuracy time base. This input allows the time base from another source to be used, instead of the counter's built-in time base. Ideally, the external time base is a high-quality one, such as one based on cesium, a primary standard. If several instruments are sharing the same time base, a distribution amplifier might be needed to maintain good signal integrity and isolation.

Internal time base out. This is the complement of the above. Often, the time base in the counter is the highest-grade one in the system, so this allows it to be sent to other instruments. When using a common time base for all instruments in a system, be aware that some counters can artificially "lock in" on a frequency and give slightly skewed readings. This is particularly true if performing a function, such as time-interval averaging. Averaging will not be done properly and the expected resolution enhancement will not be delivered if the signal being averaged is synchronous to the counter's time base. To combat this, some counters add a bit of deliberate jitter to the time base to promote better averaging.

IEEE-488. This is the most common of instrument interfaces. It is not available on all counters and is sometimes either an option or standard. This bus is also known as the *Hewlett-Packard-Interface Bus* (*HP-IB*) and the *General-Purpose Instrument Bus* (*GPIB*). Some more-recent counters are beginning to offer RS-232 interfaces. This is a response to the availability of personal computers and to the fact that IEEE-488 is becoming rarer as an interface for printers. Depending on the counter, all controls might not be programmable. In some counters, the computer can control the function selection and a few other settings and can read measurements, but not the front-end controls, such as attenuators and impedance.

Gate out. This is often a TTL-compatible or other signal that is used to display roughly where the counter's gate is opening and closing. This is not an exact indication because usually some delay occurs between when the gate actually opens and when the gate out signal shows it. Another use for this signal is to act as a trigger for some other instrument. The complement of this output is the gate in input, which is a version of external arm in.

Rear terminals. In racked systems, the signal connections are sometimes more convenient, if available, on the rear panel. As such, some counters offer an option to feed the normal front-panel signals to the rear panel instead.

Note that this might add extra capacitance to the inputs, and thus degrade the performance.

Miscellaneous specifications. Some counters have other specifications that might be useful to an application.

Math, statistics, and limits. Often, in manual operation, it is convenient to display the answer in units that are more meaningful to the application. For this reason, some counters have *add, subtract, multiply*, and/or *divide* by a constant. Usually, only two or three of these are allowed. Some counters also can subtract the previous measurement from the current one. This produces a form of dynamic difference measurement, and if, combined properly with standard deviation, a form of Allan variance can be performed, which is a frequency-stability measurement. Some instruments even provide Allan variance as a built-in function. Note that dead time is a factor in Allan variance measurements, and, if a continuous count counter or FTIA can be used, a more-accurate measurement can be made.

Some counters can also perform a limited number of statistics on the measurements. The most-common ones are *mean, sample standard deviation, minimum,* and *maximum*. The counter usually does not store every measurement in the sample set, but performs a set of running calculations on each measurement and constructs the answer at the end. Because the operations are done internally in the counter, the speed of the measurement can often be much faster than if the raw measurements were sent to a computer for calculation there.

A special feature found on just a few counters is called *smooth*. This does a running average of the measurements, plus it adjusts the display's resolution to match the stability of the signal. If the signal is very unstable, the *smooth function* truncates some of the digits automatically.

Limits are useful for unattended monitoring of a signal or for manual adjustment of a signal. The counter is given a low and high limit for a particular measurement, then it provides a visual and/or electrical indication of whether the signal is within these limits. A useful variation of this feature is to show the measurement in the form of a "meter." Manual adjustment can be tedious if the operator has to look at lots of digits. An analog type of display is easier to relate to.

Save and recall. Some counters allow several setups to be saved in memory. This is handy when repetitive testing is done, usually in manufacturing or service. Usually between 10 and 20 setups can be stored. Recalling of the setups can be either through a simple one-button control or through a more-complicated arrangement. For manual applications, the advantage of this feature is obvious. For automated applications, this feature can increase throughput because the computer need only end a simple recall command to invoke a whole new setup, instead of a series of longer commands.

19.4.2 Microwave counter specifications

A few specifications are unique to microwave counters, and some of the most common are covered here.

CW microwave

Acquisition time. Because of the way that most microwave counters operate, using a harmonic sampler technique or a YIG preselector, some time is required to determine where in the spectrum the signal is. This is called *acquisition time*. This is usually in the 50- to 250-ms range. YIG-based counters generally have longer acquisition times. If there is a general idea of where the signal already is, some counters work in a manual acquisition mode. The counter then skips the acquisition phase of the measurement and makes the measurement after properly positioning the LO to the manually specified area, taking less time <20 ms.

FM tolerance. This is largely determined by the IF bandwidth of the counter or the bandwidth of the preselector. If the input signal's FM is such that the signal occasionally falls outside of these bandwidths, the measurement will be in error. Also, because the IF or preselector is usually not exactly centered on the signal, a margin is accounted for that narrows the FM tolerance of the instrument.

In some counters, the FM rate can affect the amount of FM tolerance. If the signal is changing slowly, and with about the same time constant as the counter's acquisition phase gate time, the measurement can be fooled into thinking the frequency is higher or lower than the actual. As such, the counter might see a lower tolerance to FM when the FM rate is a low frequency. Related to this is tracking speed, which is the ability of the counter to track a changing frequency, as in the testing of a VCO. Some counters might have the ability to anticipate changes in frequency to help it track the signal better. Counters that use prescaler technology do not have a problem with FM tolerance, or with acquisition time for that matter. The trade-off is that some loss of resolution compared with a downconverter-style counter is realized, but this might not be a problem if advanced interpolators are used.

Preselection. Sometimes it is necessary to measure one signal in the presence of other signals. If this is the case, it is convenient to have a preselector, which is essentially a programmable filter. If the desired signal is larger than the other signals that are present, it might also be possible to measure it without a preselector. This is called *automatic amplitude discrimination*. Generally, a difference of about 6 to 20 dB between two signals is enough to allow the counter to measure just the stronger of the two signals. The exact behavior of this specification depends on the flatness of the counter's downconverter over frequency.

Damage level. This factor is worth considering if the counter is going to be used around transmitters and it could be accidentally connected to a large signal. Protection from signals up to 35 to 40 dBm is typical (about 8 to 10 W).

Power. Some microwave counters have the ability to measure the signal's power. This is useful for checking to see if a transmitter is within specifications or for finding faults in the cabling or connectors. Two main methods are used, true rms and detectors. The former produces much better accuracy, down to 0.1 dB, and the latter is less expensive, but has an accuracy of around 1.5 dB.

Pulsed microwave. Several unique pulse measurements can be made that are important for radar, electronic warfare, and communications applications. Pulse microwave counters have a detector built in that produces the carrier envelope of the signal. This is then used to automatically gate the counter for proper frequency measurements. It is also used to set pulse parameter measurements (see CW and pulse microwave counters in Section 19.2). However, in addition to these basic features, some enhancements and limitations are sometimes found.

Resolution. Frequency resolution on pulsed signals is usually lower than that available from a CW measurement. This is because there is less signal to measure, and the requirements are not as tight. However, sometimes it is desirable to make a CW measurement with a pulsed microwave counter. If this is the case, it is useful if the counter has the ability to switch into a CW mode to provide the 1-Hz (or better) resolutions that might be needed. For pulse measurements, one of the factors that limits resolution is the counter's IF frequency. For example, with a 50-MHz IF frequency the period is 20 ns. This means that changes in the pulse width would be seen only if they are greater than 20 ns.

Accuracy. One of the factors that limits frequency accuracy is distortion. Distortion comes from pulse distortion in the signal itself and whatever distortion the counter might add. One form of pulse distortion is video, or a dc pulling of the signal at the pulse's edge. This can be cleaned up in some counters, but there will generally be some form of distortion there. The smaller the pulse, the more pulse distortion is a factor.

Measurement time. The two biggest factors that affect measurement time are pulse-repetition interval and pulse width. This is because several pulses typically have to be averaged to get enough information to first acquire the signal and then measure it to the desired resolution level. Specific formulas in the counter's data sheet will indicate how to estimate measurement times, but, unlike a CW counter, depending on the signal parameters, the times can stretch into several seconds.

Pulse profiling. Some signals are frequency modulated, as in a chirp radar, and it is important to measure the linearity of the modulation. Even if the

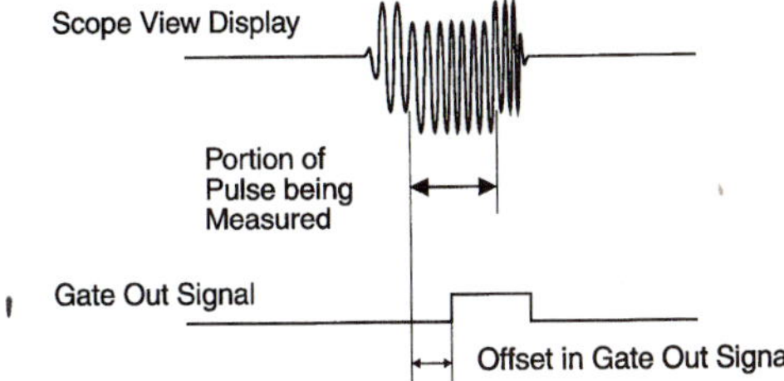

Figure 19.20 This shows how the gate-out and scope-view signals look when measuring a pulsed microwave signal.

signal is not modulated, it is often important to measure the frequency distortions at the beginning and end of the pulse. Many pulse microwave counters require external pulse generators and computers to properly gate the counter at the correct times and plot out the results; some have this function built in.

Gate display. A gate-out signal is often provided to provide an indication of where in the pulse the measurement is being made. This signal is useful if external pulse generators are being used to make a pulse profile measurement. Some counters offer scope view, which shows a picture of the pulse and the area being measured simultaneously. This has the advantage of eliminating the timing offset that the gate-out signal usually has, relative to the actual signal (Fig. 19.20).

Minimum pulse width. This specifies the shortest pulse that can be measured. This is usually specified only on pulsed microwave counters, but some CW counters have a limited ability to measure pulsed signals, too. To be practical, the counter should be able to measure down to at least a 100-ns pulse to cover most radar applications. Communication applications can get by with wider pulses.

To know where in the pulse the counter is measuring, it is convenient to have some kind of indicator. Not all pulse microwave counters will provide this. Another convenient feature is the ability to scan an RF pulse to perform a pulse profile as a built-in feature. This can be done with external programmable pulse generators and a computer, too.

19.4.3 Frequency and time interval analyzer specifications

Although many specifications are similar to those found in a counter, some are unique and need some explanation.

Sample rate. Also called *measurement rate,* this is how rapidly the instrument can make measurements. The speed needed depends on the phenomenon being measured. For example, to measure the 5-MHz jitter bandwidth of a clock circuit, the instrument must sample at 10 MHz (or faster). Sample rates in the 1-MHz range are good for moderate data-rate communications, VCO testing, and electromechanical measurements. Sample rates of 10 MHz (and higher) are better for high data rate communications, data storage, and radar applications.

What sample rate is needed? As with all signal digitizers, the Nyquist rules need to be obeyed. With an ADC system (such as a digitizing oscilloscope), the sample rate must be at least twice the frequency of the signal's bandwidth. For example, if a radio signal is at 200 MHz and the bandwidth of the modulated carrier is ±500 kHz, an ADC would have to digitize at 401 MHz (or greater) to avoid aliasing problems. This can get to be impractical with high-bandwidth signals, so the signal is usually downconverted to a lower IF signal. But, an IF of 60 MHz still requires sampling rates of greater than 120 MHz.

With this same signal, however, an FTIA might only have to digitize at 2 MHz. This is because an FTIA digitizes phase through the timing of zero crossings, not voltage. So, to determine the Nyquist frequency, look at the modulation bandwidth, not the signal bandwidth. In this example, the frequency is being modulated at a 1-MHz level, peak-to-peak. The Nyquist for this is 2 MHz, which sets the minimum sample rate for the FTIA. Particularly when the modulation rate is not known exactly, the actual necessary sample rate will probably be higher.

The difference between sample-rate requirements is much higher when measuring frequency hopping or agile signals. With an ADC-based instrument, the signal needs to be digitized at the RF level unless the instrument has the ability to follow the frequency-hopping sequence and mix down the signal into a stable IF. This means that the sample rates need to be very high. With an FTIA, the sample rates are still determined by the modulation bandwidth, so the rates can be much lower. For example, a frequency-hopping radio sends data at a 200 Kb/s rate, but hops over a frequency range of 900 to 920 MHz, a 20-MHz range. An ADC could mix this down to around a 2- to 21-MHz range and sample at about 50 Msample/s, but an FTIA would only have to sample at about 1 Msample/s (using 5 samples per bit).

To determine the required sample rate for an application, first determine what parameter is being measured and how rapidly it is being modulated. For frequency and time parameters, a sample rate of at least 2 to 10 times faster than the modulation rate is desirable, depending on the type of modulation. For example, an FSK-modulated signal might require about 6 samples per bit to capture the modulation parameters desired. If phase parameters are being looked at, the desired sample rate should be about 4 to 10 times the modulation rate. This is because of the way that the instruments interpret timing information into phase. If the device under test is being modulated with a specific known signal, it is actually possible to get by with undersampling. The analyzer or supporting-analysis software needs to know what the stimulus is in order for this to work, but the results can be much finer resolution in the measurement.

The rate that the analyzer samples can be specified by the user. Often, this can be performed by specifying the sample interval or time base (in this case, "time base" is being used the way it is used in an oscilloscope), such as a sample every 10 ms. Sampling on some analyzers can also be specified by an external signal, similar in some ways to external arming. In this situation, for example, putting in a 1-MHz signal could "tell" the analyzer to sample at a 1-MHz rate. Some counters have "sample rates" of several thousand measurements per second. Some might also build up a picture by repetitively sampling, moving the gate incrementally along the signal. This approach is OK if the phenomenon being measured is repetitive. Anything that is random (e.g., noise, jitter, and pseudorandom) needs a frequency and time-interval analyzer.

A further subtle issue is whether the samples are made continuously or not. If they are made continuously, additional information from the measurement can be derived. First, there are no measurement dead times where signal changes could be missed. Second, it is easier to relate one measurement to another. Among other things, this provides the ability to measure phase changes.

Time resolution. The amount of resolution required depends upon the application and is heavily influenced by the sample rate being used. Higher sample-rate measurements generally require finer resolution. If this holds true for the application, check to see if the instrument is making a trade-off between sample rate and resolution because some produce their best resolution only at the slower sample rates.

The crucial parameter to understand is timing resolution. The finer the time resolution, the more precise the frequency, time, or phase measurement. Resolution typically runs in the 1-ns to 10-ps range. Frequency and phase resolution can generally be calculated once the time resolution is known.

In the example given in Fig. 19.15 for a 20-MHz frequency measurement, the sample rate was set at 5 MHz. This makes the sample interval nominally 200 ns. If the FTIA's interpolator has a resolution of 100 ps, the measurement's resolution is roughly:

$$\frac{\textit{Frequency} \times (\textit{interpolator resolution})}{\textit{sample interval}} = \frac{200 \text{ MHz} \times 100 \text{ ps}}{200 \text{ ns}} = 200 \text{ kHz}$$

In the example shown in Fig. 19.16 for a phase measurement on a 20-MHz signal, the resolution is roughly:

$$\frac{360° \times (\textit{interpolator resolution})}{\textit{signal period}} = \frac{360° \times 100 \text{ ps}}{50 \text{ ns}} = 0.7°$$

To determine how much resolution is required, use the general formula:

$$\textit{Time resolution} = \frac{1}{\text{``}\textit{precision}\text{''} \times \textit{sample rate}}$$

For example, if 1-kHz resolution is desired for a 10-MHz signal, 1 part in 10^4 "precision" is required. If the signal is being frequency modulated at a 300-kHz rate, pick a sample rate of 1 MHz. So, $1/(10^4 \times 10^6$ Hz$) = 1 \times 10^{-10}$ s = 100-ps time resolution. As with counters, if the instrument has a downconverter in the front end, the resolution is enhanced; this needs to be taken into account when calculating requirements.

Memory depth. Memory depth for storing the digitized data is the third important instrument parameter. Some applications only need a short amount of memory to take a look at transient signals, such as the step response of a VCO. Other applications need a very deep memory, where probably the most extreme is the surveillance of unknown signals. The memory depth of FTIAs can range from 1000 ms on up. Deeper memory depth can also make it easier to freeze a measurement and subsequently zoom into an interesting area of the signal. With additional processing, this memory can also be used to enhance the resolution of the measurement or do a better spectral analysis.

If the desire is only to look at what can fit on a display, a memory depth of 1000 measurements (or better) is adequate. This can be the case when looking at signal changes that are controlled so that the measurements can be armed properly. For many other applications, a memory depth of 8K to 32K is nice to have. This provides the ability to generally position the measurement and zoom in for later analysis. It does not have to be as exact. The larger memory is also useful for observing widely spaced or slowly changing signals. Some applications require 256K or more memory. Such is the case for some applications in data storage, where the capture of a complete track is desired.

Arming. The arming modes for many of the analyzers can be quite sophisticated. Refer to the analyzer's detailed specifications and the application to see if the instrument allows for the measurement setup that is needed. One reason that this can get complicated is that an interaction can occur between how a set of measurements is armed and how the sample point for each individual measurement is determined (which is related to sample rate).

One feature to look for is *value trigger*. This triggers a set of measurements based on a condition being met in the signal being measured. For example, it may be required to make a set of measurements around the point that the signal changes from one frequency to another. With value trigger, this can be done by specifying that the measurements be made when the signal's frequency passes through 10 MHz, for example.

Auto-scale. Setting up an analyzer to the proper settings can sometimes be difficult, particularly if the input signal is unknown. If the analyzer has autoscale, the instrument can set the sample rate and display automatically. This can be a great time-saver, particularly in R&D situations.

Histogram. The measurements that are made can be accumulated into histogram form to examine the statistical distribution of some phenomenon. In order for this to be effective for many applications, a large number of measurements need to be accumulated. *Measurement rate, maximum measurements per bin,* and the *number of bins* are important specifications. Whether the analyzer handles the histogram through software or hardware affects some of these parameters. If the measurements are accumulated through software, the measurement rate is orders of magnitude slower than if performed via hardware.

Chapter

20

Precision Time and Frequency Sources

Leonard S. Cutler
Agilent Technologies
Palo Alto, California

John A. Kusters
Agilent Technologies
Santa Clara, California

20.1 Requirements and Definitions

Throughout most of history, people have had the need to track the events that affect their everyday life. The movement of the earth with respect to the stars provided an annual cycle. The rotation of the earth with alternating periods of light and dark provided a daily cycle. The position of the sun or the length of a shadow provided an indication of the current part of the day. As the need for more precision grew, time was expressed in terms of a burning rate, as with a candle; a material flow rate, as with water or sand through a hole; or the oscillation of a mechanical device, such as a pendulum.

None of these was sufficiently accurate for navigational needs. *Latitude* (angular distance from the equator) could be measured by knowing the angle of elevation of the North Star above the horizon. *Longitude* (the east-west angular distance from the Greenwich, England, zero meridian) was determinable only if the navigator knew the time in Greenwich. The sun travels across the sky at a predictable rate of about one degree in every four minutes. Thus, for every four minutes that the clock showing Greenwich time differs from the time determined by local noon of the sun, the observer is one degree of longitude away from Greenwich. In 1713, the British government offered an award of £20,000 for a clock that could determine longitude to within ½ degree. The prize was not claimed until 1761.

Today, emphasis on further precision dictated by space and military requirements and communication synchronization has required time standards that are even more accurate. Depending on the application, the necessary time accuracy might be picoseconds per day.

20.1.1 Time standard

A highly precise time standard consists of an ultra-stable oscillator and some form of counter or integrator that tracks the number of elapsed cycles of the oscillator. This is because the indicated time of the time standard is the integral with respect to time, properly scaled, of the frequency of the oscillator. The counter is usually arranged so that it reads out in conventional units of time. A simple example of a clock is the pendulum clock, the very best of which are only moderately good time standards. The swinging pendulum is the *oscillator* and the gear train and hands are the *counter* or *integrator*. The two requirements for a good time standard are that its oscillator has a known, constant rate of frequency and that the readout is set at some time to agree with the accepted time reference at that time.

20.1.2 Accuracy and precision

The *accuracy* of a clock is the degree of conformity of its time with respect to a standard reference. The precision of a clock is the degree of mutual agreement among a series of individual measurements of essentially identical clocks. For example, a group of clocks might have an average time in perfect agreement with a reference standard, but each member of the group might differ widely. This group would be highly accurate as a group, but very imprecise as individual units. Similarly, a group of clocks might all agree with each other very closely, but, as a group, be far from the reference standard. Each clock now exhibits high precision, but poor accuracy.

For the purposes of this chapter, *precision* implies the ability to achieve an accuracy of one part in 10^7 with respect to a defined standard, with an uncertainty or instability in the oscillator of less than one part in 10^{10}.

20.1.3 Stability

A further measure of clock performance is *stability,* or more correctly, *instability,* the spontaneous and/or environmentally caused frequency change within a given time interval, or within a given range of an environmental variable. All clocks are subject to random fluctuations in oscillator frequency, with consequent effects on the indicated time. Most also exhibit perturbations caused by aging and environmental effects. Generally, one distinguishes between systematic effects, such as frequency drift, and stochastic frequency fluctuations. Radiation, pressure, temperature, humidity, etc. can cause systematic instabilities. Random or stochastic instabilities are typically characterized in either the time domain (for example, Allan variance) or the frequency domain (phase noise). Measurements of stochastic instabilities are dependent on the measurement system bandwidth and on the sample time or integration time.

For the purposes of this chapter, *precision* implies the ability to achieve an accuracy of 1 part in 10^7 with respect to a defined standard, with an uncertainty or instability in the oscillator of less than 1 part in 10^{10}.

20.1.4 Universal time

Customarily, the unit of time in the MKS system is the second. *Universal time* (*UT*) is based on the rotation of the earth about its axis. The time basis was chosen so that, on the average, local noon would occur when the sun was on the local meridian. This assumed that the earth's rotation rate was constant and would, therefore, produce a uniform time scale. It is now known that the earth is subject to periodic, secular, and irregular variations; time based on the rotation of the earth is subject to these same variations.

Because UT was not a satisfactory time scale, a variant of UT was defined. The UT_1 time scale corrects for the position of the earth's pole and thus measures the angular position of the earth and is useful for navigation. The seasonal and other variations remain.

To compensate for these variations, the UT_2 time scale, the second variant of UT, was defined by international agreement. Corrections were applied for the seasonal variations. Annually, the non-periodic variations are examined, and, if necessary, a correction is made in the length of the second on January 31 of each year. The goal was to maintain the UT_2 time scale to within 100 ms of the actual universal time. As the length of a second changed yearly, clocks all over the world had to be adjusted each year to run at a different rate.

To solve this problem the "leap second" was invented in 1972. *UTC* (*Universal Time Coordinated*) was defined to have seconds based on atomic time, described in the following section, but with the rule that UTC is always to be within 0.9 s of UT_1. To keep this relationship, UTC is subject to stepwise corrections (leap seconds), which either add a second to or subtract a second from the last minute of the year in December or in the last minute of June. The minute in which the adjustment is made is thus either 59 or 61 s long.

20.1.5 Atomic time

Atomic time, time defined on the basis of an atomic transition, was first defined at the Twelfth General Conference on Weights and Measures meeting in Paris in October 1964. It was re-defined by the Thirteenth General Conference on Weights and Measures in October 1967. That definition, still in effect, is: "The second is the duration of 9,192,631,770 periods of the radiation corresponding to the transition between the two hyperfine levels of the fundamental state of the atom of Cesium-133." Cesium-133 was chosen for a number of reasons:

1. It is an alkali metal with a single electron outside of closed shells, so it has a useful microwave hyperfine structure and moderately low sensitivity to magnetic fields.
2. It is a heavy atom, so its thermal velocity is not high, leading to small second-order Doppler shift (described in the following section).
3. It has high enough vapor pressure at moderate temperature, so it can easily generate an atomic beam.
4. It has the lowest ionization potential of all of the elements, allowing easy detection of atoms by ionizing them on a hot wire.
5. Cesium-133 is 100% of the naturally occurring element.

The two hyperfine levels used in the definition refer to the two specific energy levels that exist in the fundamental or ground state of a cesium atom in the absence of any external fields. These levels are degenerate and the degeneracy is removed in the presence of magnetic fields. Physically, they represent states where the spin vector of the nucleus of the cesium atom is either in the same direction as the spin vector of the outer electron of the atom or in the opposite direction. With a weak magnetic field, the states that have only second-order energy dependence on the field are those with $m_F = 0$, corresponding to the total spin vector of the atom being perpendicular to the magnetic field. These states are used in all practical realizations of the definition of the second. The difference in energy between these two states also depends quadratically on the magnetic field. The populations of the two states are essentially equal, differing only by thermal energy. The assigned frequency value of 9,192,631,770 Hz is in as close agreement with earlier definitions as was experimentally possible at the time.

Atomic time is, based on present knowledge, a true uniform time scale. It is maintained through international agreement by *BIPM* (*Bureau International des Poids et Mesures*) in Paris through bimonthly analysis of data accumulated from a large number of atomic standards, mostly cesium-based, which are maintained at various sites around the world. Virtually all national standards laboratories contribute to the BIPM. In the United States, large ensembles of clocks are maintained by both *NIST* (*National Institute of Standards and Technology*) and *USNO* (*United States Naval Observatory*). Time comparisons between various clocks are obtained through a variety of measurement methods, and are routinely transmitted to BIPM for analysis. The analysis determines the apparent stability of each clock, assigns a weighting factor to the data from each clock, with the most stable clocks receiving the highest weighting, and computes atomic time from the weighted data. Because the data gathering and computing cycle occurs every two months, the correction factors reported back to the clock sites represent the state that each clock was in at the time that the data was gathered. From the correction factors accumulated over a period of time, the ensemble of clocks is updated to reflect the current best estimate of atomic time, and this is used as the national time standard for each of the countries participating with BIPM.

All other time standards are derived from or defined on the basis of UTC or atomic time.

20.2 Frequency Standards

As mentioned earlier, the basic requirement for a highly precise time standard is an ultra-stable oscillator or frequency standard. Figure 20.1 is a block diagram of a generalized atomic frequency standard. A and B are two possible hyperfine energy states of an atom separated by energy $h\nu_0$, where h is Planck's constant and ν_0 is the resonance frequency of the transition. Under normal conditions, both states are nearly equally populated. To detect resonance with externally applied electromagnetic fields, a reasonable population imbalance must be present between the states. This can be accomplished by several means, including magnetic state selection in a beam of atoms or use of optical pumping. When an electromagnetic field at the resonance frequency is applied, the population imbalance can be inverted by the atoms making a transition to the opposite state. The transition can be detected by external means and is a maximum when the applied field frequency is exactly equal to the resonance frequency, ν_0. The detected signal is used to servo a quartz crystal oscillator so that its frequency, when multiplied by the microwave synthesizer, is precisely that which provides the maximum state change.

Another type of atomic oscillator is the *MASER* (*Microwave Amplification by Stimulated Emission of Radiation*). Atoms with a pair of levels prepared

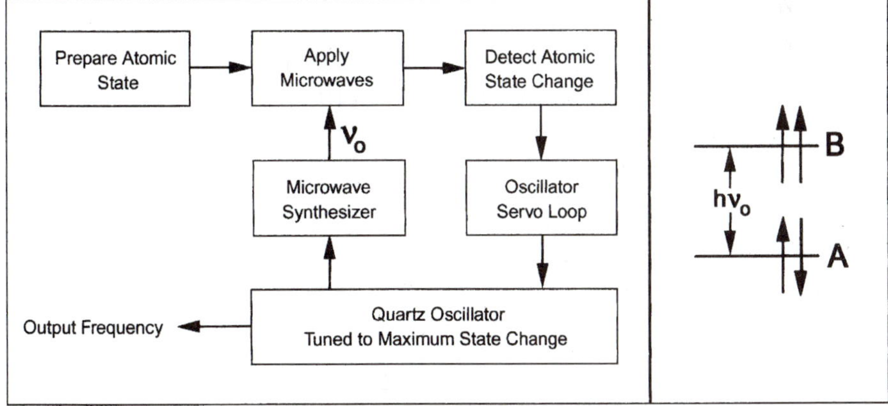

Figure 20.1 Block diagram of a generalized atomic oscillator. Levels *A* and *B* are different hyperfine energy levels. The energy difference is equal to the frequency difference times Planck's constant.

with a high population excess in the higher energy level and enclosed in a cavity resonant at the resonance frequency of the transition can produce a very pure oscillation by stimulated emission. The resonances used cover a wide range of frequencies, from RF to optical. An optical MASER is usually referred to as a *LASER* (*Light Amplification by Stimulated Emission of Radiation*).

20.3 Primary Frequency Standards

A *primary standard* is one whose value can be determined through analysis and calculations based on fundamental principles of science. Primary standards require no external reference for proper operation. All factors that affect either accuracy or stability are deducible from first principles with no experimentally determined "fudge factors" to enable agreement between theory and realization. Primary frequency standards must meet these requirements. By definition, therefore, if a primary frequency standard can be put into operation, as verified by either self-contained meter readings or indicator lights, the output frequency will be within the specified accuracy, provided that proper account is taken for all the systematic perturbing effects. Primary standards can be compared, but not calibrated. Calibration to force agreement between a primary standard and an assigned reference reduces the primary standard to a secondary standard.

Although some of the standards described here are not primary standards in the strictest sense, because they don't use the cesium atom, they have very high accuracy capability. It is entirely possible that, at some time in the future, the definition based on cesium might be replaced by, for example, Mercury-199. For this reason, these standards are described here.

20.3.1 Cesium beam standards

Because the international unit of time is defined on the basis of an atomic state transition in the undisturbed cesium-133 atom, a clock properly constructed using this transition is a primary standard.

The cesium-beam tube is a passive resonator. A typical cesium-beam tube is shown in Fig. 20.2. In the tube, heating the cesium in an oven and passing the vapor through a collimator generates a beam of cesium atoms. The beam is directed through an inhomogeneous magnetic field generated by a state selector magnet, the A magnet. This deflects atoms in the two hyperfine levels in opposite directions. The atoms in the desired state are directed through a microwave cavity, the Ramsey cavity, in a magnetically shielded region with a small, homogeneous magnetic field, the C field. In passing through the Ramsey cavity, the atoms "see" a double pulse of microwaves, which leads to several important benefits that are too detailed to pursue here. If the microwave frequency is at the atomic resonance frequency, 9.192631... GHz, and its amplitude is correct, the atoms experience a change in their hyperfine

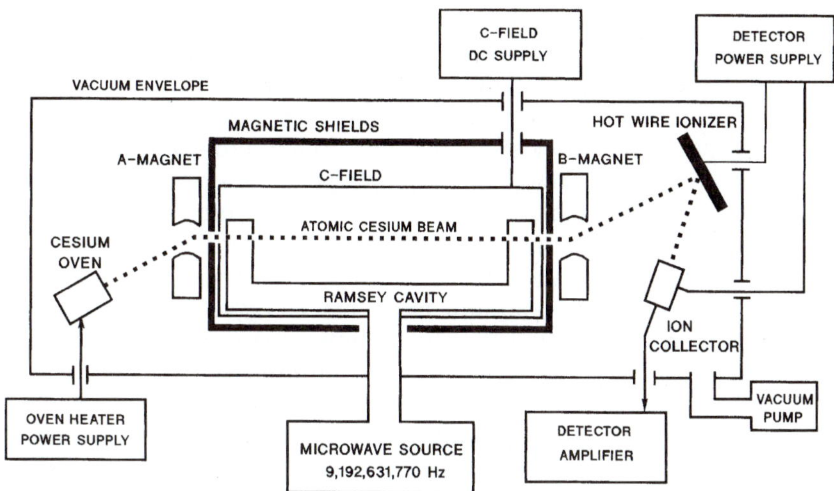

Figure 20.2 Drawing of the interior of a cesium-beam tube showing external electronic elements that control the tube.

energy state. At the output of the cavity, a second magnetic state selector, the B magnet, selects only those atoms that were originally selected by the A magnet and changed in energy state by interaction with the microwave signal. Atoms selected by the B magnet are ionized by the hot wire ionizer, then sent to a mass spectrometer and electron multiplier. When the applied microwave signal is exactly at the cesium hyperfine resonant frequency and its amplitude is correct, the output of the electron multiplier is at its maximum. To prevent attenuation of the cesium beam through collisions with background gas atoms, the tube is evacuated and is constructed with its own vacuum pump to maintain the highest possible vacuum.

The microwave source frequency is synthesized from a precision quartz oscillator, whose frequency is continuously adjusted by feedback control circuitry to keep the synthesized frequency in exact agreement with the 9.192631... GHz atomic transition frequency of the cesium-beam tube. In this manner, the crystal oscillator frequency, nominally at 5 or 10 MHz, has the same absolute accuracy as the cesium transition.

The frequency must be corrected for several systematic effects. The C field must be known to allow correction for the quadratic dependence on the magnetic field mentioned previously. If it is very homogeneous, it can be measured using some of the Zeeman microwave transitions, which have a linear dependence of frequency on magnetic field. There is also the relativistic effect caused by the velocity of the cesium beam, known as *time dilation* or the *second-order Doppler effect.* The fractional shift is $-\frac{1}{2}\, v^2/c^2$, where v^2 is the mean square velocity in the detected beam and c is the speed of light. This amounts to about -1 part in 10^{13} for the usual beam velocities. Correction for this requires knowledge of the velocity distribution in the detected beam signal. This can be determined from a careful measurement of the full lineshape of the microwave transition. Another correction is for phase shifts in the microwave cavity. This effect is complicated and is not described here.

Cesium frequency standards have come into widespread use. They provide a moderate-size, rugged, reasonably portable standard with an absolute accuracy of better than 1 part in 10^{11}. Recent advances in digital control loops and improvements in the beam tube have resulted in a commercial cesium-beam primary frequency standard that has an intrinsic accuracy of better than 1 part in 10^{12}. Its maximum frequency change from environmental effects is less than 8 parts in 10^{14}. This optimized cesium-beam standard has demonstrated warm-up time of less than 30 min to reach full operating specifications. The combination of high accuracy, environmental insensitivity, and fast warm-up is a significant advance in state-of-the-art frequency standards.

20.3.2 Laser optically pumped cesium beam standard

Optical pumping for state selection is described in a previous section. Stabilized diode lasers can perform this function very well and can also be used for detection. The latest primary standard at NIST, NIST 7, uses an optically pumped cesium beam tube. It has excellent performance.

Many schemes are available for optical pumping for state selection. The simplest is *intensity pumping,* which is very similar to that used in the rubidium standard, as described in the section on secondary frequency standards (see Section 20.4). Figure 20.3 shows the essential elements of a laser optically pumped cesium beam tube. It has the usual cesium oven, microwave cavity, and shields, but lacks the A and B magnets, the hot-wire ionizer and ion collector. The state selection laser is tuned to the allowed transition between one of the hyperfine levels in the ground state of cesium (for example, the lower level), and an optically excited state. Atoms in the beam encounter the state-selection laser beam, are pumped out of the lower level by the laser radiation, and decay by spontaneous emission back to both of the hyperfine levels. Because only the lower level is being pumped, after a few cycles of the pumping process, essentially all of the atoms will end up in the upper level and then enter the microwave cavity.

The interaction with the microwave field in this case is the same as described in the cesium section. After leaving the microwave cavity, the beam encounters the detection laser beam. This laser is tuned to a transition from the upper level to an optically excited state. The fluorescence resulting from spontaneous decay back to the ground state is detected by the fluorescence photodetector and appears as an electrical signal. If the laser is tuned to a transition involving the optically excited state, with one higher unit of total angular momentum than the upper hyperfine level, then the transition to the lower hyperfine level is forbidden and the atom repeatedly cycles between the upper level and the optically excited state, emitting a photon each cycle. This

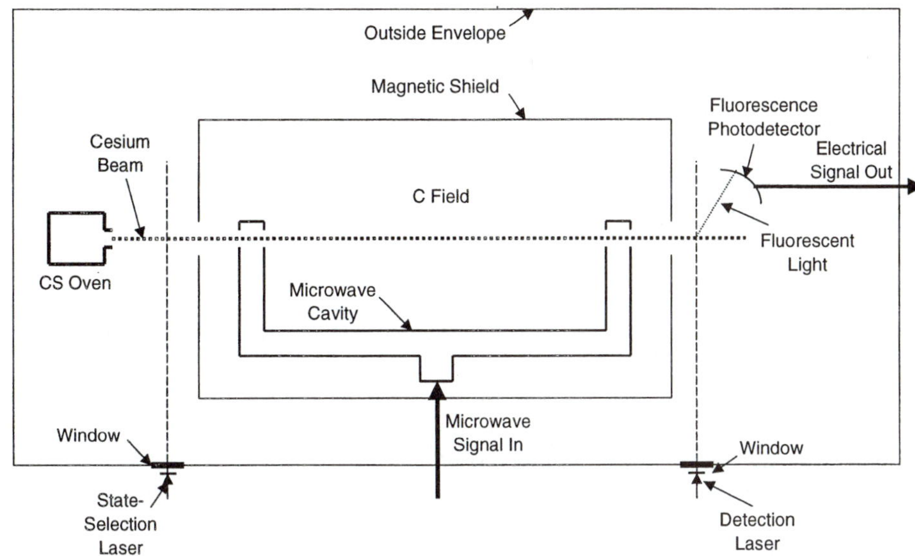

Figure 20.3 Drawing of the essential elements of a simple laser optically pumped cesium beam tube. The state selection laser beam replaces the A magnet in Figure 20.2, and the detection laser beam replaces the B magnet and hot-wire ionizer. Rather than collecting cesium ions converted from atoms, the fluorescence radiation from the atoms is collected.

greatly increases the detected signal for each atom and allows the signal-to-noise ratio to approach the shot noise limit. The ground state angular momentum for this situation is $F = 4$ and the optically excited state is $F = 5$.

A much higher fraction of the beam emitted from the oven is detected in the optically pumped beam tube, compared with the magnetic deflection tube. This improves the signal-to-noise ratio and provides the standard using the optically pumped tube a much-improved short-term stability. In addition, no strong magnetic fields are close to the region of interaction with the microwaves where the field must be weak. This improves the C-field homogeneity and, thus, the attainable accuracy. NIST 7 now has an accuracy of ± 5 parts in 10^{15}.

At least one company is working on a commercial optically pumped cesium beam tube.

20.3.3 Laser cooling

Atoms and ions can be cooled to microKelvin and below temperatures by using laser radiation in the proper way. Consider the following one-dimensional example. A laser is tuned slightly below an optical transition and the beam is directed into a cloud of atoms. Atoms travelling toward the laser source (i.e., against the beam) will "see" the laser frequency Doppler shifted toward the resonance. The atom absorbs a photon and its momentum and then spontaneously emits a photon. The absorbed momentum was opposite to the atom's momentum, so the atom loses that momentum on absorption. The spontaneously emitted photons and their momentum are, on the average, emitted in all directions, so the average momentum change with the spontaneous emission is zero. In this way, with repeated absorption and emission, atoms travelling toward the laser are slowed down close to what is called the *recoil limit.*

Another way to think about this is as follows. The laser is tuned lower in frequency than the atom's resonance frequency and the atom's spontaneous emission is at the resonance frequency. Thus, with each cycle, the atom radiates more energy than it absorbs. This energy loss must come from the kinetic energy of the atom, so it slows down.

If the laser beam is reflected back on itself so that the atoms "see" both beams, those moving away from the laser are also slowed in the same manner. This is *one-dimensional cooling.*

Consider now three mutually perpendicular laser beams, each tuned below the atomic resonance and each reflected back on itself. Now all three directions are slowed and the cooling is three-dimensional. This arrangement is called *optical molasses* and is very effective.

If one of the reflected beams in the optical molasses is replaced by a laser beam, tuned slightly higher or lower in frequency, then the cooled ball of atoms will be forced to move toward the lower frequency laser at a velocity, $v = c/2 \times \Delta f/f$, where c is the speed of light and $\Delta f/f$ is the fractional frequency difference between the lasers. For a velocity of 0.1 m/s, the required fractional frequency difference is about 6.7 parts in 10^{10}. This capability of manipulating balls of cooled atoms is extremely useful. An example is in the cesium fountain standard (described in Section 20.3.4).

20.3.4 Cesium fountain standard

Figure 20.4 shows the essential elements of a typical *cesium fountain standard.* A heated oven and collimator form a cesium beam directed to the center of a magneto-optical trap, where the optical molasses beams, described previously, intersect. The magneto-optical trap consists of a pair of coils in the Helmholtz configuration, but with their currents flowing in opposite directions. This generates a point of zero magnetic field at the center of the trap. The magnitude of the field increases in every direction from this zero field point. There are sublevels in both the ground state hyperfine levels of cesium that have energy that increases with the magnetic field. Therefore, atoms in these sublevels experience a restoring force toward the center of the trap. Atoms in states with decreasing energy are pushed out of the trap.

The trap is quite weak, so the optical molasses is used to generate a cold ball of atoms that can be retained by the trap. While the trap holds the ball, the vertical molasses beam frequencies are modified to enable pushing the ball upwards with a velocity just great enough for it to rise to the top of the

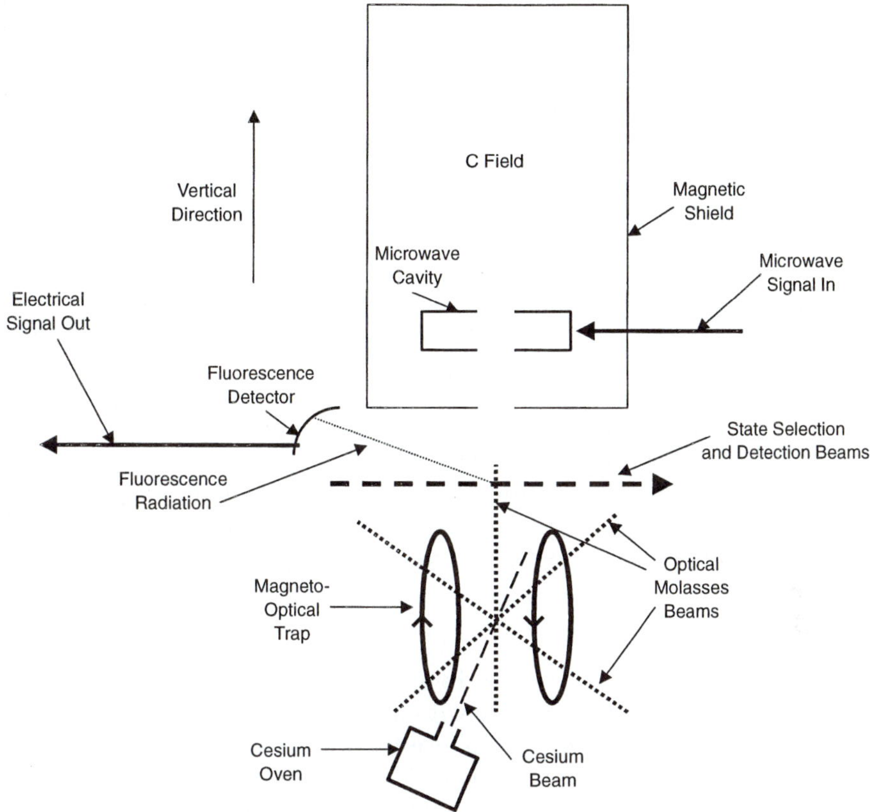

Figure 20.4 Drawing of the essential elements of a cesium fountain standard. Rather than using a thermal beam of cesium atoms, a very cold ball of cesium atoms is formed at the center of the magneto-optical trap as described in the text and tossed upward through the microwave cavity, and falls back through the cavity again under the influence of gravity. Optical state selection and detection take place through interaction with the state selection and detection laser beams.

C-field region before falling back under the influence of gravity. The molasses is turned on again and pushes the ball, which then passes through the state-selection laser beam. Here, optical pumping occurs to put the atoms in an $m_F = 0$ sublevel, the desired sublevel for the clock transition.

The ball proceeds upward through the microwave cavity, where it interacts with the microwave field in the usual way and then continues upward, slowing until it falls down through the microwave cavity and interacts with the microwave field again, thus producing Ramsey excitation. It continues down, passing through the detection laser beam, where it emits fluorescence, preferably using a transition that cycles in the same way described previously for the optically pumped cesium beam tube. The fluorescence is detected and produces an electrical signal that is processed in the usual way.

The ball is travelling slowly and takes of the order of ½ second between the microwave interactions, providing a resonance linewidth of about 1 Hz. The narrow linewidth coupled with a good signal-to-noise ratio provides good short-term stability and accuracy capabilities.

The density-dependent frequency shift caused by cesium-cesium collisions is fairly large, and is enhanced at the low temperatures. The shift can be determined by varying the density and extrapolating to zero density. Other systematic frequency shifts are similar to those in the conventional cesium beam tubes. The frequency shift caused by end-to-end phase difference in the microwave cavity is largely alleviated because the ball goes through the same cavity twice, instead of through a cavity with two separate interaction regions.

Several groups have built and are operating cesium fountain standards. The present accuracy is about 1 part in 10^{15}, limited by the systematic effects mentioned previously.

A group at Yale University is building a rubidium fountain. The advantage is that the density-dependent frequency shift is much smaller than that of cesium and this could lead to improved accuracy.

20.3.5 Trapped ion standards

Cool ions can be trapped in an electrostatic trap in an evacuated chamber so that the interaction time with an electromagnetic field can be very long, leading to very narrow resonance linewidth. The trap can be either two-dimensional or three-dimensional. Mercury-199 ions are often used because they are heavy and have the usual microwave transition in the ground state at about 40.5 GHz. They can be optically pumped with light from Mercury-202 ions or a laser.

Traps. Figure 20.5 shows the essential parts of a simple two-dimensional trap. Diagonally opposite rods are connected together and an RF voltage at a frequency of a few MHz and a dc voltage are applied between the connected pairs. This produces a linear restoring force, directed perpendicular to the rods, toward the central axis of the trap, on ions of the correct charge-to-mass ratio. The ions, if their density is low enough so that space charge effects can be neglected, will execute slow, simple harmonic motion normal to the rods of the trap, as well as rapid oscillatory motion at the RF frequency. No restoring force is along the trap parallel to the rods so that end-electrodes, at a positive voltage with respect to the quadrupole rods, are needed to confine positive ions along that trap axis. The ions are thus confined for relatively long times in an elongated cloud in the trap, provided their temperature is low enough. The interaction with the RF field and collisions between ions leads to cloud heating, so some form of cooling is needed. In addition, the velocity of the ions caused by their motion in the RF field is proportional to their distance from the axis of the trap and produces the second-order Dopp-

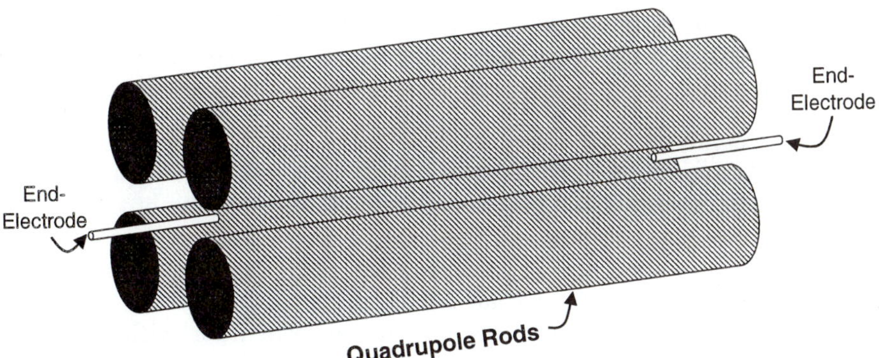

Figure 20.5 Drawing of the essential elements of a simple two-dimensional quadrupole trap. Diagonally opposite rods are connected together, and a combination of RF and DC voltage is applied between the connected pairs of rods. This produces a restoring force toward the axis of the trap for ions of the proper charge-to-mass ratio. End electrodes with a suitable voltage applied with respect to the rods provide a restoring force toward the center of the rods, preventing ion loss out the ends. An elongated cloud or a string of ions is thus trapped along the symmetry axis of the rods.

ler shift mentioned earlier. The shift must be properly averaged over the cloud and is proportional to the square of the ion cloud diameter so having a long, small-diameter cloud is important. This is a strong reason to prefer a two-dimensional trap to a three-dimensional trap. The shift can be calculated from the dimensions of the trap, the voltages, and the cloud dimensions.

Cooling techniques. In regard to cooling the ions, two methods have been used. One is to admit helium at moderately low pressure. The helium is roughly at the temperature of the apparatus and the collisions of the heavier ions with the helium atoms transfer kinetic energy to the helium on the average and cool the ions. The temperature will always be somewhat higher than that of the apparatus, but is usually low enough so that trapping times of as long as hundreds of seconds are achieved. The collision of the ions with helium atoms does produce a small frequency shift because the wave functions of the ions are perturbed during the collisions. This frequency shift can be estimated by varying the helium pressure and extrapolating to zero pressure.

The second cooling technique is *laser cooling*. Temperature as low as a few microKelvin can be reached in a high vacuum. Under these conditions the ions can be frozen into a crystalline lattice. For a small number of ions, say less than about 100, the cold ions form a string along the axis of the trap. Their velocity is almost zero and so the second-order Doppler correction is negligible. In the extreme case, only one ion is present in the trap. This is a very good example of a single quantum system, almost completely unperturbed, that makes an almost ideal frequency standard.

Complete standard. Completion of the standard requires the usual high-quality oscillator/synthesizer combination locked to the transition. The state selection is performed by optical pumping. To avoid light shift, the pumping light, as well as any cooling light, must not be on when the microwave interrogation occurs. This requires a pulse-sequenced operation.

Mercury-199 standards. A small number of helium-cooled microwave Mercury standards have been built and they have excellent performance. They have demonstrated the best short-term stability of any of the passive high-performance standards.

Work done at NIST on laser-cooled mercury ion standards promises accuracy of much better than 1 part in 10^{15}.

Mercury-199 also has an optical transition with an extremely narrow linewidth, so it would make an excellent optical frequency standard. NIST is working on this as well as the necessary RF-to-optical frequency synthesizer. This synthesizer is a very challenging project in itself.

20.4 Secondary Frequency Standards

Secondary standards require calibration to an external reference standard. These standards generally exhibit some variation in output frequency with time and the frequency varies from unit to unit much more than primary standards. The most common cause of the time variation is the aging of some parameter or element on which the frequency depends.

20.4.1 Hydrogen maser

The hydrogen maser is an active frequency standard. A drawing of the internal structure of a hydrogen maser is shown in Fig. 20.6. Molecular hydrogen

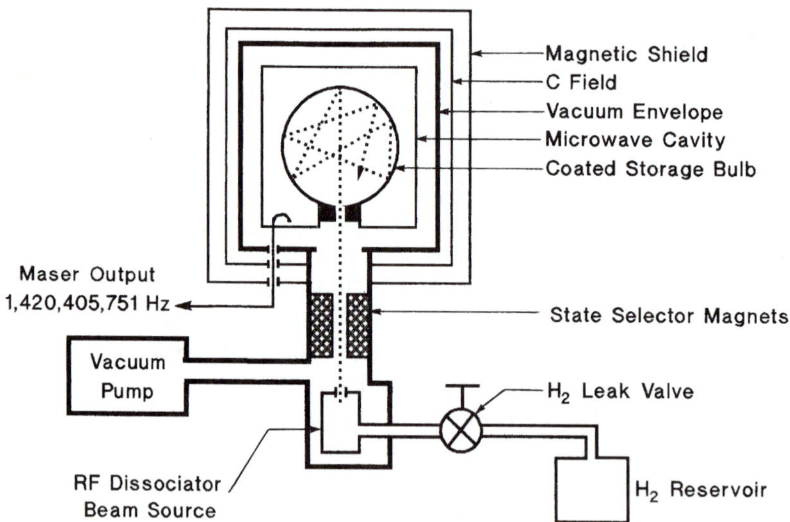

Figure 20.6 Physics package of a hydrogen maser. The storage bulb is usually made of quartz, coated with Teflon.

from the reservoir leaks into the maser through a heated palladium barrier. The hydrogen molecular bond is broken by dissociation in an electrical RF discharge. The resultant atomic hydrogen is collimated into a beam and directed through a magnetic state selector, which selects the atoms in the upper hyperfine energy level. These selected atoms are then directed into a Teflon-coated, quartz storage bulb inside of a microwave cavity. The Teflon coating allows the hydrogen atoms to collide with the wall without relaxing or changing state. However, a small phase shift results from the collisions. This leads to a frequency shift called *Wall Shift.* The existing oscillating field in the cavity stimulates the atoms to make the transition to the lower energy state and the radiation emitted is coherent with the stimulating signal. This is a self-sustaining process if both the number of atoms in the cavity and the cavity Q are high enough. The maser oscillation starts from stimulated emission by thermal noise in the cavity and builds up to the equilibrium value set by equality of the power available from the entering atoms to the power dissipated in the cavity and its load. The output frequency of a hydrogen maser is 1,420,405,751 Hz. The output signal is used to phase-lock a precision quartz oscillator/synthesizer combination operating at 5, 10, or 100 MHz. Thus, the output frequency has the same accuracy and long-term stability as the hydrogen maser signal.

To prevent losses caused by collisions between hydrogen atoms and contaminant gases and to scavenge spent hydrogen from the system, the cavity and storage bulb are kept in the highest possible vacuum, considering the hydrogen inflow, and continuously pumped with an internal vacuum pump.

Hydrogen maser accuracy is limited by frequency shifts caused by hydrogen-hydrogen collisions, the Wall Shift of the storage bulb inside the resonant cavity, and cavity mistuning. The frequency also depends on the second-order Doppler shift from the thermal velocity of the hydrogen atoms in the storage bulb. This amounts to about -1.4 parts in 10^{13} per Kelvin. The hydrogen maser is not a primary standard because the effects of the wall shift cannot be calculated from first principles. The magnitude of the wall shift in typical masers is about 1 part in 10^{11}. Current technology provides auto-tuning of the cavity to remove the effect of cavity mistuning. The resultant standard is accurate to better than one part in 10^{11}, with short-term stability superior to that of any other standard. The long-term stability of the auto-tuned maser can be very good if the Teflon coating is very clean and stable, and the temperature is held very constant.

Active hydrogen masers are larger, more expensive, and more sensitive to environmental influences than modern cesium-beam standards.

Passive hydrogen masers are also available. In these, the atoms do not oscillate. The resonance is detected by applying a microwave signal from a quartz oscillator/synthesizer combination, like the other passive standards. Although larger in size and higher in cost than a cesium-beam standard, in certain applications where short-term stability is important, a passive maser offers a slight advantage over cesium. One advantage over the active maser is that the frequency shifts caused by cavity mistuning are much smaller. Over the long term, a passive maser has the same accuracy, stability, and precision as the active maser. Its short-term stability is, however, poorer.

20.4.2 Rubidium-vapor standard

The rubidium-vapor frequency standard is an optically pumped passive standard based on the hyperfine transition in rubidium-87. A typical optical package from a rubidium standard is shown in Fig. 20.7.

In one form of the rubidium standard, a small glass bulb, the pumping lamp, filled with buffer gas and ^{87}Rb (rubidium-87), is subjected to an intense RF field, producing a plasma inside the lamp that generates light, which contains power at several discrete optical wavelengths. The most important wavelengths are 780 and 794.8 nm. The ^{85}Rb filter cell removes light corresponding to the transitions to the upper hyperfine level of ^{87}Rb because those transitions are more closely matched between the two isotopes. The remaining light, which is mainly that corresponding to the transitions to the lower state, enters the ^{87}Rb reference cell, where it interacts with rubidium atoms in the cell. The light pumps atoms from the lower level to intermediate optical states, where they spontaneously emit light and make a transition to both hyperfine levels. Because only the lower level is being pumped, the upper level becomes highly populated and, in equilibrium, the cell becomes more transparent to the incident light in the absence of microwave excitation. The light transmitted through the cell reaches a photodetector. When microwaves are applied at the hyperfine transition frequency, 6.834682613 GHz, atoms in the upper state are induced to make a transition to the lower state, where they can then be pumped. Because there are now more atoms to pump, the

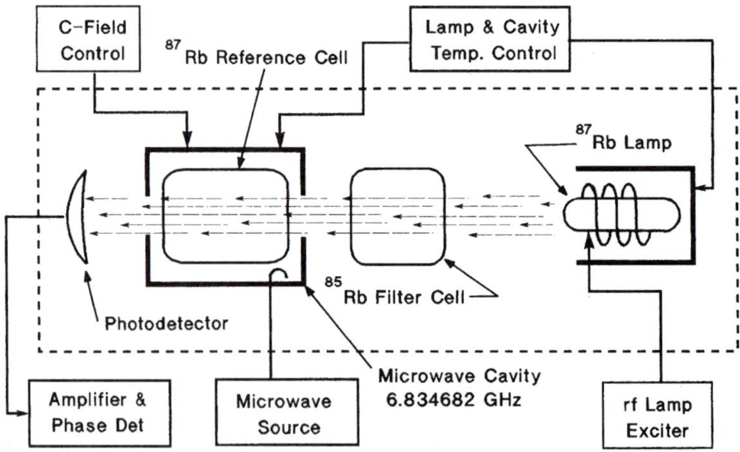

Figure 20.7 Rubidium-vapor reference diagram. Light from the rf lamp passes through a filter cell, a reference cell, and is detected by a photodetector.

cell then becomes less transparent and the photodetector current decreases. A plot of photodetector current versus microwave frequency shows a dip with the minimum occurring at the ^{87}Rb resonance frequency. The microwave frequency is synthesized from a quartz oscillator and coupled to the resonant cavity containing the rubidium-vapor reference cell.

The reference cell contains a buffer gas that has only minimal interaction with the ^{87}Rb. This effectively confines the Rb atoms and removes the Doppler broadening giving a much narrower linewidth than a pure Rb cell would have. The buffer gas does produce a frequency shift that is dependent on the gas used as well as temperature, and pressure.

Because the atoms are irradiated with both light and microwaves, a frequency shift in the microwave resonance can occur, called *light shift.*

The rubidium-vapor standard is a secondary standard because it is not self-calibrating as a result of the buffer gas-induced frequency shifts and the light shift. Once constructed, it must be compared with a more-accurate standard, such as cesium, to determine its exact frequency. Rubidium standards also exhibit aging because the shifts (mentioned previously) change with time. The aging is of the order of several parts in 10^{11} per month so the standards require periodic calibration to maintain a given accuracy specification.

Well-designed rubidium standards have very good short-term stability, usually superior to cesium-beam standards. Size, weight, and cost are significantly lower than those of any of the primary standards, making rubidium the atomic frequency standard of choice in many applications.

Because the pumping technique produces a high population in the upper state, the rubidium standard can also be operated as a maser. Some of these have been built and exhibit very good short-term stability.

Currently, several groups are working on diode laser-pumped rubidium standards. These offer very good short-term stability, as well as reduced size and power.

20.4.3 Quartz crystal oscillators

Crystalline quartz is a piezoelectric material with excellent mechanical and chemical stability. When a quartz crystal of the proper design and configuration is deformed mechanically, a voltage is induced between electrodes on the crystal or coupled to it. Conversely, when a voltage is applied to the electrodes, the crystal undergoes a mechanical deformation, which is proportional to the field strength in the crystal and its polarity. This effect makes it easy to use a quartz crystal as the frequency-determining device, based on the mechanical resonances of the crystal, in an electronic oscillator circuit. The higher-precision crystal resonators operate in a thickness shear mode, where the resonant frequency is inversely proportional to the thickness of the resonator. Overall performance of a quartz crystal depends upon the crystallographic orientation to which it was cut. The mounting structure and manufacturing techniques also strongly affect performance. Most used orientations

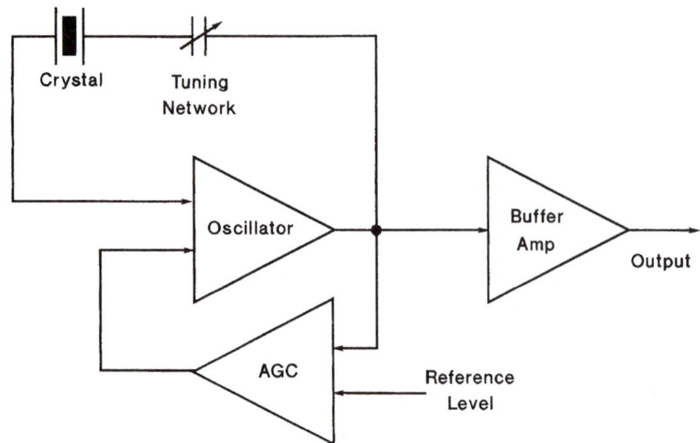

Figure 20.8 Block diagram of a quartz oscillator with tuning and AGC (automatic gain control).

are the *AT cut* and the *SC cut*. Both are known for their thermal performance because the orientations used have a frequency-vs.-temperature curve that is cubic. When operated at the point of zero slope of the frequency-temperature curve, these crystals exhibit at least first-order temperature compensation, which minimizes the dependence of the resonant frequency on temperature. Of the two, AT cuts are the most common. The SC cut is more expensive to produce, but also offers freedom from thermally generated frequency transients. SC cuts are normally used only in the highest-precision oscillators or for highly specialized military applications.

High-precision crystal oscillators are usually built inside of a temperature-controlled oven to further minimize environmental effects. A generic quartz oscillator is shown in Fig. 20.8. In the figure, the oscillator provides electronic gain with its output signal fed back through a tuning network and a quartz crystal. In this mode, the crystal and its tuning provide a filtering function that permits operation only at a discrete frequency, which is representative of the quartz crystal used. Buffering is necessary to isolate the oscillator from outside electrical loading effects. *Automatic gain control* (*AGC*) is used to ensure that the oscillator is operating at the proper crystal drive level. A change in drive level to the crystal usually results in a change in frequency.

Quartz crystal oscillators are available in frequencies below 50 kHz up to frequencies above 1 GHz for surface acoustic-wave devices. Oscillators are available in almost any configuration, size, and cost. The highest-precision oscillators exhibit aging on the order of several parts in 10^{11} per day, about 30 times greater than rubidium-vapor standards. Because the quartz crystal resonator is a mechanical device, not dependent on an atomic transition for its inherent accuracy, quartz crystal oscillators are secondary standards and must be calibrated through comparison with a more-accurate standard.

TABLE 20.1 Comparison of Frequency Standards

	Commercial quartz	Commercial rubidium	Commercial cesium	Commercial hydrogen	Laboratory cesium fountain	Laboratory LASER cooled mercury ion
Accuracy	1E-8	5E-12	1E-12	1E-13	1E-15	<1E-15
Aging per year	6E-9	2E-10	<1E-14	1E-13	<1E-15	<1E-15
Temperature stability	1E-9	2E-10	2E-14	1E-12	<1E-15	<1E-15
Stability @ 1 sec	1E-12	1E-11	5E-12	2E-12	2E-13	2E-13
Stability @ 1 day	1E-10	3E-13	3E-14	1E-14	1E-15	1E-15
Retrace	1E-10	1E-11	1E-13	1E-13	1E-15	<1E-15
Warm-up, min	5	15	30	1440		
Size, cm^3	200	1000	30000	150000		
Weight, kg	1	2	30	100		
Power, W	2	15	50	150		
Cost, $	1000	5000	60000	250000		

20.5 Comparison of Standards

Table 20.1 compares many of the features of the frequency standards. Estimates are given in a number of places. The laboratory standards are not described in terms of cost, weight, size, warm-up time, and power. By *accuracy,* in the cases of the secondary standards, it is assumed that a calibration has been performed. For example, an uncalibrated rubidium standard would have an accuracy of no better than 1 part in 10^8; uncalibrated quartz is 1 part in 10^6; uncalibrated hydrogen is about 2 parts in 10^{12}.

20.6 Time Standards

The frequency standards covered previously provide ultra-stable oscillators that permit highly precise and accurate clocks to be developed. Keeping accurate time is much more difficult than maintaining an accurate frequency standard. Correcting the frequency of a frequency standard requires setting it to agree with the frequency of an appropriate reference. However, if this same standard is driving a clock, correcting the frequency does not correct the time. Correcting the frequency allows the clock to run at the right rate but will not correct for the time errors accumulated when the frequency standard was at the wrong frequency or errors due to improper setting of the initial time.

Time is used to designate two distinct concepts. The first is associated with when an event happens, usually termed *epoch.* The second is related to time interval, the length of time between two events. The *atomic second,* as defined previously, enables us to measure time interval quite accurately, independent of other definitions or conventions. To determine the epoch of an event, time must be measured with respect to some reference. By convention and international agreement, the epoch of an event is specified by a date composed of day number, month, and year, and the time in hours, minutes, and seconds, with respect to a reference. All of these were originally based on astronomical cycles or parts of cycles, of periodic events, such as the position of the sun, as it appears in the sky and earth's movement about the sun. To keep time and to disseminate it is the responsibility of various government agencies worldwide.

20.6.1 National Institute of Standards and Technology (NIST)

NIST, formerly the *National Bureau of Standards* (*NBS*), is one of the two organizations in the United States that is primarily responsible for providing time and frequency information. NIST is a part of the Department of Commerce of the United States government. Its Time and Frequency Division is located in Boulder, Colorado. The other agency, the United States Naval Observatory (USNO), is a part of the U.S. Department of Defense and is located in Washington, DC. Roughly speaking, the NIST is responsible for the U.S. contribution to the accuracy part of the atomic second. Both organizations provide input to BIPM on the stability of the atomic second.

NIST's input is derived from a system of atomic clocks. The system consists of various primary cesium standards, developed by NIST, that are used to check the accuracy of a number of commercial atomic clocks, mostly cesium-beam clocks. The cesium-beam standards are used as system "flywheels" to maintain correct time. NIST uses several means to distribute time and frequency signals, using many different forms of communication media. The information is available to anyone who has a need to set time and/or frequency to control clocks, timing systems, calibration systems, or secondary frequency standards.

WWV and WWVH radio broadcast stations. Dissemination of time and frequency signals primarily in the western hemisphere is the responsibility of the NIST radio station WWV, broadcasting from Fort Collins, Colorado. The signal is broadcast on standard frequencies of 2.5, 5, 10, 15, 20, and 25 MHz. A companion station, WWVH, located in Kekaha, Kauai, Hawaii, broadcasts on standard frequencies of 2.5, 5, 10, 15, and 20 MHz to areas in the western Pacific.

The time signals correspond to the UTC time scale and are better than 2 parts in 10^{11} when transmitted. In transmission, the accuracy is degraded, owing to atmospheric propagation effects and to Doppler shift when the transmission path includes reflection from the ionosphere. The Doppler error can approach one part in 10^7 in the worst conditions. These signals are readily available, can be received with inexpensive receivers, and are of accuracy sufficient to set clocks to within one ms of UTC. Additional information is transmitted for UT_1 corrections.

WWVB radio broadcast station. A more accurate service provided by NIST is the WWVB signal, also broadcast from Fort Collins, Colorado. This signal is broadcast at 60 kHz and is referenced to atomic time. Commercial equipment is available that performs phase comparisons between a local reference and a signal derived from WWVB. With normal signal conditions within the United States, frequency accuracies of two to three parts in 10^{11} can be achieved in 24 h. For shorter comparison times, the best results are usually obtained when the total propagation path is in sunlight and conditions are stable. Near sunrise and sunset, there are noticeable shifts in both amplitude and phase. Time pulse modulation permits time synchronization to 100 μs or better, provided that compensation for the propagation delay is known.

UTC information is also broadcast and allows remote clocks to be set with moderate accuracy.

Other methods of time dissemination. Other services provided by NIST are (in increasing order of accuracy):

- Telephone voice messages duplicating the WWV message are obtained by dialing (303) 499–7111 in Colorado or (808) 335–4363 in Hawaii, providing accuracy to about 30 ms.
- Computer modem time transfer programs capable of synchronizing computer internal clocks to one time tick, or several milliseconds. This is available from Colorado by dialing (303) 494–4774 and from Hawaii at (808) 335–4721.
- Remote synchronization of time bases to about 1 part in 10^9.
- Common view of Loran-C stations with 24-hour continuous monitoring, using NIST-provided measurement equipment. Over a long period, overall frequency accuracy approaches parts in 10^{12}.
- Common view of GPS satellites, providing time transfer accuracies on the order of 30 to 50 ns.

20.6.2 United States Naval Observatory (USNO)

USNO provides accurate time and astronomical data. Continual observation of positions and motions of the sun, moon, planets, and principal stars, with "standard" cesium and hydrogen clocks, permits the determination of precise time. Additional observations are used in the compilation of tables and star catalogs for navigational purposes.

USNO controls the distribution of precise time and frequency, primarily for the use of the U.S. Department of Defense and its associated branches. USNO monitors a wide variety of transmitted time and frequency signals and provides routine corrections to maintain these to within the established specifications of the system. USNO uses several means to distribute time and frequency signals, using many different forms of communication media. Although designed primarily for U.S. Department of Defense use, this information is also publically available. The USNO Master Clock time is available via modem by dialing (202) 762–1594.

Loran-C. *Loran-C (Long-Range Navigation)* is a federally provided radio navigational system for civil marine use in U.S. coastal waters and is the successor to earlier radio navigational systems. Although primarily used for navigation, Loran-C transmissions are also used for time-dissemination and frequency-reference purposes. The system consists of transmitting stations arranged in local geographic groups. Each group contains a master station and two or more secondary stations. Measurements of time difference are made using receivers that achieve high accuracy by comparing zero crossings of a specified RF cycle within the pulses transmitted by the master and secondary stations within a group. The U.S. Coast Guard is responsible for the operation and maintenance of the system within the United States and in certain overseas locations. The USNO monitors and provides correctional factors to keep the system synchronized.

All Loran stations have triply redundant cesium-beam standards, which are carefully managed in time and frequency. Loran-C is one of the older of several methods of intercomparing atomic standards between international agencies, such as NIST, USNO, and the National Research Council in Canada.

All stations operate on the same frequency, 100 kHz, but differ in repetition time and time cycle. The Loran signal contains no intelligence. Thus, to determine the exact UTC time, the user must first set the clock to within about 10 ms using one of the other methods. Typical Loran-C receivers can achieve frequency accuracies of one part in 10^{12} and time accuracies of better than 1 μs after monitoring for 24 h.

Global positioning system (GPS). GPS is designed and deployed by the U.S. Department of Defense as an all-conditions, three-dimensional navigation system. GPS enables users to determine their three-dimensional position, velocity, and time anywhere in the world with unprecedented accuracy. GPS has three segments. The relationship of these three segments is shown in Fig. 20.9. The space segment presently has 27 satellites orbiting at 10,900 nautical miles, providing an orbital period of 12 h. Each satellite contains multiple atomic clocks, usually cesium and rubidium, and quartz oscillators. Each satellite broadcasts continuously on two frequencies. The first, the L1 band signal, is available to any user. This provides positional accuracy to about 100 m rms, and timing accuracy to about 300 ns rms. The second signal, broadcast in the L2 band, is available only to qualified military users and provides a higher degree of accuracy both in position and in timing.

The control segment consists of a master-control station, five monitor stations, and three ground antennas located around the world. This segment tracks the satellites, determines orbits with extreme precision, and transmits orbit definition information to each satellite. The information sent to each satellite also contains up-to-date information on the difference between the satellite clock and UTC as maintained by USNO.

The user segment determines position, velocity, and time by passive trilat-

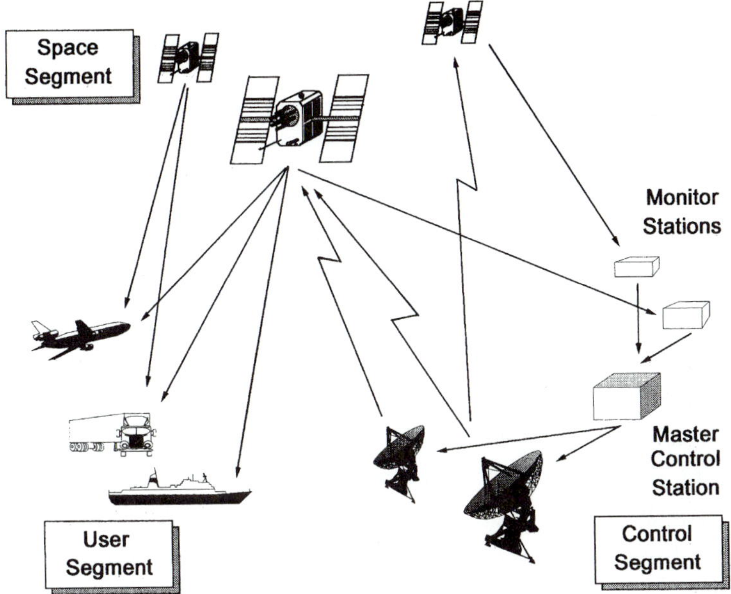

Figure 20.9 The global positioning satellite (GPS) navigation and timing system.

eration. To be fully useful, the GPS receiver must track the spread-spectrum codes transmitted by four satellites simultaneously. By the measurement of time of arrival of the signals from four satellites in view, knowing the speed of the radio signal, knowing when each signal is transmitted, and the position of each satellite, the final solution yields an accurate three-dimensional position of the receiver antenna, as well as time, as determined on the UTC time scale. The position of the satellite is known because each satellite message contains a satellite ephemeris.

With the system fully functional, at least four satellites are within sight of any ground station 24 hours per day. Under ideal conditions, frequency accuracies of parts in 10^{13} and timing accuracies of better than 100 ns are possible.

Other methods of time dissemination. Many of the services provided by NIST from Boulder, Colo., are also available from USNO. The exact particulars differ, but the basic services remain the same. In addition to those listed for NIST, USNO also provides an online computer modem service, providing current status of all of the Department of Defense timing services.

20.6.3 Other national and private time standards

In addition to NIST and USNO, national standards organizations of many other countries throughout the world provide a similar service. For example, radio transmissions similar to WWV and WWVB originate in Canada, Germany, France, United Kingdom, Switzerland, Italy, Japan, Argentina, Australia, Taiwan, Czechoslovakia, Brazil, Russia, Venezuela, China, and South Africa. Contributions to the international atomic time scale are made by at least 15 different national laboratories—each with a number of cesium-beam and/or hydrogen standards. Loran-C systems, orginally established by the U.S. Coast Guard in areas outside of the United States, are now operated by several different countries. Private Loran-C systems have been established in various areas around the world, mostly associated with marine oil exploration.

The former-USSR established GLONASS, a service similar to GPS with essentially the same goals, but with a degree of accuracy that is about one-third that of GPS. The service is maintained by URSS in Moscow.

20.7 Ensembles of Time and Frequency Standards

The majority of timekeeping services throughout the world have two or more atomic frequency standards as their source of atomic time. The primary reason for this is greater reliability. If one clock fails for any reason, a backup clock exists. Further, given that the timekeeping service has a group of clocks, improved accuracy can be obtained by combining time scales from many individual clocks.

Each primary clock is stochastic in nature. That is, the variability or instability of one clock is, in general, not related to the variability in another clock. Thus, by forming a group of clocks, each contributing to the final output, the variability of the group of clocks would be reduced through averaging. Further, by computing a weighted average, where the weighting is a function of the individual clock's stability, higher stability having a higher weight, the weighted average would represent a clock scale that is superior to the best clock in the group, and is such that even the worst clock improves the time scale.

If all clocks are essentially equal and their frequencies are normally distributed about the nominal frequency, then the accuracy and stability of an ensemble of clocks is improved by the square root of the number of clocks. Similarly, the phase noise of the group is improved by 10 log N dB, where N is the number of clocks. The variance of the time scale decreases as $1/N$, which means that the measurement time necessary to achieve a given level of accuracy is reduced by the factor $1/N$. For clocks that are not equal, the improvement is not as great, but the ensemble still yields better accuracy and stability than that of a single clock.

20.7.1 Paper clock ensembles

The clock ensemble is the basis for the BIPM time scale, and thus UTC, as covered earlier. Data from many clocks in all parts of the world are collected.

From calculations on the data, and from long-term data showing the stability of each clock with regard to the atomic second, a final value is determined that becomes the accepted definition of the atomic second. On a smaller scale, the same types of calculation are used for ensembles of clocks at NIST and at USNO.

All of these are *paper ensembles.* Calculations are performed on all of the available data to arrive at a corrected local time scale. An *average time scale* is computed from the ensemble and published after the fact to correct the real-time output of the ensemble. From the data, usually one clock is chosen as the final source of the time scale. In some facilities, this clock is adjusted to agree as closely as possible to UTC.

A paper ensemble has no output signal at the instantaneous average frequency of the ensemble because the data computation and clock adjustment cannot be performed in real time. The output clock has the short-term instability of a single clock. There is little or no redundancy for the real-time clock output.

20.7.2 Real-time clock ensembles

If the clocks can be steered in real time, it is possible to develop a hardware implementation of the paper clock. In one type of steered clock, the internal cesium servo loop contains a direct digital frequency synthesizer so that the output frequency can be changed in precise steps from the unperturbed natural frequency. The output can be steered by digital commands applied to the synthesizer through the standard's serial port. In the ensemble, data are taken continuously from all contributors to the ensemble and then used in real time to control the group of clocks.

Figure 20.10 is one embodiment of a real-time ensemble. At power-up of the ensemble, all clock outputs are at different frequencies and phase. As time progresses, each clock output is steered in frequency and phase so that eventually all clock outputs are exactly at the same frequency and phase. When this occurs, the signal out of the power summer is at maximum amplitude. The key to performance is the algorithm in the ensemble controller. Optimal performance occurs when the resultant frequency represents the average of all of the clock outputs in the ensemble, under the condition that the sum of the steering commands is zero.

In this ensemble, the output frequency is at the average of the units in the ensemble. The short-term and long-term stabilities of the output are improved by optimal combination of the individual clocks. Reliability is greatly improved since the failure of a single clock does not interrupt the output. The ensembling algorithm can assign weights to the members and adjust them continuously to maintain optimum performance. Clock data is continuously acquired and processed. The complete ensemble can be steered, if required, to synchronize to another standard. Extensions to the real-time ensemble al-

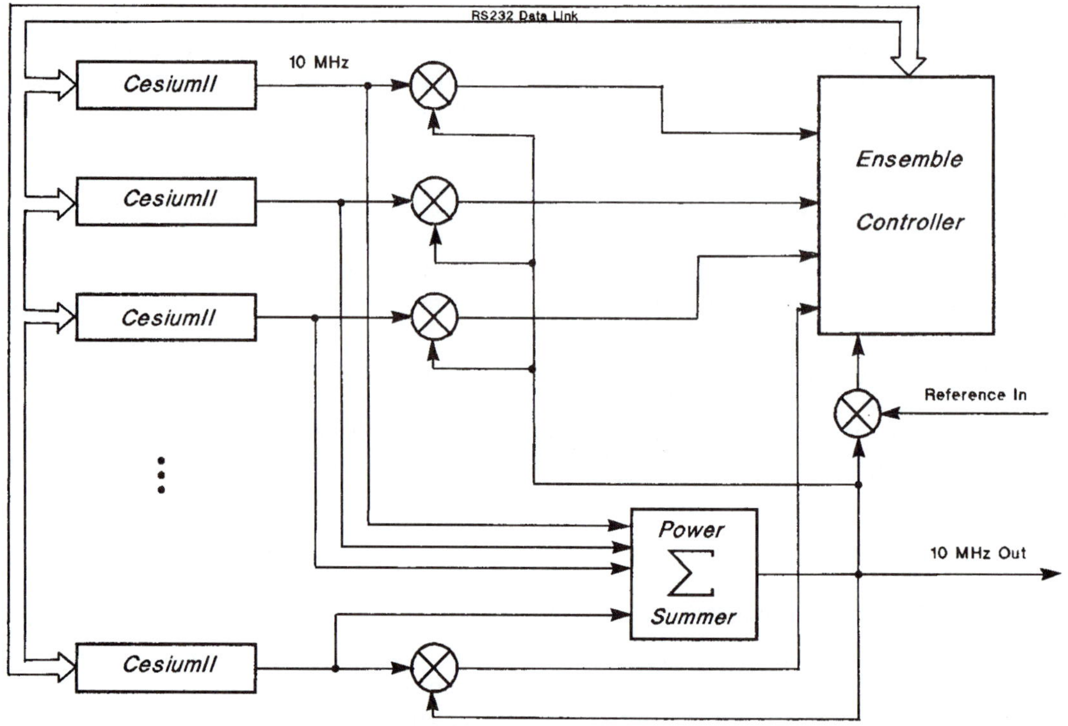

Figure 20.10 A real-time ensemble of independent cesium standards controlled such that the output of the system is the average of all of the independent standards.

gorithm permit clock removal from and installation into the ensemble of clocks with no significant perturbation of frequency.

In Fig. 20.10, all clocks are steered to a common frequency. Another approach, which also results in the same performance, permits each clock to run at its native frequency. In this case, a separate ensemble frequency source is steered to the average frequency by the ensemble controller. The data gathered in this case is the deviation between each clock and the ensemble frequency source. Although the overall advantages are the same in either case, the unsteered ensemble permits using a very low noise oscillator as the ensemble frequency source, yielding the excellent short-term stability of an oscillator while, at the same time, obtaining the accuracy and long-term stability improvements of the ensemble of cesium standards.

Another version of a real-time clock ensemble combines the virtues of the two ensembles covered previously, with cesium standards, whose outputs are steered to a common phase and frequency, and with an ensemble output that is derived from a low-noise, high-stability quartz oscillator. The combination yields an ensemble that is superior to the first ensemble in terms of the phase changes “seen” when a clock is installed or removed from the first ensemble.

The combination is superior to the second, unsteered clock ensemble, in that keeping track of the phase changes (Fig. 20.13) is not required.

20.8 Precision Time and Frequency Comparisons

The calibration of precise time and frequency sources requires a comparison between the device to be calibrated and a precision reference source. All secondary standards require periodic calibration. In most cases, the calibration is against a local reference standard. For the highest precision, comparison against a national reference standard might be required.

20.8.1 Local clocks

Local clocks are generally those time and frequency standards that are maintained at the calibration site. In some cases, where a low degree of precision is required, the time and frequency-reference signals can be transmitted from one site to another. As a result of the transmission process, the received time and frequency signals are degraded by propagation delay and induced line noise. A more-subtle effect that degrades the reference signals is the environmental performance of the transmission media. For example, propagation temperature coefficients approaching many parts per million per degree Celsius have been observed on coaxial cables.

Frequency counters. The modern electronic frequency counter is a versatile device (see Chapter 19). In its simplest mode, it provides a method to determine the frequency of any suitable signal applied to its input port. The accuracy of the measurement is directly related to the internal resolution of the counter and the accuracy of its internal frequency source. A simple test of the accuracy, resolution, and stability of the counter is to provide as an input to the counter a frequency-reference standard whose stability is known to be significantly better than the specifications of the counter. Because most high-precision sources have a frequency stability over short times that is orders of magnitude better than that of most counters, this requirement is easily met. Observation of the measurement, as displayed on the counter readout, provides the following two key parameters. First, the mean relative difference, i.e., (*standard frequency–measured frequency*)/(*standard frequency*), is a direct measure of the relative error of the counter's internal time base. Second, the instability of the readings on the counter is a measure of the instability of the internal time base and the basic counter electronics.

The performance of the frequency counter can be significantly improved in both accuracy and stability by using the reference-frequency source as an external time base for the counter. However, modern counters are still limited by their internal design to resolutions on the order of one part in 10^9. This means that most high-precision time and frequency sources cannot be evaluated adequately by direct measurement with a frequency counter.

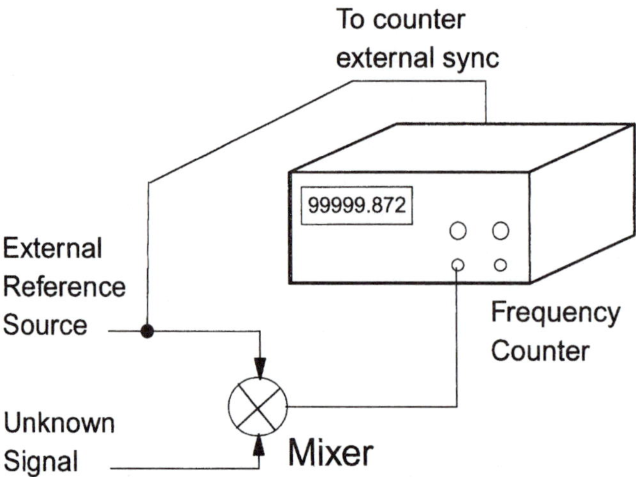

Figure 20.11 Single-mixer heterodyne frequency measurement technique.

By the use of heterodyne technique, the resolution of a frequency counter based on the reciprocal counting technique (see Chapter 19) can be greatly extended. A typical equipment setup, the "single-mixer heterodyne" technique, is shown in Fig. 20.11. In this figure, the counter internal time base has been replaced with the external frequency reference. The frequency-reference source also is applied to a diode mixer, where the signal being measured is heterodyned against the reference. The difference frequency is then measured by the counter. Reciprocal counters usually have a resolution of about one part in 10^9. For example, assume that the unknown signal is the usual if signal for an FM receiver, 10.7 MHz. Applying this directly to the counter will result in a measurement to about 0.01 Hz. After heterodyning with a 10-MHz high-stability ion source, the difference frequency will be 700 kHz. The setup shown in the figure will measure this to one part in 10^9, 0.0007 Hz. The improvement in resolution is just the ratio of the difference frequency measured by the counter to the signal frequency, in this case, 0.7/10.7, provided the counter resolution remains constant. Further resolution enhancement can be obtained if the input reference source is actually a digital synthesizer synchronized and driven by the frequency reference source. In this manner, resolution enhancement to any desired degree of precision can be obtained, limited only by the instability of either source or the synthesizer used. Overall accuracy and stability will be governed by the signal with the worst stability. Thus, unless it is known that the frequency reference source is significantly better than that being measured, the only conclusion that can be drawn from a measurement of this type is that the signal being measured is no worse than the measurement indicates, and can be much better.

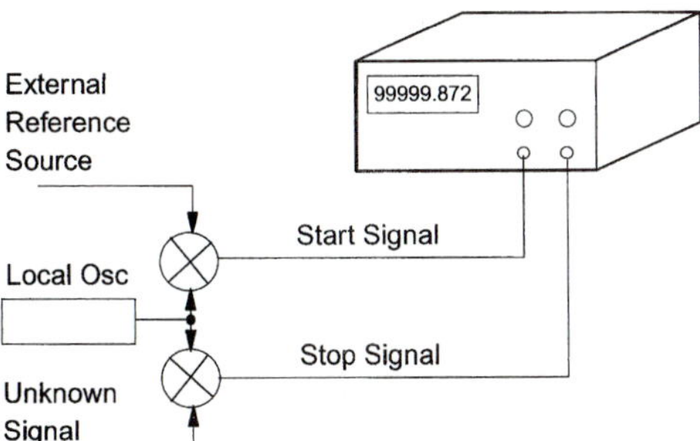

Figure 20.12 Dual-mixer time interval frequency comparison technique.

Time interval counters. Specialized time-interval counters have been developed with resolutions on the order of picoseconds. With this degree of precision, several other methods of determining the frequency of a high-precision source are available. In each case, the measurement is a comparison against a known frequency-reference source. In each case, the degree of precision is governed by the signal with the greatest amount of noise, or instability.

The first application is shown in Fig. 20.12. This technique, called the *dual-mixer time-difference technique,* is an excellent method of comparing two signals that are virtually identical in frequency. The offset frequency source used to heterodyne both channels need not be of exceptionally high quality as its instability is common to both measurement channels and will be eliminated to first order. After heterodyning, one mixer output starts the time-interval counter, and the other mixer output stops it. By tracking the counter readings over a period of time, a data plot like that shown in Fig. 20.13 results. In this figure, the measured time intervals between the start and stop signals are plotted as a function of elapsed time. The maximum time interval that can accumulate is the period of the highest frequency applied to either the "start" or "stop" ports of the counter. If a full period of time interval does accumulate, the data reduction becomes more complicated because proper one-period adjustments must be made to all of the data obtained after the

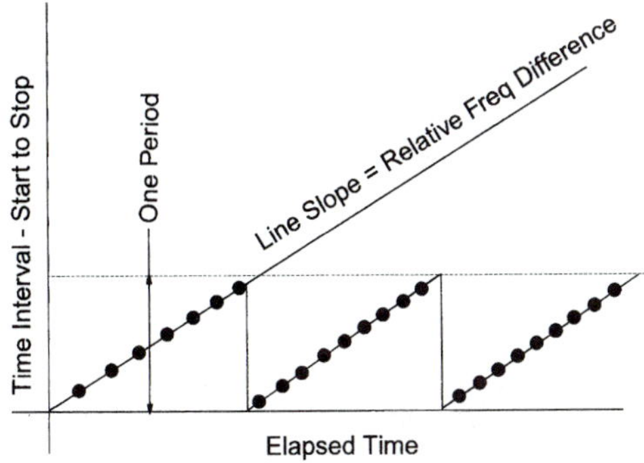

Figure 20.13 Typical data plot obtained by the dual-mixer time interval technique.

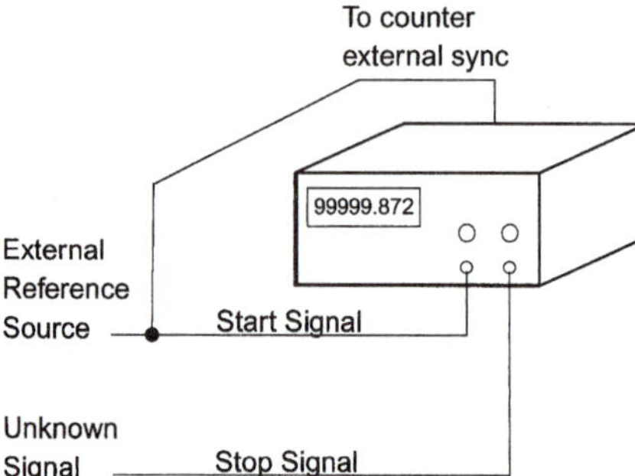

Figure 20.14 Direct time interval frequency comparison technique.

data step. In both signals are relatively stable, a determination of the unknown frequency can be determined as a relative frequency difference by computing the slope of the data. Again, as in the case of the single-mixer heterodyne system, the results indicate that the unknown frequency is no worse than the measurement indicates and might be better. A refinement of the technique shown is to measure several frequency sources, each against the other in pairs. From the data obtained, the relative acuracies and instabilities of each of the signals can be determined independently.

A simpler application is shown in Fig. 20.14. In this case, the "direct time interval" technique, both signals are applied directly to the inputs of the time-interval counter, where the start signal is the reference-frequency source and the unknown signal is very close in frequency to the reference. In this case also, a curve similar to Fig. 20.13 results. This method is used commercially to measure large numbers of cesium and rubidium standards—essentially simultaneously through continuous sampling. Because an advanced time-interval counter has a resolution of about 100 ps, a resolution of one part in 10^{10} is achieved from sampling every second. Because this method has no heterodyne advantage, data must be obtained over long periods of time. In particular, this method measures the change in time between the start signal and the stop signal. Thus, if data on the changes are accumulated over an extended period of time and are properly analyzed, the relative time difference between each standard operating as a clock can be determined by the slope of the measured data. Because the following relationship holds for all time and frequency sources:

$$\frac{\Delta f}{f} = \frac{\Delta \varphi}{\varphi} = -\frac{\Delta t}{t} \tag{20.1}$$

where $\frac{\Delta f}{f}$ is the relative frequency difference,

$\frac{\Delta \varphi}{\varphi}$ is the relative phase difference,

$\frac{\Delta t}{t}$ is the relative time difference,

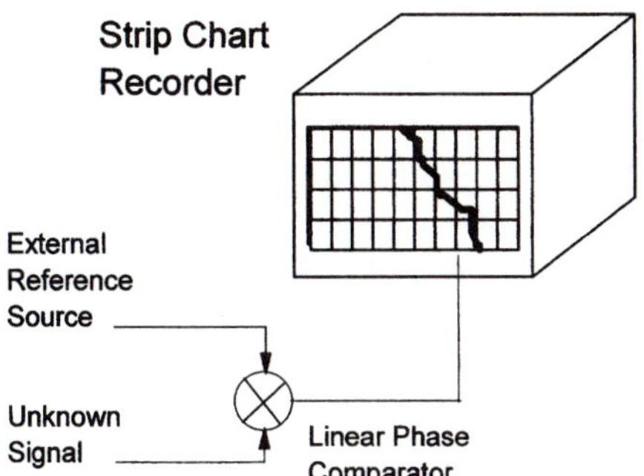

Figure 20.15 Phase plot technique.

the relative time difference between the clocks is precisely the negative of the relative frequency difference, or the relative phase difference. Analysis and experimentation indicate that if the two sources are within one part of 10^{10} of each other, data obtained over 10,000 s and properly averaged can yield accuracies of one part in 10^{13} with respect to the reference source.

Phase comparison. The *direct time interval method* is a refined form of phase comparison because knowing the relative time difference also yields the relative phase difference. A more-direct method is shown in Fig. 20.15. In this figure, a known stable signal and a signal at essentially the same frequency are compared in a linear phase comparator. The output of the phase comparator is a slowly varying voltage that is proportional to the phase difference between the two signals. Customarily, the output voltage drives a strip-chart recorder. Although different comparators have different features, customarily an output voltage of zero corresponds to zero phase difference. Full-scale output voltage corresponds to a phase difference of 360°. The strip chart records the change in phase over a given time period. *Relative phase difference* is the total change of phase observed on the strip chart divided by the total phase change accumulated by the reference signal.

As in the frequency-measurement methods covered previously, the relative accuracy and precision of the measurement will be no better than that of the least-precise source, whether or not it has been designated as the reference signal. Given the nature of the phase-comparison method, both the reference signal and the unknown signal must be quite close in frequency. For example, two signals at essentially 10 MHz that are separated by one part in 10^{10} in frequency will accumulate a total phase change of $3.6 \times 360° = 1296°$, in 1 hour. Phase changes much greater than this (i.e., frequency changes greater than one part in 10^{10} at 10 MHz) are difficult to determine on a strip-chart recording.

20.8.2 Remote clocks

A *remote clock* is used to synchronize local clocks at the calibration site. Time is transferred to the local clock using various media. Several methods were described previously as services provided by various national standards laboratories. These essentially represent available sources that feature radiowave propagation of timing signals. Each has advantages in that they are usually available continuously, and disadvantages associated with propagation delay, ionospheric effects, and other perturbations not under the direct control of the laboratory. However, in many cases, such a transmitted timing signal is adequate for the degree of accuracy and precision required by the calibration laboratory. For a higher degree of precision in time transfer, several methods have been developed.

Traveling clocks. A more-accurate method of transferring time involves physically transporting time to the calibration laboratory. Because of the need to synchronize clocks at the highest level of precision, and the need to keep the clock powered during the trip, the usual mode of transportation involves flying the clock in the passenger section of a commercial aircraft. However, cesium is an active alkali metal and is rated as an environmentally hazardous material, is spite of the protection provided by at least four levels of brazed and welded metals that contain the few grams of cesium in a typical cesium-beam tube. Prior to restrictions on the transportation of hazardous material in passenger-carrying aircraft, the usual method was to synchronize a special cesium standard, equipped with extra batteries and multisource power converters, to the national reference, then physically transport it to the local site. At the local site, the time contained on the traveling clock was transferred to the local clocks through various synchronization techniques. With current restrictions, this method is no longer in general use, but it is still used in special military applications. In the most precise applications, relativistic effects must be taken into consideration.

One-way GPS. The primary advantage of one-way GPS is that of cost and convenience. Modern GPS receivers are relatively inexpensive, do not require elaborate installation and costly antennas, can be operated essentially unattended, and produce timing signals that are sufficiently accurate for most laboratory calibration purposes.

One-way GPS is illustrated in Fig. 20.9. In this method, time is transferred from USNO via the GPS satellite system to a special receiver equipped to process timing corrections. A typical GPS receiver tracks the spread-spectrum codes from four or more satellites in view. Knowing the speed of the radio signal, the time each signal is transmitted, and the location of each satellite, the receiver creates an accurate three-dimensional position of its antenna, as well as the time determined on the UTC time scale.

The degree of accuracy is determined by many factors. Generally, the major inaccuracy is the result of artificial degradation of accuracy by the U.S. Department of Defense through a feature of GPS called *Selective Availability* (*SA*). With SA, either the timing signal, the satellite ephemeris, or both are perturbed in a random manner, resulting in a timing uncertainty of 300 ns rms. Other contributing factors are the length of cable between the antenna and the signal processor, timing delays in the antenna system and signal processor system, uncertainty in the exact satellite location, propagation delay effects caused by ionospheric perturbations, and a basic jitter of 20 to 60 ns caused by instabilities in the satellite timing systems.

In addition to the determination of position, a GPS timing receiver has another requirement. The epoch transmitted from the GPS satellite needs to be corrected before the output of the GPS receiver reflects UTC (USNO). In addition to the many corrections detailed, a further correction is needed because the timing synchronization pulse in the GPS message is not synchronized to UTC (USNO). The message transmitted by the GPS satellite also contains information about the difference between the satellite epoch and UTC.

Under ideal conditions and after determination of the appropriate correction factors, current receivers have demonstrated time-transfer accuracies of 50 to 60 ns using one-way GPS without SA. Even though the timing uncertainty with SA is 300 ns or greater, averaging the timing information from all satellites in view using a stable oscillator as a system flywheel and sufficient averaging time, time-transfer accuracies in the presence of SA of less than 100 ns have been demonstrated. Using a modern cesium-beam standard as the system flywheel and integrating for periods of several days, time-transfer accuracy of better than 10 ns can be achieved.

Common-view GPS. *Common-view GPS* is a technique that allows a highly precise transfer of frequency. This technique requires that the two sites interchanging time difference information coordinate their activities to track the same satellite(s) during specific time periods.

The basis for common-view GPS is that, if two sites are monitoring the same satellite at the same time, then to first order, the ionospheric effects mentioned previously are the same for both sites. Further, because the data obtained is referenced to the satellite clock and thus is common to both sites, actual uncertainties caused by the satellite and to SA are also eliminated.

Assume that you have two sites shown in Fig. 20.16. In this figure, the items labeled d refer to the signal delay on each path. For example, d_A refers to all of the delays and uncertainties in receiver A and d_{CA} represents all of the delays and effects in the path from clock C in the satellite to the receiver at site A.

If both sites now track the same satellite at the same time period, and later exchange the data taken, it is possible to determine, with a high degree of precision, the time difference between the two sites. If the process is routinely performed over extended periods of time, for example, weeks or months, then a high degree of agreement exists for the relative time changes between sites. This is further enhanced because the antenna delays, receiver delays, and other factors limiting the transfer of exact time with one-way GPS are now the same during each measurement cycle. Because this method retrieves the

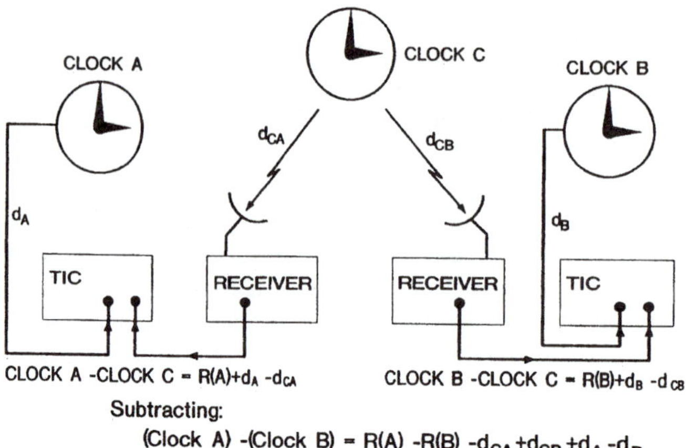

Figure 20.16 Common-view GPS. Clocks at site A and site B are compared with each other through a third clock C located in the GPS satellite.

relative difference in time from measurement to measurement, virtually all of the limiting factors in one-way GPS are now common to all measurements and are essentially eliminated.

Because the relative difference in time between sites is known then, as previously described, the relative difference in frequency is also known. This method is especially effective in transferring precise frequency from a national calibration site to any other calibration site. With care, uncertainty in relative time can be reduced to several nanoseconds. Over a period of one month, therefore, the uncertainty in relative frequency could approach parts in 10^{15} if the actual time-domain stability of the reference source at the primary calibration site permits measurement to this level.

Two-way satellite synchronization. *Two-way synchronization* via satellite uses the geosynchronous satellites commonly used for communication purposes. As a result, the equipment required for synchronization is expensive, requiring specialized satellite antennas and two-way satellite modems. The advantage of the technique is that it provides the highest degree of accuracy obtained so far. For two sites on the same continent, transfer accuracy on the order of several hundred picoseconds has been achieved.

Figure 20.17 shows a diagram of the two-way setup for the synchronization technique. Time from clock A is transmitted to the satellite. An on-board transponder in the satellite transmits clock A time to station B. The signal is received and compared through a *time-interval counter* (*TIC*) to clock B. At the same time, time from clock B is transmitted via the same satellite to station A. In Fig. 20.17, all of the individual signal and system delays are shown by the letter *d*. If all of the delays from station A to station B are lumped into a single delay (d_{AB}) and do the same for the delay (d_{BA}), then the satellite-transfer equation reduces to that shown in Fig. 20.18.

d_{SAB}

Satellite

d_{SBA}

d_{AS} d_{SA} d_{SB} d_{BS}

Antenna Diplexer and Filters

d_{TA} Transmitter Clock A

Receiver TIC

d_{RA}

Earth Station A

d_{TB} Clock B Transmitter

TIC Receiver

d_{RB}

Earth Station B

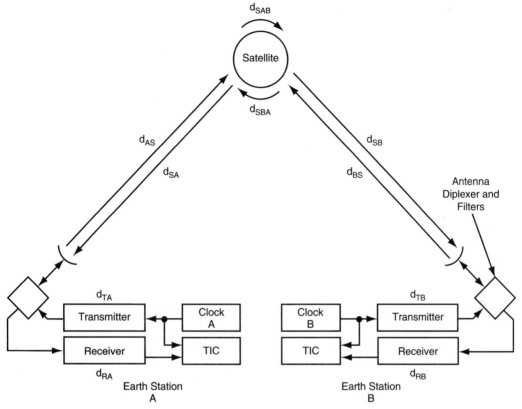

Figure 20.17 Two-way time transfer via satellite. Shown are the equipment required and the major path delays experienced in transmitting from station A to station B and vice versa.

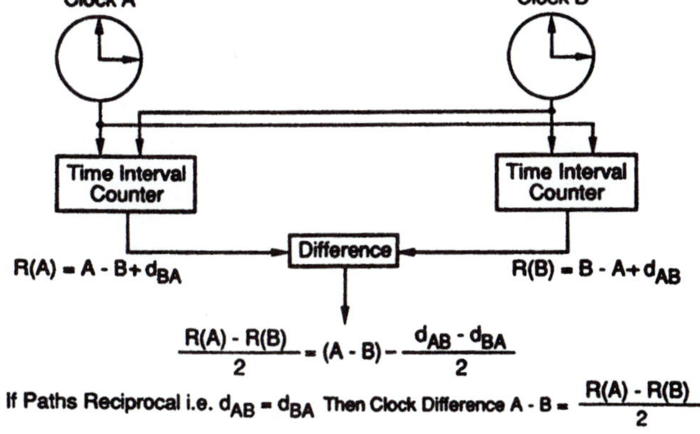

Figure 20.18 Data reduction path for two-way time transfer via satellite. $R(A)$ and $R(B)$ are the measured differences in the time interval counter.

If reciprocity can be assumed in the transmission paths (i.e., that the delay from A to B is exactly the same as from B to A), then the clock difference from station A to B is precisely as shown in the figure. Notice that to perform highly accurate time transfer from station A to station B, essentially nothing is necessary to know about either the satellite location or the location of either of the earth stations, as long as reciprocity holds.

Chapter

21

Spectrum Analyzers

Alan W. Schmidt
Agilent Technologies
Santa Rosa, California

21.1 Introduction

Spectrum analysis is a quick method to observe and measure signal levels and signal distortions. The display gives the user a visual representation of the magnitude of the Fourier transform of the input signals. The Fourier transform maps the time domain signal into the frequency domain as a set of sines and cosines. The measurement range of the signal analyzer is extremely large, exceeding 140 dB. These abilities make the spectrum analyzer a versatile instrument specially suited for the modern communications era. Spectrum analysis is essentially the examination of amplitude vs. frequency of a given signal source, antenna, or signal distribution system. This analysis can yield important information about the signals such as stability, distortion, amplitude, and types and quality of modulation. With this information, circuit or system adjustments may be made to increase efficiency or verify compliance to the many emerging regulations regarding both desired emissions and undesired emissions.

The modern spectrum analyzer has found many complex uses ranging from the R&D lab to manufacturing and field service. The signal analyzer has been a laboratory instrument of great value. The quick ability to view wide spectrum widths and then quickly zoom in to closely examine signals of interest has been highly valued by the R&D engineer. In the manufacturing area, the speed of measurements combined with the ability to access these data through computers has allowed very complex measurements to be made quickly, accurately, and repeatably.

21.1.1 Applications

Many forces are impacting the uses and needs for signal analyzers. For example, the great proliferation of high-speed digital computers has created the need for wide frequency range diagnostic instruments. These tools allow the designers to scan quickly for possible problems that may violate regulatory requirements such as those from the Federal Communications Commission (FCC).

The rapid rise in radio-frequency telecommunications has given rise to even more testing to verify regulatory requirements for transmission modes. Current requirements are strict for mobile radio telephones. These include measurements of spectrum occupancy, power levels, time domain responses, and other spurious emissions. Cable television and broadcast video services also provide opportunities for the signal analyzer to be used. Modulation bandwidth, signal-to-noise, carrier level, and harmonics are examples.

The world of radio frequency (rf) or microwave frequency use is constantly placing increasing demands upon both the end use equipment and the test equipment. Just as equipment needs for each of the end users vary, so do the requirements for the associated signal analyzers vary. Thus a complete understanding of the intended application is needed before it is possible to select an appropriate spectrum analyzer. As the measurement needs of a specific type become more intense, finding a signal analyzer specifically tailored to that application may be possible. As signal analyzers have been designed for use in specific application areas, they do not simply display raw frequency and amplitude measurements but translate those measurements into more complete solutions. Spectrum analyzers may now be obtained to assist the digital designer in diagnosing and improving the radio-frequency interference (rfi) performance of their high-speed digital systems. These analyzers "speak" the language of the application and are more easily operated and understood by electromagnetic interference engineers. There are similar examples for many other areas, such as rf mobile telecommunications and the CATV or broadcast video marketplaces.

This chapter provides basic information to describe the function, design alternatives, and use of spectrum analyzers that will provide the reader with a good basis for understanding how the spectrum analyzer works and some ability to use this information to maximize the use of the equipment in the end user application.

21.1.2 Time versus frequency domain

The usefulness of this tool may be best shown in a comparison of time domain and frequency domain analysis of a simple signal. The time domain oscilloscope provides a display of amplitude vs. time (Fig. 21.1*a*). The vertical axis of this display is representative of the amplitude of the signal, while the horizontal axis is that of time, increasing from left to right. In the spectrum analyzer, the instrument provides a display of amplitude versus frequency (Fig.

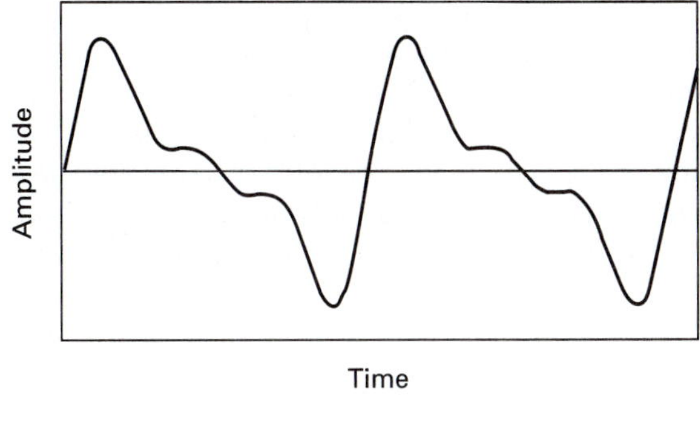

(a)

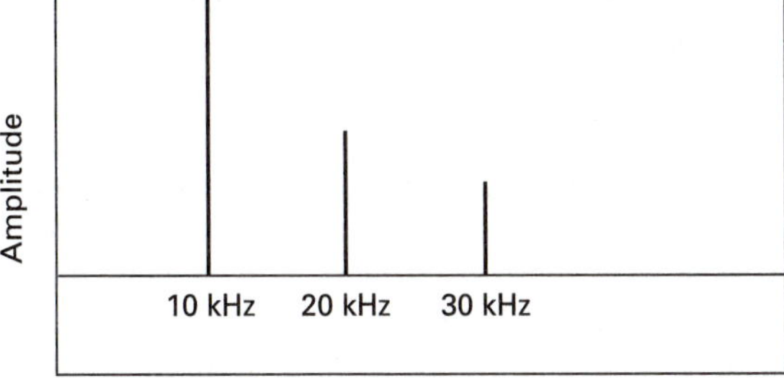

(b)

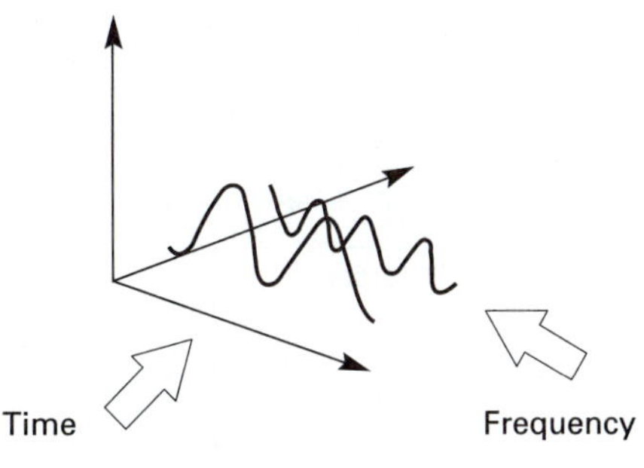

(c)

Figure 21.1 (a) The time domain view is complex and not easily understood; (b) in the frequency domain, it is easy to see the component parts of the fundamental and two harmonics; (c) a view of the mapping between time domain and frequency domain.

21.1*b*). The vertical axis of this display is also representative of the amplitude of the signal, while the horizontal axis is frequency, increasing from left to right. Figure 21.1*c* depicts the relationships of these analysis techniques. The spectrum analyzer resolves the spectral makeup of the signal and displays it in a wide amplitude and frequency range. The modern spectrum analyzer has analysis capabilities from a few hertz to well over 100 GHz in frequency and an amplitude range of well over 100 dB. This tool can rapidly display and quantify the complete makeup of the signal.

21.2 Basic Spectrum Analyzer Operation

The separation of the frequency components of signals may be attained in several ways. The most basic form of spectrum analysis, called tuned rf (TRF) analysis, is simply accomplished through the use of tunable filters where the center frequency of the filter is adjustable and is followed by broadband detectors. These TRF analyzers are somewhat difficult to construct in extremely wideband frequency ranges and lack the capability for many different resolution bandwidths. Wideband circuitry usually requires more power and is more difficult to keep stable over temperature, which leads to costly and lower-performance circuits. Since all predetection signal processing is wideband in nature, TRF analyzers are not in general use with the exception of the current similar construction found in optical spectrum analyzers.

21.3 Swept Superheterodyne Spectrum Analyzer

The swept superheterodyne spectrum analyzer (Fig. 21.2) is made up of a wideband input mixer, driven with a swept local oscillator, and the resolution

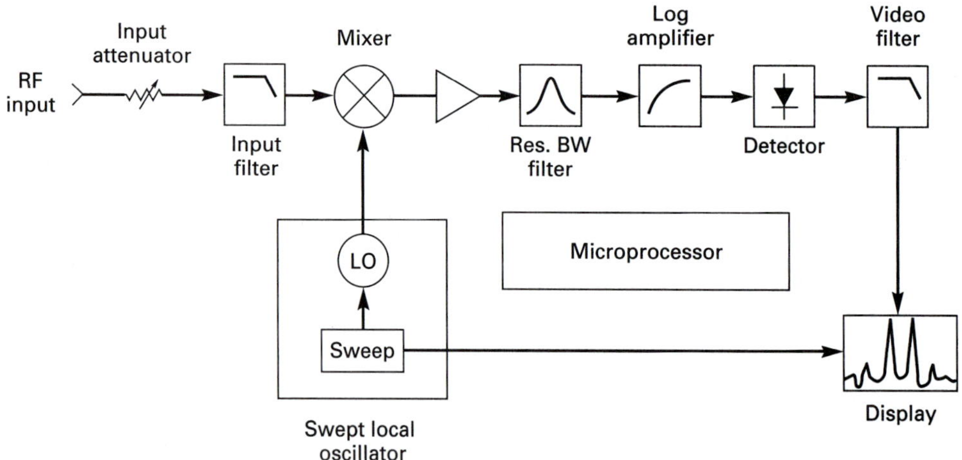

Figure 21.2 A simplified block diagram of a swept superheterodyne spectrum analyzer. Some detail has been left out in order to concentrate on these significant blocks.

capability of the instrument is determined at a fixed intermediate frequency (if). Within this fixed if one of many resolution bandwidths may be selected. This is the most common architecture for spectrum analyzers.

Real-time spectrum analyzers can be realized in two types: filter banks or fast Fourier transform types.

21.3.1 Filter bank real-time analyzers

The first is to simply have a bank of fixed tuned filters spread across the frequency range of interest and follow each filter with its own detector. This method may require an extraordinary amount of circuitry depending upon the desired frequency range and the desired frequency resolution.

21.3.2 Fast fourier transform analyzers

The second, and more common, method employs an analog-to-digital converter and performs a fast Fourier transform on the output to resolve the frequency components in the input signal. The fast Fourier transform is a numerical method of performing the Fourier transform well suited to computer methods.

The swept superheterodyne spectrum analyzer and the fast Fourier transform spectrum analyzers are by far the most common instruments used in rf and microwave work today. The swept superheterodyne instrument will be examined in some detail.

21.3.3 Block diagram

The basic block diagram of a spectrum analyzer is shown in Figure 21.2. The key point here is that a spectrum analyzer is essentially a superheterodyne receiver. The spectrum analyzer has some specific differences in that the frequency range is larger than in most receivers, the first local oscillator is able to be swept, and the major differences lie in the if structure.

Input attenuator. The first block is the input attenuator. Its purpose is to limit the power of the incoming signal to keep the rest of the instrument in its normal operating range. Most spectrum analyzers can handle 0 to −10 dBm applied to the input mixer. (The term dBm refers to decibles referenced to 1 mW.) The input attenuator is used to keep larger signals below this threshold and to optimize the measurement dynamic range of the spectrum analyzer. The attenuator itself is usually in the 0.5- to 1-W power handling range; this sets the maximum input level that the instrument can handle without damage.

Input filter. The next block is a filter for image rejection or preselection. Since mixers respond to both sums and differences of frequencies, this filter must be employed to reject the unwanted mixing product, or unwanted responses

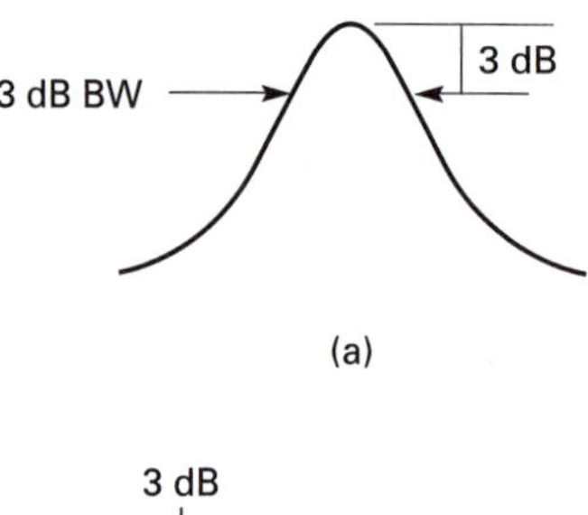

(a)

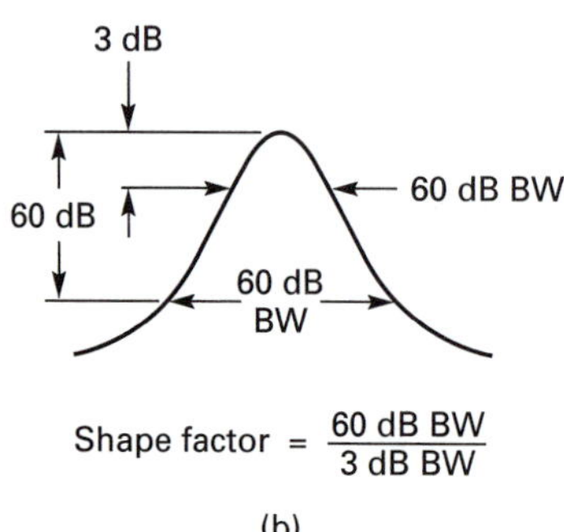

(b)

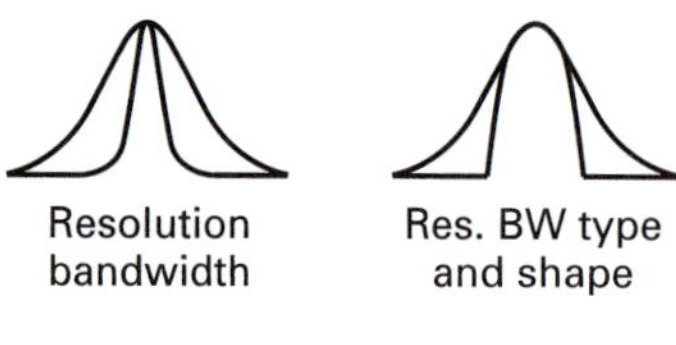

(c)

Figure 21.3 (a) The principal specification in resolution bandwidth; (b) the 60 dB/3 dB ratio and its relation to selectivity; (c) the difference in resolution bandwidth and bandwidth type.

may occur owing to signals at the image frequency. In the typical rf spectrum analyzer, this is accomplished with a low-pass filter. In most modern microwave spectrum analyzers this need is filled through the use of tracking filters. These are electronically tunable filters which continuously adjust to track the tuned frequency of the spectrum analyzer. In very high millimeter wave bands (above 50 GHz) these filters are difficult to realize physically; therefore, *unpreselected* mixers are used. This means that there will be multiple responses for one input signal because mixers respond to signals at both the sum and difference of the local oscillator. Care must be taken to identify which product is being viewed, and most signal analyzers provide a function called signal ID to facilitate this process. A marker or cursor is placed upon the unknown signal and the signal analyzer displays the true frequency of the response.

Intermediate-frequency stage (IF). The if is where the real analysis is done in the signal analyzer. The primary function is to provide a wide selection of resolution bandwidth filters. These filters are described by their 3-dB bandwidth as shown in Fig. 21.3*a*. The resolution bandwidth is one measure of the resolving power of the instrument; the narrower the filter the closer two signals may be and still be seen as separate responses. Resolution bandwidth filters are usually realized in a combination of *LC* filters, crystal filters, and digital filters. The shape factors and filter types are important factors in specifying these filters. Shape factor is a measure of how selective the filter is, usually specified as a ratio of the 3-dB/60-dB bandwidths. This impact can be seen in Fig. 21.3*b*. This ratio (3 dB/60 dB) indicates how close to a large signal contained within the 3-dB bandwidth a signal one million times smaller (−60 dB) can be resolved. The filter type has a significant impact upon the performance of the spectrum analyzer. While some filter types such as Butterworth or Chebyshev filters have superior selectivity (ability to separate signals), and gaussian and synchronously tuned filters have better time domain performance (better swept amplitude accuracy), the end application will play a significant part in which filter type is best. Superior shape factor performance will allow for better resolution of closely spaced signals. Better time domain performance (no overshoot) allows for faster sweep speeds with good amplitude accuracy. Figure 21.3*c* shows how different resolution bandwidths and different filter types may impact resolving signals. There are other factors which impact resolution which have to do with local oscillator stability, and those items are discussed along with the oscillator. Step gain amplifiers with accurate amplification settings serve to allow precise adjustment of the spectrum analyzer's sensitivity and measurement range. These amplifiers provide very precise steps in instrument gain, usually adjustable in at least 1-dB or finer steps with a range of over 50 dB.

Log amplifier. The log amplifier processes the incoming signal in a logarithmic fashion which allows a large range of incoming signals to be measured and compared. One technique to achieve this compression is to build an amplifier whose gain varies with signal amplitude. At small signal levels the gain may be 10 dB, while at larger amplitudes, the gain drops to 0 dB. Cascading several stages of this type of amplifier will be necessary to obtain the desired log range. Log amplifiers usually have ranges on the order of 70 dB to in excess of 100 dB. Along with the log range the fidelity (how closely the log compression truly fits a log curve) is a significant factor to consider. This error will factor directly into the amplitude error of the measurement.

Detector. The detector in most basic spectrum analyzers is a linear envelope detector, similar to those found in AM radios. With the signal already having been compressed with the log amplifier this linear detector yields a great range without large linear range requirements placed on the detector. Some analyzers have taken a different approach. In these instruments large-range linear detectors like synchronous detectors are used and they are followed by direct current log amplifiers to still yield the 80- to 100-dB displays.

Video filters. There are also video filters which follow the envelope detectors. These filters allow some postfiltering or averaging of the detected output.

Video filters are generally set to the same bandwidth or larger than the resolution bandwidth unless the measurement needs averaging. Averaging may be needed if there is a combination of noise and signal, where the random noise will average out while the signal remains.

Analog-to-digital converter. Finally there is the analog-to-digital converter (ADC). Ideally an ADC with a large number of bits (16 or greater) operating at speeds greater than the largest-resolution bandwidth would be used. Since in some cases these are too power-consuming or are just not available, additional circuitry is needed to capture the desired information. In many cases a system of analog and digital peak detectors is used to record the largest values in a given time or span. In order to give a feeling of real signal or noise bandwidth, various combinations of peak and minimum detectors are used and various algorithms are used to choose which detector to display in order to approximate the real analog signal amplitude changes. One method is to detect that the signal has both risen and fallen in a measurement period, if so, then alternately display the peak and minimum values.

Swept local oscillator. The swept local oscillator is a key element in the total signal analyzer. The stability and spectral purity of the swept local oscillator can be a limiting factor in many performance areas. Residual FM is a measure of local oscillator stability. The ideal local oscillator would be exactly stable and have no *frequency modulation.* In the normal signal analyzer with very narrow resolution bandwidths, a few hertz of FM can cause the signal to smear as seen in Fig. 21.4*a*. The stability of the local oscillator may set the minimum resolution bandwidth, which can be useful, and not have the resulting jitter impair the measurement. The required stability may be obtained in many ways: discriminator loops, frequency lock loops, or phase locked loops. Each of these techniques has merit and should be matched by the rest of the instrument in appropriate resolution bandwidths. Even with very frequency-stable local oscillators, there is remaining instability. This is called phase noise or phase noise sidebands. The impact of phase noise can be to block observation of close-in signals which might have otherwise been observable if we were to consider only bandwidth and shape factor; see Fig. 21.4*b*. A significant application of modern signal analyzers is to make direct phase noise measurements of other devices. This is obviously, in that case, an important factor.

21.3.4 Microprocessor

Perhaps the most important item in any modern instrument is the microprocessor and its associated instructions. This processing power orchestrates all of the hardware in the instrument to ensure that measurements are accurate. Older spectrum analyzers required users to maintain the integrity of the measurement by adjusting the resolution bandwidth, sweep time, and fre-

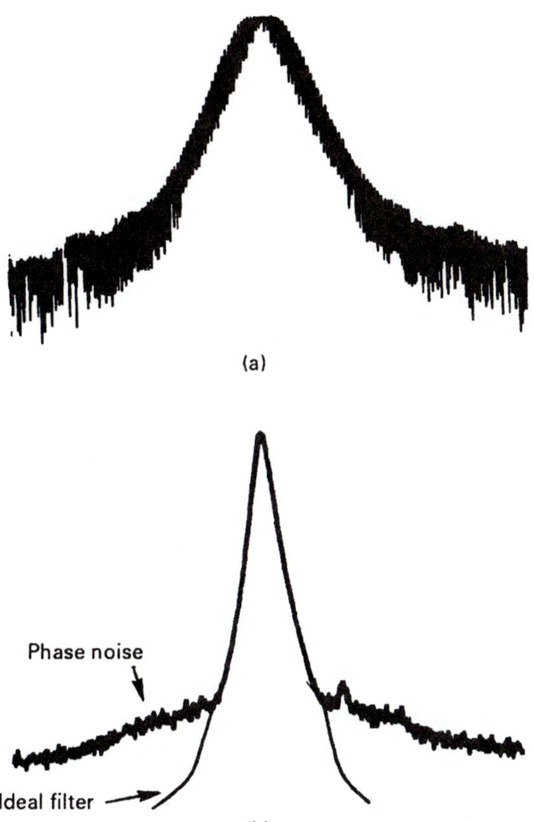

Figure 21.4 Local oscillator impact on resolution bandwidth. (a) The impact of residual FM or jitter, causing the signal to smear in frequency; (b) the impact of phase noise and the limits it places on filter selectivity.

quency span controls to maintain calibration. This coordination is required to ensure that the input signal is swept through the if filter slowly enough to allow full-amplitude response. If the signal is swept too rapidly, the response will appear delayed and lower in amplitude, as shown in Fig. 21.5. The relationship of sweep time, resolution bandwidth, and frequency span is given by

$$\frac{\text{Sweep time} \propto \text{frequency span}}{(\text{Resolution bandwidth})^2} \tag{21.1}$$

where sweep time = time required to sweep the frequency span
frequency span = total frequency change during the sweep
resolution bandwidth = resolution bandwidth used

The sweep time is proportional to the sweep time divided by the resolution bandwidth squared. Thus a reduction from a 10-kHz resolution bandwidth to 1-kHz resolution bandwidth, while keeping the span width constant, will cause the sweep time to be increased by a factor of 100. Here the microproces-

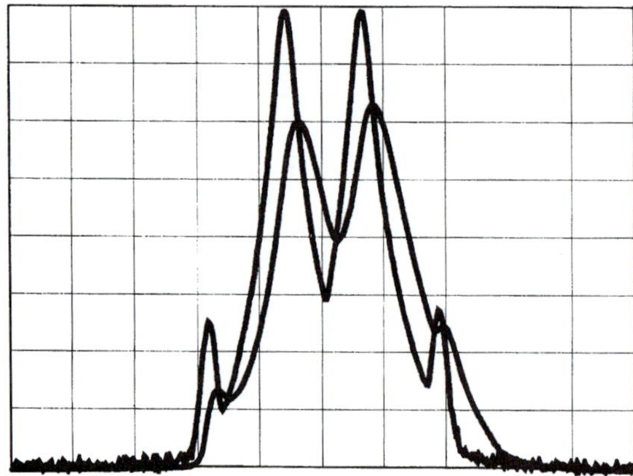

Figure 21.5 The impact of sweeping the local oscillator too fast. The filter does not have time to reach full amplitude, causing the drop in amplitude and time delay through the filter and the shift to the right.

sor can track all instrument settings and make adjustments to remain calibrated.

Much of the instrument is also calibrated, corrected, or aligned by the microprocessor. Calibration information may be stored in the read-only memory of the instrument to be accessed later in the actual measurement to remove errors due to the hardware. An example of this capability is the correction of frequency response errors due to the input filter and mixer response. Accurate measurements of the hardware's performance are taken and stored in the memory of the instrument. Then as measurements are made corrections are added or subtracted to correct for the hardware response. Many other sets of correction data can be measured, stored, and corrected such as attenuator accuracy and log amplifier accuracy. Also markers and various functions now clearly an expectation of a basic instrument are provided through this central microprocessor. In the case of markers, the microprocessor will handle reading the data stored at a given frequency and displaying this information to the user in appropriate resolution. The inclusion of the microprocessor also gives the remote capability needed in today's complex and competitive world. This allows the instrument to be assembled into more complex systems orchestrating many instruments through the power of a measurement controller. These systems are then put to high-speed use in R&D, manufacturing, service, and field operations.

21.3.5 Dynamic range considerations

Dynamic range is a major factor in the selection of the spectrum analyzer. This key specification generally is stated in the instrument's ability to measure two signals simultaneously. In this area sensitivity, harmonic distortion, and third-order distortion are key elements.

Noise level. The ability of a signal analyzer to detect a signal is usually stated as displayed average noise level. The noise level displayed is measured by placing the analyzer in a high-gain state and using the narrowest-resolution bandwidth available. Since noise level is a function of bandwidth it is defined by

$$\text{Noise-level change (dB)} = 10^{\frac{\log(\text{Res BW2})}{\text{Res BW1}}} \tag{21.2}$$

Therefore, changing the resolution bandwidth from 10 to 1 kHz will result in a drop of 10 dB displayed noise.

$$\begin{aligned}\text{Noise-level change (dB)} &= 10^{\frac{\log(1\,\text{kHz})}{10\,\text{kHz}}} \\ &= -10\ \text{dB}\end{aligned} \tag{21.3}$$

where noise-level change (dB) = observed measurement level change
Res BW1 Res BW2 = different resolution bandwidths used

This effect is shown in Fig. 21.6*a*. Note these measurements should be made with the input attenuator set to 0 dB for maximum sensitivity. This specification should give an accurate measure of how small a signal an analyzer can measure. If the signal power level is equal to the noise level, the two powers will yield a 3-dB response in the noise floor as shown in Fig. 21.6*b*. With the sensitivity specification we can construct a signal-to-noise graph for each resolution bandwidth as shown in Fig. 21.7. With this graph we can see that the largest signal-to-noise occurs with the largest signal level input. If this were the only concern, the largest dynamic range would be obtained with large input signals. In reality it's not quite that simple.

Distortion products. In a spectrum analyzer, there are nonlinearities which produce distortion products. Examples are harmonic distortion and third-order distortion. There are actually many orders of products, but usually second- and third-order are the most prominent. The following is an example of how a signal analyzer behaves. One key subject will be the ability to determine if the observed distortion is caused by the device under test or internally in the analyzer. Many times in making signal analyzer measurements there is a need to determine or separate internally (in the spectrum analyzer) generated distortions from those generated in the device under test. The input attenuator plays a large role in this task. In order to test for overload, it is necessary to increase the attenuator. If the displayed signal reads the same amplitude, there was no gain compression and it is safe to return to the original attenuation level. If, however, there is a difference, the attenuator should be once again increased until the level readings are constant. Gain compression or overload occurs when a large signal causes the measuring instrument to indicate a lower level than it should. In the case of distortion products these results are even more dramatic. The reason is that internally generated distortion products rise faster than the signal level change. For illustration purposes third-order products change at three times the rate of the signal level. Thus for every 1-dB change in signal level, the distortion products will

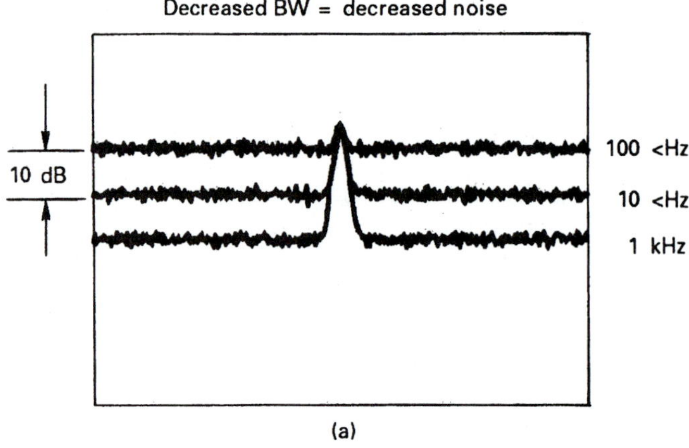

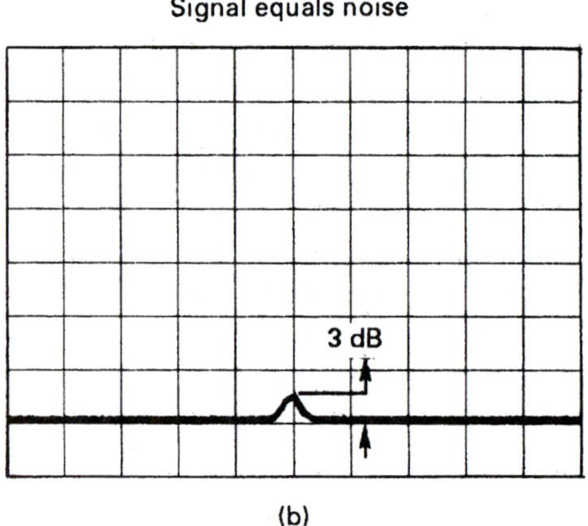

Figure 21.6 (a) The 10-dB drop in noise level for a factor of 10 change in resolution bandwidth. The gain in signal-to-noise is apparent. (b) The impact of measuring a signal at the same level as the noise. The two powers combine to yield a 3-dB bump in the noise.

rise by 3 dB. This is illustrated in Fig. 21.8*a*. The method to check for internally generated products is the same as that of compression. Simply change the input attenuator and observe the displayed level of the distortion products for change; if they are constant, the measurement is one of the device under test.

In a similar manner, second-order products must be measured and tested. The second-order products, however, move 2 dB for every 1-dB change in level. The input attenuator may be used to determine whether the distortion is produced in the analyzer or the device under test.

As with signal-to-noise, a graph can be constructed to indicate signal-to-

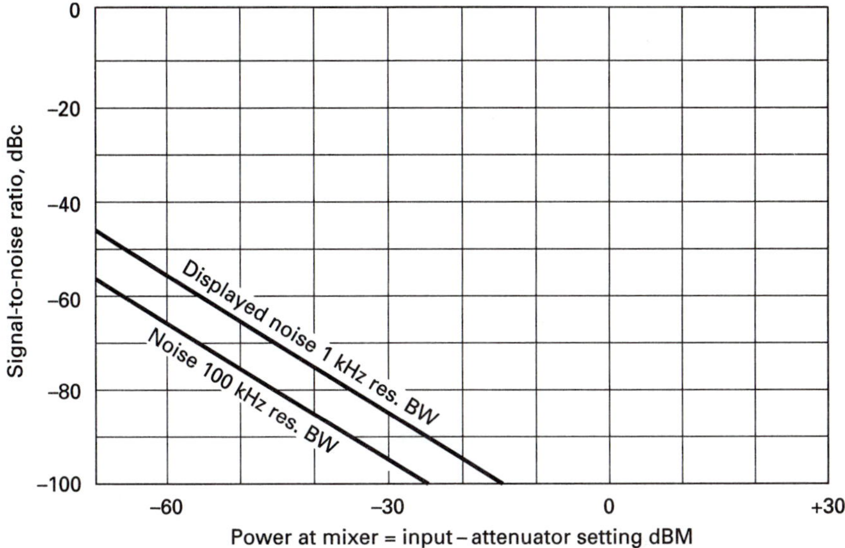

Figure 21.7 Signal-to-noise graph for two bandwidth settings. Note the 10-dB difference in the level because of the factor of 10 difference of 100 Hz to 1 kHz.

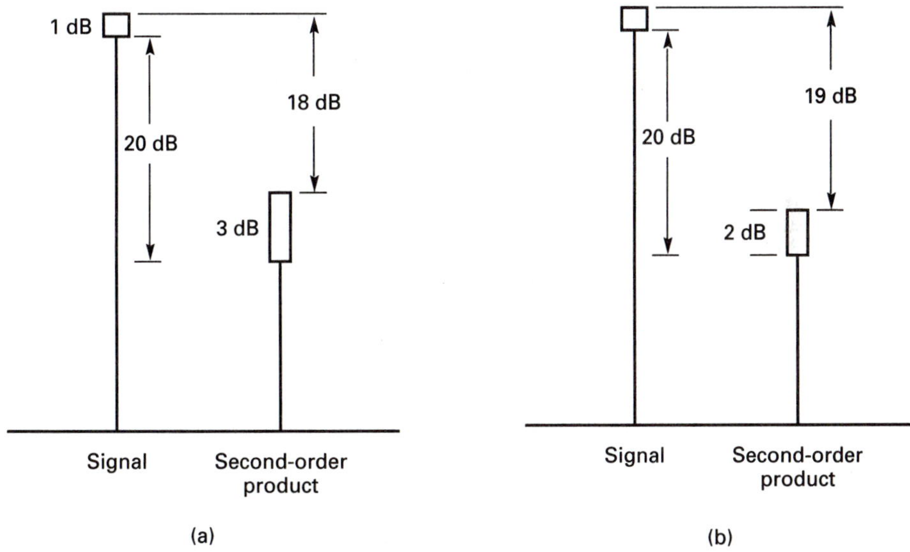

Figure 21.8 The relationship of distortion products to signal level is increased 1 dB. (a) The third-order products grow by 3 dB, thereby reducing the difference to the carrier by 2 dB; (b) the second-order products grow by 2 dB, decreasing the difference to the carrier by 1 dB. This effect is present in both the device under test and the measuring instrument. By changing the input attenuator, one can determine if the products are generated in the spectrum analyzer or come from the device under test.

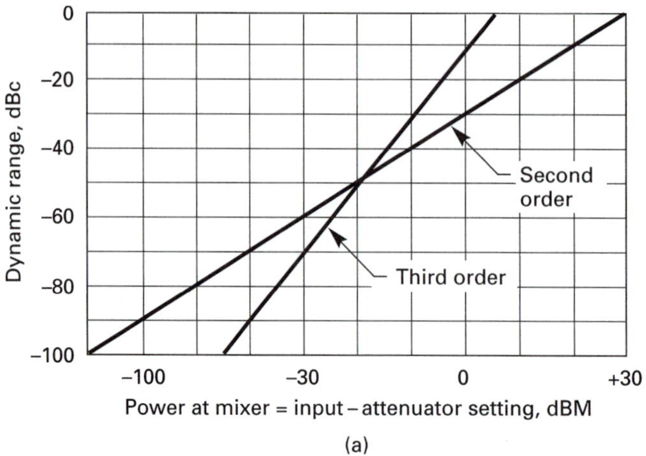

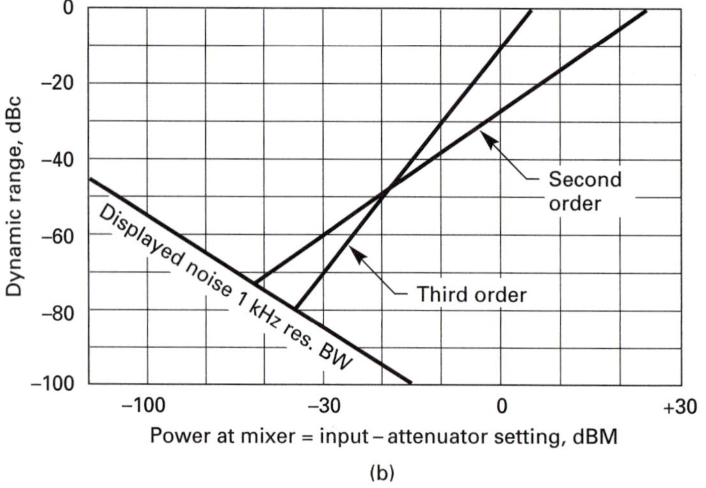

Figure 21.9 (a) The dynamic range of curves for third- and second-order products can be seen, with the third-order products having a slope of 2 and the second-order products a slope of 1; (b) when combining the distortion curves with the noise floor, the dynamic range of the spectrum analyzer can be seen. Optimum mixer levels can be found where the distortion products interesect with the noise floor.

distortion range (see Fig. 21.9*a*). In this case the vertical axis is in dBc (dB relative to carrier power), and the horizontal axis is power incident on the input mixer of the analyzer (input level-attenuator setting). In the case of third-order distortion, specifications are usually for mixer levels of −30 dBm. (Some manufacturers specify total input power, while some specify each tone at a level. Be sure to check this for the potential 3-dB difference.) Starting from the specified level one can extend in either direction with a slope of 2. (With a 1-dB change producing a 3-dB drop in distortion, the difference will increase or decrease by 2.) The same line may be plotted for second-order distortion, but with a slope of 1, and be added to the distortion graph. As we can see from this graph, the highest maximum distortion difference comes with the lowest input signal level. If the distortion graphs are combined with the sensitivity graph, the result is seen in Fig. 21.9*b*.

Now there is an obvious trade-off to be made. The maximum signal-to-noise occurs with the highest input signal levels, while the largest distortion-free range occurs with the lowest input level. The point here is to understand the measurement and adjust accordingly. For instance, if the measurement is third-order distortion, the optimum input level will be that level where the displayed noise and the distortion products are equal. This occurs at the intersection of the signal-to-noise line and the third-order distortion line. For most accuracy, and dynamic range, that is the input mixer level where the measurement should be made.

To adjust to cover all distortion products, simply choose that input level of the highest intersection with the noise floor and distortion lines. Many modern signal levels allow the user to set the mixer level to a specified level, and the microprocessor will adjust the input attenuator to obtain the needed distortion-free range. Often these graphs are not provided by the manufacturer, yet taking the time to construct them may well be worth it.

21.4 Auxiliary Equipment

Several pieces of auxiliary equipment greatly enhance the capabilities of the spectrum analyzer. These consist of tracking generators, preselectors, and high-impedance probes.

21.4.1 Tracking generator

The tracking generator can transform the spectrum analyzer into a very high dynamic range scalar network analyzer. The tracking generator combines the swept first local oscillator signal with a fixed oscillator of the frequency of the first if. The net result is that a signal is generated which exactly matches the tuned frequency of the spectrum analyzer. With this setup the frequency response of any device placed between the spectrum analyzer and the tracking generator may be measured. The total measurement range will be from the maximum output power of the tracking generator to the sensitivity limit of the spectrum analyzer. This may range from +10 to −140 dBm.

21.4.2 Tracking preselector

The tracking preselector allows the analyzer to be used in an environment which has many signals present. The large number of signals could provide a total signal power (the sum of all the individual signals) that is too large for the input mixer to handle. In this case employing a tracking filter which stays aligned with the tuned frequency of the analyzer can filter out much of the signal clutter and allow the input mixer to function normally.

21.4.3 High-impedance probe

The input impedance of the swept spectrum analyzer is usually 50 to 75 Ω to match up with the majority of rf and microwave devices. There are times when a very high impedance input is more desirable. The probing of a circuit or integrated circuit, for example, usually needs high impedances. There are active probes consisting of a high input impedance amplifier with a 50-Ω output impedance to facilitate this type of measurement and retain high sensitivity.

Chapter

22

Lightwave Signal Sources

Waguih Ishak
Agilent Technologies
Palo Alto, California

22.1 Introduction: Fiber-Optic Communications Systems

Fiber-optic communications developed very rapidly in the past two decades. Many systems have been installed and many others are planned in the United States, Europe, and Japan. These systems clearly compete very well with the traditional communications systems as a cost-effective means for information exchange. Fiber-optic systems typically operate at hundreds of megabits per second, and new systems operating at 2.4 Gbit/s are now being installed. At the same time, research laboratories around the world are developing multi-Gbit/s components and systems with potential for terabit per second communications links in the late 1990s.

In addition to the increase in the data rates over fiber-optic networks, the performance of the devices, components, and subsystems used in such networks is improving at a very high rate. For example, a semiconductor laser needed in an optical amplifier system must have a mean time between failures (MTBF) of more than 100,000 h. If the amplifier is used in the submarine cable (transatlantic or transpacific), the laser must be reliable to withstand severe operating conditions (temperature, humidity, pressure, etc.). In addition to developing high-performance components, the trend continues toward lowering the effective cost per bit and mile of information. As a result, the designers of lightwave devices, components, and subsystems are faced with a challenge. They need to maximize the performance of each system building block, minimize the adverse interactions between these blocks, and at the same time design for manufacturability and cost-effectiveness. In order to help the designers achieve these goals, new measurements and character-

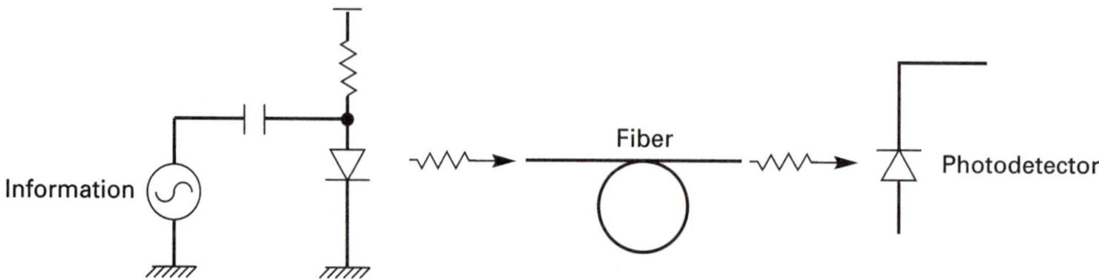

Figure 22.1 A block diagram of a fiber-optic communication system.

ization techniques are needed in the area of lightwave communications test instrumentation.

The small size, large bandwidth, and very low attenuation of optic fibers make them attractive as alternatives to conventional copper cables in telecommunications applications such as telephone and CATV systems and in data communications applications such as computer interconnects systems.

A basic communications system consists of a transmitter, a receiver, and an information medium as shown in Fig. 22.1*a*. The transmitter is a system capable of generating the information to be sent to the receiver. The information can be in analog or digital format. The transmission medium carries the information over some distance and delivers it to the receiver. The receiver interprets the information and transforms it into an accessible form. In the case of the fiber-optic communications system, the transmitter is an optical source (laser or light-emitting diode, LED) which is modulated with the information to be transmitted. Optical sources can be modulated internally (by varying their operating currents) or externally (by using an external optical modulator following the optical source). The transmission medium is an optic fiber and the receiver is a photodetector, as shown in Fig. 22.1.

An actual fiber-optic communications system uses many more electrical and optical components than mentioned above. Examples include optical isolators, optical polarization controllers, fiber couplers, optical filters, optical connectors, electronic demodulators and amplifiers, and electrical signal generators.

Types of fiber-optic communications systems. State-of-the-art-fiber-optic communications systems use semiconductor lasers as the optical source. If the data rate is below a few Gbit/s, the lasers can be modulated by varying their bias current. However, if the data rate is in excess of 10 Gbit/s, an external optical modulator (such as a lithium niobate Mach-Zehnder modulator) is coupled to the laser source. The fiber can span large distances (for example, the transatlantic fiber link is more than 6000 km) and repeaters will be needed to boost the signal level above the noise. In most of the installed fiber systems, these repeaters are electronic repeaters which transform optical signal to an electrical signal using a photodetector (O/E converter), amplify the electrical signal using electronic amplifiers, and then convert the electrical signal back to an optical signal using a laser (E/O converter). These repeaters limit the data rate on the fiber because of the limited electronic amplifier bandwidths. At the end of the communications channel, the receiver consists of a photodetector, which converts the optical signal to electrical signal, followed by an amplifier to boost the signal level and a demodulator to decode the information. In such a system, the information determines the data rate and the laser wavelength determines the carrier frequency. For example, a typical system will use a laser operating at 1.3 μm wavelength corresponding to a carrier frequency of about 176 THz (176,000 GHz), and the information will modulate the laser at a rate of 560 MHz.

The invention of optical fiber amplifiers (erbium-doped fiber amplifiers, EDFAs) has changed the way fiber-optic communications systems are configured. As will be described later, commercially available EDFAs can boost an optical signal level by 30 dB or more over a significantly wide band (1530 to 1570 nm). The use of EDFAs as signal boosters at the transmitter and as preamplifiers across the fiber and at the receiver should eliminate many of the repeaters needed in the conventional systems described above. This is significant because these repeaters do not allow the option of varying the data rate over the communications system. However, the EDFAs, with their extremely wide band, will make the fiber-optic communications system upgradable without major changes in the configuration. In addition, since the EDFA is a small (few inches cubed) system which can be easily coupled to the transmission fiber, it will be possible to use a large number of EDFAs in the system, allowing longer transmission distance. Recently, AT&T demonstrated a 9000-km undersea system using 300 EDFAs uniformly spaced along the fiber length.

Another development is the emergence of wavelength division multiplexing (WDM) as a means for increasing the capacity of fiber-optic communications systems. In WDM systems, the transmitter consists of several laser sources operating at different wavelengths. The laser outputs are modulated and multiplexed for transmission over the same optical fiber. At the receiver, a demultiplexer is used to separate the channels for the appropriate photodetector circuit as shown in Fig. 22.2. In a typical system, single-mode semiconductor lasers are used with wavelength separation of a few nanometers (corresponding to hundreds of GHz) and EDFAs are used as signal boosters along the fiber. In future systems, it is expected to use channel separation of a few GHz.

Coherent fiber-optic communications systems (Fig. 22.3) are useful for applications requiring high sensitivity and enhanced selectivity between the various transmitted channels. These systems, which are similar to the heterodyne radio communications systems, consist of a transmitter which is either frequency- or phase-modulated using frequency shift keying (FSK) or phase shift keying (PSK) techniques. The modulated signal is transmitted over the fiber and then received using a heterodyne, or homodyne, receiver. The receiver requires an optical local oscillator which is used to mix with the incom-

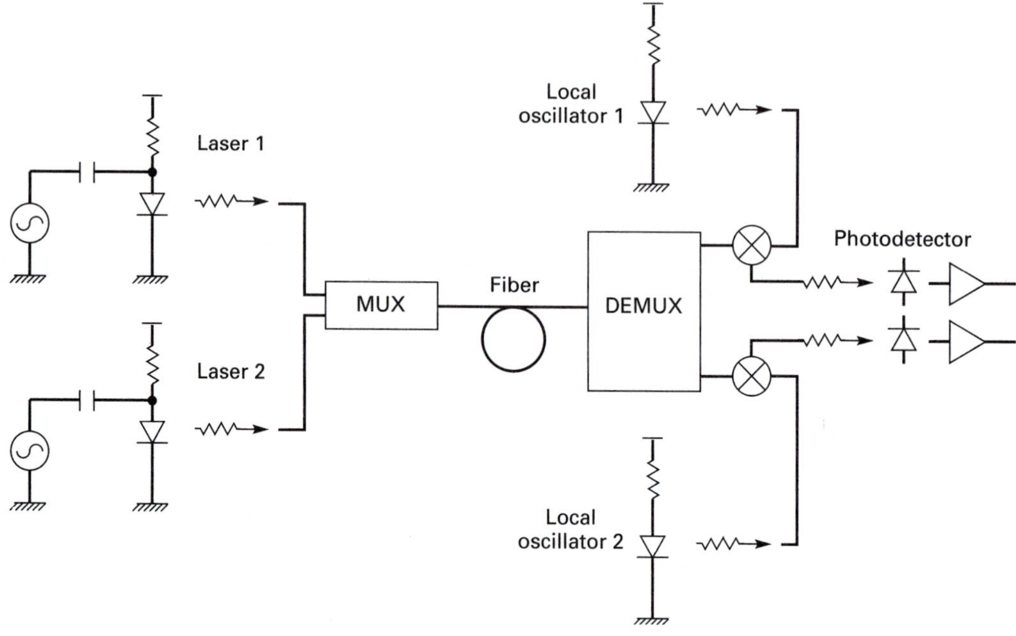

Figure 22.2 A block diagram of a WDM fiber-optic communication system.

ing signal to subtract the carrier frequency. The detected signal is enhanced by the optical power from the local oscillator, hence the improvement in sensitivity.

As the technology advances, however, and as the trend toward lowering the effective cost per bit and kilometer of information continues, the designers of fiber systems need to maximize the performance of each system building block and minimize adverse interactions in order to optimize the system's overall performance. New techniques and tools are needed to help design, manufacture, test, and support these systems and components.

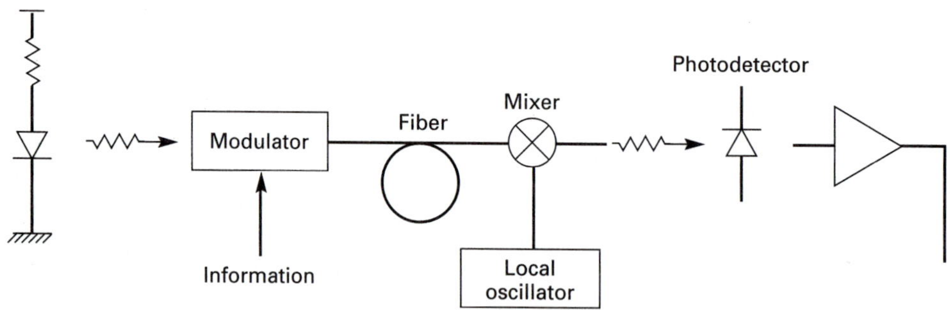

Figure 22.3 A block diagram of a coherent fiber-optic communication system.

One of the most important lightwave test instruments is an optical source capable of producing a high-quality optical signal with enough power and wavelength tuning range to test optical components and subsystems. It is highly desirable, by design and test engineers, to have an optical source instrument which can be programmed (manually or under computer control) to produce a stimulus optical signal, tunable over a certain wavelength range (typically centered around 850, 1330, or 1550 nm). Design and test engineers use such an instrument as the input source to their device (or subsystem) under test where the output from the device (or subsystem) under test is connected to an optical receiver (such as a power meter) to measure insertion and return loss vs. wavelength. The accuracy and stability of the signal wavelength generated by an optical source instrument are very important in optical device characterization. Therefore, such an instrument must contain appropriate circuitry to ensure stable (typically within 0.01 nm over several hours) and accurate (typically better than 0.01 nm) optical signals.

In its simplest configuration, an optical signal source can consist of a single-frequency semiconductor laser (such as a distributed feedback laser, DFB) with the appropriate circuitry to control the injection current to the laser. In addition, the laser can be thermally stabilized by controlling its temperature. Although this optical source can be useful on an optical bench, it is severely limited by the narrow tuning range of DFB lasers (typically a few nanometers). Moreover, the linewidth of the optical signal from such a source is in the order of a few MHz, which makes it inappropriate for some test applications, such as noise measurements which require an optical signal with less than 100-kHz linewidth.

To increase the tuning range of the DFB-based optical sources, a bank of lasers, each emitting at a different wavelength, can be used. While it is possible to cover a wide tuning range utilizing this technique, it is difficult to avoid tuning gaps. In addition, different DFB lasers will have different output power and different linewidth, resulting in a complex instrument design to compensate for the various laser outputs.

This chapter focuses on tunable external cavity lasers (ECLs) as the most general class of optical signal generators. It is possible to build ECL systems with tuning ranges in excess of 100 nm, high output power (>1 dBm), and very narrow linewidth (<100 kHz).

22.2 External Cavity Laser Fundamentals

External cavity lasers have been known as tunable sources for a long time and are often used in lab measurements. An ECL simply consists of a semiconductor laser chip placed in the external cavity of a diffraction grating as shown in Fig. 22.4. The semiconductor laser chip is a Fabry-Perot laser diode in which the internal laser resonator is disabled by an antireflection coating on one laser facet. The resonator is then built by adding an external reflector,

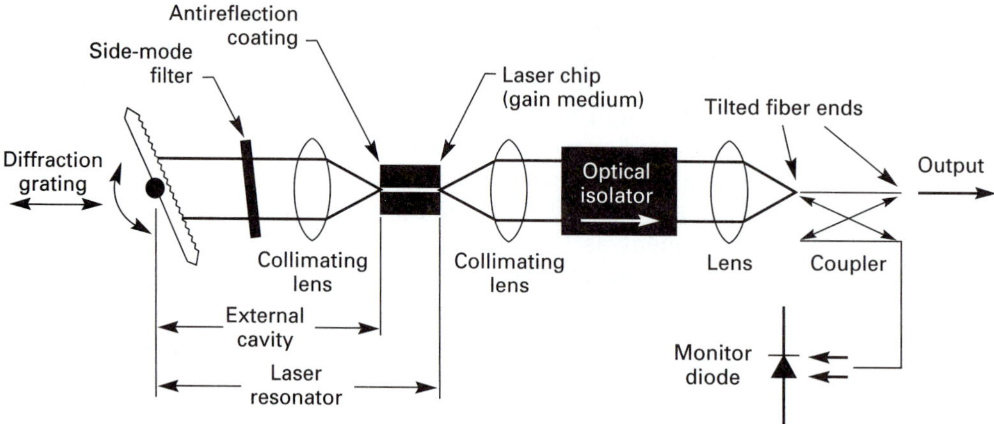

Figure 22.4 External cavity laser configuration.

the diffraction grating. The grating acts both as a plane mirror and as a wavelength-selective element.

The performance of the ECL is determined to a large extent by the external cavity formed between one surface of the semiconductor laser chip and the surface of the diffraction grating. The laser output power, single-mode operation, tuning linearity, and wavelength stability are strongly related to the components used in the external resonator. Since most of these components are temperature-dependent, the entire cavity and the laser chip must be temperature-stabilized.

Most ECLs produce an output spectrum which is multimode. While this may be useful in some component testing applications, it is highly desirable to have a single-mode output. Several techniques have been proposed to address this issue including adding a side-mode filter, which is a very narrow band wavelength filter that increases the wavelength selectivity compared to just a diffraction grating alone. However, the operation of the ECL requires the side-mode filter to be synchronously tuned with the diffraction grating, an operation performed by the control circuitry of the ECL.

Figure 22.5 explains the principle of operation of a single-mode ECL. The grating is tuned by rotation, so the wavelength where the reflection is maximum is dependent on the angle of incidence of the emitted laser beam. The external cavity laser resonator with its comblike filter characteristic will allow a large number of possible lasing wavelengths. These possible lasing modes are called "cavity modes." The spacing between two modes is determined by the resonator length and can be easily calculated. The cavity length is usually chosen to meet other requirements within the system (such as sizes and characteristics of the other optical components, the desired ECL linewidth, and the system size limitations).

The diffraction grating filter bandwidth is determined by the optical beam

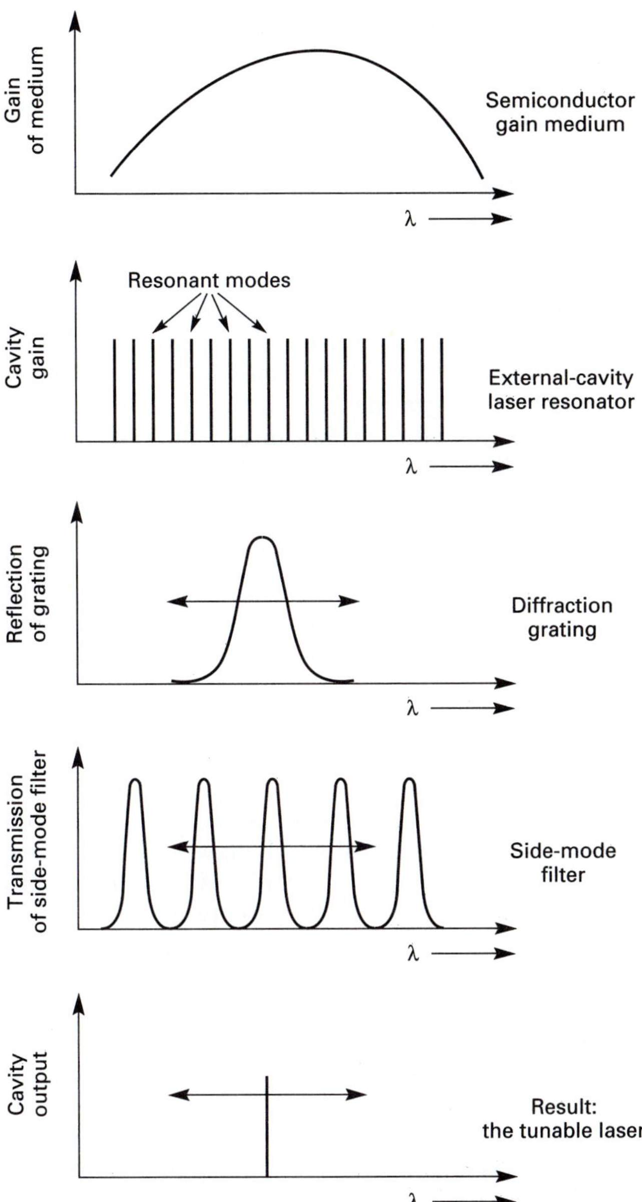

Figure 22.5 Principle of operation of external cavity lasers.

diameter, the mechanical layout of the filter, and the spacing of the rulings imprinted into the grating. From Fig. 22.5, it is easy to see that with just a diffraction grating, a more or less undefined mode is selected. Looking at the external cavity mode spacing and the filter characteristic of the diffraction grating, it is clear that a single-mode output is not guaranteed at all wavelengths because several modes have comparable gain conditions and are able to lase. The situation can be improved if the semiconductor laser chip has a flat and smooth (ripple-free) gain characteristic. Although good antireflection coatings can help achieve such characteristics, it is impractical to cover a very wide range greater than 100 nm of tuning using this solution.

To improve the wavelength selectivity of the external cavity, the side-mode filter is added to the cavity. The filter should be at least 10 times more selective than the stand-alone grating and should have nearly no feedback loss and easy wavelength tunability. This can be achieved by using a solid-glass Fabry-Perot etalon with a typical bandwidth of 0.1 nm (compared to 1 nm for the stand-alone grating).

Adjustment of the external cavity to produce a specific output wavelength consists of selecting the appropriate angles for both the grating and the side-mode filter. To allow continuous tuning, the reflection curve of the diffraction grating, the transmission maximum of the filter, and the resonant modes of the external cavity laser resonator are all shifted by the same wavelength increment.

22.3 Design Challenges

External cavity laser sources are very sensitive to external influences such as temperature fluctuations. Temperature changes cause changes in the optical length of the resonator, resulting in a change in the output frequency of the laser. A wavelength stability of 100 MHz (centered around the optical carrier frequency of about 230 THz) translates into a maximum of 1 pm change in the length of the resonator. A combination of careful optomechanical design of the resonator and a very accurate calibration procedure is needed to achieve this stability.

Another challenge is the high-reliability requirement for optical signal generators. Since the most critical component inside the laser cavity is the semiconductor laser chip, precautions such as hermetically sealing the semiconductor laser chip must be taken to ensure a long lifetime for the signal generator. In addition, the laser cavity must be kept in a humidity-resistant environment to avoid condensation on the optical surfaces of the optical components inside the laser cavity which can dramatically deteriorate the performance of the signal generator.

An important characteristic of optical signal generators is their ability to produce fast tuning speeds. For example, in a manufacturing environment, it is highly desirable to test optical components by stepping the wavelength of the optical signal generator at speeds of 5 to 10 nm/s. This means that within

100 to 200 ms the new wavelength and power level must be settled and stabilized at the output of the signal generator. In order to achieve such a tuning speed, a clever design of the motor driver circuitry is needed. To rotate the grating to change the wavelength of the laser, a precise stepper motor with a control board is used. Microstepping (about 10,000 microsteps per revolution) provides a mechanical resolution of 50 nm, and a precision switch is used to provide an absolute reference point needed to bring the motor to the right position after a power shutdown.

Another motor is needed to drive the single-mode filter. To ensure single-mode operation for the signal generator, the filter must be well synchronized with the grating drive. A dc motor with an encoder can be used to drive the filter. The control board for the motor receives its control signals from a central microprocessor board, which also sends control signals to the stepping motor of the grating.

22.4 Specifications

Wavelength range. The tuning range of an external cavity laser is ultimately limited by the width of the gain curve of the semiconductor laser chip (upper trace in Fig. 22.2). In order for the ECL to lase, the semiconductor laser must supply enough gain to overcome the losses of the other components in the cavity (lenses, reflections from grating, side-mode filter). In the near-infrared region (1.3 and 1.55 μm), typical wavelength ranges are 20 to 80 nm.

Wavelength accuracy. This parameter is determined by the accuracy of the grating angle and hence by the algorithm used to rotate the grating. It is limited by the resolution of the stepping motor used to rotate the grating. A typical accuracy number is ± 0.1 nm.

Wavelength stability. This parameter is defined as the change in the output wavelength over a period of time (typically 1 h). It is a function of the stability of the components used inside the cavity (semiconductor laser, lenses, side-mode filter, etc.). A typical number is 0.02 nm/h.

Wavelength resolution. The wavelength resolution is determined by the resolution of the angle rotation of the grating, and hence the resolution of the stepping motor. Using stepping motors with more than 100,000 steps per revolution, it is possible to achieve wavelength resolution of 0.015 nm.

Tuning speed. This parameter is defined as the time it takes to change the wavelength output of the ECL by a certain amount (typically 1 nm). Typical numbers range from 1 s (for slow lasers) to 250 ms (for fast lasers).

Side-mode suppression ratio. The ratio (in dB) between the power of the fundamental mode and the power of the highest spurious mode. Using side-mode filters, it is possible to obtain more than 50-dB suppression.

Linewidth. This parameter defines the width of the laser spectrum and is one of the major advantages of external cavity lasers. By extending the laser cavity length, lower linewidth values are possible. For a cavity 10 cm long, the linewidth is about 100 kHz.

Output power. This parameter depends on the output power of the semiconductor laser diode and the losses of the other components used inside the cavity. Typically, external cavity lasers are capable of producing more than 100 μW of output power.

Power stability. This parameter is defined over a certain time period (typically 1 h). It is a very important parameter since it is always desirable to generate a stable output for long tests. Typical numbers range from ± 0.05 to ± 0.1 dB.

Power flatness vs. wavelength. This is one of the most important parameters of a laser since most component testing requires constant output power. A good ECL will have better than ± 0.2 dB flatness over the tuning wavelength range.

Chapter

23

Lightwave Signal Analysis

Waguih Ishak
Agilent Technologies
Palo Alto, California

23.1 Introduction

High-capacity digital transmission systems and analog-modulated microwave frequency systems based on fiber-optic (FO) systems have emerged as very competitive to conventional communication systems. A variety of lightwave components have been developed to support these high-speed systems. Most notable among these components are semiconductor lasers, specifically single-frequency distributed feedback (DFB) lasers, and high-speed pin photodetectors.

The design of the communications systems depends heavily on the characteristics of the lightwave signal propagating through the various components of the system. A lightwave signal is an electromagnetic wave in the wavelength range between 200 and 2000 nm, i.e., a frequency range between 150 and 1500 THz. This signal is generated by the laser in the transmission section of the system. It is critical to the success of the system to use high-quality lasers (high power, large modulation bandwidth, low noise, etc.). In addition, as this signal propagates through the system, its characteristics (power, polarization, etc.) change. Accurate methods of measuring the characteristics (power, modulation speed, spectrum, linewidth, state of polarization, etc.) of a lightwave signal or, alternatively, analyzing the lightwave signal are therefore essential for the design of high-capacity fiber-optic transmission systems.

23.2 Light Sources

The term *laser* stands for "light amplification by stimulated emission of radiation." We focus the discussion in this section on semiconductor laser diodes

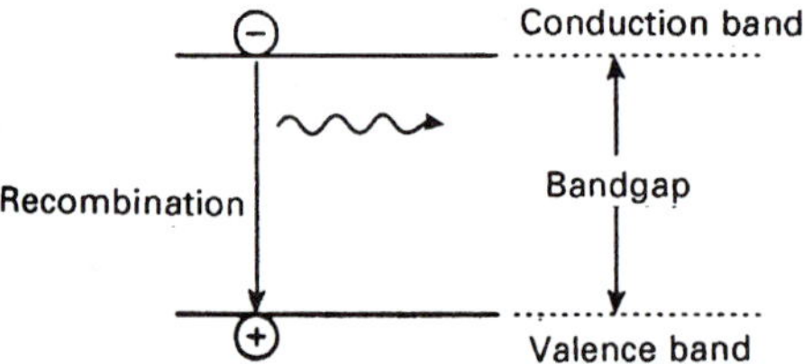

Figure 23.1 Spontaneous emission from semiconductor lasers.

(LDs) which are now being routinely used in FO systems because of their small size, high efficiency and reliability, and excellent control of wavelength, power, and spectral characteristics. Because the light output from a laser is coherent, LDs are used for applications requiring high data rates (in excess of 100 Mbit/s). On the other hand, the incoherent nature of the output signal from an LED limits their applications to data rates less than 100 Mbit/s. While the structure and the fabrication process of LEDs and LDs are very similar, their light output characteristics are quite different.

23.2.1 Light-emitting diodes (LEDs)

LEDs produce light with a wide spectral width, and when used in FO communications systems, they can be modulated at frequencies up to 100 MHz. LEDs have the advantages of low temperature sensitivity and no sensitivity to back reflections. Furthermore, LEDs produce incoherent light output which is not sensitive to optical interference from reflections.

LEDs generate light by spontaneous emission. This occurs when an electron in a high-energy conduction band changes to a low-energy valence band, as shown in Fig. 23.1. The energy lost by the electron, the bandgap energy E_g is released as a photon, the entity of light. The energy of the released photon is equal to the energy lost by the electron, and the wavelength of the emitted photon is a function of its energy. Because different materials have different orbital states which determine the energy levels of the various electrons, the wavelength of the emitted photon is determined by the material used to make the LED.

The wavelength of the emitted photon is given by

$$\lambda = \frac{hc}{E_g} = \frac{1.24\mu m}{E_g(\text{eV})} \tag{23.1}$$

where h = Planck's constant = 6.62×10^{-34} W/s^2
c = speed of light = 2.998×10^8 m/s
E_g = bandgap of material, joules

For a semiconductor material with $E_g = 0.9$ eV, the wavelength of the emitted photons will be about 1.38 μm. The most commonly used materials for

LEDs are gallium arsenide (GaAs) with $E_g = 1.42$ eV and gallium phosphide (GaP) with $E_g = 2.24$ eV. By adding other materials, such as aluminum or indium, to the GaAs or the GaP, it is possible to tailor the bandgap energy to achieve any wavelength in the 0.5- to 2.0-μm range.

With appropriate n and p doping, the above materials can be used to form a simple pn diode which can function as an LED. By forward biasing the pn junction of the diode, conduction-band electrons are generated. For a better confinement of the output optical power, a double heterostructure (DH) LED is used. In a DH-LED, the junction is formed by dissimilar semiconductor materials with different bandgap energy and refractive index values and the free charges are confined to recombine in a narrow, well-defined semiconductor layer (called the active layer).

The spectrum of an LED has a broad distribution of wavelength centered around a wavelength calculated from the above equation. The spectral width is often specified as the full width at half maximum (FWHM) of the distribution. Typical values for the FWHM range from 20 to 100 nm.

23.2.2 LED parameters

Total power. The summation of the power at each trace point, normalized by the ratio of the trace point spacing and resolution bandwidth. This normalization is required because the spectrum of the LED is continuous, rather than containing discrete spectral components (as a laser does).

$$\text{Total power} = \sum_{i=1}^{N} P_i \frac{\text{trace point spacing}}{\text{resolution bandwidth}} = P_0 \tag{23.2}$$

Mean (FWHM). These wavelengths represent the center of mass of the trace points. The power and wavelength of each trace point are used to calculate the mean (FWHM) wavelength.

$$\text{Mean (FWHM)} = \bar{\lambda} = \sum_{i=1}^{N} P_i \frac{\text{trace point spacing}}{\text{resolution bandwidth}} \frac{\lambda_i}{P_0} \tag{23.3}$$

Sigma. An rms calculation of the spectral width of the LED based on a gaussian distribution. The power and wavelength of each trace point are used to calculate sigma.

$$\text{Sigma} = \sigma = \sqrt{\sum_{i=1}^{N} \frac{\text{trace point spacing}}{\text{resolution bandwidth}} (\lambda_i - \bar{\lambda})^2 P_0} \tag{23.4}$$

FWHM (Full Width at Half Maximum). Describes the spectral width of the half-power points of the LED, assuming a continuous, gaussian power distribution. The half-power points are those where the power spectral density is one-half that of the peak amplitude.

$$\text{FWHM} = 2.355\sigma \tag{23.5}$$

3-dB width. Used to describe the spectral width of the LED based on the separation of the two wavelengths that each have power spectral density equal to one-half the peak power spectral density. The 3-dB width is determined by finding the peak of the LED spectrum and dropping down 3 dB on each side.

Mean (3 dB). The wavelength that is the average of the two wavelengths determined in the 3-dB width measurement.

Peak wavelength. The wavelength at which the peak of the LED's spectrum occurs.

Density (1 nm). The power spectral density (normalized to a 1-nm bandwidth) of the LED at the peak wavelength.

Distribution trace. A trace can be displayed that is based on the total power, power distribution, and mean wavelength of the LED. This trace has a gaussian spectral distribution and represents a gaussian approximation to the measured spectrum.

23.2.3 Fabry-Perot (FP) laser diodes

Lasers are capable of producing high output powers and directional beams. When used in fiberoptic communication systems, semiconductor lasers can be modulated at rates up to about 10 GHz. However, lasers are sensitive to high temperature and back reflections. Additionally, the coherent emitted light is sensitive to optical interference from reflections. Two different types of semiconductor lasers are commonly used in FO communications systems: Fabry-Perot (FP) and distributed feedback (DFB) lasers.

The Fabry-Perot laser is different from an LED in that it generates light by stimulated emission, where photons trigger additional electron-hole recombinations, resulting in additional photons as shown in Fig. 23.2. A stimulated

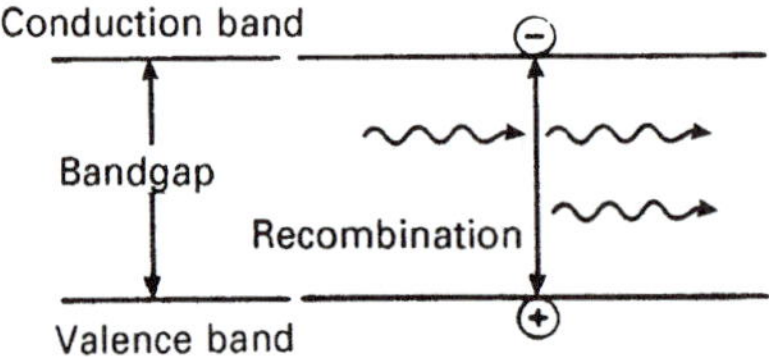

Figure 23.2 Stimulated emission from semiconductor lasers.

photon travels in the same direction and has the same wavelength and phase as the photon that triggered its generation. Stimulated emission can be thought of as amplification of light. As one photon passes through the region of holes and conduction band electrons, additional photons are generated. If the material is long enough, enough photons can be generated to produce a significant amount of power at a single wavelength.

An easier way to build up power is to place a reflective mirror at each end of the region where the photons multiply so that they can travel back and forth between the two mirrors, building up the number of photons with each trip. These mirrors form the resonator needed for the operation of the laser. Additionally, in order for the laser action to occur, a greater number of conduction band electrons than valence band electrons must be present. Called population inversion, this is achieved by forcing a high current density into the active region of the laser diode.

The possible wavelengths produced by the resonator are given by

$$f_{\text{res}} = \frac{mc}{2ln} \tag{23.6}$$

where m = integer
c = speed of light
l = length of resonator
n = refractive index of laser cavity

The mode spacing, which is the separation between the different wavelengths, is determined as

$$\text{Mode spacing} = \frac{c}{2ln} \tag{23.7}$$

23.2.4 FP laser parameters

Total power. The summation of the power in each of the displayed spectral components or modes that satisfy the peak excursion criteria (defined below).

$$\text{Total power} = \sum_{i=1}^{N} P_i = P_0$$

Mean wavelength. Represents the center of mass of the spectral components on screen. The power and wavelength of each spectral component are used to calculate the mean wavelength.

$$\text{Mean wavelength} = \sum_{i=1}^{N} \frac{P_i \lambda_i}{P_0} = \bar{\lambda} \tag{23.8}$$

Sigma. An rms calculation of the spectral width of the FP laser based on a gaussian distribution.

$$\text{Sigma} = \sigma = \frac{\sum_{i=1}^{N} P_i(\lambda_i - \lambda)^2}{P_0} \tag{23.9}$$

FWHM. Describes the spectral width of the half-power points of the FP laser, assuming a continuous, gaussian power distribution. The half-power points are those where the power spectral density is one-half that of the peak amplitude.

$$\text{FWHM} = 2.355^{*}\sigma \tag{23.10}$$

Mode spacing. The average wavelength spacing between the individual spectral components of the FP laser.

Peak amplitude. The power level of the peak spectral component of the FP laser.

Peak wavelength. This is the wavelength at which the peak spectral component of the FP laser occurs.

Peak excursion. The peak excursion value (in dB) can be set by the user and is used to determine which on-screen responses are accepted as discrete spectral responses.

23.2.5 Distributed feedback (DFB) lasers

DFB lasers are similar to FP lasers, except that all but one of their spectral components are significantly reduced. Because its spectrum has only one line, the spectral width of the DFB laser is much less that of an FP laser. This greatly reduces the effect of chromatic dispersion in fiberoptic systems, allowing for greater transmission bandwidths.

The distributed feedback laser utilizes a grating, a series of corrugated ridges, along the active layer of the semiconductor, as shown in Fig. 23.3. Rather than using just the two reflecting surfaces at the ends of the diode, as a Fabry-Perot laser does, the distributed feedback laser uses each ridge of the corrugation as a reflective surface. At the resonant wavelength, all reflections from the different ridges add in phase. By having much smaller spacings between the resonator elements, compared to the Fabry-Perot laser, the possible resonant wavelengths are much farther apart in wavelength, and only one resonant wavelength is in the region of laser gain. This results in the single laser wavelength.

The ends of the diode still act as a resonator, however, and produce the

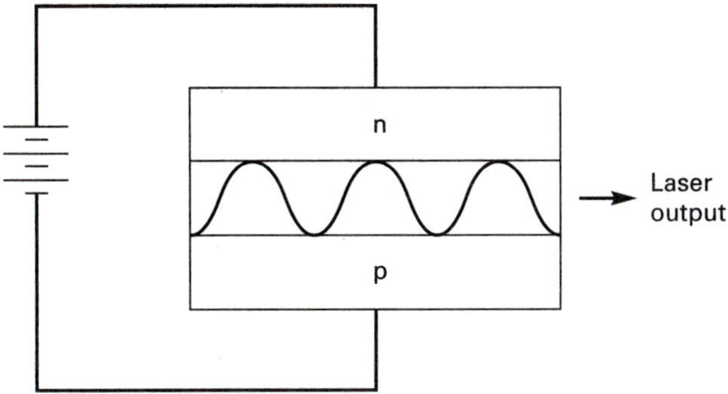

Figure 23.3 Distributed feedback laser.

lower-amplitude side modes. Ideally, the dimensions are selected so that the end reflections add in phase with the grating reflections. In this case the main mode will occur at a wavelength halfway between the two adjacent side modes; any deviation is called a mode offset. Mode offset is measured as the difference between the main-mode wavelength and the average wavelength of the two adjacent side modes.

The amplitude of the largest side mode is typically between 30 and 50 dB lower than the main spectral output of the laser. Because side modes are so close to the main mode (typically between 0.5 and 1 nm) the dynamic range of an optical spectrum analyzer determines its ability to measure them. Dynamic range is specified at offsets of 0.5 and 1.0 nm from a large response. Hewlett-Packard optical spectrum analyzers, for example, specify a dynamic range of −55 dBc at offsets of 0.5 nm and greater and −60 dBc at offsets of 1.0 nm and greater. This indicates the amplitude level of side modes that can be detected at the given offsets.

23.2.6 DFB laser parameters

Peak wavelength. The wavelength at which the main spectral component of the DFB laser occurs.

Side-mode suppression ratio (SMSR). The amplitude difference between the main spectral component and the largest side mode.

Mode offset. Wavelength separation (in nanometers) between the main spectral component and the SMSR mode.

Peak amplitude. The power level of the main spectral component of the DFB laser.

Stop band. Wavelength spacing between the upper and lower side modes adjacent to the main mode.

Center offset. Indicates how well the main mode is centered in the stop band. This value equals the wavelength of the main spectral component minus the mean of the upper and lower stop band component wavelengths.

Bandwidth. Measures the displayed bandwidth of the main spectral component of the DFB laser. The amplitude level, relative to the peak, that is used to measure the bandwidth can be set by the user.

Peak excursion. The peak excursion value (in dB) can be set by the user and is used to determine which three on-screen responses will be accepted as discrete spectral responses. To be counted, the trace must rise and then fall by at least the peak excursion value about a given spectral component. Setting the value too high will result in failure to count small responses near the noise floor.

23.3 Optical Power Meters

Power is the most basic and common measurement for optical signals. It is commonly measured using a power meter utilizing a photodetector with appropriate bias circuits. The strongest concern in power measurements is accuracy.

23.3.1 Types of optical power meters

If the signal to be measured is a parallel beam of light, a large-area (about 5 mm in diameter) photodetector is usually used. The large area of the photodetector can capture the several modes (from a fiber cladding, for example) composing the optical signal. However, it is important to know the wavelength of the signal before an accurate estimate of the power can be made. This is because photodetectors exhibit a wavelength-dependent responsivity. The disadvantage of this method is the presence of a large dark current (current generated in the photodetector with no signal present) for the large-area photodetector.

Another method of measuring optical power is to focus the optical signal into a tiny spot (less than 1 mm) on a small-area photodetector. The advantages are higher sensitivity and lower cost. However, it is difficult to calibrate such a system, and small-area photodetectors usually result in nonuniform signal distribution.

Wavelength-independent detectors (such as pyroelectric radiometers and thermopiles) can be used with parallel beam signals to measure the optical power. This method is indirect and is based on measuring the temperature change of the detector as the optical signal falls onto it. One major advantage is the wavelength insensitivity, but the sensitivity is inferior relative to the photodetector-based methods.

A problem common to all methods is reflection. Multiple reflections from

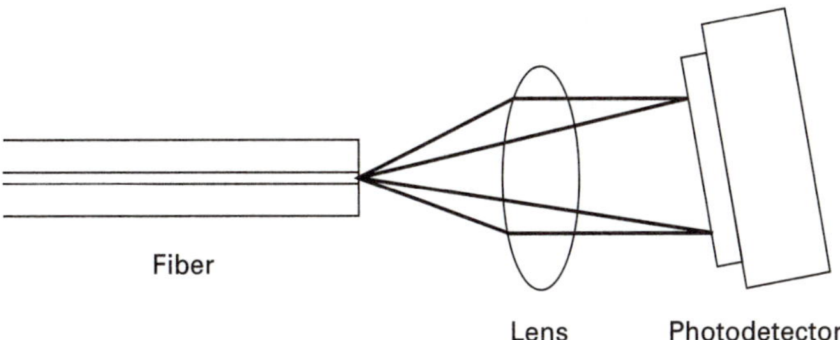

Figure 23.4 Power meter block diagram.

surfaces within the power meter can cause severe inaccuracies, and commercial power meters must provide means of eliminating or correcting for this effect. A simple yet effective method to solve this problem is to tilt the photodetector surface relative to the normal to the plane of incidence and to use antireflection coatings as shown in Fig. 23.4. An antireflection coating usually consists of several layers of thin dielectric films. The thicknesses and the dielectric constants of the various films are adjusted so that reflections from the films will add constructively in the transmission direction (the direction of the incident optical signal) and destructively in the reflection direction, thus eliminating reflected signals from reaching the device under test.

23.3.2 Accuracy of optical power meters

A general-purpose power meter can measure both absolute and relative optical power. The critical specification in absolute optical power measurements is the uncertainty of the displayed power. The more accurate the measurements of power, the higher the confidence in component verification during incoming inspection, production, installation, and maintenance. In these areas, a more accurate power meter will increase yield and reduce cost. In order to achieve accurate absolute power measurements which are traceable to national standard laboratories, the detector noise and the postamplifier drift must be minimized. This can be achieved by mounting the photodetector and the transimpedance amplifier on a thermoelectric cooler inside a hermetically sealed package. By keeping the temperature of the photodetector constant, the error caused by the variation of the quantum efficiency of the photodetector is almost eliminated. This results in a low-noise power measurement. In addition, by cooling the detector, the noise equivalent power is kept to a minimum. Furthermore, by keeping the temperature of the amplifier constant, the drift is minimized.

Relative power measurement is useful in determining the insertion loss of optical components. For example, in determining the insertion loss of fiber connectors, the light out of one connector is taken as the reference; then the connectors are mated and the measurement is taken at the far end of the second fiber. In order to detect very small changes in loss (for example, when testing very low loss components), very high power meter resolution is required. State-of-the-art power meters have 0.001 dB resolution. The key to achieving high resolution is the design of the photodetector and its associated circuity (bias and amplifier).

On the other extreme, a high dynamic range power meter is needed to measure high losses (for example, in characterizing long fibers). The linearity of the detection circuit determines the limit on the dynamic range. It is crucial to make sure that the gain of the postamplifier chain is matched to avoid nonlinearities.

23.3.3 Specifications of optical power meters

Wavelength range. The range of optical wavelength over which the rest of the specifications are valid. This parameter is dependent upon the type of photodetector used inside the power meter. For silicon detectors, the wavelength range is between 450 and 1000 nm. For indium gallium arsenide detectors, the wavelength range is 800 to 1700 nm.

Power range. The minimum and maximum detectable power levels. Typical range from −100 to +3 dBm.

Resolution. The display resolution in dB or watts. Typically 0.001 dB or 10 pW.

Noise. The internal noise generated by the power meter. Typically 1 to 50 pW.

23.4 Lightwave Signal Analyzers

Lightwave signal analyzers are useful for measuring the optical signal strength and distortion, modulation depth and bandwidth, intensity noise, and susceptibility to reflected light. However, these analyzers are quite different from the optical spectrum analyzers discussed later in this chapter. The optical spectrum analyzers show the spectral distribution of average optical power and are useful for observing the modes of multimode optical sources or the side-lobe rejection of single-mode sources. Their measurement resolution is typically 0.1 nm or about 18 GHz at a wavelength of 1300 nm. On the other hand, the lightwave signal analyzer displays the total average power and the modulation spectrum but provides *no* information about the wavelength of the optical signal. This distinction is shown in Fig. 23.5.

23.4.1 Block diagram

A simplified block diagram for a high-speed lightwave signal analyzer is shown in Fig. 23.6. The input lightwave signal passes through an optical

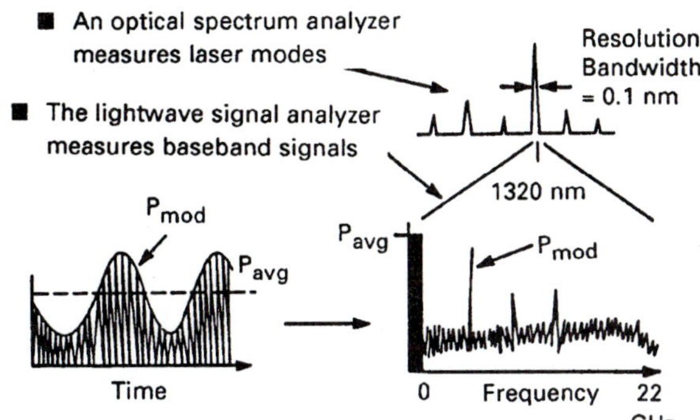

Figure 23.5 Differences between a lightwave signal analyzer and optical spectrum analyzer.

attenuator to prevent overloading of the front end. The high-speed photodetector converts the optical signal to an electrical signal. The time-varying component of the photocurrent, which represents the demodulated signal, is fed through the preamplifier to the input of a conventional microwave spectrum analyzer. The dc portion of the photocurrent is fed to a power meter circuit.

23.4.2 Measured parameters

A key parameter in any lightwave system is the modulation bandwidth of the optical source of the system. Current-modulated semiconductor lasers have bandwidths approaching 20 GHz. A high-speed lightwave signal analyzer can measure the intensity modulation frequency response of these lasers provided the photodetector and the microwave spectrum analyzer (shown in Fig. 23.6) have wider bandwidth than the device under test. Another key parameter, for a number of reasons, is the laser noise spectrum. It obviously impacts the signal-to-noise ratio in a fiber-optic transmission system. Furthermore, it can be shown that the intensity noise spectrum has the same general shape as

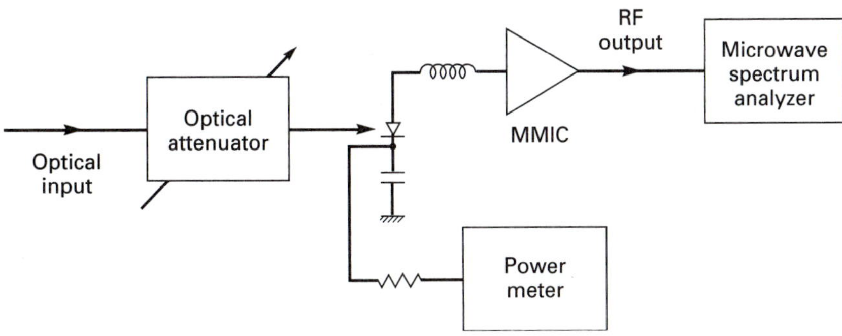

Figure 23.6 Lightwave signal analyzer block diagram.

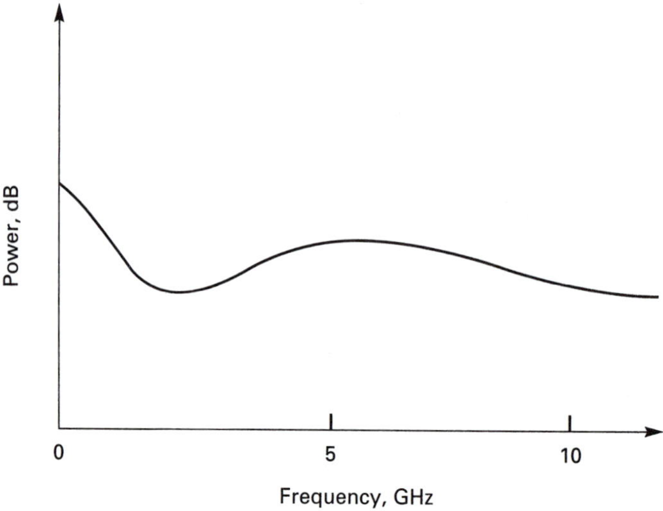

Figure 23.7 Relative intensity noise of a semiconductor laser.

the intensity modulation response and can be used as an indicator of potential modulation bandwidth. Again, a high-speed lightwave signal analyzer can be very useful in measuring the laser noise spectrum.

The laser intensity noise spectrum can be generally affected by both the magnitude and the polarization of the optical power that is fed back to the laser. This is called reflection-induced noise and is typically caused by reflections from optical connectors. This reflected power upsets the dynamic equilibrium of the lasing process and typically increases the amplitude of the intensity noise. It can also induce a ripple on the spectrum. The lightwave signal analyzer is ideal for measuring the laser intensity noise. It should be mentioned here that while it is possible to use an optical time domain reflectometer to measure the magnitude of an optical reflection, the lightwave signal analyzer is the only instrument that can measure the effect of these reflections on the noise characteristics of the laser under test.

Perhaps one of the most important parameters of laser sources to be measured by a lightwave signal analyzer is the relative intensity noise (RIN). It is the ratio of the optical noise power to the average optical power and is an indication of the maximum possible signal-to-noise ratio in a lightwave system, where the dominant noise source is the laser intensity noise. Figure 23.7 shows an example of the relative intensity noise measurement of a high-speed semiconductor laser.

23.5 Optical Spectrum Analyzers

Optical spectrum analysis is the measurement of the optical power as a function of the wavelength. This is an important measurement for optical sources such as lasers and LEDs. The spectral width of a light source is an important parameter in fiber-optic communications systems because of chromatic dispersion, which occurs in the fiber and limits the modulation bandwidth of the system, as described earlier. The effect of chromatic dispersion can be seen in the time domain as pulse broadening of a digital waveform. Since chromatic dispersion is a function of the spectral width of the light source, narrow spectral widths are desirable for high-speed communications systems.

Because there are many different types of light sources, with varying spectra, there is a need to have an instrument capable of measuring the spectrum of single-mode lasers (such as a DFB or a DBR laser) or a multimode laser (such as a FP laser). This instrument is called an optical spectrum analyzer. It is probably the most common instrument used by researchers and manufacturing engineers who are involved in optical measurements.

Optical spectrum analyzers (OSAs) can be divided into three categories: diffraction-grating–based and two interferometer–based architectures, the Fabry-Perot and Michelson interferometer–based OSAs. Diffraction-grating–based OSAs are capable of measuring spectra of lasers and LEDs. The resolution of these instruments is variable, typically ranging from 0.1 to 10 nm. Fabry-Perot (FP) interferometer–based OSAs have a fixed, narrow resolution, typically specified in frequency, between 100 MHz and 10 GHz (i.e., 0.01 to 0.1 nm at the 1.55-μm wavelength range). This narrow resolution allows them to be used in measuring laser chirp but can limit their measurement spans much more than the diffraction-grating–based OSAs. Michelson-interferometer–based OSAs display the resolution by calculating the Fourier transform of a measured interference pattern.

23.5.1 Block diagram

The basic block diagram of an optical spectrum analyzer is shown in Fig. 23.8. The incoming optical signal passes through a wavelength-tunable opti-

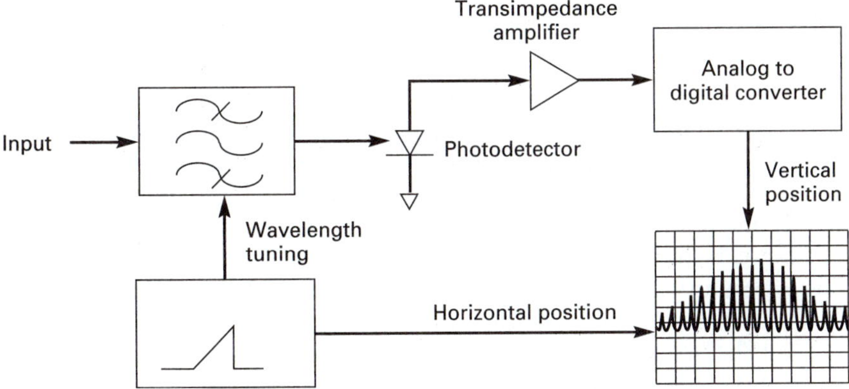

Figure 23.8 Simplified block diagram of an optical spectrum analyzer.

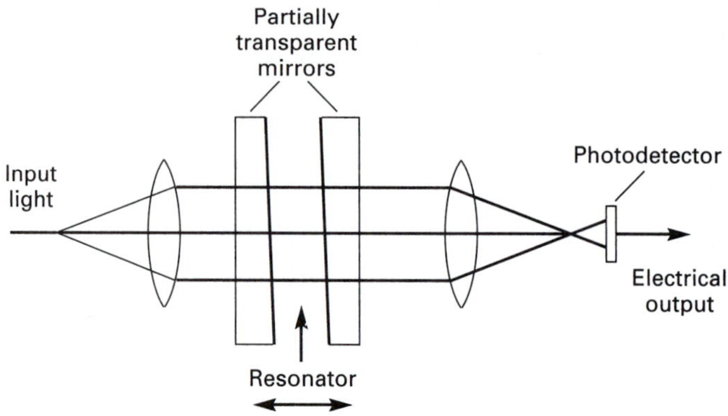

Figure 23.9 Fabry-Perot–based optical spectrum analyzer.

cal filter (monochromator or interferometer) which resolves the individual spectral components. The photodetector then converts the optical signal into an electric signal proportional to the incident optical power. An exception to this description is the Michelson interferometer, which is not actually an optical filter.

The current from the photodetector is then converted to a voltage by the transimpedance amplifier and then digitized using an A/D converter. The signal is applied to the display as the vertical data and a ramp generator determines the horizontal location of the trace as it sweeps from left to right. This ramp is used to tune the optical filter so that its resonance wavelength is proportional to the horizontal position. A trace of optical power vs. wavelength results. The displayed width of each of the modes of the optical signal under test is a function of the spectral resolution of the optical filter. Therefore, the resolution of the OSA is mainly determined by the characteristics of the tunable optical filter.

23.5.2 FP interferometer–based optical spectrum analyzers

The FP interferometer, shown in Fig. 23.9, consists of two highly reflective, parallel mirrors that act as a resonant cavity which filters the incoming optical signal. Because the spacing between the mirrors is essentially fixed, the resolution of an FP interferometer–based OSA is typically fixed. The wavelength of the cavity is varied by slightly changing the spacing between the mirrors from its nominal value by a very small amount.

The advantage of the Fabry-Perot interferometer is its very narrow spectral resolution, which allows it to measure laser chirp. The major disadvantage is that at any one position multiple wavelengths will be passed by the filter. (The spacing between these responses is called the free spectral range.) This problem can be solved by placing a monochromator (a wavelength-tunable narrowband optical filter) in cascade with the Fabry-Perot interferometer to

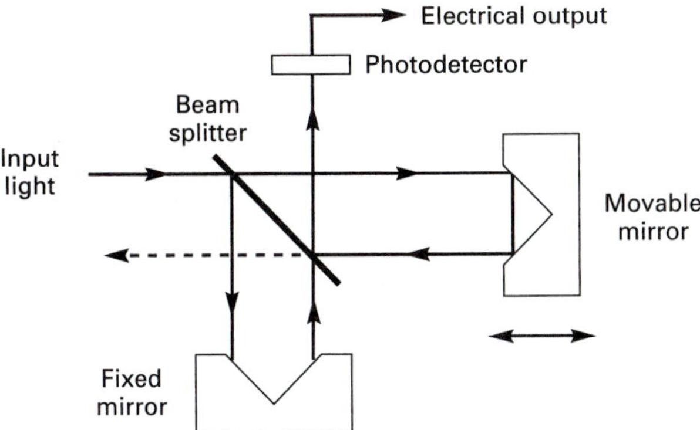

Figure 23.10 Michelson-interferometer–based optical spectrum analyzer.

filter out all power outside the interferometer's free spectral range about the wavelength of interest.

23.5.3 Michelson-interferometer–based optical spectrum analyzers

The Michelson interferometer, shown in Fig. 23.10, is based on creating an interference pattern between the signal and a delayed version of itself. The power of this interference pattern is measured for a range of delay values. The resulting waveform is the autocorrelation function of the input signal. This enables the Michelson-interferometer–based spectrum analyzer to make direct measurements of coherence length, as well as very accurate wavelength measurements. Other types of optical spectrum analyzers cannot make direct coherence-length measurements.

To determine the power spectra of the input signal, a Fourier transform is performed on the autocorrelation waveform. Because no real filtering occurs, Michelson-interferometer–based optical spectrum analyzers cannot be put in a span of zero nanometers, which would be useful for viewing the power at a given wavelength as a function of time. This type of analyzer also tends to have less dynamic range than diffraction-grating–based optical spectrum analyzers.

23.5.4 Diffraction-grating–based optical spectrum analyzers

Monochromators. The most common optical spectrum analyzers use monochromators as the tunable optical filter. In the monochromator, a diffraction grating (a mirror with finely spaced corrugated lines on the surface) separates the different wavelengths of light. The result is similar to that achieved with

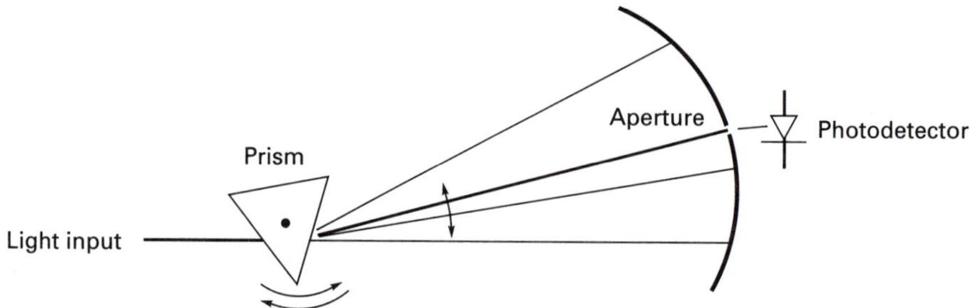

Figure 23.11 Prism-based optical spectrum analyzer.

a prism. Figure 23.11 shows what a prism-based optical spectrum analyzer might look like. The prism separates the different wavelengths of light, and only the wavelength that passes through the aperture reaches the photodetector. The angle of the prism determines the wavelength to which the optical spectrum analyzer is tuned, and the size of the aperture determines the wavelength resolution. Diffraction gratings are used instead of prisms because they provide a greater separation of wavelengths, with less attenuation. This allows for better wavelength resolution.

A diffraction grating is a mirror with grooves on its surface, as shown in Fig. 23.12. The spacing between grooves is extremely narrow, approximately equal to the wavelengths of interest. When a parallel light beam strikes the diffraction grating, the light is reflected in a number of directions.

The first reflection is called the zero-order beam ($m = 0$), and it reflects in the same direction as it would if the diffraction grating were replaced by a plane mirror. This beam is not separated into different wavelengths and is not used by the optical spectrum analyzer.

The first-order beam ($m = 1$) is created by the constructive interference of reflections off each groove. For constructive interference to occur, the path-length difference between reflections from adjacent grooves must equal one

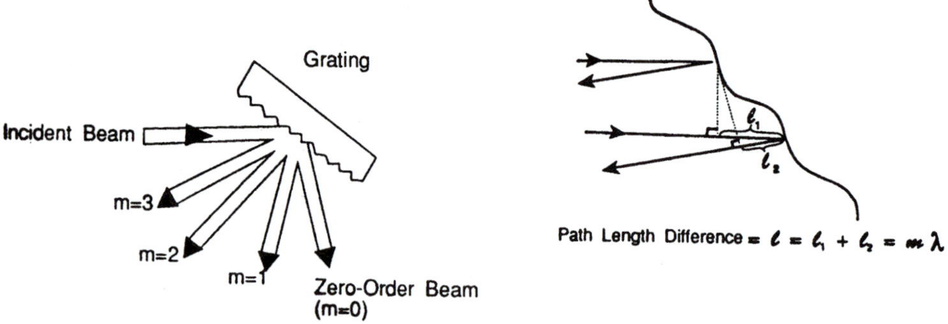

Figure 23.12 Diffraction grating principle of operation.

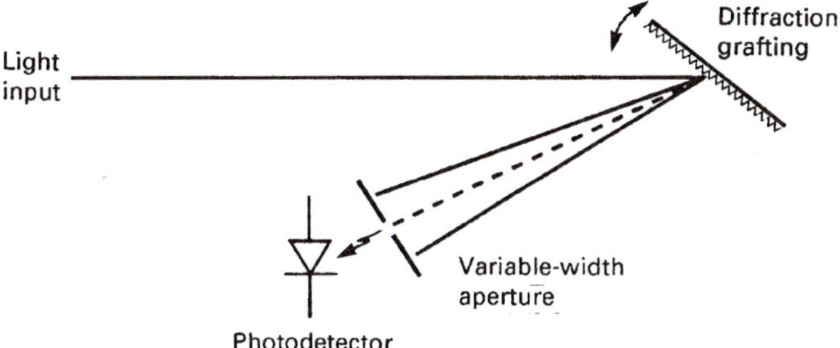

Figure 23.13 Diffraction-grating–based optical spectrum analyzer.

wavelength. If the input light contains more than one wavelength component, the beam will have some angular dispersion; that is, the reflection angle for each wavelength must be different in order to satisfy the requirement that the pathlength difference of adjacent grooves is equal to one wavelength. Thus the optical spectrum analyzer separates different wavelengths of light.

For the second-order beam ($m = 2$), the path-length difference from adjacent grooves equals two wavelengths. A three-wavelength difference defines the third-order beam, and so on. Optical spectrum analyzers utilize multiple-order beams to cover their full wavelength range with narrow resolution.

Figure 23.13 shows the operation of a diffraction-grating–based optical spectrum analyzer. As with the prism-based analyzer, the diffracted light passes through an aperture to the photodetector. As the diffraction grating rotates, the instrument sweeps a range of wavelengths, allowing the diffracted light—the particular wavelength depends on the position of the diffraction grating—to pass through to the aperture. This technique allows the coverage of a wide wavelength range.

Diffraction-grating–based optical spectrum analyzers contain either a single monochromator, a double monochromator, or a double-pass monochromator. Figure 23.14 shows a single-monochromator-based instrument. In these instruments, a diffraction grating is used to separate the different wavelengths of light. The second concave mirror focuses the desired wavelength of

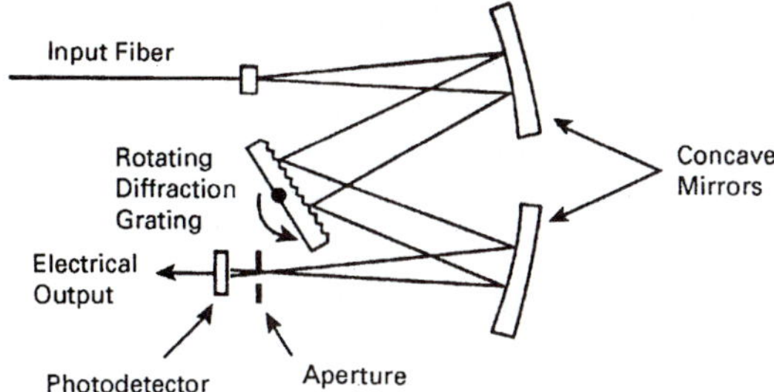

Figure 23.14 Single-pass grating-based optical spectrum analyzer.

light at the aperture. The aperture width is variable and is used to determine the wavelength resolution of the instrument.

Double monochromators. Double monochromators, as shown in Fig. 23.15, are sometimes used to improve on the dynamic range of single-monochromator systems. Double monochromators are equivalent to a pair of sweeping filters. While this technique improves dynamic range, double monochromators typically have reduced span widths due to the limitations of monochromator-to-monochromator tuning match; double monochromators also have degraded sensitivity due to losses in the monochromators.

The double-pass monochromator provides the dynamic-range advantage of the double monochromator and the sensitivity and size advantages of the single monochromator. Figure 23.15 shows the double-pass monochromator. The first pass through the double-pass monochromator is similar to conventional single monochromator systems. In Fig. 23.15, the input beam (1) is collimated by the optical element and dispersed by the diffraction grating. This results in a spatial distribution of the light, based on wavelength. The diffraction grating is positioned such that the desired wavelength (2) passes through the aperture. The width of the aperture determines the bandwidth of wavelengths allowed to pass to the detector. Various apertures are available to provide resolution bandwidths of 0.08 and 0.1 to 10 nm in a 1, 2, 5 sequence. In a single-monochromator instrument, a large photodetector behind the aperture would detect the filtered signal.

The system shown in Fig. 23.15 is unique in that the filtered light (3) is sent through the collimating element and diffraction grating for a second time. During this second pass through the monochromator, the dispersion process is reversed. This creates an exact replica of the input signal, filtered by the aperture. The small resultant image (4) allows the light to be focused onto a fiber which carries the signal to the detector. This fiber acts as a second aperture in the system. The implementation of this second pass results in the high sensitivity of a single monochromator, the high dynamic range of a double monochromator, as well as polarization insensitivity (due to the half-wave plate).

23.5.5 Specifications of optical spectrum analyzers

Wavelength tuning. This parameter is controlled by the rotation of the diffraction grating. Each angle of the diffraction grating causes a corresponding wavelength of light to be focused directly at the center of the aperture. The diffraction-grating angle must be precisely controlled and very repeatable over time to assure accurate tuning.

Wavelength repeatability. This parameter specifies the wavelength tuning drift in a 1-min period. This is specified with the OSA in a continuous sweep mode and with no changes made to the tuning.

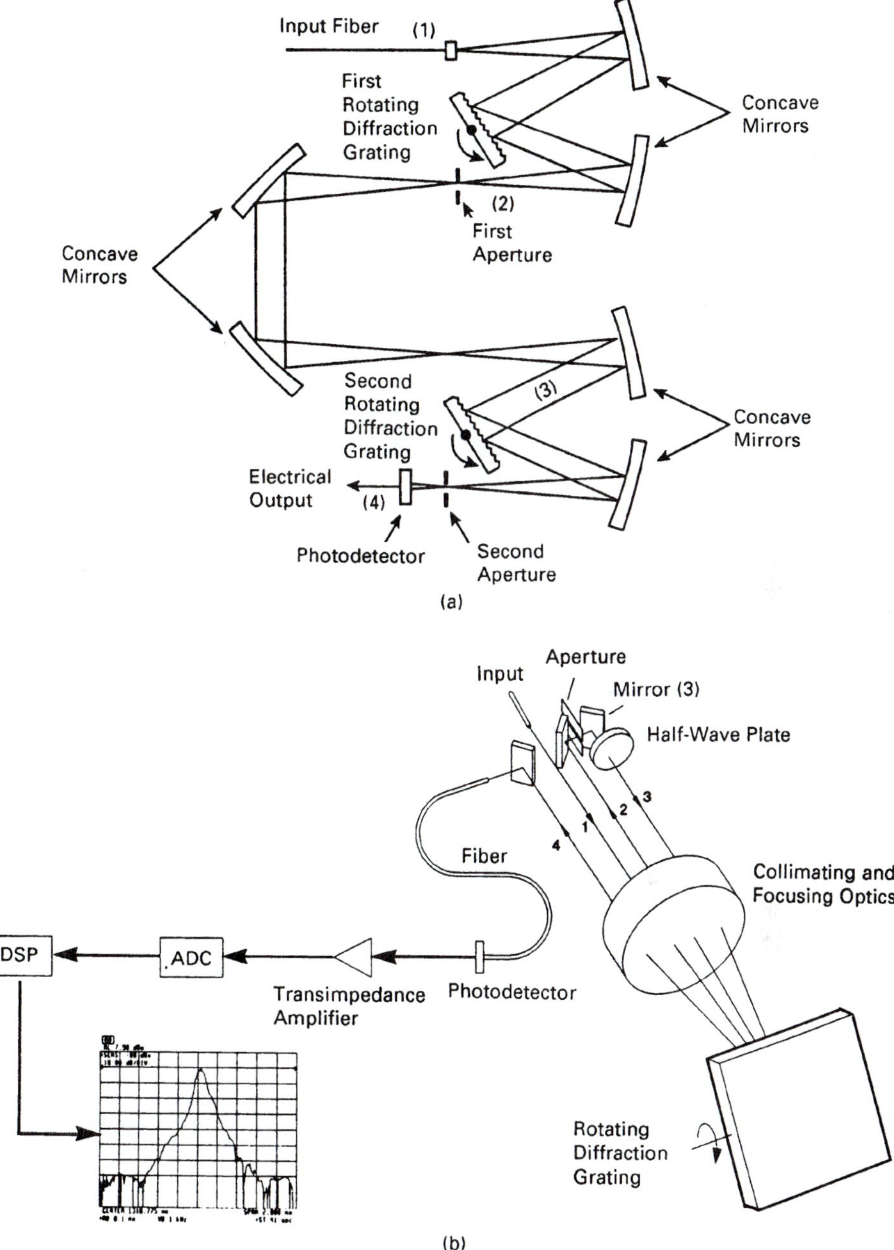

Figure 23.15 Double-pass grating-based optical spectrum analyzer. (a) Schematic drawing; (b) implementation in instrumentation.

Wavelength resolution bandwidth. This parameter determines the ability of the OSA to display two signals closely spaced in wavelength as two distinct responses. It is determined by the bandwidth of the optical filter and is defined as the filter bandwidth at the half-power level, referred to as the FWHM. Typical values for the wavelength resolution bandwidth vary from 0.08 to 5 nm.

Dynamic range. This parameter is commonly specified at 0.5- or 1.0-nm offsets from the main response. A −60 dB dynamic range specification at 1 nm and greater indicates that the OSA's response to a purely monochromatic signal will be −60 dBc or less at offsets of 1.0 nm or greater.

Sensitivity. This parameter is defined as the minimum detectable signal or, more specifically, 6 times the rms noise level of the instrument. This is one difference between OSAs and rf and microwave spectrum analyzers, whose sensitivities are defined as the average noise level.

Tuning speed. This parameter depends on the sensitivity scale used and the area of the photodetector.

Polarization insensitivity. This parameter represents how much reflection loss results when light signals of different polarization states fall onto the diffraction grating.

23.6 Linewidth Measurements

In high-speed fiber-optic communications systems, and to minimize the transmission penalties resulting from the dispersion in long fiber links, high-quality lasers are needed. These lasers should operate in a single longitudinal mode (i.e., single-frequency oscillation) with minimal dynamic linewidth broadening (i.e., frequency chirp) under modulation. In coherent and wavelength division multiplexing communications systems, the lasing linewidth becomes an important determinant of system performance. It is therefore important to be able to measure the linewidth of single-frequency lasers accurately.

A simple block diagram capable of measuring the linewidth of a wide variety of laser sources is shown in Fig. 23.16 and is called a delayed self-

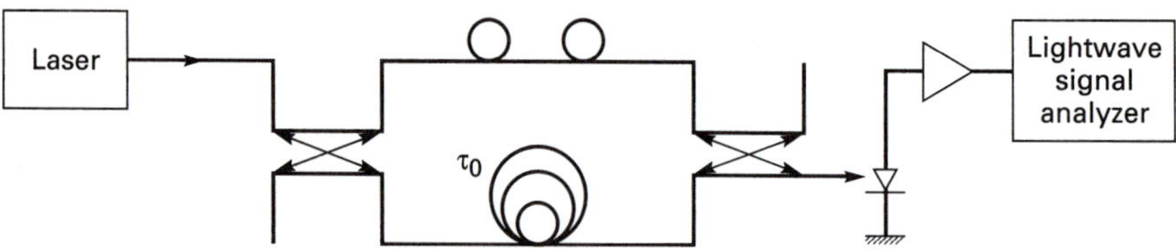

Figure 23.16 Delayed self-homodyne linewidth measurement system.

homodyne measurement system. A frequency discriminator is used to convert the optical phase (or frequency) deviation of the signal under test into intensity variation, which can be detected using the high-speed lightwave signal analyzer. The frequency discriminator is essentially an unbalanced fiber-optic interferometer with a directional coupler at the input. The directional coupler splits the incoming optical signal into two equal parts. The two signals then travel along separate fiber paths where they experience a differential delay. The two signals are then recombined using another directional coupler at the output of the interferometer. Since the fiber-optic interferometer does not preserve the polarization state, a polarization controller is added to one arm of the interferometer. By adjusting the polarization controller, it is possible to maximize the interference between the two signals at the output of the interferometer.

The minimum linewidth which can be measured using the circuit shown in Fig. 23.16 is dependent on the amount of delay in the lower arm of the discriminator. For a 730-m-long fiber (corresponding to about 3.5 μs of delay), the minimum linewidth will be about 225 kHz, far below typical linewidths of semiconductor lasers.

To measure the linewidth of a single-frequency laser, such as a DFB laser, an isolator is used to couple the laser to the discriminator. The isolator is used to reduce the perturbations of the laser by the reflections from the fiber or the optical interfaces within the interferometer. After passing through the isolator, the signal is then applied to the discriminator. The two signals arriving at the output of the interferometer (inside the discriminator) are uncorrelated and are recombined by the directional coupler. Therefore, the output signal is the sum of two signals with equal linewidth and center frequency. It is possible to prove that this output signal is the autocorrelation function of the two signals (in the two arms of the interferometer) and for gaussian line shapes, this autocorrelation function has a linewidth equal to $\sqrt{2}$ times that of the original signal. By measuring the linewidth of the output signal (on the screen of the lightwave signal analyzer) and dividing it by $\sqrt{2}$, the linewidth of the signal under test can be determined.

23.7 Lightwave Polarization Analyzers

Polarization is a fundamental property of light that dramatically affects the performance of many optical components. An optical isolator, for example, would be impossible to design and manufacture without paying attention to the various polarization conditions within the isolator assembly. The same is true for other polarization-sensitive optical components such as polarization-maintaining fiber, lasers, beam splitters, modulators, interferometers, retardation plates, and polarizers and polarization adjusters. It is important to note, however, that polarization effects may be undesired or desired. For example, simple movement of fiber-optic cables often changes optical power.

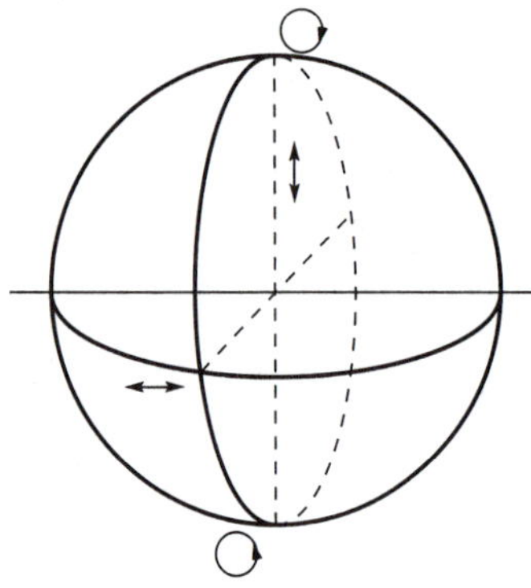

Figure 23.17 Poincaré sphere.

Or, for systems which require interference, polarization is a key parameter to control.

The electric field of a lightwave signal can be resolved into two components: horizontal and vertical. In polarized light, the field components combine according to their relative magnitude and phase and create characteristic polarization states that can be grouped into linear, circular, and elliptical states of polarization. The state of polarization can be graphically described as a point on a sphere, called a Poincaré sphere as shown in Fig. 23.17. On a Poincaré sphere, all linear states exist on the equator of the sphere; right-hand circular is on top, left-hand circular is on bottom, and all other locations are elliptical. A coordinate on a Poincaré sphere is a vector, called a Stokes vector, which has four elements called the Stokes parameters. These parameters are a numerical method of describing the state of polarization (SOP). The degree of polarization (DOP) is the ratio of polarized light to the total power. Many polarization-dependent properties are affected by the DOP.

23.7.1 Block diagram

A block diagram of a lightwave system capable of measuring the SOP and DOP of a lightwave signal is shown in Fig. 23.18. The incoming light is divided into four paths and each path is processed separately. They are each passed through polarizing and retarding elements, and their respective intensities are measured. These signals are processed by a computer to arrive at the average optical power and the horizontal, 45°, and circular polarization components of the incoming signal, as defined by the Stokes parameters. From these four components, the SOP of the incoming signal is computed.

23.7.2 Polarization-dependent loss, PDL

Two of the most important measurements which can be done using a polarization meter are polarization-dependent loss (PDL) and polarization-mode dispersion (PMD). Many optical components such as couplers, isolators, fiber

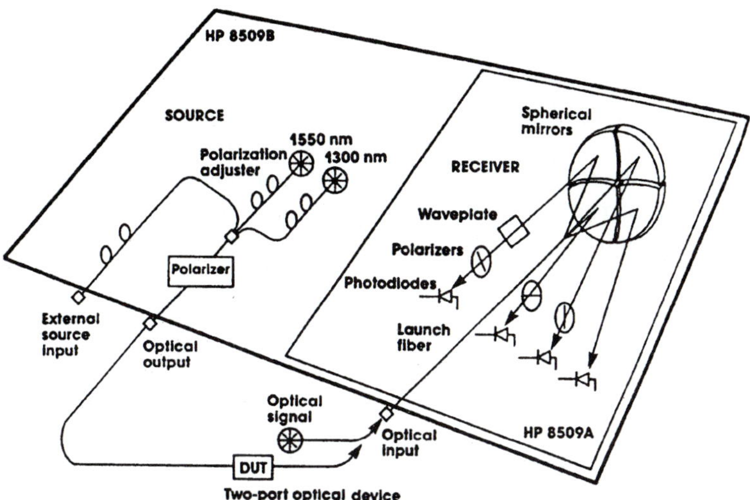

Figure 23.18 A block diagram for a polarization measurement system.

amplifiers, and connectors will have loss or gain, which is a function of the state of polarization incident to them. Their PDL, sometimes called polarization sensitivity, is a parameter which must be measured before such a component can be utilized in an actual system. By varying the state of polarization of an input signal, the polarization meter will then measure the output power from the device under test as a function of the input power and input state of polarization. The input state of polarization should be varied in such a way to cover most of the Poincaré sphere to make sure that worst-case performance is determined.

23.7.3 Polarization-mode dispersion, PMD

In very long, high-speed fiber-optic transmission systems, PMD can be the limiting factor because light traveling through a device in different polarization mode experiences different propagation delays. In any component exhibiting PMD, two polarization states exist that represent the fastest and slow-

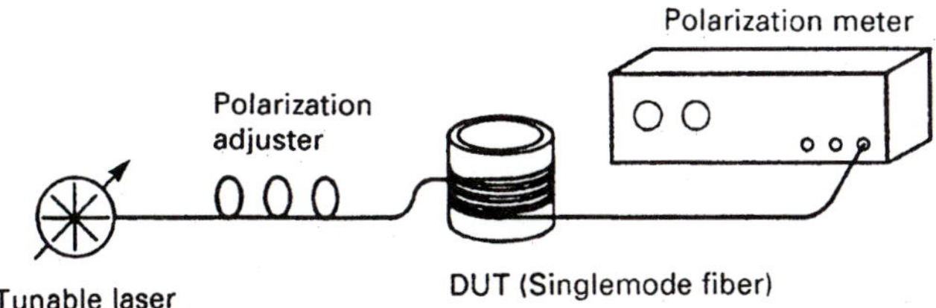

Figure 23.19 Polarization-mode dispersion measurement block diagram.

est propagation. By determining those two states, the PMD value can be calculated. One way of measuring PMD is to change the wavelength of the source incident on the device under test and monitor the output state of polarization. Because of the delay difference due to PMD, as the wavelength of an optical signal is varied, a fixed input state of polarization is transformed into a continuous range of output polarizations. These output states are displayed as an arc on the Poincaré sphere. The delay difference is directly proportional to the measured arc angle as shown in Fig. 23.19.

Chapter

24

Lightwave Component Analyzers

Waguih Ishak

Agilent Technologies
Palo Alto, California

24.1 Introduction

As the technology of fiber-optic communication systems advances to meet the demand for higher data rates, the trend toward lowering the effective cost per bit per mile will continue. The component designer will strive for lower cost, and better design and testing tools will be crucial to achieving this goal. Therefore, optical component analyzers play a significant role in the testing and debugging of the key components utilized in communications systems.

There are three basic classes of component analyzers: First, modulation domain component analyzers are used to measure the response of optical components as a function of the frequency of modulation of the optical carrier. Second, wavelength domain component analyzers are used to measure the parameters of optical components as a function of the optical wavelength. Third, reflection domain analyzers are used to measure the amplitudes and locations of the reflections from various parts within the optical component under test. The three types complement each other, and in order to do a complete component analysis of all components within a communications system, the three types are needed by the design, manufacturing, and test engineers.

In modulation domain component analyzers, a light signal (from a laser source) is modulated at a rate controlled by the instrument and then applied to the device under test. The output is then received by the analyzer, demodulated, and processed to compute the scattering parameters (S parameters). The modulation frequency is varied, the S parameters are remeasured, and this process continues over the desired range of modulation frequency. The results (S parameters), as a function of the modulation frequency, are then displayed on the screen. This technique is similar to electrical network analyzers except that in optical analyzers the carrier frequency is more than four orders of magnitude higher than the electrical carrier frequency.

In wavelength domain analyzers, a wavelength-tunable light signal (from a laser source) is applied to the device under test and the output power is received by the instrument. The insertion loss of the device is then calculated as the difference between the input and the output power levels. The process is repeated as the wavelength of the optical signal is tuned, and the insertion loss is plotted as a function of the wavelength.

In reflection domain analyzers, an optical signal (from a low-coherence source such as an LED) is first split into two beams (through an interferometer). One beam goes to the device under test and the second beam is applied to a reflecting surface (such as a mirror) which is translated along the beam direction. The reflections from the moving mirror add coherently with any reflections from the device under test and show as a peak on the display (as a function of the position of the mirror).

24.2 Modulation Domain Component Analyzers

In ultrahigh-speed communications systems, the modulated signals contain spectral components reaching well into the microwave region. In digital systems, where the information source is an encoded bit stream of high and low states, the information-carrying signal is usually a pulse sequence. The required bandwidth to transmit this sequence is derived from the pulse width of the signal, or conversely, the system's bandwidth defines the pulse shape, rise time, and overshoot of the applied time domain signal. With multigigabit systems, the signal energy is no longer propagated by electrons in conductors but rather in electromagnetic fields along transmission lines. Signal scattering caused either by mismatches (reflections) or by multipath transmission significantly degrades system performance.

Modulation domain component analyzers (also called high-speed component analyzers) are very effective in measuring high-speed devices in terms of their modulation frequency response (bandwidth), microwave impedance, optical return loss, propagation time or equivalent electrical length, and microwave and optical interactions with other high-speed devices. Typical devices are shown in Table 24.1.

The high-speed lightwave component analyzer measurement concept is shown in Fig. 24.1. The analyzer measures a modulation transfer function of a device under test, providing the amplitude and phase characteristics. The information source is a sine-wave signal incident to the transmitter. The sine-wave modulation signal is placed on the light carrier to create an amplitude-modulated light signal, which is transmitted through an optical device such as an optical fiber. The receiver demodulates the modulated signal after it has been operated upon by the device under test. The signal processing unit forms the complex ratio of the device's input and output (i.e., the modulation

TABLE 24.1 Types of Lightwave Devices

	Electrical (rf) output	Optical (modulated) output
Electrical (rf) input	Electrical-to-electrical devices Amplifiers Coaxial cables Passive components Repeater links	Electrical-to-optical devices Laser diodes and LEDs Optical sources Optical modulators
Optical (modulated) input	Optical-to-electrical devices PIN Photodiodes Avalanche photodiodes Optical receivers	Optical-to-optical devices Optical fibers Passive components Optical modulators Regenerators

transfer function amplitude and phase) and applies the appropriate corrections for the device being measured.

24.2.1 Modes of operation

The analyzer can be used in one of four different modes: (1) The "O/O" mode refers to optical-to-optical mode in which the input and output signals to and from the device under test are both optical signals. This mode is used to test optical fiber, modulators, attenuators, and isolators. (2) In the "O/E" mode, which is the optical-to-electrical mode, the input signal is optical and the output signal is electrical. This mode is useful in testing O/E converters such as photodetectors. (3) The "E/O" mode, electrical-to-optical, is used to test lasers and optical sources in which the input signal is electrical and the output is optical. (4) As for the "E/E" mode, the input and output signal are both electric and the analyzer in that mode is simply a conventional electric component analyzer.

A useful feature for lightwave component analyzers is the ability to com-

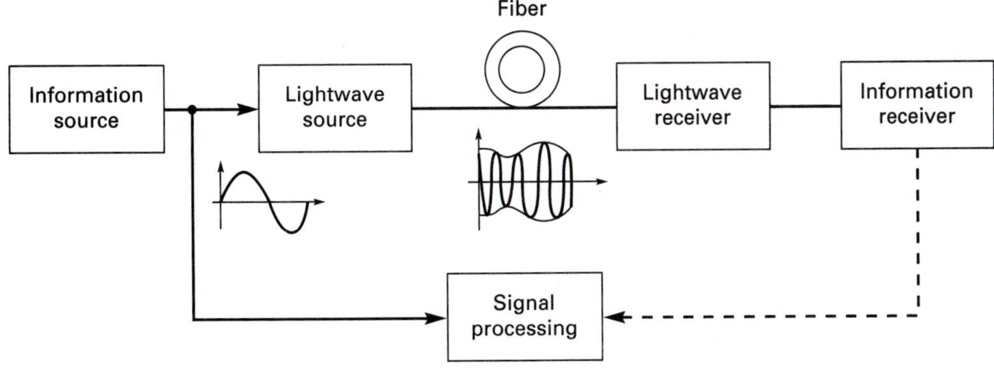

Figure 24.1 Lightwave component analyzer schematic.

TABLE 24.2 Lightwave Component Analysis Measurements

Transmission	Reflection
Insertion loss or gain	Return loss
Frequency response Modulation bandwidth Flatness Slope responsivity	Impedance (electrical and rf)
Time domain analysis Rise time Pulse dispersion	Reflectometry Electrical time domain Optical frequency domain
Delay	Delay
Length	Length
Insertion modulation phase	
Group delay	
Dielectric constant, refractive index	
Reflection sensitivity (for electrical-to-optical devices)	

pute time domain information. From the measured frequency domain data, a suitable Fourier transform (usually FFT) can be applied to produce the time domain data. The algorithms needed to perform the FFT are stored in the analyzer electronic memory.

24.2.2 Basic transmission and reflection measurements

Table 24.2 lists the basic transmission and reflection measurements most useful for the design, evaluation, and characterization of high-frequency components.

As shown in Fig. 24.1, the heart of the component analyzer consists of a modulated lightwave source and an optical receiver. The lightwave source produces light that is intensity modulated at microwave frequencies. There are several optical modulation techniques which can be used inside the optical block of the component analyzer. Modulation can be achieved by direct current injection modulation of the laser diode, by external optical modulators, or by heterodyning methods. Direct laser modulation is a simple technique for low (<5 GHz) rates. While direct modulation of laser diodes up to 20 GHz has been achieved in research labs, the approach suffers from signal-induced degradation of the optical spectrum. Similarly, optical heterodyne techniques, in which two highly stable single-line lasers, slightly offset in frequency, are combined to produce a beat signal, can achieve the desired modulation bandwidth. Typically, the lasers, which are temperature or electrically tuned, require precise control of their parameters and long measurement times. In addition, modulation phase is not well defined.

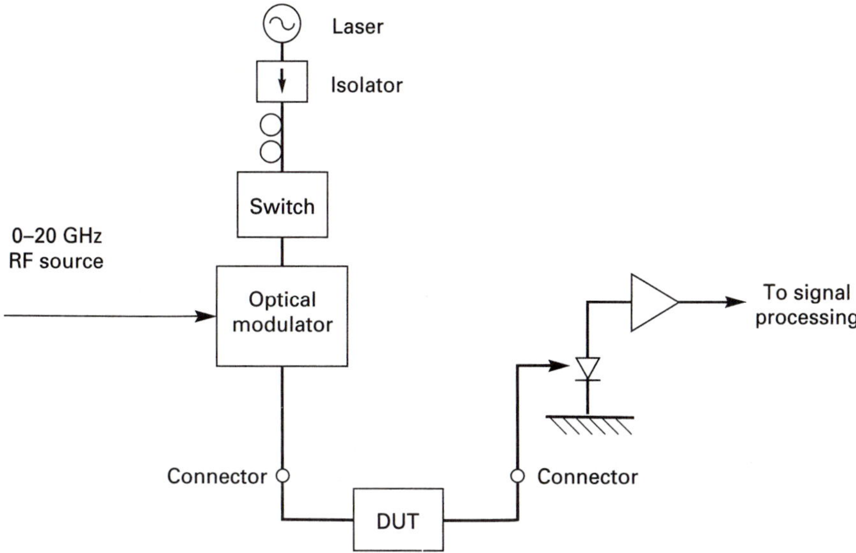

Figure 24.2 High-speed lightwave component analyzer block diagram.

24.2.3 External modulation

As for external modulation, the light output of a high-power semiconductor laser diode is intensity modulated by a Mach-Zehnder modulator (such as a lithium niobate traveling-wave, electro-optic modulator). This allows the use of a variety of laser sources, such as Fabry-Perot, DFB, or external cavity lasers, with the same modulator. The drawbacks to this technique over direct modulation are the added components, additional optical loss, and the need for polarization control.

Figure 24.2 shows the optical circuitry of a high-speed lightwave component analyzer using an external modulator. The source of light is an unmodulated (CW) semiconductor laser diode. The light is directed through an optical isolator to protect the laser from the destabilizing influence of reflected light. Next, the light is passed through a polarization controller, which is rotated to create the proper state of polarization required at the input of the optical modulator. The modulator usually operates at maximum efficiency when its light input is linearly polarized and the E-field is in the plane of the modulator circuit. After exiting the modulator, the laser light is routed out of the front panel of the instrument through an optical connector. After passing through the device under test, the optical signal enters the analyzer through another connector and then goes to the receiver circuit. The receiver consists of a sensitive photodetector followed by an amplifier and other electronic circuits.

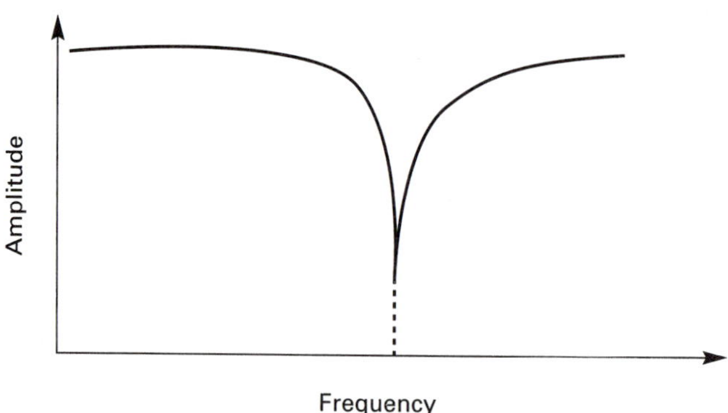

Figure 24.3 Chromatic dispersion.

24.2.4 Measurement examples

We will discuss three examples for important, unusual lightwave measurements for which the lightwave component analyzer is used.

Chromatic dispersion fiber measurement. Chromatic dispersion in standard single-mode optical fiber crosses through the zero dispersion point in the 1300-nm band and reaches values of approximately 16 ps/km-nm at 1550 nm. This dispersion causes undesired effects for 1550-nm fiber circuits. Consider an intensity-modulated signal with upper and lower sidebands equally spaced from the carrier. As the signal propagates along the fiber, each component travels at a slightly different propagation velocity, resulting in time delays from one component to another. A situation can exist in which the AM component vanishes and nulls occur in the baseband AM response of the fiber. By connecting a fiber (or a fiber circuit) as a device under test for the lightwave component analyzer and using a 1550-nm modulated laser source, the transmission characteristic of the device can be displayed on the screen if the modulation frequency is swept. From the frequency of the null, the chromatic dispersion can be calculated using closed form expressions. An example of the transmission characteristic of a single-mode fiber at 1550 nm is shown in Fig. 24.3.

Laser FM response measurement. For high-speed FSK modulated systems, there is interest in measuring the FM response of DFB lasers as a function of bias current. An optical frequency discriminator is usually used to convert the FM signal into an AM signal, which is then analyzed using the analyzer detection circuitry. In this measurement, the DFB is the device under test and the analyzer is used in the "E/O" mode. An example of the AM and FM response of a DFB, as measured by the analyzer, is shown in Fig. 24.4.

Receiver optical launch measurement. The launch circuit of an optical receiver is designed for low overall return loss. In addition, it is useful for the designer

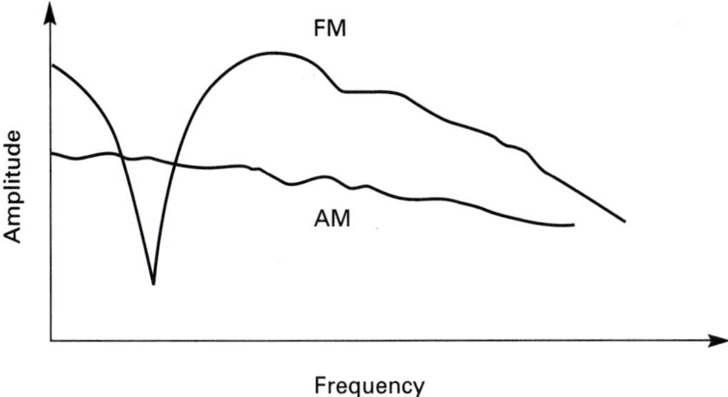

Figure 24.4 AM and FM laser response.

to find out the location and the return loss values for each individual reflection inside the components used to build the launch. By using the time domain capability of a lightwave component analyzer, selective reflection measurements are easily made on a wide variety of optical components. An optical mismatch can be measured by separating the incident test signal from the reflected one by means of a directional coupler. The ratio of the reflected to the incident signals yields the reflection coefficient magnitude at the optical frequency and can be measured by an optical power meter. However, using the lightwave component analyzer technique and applying a modulated test signal has the advantage that the origin of the reflecting signal can be determined by computing the propagation delay and distance from the modulated phase shift of the reflected signal. Furthermore, it is possible to apply the same technique to multiple reflections, thus separating them in the distance or time domain.

Figure 24.5 shows a selective reflection measurement of a photodiode re-

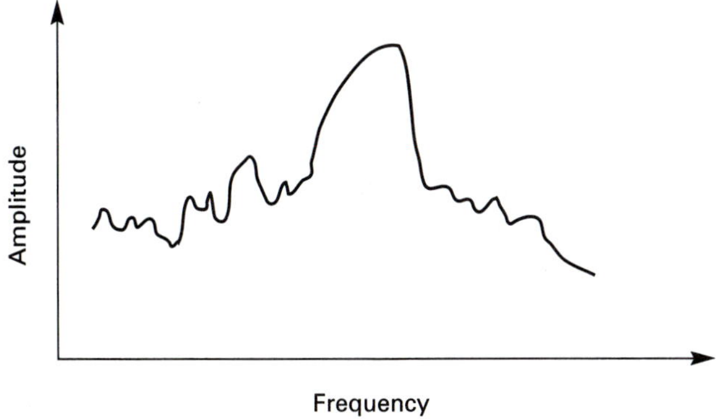

Figure 24.5 Optical reflection measurements of a lightwave receiver.

ceiver optical launch using a GRIN lens. The lower peak on the left is due to the interface between the fiber and the first surface of the GRIN lens. The higher peak to the right is due to a combination of the reflection from the second surface of the GRIN lens and the photodiode surface. The two peaks are separated by 14 mm, which corresponds to the length of the GRIN lens.

24.2.5 Specifications of high-speed component analyzers

Frequency range. This parameter defines the range of modulation frequencies of the instrument and is limited by the bandwidth of the external optical modulator. It can be as high as 40 GHz.

Frequency resolution. This parameter defines the resolution by which the modulation frequency can be stepped within the frequency range. It is limited by the rf circuitry which drives the external modulator and can be as low as 1 Hz.

Output power. This parameter defines the output power level from the optical port of the analyzer. It is limited by the amount of power available from the semiconductor laser source and the losses of the optical components (isolator, modulator, etc.). It can be as high as +5 dBm.

Dynamic range. This parameter is defined as the difference (in dB) between the highest output power from the analyzer (to the device under test) and the received optical power (after passing through the device under test). It simply defines the highest insertion loss that can be measured by the analyzer and is limited by noise considerations. This parameter can be as high as 50 dB.

24.3 Wavelength Domain Component Analyzers

An important measurement for optical components is to determine the dependence of various parameters (such as insertion loss, return loss, and group delay) on wavelength. This is particularly important for components used in wideband and wavelength division multiplexing communications systems. An example of such components is the erbium-doped fiber amplifier (EDFA). This device can amplify optical signals over a wide range of about 40 nm centered around 1550 nm. Because of the excellent performance of EDFAs (high gain, low polarization effects, etc.), new installations for fiber links will use direct optical amplification, which will eliminate the need for electronic repeaters and therefore can be easily upgraded to higher transmission speeds as the need arises.

The development of the EDFA technology resulted in a strong need for component analyzers to characterize the amplifier and the components used to build the amplifier (such as couplers, filters, isolators, and active fibers) as a function of wavelength. Two block diagrams depicting such component

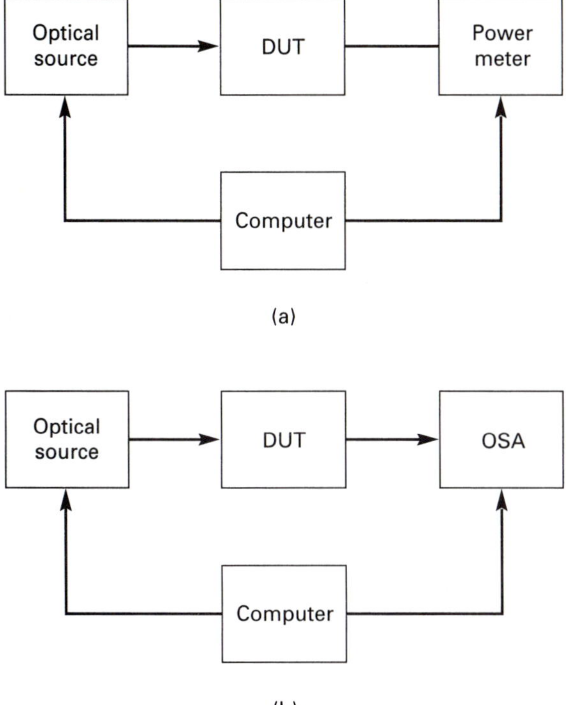

Figure 24.6 Wavelength domain lightwave component analyzer block diagram: (*a*) using a power meter; (*b*) using an optical spectrum analyzer.

analyzers are shown in Fig. 24.6*a* and *b*. The simplest system uses a tunable laser source (such as the external cavity laser described in Chap. 22) and a power meter (such as the one described in Chap. 23). As shown in Fig. 24.6*a*, the component under test is connected between the laser source and the power meter. An optical signal from the source is applied to the input of the component under test and the output from the component is applied to the input of the power meter. The wavelength of the input signal is varied over the tuning range of interest and the output power is measured at each wavelength. A computer can be used to record the output power versus the input signal power level and wavelength. From these data, the insertion loss of the component under test can be plotted as a function of the wavelength. Such a system is simple to implement and use but is limited to one measurement, insertion loss versus wavelength. In addition, because the power meter does not have a front-end optical filter, the noise level of the measured power will be relatively high and the dynamic range of the component analyzer (shown in Fig. 24.6*a*) will be low. This disadvantage severely limits the use of this block diagram as a general optical component analyzer.

If the power meter in Fig. 24.6*a* is replaced by an optical spectrum analyzer, the system dynamic range will be improved considerably because of the optical filter built into the OSA. The controller shown in Fig. 24.6*b* will

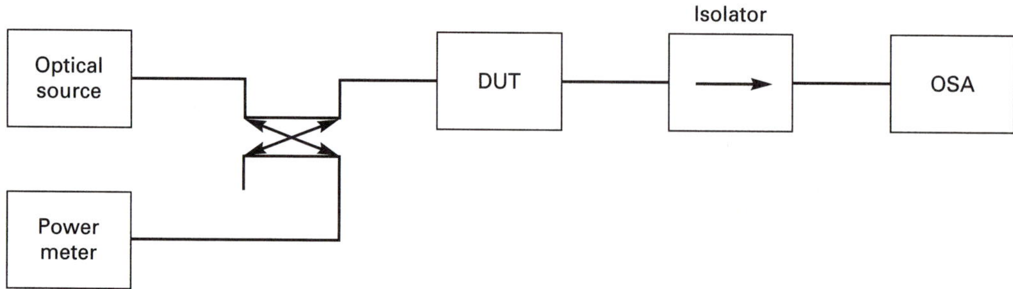

Figure 24.7 Lightwave component analyzer block diagram.

synchronize the laser source and the OSA so that the OSA is tuned to the wavelength of the laser source signal. Therefore, the noise associated with the measurement will be at a very low level and components with high insertion losses can be easily measured using this technique.

A practical component analyzer will include additional optical elements to improve the performance of the systems shown in Fig. 24.6*a* and *b*. As shown in Fig. 24.7, an optical coupler is used to accurately measure the power level of the input signal (the output of the optical source) and an optical isolator is used at the input of the optical spectrum analyzer to protect the optical source from reflections.

24.4 Reflection-Based Component Analyzers

24.4.1 Techniques

A number of techniques exist for measuring reflections. Optical time domain reflectometry (OTDR), discussed in detail in Chap. 25, is commercially the most successful and is suitable for finding faults in long lengths (meters or kilometers) of fibers. A second reflectometry technique is optical frequency domain reflectometry (OFDR), as described earlier. This technique is suitable for testing reflections within devices but limited in resolution by the maximum modulation frequency of the instrument. For example, a 20-GHz high-speed lightwave component analyzer will have a resolution limit of about 5 mm. A third technique, based on using a power meter, is to measure the continuous-wave (CW) reflected light from the device under test. While strictly not a reflectometry technique because it does spatially resolve the reflections, it has been widely used to measure the ratio of total reflected power to total incident power. The technique of white-light interferometry, described in this section, offers greatly improved performance over the above-mentioned techniques in terms of sensitivity and resolution.

24.4.2 White-light measurement techniques

The white-light measurement technique uses optical interference and is based on the use of a Michelson interferometer, which has been in existence

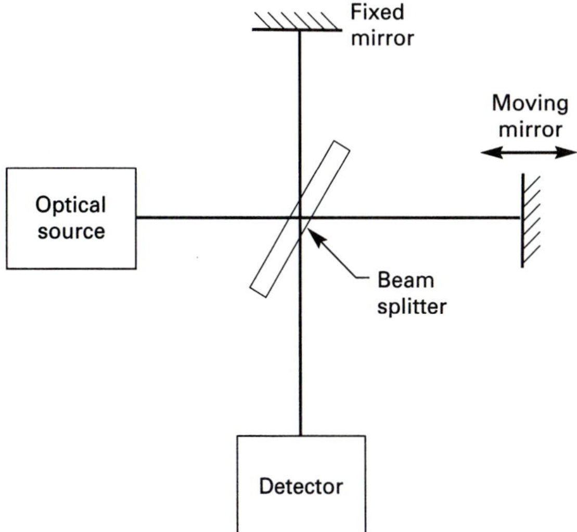

Figure 24.8 Michelson interferometer.

for over 100 years. A simple block diagram for an open-beam Michelson interferometer is shown in Fig. 24.8. One of the uses of the Michelson interferometer is the measurement of the coherence length of various optical sources. This concept can be generalized to build a high-resolution, high-sensitivity component analyzer capable of measuring reflections, with micron resolution, within optical devices and subsystems. This analyzer will be referred to as a low-coherence reflectometer.

A block diagram of a fiber-optic, low-coherence reflectometer is shown in Fig. 24.9. The use of fiber optics offers advantages in the manufacturing of the instrument and is compatible with the fiber-optic pigtailed devices it is designed to measure. An LED (or another low-coherence source) is coupled

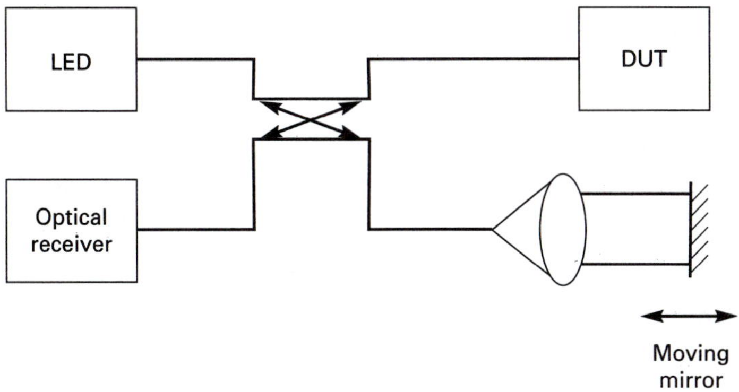

Figure 24.9 Low-coherence reflectometer.

into the input of a fiber-optic interferometer and is divided equally between the two arms of the interferometer by a 3-dB coupler. The device under test is connected to one arm of the interferometer, and a moving mirror is connected to the other arm. The reflected powers from each arm are then added together using the 3-dB coupler and sent to a detector which generates a photocurrent that is proportional to the square of the incident optical field. When the path-length difference between the moving mirror and the device under test reflection is longer than the coherence length of the source, the returning optical phases are uncorrelated and the output from the photodetector will be zero. For path-length differences within the coherence length of the source, the input to the photodetector is a beat frequency equal to the Doppler frequency shift of the moving mirror. The photodetector will respond to this Doppler shift, and its output allows the determination of the magnitude and the location (with respect to the movable mirror) of any reflection sites along the test arm of the interferometer. Because LEDs have coherence length in the order of tens of microns, this method allows the detection of reflections within the device under test at a very high resolution (tens of microns or even less).

24.4.3 Practical considerations

For a practical design of an instrument based on the interferometric technique mentioned above, several physical phenomena must be considered:

Signal-to-noise ratio. Reflection sensitivity is determined by the amount of noise that passes onto the photodetector. This noise comes from three sources: thermal receiver noise, optical shot noise, and optical intensity noise. In most cases in which a reasonable source power can be obtained, the shot noise or intensity noise will dominate the receiver noise. As the source power increases, the shot noise (which grows linearly with the optical power) will be dominated by the intensity noise (which grows quadratically with the optical power). When this occurs, reflection sensitivity is not improved as the source power is increased. Modifications to the basic Michelson interferometer configuration (for example, by using reference power attenuation) are possible to overcome this difficulty and overcome the effect of intensity noise.

Polarization dependence. The magnitude of the optical reflectivity is obtained from the strength of the interferometric beat signal which depends on the relative polarization states of the signals reflecting from both arms of the interferometer. If the two signals are in the same polarization state, the interference produces a strong signal. On the other hand, if they are orthogonally polarized, the signal becomes zero. Therefore, the effect of polarization must be considered.

To remove the dependence of the detected signal on the polarization state of the test arm reflection, a polarization diversity receiver can be used as

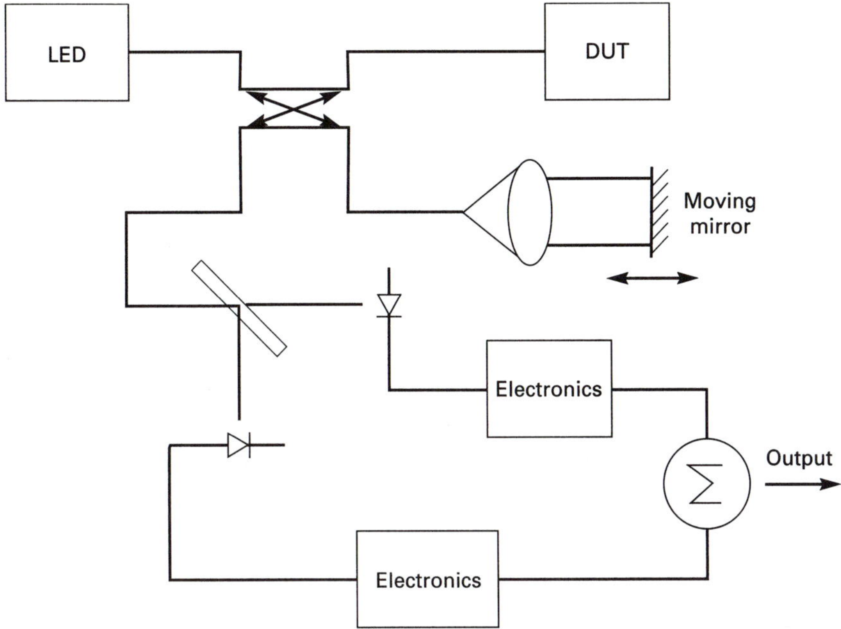

Figure 24.10 Polarization-diversity receiver-based reflectometer.

shown in Fig. 24.10. A polarizing beam splitter is used to split the incoming light signal into two orthogonal polarization components and direct them to two photodetectors. This technique is very powerful and is usually used in communications systems.

Dispersion effects. Since the velocity at which an optical pulse travels along a fiber depends on its wavelength, different colors contained in a light pulse propagate at different speeds causing chromatic dispersion which produces broadening of the optical pulse. This is particularly significant in the case of the LED-based reflectometer because of the broadband source spectrum needed to achieve high resolution. A mismatch between the chromatic dispersion properties of the two arms of the interferometer will result in the broadening of the beat signal and a reduction in its peak magnitude.

To overcome this problem, a good knowledge of the dispersion properties of the components used in both arms of the interferometer is essential. When these properties are understood, they can be modeled and stored in the instrument memory to cancel this effect as the measurement takes place.

An example of the capability of white-light interferometers as a powerful tool to characterize optical components is shown in Fig. 24.11. A packaged laser containing a single-mode optical fiber, a window, a lens, and the laser chip mounted on a heat sink is used as the device under test. The display shows seven strong reflections from the seven surfaces indicated inside the

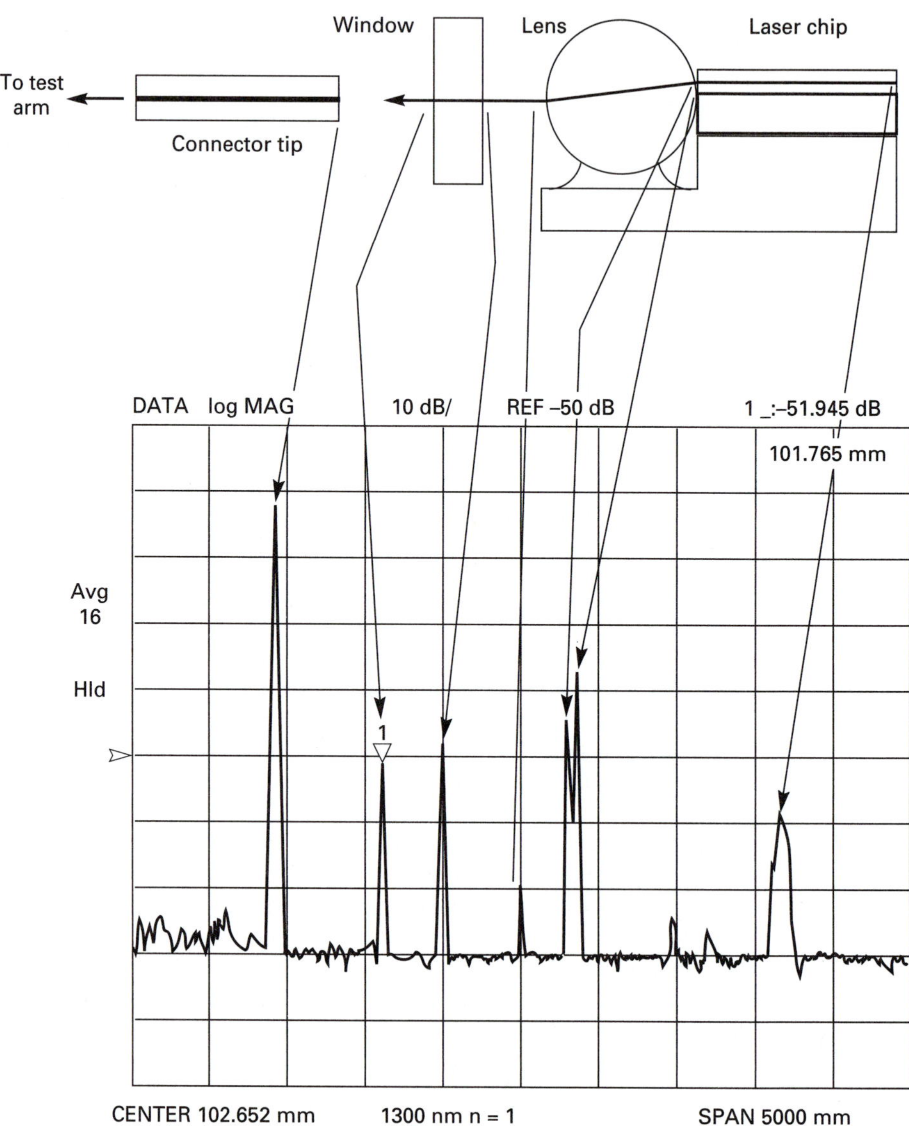

Figure 24.11 Measuring reflections from a laser module using low-coherence reflectometer.

laser package. Measurements of this type are nondestructive and valuable to both the component designer who wants to optimize each element of the component and the process engineer who wants to monitor the components' quality and assembly process consistency.

24.4.4 Specifications of reflection-based component analyzers

Dynamic range. This parameter is also called return loss range. It is limited by the intensity noise of the instruments. Typical value: 70 dB.

Amplitude accuracy. This parameter is also called return loss accuracy. It is defined as the accuracy in the amplitude measurements of the reflections in dB and is limited by the detection circuit elements. Typical value: ±2 dB.

Spatial resolution. This parameter is defined as the smallest distance in air between two reflections which is resolvable by the instrument. It is also defined as the 3-dB bandwidth of a reflection peak. Typical value: 25 μm in air (equivalent to 17 μm in a glass fiber).

Spatial accuracy. This parameter is defined as the accuracy in the distance measurement and is limited by the resolution and accuracy of the moving mirror mechanical driving assembly (motor). Typical value: 10 to 100 μm.

Chapter

25

Optical Time Domain Reflectometers

Waguih Ishak
Agilent Technologies
Palo Alto, California

25.1 Introduction

The development of high-power, narrow-linewidth semiconductor lasers and the invention and development of erbium-doped fiber amplifiers have impacted fiber-optic communications links in two important areas. First, new network configurations have emerged such as wavelength division multiplexing schemes. Second, long fiber links between repeaters are becoming more and more common. For example, the systems installed under the Mediterranean will use a 150-km repeaterless link. As these systems have become available, the demand on testing and monitoring the network has increased and the need for higher sensitivity, resolution, and SNR test equipment has also increased.

The optical time domain reflectometer (OTDR) is the most commonly used instrument to test a fiber-optic link. An OTDR is an instrument that characterizes optical fiber by launching a probe signal into the fiber under test and detecting, averaging, and displaying the return signal. The distance to a given feature (such as a splice, a bent, or a kink) is determined by simply timing the arrival of that part of the return signal at the detector.

25.2 OTDR Fundamentals

25.2.1 Block diagram

A block diagram of a generic OTDR is shown in Fig. 25.1. An electronic probe signal generator is used to modulate the intensity of a laser. In a conventional OTDR, the probe signal is a single square pulse. In field portable instruments, a semiconductor laser is used as the source. The output of the laser is

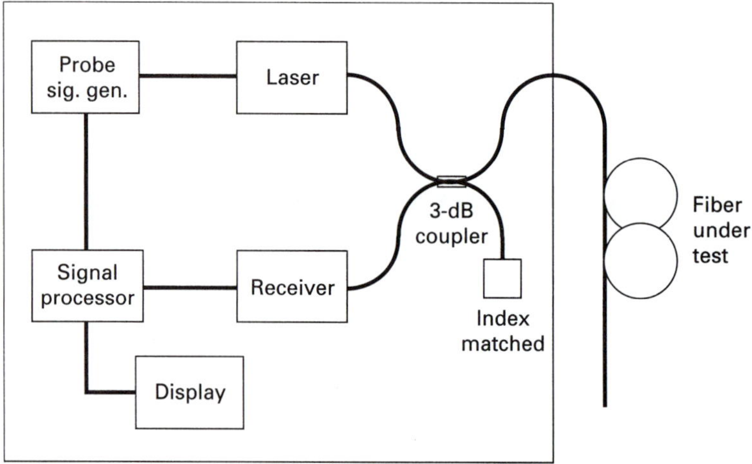

Figure 25.1 Optical time domain reflectometer block diagram.

coupled into the fiber under test using a beam splitter or, as shown in Fig. 25.1, a 3-dB fiber directional coupler. In the case of a fiber coupler, the unused output of the directional coupler must be terminated by a method such as refractive index matching in order to prevent the probe signal from being reflected directly back into the receiver. The return signal from the fiber under test is coupled to the receiver via the beam splitter or coupler. It is important that this beam splitter be polarization-independent; otherwise polarization variations in the backscattered signal will appear to be changes in the power. The amplified signal is then processed—usually by averaging—before being transferred to the display.

25.2.2 The backscattering impulse response

The main object of the OTDR measurement is to determine the backscattering impulse response of the fiber under test. This is strictly defined as the response of the fiber to a probe signal which is an ideal delta-function impulse, as detected by an ideal receiver at the same port into which the probe was injected. Since the physical processes that generate it are all linear, the backscattering impulse response is indeed a true impulse response, and the fiber can be treated as a linear system or network. In practice, however, the signal that is displayed is not the backscattering impulse response itself but rather a smoothed version of it. The smoothing is due to the effects of the "nonzero" duration of the probe pulse and the receiver response and will be described in more detail at the end of this section.

Figure 25.2 illustrates what a typical measurement might look like. The measured response is plotted on a logarithmic scale (in dB). Although the data are acquired as a function of time, they are displayed as a function of

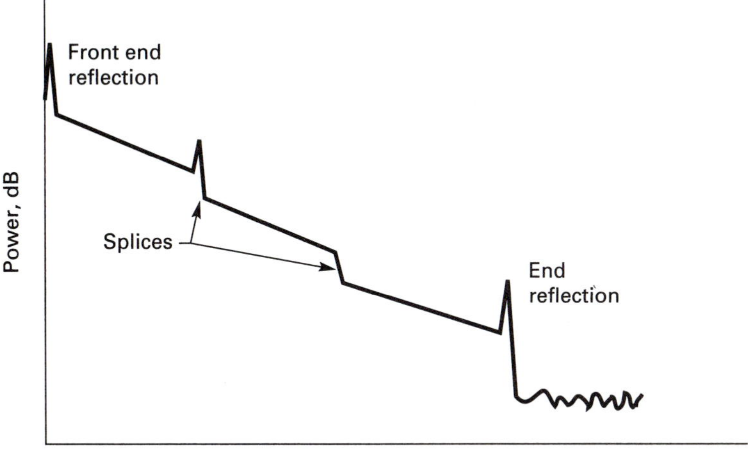

Figure 25.2 Typical display from a time domain reflectometer.

distance using a conversion factor which approximately equals 10 μs/km, the round-trip propagation delay of light in fiber. An ideal fiber without breaks, splices, or reflections would be displayed as a perfectly straight line, indicative of the exponential decay due to propagation loss. The measured response typically exhibits two kinds of features: positive spikes that are due to reflections, and relatively smooth exponentially decaying signals that are due to Rayleigh backscattering. These backscattering curves can exhibit discontinuities, which may be either positive or negative, depending on the physical causes of that particular feature. All of the features that comprise the backscattering impulse response (shown in Fig. 25.3) are derived from either of two physical effects: Fresnel reflections and Rayleigh scattering.

25.2.3 Fresnel reflections

Just as an electrical TDR measures mismatches and discontinuities in the electrical impedance, an optical TDR measures discontinuities and mismatches in the optical refractive index. Such a mismatch results in what is commonly called a Fresnel reflection. Fresnel reflections commonly occur at connectors, mechanical splices, and unmatched ends of a fiber cable. If the interface is between fiber and air, as it might be at a junction where an air gap exists or at the end of a "perfectly" broken fiber (a clean break perpendicular to the optical axis), the reflection should be approximately 3.4 percent (−14.7 dB) of the incident power.

In practice, however, reflections having just this magnitude are not often seen. For example, the measured reflections at glass-air interfaces can be much smaller if the fiber is dirty or not cleanly broken in the vicinity of the

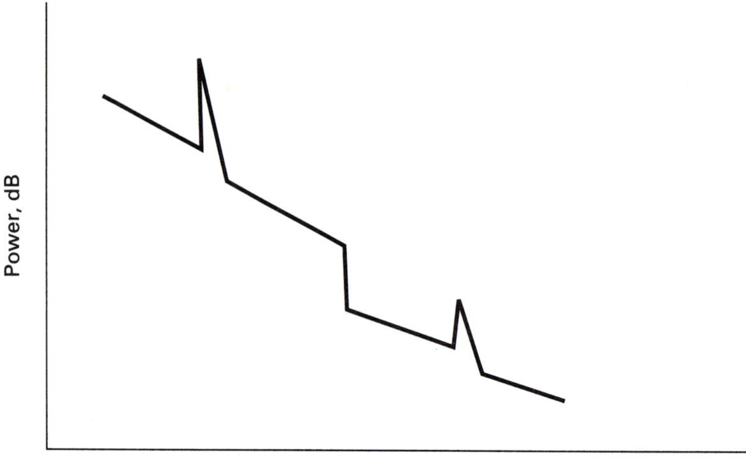

Figure 25.3 Reflection situation.

core. Even a clean, flat fiber end can exhibit low reflection if it is angled with respect to the optical axis. In these cases, the light is reflected or scattered at an angle that is too large for it to be captured in the fiber core and guided back to the input. For example, if the end of a single-mode fiber is cleaved at an angle of only 6° the reflected power that is guided back to the input is 69 dB down from the incident power.

On the other hand, conditions can exist that actually enhance Fresnel reflections or that make such reflections difficult to suppress. For example, a gap at a connector joint may form an etalon between the fiber faces which, if the spacing is right, can lead to a resonant enhancement of the reflection. Also, the polishing of a fiber end appears to cause its refractive index to increase (perhaps by more than 50 percent) in a thin layer near its surface. This can lead to an index discontinuity within the fiber itself, resulting in the presence of a reflection regardless of the condition of the end surface.

The ability of OTDRs to measure the magnitude of Fresnel reflections will be increasingly important in the future as more and more narrow linewidth laser sources are installed in coherent and/or wavelength multiplexed transmission systems. The reason is that these sources exhibit serious spectral instabilities when large reflections are present. Occasionally, other situations may arise wherein it is important to know the magnitude of Fresnel reflections. In these cases, an OTDR's ability to measure refections will prove to be valuable.

In most cases, however, the magnitudes of Fresnel reflections in the backscattering impulse response have little or no significance. This may seem surprising since reflection peaks appear to be the most dramatic features of the measured signal. However, the presence or absence of a reflection peak at a connector or splice does not by itself give an indication of the quality of the joint. That and other information are best determined by observing changes in the level and slope of the part of the response that is due to Rayleigh scattering.

25.2.4 Rayleigh scattering

Whereas electrical TDRs measure only discrete reflections, OTDRs also measure a low-level background of backscattered light, Rayleigh scattering, whose decay is ideally proportional to that of the probe signal that travels along the fiber. In principle this allows the propagation loss of a fiber segment to be measured by observing the slope of the backscattering impulse response. In practice, this is usually the case; however, some care must be taken in making this interpretation if a variation in backscattering parameters as a function of distance is suspected (i.e., if the fiber does not scatter uniformly along its length).

Rayleigh scattering is the dominant loss mechanism in single-mode fiber. It occurs because of small inhomogeneities in the local refractive index of the fiber, which reradiate the incident light in a dipolar distribution. In single-mode fiber, only a very small fraction of the scattered light falls within the small capture angle of the fiber and is guided back to the input.

One of the main challenges of an OTDR measurement is that the backscattered signal is very weak. The dependence of Rayleigh scattering on wavelength is ¼, and has motivated fiber system designers to use longer transmission wavelengths: typically 1.3 μm (the wavelength of minimum dispersion) or 1.55 μm (the wavelength of minimum loss). Unfortunately, at these wavelengths there is little scattering to be captured, guided in the backward direction, and detected by an OTDR. For example, the detectable backscattered power generated by a 1-s probe pulse is roughly 2000 times smaller (−33 dB) than what would be reflected by a 3.5 percent Fresnel reflection at the same location.

Not only must an OTDR be able to measure these small signal levels, it must also be able to detect small changes in them on the order of 0.05 dB (<1 percent). For example, connectors and splices in many fiber transmission systems are required to have insertion losses on the order of 0.1 dB. If the total allowable loss is exceeded, the faulty junction(s) must be identified. In an OTDR measurement, a lossy junction would appear as an abrupt lowering of the backscattering level. Such a discontinuity could be caused by a misalignment of the fibers at a splice or connector, where core offsets of as little as 1 μm can result in substantial losses. Another source of loss might be a bend in the fiber which is sharper than its allowable bend radius, causing some of the guided light to be radiated into the cladding and lost. In this case, the OTDR would also measure a discontinuity. Finally, in the case of a catastrophic fault, such as a break in the fiber, the backscattering impulse response would abruptly drop into the noise.

On rare occasions, when a fiber is joined to a lossier fiber, the measured backscattering level can actually increase abruptly, followed by a more rapid decay. This points out once again that the backscattering impulse response does not always precisely indicate what happens to a forward-traveling signal (such a signal would certainly not experience gain at the junction!). However, in the great majority of cases, and with proper interpretation, the backscattering impulse response measured by an OTDR can present an accurate and useful picture of the propagation conditions along a fiber-optic cable.

25.3 The Complementary Correlation OTDR

25.3.1 Spread-spectrum techniques

Spread-spectrum techniques such as correlation offer the possibility of providing measurements with improved signal-to-noise ratios, SNRs, without sacrificing response resolution. Such techniques are commonly used in radars and other peak-power limited systems where increases in the transmitted energy would otherwise result in degraded resolution.

One way of applying correlation to OTDR measurements is to correlate (*) the detected signal $s(t)$ with the probe signal $p(t)$:

$$s(t) * p(t) = [p(t) * r(t) * f(t)] * p(t) = [p(t) * p(t)] * [f(t) * r(t)] \tag{25.1}$$

where $r(t)$ = impulse response of receiver
$f(t)$ = backscattering impulse response of fiber

To the extent that the autocorrelation of the probe signal approximates a delta function, the fiber backscattering response $f(t)$ can be accurately recovered, subject (as always) to the response of the receiver.

$$[p(t) * p(t)] * [f(t) * r(t)] \sim d(t) * [f(t) * r(t)] = f(t) * r(t) \tag{25.2}$$

In this case, the duration of the autocorrelation of the probe signal determines the response resolution and not the duration of the probe signal itself, which may be long and energetic.

25.3.2 Complementary code OTDR, CCOTDR

The CCOTDR described here realizes the full advantage of correlation by probing and correlating with pairs of probe signals that have complementary autocorrelation properties. These probe signals are the complementary codes, which were first introduced by M. J. E. Golay in the late 1940s as a method of improving the performance of multislit spectrometers. Golay codes have

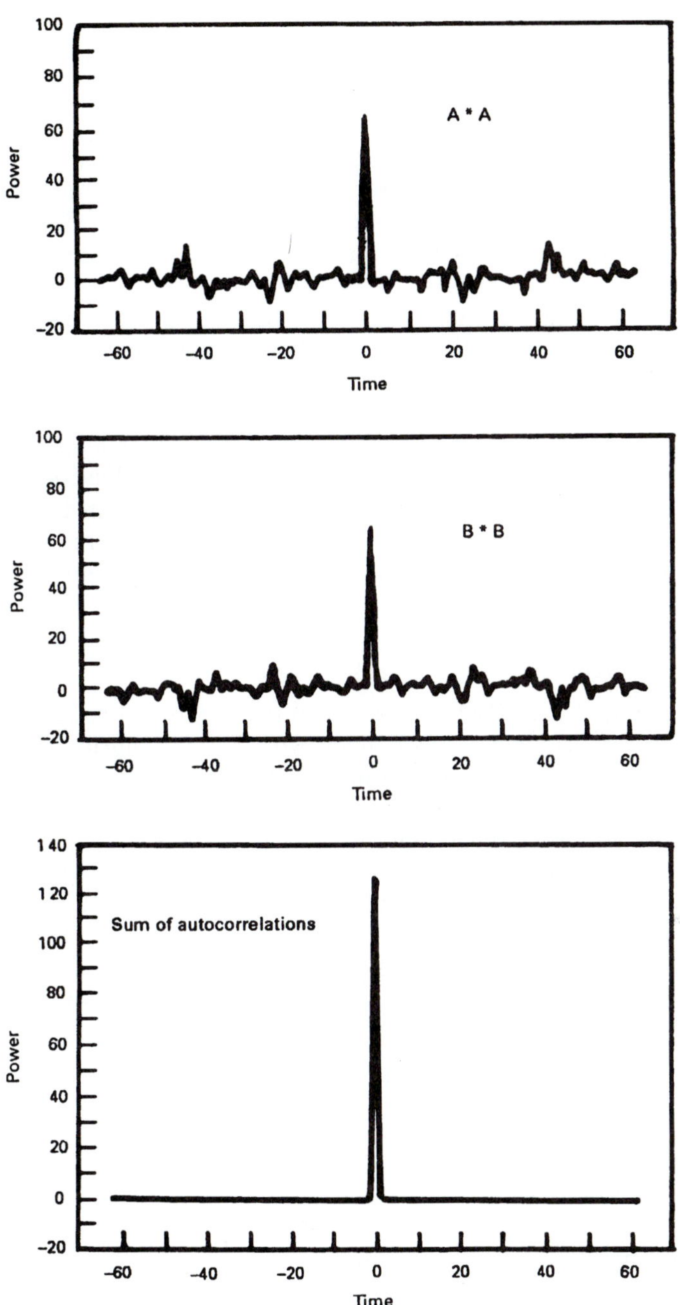

Figure 25.4 Golay code.

the following property:

If A and B are an L-bit complementary code pair, then

$$(A * A) + (B * B) = 2L\delta(t) \tag{25.3}$$

The unique autocorrelation properties of the Golay codes are shown graphically in Fig. 25.4. The upper two plots show the individual autocorrelations of each one of a 64-bit complementary code pair. Each of the individual codes consists of a pattern of 1's. The value of each of the autocorrelation peaks is equal to the number of bits in the individual code. Each of the individual autocorrelations also exhibits side lobes that are up to 10 percent of the peak height. However, when the autocorrelations are added together, the peaks add together to a value of $2L$ and the side lobes cancel exactly!

It is this contribution of all of the bits to the autocorrelation peak along with the complete cancellation of the side lobes that allows the correlation technique to work in practice. In designing a practical system of this kind, it is essential to work with an autocorrelation function that is perfect in principle, since finite side lobes will always exist in a real, nonideal system. Using complementary codes, the side lobes in a real system can be low enough so that the full advantage of correlation can be realized.

25.3.3 CCOTDR block diagram

A block diagram of a CCOTDR is shown in Fig. 25.5. A code generator is used to drive a semiconductor laser diode with the appropriate Golay codes and their 1's complements. The output power from the laser is coupled into the fiber under test through a directional coupler whose second output was index matched to suppress reflections. A portion of the return signal is coupled to the receiver via the same directional couple. The amplified signal is digitally sampled, averaged and processed to reconstruct the fiber backscattering impulse response.

Figure 25.6 summarizes the superior performance of a CCOTDR as compared with a conventional OTDR. The top curve shows a measurement made

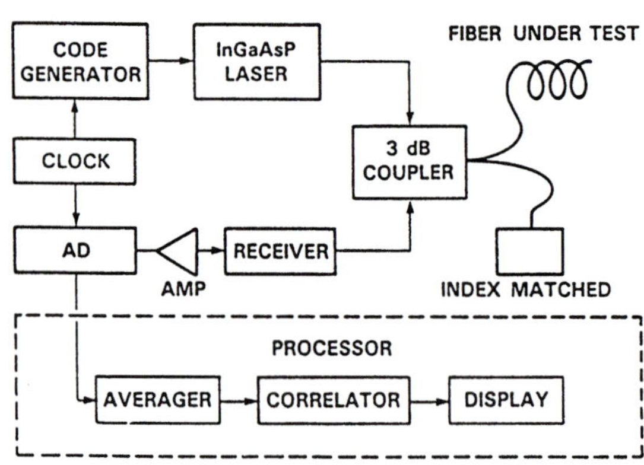

Figure 25.5 Correlation optical time domain reflectometer block diagram.

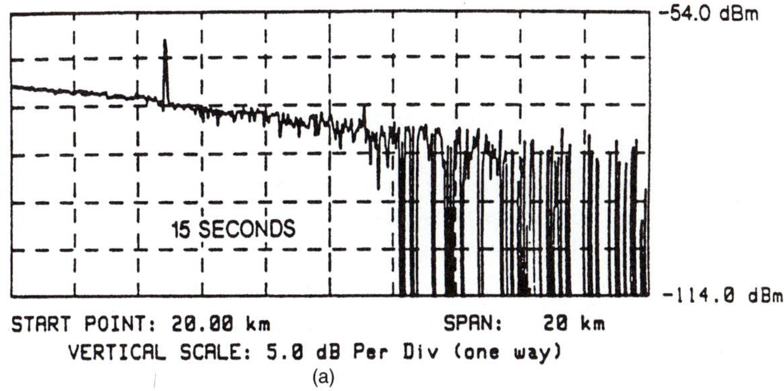

(a)

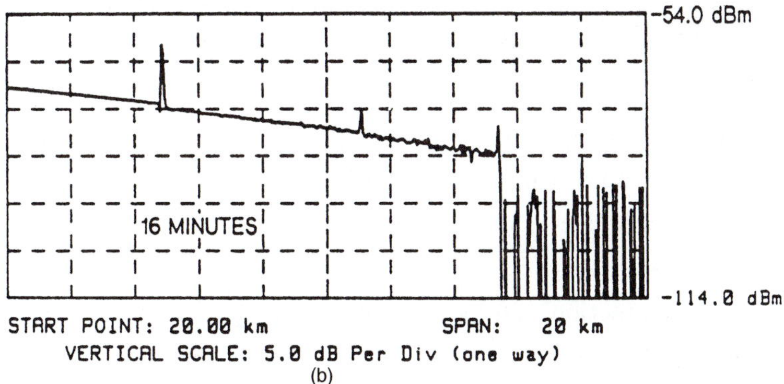

(b)

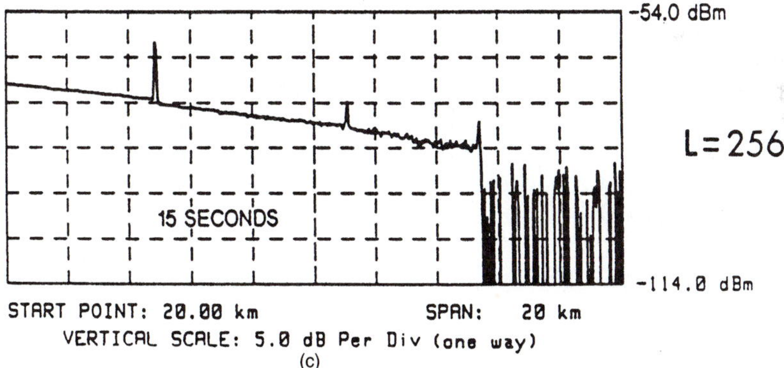

(c)

Figure 25.6 Comparison between (a) conventional OTDR with 15 s averaging time; (b) conventional OTDR with 16 min averaging time; and (c) correlation OTDR with 15 s averaging time.

using a conventional OTDR after 15 s of averaging time. The middle curve shows the same measurement after 16 min of averaging. The fiber end can be clearly identified from this curve. In addition, the curve is free from noise and a second reflection can be easily seen at 31 km. The bottom curve shows the CCOTDR results after 15 s of averaging. The curve is *identical* to the middle curve which shows the speed advantage of the CCOTDR. Comparing the top and the bottom curves (which were obtained after 15 s of averaging time), the CCOTDR is clearly superior in dynamic range.

25.4 OTDR Applications

An OTDR measurement of a fiber's backscattering impulse response contains information that can be useful for many different purposes. Indeed, OTDRs are used in a wide variety of applications. Some of these applications are listed below.

- Fiber length determination
- Fiber characterization
- Active splicing during installation
- Splice and connector maintenance
- Detection and location of catastrophic faults
- Security: detection and location of tap
- Component testing
- Network testing
- Sensing: interrogation of optical sensors

25.5 OTDR Specifications

The performance of an OTDR can be specified using a set of parameters that describe the quality of the measurements. Figure 25.7 depicts some of the key features of a backscattering impulse response that can be used to define and describe these key performance parameters.

Dynamic range. The dynamic range, expressed in optical dB (dB_{opt}), is

$$1\ dB_{opt} = 2\ dB_{elect} \tag{25.4}$$

(The reason for this relationship can be mathematically traced to the fact that optical power is proportional to the photocurrent from the photodectector while electrical power is proportional to the square of the current in the electric circuit.) Equation 25.4 is a measure of the range of optical power levels over which useful measurements of the backscattering impulse response can be made. It is usually defined as the range from the initial backscattering

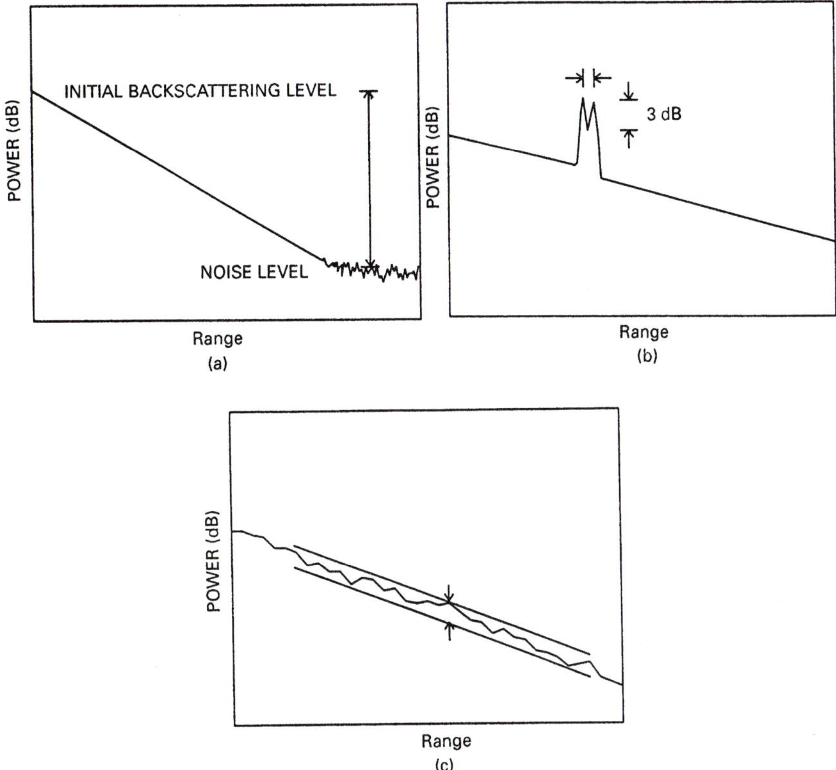

Figure 25.7 Parameters of OTDR: (a) dynamic range; (b) response resolution; and (c) amplitude sensitivity.

level to the noise as shown in Fig. 25.7*a*. Two different terms have evolved to describe the dynamic range. Two-way dynamic range (TWDR) is the dynamic range that is actually measured by the instrument, and corresponds to the difference in actual optical power levels. One-way dynamic range (OWDR) is equal to half the TWDR. It is commonly used to describe the performance of the OTDR because it allows users to estimate how far into the fiber they can see by dividing the OWDR by the propagation loss of the fiber. For example, an OTDR with a OWDR of 20 dB can see up to a distance of about 57 km using a fiber with propagation loss of 0.35 dB/km.

Response resolution. The two-point response resolution is a measure of how closely two faults can be measured and distinguished from one another. As shown in Fig. 25.7*b*, the resolution is defined as the minimum separation between two equal peaks such that the measured signal drops by some arbitrary fraction of the peak height. This parameter is directly proportional to the probe pulse width.

Amplitude sensitivity. This parameter specifies the minimum amplitude variation that can be detected at a certain location on the measured signal as shown in Fig. 25.7*c*. It is limited by noise.

Linearity. The measurement linearity can be expressed in terms of the deviation of the amplitude sensitivity envelope from a straight line when an ideal fiber having a perfect exponential decay is measured. It is expressed in dB/dB. Deviations from linearity can be caused by nonideal receiver or analog-to-digital converter circuits.

Measurement time. The measurement time required to achieve a specific dynamic range capability is a basic and essential specification of the performance of the OTDR.

Range resolution. This parameter is a measure of the uncertainty of the absolute position of a given feature and is usually directly related to the sampling period.

Chapter

26

Impedance Measurement Instruments

Yoh Narimatsu

Agilent Technologies
Kobe, Japan

26.1 Introduction to Impedance Measurements

Impedance (admittance for its reciprocal value) is a fundamental parameter associated with electronic materials, components, and circuits (these are inclusively called *devices* in the later sections). When a current flows through a device, the device gives an opposition to the current flow; the degree of opposition is called *resistance* if the current is a *direct current* (*dc*), and *impedance* if it is an *alternating current (ac)*. Ohm's law governs the relationship between the resistance, the current, and the resultant voltage drop across the *device under test* (*DUT*) for dc cases, and can be mathematically extended to deal with ac cases. The DUT impedance can, therefore, be expressed by the ratio of the voltage across the DUT to the current flowing through it. However, impedance is a more fundamental property inherent to a device; it can be accurately determined and maintained.

The many impedance- or admittance-measuring methods vary in DUT type, cost, accuracy, and test conditions. For many years, a variety of bridges, most of which were manually operated, were used for medium- to high-accuracy measurements. Nowadays, however, digital impedance meters (names vary widely), which automatically perform impedance measurements, have also been widely accepted, owing to their ease of use (less susceptible to human error), relatively high accuracy, and convenience for systemization.

26.2 Basic Definitions

When a direct current is applied to a DUT, the relationship between parameters is expressed by Ohm's law:

$$R = \frac{E}{I} \tag{26.1}$$

where the resistance (more specifically, dc resistance), R, is the degree of opposition that the DUT offers to the current, I, and E is the resultant voltage drop across the DUT.

The same description holds true if an alternating current flows through the DUT. In this case, however, an extended version of Ohm's law should be applied because ac voltage or current consists of two independent elements, magnitude and phase. Alternating current can be conveniently dealt with if voltage, current and impedance are each expressed as complex numbers. Then, DUT impedance Z is derived with the following equation, where each value is a complex number:

$$Z = \frac{E}{I} \tag{26.2}$$

A complex impedance (Z) can be further expressed either in the rectangular coordinate form $Z = R + jX$ or in the polar form $Z = |Z|e^{j\theta}$. Similarly, admittance $Y(=1/Z)$ can be expressed as either $Y = G + jB$ or $Y = |Y|e^{j\theta} = |1 \neq Z \neq e^{-j\theta}$. These expressions are summarized in Table 26.1 and illustrated in Fig. 26.1. In the figure, an impedance and its associated admittance are shown together in the same complex plane. The unit for impedance, resistance, and reactance is the ohm (Ω), and the unit for admittance, conductance, and susceptance is the siemens (S). Generally, impedance is convenient for calculations when devices are connected in series, whereas admittance is convenient when they are connected in parallel. If alternating current I with

TABLE 26.1 Definitions of Impedance and Admittance

Impedance Z	Admittance Y (= $1/Z$)
R X (series circuit)	G B (parallel circuit)
$Z = R + jX$ $= \|Z\|e^{j\theta}$	$Y = G + jB$ $= \|Y\|e^{j\phi}$ $= \|1/Z\|e^{-j\theta}$
Where,	Where,
R = ac resistance	G = ac conductance
X = reactance	B = susceptance
$\|Z\| = \sqrt{R^2 + X^2}$ = impedance magnitude	$\|Y\| = \sqrt{G^2 + B^2} = 1/\sqrt{R^2 + X^2}$ = admittance magnitude
$\theta = \tan^{-1}(X/R)$ = impedance phase angle	$\phi = \tan^{-1}(B/G) = -\tan^{-1}(X/R)$ = admittance phase angle
$R = \frac{G}{G^2 + B^2}$ $X = -\frac{B}{G^2 + B^2}$	$G = \frac{R}{R^2 + X^2}$ $B = -\frac{X}{R^2 + X^2}$

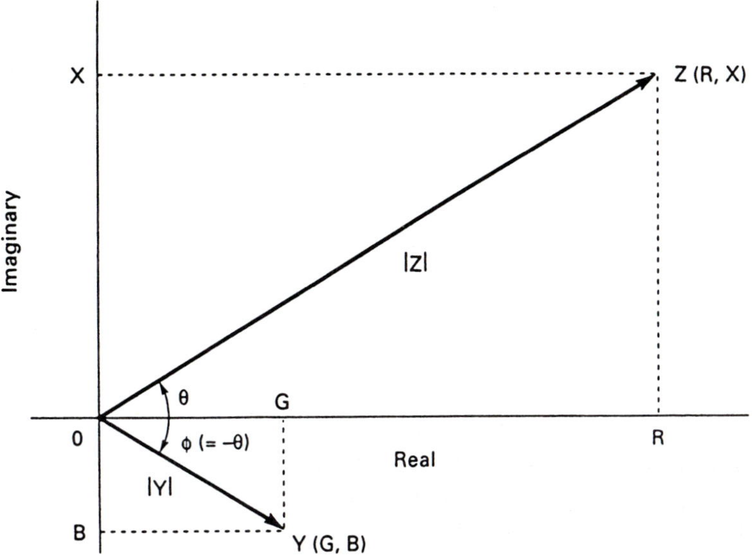

Figure 26.1 An impedance vector and its associated admittance vector on the complex plane.

frequency f is applied to an inductor of inductance L (a pure or lossless inductor is assumed here), its degree of opposition to the current is proportional to L and to the angular frequency $\omega(=2\pi f)$ of the current. The phase of the resultant voltage (E) will lead by 90° (or $\pi/2$ radian) relative to I. This can be denoted as follows because a 90° phase lead is mathematically expressed by multiplying by the imaginary unit j:

$$E = j\omega LI \tag{26.3}$$

If E in Eq. 26.2 is substituted by Eq. 26.3, the impedance Z of the inductor becomes:

$$Z = j\omega L \tag{26.4}$$

By comparing Eq. 26.4 with the general form $Z = R + jX$, it can be seen that ωL is the reactance given by inductance L at angular frequency ω. The reactance provided by an inductance is called the *inductive reactance,* X_L.

Likewise, a pure capacitance (C) provides the capacitive reactance (X_c) of $1/\omega C$, and the phase of the voltage across the capacitor lags by 90°, relative to the applied current. Thus, impedance Z of the capacitor becomes:

$$Z = -j\frac{1}{\omega C} = -j\frac{1}{2\pi fC} \tag{26.5}$$

The quality factor (Q) is defined for a device of impedance Z, as follows. Here, a resistance element is involved, in addition to a reactance element. That is, the device impedance is no longer pure:

$$Q = \frac{X}{R} = \tan\theta \tag{26.6}$$

Because a reactance element stores energy, but a resistance element dissipates it, a large Q of a device implies it can store energy with little loss. Also, a large Q for an inductor or capacitor implies that the degree of its "purity" is high.

The dissipation factor (D) is also defined. It is the reciprocal value of Q; thus, it indicates the size of the energy loss, when compared with the stored energy in a device:

$$D = \frac{1}{Q} = \frac{R}{X} = \cot\theta \tag{26.7}$$

A similar description can be applied to admittance, and the related expressions are listed in Table 26.2.

TABLE 26.2 Impedance, Admittance, and Their Related Definitions for *RL* and *RC* Combinations

	Impedance expression	
Circuit	R_s L_s (series)	R_s C_s (series)
Impedance	$Z = R_s + j\omega L_s$	$Z = R_s - j\dfrac{1}{\omega C_s}$
	$= R_s + jX_L$	$= R_s - jX_C$
Quality factor	$Q = \dfrac{X_L}{R_s} = \dfrac{\omega L_s}{R_s}$	$Q = \dfrac{X_C}{R_s} = \dfrac{1}{\omega C_s R_s}$
Dissipation factor	$D = \dfrac{1}{Q} = \dfrac{R_s}{X_L} = \dfrac{R_s}{\omega L_s}$	$D = \dfrac{1}{Q} = \dfrac{R_s}{X_C} = \omega C_s R_s$
	Admittance expression	
Circuit	$R_p(=1/G_p)$, L_p (parallel)	$R_p(=1/G_p)$, C_p (parallel)
Admittance	$Y = G_p - j\dfrac{1}{\omega L_p}$	$Y = G_p + j\omega C_p$
	$= G_p - jB_L$	$= G_p + jB_C$
Quality factor	$Q = \dfrac{B_L}{G_p} = \dfrac{R_p}{\omega L_p}$	$Q = \dfrac{B_C}{G_p} = \omega C_p R_p$
Dissipation factor	$D = \dfrac{1}{Q} = \dfrac{G_p}{B_L} = \dfrac{\omega L_p}{R_p}$	$D = \dfrac{1}{Q} = \dfrac{G_p}{B_C} = \dfrac{1}{\omega C_p R_p}$

26.3 Characteristics of Electronic Components

26.3.1 Equivalent circuits

No actual device consists of a single element of inductance, capacitance, or resistance in a strict sense. Even an air dielectric capacitor, which is usually considered to be a very pure capacitor, has the lead inductance, and losses caused by the lead resistance, electrode surface finish, contamination, eddy current, etc. For actual applications, however, it can be treated as a pure capacitor if its capacitance value is less than a few hundred pF and the operating frequency is less than several kHz.

For quantitative evaluation of a device, it is convenient if its characteristics can be approximated by a combination of ideal elements near the operating condition concerned. The resultant combination is called an *equivalent circuit.* Such an equivalent circuit does not necessarily reveal the physical structure of an actual device. For the air capacitor example covered earlier, its equivalent circuit might be a pure capacitor in the low-frequency region, and even a series combination of a resistor and an inductor at a few hundred MHz, above the series resonant frequency of the capacitor. In general, a complex impedance at a given frequency is expressed by two quantities, a real part and an imaginary part, and can be synthesized by a combination of a resistance (or conductance) and either an inductance or a capacitance. For example, if a device has an impedance of $1000 + j1000\ \Omega$ at 1 kHz, its equivalent circuit can be that of either Fig. 26.2A or 26.2B, and the two circuits exhibit

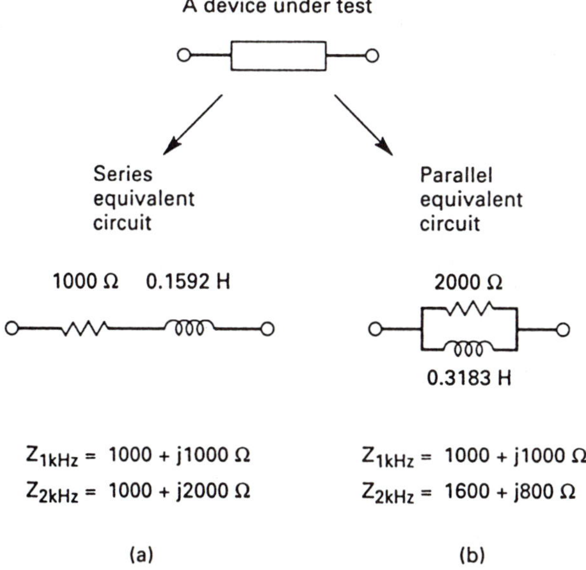

Figure 26.2 Series (a) and parallel equivalent circuits of an inductor (b).

TABLE 26.3 Parameter Conversion between Series and Parallel Equivalent Circuits

A combination of inductance and resistance	
$L_s = \frac{Q^2}{1+Q^2} L_p = \frac{1}{1+D^2} L_p$	 R_s, L_s (series)
$R_s = \frac{1}{1+Q^2} R_p = \frac{D^2}{1+D^2} R_p$	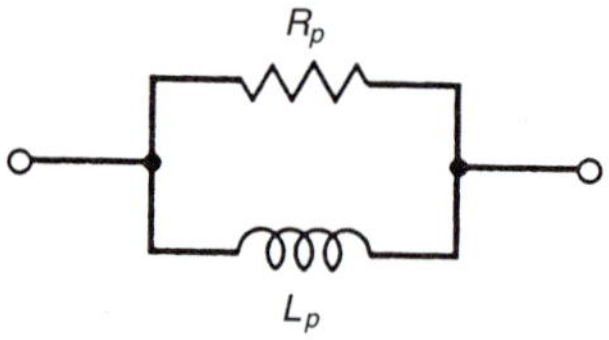
$L_p = \left(1+\frac{1}{Q^2}\right) L_s = (1+D^2) L_s$	R_p, L_p (parallel)
$R_p = (1+Q^2) R_s = \left(1+\frac{1}{D^2}\right) R_s$	
A combination of capacitance and resistance	
$C_s = (1+D^2) C_p = \left(1+\frac{1}{Q^2}\right) C_p$	 R_s, C_s (series)
$R_s = \frac{D^2}{1+D^2} R_p = \frac{1}{1+Q^2} R_p$	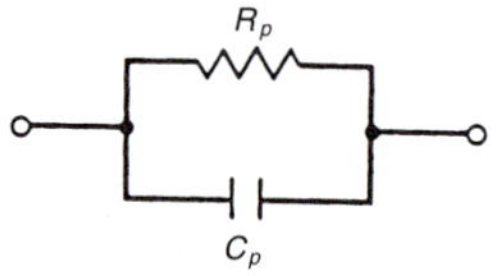
$C_p = \frac{1}{1+D^2} C_s = \frac{Q^2}{1+Q^2} C_s$	R_p, C_p (parallel)
$R_p = \left(1+\frac{1}{D^2}\right) R_s = (1+Q^2) R_s$	

exactly the same characteristics at this frequency. If the device, however, provides an impedance of $1000 + j2000\ \Omega$ at 2 kHz, it can be said that the equivalent circuit better represents the actual characteristics of the device. If the configuration of an equivalent circuit is close enough to the actual structure of a device (often with many elements), it can exhibit a good match with actual characteristics of the device. In reality, however, equivalent circuits with only a few elements are frequently used for simplicity and convenience reasons.

For equivalent circuits with two elements, parameter conversion equations are listed in Table 26.3.

26.3.2 Parameter dependence

With actual devices, their inductance, capacitance, and resistance elements are subject to change, depending on operating conditions, such as frequency, temperature, applied voltage and current, and time passage, for example. Therefore, it is important to remember that a test result of a device for a given test condition might not hold true at another condition. In other words, it is important to measure the characteristics of a device with all parameters as close as possible to the actual operating condition.

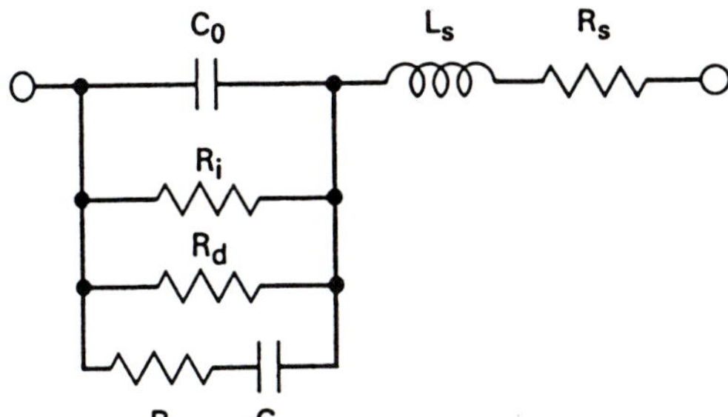

Figure 26.3 An equivalent circuit of a ceramic capacitor.

Figure 26.3 illustrates an equivalent circuit of a ceramic dielectric capacitor with major parasitic elements included, where, C_0 = nominal capacitance at low frequencies, L_S and R_S = series inductance and resistance mainly caused by leads and electrodes, R_i = insulation resistance, R_d = dielectric loss of ceramic material, and C_a and R_a = dielectric absorption capacitance and resistance.* Owing to the series inductance, the equivalent series capacitance becomes larger than the nominal capacitance value (C_0) near the series resonance frequency. The dielectric material is frequency-dependent, too, and its loss becomes larger as the operating frequency increases. Usually, a material with a high dielectric constant, useful to make a capacitor size smaller, has a large dependence on temperature (i.e., its capacitance value varies with change in ambient temperature), and its temperature coefficient might not be constant over temperature. It usually has a large dependence on the applied signal level, too.

A similar description can be applied to an inductor with a magnetic material core, shown in Fig. 26.4, where, L_0 = nominal inductance at low frequencies, R_s = series resistance, C_i and R_i = parasitic capacitance and associated dielectric loss, and R_p = hysteresis and eddy current loss for core material.

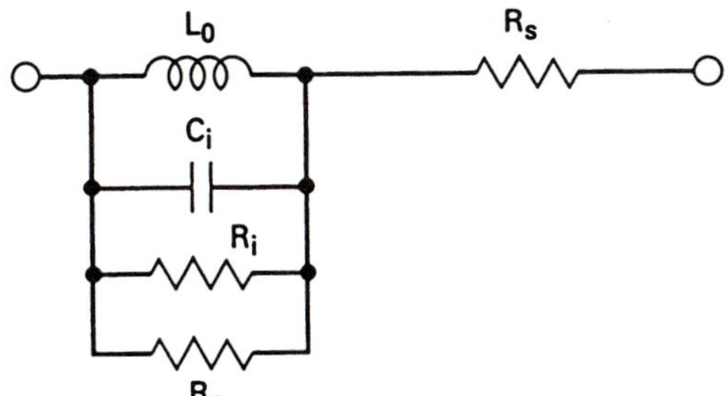

Figure 26.4 An equivalent circuit of an inductor with a core.

* The dielectric absorption is explained by the slow diffusion of electric charges into the dielectric material, and it sometimes has a very large time constant of the order of hours or more.

Because an inductor is usually realized by several layers of wire windings around a magnetic core material, parasitic capacitance between layers and windings tends to be introduced. Frequency dependence of an inductor is mainly determined by the parasitic capacitance and the magnetic material property. Wire resistance, skin effect, and losses of magnetic material are the major sources of the loss of an inductor, and each of them has frequency and temperature dependence, too. The property of the magnetic material is also sensitive to the magnetic field strength. Therefore, the inductance of an inductor with a magnetic core varies, depending on the applied signal level, which is converted into the magnetic field strength.

26.4 Impedance Measuring Techniques

A wide variety of techniques are used to measure the impedance of a device, and a suitable technique should be chosen. Although it is impossible to list all the techniques currently available, the representative techniques often used are explained here. Each technique also has variations, depending on actual applications.

26.4.1 The voltmeter-ammeter method

Figure 26.5 shows the basic connection for measuring the dc resistance of a device with the voltmeter-ammeter method. The resistance can be derived by directly applying Ohm's law:

$$R_x = \frac{E}{I} \tag{26.8}$$

where E (the reading of the voltmeter, V) is the voltage across the device and I (the reading of the ammeter, A) is the current flowing through the device.

The dc supply (V_s) can be either a voltage or a current source. If an ac supply and appropriate ac voltage and current meters are used, the magnitude of the device impedance can be obtained. Although this method is easy

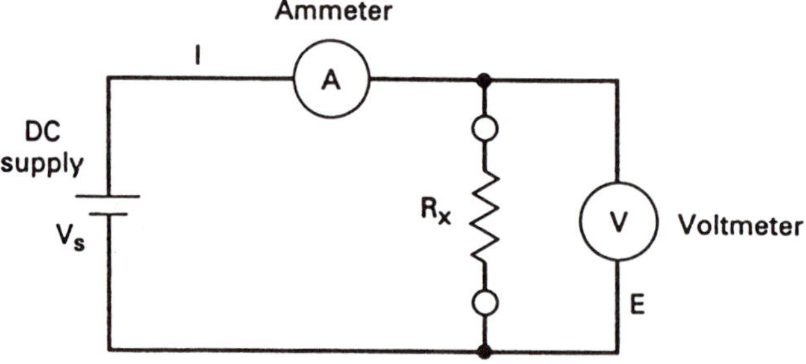

Figure 26.5 A dc resistance measurement with the voltmeter-ammeter method.

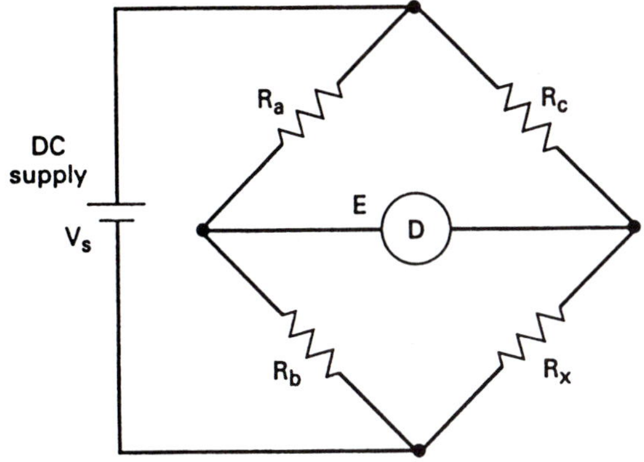

Figure 26.6 The Wheatstone bridge.

to understand and to implement, errors associated with the meters (nonlinearity, internal resistance, etc.) directly affect the measurement results. Therefore, this is mainly used for low- to medium-accuracy applications (no better than 1 percent).

26.4.2 The bridge method

A variety of bridges have been implemented depending on the applications. The word *bridge* was originally used for a null detector bridging the two points of balancing arms. But it is now widely used to indicate impedance-measuring circuits employing null detection and balancing arms. Generally, a bridge can yield relatively good measurement accuracy because it is primarily determined by the stable passive components forming the balancing arms. Figure 26.6 is the Wheatstone bridge for dc resistance measurement with a dc voltage source, V_s. Assuming that $E = 0$ (the reading of null detector D is 0), resistance R_x is derived as:

$$R_x = \frac{R_b R_c}{R_a} \tag{26.9}$$

Neither V_s nor E explicitly appears in this equation; that is, neither the variation of the power supply nor the nonlinearity of the voltmeter becomes the primary source of the measurement error. The same description can be applied to the bridges used for ac impedance measurements. For ac bridges, the balancing arms always consist of one or more reactance elements, in addition to resistance elements.

26.4.3 The resonance method

A combination of inductance and capacitance resonates at a certain frequency. If one of the values of the combination and the resonance frequency

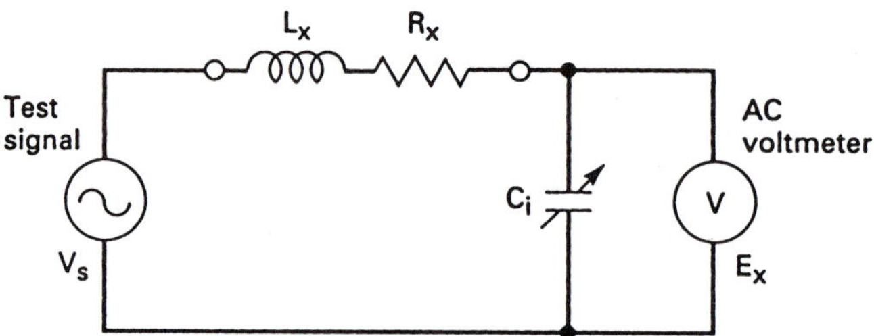

Figure 26.7 The principle of the Q meter.

are known, the other value can be calculated. This is the principle of the resonance method. The Q meter is a direct implementation of this method, and Fig. 26.7 gives its basic circuit diagram. In addition to the manually operated Q meter, the automatic Q meter, with which the tuning process is automatically performed is available.

The Q meter typically has a built-in variable capacitor, and an inductor is externally connected. The resonance frequency (f) of the combination circuit can be calculated as:

$$f = \frac{1}{2\pi\sqrt{L_x C_i}}$$

So, if f and C_j are known, L_x is calculated as:

$$L_x = \frac{1}{(2\pi f)^2 C_i} \tag{26.10}$$

The Q value of the inductor is also obtained as:

$$Q = \frac{|E_x|}{|V_s|} = \frac{\omega L_x}{R_x} \tag{26.11}$$

26.4.4 The automatic bridge method

Many bridges, such as the Wheatstone bridge described in Section 26.4.2, are manually operated (i.e., iterative balancing manipulation is needed to reach the final null, balance, condition). In recent years, however, many automatic bridges have been developed to eliminate the time-consuming manual operation and associated measurement errors.

Figure 26.8 is an example of the automatic impedance bridge. In this block diagram, the reference resistor (R_r) is connected to the null point (P) and the residual voltage there is amplified to be fed back to the other end of the

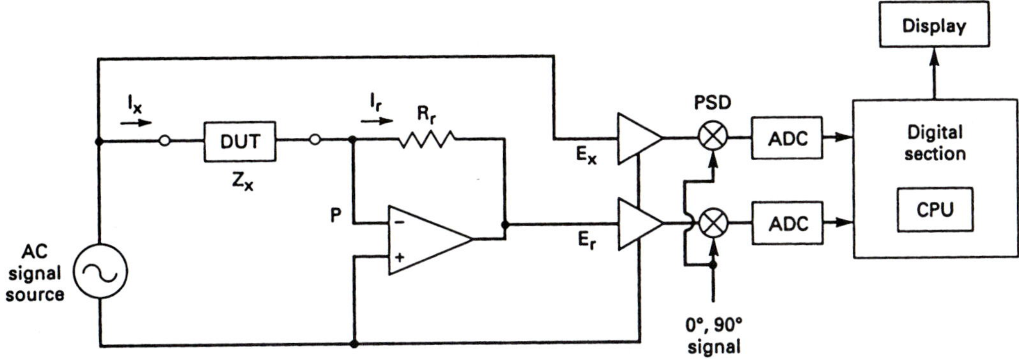

Figure 26.8 An example of the automatic impedance bridge.

resistor to establish the null condition. Once the null condition is achieved, the DUT impedance (Z_x) is calculated as follows:

$$E_r = -I_r R_r$$
$$E_x = I_x Z_x$$
$$I_r = I_x$$

Then:

$$Z_x = -R_r \frac{E_x}{E_r} \tag{26.12}$$

One or more microprocessors are used to control the internal functional blocks and converting the measured data into the desired final outputs. Modern automatic bridges are also equipped with communication paths with external devices, such as a computer and a device handler for systemization.

26.4.5 Specifications of impedance measuring instruments

Fundamental items found in the specifications of impedance measuring instruments are listed here.

Impedance range. The *impedance range* specifies the limits of DUT impedance, within which an instrument can make measurements. The range of *R*, *C*, *L*, and other parameters is often specified as well. Measurement performance of an instrument is usually degraded near the boundary of its specified range.

Measurement accuracy. The *measurement accuracy* specifies how close the measured impedance value is to the true value. To express the measurement accuracy, the uncertainty of the measured value is usually indicated in the basic form of percent of measured value plus offset value. The percent term becomes significant if the measured value is near the full scale of a measuring instrument; the offset term becomes significant if it is close to zero. Because the accuracy of an instrument is determined by many parameters, such as DUT impedance itself, test frequency, and test signal level, it is sometimes expressed with a complicated formula, which is the function of these parameters.

Test signal level. This specifies the test signal level applied to a DUT at measurement. This specification becomes important for DUTs, which have impedance that varies with signal level. Also, in some situations the test signal level is specified by the industry standard for commercial transaction purposes.

Measurement time. The measurement time for an automatic impedance measuring instrument is a period of time for which a DUT is being measured. This specification is important at the production line environment of component manufacturers, for example, where the test throughput is of primary concern.

A simple measurement time specification often specifies only the period of each consecutive measurement with a DUT continuously connected to the instrument. The measurement time, however, becomes longer because of additional rebalancing (settling) time if the DUT is disconnected from the instrument each time after measurement. This is usually the case at the production line, and the measurement time specification should, therefore, be carefully examined for actual applications.

Operating temperature range. The operating temperature range is the range of ambient temperature within which an instrument can be used. Because every instrument has temperature dependence, its accuracy is somewhat degraded if the operating temperature is different from its specified (calibration) temperature, usually 23°C. The degree of accuracy degradation is usually specified, too.

26.5 Connection and Guarding

26.5.1 Multiterminal connections

The two-terminal connection. As covered in the previous sections, impedance is usually defined between two terminals of a device. Therefore, it is natural to connect it to an impedance measuring instrument with two test leads, as

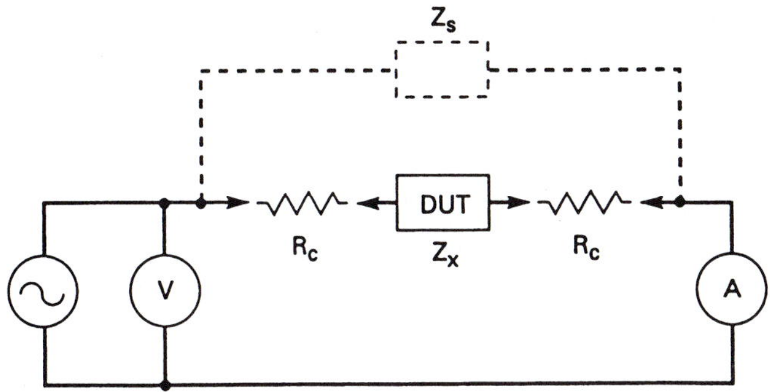

Figure 26.9 The two-terminal connection.

shown in Fig. 26.9. Although this connection is simple to implement and easy to understand, it has some potential measurement errors. First, a stray or residual impedance (Z_s) might exist between the two test leads. If a stray impedance exists in parallel with a device, the measured impedance becomes their parallel combination value, which is not the desired result. Second, a contact resistance (R_c) appears between the device terminal and test lead at each of the contact points, and it is placed in series with the device. The contact resistance varies from 10 Ω to as large as several ohms, and is unstable. As a result, it introduces a significant amount of error if the device impedance is relatively small, such as less than several hundred ohms.

The three-terminal connection. To solve the first problem, the three-terminal connection is used. The third terminal and its associated structure, called the *guard terminal,* is used to prevent the stray impedance from causing error (Fig. 26.10). The stray impedance is now placed in parallel with the signal source (Z_a) and the ammeter (Z_b). The leakage current caused by Z_a does not

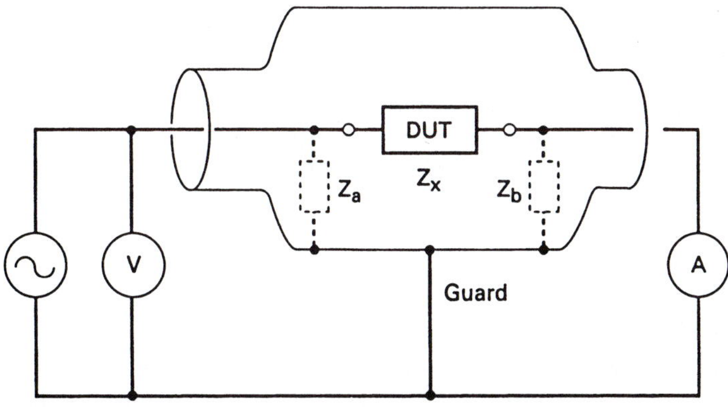

Figure 26.10 The three-terminal connection.

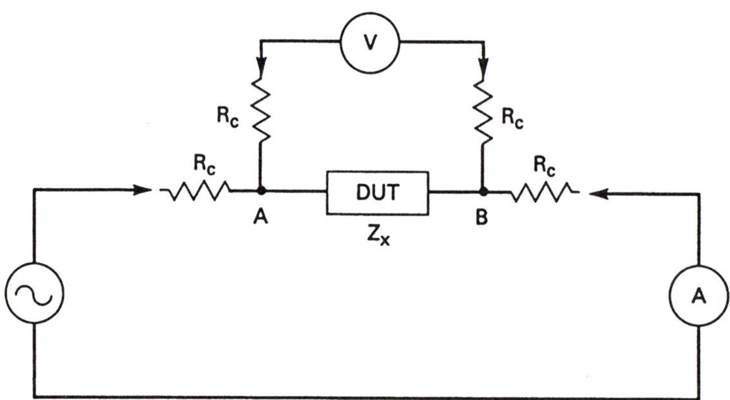

Figure 26.11 The four-terminal connection.

flow through the ammeter and the leakage current caused by Z_b becomes small because the internal impedance of the ammeter is small and the resultant voltage drop across it becomes small, too. Thus, only the current flowing through the DUT is measured by the ammeter. Usually, the stray impedance is either capacitive, resistive, or a combination of both, and such a guard terminal is effective when DUT impedance is relatively high, above several kΩ.

The four-terminal connection. To deal with the second problem, namely, to eliminate the measurement ambiguity caused by the contact resistance, the four-terminal connection is used. The current supply leads and the voltage sensing leads are independently provided, as shown in Fig. 26.11. With this connection, DUT impedance is defined across the voltage sensing points (A-B) because of no voltage drop through the voltage-sensing leads, and the effect of the contact resistance can be completely eliminated. Note that the input impedance of the voltmeter should be much larger than the device impedance to be measured so that it can be neglected.

The five-terminal connection. The three-terminal and the four-terminal connections can be combined, resulting in the five-terminal connection. With this connection, benefits from both connections can be obtained with a drawback of having a somewhat-complicated configuration.

26.5.2 The four-terminal-pair connection

When an alternating current flows through a conductor, ac magnetic flux is generated around it. An ac voltage is induced at another conductor if the magnetic flux is coupling with it. With the four-terminal connection covered in Section 26.5.1, there is a possibility of an error voltage being induced at

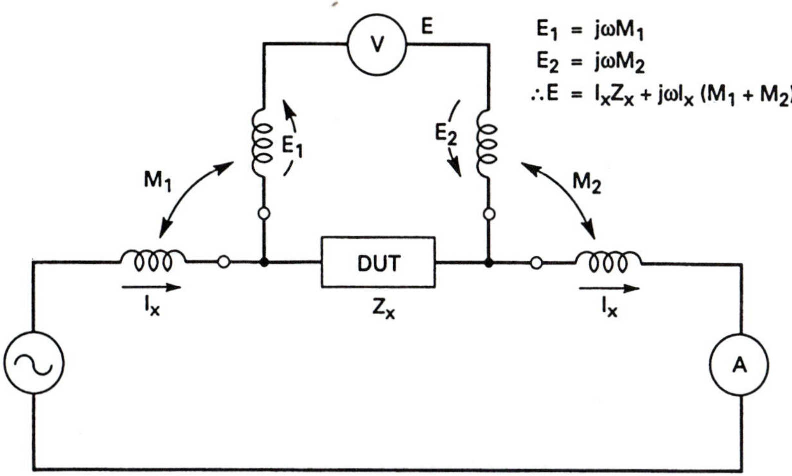

Figure 26.12 Error voltage induction with magnetic coupling.

the voltage-sensing leads with this mechanism (Fig. 26.12). The higher the test frequency and the larger the current in the test leads, the larger the error voltage that tends to be introduced. One solution for this undesirable voltage is to twist the two current supply leads together and also do the same for the voltage-sensing leads. If the current supply leads are twisted together, the current flowing in each lead is of opposite direction and the generated magnetic flux by each current cancels out. Also, each of the generated error voltages at the twisted voltage-sensing leads has opposite polarity and again

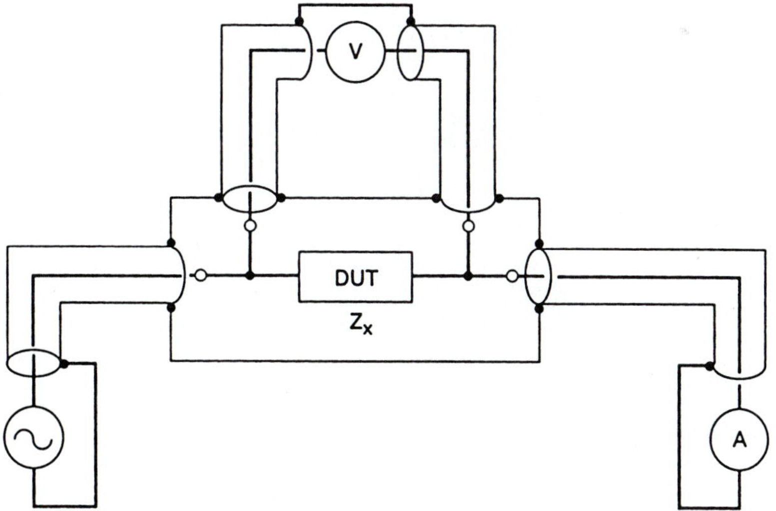

Figure 26.13 The four-terminal-pair connection.

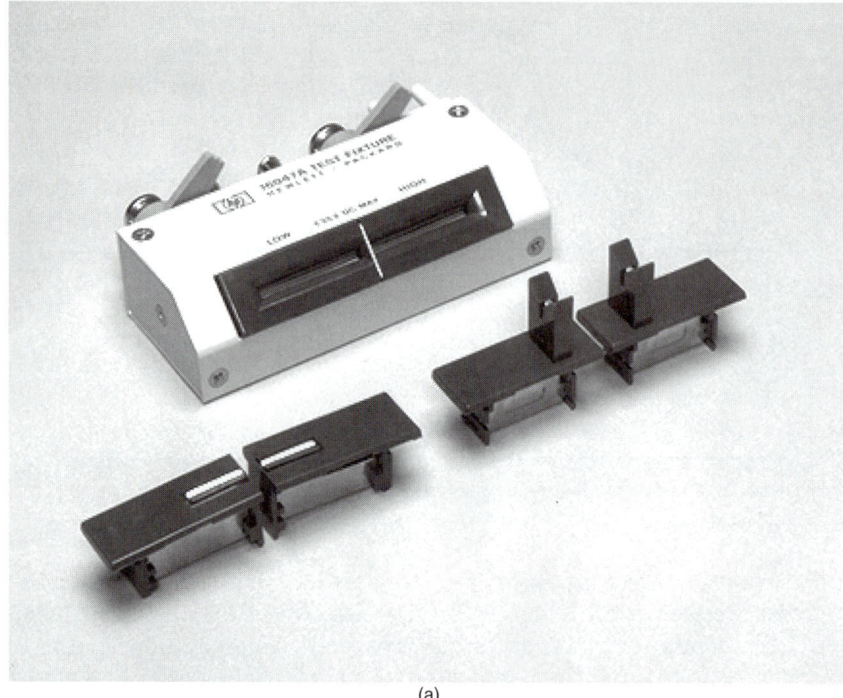

(a)

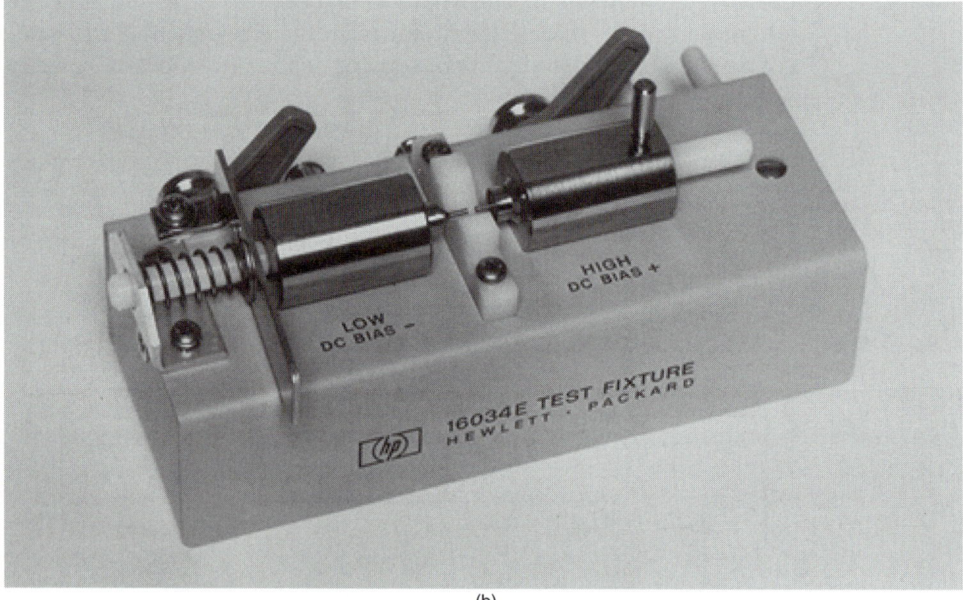

(b)

Figure 26.14 Examples of lower-frequency test fixtures: Leaded device test fixture (a), SMD/chip device fixture (b), SMD/chip device tweezer fixture (c), dielectric material test fixture (d).

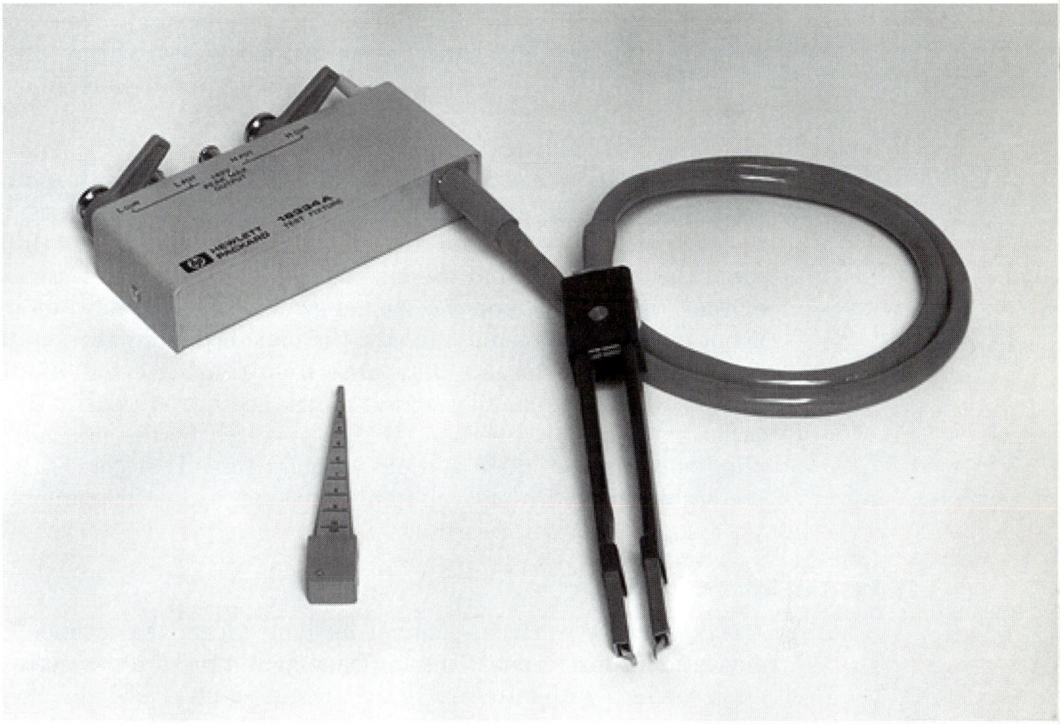

(c)

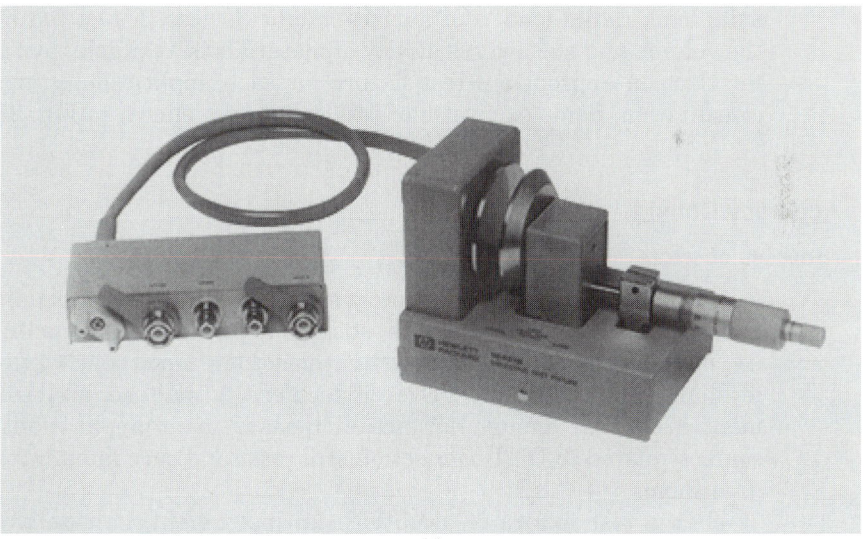

(d)

Figure 26.14 *(Continued)*

cancels out. Twisting the test leads, however, is not always practical to implement and the degree of cancellation becomes worse as the test frequency becomes higher. In such a situation, the four-terminal-pair connection provides a better solution.

Figure 26.13 illustrates the basic configuration of the four-terminal-pair connection, where coaxial cable is used for the test leads and the outer conductor of each test lead is tied together at the DUT end. With this structure, the current flowing through the inner conductor returns via the outer conductor of the same lead, and the magnetic flux in each conductor is canceled out and not radiated outside the coaxial structure. Thus, unwanted error voltage is not introduced. Needless to say, the four-terminal nature of this configuration prevents the contact resistance from causing error. Another benefit of this structure is a smaller stray inductance introduced by the test leads in series with the DUT because it is suppressed by the mutual magnetic coupling between the inner and outer conductors. Therefore, this connection is preferably used with modern impedance-measuring instruments.

26.5.3 Test fixtures

A *test fixture* is a crucial element for impedance measurement to interface a device to an instrument. Having easy mounting of a device to a test fixture often conflicts with minimizing the amount of induced error. So, various types of test fixtures have been designed, depending on applications. A variety of electrode shapes are also available for accepting various devices, such as axial-lead, radial-lead, and surface-mount devices. A test fixture to measure the volume and surface resistivity of materials is available, too. It is advisable to select an appropriate test fixture for each application with the supplier's consultation. Some examples of test fixture are shown in Fig. 26.14.

26.6 Accuracy Considerations

26.6.1 Sources of errors

If an impedance measuring instrument is very strictly examined, it might have potential sources of errors, such as dependence on temperature, humidity, or line voltage; internal system noise; and nonlinearity with test signal level and DUT impedance. Even if such errors that are intrinsic to an instrument are small enough for actual measurement applications, other error sources related to DUT connection still exist and care must be taken to eliminate them.

When a test fixture is used with an instrument, it usually has a certain amount of stray impedance or admittance—even without a device inserted. Its form might be stray capacitance and conductance in parallel with the DUT, and/or stray inductance and resistance in series. Such stray components become on offset error and can be mathematically subtracted from the measurement results on which they are superimposed. Most instruments are equipped with stray impedance compensation (cancellation) capability.

Another error source is associated with test-lead extension at ac measurements. When an ac signal travels through a test lead, its magnitude and phase vary as transmission-line theory indicates. Errors caused by this effect become larger as the lead length becomes longer and the test frequency becomes higher. Compensation of this type of error is more difficult than compensation for stray impedance because it is mathematically equivalent to multiplying the measured values by complex compensation coefficients. Again, some modern instruments have built-in compensation capability for extended test leads.

26.6.2 Compensation and calibration

As shown in the previous section, *compensation* usually indicates the means to eliminate error sources given by the external conditions. On the other hand, *calibration* usually indicates the procedure used to adjust an instrument using appropriate impedance standards so that it can measure devices with specified accuracy. Therefore, compensation and calibration are sometimes dependent. However, the sources of measurement errors should be always clearly identified and each of them must be appropriately implemented to obtain reliable measurement results.

26.6.3 Impedance standards and traceability

Many impedance standards are available to calibrate an impedance-measuring instrument. Their primary parameter is resistance, capacitance, or inductance, and sometimes a secondary parameter (such as D, Q, or time constant is indicated together). "Open" and "short" standards are often used in a calibration process, too. Although the resistance standard (including open and short) is the only possible standard for dc resistance calibration, capacitance and inductance standards are also used for ac impedance calibration. Because stable and low-loss inductors are difficult to realize, capacitance standards are preferred for precision applications.

Because an impedance-measuring instrument is calibrated with respect to the values of impedance standards, it is essential to provide a reliable value to a standard. This is achieved with the concept of traceability: A value given to a standard can be traced back to upper-level standards, and eventually to the national standards. It is apparent that a lower-level standard has larger uncertainty. When an impedance value cannot be directly traced back to the same type of impedance standard, it can be substituted by a combination of other traceable quantities and conversion theory. One example is a coaxial line with air dielectric material. A combination of physical dimension measurements and microwave transmission-line theory allows a coaxial line to be used as an impedance standard.

26.7 Impedance Measuring Instruments

To become more familiar with the popular impedance-measuring instruments, internal structure and features of typical instruments will be discussed. Although the guard structure is almost always provided with an impedance-measuring instrument, it is not explicitly shown, unless otherwise indicated to simply illustrate the basic operation principle of each instrument.

26.7.1 Multimeters

Dc resistance-measuring capability is usually provided as one of the functions of a multimeter; a variety of which are currently available, ranging from simple analog circuit testers to precision digital multimeters. Although the dc resistance measurement circuits in a multimeter have some variation, they are all essentially based on the voltmeter-ammeter method. Figure 26.15A shows a circuit used in an analog circuit tester with measurement accuracy of 3 percent of the meter scale, and Fig. 26.15B shows a dc resistance measurement block diagram for a digital multimeter with a basic accuracy of 50 ppm of the reading. Note that some digital multimeters have the ability to select between the two-terminal and the four-terminal connection.

26.7.2 The universal bridge

An impedance-measuring tool containing several types of manual bridges, sharing the key bridge elements, is available and is sometimes called the *universal bridge*. Typically, it consists of resistance-, capacitance-, and induc-

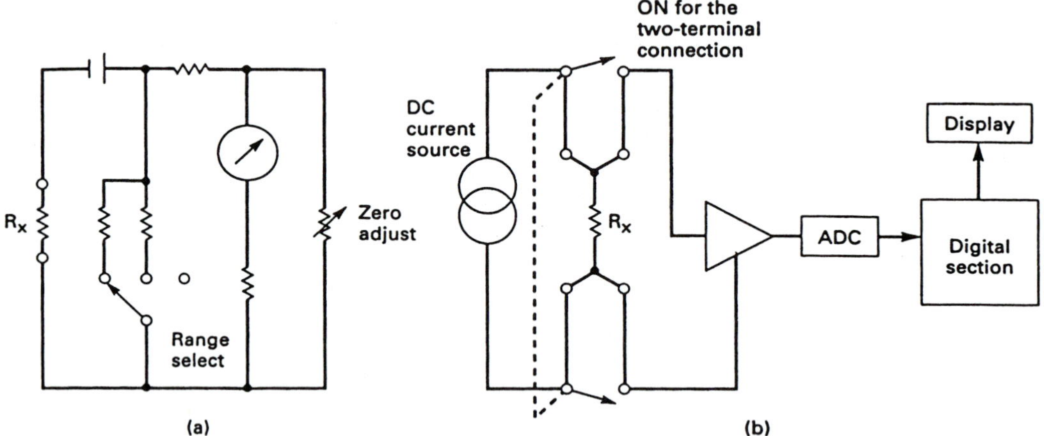

Figure 26.15 Dc resistance-measurement circuits: A part of an analog multimeter (a) and a part of a digital multimeter (b).

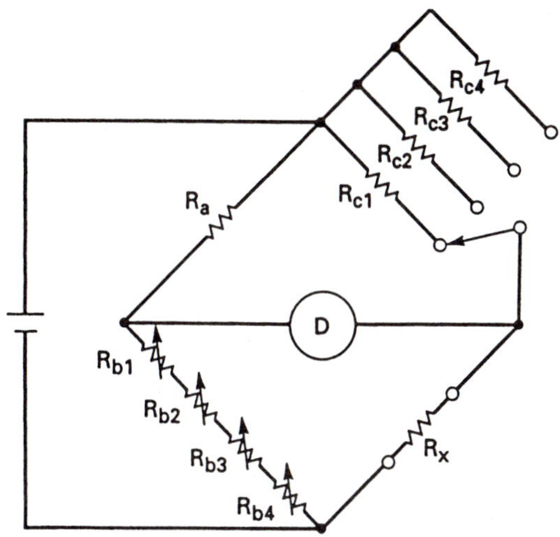

Figure 26.16 A Wheatstone bridge implementation.

tance-measuring bridges, one of which can be chosen and configured, depending on the DUT type.

For a resistance measurement, the Wheatstone bridge configuration is used and its operation principle is explained in Section 26.4.2. Strictly speaking, because it consists of only resistors for the balancing arms, it can only be used for dc resistance measurement. However, if the measurement frequency is low enough, such as less than a few kHz, the reactance elements associated with the circuit are negligibly small, so the ac resistance of a DUT can be measured, too. This is convenient in such cases when electrochemical reaction occurs if a dc voltage is applied to a device, or when a thermal electromotive force prevents the bridge from making accurate dc measurements.

To expand the measurement range, the Wheatstone bridge in actual use is modified from its original form, as shown in Fig. 26.16, for example. In this configuration, R_b is a step-variable resistor whose value (reading of the step dial) is in proportion to the value of R_x, and R_c is a range multiplier to extend the measurement range (four decades, in this case). This modification is also applicable to the capacitance- and inductance-measuring bridge configuration.

For capacitance measurement, two types of capacitance bridges are provided and one of them is chosen to measure either the equivalent series capacitance or equivalent parallel capacitance. They are called the *series* and *parallel resistance bridges,* shown in Fig. 26.17. If the dissipation factor (D) of a capacitor under test is small enough (less than 0.03, for example), the difference between the capacitance values, depending on the choice of equivalent circuit is small, but becomes significant as D increases (see Table 26.3). Therefore, the selection of bridge circuit, based on some insight about the device structure, becomes important for large-loss capacitor measurements.

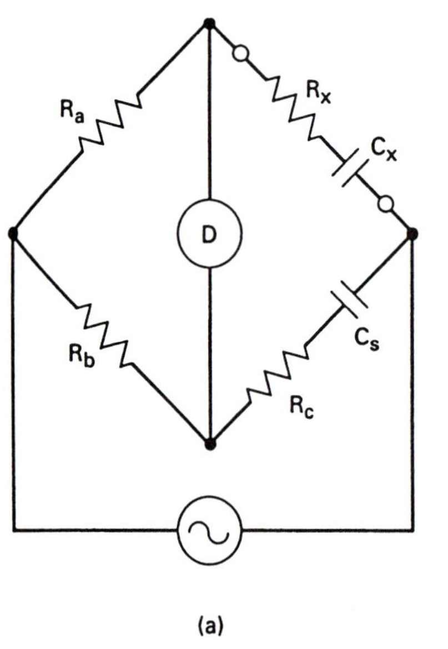

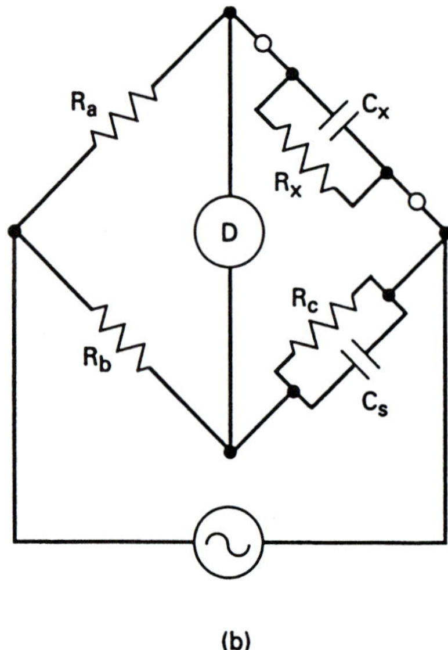

Figure 26.17 Capacitance-measuring circuits for the universal bridge: The series resistance bridge (a) and the parallel resistance bridge (b).

The balance condition for the series resistance bridge is:

$$C_x = C_s \frac{R_b}{R_a} \tag{26.13}$$

$$R_x = R_c \frac{R_a}{R_b} \tag{26.14}$$

$$D = \omega C_s R_c \tag{26.15}$$

For the parallel resistance bridge, the same equations can be used for C_x and R_x, but the dissipation factor is:

$$D = \frac{1}{\omega C_s R_c} \tag{26.16}$$

To measure inductance, the Maxwell-Wien bridge and the Hay bridge configurations are used. The Maxwell-Wien bridge is chosen for an equivalent series inductance measurement and the Hay bridge for an equivalent parallel inductance measurement. The basic circuits of these bridges are shown in Fig. 26.18.

The balance condition of the Maxwell-Wien bridge is:

$$L_x = C_s R_a R_b \tag{26.17}$$

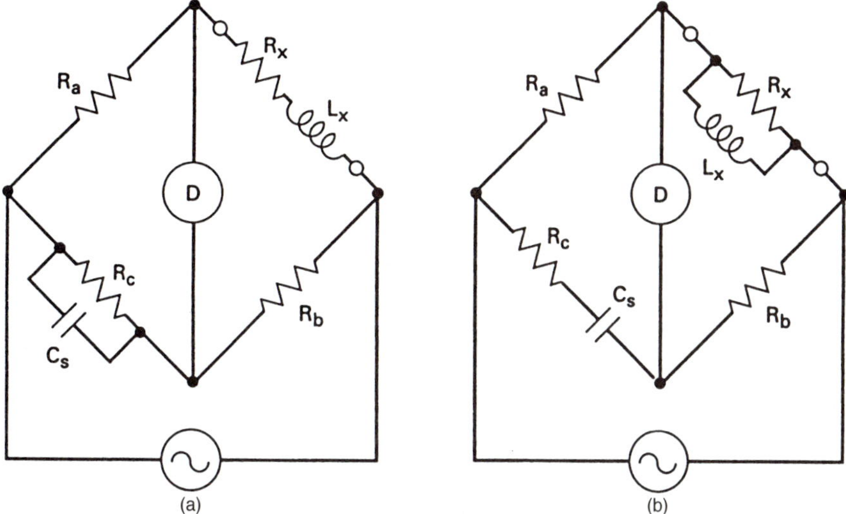

Figure 26.18 Inductance-measuring circuits for the universal bridge: The Maxwell-Wien bridge (a) and the Hay bridge (b).

$$R_x = \frac{R_a R_b}{R_c} \tag{26.18}$$

$$Q = \omega C_S R_c \tag{26.19}$$

For the Hay bridge, the same equations can be used for L_x and R_x, but the Q factor is:

$$Q = \frac{1}{\omega C_S R_c} \tag{26.20}$$

An example of a commercially available universal bridge with an internal 1-kHz signal source has a basic measuring accuracy of 0.1 percent of the reading with four significant figures.

26.7.3 The Kelvin bridge

The Kelvin bridge is a four-terminal connection bridge that is frequently used to measure dc resistances as low as 100 $\mu\Omega$ with accuracy of better than 1 percent. Its circuit schematic diagram is shown in Fig. 26.19. The Kelvin bridge is usually designed in such a way that R_a and R_u, and R_b and R_v are ganged (i.e., $R_b/R_a \approx R_v/R_u$). If each contact and lead resistance is small enough, compared with each resistor value, the balancing condition becomes:

$$R_x = R_s \frac{R_b}{R_a} + \left(\frac{R_b}{R_a} - \frac{R_v}{R_u}\right) \frac{R_u R_w}{R_u + R_v + R_w} \tag{26.21}$$

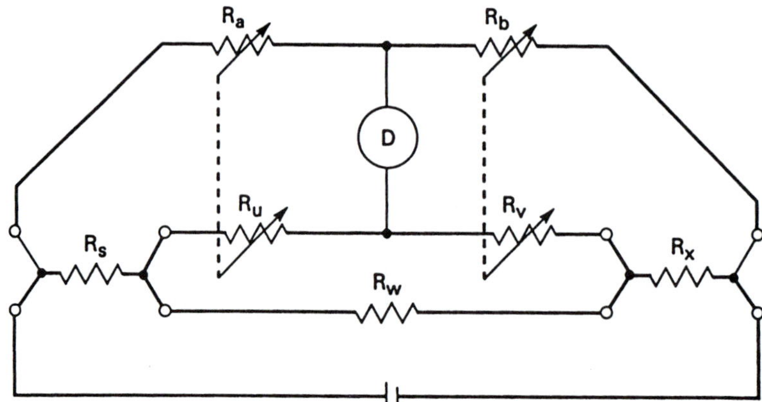

Figure 26.19 The Kelvin bridge.

If R_w is reasonably small, the second term of Eq. 26.23 becomes small, compared with the first term—even if the error of ganging is in the order of a few percent. Then it can be simplified to:

$$R_x = R_s \frac{R_b}{R_a} \tag{26.22}$$

Because this bridge has the second ratio arms, R_u and R_v, it is sometimes called the *Kelvin's double bridge.*

26.7.4 Milliohmmeters

A milliohmmeter with the four-terminal connection is designed so that a small resistance can be easily measured in various applications. Examples of a small resistance include the contact resistance of a connector and a switch, the internal resistance of a battery, and the series resistance of a fuse. Because the contact resistance of a delicate contact sometimes has dependence on the test-signal level, some milliohmmeters are equipped with a test-signal limiting circuit. Without this circuit, a large signal might appear across the test leads and "clean up" the contact contamination of a delicate device upon connection, resulting in a measured value of the contact resistance that is smaller than it should be. Both dc and ac milliohmmeters are commercially available. A dc meter is less expensive, owing to its simpler structure, but an ac meter can make measurements under the presence of a dc error signal in such cases as a battery measurement and a contact resistance measurement with thermal electromotive force presence.

Figure 26.20 is an example of the ac digital milliohmmeter. A *phase-sensitive detector* (*PSD*) is used to extract only the resistance element of a DUT—even if a large reactance element exists. The series resistance measurement of transformer windings is an example. A constant-current ac source is pro-

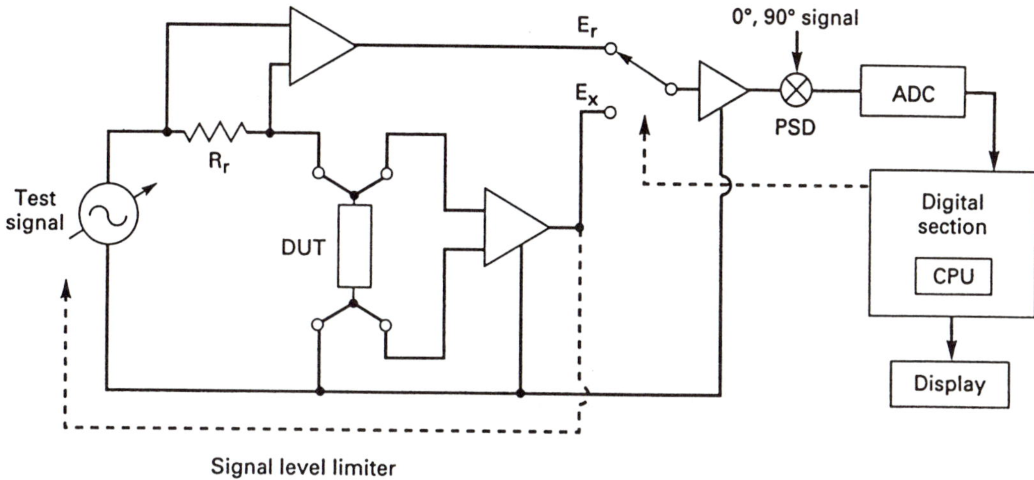

Figure 26.20 The ac milliohmmeter.

vided; thus, the output voltage from the PSD is proportional to the resistance element (R_x) of the DUT. With the test-signal limiting circuit, the signal voltage across the test leads is limited at the level of not more than 20 mV. Such a meter can offer a basic measurement accuracy of 0.4 percent of the reading.

26.7.5 High-resistance meters

An example of the high-resistance meter is shown in Fig. 26.21. A high-resistance meter typically consists of a high-voltage dc source and a current-voltage converter with the usual guard structure. Also, an electrode contact check mechanism is usually incorporated to distinguish a high DUT resistance from open or insufficient contact. Because the current to be detected is very small (sometimes in the order of nanoamperes or less) and noisy, enhanced noise-rejection capability is provided inside of the meter to obtain stable readings at the display. In this example, the ammeter section is completely isolated using a separate power supply, and the measured data and the control signal are transferred through an optical fiber cable between the ammeter section and the digital section. A typical high-resistance meter can offer a basic measurement accuracy of 0.6 percent of the reading up to 10^{16} Ω.

Some meters also have a timer function, with which the resistance of a device can be measured at a specified time after applying a test voltage to a DUT. This is convenient because many high-resistance and insulation materials have a large time constant and absorption current, and measured resistance values have time dependence.

Because a high dc test voltage is applied to a DUT, safety considerations

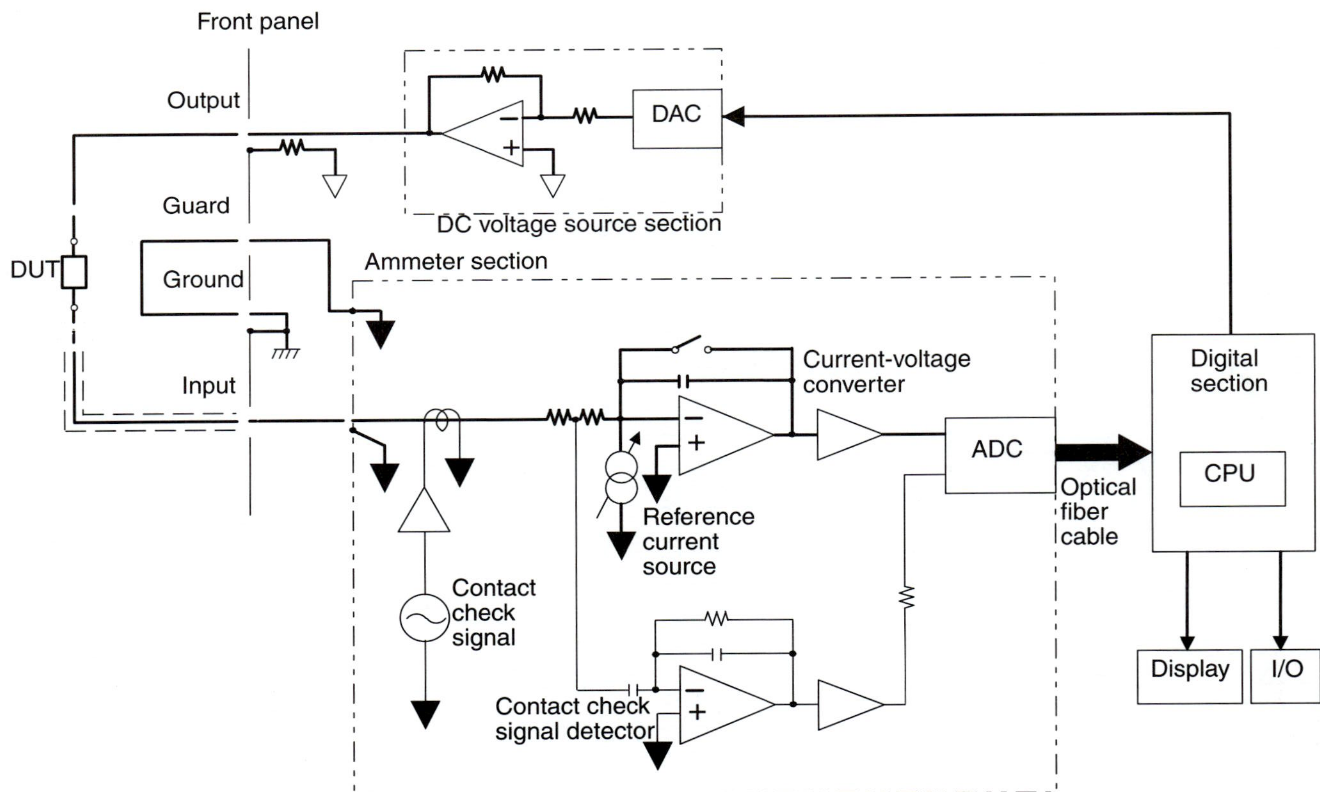

Figure 26.21 The high-resistance meter with contact check function.

should be given to the measurement system (high-resistance meter and test fixture). Provision of some interlock mechanism is advisable so that the operator can be kept from touching places where high voltage is exposed during operation.

26.7.6 Transformer-ratio-arm bridges

With a well-constructed transformer, an ac voltage or current ratio can be precisely determined by the ratio of wire windings; thus, accurate DUT values can be read by counting the number of windings of the transformers. The transformer-ratio-arm bridge implementations have many variations. An example is shown in Fig. 26.22, where the combination of the transformer taps, several internal reference capacitors, and resistors are used to cover a wide range of DUT values. Because a transformer cannot be considered ideal at higher frequencies, a typical transformer-ratio-arm bridge accuracy is specified at 1 kHz, and error terms in proportion to the measurement frequency squared are added above 1 kHz. A typical transformer-ratio-arm capacitance bridge has basic measurement accuracy of 0.01 percent of the reading, which is mainly limited by the uncertainty of the internal standard capacitors.

A microprocessor-controlled transformer-ratio-arm capacitance bridge is available for automated applications. Its test frequency can be chosen from 50 Hz up to 20 kHz and automatic balancing is realized with all of the previously

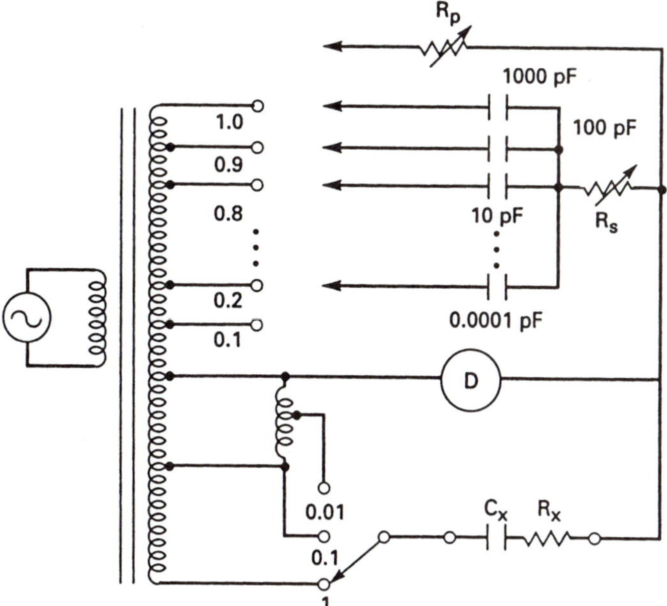

Figure 26.22 A transformer-ratio-arm bridge.

mentioned features preserved. Its basic accuracy is specified as 5 ppm of the reading at 1 kHz after calibration.

Because of their inherent high precision and long-term stability, transformer-ratio-arm bridges are mostly used in the standard lab environments.

26.7.7 Capacitance meters

In addition to the transformer-ratio-arm bridges previously described, many digital capacitance meters are commercially available. Figure 26.23 is an example of 1-MHz digital capacitance meter. In this meter, 1:1 ratio transformers are used to implement an octave ranging sequence, instead of the popular

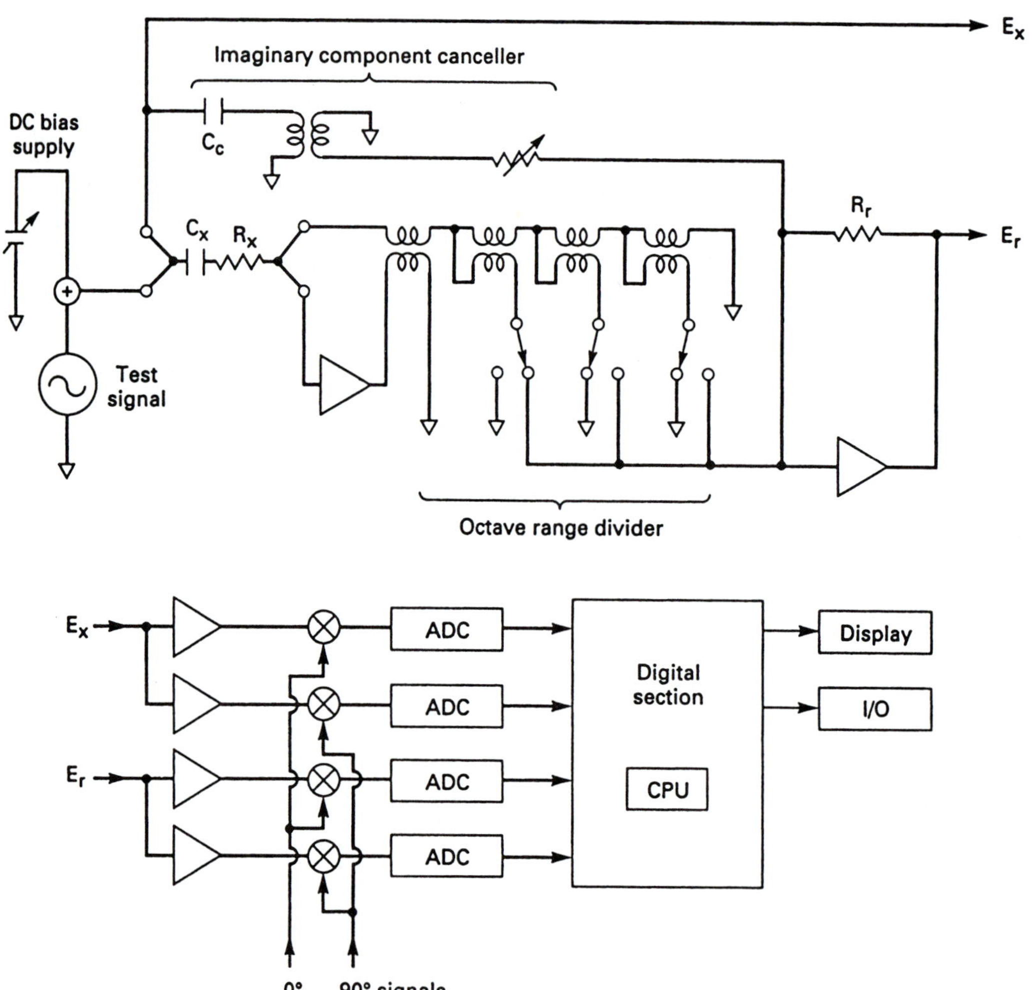

Figure 26.23 The 1-MHz digital capacitance meter.

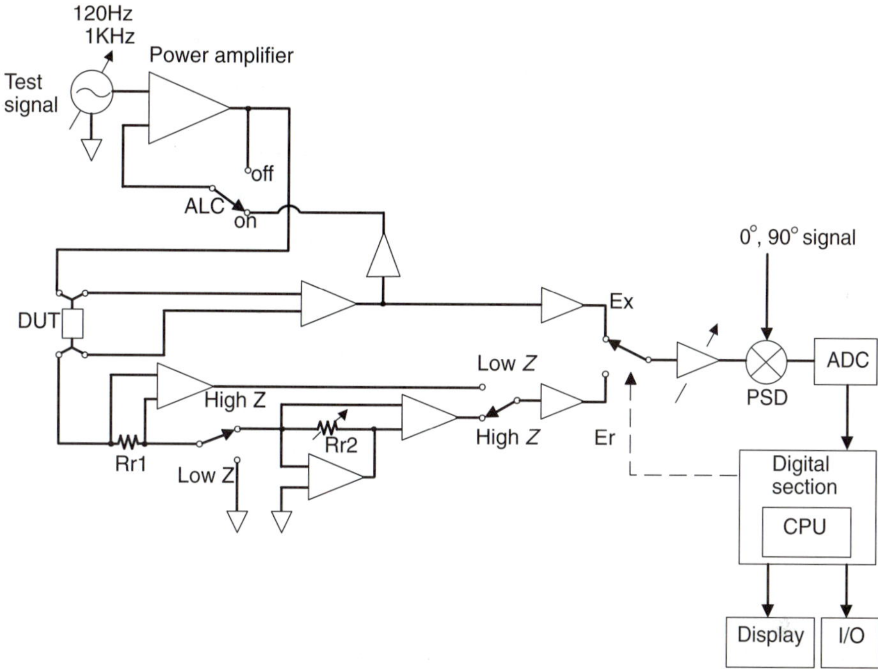

Figure 26.24 The high-capacitance meter with ALC.

decade-ranging sequence to realize good linearity and to guarantee long-term stability. Also, the imaginary component signal-canceling circuit is provided to utilize the full dynamic range of the *analog-to-digital converters* (*ADC*). This meter can offer a basic measurement accuracy of 0.05 percent of the reading after performing the proper calibration procedure.

Another example of the digital capacitance meter is shown in Fig. 26.24. It is designed to measure relatively high values of capacitors. With a high-capacitance DUT, the measured impedance becomes small and the signal level across it, therefore, tends to become small. Because the industry standard often requires a certain level of test signal across DUT (even under such a condition), 1 V for 10 μF at 1 kHz, for example, a powerful test-signal source and an *automatic signal-level control circuit* (*ALC*) are provided. Its accuracy is specified as 0.3 percent of the reading at a typical measurement condition.

26.7.8 LCR meters

Many digital LCR meters are commercially available these days. They usually measure DUT impedance, and the internal microprocessor converts the measured impedance values into user-specified parameter values such as *L, C, R, D,* and *Q*. Unlike the dedicated meters, such as the digital capacitance

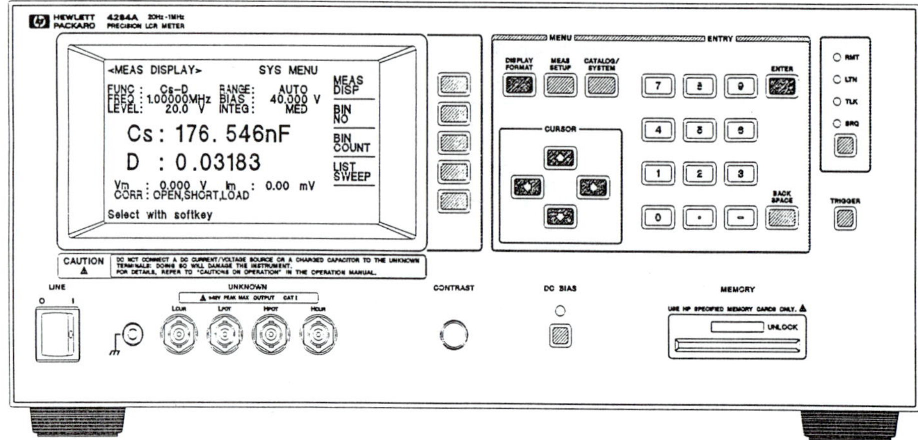

Figure 26.25 Typical digital LCR meter front panel.

meter covered in the previous section, a typical digital LCR meter can cover a variety of DUT types with somewhat moderate measurement impedance range. A typical digital LCR meter front panel is shown in Fig. 26.25. A test fixture is attached to the LCR meter via the four BNC connectors located at the lower left area of the front panel.

A schematic diagram example of inductance measurement with a digital LCR meter is shown in Fig. 26.26. It can supply a constant current/voltage ac test signal of a specified level to the inductor under test. It can also supply a dc bias current up to 40 A with the bias current source unit. Because the permeability of a magnetic core material varies according to the applied magnetic flux density, it is necessary to evaluate the inductor near the actual operating condition. When an inductor is used as a choke inductor in a switching power supply, for example, its inductance significantly decreases under a large signal level and even saturates (inductance becomes almost zero) in extreme cases.

Recently, many *surface-mount electronic devices* (*SMDs*) are popularly used to reduce the size of electronic products. As the actual operating frequency of such devices becomes higher, with the cellular phones and BS tuners, for example, it is necessary to evaluate them at high frequencies. A commercially available digital LCR meter can measure devices up to 1 GHz. To make measurements at high frequencies, a proper test fixture should be chosen and an appropriate calibration procedure should be performed before connecting a DUT to minimize the measurement uncertainly introduced by the unwanted parasitic elements associated with the test fixture.

Because it is difficult to obtain accurate inductance standards, stable resistors and capacitors are usually used as the standards for calibrating the digital LCR meter. Capacitance is treated as negative inductance and its polarity is converted to positive by the microprocessor in the inductance meter to

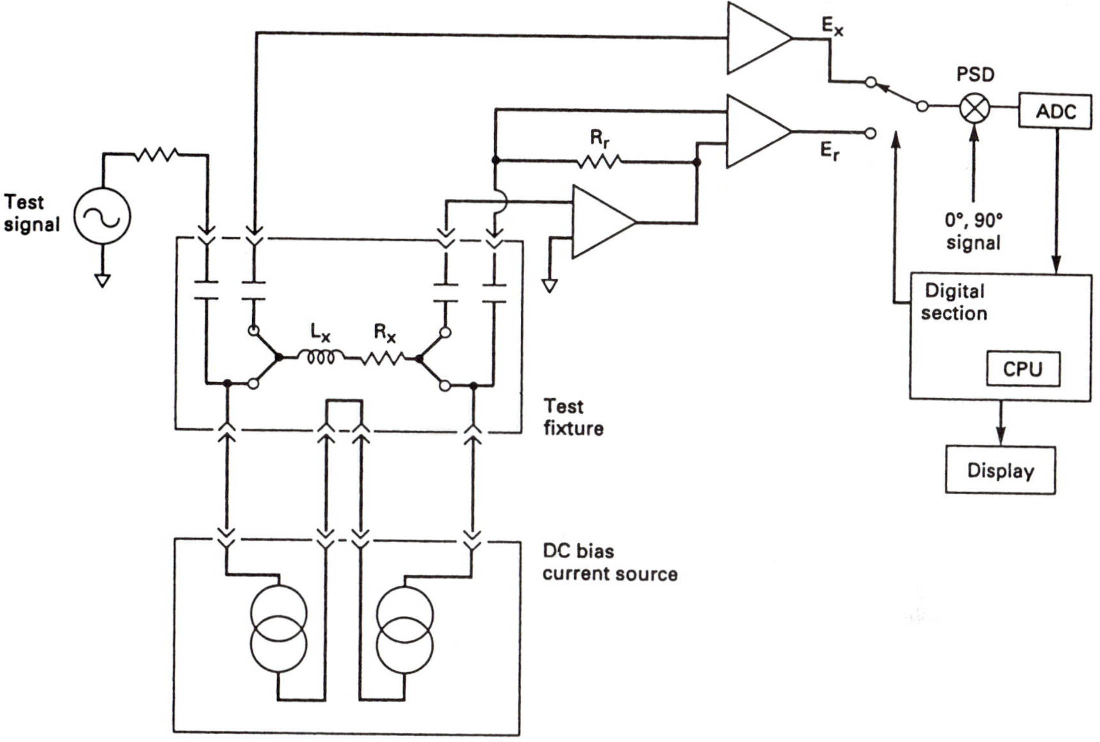

Figure 26.26 Inductance measurement with the digital LCR meter and the dc bias current source.

guarantee the accuracy—even for inductance measurements. Thus, a typical 1-MHz digital LCR meter can offer a basic measurement accuracy of 0.05 percent of the reading, whereas the 1-GHz LCR meter provides a basic measurement accuracy of 1 percent of the reading after proper calibration.

26.7.9 Impedance analyzers

In some cases, the impedance value of a device is more important than its *L, C,* or *R* value, and the impedance variation over frequency or other parameters is of primary concern. Examples include evaluation of a bypass capacitor, an *electromagnetic interference* (*EMI*) choke inductor, and the input-output impedance of a filter. A dielectric material evaluation over frequency and temperature variation is extremely important to design a stable dielectric resonator and filter. An impedance analyzer is usually equipped with a sweep function of frequency and other parameters, so the impedance of a device can be evaluated over a specified range of a parameter.

An example of the impedance analyzer covers the test-frequency range from 100 Hz to 40 MHz with the four-terminal-pair connection and up to 100

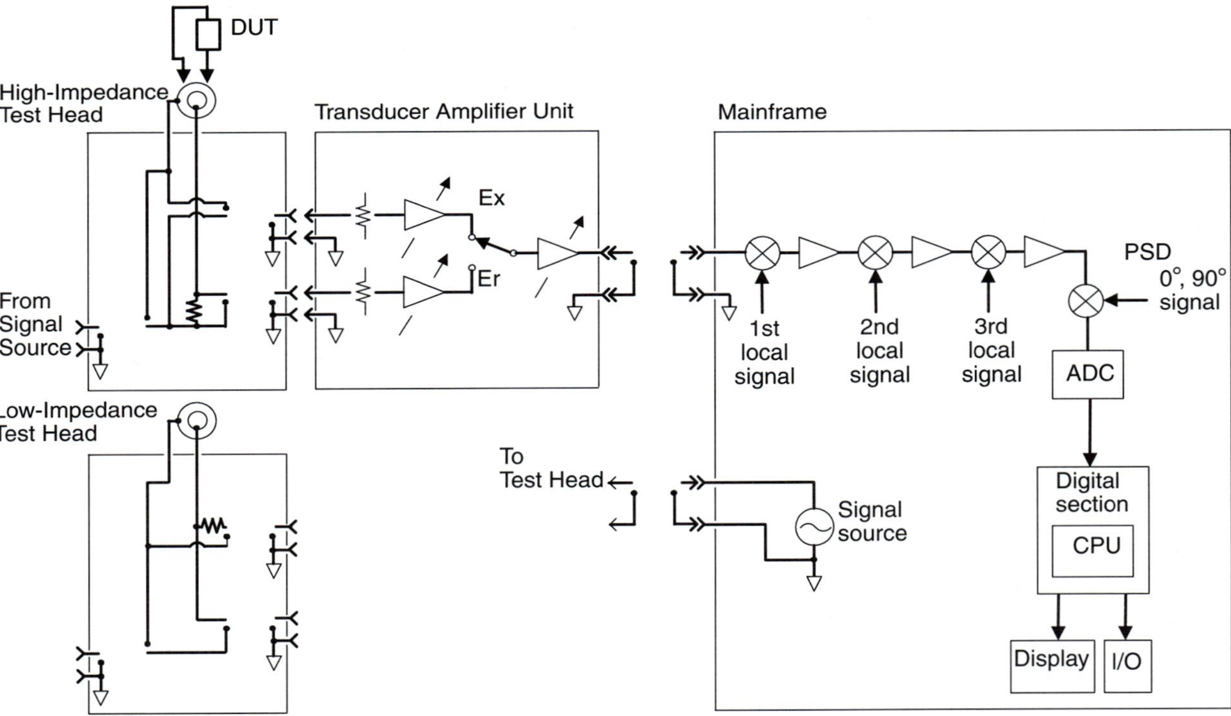

Figure 26.27 The 1.8-GHz impedance analyzer with two test head configurations for wide range of DUT impedance.

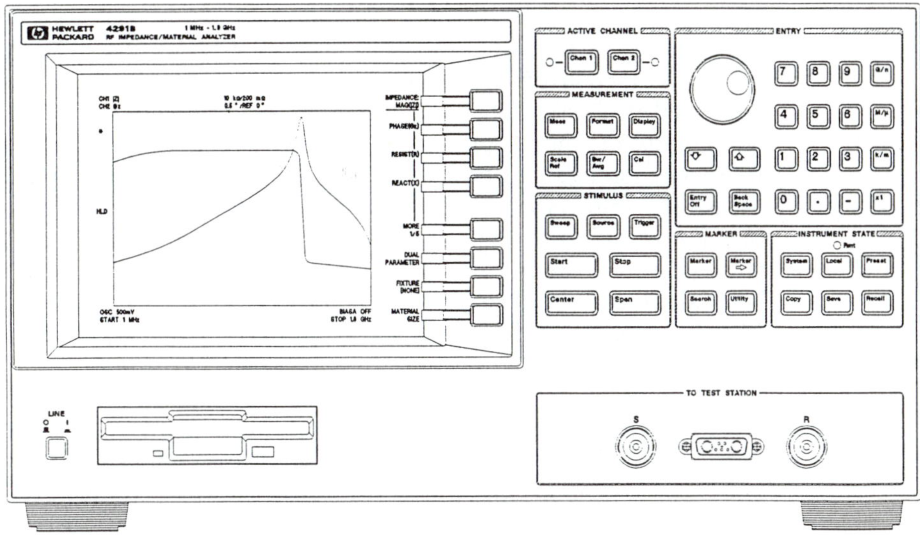

Figure 26.28 An example of impedance analyzer front panel.

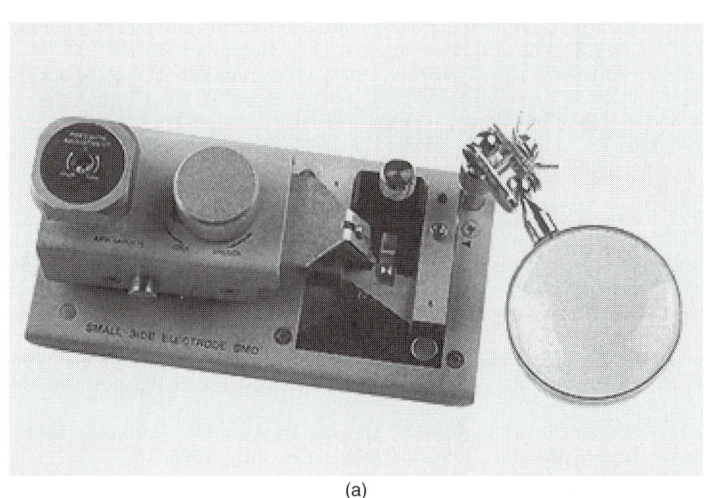

(a)

(b)

Figure 26.29 Examples of high-frequency test fixtures: SMD test fixture (A) and dielectric material test fixture (B).

MHz with an impedance probe; the basic measurement accuracy is 0.17 percent of the reading with the former configuration and 1.5 percent with the latter. Its measurement principle is based on the automatic bridge method described in Section 26.4.4. As an example of device analysis, the variation of magnitude and/or phase of the device impedance relative to the test-frequency change can be graphically displayed. The internal microprocessor allows the impedance values to be converted into *L, C, R,* and other values, too. For simple devices (such as a capacitor, an inductor, or a resonator), the equivalent circuit analysis function provides an estimation of the device's equivalent circuit element values, based on the frequency characteristics. This function is useful to evaluate the parasitic elements associated with a device.

Figure 26.27 illustrates another example of the impedance analyzer up to 1.8 GHz with basic accuracy of 0.8 percent of the reading. Two types of test heads are provided and appropriately selected, depending on DUT impedance. Dc bias can be supplied to the device via the test head. The analyzer has the capability of controlling an external oven for evaluation of DUT material property over temperature variation up to 200°C. The analysis results are graphically displayed on the screen, as shown in Fig. 26.28.

Figure 26.29 shows two examples of test fixtures used with the analyzer to evaluate devices at high frequencies. At high frequencies, such as 1.8 GHz, a significant amount of error caused by inappropriate DUT connection would be introduced, and proper use of a test fixture with consultation by its supplier is strongly recommended.

Bibliography

1. *HP 42851A Precision Q Adaptor Operation Manual (Second Edition)*, Hewlett-Packard Co., December 1995.
2. Cutkosky, R. D., "Techniques for Comparing Four-Terminal-Pair Admittance Standards," *Journal of Research of the N.B.S.,* Vol. 74C, Nos. 3 and 4. July-December 1970.
3. *HP 16380A Standard Air Capacitor Set Operation and Service Manual,* Hewlett-Packard Co., 1990.
4. *HP 4338B Milliohm Meter Operation Manual (Third Edition)*, Hewlett-Packard Co., June 1998.
5. *HP 4339B High Resistance Meter Operation Manual (Third Edition)*, Hewlett-Packard Co., June 1998.
6. *Type 1615 Capacitance Bridge-Type 1620 Capacitance Measuring Assembly,* QuadTech Inc., December 1992.
7. *Model 2700A Multi-frequency Ultra-precision Capacitance/Loss Bridge Preliminary Specifications,* Andeen-Hagerling Inc., 1998.
8. *HP 4278A 1kHz / 1MHz Capacitance Meter Operation Manual,* Hewlett-Packard Co., December 1996.
9. *HP 4268A High Capacitance Meter Operation Manual,* Hewlett-Packard Co., December 1998.
10. *HP 4284A Precision LCR Meter Operation Manual (Fifth Edition)*, Hewlett-Packard Co., December 1996.
11. *HP 42841A Bias Current Source Operation Manual,* Hewlett-Packard Co., December 1996.
12. *HP4286A RF LCR Meter Service Manual,* Hewlett-Packard Co., July, 1995.
13. *HP 4194A Impedance / Gain-Phase Analyzer Operation Manual,* Hewlett-Packard Co., December 1996.
14. *HP 4291B RF Impedance / Material Analyzer Service Manual,* Hewlett-Packard Co., May 1998.

Chapter

27

Semiconductor Test Instrumentation

James L. Hook
Agilent Technologies
Santa Clara, California

27.1 Introduction to Semiconductor Test Instrumentation

At the beginning of the semiconductor age, devices were simple two- or three-pin components, such as diodes and transistors, most of which could be readily tested with general-purpose electronic instruments. Today, after 50 years of ever-accelerating technical advancement, semiconductor devices are among the most complex creations of mankind. A modern microprocessor, for example, requires the coordinated contributions of thousands of technologists working together for several years to design and manufacture the many millions of devices per year that are routinely used in computers, communications systems, automobiles, and games.

Along with the growth in semiconductor complexity, it was necessary to invent new test instruments, and combinations of instruments (referred to as *semiconductor test systems*), capable of coping with the testing needs of ever-increasing performance and variety of devices. This chapter describes the most common instruments and test system types in use in the design and manufacture of semiconductors today. Of necessity, the level of detail must be restricted while examining this area because the variety of instruments and systems on the market today would require volumes, if described in detail. In the following sections the term *Semiconductor Test Equipment, STE* is used to refer to semiconductor test instruments and systems as a whole.

27.1.1 Where semiconductor test equipment is used

STE is used at several design and process steps that are key to bringing semiconductor devices to market. This process beings in the R&D lab, where device design and fabrication process teams have completed the first units of a new device type in the form of a silicon wafer containing hundreds of devices in an array. Characterization testing must be performed on the devices to determine if the devices have been properly designed and fabricated and will perform as expected. This characterization step is accomplished using an appropriate test system that performs a series of functional and performance tests on the devices and captures the results for analysis. This build/characterize cycle is usually repeated several times as problems are found and fixed until the device is deemed ready for *manufacturing release*—a term that implies that the devices can be manufactured in volume with high enough yield to be profitable in the marketplace.

Once in the manufacturing phase, STE is used at the end of several process steps to be sure that the fabrication process is proceeding under control, and then after the wafers are finished, to verify that the devices are fully functional and perform to specifications before delivery to the end user. The STE used in the fabrication process is similar for all types of devices because the parameters being measured include simple resistance, capacitance, breakdown voltage, leakage current, and gain. These tests are performed on special test cells embedded in each wafer along with the devices being manufactured. Test systems used during the wafer-fabrication process are referred to as *Parametric Test Systems* because they measure very basic semiconductor parameters on the wafers.

Once the wafers have completed the fabrication process, everything changes. Now, instead of being able to use a single type of equipment for testing all types of wafers, the STE used must be capable of testing the particular type of device fabricated on each wafer. Because this could range from simple transistors to complex microprocessors, a wide variety of STE types are in use today. Figure 27.1 shows the typical testing steps used in semiconductor manufacturing. In addition to the differences between test systems for different types of devices, often significant differences occur among the wafer-sort test systems and final test systems used to test the same parts. These differences are driven by differing test requirements at each test stage, as well as test throughput requirements, device yield (percent of good devices at each test stage) and overall cost-of-test considerations. Wafer-sort test systems often have lower performance and lower cost than final test systems for the same device. However, because wafer sort often involves more lengthy tests to thoroughly test device functionality, in the end, wafer sort and final test frequently have about the same economic impact on the overall cost of test. In the hyper-competitive semiconductor marketplace of today, overall cost of test is the most optimized parameter for semiconductor test systems.

27.1.2 Instrumentation components of modern semiconductor test systems

A semiconductor test system can be described as a tightly integrated group of specialized instruments whose purpose is to economically test the function-

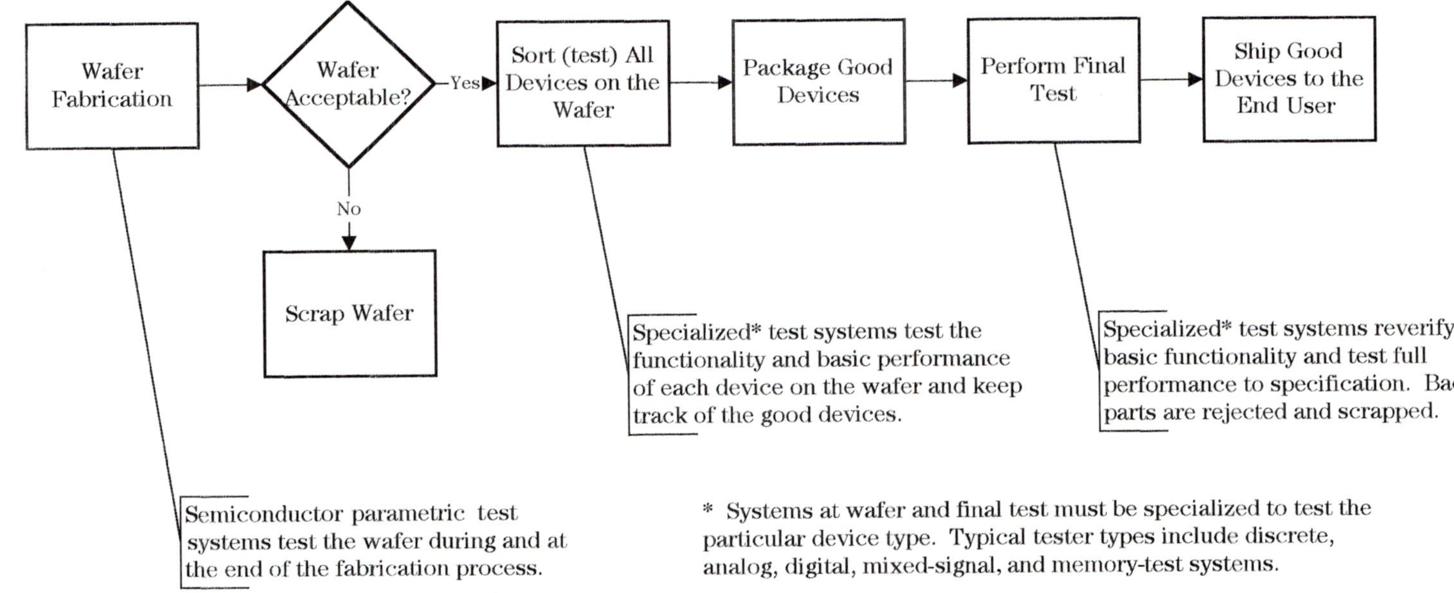

Figure 27.1 Typical testing steps used in the semiconductor-manufacturing process.

TABLE 27.1 Instruments Used in Semiconductor Test Systems

Instrument type	Main use in semiconductor test systems	Parametric	Discrete	Analog	Digital	Mixed-Signal	Memory
General-Purpose Instruments							
Waveform Generator	Generate analog stimulus waveforms	•	•	•		•	
Signal Processors	Analyze analog waveforms (DSP-based)			•		•	
Power Supply	Provide power for test system	•	•	•	•	•	•
Oscilloscope	Capture and analyze device waveforms		•	•		•	
RF Signal Generator	Generate RF stimulus waveforms		•	•		•	
Time/Freq. Source	Time reference for system and DUT					•	
Spectrum Analyzer	Analyze waveform frequency content			•		•	
Lightwave Source	Generate signals for lightwave devices		•				
Lightwave Analysis	Analyze signals from lightwave devices		•				
Logic Analyzer	Capture and display digital waveforms				•	•	
Protocol Analyzer	Analyze communications protocol			•		•	
Instruments Unique to Semiconductor Test							
Parametric Measurement Unit	Force/measure voltage and current	•	•	•	•	•	•
Pin Electronics	Stimulate/measure digital input/outputs			•	•	•	•
Sequencer/APG	Generate sequences of digital vectors			•	•	•	•

ality and performance of a particular set of semiconductor device types at one or more stages of device design or manufacturing. Many of the specialized instruments integrated within typical semiconductor test systems have capabilities similar to general-purpose instruments available on the market. However, in virtually all cost-sensitive semiconductor test situations, general-purpose instruments end up being too costly, too large, or too slow to satisfy the overall cost-of-test requirements. Thus, even common electronic test instruments must be redesigned for integration within test systems to achieve the overall system cost and performance goals.

In addition to the specialized versions of standard test instruments included in semiconductor test systems, three instrument types are unique to semiconductor testing. These instruments have evolved over the last three decades as the requirements for semiconductor testing became more and more specialized. Table 27.1 summarizes both the specialized versions of standard electronic instruments, as well as the unique semiconductor test instruments typically found in the various types of semiconductor test systems today (see Section 27.4 for a brief description of test-system types). Because the standard electronic instruments mentioned in Table 27.1 are thoroughly described in other chapters of this book, this chapter describes only the unique semiconductor test instruments in detail.

27.2 Electronic Instrumentation Unique to Semiconductor Testing

Three unique instrument types have evolved to meet particular semiconductor test needs: the *Parametric Measurement Unit* (*PMU*), the *Digital Pin Electronics* (*PE*), and the *Sequencer/APG.* The PMU is used in every type of test system while the PE and Sequencer/APG are found only in test systems that must test digital components. These three semiconductor test instrument types are described in some detail in the following sections.

27.2.1 The parametric measurement unit (PMU)

The PMU is the instrument used to perform most types of dc stimulus and measurement functions on semiconductor devices today. Common tests performed on devices with PMUs include input/output leakage, output drive voltage/current capability, power supply voltage/current measurement, etc.

At the most basic level, the PMU is a simple instrument that is capable of forcing either a voltage or current and simultaneously measuring the resulting current or voltage, respectively, on whatever load the PMU is connected to (a *device under test, DUT,* or another instrument). Because PMUs are called upon to do a wide range of stimulus and measurement functions, several PMUs often have quite different features and performance levels within a test system. In fact, it is common to have three distinct types of PMUs within a system each optimized for making a specific class of measurements efficiently. First, there is always a main PMU with several voltage and current ranges that is used to make the most precise measurements on the DUT. Next are special PMUs, optimized to supply power to the DUT, which must be capable of driving large capacitive loads and supplying up to several amps of current on the highest power devices. Finally, small PMUs are associated with each tester pin, or channel, to allow input/output leakage and other low-power measurements to be made on many DUT pins simultaneously to facilitate high throughput. Altogether, it is not uncommon for a single large digital test system to include over 1000 PMUs making the PMU one of the most common instruments in the semiconductor test industry. PMUs are often referred to by other names, including *Source/Measurement Unit* (*SMU*), *Voltage/Current Supply* (*VI*), *Force/Measure Generator* (*FMG*), *Device Power Supply* (*DPS*), etc.

PMU designs vary widely because of the range of measurements that they are required to perform. For example, Parametric Test Systems used in wafer production testing are called upon to make measurements down into the low femtoamp and microvolt ranges; in standard digital test systems, nanoamp and millivolt measurements are adequate. Discrete power-transistor test systems, on the other hand, require measurements into the kilovolts and tens of amps.

Before looking at how a PMU operates in detail, it is helpful to understand how PMUs are used for making basic measurements. Although it is somewhat of an oversimplification, the PMU basically operates by attempting to force a voltage (force voltage) on its load while monitoring its output current. If the output current ever reaches a preprogrammed limit value (I_{limit}), the PMU will lower its output voltage, as required, to keep the output current at I_{limit}. Therefore, as long as I_{limit} has not been reached, the PMU acts as a voltage source. However, if I_{limit} is reached, the PMU switches modes and acts as a current source. Then, if the load impedance changes so that the current decreases below the I_{limit}, the PMU switches back into the voltage source mode. Based on an understanding of the load and the required measurement parameters, the test programmer can readily use the PMU as a power supply, an input/output leakage measurement instrument, a device output-current drive-measurement instrument, a voltmeter, an ohmmeter, or an ammeter. Figures 27.2A, 27.2B, and 27.2C show how the PMU is used to make a number of common measurements. Assume that a DUT requires a +5-V V_{cc} supply, capable of sourcing 100 mA, as shown in Fig. 27.2A. The PMU can be programmed to supply V_{cc} to the part by setting the force voltage to +5 V and the I_{limit} to 200 mA (greater than the maximum required, but low enough to protect the fixture from damage if something goes wrong). When enabled, the PMU forces +5 V onto the device V_{cc} pin. During the entire test, the PMU continually monitors the load current in case the device fails and becomes a short circuit, for example. If this happens, once the current reaches 200 mA the PMU switches into current source mode and limits the current to 200 mA—even if it has to lower the force voltage to zero volts in the process. By placing the DUT into various operating modes and measuring the resulting V_{cc} current, the PMU can verify all device power specifications.

Figure 27.2B shows how a PMU is used to measure device input leakage. Assume that a CMOS IC has an input leakage specification of 1 μA (a maximum at +5 V). To verify that the input meets this specification, the PMU is programmed to force +5 V on the device input with I_{limit} set to full scale on a current range that can measure 1 æ A accurately (a 2–μA range is assumed in Figure 27.2B). To make the measurement, the PMU forces +5 V, waits for the voltage to settle, and then measures the actual leakage current flowing into the DUT input. If this current is ó1 μA, the device passes the test.

Figure 27.2C shows how a PMU can be used to measure a device output drive current capability. Assume that a DUT output is specified to drive at least 3.0 V at 2.0 mA into a load. The PMU would be programmed to force 0.0 V on the DUT output pin with the I_{limit} set to 2 mA. The test is then run by placing the DUT output into a mode where it is attempting to drive a "1" (high output). When the PMU is turned on, it will attempt to drive 0.0 V onto the DUT output. Because the DUT is attempting to put out a higher voltage, it will drive current into the PMU until it reaches the programmed I_{limit} of

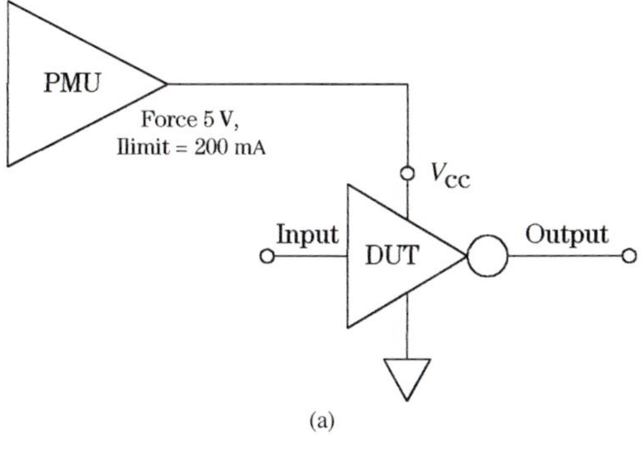

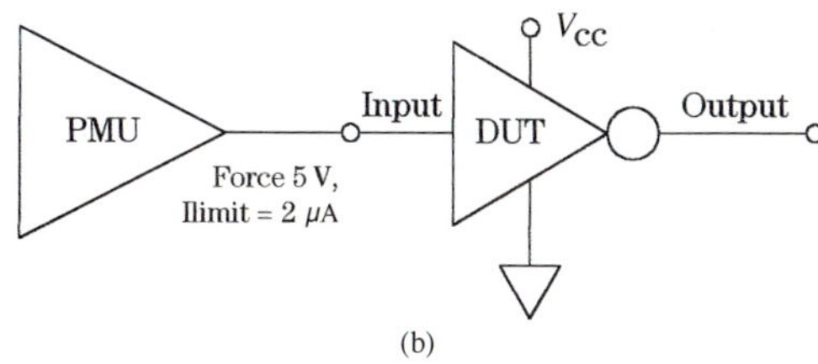

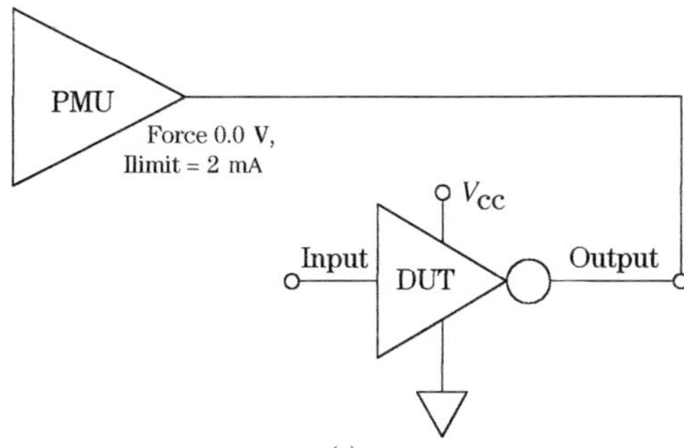

Figure 27.2 The PMU as a device power supply (a), as a leakage test instrument (b), and as an output-drive test instrument (c).

2 mA. When this happens, the PMU switches into the current source mode and its output voltage moves to whatever drive voltage the DUT is capable of sourcing 2 mA into. The PMU output voltage is then read and if it is greater than 3.0 V, the device passes the test.

Based on these examples, it is easy to explain how the PMU can also operate as a voltmeter, an ammeter, or an ohmmeter. Voltmeter operation is achieved by setting the I_{limit} to a low value on its lowest current range and its force voltage to 0.0 V. Then, when the PMU is connected to any active signal source, it will always operate in the current source mode, but at such a low current that it does not "load" the node being measured significantly. Measuring the PMU output voltage at this time essentially yields a voltage reading just as if a voltmeter was connected to the signal source. Operation as an

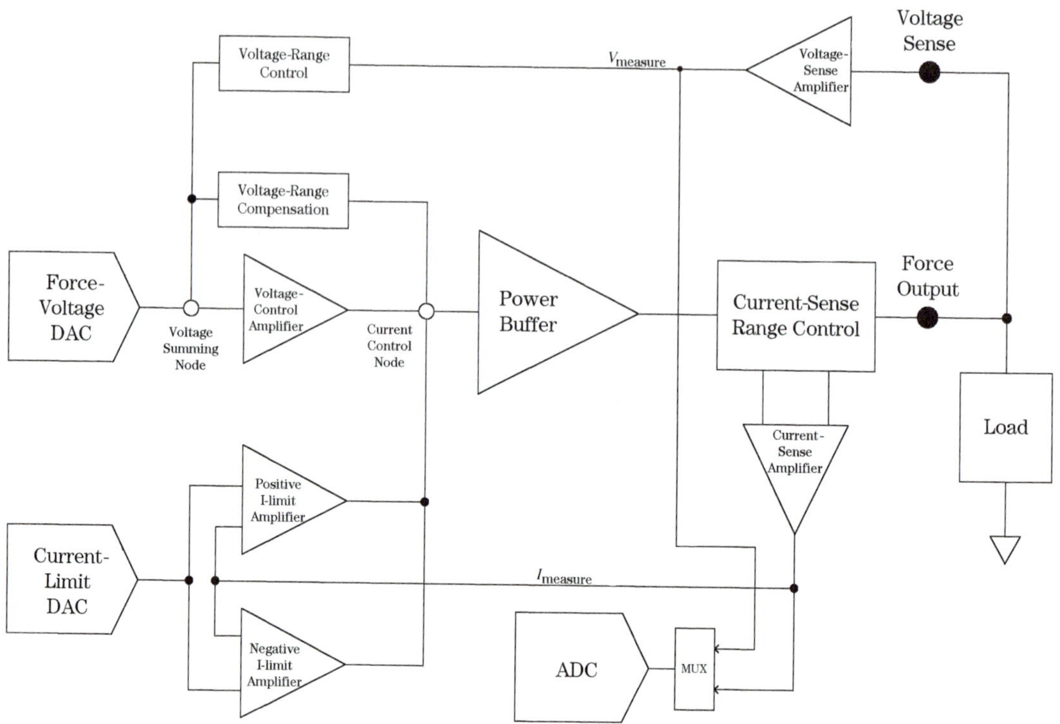

Figure 27.3 The PMU block diagram.

ammeter is accomplished by forcing a desired voltage on a load, with an adequately high I_{limit}, and then measuring the resulting source/sink current. The PMU can be readily used as an ohmmeter, if the load is a resistance, by simply forcing a known voltage on the load, measuring the resulting output current and then using Ohm's Law to calculate the resistance using the equation $R = V/I$.

Figure 27.3 shows the simplified block diagram of a classic PMU used in semiconductor test systems. Several control circuits determine the voltage and current levels that the PMU will use to make the desired measurements: the force-voltage DAC, the current-limit DAC, the voltage-range control, and the current-sense range control. These functions are typically set before each measurement and then the PMU is enabled to perform the actual forcing and measuring functions.

The voltage range control and the force voltage DAC work together to determine the voltage that the PMU will attempt to force on the load. The voltage-range control determines the full-scale range over which the PMU will operate and the force-voltage DAC determines the actual forcing voltage within that range. Having multiple ranges allows the instrument to achieve much finer resolution (tens of microvolts) when used on a ±1-V range, for example, than would be possible on an instrument with only a single ±15-V range. This optimizes the force or measures accuracy, depending on measurement need. Typical full-scale ranges of ±1 V to ±15 V are common in IC test systems while programmable ranges in the kilovolts are often required in high-voltage discrete semiconductor test systems. The voltage range control circuitry consists of resistive dividers or, for low-voltage ranges, voltage amplifiers, that change the weighting factor of $V_{measure}$ going into the voltage-summing node. The voltage-range-compensation circuitry is closely associated with the voltage-range control and includes feedback networks that determine the voltage-control amplifier frequency compensation to be used during a test. Normally, a separate compensation network is associated with each voltage range to optimize the overall PMU settling time and stability for the selected range and expected loads.

Once the voltage functions are programmed, the current-limit functions are established, based on the current requirements of the measurement. This is accomplished using the current-sense range control and the current-limit DAC. As with the voltage-control functions, the current-sense range control determines the full-scale current that can be forced or measured and the current-limit DAC determines the actual current limit value within that range. The current-sense range control consists of resistors and switches, through which the PMU output current flows. The current-sense amplifier senses the voltage across the selected range resistor and sends the resulting $I_{measure}$ signal, which is proportional to the current flowing through the resistor, to the I_{limit} amplifiers, where it is used to control the PMU output voltage if the programmed current limit is reached. The $I_{measure}$ signal is also connected to the *analog-to-digital converter* (*ADC*) so that the actual PMU output current can be read by the test program. Most PMUs have several current ranges so that currents from a few nA to an amp (or more) can be forced or measured. This is necessary because DUT inputs typically have very low leakages to be measured while the power supply pins might require hundreds of milliamps and the PMU is called upon to make both types of measurements.

Before explaining how the PMU performs measurements, it is necessary to briefly explain the functions of the other circuit blocks in Fig. 27.3. The voltage-control amplifier is a summing amplifier that sums the force-voltage DAC, the $V_{measure}$ feedback coming from the voltage range control and the internal voltage feedback coming from the voltage range compensation circuitry. While continually summing these three inputs, it moves the output to the current-control node in such a way as to attempt to make $V_{measure}$ (as divided or multiplied by the voltage-range control) exactly equal to the voltage from the force-voltage DAC. The power buffer's role is to provide the voltage and current amplification needed to drive the load over the full operating range of the PMU. The output of the power buffer is connected to the current-sense range control so that the PMU's output current goes through the selected-range resistor on its way to the load. As explained earlier, a voltage proportional to the output current, $I_{measure}$, is generated by the current-sense amplifier and is used to control the current limit and provide output-current measurement capability via the PMU ADC circuitry. From the current-sense range control, the PMU output goes through relays and cables (not shown in the diagram) to the load.

Because the current-sense range control resistors, the relays and the cables all produce voltage drops as the output current flows to the load, it is necessary to sense the actual voltage at the load in order to achieve accurate voltage forcing. This is accomplished by connecting the voltage-sense amplifier directly to the load via the voltage-sense input. The voltage-sense amplifier has very high input impedance, so it is able to sense the load voltage without loading the output significantly itself. The output of the voltage-sense amplifier, $V_{measure}$, provides feedback to the voltage-control amplifier, after being scaled by the voltage-range control, and also connects to the measurement ADC to allow the actual output voltage to be measured, regardless of the PMU forcing state.

The I_{limit} amplifiers perform the important function of limiting the PMU output current if the load should fail or when the user is attempting to force a known current into the load. During measurements, the positive I_{limit} amplifier continually compares the $I_{measure}$ signal against the current-limit DAC voltage. If the PMU is sourcing current, and $I_{measure}$ is less than the DAC voltage, the positive I_{limit} amplifier does nothing. However, if $I_{measure}$ becomes greater than the current-limit DAC voltage, the positive I_{limit} amplifier becomes active and pulls down on the current-control node, overpowering the voltage-control amplifier, and lowering the PMU output voltage so that the output current stays precisely at the programmed current limit value. The negative I_{limit} amplifier works identically, except that it operates only when the PMU is sinking current. If the load later changes impedance so that it can support the full force voltage without exceeding the current limit, both I_{limit} amplifiers get out of the way and let the voltage control amplifier take over again.

Finally, the ADC and its input multiplexer (MUX) allow the instrument to read the actual PMU output current or voltage at the load during the measurement. This data can be read back into the test controller and compared against test limits to determine if the measurement passed or failed. The data can also be logged into a test result file for later analysis.

To illustrate more clearly how all of the above circuits work together during a typical PMU measurement, it is helpful to consider a simple example. Assume that the PMU load is a resistor that is specified to be 1 kΩ ±5%, and that you want to verify that the resistor is within specification with 5 V across it. A typical PMU with a 15-V voltage range and a 20-mA current range could be programmed to execute the sequence of operations below and return the actual measured resistance value to the test program as follows:

1. Set the voltage-range control to select the 15-V range.
2. Set the force-voltage DAC to +5 V.

3. Set the current-sense range control to select the 20-mA range.
4. Set the current-limit DAC to 10 mA (5 mA is typical, 10 mA keeps the current limit out of the way, unless the resistor shorts out during the test).
5. Enable the PMU: the voltage-control amplifier sums the force-voltage DAC and $V_{measure}$ signals, determines that the load voltage is too low, and begins to increase the voltage on the current-control node. The power buffer amplifies this voltage and begins slewing the output voltage toward +5 V. As the output voltage increases, the current through the load increases toward 5 mA. When the output voltage finally reaches exactly +5 V, as sensed by the voltage-sense amplifier, the voltage-control amplifier stops increasing the output voltage and the force voltage remains stable at +5 V.
6. The system controller reads the $I_{measure}$ voltage and converts it to a current value, based on the selected current range (for example, 5.1 mA) and also reads $V_{measure}$ and finds it to be 5.0 V.
7. The system controller divides 5 V by 5.1 mA (Ohm's Law) and determines that the resistor is 980.4 Ω, well within the ±5% specification (950 Ω to 1050 Ω).
8. The PMU is then disabled and the measurement is complete.

As occasionally happens during semiconductor testing, the load might be the on-resistance of an active device and this device might fail during the test and become a short circuit. If this had happened in this example, the PMU would have behaved differently from step 5 onward, as follows:

5. Enable the PMU: the voltage-control amplifier sums the force-voltage DAC and $V_{measure}$ signals, determines that the load voltage is too low and begins to increase the voltage on the current-control node. The power buffer amplifies this voltage and begins slewing the output voltage toward +5 V. As the output voltage increases, the current through the load increases toward 5 mA. However, assume that at 3.5 V, the load fails and becomes a low-resistance short circuit. The $I_{measure}$ signal would quickly detect that the output current is now >10 mA and that the positive I_{limit} Amplifier pulls down on the current-control node until the output voltage comes down to a low enough voltage to result in exactly 10 mA of output current.
6. The system controller reads the $I_{measure}$ voltage and converts it to a current value, based on the selected current range, and finds that the current is 10.0 mA. It also reads $V_{measure}$ and finds that it is 0.01 V (for example).
7. The system controller divides 0.01 V by 10.0 mA (Ohm's Law) and determines that the resistor is 1.0 Ω, well outside of the ±5% specification (950 Ω to 1050 Ω).
8. The PMU is then disabled and the measurement is complete.

In summary, the PMU is an extremely versatile dc measurement instrument that is widely used in semiconductor test systems today. In fact, in addition to its usefulness for making a wide variety of measurements on the DUT, the PMU is often used to perform system dc calibration and diagnostic measurements on other instruments in the test system itself.

27.2.2 The pin electronics instrument (PE)

During the early years of semiconductors, all devices were analog in nature and could be tested using PMUs, plus traditional waveform generation and measurement instruments. However, with the development of digital circuitry, test systems had to cope with a new class of input/output function: the digital pin. Although inside the IC, the digital pin was essentially an analog circuit, digital input/output pin specifications quickly standardized around product families, referred to by names such as *Transistor-Transistor-Logic* (*TTL*) and *Emitter-Coupled Logic* (*ECL*). Unlike device analog pins, which were stimulated or measured using ac or dc voltages and/or currents, on digital pins the drive voltages and currents became of secondary interest compared to the timing and sequence of digital states going into or coming out of the pin. A new type of instrument, referred to here as the *Pin Electronics Instrument* (*PE*), began to provide test capability for digital pins. Today, in a world dominated by digital circuitry, the PE instrument has become the most complex and highly specialized instrument within semiconductor test systems. Because one PE instrument is required for controlling each device input and output pin, test systems used to test digital and memory devices now include hundreds of PE instruments.

Functional and performance requirements for PEs are derived largely from the particular type of digital devices being tested. For example, a tester targeted for testing nonvolatile memory devices at the wafer level can have relatively crude timing, operate at a few megahertz, and present a sequence of algorithmically generated digital states (vectors) to the device under test. At the other extreme, a state-of-the-art microprocessor tester must be capable of presenting billions of complex digital states to several hundred pins simultaneously at rates of several hundred megahertz with timing precision measured in the tens of picoseconds. Clearly, one PE cannot cost effectively satisfy all requirements.

Modern PEs are a collection of special-purpose circuits that, together, are capable of performing all stimulus and measurement functions on digital inputs, outputs, or I/O pins. These functions include testing for dc input leakage, output leakage, and output drive capability, as well as providing the high-frequency formatted and timed input waveforms to input pins and comparing device output pins against the expected response. Typical PE instruments include the superset of all digital stimulus and measurement circuits so that any tester pin or PE channel can be connected to any DUT digital

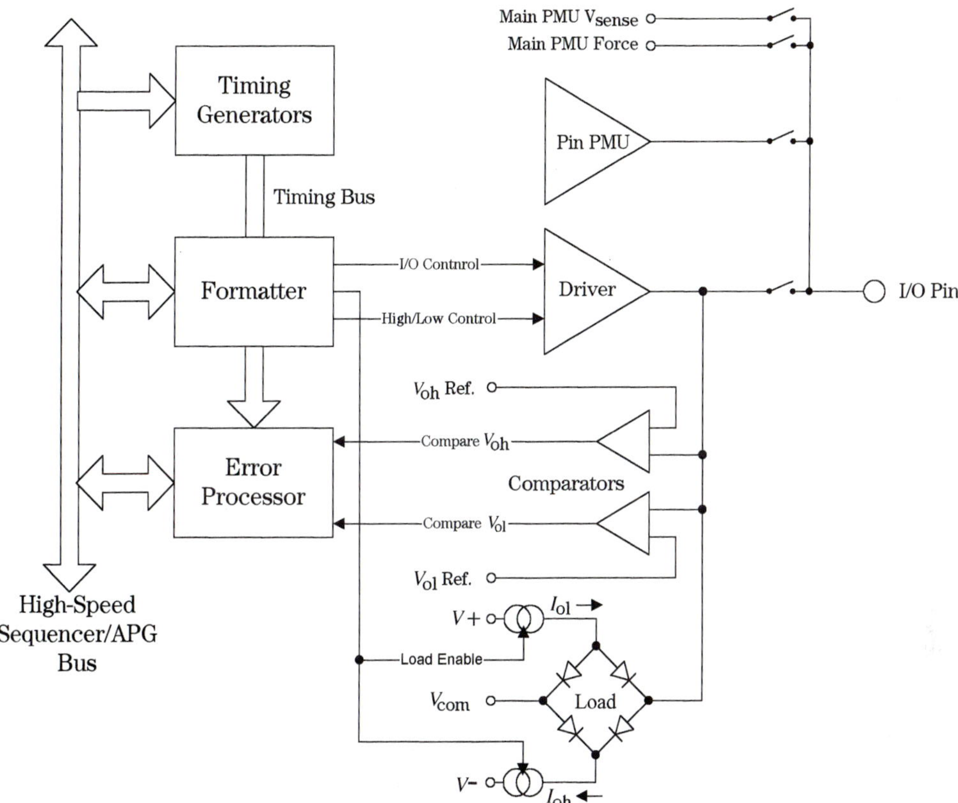

Figure 27.4 Pin electronics block diagram.

input, output, or I/O pin without regard to the type of function that the DUT pin represents. This capability greatly simplifies the construction of DUT interface fixturing.

Figure 27.4 shows the block diagram of a typical PE instrument found in digital, memory, and mixed-signal semiconductor test systems. The PE instrument interfaces between the high-speed test *Sequencer/Algorithmic Pattern Generator* (*APG*) and the device under test with one PE connected to each device digital pin. Each PE instrument includes seven basic blocks—each of which performs a unique function during dc or high-frequency digital test sequences.

For all the complexity of the PE instrument, it has only one measurement terminal, the I/O pin. The I/O pin is typically connected directly to a DUT pin through the tester DUT interface, which consists of coaxial cables, connectors, spring contacts, and specialized printed circuit boards that interface to device-handler contactors or wafer probe cards. The I/O pin terminal can connect the DUT to three measurement circuits through relays or solid-state switches: the external main system PMU, pin PMU, and the digital *Driver/Comparator/Load (DCL)* circuits. As shown in Fig. 27.4 and explained in the previous section, the main system PMU can be connected directly to the DUT pin for making precise dc voltage, current, or impedance measurements during the dc section of a device test. Alternatively, when test requirements allow, and throughput is more of an issue than measurement accuracy or versatility, the DUT pin can be connected to the pin PMU for performing leakage and other low current measurements. The pin PMU is simply a PMU with limited current capability that has been packaged within the PE instrument to allow low-power dc measurements to be made on many pins in parallel as a means of improving test throughput. This capability can become extremely important when dc measurements need to be made, for example, on each cell of a flash memory device during a test. With many millions of cells to test, serializing the measurements by using the main system PMU would extend test times tremendously because each measurement could require a millisecond or more to complete.

Once the dc portions of a test program have been completed, the PMU and pin PMU circuits are disconnected and the DCL circuits are then connected to the DUT for the high-frequency digital tests. The driver converts the digital signals from the formatter to the input levels specified for the specific device test. TTL-compatible devices, for example, might require input levels of 2.0 V for the input high voltage (V_{ih}) and 0.8 V for the input low voltage (V_{il}). The V_{ih}/V_{il} levels are usually set to several different sets of values during various parts of a test to be sure that the device functions properly under all specified voltage conditions. Once the V_{ih} and V_{il} levels have been programmed, the driver simply switches between V_{ih} and V_{il}, as commanded by the formatter section. Because many devices have I/O pins capable of acting as an input one instant and an output the next, the drivers must be capable of being turned off, or put into a high-impedance mode, when the device I/O pin is acting as an output to avoid drive conflicts (this on/off control function is shown as the I/O control signal in Fig. 27.4). Most test system drivers have a 50-Ω output impedance to allow driving through 50-Ω coaxial cables in the DUT interface while maintaining high signal fidelity (controlled rise/fall times and minimal overshoot).

The comparators are used to sense the state of the DUT output pins during digital test sequences. The comparators consist of two level comparator circuits, one comparing the DUT output against the programmed output low-voltage reference (V_{ol}) and the other comparing the DUT output against the programmed output high-voltage reference (V_{oh}). The comparators determine if the device is high, low, or in-between (tri-state) on a cycle-by-cycle basis. The outputs of the comparators connect to the error processor circuitry, which digitally compares the state of the V_{oh}/V_{ol} comparators, at a particular instant in time, against the expected state from the sequencer/APG. As with the drivers, the comparator reference voltages are often programmed to several different sets of values during a test sequence to be sure that the device outputs function properly under all test conditions. The comparators are also used to test device I/O pins when they are in the output mode. Comparators present a high input impedance load to the device output (1 MΩ + 50 pF typical) and thus, unlike the drivers and load, they can be left connected throughout the entire digital test sequence.

The load circuit, sometimes referred to as a *programmable load,* presents a current load to the device output during digital test sequences. This allows the device output to be properly (and conveniently) loaded with the specified high- and low-state current loads, I_{oh} and I_{ol}, respectively, during test sequences without the need for external load components. The load, if used, is only enabled during cycles when the device output is being tested with the comparators because leaving it enabled all the time would load the Driver output, resulting in V_{ih}/V_{il} level errors when the PE is in the drive mode. The circuit shown in Fig. 27.4 is a classic diode bridge commutating load circuit. To use the load, the user must first program the proper I_{oh} and I_{ol} current levels into the two current sources and set V_{com} to a voltage in-between the DUT V_{oh} and V_{ol} output levels. When the device is driving the load with a voltage less than V_{com}, the device must sink the I_{ol} current. Conversely, when driving the load with a voltage higher than V_{com}, the device must source the I_{oh} current. As the device transitions from high to low, or low to high, the diode bridge automatically switches from sinking to sourcing current as the device output voltage passes through the V_{com} level. I_{oh} and I_{ol} current values typically range from 100 μA to >20 mA, depending on the device technology.

During test execution, the DCL circuits are controlled individually per pin on a cycle-by-cycle basis. That is, some channels are driving inputs, other channels are comparing, and still others are inactive—and the usage changes every device cycle, as needed. The timing generators begin this control process by generating precise timing markers, relative to the beginning of a device or tester cycle, which are then used by the formatter and error processor to control when drivers transition, when comparators are strobed, and when drivers and loads are enabled or disabled. Timing generators have programming resolutions and accuracies that range from a few picoseconds to a nanosecond, or more, depending on the application. Most test systems use multiple independent timing generators for each pin to allow for greater per-pin flexibility and improved timing calibration.

The formatter circuits combine data state information and formatting commands from the high-speed sequencer/APG with preprogrammed timing markers from the timing generators to generate digital waveforms that are then used to control the drivers. Each formatter generates the waveform for a single pin that looks like the timing diagram in the device datasheet, except that the waveforms are strictly logic levels, that is, the V_{ih} and V_{il} levels have not yet been applied. Over the years, many formats have become standardized and are referred to using such industry-standard names as *Return-to-One* (*RO*), *Return-to-Zero* (*RZ*), *Non-Return-to-Zero* (*NRZ*), *Surround-by-Com-*

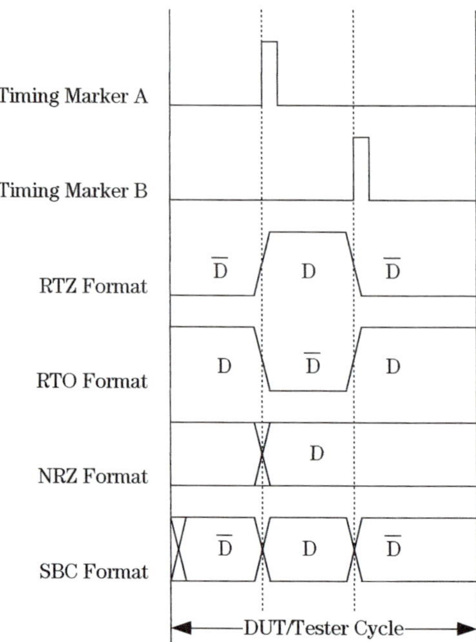

Figure 27.5 Industry-standard-format waveforms.

plement (*SBC*), etc. Figure 27.5 shows typical waveforms for these common formats. The "return" formats generate low-true and high-true pulses (two transitions per device cycle) that are used to generate clocks and chip enables, but the "nonreturn" formats generate single transition-per-cycle signals that are used mainly for buses. Specialized formats, such as SBC, guarantee a transition at each timing marker by automatically transitioning to the complement of the desired input data at the beginning of the device cycle. This format results in a transition to the true data state at the device input setup time and another transition to the complement of the data state at the device hold time, thus testing both setup and hold time in the same cycle. Most formatters are capable of switching from one standard format type to another on a cycle-by-cycle basis so that complex series of device cycle formats can be executed without pauses ("dead cycles") in the test execution. The formatter also controls the load enable function so that the driver and load can be accurately synchronized with each other and with the comparators within a device cycle to avoid drive conflicts on device I/O pins.

As explained in the comparator description, the error processor uses data from the sequencer/APG to determine the expected state of the device output at specific points in time and then strobes the comparators to see if the device output is indeed in the proper state. Immediately after each comparator strobe, the error processor determines if the device state matches what was expected and then initiates appropriate actions, depending on the result. If the device output state matches what was expected, testing typically continues as planned. If an output is in the wrong state, an error flag is generated and sent to the high-speed sequencer/APG, where it can be ignored, used to trigger a branch in the program, or to abort the test and return a FAIL result.

In some test cases, failure data is captured throughout a high-speed test sequence and then analyzed at the end to see if the device passed or failed. This is particularly common in mixed-signal device testing, where the results of many digital comparisons must be considered together to determine if the end result is good or bad. For example, a *Fast Fourier Transform* (*FFT*) routine can be used to analyze an array of 10,000 digital words to determine the frequency and phase content of a complex waveform digitized by an ADC. Also, most high-density memory devices contain redundant rows and columns within their memory arrays that can be used to electrically replace defective cells. On these devices, the error processor sends failure data to special error-capture hardware circuits in the sequencer/APG which is then analyzed at the end of the test to determine if the device is repairable and, if so, how to effect repairs.

Although this has been a brief and very high-level description of the PE instrument, actual implementations are quite complex. Test systems of the past often implemented PE instruments using groups of modules integrated into one or more PE boards per test site. Today, with the push to higher and higher performance levels, PE functions have been integrated into custom chip sets, and, in some cases, single, large ICs, incorporating multiple PE channels. PE performance levels are the dominant factor in overall system cost because the PEs are a relatively high percentage of overall digital test system content. In fact, system costs are often measured in $K/pin as a way of comparing relative costs. Low-end systems used in wafer-sort applications can cost as little as $1K/pin for typical configurations. At the other extreme, the highest-performance systems can cost more than $6K/pin, driving total system costs into the multi-million dollar price range.

27.2.3 The sequencer/APG instrument

Most high-technology products on the market today include one or more high-performance computing devices tightly integrated within the product architecture that control all aspects of the product functionality. This is especially true of semiconductor test systems, where typically three or more high-performance processors control the various aspects of the test system. The first processor is usually referred to as the *system controller*; it consists of an off-the-shelf high-performance workstation, running either the Unix or Windows/NT operating system. The system controller is responsible for all high-level activities within the test system, including the user interface, networking interfaces, and high-level test program control. The second processor is often an embedded microprocessor that controls the detailed, step-by-step, execution of test programs by coordinating the activities of the system resources. This processor is often referred to as the *test site controller* and is

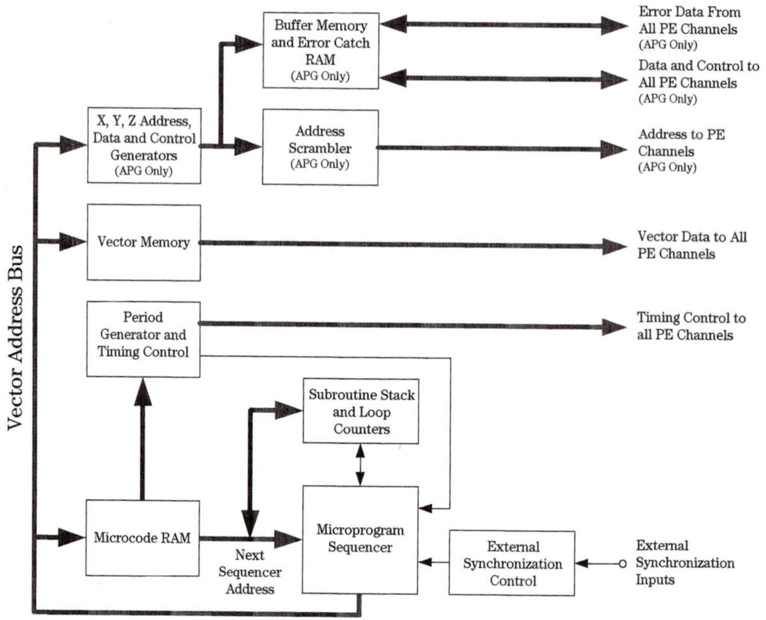

Figure 27.6 Sequencer/APG block diagram.

described later in this chapter. In highly parallel test systems, it is common to find individual test site controllers integrated into each device test site. In such cases, as many as 32 test site controllers might be contained within a large test system.

The third type of processing device integrated within semiconductor test systems is a highly specialized processor, usually referred to as the *sequencer* or *Algorithmic Pattern Generator (APG)*. The purpose of the sequencer/APG is to control the high-frequency digital test sequences that are executed on the DUT when verifying device functionality and performance. At first, one might ask why such a specialized device is necessary when off-the-shelf microprocessors are available with tremendous processing power. Basically, the answer is that simple, off-the-shelf processors are aimed at solving general-purpose computing problems and don't come close to having the proper feature set, performance level, or sheer control capability needed to run a large semiconductor system test sequence. The truth of this statement will become apparent when examining the role that the sequencer/APG plays in testing semiconductor devices.

Figure 27.6 shows a high-level block diagram of a sequencer/APG used in test systems to test devices that include a combination of logic and memory in the same IC; microcontrollers and flash memory devices are typical examples. Beginning with the microcode RAM, microprogram sequencer, and subroutine stack and loop counters, the components are typical of most computing devices. Basically, the microcode RAM stores a sequence of control instructions that are executed by the microcode sequencer during test execution. These instructions tell the sequencer what to do on a cycle-by-cycle basis and include functions like "advance" to the next instruction, "call" or "return" from a subroutine, "increment" or "decrement" from a loop counter, "jump" and "stop." The output of the microprogram sequencer is the address (often referred to as the *vector address*) of the next instruction to be executed. This vector address is then bussed to all other sequencer/APG functions via the vector address bus to control what happens next and to maintain synchronization between all test resources throughout the entire test sequence. The subroutine stack and loop counters, when used with sequencer conditional branching functions, allow the microprogram sequencer to generate complex control sequences, often billions of cycles long, that control the rest of the sequencer/APG and PE components. Most microprogram sequencers have the ability to synchronize their operation to external inputs via an external synchronization control circuit. This can take the form of either waiting until some set of conditions occur and then beginning the test (referred to as *match mode*) or, in the most extreme cases, using an internal phase-locked loop to synchronize tester operation to an external clock coming from the device. At the date of this writing, the most-complex sequencer/APGs are capable of operating at frequencies greater than 1 GHz and generating control words thousands of bits wide (directly related to the number of PE channels), making their performance many times higher than off-the-shelf processors could possibly achieve.

Although the sequence-control components described are typical of those found in most processors, after that, the sequencer/APG begins to deviate from traditional computing machines. The concept of a device cycle with many timing events synchronized to the device cycle is foundational in digital device specifications and, thus, in test systems. Because the DUT and the test system typically operate synchronously, the sequencer/APG includes a period generator and timing control function that generates the basic control signals that determine the tester/DUT cycle time on a cycle-by-cycle basis. Most devices to be tested operate in many different modes, each with a unique timing cycle definition. A common example is a typical DRAM, which has several different modes or cycle types (read, write, read-modify-write, refresh, extended data out, and various combinations and variations of these) that are switched between on a cycle-by-cycle basis. Before test execution begins, the period generator and timing control is loaded with the period definitions for each cycle type to be implemented for the test. Then, once high-speed execution begins, the period generator and timing control circuitry (typically, counters and fast static RAMs) receive cycle-by-cycle control codes from the microcode RAM that specify which one of the predefined cycle definitions to use for that particular cycle. One of the outputs from the period generator and timing control circuit is a synchronization pulse that signals the end of one period to the microprogram sequencer and triggers it to start the next cycle. The other outputs consist of clocks and other timing-control signals that are sent to all PE channels to provide synchronization and cycle-by-cycle formatting control. Typical period generators have the ability to generate cycle times ranging from a few microseconds to the minimum cycle time of the test system so that the test sequence will run precisely at the frequency specified for the DUT. Although simple in concept, timing generation and control across all tester pins is the most challenging aspect of high-performance semiconductor test system design and achieving the desired timing accuracy is frequently the main consideration when selecting a particular system architecture. This is the direct result of the difficulty of synchronizing many timing edges within tens or hundreds of picoseconds across hundreds of PE channels under a variety of test conditions. In fact, the newest architectures go to the extreme length of duplicating the sequencer/APG for each individual pin so that only one or two synchronization signals need to be shared across all PE channels.

Once the program sequence and timing control are specified for a test sequence, the remainder of the work performed by the sequencer/APG involves generating test vectors (cycle-by-cycle data states) that are bussed to the PEs to control device stimulus and response functions. A classic digital test system (without the APG functions) uses a large vector memory to store the data needed to control each individual pin on a cycle-by-cycle basis. During test execution, the vector address bus selects a particular word out of the vector memory to provide the data and format control for the PE channels. Typical vector memories are configured with four bits per PE channel to control drive/compare and format functions and with depths of from 64K words, for low-end digital testers, to several megawords for high-end systems. Simple calculations performed using these numbers shows how large the vector memory can become—a 1000-pin system with 4M × 4 bits of vector memory per channel can require up to two gigabytes of high speed RAM to store the vectors used in a single device test program.

The sheer volume of data to be stored becomes a significant design and cost issue for digital test systems and results in another big challenge—how does a test engineer generate all the bits needed to test a complex device? Although the vector data for small devices can be readily generated by manual processes, as devices become more and more complex, this becomes impossible. Fortunately, advances in computer-aided design and engineering (CAD and CAE) over the last 20 years have resulted in software tools that do much of the work of generating test vectors starting with the device design database itself. This allows the device to be tested via simulation software before the silicon is available and then facilitates generation of much of the test program that runs on the tester to test the actual device. Without these software tools, neither the devices nor the test programs could be completed and brought to market as rapidly as they are today.

For test systems designed to test pure memory devices (DRAMs, SRAMs, etc. with only minor supporting logic) the vector memory can be eliminated.

Because the data and sequences needed to test memory devices are highly algorithmic, a programmable *Algorithmic Pattern Generator* (*APG*) can be used to generate long sequences of data states, instead of storing large volumes of pseudo-random data in a vector memory. This greatly reduces the complexity and cost of a memory test system, when compared to a digital test system, and, in the process, makes program generation much simpler. Most APGs include the blocks of circuitry shown in Figure 27.5 in the boxes labeled *APG only*. Because memory devices typically have two or three address fields (rows, columns, and sometimes multiple subarrays), a data I/O bus and various control signals, the APG includes multiple word generators that generate the data states independently for each group of inputs and outputs. Classic memory test systems include X, Y, and Z address field generators, a device data I/O generator and a field for such control bits as Write Enable, Row Address Strobe, Column Address Strobe, etc. Each of these generators executes an instruction every cycle, as selected from a microcode memory within the generator by the vector address bus. Typical instructions executed by the generators include increment, decrement, complement, and other logical functions that together create the sequence of data states needed to test the device. These sequences are somewhat standardized (many have industry standard names, such as *March, Checkerboard, Moving Inversion, Galpat,* etc.) and are highly algorithmic allowing a relatively small program (often <50 microcode instructions) to generate the many millions of device cycles needed to test the millions of memory cells in the device.

APGs have several additional features designed to cope with specific test issues (Fig. 27.6), namely an address scrambler and a buffer memory/error catch RAM. Each of these functions solves a particular problem associated with testing one type of memory or another. The address scrambler, for example, solves the problem of how to topologically unscramble the internal rows and columns of the device under test (they typically end up being scrambled, arranged in pseudo-random order, during the IC layout process as part of various design tradeoffs) so that the address sequence through the device under test and the APG are topologically identical. This is important because many test algorithms stress individual memory cells by performing operations on adjacent cells. Without an address scrambler, the cell being stressed would most likely not be physically surrounded by its logical adjacent cells. The address scrambler circuit is a simple RAM-based lookup table that maps each logical address for a single address generator (X or Y) to the topologically correct physical address for the device under test.

The buffer memory/error catch RAM is a crucial component for testing ROMs that have already been loaded with a random pattern (like video game memory) or for testing repairable memory devices at wafer sort. In the case of a ROM, the data image of the ROM is loaded into the buffer memory, which then supplies the data that tells the error processor what to expect when reading the memory under test during test execution. Of course, the buffer memory must be large enough to store all of the data contained within the device under test (typically, somewhere between 8 Mbits and 128 Mbits).

Memory devices that are repairable (include redundant rows and columns to replace defective cells in the memory array), drive the need for a large memory array inside the tester to capture the exact failure location of all bad cells discovered during testing. In this case, the buffer memory (now referred to as the *error catch RAM*) operates in an error-catch mode during the high-speed test-pattern execution and accumulates a bit-by-bit failure image of the device under test. At the end of the test, the failure pattern can be analyzed and, if the number and arrangement of the failed bits is within repair guidelines, the defective rows/columns can be electrically replaced with the redundant rows and columns, thus completely repairing the array. This approach is very common in the highest-density memory devices (DRAMS, flash memory, etc.) because manufacturing yield can be significantly improved if a relatively small number of random silicon defects per device can be repaired by replacing them with redundant rows and columns.

As mentioned earlier, test systems designed primarily to test pure logic devices traditionally omit the APG functions, but systems designed to test pure memory devices traditionally omit the vector memory. However, over the last few years, the line between pure logic and pure memory devices has blurred significantly and many systems now include both vector memory and APG functionality. This is driven by the trend to include relatively large memory arrays within microcontroller and microprocessor designs, and, conversely, by the trend to include complex logic functions (memory controllers) inside of memory ICs to achieve improved memory interface performance. These ICs require the use of sophisticated logic and memory test functions to be adequately and efficiently tested.

Now it's necessary to include information concerning the implementation of the sequencer/APG instrument in actual test systems. There is a basic difference in the architecture of digital vs. memory test systems that complicates the design of systems that must include both vector memory and APG functions. Digital test systems that use large vector-memory arrays perform best when designed using a "per-pin" architecture; that is, virtually all tester functions are duplicated and dedicated on each PE channel (pin), including the sequencer, timing controls, vector memory, formatters, and driver/comparator/loads. This approach results in the highest performance and greatest flexibility because each channel primarily interacts within itself, rather than with all other pins on a cycle-by-cycle basis. APG-based memory test systems are, on the other hand, typically implemented with the APG, including the sequencer, timing controls, address and data generators, address scramblers, buffer memory/error-catch RAM, etc., as shared resources which must communicate across all channels on a cycle-by-cycle basis. This sharing of resources across many pins, while reducing system cost, significantly complicates system design and is the main reason why memory test systems are typically limited to lower maximum frequency of operation (100

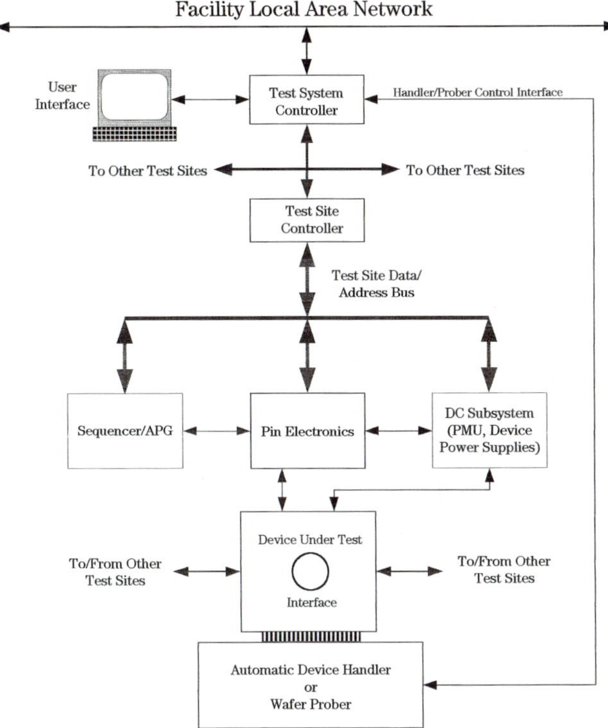

Figure 27.7 Semiconductor test-system block diagram.

MHz or less), compared with digital test systems, which can operate at frequencies of several hundred megahertz to one gigahertz. Systems that include both functionalities use very inventive hardware and software architectural approaches to solve this problem, but in the end, some compromises must be made.

27.3 Semiconductor Test System Architecture Overview

Now that semiconductor test instruments have been covered in some detail, it's time to step back and see how these instruments are combined to implement actual test systems. Because a very wide range of semiconductor test system configurations are in use today, this section is restricted to a general, high-level description that illustrates the most important semiconductor test system architectural concepts.

Figure 27.7 shows a test-system architecture typical of modern digital/memory test systems used for production testing. To begin with, virtually all production test systems are connected to the user's *Local Area Network* (*LAN*), which provides a convenient way to access test programs and send test results to shared network servers. This is usually accomplished by connecting the test system controller, typically an off-the-shelf workstation running the Unix or Windows/NT operating system, to the LAN. The test system controller includes multiple interfaces, including a keyboard and monitor for interacting with the user, a high-speed interface to the test sites (a collection of hardware instruments and software that implement a logical test system to perform the actual testing), and interfaces to the mechanical handling equipment (a packaged device handler or wafer prober). The test system controller is the highest-level processor in the test system and it essentially supervises all tester activities, as commanded from the user interface, or in some cases, from commands originating at another workstation connected to the LAN. In production test applications, it "tells" the test sites to start testing and records the test results at the end of test. In engineering applications, it can also perform data analysis and display functions on the test results using a wide variety of program debug and device troubleshooting tools.

The test site controller can be implemented as a physical hardware processor or as a software process that runs on the test-system controller, depending on the system architecture. In either implementation, it is responsible for conducting the detailed step-by-step test process necessary to complete testing of the device under test. This is accomplished by having the test-site controller execute a test program that has been written by a test engineer, who is very familiar with both the operation of the device and the test system. During test execution, the test-site controller sends commands to all other test-site components, via the test-site data/address bus, causing the site instruments to make measurements and run high-speed test sequences until the device is fully tested and ready to be "binned out." Although most devices are not simply dumped into a "bin" at the end of test today, the concept of logical bins has been preserved from the past. At the end of the test sequence, each device is assigned to a bin, indicating that it either passed or failed the test sequence. For most parts, there are multiple "pass" bins that reflect different levels of acceptable performance (thus, different price levels) and multiple "fail" bins that are all rejected parts, but have failed in a different way. All of the failed bin parts are thrown out, but the fail-bin statistics are gathered to provide feedback to the manufacturing process as part of the overall quality-assurance and yield-improvement strategy. Acting under the control of the test-site controller, the system instruments (PMUs, PEs, sequencer/APG, and possibly other system instruments, depending on system type) generate stimulus signals and measure responses from the DUT via a complex mechanical assembly, usually referred to as the *Device Under Test interface* (*DUT interface*). The DUT interface typically includes many interconnect components (wires, connectors, spring contacts, printed circuit boards, etc.) that connect the instruments to the pins of the DUT. The last stage of interconnect is usually a custom-handler interface assembly with DUT-specific contacts for the packaged part being tested, or a DUT-specific wafer probe card that contacts the small metal pads (device terminals) of the devices on the wafer. The DUT interface usually consolidates signals from multiple test sites into one interface connected to a single handler or wafer prober. This reduces cost and complexity by allowing one DUT interface to connect from eight to 16 sites (more, in some cases) to a single handler or prober. It is also common for one large test system to utilize two DUT interfaces connected to two handlers or two probers to optimize the test capacity per square foot of floor space.

One final issue concerns the implementation of test sites within large test systems: the virtual site vs. truly independent site approaches. Users often need to use the same test system for testing high pin count devices during one period of time and smaller pincount devices at other times. This results in a cost/configuration problem. For example, if the large pin-count device requires 512 pins and the small pin-count device requires only 32 pins, the user must buy an expensive 512-channel tester for the larger device. Then, when the user is testing the smaller device, 480 tester channels go unused—a tremendous waste of costly resources. Even when all devices to be tested have a low pin count, it might be advantageous to test many devices in parallel on a single high-pin-count test system to reduce the per-site tester overhead (chassis, cooling components, test system controllers, etc.) and the number of handlers or probers to be purchased. The problem of testing many low pin-count devices on a single large tester has been solved by creating many small "virtual sites" out of a large single site. In the previous example, the 512 channels would be divided into 16 groups of 32-channel "virtual" sites and 32 of the low pin-count devices could be cost-effectively tested in parallel on the larger system. This approach has been successfully used in many cases where the test sequence is highly similar for all 32 devices; that is, only minor conditional branching or test-flow differences across all 32 devices. If, for some reason, each device requires a different test flow, this scheme breaks down because the devices will end up waiting on each other during the test flow and the overall test time gets dramatically longer.

This latter case is typically true on devices, such as programmable nonvolatile memories (flash, EEPROM, etc.), where erase and programming sequences are slow and must be controlled individually on a device-by-device basis. On these devices, using the virtual site approach results in poor test throughput and significantly complicates the control aspects of testing many devices in parallel. As a result, test systems used for these types of devices typically implement multiple physically independent test sites so that each device is tested individually, thus making the test program relatively straightforward to write, and achieving the highest possible throughput. This approach does result in additional system hardware overhead per test site, but the incremental cost is more than compensated for by the much higher test throughput achieved.

27.4 Common Semiconductor Test System Configurations

No description of semiconductor test systems would be complete without a few brief comments on the major classes of test systems utilized by the semiconductor industry today. The variety of test systems available on the market is driven by the competitive market pressures on semiconductor manufacturers to bring ever higher-functionality and higher-performance ICs to market at the lowest possible prices. This forces all aspects of semiconductor manufacturing to be highly optimized including device test. Thus, for each major class of devices, and sometimes for different device architectures within a common classification of devices, specific highly optimized test systems have been designed to achieve the required cost of test needed to successfully compete. Without going into great detail, the following sections outline the salient differences between the major types of semiconductor test systems in common use today, starting with the simplest systems and moving toward the latest and most-complex systems coming to market at the beginning of the 21st century.

27.4.1 Parametric test systems

Parametric test systems are utilized in the wafer-fabrication process for virtually all semiconductor device types—from diodes to microprocessors. Their main use is to monitor the fabrication process so that high yield is realized when the wafers are completed. Parametric test systems are usually small and on the low end of the STE price spectrum. They include a variety of PMUs, simple waveform generators, and simple waveform measurement instruments that can measure various parametric-level parameters, such as resistance, capacitance, gain, etc., on test cells designed into each wafer. Special low-leakage relay matrices are used to connect the instruments to the test cells to facilitate readings down into the microvolt and femtoamp ranges.

27.4.2 Discrete semiconductor test systems

Discrete test systems were the first specialized semiconductor test instrument-based systems developed in the early days of semiconductor manufacturing when device types consisted mainly of diodes and transistors. In many ways, discrete test systems are similar to parametric test systems, except that they typically have much higher voltage- and current-test capabilities to test higher-power devices. High-power capability is largely achieved by extending the voltage and current ranges of the PMUs and making measurements using pulsed voltage/current techniques to limit dissipation. Of course, many specialized test capabilities—designed into discrete test systems used for testing the highest-power discrete devices—were voltages of several kilovolts and currents up to or greater than 1000 A are required. This is especially true when the devices under test are to be used in high-power switching applications, such as motor controls.

27.4.3 Analog test systems

Analog test systems came into existence when diodes and transistors began to be combined into simple analog circuits, such as op amps, comparators, and voltage regulators. Over time, these systems became more and more complex, as such devices as *Analog-to-Digital Converters* (*ADCs*), *Digital-to-Analog Converters* (*DACs*), and various amplifiers came into existence. Analog test systems include PMUs, waveform generators, waveform digitizers, and specialized instruments needed to test op amps and converters efficiently. Most analog test systems also connect the test system instruments to the DUT through a relay matrix, similar to those found in parametric test systems. Although not capable of performing meaningful testing on digital parts, analog test systems often include a simple sequencer that is capable of controlling the relatively few digital inputs/outputs on DACs and ADCs at low frequency. Analog test systems, although not a large share of the semiconductor test market, do play an important role in testing op amps, comparators, DACs, ADCs, and analog-switch and power-management ICs used in a wide variety of state-of-the-art computer and communications products today.

27.4.4 Digital test systems

Along with the digital age in the 1960s came a need for digital IC test systems that has grown steadily for well over three decades now. Early digital device families (such as Digital-Transistor-Logic devices, which we would consider to be primitive today) were largely tested on analog test systems, supplemented with custom test adapters. As digital technology accelerated, specialized digital test systems began to emerge and gradually became the largest semiconductor test market segment—a position that has been eclipsed by memory test systems as a result of the *Personal Computer* (*PC*) revolution. Specialized pin electronics and sequencer instruments evolved as pin-count requirements accelerated and device tests grew more and more complex. Today, as mentioned earlier in this chapter, digital test systems are the largest, and most costly, of the semiconductor test systems, driven by the simultaneous need for high pin count and high performance.

In recent years, new classes of digital-intensive mixed-signal devices have emerged as a byproduct of the personal computer revolution. These devices are often referred to as *multimedia* devices because they turn digital data into the analog signals that drive the graphics displays and speakers in workstations, PCs, and games. Because of the digital-intensive nature of these devices, and their evolution alongside other PC chipsets, they are usually tested on digital systems that have been enhanced through the addition of high-performance analog test instruments (mainly *Digital Signal Processing-*

based (*DSP*) waveform generation and measurement instruments) rather than being tested on classic mixed-signal test systems (see Section 27.4.6).

27.4.5 Memory test systems

Specialized memory test systems began to evolve from digital test systems during the mid-1970s as memory devices grew in size and complexity. In fact, the first memory test systems were a lot like digital test systems, with the sequencer replaced by the algorithmic pattern generator—a trait that largely continues to this day. The invention of the first multiplexed address DRAM (4K bits) helped fuel the computer revolution, which has driven electronics growth so explosively over the last two decades. Computers of all sorts need ever more memory to run their programs. This generates the demand for billions of memory ICs per year to be manufactured and tested on memory test systems, resulting in memory test systems becoming the largest segment of the semiconductor test market. Modern memory test systems are still a lot like their earlier ancestors, except that new technology allows for many (16 to 32) high-performance test sites to be packaged in the same space as a single test site of the 1970s.

Memory testers are in the middle of an architectural revolution today, driven by the need for higher communication bandwidth between the newest microprocessors and their program memory, as well as the need for embedded NVM arrays within microcontrollers. The need for higher bandwidth is driving memory testing more toward digital performance levels as specialized memory interface schemes are developed to increase the data-transfer rate needed for the next generation of microprocessors. Meanwhile, special testing challenges associated with NVM arrays in embedded controller devices are driving the need for more completely integrated memory/logic test capabilities within the same test system. It is likely that these conflicting requirements will be resolved with new and innovative hardware/software system architectures over the next few years.

27.4.6 Mixed-signal test systems

One look at the block diagram of a state-of-the-art mixed-signal test system quickly communicates the incredible complexity of instrumentation required to test today's most complex *mixed-signal ICs* (ICs that contain both significant digital and analog circuitry). Early mixed-signal test systems evolved from analog test systems through the addition of high-end ac signal generation and measurement instruments that were DSP based. These instruments were needed to efficiently test new classes of telecommunications devices, such as modems, which not only needed digital and analog test support, but had many specifications dealing with frequency-domain parameters related to industry standards. As the communications revolution expanded beyond traditional telephone line-based systems into local-area networks, cellular phone systems, ISDN, fiber optics, satellite links, the Internet, etc., electronic assemblies containing hundreds of discrete logic and analog components were compressed into single ICs. Inside these ICs, digital processors implemented the complicated protocols associated with these communications advances, generated the complex signals needed to drive the line (or antenna), and executed the algorithms needed to extract the signal of interest from a sea of other signals, noise, and signal distortions caused by imperfect transmission lines.

Mixed-signal test system complexity results from the need for tightly synchronized analog, digital, and DSP functions that push the state of the art in performance and functionality in all areas simultaneously. A typical test on a mixed-signal IC that interfaces to a high-frequency telephone line, for example, involves a DSP-based waveform generator that generates a waveform that contains the data, emulates a particular protocol, and introduces simulated transmission-line signal defects (noise, reflections, impedance mismatches, etc.); getting that signal through the DUT interface without distortion; providing the digital control to the DUT that causes it to capture and process the input signal; capturing the resulting digital output sequence from the device; and processing this digital output using DSP to see if the DUT correctly extracted the data of interest from the input signal. Many variations of this type of test must be performed on the DUT in order to test all device transmit and receive functions for all supported protocols.

The rapidly expanding cellular telephone marketplace is driving the need to integrate even more technology into mixed-signal ICs—namely, the RF section. As semiconductor-manufacturing processes gradually become capable of implementing high-performance analog, digital, and RF circuitry within the same IC, mixed-signal test systems are faced with the challenge of adding new RF hardware and software components to their already highly complex system configurations to test the DUT RF circuitry.

27.5 Continuing Trends in Semiconductor Test Instruments and Systems

As mentioned in the description of the various specialized semiconductor test system types on the market, semiconductor testing is continuing to evolve to meet the needs of technical advances in the design and fabrication of ICs. Several trends in instrumentation, system architecture, and system implementation are visible and are likely to continue. They include:

1. Test instrumentation (PMUs, PEs, and sequencer/APGs) will continue the trend of being integrated into single ICs and even combined (especially the PE and sequencer/APG) into single, very large, ICs to allow a higher number of test sites to be integrated into a limited amount of floor space. This trend is also being pushed by the need for higher frequency of operation and improved timing accuracy on digital, memory, and mixed-signal test systems.

2. Test system architecture and implementation will be increasingly tied directly to specific device architectures—especially on the high-volume commodity devices, such as DRAMs and flash memories, to achieve the required performance levels at an acceptable cost of test. Because of this, test-system availability will often become part of the crucial path in bringing new part types to market. This trend is already having an impact on the test-system industry, where large installed bases of test systems quickly become obsolete (as far as being able to test the latest new devices is concerned), thus creating opportunities for new competitors to enter the market with newer, more focused, test systems.
3. Because of the need to integrate many test sites into a small footprint to improve *test-cell efficiency* (the collection of tester, interfaces, and handling equipment needed to perform production testing), an increasing trend is for test-system manufacturers to work directly with the supplier of other test-cell components to integrate an overall test cell that maximizes performance and reliability while reducing floor-space requirements.
4. Another fallout of the high level of integration is the move toward liquid-cooled electronics as a way to improve density and reliability, as well as provide the thermal stability needed to achieve the highest performance levels. Once used mainly in the realm of super computers, water cooling has already become a common approach in the highest-performance digital test systems and will likely move into other test systems over the next few years.

Chapter

28

Network Analyzers

Daniel R. Harkins
Agilent Technologies
Santa Rosa, California

28.1 Introduction

Electronic circuits operating at high frequency present some unique challenges to proper characterization. At high frequencies the wavelengths of operation become similar in dimension to the physical properties of circuit elements. This results in circuit performance that is distributed in nature. Rather than describing the voltage and current at a specific circuit node it is more appropriate to describe how waves in a transmission medium respond to a component in their path. Network analyzers are a class of instruments that have been developed to characterize radio-frequency (rf) components accurately and efficiently as a function of frequency. Network analysis is the process of creating a data model of the transfer and/or impedance characteristics of a linear network through stimulus-response testing over the frequency range of interest. At frequencies above 1 MHz, lumped elements actually become "circuits" consisting of the basic elements plus parasitics like stray capacitance, lead inductance, and unknown absorptive losses. Since parasitics depend on the individual device and its construction they are almost impossible to predict. Above 1 GHz component geometries are comparable to a signal wavelength, intensifying the variance in circuit behavior due to device construction. Network analysis is generally limited to the definition of linear networks. Since linearity constrains networks stimulated by a sine wave to produce a sine-wave output, sine-wave testing is an ideal method for characterizing magnitude and phase response as a function of frequency. This chapter discusses the key parameters used to characterize rf components, the types of network analyzer techniques used to make measurements, and considerations to be made in obtaining the most accurate results.

28.2 Component Characteristics

Rf (frequencies less than 3 GHz) or microwave (frequencies in the 3- to 30-GHz range) energy can be likened to a light wave. Incident energy on a device under test (DUT) (for example, a lens) is either reflected from or transmitted through the device (Fig. 28.1). By measuring the amplitude ratios and phase differences between the two new waves it is possible to characterize the reflection (impedance) and transmission (gain) characteristics of the device.

28.2.1 Reflection and transmission

There are many terms used to describe these characteristics. Some use only magnitude information (scalar) and others include both magnitude and phase information (vector). If an incident wave on a device is described as V_{INCID}, the ratio of V_{INCID} and I_{INCID} is called the transmission system characteristic impedance Z_0, and a device terminating a transmission system has an input impedance called a load impedance Z_L, then key device characteristics can be defined as:

Reflection terms:

$$\Gamma = \frac{V_{\text{REFLEC}}}{V_{\text{INCID}}} = \frac{Z_L - Z_0}{Z_L + Z_0} \tag{28.1}$$

where Γ = device reflection coefficient
V_{INCID} = incident wave on a test device
V_{REFLEC} = reflected wave from a test device
Z_0 = transmission medium characteristic impedance
Z_L = impedance of test device

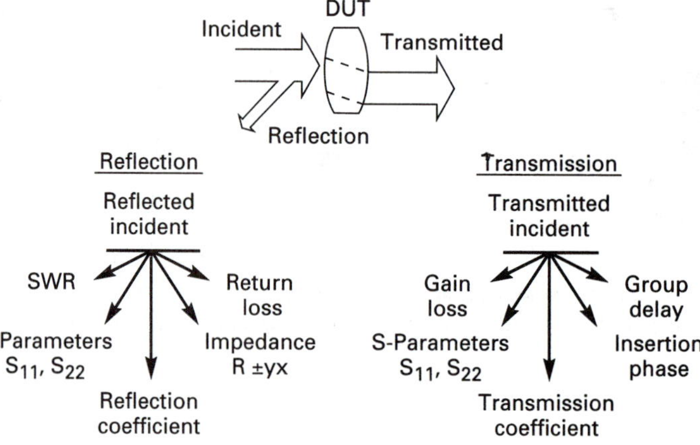

Figure 28.1 Reflection and transmission characteristics of waves incident on a device under test.

$$\rho = |\Gamma| \tag{28.2}$$

where ρ = magnitude of reflection coefficient
Γ = complex reflection coefficient

$$\text{SWR} = \frac{1+\rho}{1-\rho} \tag{28.3}$$

where SWR = standing-wave ratio of current or voltages on a transmission medium
ρ = magnitude of reflection coefficient

$$Z_L = \frac{1+\Gamma}{1-\Gamma} Z_0 \tag{28.4}$$

where Z_L = load complex impedance
Γ = complex reflection coefficient
Z_0 = transmission medium characteristic impedance

$$\text{Return loss (dB)} = -20 \log \rho \tag{28.5}$$

where ρ = magnitude of reflection coefficient

Transmission terms:

$$\text{Transmission coefficient} = \frac{V_{\text{TRANS}}}{V_{\text{INCID}}} \tag{28.6}$$

where V_{INCID} = incident wave on a device under test
V_{TRANS} = transmitted wave through a device under test

$$\text{Insertion loss (dB)} = 20 \log \left(\frac{V_{\text{TRANS}}}{V_{\text{INCID}}} \right) \tag{28.7}$$

where $|V_{\text{INCID}}|$ = magnitude of incident wave on a device under test
$|V_{\text{TRANS}}|$ = magnitude of transmitted wave through a device under test

$$\text{Gain (dB)} = 20 \log \left(\frac{V_{\text{TRANS}}}{V_{\text{INCID}}} \right) \tag{28.8}$$

where $|V_{\text{INCID}}|$ = magnitude of incident wave on a device under test
$|V_{\text{TRANS}}|$ = magnitude of transmitted wave through a device under test

$$\text{Insertion phase} = {<}V_{\text{TRANS}} - {<}V_{\text{INCID}} \tag{28.9}$$

where $<V_{\text{INCID}}$ = vector relative phase angle of incident wave on a device under test
$<V_{\text{TRANS}}$ = vector relative phase angle of transmitted wave through a device under test

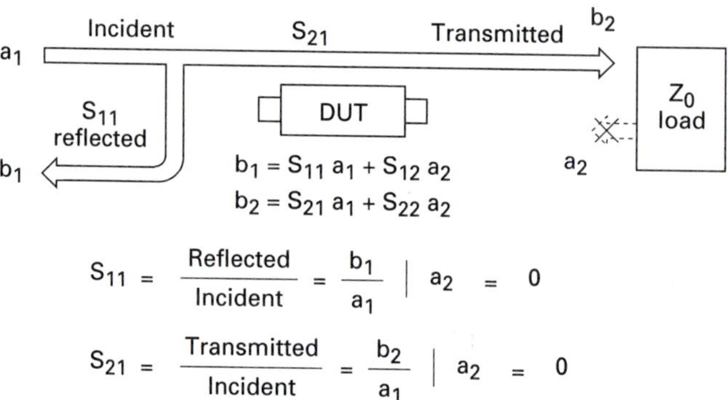

$$S_{11} = \frac{\text{Reflected}}{\text{Incident}} = \frac{b_1}{a_1} \Big| \, a_2 = 0$$

$$S_{21} = \frac{\text{Transmitted}}{\text{Incident}} = \frac{b_2}{a_1} \Big| \, a_2 = 0$$

Figure 28.2 Scattering parameter measurements of a device under test (forward measurement waves shown).

28.2.2 Scattering (*S*) parameters

Many component measurements are of two-port networks such as amplifiers, filters, and cables. These component characteristics are typically used to determine how a particular device would contribute as a part of a more complex system. To provide a method to model and analyze full two ports in the rf environment, scattering parameters (*S* parameters) were defined (see Fig. 28.2). This is a characterization technique similar to lower-frequency *Z* or *Y* modeling except that it uses incident, transmitted, and reflected waves to characterize the input and output ports of a device as opposed to using voltage and current terms which are impossible to measure at high frequencies. The *S* parameter terms relate to other characterizations with certain conditions. For instance, S_{11} is equivalent to a device input reflection coefficient Γ_{IN} under the condition the device has a perfect Z_0 match on the output. *S* parameter characterization of devices plays a key role in the ability to measure, model, and design complex systems with multiple components. *S* parameters are also defined such that they can be measured with a network analyzer.

28.3 Network Analysis System Elements

A network analyzer measurement system can be divided into four major parts: a signal source providing the incident signal, signal separation devices to separate the incident, reflected, and transmitted signals, a receiver to convert the microwave signals to a lower intermediate (if) signal, and a signal processor and display section to process the if signals and display detected information (Fig. 28.3).

28.3.1 Signal source

The signal source (rf or microwave) produces the incident signal used to stimulate the test device. The test device responds by reflecting part of the inci-

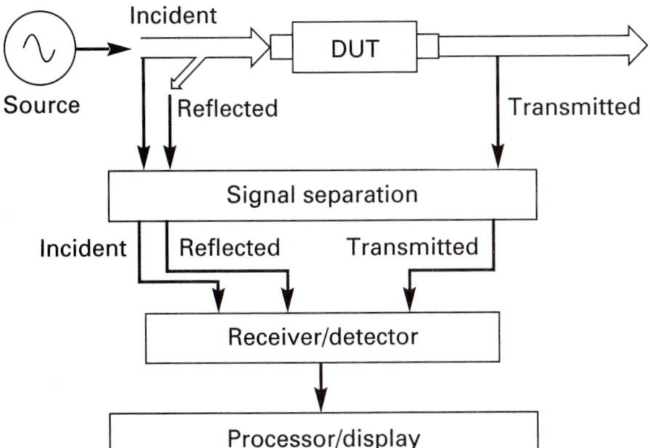

Figure 28.3 Key elements of a network analyzer measurement system.

dent energy and transmitting the remaining part. By sweeping the frequency of the source the frequency response of the device can be determined. Frequency range, frequency stability, signal purity, and output power level and level control are factors which may affect a measurement. There are basically two types of sources used for network analyzer measurements, sweep oscillators and synthesized sweepers (including synthesized signal generators). The sweep oscillator is low in cost, but its frequency accuracy and stability is much less than that of a synthesizer. If a device response changes significantly over the residual FM spectrum width of a sweep generator (such as a crystal filter characteristics), a more stable source such as a synthesizer or synthesized sweeper should be used. Also, if a device's phase response changes rapidly with frequency (i.e., electrically long devices such as long cables), it is important to use a stable frequency source such as a synthesizer to avoid drift.

28.3.2. Signal preparation

The next step in the measurement process is to separate the incident, reflected, and transmitted signals. Once separated, their individual magnitude and/or phase differences can be measured. This can be accomplished through the use of directional couplers, bridges, power splitters, or even high-impedance probes. Figure 28.4 identifies potential transmission measurement configurations. A directional coupler is a device that consists of two coupled transmission lines that are configured to couple energy to an auxiliary port if it goes through the main port in one direction and not in the opposite direction. Directional couplers usually have relatively low loss in the mainline path and thus present little loss to the incident power. In a directional coupler structure (Fig. 28.5) the coupled arm samples a signal traveling in one direction only. The coupled signal is at a reduced level and the amount of reduced level is called the coupling factor. A 20-dB directional coupler means

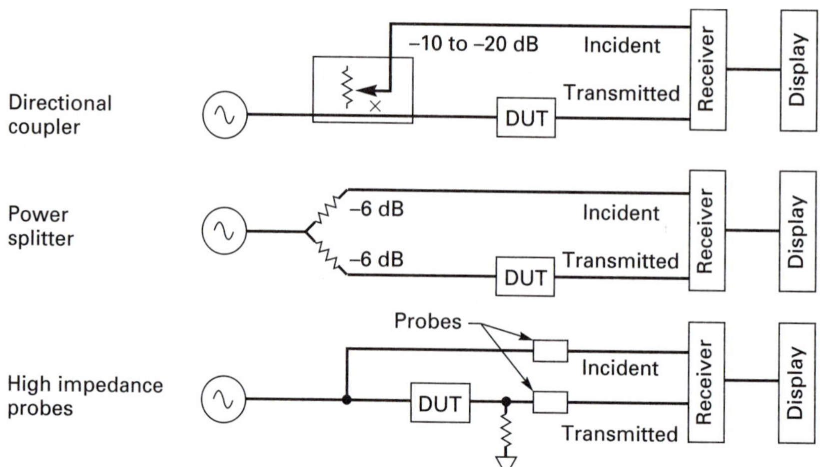

Figure 28.4 Transmission measurement configurations.

that the coupled port is 20 dB below the input, which is equivalent to 1 percent of the incident power. The remaining 99 percent travels through the main arm. The other key characteristic of a directional coupler is directivity. Directivity is defined as the difference between a signal detected in the forward direction and a signal detected in the reverse direction. Sources of imperfect directivity are signal leakage, internal coupler load reflections, and connector reflections. A typical directional coupler will work over several octaves with 30 dB directivity. The two-resistor power splitter (Fig. 28.6) is used to sample either the incident signal or the transmitted signal. The input signal is split equally between the two arms, with the output signal (power) from each arm being 6 dB below the input. A primary application of the power

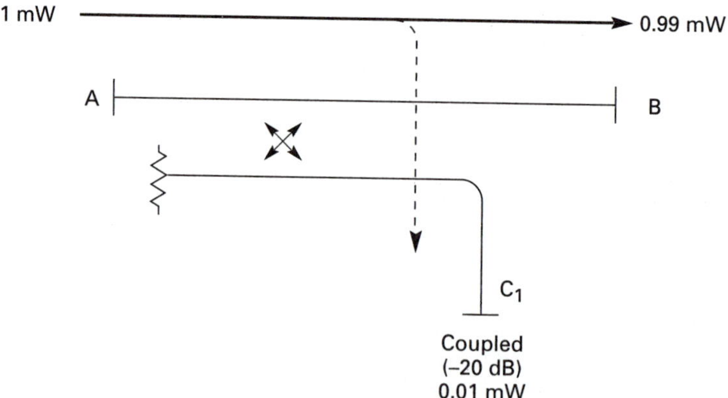

Figure 28.5 Directional coupler coupling characteristics.

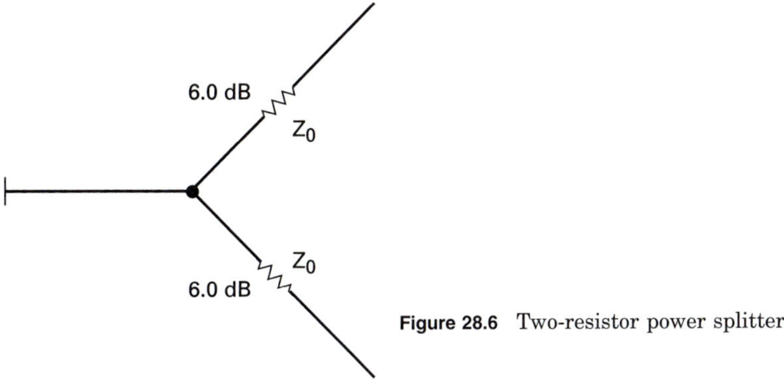

Figure 28.6 Two-resistor power splitter.

splitter is for producing a measurement with a very good source match. If one side of the splitter output is taken to a reference detector and the other side goes through a device under test to a transmission detector, a ratio display of transmitted to incident has the effect of making the resistor in the power splitter determine the equivalent source match of the measurement. All other influences on source match prior to the splitter are ratioed out. Power splitters are very broadband, have excellent frequency response, and present a good match at the test device input.

In environments different from the typical 50 or 75 Ω, measurements can be made with high-impedance probes. It is important that the probe impedance be large with respect to the circuit impedance so unnecessary loading does not occur.

Figure 28.7 illustrates the reflection measurement configuration. Reflection measurements require a directional device. Separation of the incident and reflected signals can be accomplished using either a dual directional coupler or a bridge. The essential difference is in the power levels involved. A direc-

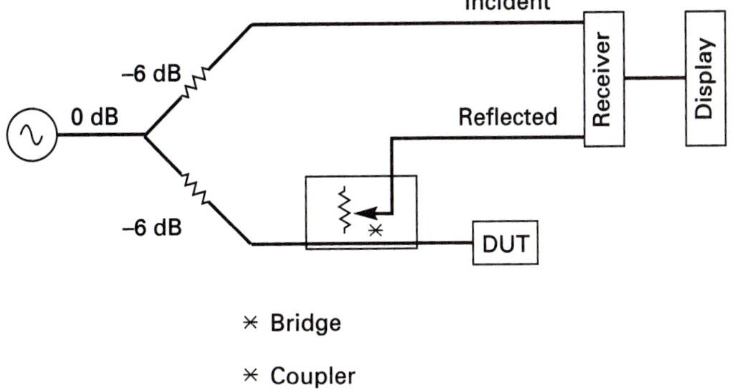

Figure 28.7 Reflection measurement configuration.

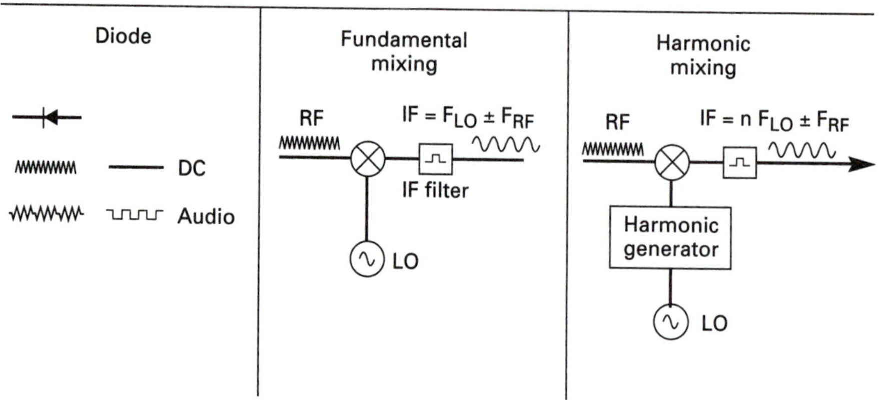

Figure 28.8 Network analyzer receiver techniques.

tional coupler has less main arm loss whereas a bridge tends to have better response over a broad frequency and for that reason is more commonly used.

28.3.3 Receiver

The receiver provides the means of converting the rf or microwave signals to a lower if or dc signal to allow for accurate detection. There are basically three receiver techniques used in network analysis (Fig. 28.8). The simplest technique is the use of a diode detector as a broadband sensor that converts all incident rf energy to a dc signal that is proportional to the power incident on the diode. The other two types of receivers are broadband tuned receivers that use either a fundamental mixing or harmonic mixing input structure to convert an rf signal to a lower-frequency if signal. The tuned receivers provide a narrowband-pass intermediate-frequency (if) filter to reject spurious signals and extend the noise floor of the receiver. Receivers using the broadband diode detection are used in scalar network analyzers and the tuned receiver technique is used in vector network analyzers. The scalar measurement system is the most economical measurement as well as the simplest to implement. The vector measurement systems (tuned receivers) have the highest dynamic ranges, are immune from harmonic and spurious responses, can measure phase relationships of input signals, and provide the ability to make more complex calibrations that lead to more accurate measurements.

28.3.4 Processor display

Once the rf has been detected, the network analyzer must process the detected signals and display the measured quantities. Network analyzers are multichannel receivers utilizing a reference channel and at least one test channel. Absolute signal levels in the channels, relative signal levels (ratios) between the channels, or relative phase difference between channels can be measured depending on the analyzer. Relative ratio measurements are usually made in dB, which is the log ratio of an unknown signal (test channel) with a chosen reference signal (reference channel). This allows the full dynamic range of the instrumentation to be used in measuring variations of both high- and low-level circuit responses. For example, 0 dB implies the two signal levels have a ratio of unity while ±20 dB implies a 10:1 voltage ratio between two signals. All network analyzer phase measurements are relative measurements, with the reference channel signal considered to have zero phase. The analyzer then measures the phase difference of the rest channel with respect to the reference channel.

28.4 Measurement Accuracy

The specified accuracy of any network analyzer measurement is the result of many factors that must be considered about both the device being tested and the particular network analyzer system being used for the measurement. Figure 28.9 is a diagram of the factors that must be considered to determine the

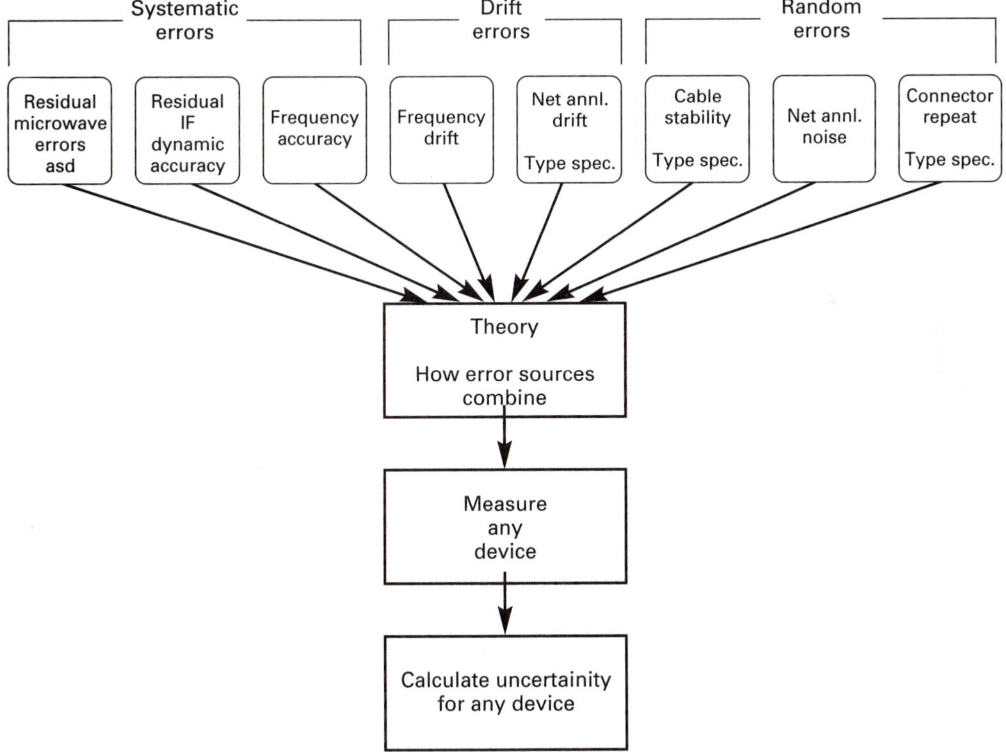

Figure 28.9 Error sources contributing to measurement uncertainty and specified network analyzer performance.

level of uncertainty in any specific measurement. As long as each of the error sources can be determined and a theoretical model of the measurement technique is understood, the accuracy of the final result is assured. The resultant uncertainties are not only a function of the measurement system but are also a function of the parameters of the device under test.

28.4.1 Uncertainty

The models and analysis techniques used to calculate the uncertainty of any measurement end up with an expression that is a function of the error sources in the measurement. A typical expression for the resulting uncertainty is of the form

$$\text{Mag uncert} = \text{Systematic} + [\text{Random}^2 + (\text{Drift \& stability}^2)]^{1/2} \quad (28.10)$$

where Mag uncert = uncertainty of a magnitude measurement
Systematic = systematic measurement system errors
Random = random measurement system errors
Drift & stability = drift characteristics of a device or test system

In this form the systematic errors add in a worst-case manner and the random, drift, and stability errors are characterized in a root sum squared (RSS) fashion as indicated in the second term of the expression. Understanding the source and magnitude of these errors for a particular test system will be key to determining the quality of the measurement being made.

Systematic errors are those that don't change after a calibration and are stationary during the measurement. Systematic errors relate to how well the actual test system duplicates an ideal stimulus-response test environment. Therefore, the residual directivity of the test system, the tracking of a through response, and the equivalent source and load matches are real system measurement characteristics that can result in errors. A simple example is that of equivalent source match. If the test system source is not a perfect match, the reflected wave of a device under test will re-reflect off the nonideal source and return as a second incident wave to the device under test. The result is a possible measurement error due to the multiple re-reflections between the test source and the device being tested. This error is definitely a function of the magnitude of the two mismatches causing the multiple reflections. There are a number of different techniques that can be used to improve this class of systematic errors. In the case of equivalent source match, adding a pad at the test port or doing a one-port system calibration are both techniques that can improve the apparent mismatches. Another class of systematic errors relates to the detection processing for a particular network analyzer system. Dynamic accuracy is a key consideration in most measurement system configurations. Dynamic accuracy relates to the ability of a receiver to accurately detect a signal over a large amplitude range. In determining the range of signals a receiver can characterize, the largest input signal is usually limited by compression in the input receiver structure, and the smallest signal that can be detected is limited by the noise floor of the receiver or signal crosstalk and leakage through undersirable paths in the measurement hardware. The inherent linearity of the signal detection scheme is dependent on the type of receiver detector used. Calibration techniques exist that address various systematic errors. The choice of what calibration should be used for a particular test depends on the device under test characteristics and on the particular network analysis system being used for the measurement.

28.4.2 Random errors

The second source of errors is those that can be classified as random. Key contributions to random measurement errors are noise sources, connector repeatability, and cable stability. In any system there are a number of noise sources. The system sensitivity is determined to be the noise of the receiver front end down converter or detector. The source spectral purity and (if used) receiver local oscillators can add noise to the data stream. Receiver structures that provide variable detection bandwidth and data averaging offer some methods to reduce noise errors. Vector network analyzer receivers offer this capability. The receiver if bandwidth can be set by the user to trade off sensitivity for receiver sweep speed.

Connector repeatability can vary significantly depending on the quality of the connector system being used. In each connector standard there are varying grades of quality to the parts being used. They are usually classified as commercial, instrumentation, or precision quality connectors. The part cost, tolerances, and rf performance vary accordingly. In any connector-class precision connector repeatability could be greater than 60 dB while commercial-grade connector repeatability could be less than 30 dB. In any particular situation, connector repeatability can be characterized by making multiple connections and measuring the resulting differences in the data. The analysis should be done with a large sample base and characterized statistically.

Cables are a major source of error. If they are not moved after calibration, the error can often be very small, but this is not the typical use of the system. Typically the transmission phase error will be larger than the magnitude error. Hard line cables tend to be more stable if the measurement requires very little movement. But if the cables must be moved often, then a high-quality flexible cable is a must.

Drift and stability represent changes that occur in a system over time and temperature. Typical sources of this type of error can be attributed to receiver downconversion and detection if changes with temperature. Many of the ratio portions of a network analyzer measurement system help common mode out potential sensitivities to drift. The most appropriate method to address these sources of error is to use the most stable hardware to start with and then to recalibrate the measurement frequently enough to avoid problems in the specific measurement environment.

28.5 Scalar Network Analysis

The most unique components in a scalar network analyzer system are the diode detectors used as the rf power sensing device. This results in very economical broadband magnitude measurements of rf characteristics.

28.5.1 Diode detectors

Diode detectors convert the rf signal to a proportional dc voltage. If the signal is amplitude-modulated, the diode recovers the modulation. Diode detectors can be very broadband (10 MHz to 50 GHz), have fast response times, and have a dynamic measurement range up to 76 dB. Typical detector return loss is 20 dB.

Diode detectors have a square-law region over which the voltage out is proportional to the power in (Fig. 28.10). This is called the square-law region, since voltage out is proportional to the square of voltage in. Above a certain power level the response becomes linear. The scalar network analyzer receiver has the ability to compensate for this detector characteristic change to extend the allowable dynamic range.

Diode detection schemes use either dc detection or ac detection. Dc detection produces a dc signal that is proportional to the power incident on the diode. The diode output is read directly by the analyzer, making the analyzer,

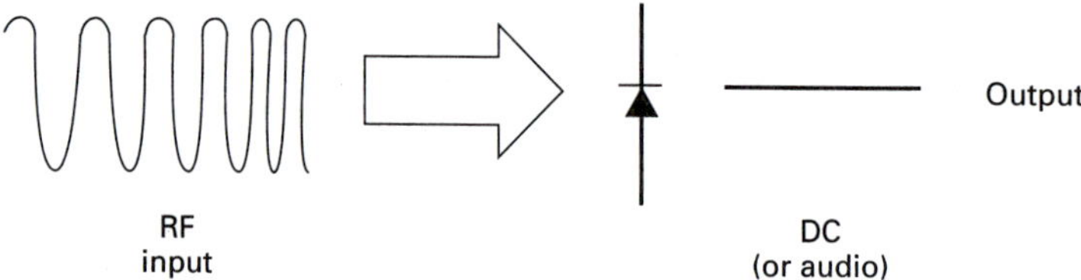

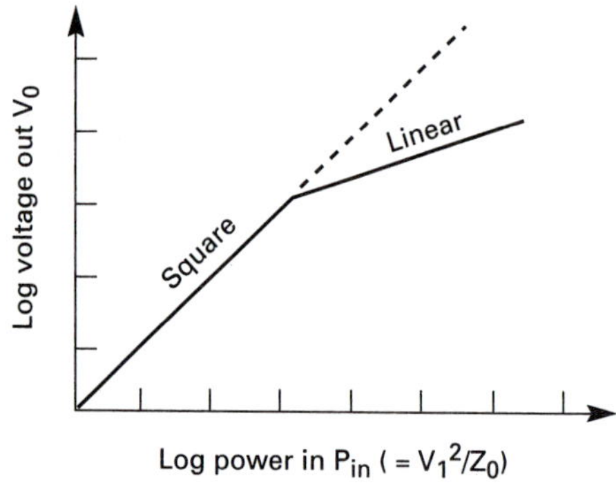

Figure 28.10 Diode detector characteristics.

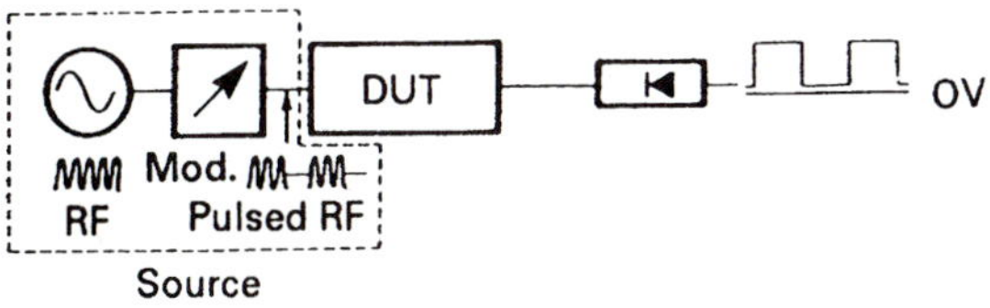

Figure 28.11 Block diagram of square-wave modulated ac detection. Diode converts pulsed rf into square wave.

in this case, a fancy voltmeter with a logarithmic response. Ac detection also produces a signal proportional to the power incident, but the rf source is square-wave amplitude-modulated, which results in a square-wave output of the detector (Fig. 28.11). Ac detection provides benefits over dc detection by avoiding diode detector dc drift and temperature sensitivities. Also, ac detection is not affected by signals at the input of a detector that are not modulated. Ac detection does require rf-source modulation which is sometimes difficult to achieve and can possibly affect the device under test performance.

The scalar analyzer receiver accommodates multiple (up to four) detector inputs. Its low-frequency circuitry provides detector compensation to account for square-law to linear detection as well as temperature compensation to ensure maximum dynamic range and detector dynamic accuracy.

In using broadband detectors, attention must be paid to the fact that they can respond to all signals present at the input ports in the frequency range of the detectors. In each measurement situation, attention must be given to source harmonic levels and spurious signals. If the signals detected are in the square-law region of the detector, unwanted signals add in a power sense. In the linear portion of the detector range, unwanted signals add uncertainty in a linear sense.

The directional bridge in a scalar analyzer system is a combination of detector and signal separation device. The directional bridge works much like a Wheatstone bridge (Fig. 28.12). If all four arms are equal in resistance (i.e., test port = 50 Ω) a voltage null is measured. If the test port load is not 50 Ω, then the voltage across the bridge is proportional to the mismatch (deviation

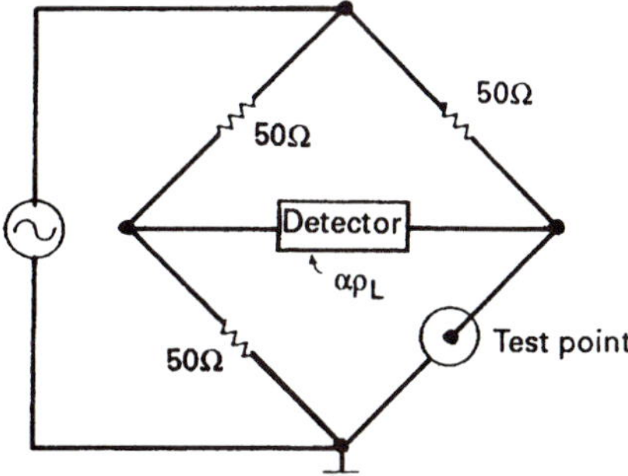

Figure 28.12 Scalar analyzer directional bridge.

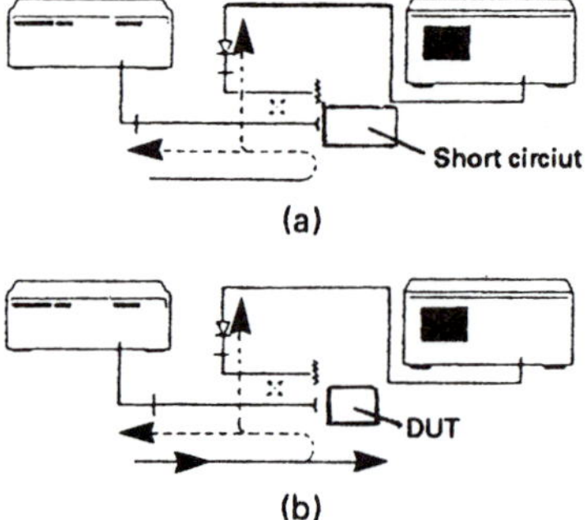

Figure 28.13 Reflection measurement. (a) Calibration configuration; (b) measurement configuration.

from 50 Ω) of the device under test (DUT). Directional bridges have very broadband characteristics with very good (40 dB) directivity performance. One trade-off is that they have 6 dB insertion loss in the incident rf path. This can affect how much incident power is available at a DUT, which can then limit the dynamic range of a transmission measurement.

28.5.2 Reflection measurements

An analysis of the signal flow paths in the basic reflection measurement configuration (Fig. 28.13) results in a reflection uncertainty expression that accounts for any uncertainties that would be introduced in the process of measurement calibration and then DUT characterization. Figure 28.14 shows how the uncertainties of a reference calibration and the actual measurement can add to the total measurement uncertainty. A simplified reflection uncertainty equation for this situation can be given by

$$\Delta\rho = A + B\rho_L + C{\rho_L}^2$$

where $\Delta\rho$ = reflection magnitude uncertainty
A = directivity
B = calibration error, frequency response, display and instrument errors
C = effective source match
ρ_L = reflection coefficient of DUT

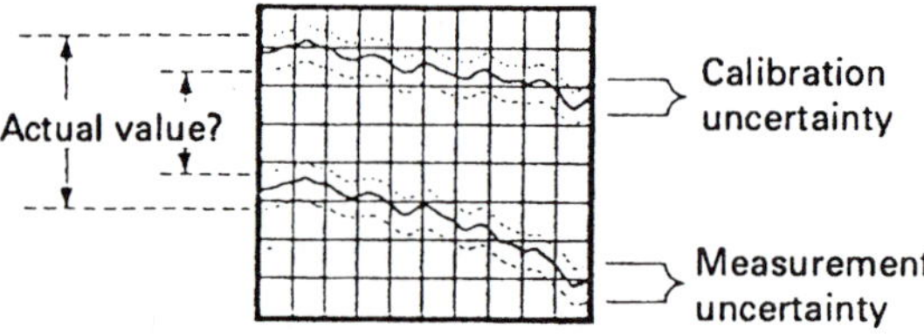

Figure 28.14 Reflection measurement uncertainty.

The first factor is the directivity term of the signal-separation device. As was discussed in the definition of directivity, this is equivalent to a portion of the incident energy leaking directly to the reflected signal detector. It is independent of the reflection term being measured. With high reflection terms, directivity is not a major concern, but with high return loss test devices it can be a major source of error. It is important to pick a signal-separation device with a directivity large compared with the return loss of the reflections being measured. Also, care must be taken to adding adapters at the test port output of the reflection device because the connector match of the adapter can be a limiting factor on the system directivity.

The B term in the uncertainty equation relates to errors in the direct path of the incident wave on the DUT and the return path of the reflected wave to the reflected detector. The frequency response portion of this term can be calibrated out by doing a normalization measurement with a short circuit ($\rho = 1$) on the test port. This reduces frequency response errors but will not account for display and instrumentation error such as receiver dynamic accuracy.

The final uncertainty term is the effective source match term C. This is a result of the reflected wave re-reflecting off the nonperfect source structure and being measured as another incident wave. This uncertainty term is a potential problem when the DUT has a ρ_L close to unity. The equivalent source match can be improved by using a power splitter-ratio measurement configuration, by improving the output source leveling of the stimulus, or by putting pads and attenuators in the stimulus path.

Another calibration technique is used in scalar systems that reduces the error due to the sum of directivity and source match errors. By averaging the short- and open-circuit responses, the calibration error due to the sum of the directivity and source match errors can be removed. The open and short average tends to average out calibration error, thus making $B = 0$. Figure 28.15 shows the impact of open and short averaging on a calibration.

A very useful tool available from most instrumentation suppliers is the reflectometer and mismatch error calculator. It is a simple device used to relate directivity, mismatch, SWR, and uncertainty. Directivity can be converted directly to a linear term equivalent to the A-error term in a reflection coefficient measurement. In addition conversion can be directly made between reflection coefficient, return loss, and standing wave ratio (SWR). Ripple due to multiple mismatches can also be calculated. This is of value in transmission calculations as well as the equivalent source match considerations for reflection uncertainty.

28.5.3 Transmission measurements

A scalar analyzer transmission measurement is done by first calibrating to a through reference connection and then substituting a DUT for the through path. The uncertainty of the resulting transmission magnitude measurement

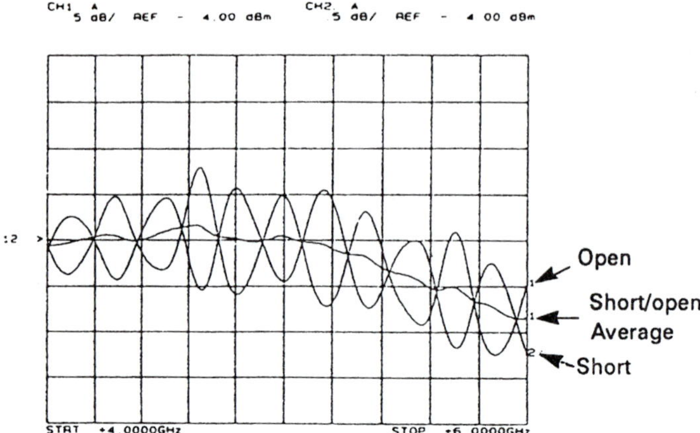

Figure 28.15 Scalar analyzer open and short averaging characteristics.

is the sum of the calibration measurement uncertainty and the device measurement uncertainty. The dominant sources of error are source and detector mismatches. Frequency response errors are eliminated through normalization, but the through calibration has an uncertainty due to the multiple reflection between the source match and detector load. When the DUT is inserted, similar uncertainties occur between the source match and DUT input match as well as the DUT output match and the detector match. Figure 28.16 graphically shows how these mismatches interact. Figure 28.17 shows how the actual multiple mismatch uncertainty adds. The "mismatch error limits" portion of a reflectometer slide rule provides a simple way to calculate these values.

Padding source or detector, improving source match by leveling techniques, or using a power splitter-ratio measurement are all techniques that can be employed to improve the equivalent source match or detector match of the system and thus reduce the measurement uncertainties.

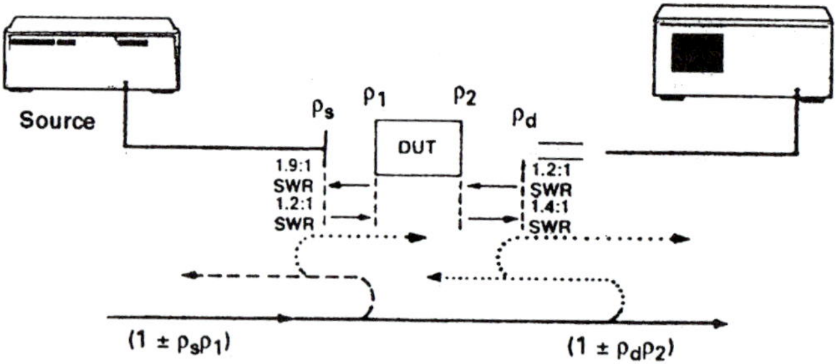

Figure 28.16 Transmission measurement mismatch uncertainty.

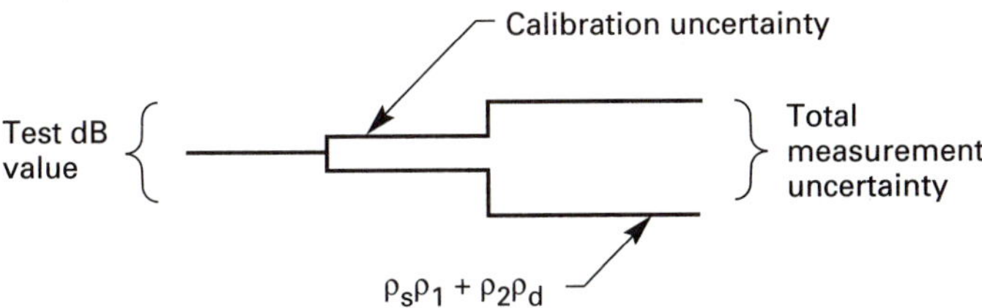

Figure 28.17 Mismatch uncertainty model where ρ_s = source reflection, ρ_1 = DUT input reflection, ρ_2 = DUT output reflection, and ρ_d = detector reflection.

28.5.4 Special applications

While the main application of scalar network analyzers is frequency domain characterization of linear networks, there are some additional applications they do address. For example, it is possible to use a frequency domain measurement to calculate return loss as a function of distance down a transmission structure. By applying Fourier transform analysis techniques to frequency domain information it is possible to obtain a time domain simulation that can be scaled in distance for a particular transmission medium propagation velocity. The resulting output is a useful tool for fault location analysis of transmission structures.

With some scalar network analyzer systems, detectors are individually characterized for frequency response and dynamic accuracy. This results in the situation that the detectors can measure power with power meter accuracies. With such accuracy, scalar analyzers become very useful in measuring compression and amplitude-sensitive characteristics of active devices such as amplifiers. Also, with excellent power measuring capability and available techniques such as ac detection, scalar analyzers are seeing increasing applications in measuring translated frequency devices such as mixer components or communications up and down converter system components.

28.6 Vector Network Analysis

Basic measurement characteristics. The vector network analyzer configuration is distinguished from the scalar network analyzer primarily by the complexity of the receiver structure and the information extrapolated from the detectors (Fig. 28.18). The signal-separation devices consist of splitters, couplers, and/or bridges. The signal-handling components and appropriate rf switching are usually configured together in a "test set" portion of the measurement system. This is done because of the need for frequent, repeatable, and accurate signal switching with the complex calibrations used.

28.6.1 Tuned receivers

The receiver portion of a vector network analyzer system is a fundamental or harmonic mixing multichannel receiver that is tuned in such a fashion that

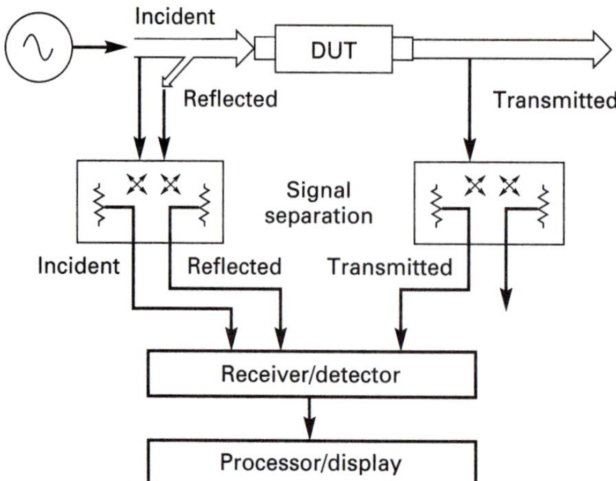

Figure 28.18 Vector network analyzer measurement.

it tracks the frequency of the source. The receiver converts broadband swept rf signals down to an if frequency that is fixed and independent of the rf test frequency. The if frequency is low enough that precision detection circuitry can be used to determine the magnitude of the signal in each receiver channel and the phase relationship of any two receiver channels. The result is a wide dynamic range (100 dB+) spurious free multichannel receiver that can characterize vector quantities (such as reflection coefficient and gain) between any of its multiple inputs. The ability to measure phase characteristics gives the vector network analyzer the capability to make characterization of complex impedance and phase delay characteristics of devices under test. The ability to measure vectors and do complex calculations allows the measurement system to perform complex calibrations by measuring well-known standards and calculating correction factors to apply to measured data of a DUT. The detectors' ability to do complex vector manipulation on measured data allows the system to improve the quality of measurement significantly and reduce the uncertainties associated with measured results. The detector also has the ability to manipulate error-corrected data to present information in many different display formats from rectilinear phase or magnitude as a function of frequency to polar displays of vector quantities.

Much like the characteristics of the scalar detectors, vector network analyzer receivers have limitations to their performance. The receivers are expected to be linear in their conversion characteristics. Therefore, there are maximum allowable input signals to each receiver channel before compression or limiting begins to take place. At low signal levels receiver sensitivity and accuracy is limited by noise and by low-level leakage of signals (crosstalk) not part of the measurement path. Each network analyzer system, including appropriate signal-separation test set, must be carefully understood from a signal level standpoint to obtain optimum performance from the system. Care

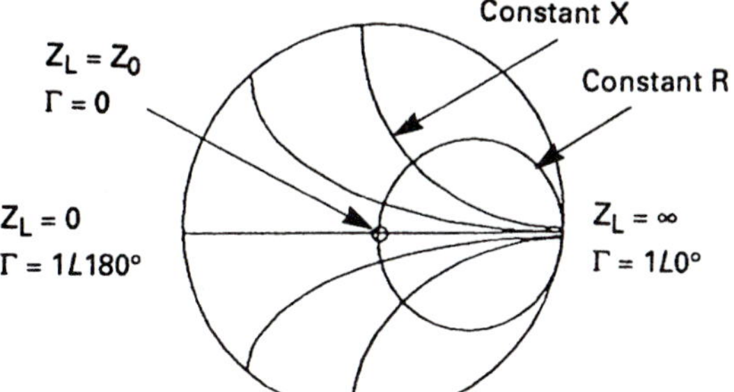

Figure 28.19 Impedance measurements displayed on a Smith chart.

must be taken to maintain optimum incident test signal and receiver input signal magnitudes.

28.6.2 Reflection measurements

With the ability to measure vector quantities, it is possible to measure the ratio of a device reflected signal to an incident signal. This ratio is the complex number representation of reflection coefficient Γ_L. The vector network analyzer can display either the magnitude or the phase of the reflection coefficient as a function of frequency. It can also display a polar display of both the magnitude and phase of the reflection coefficient. Since each specific reflection coefficient relates to a unique impedance, it is possible to relate the vector reflection coefficient to impedance. Figure 28.19 is an outline of a common polar display called a Smith chart. It is a polar display of reflection coefficient with overlays of constant-impedance lines normalized to a characteristic impedance Z_0. The Smith chart polar display becomes a very effective analytic tool at evaluating the input impedance of a device under test. All positive real resistance values transform to points inside the unity reflection coefficient circle.

Figure 28.20 is a model of the reflection measurement configuration. Expressions for the measured reflection coefficient (S_{11}) and the actual reflection coefficient show that measurement uncertainty is affected by directivity, tracking, and source match terms. In this situation one can do a better job than just taking care of the best possible raw performance by taking several

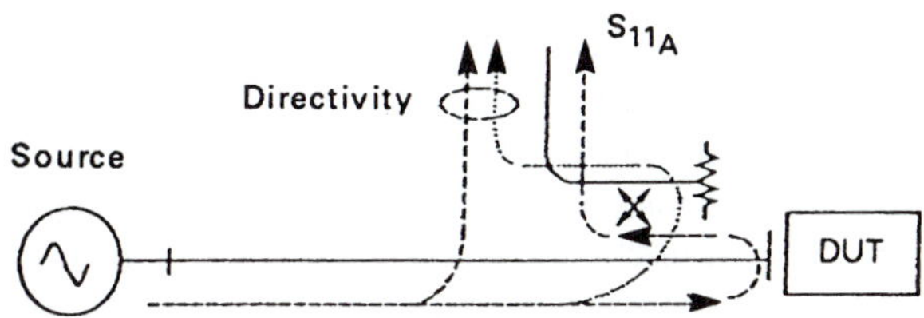

Figure 28.20 Reflection measurement signal paths.

TABLE 28.1 Typical One-Port Accuracy Enhancement Performance for Response and One-Port Calibration Techniques

	Typical after response calibration, dB	Typical after one-port calibration (S_{11} one-port), dB
Directivity D	30	50
Source match M_S	16–20	>40
Reflection track T_R	±1.5	±0.05

$\Delta S_{11} = S_{11M} - S_{11_A} = D + S_{11_A} \times T_R + M_S \times S_{11_A}^2$ (7-mm connector)

calibration steps. The first step is to put a short circuit on the test port and to normalize the measured data to a reflection of a short circuit. ($\Gamma = 1$ at 180°). Called a response calibration, this removes the frequency response errors in the measurement system.

A more complex approach is to do a one-port calibration. This procedure requires measuring several different devices to extrapolate the error terms of the reflection measurement. The first device measured is a precision load. The measured data are the directivity term of the error model. Next an open circuit and a short circuit are both measured. From these two sets of measurements it is possible to extract the source match and frequency response errors of the system. After the calibration the network analyzer detector stores error terms that are used to convert measured data into the corrected display of the DUT reflection characteristics. The "ideal" calibration standards correct the measurement only to the degree that they are ideal. For example, at very high frequencies it is difficult to build a perfect fixed termination, so high-frequency calibration devices include sliding terminations. With a sliding termination the directivity vector at a given frequency can be determined by sliding the load on an airline to create a "circle" of data points. The center of this circle is the directivity vector at that frequency. Table 28.1 is an example of the typical performance one can expect for an rf network analyzer in a 7-mm connector system with the two different calibration approaches.

28.6.3 Transmission measurements

In making measurements of two-port devices it is usually of interest to measure both the reflection and transmission characteristics of the unknown device. This presents some interesting interactions which must be considered. Figure 28.21 represents the measurement situation. The first observation is that the load match of the measurement system at the output of the DUT will impact the input match of the device. In the transmission path frequency response, source mismatch interactions, load mismatch interactions, and crosstalk are factors that will impact the accuracy of a measurement. As in

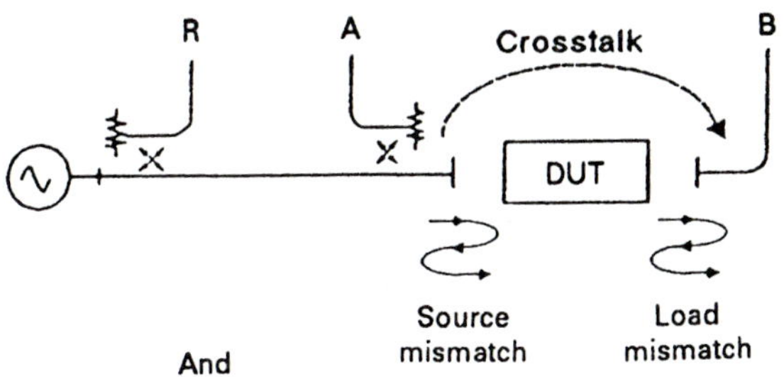

Figure 28.21 Transmission measurement signal paths.

the reflection measurement situation, it is possible to use measurements of known standard devices to calibrate the total measurement system by systematically calculating error terms that represent the characteristics of the particular test system.

Figure 28.22 is a flow graph and expression of the transmission measurement model. Notice the interaction of source match M_s with S_{11A}, load match M_L with S_{22A}, the transmission frequency response error T_t, and the crosstalk of leakage (c) error. This model also indicates that in order to accurately extract S_{21A} data, accurate S_{11}, S_{12}, and S_{22} information must be known.

A full two-port measurement calibration is used to mathematically remove the effects of these errors. A one-port reflection calibration is used to characterize source match, a through connection to characterize transmission frequency response and load match, and an isolation calibration to determine transmission leakage or crosstalk. A complete full two-port measurement model includes a model for both forward and reverse direction measurements. To accurately measure a single S parameter, it is necessary to measure all of them. Table 28.2 is an example of the measurement system improvements gained by doing a full two-port calibration as compared with a simple response calibration for a typical test system with 7-mm test ports and a 7-mm set of calibration standards.

The calibration technique discussed above used a traditional open/short/

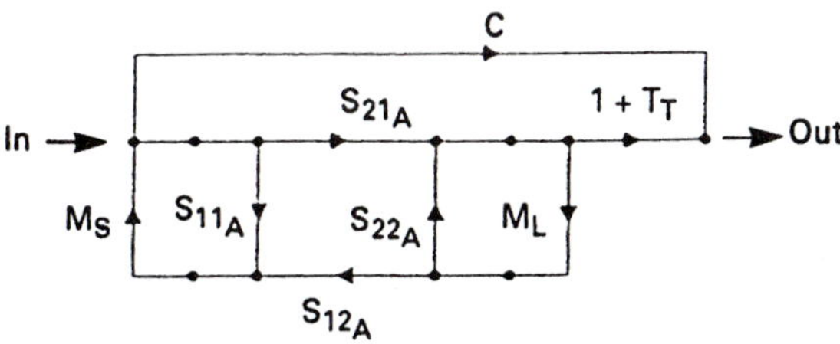

$$S_{21_M} = C + \frac{S_{21_A}(1 + T_T)}{1 - M_S S_{11_A} - M_L S_{22_A} - M_S M_L S_{21_A} S_{12_A}}$$

Figure 28.22 Transmission measurement flow graph.

TABLE 28.2 Typical Two-Port Calibration Compared with a Response Calibration (7 mm)

	Typical response measurement calibration, dB	Typical after full two-port calibration, dB
D	30	50
M_S	16–20	>40
T_R	±1.5	±0.05
T_T	±0.02	±0.03
M_L	16–20	40
C	90	>100

$\Delta S_{21} \cong C + T_T \cdot S_{21_A} + M_S \cdot S_{11_A} \cdot S_{21_A} + M_L \cdot S_{22_A} \cdot S_{21_A}$

D = directivity, M_S = source match, T_R = reflection tracking, T_T = transmission tracking, M_L = load match, and C = crosstalk.

load/through set of standards. However, it is not always simple to manufacture these standards, especially in a noncoaxial media. Fixed broadband loads are difficult to manufacture, so for greater precision a sliding load can be used in place of a fixed load. At millimeter-wave frequencies in waveguide structures offset loads and fixed loads are used to establish the center point of the directivity vector. An offset short is used in waveguide structures, since an open-circuit standard is not possible. The through/reflection/line (TRL) calibration calculates the same 12 error terms (directivities of two-ports, forward and reverse match for two-ports, forward and reverse tracking for transmission and reflection, and forward and reverse isolation) as the other approaches, but it uses a through connection, a high but unknown reflection, and a line (whose reference impedance is the Z_0 reference) to extract the calibration information. The TRL calibration method offers the advantage of simplicity and accuracy over other calibration techniques. It is especially useful in unusual transmission-line environments such as microstrip where standards are difficult to achieve. Table 28.3 gives an idea of the types of accuracy that can be achieved with the calibration standards available in the 7-mm

TABLE 28.3 Accuracy Comparison of Different Calibration Methods with 7-mm Standards

Residual errors	Open short load, fixed, dB	Open short load, slide, dB	Open short offset, dB	TRL, dB
Directivity δ	−40	−52	−60	−60
Match τ	−35	−41	−42	−60
Tracking μ	±.1	±.047	±.035	±0

Comparison (7 mm) 18 GHz.

TABLE 28.4 Summary of Accuracy Enhancement Techniques

Measurement calibration	When to use	Errors removed
Response	Transmission measurement Reflection measurement When the highest accuracy not required	Frequency response only
S_{11} one-port	Reflection measurement: highest accuracy for one-port devices (could be used for well-matched two-port devices)	Directivity Source-match Frequency response
Full two-port	Transmission measurement Reflection measurement Highest measurement accuracy for two-port devices	Directivity Source, load match Isolation Frequency response

connector environment for the various calibration techniques. Table 28.4 is a summary of the various accuracy enhancement options. The data sheet for any particular network analyzer system presents graphical plots of the system measurement capability as a function of the system performance, the calibration technique, and the DUT parameter being measured. Figure 28.23 is a sample plot of S parameter measurement accuracy for an rf network analyzer using full two-port calibration and in several connector environments (type N, 3.5 mm, 7 mm).

28.6.4 Special considerations

As vector network analyzers have evolved their internal computation and control capabilities, several new measurement capabilities have evolved. Vector network analyzers can determine the group delay of a device by calculating the slope of the measured phase information. Group delay is a term used to describe how linear the phase of a device is and thus how much distortion it will potentially contribute to communication systems. Group delay T_G has a dimension of time.

One of the most powerful analytical tools is a built-in time domain function. The time domain function performs a digital Fourier transform on measured frequency data and displays a simulated impulse or step response to the measured device in the time domain. The time domain display can be used to analyze step and impulse characteristics in the transmission mode and to simulate a time domain reflectometer mode in reflection characterization. This simulation can give a great deal of insight into test device properties. Figure 28.24 is a network analyzer display of the simulated impulse response on the frequency domain measurement of a surface acoustic wave device. This display shows the through transmission, the rf feedthrough, and the multiple

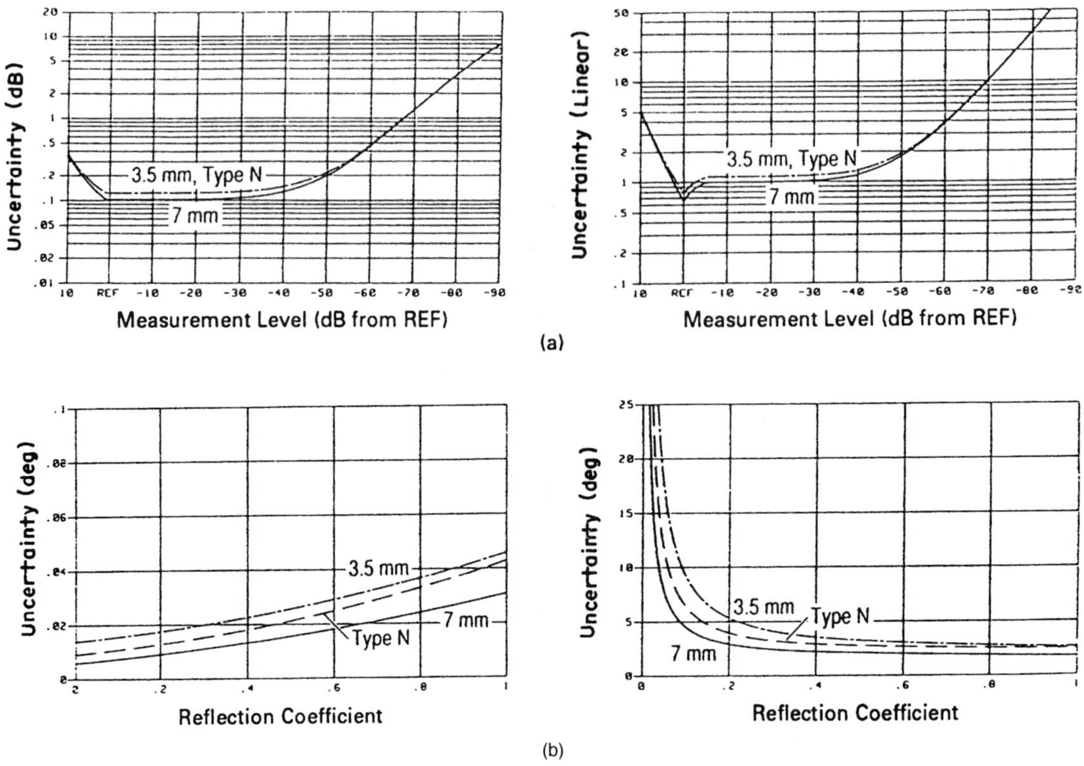

(a)

(b)

Figure 28.23 Performance characteristics of an rf vector network analyzer with two-port calibration. (a) Transmission measurements; (b) reflection measurements.

reflection triple travel path. By "gating" out all the time domain data except the through transmission, it is possible for the analyzer to transform back into the frequency domain and display a frequency response of just the through path of the device.

Vector network analyzers have also incorporated calibration techniques that allow the test set ports to be calibrated for power out and power incident

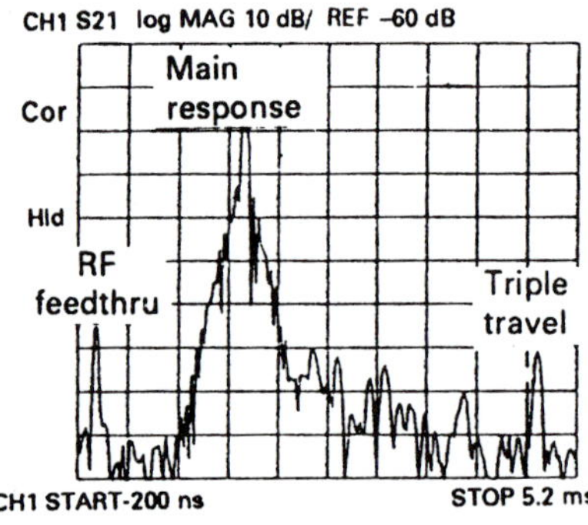

Figure 28.24 Simulated time domain impulse response of a surface acoustic wave device.

to levels approaching that of power meter accuracy. With this capability the characterization of active devices as a function of power became a new application area. Gain compression in amplifiers is an example of very important device considerations for power-sensitive components. In addition some vector network analyzer block diagrams have the ability for the receiver to track harmonics or the source as well as the fundamental input frequency. This allows swept characterization of harmonic content. In addition to harmonics, some systems allow the ability to offset the receiver from the source frequency such that frequency offset measurements can be implemented. This is useful for measuring frequency converters or mixers.

Chapter

29

Logic Analyzers

Steve Warntjes
Agilent Technologies
Colorado Springs, Colorado

29.1 Introduction to the Digital Domain

The advent of digital circuits dramatically changed the concerns of engineers and technicians working with electronic circuits. Ignoring for a moment digital signal quality or signal integrity, the issues switched from the world of bias points and frequency response to the world of logic ones, zeros, and logic states (see Fig. 29.1). This world has been called the "data domain."

Using off-the-shelf components virtually guarantees correct values of voltage and current if clocks are kept to moderate speeds (less than 50 MHz) and fan-in/fan-out rules are observed. The objective for circuit verification and testing focuses on questions of proper function and timing. Although parametric considerations are simplified, the functional complexity and sheer number of circuit nodes is increased tremendously. Measurements to address these questions and to manage the increased complexity are the forte of the "logic analyzer." Logic analyzers collect and display information in the format and language of digital circuits.

Microprocessors and microcontrollers are the most common logic-state machines. Software written in either high-level languages, such as C, or in the unique form of a processor's instruction set (assembly language) provide the direction for these state machines that populate every level of electronic products. Most logic analyzers can be configured to format their output as a sequence of assembly processor instructions or as high-level language source code. This makes them very useful for debugging software. For real-time or time-crucial embedded controllers, a logic analyzer is a superb tool to both trace program flow and measure event timing.

Because logic analyzers do not affect the behavior of processors, they are

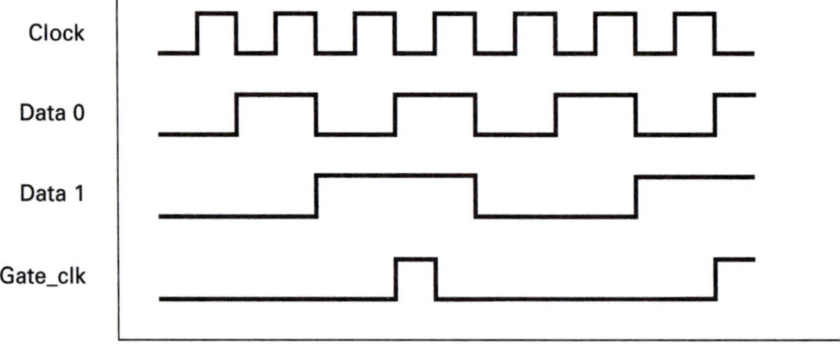

(a)

I = 1
I = 0
State 0
O = 0
E = 0
S = 1
I = 1
State 1
O = 0
E = 0
I = 0
State 2
O = 1
E = 0
I = 1
I = 0
S = 0
I = 1
S = 1
I = 1
State 4
O = 0
E = 0
S = 1
I = 0
State 3
O = 0
E = 1
S = 0

(b)

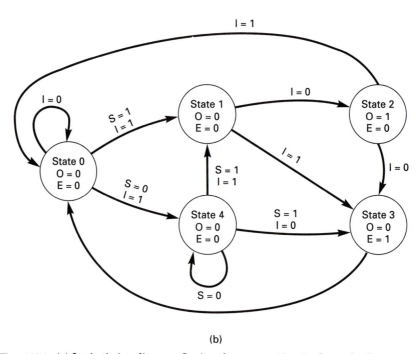

Figure 29.1 (a) Logic timing diagram. Logic value versus time is shown for four signals. (b) Logic state diagram. Inputs I and S control transitions from state to state. O and E are outputs set to new values upon entry to each state.

excellent tools for system performance analysis and verification of real-time interactions. Data-stream analysis is also an excellent application for logic analyzers. A stream of data from a digital signal processor or digital communications channel can be easily captured, analyzed, or uploaded to a computer.

29.2 Basic Operation

The basic operation of logic analysis consists of acquiring data in two modes of operation: asynchronous timing mode and synchronous state mode. Logic analysis also consists of emulation solutions used to control the processor in the embedded system under development and the ability to time correlate high-level language source with the captured logic analyzer information.

29.2.1 Asynchronous mode

On screen, the asynchronous mode looks very much like an oscilloscope display. Waveforms are shown, but in contrast to an oscilloscope's two or four channels, there are a large number of channels: eight to several hundred. The signals being probed are recorded either as a "one" or a "zero." Voltage variation (other than being above or below the specified logic threshold) is ignored, just as the physical logic elements would do. In Fig. 29.2, an analog waveform is compared with its digital equivalent. A logical view of signal timing is cap-

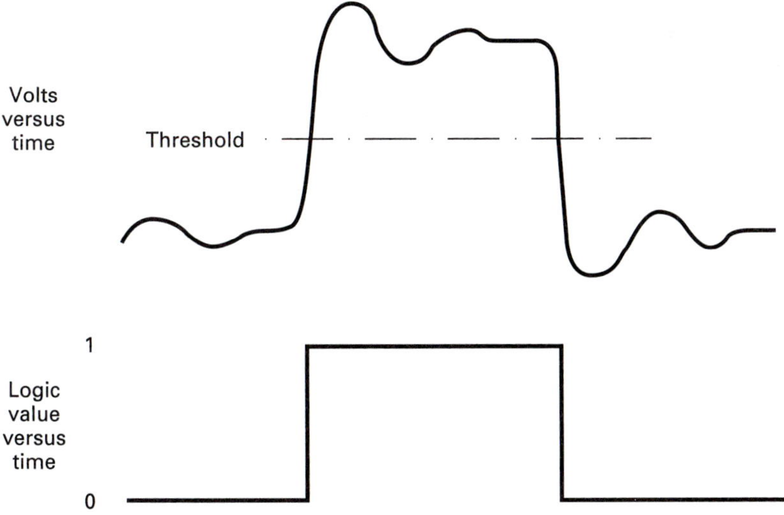

Figure 29.2 Analog versus digital representations of a signal.

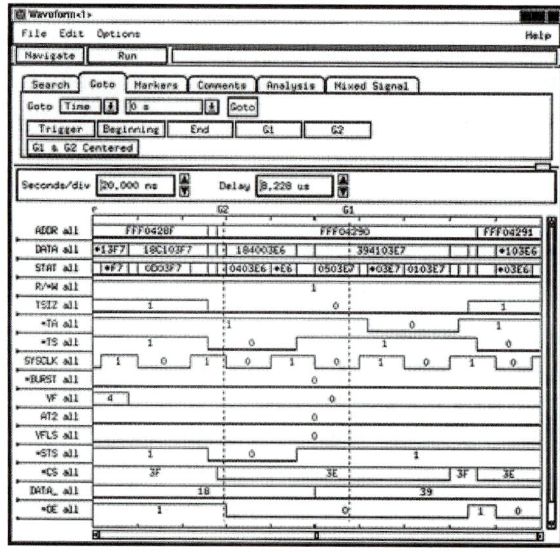

Figure 29.3 Timing mode display. Display and acquisition controls are at the top. Waveforms are displayed on the bottom two-thirds of the display. Note the multibit values shown for "A8–15."

tured. As with an oscilloscope, the logic analyzer in the timing mode provides the time base that determines when data values are clocked into instrument storage. This time base is referred to as the "internal clock." A sample logic analyzer display showing waveforms captured in timing mode is shown in Fig. 29.3.

29.2.2 Synchronous mode

The synchronous state mode samples the signal values into memory on a clock edge supplied by the system under test. This signal is referred to as the

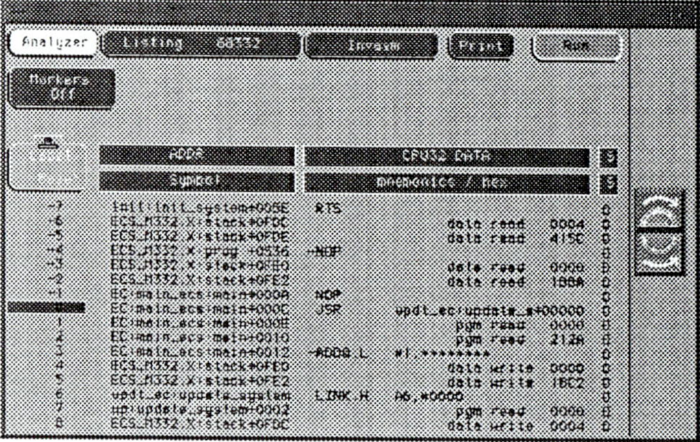

Figure 29.4 State mode display. Listing shows inverse assembly of microprocessor bus cycles.

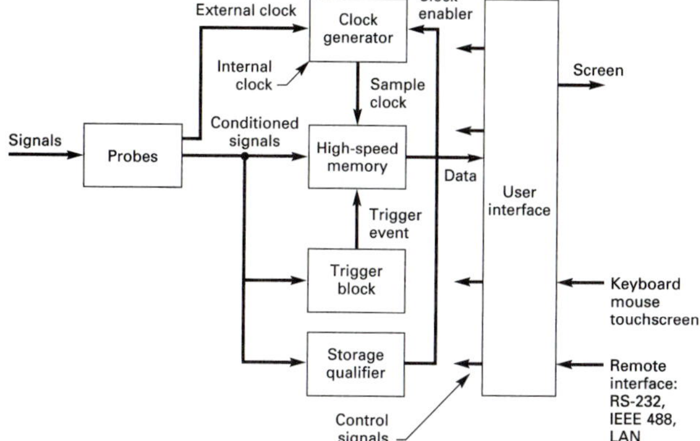

Figure 29.5 Logic analyzer block diagram.

"external clock." Just as a flip-flop takes on data values only when it is clocked, the logic analyzer samples new data values or states only when directed by the clock signal. Groupings of these signals can represent state variables. The logic analyzer display shows the progression of states represented by these variables. A sample logic analyzer display showing a trace listing of a microprocessor's bus cycles (state mode) is shown in Fig. 29.4.

29.2.3 Block diagram

An understanding of how logic analyzers work can be gleaned from the block diagram in Fig. 29.5. Logic analyzers have six key functions: The probes, high-speed memory, the trigger block, the clock generator, the storage qualifier, and the user interface.

Probes. The first function block is the probes. The function of the probes is to make physical connection with the target circuit under test. To maintain proper operation of the target circuit, it is vital that the probes not unduly load down the logic signal of interest or disturb its timing. It is common for these probes to operate as voltage dividers. By dividing down the input signal, voltage comparators in the probe function are presented with the lowest possible voltage slew rate. Higher-speed signals can be captured with this approach. The voltage comparators transform the input signals into logic values. Different logic families (i.e., TTL, ECL, or CMOS) have different voltage thresholds, so the comparators must have adjustable thresholds.

High-speed memory. The second function is high-speed memory, which stores the sampled logic values. The memory address for a given sample is supplied internally. The typical memory depth is hundreds of thousands of samples. Some analyzers can store several megasamples. Usually the analyzer user is interested in observing the logic signals around some event. This event is

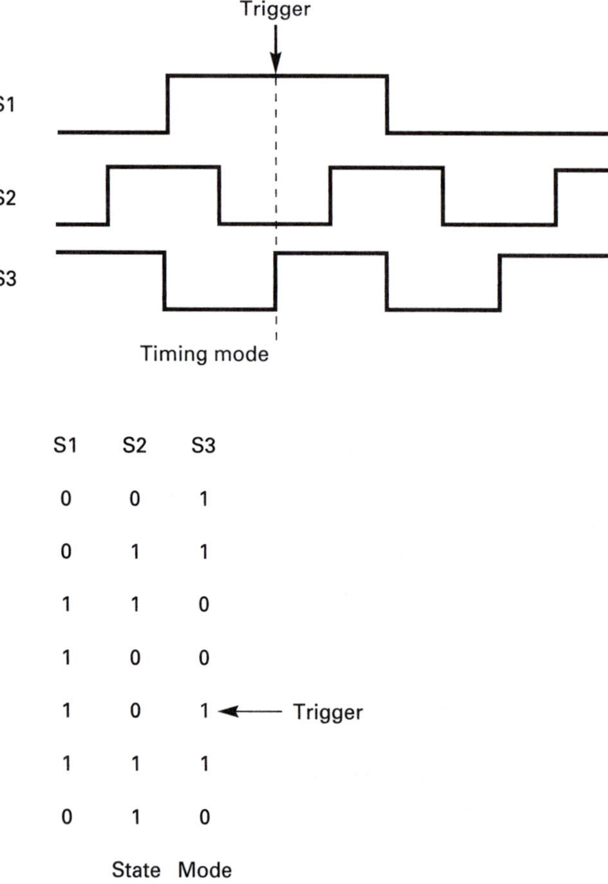

Figure 29.6 Example of trigger pattern showing match found with timing mode data and then state mode data. Trigger pattern is "1 0 1" for input signals S_1, S_2, and S_3.

called the "measurement trigger." It is described in the next functional block. Samples have a timing or sequence relationship with the trigger event, but are arbitrarily placed in the sample memory, depending on the instantaneous value of the internally supplied address. The memory appears to the user as a continuously looping storage system.

Trigger block. The third functional block is the trigger block. Trigger events are a user-specified pattern of logical ones and zeros on selected input signals. Figure 29.6 shows how a sample trigger pattern corresponds with timing and state data streams. Some form of logic comparator is used to recognize the pattern of interest. Once the trigger event occurs, the storage memory continues to store a selected number of posttrigger samples. Once the posttrigger store is complete, the measurement is stopped. Because the storage memory

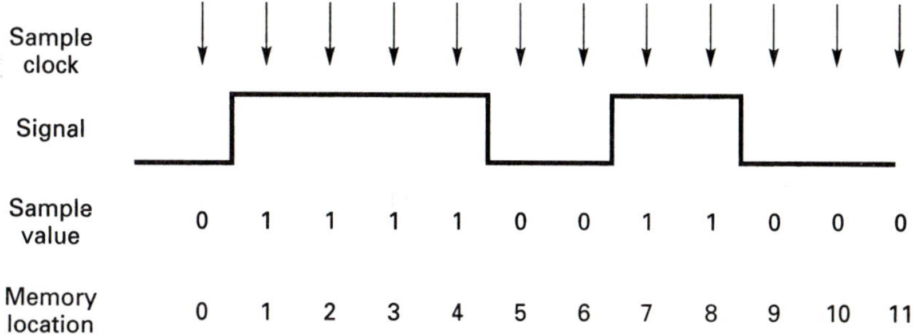

Figure 29.7 Continuous storage mode. A sample value is captured at each sample clock and stored in memory.

operates as a loop, samples before the trigger event are captured, representing time before the event. Sometimes this pretrigger capture is referred to as "negative time capture." When searching for the causes of a malfunctioning logic circuit, the ability to view events leading up to the problem (i.e., the trigger event) makes the logic analyzer extremely useful.

Clock generator. The fourth block is the clock generator. Depending on which of the two operating modes is selected, state or timing, sample clocks are either user supplied or instrument supplied.

In the state mode, the analyzer clocks in a sample based on a rising or falling pulse edge of an input signal. The clock generator function increases the usability of the instrument by forming a clock from several input signals. It forms the clocking signal by "OR"ing or "AND"ing input signals together. The user could create a composite clock using logic elements in the circuit under test, but it is usually more convenient to let the analyzer's clock generator function do it.

In timing mode, two different approaches are used to generate the sample clock. Some instruments offer both approaches, so understanding the two methods will help you get more from the instrument. The first approach, or "continuous storage mode," simply generates a sample clock at the selected rate. Regardless of the activity occurring on the input signals, the logic values at the time of the internal clock are put into memory (see Fig. 29.7).

The second approach is called "transitional timing mode." The input signals are again sampled at a selected rate. The clock generator function only clocks the input signal values into memory if one or more signals change their value. Measurements use memory more efficiently because storage locations are used only if the inputs change. For each sample, a time stamp is recorded. Additional memory is required to store the time stamp. The advantage of this approach over continuous storage is that long-time records of infrequent activity or bursts of finely timed events can be recorded (see Fig. 29.8).

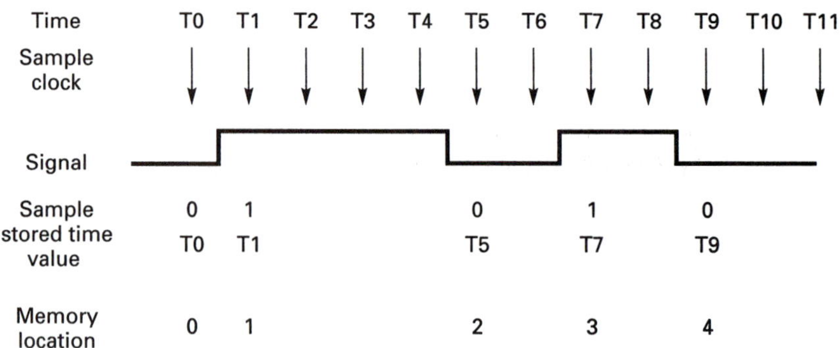

Figure 29.8 Transitional storage mode. The input signal is captured at each sample clock, but is stored into memory only when the data changes. A time value is stored at each change so that the waveform can be reconstructed properly.

Storage qualifier. The fifth function is the storage qualifier. It also has a role in determining which data samples are clocked into memory. As samples are clocked, either externally or internally, the storage qualifier function looks at the sampled data and tests them against a criterion. Like the trigger event, the qualifying criterion is usually a one-zero pattern of the incoming signal. If the criterion is met, then the clocked sample is stored in memory. If the circuit under test is a microprocessor bus, this function can be used to separate bus cycles to a specific input/output (I/O) port from instruction cycles or cycles to all other ports.

User interface. The sixth function, the user interface, allows the user to set up and observe the outcome of measurements. Benchtop analyzers typically use a dedicated keyboard and either a cathode-ray tube (CRT) or a liquid crystal display (LCD). Many products use graphic user interfaces similar to those available on personal computers. Pull-down menus, dialog boxes, touch screens, and mouse pointing devices are available. Logic analyzers are used sporadically in the debug process, so careful attention to a user interface that is easy to learn and use is advised when purchasing.

Not all users operate the instrument from the built-in keyboard and screen. Some operate from a personal computer or workstation. In this case, the "user interface" is the remote interface: IEEE-488 or local area network (LAN). Likewise, the remote interface could be the user's Web browser of choice if the logic analyzer is Web enabled.

29.2.4 Emulation solutions

An emulation solution is a connection to the processor in the embedded system used to control the program execution. The connection is usually on the order of five to 10 processor signals. These signals control the program execution, including the state of the processor, such as running, reset, or single

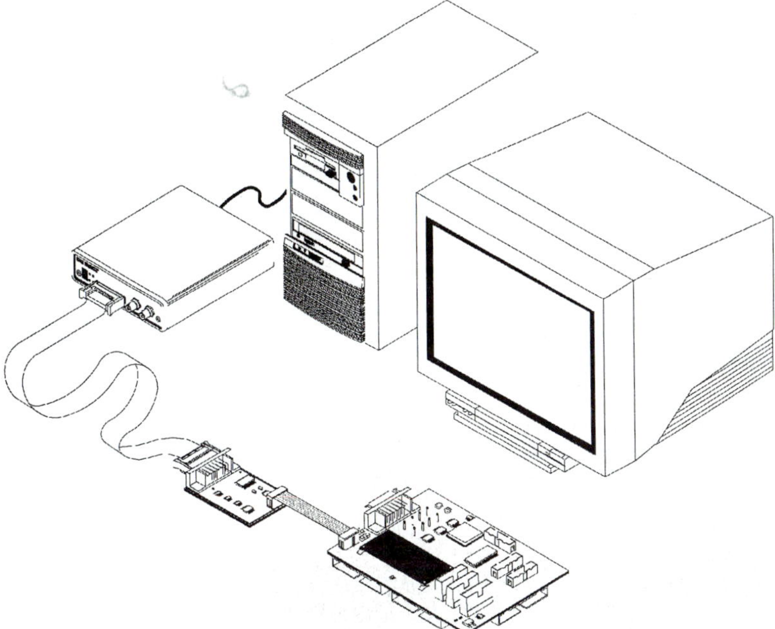

Figure 29.9 Example emulation solution with a PC connected to a customer target to control the processor under development.

stepping, and also the ability to quickly download the processor's executable code and examine/modify processor memory and registers. This processor control is usually accomplished through dedicated on-processor debug resources. Emulation solution allows you to perform software and hardware debugging, as well as time correlate software and hardware activity (see Fig. 29.9).

29.2.5 High-level language source correlation

High-level language source correlation provides a real-time trace or acquisition of processor address, data, and status information linked to a software high-level source language view. This information is then time correlated to activity captured by the rest of the logic analyzer's acquisition modules, such as oscilloscope channels. Symbols from the user's software program can also be used to specify trigger conditions and are listed in the analyzer display (see Fig. 29.10). This feature uses the information provided in the object file from the customer's compiler to build a database of source files, line numbers, and symbolic information. The HLL Source correlation is a nonintrusive tool that does not typically require any major changes in the software compilation process.

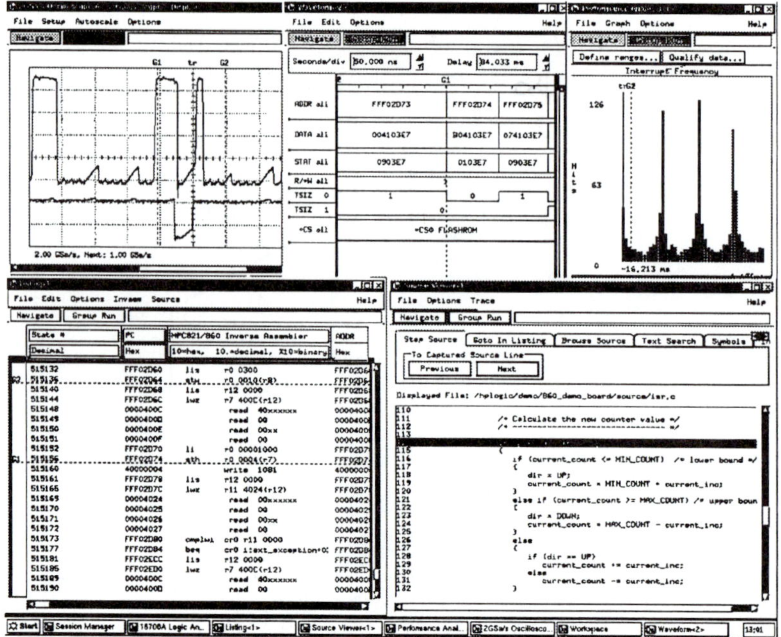

Figure 29.10 Correlated high-level language display. Lower right is the high-level source code; lower left, the assembly code listing. Upper left is a time-correlated oscilloscope display; upper middle, a timing waveform display; upper right, a performance analysis snapshot.

29.3 Using the Key Functions

Basic proficiency centers around a solid setup of the logic analyzer and piloting of the trigger function.

29.3.1 Setting up

Basic setup involves configuring the analyzer into one or more machines, attaching channel probes, selecting the clock source, grouping channels into labels, and specifying the display format for these labels.

1. Determine if the instrument is to act as one logic analyzer (i.e., one machine) or multiple machines. Typically, the instrument is configured as one machine to watch the system timing or analyze a specific processor. In some cases, the instrument is configured as two or more machines. For example, one part of the instrument could be configured as a state machine to watch processor bus activity; the other part of the machine could be configured as a timing machine to watch I/O signals. The many possible variations are based on the specific needs of the customer.

   There is a great deal of variation between different vendors and logic

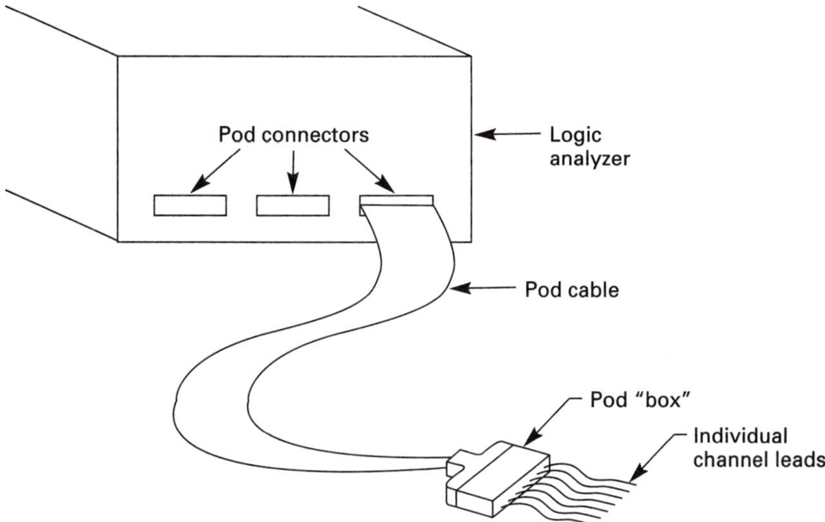

Figure 29.11 Logic analyzer probes. Individual channels are grouped into "pods."

analyzer models; however, the analyzer is usually configured in a "system" menu. In addition to selecting whether the analyzer will operate as one or more machines, each machine is set to state or timing mode. Input channels are also assigned to each machine.

2. Connect probes to the target circuit. The large number of channels involved in a logic analyzer measurement is a challenge to connect physically. Because of the large number of channels, unique probe cables spanning from instrument to target signal are not used in logic analyzers for each channel. Instead of having individual leads, eight or more channels are grouped into "pods." Each pod connects to the instrument as a single cable. At the end of each pod, the channels flare out for individual connection to the target circuit. In some cases, a small box is at the pod's end to facilitate connection. The name *pod* refers to this small box (see Fig. 29.11).

   The physical grouping of channels into pods is the default way of identifying channels. A 32-channel analyzer might have two pods of 16 channels each. A given channel can be identified as a "pod number, bit number." The first channel would be pod 1, bit 0 (digital counting convention starts with zero). The next channel would be pod 1, bit 1. The eighth channel would be pod 1, bit 7. The thirty-second channel would be pod 2, bit 15.

   The probes are typically connected to the target system in one or more of the following mechanisms. The first way is to simply connect the individual channel leads to the target system. This is typically accomplished with chip clips or grabbers, which are attached to the package lead. With the advent of surface mount components, this has become difficult and

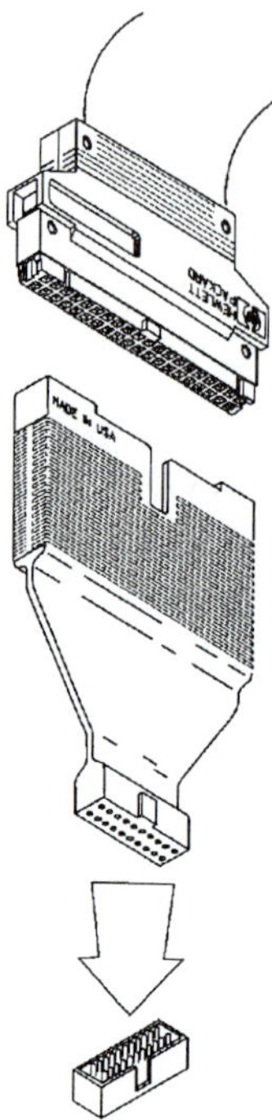

Figure 29.12 An example connector mounted on the target system, termination adapters and logic analysis cable.

sometimes impossible; attachment in this case is accomplished via special surface mount wedge probing techniques or accessories (see Fig. 29.12). The second connection scheme involves plugging the pod cable directly into the target circuit. This usually involves some probe termination adapters or special circuits on the customer's target system. Customer help via changes to their design and/or board layout will continue to be a viable solution as chip interconnect geometries continue to shrink.

The third and common solution is to connect through a probing adapter or analysis probe. Analysis probes are probing adapters made to attach to specific integrated circuits in specific packages. They also may have circuitry to simplify the clocking or enhance the captured data. Analysis probes are the cleanest, easiest way to probe the target circuit.

3. Assign labels to connected channels. Referring to a channel as pod 1, bit 3 is not easy to remember, so user-defined labels are assigned to channels. In the last example, a "memory strobe" was mentioned. In the format menu, you can create a label called "MEM_STB" and associate it to channel pod 1, bit 0. The microprocessor's address bus could consist of 16 channels. The probe hookup could be to channels pod 2, bit 0 through pod 2, bit 15. Create a label called "ADDRESS" and assign it to the selected channels.

   In this way, all the available channels of the analyzer are assigned.

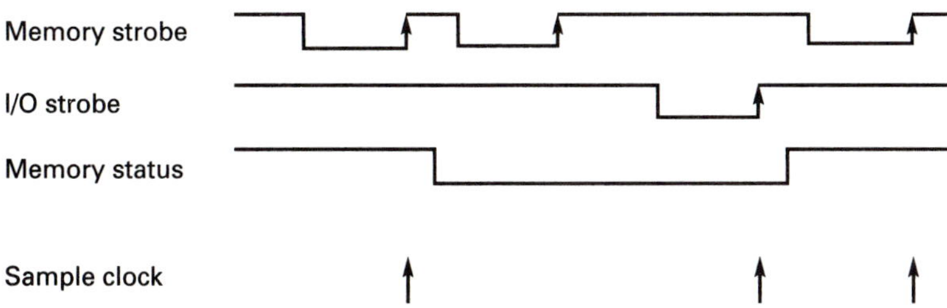

Figure 29.13 State mode clock generation. More useful sample clocks can be created by "AND"ing or "OR"ing signals together. Example shows the rising edge of "MEMORY_STROBE" generating a sample clock only when "MEMORY_STATUS" is high. Another sample clock is generated on the rising edge of "I/O_STROBE" only when "MEMORY_STATUS" is low.

Some channels might be left over, not connected to target signals nor assigned to any labels.

For the 32-channel analyzer of the example, label assignments end up as:

Label	Channels
ADDRESS	pod 2, bits 0 through 15
DATA	pod 1, bits 8 through 15
MEM_STB	pod 1, bit 0
IO_STB	pod 1, bit 1
MEM_STAT	pod 1, bit 2

Note that the latter three labels are single-channel or single-bit values. "ADDRESS" refers to 16 channels or bits of data. "DATA" refers to 8 channels or bits of data.

4. Select a clock. If the analyzer is to be operated in asynchronous timing mode, the internal clock must be selected and set to an operating rate. Because this rate determines the time-interval accuracy of the measurement, it is customary to specify clock rate by the clock period, rather than by frequency. A 100-MHz sample rate would be selected as a 10-ns period. In general, the fastest sample period should be selected unless the total sample record (i.e., sample period times memory depth) is shorter than the duration of the signal behavior being observed. See Sec. 29.4.1 for more information on sample rate. If operating in synchronous state mode, one or more input channels must be selected as the clock source. For most analyzers, only certain input channels or certain pods can be clock sources. The selection of clocks typically will be done in a menu called the "format specification." One or more channels are selected, and either their rising or falling edge (or both) is picked as the sampling event. If the analyzer is capable of having clock qualifiers, one or more channels can be selected to qualify another channel's edge by being a logic one or zero, as specified.

   For example, if a microprocessor bus is being probed, a sample could be generated on the rising edge of a memory strobe (pod 1, bit 0) ANDed with

	Address (hex)	Data (dec)	Status (bin)
TRIGGER	FD00	36	111
TRIG +1	FD01	80	101
TRIG +2	FD02	57	001
TRIG +3	010A	32	111
TRIG +4	FD20	63	101
TRIG +5	FD21	41	010

Figure 29.14 State display using hexidecimal, decimal, and binary data formats.

a logic one on memory status (pod 1, bit 2) OR on the edge of an I/O strobe (pod 1, bit 1) ANDed with a logic zero on memory status (pod 1, bit 2). This is shown in Fig. 29.13.

5. Select a display mode for the labels. Two fundamental options are to view the data as graphic waveforms or as text listings. Usually, this choice falls out of the mode (timing or state) in which the analyzer is operating. Timing mode is usually best viewed as graphic waveforms; in the state mode, the output is best viewed as a listing of numbers or state symbols. The state symbols could be microprocessor instructions mnemonics. The choice between graphics and a listing is usually made by selecting the analyzer's "trace" menu or its "listing" menu.

If the choice is a listing, selections can be made back in the format menu, where the labels were defined. For each label, the user can select a number base (i.e., binary, decimal, hex, etc.). For labels referencing a single channel, this feature provides little value, but for multichannel labels, such as ADDRESS or DATA, the readability is greatly improved. Examples of different numeric bases for labels are shown in Fig. 29.14.

Besides numeric formatting, a set of symbols can be defined to represent each value of a label. Figure 29.15 shows a sample translation table and state listing. The state listing shows how numeric values of the captured inputs are replaced in the listing with more readable labels. Using these features requires effort in the instrument setup, but can dramatically reduce the time to interpret measurement results.

If the analyzer is probing a microprocessor, a display postprocessor called an "inverse assembler" can be turned on. For simple microprocessors, inverse assembly is a simple mapping of symbols to each value of the data bus. More complicated processors have such features as instruction prefetching, caching, and memory management. The inverse assembler for a processor such as this is very sophisticated. It must use all labels for a given state (i.e., address, data, and status) plus include knowledge of prior states to do a decent job of representing instruction execution.

Translation table	
Status	Assigned labels
000	"DATA - WRITE"
001	"DATA - READ"
010	"DATA - WRITE"
011	"DATA - READ"
100	—
101	"CODE - END"
110	—
111	"CODE - BEG"

(a)

State listing	
	Status
TRIGGER	CODE - BEG
TRIG +1	CODE - END
TRIG +2	DATA - READ
TRIG +3	CODE - BEG
TRIG +4	CODE - END
TRIG +5	DATA - WRITE

(b)

Figure 29.15 Example symbol translation table and state listing showing the use of translation. (*a*) Symbolic labels are defined in the translation table for the eight possible values for the signal label "status." (*b*) The state listing shows an example of how the numeric values for "status" in Fig. 29.14 would be replaced by symbols.

29.3.2 Triggering on the problem

The trigger function searches for an event around which a picture of signal timing or state activity is captured. For some measurements, the objective might simply be finding the event or not finding it. For other measurements, the trigger event is the reference point for observing activity.

Whether it's a state or timing measurement, all trigger events are formed from simple trigger resources: patterns and counters. Sophisticated and complex triggers can be formed by using a third resource referred to as the "trigger sequencer" or "trigger state machine." Don't be intimidated by all of the "flexibility"; most measurements use very simple triggers.

Triggers are typically programmed in a logic analyzer's "trigger menu." Figure 29.16 shows an example trigger menu where the building blocks (patterns, counters, and sequencing) have been combined to form a complete trigger specification.

For a group of input signals, a pattern is defined by selecting either a logic "one," "zero," or "don't care" for each signal in the group. Selection of "don't care" for a signal effectively removes that signal from consideration when judging whether a pattern is true. The most common measurement is to set up one pattern and trigger the measurement when that pattern is true. Figure 29.17 shows two patterns and how several test cases match or don't match these patterns. Most analyzers provide 4 to 16 pattern resources. As described under "Setting Up" (Sec. 29.3.1), logic circuits involving microprocessors usually will define signal groups or labels for the processor's address bus, data bus, and status lines. A trigger pattern involves each of these labels and others defined to include all signal inputs.

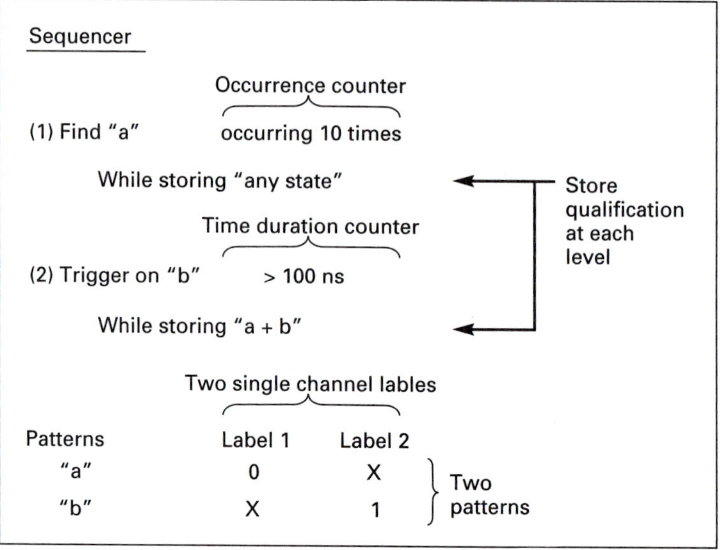

Figure 29.16 Trigger menu example. The three basic elements of a trigger—patterns, counters, and sequence levels—are combined to form a complete specification. Two patterns, *a* and *b*, are formed by selecting "0," "1," or "Don't care" values for input labels "Label 1" and "Label 2." Pattern *a* is used in the first sequence level, along with an occurrence counter. When *a* occurs 10 times, the sequencer advances to level 2. There, pattern *b* happening for 100 ns (a time-duration counter) will trigger the analyzer. Programming of the storage-qualifier function is also shown. After finding pattern *a* in level 1, all states are stored. Then, after the trigger is found, only states matching pattern *a* or *b* will be stored.

		Address (hex)	Data (hex)	MEM_STB (bin)	IO_STB (bin)	MEM_STAT (bin)
Pattern 1		FXXX	6X	1	0	X
Pattern 2		010X	XX	1	X	X
X = don't care						
Pattern match	Example					
Yes	#1	F10A	68	1	0	0
No	#2	E501	62	1	0	0
Yes	#3	0101	EA	1	1	1
No	#4	0105	6F	0	1	1

Figure 29.17 Pattern examples with "Don't cares." Patterns 1 and 2 are defined to include "Don't cares." Both patterns are compared with four examples seeking a match. Pattern 1 only matches example 1. Pattern 2 matches examples 1 and 3.

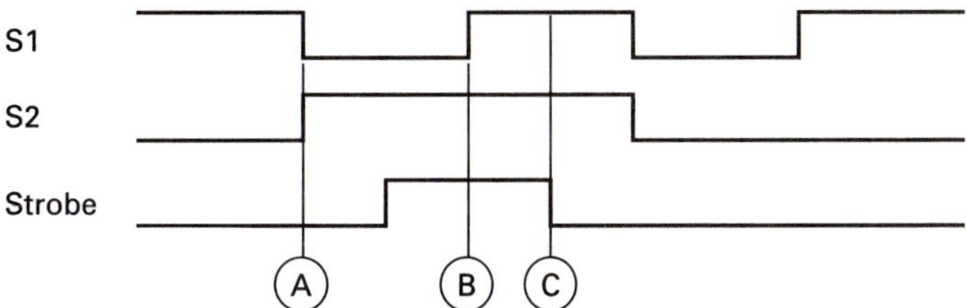

Figure 29.18 Triggering with an edge event. Point C is found by selecting a high-to-low edge for "STROBE," (i.e., $S_1 = 1$, $S_2 = 1$, STROBE = high-to-low edge). Without the edge event, points A and B would be found. If $S_1 = 1$, $S_2 = 1$, STROBE = 0, then point A. If $S_1 = 1$, $S_2 = 1$, STROBE = 1, then point B.

Ranges. A special type of "pattern" usually unique to state mode is the range. "Ranges" have the form of "beginning value/ending value." "Values" are numbers that can be programmed as "one/zero" patterns. "Don't cares" do not make sense, so they are not allowed. A "range pattern" is considered true when compared with any input that is between the beginning and ending values, including those values. The range is very useful when looking at processor address and data buses. A software problem called the "wild pointer" can be solved easily with a range pattern. Wild pointers are calculated addresses for a data access. Because the calculation is bad, the access occurs to unexpected locations that may not repeat. The range pattern opens a wide net to catch the errant access.

Edge events. In timing mode, the pattern resources operate the same as they do in state mode. Not in common is the "edge event." On one or more signals within a label, a low-to-high or high-to-low transition (or both) can be specified. If edge events are used together with a pattern, the trigger criterion is considered true when all "zero/one/don't care" criteria are met and when the signals with an edge specification transition in the correct way. The benefit of using an edge over a common pattern can be seen in Fig. 29.18. The desired trigger point (C) can only be achieved by selecting the high-to-low edge of signal STROBE. Otherwise, points A or B would be selected.

Glitch events. Another criterion unique to timing mode is a "glitch event." As observed in a logic circuit, glitches are very narrow pulses caused by timing races or incomplete switching brought on by timing violations. The chance of such a narrow pulse (e.g., 1 ns wide) being captured on a timing sample is very small. A glitch-triggering resource looks for two signal edges happening between samples (see Fig. 29.19). As long as the pulse is wide enough to meet the minimum specification for the glitch trigger, such problems can be located easily. Analyzers with glitch capability usually can store and display such events as they occur in the sample stream. The glitch capability is sometimes used as an undersampling indicator in continuous storage mode. The sam-

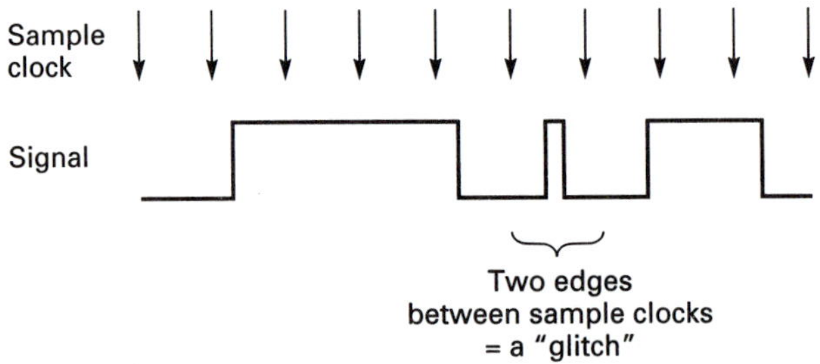

Figure 29.19 Glitch triggering.

pling rate can be reduced to stretch out the total recorded time. The appearance of glitches means that the sample rate has been reduced too much.

Runt trigger. Another unique criterion in the timing mode is the "runt trigger." In digital logic families, two thresholds are used to discriminate between logic ones and logic zeros. A "zero" is defined to be less than 0.8 V, and a "one" is greater than 2.0 V. The middle region between 0.8 and 2.0 V is the switching zone used to improve noise margin. A runt pulse crosses one threshold, but not the other. Figure 29.20 shows an example waveform for a runt pulse.

29.3.3 Counter resources

Counter resources can count either pattern occurrences or time duration. The trigger event can be defined as the first occurrence of a pattern or a specified repeat of the pattern. When operated in timing mode, a pattern repeat count has the effect of setting a minimum time duration. For this to work properly, of course, the counter must reset if a nonmatch occurs before the desired count is reached. A time counter directly qualifies a pattern's duration in timing mode or the time between patterns in the state mode. Usually, time-duration counters have more resolution, more flexibility, and are more straightforward to use compared with occurrence counters used for this purpose.

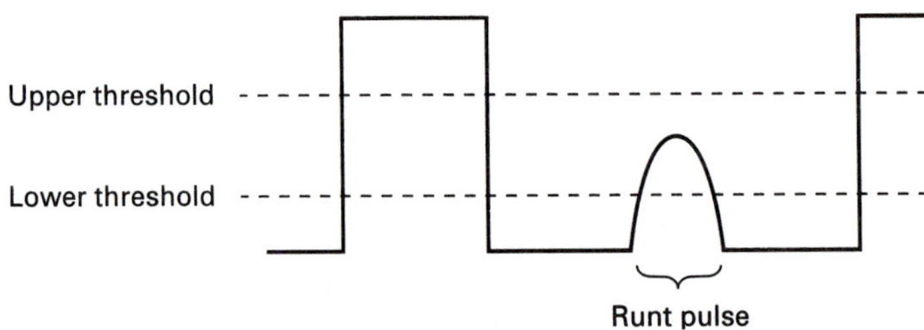

Figure 29.20 Runt pulse. The runt pulse of this example crosses the lower threshold, but falls back before the upper threshold.

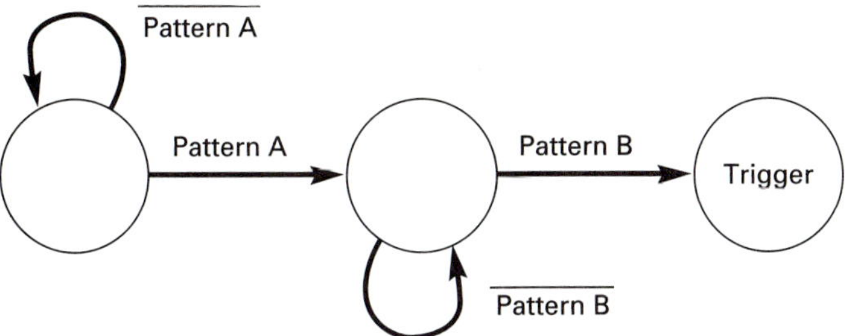

Figure 29.21 State flow diagram corresponding to a sequenced trigger. Bubbles represent discrete states. True or false values for pattern *A* and pattern *B* are inputs that control which path is taken to transition from one state to the next.

29.3.4 Sequencing

It is the nature of digital circuits and state machines for simple patterns to occur repetitively and not be unique unless referenced to a unique event. This is why counters are so useful in positioning the trigger event. Equally essential is the trigger sequencer. The trigger sequencer is itself a state machine that can be programmed to follow state flow in the target circuit or to find a desired circuit activity sequentially. In its simplest form, the trigger condition is described as "find pattern A, then find and trigger on pattern B." On a state flow diagram, this looks like Fig. 29.21.

This feature can be essential when tracing a software subroutine or function called from many places. A simple pattern would trigger on entry to the function, but triggering on a specific call of the function requires a sequenced trigger. Figure 29.22 shows the dilemma. A function calling another function

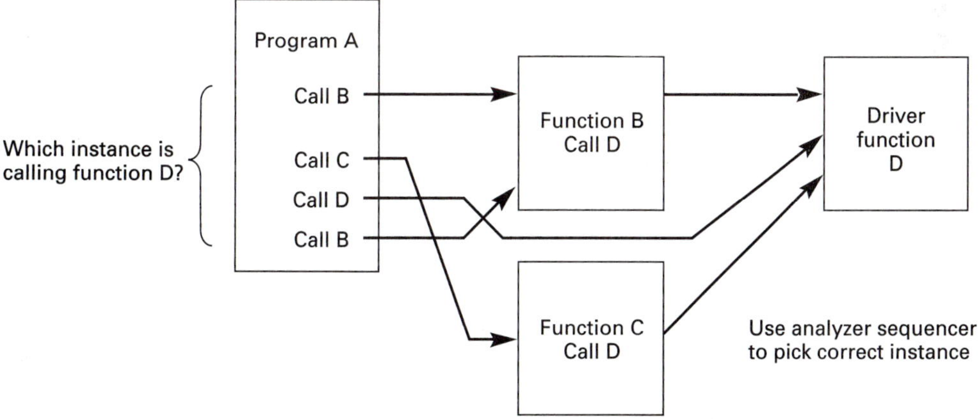

Figure 29.22 Four software functions or subroutines are shown which call each other. The call to D resulting from the second call to function B (from program A) can be selected by "Find A, then find 2nd occurrence of B, then trigger on D. Restart sequence on any return to A."

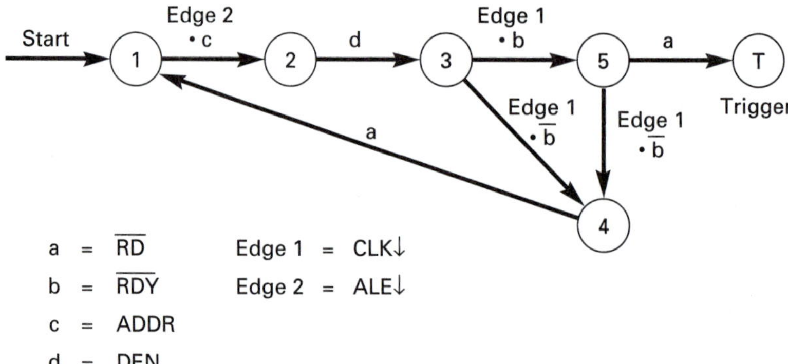

Figure 29.23 State flow diagram for using sequencer to catch invalid memory wait cycle. Six input signals involved in a memory bus cycle are selected as patterns *a, b, c,* and *d* and edges "edge1" and "edge2." Combined in Boolean expressions, these patterns direct the sequencer through the various states, eventually triggering if an invalid memory cycle occurs.

that calls another function is quite common. A trigger sequencer with multiple levels is key to tracking down a state in such an instance.

The power of the trigger sequencer is really unleashed when multiple paths can be taken at each sequence level. For example, "Starting at level 0, find pattern A and go to level 1. If you find pattern B, go to level 2 and trigger, but if you find pattern C instead, return to level 0 and start over."

Logic analyzers commonly offer sequencers with tens of levels or discrete states. Often several paths are available out of each state. One of these paths is usually a global reset, which returns to the initial state from any of the sequence levels. Figure 29.23 shows a complex state sequence and Fig. 29.24 shows how a trigger specification could be programmed to follow it.

In timing mode, the sequencer could be programmed to follow a multisignal protocol, such as that used in a microprocessor direct memory access (DMA) cycle.

29.3.5 Triggering between multiple machines

As covered in Sec. 29.3.1, most instruments can be configured as multiple machines. The different machines can interact with each other to form interactive trigger specifications. For example, one machine configured in state mode could be tracking the state behavior of the target circuit and locate a problem. The second machine configured in timing mode would be triggered by the state machine to capture the signal behavior of the target circuit. In this way, the root cause of the problem could be determined.

In addition to a specific logic analyzer or set of logic analyzer channels configured as multiple machines (both a timing and a state analyzer), oftentimes logic analyzers contain additional capability, such as digital pattern generators and/or oscilloscope acquisition channels. The addition of this func-

(1) Find "c * edge 2" 1 time

(2) Then find "d" 1 time

(3) Then find "~b * edge 1" 1 time
Else on "b * edge 1" go to level 5

(4) Then find "no state" 1 time
Else on "a" go to level 1

(5) Trigger on "a" 1 time
Else on "~b * edge 1" go to level 4

Patterns/Labels	RDY	RD	DEN	ADDR	CLK	ALE
"a"	X	0	X	XXXX	X	X
"b"	0	X	X	XXXX	X	X
"c"	X	X	X	1 2 3 4	X	X
"d"	X	X	0	XXXX	X	X
"edge 1"	X	X	X	XXXX	↓	X
"edge 2"	X	X	X	XXXX	X	↓

Figure 29.24 Sequencer programming to follow the state flow of Figure 29.23. Using the programming commands of the example trigger menu, a trigger specification is shown which implements the invalid memory wait cycle state machine.

tionality coupled with traditional logic analyzer acquisition and the ability to time-correlate and cross trigger makes an extremely flexible instrument for debugging embedded systems.

Traditionally, with multiple machine logic analyzers, you could look at both the synchronous (state) and asynchronous (timing) view of a single channel by probing the node under question with two separate probes—double probing. This is often useful to help determine the root cause of a problem or to help with cross triggering, but causes more probing difficulty. Another method to solve this problem is dual analysis per pin (DAPP). DAPP allows the user to see the traditional logic analyzer view of the channel (synchronous or asynchronous) and adds a higher-speed asynchronous (timing) analyzer that captures information concurrently.

29.4 Instrument Specifications/Key Features

Key parametric specifications include maximum sample rate for internal (asynchronous) and external (synchronous) clocks, setup and hold or capture aperture, and probe loading. Key functional specifications include the number of channels, memory depth, the number of trigger resources, the availability of analysis probes, inverse assemblers, emulation solutions, and high-level language support, time stamps, and correlation between measurement modules.

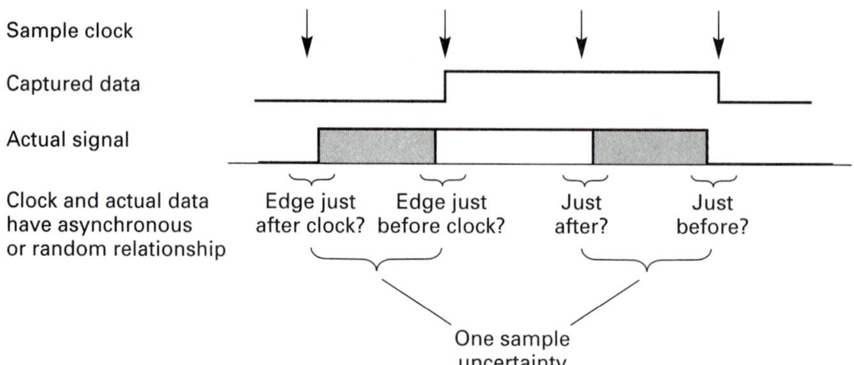

Figure 29.25 Time-interval uncertainty of the timing mode. The analyzer's captured data represent the actual signal only at the time of the sample clock. The actual signal as shown could have gone high almost a full sample period before it was recorded as high. The same uncertainty exists when the signal goes low; it could have gone low almost a full sample period before it was recorded.

Local-area network (LAN) interfaces have emerged as crucial links in research and development (R&D) to tie logic analyzers to project software databases and to other tools on the personal computer. Connectivity to the personal computer via LAN or compatible floppy disk drive is used to interface to word processors/spreadsheets and to save product configuration files/measurement results for future use. Web interfaces are also useful to provide personal computer connectivity. Data and control interfaces for IEEE-488 and RS-232 have been important in manufacturing applications for data access and control.

29.4.1 Sample rate

The sample rate determines the minimum time interval that can be resolved and measured in the timing mode. The relationship of the sample clock to an input signal transition is completely random. The transition can occur just after a preceding clock edge, or it could happen just before the clock that did capture it. This is shown graphically in Fig. 29.25.

The uncertainty of placing a signal transition is a full sample period. Two edges of the same signal can be measured to an accuracy of two sample periods. Measuring a transition on one signal versus a transition on another signal can be done to an accuracy of two sample periods plus whatever skew exists between the channels. "Channel skew" is the difference in path delay between two channels of the analyzer. Channel skew is usually specified by the analyzer vendor and will typically be one half of the minimum sample period.

For state measurements, the sample rate determines the maximum clock rate that can be measured in the target state machine. Historically, the re-

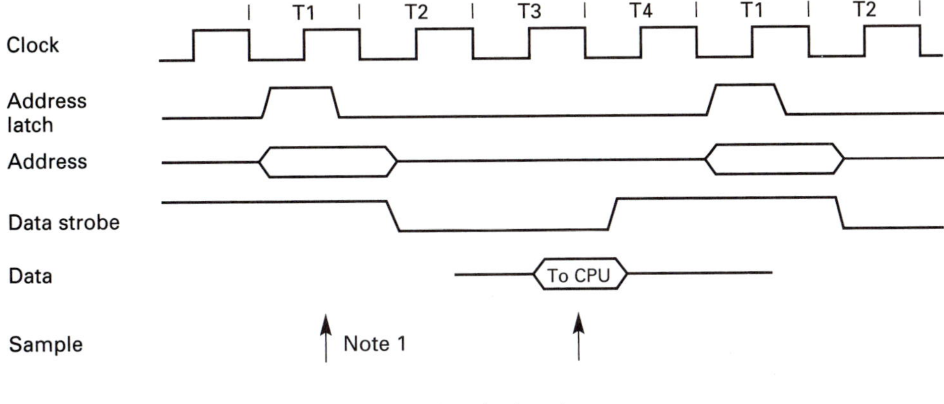

Figure 29.26 Processor bus cycle with latched address. Note 1: Data and address cannot be captured with a single sample.

quired sample rate for processor buses is one half to one fourth of the processor's clock; however, as processor data-transfer rates continue to increase, some processors transfer data at or above the processor clock rate. For the processor in Fig. 29.26, bus cycles take up four processor clocks. Figure 29.27 shows a more-sophisticated processor that uses two clocks for most bus cycles, but can operate with only one clock in a burst mode. These processors and certain reduced instruction set computer (RISC) processors require state sample rates that are equal to the processor's clock. Careful thought to the maximum bus rate, not the processor clock rate, should be given when determining the need for maximum state sample rate.

29.4.2 Setup and hold

Closely related to state sample rate are the setup-and-hold time specifications for the analyzer. Some analyzers have a selectable "capture aperture." Both specifications refer to the time interval relative to a clock edge when the data must be stable for accurate capture. With any synchronous logic circuit, the occurrence of a clock causes the output of a flip-flop or register to transition to its next state value. The transition must be complete and stable a specified time before the next clock. Likewise, the inputs to the flip-flop must be held for a specified time after the clock. A similar situation exists for microprocessor buses. The output of memories changes as the next address is presented. The memory data must be stable before they are sampled by the processor and held after the clock the required amount.

Like flip-flops, registers, and memory elements, a logic analyzer also needs stable data at a specified time before the external sample clock. This is the "setup time specification." The "hold specification," likewise, must be met, but it is typically zero time. For setup time, a number of vendor attitudes are

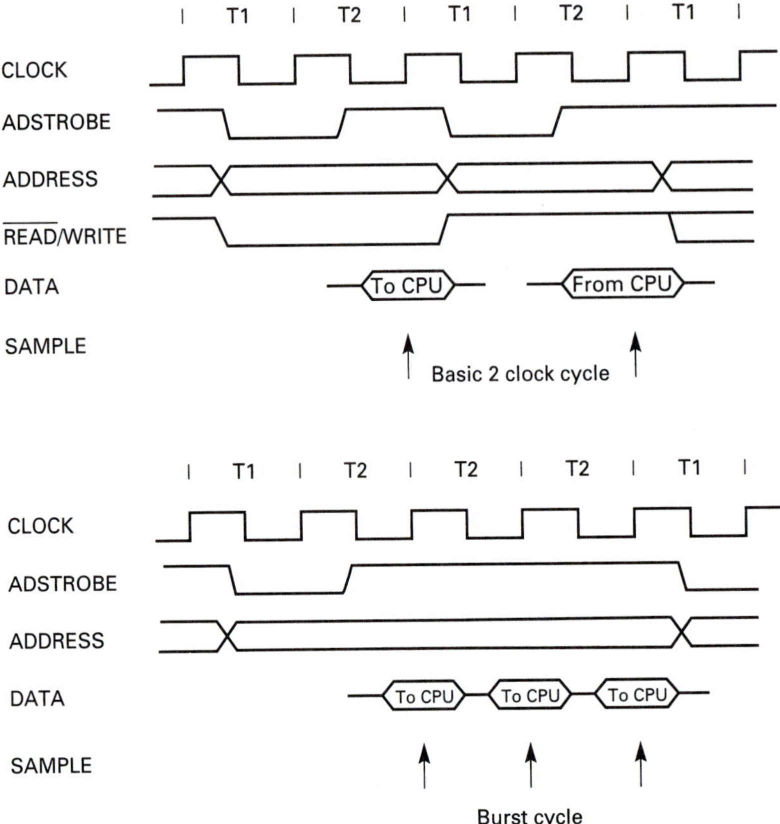

Figure 29.27 High-speed processor bus cycles. Both basic and burst-mode cycles are shown.

represented in setup specifications. From the perspective of the instrument user, the analyzer should have the same characteristics as the user's circuit. At high clock rates, a very short setup time is difficult for vendors to achieve. It is common for the setup time to be in the range of one half to one third of the clock period at maximum sample rate. Users planning on using the analyzer at its maximum state speed should check this specification to be sure their circuit will present acceptable setup times. Some analyzers have a selectable aperture in which the state data are sampled. This can be an extremely useful feature for checking or characterizing the target system's own setup and hold margins.

29.4.3 Probe loading

The last parametric specification is "probe loading." Logic analyzer probes are usually specified with static resistive and capacitive loading values. Obvi-

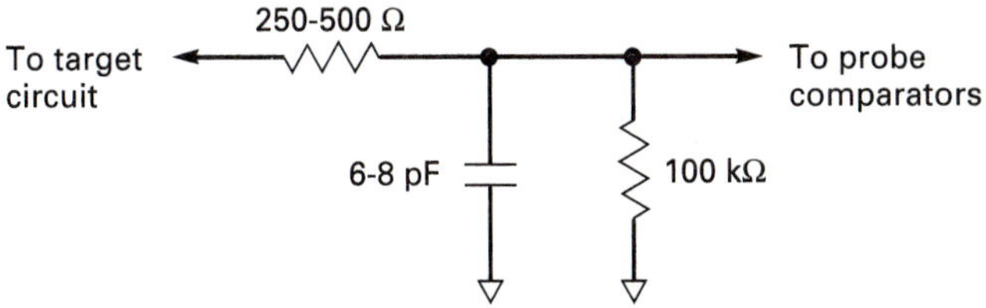

Figure 29.28 Equivalent circuit for logic analyzer probe.

ously, the target system should not be perturbed by probe loading. However, the effects of these static loads cannot be interpreted for logic analyzers as easily as they might be for an oscilloscope. Logic analyzers are sensitive to edge placement, so signal rise time and delay are more important than pulse response. Latitude is taken in optimizing the probe-comparator network.

Most logic analyzers have probe specifications of 100 kΩ and 6 to 8 pF. In series with this shunt reactance is a 250- to 500-Ω resistance (see Fig. 29.28). The capacitance is largely isolated from the target system by the series resistance. As logic analyzer sample rates continue to increase, the effect of the probe capacitance on the target system also increases. Consequently, logic analyzer probing systems have been developed to reduce the probe capacitance in the target system. These more-expensive probing systems have capacitance in the 2- to 5-pF range.

Analyzers having greater than 1-GHz sample rates supply SPICE models for their probes so that the true impact on signal integrity can be evaluated and designed to allow the system under test to be measured.

29.4.4 Channel count

The key functional specification is "channel count." The number of input channels, along with maximum sample rate, drives the cost of a logic analyzer. Selecting the appropriate number of channels for a current project, as well as future needs, is very important. The use model for debugging a problem in timing mode often involves an incremental hunt-and-probe approach. A minimum count of 32 channels is recommended. Thirty-two channels is also the minimum for probing an 8-bit microprocessor: 16 channels for addresses, 8 channels for data, and about 8 channels for status and random circuit points. For 16-bit processors, the minimum is closer to 56 channels: 24 for addresses, 16 for data, and 16 for status and random probes. For 32-bit processors, the number is 80 channels: 32 for addresses, 32 for data, and 16 for status. For analyzers that can be configured to operate as two simultaneous machines (e.g., one as a state analyzer and the other as a timing analyzer), 8 to 32 channels should be added to the preceding recommendations.

29.4.5 Memory depth/maximum samples

Memory depth is directly related to the maximum time window captured in timing mode or the total number of states or bus cycles captured in state mode. Often, depth is traded against sample rate, channel count, or the availability of instrument resources, such as time stamps for state samples. Memory can sometimes be cascaded or interleaved for more depth, but at the expense of channels or time stamps.

The high cost of memory components makes it uneconomical to record directly into logic analyzer memory at 100 MHz to 2 GHz. The data are usually decelerated or fanned out to multiple banks of memory. Memory banks can be interleaved together to achieve maximum sample rates. This also provides "double depth" memory. If the interleaving is switched off, the sample rate must drop along with a decrease in memory depth. The remaining memory can be used to create additional channels.

As a general rule, more memory depth is better. Most analyzers provide from hundreds of thousands to millions of samples. Capturing a packet of data, such as a single scan line from a television or computer monitor, might require a certain analyzer depth. A common measurement is to trace a processor "crash." A crash occurs when a processor fetches an instruction stream that directs it to a portion of memory that has not been loaded with a valid program. In such a case, the processor will fetch an invalid instruction or one intended to literally stop the processor's execution. Once the crash is detected, deep memory can be searched backward to the root cause of the crash. A third purpose of deep memory is to trace software written in a high-level programming language. Ten or more bus cycles might be needed to capture a single high-level instruction. A useful snapshot will require the capture of several hundred high-level instructions, meaning several thousand bus samples.

Transitional storage mode for timing measurements is an excellent feature to optimize memory needs. Samples are only stored when input signals change values. High-resolution pictures over wide time windows can be captured. The tradeoff here is the maximum sample rate and richness of trigger features, which can be achieved in this mode relative to the simpler continuous storage mode.

29.4.6 Trigger resources

The key to finding obscure or subtle problems is in the number and sophistication of trigger resources. Most important are the number of patterns and some number of sequencer levels.

"Sequence levels" refer to the number of states that the trigger block can be programmed to step through while finding the trigger event or while controlling storage qualification. For example, a three-level sequencer could be set to first find pattern 1 and then find pattern 2 before finally finding pattern 3 and triggering.

It is common for instruments, except the highest-sample-rate machines, to offer 8 to 16 simultaneous patterns and a like number of sequence levels. From a user-interface perspective, 20 or 30 patterns can sometimes be defined and kept ready for a measurement. The instrument hardware is usually more limited than that, allowing less than 16 simultaneous patterns.

Usually, a counter is associated with each level of the sequencer. Range patterns (see Sec. 29.3.2) are expensive to implement. Unless the analyzer has been specifically designed for software tracing, there will usually only be one or two range patterns. Rich trigger capabilities can be more effective than capturing a long record with deep memory and then having to wade through all of the data.

29.4.7 Analysis probes/inverse assemblers/emulation solutions and high-level language support

A key benefit to using a logic analyzer is looking at many channels of data simultaneously. Hooking up all those channels (tens to hundreds) is a large barrier to using a logic analyzer. Prewired probes, called "analysis probes," can effectively overcome this barrier. Analysis probes are available for many processors and standard buses, such as the PC PCI bus and the standard serial bus, RS-232. Many processor package types, such as leadless chip carriers (LCC), quad flat packages (QFP), and ball grid arrays (BGA), are available. Purchasing an analysis probe is well worth the money. Not all vendors support all processors or all package variations. Careful attention should be paid to this limitation when selecting a logic analyzer vendor.

Close companions to analysis probes are inverse assembler software packages. These software packages usually run on the logic analyzer and are used to format the captured logic analyzer information into processor instructions. This greatly enhances the readability of the instrument. In early processors, the execution and bus interface units were tied directly, and the correlation of bus activity to execution was straightforward. Processors using pipelining, caching, and memory translation cause poor correlation of execution and bus activity (that the logic analyzer observes). Analysis probes for these processors incorporate special hardware and/or use special information emitted by the processor to overcome these problems.

The availability of emulation solution is often tied to both the logic analyzer vendor and the processor vendor. For the logic analyzer vendor to offer an emulation solution, the processor-specific debug capability must exist inside the processor provided by the processor vendor. Key specifications for emulation solutions consist of the download rate and the ability to access/control processor-specific functions. Although an emulation solution has the advantage of being processor package independent, users must usually put the processor control connector on the system under development. Another key feature is the ability to stop the processor when an event of interest is found. The ability to quickly stop the processor, in this case, is a function of the emulation solution implementation by both the logic analyzer vendor and the processor vendor.

The ability of the logic analyzer to display high-level language source correlation is usually a function of the specific language tools used to generate the machine code that the processor executes. These language tools generate files that contain the source correlation information and must be in a format readable by the logic analyzer. The degree to which source correlation is supported is also a function of the language tools/logic analyzer cooperation. Various degrees of source level support can be provided. The most basic is high-level language function names and variable support; the most advanced is high-level language source lines displayed in the logic analyzer window. Given the large number of software tools available for processors, users should be careful to ensure that the language tools they plan to use are fully supported by the logic analyzer.

29.4.8 Nonvolatile storage

Instrument setup involves channel selection and the creation of data labels and trigger specs. The ability to save these setups, along with measurement data in nonvolatile storage, is crucial to productivity. Flash memory and disk drives are the two most common options available.

Closely related to setup is the documentation of setups and measurement results. The connection of the analyzer to a printer or plotter is a good way to output this information, but transfer of the data to a personal computer via a compatible disk drive is quickly becoming the preferable route. In this case, popular documentation packages, database programs, or spreadsheets can use the information directly. In addition to the media compatibility between PC and instrument, the data format compatibility with word processors or graphics packages must be considered.

29.4.9 Time stamps

When operating in the state mode, the timing of clocks is determined by the circuit under test. Sometimes the clocking is sporadic, or store qualification is engaged, which obliterates any implied timing from a regularly paced state clock. The ability of a logic analyzer to store a time value along with the captured data can be very useful. The stored time value is called a "time stamp."

The time stamp is especially useful with an instrument split into state and timing machines. It enables the user to correlate the state and timing measurements. Modular instruments that can host multiple logic analyzers or other measurement modules, such as pattern generators or oscilloscopes, sometimes are able to time stamp and correlate events from each of the modules.

29.5 Getting the Most from a Logic Analyzer

The attributes of a logic analyzer that make it useful—logic domain data acquisition, presentation, large channel count, and multiple domain (scope to high-level language) time correlation—can also make the instrument intimidating and hard to use. The recommendations that follow will help you use the capability of the logic analyzer and reduce the intimidation factor.

29.5.1 Recommendations to improve usability

- *Plan the use of input channels.* Multibit buses (such as address, data, and status) should all be assigned to adjacent channels. Hold out some channels on a free pod for "hunt and probe" timing measurements. First choice is for these channels to be the default setup. When the instrument is first turned on, check to see the default channel assignment. Usually, the instrument will default to a timing mode measurement. Identify which pod (probably, pod 1) contains the selected channels. As you move these exploratory channels along, you might find that you've probed a multibit bus. You might want to add this signal and its companions to your suite of input channels. If you took the time to plan ahead, you'll have already reserved channels for all the other bits. If a split timing and state measurement looks promising, it will be easy to incrementally set up the machine because usable pod configurations will have been reserved.
- *Use analysis probes or specially designed package clips to mass-connect the probes.* Little value is added by the user to hook up all these lines, and the frustration can be overwhelming when general-purpose clips fall off.
- *Analysis probes usually come with preconfigured setups, including labels and symbols.* They also come with inverse assemblers or data formatters so that information can be presented in standard mnemonics.
- *Print and save configurations containing probe setup—especially when using flying leads.* Annotate the setup printed or saved with probing locations.
- *Take the time to create labels and multibit symbols.* After the potential for a probing snarl, the next opportunity for confusion will be when making sense of the display. Labels and symbols keep the instrument tracking your thought process and avoid mistakes of interpretation.
- *Frequently store setups, labels, and symbols on your storage.* Store your "hunt and probe" setup, as well as snapshots of your incremental measurements—especially complicated triggers or split machine configurations. These stored setups and measurements will allow you almost single-button recall of your thoughts during the troubleshooting process.
- *Invest in hardcopy output for your analyzer and use it.* Manual notes and sketches of waveforms or trace listings are unnecessary and time consuming. Directly output to a printer or use a LAN connection to transfer machine-readable data to your PC or workstation.

- *Invest time to learn how to set triggers and storage qualification beyond a simple single edge or pattern.* Most users never get beyond these simple triggers and consequently consume large amounts of time analyzing logic analyzer information. Understanding powerful concepts, such as sequencing and storage qualification, can find tough problems in minutes that would take hours to understand by analyzing the acquired data.
- *Invest in an instrument that leverages already familiar user interface paradigms.* This usually means a windowed interface with keyboard and mouse. Logic analyzers are usually used intensely for bursts of time and then left unused; investing in an instrument that will be easy to learn will increase productivity.

Chapter

30

Protocol Analyzers

Steve Witt

Agilent Technologies
Colorado Springs, Colorado

30.1 Introduction

What is a protocol? Why does it need analysis? Don't networks just work? Why are there so many different protocols, standards, and networking technologies? The field of computer networking is complex and becoming increasingly so.

Existing computer networks consist of many different computer systems, applications, and network topologies. The capital investment in cabling and transmission infrastructure is massive. The number of users demanding access to computer networks is ever increasing and these users are demanding more bandwidth, increased performance, and new applications. A constant stream of new equipment and services are being introduced in the marketplace. In this complex environment, computer networking is made possible only by equipment and services vendors adhering to standards covering protocols, physical connectors, electrical interfaces, topologies, and data formats. Protocol analysis is used to ensure that the products implemented according to these standards behave to specification.

30.1.1 Protocol definition

Generally speaking, a *protocol* is a code or a set of rules that specifies the correct procedure for a diplomatic exchange. In terms of computer networks, a *protocol* is a specific set of rules, procedures, and conventions that define the format and timing of data transmission between devices connected to a computer network. Protocols are defined so that devices communicating on a computer network can exchange information in a useful and efficient manner.

Protocols handle synchronization, addressing, error correction, header and control information, data transfer, routing, fragmentation and reassembly, encapsulation, and flow control. Protocols provide a means to exchange information so that computer systems can provide services and applications to end users. The wide range of protocols in use is due to the wide range of transmission medium in use, the wide range of applications and services available to end users, and the numerous independent organizations and vendors creating protocol standards.

Communication via a computer network is possible only with an agreed-upon format for the exchange of information and with a common understanding of the content of the information being exchanged. Therefore, protocols must be defined by both semantic and syntactic rules. *Semantics* refers to the meaning of the information in the frame of data, including control information for coordination and error handling. An example of the semantic information in a frame is a request to establish a connection, initiated by one computer and sent to another computer. *Syntax* refers to the structure, arrangement, and order of the protocol, including data format and signal levels. An example of the syntax of a protocol is the relative position of a protocol field in the frame, such as the network address. Protocol analysis is concerned with both the syntax and the semantics of the protocols.

30.1.2 Protocol standards

Communication between devices connected to computer networks is controlled by transmission and protocol standards and recommendations. These standards are necessary for different vendors to offer equipment and services that interoperate with one another. Although standards can be defined and implemented in the private sector by computer and network component vendors, such as Cisco Systems, Hewlett-Packard, and IBM, most standards and recommendations are created by organizations, including, but not limited to, ANSI (American National Standards Institute), CCITT (Consultative Committee on International Telegraphy and Telephony), ETSI (European Telecommunications Standards Institute), IEEE (Institute of Electrical and Electronic Engineers), ISO (International Standards Organization), and the ITU (International Telecommunications Union). The ATM Forum and the IETF (Internet Engineering Task Force) are technical working bodies that develop standards for networking products.

30.1.3 The OSI reference model

The International Organization for Standardization (ISO), located in Geneva, Switzerland, is responsible for many of the international standards in computer networking. The ISO defined a model for computer communications networking, called the *Open Systems Interconnection Reference Model.* This model, commonly called the *OSI model,* defines an open framework for two computer systems to communicate with one another via a communications network. The OSI model, see Fig. 30.1, defines a structured, hierarchical network architecture.

The OSI model consists of the communications subnet, protocol layers 1 through 3, and the services that interface to the applications executing in the host computer systems (protocol layers 4 through 7). The combined set of protocol layers 1 through 7 is often referred to as a *protocol stack.* Layer 7, the applications layer, is the interface to the user application executing in the host computer system. Each layer in the protocol stack has a software interface to the application below it and above it. The protocol stack executes in a host computer system. The only actual physical connection between devices on the network is at the physical layer, where the interface hardware connects to the physical media. However, there is logical connection between the two corresponding layers in communicating protocol stacks. For example, the two network layers (layer 3 in the OSI reference model) in two protocol stacks operate as if they were communicating directly with one another, when, in actuality, they are communicating by exchanging information through their respective data-link layers (layer 2 in the OSI reference model) and physical layers (layer 1 in the OSI reference model).

Current network architectures are hierarchical, structured, and based in some manner on the OSI reference model. The functionality described in the OSI reference model is embodied in current network architectures, albeit at different layers or combined into multiple layers.

30.1.4 Network troubleshooting tools

Two broad categories of products are used to implement and manage computer networks: those that test the transmission network and those that test the protocol information transferred over the transmission network. Testing the protocols is commonly referred to as *protocol analysis* and it can be accomplished with several different types of products including:

- *Network management systems.* Comprehensive, integrated network-wide systems for managing and administrating systems and networks. Protocol analysis is one of many applications performed by network management systems. Network troubleshooting is performed by acquiring network data from devices on the network and from instrument probes distributed through the network.
- *Distributed monitoring systems.* Performance monitoring and troubleshooting applications that are implemented with instrument probes or protocol analyzers that are distributed throughout the network. The probes and analyzers are controlled with a management application running on a workstation or PC.
- *Protocol analyzers.* Specialized instrumentation dedicated to protocol analysis. Protocol analyzers are used to troubleshoot network problems and to monitor the performance of networks.

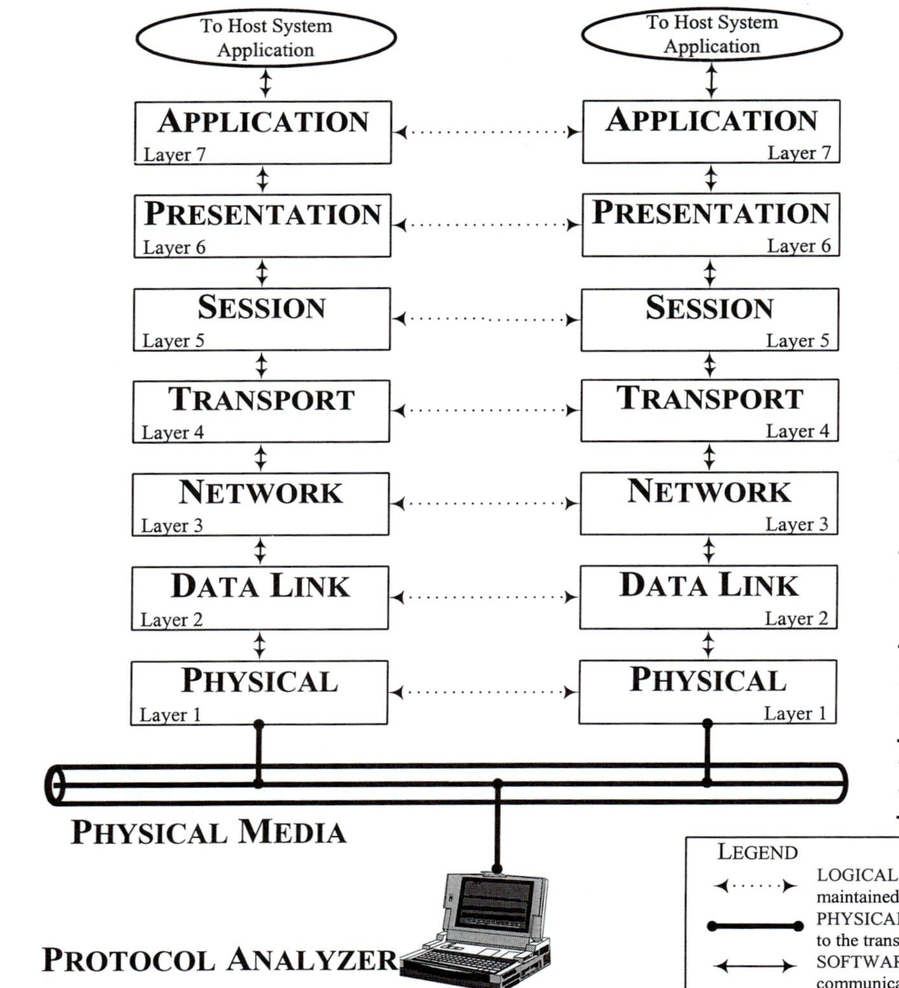

Figure 30.1 The OSI Reference Model. The OSI Reference Model defines an internationally accepted standard for computer network communications. Host system applications exchange information in a hierarchical and structured manner.

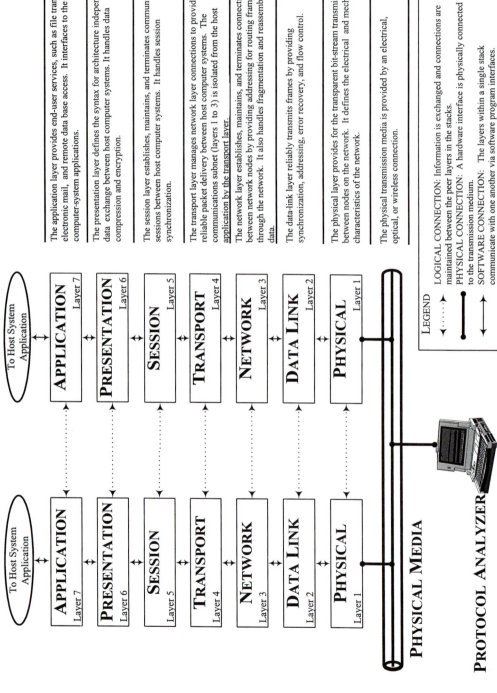

- *Hand-held test tools.* Special-purpose tools that are very small, lightweight, and usually battery operated. They perform a variety of measurements, such as continuity tests, transmission tests, and simple protocol analysis measurements (such as simple statistics and connectivity tests).

30.1.5 The need for protocol analysis

In order for two applications running on two different computer systems to communicate with one another (e.g., a database application executing on a server and a client application performing database queries), meaningful information must be continually, efficiently, and correctly exchanged. This requires a physical connection to exist—either twisted-pair copper wire, coaxial cable, optical fiber, or wireless (e.g., radio transmission). The physical characteristics and specifications of the transmission media must be standardized so that different computer systems can be electrically connected to one another. The bit streams exchanged over the physical media must be encoded so that the analog stream of information can be converted to digital signals. In order for two people to effectively communicate, they must speak the same language. Similarly, for two computer systems to communicate, they must speak the same "language." Therefore, the bit stream must conform to a standard that defines the encoding scheme, the bit order (least-significant bit first or most-significant bit first), and the bit sense (a high value defined either as a one or a zero). Any errors in the transmission must be detected and recovered and, if necessary, the data must be retransmitted. A protocol analyzer is used to examine the bit stream and ensure that it conforms to the protocol standards that define the encoding schemes, bit sequences, and error conditions.

Once a bit stream can be transmitted and received, the physical communication is established and the exchange of information can be accomplished. Information is exchanged in logical units of information. A protocol frame, packet, message, or cell (in this chapter, the term *frame* is used to mean any or all of these) is the logical unit of information transmitted on the physical infrastructure of a computer network. Depending on the type of network and protocols, these frames of data are either fixed in size, such as the 53-byte cells used by ATM networks, or they can be variable in size, such as the 64- to 1518-byte frames used by Ethernet networks. The most fundamental aspect of protocol analysis is the collection and analysis of these frames.

Networks usually have more than one path connecting different devices. Therefore, the frames in which the data is contained must be addressed properly so that they can traverse single or multiple routes through the network. Fragmentation and reassembly issues must be properly handled—frames are often disassembled and reassembled so that they can be of a proper size and can be encapsulated with the proper header information to ensure that the intermediate and end devices in the network can properly manipulate them. The network must also handle error conditions, such as nodes that stop responding on the network, nodes that transmit errored frames or signals, and nodes that use excessive bandwidth. A protocol analyzer is used to examine the addresses of the frames, check fragmentation and reassembly, and investigate errors.

Connections are established so that communication is efficient. This prevents the communication channel from being redundantly set up each time that a frame is transmitted. This is similar to keeping a voice line open for an entire telephone conversation between two people, rather than making a phone call for each sentence that is exchanged. To ensure this efficiency, connections or conversations are established by devices on the network so that the formal handshaking doesn't have to be repeated for each information exchange. Protocol analysis includes scrutinizing protocol conversations for efficiency and errors. The data that is transferred to the host system application in the frames must conform to an agreed-upon format, and, if necessary, it must be converted to an architecture independent format so that both computer systems can read the information. Each time a user enters a command, downloads a file, starts an application, or queries a data base, the preceding sequence of processes is repeated. A computer network is continuously performing these operations in order to execute an end user's application. Protocol analysis involves examining the formatted data that is exchanged between host-system applications.

30.1.6 Protocol analysis

A *protocol analyzer* is a dedicated, special-purpose computer system that acts as a node on the network, but, unlike a typical node on the network, it monitors and captures all of the network traffic for analysis and testing. Protocol analyzers provide a "window into the network" to allow users to see the type of traffic on the network. Many problems can be solved quickly by determining what type of traffic is or is not present on the network and if protocol errors are occurring. Without a "window into the network," network managers must rely on indirect measures to determine the network behavior, such as observing the nodes attached to the network and the problems the users are experiencing.

The term *protocol analyzer* was created in the early 1980s to describe a new class of products dedicated to testing serial data-communication networks. These early products provided a set of features that focused on examining the communication protocols and ensuring that they conformed to the standards. This class of products has evolved to include support for all types of computer and communications networks, architectures, and protocols. They provide capabilities to compare data frames with the protocol standards (protocol decodes), load and stress networks with traffic generation, and monitor network performance with statistical analysis. The measurement capabilities have expanded to focus on a broad set of applications, such as network troubleshooting, network performance monitoring, network planning, network security, protocol conformance testing, and network equipment development. These new applications go far beyond simply examining network traffic with protocol decodes.

As new networking technologies are introduced and higher-performance network components are delivered to the marketplace, the constant pressure is to drive costs down and increase performance. Although new network architecture designs are inherently more reliable, the driving factors of cost and performance prohibit building extensive test capability into the network itself. Therefore, additional testing tools, such as network-management systems, distributed monitoring systems, and dispatched protocol analyzers are necessary to guarantee that network down time is kept to a minimum. All of these tools use protocol analysis to implement the necessary testing capabilities.

30.2 Protocol Analyzers

The term *protocol analyzer* is typically used to describe a class of instruments that are dedicated to performing protocol analysis. Protocol analyzers are implemented in one of three ways:

- Portable protocol analyzers
- Embedded protocol analyzers
- High-end protocol test sets

Each of these products has unique advantages and disadvantages (see Table 30.1) making it the suitable choice for specific applications.

30.2.1 Dispatched portable protocol analyzers

The term *protocol analyzer* most commonly refers to the portable instruments that are dispatched to trouble sites on the network (i.e., the crucial links or segments that are experiencing problems). The emphasis in these products is portability—products that are light weight, rugged, and include the maximum amount of functionality. Network troubleshooters require a product that connects to any point in an internetwork, regardless of the physical implementation of the network. Therefore, most popular protocol analyzers are able to accommodate multiple network interface modules. To avoid multiple trips back to the office to get additional equipment, network troubleshooters require a "one-handle solution," a product that integrates as much test capability as possible into one portable product.

Portable protocol analyzers are most commonly used for network troubleshooting in installation and maintenance applications. Such products focus on installing networks and troubleshooting network problems. Installing networks requires the network engineer to stress the network or network devices using scripts that emulate setting up logical connections, placing calls, and

TABLE 30.1 Protocol Analyzer Products

Product	Description	Application	Advantages	Disadvantages
Dispatched Portable Protocol Analyzers	■ Integrated, stand-alone unit ■ Lightweight, portable, rugged package ■ Built upon an integrated PC system ■ Provides "One Handle" solution ■ Supports multiple network interfaces	■ Troubleshooting ■ Installation ■ Performance monitoring	■ Good price performance ■ Portability ■ Accommodates many different network interfaces ■ Turnkey solution: hardware and software configured at the factory ■ Suitable for simple development applications	■ Limited multi-port capability because of the size of the package ■ Limited capability for R&D applications
Embedded Protocol Analyzers	■ "Cards and software" (line-interface hardware and application software) for each network interface ■ Built on standard computing platform: PC or UNIX workstation	■ Troubleshooting ■ Performance monitoring	■ Lower relative cost ■ Takes advantage of an existing PC or workstation ■ Additional cards and software can be added for additional network interfaces	■ Limited to a specific physical site unless built in a portable PC ■ Requires user to set up and configure hardware and software
High-End Protocol Test Sets	■ Built on standard computing platform: PC or UNIX workstation ■ Standard card cage allows user to configure the product to the application	■ Conformance testing ■ Stress testing ■ Traffic generation	■ Sophisticated measurement set ■ Programmable measurements ■ Modular to allow upgrade of test capability ■ Many software applications ■ Supports new, emerging network technologies	■ Higher relative cost ■ Limited portability ■ Complex capabilities are typically more difficult to use

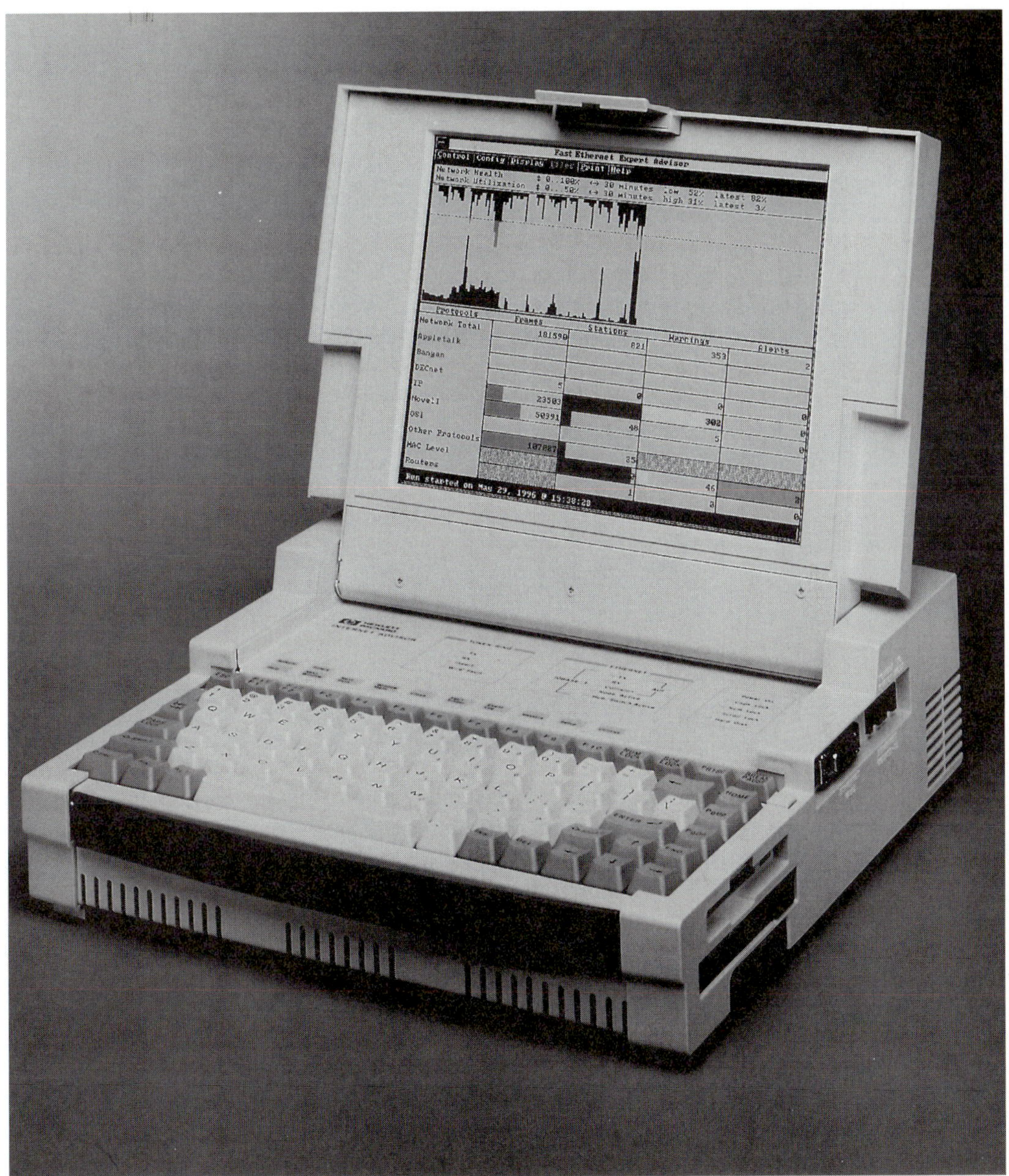

Figure 30.2 A portable protocol analyzer.

Figure 30.3 An embedded protocol analyzer.

creating high-traffic scenarios. Troubleshooting network problems requires that the network engineer have extensive protocol decodes available to examine whatever frames are present on the network. Protocol statistics and expert analysis are used to identify network errors.

Figure 30.2 shows a typical dispatched portable protocol analyzer that is light weight, rugged, integrates a standard PC, and will accommodate modules for different network interfaces.

30.2.2 Embedded protocol analyzers

The flexibility and portability of a portable protocol analyzer put it at a price premium above the protocol analyzer solutions that are implemented as embedded systems. Embedded protocol analyzers are implemented as software applications that run in a computer platform, such as a UNIX workstation or a desktop PC. The line interface is provided by a special-purpose analysis and acquisition system or a commercially available standard network interface card. Embedded protocol analyzers, such as portable protocol analyzers, are used to test established networking technologies. The functionality of an embedded protocol analyzer is usually focused on performance-monitoring applications, making use of the statistical measurement capability.

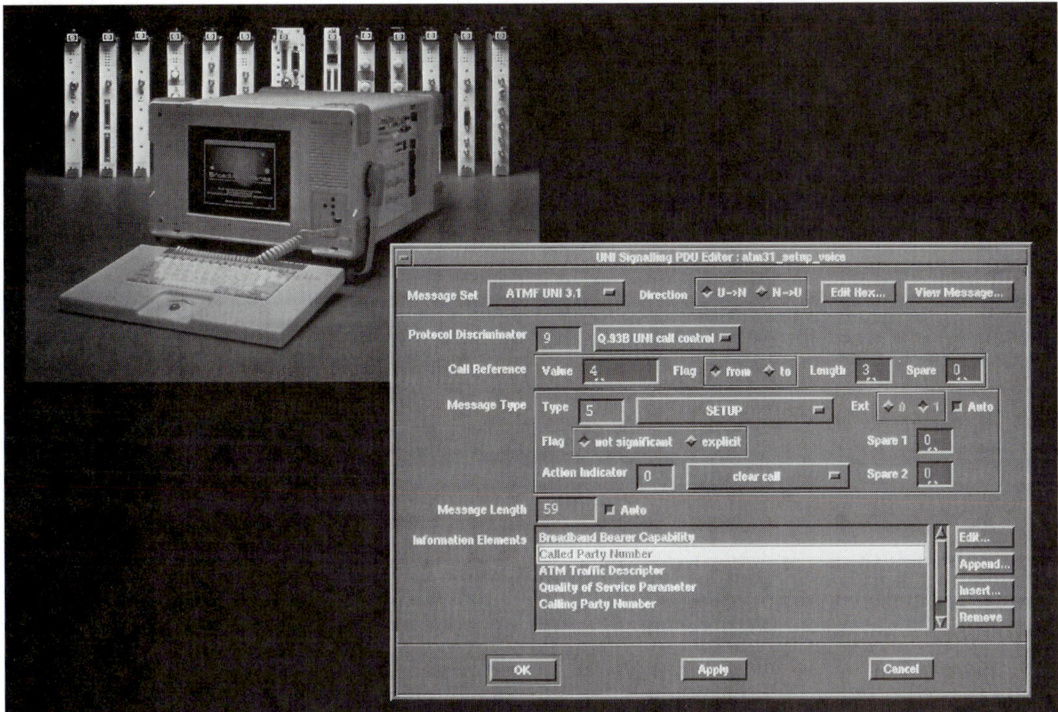

Figure 30.4 A high-end protocol test set with the protocol configuration screen.

Because embedded protocol analyzers take advantage of an existing computer platform, they are usually the lowest-cost alternative for protocol analysis. But lower cost comes with some disadvantages, such as requiring the user to install and configure the hardware and software. They also do not provide a portable, one-handle solution with multiple-line interfaces. However, these products are usually used by a network manager to monitor a specific crucial segment or link, so a product that is dedicated to a single network interface is an acceptable trade-off.

Figure 30.3 shows a set of PC cards that are used with a software application to implement an embedded protocol analyzer.

30.2.3 High-end test sets

High-end protocol test sets are used in development applications for newly emerging network technologies. R&D engineers developing network equipment, designers of new networks, and network engineers, who are commissioning new equipment all require power, performance, and flexibility in protocol-analysis tools. Therefore, the testing functions in a high-end test set emphasize traffic generation, simulation, emulation, stimulus/response testing, and programming capability.

As the new network technologies are deployed, this class of product is used for installation and maintenance until portable or embedded solutions become available. Products in this category are more expensive than the portable-protocol analyzers and embedded-protocol analyzers used for installation and maintenance applications. High-end test sets are usually offered in a card cage, such as VXI, which allows the user to configure the exact network interfaces and measurements required for the application, and to analyze multiple-network connections simultaneously (multiport analysis). Figure 30.4 shows a typical high-end protocol test set offering a UNIX-based computing platform with slide-in modules for the line interfaces and analysis and acquisition systems.

30.3 Protocol Analysis

A network fault is any degradation in the expected service of the network. Examples of service degradation include an excessively high level of bit errors on the transmission medium, a single user monopolizing the network bandwidth, a misconfigured device, or a software defect in a device on the network. Regardless of the cause, the network manager's fundamental responsibility is to ensure that such problems are fixed and that the expected level of service is restored. To accomplish this, the network manager must troubleshoot the network and isolate faults when the inevitable problems occur. Many of the difficult to detect, intermittent problems can only be recreated if the network is stressed by sending frames. Many of the network problems are caused by the constantly changing configurations of the devices attached to the network, so the network manager must manage the configuration of the network devices. In order to ensure the bandwidth and performance that users demand, the network manager must monitor the performance of the network and plan accordingly for future growth. Network managers are also concerned with the security of their networks and use protocol-analysis tools to monitor for illegal or unauthorized frames on network segments.

30.3.1 Protocol analysis applications

Fault isolation and troubleshooting. Most network problems (e.g., network down time) are solved by following a rigorous troubleshooting methodology. This methodology, not unique to network troubleshooting, consists of observing that a problem has occurred, gathering data about the problem, formulating a hypothesis, and then proving or disproving the hypothesis. This process is repeated until the problems are resolved. Protocol analysis is used in the network troubleshooting process for first observing that a problem has occurred, and next gathering data (using protocol analysis measurements, such as protocol decodes and protocol statistics, as described in Section 30.3.2). The user then formulates a hypothesis and uses the protocol-analysis measurement to confirm the cause of the problem, ultimately leading to a solution. The protocol analyzers can then be used to confirm that the problem has indeed been repaired.

Performance monitoring. Determining the current utilization of the network, the protocols in use, the errors occurring, the applications executing, and the users on the network is crucial to understanding if the network is functioning properly or if such problems as insufficient capacity exist. Monitoring performance can be used over short time periods to troubleshoot problems or it can be used over long time periods to determine traffic profiles and optimize the configuration and topology of the network.

Network baselining. Every network is unique: different applications, distinct traffic profiles, products from numerous vendors, and varying topologies. Therefore, network managers must determine what is normal operation for their particular network. A network baseline is performed to determine the profile of a particular network over time. A *profile* consists of statistical data, including a network map, the number of users, protocols in use, error information, and traffic levels. This information is recorded on a regular basis (typically, daily or weekly) and compared to previously recorded results. The baselining information is used to generate reports describing the network topology, performance, and operation. It is used to evaluate network operation, isolate traffic-related problems, access the impact of hardware and software changes, and plan for future growth.

Security. Networks are interconnected on a global scale; therefore, it is possible for networks to be illegally accessed. Illegal access can be knowingly performed by someone with criminal intent or it can be the result of an errored device configuration on the network. Protocol-analysis tools, with their powerful filtering, triggering, and decoding capabilities, can detect security violations.

Stress testing. Many errors on networks are intermittent and can only be recreated by generating traffic to stress network traffic levels, error levels, or by creating specific frame sequences and capturing all of the data. By stress testing a network and observing the results with protocol statistics and decodes, many difficult problems can be detected.

Network mapping. Networks continually grow and change, so one of the big challenges facing network engineers and managers is determining the current topology and configuration of the network. Protocol-analysis tools are used to provide automatic node lists of all users connected to the network, as well as graphical maps of the nodes and the internetworks. This information is used to facilitate the troubleshooting process by quickly being able to locate users. It is also used as a reference to know the number and location of network users to plan for network growth.

Connectivity testing. Many network problems are the result of not being able to establish a connection between two devices on the network. A protocol analyzer can become a node on the network and send frames, such as a PING, to a device on the network, and determine if a response was sent and the response time. A more sophisticated test can determine, in the case of multiple paths through the network, which paths were taken. In many WANs, connectivity can be verified by executing a call-placement sequence that establishes a call connection to enable a data transfer.

Conformance testing. *Conformance testing* is used to test data-communications devices for conformance to specific standards. These conformance tests consist of a set of test suites (or scenarios) that exercise data communications equipment fully and identify procedural violations that will cause problems. These conformance tests are used by developers of data-communications equipment and by carriers to prevent procedural errors before connection to the network is allowed. Conformance tests are based on the applicable protocol standard.

30.3.2 Protocol analysis measurements

Protocol analyzer functionality varies, depending upon the network technology (e.g., LAN vs. WAN), the targeted user (e.g., R&D vs. installation), and the specific application (e.g., fault isolation vs. performance monitoring). *Protocol analysis* includes the entire set of measurements that allow a user to analyze the information on a computer communications network. However, no single product provides all of the following measurements for all networking technologies.

- Protocol decodes
- Protocol statistics
- Expert analysis
- Traffic generation
- Bit-error rate tests
- Stimulus/response testing
- Simulation

Because protocol analysis requires associating the results of different measurements, these measurements are typically made in combination. Thus, protocol analyzers include different sets or a combination of these measurements. A good user interface combines pertinent information for the user, combining the measurement capability.

Protocol decodes. *Protocol decodes,* also referred to as *packet traces,* interpret the bit streams being transmitted on the physical media. A protocol decode actually decodes the transmitted bit stream. The bit stream is identified and broken into fields of information. The decoded fields are compared to the expected values in the protocol standards, and information is displayed as values, symbols, and text. If unexpected values are encountered, then an error is flagged on the decode display. Protocol decodes follow the individual conversations and point out the significance of the frames on the network by matching replies with requests, monitoring packet-sequencing activity, and flagging errors. Protocol decodes let the user analyze data frames in detail by presenting the frames in a variety of formats.

Decodes are typically used to troubleshoot networks. The ability to troubleshoot a problem is directly related to the ability to capture the frames from the network. Protocol conversations are based on sequences of frames. If a protocol analyzer misses frames, then protocol conversations are not recorded accurately and the troubleshooter cannot determine which is at fault, the protocol analyzer or the network. It is, therefore, essential to capture all of the frames without dropping or missing any.

Because protocol analysis can be performed in one of two ways, real time or post process (refer to Section 30.4.1), most analyzers provide a method to capture and store frames for subsequent decoding. Some analyzers also provide the capability to decode frames as they are captured (in real time). This allows the user to display data traffic as it occurs, eliminating the need to stop the testing to find out what is happening on the network.

Network problems are isolated by starting at an overview level and then focusing in on the network traffic in increasing detail. Because a protocol stack is essentially a virtual network operating on the physical medium, and many such virtual networks can coexist on one physical medium, protocol decodes can display all of the network traffic or the protocol decode can be filtered to isolate a single protocol stack or a particular protocol.

Protocol decodes are typically displayed in three formats: a summary decode (Fig. 30.5A), a detailed decode (Fig. 30.5B), and a data decode. These examples show a frame captured from the Internet. The http protocol is executing over TCP/IP on a 10BASE-T network. The summary decode shows a consolidated view of the frame information. The contents of each frame are summarized into one line on the display. This is used to quickly determine the type of traffic being transmitted on the network and to determine the upper-layer protocol conversations that are occurring.

The summary decode uses an underline cursor to aid in viewing a specific frame, as well as to indicate the frame that is examined in further detail, using a "drill down" user interface. In the example of Fig. 30.5A, frame number 4624 is underlined. The summary decode provides a reference number for the frame, a time stamp for the specific frame, the source and destination addresses (in this case, the MAC addresses), and the description field of the information contained in the frame, including encapsulated protocols, address and control information, and error information.

The detailed decode (Fig. 30.5B), breaks out all of the protocols in the

Network Stack Summary Decode

Control | Config | Actions | Format | Other displays | Print | Help

Frame	Time	Source	Destination	Description
4618	04.171544	Becky's PC	Cisco----01-FE-1D	TCP S=1120 D=808E
4619	04.250532	HP-------68-38-E4	Broadcast	SAP R HP28699A_08E
4620	04.260417	HP-------8B-4B-9E	HP-------10-DC-07	NETB Session data
4621	04.263308	HP-------10-DC-07	HP-------8B-4B-9E	SMB R Write
4622	04.265490	HP-------8B-4B-9E	HP-------10-DC-07	SMB C F=8162 Write
4623	04.266372	HP-------8B-4B-9E	HP-------10-DC-07	NETB Session data
4624	04.308424	Cisco----01-FE-1D	Becky's PC	HTTP R PORT=1122 (
4625	04.313735	Cisco----01-FE-1D	Becky's PC	HTTP R PORT=1122 H
4626	04.316341	Cisco----01-FE-1D	Becky's PC	HTTP R PORT=1122 H
4627	04.317574	Becky's PC	Cisco----01-FE-1D	TCP S=1122 D=808E
4628	04.317964	Cisco----01-FE-1D	Becky's PC	HTTP R PORT=1122 H
4629	04.319472	Cisco----01-FE-1D	Becky's PC	HTTP R PORT=1122 H
4630	04.320265	HP-------0B-2B-79	HP#------00-00-01	Ethernet TYPE=8005
4631	04.321837	Becky's PC	Cisco----01-FE-1D	TCP S=1122 D=808E
4632	04.323591	HP-------10-DC-07	HP-------8B-4B-9E	TCP S=2905 D=440E
4633	04.324065	HP-------8B-4B-9E	HP-------10-DC-07	NETB Session data
4634	04.324947	HP-------8B-4B-9E	HP-------10-DC-07	NETB Session data
4635	04.329378	HP-------10-DC-07	HP-------8B-4B-9E	TCP S=2905 D=440E
4636	04.329796	HP-------8B-4B-9E	HP-------10-DC-07	NETB Session data
4637	04.336286	HP-------10-DC-07	HP-------8B-4B-9E	SMB R Write
4638	04.338457	HP-------8B-4B-9E	HP-------10-DC-07	SMB C F=8162 Write
4639	04.339338	HP-------8B-4B-9E	HP-------10-DC-07	NETB Session data
4640	04.340220	HP-------8B-4B-9E	HP-------10-DC-07	NETB Session data
4641	04.343099	HP-------36-CB-34	Cisco----01-FE-1D	SNMP Trap
4642	04.346605	HP-------10-DC-07	HP-------8B-4B-9E	TCP S=2905 D=440E

(a)

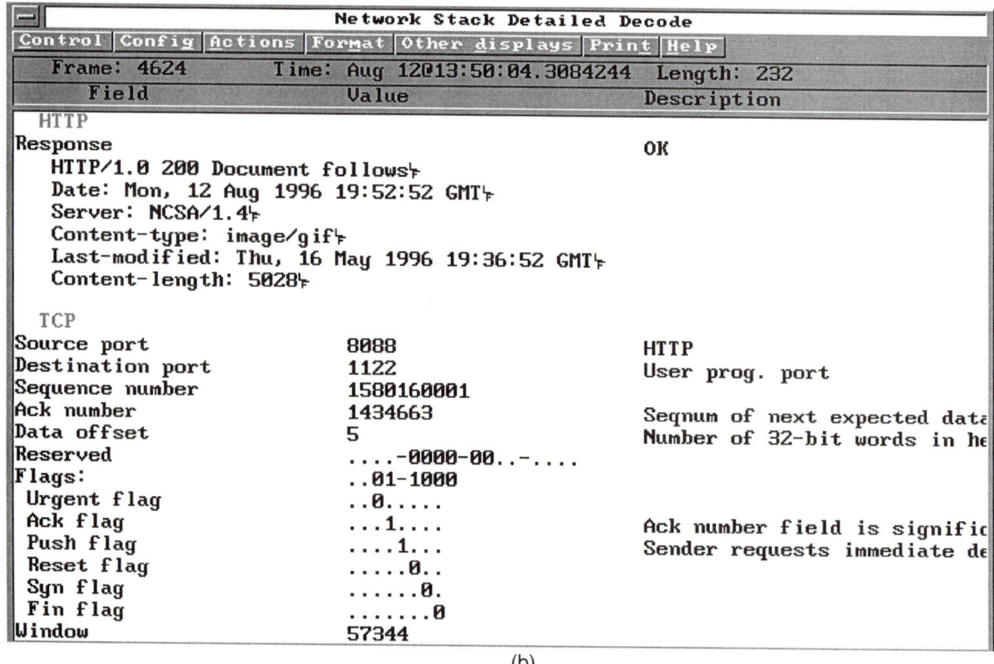

Network Stack Detailed Decode

Control | Config | Actions | Format | Other displays | Print | Help

Frame: 4624 Time: Aug 12@13:50:04.3084244 Length: 232

Field	Value	Description
HTTP		
Response		OK
HTTP/1.0 200 Document follows		
Date: Mon, 12 Aug 1996 19:52:52 GMT		
Server: NCSA/1.4		
Content-type: image/gif		
Last-modified: Thu, 16 May 1996 19:36:52 GMT		
Content-length: 5028		
TCP		
Source port	8088	HTTP
Destination port	1122	User prog. port
Sequence number	1580160001	
Ack number	1434663	Seqnum of next expected data
Data offset	5	Number of 32-bit words in he
Reserved	....-0000-00..-....	
Flags:	..01-1000	
Urgent flag	..0.....	
Ack flag	...1....	Ack number field is signific
Push flag	....1...	Sender requests immediate de
Reset flag	.....0..	
Syn flag	......0.	
Fin flag	0	
Window	57344	

(b)

Figure 30.5 (a) Ethernet summary decode, (b) ethernet detail decode.

underlined frame and identifies all of the fields in each protocol. Each field in the protocol layer is described and all parameters are identified with a textual description, as well as the actual field values. This decode view is used to troubleshoot specific problems within a protocol or in a transaction between devices on a network.

Each frame that is captured by a protocol analyzer from the network under test is provided a time stamp that is generated by an internal clock in the protocol analyzer. The time stamp is used to determine the particular time at which the frame was received, allowing a user to make crucial timing measurements, such as determining if a protocol time-out has occurred. Many protocols are based on receiving specific responses within a specified amount of time. In Fig. 30.5B, the top of the display shows that frame number 4624 was captured on Monday, August 12 at 13:50:04.3084244. An abbreviated time stamp is shown in the summary decode. Typical network troubleshooting involving timing is performed at the microsecond or millisecond resolution.

Captured frames are provided with a sequential number for convenient reference by the user. The frame numbers are not derived from the network under test and they are not part of the transmitted information on the network. They provide a convenient, but arbitrary, reference used to make it simpler to identify a particular frame from the thousands of frames captured in a single-trace capture.

Protocol statistics. An average high-speed computer network (such as a 10 M bit per second Ethernet) handles thousands of frames per second. It takes only seconds to fill the capture buffer of a protocol analyzer. If you are looking for a specific event on the network, then you can use capture filters to filter the data going into the capture buffer, or you can use triggers to stop the capture buffer around a specific event on the network and invoke protocol decodes to troubleshoot the problem. But this approach relies on knowing the cause of the trouble. In many cases, this is not known. Also, although decodes pinpoint specific problems, they do little to identify trends or provide performance information; thus, the need for protocol statistics exists.

Protocol statistics reduce the volumes of data captured into meaningful information. Protocol statistics provide valuable insight for determining the performance of a network, pinpointing bottlenecks, isolating nodes or stations with errors, and identifying stations on the network. Protocol analyzers keep track of hundreds of different statistics. The statistical information is displayed as:

- Histograms
- Tables
- Line graphs vs. time
- Matrices

TABLE 30.2 Types of Statistical Measurements

Statistical measurement	Description	Examples
Application	Ascertain the usage distribution of the applications running on the network. Used to determine how a network is being utilized and provide information for optimizing the network.	—Database applications —Graphics applications —E-mail applications —Internet access
Destination Addresses	Determine the type of destination address contained in a frame. Networks with high levels of broadcast traffic sent to all nodes on the network typically experience performance problems.	—Unicast address —Multicast address —Broadcast address —Source routing on a Token-Ring network
Error	Detect errors for the entire network, a particular node, or a logical channel. Devices creating errors affect the performance of the entire network and can create problems on other devices.	—Transmission errors —FCS errors —Protocol errors —Invalid frames
Frame Size	Calculate the average, minimum, and maximum frame sizes on the network. Used to determine if the network is being utilized with high data throughput (large frames) or if it is being overloaded with high overhead (many small frames).	—Frame size distribution —Average, minimum, and maximum frame sizes —Illegal frame lengths
Node Discovery	Automatically discover the nodes and stations attached to the network and identify their MAC addresses, network address, and alias or friendly name.	—Create a network node list —Identify interconnect devices (e.g., routers) —Identify new nodes —Identify aged nodes
Number of Nodes	Calculate the number of devices using the network.	—Current, average, minimum, and maximum count
Protocol	Ascertain the protocol and protocol stack distribution for the traffic running on the network. Used to determine how a network is being utilized and provide information for optimizing the network. Determine protocol usage, which stations are using bandwidth, and the respective amount of control protocols vs. data-transfer protocols.	—Distribution of protocols —Distribution of protocol stacks —Utilization of specific protocols (e.g., network layer protocols)
Utilization	Calculate the amount of traffic on the network vs. the theoretical bandwidth available on the network. Performance problems and network failures typically increase as the network utilization approaches its theoretical limit.	—Utilization (percent vs. time) —Current, average, minimum, and maximum throughput

Major classifications of statistical measurements are applicable to the full range of networks and protocols. Table 30.2 describes the major types of statistical measurements that can be made on networks. This is a list of what can be measured. Table 30.3 describes how it can be measured. The statistical-measurement data can be filtered by the criteria listed, with these measurements made by combining the entries listed in these two tables. For example, statistics can be filtered by a particular node, allowing error statistics

TABLE 30.3 Methods for Filtering Statistical Data

Filter criteria	Description	Examples
Connection	Track the performance of conversation pairs of MAC addresses, network addresses, or subnet types. Track errors and bandwidth utilization by connection pair.	Throughput statistics between: —File server and a router —Client and server —Two routers
Logical Channel	Track the performance of the virtual circuits established in a switched network. These statistics are good for isolating problems to a particular device or determining the performance of the switched network.	Throughput statistics for: —X.25 Logical Channel Number (LCN) —Frame Relay Data Link Connection Identifier (DLCI)
Network	Statistics are summed for all of the traffic observed on the network. These statistics give a good overall indication of network performance and can provide a network baseline.	Utilization (percent vs. time) —Total node count —Total error count —Average throughput —Average frame size
Node	In a LAN environment, statistics are sorted on a per-node basis. These statistics are good for isolating problems to a particular node or for identifying the crucial nodes, such as routers, that have high bandwidth.	—Frames received per node —Frames transmitted per node —Bytes received per node —Bytes transmitted per node
Top Talkers	Identify the nodes on the network that are transmitting the most traffic. Network problems are often a function of traffic. Top talkers is also useful for identifying key nodes, such as routers and servers.	—Nodes generating most frames —Nodes generating most IP traffic —Nodes generating most file transfers

to be collected for that node to determine if the specific node is operating properly. Figure 30.6 shows a summary statistics display for analyzing the performance of a 10BASE-T network. It is similar to the types of measurements made on most networks. The upper left part of the display provides a history of the overall network utilization. In this case, the network utilization has ranged from 0 to 20% utilization over the last three minutes. Typical statistics measurements can be set up to display the last several minutes of traffic or, with coarser granularity, more than one week of data. This statistics display also shows gauges that indicate instantaneous and cumulative counts of "collisions" and "errors." The current "node count" and "bytes/frame" are displayed. The percent of multicast vs. unicast vs. broadcast traffic is displayed in the destination addresses graph. The percentage of different protocols on the network is shown in the protocols graph. And, finally, the average throughput (in average frames/second) for crucial nodes on the network is displayed in the selected nodes graph. Each of these measurements has a user-defined threshold so that visual alarms can be displayed.

Expert analysis. Troubleshooting computer networks is complicated by the wide range of network architectures, protocols, and applications that are simultaneously in use on a typical network. Expert analysis reduces thousands

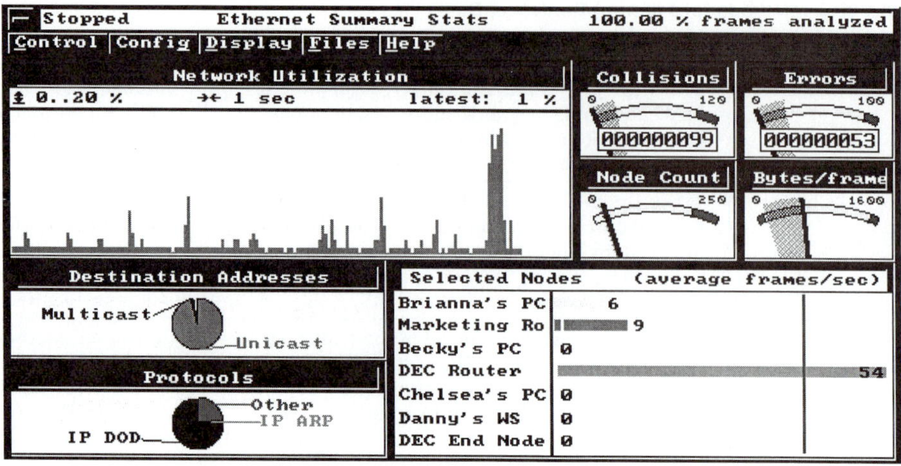

Figure 30.6 Protocol statistics.

of frames to a handful of significant events by examining the individual frames and the protocol conversations for indications of network problems. It watches continuously for router and bridge misconfigurations, slow file transfers, inefficient window sizes, connection resets, and many other problems. And it does this in real time for each protocol stack running on the network, as the network events occur. Thus, data is transformed into meaningful diagnostic information.

The typical network engineer will have expertise in certain protocol stacks and certain applications, but no single person has expertise in all of the applications and protocols. Therefore, expert analysis combines the knowledge of many networking experts into a single protocol analyzer. Expert analysis is performed by using the analysis capability (data capture, filters, triggers and actions, statistics, etc.) to monitor the network traffic. Figure 30.7 is an example of one of the many expert-analysis screens that simplify network troubleshooting. Rather than setting filters, collecting frames, decoding the frames, and correlating sequences of frames, the expert analysis screen indicates events occurring on the network, their severity, and appropriate additional information to troubleshoot the problem. The events are categorized as either *Normal, Warning,* or *Alert.* Normal events, such as the OSPF router identified take note of normal occurrences that can be of particular interest. Warning and alert-level events indicate an increasing level of severity that can suggest degraded network performance or catastrophic network failures. The highlighted event in Fig. 30.7 shows that two nodes, one with IP address 15.42.144.11 and one with IP address 15.6.74.53, are experiencing excessive TCP retransmissions. This is a warning level event, meaning that the network performance is being adversely affected. Most expert analysis applications provide context sensitive online documentation that provides trouble-

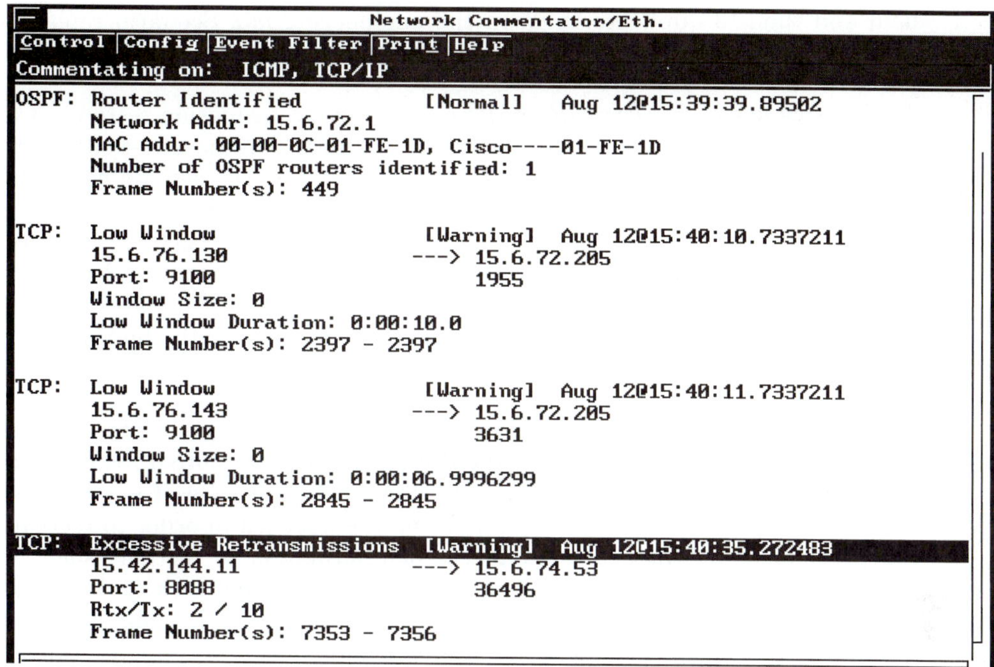
Network Commentator/Eth.
Control | Config | Event Filter | Print | Help
Commentating on: ICMP, TCP/IP

```
OSPF: Router Identified               [Normal]   Aug 12@15:39:39.89502
      Network Addr: 15.6.72.1
      MAC Addr: 00-00-0C-01-FE-1D, Cisco----01-FE-1D
      Number of OSPF routers identified: 1
      Frame Number(s): 449

TCP:  Low Window                      [Warning]  Aug 12@15:40:10.7337211
      15.6.76.130                    ---> 15.6.72.205
      Port: 9100                            1955
      Window Size: 0
      Low Window Duration: 0:00:10.0
      Frame Number(s): 2397 - 2397

TCP:  Low Window                      [Warning]  Aug 12@15:40:11.7337211
      15.6.76.143                    ---> 15.6.72.205
      Port: 9100                            3631
      Window Size: 0
      Low Window Duration: 0:00:06.9996299
      Frame Number(s): 2845 - 2845

TCP:  Excessive Retransmissions       [Warning]  Aug 12@15:40:35.272483
      15.42.144.11                   ---> 15.6.74.53
      Port: 8088                            36496
      Rtx/Tx: 2 / 10
      Frame Number(s): 7353 - 7356
```

Figure 30.7 Expert analysis.

shooting suggestions. They also provide drill-down capability to access additional measurements. In this particular case frame, numbers 7353 to 7356 in the capture buffer provide more information about the problem.

Traffic generation. Many network problems are very difficult to diagnose because they occur intermittently, often showing up only under peak load. Traffic generators provide the capability to simulate network problems by creating a network load that stresses the network or a particular device sending frame sequences on to the network. In an ATM network, a traffic generator can be used to create cell-loss and cell-delay measurements. They are also used to stress a network or a particular device on the network under various loads. They can be used to increase the background traffic level on an in-service network, or they can be used to stress test an out-of-service network or specific device. In either case, the user can measure the response of a device or an entire network under an increased load.

A powerful feature of traffic generation is *scenario replay,* the ability to capture a buffer of live network data and then use the protocol analyzer to retransmit the data in either its original form or as modified by the user. The user can define the frame, can cut and paste the frame from one already captured, or send a captured file. This is used to capture live problems in the field and then to duplicate them in the engineering lab. Scenario replay is used for field service, as well as R&D applications.

Bit-error-rate tests. *Bit-Error-Rate (BER) tests* are transmission tests used to determine the error rate of the transmission media or the end-to-end network. Although advanced BER measurements reside in the domain of sophisticated transmission test sets, protocol analysis, particularly in a WAN or ATM environment, often requires a verification of the media. BER tests are performed by transmitting a known bit pattern onto the network, looping it back at a point on the network, and receiving the sequence. The bit-error rate is calculated as a percentage of the bits in error compared to the total number of bits received.

Stimulus/response testing. Although many networking problems can be quickly solved with decodes and statistics, many of the more-difficult problems cannot be solved in a nonintrusive manner. In this case, it is necessary to actively communicate with the devices on the network in order to recreate the problem or obtain necessary pieces of information to further isolate a problem. So, the user can actively query or stress the network and observe the results with decodes and statistics. For example, observing Ethernet frames with the same IP address and different MAC addresses might indicate a duplicate IP address problem, but it might also just be the result of a complex router topology. To determine the node's identity (router or true duplicate IP address), the user can search through thousands of captured frames with decodes, but by sending an *Address Resolution Protocol (ARP)* frame to the node on the network and comparing the results to the addresses of the routers on the network, determined with node-discovery statistics measurements, the user can quickly isolate duplicate IP addresses.

Typical active tests include:

- Connectivity tests
- Installation tests
- Network path analysis
- Adapter status commands
- Address resolution tests
- MIB (Management Information Base) queries

Simulation. In the context of protocol analysis, *simulation* can take two forms: *protocol simulation* and *protocol emulation.* Protocol simulation allows the user to send strings of data containing selected protocol headers along with the encapsulated data. In this way, the operation of a network device can be simulated for the purpose of testing a suspected problem or for establishing a link to confirm operation. Protocol emulators are software that controls the operation of the protocol analyzer automatically. For example, in the case of X.25, a protocol emulator takes over the functions of bringing up the link at layers one, two, and three, and automatically maintaining the connection. These functions are specified by the user and the operation is then placed under programmatic control. Typically, simulation tests are executed on top of an emulation program that is automatically handling the lower layers of the protocol stack (e.g., emulating a frame-relay circuit while simulating a PING).

30.3.3 Protocol-analysis measurement options

Protocol analysis consists of using measurements like those described in the previous section to isolate network problems or monitor network performance. The protocol-analysis measurement options described in this section are a set of orthogonal capabilities to the measurements described in the previous section. For example, a protocol analyzer can gather statistics on the network traffic, but a more-effective troubleshooting approach is created by setting a capture filter and running the statistics so that only the data between a file server and a router is analyzed.

Data capture. The most fundamental attribute of protocol analysis is the ability to capture the traffic from a live network and to store this data into a capture buffer. The captured data can then be analyzed and reanalyzed at the user's discretion. The capture buffer is either part of the main memory of a PC or it is a special RAM buffer dedicated to capturing frames. Information stored in the buffer includes all of the frames on the network, a time stamp indicating when they were captured, and network-specific status information necessary to reproduce the exact state of the captured traffic. Because protocol analysis is used primarily for fault isolation, it is, by definition, used when the network is either experiencing heavy traffic loads or is experiencing errors. It is also necessary to analyze sequences of frames because most protocols are state dependent. Therefore, it is essential that all traffic on the network is captured, regardless of the level of utilization, or whether the frames are errored or damaged. Once data is captured in the capture buffer, it can be repeatedly examined for problems or events of significance. Search criteria, such as filter criteria, are based on address, error, protocol, and bit patterns.

Data logging. Many network troubleshooting sessions can be spread over hours and days. The capture buffer on a typical protocol analyzer is filled in seconds on a high-speed network. Therefore, data-logging capabilities are crucial for setting up long troubleshooting sessions. The user can specify a file name and a time interval for recording crucial information. This information is then regularly stored to hard disk and can be examined by the user at a later time. Information that is typically stored to disk includes frames that match a user-specified filter, statistics results, or the results of a programmed test.

Filtering. The key to successfully troubleshooting network problems is based on eliminating the unnecessary information and focusing on the crucial information that is essential to solving the problem. Computer networks process thousands of frames per second. A protocol analyzer can quickly fill a capture buffer with frames, and a user can sift through protocol decodes searching for the errors. But this task is time-consuming and tedious. The most-powerful function of protocol analysis is the ability to filter the data on the network in order to isolate problems.

The function of a filter is very similar to that of a trigger (sometimes called a *trap*). Specific filter patterns are set by the user of a protocol analyzer and these filters are then compared with the data from the network. Filters range from simple bit-pattern matching to sophisticated combinations of address and protocol characteristics. Fundamentally, the two types of filters are *capture filters* and *display filters*. Capture filters are used to either include or exclude data from being stored in a protocol analyzer's capture buffer. Capture filters make it possible to only collect the frames of interest by eliminating extraneous frames. This effectively increases the usage of the capture buffer. Rather than a capture buffer that only contains six error frames out of 40,000 captured frames in the buffer, the data is filtered so that only errored frames are in the buffer. More frames of interest can be captured and they can be located more quickly. A disadvantage of capture filters is that it is necessary for the user to know what to filter on (i.e., to have some idea of what problem to investigate). A second disadvantage is that the frames that were filtered out might contain the sequence of events leading up to the errored frame.

In many situations, the source of a network problem is not known; therefore, it is necessary to capture all of the frames on the network and use display filters to repeatedly filter the frames. Because all of the frames are stored in the capture buffer, the frames can be played back through the display filters. Display filters act upon the frames once they have been captured. Frames can be selected for display by measurements, such as protocol decodes.

Filter conditions can be combined to form more-powerful filter criteria. Typically, as the troubleshooting process progresses, the user discovers more and more information about the network problem. As each new fact is discovered, it can be added to the filter criteria until, finally, the problem is identified. For example, to isolate a faulty Ethernet network interface card, it is necessary to filter on the MAC address of the suspicious node and bad *Frame Check Sequence* (*FCS*) simultaneously.

Table 30.4 summarizes the different types of filter criteria. Typically, the same criteria can be used by either display filters or by capture filters.

Triggers and actions. In order to troubleshoot network problems, it is often necessary to identify specific frames or fields in frames. Triggers are used to detect events of significance to the user and then initiate some action. Trig-

TABLE 30.4 Filter Criteria

Frames containing the specified filter criteria are isolated to the exclusion of all other traffic on the network. Frames matching the filter criteria can be either included or excluded. This allows the user to focus on specific frames or to exclude frames that are extraneous to the troubleshooting process. Filters can be used in combination with one another to form more powerful filters.

Filter class	Filter type	Filter criteria
Address	MAC Layer Source Address	Frames with the specific source address of the device sending the frame.
	MAC Layer Destination Address	Frames with the specific destination address of the device specified to receive the frame.
	Network Layer Source Address	Frames with the specific source network address of the system in an end-to-end network that originated the frame.
	Network Layer Destination Address	Frames with the specific destination network address of the system in an end-to-end network that is specified to receive the frame.
	Broadcast	Frames with the destination address equal to a broadcast address (frames to be sent to all devices on the network).
	Multicast	Frames with the destination address equal to a multicast address (frames to be sent to a group of devices on the network).
	LCN, Logical Channel Number	Frames containing a specific LCN. The LCN identifies one of many virtual circuits established in a packet-switched X.25 network.
	DLCI, Data Link Connection Identifier	Frames containing a specific DLCI. The DLCI identifies one of many virtual circuits in a packet-switched frame-relay network.
	VPI/VCI, Virtual Path Identifier/Virtual Channel Identifier	Frames containing a specific VPI/VCI pair. The VPI/VCI pair is used to identify a specific virtual circuit in an ATM network.
Error	FCS (Frame Check Sequence)	Frames experiencing an FCS error (indicating that a frame was not properly received).
	All Protocol Errors	Frames experiencing any protocol errors.
	Specific Protocol Error	Frames containing a specific protocol error (e.g., an illegally long frame).
	All Transmission Errors	Frames experiencing any transmission errors.
	Specific Transmission Errors	Frames experiencing a specific transmission error (e.g., an alignment error).
	Invalid Frames	All invalid frames.
Protocol	Specific Protocol	Frames containing a specific protocol type (e.g., IP).
	Specific Protocol Stack	Frames containing the protocols specific to a particular protocol stack (e.g., TCP/IP).
	Specific Protocol Layer	Frames containing protocols specific to a particular layer (e.g., the network layer).
	Specific Protocol Field	Frames containing a specific protocol field (e.g., a TCP destination port of a specified value).
	Specific Frame Types	Frames of a specific frame type (e.g., an ICMP redirect frame).
Bit Pattern	Position dependent bit pattern	Frames containing a hexadecimal pattern at a fixed offset from the beginning of the frame.
	Position independent bit pattern	Frames containing a hexadecimal pattern anywhere in the frame.

gers and filters operate the same way in terms of recognizing conditions on the network. The parameters for setting trigger criteria are the same as the filter types specified in Table 30.2. The trigger is a key capability of protocol analyzers because it allows the automatic search of a data stream for an event of significance, resulting in some action to be taken. Possible trigger actions include:

- Visual alarm on the screen
- Audible alarm
- Start capturing data in the capture buffer continuously
- Start capturing data, fill the capture buffer, and stop
- Position the trigger in the capture buffer and stop capturing data
- End the data capture
- Increment a counter
- Start a timer
- Stop a timer
- Make an entry in the event log
- Start a specific measurement
- Send an SNMP trap
- Log data to disk

30.4 Protocol Analyzer Implementations

Because a variety of requirements are necessary for protocol analysis, there are many different product implementations. Many of the differences between products represent nothing more than the preferences or innovations offered by individual vendors. But there are some fundamental trade-offs in the implementation of a protocol analyzer that have significant impact on the feature set and cost of the product. These trade-offs include:

- Real-time analysis vs. post-processing
- Passive testing vs. intrusive testing
- Dedicated hardware acquisition system vs. software acquisition system
- Single port vs. multi-port

30.4.1 Real-time analysis vs. post-processing

Network troubleshooting can be performed in real-time as problems occur, or it can be done in a post-process mode by capturing a buffer of data for subsequent analysis.

Real-time analysis. The most intuitive way to use a protocol analyzer is in real-time, making measurements on the live network as the traffic is occurring. However, extra processing power is required to keep up with real-time network events. Real-time analysis is performed to troubleshoot catastrophic network failures as they occur. The network engineer will interact with the protocol analyzer to gather information, query the network, and solve the problem. Real-time analysis is also used to gather performance monitoring information over long periods of time that would typically overrun a capture buffer.

Post-process. In post-process mode, data is captured and analyzed offline. This allows the user to closely scrutinize the collected data. Post-process is typically used to solve intermittent problems. The most common scenario is to set a specific capture trigger that will cause a capture buffer full of data to be stored. The capture buffer is then studied with protocol decodes, display filters, statistics, or expert analysis measurements to track down the intermittent problem. As a general rule, post-process solutions require less processor bandwidth than real-time analysis applications because the analysis does not need to keep up with real-time network events.

30.4.2 Passive testing vs. intrusive testing

All protocol analyzers must electrically connect to the network under test. The manner in which they connect is dictated by the specifications of the

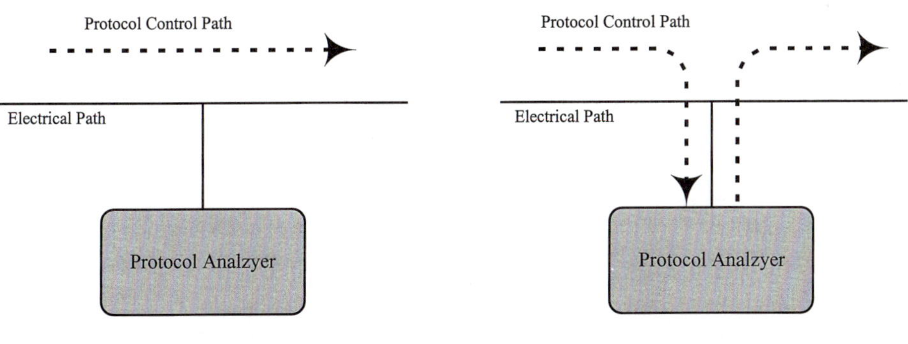

Figure 30.8 Passive vs. intrusive testing. A protocol analyzer must electrically connect to the network under test in order to monitor network traffic. However, the analyzer can be an active or passive participant in the protocol control on the network. It can passively monitor the traffic without perturbing the protocol of the network, as shown in the passive testing mode. However, many troubleshooting and network installation situations require that the protocol analyzer actively participate in the protocol on the network and generate network traffic, as shown in the intrusive testing mode.

TABLE 30.5 Comparison of Analysis and Acquisition Systems

Software analysis and acquisition systems are implemented as applications in the computer platform (PC or UNIX). Hardware systems are implemented as separate processors and circuitry dedicated to analysis and acquisition.

Attribute	Comments	Software solution	Dedicated hardware solution
Cost	Hardware solutions are more costly because of the material cost of the dedicated processor and circuitry.	★★★★	★★
Size	Hardware solutions are implemented as computer cards or modules. Software solutions are built as applications on the hard disk of the computing platform that acquire data from the network by using a standard network interface card.	★★★	★★
Performance	The computing platform shares its processing power with the software solution while the hardware solution offloads the computing platform.	★	★★★★
Full line rate acquisition	Hardware solutions capture traffic at 100% network utilization while software solutions are limited to partial capture on most networks, often discarding frames.	★	★★★★
Capture all errored frames	The standard network interface cards used with software solutions ignore many errored frames, which are of particular interest in network troubleshooting. Hardware solutions are built with special circuitry to capture these frames.	★	★★★★
Triggering	The number of triggers and the sophistication of the trigger conditions is far greater in a hardware solution that has special circuitry to implement triggers.	★★	★★★★
Filtering	The number of filters and the sophistication of the trigger conditions is far greater in a hardware solution that has special circuitry to implement filters.	★★	★★★★
Statistics	Network statistics require examining and sorting network traffic in real time. The processing power of a hardware solution offers more-sophisticated real-time statistics.	★★	★★★★
Expert analysis	Expert analysis requires capturing all the network frames and following complex frame sequences. The capture, filter, and trigger facilities of a hardware solution are superior to those of a software solution.	★	★★★★
Simple troubleshooting	Simple troubleshooting involves observing network traffic and looking for connections, data transfer, and utilization. This can be performed well with either type of solution.	★★★	★★★★
In-depth troubleshooting	In-depth troubleshooting requires excellent data transfer, triggering, filtering, and the ability to examine errored frames. Hardware solutions provide a much more robust feature set.	★	★★★★

TABLE 30.5 Comparison of Analysis and Acquisition Systems (*Continued*)

Attribute	Comments	Software solution	Dedicated hardware solution
Performance monitoring	Performance monitoring is a function of the statistical measurement capability of the product and is greatly enhanced by the processing power of a dedicated hardware solution.	★★	★★★★
Established network technologies	Mature technologies are readily available on standard PC cards and PCMCIA cards for integration into a software solution. These same technologies are also available in custom hardware solutions.	★★★★	★★★★
Newly emerging technologies	Emerging technologies are sometimes available on standard network interface cards that can be integrated into a software solution. However, the test solutions for these new technologies usually must be implemented as custom hardware solutions.	★★	★★★★

★★★★ Excellent; ★★★ good; ★★ average; ★ poor.

physical layer of the network being tested, and this might be a passive or intrusive electrical connection. However, in terms of the logical or protocol connection, the protocol analyzer can be a passive monitor or it can actively participate in the protocol and perform intrusive testing. Figure 30.8 shows the logical flow of the protocol control in a passive and an intrusive situation.

Passive testing. Many problems can be discovered by using passive measurements, such as statistics and decodes, that monitor the network under test. Passive tests do not alter or participate in the protocol on the network. An advantage of passive testing is that it does not contribute to problems that might be occurring on the network, nor is any traffic load added to the network.

Intrusive testing. Some troubleshooting scenarios require that the network be stressed or loaded in order to cause problems. A protocol analyzer can act as a node on the network and intrusively stimulate the network under test. Traffic generation, *Bit-Error Rate* (*BER*) tests, stimulus/response tests, and simulation are intrusive measurements. A protocol analyzer that is designed to implement intrusive testing includes circuitry to maintain a transmit path in addition to a receive path. Unlike a normal node on a network that is either transmitting or receiving data, a protocol analyzer must always maintain its receive path so that it can observe all of the data on the network, including data that the protocol analyzer itself transmits.

30.4.3 Dedicated hardware acquisition system vs. software acquisition system

The distinguishing feature between two broad categories of protocol analyzers is the method of implementation for the analysis and acquisition systems. An analysis and acquisition system can be built as a software application in a PC or it can be implemented with a dedicated system using a special-purpose microprocessor and specialized hardware. Table 30.5 describes the advantages and disadvantages of the two implementations.

Software acquisition system. Low-cost analyzers are often implemented using the same commercial network-interface cards that are commonly used in PC systems or workstations. The acquisition and analysis functionality is then implemented in software as a PC application. The acquisition system is, therefore, limited in performance by the computing platform, which must also execute the user interface and measurement software. Commercial network interface cards are not typically designed to pass all data and all errored frames to the host; therefore, their suitability to protocol analysis is restricted.

Dedicated hardware acquisition system. In most high-performance protocol analyzers, the data-analysis and acquisition system is implemented in dedicated hardware, often with special-purpose processors that are specially designed to facilitate manipulating and examining data at high speeds. This allows computationally intensive measurements to be executed, and it also allows multiple measurements to be executed simultaneously. Simultaneous mea-

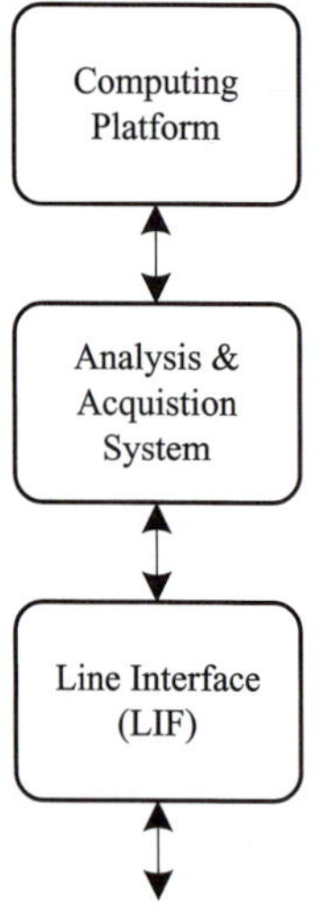

Figure 30.9 Single-port protocol analyzer architecture. A single-port protocol analyzer allows testing of a single network connection. A single line interface and a single analysis and acquisition system are dedicated to the network under test.

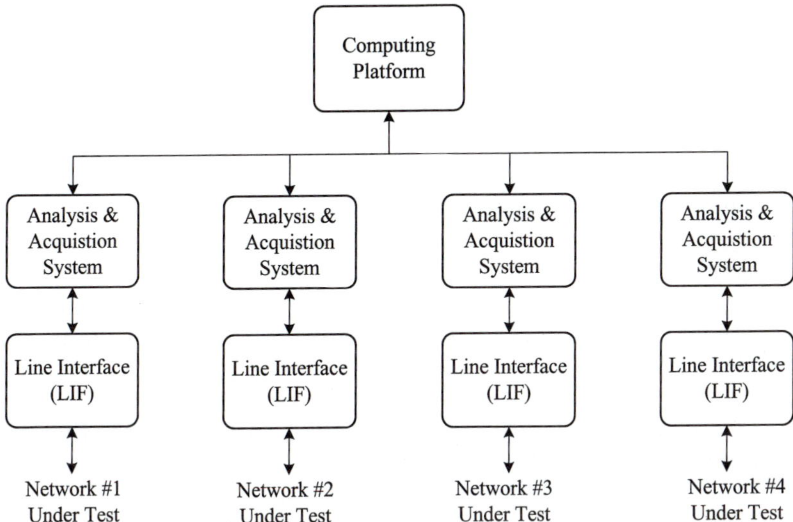

Figure 30.10 Multi-port protocol analyzer architecture. Most multi-port protocol analyzers allow for simultaneous testing of two to eight different network connections. In this example, a four-port architecture demonstrates that each network connection requires its own line interface and its own analysis and acquisition system, while sharing a computing platform.

surement execution is required for most troubleshooting scenarios. An example of a high-performance measurement is "top talkers" on a 100BASE-T LAN (where data is transferred at 100M bits per second). The top talkers measurement displays the nodes on the network that are transmitting the most traffic. This requires that the protocol analyzer process each frame on the network in real-time, examining its source address and keeping a sorted data base of the addresses.

30.4.4 Single port vs. multi-port

Most networking troubleshooting applications require that a protocol analyzer only connect to one point on the network. However, in certain situations, it is required to connect a protocol analyzer to more than one point on the network. Typical scenarios usually involve the interconnect devices on the networks, such as bridges, routers, hubs, and switches. These devices have multiple network connections and they perform transformations on the protocol information that flows through them. Multi-port protocol analyzers will typically support from two to eight simultaneous test ports.

Single port. *Single-port analysis* is used to describe a protocol analyzer that is using one, and only one, line interface to test the network under test. Figure 30.9 shows the block diagram for a single-port protocol analyzer. There is a dedicated analysis and acquisition system and a dedicated computing system for the single line interface. Most protocol analysis and network troubleshooting is performed in a single-port configuration.

Multi-port. *Multi-port protocol analysis* is used to simultaneously test multiple points on the network under test. The different points could be the same physical network technology or they could be different network technologies. For example, a two-port analyzer could be used to test the latency of frames being processed by a router that routes T-1 traffic onto an 10BASE-T Ethernet network. Multi-Port protocol analyzers are significantly more complex and costly than single-port protocol analyzers with each line interface requiring a dedicated analysis and acquisition system. They are most commonly implemented with additional dedicated hardware that allows the multiple ports to handle all of the data, and to provide time synchronization between the ports so that captured data can be correlated.

Figure 30.10 shows the block diagram for a multi-port protocol analyzer that is constructed with a line interface and an analysis and acquisition system dedicated to each port.

30.5 Basic Protocol Analyzer Architectures

Any protocol analyzer has three main components:

- Computing platform
- Analysis and acquisition system
- Line interface

30.5.1 Computing platform

The computing platform is a general-purpose processing system (typically a PC or a UNIX workstation). The computing platform executes the user interface for the product and controls the measurements that are executed in the analysis and acquisition systems. It is very common for other applications to be run in conjunction with a protocol analyzer. These include spreadsheet applications for analyzing data and measurement results, network-management software, and terminal-emulation applications that log into computer systems on the network. Therefore, computing platforms are usually based on an industry-standard system that allows a user to openly interact with the application environment.

30.5.2 Analysis and acquisition system

Fundamentally, a protocol analyzer acquires data from the network and then analyzes that data. Thus, the analysis and acquisition system is the core of a

TABLE 30.6 Protocol Analyzer Selection Criteria

Selection criteria	Common attributes	Typical applications
☑ Package	Portable, <20 lbs (9 kg)	The protocol analyzer is dispatched to the troubleshooting location.
	Transportable, 20 to 50 lbs (9 to 22 kg)	The protocol analyzer is moved infrequently or is used on a cart.
	Stationary, >50 lbs (22 kg)	The protocol analyzer is used in a fixed location, such as a lab bench or rack.
☑ Operation	Floor	The protocol analyzer is used in wiring closets and behind racks of equipment.
	Rackmount	The protocol analyzer is used in a test equipment bay or in a rack of network equipment.
	Table	The protocol analyzer is used in an office or lab environment.
☑ Power	Ac	Typical operation for most test equipment.
	Battery	Needed for portability or in places where ac power is not available.
☑ Computing platform	PC with Microsoft Windows	Preferred by users familiar with PC systems and used for running other PC applications.
	UNIX Workstation	Preferred by users familiar with UNIX systems and used for running other UNIX applications.
	Vendor proprietary	Typically unique to a specific test application.
☑ Display system	■ Resolution (e.g., VGA and SVGA) ■ Color	The quality of the display greatly impacts usability, but also has a big impact on the cost of the product.
☑ User interface	■ Ease of use ■ Use of color ■ Graphical	The protocol analyzer should be intuitive and as easy to use as possible.
☑ I/O	■ Serial ■ Parallel ■ LAN ■ PCMCIA ■ HP-IB	Used for printing measurement results and reports. It is also used for remote control of the product. PCMCIA provides the greatest flexibility because it can handle various interface cards.
☑ Capture performance	All frames Errored frames	Acquiring and analyzing all of the frames on the network, regardless of utilization or error levels, is essential to in-depth troubleshooting applications and performance.
☑ Analysis and acquisition system	Software	Used for monitoring and simple troubleshooting applications where a low-cost solution is preferred.
	Dedicated hardware	Used for performance monitoring, in-depth troubleshooting, and installation where a full-featured, full-capture-rate solution is required.
☑ Number of test ports	Single port	Used for most troubleshooting situations where testing is focused on the segment or link.
	Multiport	Used to test interconnect devices (bridges, routers, switches, etc.) in internetworking situations.

TABLE 30.6 Protocol Analyzer Selection Criteria (*Continued*)

Selection criteria	Common attributes	Typical applications
☑ Network interfaces	■ Ethernet ■ Token-passing bus ■ Token ring ■ VG ANYLAN ■ Fast Ethernet ■ FDDI ■ MAN (Metropolitan Area Network) ■ ITU-T G.804 PDH ■ T1.105 SONET ■ G.708 SDH ■ TAXI ■ Fiber channel ■ IBM block encoded ■ HSSI, High-speed serial interface ■ V-series	The ability to connect the protocol analyzer to the networks on your site is the most crucial criteria to consider when purchasing a protocol analyzer. This is determined by the network interfaces offered by the vendor.
☑ Protocol stacks	■ AppleTalk ■ ATM ■ Banyan VINES ■ DECnet ■ Frame Relay ■ IBM/SNA ■ ISDN ■ ISO ■ LAN Manager ■ MAP ■ Novell NetWare ■ Sun ■ TCP/IP ■ X Windows ■ X.25 ■ XNS	The protocol stacks supported by the protocol analyzer determine if you can test the protocols running on your network. The protocol stacks (including all of the individual protocols in the protocol stacks) should be supported by all of the measurement capability and the measurement options in the protocol analyzer.
☑ File import/export	■ ASCII ■ CSV ■ Vendor Specific	Data captured by a protocol analyzer is imported into other applications, such as database programs and spreadsheet applications. This is accomplished through standard interchange formats.
☑ Measurements	Protocol decodes	Used for in-depth troubleshooting.
	Statistics	Used for troubleshooting and performance monitoring.
	Expert analysis	Used for in-depth troubleshooting.
	Traffic generation	Used to stress-test existing networks and to install new networks.
	BER testing	Used to verify the transmission network.
	Stimulus/response testing	Used to query the network under test.
	Simulation/emulation	Used to create installation tests for new networks and equipment.
	Programming language	Used to create specific network scenarios, usually for development applications or recreating network problems.

TABLE 30.6 Protocol Analyzer Selection Criteria (*Continued*)

Selection criteria	Common attributes	Typical applications
☑ Measurement options	Display filtering	Used in troubleshooting applications to pinpoint specific frames.
	Capture filtering	Used in troubleshooting applications to reduce the amount of data captured by the analyzer.
	Triggers and actions	Used in troubleshooting applications to isolate intermittent failures.
	Timers and counters	Used to create simple statistics or to verify protocol timing sequences.
☑ Connectivity	Remote control	Used to control a protocol analyzer in a remote location via a LAN or modem connection.
	SNMP manageable	Used to query SNMP network devices or to provide SNMP information to queries from other network devices. This typically requires that the protocol analyzer be built with a MIB (Management Information Base).
	Terminal emulation	Allows the user to log on to other systems connected to the network. It is used to test connectivity and also to access applications available on other systems.
	Network management systems	The protocol analyzer can be invoked by Network Management Systems.
☑ Operating specifications	■ Safety regulations ■ Electromagnetic compatibility ■ Temperature	These specifications vary by country and by the requirements of the application in which the protocol analyzer will be used.
☑ Reliability	■ MTBF (Mean Time Between Failures) for the hardware ■ Software defects and upgrades	The more important the network being tested, the higher reliability required of the protocol analyzer.
☑ Installation	■ Factory setup ■ User setup	Most protocol analyzers are shipped completely configured as turnkey solutions. However, embedded protocol analyzers require the user to install the hardware and software.
☑ Support	■ Phone (help desk) ■ Bulletin boards ■ World Wide Web site ■ Technical consultants	The particular type of support required depends on the specific network application, the preference of the user, and the geographic location.
☑ Training	■ Classes ■ Computer-based training ■ Tutorials ■ Application notes	The particular type of training required depends on the specific network application, the preference of the user, and the geographic location.
☑ Help Systems	■ On-line help ■ Documentation	Using a protocol analyzer requires help systems to describe the operation of the product, the network interfaces, and the protocols supported.

protocol analyzer. This system is essentially responsible for transferring data from the line interface to the capture buffer, ensuring that all of the error conditions, the protocol state information, and the protocol data is correctly stored and time stamped. During real-time and in post-process mode, the triggers and actions, the timers and counters, and the protocol followers are executed in the analysis and acquisition system. Additionally, the measurements are typically executed in a distributed fashion between the computing platform and the analysis and acquisition system. In low-cost, software-based analyzers, the analysis and acquisition-system functions are performed by the computing platform. In high-performance protocol analyzers a dedicated processor and special-purpose hardware are used to implement the processing required by the analysis and acquisition functions.

30.5.3 Line interface

The physical hardware and firmware necessary to actually attach to the network under test are implemented in the line interface. Additionally, the line interface includes the necessary transmit circuitry to implement simulation functions for intrusive testing. The function of the line interface is to implement the physical layer of the OSI Reference Model and provide framed data to the analysis and acquisition system.

30.6 Protocol Analyzer Selection Criteria

When selecting a protocol analyzer, it is important to recognize that there is a wide range of products on the market. The user must first determine how the product will be used: the information in this chapter describes the functionality that is available. Table 30.6 lists the major categories of product features that can be used to evaluate protocol analyzers. No single product incorporates all of the capabilities outlined, so it is important to look for the combination of selection criteria that meets your needs.

Chapter

31

Bit Error Rate Measuring Instruments: Pattern Generators and Error Detectors

Hugh Walker
Agilent Technologies
South Queensferry, Scotland

31.1 Introduction

Digital circuits, switches, transmission systems, and storage devices are currently some of the highest growth areas in electronics. Fueled by the almost limitless bandwidth of fiberoptic cables, digital telecommunications is transforming the telephone industry and creating the rapid convergence of voice, data, and video communications. In the space of a few years in the 1990s, the speed of digital systems has increased from hundreds of megabits per second (Mbits/s) to tens of gigabits per second (Gbits/s).

The fundamental measure of performance or quality in these new digital systems is the probability of any stored or transmitted bit being received in error. With the latest equipment, the probabilities are very low, being on the order of 10^{-12} or less. However, it is still necessary to measure the performance of these systems and in particular to analyze the margins of safety available and to explore potential weaknesses which could lead to degraded performance later. This is the purpose of digital pattern generators and error detectors, sometimes referred to as "bit error rate test sets," or BERTS.

31.2 Sources of Errors

Errors in digital systems arise as a result of several distinct practical effects. When viewing a random digital signal on an oscilloscope, the common eye diagram shown in Fig. 31.1 is displayed. To obtain the eye diagram display

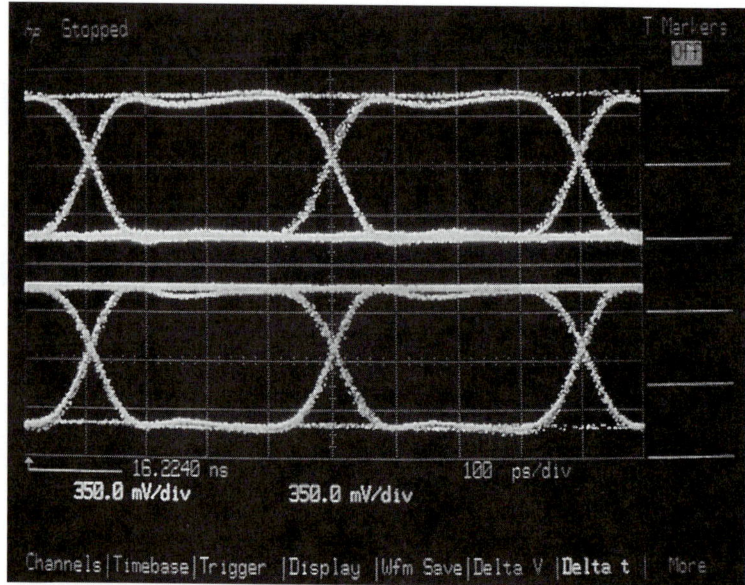

Figure 31.1 The eye diagram of a random digital signal displayed on a digital sampling oscilloscope. Over successive sweeps, the random bit patterns build up a composite picture of all the possible pattern sequences and transitions. The large open area in the center of the pulse is called the "eye opening," with the one value at the top and the zero value at the bottom. The distance between the top and the bottom at the center of the pulse is called the "eye height," while the distance between the transitions is called the "eye width." The eye diagram is a useful qualitative measure of a digital system's performance. An engineer can spot immediately if the eye is degraded by noise, timing jitter, pulse degradation, or intersymbol interference (ISI). As explained in the text, ISI is pattern-dependent, so different test patterns will create different eye diagrams.

on an oscilloscope, the sweep is triggered using the data clock source, and the time base is adjusted so that, say, two or four bit periods are displayed. On each successive sweep, the random bit patterns (and transitions) build up on the display, either through the persistence of the screen phosphor or through digital storage. With random data, all possible combinations of bit sequences will be explored, so the eye diagram then shows the extent of pulse distortion that might occur. The eye diagram is important to designers and testers of digital circuits because it shows at a glance the quality of a system.

The displayed eye diagram exhibits an open area in the center of the pulse separating the '1' level from the '0' level. This is termed the "eye opening." The wider this gap, the lower is the probability of a '1' being confused with a '0' and vice versa. The space between adjacent vertical transitions at the edges of the pulse is termed the "eye width." The wider this gap, the more tolerant the system will be of the point at which the digital signal is sampled to determine the instantaneous binary value. Errors occur either when the eye opening is partially closed or when the relative sampling instant is displaced by timing jitter (as described below). The following subsections discuss sources of errors.

31.2.1 Excessive noise

Excessive noise generally arises from thermal noise or crosstalk in sensitive circuits which, by addition to the binary signal, closes the eye opening and creates a "fuzziness" on the eye diagram display. Examples of noise-limited systems are an optical receiver operating near its minimum light level or a digital radio or satellite receiver operating with a low signal attenuated by fading. A characteristic of noise-limited systems is that the errors occur randomly according to a Poisson distribution and are not influenced by the transmitted bit pattern.

31.2.2 Intersymbol interference

Intersymbol interference (ISI) is caused by short-term storage effects in a digital circuit. At any time, the received signal represents not only the current digital value but also the residues of previous digital values (determined by the system's impulse response). With a random digital signal, these residual appear on the eye diagram rather like thermal noise. However, the levels are not random but are determined by the preceding digital pattern. The purpose of the pattern generator is to simulate a wide range of digital sequences and stress patterns to explore the effects of ISI. This is sometimes referred to as the "pattern dependency" of a system.

31.2.3 Timing jitter

Timing jitter causes a horizontal displacement of the eye diagram and so reduces the eye width. One of the best definitions of timing jitter has been provided by the CCITT (International Telephone and Telegraph Consultative Committee): "Timing jitter is the short-term variation of the significant instants of a digital signal from their ideal positions in time." The significant instant might be the rising or falling edge of a pulse. The effect of jitter can be seen in the diagram in Fig. 31.2. At certain points in time, the pulse is significantly offset from its correct position. If this offset becomes large, then there will be an error when we try to sample and decode the digital signal. The disturbance or offset of the timing instant is usually measured in unit intervals (UIs) peak to peak, where the unit interval is equivalent to one bit period.

Jitter arises from a variety of causes, including superimposed noise and crosstalk affecting the trigger point of logic decision circuits. Another important source of jitter is clock recovery circuits which derive the reference sampling clock from a received data stream. Depending on the pattern density, the clock recovery circuit (usually a tuned tank circuit or phase-locked loop) may drift toward its natural frequency and lose or accumulate phase shift

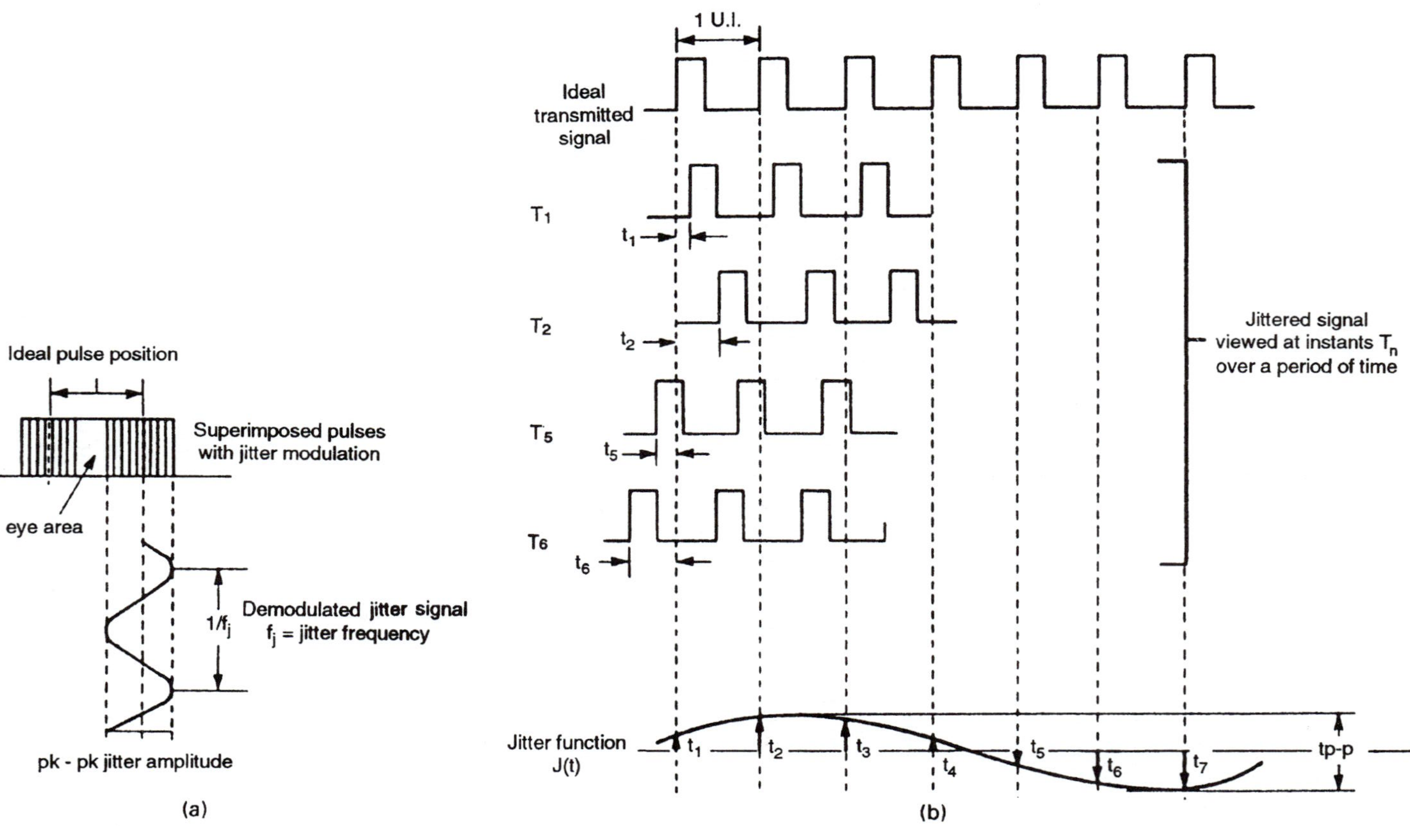

Figure 31.2 Timing jitter disturbs the pulse from its ideal position in time, and the perturbations cause a narrowing of the eye area, as shown in (a). Looked at in real time (b) at instants T_1, T_2, T_3, and so on, one can see that the bit pattern is displaced from the ideal positions in time. The instantaneous offsets t_1, t_2, t_3, and so on from the ideal positions form the jitter function $J(t)$. If jitter becomes excessive, the eye opening will be closed sufficiently to cause errors when sampling the data. Sampling is usually timed to occur at the center of the eye, at the point of greatest eye height.

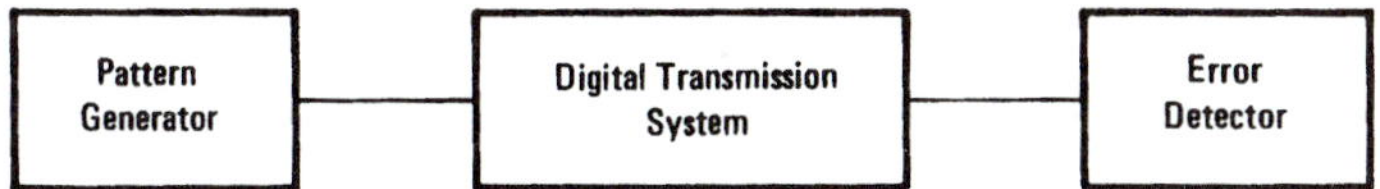

Figure 31.3 To test a digital transmission system, a pattern generator is connected to the input of the system under test, and an error detector is connected at the output.

relative to the incoming data. In analyzing the source of errors, it is often necessary to check jitter levels on reclocked data and to check a system's tolerance to specific levels of superimposed jitter. These measurements are usually carried out with a dedicated telecommunications jitter test set, a digital sampling oscilloscope, or a time-interval analyzer.

31.3 Error Measurements

To test a digital system, the input is stimulated by a test pattern. Usually this is a pseudo-random binary sequence (PRBS), although other specific stress patterns (referred to as "userdefined word patterns") may be used to explore limits of performance. Typical stress patterns might include long runs of "zeros" to test clock recovery or patterns with alternate periods of high and low "mark density" (or "one" density) to check storage effects in optical transmitters and receivers which may respond to average mean level in a data signal.

For telecommunications and data communications transmission systems, the object is to simulate the random traffic experienced under normal operating conditions. The problem with a truly random signal is that the error detector will have no means of knowing the actual bit values that were transmitted and therefore no way of detecting errors. Instead, a pseudo-random signal is used, which means that it has virtually all the statistical characteristics of a true random signal and appears as such to the item under test while in fact being completely deterministic and therefore predictable by the error detector. A range of maximal length PRBS patterns has been specified for this, as described later. At the error detector (Fig. 31.3), the output of the system under test is compared bit by bit with a locally generated, error-free reference pattern.

The probability of error in any transmitted bit is a statistical property and has to be treated as such. Any attempt to measure this over a given time period can be expressed in various ways, the most common of which is

$$\text{Bit error ratio (BER)} = \frac{\text{number of errors counted in the averaging interval}}{\text{total number of transmitted bits in the averaging interval}}$$

Clearly, the result will have a statistical variance from the long-term mean error ratio dependent on the size of the sample taken from the population—in this case the number of errors counted. Three methods of computing BER are

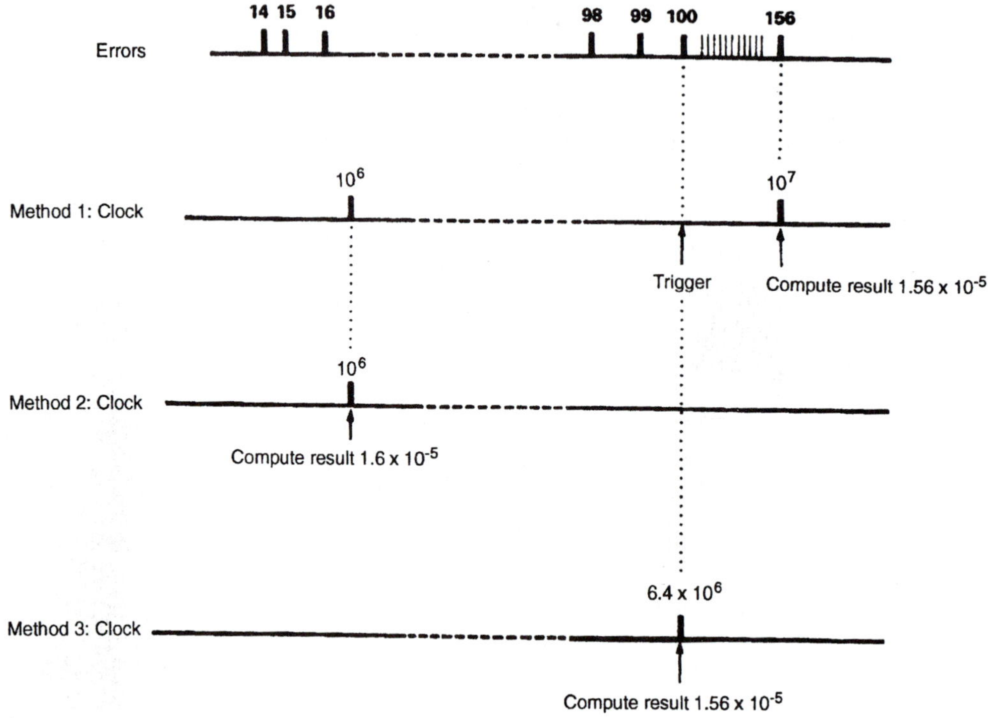

Figure 31.4 Three methods of computing BER. Method 1, used on early BER testers, simply defines the measurement time by the number of clock periods (10^6, 10^7, and so on). The accumulated error count can then easily be converted to BER; however, the measurement period varies according to the bit rate. Method 2 defines the measurement period in seconds, minutes, and hours, and a microprocessor is used to calculate BER from the accumulated error count and clock count in that period. The advantage is that the measurements are consistent with error-performance standards. Method 3 determines the measurement period as that required to accumulate sufficient errors for a statistically significant result, e.g., 100 errors. This may lead to very long measurements at low BER values.

in general use, and these are illustrated in Fig. 31.4. The first method, common on early test sets, simply counted the number of clock periods to provide a time base or averaging interval. This could be implemented easily using discrete logic decade dividers. Now that microprocessors are available, more convenient gating periods are used. In the second method, a timed gating period of, say, 1 second, 1 minute, or 1 hour is used, and a calculation of BER is made from the accumulated totals. The advantage of this method is that it provide results which are compatible with the error-performance criteria discussed later. The third method determines the gating period by counting sufficient errors (typically 100 or more) for statistically reliable results. Again, the processor calculates BER from the accumulated totals. However, this method can lead to very long gating periods with low BER values. For example, a system running at 100 Mbits/s with a BER of 10^{-12} would take nearly 12 days to accumulate 100 errors.

The most commonly used approach is method 2, which calculates BER after a fixed, repetitive gating period. In this case, the variance in the result will be continuously changing, so it is normal to give some kind of warning if the variance exceeds generally acceptable levels. The most widely accepted level is 10 percent, i.e., an error count of at least 100 errors. In practical digital transmission systems, particularly those using radio propagation, the BER can vary substantially over time. In this case, the long-term mean value only provides part of the story. Communications engineers are also interested in the percentage of time when the system under test is unacceptably degraded. This is called "error analysis" or "error performance," which is discussed in Sec. 31.3.3.

31.3.1 Test patterns

As alluded to earlier, the choice of test pattern is usually made between a PRBS to simulate traffic and specific word patterns to examine pattern-dependent tendencies or critical timing effects. With a PRBS, the choice of binary sequence and the resulting spectral and run properties are important.

These properties may be summarized:

- Sequence length, in bits
- Shift register feedback configuration defining binary run properties
- Spectral line spacing, which depends on bit rate (see Fig. 31.5)

PRBS patterns have been standardized by the CCITT for testing digital transmission systems (Recommendations 0.151, 0.152 and 0.153). The most

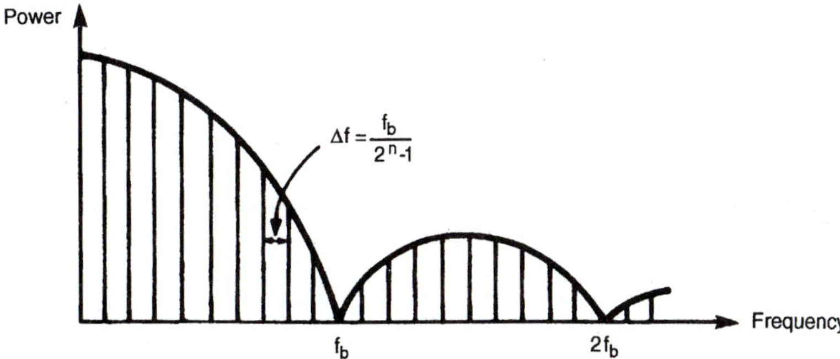

Figure 31.5 This is the ideal spectrum of a binary nonreturn to zero (NRZ) PRBS. It has a line spectrum in which the line spacing is determined by the bit rate (f_b) and the sequence length $2^n - 1$ (see Table 31.1).

TABLE 31.1 Relationship Between Sequence Length and Bit Rate with Corresponding Spectral Line Spacing for Some Standard Telecommunications Transmission Rates

Bit rate (f_b), kbit/s	Sequence length (n)	Polynomial	Spectral line (f_b/n), Hz
1,544	$2^{15} - 1$ bits	$D_{15} + D_{14} + 1 = 0$	47.1
2,048	$2^{15} - 1$ bits	$D_{15} + D_{14} + 1 = 0$	62.5
34,368	$2^{23} - 1$ bits	$D_{23} + D_{18} + 1 = 0$	4.1
44,736	$2^{15} - 1$ bits	$D^{15} + D_{14} + 1 = 0$	1365.3
139.264	$2^{23} - 1$ bits	$D_{23} + D_{18} + 1 = 0$	16.6

commonly used patterns in digital transmission testing are summarized in Table 31.1.

Table 31.1 shows some examples of CCITT-specified sequences for several standard telecommunications bit rates. Note that the longer sequences give closer spectral line spacing, and typically, the higher the operating bit rate, the longer is the required sequence to simulate real data traffic. For tests in the gigabit per second range, some test sets now provide a $2^{31} - 1$ sequence length.

Adequate (i.e., close enough) spectral line spacing is important when testing systems containing relatively narrow band (high Q) clock timing recovery circuits in order to see the jitter contribution of these and its effect on error performance. The shift register configuration is defined by a polynomial of the type shown in Table 31.1. The letter D stands for delay, and for example, $D_{15} + D_{14} + 1 = 0$ means that the outputs of the fifteenth and fourteenth stages of the shift register are connected to an exclusive-OR gate, the output of which drives the first shift register stage, as shown in Fig. 31.6. This basic circuit arrangement generates a sequence with a maximum run of ones rather than zeros. It is common to invert the output to generate a maximum run of zeros, since this may create more stringent conditions for a clock recovery circuit. The simple three-stage PRBS generator shown in Fig. 31.7 with the truth table helps to explain the operation of the feedback shift register. This has a sequence length of $2^3 - 1$, or 7, bits.

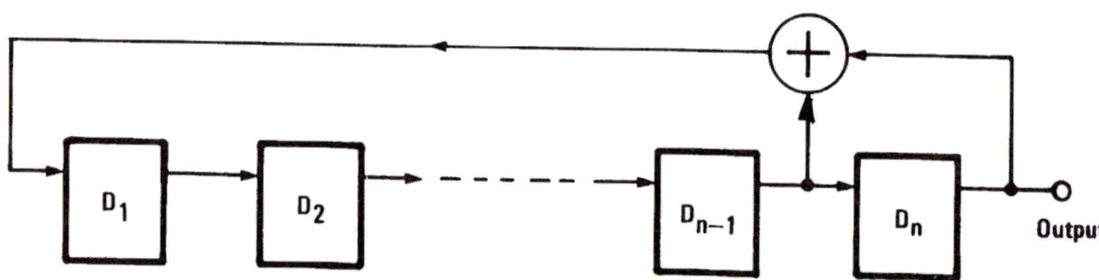

Figure 31.6 The PRBS pattern generator consists of a shift register with feedback taps connected to an exclusive-OR gate. The output of the shift register may be inverted to create a signal with the maximum run of zeros. The polynomial which defines the PRBS determines where the feedback connections are made. The expression for this diagram would be $D_n + D_{n-1} + 1 = 0$.

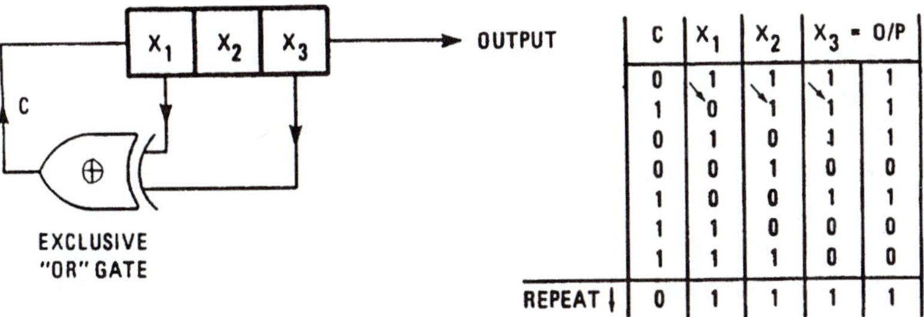

	C	X_1	X_2	X_3	= O/P
	0	1	1	1	1
	1	0	1	1	1
	0	1	0	1	1
	0	0	1	0	0
	1	0	0	1	1
	1	1	0	0	0
	1	1	1	0	0
REPEAT ↓	0	1	1	1	1

Figure 31.7 The simple three-stage PRBS generator demonstrates how the feedback shift register creates the pseudo-random sequence. The output of the exclusive-OR gate provides the input to the shift register. The truth table shows how the bit pattern steps through the shift register with each clock pulse.

The choice of shift register configuration affects the run properties of the PRBS—an example is shown in Fig. 31.8 for the $D_{15} + D_{14} + 1 = 0$ polynomial. This graph shows that there are periods in the sequence with a low number of ones relative to zeros, which is more stressful to clock recovery circuits. Run properties affect timing jitter in terms of the length of zero blocks over which phase error is accumulated by the timing recovery circuits. This leads to pattern-dependent jitter and, if not controlled, to errors.

Figure 31.8 This graph shows the running digital sum for a commonly used PRBS. Notice the initial steep descent of the graph, indicating a predominance of ones or zeros (depending on inversion) for the initial period of the sequence.

Some test sets allow users to take a standard PRBS and modify it using zero substitution and variable mark density. A maximal length (m-sequence) PRBS usually has a maximum run of zeros equal to one less than the degree of the defining polynomial. This can be extended by deliberately overwriting subsequent bits with zeros to any extent the user requires. Zero substitution is useful for checking the limits of a clock recovery circuit. Variable mark density allows the user to change the density of ones in the pattern from the regular 50 percent of a PRBS pattern. A test set may allow the ratio to be varied from 12.5 to 87.5 percent. The advantage of modifying a standard PRBS is that many of the "random" traffic-simulating characteristics are retained while the user explores the limitations of a system under fairly realistic conditions.

In addition to PRBS patterns, most test equipment now provides at least some facilities for the user to program a repetitive-word pattern. At the simplest level, this could be a particular pattern with minimum or maximum density of marks—for example, the 3 in 24 pattern used in North America for testing T1 circuits at 1.544 Mbits/s. This pattern complies with the minimum mark density of 12.5 percent while providing the longest allowed run of 15 zeros.

However, with the increasing complexity of telecommunications equipment, there is also the need to simulate frame structures and even specific data messages. Indeed, many systems today will not respond to an unstructured, unframed PRBS. Framing is used by telecommunications equipment to identify individual channels which are time division multiplexed (TDM) together in a serial data stream. To mark the beginning of a sequence of multiplexed channels, the telecommunications equipment transmits a fixed frame word or pattern of bits which is recognized at the receiving end. If the frame word pattern is not present, the receiving equipment will assume a fault and transmit alarms.

Pattern generators, therefore, will increasingly require large pattern memories which can be programmed from a computer to simulate a wide range of framed test signals and, of course, to simulate specific malfunctions and errors in those framed test signals. Figure 31.9 shows a typical range of patterns downloaded from disk available on a gigabit per second pattern generator and error detector.

31.3.2 Error detection

The basic error-detection process involves comparing the incoming data stream from the item under test with a reference pattern in an exclusive-OR gate. When both data patterns are identical and synchronized, the exclusive-OR gives zero output. When a difference exists, i.e., when an error is present in the received stream, an output is generated as shown in Fig. 31.10. This simple circuit will detect all logic errors. An additional feature allows errored ones and errored zeros to be counted separately. Error mechanisms in some

```
          16:22:39 FEB 1, 1993                                         MENU
DISC      HP 70842B ERROR DETECTOR (Patterns)                  (0,17)  CURRENT
PATT 6                                                                 PATTERN
          Current Pattern INACTIVE  Length:          127
DISC      2^7              from Patt  4                                INTERNL
PATT 7    Patt.  1:                 Length:        8,192               PATT 1
          Patt.  2: 2^11            Length:        2,048
DISC      Patt.  3: SONET STS-48    Length:        1,152               INTERNL
PATT 8    Patt.  4: 2^7             Length:          127               PATT 2
          Patt.  5: SONET STS-12    Length:       77,760
DISC      Patt.  6: SONET STS-48    Length:      311,040               INTERNL
PATT 9    Patt.  7: CID STM-4       Length:       20,720               PATT 3
          Patt.  8: CID STM-16      Length:       22,440
DISC      Patt.  9: SDH STM-4       Length:       77,760               INTERNL
PATT 10   Patt. 10: SDH STM-16      Length:      311,040               PATT 4
          Patt. 11: FDDI Jitter     Length:        1,280
          Patt. 12: FDDI Wander     Length:       90,000
DISC                                                                   DISC
PATT 11                                                                PATT 5

DISC                                                                   CANCEL
PATT 12                                                                EDIT
```

Figure 31.9 An example of stored word patterns in a BER test set. The list includes some simple PRBS patterns, SONET/SDH frame patterns, and test patterns for FDDI LAN applications.

devices, such as laser transmitters, lead to predominance of one polarity over the other.

A further refinement available on some instruments allows the user to identify the exact bit within a long word pattern which is in error. Often error generation is systematic; i.e., it is caused by inherent deterministic impairments in the equipment such as intersymbol interference, storage effects, and delay-line reflections. A particular sequence of bits in the pattern creates a high probability of error in a subsequent bit. Errored bit identification helps the user to focus on the portion of pattern which creates problems and then to investigate the causes. Counting bits from the start of the pattern as a reference point, the instrument locates the position of an errored bit and identifies it. On successive repeats of the pattern it calculates the "bit" BER, which, if the cause is systematic, will be far higher than the average BER for the pattern. Of course, if the error is purely random, then on average the

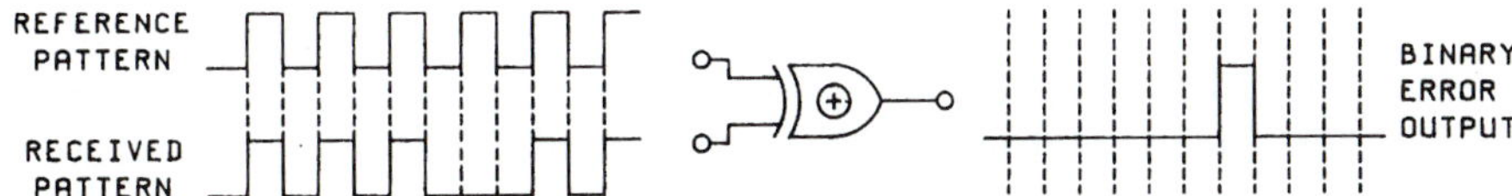

Figure 31.10 Error detection using an exclusive-OR gate simply involves comparing bit by bit the received pattern with the reference pattern. When a difference is present, the exclusive-OR gate gives an output.

"bit" BER will be similar to the overall average. This capability, in conjunction with a digital sampling oscilloscope, is very useful to an engineer designing equipment and trying to locate the often complex source of errors. For example, reflections in transmission lines often give rise to pulse degradation many bits delayed from the launch pulse. By monitoring a frequently errored bit, the engineer can adjust the word pattern to explore the effects of ISI and reflections.

The reference pattern for error detection could be supplied locally from the pattern generator, but normally, a separate reference pattern generator is provided in the error detector so that the transmit and receive portions of the test equipment can be separated—an important requirement for end-to-end testing of telecommunications links.

Synchronization. This raises the question of synchronization of the two patterns to be compared before error detection can commence. In order to synchronize rapidly yet remain in synchronism at high error rates or during large error bursts, a "sync criterion" BER has to be established which has variable gate times over which a test for synchronism is made.

The sync criterion may be expressed as

$$\text{Sync gain} = \text{sync loss}$$
$$x/n = X/N$$

where x and X = error counts
n and N = total bit counts
$x << X$
$n << N$

(*Note:* To avoid oscillation, it is normal to make the sync loss BER greater than the sync gain BER, while the gate time for sync gain is much less than for sync loss. The longer period for sync loss ensures that the error detector is not thrown out of lock by a burst of errors. For example, one test set has a synch loss criterion of >20,000 errors in 500,000 bits and a sync gain criterion of <4 errors in 100 bits.)

The normal method of achieving synchronization (see Fig. 31.11) is to open the feedback loop in the reference pattern shift register and feed the input data signal into the register until it is full, close the feedback loop, and test for sync. Clearly, two PRBS patterns out of sync have a BER of approximately 0.5, so the sync criterion must be lower than this. Some error detectors allow the user to set the sync gain/loss criteria, for example, by setting the threshold BER in a particular gating period. Since measurements are invalid during resynchronization, it is desirable to minimize these periods.

In-service measurements. Finally, as mentioned earlier, most of the communications equipment in use operates with a standard frame structure. Framing

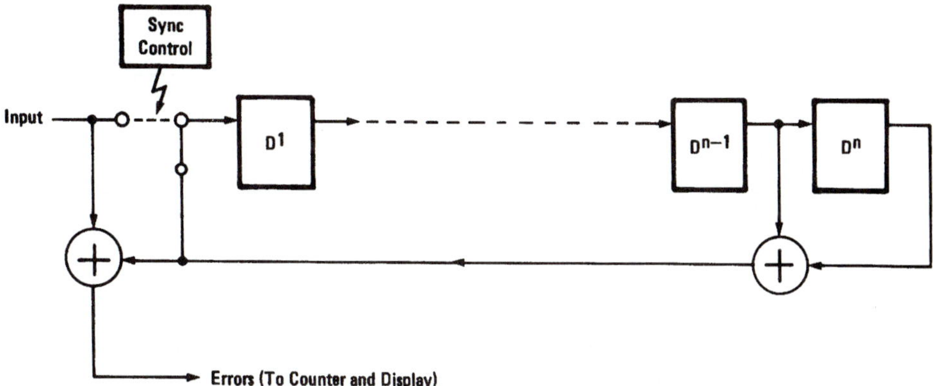

Figure 31.11 To obtain pattern synchronization in a closed-loop error detector, the feedback loop on the reference pattern generator is temporarily opened, and a sample of the input bit pattern is fed into the shift register. The loop is then closed, and a synchronization check is made by checking the output of the error detector. A BER of approximately 0.5 indicates a loss of synchronization. In this case, the process is repeated until synchronization is obtained. Some test sets allow the criteria for sync gain and loss to be adjusted for a particular application.

allows multiplexed channels to be identified in a high-speed data stream. Frame structures also carry information on system status and error detection, e.g., using cyclic redundancy checksum (CRC) or parity. The frame word itself provides a constant repetitive pattern even when operating in live traffic and so provides a simple way of monitoring errors. Some error detectors, specifically designed for maintenance of digital communications systems, lock onto the frame pattern and report on frame and code errors while the system is "in service" in much the same way as logic errors in a PRBS are counted in an "out-of-service" test.

31.3.3 Error analysis

In practical systems, the BER can vary substantially over time (Fig. 31.12) due to propagation effects in radio and satellite systems, electrical interference, and random traffic patterns. For example, a system may have a satisfactory long-term BER level while being subjected to short bursts of intense errors interspersed with long periods of error-free operation. If the system is subjected to high BER for periods of more than a few seconds, it is considered to be unusable and is described as "unavailable." In commercial communications, customers want to know what grade of service a network operator guarantees to provide because this determines the price paid for the service. This is generally classified as the percentage of time a service meets or exceeds expectations.

To meet this measurement requirement, it is necessary for the error detector to measure BER and count errors and then to classify the results as percentages of the total elapsed measurement period that certain thresholds are

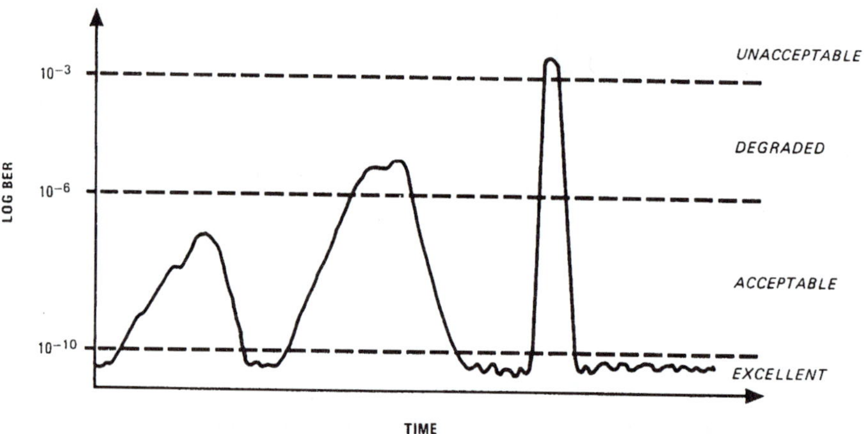

Figure 31.12 Error performance. In a practical system, particularly those using radio propagation, the BER may vary a great deal over time. One could measure the average long-term BER, but communications engineers find it more useful to categorize the performance into bands and then express the error performance in terms of the percentage of time each threshold is exceeded.

exceeded. This is called "error analysis." The industry standard specification is CCITT Recommendation G. 821 shown in Table 31.2.

G.821 defines how error performance parameters are calculated in accordance with the flow diagram shown in Fig. 31.13. The total measurement time is divided into 1-s periods, and unavailable time is subtracted to obtain the available time on which the G.821 parameters are calculated.

A period of unavailable time begins when the BER in each second is worse than 10^{-3} for a period of 10 consecutive seconds. These 10 s are then considered to be unavailable time. A new period of available time begins with the first second of a period of 10 consecutive seconds each of which has a BER better than 10^{-3}.

During available time, any second containing one or more errors is logged

TABLE 31.2 CCITT Recommendation G.821: Error Performance of an International Digital Connection Forming Part of an ISDN*

Performance classification	Objectives
Degraded minutes (DM)	Fewer than 10 percent of 1-minute intervals to have a bit error ratio (BER) worse than 1E-6
Severely errored seconds (SES)	Fewer than 0.2 percent of 1-second intervals to have a bit error ratio (BER) worse than 1E-3
Errored seconds (ES)	Fewer than 8 percent of 1-second intervals to have any errors—equivalent to 92 percent error-free seconds

*Measured over a period T_L (e.g., 1 month) on a unidirectional 64-kbit/s channel of the hypothetical reference connection (HRX) of 27,500 km.

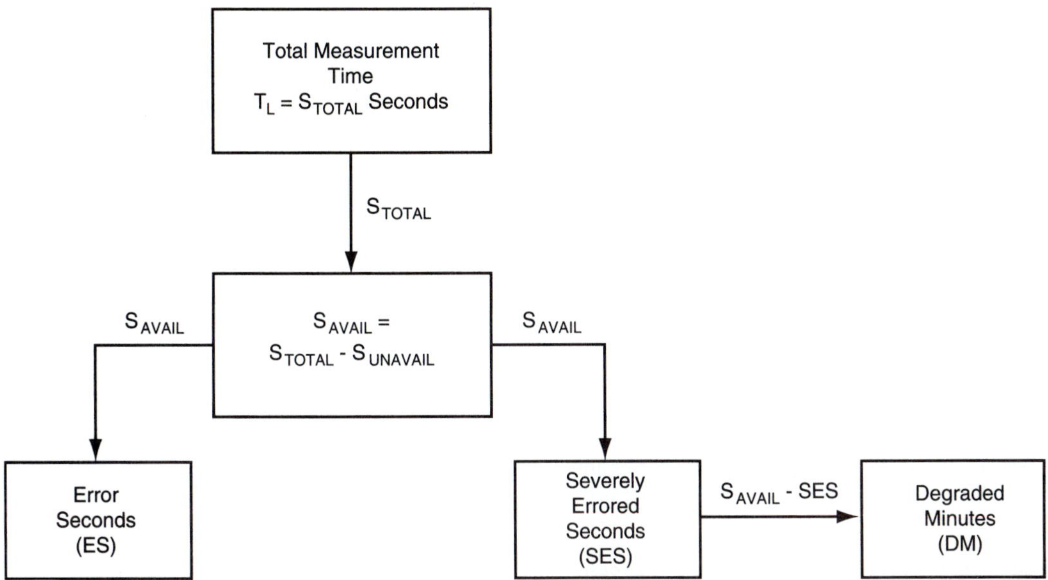

Figure 31.13 This diagram shows how the error detector classifies the error measurements according to the G.821 criteria. From the total measurement time T_L, the tester subtracts periods of unavailable time (periods of 10 s or more when the BER is worse than 10^{-3}). From the remaining available time (S_{AVAIL} seconds), errored seconds are accumulated, and simulataneously, any seconds with the BER worse than 10^{-3} are accumulated as severely errored seconds (SES). The remaining nonseverely errored second periods are grouped together in 60-s blocks, and any that have an average BER of worse than 10^{-6} are classified as degraded minutes (DM).

as an errored second (ES). Any period of 1 s with a BER exceeding 10^{-3} is classified as a severely errored second (SES) and is subtracted from available time. The remaining seconds are grouped into 60-s periods. Any of these 1-min periods with a mean BER exceeding 10^{-6} are classified as degraded minutes (DM).

Almost all BER test sets now incorporate the basic G.821 analysis outlined above. When the original G.821 recommendation was formulated, communications engineers were principally concerned with the quality of voice and low-speed data transmission based on channel bit rates of 64 kbits/s. Later, standards committees investigated the appropriate error-performance criteria for high-speed data and video services. For example, there is an Annex D to the G.821 recommendation which suggests algorithms for adapting the standard to bit rates above 64 kbits/s. Some test sets incorporate the Annex D recommendation.

The original G.821 analysis is based on a minimum gating period or errored/error-free interval of 1 s. For higher-speed data and video applications, a shorter interval might be appropriate. Some test sets now additionally provide analysis based on errored/error-free intervals of deciseconds, centiseconds, and milliseconds.

When planning to use a BER test set in installation and maintenance of

```
Link        :                              Instrument   : HP3784A
                                           Tx Interface: 75ohm  TERM HDB3
Tx Location:                               Tx Clock     : STD RATE 2MHz    +0ppm
                                           Tx Pattern   : PRBS 15 ZERO SUB 000
Rx Location:                               Rx Interface: 75ohm TERM  HDB3
                                           Rx Clock     : 2MHz
                                           Rx Pattern   : PRBS 15 ZERO SUB 000
                                           Rx Hit Thres: 0.500UIP

Fri, Oct 21, 1988 11:05:44 START OF MANUAL GATING PERIOD
Bit Error Intervals
and Alarm Conditions.
21-Oct-88 11:06:49 Bit Errors this Second........ 1
21-Oct-88 11:06:50 Bit Errors this Second........ 2
21-Oct-88 11:06:51 Bit Errors this Second........ 1
21-Oct-88 11:07:02 Bit Errors this Second........ 2
21-Oct-88 11:07:03 Bit Errors this Second........ 4
21-Oct-88 11:07:04 Bit Errors this Second........ 4
21-Oct-88 11:07:16 Bit Errors this Second........ 2
21-Oct-88 11:07:17 Bit Errors this Second........ 3
21-Oct-88 11:07:18 Bit Errors this Second........ 5
21-Oct-88 11:07:19 Bit Errors this Second........ 5
21-Oct-88 11:07:20 Bit Errors this Second........ 5
21-Oct-88 11:07:21 Bit Errors this Second........ 4
21-Oct-88 11:07:29 Bit Errors this Second........ 2

Fri, Oct 21, 1988 11:07:45 STOP, ELAPSED TIME 00 Days 00 Hours 02 Mins 01 Secs

SUMMARY:
Bit Error Result
    Bit Error Count................ 40 Bit Error Ratio........ 1.6E-07
    Bit Error Secs................. 13 Bit Error Free Secs........ 108
Bit Error Analysis
    Availability............ 100.00%  Number of Error Seconds with N errors:
    Unavailability.......... 0.0000%    1 error................. 2
    Severely Errored Seconds. 0.0000%   2-10 errors........... 11
    Error Secs.............. 10.744%    >10 errors............. 0
    Degraded Minutes......... 0.0000
    Error Bursts>100 Errors........ 0  L.T.Mean.E.Ratio...... 1.6E-07
Code Error Result
    Code Error Count................ 0 Code Error Ratio............. 0
    Code Error Secs................. 0 Code Error Free Secs....... 121
Alarm Durations (seconds)
    Power Loss......... 0.0  Tx Clock Loss....... N/A  Rx Clock Loss....... N/A
    Rx Data Loss....... 0.0  Sync Loss........... 0.0  AIS Secs............ 0.0
    Jitter Loss........ 0.0  Slip Secs........... 0.0
```

Figure 31.14 The modern BER test set generates a large amount of information on errored events and calculates error performance measurements. This sample printout shows the type of data available.

telecommunications or data communications links, G.821 analysis is essential. For laboratory and manufacturing applications, it is less important, and one may consider pattern flexibility and error position identification more valuable.

31.3.4 Data logging

Since error performance measurements may need to run for several hours or days to accumulate statistically significant results, most of the time the test sets will be left unattended. Hence the error detector must have a means to log measurement data and error events/alarms for later analysis. Furthermore, long-term tests may be affected by power supply interruptions. Data logging protects valuable test data which would otherwise be lost. When power is reinstated, the instrument can be designed to recover from the interruption automatically and recommence the test without operator involvement.

Usually, data logging is provided by outputting results and events to a printer as they happen or by storing information in nonvolatile memory for analysis and display later. An example printout is shown in Fig. 31.14.

Long-term tests may produce a large amount of measurement data, which is time-consuming to analyze. Graphic display of results, as shown in the example in Fig. 31.15, helps the user quickly identify the periods of interest.

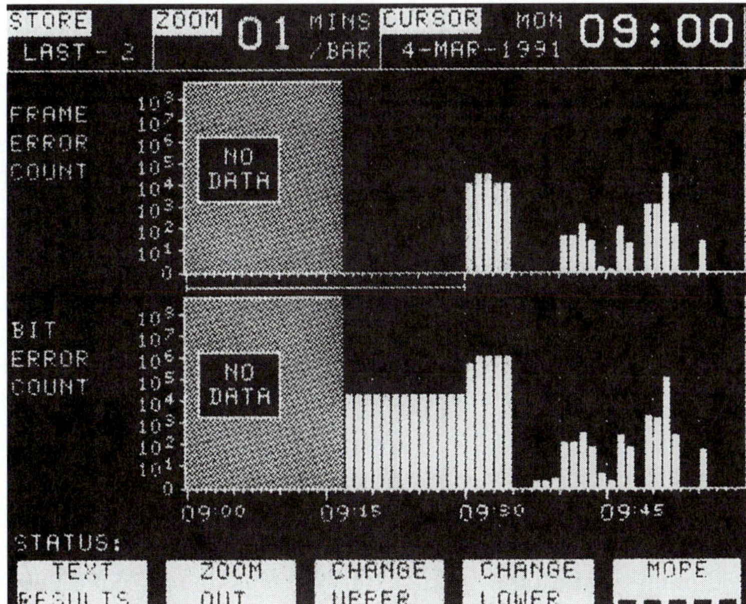

Figure 31.15 An additional feature on some BER test sets is the graphic display of results. In this example, a histogram display shows how bit error count and frame error count are varying with time.

Note that the results are time stamped, allowing error/alarm events to be correlated with system malfunctions.

31.4 Bit Error Rate (BER) Instrument Architecture

A BER tester consists of a pattern generator and error detector, often combined in a single instrument, though sometimes separate. Applications of BER testers can be divided into two broad categories:

- Telecommunications testing
- General laboratory/production testing of devices and systems

In both cases the measurements and modes of operation are similar; the main difference is in the electrical interfaces required.

If the BERT is used for testing telecommunications equipment and systems either in the field or in the factory, it must have coded interfaces according to the CCITT Recommendation G.703 (or the equivalent North American ANSI T1 standards). As mentioned earlier in this chapter, the G.703 interface specification defines digital codes (such as HDB3, B3ZS, and CMI) which ensure that a minimum number of transitions occurs in the data stream even with long runs of ones or zeros. This characteristic ensures that a timing clock can be reliably recovered in the receiving equipment. A single output or input connection is all that is required when connecting the test set, and the user does not need to be concerned about the relative phasing of parallel clock and data signals. G.703 also defines other important interface characteristics such as specific bit rates (see Table 31.1), voltage levels, pulse shape, impedance, frame structure, and so on. Put simply, if the test set complies with the relevant interface standard, it is ready to connect to telecommunications equipment. This is also true for the new generation of synchronous optical network (SONET) equipment for which standards now exist up to 2.488 Gbits/s.

For more general use, conventional clock and binary NRZ (nonreturn to zero) data are necessary. With higher-speed application, clock/data phasing becomes critical, so the test set should have a means of adjusting the clock timing for optimal sampling of the eye diagram. Two examples of BER testers now follow to explain some of these concepts.

31.4.1 Telecommunications BER tester with coded interfaces

A pattern generator and an error detector for telecommunications applications are shown in the block diagrams in Figs. 31.16 and Fig. 31.17. This serial implementation is suitable for bit rates up to around 200 Mbits/s [typical maximum rates are 44.736 Mbits/s (DS3) in North America, and 139.264 Mbits/s outside North America].

Referring to Fig. 31.16, the PRBS and word generator is clocked either from

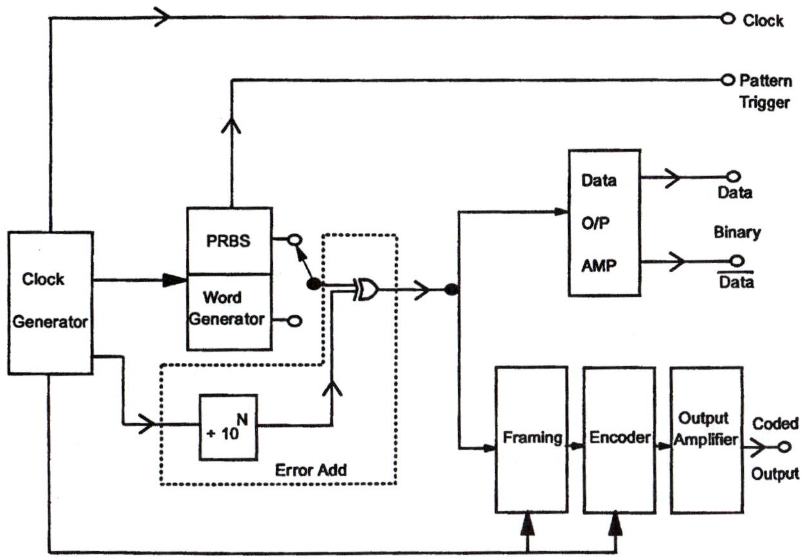

Figure 31.16 Pattern generator block diagram for a telecommunications BER test set.

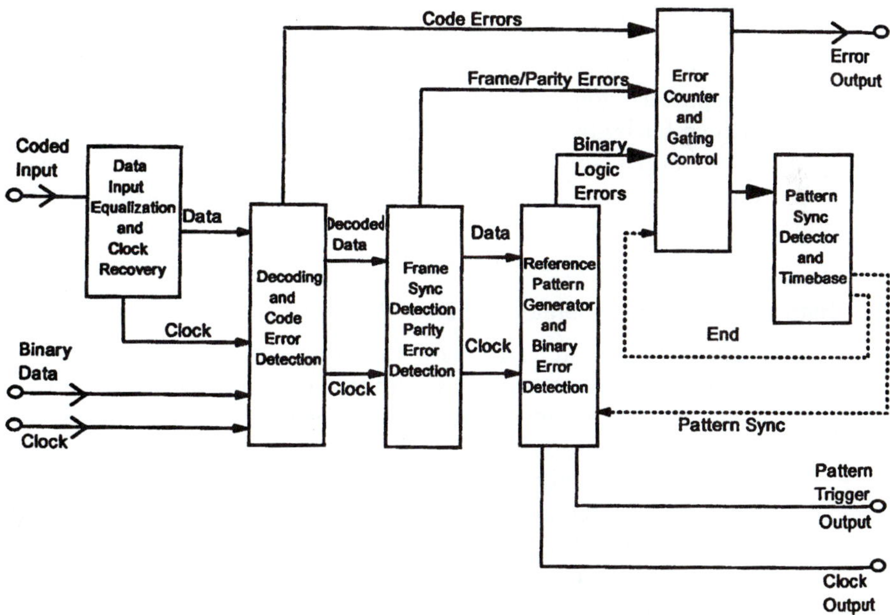

Figure 31.17 Error detector block diagram for a telecommunications BER test set.

a fixed-frequency clock source (to G.703) or from a synthesizer to provide a variable clock rate. Most telecommunications applications require a few specific clock frequencies and the ability to provide small offsets of ±15 to ±50 ppm. The PRBS and word generator circuit usually provides a trigger pulse to signify repetition of the pattern. The output from this circuit drives either the binary data output amplifier, which provides DATA and DATA (bar) with an accompanying clock signal, or the coded data output circuitry. This may add framing to the signal (a fairly standard requirement in North America but not essential elsewhere) and then adds the appropriate interface code for effective clock recovery. The output amplifier provides a signal conforming to the electrical interface specification, which may require a pseudo-ternary alternate-mark inversion signal. This means that alternate marks or ones are encoded as positive and negative pulses in order to provide a near-zero average dc level.

Error can be added to the pattern by an exclusive-OR gate which is controlled by single shot or repetitive pulses from the clock generator. The decade divider sets a BER of 10^{-N}.

The companion error detector, shown in Fig. 31.17, receives the standard coded-interface signal, recovers the clock, and strips off the coding to provide binary data and clock signals. In this process it detects any violations of the interface code algorithm and sends signals to the error counter. This provides the first level of in-service error detection.

For an instrument equipped to deal with framed signals, the receiver then locks onto any framing present, checks for frame errors, and decodes any embedded alarm signals, parity, or CRC bits, thus providing a further level of in-service measurement.

Finally, the binary data and clock are fed to the error detector and reference pattern generator (as described in Sec. 31.3.2), which checks the received pattern bit by bit for logic errors. A time base controls the measurement gating for single-shot, repetitive, and manual gating. The error counts accumulated are processed to provide BER and error performance analysis (see Sec. 31.3.3).

31.4.2 High-speed pattern generator and error detector

Figures 31.18 and 31.19 show the architecture for a 3-Gbit/s pattern generator and error detector. Because of the high bit rate, it is not practical to implement the PRBS and word generation directly in serial form. Instead, the patterns are generated as parallel 16-bit words at a maximum rate of 200 Mbits/s, where shift registers and high-capacity memory can be implemented using bipolar technology—the high-speed circuitry would normally use gallium arsenide ICs. The high-speed multiplexer (using a pyramid of single-pole double-throw switches shown in Fig. 31.20) converts the parallel data to a serial stream at rates up to 3 Gbits/s.

The clock input is generated by a frequency synthesizer. The clock output amplifier is driven through a fixed delay line, and the pattern generation and

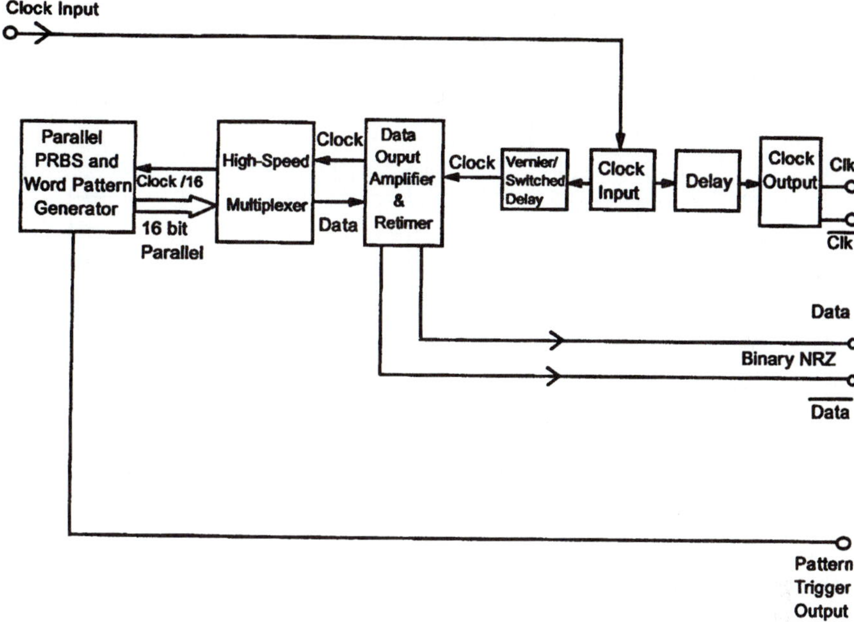

Figure 31.18 High-speed pattern generator block diagram.

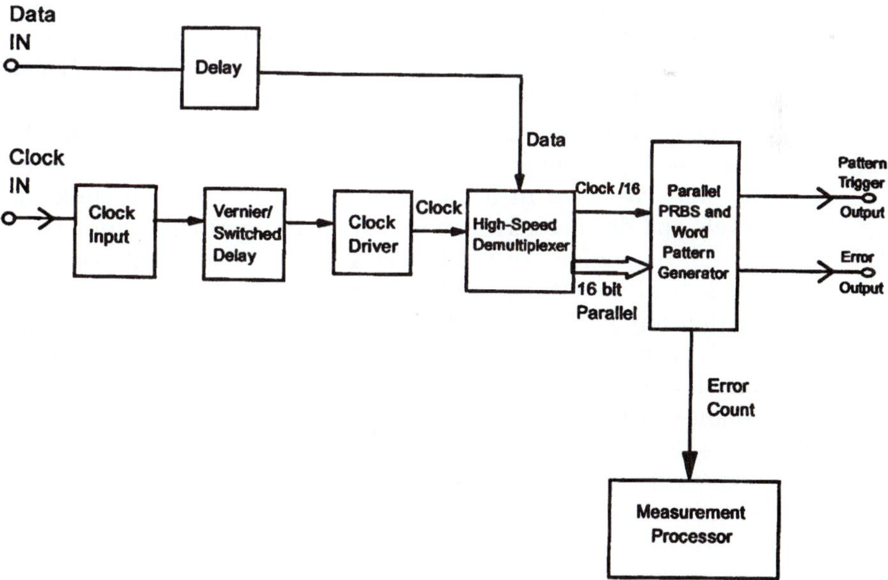

Figure 31.19 High-speed error detector block diagram.

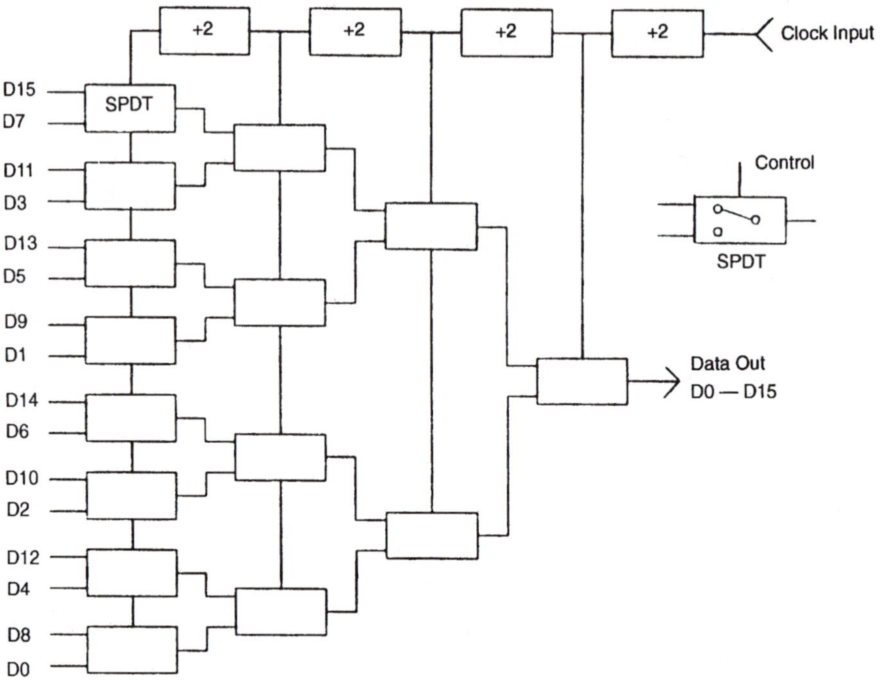

Figure 31.20 Pyramidal multiplexer configuration. Here, 16-bit data from the parallel gate array are fed to the first eight switches. The output of these switches is fed to the next four switches and so on until a single serial data stream is generated. The control of switch positions for each level is derived by dividing the clock rate by 2 at each stage.

output amplifier are clocked through switched and vernier delay circuitry so that clock/data phase can be varied both positively and negatively. The switched delays are 250, 500, and 1000 ps, while the vernier provides 0 to 250 ps in 1-ps increments.

The retimer associated with the output amplifer reclocks the data through a D-type flipflop to maintain minimum jitter. Since this type of test set would normally be used in a laboratory environment, the clock and data output levels and dc offsets can be varied for the particular application.

The companion error detector, shown in Fig. 31.19, uses a similar parallel implementation. Clock and data inputs pass through fixed and vernier delays so that any arbitrary clock/data phase can be adjusted for optimal sampling in the error detector. In fact, by adjusting the decision threshold and clock phasing under control of the internal processor, the error detector operating conditions can be optimized automatically. The high-speed demultiplexer converts the serial data stream to 16-bit parallel words along with a divide-by-sixteen clock signal. The parallel implementation reference pattern generator synchronizes with the incoming data and makes a bit-by-bit comparison. Any errors that occur are counted by one of two counters. One counter counts errors, while the other counter is being read in succession. The measurement processor provides error performance analysis down to 1-ms resolution.

31.5 Bit Error Rate Measuring Instrument Specifications

This section describes the major specifications of BER testers and the relevance for different applications, principally telecommunications testing and general-purpose device and system testing.

31.5.1 Patterns

The pattern generator and matching error detector will provide a range of PRBS patterns. The range might be quite limited for a dedicated tester aimed at maintenance of 1.5-Mbit/s (T1) or 2-Mbit/s (E1) links or may be wide for general-purpose applications.

For international application, the patterns should conform to CCITT Recommendations 0.151, 0.152, and 0.153, as shown in Table 31.3. For North American applications, you also may find the QRSS (quasi-random signal source), which is a $2^{20} - 1$ PRBS, but with the maximum run of zeros limited to 15. If the tester uses standard PRBS patterns, the user can be almost certain it will interwork with other models of BER tester, since the defining polynomials will be the same. When testing telecommunications links, one also can be sure of measuring them in accordance with the international standard in case of a dispute between vendor and customer.

For general laboratory use, a wider range of PRBS patterns may be useful. If not covered by the CCITT recommendations, it is usual to quote the polynomial used. For example,

$$2^{31} - 1 \quad D31 + D28 + 1 = 0$$
$$2^{10} - 1 \quad D10 + D7 + 1 = 0$$
$$2^{7} - 1 \quad D7 + D6 + 1 = 0$$

As mentioned in Sec. 31.3.1, for general-purpose applications, it is useful to be able to modify the PRBS pattern to add arbitrarily long runs of zeros or to vary the mark density, i.e., the ratio of ones to zeros.

Programmable user word patterns are becoming increasingly important as

TABLE 31.3 Test Pattern versus Bit Rate (0.151, 0.152, and 0.153)

Bit rate	Test pattern
50 bits/s to 168 kbits/s (0.153)	$2^9 - 1$, $2^{11} - 1$, $2^{20} - 1$
64 kbits/s (0.152)	$2^{11} - 1$
1.544 Mbits/s	$2^{15} - 1$, $2^{20} - 1$
2.048 Mbits/s	$2^{15} - 1$
8.448 Mbits/s	$2^{15} - 1$
34.368 Mbits/s	$2^{23} - 1$
44.736 Mbits/s	$2^{15} - 1$, $2^{20} - 1$
139.264 Mbits/s	$2^{23} - 1$

a means of simulating extreme data patterns found in live traffic and to simulate and capture frame structures. For maintenance of telecommunications networks, a few standard word patterns are all that is required. In North America it is now common to provide a range of word patterns in a T1 tester for 1.544 Mbits/s. These might include all 1s/0s, 1:1 pattern density, 1:7 density, 3 in 24, and 55-octet (8-bit) user word.

For general-purpose applications, particularly at high bit rates, very much larger pattern memories are quite common. For example, a 3-Gbit/s test set (HP 71603B) has a pattern memory of 4 Mbits so that complete frame structures for SONET/SDH systems and ATM systems can be simulated. With very large pattern memories, you should consider how they will be programmed, because it is impractical to program them by hand using the keyboard and display. Some test sets provide a floppy disk drive for downloading the word pattern. These disks can be programmed on a PC, and the test equipment manufacturer may well supply a disk with a range of standard patterns already stored.

When reviewing the large pattern memory capability, one should investigate how the memory is arranged. Very large memories will be arranged as parallel bit maps, which may limit the resolution the pattern can have. In other words, the stored pattern must be an integer number of smallest steps in the memory. Some frame structures may require the repetition of several frames in order to make up an integer number of steps. For example, one pattern generator offers 1-bit resolution for pattern lengths from 1 bit to 32 kbits, rising to 128-bit steps for pattern lengths of 2 to 4 Mbits. Using the 2- to 4-Mbit range, your pattern would have to be an exact multiple of 128 bits.

Finally, some pattern generators and error detectors allow switching between two stored patterns in order to investigate how a device under test responds to varying patterns. A gating input is usually provided to control the switching.

31.5.2 Clock and data inputs and outputs

The requirements in this area will depend on the intended application of the tester. For testing telecommunications equipment and systems, the main requirement is standard coded data interfaces (clock signals are not generally required). For each of the standard telecommunications bit rates, the interface should conform to a recognized telecommunications standard such as CCITT G.703 or ANSI T1.403 or ANSI T1.102/T1.105. The error detector also may have high-impedance inputs to allow bridging measurements, and the available sensitivity of the input should be able to handle levels of −20 to −30 dB below standard to cope with protected test points on equipment. For example, the input sensitivity might be quoted as +6 to −30 dBdsx, where dsx stands for digital crossconnect—the usual test reference point in a telephone exchange or central office. Unless specified otherwise, one can assume that the tester meets the standard jitter specifications at the interface (CCITT Recommendations G.823 and G.824), since this is included in G.703.

For general applications, more flexible and detailed specifications are needed. In most applications, both data and clock will be required at generator and error detector. One should check that the inputs and outputs can interface with standard logic levels such as ECL or TTL. Whereas telecommunications equipment normally operates with 75-Ω impedances, high-speed general circuit applications will probably require 50-Ω connections. A general-purpose test set will probably have programmable output levels and offsets, e.g., 0.25 to 2 V p-p with 10-mV resolution. Both data and inverted-data outputs are usually provided.

In high-speed testers, check the jitter specification on the clock and data outputs. This is usually specified in picoseconds (e.g., 10 ps rms) when transmitting a PRBS. Wave shape is also important and is defined by the transition times and overshoot. A poor wave shape from the pattern generator may limit the testing and hence the quality of development work.

One final point that is sometimes overlooked in digital systems is the return loss or impedance matching at the tester input and output. Compared with analog systems, return loss is not quite so important; however, anything less than 10- to 15-dB return loss may cause problems with transmission-line reflections and degraded pulse shape. The G.703 interface specification gives recommended return loss values for telecommunication systems.

31.5.3 Framing

This is mainly applicable to telecommunications equipment testing. As with coded-data interfaces, framing is governed by the appropriate standard such as CCITT Recommendations G.704 and G.706. When framing is included in the tester, a wider range of measurements is possible, including in-service monitoring. In some cases with equipment in North America, framing at T1 (1.544 Mbits/s) and DS3 (44.736 Mbits/s) is mandatory, as it is also with the new generation of SONET/SDH transmission equipment.

General-purpose testers with large pattern memories also can provide frame simulation. However, the extra flexibility provided by this approach is not required in dedicated applications.

31.5.4 Error insertion

This facility provided in the pattern generator allows single errors to be inserted in a PRBS or word pattern or a specified BER to be generated. This is a useful check of the error detector, but of course, errors in a pseudo-random data stream will not be detected by the equipment under test. If, however, the data stream is structured either by using the pattern memory or by built-in frame capability, then errors inserted in the frame will be detected by the equipment under test.

This is one of the advantages of a framed tester for telecommunications

TABLE 31.4 Clock Tolerance at Hierarchical Interfaces

Clock rate	Tolerance
PDH	
64 kbits/s	±100 ppm
1.544 Mbits/s (DS-1)	±50 ppm
2.048 Mbits/s (E1)	±50 ppm
8.448 Mbits/s	±30 ppm
34.368 Mbits/s	±20 ppm
44.736 Mbits/s (DS-3)	±20 ppm
139.264 Mbits/s	±15 ppm
SONET/SDH	
51.84 Mbits/s (STS-1)	<4.6 ppm
155.52 Mbits/s (STS-3, STM-1)	<4.6 ppm
622.08 Mbits/s (STS-12, STM-3)	<4.6 ppm

applications, since it allows the built-in alarms and monitors on the equipment under test to be checked. The error detector should have appropriate measurements such as code and frame errors in addition to standard logic errors in data.

31.5.5 Clock rates

For telecommunications applications, a few standard bit rates are all that is required. Table 31.4 shows the most common rates and the required clock accuracy to meet the interface specifications.

For general applications, a synthesized clock source is more useful. Sometimes this is built-in; sometimes an external synthesizer is required, in which case a clock input is specified on the pattern generator. The clock input should be compatible with the synthesizer output, which will probably be a sine wave. For best jitter performance, the synthesizer should have low single-sideband phase noise—e.g., −130 to −140 dBc/Hz.

31.5.6 Error measurements

All test sets will provide the basic error measurements such as error count and bit error ratio in a timed gating period. If the test set operates only with PRBS or word patterns, then the errors detected will be simply logic errors from bit-by-bit comparison with the reference pattern generator. If the tester has framed data capability, then errors also can be detected in frame words and in coding. The range of error types might include the following:

Out-of-service measurements

Logic errors or bit errors (binary reference pattern comparison)

In-service measurements

Frame errors

CRC4 or CRC6 code errors

Remote end block errors (REBE) or far end block errors (FEBE)

Parity errors

Interface code errors or bipolar violations

As mentioned in Sec. 31.4.1, an error detector capable of operating with framed signals can derive one or more of the in-service measurements listed above. These may or may not be important in your application, but increasingly, measurements on operational telecommunications equipment require this facility. Frame errors are detected by checking for any errored bits in the periodic frame alignment signal (FAS). Cyclic redundancy checksum (CRC4 and CRC6) is calculated on blocks of live data, and the remainder is transmitted as part of the frame structure. At the receiving end, the error detector decodes these data and compares them with the locally calculated CRC remainder. A discrepancy indicates one or more errors in the data block. This is a powerful method of in-service error checking which is becoming universally accepted internationally (CRC4) and in North America (CRC6). In North American DS3 systems and in the new generation of SONET equipment, a similar in-service check is provided by parity error checks on the data bits in the frame. A further enhancement is REBE or FEBE, whereby block error information detected by CRC or parity is sent back to the transmitting end by means of the frame structure. Lastly, the interface code itself can be checked for errors.

Review the specification of the error detector for how many error types are available and for compliance with the international or North American standards.

Error analysis will normally be included based on CCITT Recommendation G.821, described in Sec. 31.3.3. This is based on the 1-s error-free interval. Some test sets also provide finer resolution of the interval down to 1 ms, which can be useful in research work and field trials. Some test sets incorporate G.821 Annex D additions and possibly analysis according to the new CCITT Recommendation M.550 for maintenance of telecommunications links.

Lastly, look at the data-logging capabilities provided. Long-term tests result in a lot of data, so a means for storing them and displaying them graphically can be a valuable facility.

Chapter

32

Microwave Passive Devices

Frank K. David

Agilent Technologies
Santa Rosa, California

32.1 Introduction

The range of frequencies defined as "microwave" is 3 to 30 GHz, where a "gigahertz" (GHz) is 1 billion (1×10^9) cycles per second. The frequency range below this is called "radio frequency" (rf) and spans 300 kHz to 3 GHz; the range above it is called "millimeter wave" and spans 30 to 300 GHz.

The definition of a microwave device is not set by the frequency range of use; it is set by the design techniques used to create it and the manner in which it is applied. Microwave devices have been created and used in the rf range as low as 200 MHz and in the millimeter frequency range all the way to 300 GHz.

Microwave passive devices are used inside microwave measurement instruments, and they are used to combine instruments to create more complex measurement systems. In all cases, these devices will split, combine, filter, attenuate, and/or shift the phase of a microwave signal as it propagates through a particular transmission system. This chapter describes the most useful and prevalent transmission systems for microwave signals and the passive devices that are most effectively achieved in each system.

32.2 Coaxial Transmission Lines

A "coaxial transmission line," or "coaxial cable," is used to transport electromagnetic power from one place to another (one device to another) with minimum loss of power and maximum isolation from other electromagnetic signals that may be present.

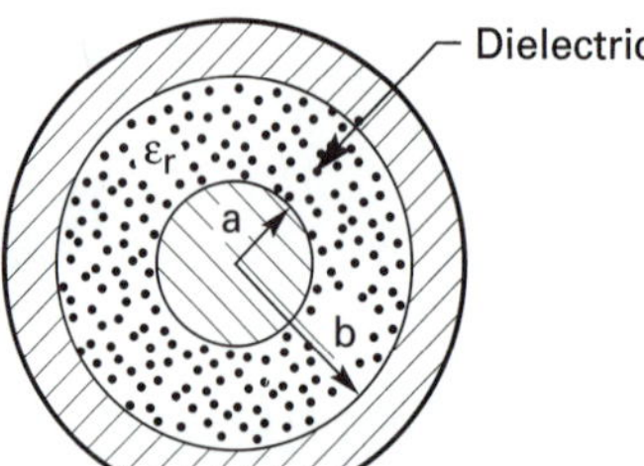

Figure 32.1 Coaxial line cross section.

32.2.1 Physical characteristics

A coaxial cable is the concentric arrangement of two cylindrical metal conductors separated by a dielectric that fills the space between the metal surfaces; a cross section is shown in Fig. 32.1. In cases where extremely low loss or very high power handling is required, the dielectric between the conductors is air, or a pressurized inert gas, with periodic ceramic or plastic supports to maintain alignment of the center and outer conductors. However, for most coaxial cables, flexibility is important, and this is achieved by using a dielectric that is a low-loss plastic material that supports the center conductor in a continuous manner the entire length of the cable, a center conductor comprised of many strands of a smaller-diameter wire, and an outer conductor that is a braid of very small-diameter wire or a thin foil of metal. For protection, and to give increased ruggedness to the cable, a nonconductive material, called a "jacket," covers the outer conductor.

32.2.2 Electrical characteristics

The electric and magnetic fields, designated by E and H, respectively, exist in the dielectric between the center and outer conductors, as shown in the cross section in Fig. 32.2. This field pattern, called a "mode of propagation," has all E and H lines lying in a plane that is transverse to the direction of propagation. For example, in Fig. 32.2, the field lines lie in the plane of the paper, and the power would be propagating perpendicularly out of the paper. This is called the "transverse electromagnetic mode" (TEM), and it has the useful characteristics that the phase velocity and characteristic impedance

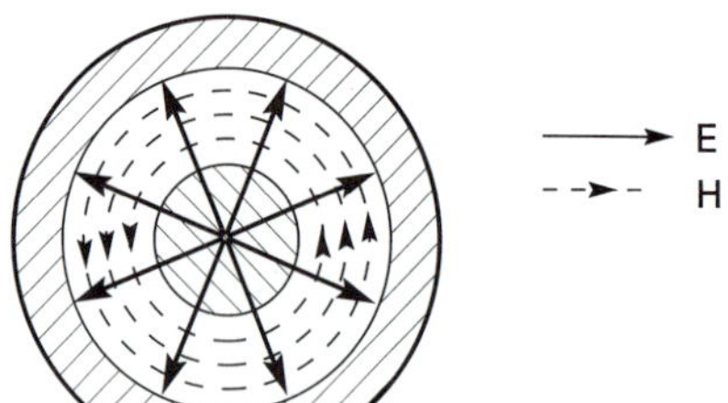

Figure 32.2 Coaxial line field patterns.

are independent of frequency from virtually direct current to the onset of the first higher-order (non-TEM) mode on a coaxial line.

The phase velocity for a TEM mode is given by

$$v = c/\sqrt{\varepsilon_r} \tag{32.1}$$

where ε_r = relative permittivity of the cable's dielectric
$\varepsilon_r = \varepsilon/\varepsilon_0$ = dielectric permittivity/permittivity of free space
c = velocity of light in free space

Some other related quantities are

λ = wavelength

$\lambda = v/F$ where frequency F is in hertz (cycles per second)

θ = the electrical length of a line of physical length L

$\theta = 2\pi F\sqrt{\varepsilon_r}L/c$ radians

When the electrical length of a line is 2π radians at a particular frequency, it is one wavelength long.

In a TEM transmission system, signals of all frequencies propagate along the line at the same velocity. This is an important characteristic for a pulse (which is comprised of many harmonically related frequencies), which will propagate along the line without distortion of its rising or falling edges. The characteristic impedance is given by

$$Z_0 = E/H = V/I = (60/\sqrt{\varepsilon_r})\ln(b/a)\,\text{ohms} \tag{32.2}$$

where V = voltage difference between inner and outer conductors
I = current flowing on the inner (or equally on the outer) conductor
a = radius of the inner conductor
b = radius of the inner surface of the outer conductor

This is one of the nice features of a TEM transmission line: the ratio of the E and H fields that propagate together in one direction (the definition of a transmission line's characteristic impedance) is the same as the ratio of the voltage and current, and these ratios are related to the geometry of the line's cross section.

32.2.3 Power handling and loss

There are two factors that will limit how much power can propagate along a coaxial line: resistive heating of the conductors and arcing between the inner and outer conductors. The resistive heating is always concentrated in the inner conductor because it has substantially less surface area than the outer conductor, and for all conductors at rf and microwave frequencies, the current is concentrated on the surface due to a phenomenon called "skin effect" (dis-

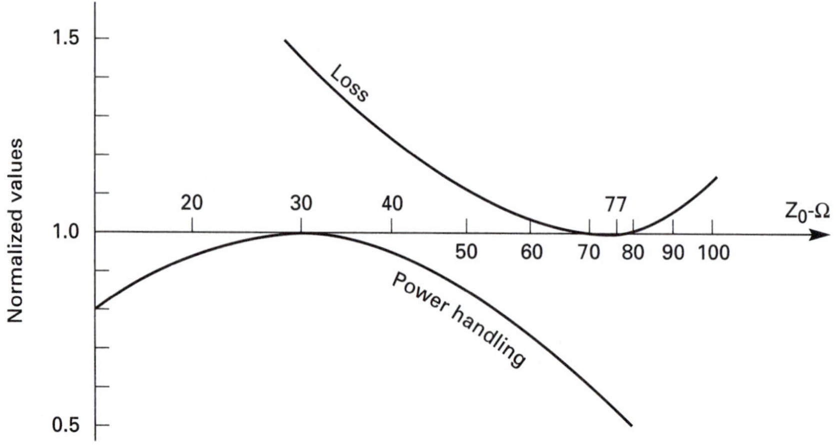

Figure 32.3 Power and loss versus Z_0.

cussed in the section on waveguides). The solution for reducing the resistive heating is to increase the surface area of the inner conductor by increasing its diameter. Doing this will lower the Z_0 of the line, which is not a problem by itself, but it will move the center conductor closer to the outer conductor, and this increases the electric field strength between the conductors, which can cause dielectric breakdown and arcing. It would seem that there might be some ratio of outer to inner conductor that would balance the center conductor heating and dielectric breakdown and give maximum power-handling capability. There is; it is the ratio of b/a that gives a Z_0 of 30 Ω.

The power loss along a coaxial line is a function of the resistive loss in the conductors, loss in the dielectric, and the relative value of these loss terms compared with the Z_0 of the line. In the vast majority of cases, the loss in the dielectric is very small compared with the resistive loss, and with this being the case, it can be calculated that the coaxial geometry that gives the minimum loss per unit length gives a Z_0 of 77 Ω.

The relative variation of power handling and loss versus Z_0 is shown in Fig. 32.3, and this figure shows that the "industry standard" of 50 Ω is a compromise between maximum power-handling capability and minimum loss. A commonly used impedance level, such as 50 Ω, allows for some standardization of cable sizes and connectors to join cables or insert other devices into the coaxial transmission system. This is a necessary feature to facilitate the interconnection of devices and/or instruments.

32.3 Coaxial Passive Devices

A great number of devices have been realized in coaxial structures, but the physical constraints make some devices very easy to realize and others very difficult. This section will examine some of the devices that are relatively

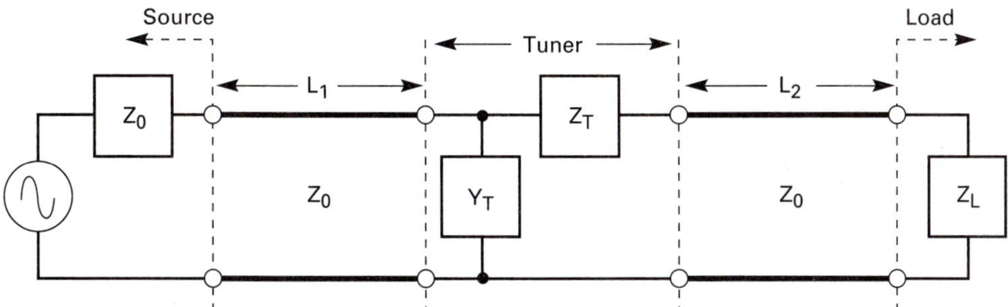

Figure 32.4 Load matching with a tuner.

easy to realize and that perform very well in a coaxial structure, and because of this, they are often used in or with measurement instruments.

32.3.1 Tuners

A "tuner" is a device that can be inserted into a transmission line (usually via connection with coaxial connectors) between a source and a load to provide a way to "tune out" any reflection from the load and hence allow maximum power transfer from the source to the load-the term "load" always refers to some physical component that absorbs or radiates the power generated by the source. A typical connection of a tuner is shown in Fig. 32.4.

As seen from the location of the source, the tuner creates a reflection in the transmission path that cancels (compensates) the reflection of the load. The tuner is properly adjusted when no power is reflected back into the source. Once a tuner is adjusted to provide a match, the frequency cannot change by more than 2 to 3 percent before the tuner needs to be readjusted; thus tuners are used in narrow-band applications.

As shown in Fig. 32.4, the tuner can be represented by a shunt admittance Y_T. This admittance has only a reactive element which is called "susceptance" (it has no loss; it only stores electrical or magnetic energy), and this element is adjustable—this is the "tuning."

However, this single element called a "single-stub tuner" cannot match every load unless the distance L_2 shown in Fig. 32.4 can be adjusted, and the most practical way to do this is to use a slide-screw tuner, as shown in Fig. 32.5. In this case, the metal conductor that functions as the center conductor of the shunt connection is not connected directly to the main line's center conductor; it can be positioned very close, capacitively coupling it to the center conductor, as shown in Fig. 32.6, which is a cross section through the micrometer-adjusted stub. The mechanical tuning of this capacitive coupling has the effect of changing the value of Y_T. This adjustment, and the adjustment of sliding the stub along the main line to adjust L_2, allows this single-stub tuner to match any load impedance.

The main transmission line for the slide-screw tuner is a variation of a

Figure 32.5 Slide-screw tuner.

coaxial line; it is often referred to as a "slab line" (see Fig. 32.6). Even though the outer conductor is not concentric with the center conductor and the "slabs" are open at the ends (far from the center conductor), this line is very nearly TEM in performance and very convenient for introducing a shunt element that can be moved along the line.

32.3.2 Filters

A "filter" is a device that will allow one range of frequencies to pass through it and will block another range. Four categories are shown in Fig. 32.7: low-pass, high-pass, bandpass, and bandstop. The frequency range that passes through the filter is called the "passband," and the range that is blocked is the "stopband."

Ideally, all the power would pass through the filter in the passband and none of the power would pass through in the stopband. In practical filters,

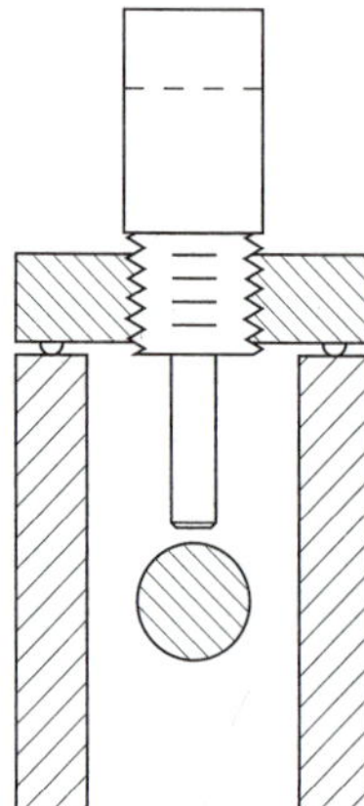

Figure 32.6 Slide-screw tuner: cross section through micrometer-adjusted stub.

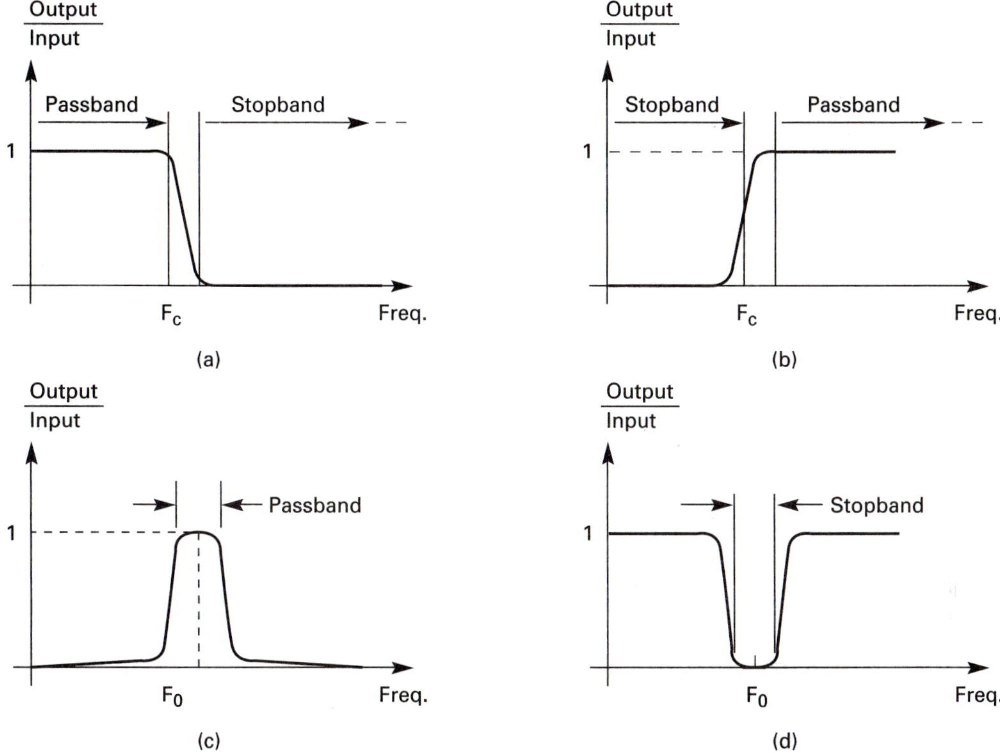

Figure 32.7 Filter response. (*a*) Low-pass; (*b*) high-pass; (*c*) bandpass; (*d*) bandstop.

between 80 and 95 percent of the input power shows up at the output in the passband (the power loss is usually due to conductor loss in the filter elements), and some small amount shows up at output in the stopband.

When the filter blocks the power from propagating through it, it does not absorb the power; rather, it reflects it back to the source. Thus the filter's transmission characteristics are a result of its reflection characteristics.

In coaxial structures, there are four ways to create a low-pass or bandpass filter: coupled lines, lumped-element, stepped-impedance, and semi-lumped-element. Coupled lines can create bandpass filters with very low passband loss and very rapid (with frequency) transitions from passband to stopband, but they are very bulky, relatively expensive, and somewhat fragile. Lumped-element filters can be very small, but they are limited to lower frequency ranges, and some of the elements are very difficult to create and integrate into the coaxial structure. The remaining two approaches lend themselves to coaxial implementations that are easy and cheap and which work almost exactly as predicted. Cross sections through the filter in the direction of propagation are shown in Fig. 32.8*a* and *b* for the stepped-impedance and the semi-lumped-element filters, respectively. Both types are comprised of alternating

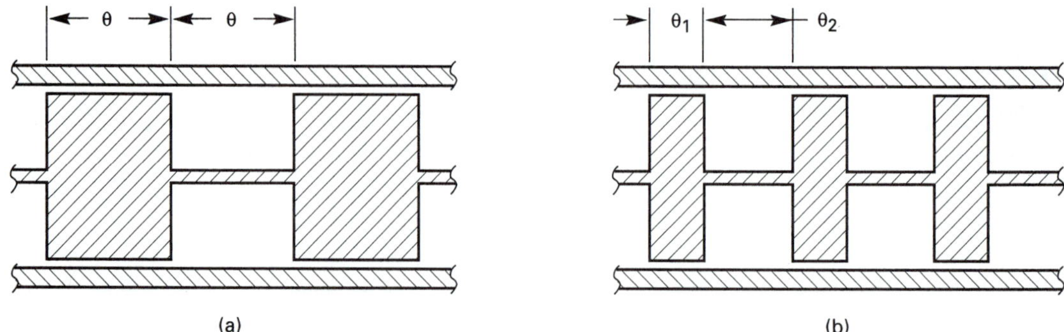

Figure 32.8 Filter cross section. (*a*) Stepped-impedance; (*b*) semi-lumped-element.

low- and high-impedance sections of coaxial line, and their power transmission versus frequency is shown in Fig. 32.9.

The stepped-impedance filter is also called a "half-wave filter" because the electrical length of the sections is one-half wavelength at the center frequency of the passband. As Fig. 32.9*a* shows, this can be used for a low-pass filter, but it performs much better as a bandpass filter when $\theta = \pi$.

The semi-lumped-element filter has electrical lengths for both the high- and low-impedance sections that are shorter than one-eighth wavelength, or $\theta < \pi/4$; it uses these short lengths of line to approximate lumped elements. The low-impedance line approximates a shunt capacitor, and the high-impedance line approximates a series inductor. This makes an excellent low-pass filter, where the first spurious response is at five times the low-pass cutoff frequency, as shown in Fig. 32.9*b*.

32.3.3 Couplers

When two transmission lines are brought together such that the two center conductors share the same outer conductor, or "ground plane," power flow on one transmission line will transfer, or couple, some power onto the other

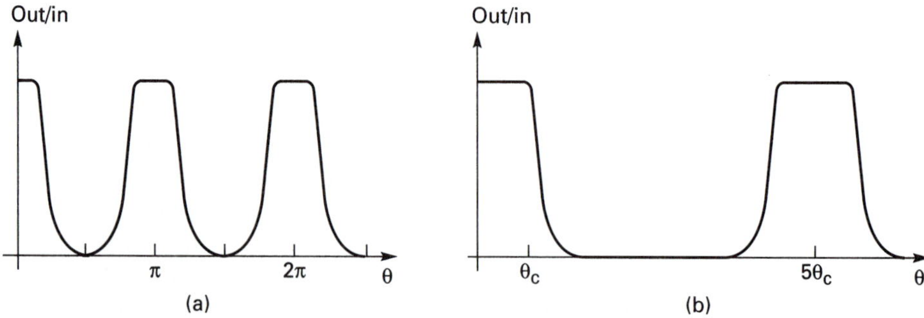

Figure 32.9 Filter passband/stopband response. (*a*) Stepped-impedance; (*b*) semi-lumped-element.

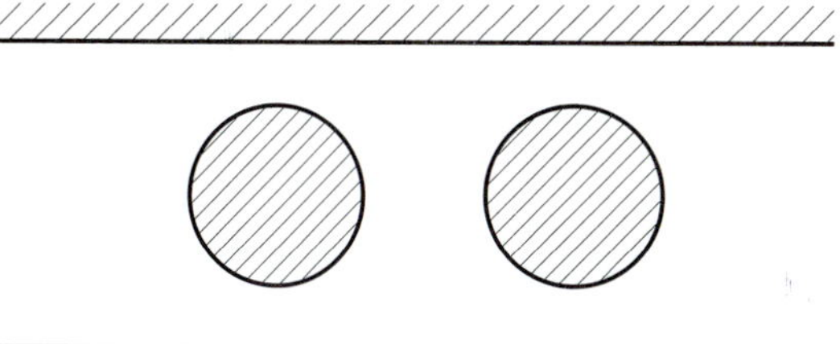

Figure 32.10 Coupled-line cross section.

transmission line. A cross section of two coupled lines is shown in Fig. 32.10. This coupling comes about because the presence of electric charge on one center conductor will induce (opposite) charge on the other center conductor, and the movement of charge, or current, on one center conductor will induce the (opposite) movement of charge on the other center conductor. One structure used for coaxial directional couplers is the "slab line" (described in the section on tuners). This arrangement of cylindrical (or sometimes rectangular) conductors between parallel plates is very convenient for positioning center conductors close to each other to allow power to couple from one to the other.

A plan view of the two center conductors of a coupled section is shown in Fig. 32.11. The numbers 1 through 4 in the figure refer to "ports." For a TEM structure, a "port" is a coaxial connector or even an imaginary reference plane perpendicular to the flow of power along two conductors (two wires, the inner and outer conductors for a coaxial cable, the center conductor and the two slabs of a slab line, etc.). Power can flow into or out of a port, or it can do both simultaneously.

In addition to coupling, or tapping off, of power from one transmission line to another, couplers display a very useful characteristic called "directivity," where the coupled power exits at one coupled port but not the other. Refer-

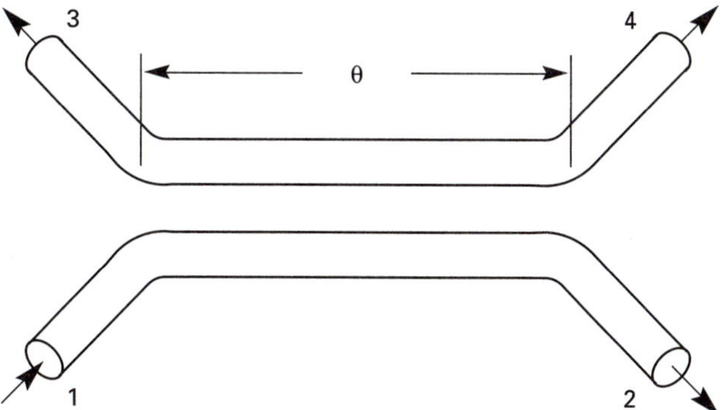

Figure 32.11 Coupled-line plan view.

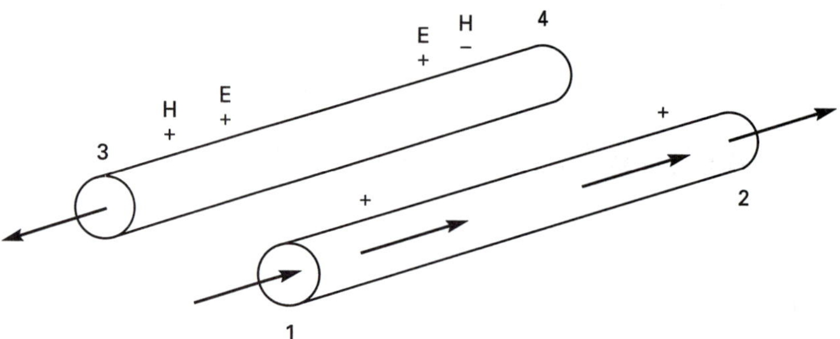

Figure 32.12 *E*- and *H*-field coupling between two short sections of line.

ring to Fig. 32.11, the power flowing into port 1 and along the main arm of the coupler 1–2 couples to the coupled arm 3–4 such that all the coupled power exits at port 3 and none at port 4; port 4 is "isolated." Similarly, power into port 2 will cause all the coupled power to exit at port 4 and none at port 3. If there is a reflection on the main line such that power is simultaneously flowing into and out of port 1, the coupled power that shows up at port 3 is still entirely due to the power flowing into port 1. This ability to direct power in the coupled arm based on the direction of propagation of power in the main arm is what makes this device so useful. In real couplers this isolation is not infinite; some power shows up at the isolated port.

The significant parameters that describe a coupler are its coupling factor, directivity, and power split; for power input at port 1,

$$\text{Coupling factor}\, k = -10 \times \log\left(\frac{\text{power out 3}}{\text{power into 1}}\right) \text{dB} \tag{32.3}$$

$$\text{Directivity}\, D = -10 \times \log\left(\frac{\text{power out 4}}{\text{power out 3}}\right) \text{dB} \tag{32.4}$$

$$\text{Power out port 2} = (\text{power in port 1}) - (\text{power out port 3}) \tag{32.5}$$

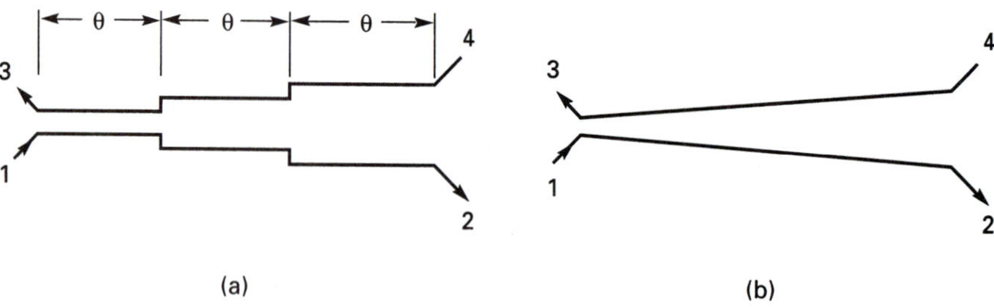

Figure 32.13 Broadband couplers. (*a*) Multisection; (*b*) tapered.

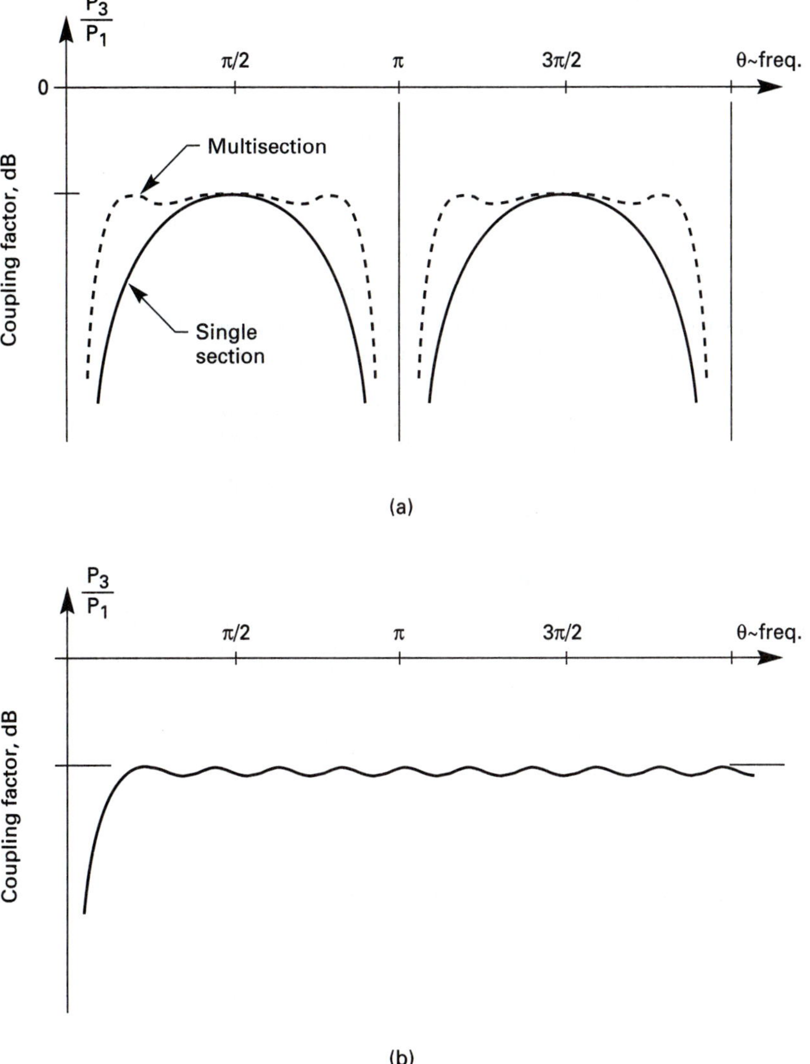

Figure 32.14 Coupling versus frequency. (*a*) Multisection; (*b*) tapered.

The coupling factor is primarily a function of how close the lines are to each other (with respect to how close they are to the ground planes); the closer they are, the stronger (the tighter) is the coupling. In practical couplers, this can range from a tight coupling of 3 dB to a loose coupling of 30 dB. The directivity is not related to the coupling; it is inherent in the propagating wave phenomenon that transfers power from one conductor to another. A good coaxial instrumentation coupler can achieve directivities of 30 to 40 dB over a wide frequency range.

The propagating wave phenomenon that causes directivity is due to the fact that power is transferred from one conductor to another by a combination of coupling from the E and H fields. A very short section of coupled lines is shown in Fig. 32.12, where the power enters at port 1. The E field on conductor 1–2 induces a voltage on conductor 3–4 that is symmetrical at port 3 and 4. The H field from conductor 1–2 induces a voltage on conductor 3–4 that is asymmetrical at ports 3 and 4. The net result is that the induced voltages sum at port 3 and cancel at port 4; this is shown by the "plus or minus" notation for the E- and H-induced voltages on conductor 1–2. Note that for the coupled lines, the main-line power comes in at port 1 and the coupled power goes out from port 3; this is typical for TEM couplers, and it is why they are sometimes referred to as "backward wave" couplers (as opposed to waveguide couplers that are "forward wave" couplers).

For a single-section coupler, the coupling coefficient, as defined above, varies with frequency; it is a maximum at a frequency where the electrical length of the coupled section is one-quarter wavelength; for lower or higher frequencies, the coupling becomes less. The frequency range over which the coupling is acceptably close to the maximum is called the coupler's "bandwidth." A single-section coupler has a half-power bandwidth that covers a 2 : 1 frequency range (one octave). This bandwidth is acceptable for many applications, but some applications require the coupler to work over a 10 : 1 frequency range (one decade). To do this, the solution is to use multisection or tapered-section couplers, as shown in Fig. 32.13. The coupling versus frequency for the single- and multisection couplers is shown in Fig. 32.14*a*, and for the tapered section in Fig. 32.14*b*.

32.4 Planar Transmission Lines

Coaxial structures have low loss and TEM characteristics, but it is very difficult to integrate any active devices into them. The only success has been to place the diode or transistor chip in a package that can be inserted into a coaxial line. This nicely solves the physical part of integration, but the package adds significant parasitic inductances and capacitances such that the frequency performance of the final assembly is limited. With the goal of having a flat surface for attachment of active devices in chip form, it would seem that flattening the coaxial cable's center conductor and "unwrapping" its outer conductor to gain access to the flattened center conductor would be the solution. It is a good solution, and it has led to a class of "TEM-like" planar transmission lines.

32.4.1 Physical characteristics

There are many variations of planar lines; some of the more prevalent ones are shown in Fig. 32.15. The name for each line is, respectively, "stripline," "microstrip," "suspended microstrip," and "coplanar waveguide" (CPW). The

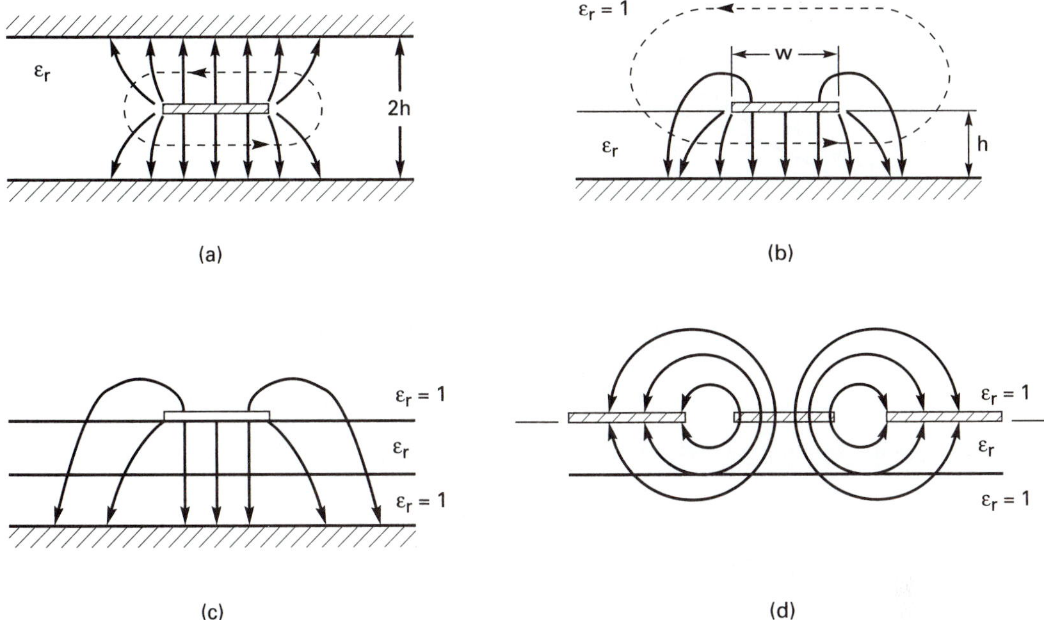

Figure 32.15 Planar line cross sections. (*a*) Stripline; (*b*) microstrip; (*c*) suspended microstrip; (*d*) coplanar waveguide.

figure also shows the *E*-field pattern and one line of the *H*-field pattern (for power propagating out of the plane of the paper).

In all cases, the "center conductor" (a carryover from coaxial terminology) is supported and separated from the ground plane by a low-loss dielectric. The center conductor is usually etched from a thin layer of copper laminated to the dielectric.

32.4.2 Electrical characteristics

Stripline is the most similar to coaxial cable; in fact, it is like slab-line cable with a flat center conductor and a uniform dielectric to contain all the field lines; this gives it true TEM characteristics. The drawback of this configuration is the lack of access to the center conductor and access from the center conductor to the ground planes.

Microstrip and suspended microstrip allow very easy center conductor access, but access from the center conductor to the ground plane can be difficult (but not impossible) if the dielectric is a hard, brittle material (like a ceramic). These configurations have field lines that go through the air and the dielectric; this creates a condition that does not allow true TEM propagation. However, the deviation is not severe, and these configurations are effectively analyzed and used as TEM up to a frequency of 30 GHz or more. This situation

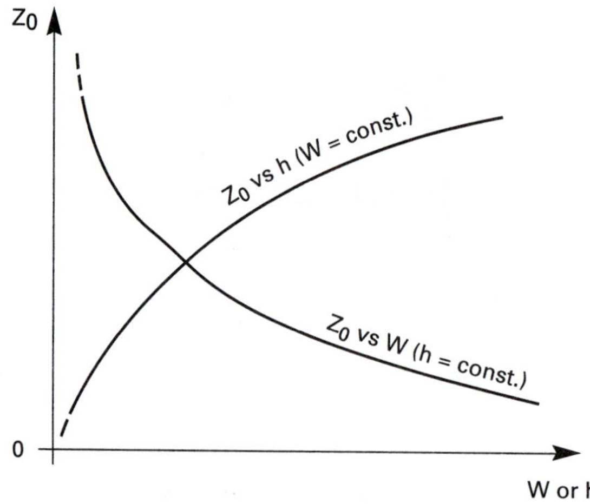

Figure 32.16 Z_0 versus line geometry.

leads to the label "quasi-TEM." Suspended microstrip, because most of the field path length is in one dielectric (air), is the closest to TEM of the two.

Coplanar waveguides allow complete access because all the conductors in this transmission system lie in the same plane, but they are the least TEM-like of the group. In addition to the problem of mixed dielectrics for the field lines to go through, there are field components that do not lie in the plane transverse to the direction of propagation. However, up to about 30 GHz, this configuration will perform as though it were TEM for all but the most precise components (narrow-band filters and tuning elements).

The expression for Z_0 for any of these planar structures is not a simple, closed-form mathematical expression, as it is for coaxial cable. Primarily, this is due to the flat center conductor having a very nonuniform distribution of charge across its width W, the infinite extent of the ground plane(s), and the mixed dielectric in the case of microstrip, suspended microstrip, and CPW. However, the relation of Z_0 to strip width W and ground plane spacing h has an expected relation, as shown qualitatively in Fig. 32.16.

The important feature to note on the curves for Z_0 versus W or h is the very slow change in Z_0 for the higher values of these parameters. This behavior places a practical limit of 20 to 130 Ω that can be achieved by these structures for dielectrics with $4 < \varepsilon_r < 10$.

32.4.3 Power handling and loss

Compared with coaxial cable, all these planar structures have reduced power-handling capability and higher loss. This is caused by the nonuniform charge distribution across the strip's width; most of the charge is concentrated on the edges. This concentration increases the effective resistance of the strip and the electric field strength at the edges, which causes dielectric breakdown and arcing at high power. Since these structures are best used for the integration of multiple functions in a small space, the high power limitation is not a problem. This loss becomes a problem when a section of line is used as part of a filter, but in all but the most demanding requirements, the loss is acceptable.

The quasi-TEM structures have another loss mechanism that results from their mixed dielectric. A small amount of power exists in a non-TEM due to the air-dielectric interface, and this power tends to radiate into the air as the frequency increases. This becomes a significant factor above 30 GHz.

32.5 Planar Passive Devices

Virtually every passive device that can be created in a coaxial structure can be created in a planar structure. In most cases (couplers being a notable exception), the planar devices work as well as the coaxial devices; in a few cases, they work better. In all cases, the planar devices are much easier and cheaper to construct, and they are much more rugged.

32.5.1 Directional couplers

A coupled-line directional coupler, similar to that described in the slab line, can be created easily in a planar structure. Figure 32.17 shows a cross section of the *E*-field patterns for the even and odd modes, respectively, for a microstrip coupler.

This coupler is small (due to the dielectric constant of the substrate), extremely rugged, and cheap to make. The only problem with a coupler in a quasi-TEM planar structure is the mixed dielectric. As Fig. 32.17 shows, the even and odd modes have field patterns that occupy different amounts of air and dielectric. This causes the two modes to have different velocities of propagation as they move along the coupled lines, and this degrades the directivity to the point that, over an octave or more in frequency, 20 dB is considered good performance.

32.5.2 Power splitters

All the power splitters described in this section could be and have been constructed in coaxial conductors, but this implementation is mechanically com-

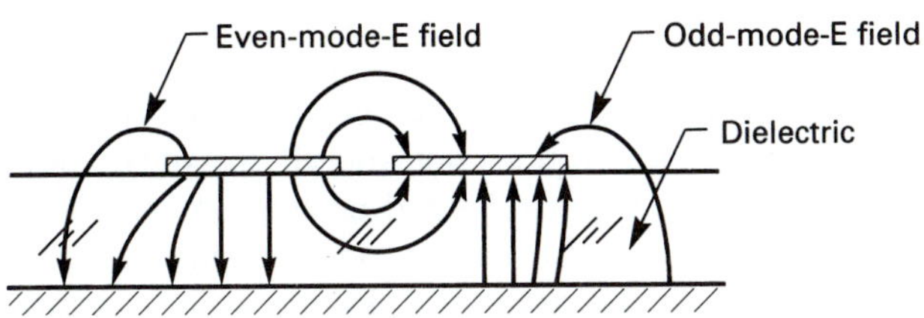

Figure 32.17 Planar coupled-line cross section.

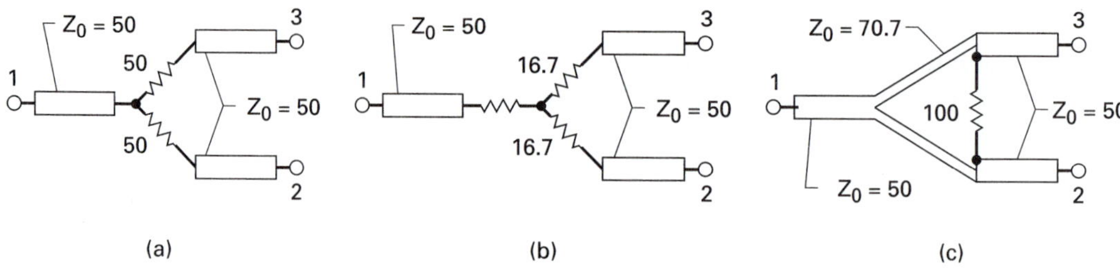

Figure 32.18 Equal phase power splitters. (*a*) Two-resistor; (*b*) three-resistor; (*c*) Wilkinson.

plex, fragile, and expensive, whereas the planar implementation is easy, rugged, and cheap. This useful device, which is rarely created with coxial conductors, is used frequently in planar structures.

All power splitters do what the name implies: Power comes in one port and splits to two (or more) output ports. The output ports can have equal or unequal amplitude and phase relative to each other; this section will describe the case of power splitting into two paths with equal amplitude and equal and unequal phase. Technically, a 3-dB coupler (described in the preceding section) meets the definition of a power splitter, but this section will be further limited to splitting power by splitting conductor paths, not by field coupling.

Power splitters with outputs of equal phase can be of two types: resistive, as shown in Fig. 32.18*a* and *b*, and "3-dB," or "Wilkinson" as it is often referred to after its inventor, in Fig. 32.18*c*. The figure shows a plan view of planar transmission lines connecting the input port (1) and the output ports (2 and 3) to the resistor values shown. With the exception of the 70.7-Ω line in Fig. 32.18*c*, the length of the lines is not important. The length of the resistors is very important; they must be less than 5 percent of a wavelength at the highest frequency of operation if they are to behave like a resistor and not a resistor plus an inductor.

The resistive splitters are exceedingly broadband; they can operate from direct current to a frequency where the transmission system has higher-order modes (this can cover several decades of frequency), and they have 6 dB of loss from the input to an output port. The two-resistor splitter is used when one output port is used for leveling the input signal or for a reference signal for a ratio measurement. The three-resistor splitter is used for general splitting of one signal or the combining of two signals. The Wilkinson splitter does not have any resistive loss as it splits the input signal; it has the minimum loss that a splitter can have—each output port is 3 dB below the input. The resistor connecting ports 2 and 3 does not absorb any of the power that comes in from port 1, but it does absorb reflected power that comes back in ports 2 or 3. This provides some isolation between one output port and another and between an output port and the input. To provide an impedance match for all the ports of the splitter (when the input and outputs are terminated in 50 Ω),

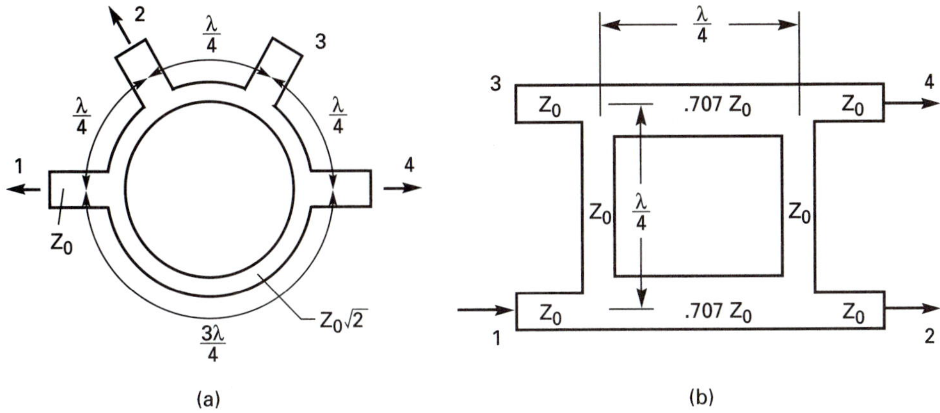

Figure 32.19 Hybrid junctions. (*a*) Rat-race; (*b*) branch-line.

the section of transmission line connecting the Y junction with the output ports is a quarter-wave impedance transformer that converts the 50 Ω of an output port to 100 Ω at the junction; thus the output ports appear as two 100-Ω impedances in parallel at the junction, and the input is then matched to 50 Ω. However, the impedance transformation is correct only at the frequency when the 70.7-Ω line is one-quarter wavelength long; it is still very usable for frequencies 20 percent below and above this frequency. For broader bandwidth applications, multisection impedance transformers are used between the Y junction and ports 2 and 3.

Power splitters with outputs of unequal phase (usually 180° out of phase) are referred to as "hybrid junctions." Figure 32.19 shows two implementations, called the "rat-race" and the "branch-line couplers," respectively. It is very easy to create these devices in any of the planar structures described.

The rat-race coupler has power entering at port 1, splitting evenly, and exiting at ports 2 and 4; port 3 is isolated. If ports 2 and 4 are terminated with 50 Ω, then port 1 is also matched to 50 Ω. The phase of the signal at port 2 is 180° different from that at port 4. The equal splitting of power and the phase difference are achieved only for a narrow range of frequencies, since the operation of this device is based on two signals combining at an output port with constructive or destructive (for the isolated port) phase delays, and these phase delays vary rapidly with frequency. This limits this device to 10 to 20 percent bandwidth.

The branch-line coupler has the characteristics of the rat-race coupler described above, except that a single section (as shown) has even less bandwidth than the rat-race coupler. The bandwidth can be extended by using a multisection version with three or more branches between the main lines.

There is one other power splitter that is harder to implement than the rat-race or the branch-line coupler, but it provides a very broadband 180° out-of-phase pair of outputs. It too is referred to as a "hybrid junction," and it is

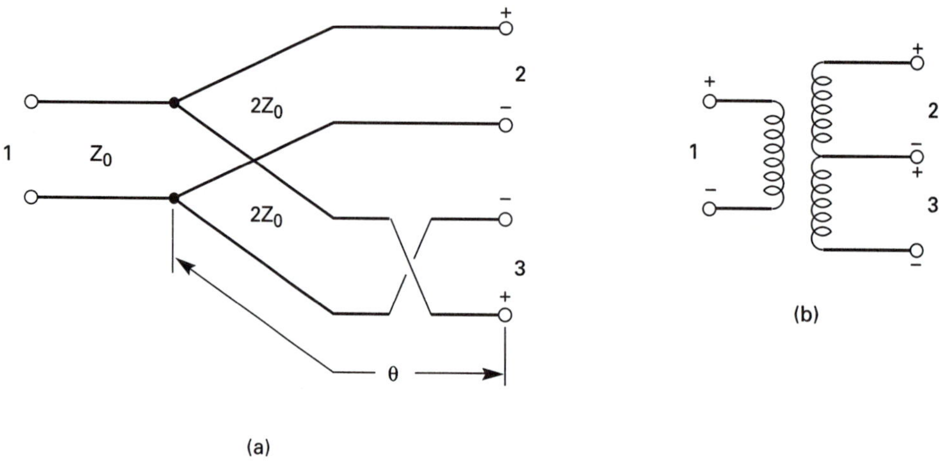

Figure 32.20 Broadband hybrid junction. (*a*) Schematic; (*b*) equivalent circuit.

shown in Fig. 32.20*a*. Swapping conductors in one arm gives a very good 180° phase shift as long as the crossover section is very short compared with a wavelength. For frequencies where $\pi/8 < \theta < 3\pi/8$, this junction can be represented by the equivalent circuit shown in Fig. 32.20*b*.

32.5.3 Filters

All the filter types that could be created in coaxial cable can be created in any of the planar structures: coupled lines, lumped-element, and stepped-impedance half-wave and semi-lumped-element lines. Unlike coaxial cable, coupled-line filters are easy, rugged, and cheap to construct in planar structures. A plan view of a coupled-line bandpass filter is shown in Fig. 32.21*a*, where half-wave, open-ended resonators couple to the input and output lines and each other with quarter-wave coupling sections. The quarter- and half-wave lengths occur at the center frequency of the filter. These filters have a passband response at frequencies when the resonators are odd multiples of a half-wavelength, as shown in Fig. 32.21*b*.

There is a form of a coupled-line filter that has some very useful characteristics; it is called a "directional filter," and it is shown in Fig. 32.22. The full-wave resonant ring in the middle is coupled to the input and output circuits through a quarter-wave directional coupler, and the input and output lines are terminated with an impedance equal to that of the transmission line.

At frequencies below the first resonance of the ring (the frequency where the electrical length of the ring is one wavelength), most of the input power goes into the input line's termination. Very little power is transferred into the ring because the coupling factor is low and the ring is nonresonant and does not transfer power to the output circuit. When the frequency reaches the resonant frequency of the ring, the coupling is most efficient, and the ring's

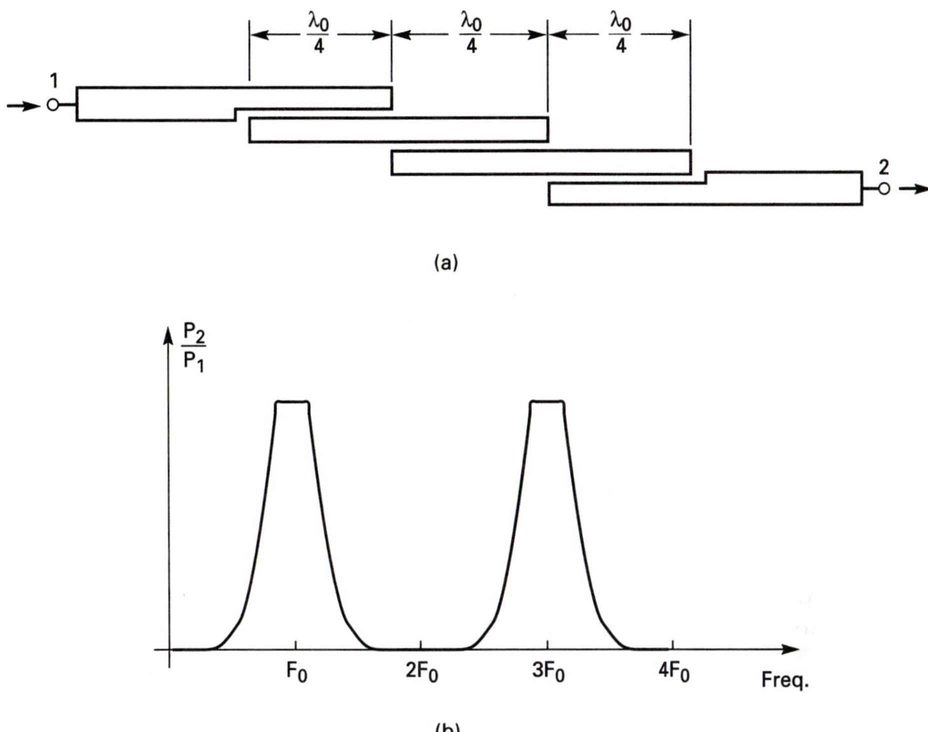

Figure 32.21 Coupled-line filter. (*a*) Plan view; (*b*) response versus frequency.

resonance transfers power to the output circuit with little loss. Because the output circuit is a directional coupler, the power goes to the output port and not the output line's termination. At frequencies just above resonance, the operation is the same as below resonance; the input power goes into the input line's termination and not the output port. At a frequency of twice the first

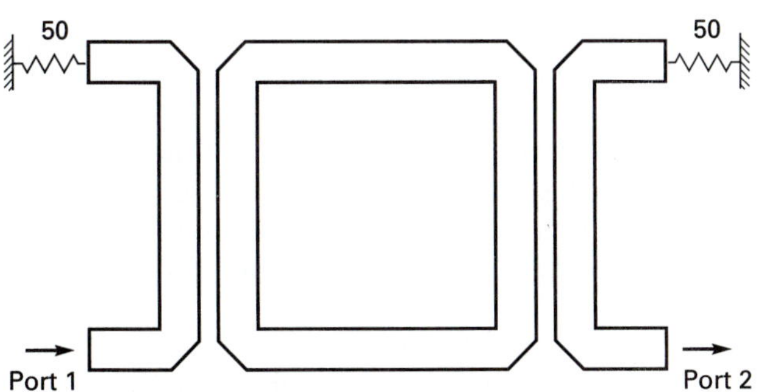

Figure 32.22 Directional filter.

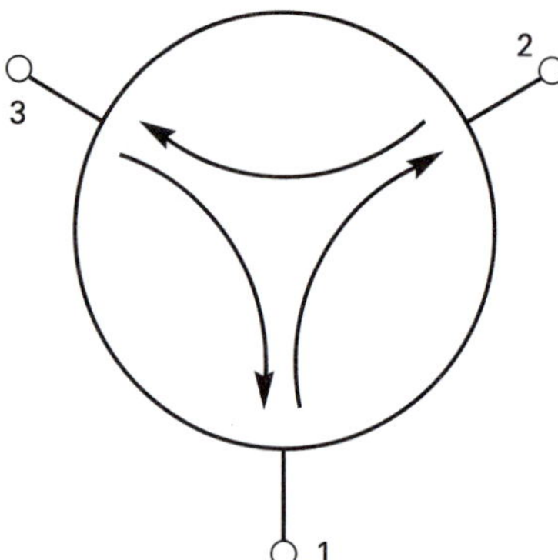

Figure 32.23 Circulator.

resonance of the ring, the ring will resonate and would transfer power with good efficiency, but the input and output couplers are one-half wavelength at this frequency, so they will not transfer power into or out of the ring; this helps in extending the stopband, giving the same response as shown in Fig. 32.21*b*. The unusual and useful feature of this filter is that it provides a match at both input and output ports for both the passband and the stopband; this can be contrasted to all the filters previously discussed, which reflected all the power back to the source for frequencies in the stopband.

32.5.4 Circulators and isolators

A "circulator" is a three-port device that is most readily implemented in a planar structure, and it differs from all the devices discussed so far in that it is "nonreciprocal"; i.e., the performance of the device is significantly different depending on whether power is flowing through it in the forward or reverse direction. As Fig. 32.23 shows, power flowing into port 1 goes to port 2 (and not to port 3), but power flowing into port 2 does not go back to port 1; it goes to port 3. This "magic" is caused by the presence of a ferrite material above and below the Y junction of the planar conductors from and to the three ports. When this material is magnetically biased by permanent magnets, it interacts with the propagating magnetic field to shift the field's direction and selectively route it from one port to another. This routing of power and the maintenance of an impedance match at all ports can be achieved over an octave of frequency.

An "isolator" is a circulator with an impedance-matched load (a termination) on one port; e.g., if port 3 was terminated, power from port 1 would go to port 2, and any reflections from port 2 would be dissipated in port 3's termination and never reach port 1. This effectively isolates port 2 from port 1.

There is an isolator that is not based on a circulator, and it is called a "peripheral-mode isolator." It is a two-port device that uses a magnetically biased ferrite to shift the forward-propagating magnetic field to one side of the transmission structure, where there is low loss, and to shift the reverse-propagating magnetic field to the other side of the structure, where there is high loss. This can give a forward-to-reverse power ratio of 20 dB or more, and it can operate over two to three octaves of frequency range.

32.6 Waveguides

Although any transmission line effectively guides electromagnetic energy, the term "waveguide" usually refers to hollow pipes used as transmission lines. The most commonly used waveguide has a rectangular cross section, as shown in Fig. 32.24*a*, but square and round guides are also used (Fig. 32.24*b* and *c*). For special applications, irregular shapes may be used, such as the ridged guide shown in Fig. 32.24*d*.

Waveguides are made of metal, and since the electromagnetic energy travels inside the pipe, they are shielded transmission lines. There is very little penetration into the surface of the metal because of the "skin effect" (this is the tendency of rf currents to flow nearer the surface of a conductor as the frequency is increased). The current density is greatest near the surface and decreases exponentially with depth into the metal. The "skin depth" is the depth of penetration at which the current density is $1/e$ of its value at the surface. The skin depth is proportional to the square root of the conductivity of the material and inversely proportional to the square root of the frequency. With an excellent conductor such as copper or silver, the skin depth at 10 MHz is less than a thousandth of an inch, and at microwave frequencies, it is negligible. Thus the wall thickness of a waveguide is determined by mechanical factors rather than electrical problems. Losses in the waveguide are a function of the conductivity of the surface metal, and therefore, the surface metal should be a good conductor. For considerations of economy, ease of fabrication, or light weight, aluminum and brass are commonly used for waveguides. These materials are then plated inside the tube with a few thousandths of an inch of silver or rhodium to increase the conductivity.

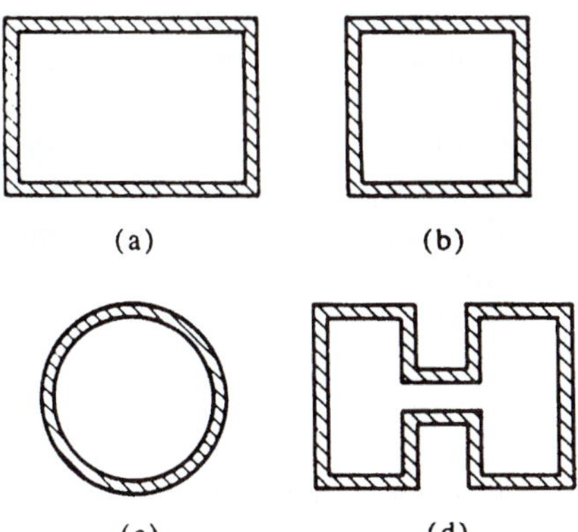

Figure 32.24 Waveguide cross sections. (*a*) Rectangular; (*b*) square; (*c*) round; (*d*) ridged.

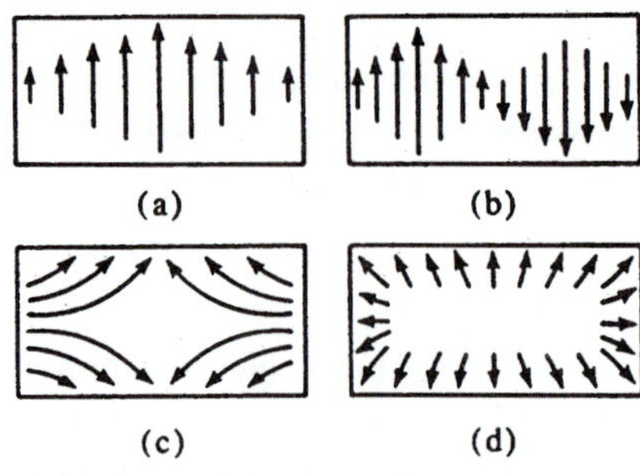

Figure 32.25 Modes in rectangular waveguides. (*a*) TE_{10}; (*b*) TE_{20}; (*c*) TE_{11}; (*d*) TM_{11}.

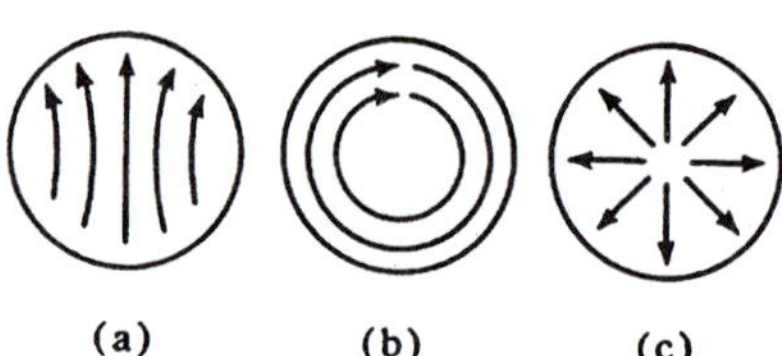

Figure 32.26 Modes in round waveguides. (*a*) TE_{11}; (*b*) TE_{01}; (*c*) TM_{01}.

32.6.1 Modes in waveguides

The electromagnetic wave in a waveguide consists of an electric vector and a magnetic vector which are perpendicular to each other. There are many different possible patterns or arrangements of these vectors, and these patterns are called "modes." In general, there are two kinds of modes. In one set of modes, called "transverse electric (TE) modes," the electric vector is *always* transverse or perpendicular to the direction of propagation. In the second set, "transverse magnetic (TM) modes," the magnetic vector is *always* transverse to the direction of propagation. In all modes, the electric field must always be perpendicular to the waveguide wall at the surface, and the magnetic field at the wall surface is always parallel to the wall.

The modes in rectangular waveguides are designed further by two subscripts. The first subscript indicates the number of half-wave variations of the *electric* field (in both TE and TM modes) across the wide dimension of the waveguide, and the second indicates the number of half-wave variations of the electric field across the narrow dimension. Some typical electric field patterns are shown in Fig. 32.25. In each case, the magnetic field is perpendicular to the electric field. The length of the arrow indicates the relative amplitude of the field at that point.

In round waveguides, two subscripts are also used to designate mode; the first is related to the number of circumferential variations in the field, and the second relates to the number of radial variations. The three most common modes in round guides are shown in Fig. 32.26.

32.6.2 Cutoff wavelength

There is a lower limit to the frequency which can be transmitted through a fixed size and shape of waveguide. The "cutoff wavelength" is the largest wavelength corresponding to the lowest frequency that can be propagated. For a rectangular guide, the cutoff wavelength λ_c is given by

$$\lambda_c = \frac{2}{\sqrt{(m/a)^2 + (n/b)^2}} \tag{32.6}$$

where m and n = subscripts, such as TE_{mn} or TM_{mn}
a = wide dimension of the guide
b = narrower dimension

Thus, for the TE_{10} mode in a rectangular guide, the cutoff wavelength is $2a$, or twice the wide dimension. For the TE_{01} mode in a square or rectangular guide, $\lambda_c = 2b$.

From Eq. 32.6 it is evident that the TE_{10} mode has the highest cutoff wavelength. Therefore, for a given frequency, it is possible to select waveguide dimensions such that only the TE_{10} mode will propagate, and for all other modes that frequency will be *below cutoff.* The TE_{10} mode is consequently called the "dominant mode," and all other modes are "higher modes."

Frequencies below cutoff are attenuated very highly in a waveguide, and to all intents and purposes, they do not propagate. A discontinuity in a guide may excite higher modes, but they die out quickly if they are below cutoff. If two or more modes were *above* cutoff and propagated together, they could interact in an undesirable manner. For this reason, waveguide sizes are chosen to cover specific frequency bands so that only the dominant mode can propagate at these frequencies.

For the three modes shown in Fig. 32.26 for a round waveguide, there are also cutoff wavelengths:

For the TE_{11} mode, $\lambda_c = 1.706d$

For the TE_{01} mode, $\lambda_c = 0.820d$

For the TM_{01} mode, $\lambda_c = 1.306d$

where d is the inside diameter of the round guide.

Since the TE_{11} mode has the largest cutoff wavelength, it is the dominant mode in circular waveguides, and with proper choice of diameter, it is the only mode that will propagate. However, if the waveguide is bent, this mode could propagate in the wrong polarization, i.e., with the electric vector rotated from the vertical. The other two modes shown in Fig. 32.26 are symmetrical and are not affected by rotation of the axis. In fact, the TM_{01} mode is used in rotary joints where symmetry is important. The TE_{01} mode has the lowest attenuation of all modes in a round waveguide and is thus useful in long waveguide runs.

32.6.3 Guide wavelength

The wavelength of a signal propagating in a waveguide is greater than the wavelength of the same signal in free space. It is a function of the cutoff wavelength, which in turn depends on the dimensions of the guide. The exact relationship is

$$\frac{1}{\lambda_g^2} = \frac{1}{\lambda_0^2} - \frac{1}{\lambda_c^2} \tag{32.7}$$

where λ_g = wavelength in the guide
λ_0 = wavelength in free space
λ_c = cutoff wavelength

Equation 32.7 applies to all modes in all cross sections of waveguides.

It should be noted that if $\lambda_0 = \lambda_c$ in Eq. 32.7, then the right-hand side is zero, and λ_g must be infinite. If $\lambda_0 > \lambda_c$, the right-hand side is negative, and λ_g must be imaginary. This would indicate that there can be no propagation in the guide when the free-space wavelength exceeds the cutoff wavelength.

32.6.4 Velocity of propagation

The velocity of propagation of a signal in free space is equal to the speed of light, designated c. Since the velocity of propagation equals the product of wavelength and frequency,

$$c = \lambda_0 F \tag{32.8}$$

In a wavelength, the velocity of propagation V_p is also equal to the product of wavelength and frequency. That is,

$$V_p = \lambda_g F \tag{32.9}$$

Since λ_g is greater than λ_0, it must follow that the velocity of propagation in a waveguide exceeds the speed of light, which is contrary to physical principles. However, the guide wavelength is the length of one cycle in the guide, and the velocity V_p is the velocity of the phase of the signal. No intelligence or modulation travels at this velocity. Therefore, V_p is called the "phase velocity."

If the signal is modulated, the modulation will travel at a slower rate in the guide. That is, the modulation keeps slipping backward compared with the phase of the carrier. The velocity of the modulation envelope is called the "group velocity" and is designated V_g. The group velocity is reduced from the speed of light by the same ratio that the phase velocity exceeds the speed of light. Therefore,

$$V_p V_g = c^2 \tag{32.10}$$

32.6.5 Loss

Waveguides have lower attenuation than coaxial lines, but they do have measurable loss. Attenuation in an empty (or air-filled) guide is copper loss as a result of currents in the waveguide walls. If the waveguide contains a dielectric material, the dielectric loss will be added to the copper loss.

The waveguide attenuation depends on the resistivity of the metal, as might be expected, but more important, the resistance of the walls increases as the frequency is increased. This is caused by the skin effect, since at higher frequencies the thickness of metal carrying the current is decreased. In addition, the larger the guide, for a specific frequency, the lower will be the attenuation. This means that the attenuation is greatly increased at higher frequencies, since smaller waveguides are used, and the skin effect increases the resistance. Typically, at 3 GHz, the attenuation in a standard waveguide for that frequency is about 0.6 dB/100 ft; at 10 GHz, it is about 5 dB/100 ft; and at 25 GHz, it is about 15 dB/100 ft in a brass guide and 10 dB/100 ft in a silver guide. Thus attenuation is usually negligible at frequencies below 3 GHz, and it may be neglected in short runs at frequencies up to 10 GHz. However, at higher frequencies, it is an important consideration.

The TE_{01} mode in a round waveguide is exceptional in that its attenuation decreases as frequency is increased. Thus, at frequencies above 25 GHz for long, straight runs, this mode is sometimes used, but care must be taken to prevent lower modes such as the TE_{11} from being excited.

32.6.6 Characteristic impedance

In a conventional open-wire line to a coaxial line, the characteristic impedance is determined by the physical dimensions of the line and is independent of frequency. This is not true of waveguides. An expression for the characteristic impedance of a rectangular waveguide carrying any TE mode is

$$Z_0 = 377 \frac{b}{a} \frac{\lambda_g}{\lambda_0} \tag{32.11}$$

For TM modes, the characteristic impedance is

$$Z_0 = 377 \frac{b}{a} \frac{\lambda_0}{\lambda_g} \tag{32.12}$$

The wide and narrow dimensions of the waveguide are a and b, respectively. For square or round guides, $a = b$. It should be noted also that since λ_g is greater than λ_0, the characteristic impedance for a TE mode is greater than that for a TM mode. The ratio λ_g/λ_0 is not constant with frequency, and therefore, Z_0 must vary with frequency.

From Eqs. 32.6 and 32.7, it is evident that for the TE_{10} mode in a rectangular waveguide, the guide wavelength λ_g is independent of b, the narrow dimension of the guide. Therefore, from Eq. 32.11, the characteristic impedance for the TE_{10} mode is directly proportional to the dimension b.

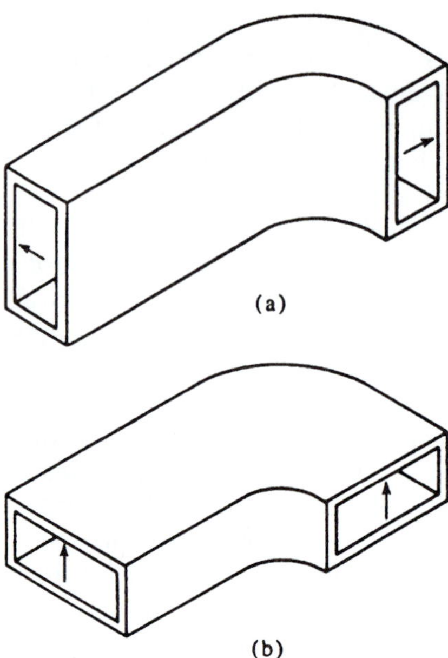

Figure 32.27 Waveguide bends. (*a*) *E*-plane bend; (*b*) *H*-plane bend.

32.6.7 Dielectric

In some applications, the waveguide is filled with a dielectric material in order to reduce the size or increase the power-handling capability of the system. Since all materials exhibit some loss at microwaves, the dielectric loss must be added to the copper loss. The dielectric loss is appreciable, and for this reason, dielectrics are used only in short lengths of waveguide.

If free space were filled with a dielectric material, the "free-space wavelength" (i.e., the wavelength in the dielectric) would be reduced by the square root of the dielectric constant. When a waveguide is filled with a dielectric, the cutoff wavelength is unchanged and is still given by Eq. 32.6, but in determining λ_g from Eq. 32.7, the value of λ_0 must be this reduced value of the free-space wavelength. Also, in determining whether a waveguide is below cutoff for a specified frequency, the dielectric free-space wavelength must be used.

The dielectric constant can be included in Eqs. 32.7, 32.11, and 32.12. Equation 32.7 becomes

$$\frac{1}{\lambda_g^2} = \frac{\varepsilon_r}{\lambda_0^2} - \frac{1}{\lambda_c^2} \tag{32.13}$$

For TE modes, the characteristic impedance becomes

$$Z_0 = \frac{377}{\sqrt{\varepsilon_r}} \frac{b}{a} \frac{\lambda_g}{\lambda_0} \tag{32.14}$$

For TM modes, it is

$$Z_0 = \frac{377}{\sqrt{\varepsilon_r}} \frac{b}{a} \frac{\lambda_0}{\lambda_g} \tag{32.15}$$

32.6.8 Bends and turns

Waveguides may be bent or twisted without causing reflection or introducing loss as long as the cross section of the guide is kept uniform. When the waveguide is bent, the bend is designated an "*E*-plane bend" if the direction of the *E* field is changed. Otherwise, it is an "*H*-plane bend," since then the magnetic lines or *H* field must be bent. Waveguide bends are shown in Fig. 32.27. These are 90° bends, but bends may be made with the arms at any angle to each other as long as the bend is made uniformly and smoothly.

32.7 Waveguide Passive Devices

32.7.1 Power dividers

The simplest power divider is a three-port junction called a "waveguide tee." There are two types, as shown in Fig. 32.28. The designations "*E* plane" and "*H* plane" are determined in the same manner as they were for the bends in Fig. 32.27. In both tees, the input arm is the leg of the tee which is perpendicular to the other two legs. (It is also common to refer to them as "arms.") The other two are called "side" legs or arms.

The *H*-plane tee in Fig. 32.28*a* is also called a "shunt tee" because the two outputs are in shunt with one another across the input. Thus the input sees an impedance only half that of the main line. In addition, there are fringing fields caused by the abrupt junction which add a reactive element to the im-

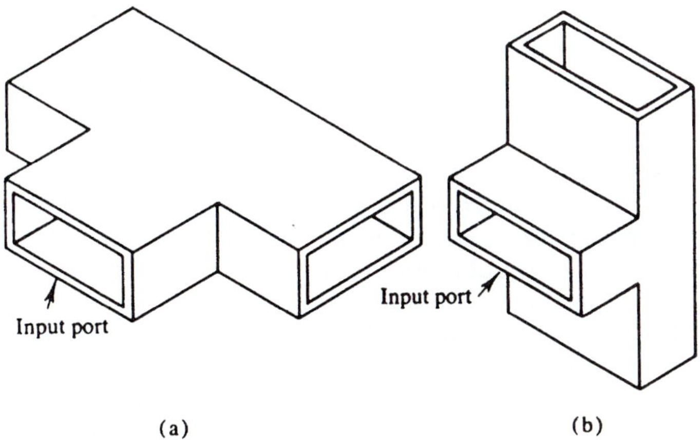

Figure 32.28 Waveguide tees. (*a*) *H*-plane tee; (*b*) *E*-plane tee.

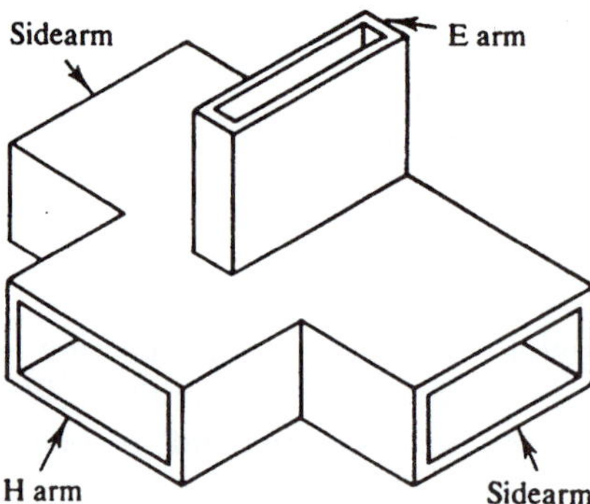

Figure 32.29 Hybrid tee.

pedance. The input arm can be matched to this impedance, however, by a reactive element. When this arm is matched, power entering splits equally and *in phase* in the other two arms.

Since it is impossible to match all three parts of a three-part junction, the sidearms are not matched. If power enters a sidearm, a large part of it will be reflected back toward the source. However, if two signals of equal power and phase are fed into both sidearms, they will add, and the sum will appear at the input arm with no reflections. That is, the reflection in a sidearm is equal to and out of phase with the signal coming from the other sidearm, so the two cancel.

The E-plane tee of Fig. 32.28*b* is called a "series tee" because the two outputs are in series across the input. Thus the input sees twice the impedance plus the effect of fringing fields. Again, it may be matched with a reactive element. When the input arm is matched, power splits equally and *out of phase* in the other two arms. By reciprocity, if two signals of equal amplitude but out of phase are fed into the sidearms, their sum appears at the input arm.

The "hybrid tee" (Fig. 32.29) is a four-port junction which is a combination of an E-plane tee and an H-plane tee. When the E arm and the H arm are matched, the other two arms are also matched automatically, and the tee exhibits unusual properties which have caused it to be called a "magic tee." If power is fed into the H arm of a magic tee, it divides equally and in phase in the sidearms, just as in an H-plane tee. There is no coupling to the E arm, since the E arm and the H arm are cross-polarized. Also, if power is fed into the E arm, it divides equally and out of phase in the sidearms, as in an E-plane tee, with no coupling to the H arm. If power is fed into a sidearm, it divides equally in the E arm and the H arm, with no coupling to the other sidearm. If signals are fed into both sidearms, even signals of different amplitude and phase, the algebraic sum of the signals appears in the H arm, and the algebraic difference appears in the E arm. Thus the H arm is also called the "sum arm," and the E arm is called the "difference arm."

Any four-arm junction which exhibits the same characteristics as a magic

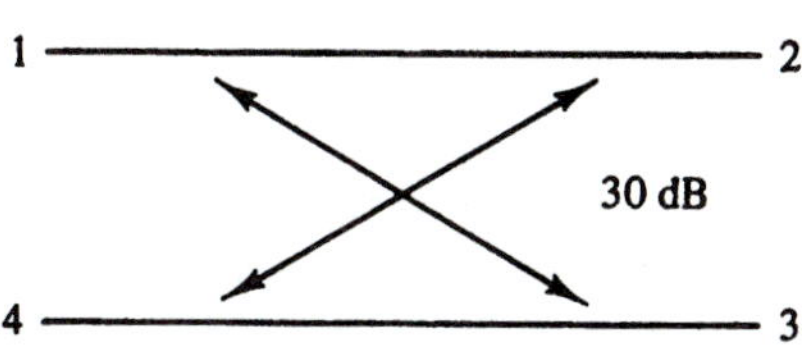

Figure 32.30 Directional coupler symbol.

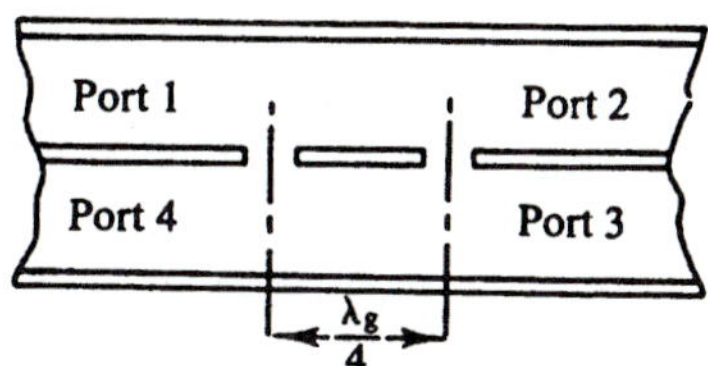

Figure 32.31 Sidewall coupler.

tee is called a "hybrid" or a "hybrid junction." The important characteristic is division of an input signal between two ports and isolation of the remaining port.

The output arms of a power divider must be matched to their loads, or reflections will upset the power division. For example, if the H-plane tee in Fig. 32.28a has one sidearm matched but a mismatch in the other, there will be a reflection from the mismatch which reenters the junction. Part of this reflected signal will cross to the other sidearm, where it will combine vectorially with the signal there. Thus the power in the matched arm may be more or less than half the input power depending on the phase of the reflected signal. In the mismatched arm, the power is always less than half the input, since some power is reflected back to the junction. The same thing holds true for the E-plane tee and the septum power divider. In the magic tee, a reflection in one sidearm will not affect the power in the other sidearm, since the two are isolated from each other.

32.7.2 Directional couplers

The schematic symbol for a waveguide directional coupler is shown in Fig. 32.30. With the ports numbered as shown, the parameters defined in Eqs. 32.3, 32.4, and 32.5 apply, but the numbering of the ports in relation to their physical location is different for TEM and waveguide couplers. Note that for power entering at port 1, coupled power exits at port 3; this is a "forward-wave" coupler.

Waveguide couplers also differ from TEM couplers in the electromagnetic phenomena that cause directivity. TEM couplers use the E- and H-fields' symmetrical and asymmetrical coupling (respectively) to create the phasing of the coupled signal that adds in one direction and cancels in the other. Waveguide couplers use two or more holes between the coupled guides to achieve phase addition and cancellation. An explanation of the operation is shown in Fig. 32.31 for a sidewall coupler. The two waveguides are joined so that they have one common sidewall with two identical holes through the common wall that are spaced a quarter of a guide wavelength apart. When a signal is incident at port 1, each hole couples some energy into the auxiliary guide, and the energy from *each* hole flows equally toward ports 3 and 4. The energy traveling from port 1 to port 3 travels the same distance through *either* hole, and thus the two coupled signals are in phase and add in this direction. However, in going from arm 1 to arm 4, one of the coupled signals must travel half a guide wavelength more than the other. Therefore, in the direction of arm 4, the two coupled signals cancel. It also should be noted that each hole is a discontinuity which sends a reflection back to arm 1. However, since one path from arm 1 to the hole and back is half a wavelength different from the other, the two reflections cancel, resulting in a good match.

In order to achieve tighter coupling, additional holes may be used. In general, the match, directivity, and bandwidth are all improved by increasing the number of holes.

32.7.3 Attenuators

In waveguides, both fixed and variable attenuators use a "resistance card," or "lossy card," to absorb power from the electric field of the dominant mode. This card is usually a thin ceramic (glass) or plastic card that has resistive material coated on one or both sides, and the card is placed in the guide such that the surfaces are parallel to the E field. The input and output ends of the card are tapered to reduce the reflections caused by the card's presence in an otherwise empty guide. The amount of attenuation is a function of the length of the card (along the direction of propagation), the sheet resistance on the card, and the location of the card relative to the position of the maximum E field. Most variable attenuators have a fixed length and resistivity for the card; the variable attenuation is achieved by positioning the card in areas of different field strength.

One form of a variable attenuator that is relatively easy to implement in rectangular waveguides is shown in Fig. 32.32. This attenuator makes use of the fact that the E field in a waveguide (for the dominant mode) is maximum at the center and zero at the walls. A lossy card is placed inside the waveguide and is controlled by thin horizontal metal rods, which have negligible effect on the match because they are perpendicular to the E field. When the card is against a sidewall of the waveguide, where the E field is a minimum,

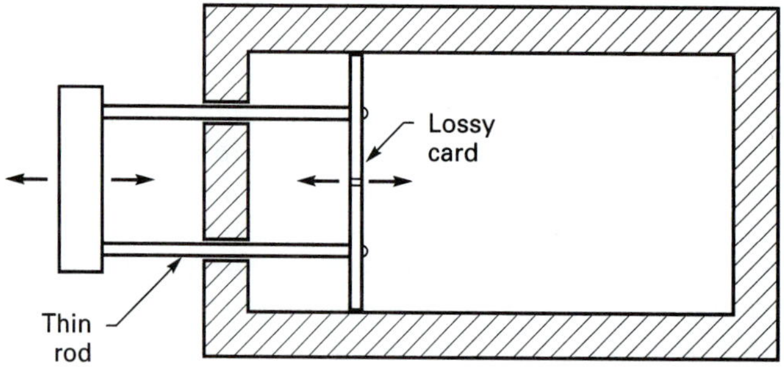

Figure 32.32 Variable attenuator, cross section through adjusting mechanism.

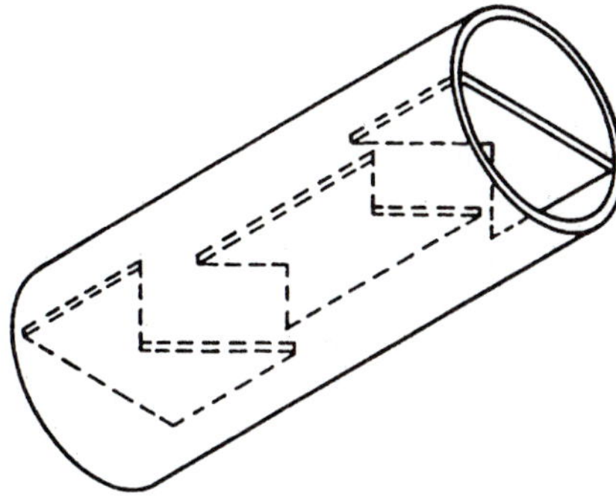
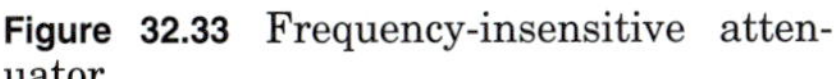

Figure 32.33 Frequency-insensitive attenuator.

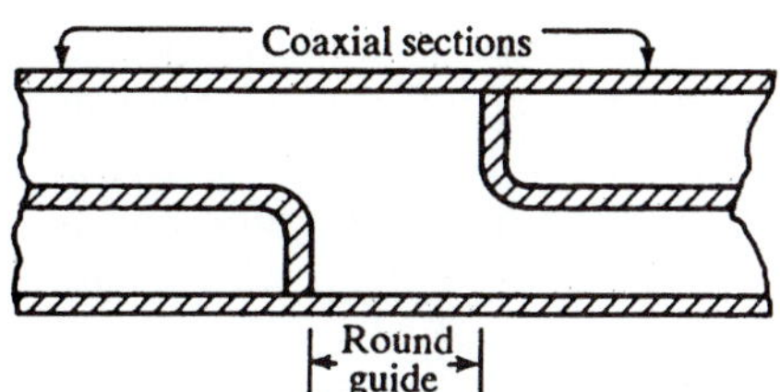

Figure 32.34 Cutoff attenuator in coaxial line.

it causes little attenuation. As it is moved from the sidewall to the center, the attenuation increases. The movement may be controlled by a calibrated knob on the outside of the waveguide, so the amount of attenuation for each position may be read directly on a dial. However, the amount of attenuation and the phase shift associated with it vary with frequency, so for each frequency a new calibration would have to be created.

A frequency-insensitive variable attenuator can be built in a round waveguide as shown in Fig. 32.33. The attenuator contains three cards of lossy material. The input and output cards are fixed with their planes perpendicular to the E field in the guide; in this position, they have minimum attenuation to the dominant mode with perpendicular polarization, but they have very high attenuation to any other polarization of the mode that may exist. The center card can be rotated about the axis of the round guide such that it can have any angle from 0 to 90° relative to the direction of the E field at the guide's center, giving maximum to minimum attenuation, respectively. The loss depends only on the angle of rotation of the center card and is independent of frequency. This attenuator is often connected to a rectangular waveguide through rectangular-to-circular adapters.

A special form of attenuator which contains no lossy material can be built around a section of waveguide which is below cutoff. Such a device is called a "cutoff attenuator."

Earlier it was pointed out that a signal cannot be propagated in a waveguide if the wavelength of the signal is less than the cutoff wavelength of the guide. In fact, if a signal below cutoff is fed into a waveguide, *most* of it will be reflected as if the waveguide were an open circuit. Some of the signal will enter the guide and be attenuated rapidly. If the wavelength of the signal is much greater than the cutoff wavelength, the attenuation is

$$\alpha = \frac{2\pi}{\lambda_c} \tag{32.16}$$

where α = attenuation, nepers per unit length
λ_c = cutoff wavelength

Equation 32.16 shows that the attenuation is only a function of the length of the cutoff section and is independent of frequency.

A practical cutoff attenuator consists of a section of round guide placed in a coaxial line, as shown in Fig. 32.34. The coaxial inner conductor is stopped at the section of round guide and is bent and attached to the outer conductor to form a coupling loop which will excite the TE_{11} mode. However, with the guide below cutoff, the TE_{11} mode is attenuated. Some signal reaches the far end of the cutoff section, where another loop couples the TE_{11} mode back to a coaxial mode. From Eq. 32.16, the attenuation is proportional to the spacing between the two coupling loops. In variable form, the spacing can be varied by turning a calibrated knob.

32.7.4 Variable phase shifter

When a waveguide is filled with a low-loss dielectric material, the guide wavelength is reduced, as is indicated by Eq. 32.13. Thus a given length of guide will contain more wavelengths when filled with dielectric. In effect, the phase shift through the guide has been increased. When a thin slab of dielectric is placed in the guide, it also increases the phase shift, but not as much as filling the guide completely. As with a variable attenuator, the effect of the dielectric slab is greatest near the center of the waveguide and minimal near the sidewall. Therefore, a variable phase shifter can be constructed in the same manner as a variable attenuator, by replacing the lossy material with a low-loss dielectric. In Fig. 32.33, for example, if the lossy resistance card is replaced by a piece of polystyrene or other low-loss dielectric material, the result is a variable phase shifter. The dial controlling the movement of the slab could be calibrated in degrees rather than decibels.

32.7.5 Resonant cavities

If a short circuit is placed across a waveguide, an incident signal will be reflected completely. If now a short circuit with a small hole in the shorting plate is instantaneously placed a half guide wavelength back toward the source, the reflected signal will be trapped and reflected again and travel in the original direction. Most important, the second reflection will cause the twice-reflected signal to be *in phase* with a new incident signal that enters through the small hole in the shorting plate. This section will then support a signal which bounces back and forth between the two shorting plates. The closed section is called a "resonant cavity," or simply a "cavity," and is analogous to a resonant "tank" circuit at lower frequencies. Basically, a resonant cavity consists of a section of transmission line and two large discontinuities which are so spaced that reflections from them are in phase with the incident signal. Any kind of transmission line, including coaxial lines, as well as waveguides can be used. It should be noted that the discontinuities or shorting plates can be spaced any number of half-wavelengths apart. There also must be some means of coupling an incident wave to the cavity, i.e., some means of getting the signal inside, and this is discussed under "Coupling."

Just as several modes exist in a waveguide, so too are there several cavity modes. Cavity modes are designated by a third subscript which indicates the length of the cavity in half-wave-lengths in the guide. Thus a rectangular cavity mode designated TE_{101} would mean a TE_{10} mode in a rectangular guide with the shorts spaced one half-wavelength apart. The TM_{112} mode would indicate a TM_{11} mode in the waveguide with a cavity length two half-wavelengths long. In a round guide, the TE_{111} mode indicates the dominant TE_{11} waveguide mode in a cavity half a guide wavelength long. In cavities in round guides, there is also another set of TM modes in which the electric field is always parallel to the axis of the cylinder. The third digit is always zero for one of these special TM modes.

32.7.6 Coupling

As has been noted, there must be some opening in the cavity to get the signal in and out. When there is only one coupling mechanism used only to get the signal into the cavity, the cavity is called an "absorption cavity." When an output coupling is also used, the cavity is a "transmission cavity."

In a rectangular cavity formed by placing suitably spaced platees across a waveguide, coupling can be accomplished by small openings in the shorting plates, as shown in Fig. 32.35. The round hole in Fig. 32.35*a*, if small enough,

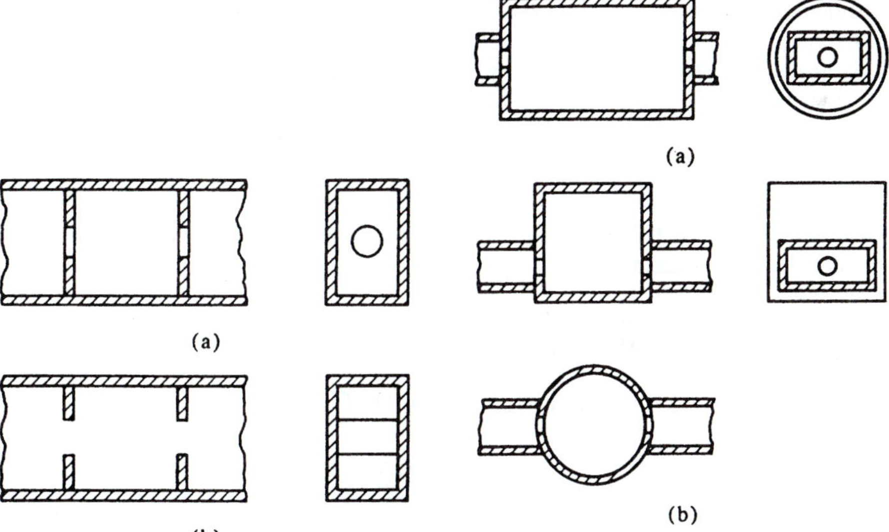

Figure 32.35 Coupling through apertures in rectangular cavities.

Figure 32.36 Aperture coupling in round cavities. (*a*) End-wall coupling; (*b*) round-wall coupling.

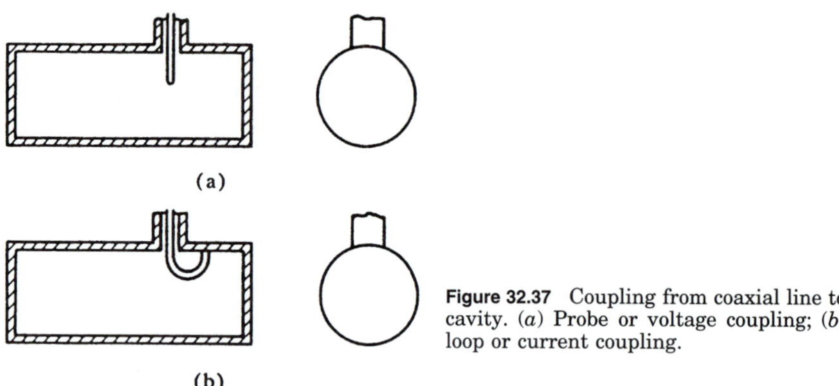

Figure 32.37 Coupling from coaxial line to cavity. (*a*) Probe or voltage coupling; (*b*) loop or current coupling.

permits some power to enter and still has little effect on the operation of the cavity. As the hole size is increased, the resonant frequency of the cavity decreases, and Q decreases. As with a resonant-tank circuit at lower frequencies, at resonance, the transmitted signal passes through the cavity with little loss, since the two discontinuities produce reflections that cancel each other.

Coupling into a cylindrical cavity in a round-waveguide transmission line can be accomplished simply by using a round hole in the shorting plate, as with rectangular guides. In general, a cylindrical cavity can be built with a higher Q than is achievable with a rectangular guide, so it is frequently desirable to use a cylindrical cavity in conjunction with a rectangular guide. Again, a round hole may be used, and two possible configurations are shown in Fig. 32.36.

A cavity is sometimes used in conjunction with coaxial lines, and it becomes necessary to couple from the coaxial line to the cavity. Two methods of coupling a cylindrical cavity to a coaxial line are shown in Fig. 32.37. In Fig. 32.37*a*, the inner conductor of the coaxial line is simply extended into the cavity to form a coupling probe. In Fig. 32.37*b*, the inner conductor is looped around and fastened to the inside wall of the cavity. Both methods also may be used to couple coaxial lines to rectangular cavities.

32.7.7 Cavity *Q*

As is true with a resonant-tank circuit, the Q of a cavity is

$$Q = 2\pi F \max \frac{\text{energy stored}}{\text{power loss}} \tag{32.17}$$

In a resonant cavity, the power loss is copper loss in the walls and is proportional to the skin depth. The energy stored depends on the ratio of volume to surface area. Thus a good approximation for the Q of a cavity is

$$Q = \frac{V}{A\delta} \tag{32.18}$$

where V = volume
A = surface area
δ = skin depth

For a rectangular cavity, maximum Q occurs when all three dimensions are equal. For a round cavity, maximum Q occurs when the diameter and length are equal. Q can be increased by polishing the inside of the cavity and plating it with a highly conductive material such as silver or copper.

The resonance curve of a cavity is identical to the resonance curve of an LC tank. The distance between the half-power frequencies is termed the "3-dB bandwidth" of the cavity and is designated ΔF. The Q of a loaded cavity is given by

$$Q = \frac{F_0}{\Delta F} \tag{32.19}$$

where F_0 is the resonant frequency. The loaded Q includes the effect of the coupling mechanisms on the intrinsic Q of the cavity.

32.7.8 Transitions

Most systems are not restricted to one type of transmission line but may consist of a combination of different sizes or shapes of waveguides as well as sections of coaxial line. It is necessary, then, to provide transitions, or "transducers," which couple signals from one type of line to another without loss due to mismatch or lossy elements.

A common type of transition between two different sizes of waveguide is a

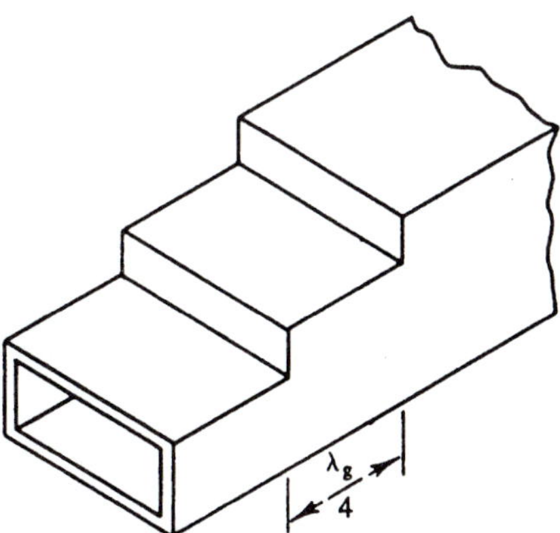

Figure 32.38 Quarter-wave transformer.

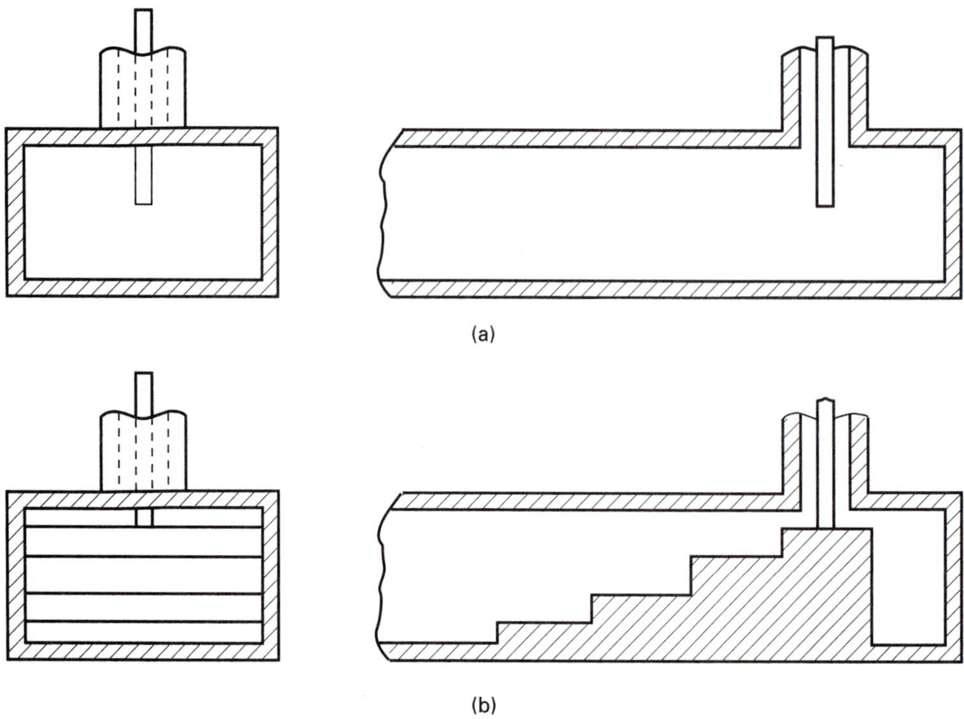

Figure 32.39 Coaxial-to-waveguide transitions. (*a*) Probe; (*b*) stepped-impedance.

linear taper from one to the other. The two guides may differ only in their narrow dimensions, in which case the taper is in that dimension only. This is simply a gradual change between two impedances, since the impedance of a guide is proportional to its height. If the taper is four or more wavelengths long, the VSWR will be under 1.05, indicating a good impedance transformation. If it is desired to make the transition in a shorter distance, a quarter-wave transformer can be used, just as at lower frequencies, although it is more easily done with waveguides. This is simply a section of guide which is one-quarter of a guide wavelength long and which has a characteristic impedance which is the square root of the product of the other two characteristic impedances. That is,

$$Z_0 = \sqrt{Z_1 Z_2} \tag{32.20}$$

where Z_0 = characteristic impedance of quarter-wave matching section
Z_1 = characteristic impedance of input line
Z_2 = characteristic impedance of output line

A typical quarter-wave transition is shown in Fig. 32.38. This will be a good match only over a narrow frequency band, since the transition is exactly a quarter-wavelength at only one frequency. Somewhat broader bandwidth can be achieved by using two or more quarter-wave sections, in each case matching to some intermediate impedance. Tapers also may be used between rectangular and round guides, but in general, they present a difficult fabrication problem.

Transitions between coaxial lines and waveguides are similar to the couplings between coaxial lines and cavities shown in Fig. 32.39. The probe coupling is simply a probe formed by extending the inner conductor of the coaxial line, as shown in Fig. 32.39*a*. The stepped-impedance transition, shown in Fig. 32.39*b*, has the center conductor of the coaxial line directly connected to the "top step" of the stepped-impedance transformer. This top step is a quarter-wave section of reduced-height waveguide with an impedance given by Eq. 32.11 that matches this impedance to the impedance of the coaxial input (usually 50 Ω). The other steps match this impedance to the impedance of the full height waveguide. The backside of the waveguide is full height with a short circuit positioned to present a high impedance at the location of the coaxial center conductor.

Chapter

33

Impedance Considerations

K. D. Baker and D. A. Burt

Department of Electrical Engineering
Utah State University, Logan, Utah

33.1 Introduction

"Electrical impedance" may be defined as an apparent opposition to the flow of current in an electric circuit. Quantitatively, impedance is the ratio of the voltage across a circuit to the current flowing in the circuit (a generalized form of Ohm's law):

$$Z = \frac{V}{I} \tag{33.1}$$

where Z = impedance, Ω
V = voltage across Z
I = current through Z

An understanding of the impedance concept is important for proper use of nearly all electrical measuring instruments and the correct interpretation of the measured values. Whenever a measuring instrument is connected to another device, attention should be given to the effect of the added loading (or change in the total impedance attached to the device) on the operation of the device, particularly with respect to the accuracy of the measured values.

Ideally, the application of a test instrument to a device under test would not disturb the operation of the device, and the measured values would be identical to those values in the device when the instrument was not connected. Unfortunately, test instruments require extraction of energy from the device under test (by drawing a current) and thus modify the operation to varying degrees depending on the impedance levels of both the test instrument and the device under test. For example, almost all electronic circuits will be affected by a voltage measurement using a common meter-type low-impedance multimeter, resulting in voltage readings in a transistor circuit typically being 10 to 20 percent low. On the other hand, high-impedance electronic digital multimeters (DMM) will read values that are not appreciably lower than the undisturbed circuit values in all but the highest-impedance circuits such as field-effect transistor circuits.

In cases involving signal sources, such as signal generators and pulse generators, the output-signal amplitude of the instrument is calibrated only when a specified impedance is connected to its output terminals. Any other value of impedance will result in magnitude errors unless corrections are made for the impedance level involved. Additionally, unless the impedance of the source at its output terminals is made equal to that of the interconnecting cable and of the load (achieving this condition is called "impedance matching"), troublesome signal reflections may occur, also resulting in calibration errors and in many cases spurious signals. Similar considerations are important in connecting output devices such as loudspeakers, indicating equipment (such as meters and recorders), and wattmeters.

Impedance-related problems of measurements are complicated by the fact that the impedance levels depend on the frequency and waveform of the signals. Often, an instrument that does not affect the test device appreciably at low audio frequencies will be completely useless at a few hundred kilohertz because of drastic loading at these higher frequencies.

In the discussion that follows, an introduction to the impedance concept will be followed by specific problems and cases related to giving due consideration to the impedance aspects for proper measurement techniques.

33.2 The Impedance Concept

The current flowing in an electric circuit as a result of a given impressed voltage is determined by the circuit impedance; the higher the impedance, the more the circuit "impedes" the flow of current and hence the lower the value of current. A conceptual aid to the uninitiated reader is the analogy of the electrical system to the flow of water through a pipe system. The amount of water that flows through a pipe depends on two factors: the water pressure at the input and the critical characteristics of the pipe such as size, shape, and smoothness. The amount of water that flows is analogous to the flow of electrons, or the current, through the electric circuit; the forcing pressure is analogous to the applied voltage, and the properties of the pipe that restrict or impede the water flow are analogous to the electrical impedance. Thus, for a given pressure, we can vary the water flow by varying water-system impedance by the setting of a valve. Similarly, for a given voltage, current can be controlled by varying the circuit impedance. The impedance of electric circuits for the case of dc sources is simply the equivalent series resistance R in ohms, defined by the ratio of the applied voltage V in volts to the resulting current I in amperes. Just as the current in a circuit that flows because of an impres-

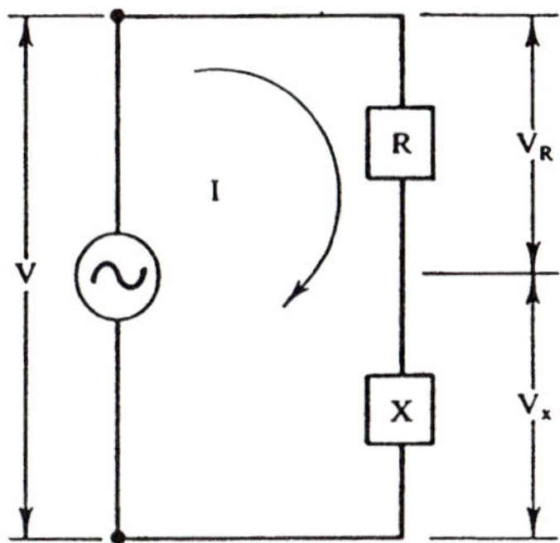

Figure 33.1 Simple series circuit with resistance R and reactance X.

sed dc voltage is determined by the dc resistance, the current due to a time-varying voltage is determined by the circuit impedance presented to the time-varying signal (which involves circuit inductance and capacitance in addition to resistance). Hence the generalized definition of impedance Z is the ratio of the voltage V across a circuit to the current I flowing in the circuit (Eq. 33.1), including time-varying effects. The commonly accepted units of impedance are ohms, just as in the case of resistance for dc circuits.

For the important, common case of sinusoidally varying voltages (ac signals), the impedance Z of a circuit can be visualized as a complex number which is the sum of series resistance R and reactance X (see Fig. 33.1):

$$Z = R + jX \tag{33.2}$$

where the j preceding the reactance is literally $\sqrt{-1}$ and indicates that the current and voltage are 90° out of time phase in the reactance.*

33.2.1 Impedance of a series circuit

The total applied voltage V in Fig. 33.1 must appear across the two series elements R and X. This simple circuit is denoted as a series circuit because there is only one path for the current I, which must necessarily flow through both R and X. By Ohm's law, the voltage V_R across R must be IR and the voltage V_X across X must be IX. The total voltage across R and X, however, is not the simple sum of magnitudes $|V_R| + |V_X|$, because although the voltage across the resistance is in phase with the current, this is not true in the reactance, where a 90° phase relationship exists, as shown by the voltage and

* Mathematically speaking, the resistance R is the real part and X is the imaginary part of the complex impedance Z.

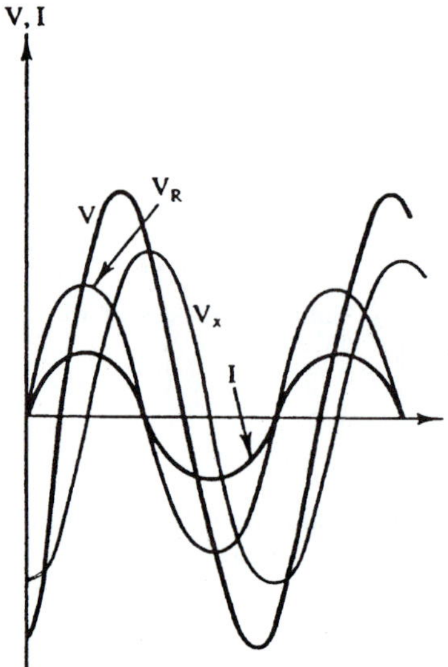

Figure 33.2 Time plot of current I, voltage across the resistance V_R, voltage across the reactance V_X, and total circuit voltage V for the simple series circuit.

current plot in Fig. 33.2. As can be seen, the magnitude of the voltage across the circuit is not merely the sum of the magnitudes of the two voltage components but depends on the phase relationship as well. The instantaneous circuit voltage at any time is equal to the sum of the instantaneous component voltages (V_R and V_X) at the same time.

The impedance relationship $Z = R + jX$ can then be thought of as a basic definition of two components of impedance: a resistance R having the voltage in phase with the current and a reactance X having a voltage across it that is leading the current by 90°. (If the reactance is negative, then the voltage will lag the current through the reactance.) The impedance $R + jX$ can be visualized graphically as shown in Fig. 33.3 by plotting R along the horizontal axis (real axis) and X along the vertical axis (imaginary axis). X is plotted 90° with respect to R to portray the 90° phase relationship designated by the j

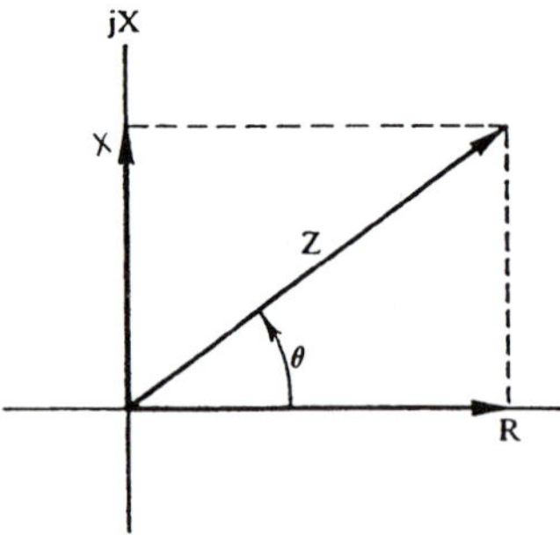

Figure 33.3 Complex impedance diagram of series circuit. Resistance R is plotted on the horizontal axis and reactance X is plotted on the vertical axis, with positive reactance plotted upwards. The length of the arrow represents the magnitude in ohms. The angle θ is the impedance phase angle.

preceding the X. Mathematically, since j is $\sqrt{-1}$, it plots 90° on the polar plot (by mathematical convention, angles are measured counterclockwise from the real axis). The total impedance is a sum of the two components R and X, which are perpendicular. Thus, rather than merely forming the algebraic sum, the complex sum must be formed as shown in Fig. 33.3.

This sum is represented by the length of the resultant arrow and the phase angle θ. This polar form of impedance can be written as

$$Z = |Z| \angle \theta \tag{33.3}$$

where $|Z|$ = ratio of magnitude of V to I
$\angle \theta$ = angle in electrical degrees that current lags voltage

For passive circuits, θ will be between -90 and $+90°$. Negative angles mean that the current leads the voltage as a result of a negative reactance (reactance can be positive or negative). The two forms of the impedance Z are equivalent and can be transformed by fundamental, complex function operations:

$$|Z| = \sqrt{X^2 + R^2} \tag{33.4}$$

$$\theta = \text{angle whose tangent is } \frac{X}{R} \tag{33.5}$$

If the impedance is stated in polar form ($|Z| \angle \theta$), the inverse operation to resistive and reactive components is

$$R = |Z| \cos\theta \tag{33.6}$$

$$X = |Z| \sin\theta \tag{33.7}$$

There are only three basic types of passive circuit elements: resistance, capacitance, and inductance. The resistance, of course, is distinguished by voltage and current that are in phase ($\theta = 0$), but capacitance results in a negative reactance ($\theta = -90°$), and inductance gives a positive reactance ($\theta = +90°$). The total circuit reactance X is related to the total equivalent series inductance L and capacitance C by the relations

$$X = X_L - X_C \tag{33.8}$$

$$X_L = 2\pi f L \tag{33.9}$$

$$X_C = \frac{1}{2\pi f C} \tag{33.10}$$

where f = frequency of the applied sinusoidal voltage in hertz
L = inductance in henrys
C = capacitance in farads

Figure 33.4 Series RL circuit.

Example 33.1: Impedance calculation for a series RL circuit. It is desired to find the impedance of a series circuit shown in Fig. 33.4 consisting of a 100-Ω resistor and 10-mH coil at a frequency of 1 kHz.

$$Z = R + jX \qquad R = 100\ \Omega$$
$$X = 2\pi fL = (2\pi)(1000)(0.01) = 62.8\ \Omega$$
$$Z = 100 + j62.8\ \Omega \qquad \text{or} \qquad 118.2\angle\theta$$

where θ = arctan 62.8/100, or approximately 32.2°. The impedance diagram is shown in Fig. 33.5.

Example 33.2: Impedance calculation for a series RLC circuit. It is desired to find and plot in polar form the impedance of a series RLC circuit (Fig. 33.6) having values of

$$R = 50\ \Omega \qquad L = 0.1\ \text{H} \qquad C = 10\mu\ \text{F} \qquad f = 100\ \text{Hz}$$
$$X_L = 2\pi fL = (2\pi)(100)(0.1) = 62.8\ \Omega$$
$$X_C = \frac{1}{2\pi fC} = \frac{1}{(2\pi)(100)(10^{-5})} = 159\ \Omega$$
$$X = X_L - X_C \approx -96\ \Omega$$

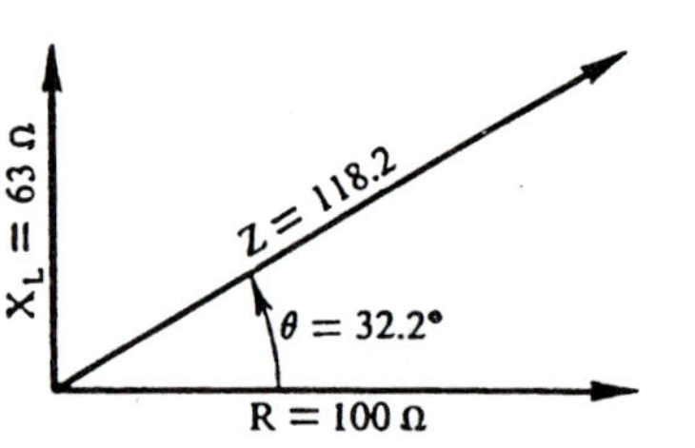

Figure 33.5 Impedance diagram of Example 35.1.

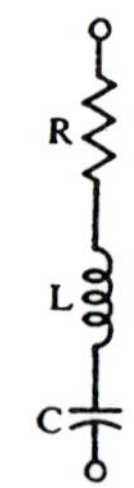

Figure 33.6 Series RLC circuit.

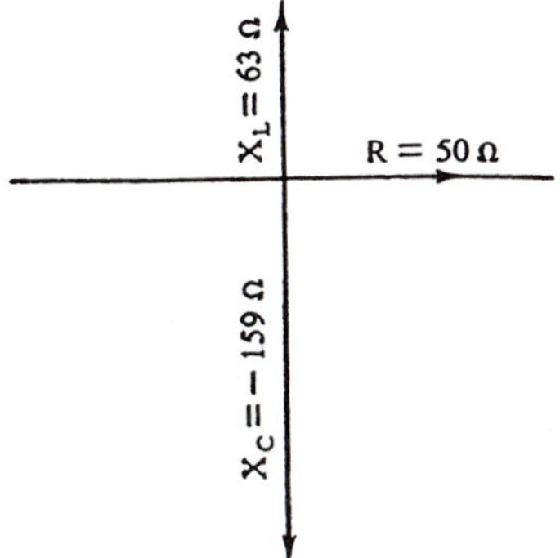

Figure 33.7 Plot of individual impedance elements for Example 33.2.

$$Z \approx 50 - j96\ \Omega$$

$$|Z| = \sqrt{R^2 + X^2} = \sqrt{(50^2) + (96)^2} = 108\ \Omega$$

$$\theta = \arctan(-96/50) = -62.5°$$

$$Z = 108\angle - 62.5°\Omega$$

The individual impedance components are shown in Fig. 33.7, and the total impedance is shown in Fig. 33.8.

Example 33.3: Calculation of current in a series circuit. The series *RLC* circuit shown in Fig. 33.9 has a 1-kHz signal of 1 *V* rms* applied across it. It is desired to find the resulting current. The total series impedance is

$$Z = R_1 + R_2 + j(X_L - X_C)$$

where $X_L = 2\pi f L = (2\pi)(1000)(10^{-2}) = 62.8\ \Omega$

$$X_C = \frac{1}{2\pi f C} = \frac{1}{2(\pi)(1000)(10^{-6})} = 159\ \Omega$$

$$Z = 20 + 30 + + j(62.8 - 159) = 50 - j96 = 108\angle{-62.5°}$$

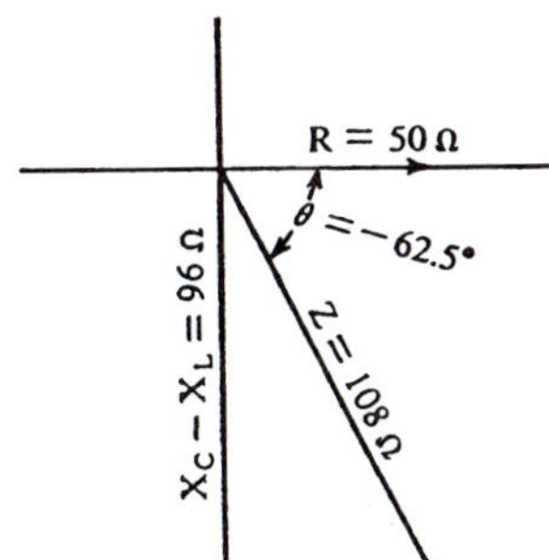

Figure 33.8 Equivalent impedance plot for Fig. 33.7.

* The root mean square (rms) value of a sinusoidally varying current or voltage is the effective value, which turns out to be 70.7 percent of the peak value.

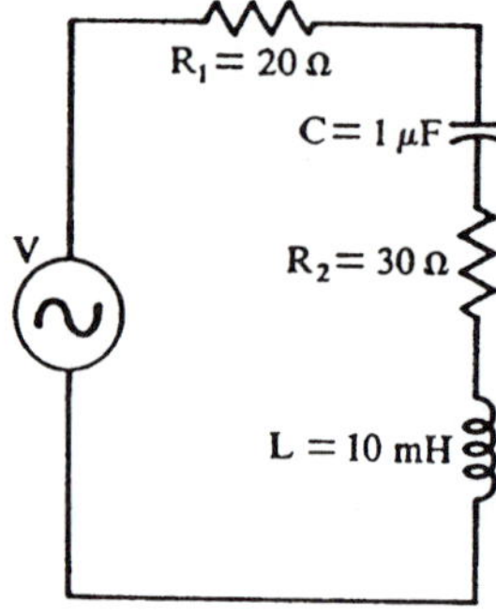

Figure 33.9 Series *RLC* circuit of Example 35.3.

The current, therefore, is

$$I = \frac{V}{Z} = \frac{1}{108\angle - 62.5^\circ} = 0.0092\angle 62.5^\circ \text{A}$$

The magnitude of the current is 0.0092 A (9.2 mA), and the phase angle indicates that the current leads the voltage by 62.5°, as illustrated in Fig. 33.10.

33.2.2 Impedance of parallel circuits

In contrast to the series circuit having the same current flowing through the circuit elements, the parallel or shunt circuit illustrated in Fig. 33.11 is characterized by a common voltage across the elements. The impedance of this parallel circuit can be found in equivalent series form, or as it is often said, the impedance can be combined by the expression

$$\frac{1}{Z_p} = \frac{1}{Z_1} + \frac{1}{Z_2} \qquad \text{or} \qquad Z_p = \frac{Z_1 Z_2}{Z_1 + Z_2} \tag{33.11}$$

where Z_p is the equivalent impedance of the two impedances Z_1 and Z_2 in

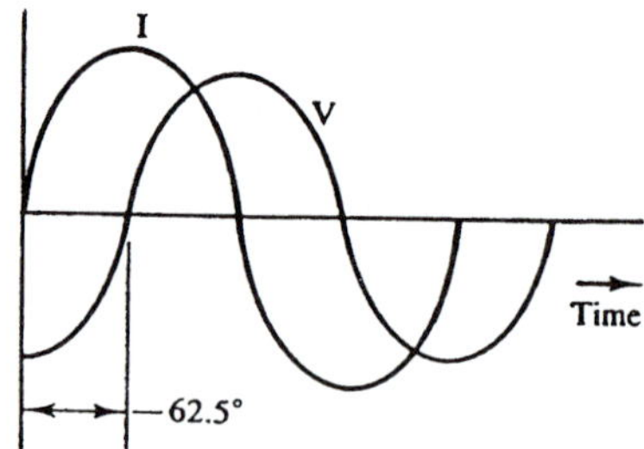

Figure 33.10 Plots of current and voltage for Example 33.3.

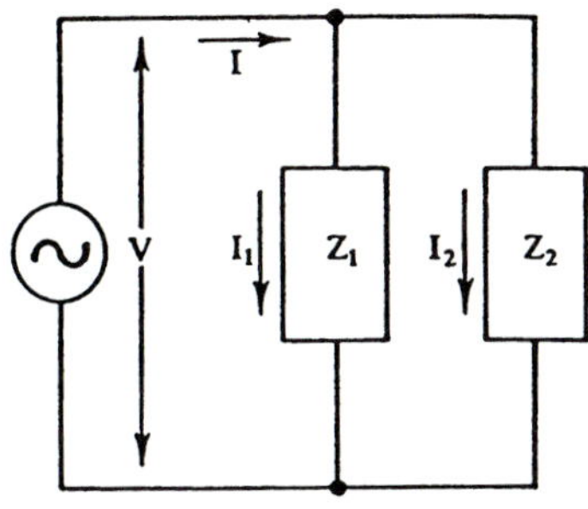

Figure 33.11 Simple parallel circuit.

parallel. The combination of more than two shunt-connected elements can be analyzed by repeated application of the two-element combination (Eq. 33.11).

Example 33.4: Calculation of impedance of a parallel RC circuit. It is desired to find the equivalent impedance of the parallel combination of a 1000-Ω resistor and a 10-pF capacitor at 10 MHz.

$$Z_p = \frac{Z_1 Z_2}{Z_1 + Z_2}$$

where, in this case,

$$Z_1 = 1000\ \Omega$$

$$Z_2 = -jX_C = -\frac{j1}{2\pi fC} = -\frac{j1}{2\pi \times 10^7 \times 10^{-11}} = -j1590\ \Omega$$

$$Z_p = \frac{(1000)(-j1590)}{1000 - j1590} = \frac{1.59 \times 10^6 \angle -90°}{1880 \angle -57.8°} = 845 \angle -32.2°\ \Omega$$

$$R = |Z| \cos\theta = 845 \cos(-32.2°) = 716\ \Omega$$

$$X = |Z|(\sin\theta = 845 \sin(-32.2°) = -450\ \Omega$$

$$Z_p = 716 - j450\ \Omega$$

33.3 Input and Output Impedance

33.3.1 Input impedance

The "input impedance" of an electric device is the ratio of voltage applied to the input terminals to the current flowing into the input terminals (see Fig. 33.12):

$$Z_{in} = \frac{V_{in}}{I_{in}} \tag{33.12}$$

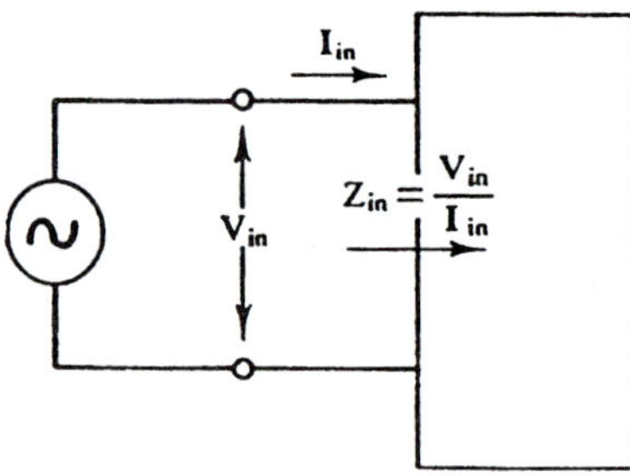

Figure 33.12 Input impedance of a device.

where Z_{in} = input impedance
V_{in} = voltage across the input terminals
I_{in} = current flowing into the input terminals

In the case of dc applied voltages, the input impedance will be a resistance; however, for the ac-signal case, the general impedance including the phase relationship must be utilized. Only in the special case where the input current happens to be in phase with the applied voltage will the input impedance be a pure resistance (reactance equals zero). A low-input-impedance device will draw more current from a given applied voltage source than will a high-input-impedance device. The low-impedance device is said to "load" the source more heavily than the high-impedance device. It then follows that the input impedance of an instrument will determine the degree that the operation of any device is changed by the application of the test instrument. Accordingly, before connecting any test instrument, consideration should be given to the input impedance of the instrument and the consequences of connecting it to the circuit.

33.3.2 Output impedance

The "output impedance" of a device is the equivalent source impedance seen by the load. The output impedance has meaning only for an active device at a pair of terminals considered as a signal source. The term "equivalent impedance" implies that the device can be represented by an equivalent circuit* (Thevenin's equivalent circuit), such as that shown in Fig. 33.13. Here the voltage source is the unloaded output voltage, and Z_{out} is the impedance looking back into the device with all active sources replaced by their internal impedances.

* The term "equivalent circuit" means that the circuit has the same voltages and currents as the original circuit, although the actual impedance elements may be much different.

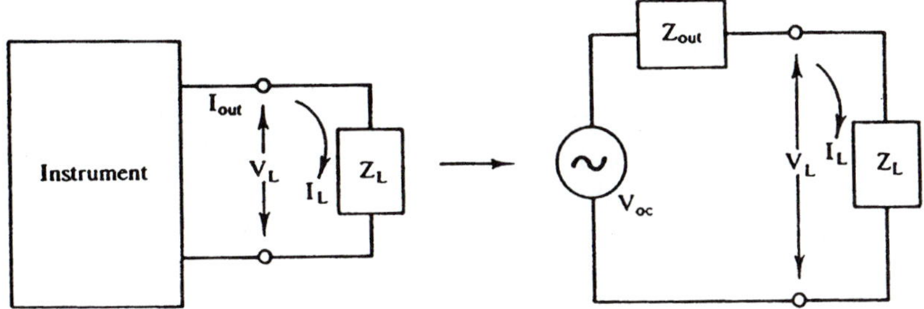

Figure 33.13 The instrument can be replaced by the equivalent circuit for determining currents and voltages.

The output impedance can be stated in terms of the unloaded output voltage of the device:

$$Z_{out} = \frac{V_{oc} - V_L}{I_L} \tag{33.13}$$

where Z_{out} = output impedance
V_{oc} = open-circuit voltage of the source
V_L = load voltage
I_L = load current

This expression conveys the idea that the drop in the output voltage ($V_{oc} - V_L$) as the device is loaded is determined directly by the output impedance.

The output impedance might be considered as a measure of the susceptibility of the device to loading; i.e., the lower the output impedance, the less effect the load will have on the output voltage. This can be seen by noting that the voltage drop across the output impedance for a given current will be less for lower output impedance; hence the output voltage will be higher.

33.3.3 Determination of input and output impedances

Input and output impedances of electronic devices may be determined by measuring the ratio of the appropriate voltage to current, or they may be calculated if the internal parameters of the instruments are known. Usually, however, this will not be necessary, since the instruction manual accompanying instruments used for measurements will list the input and output impedances which, at the design frequencies, approximate pure resistances. Statements such as "impedance of 50 Ω" and "50-Ω output impedance" imply that the reactance is essentially zero over the normal operating frequency range.

In some applications, the input and output impedances may not be adequately known. These parameters usually can be measured with the appropriate instruments. In the case of input impedance, the definition is given in Eq. 33.12. Thus, if the input voltage and current can be determined at their respective phase angles, the input impedance is specified. The most convenient method is to use an impedance bridge or other impedance-measuring instrument and actually measure impedance directly with the instrument activated and at the desired frequencies.

The output impedance also can be measured by an impedance bridge when the device can remain active but suppress its output so that it does not affect the bridge. If the output impedance is nearly a pure resistance, its magnitude can be determined simply by noting the open-circuit output voltage, loading it with a known resistor, and then measuring the loaded output voltage:

$$R_{out} = \frac{R_L(V_{oc} - V_L)}{V_L} \tag{33.14}$$

where R_{out} = magnitude of the output resistance
R_L = magnitude of the load resistance
V_{oc} = open-circuit voltage
V_L = loaded voltage

When this method is used, care should be exercised not to exceed the output capabilities of the instrument.

33.4 Effects of Instrument Input and Output Impedances on Measurements

In most measurement situations it is desirable that the test instrument not draw any appreciable power from the device being measured. This section will cover low-power measurements. The measurements involving intentional power transfer will be discussed in a following section. In the following discussion for low insertion power measurements, the measuring device can be connected across the circuit (shunt-connected) or inserted in the circuit (series-connected).

33.4.1 Shunt-connected instruments (voltmeters, oscilloscopes, logic probes, etc.)

For measurement or display of voltages in a circuit, a test instrument such as a voltmeter or oscilloscope is normally connected across a circuit in shunt (parallel) with the elements, as illustrated in Fig. 33.14. In this case, the loading effect of the test instrument due to its finite input impedance must be taken into account. The voltage at point A with respect to point B will be modified by the current I_v drawn by the voltmeter. Only if this current is negligible compared with the circuit current I_c will the voltmeter read the same value that existed at A before the meter was connected. Ideally, for these instruments, the input impedance should be infinite, i.e., should appear as an open circuit drawing no current and hence presenting no loading to the

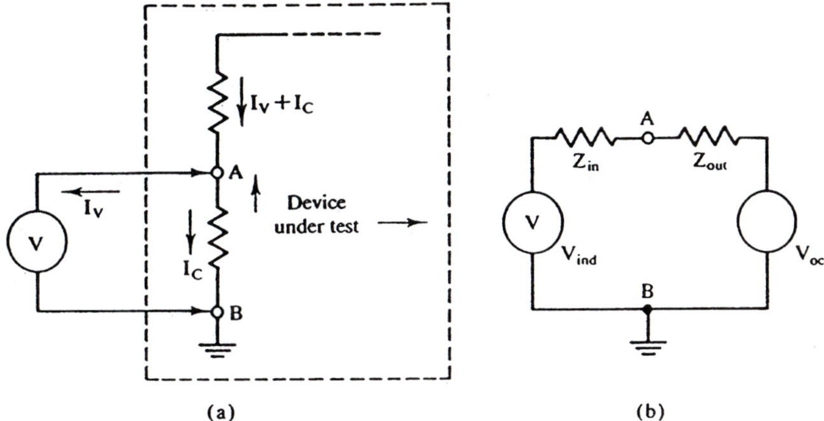

Figure 33.14 (*a*) Voltmeter connected in shunt across circuit points *AB*. (*b*) Equivalent circuit of test setup of (*a*).

circuit. Although this condition is impossible to achieve, the loading effect of the instrument usually can be neglected if it is determined that the input impedance of the test instrument is very large compared with the output impedance of the device under test (usually a factor of 20 or more depending on the desired accuracy). Otherwise, corrections must be made for the loading effects. The voltage will be reduced from the unloaded output voltage of the device under test by a "voltage-divider" action. Referring to the equivalent circuit of Fig. 33.14*b*, the unloaded voltage will be related to the indicated voltage according to the relation

$$V_{ind} = \frac{V_{oc} Z_{in}}{Z_{in} + Z_{out}} \tag{33.15}$$

where V_{ind} = indicated voltage
V_{oc} = open-circuit (unloaded) voltage
Z_{in} = input impedance of the test instrument
Z_{out} = equivalent output impedance of the circuit under test

It should be borne in mind that Z_{out} and Z_{in} will, in general, be frequency-dependent because of their reactance, so the indicated voltage also will depend on frequency. Generally, the shunting effect of Z_{in} becomes worse at higher frequencies (because of the input capacitance of the instrument),* giv-

* In the specifications of most test instruments, the input resistance and an effective shunting capacitance are given. The equivalent series impedance Z_{in} can be calculated by Eq. 33.11:

$$Z_{in} = \frac{Z_1 Z_2}{Z_1 + Z_2} = \frac{(R_{in})(-jX_{in})}{R_{in} - jX_{in}}$$

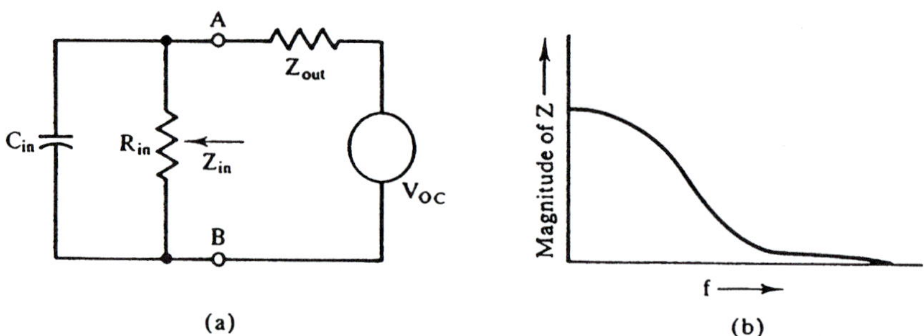

Figure 33.15 (*a*) Equivalent circuit of test setup where Z_{in} includes a shunt capacitance. (*b*) Impedance magnitude of Z_{in} versus frequency, illustrating that loading becomes more severe at higher frequencies.

ing a "high-frequency rolloff" or a reduction in impedance and hence more loading at the higher frequencies (see Fig. 33.15). Also, as a consequence of the shunt capacitance, sharply changing nonsinusoidal waveforms are rounded off because of the finite time it takes to charge the capacitance. This concept is illustrated in Fig. 33.16 for a squarewave source. In addition to the amplitude reduction due to instrumental loading, if the shunting due to the test instrument is severe, it may cause permanent damage to the circuit or device under test.

The loading effect of test instruments can be reduced at the expense of sensitivity by the use of a voltage-divider probe. A commonly used 10:1 probe is illustrated in Fig. 33.17. This device gives a 10:1 increase in impedance level but also attenuates signals by a factor of 10. The probe must be properly adjusted.

Example 33.5: Loading effect of a voltmeter. The voltage across a circuit having an output impedance of 100 kΩ is measured with a 20,000-Ω/V voltmeter. It is desired to find the percentage error in the reading due to the loading effect of the voltmeter on the 10-V range.

The input impedance of the voltmeter on the 10-V range is Z_{in} = (20,000 Ω/V) × (10 V) = 200 kΩ. The indicated voltage will be (Eq. 33.15)

$$V_{ind} = \frac{V_{oc}Z_{in}}{Z_{in} + Z_{out}} = V_{oc}\frac{200\text{ k}\Omega}{300\text{ k}\Omega} = 0.67V_{oc}$$

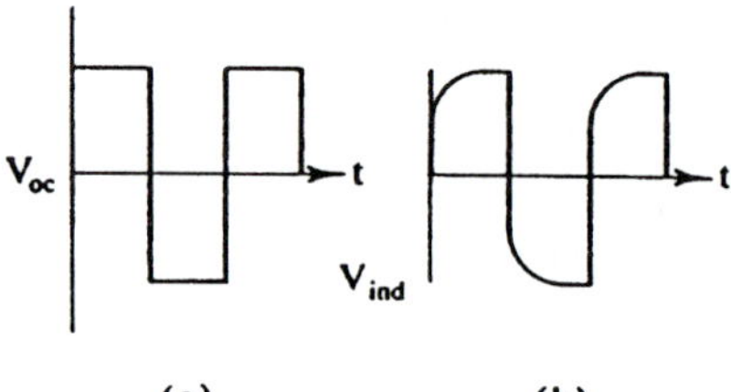

Figure 33.16 (*a*) Squarewave on circuit before instrument is connected. (*b*) Voltage is rounded because of shunt capacitance of the instrument.

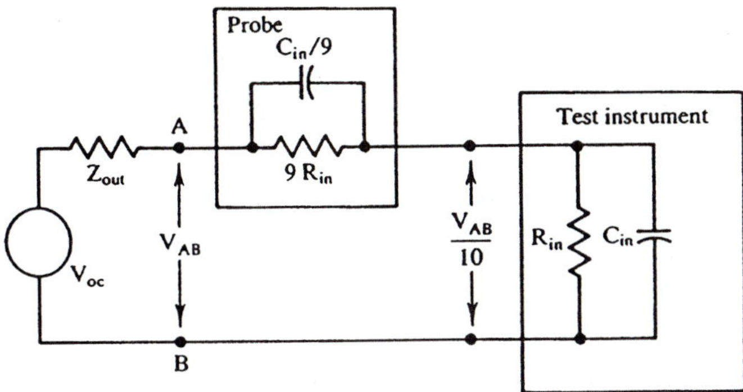

Figure 33.17 A 10:1 voltage-divider probe for increase of input impedance of shunt-connected instruments.

or 67 percent of the true, unloaded voltage value; therefore, the indicated reading (V_{ind}) will be 33 percent low.

The loading problem illustrated in Example 33.5 is typical of cases where a common-type voltmeter such as a small portable multimeter is used. Voltmeters of this type will usually seriously modify the normal in-use voltages in electronic circuits; however, they will not affect low-impedance circuits, such as power supplies. Voltmeters such as electronic voltmeters are available and should be used on high-impedance circuits to decrease this loading problem.

Example 33.6: Effect of loading by oscilloscope at two different frequencies. An oscilloscope having an input resistance of 1 MΩ shunted by a 50-pF capacitance is connected across a circuit having an effective output impedance of 10 kΩ. If the unloaded circuit voltage is a 1-V-peak, 100-kHz sine wave, what will be the voltage indicated on the oscilloscope? Repeat for a 1-MHz sine wave.

The equivalent circuit for the system is shown in Fig. 33.18. The input impedance of the oscilloscope is a 1-MΩ resistor shunted by the capacitive reactance of the 50-pF capacitor. At 100 kHz,

$$X_c = \frac{1}{2\pi fC} = \frac{1}{2\pi \times 10^5 \times (50 \times 10^{-12})} = 32{,}000\ \Omega$$

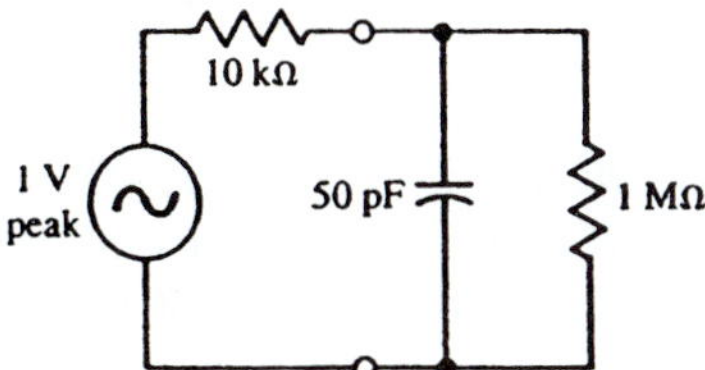

Figure 33.18 Equivalent circuit of Example 33.6.

The parallel combination of R and $-jX_c$ will be

$$Z_{in} = \frac{R(-jX_c)}{R - jX_c} = \frac{10^6 \times (-j3.2 \times 10^4)}{10^6 - j3.2 \times 10^4} = -j3.2 \times 10^4 \ \Omega$$

The indicated voltage V_{ind}, then, will be

$$V_{ind} = \frac{V_{oc} Z_{in}}{Z_{in} + Z_{out}}$$

$$= 1.0 \times \frac{-j3.2 \times 10^4}{-j3.2 \times 10^4 + 10^4} = \frac{3.2 \times 10^4 \angle -90°}{3.4 \times 10^4 \angle -73°} = 0.94 \angle -17° \text{V}$$

This means that the magnitude of the indicated voltage will be 0.94 V (6 percent below the true value) and shifted in phase by 17° lagging behind the unloaded value as shown.

For the frequency of 1 MHz, the shunting reactance of the oscilloscope will be

$$X_c = \frac{1}{2\pi \times 10^6 \times 50 \times 10^{-12}} = 3200 \ \Omega$$

giving

$$Z_{in} = \frac{R(-jX_c)}{R - jX_c} = \frac{10^6(-j3.2 \times 10^3)}{10^6 - j3.2 \times 10^3} \approx -j3200 \ \Omega$$

$$V_{ind} = 1 \times \frac{-j3.2 \times 10^3}{10^4 - j3.2 \times 10^3} = \frac{3.2 \times 10^3 \angle -90°}{1.0 \times 10^4 \angle -18°} \approx 0.32 \angle -72° \text{V}$$

In this case, the observed voltage is less than a third of the unloaded value and lags by 72°. This typical case illustrates the increased shunting problem as the frequency increases. The severe loading illustrated in this example could be lessened by the use of a voltage-divider probe. A 10:1 voltage-divider probe on the oscilloscope will result in an input impedance typically of 10 MΩ shunted by 5 pF. Use of the probe will give an input impedance, then, of about $-j32{,}000\ \Omega$ at 1 MHz, and the indicated voltage will be $0.94\angle -17°$ V, the same as at 100 kHz without the probe. Account has to be taken of the factor-of-10 voltage reduction in the probe. This is easily done by increasing the oscilloscope gain by a factor of 10.

Example 33.7: Effect of loading on squarewave. Consider the effect of the oscilloscope on the circuit of Example 33.6 if the unloaded circuit voltage were a 1-V peak, 250-kHz square-wave. The equivalent circuit of the system will be identical to Example 33.6; however, the problem cannot be solved by the sine wave impedance techniques used previously. In this case, sharp edges will be

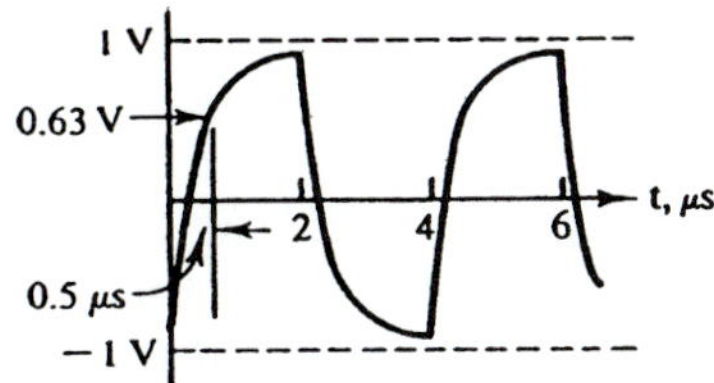

Figure 33.19 Rounding of squarewave by oscilloscope in Example 33.7.

rounded as shown in Fig. 33.19, and a more meaningful concept is that of the rise time of the sharply changing waveform. An idea of the rise time is easily derived in terms of the time constant of the circuit:

$$\tau = RC_{sh} \tag{33.16}$$

where τ = circuit time constant
R = resistance through which the capacitor is charged
C_{sh} = shunt capacitance

If the input resistance of the oscilloscope is high compared with R_{out}, R will be essentially R_{out}. Mathematically, the time constant is the time required for the voltage waveform to change 63 percent of the total ultimate change. In about five time constants, the voltage will be essentially its final value. In the case here,

$$\tau = 10^4 \times 50 \times 10^{-12} = 5 \times 10^{-7}\ \text{s} = 0.5\ \mu\text{s}$$

The 10 to 90 percent rise time is 2.2τ or, in this case, 1.1 μs. The resulting waveform will be as shown in Fig. 33.19.

Example 33.8: Severe detuning of a tuned circuit due to an oscilloscope. As an example of circuit performance deterioration due to shunting effects of test instruments more severe than those considered in the foregoing examples, consider the effect of connecting an oscilloscope across a tuned circuit consisting of a parallel 20-pF capacitor and 50-μH coil. The oscilloscope has input impedance consisting of a 1-MΩ resistance shunted by 50-pF capacitance.

Before the oscilloscope is connected, the tuned circuit resonates at a frequency of

$$f_0 = \frac{1}{2\pi\sqrt{LC}} = \frac{1}{2\pi\sqrt{50 \times 10^{-6} \times 20 \times 10^{-12}}} \approx 5\ \text{MHz}$$

Application of the oscilloscope makes the total capacitance across the coil 70 pF and shifts the resonant point to

$$f_0 = \frac{1}{2\pi\sqrt{50 \times 10^{-6} \times 70 \times 10^{-12}}} \approx 2.7\ \text{MHz}$$

Thus, if the circuit were a tank circuit of a 5-MHz amplifier, its gain at 5 MHz would drop off to nearly zero because the input capacitance of the oscilloscope shifted the resonant peak to 2.7 MHz. If the tuned circuit were in an oscillator circuit, it would shift the oscillations to 2.7 MHz or perhaps cause it to stop oscillating if it did not have sufficient drive at the new frequency.

The detuning of the resonant circuit could be lessened considerably by use of a voltage-divider probe. However, radio-frequency (rf) circuits are always difficult to measure without detuning effects, with the problem becoming worse as the frequency is increased.

33.4.2 Series-connected instruments (ammeters, etc.)

In a few measurement situations, rather than connecting the instrument in shunt with the circuit, it is necessary to break the circuit and insert the instrument in series. A notable example is the ammeter for measurement of current, as illustrated in Fig. 33.20. In this case, the idealized value of instrument input impedance Z_A is opposite from the shunt-connected instrument described above; i.e., the ammeter should appear as a short circuit ($Z_A = 0$). This can be seen from the equivalent circuits in Fig. 33.20. In order for the circuit to be undisturbed when the ammeter is added (Fig. 33.20*b*), the current I and voltage V_{in} should be the same as in Fig. 33.20*a*. This means that the voltage drop across the ammeter V_A must be negligible, which in turn means that the series impedance introduced by the ammeter must be negligible. All practical ammeters will have, of course, a nonzero input impedance,

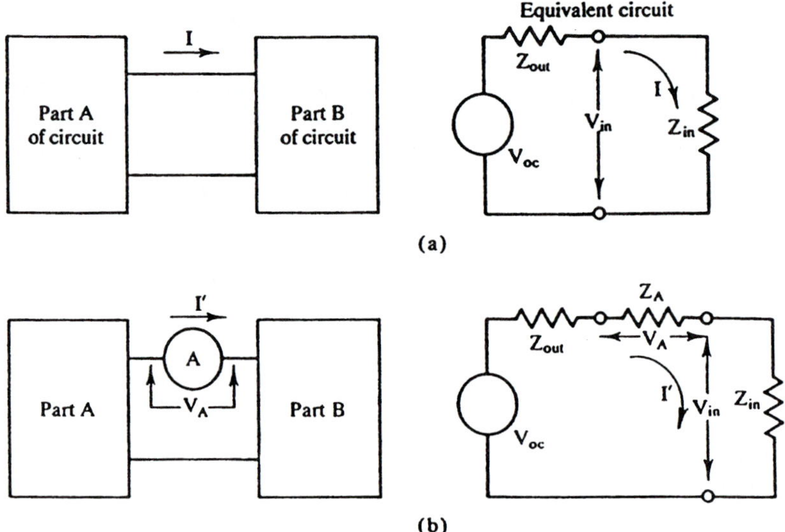

Figure 33.20 Breaking circuit between points A and B and inserting ammeter in series to measure current.

so in a measurement situation care must be taken to keep the ammeter impedance very much lower (say, a factor of 20) than the circuit impedance. The actual factor required depends on the desired accuracy, and it should be borne in mind that corrections can be made for the impedances involved in most cases.

It should be noted here that many ammeters are of the "clamp-on" type, where the instrument probe opens a jaw that then is closed around a wire to measure the current. These instruments act as current transformers and so will inductively couple into the circuit under test rather than directly insert an impedance as described above.

33.5 Impedance Matching

33.5.1 Matching output impedance to load

Typical cases where it is important to consider impedance matching include most applications involving waveform generators (pulse generators, rf transmitters, etc.) which utilize a transmission line, commonly a coaxial cable, to transfer the energy from the source to the input of the device being excited. In addition, there are many relatively low-frequency cases such as audio amplifiers driving loudspeakers and other electromechanical transducers where impedance matching is utilized for achieving high power transfer, proper damping, etc.

If the source has a purely resistive output impedance of R Ω, an impedance match is achieved by making the load or the input impedance of the following device also equal to R Ω. (Neglect, for the moment, the effect of any interconnecting cables, which will be discussed below.) This matched condition can be shown as being necessary for maximum power transference to the load.

If the output impedance of the source contains some reactance ($Z_{out} = R_o + jX_o$), two so-called matched conditions are defined. The first, an image match, occurs where the load impedance is also equal to $R_o + jX_o$. The second case, which is usually more desirable, achieves maximum power transfer, a so-called conjugate or power match, by making the load resistance equal to that of the source and the load reactance equal in magnitude to the source reactance but opposite in sign; i.e., $Z_L = R_o - jX_o$, as shown in Fig. 33.21. Sum-

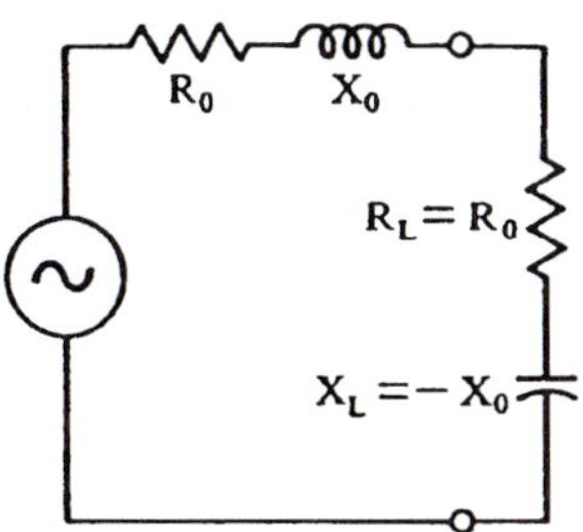

Figure 33.21 Conditions for maximum power transfer from generator to load R_L.

ming the total impedance around this circuit, X_o and X_L cancel to zero, leaving only R_o. This condition is equivalent to making the circuit series resonant, and thus the resistances R_o and R_L are matched, resulting in maximum power transfer.

33.5.2 Instrument interconnections

When interconnecting for rf or other ac signals, open wires are seldom used because they radiate, thereby losing energy and producing stray fields, and these systems are also subject to interference generation and pickup (see Chap. 34). Also, the capacitive effects between the wires and surrounding objects may produce serious degrading effects. Usually, a system of constant-impedance connectors and transmission lines such as coaxial cables or waveguides is used to transfer the energy more efficiently and predictably.* The effect of these transmission lines on the impedance must be considered for proper use of instruments.

In audio-frequency applications, transmission lines as described above are not necessary, and twisted-pair interconnections are common. At these low frequencies, the effect of the wires on the impedances is small, and output and load impedances can be reasonably well matched, neglecting the interconnecting wires.

33.5.3 Effects of transmission lines—matched conditions

All transmission lines have a basic parameter known as "characteristic impedance" Z_0 that determines impedance levels. This characteristic impedance can be defined as the input impedance to a hypothetical infinitely long section of the transmission line. An interesting way to understand this concept is to consider feeding a signal into one end of a line that extends to infinity. Since an electric signal has a finite propagation time, any signal starting at the input end of the line would never reach the other end. Thus the generator would never "know" what the load is on the other end of the line. The question, then, is, "If the generator has a known output voltage, what is the current that starts down the line?" This current is determined by the wire inductance and shunt capacitance of the line and is known as the "characteristic impedance" of the line. Over the most useful frequency range of transmission lines, this impedance is very nearly a constant resistance of relatively small magnitude, usually in the range of 25 to 600 Ω.

The most commonly encountered transmission systems have 50-Ω characteristic impedances and hence are called "50-Ω systems," e.g., common coaxial

* Normally, a coaxial cable is used for rf connections at frequencies to approximately 1 GHz. At frequencies above this, the coaxial cable becomes too lossy for long distances, and a hollow pipe known as a "waveguide" is utilized.

cables such as RG8/U and RG58/U. Also frequently encountered are 75- and occasionally 93-Ω systems. The input impedance of a transmission line of characteristic impedance Z_0, if terminated with an impedance Z_0, also will be Z_0. For example, the input impedance of a 50-Ω line with a 50-Ω resistor connected on the opposite end will be 50 Ω regardless of the length of the line.*

Most signal sources which have high-frequency components have output impedances that approximate 50-Ω resistance. If such a device is to be connected to a 50-Ω load, a matched system is simply achieved by use of a 50-Ω interconnecting cable. In other cases, it may be desirable to transform the load and/or source impedance to values that will match common transmission lines by use of series reactance, impedance transformers, or other impedance-matching techniques.

The transit of electric signals down a transmission line takes a finite time; hence this delay time for signal propagation must be considered. For most coaxial cables, the velocity of propagation is approximately two-thirds that of light in free space, giving a value of about 2×10^8 m/s. This means that a propagation time of about $0.16\ \mu s = 160$ ns per 100 ft of cable. (It takes 160 ns for a signal to travel 100 ft down this cable.)

33.6 Effects of Impedance Mismatch

Unmatched systems present many difficulties. The input impedance of a transmission line terminated in an impedance other than its characteristic impedance is, in general, a much different value from the terminating impedance. The transmission line transforms the impedance in a complex manner that is a function of the line length, operating frequency, and the degree of mismatch on the line. The impedance transformation is due to the fact that all the energy incident on the load will not be absorbed as in the matched case; part of the energy will be reflected back up the line toward the source, resulting in standing waves on the line. As a consequence, the load on the generator is not known unless measured or calculated, the power to the load will be less than maximum, and the standing waves on the line will increase the line loss. For pulse sources, the reflections back and forth on an unmatched line system may be particularly bothersome.

The reflections are caused by the finite transit time down the line. When the generator sends a pulse down the line, it sends it into the characteristic impedance of the line. Thus the power sent down the line is independent of the load on the end of the line. When the pulse reaches the end of the line, if

* Actually, there is a small reactance associated with the power loss in the cable that is usually negligible if the lines are relatively short and used well below the maximum usable frequency. The power loss in the cable (attenuation) may be important for long cables or high frequencies. This loss is usually given in reference books in decibels per 100 ft at various frequencies.

there is a mismatch, some of the power will be reflected, and a pulse will then be propagated back up the line. In the extreme case where the generator is also mismatched, there will be a large number of reflections or bounces back and forth on the line, causing multiple pulses on the output for each single input pulse.

A section of line shorter than a quarter wavelength which is unterminated (essentially open-circuited) and connected across a circuit will appear nearly as a capacitance shunting the circuit. (The capacitance of unterminated cables is listed in reference books; e.g., RG58/U has 28.5 pF/ft.) This effective shunting capacitance may have a detuning or other deleterious effect.

A final problem associated with unmatched systems is the fact that the calibrated output of signal generators is normally specified for a matched-load condition, i.e., when a resistance equal to the generator output impedance is connected to its output terminals. Under these matched conditions, equal voltage will appear across the load, and the internal impedance of the generator and the calibration should be correct. This means that if the generator is essentially unloaded, its output voltage will be twice the listed value for matched output. If the effective load is smaller in magnitude than Z_0, the voltage at the output terminals will be less than the calibrated value. The corrections for unmatched values of impedance are the same as those of Eq. 33.15, and again, extreme loads (approaching short circuit) may distort the waveform.

Chapter

34

Electrical Interference

D. A. Burt and K. D. Baker
Department of Electrical Engineering
Utah State University, Logan, Utah

34.1 Introduction

In a communications or measurement situation, any signal disturbance other than the desired signal is termed "interference." These extraneous signals, which hinder the measurement of the desired signals, assume a variety of forms and can find many devious ways of getting into or out of electronic equipment. This chapter is devoted to describing these unwanted conditions and means of reducing or eliminating their effects.

34.2 Interference and Noise Terminology

atmospheric noise or interference Radio-wave disturbances originating in the atmosphere, principally because of lightning discharges, also called "atmospheric" or simply "sferics."

common-mode interference Conducted interference caused by voltage drops across wires (usually grounds) common to two circuits or systems.

conducted interference Interference caused by direct coupling of extraneous signals through wires, components, etc.

cosmic noise Interference caused by radio waves emanating from extraterrestrial sources.

coupling The transfer of power between two or more circuits or systems.

crosstalk Electrical disturbances in one circuit as a result of coupling with other circuits.

electromagnetic interference (EMI) A general term for electrical interference throughout the frequency spectrum from subaudio up to microwave frequencies.

electrostatic induction Signals coupled to the measuring circuit through stray capacitances, also commonly called "capacitive pickup."

hum Electrical disturbance at the ac power-supply frequency or harmonics thereof.

impulse noise Noise generated in a discrete energy burst (not of random nature) which has an individual characteristic wave shape. (This is the usual type of noise generated by rotating machinery such as a dc motor or generator and switching regulators.)

interference Extraneous signals, noises, etc. which hinder proper measurements in electronic systems.

magnetic induction Interference coupled to the measuring circuit by magnetic fields.

noise Unwanted signals, commonly used to identify statistically random disturbances.

radiated interference Interference transmitted from a source to another remote point with no apparent connection between the points.

radio-frequency interference (RFI) Electromagnetic interference (EMI) in the frequency band normally used for communications (approximately 10^4 to 10^{12} Hz).

random noise Irregular signal whose instantaneous amplitude is distributed randomly with respect to time; also called "gaussian noise" because mathematically, the distribution follows a normal or gaussian curve.

signal-to-noise ratio A ratio of signal level to noise level. (This is sometimes expressed in voltage and sometimes in power; how it's used in a specific case is determined by context. The higher the signal-to-noise ratio, the less is the importance of the interference.)

static Radio interference (crackling sound) detectable as noise in the audio stages of a radio receiver.

thermal noise Random radio-frequency noise generated by thermal agitation of electrons in a resistor.

white noise An electric signal whose frequency spectrum (power) is continuous and uniform.

34.3 Instrument Noise

Internal noise is generated in all electronic equipment and limits the ultimate measurement sensitivity that can be achieved. Depending on the instrument bandwidth and input configuration, there is a theoretical noise due to thermal agitation given by

$$V_n = 2\sqrt{kTBR} \tag{34.1}$$

where V_n = noise voltage rms
k = Boltzmann's constant, 1.38×10^{-23} J/K
T = absolute temperature, K

B = system bandwidth, Hz
R = input resistance, Ω

Reduction of the noise beyond this theoretical value is not possible without changing the basic design of the instrument. As the instrument's sensitivity approaches this theoretical limit, the signal becomes masked by the noise.

Since noise introduced at the input of an amplifier, called "transducer" or "detector noise," will be amplified by succeeding stages along with the signal, the noise associated with the input stage usually will determine the overall signal-to-noise ratio and therefore will be the limiting parameter for ultimate useful sensitivity or the minimum detectable signal of an instrument.

Most instruments are designed to maintain a noise level below the sensitivity at which the instrument is intended to operate. Therefore, internal noise of the instrument is not evident at the output. This fact, however, determines the minimum measurable input signal. Operation can be extended to lower signal levels by additional amplification of the output down to the limit where the signal approaches the noise level, as determined at the input of the instrument.* Once this limit is reached, further sensitivity cannot be achieved without reducing the input noise level. Only a few means are available to minimize the noise limitation introduced by the input characteristics of instrumentation. These are as follows:

1. Optimize the instrument noise figure to approach as nearly as possible the theoretical noise (Eq. 34.1). This can be accomplished in some cases by critical adjustment of the input circuitry (tuning, impedance level, etc.) and careful selection of low-noise input devices, i.e., diode or transistor.
2. Use a low-noise preamplifier that has lower inherent noise preceding the instrument.
3. Reduce system bandwidth or frequency response by filtering (if the nature of the desired signal will not be too seriously impaired).
4. Cool the input circuitry to reduce thermal-agitation noise.

34.4 Radiated Interference

All electric equipment can radiate interfering signals, and with the proliferation of electric equipment, this radiation is becoming more and more serious. A radiated signal can be classified into one of three types: electric or high-impedance, magnetic or low-impedance, and plane wave. The electric and magnetic coupling is most important when the radiating source is on the or-

* The use of external amplifiers may complicate the calibration of the measurement system.

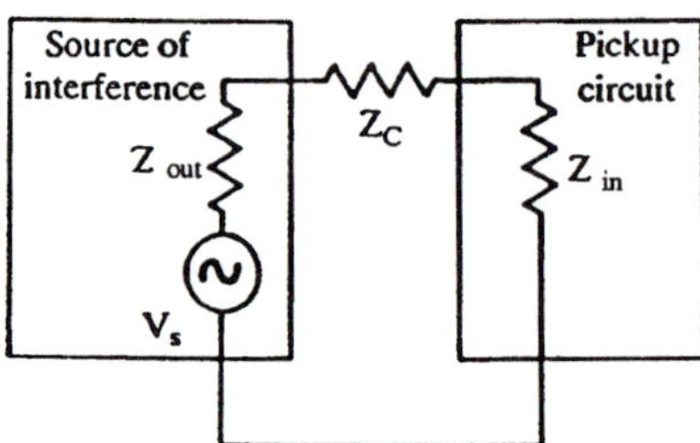

Figure 34.1 Equivalent circuit illustrating interference coupling.

der of a few wavelengths or less away. This is called the "near-field region." At larger separations, the electric and magnetic fields become negligible, and plane-wave propagation takes over. This region is called the "far field."

34.4.1 Electric or high-impedance interference

This interference is essentially capacitive coupling from sources that have high potentials and little current flow. The characteristic coupling from these sources is directly proportional to the frequency of the signal and inversely proportional to the separation distance from the source.

The coupling of energy from the source to the pickup circuit can be described in terms of an impedance Z_C coupling the two together, as illustrated in Fig. 34.1. The actual energy coupled between the two circuits is determined not only by the coupling impedance but also by the source voltage V_s, the output impedance of the source Z_{out}, and the input impedance of the pickup point Z_{in} (see Chap. 33). It is difficult to assign exact numbers to the value of coupling impedance, since the exact configuration must be known. However in general, for frequencies below about 10 MHz and at distances from inches to several feet, this impedance will be at least several thousand ohms, but its value decreases as the frequency increases.

Some of the more obvious ways to reduce the effects of this interference are to maintain low-voltage and low impedance levels at both the source and the pickup points, require large spacing, keep all radiating and pickup areas as small as possible, and limit the high-frequency response of the system. Since these methods may not be adequate, shielding is also used. Proper shielding will be discussed later in this chapter.

34.4.2 Magnetic or low-impedance interference

Magnetic interference is coupled into a circuit by the transformer action of stray magnetic fluxes. Sources for magnetic coupling are usually a loop or wire carrying a significant amount of current. The coupling between the source and receiver is inversely proportional to the frequency and distance. For frequencies below 1 MHz and at distances ranging from inches to several feet, the coupling impedance can be on the order of a few ohms. Some methods to reduce these types of signals are to use low currents and high imped-

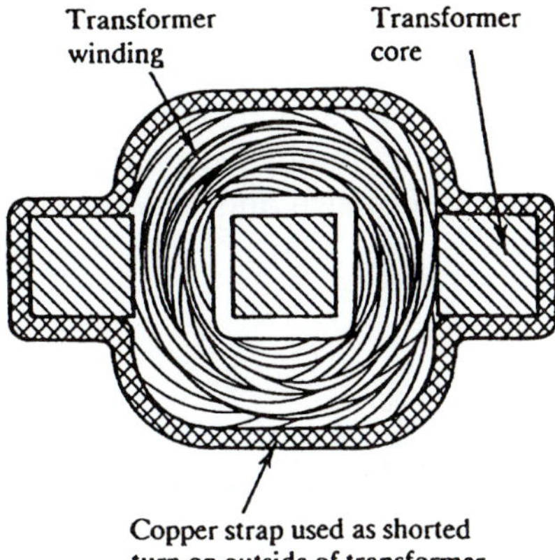

Figure 34.2 Shorted turn on outside of transformer to attenuate stray magnetic fields.

ance levels, require large spacing, keep current-carrying wires short, and reduce the low-frequency response of the system. It is interesting to note the paradox that exists in considering the ways to reduce electric and magnetic noise. Some techniques that reduce the effects of one type of interference actually increase the other type. However, both types of noise may be reduced by increasing distance and by shielding.

Magnetic interference is the most difficult to shield against; therefore, it is often easier to control it at the source. When attempting to eliminate stray magnetic fluxes, consider the principle that the magnetic flux produced in any closed path is proportional to the sum of the current enclosed by that path. Simply stated, this means that if there is a device that draws sufficient current to produce interfering fields, the power and return leads should be designed to run side by side in a twisted pair or a coaxial power cable. This simple procedure requires that the current which flows to the unit must flow back the same way so that the sum of the current inside a closed path around the two wires is zero, thereby reducing the external magnetic fields.

Another common source of magnetic interference is the stray flux that may emanate from power transformers. A common method to minimize these fields is to provide a shorted turn around the outside of the transformer core. This shorted turn must have low impedance and be wrapped in the same direction as the transformer windings, as shown in Fig. 34.2. Thus leakage fluxes outside the transformer will be attenuated by the short circuit caused by this shorted turn.

The physical proximity is of utmost importance in determining the level of interference. In some cases, the interference can be reduced to a tolerable level only by physical separation and by proper orientation of wires.

34.4.3 Plane-wave interference

Radiated electromagnetic fields approximate a plane wave in the far field (all radio propagation is due to plane-wave propagation). The magnitude of the coupling impedance between the source and the receiver depends on the sending and receiving configurations and their separation distance. At frequencies above 1 GHz, practically all coupling is due to plane-wave propagation. Rejection of this interfering signal is again accomplished mainly by requiring the radiating and pickup points to be small in size and widely separated, as well as by the use of shielding.

34.5 Shielding

One of the most important methods of eliminating interference is the use of shields at either the source or the receiver or both. This technique is to enclose the device in a conductive shield which must be designed to reflect and/or absorb electromagnetic energy that attempts to penetrate it. The design of the shield itself usually depends on the frequencies and types of interference that must be contained or excluded.

34.5.1 Plane-wave and electric-field shielding

When a plane wave or high-impedance electric signal impinges on a surface of high conductivity, the signal is essentially all reflected. Shielding then is effected by enclosing the object to be shielded in a completely closed conductive surface. For this surface to be effective, it must have high conductivity, and there can be no discontinuity or holes which approach a significant part of the wavelength of the signal that is to be rejected. This might appear to require a completely welded silver case. Because of skin effect (see Chap. 36), however, a silver plating will produce almost the same effective conductivity as would solid silver for most frequencies of concern. For very high frequencies, some critical instruments will have silver-plated chasses and shields; however, in normal applications, most conductive metals will produce satisfactory results. To avoid heavy, bulky, solid-metal boxes, metallic-wire mesh is often used as a shield material. Mesh having about 50 percent solid area and 60 or more strands per wavelength is nearly as effective as solid metal.

Most electronics are constructed on printed circuit boards, which can be single- or multilayer boards. A good practice is to provide one layer of the board as a solid copper sheet connected as a ground plane with holes where wires need to feed through. This ground plane can then act as a shield between components mounted on opposite sides of it and also as a low-impedance ground point for all the electronics on the board.

In most instruments, the outer enclosure is made of metal and designed as the final shield. With the exception of floating inputs, this enclosure will be connected directly to the input signal return or to the ground of the instrument. If the shield is maintained at the same potential as the ground point of the amplifier input, the interfering signal will be shunted out and will not

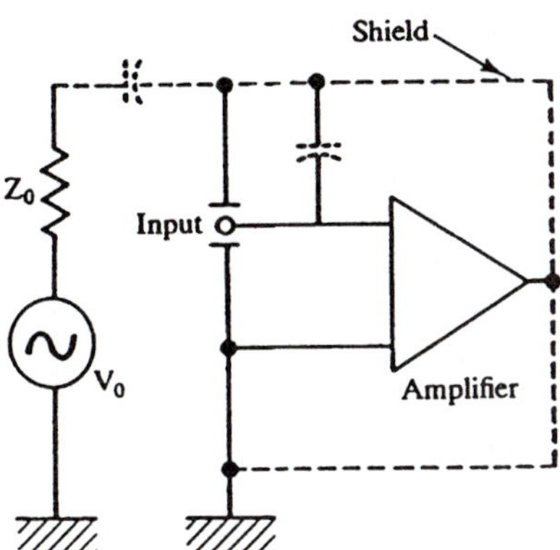

Figure 34.3 Proper connection of shield to eliminate the effects of internal shunt capacitance to the shield.

affect the input. This can be seen by referring to Fig. 34.3, which shows an amplifier as an example. As can be seen, the interfering source will be coupled to the shield through stray capacitance, as well as the amplifier input. As long as the shield is maintained at the same potential as the amplifier input, there will be no energy flow between the shield and the amplifier input. In effect, the interfering signal coupled to the shield will be shunted out.

The shields of interconnecting cables are basically an extension of the cases around instruments that are being connected together. Their primary purpose is to guard the cable and minimize conducted interference into the circuits from external radiation sources.

34.5.2 Magnetic shielding

Shielding to eliminate magnetic fields is more difficult than shielding for other types of interference because of the low coupling-impedance level. These fields may be static fields such as the earth's magnetic field or the time-varying fluxes that emanate from current loops in the near field. As stated previously, it is better to eliminate as much of the interference at the source as possible. Once a magnetic field has been reduced to a reasonable level, it becomes practical to use magnetic shielding for further reduction. The basis for magnetic shielding is to provide a low-reluctance path that will attenuate the field external to the sensitive device being shielded or, conversely, to reduce magnetic flux emanating from the instrument. The procedure then requires the use of metal for shielding with high magnetic permeability such as transformer iron or Mumetal. The proper permeability and thickness must be selected, however, according to the strength of the field. If the permeability is too high for the cross-sectional area, the metal and magnetic field strength may saturate, resulting in a poor shield. A common procedure is to use a shield built up in layers, with an air gap or nonmagnetic material between each layer, the lowest permeability metal being on the interference source side of the shield to avoid saturation. Depending on the strength of the field, many layers of shielding may be required. If the object to be shielded is large, this type of shielding can become both expensive and heavy quite rapidly.

Magnetic shields must be preformed without seams and then heat treated to achieve their maximum capabilities. Any type of discontinuity can distort the magnetic path and cause leakage from that point. Bending metal even once can change its characteristics sufficiently to cause such irregularities.

General shielding is normally designed to eliminate electric fields and decrease magnetic fields to some degree. For this reason, most instrument enclosures are made from mild steel. This gives a higher penetration loss than nonmagnetic materials but does not have quite the reflectivity of the higher-conductivity materials such as copper.

34.5.3 Holes in shields

Instrument setups normally require holes through the shield for ventilation. As a rule, for good interference rejection, all uncovered holes should be less than ⅛ in in diameter. If larger holes are necessary, they should be covered with a fine wire mesh. In the case where it is necessary to have a large hole with no screen, the hole should be designed with protruding sleeves, as shown in Fig. 34.4. This type of design acts as a waveguide beyond cutoff and will attenuate all signals below the cutoff frequency of the waveguide. The equations giving the cutoff frequency and attenuation are given in Fig. 34.4. If the sleeve is made three times as long as the diameter, an attenuation of approximately 100 dB can be achieved for frequencies one-fifth the cutoff frequency or below. At frequencies above the cutoff frequency, however, the sleeve becomes a waveguide and will allow signals to enter directly into the instrument. In a typical case, a sleeve diameter of 2 cm and a length of 6 cm

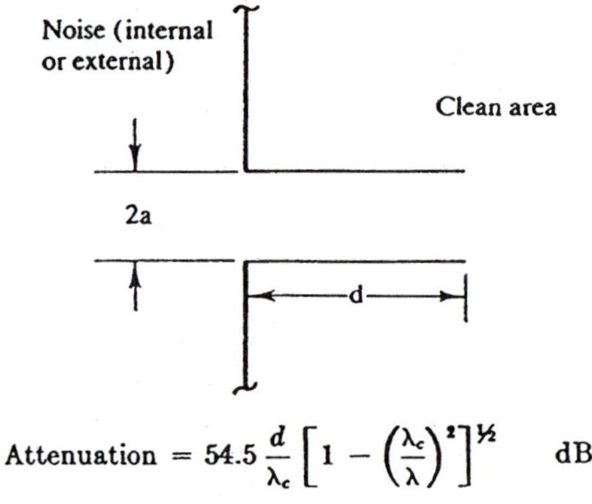

$$\text{Attenuation} = 54.5 \frac{d}{\lambda_c} \left[1 - \left(\frac{\lambda_c}{\lambda} \right)^2 \right]^{1/2} \quad \text{dB}$$

where λ_c = waveguide cutoff wavelength = $3.412a$
λ = wavelength of interfering signal
d = length of aperture, m
a = radius of aperture, m

Figure 34.4 Design for sleeve to provide a large hole in a shield.

will have a cutoff frequency of about 9 GHz and an attenuation of approximately 100 dB below 2 GHz.

34.5.4 Shield bonds

At joints in the shield and where access doors are necessary, continuous electrical and mechanical bonds are required. Therefore, the joints should be permanent where possible (by welding or brazing) or semipermanent with good, clean, machined metallic surfaces bolted together with a radio-frequency (rf) gasket between them. The rf gasket will ensure continuous low-impedance contact throughout the joint.

Bonding straps of copper or braid may be used where direct bonds are not possible, and these function reasonably well at lower frequencies. However, this type of bond is never as good as a direct bond. In general, since a low-impedance path between points is required, it can be accomplished only when the path length between the points is short in comparison with the wavelength being rejected.

34.6 Conducted Interference

Any wires that must penetrate the shield enclosure provide a path for conducting an interfering signal through the shield, thereby conveying the noise directly or, once the shield is penetrated, by radiation into the measuring system. This conducted interference can be reduced by proper filtering.

In some cases, the filtering can be accomplished satisfactorily by a simple bypass capacitor to the case. The most effective type of bypass capacitor is a feedthrough design. Most electronic components are mounted on printed circuit boards. These boards can have one layer used as a ground plane. It is a good practice to use bypass capacitors with as short leads as possible between this ground plane and each of the major integrated circuit chip supply voltages. Simple ferrite beads placed on wire leads as they enter the enclosure will help reduce the interference.

34.6.1 Filters

Many filter designs are available for interference reduction. The choice of filter types is determined by the frequency and characteristics of the desired signals as well as the interference to be rejected. Accordingly, filters are classified as low-pass, high-pass, or bandpass depending on the range of desired frequencies to be transmitted. The basic idea of the filter is to provide a high series impedance and a low-impedance shunt for the interfering signal. The filter should affect the desired signals as little as possible.

Low-pass filters are used in primary power leads to reduce high-frequency disturbances, the most common type of conducted interference. A simple low-

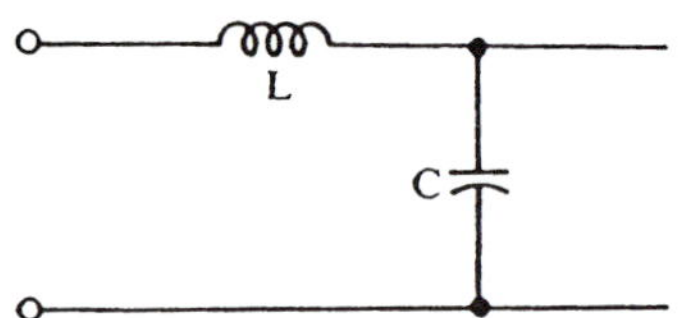

Figure 34.5 Simple LC low-pass filter.

pass filter is illustrated in Fig. 34.5, which uses a series inductor for high impedance at high frequencies and a capacitor to shunt the high frequencies to ground. Because of this low-pass characteristic of the filter, it introduces small insertion loss for the power currents and high insertion loss for higher-frequency interference currents.

Additional filter types using only passive components are shown in Fig. 34.6. For high-frequency signal circuits, the low-frequency interference components can be reduced by the high-pass filter. The bandpass filter can be used to selectively pass a narrow-frequency-range signal, rejecting all other frequencies. This filter is particularly effective in suppressing undesired harmonics of the signal. A similar filter known as a "bandstop" or "blocking filter" (sometimes referred to as a "trap") can selectively attenuate a particularly bothersome interference of specific frequency.

If application of the simple filter types just described is insufficient, two or more filters can be placed in series, or more sophisticated types of filters can be utilized. In particular, the use of active filters utilizing operational amplifiers can provide much more off-band rejection and steeper skirts on the passband curve.

Care must be taken in installing filters or their effectiveness in reducing the interference may be impaired or negated. It is important that no part of the wire be exposed within the shielded area prior to the filter, for this wire can radiate or receive noise and seriously reduce the effect of the filter. The proper method of applying filters is illustrated in Fig. 34.7. The filter should be placed as near to the source of interference as possible and must be shielded and grounded properly to provide a low-impedance shunt for bypassing the interfering signal current without allowing it to couple energy into signal circuits. Attention must be paid to inductive and capacitive coupling of interfering signals to other coils and leads by physical isolation, particularly of the input leads to the filter and cleared output leads.

34.7 Common-Mode Interference

Common-mode interference is the undesirable response of an instrument to a signal that is introduced by the inability to provide the same ground potential to an instrument and its signal source. This can be caused by long cables

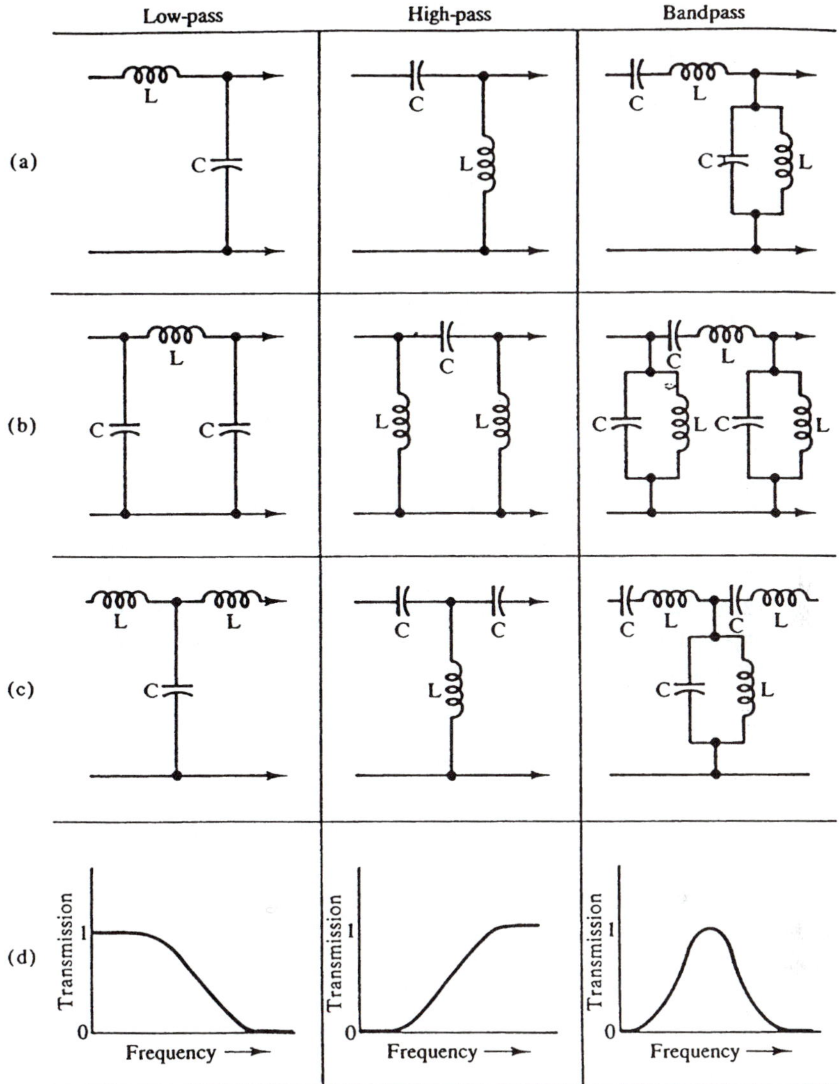

Figure 34.6 Simple low-pass, high-pass, and bandpass filters. (a) L section; (b) π section; (c) T section; (d) the transmission factor as a function of frequency is a measure of the effectiveness of the filter. Ideally, it would have a value of unity for desired frequencies and zero for interfering frequencies.

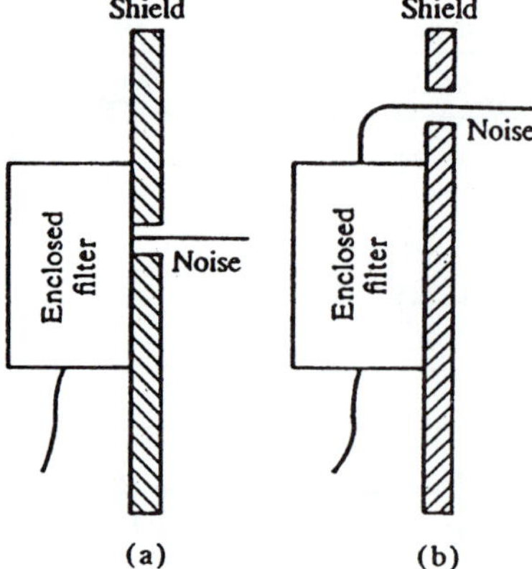

Figure 34.7 Placement of input filter for any wires penetrating a shield. (a) Correct; (b) incorrect.

picking up signals, by electrostatic or magnetic coupling, or just by poor grounding, resulting in ground loops, as discussed in Chap. 35. An illustration of the common-mode interference problem is shown in Fig. 34.8*a*, illustrating an amplifier with input impedance Z_{in} connected by a cable to a distant transducer. The transducer ground *A* and amplifier ground *B* are widely separated and may, therefore, not be at the same potential. This circuit can be represented by the equivalent circuit shown in Fig. 34.8*b*, where the differ-

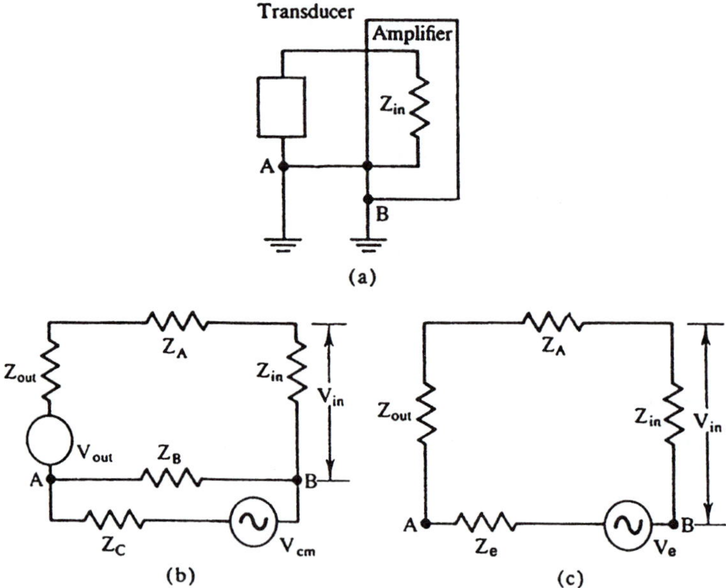

Figure 34.8 Common-mode interference. (a) Typical amplifier with remote transducer; (b) equivalent circuit of system in (a); (c) simplified equivalent circuit of (a).

ence potential between the output-transducer ground A and the amplifier ground B has been represented by a voltage source V_{cm} and series resistance Z_C. The transducer is represented by a voltage source V_{out} with an output impedance Z_{out}. The cable signal-conductor and return-lead impedances are represented by Z_A and Z_B, respectively.

In order to simplify the analysis, assume for the moment that the transducer voltage V_{out} is zero. Now the interference voltage V_{in} due to the common-mode voltage can be readily determined from the equivalent circuit. The portion of the circuit containing the voltage source V_{cm} and impedances Z_C and Z_B can be replaced by a second equivalent circuit (Thevenin's equivalent), as shown in Fig. 34.8*c*. The equivalent voltage source will be

$$V_e = V_{cm} \frac{Z_B}{Z_B + Z_C} \tag{34.2}$$

where V_e = equivalent interference source voltage
V_{cm} = common-mode voltage
Z_C = common-mode source impedance
Z_B = line impedance

and the equivalent output impedance at point AB is

$$Z_e = \frac{Z_B Z_C}{Z_B + Z_C} \tag{34.3}$$

where Z_e is the equivalent impedance at AB.

The voltage V_{in} now is simply the fraction of V_e that appears across Z_{in}:

$$V_{in} = \frac{V_e Z_{in}}{Z_{in} + Z_{out} + Z_A + Z_e} \tag{34.4}$$

where V_{in} = interference voltage at the amplifier input
Z_{in} = amplifier input impedance
Z_{out} = transducer output impedance

If typical values are substituted into these equations, the seriousness of the problem can readily be seen. Assuming Z_{in} = 10 MΩ, Z_{out} = 1000 Ω, Z_A = Z_B = 10 Ω, and Z_C = 1 Ω, the following results are obtained:

$$V_e = V_{cm} \qquad \approx V_{cm}$$

$$Z_e = \qquad \approx 1\Omega$$

$$V_{in} = V_{cm} \frac{10^7}{10^7 + 10^3 + 10 + 1} \approx V_{cm}$$

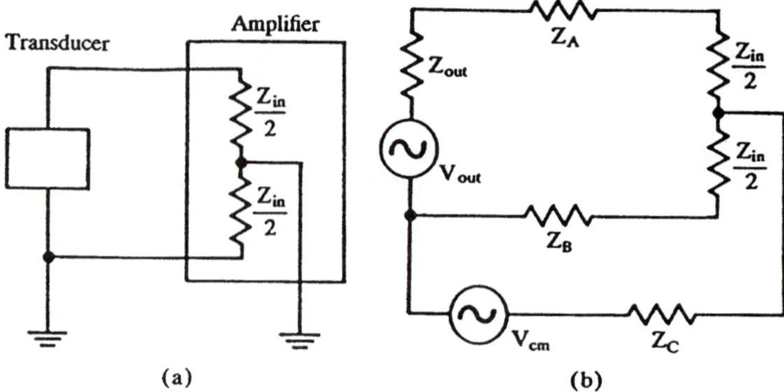

Figure 34.9 Balanced input to ground amplifier. (*a*) Typical connection for balanced input to ground amplifier; (*b*) equivalent circuit of (*a*).

It is completely possible for V_{cm} on a long cable to be on the order of 1 V. If sensitive (millivolt) readings are to be made, they are obviously impossible with this type of connection.

34.7.1 Reduction of common-mode interference by use of differential inputs

By the use of proper connections to differential inputs, the common-mode problem can be reduced. An instrument has a "differential input" if it responds to the algebraic difference between two separate input signals. Three commonly used types of differential inputs are illustrated in Figs. 34.9 through 34.11. Consider first the balanced input amplifier shown in Fig. 34.9*a* and its equivalent circuit in Fig. 34.9*b*. In this case, neither input to the amplifier is grounded, and the input impedance Z_{in} can be visualized as split and balanced to ground. As long as the input impedance is large com-

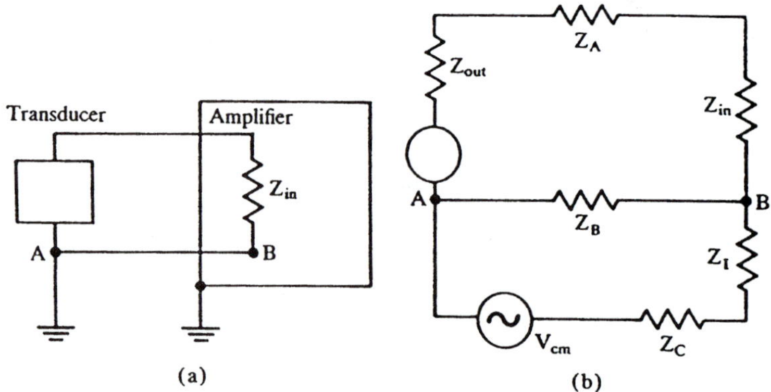

Figure 34.10 Floating-input amplifier for reduction of common-mode interference. (*a*) Typical floating-input amplifier; (*b*) equivalent circuit of (*a*).

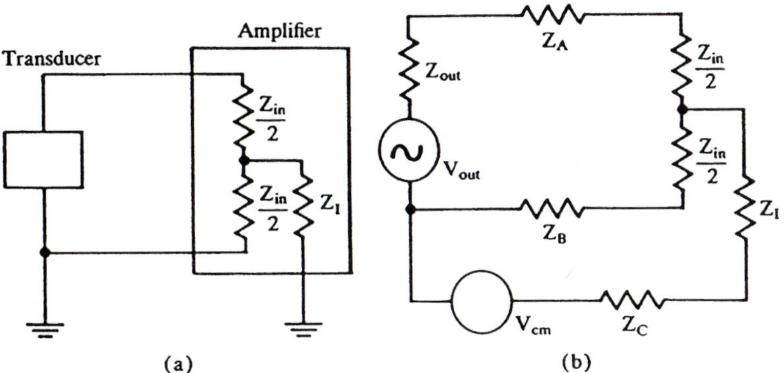

Figure 34.11 Balanced differential amplifier with isolation to ground. (*a*) Typical connection for balanced differential amplifier; (*b*) equivalent circuit for (*a*).

pared with all other impedances, it is obvious that V_{cm} will essentially appear equally at each input with respect to ground; therefore, the common mode will cancel. If it were possible to make the system perfectly balanced by using a balanced-to-ground transducer or making $Z_{out} + Z_A = Z_B$, it is theoretically possible for the amplifier output to be independent of V_{cm}. However, in practical cases, this is not completely possible.

The floating-input amplifier shown in Fig. 34.10 differs from the balanced-input case of Fig. 34.9 in that the inputs are isolated from ground with no intentional impedance connections to the ground system of the amplifier. Although the inputs are floating, in practice, both will have some finite leakage to ground. Normally, this leakage will be unbalanced to the extent that the leakage on one input terminal may be negligible compared with that on the other. The equivalent circuit for this case is shown in Fig. 34.10*b*, where Z_I represents the leakage impedance. This equivalent circuit can be simplified to give a Thevenin's equivalent circuit identical to Fig. 34.8*c*. However, the equivalent voltage will be

$$V_e = V_{cm} \frac{Z_B}{Z_B + Z_C + Z_I} \tag{34.5}$$

where V_e = equivalent interference voltage
V_{cm} = common-mode voltage
Z_B = line impedance
Z_C = common-mode source impedance

It can be seen then that if Z_I can be maintained very large, V_e can be reduced to the point that the amplifier input voltage will be insignificant, since V_{cm} will appear almost entirely across Z_I. It should be pointed out that in practical cases, Z_I will consist of the actual amplifier-leakage impedance in parallel with insulation and line impedance to ground. Any leakage on lines, such as

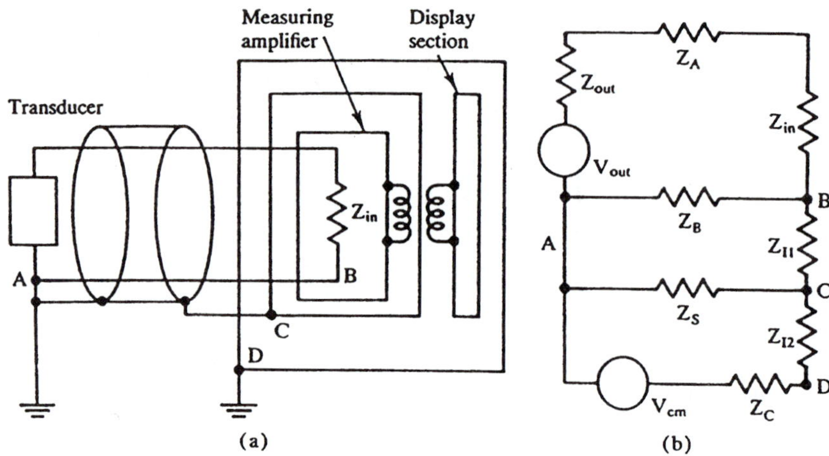

Figure 34.12 Connection for guarded-section amplifier. (*a*) Typical connection for guarded-section amplifier; (*b*) equivalent circuit of (*a*).

that due to contamination or condensation, could reduce this impedance and cause difficulty.

The use of an amplifier that has a balanced floating input to ground, as shown in Fig. 34.11, combines the advantages of both prior systems and diminishes the disadvantages.

34.7.2 Guarding

"Guarding" is a technique that is often used to reduce common-mode interference as well as to reduce the effects of shunt capacitance and leakage.* The guarded connection for common mode rejection is given in Fig. 34.12. This technique is essentially a floating input with a guard system for making the leakage negligible. This isolation is achieved by enclosing the measuring or input section of the instrument in a shield that is isolated from ground. Also, the guarded system effectively utilizes a cable shield. The equivalent circuit of this system is shown in Fig. 34.12*b*. Z_s represents the cable shield impedance, Z_{I1} represents the isolation impedance from the amplifier from input *B* to the guard shield *C* and ground point *D*. The common-mode voltage due to any potential difference between grounds *A* and *D* will be dropped across Z_{I2} because of the low-impedance path through the shield. Input *B* is further isolated by isolation impedance Z_{I1}; thus the interference input to the amplifier is essentially nonexistent.

* Application of the guarding technique for the cancellation of leakage-impedance effects is discussed in Chap. 36.

If it is not necessary to connect the source transducer to ground, an alternate solution is to leave the source ungrounded and connect its shield only to the cable shield of a differential unguarded amplifier. This will give the same results as the guarded amplifier if the impedance to ground at the transducer is maintained at a value comparable with the impedance Z_{I2} of the guarded amplifier.

34.7.3 Isolation

The methods discussed in this chapter all require significant isolation between sections of the electronics. There are several ways to accomplish this; however, the most effective way is by the use of optical isolators. Numerous devices capable of providing the required isolation are available. The best way to use them is to convert the signals to digital form and then pass them through the isolator. This data transfer will then not degrade the information beyond that caused by the analog-to-digital converter. At the present time, optical isolators do have a bit-rate limitation (bit rates that can be up to tens of megahertz). In those circuits which require higher bit rates, rf techniques can be used. Once again, high-speed matched and balanced drivers and receivers are available. In most instances, the isolation impedance is limited by stray capacitance. Therefore, its effectiveness in reducing common-mode interference will be less at the higher frequencies.

34.8 Crosstalk

In most practical instrumentation systems, there will be different signals in different channels. Any time the signal from one channel appears in a second adjacent channel, it can be termed "crosstalk." The presence of this type of interference can be verified readily if the suspected interfering channel can be deactivated.

One instance of crosstalk occurs when the input of an instrument has a number of different input signals that can be selected, as shown in Fig. 34.13*a* for two signals. In this case of two signals, when V_1 is selected as illustrated, V_2 is the unwanted signal; however, the stray leakages in the switch can couple this voltage into the amplifier. An equivalent circuit for this is shown in Fig. 34.13*b*, and the voltage that will appear on the amplifier's input due to V_2 is (for large Z_{in})

$$V_{in} = \frac{V_2(Z_1 + Z_C)}{Z_1 + Z_C + Z_2 + Z_o} \tag{34.6}$$

where V_{in} = voltage coupled to input of amplifier
Z_o = impedance of R_o and C_o in parallel
R_o = leakage resistance of open switch
C_o = stray capacitance of open switch

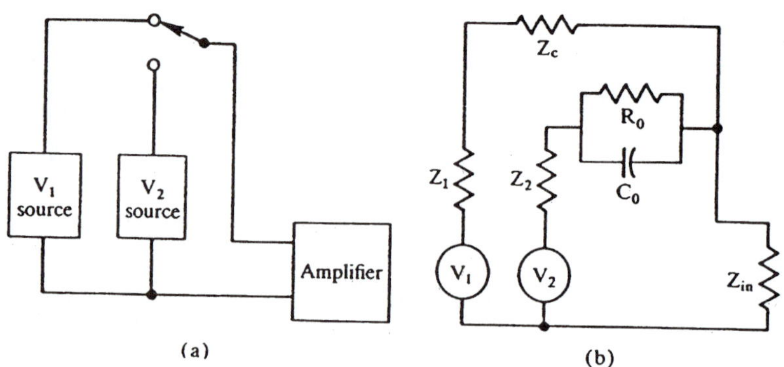

Figure 34.13 Crosstalk in switch. (*a*) Typical switched-input amplifier; (*b*) equivalent circuit of (*a*).

Z_1 = source 1 impedance
Z_2 = source 2 impedance
Z_C = contact impedance of closed switch
Z_{in} = input impedance of the amplifier

At very low frequencies $Z_o \approx R_o$, and at high frequencies $Z_o \sim 1/\omega C_o$. From Eq. 34.6, it can be seen that to maintain minimum crosstalk, Z_1 should be minimized and Z_o should be as large as possible (usually Z_C and Z_2 will be negligibly small).

Very often crosstalk will appear between two separate amplifiers where an input on one amplifier also will produce an output on the second amplifier. The same arguments hold in this case as in the switched-input amplifier. Any stray capacitance or resistance anywhere from the first amplifier to the second will cause coupling. Often the problem will be caused by something in common being shared by the two amplifiers, such as a common power supply or using the same power or ground wire between them. The obvious solution to this problem is to require complete shielding between the two amplifiers and to utilize separate power supplies. Economics do not always make this possible. Therefore, leads common to the two amplifiers should not be allowed unless absolutely necessary. Situations involving common leads require that the leads be short and the conductors large enough that their impedance is insignificant. A common power supply then should have a low output impedance and separate leads taking power to each amplifier. These leads also can be decoupled by use of appropriate filters.

34.9 Contact Potential

When two conductors of dissimilar materials are connected together, a voltage is generated at the junction which can be a source of contact static-offset interference. This voltage is a function of the temperature of the junction. In

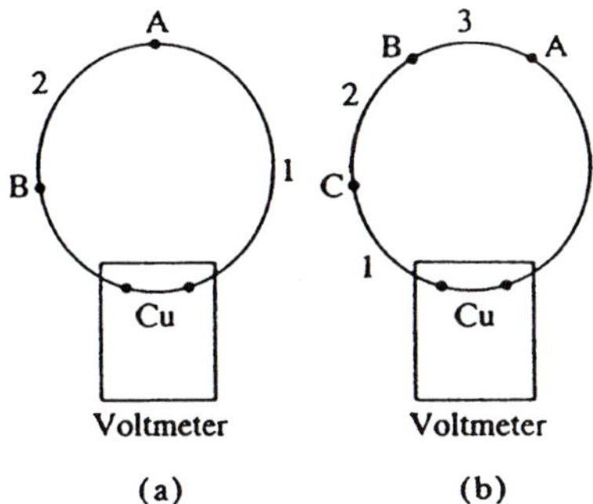

Figure 34.14 Contact potential by thermocouple action. (*a*) Circuit with two dissimilar conductors; (*b*) circuit with three dissimilar metals.

any realizable system where dissimilar metals are used in a circuit, more than one junction will always be involved. As shown in Fig. 34.14*a*, most voltmeters will have copper wires (conductor 1) so that if a junction *A* is made with conductor 2 and the copper wire, the same junction in the reverse must be made by conductor 2 and the other copper meter lead at *B*. Thus, if the two junctions are at the same temperature, no net voltage will be produced, since the voltages generated at the two junctions will cancel each other. If, as in Fig. 34.14*b*, a third homogeneous conductor 3 is inserted, as long as the temperatures of the three junctions are the same, the sum of the contact potentials around the loop will still be equal to zero. As can be seen, then, regardless of the number of junctions made, as long as all junctions are uniform and are at the same temperature, no voltage will be produced. Problems can develop, however, if different points in the circuit are at different temperatures. In most instances, any problems can be forestalled by merely using the same conductor material throughout the system. When the thermoelectric effect is being exploited to measure temperature, one junction is held at a constant known temperature. The net voltage difference will then be a function of the temperature difference.

34.10 Instrument Interconnections

Any realizable system of instruments requires interconnecting wires or cables. Because of their length, these wires and cables are prone to pick up noise. For instance, a cable may be of the proper length to be a receiving antenna for an rf signal, or the cable may form a large loop that will have large circulating currents flowing in it because of magnetic coupling. The type of cabling used and the way it is connected, therefore, depend largely on the type of interference to be rejected. For instance, if the coupling is magnetic, all cable shields should only be grounded on one end. If it is rf interference, the cable should be grounded at intervals much less than a wavelength of the interfering signal.

In severe cases, the cable may be double shielded or shielded with a twisted pair of wires inside so that the outer shield can be grounded in several places to reduce rf pickup and the inner shield can be left open on one end of the twisted pair to reduce magnetic pickup. In general, in all low-level circuits, a properly shielded cable must be used to minimize interference pickup. Conversely, high-level circuits should use the proper shielded cable to minimize radiated interference.

The physical location and routing of cables must be engineered properly to minimize the possibility of pickup. Low-level-signal cables and wires must not run near or parallel to high-level-signal or power cables. When low-level and high-level cables must be in proximity, they should cross at right angles to each other.

Any cabling in a system that produces a loop should be routed to minimize the area of the loop and thus reduce magnetically induced interference. Another method of effectively reducing magnetic interference is to twist the wires tightly together (twisted pair). This produces many small opposing loops which cancel the net magnetic pickup or transmission.

34.11 General Characteristics and Source of Interference

Hum. This audio-frequency noise is prevalent because of ac power transmission and distribution systems. In the United States, the fundamental ac frequency is 60 Hz; hence this frequency and its harmonics will characterize the hum. The magnitude of this interference at any location depends primarily on the proximity of power transformers, fixtures such as fluorescent lights, and all power wires. In general, however, this type of interference will be appreciable in nearly every measurement situation in varying degrees of severity. One notable exception is that of measurements made in space.

Impulse noise. This type of noise is normally generated any time the magnitude of a current is changed abruptly. The frequency spectrum is broad; therefore, it is very difficult to filter and is normally more easily controlled at the source. Common sources of impulse noise are dc and universal motors and generators (brush-type machines), automobile ignition systems, relays, dc-to-dc converters, switching regulators, etc. These sources may transfer energy to the measuring setup by radiation or conduction mechanisms.

Atmospheric noise. This noise is generated in the atmosphere primarily by lightning discharges and is, therefore, dependent on time of day, geographic location, and weather. Its frequency covers a broad spectrum and is characterized by a crackling sound (static) in an AM receiver.

Cosmic noise. There are many sources in space which transmit a wide band of random noise. One of the primary sources is the sun.

Thermal noise. This signal is generated by the thermal agitation of electrons in a resistance. It is a broadband random signal that defines the theoretical limit of sensitivity of an instrument.

Continuous interference. A continuous type of interference may be narrowband, as in the case of commercial radio, television, and amateur radio transmitters, rf conduction-induction heaters, etc., or relatively broadband, as in diathermy machines, x-ray machines, arcs and corona from high-voltage power lines, and other sources of arcs such as fluorescent and mercury vapor lighting fixtures, neon signs, and rotary machines with brushes or slip rings. The interference may be conducted or radiated.

Mechanically generated interference. The movement of conductors in a magnetic field is a source of interfering signal. Noise is also generated by flexing of coaxial cables and production of static electricity due to friction. Another class of noise known as "microphonics" is generated by vibration and shock, particularly of electron tubes such as multiplier phototubes.

34.12 Specific Interference-Source-Reduction Techniques

In most instances, reduction of noise and interference is most successful if sensitive instruments can be physically removed as far as possible from the interfering sources and steps are taken to reduce the interference at the source. Some specific examples of these interference-reduction techniques are as follows:

Relays, Controllers, Switching Devices

1. Shunt relay or switch contacts with a capacitor to reduce current surges. In general, a current-limit resistor should be placed in series with the capacitor to prevent deterioration of the switching contacts.
2. Enclose switching devices in a shield.

DC-DC Converters, Inverters, Switching Regulators, Light Dimmers

1. Shield the unit.
2. Use low-pass filters on all leads passing through the shield (at minimum, use feedthrough capacitors).
3. Use Faraday shields on transformers.

Arc and Gaseous Discharge Devices—Neon Signs, Fluorescent Lighting Fixtures, Arc Welders

1. Install bypass capacitors on lines.
2. Use special conductive coatings over glass.

3. Use shield cases.
4. Substitute incandescent lights for fluorescents.

Mechanically Induced Interference (Microphonics)

1. Support cables to reduce movement.
2. Use low-noise cables.
3. Use vibration- and shock-damping mounts.
4. Select low-noise devices and hand pick for lowest noise.

Electric Motors and Generators

1. Use a good shielded housing.
2. Bond housing to ground.
3. Use bypass capacitors at brushes.
4. Use a feedthrough capacitor at the armature terminal.
5. Shield terminals and interconnecting wiring.
6. Keep brushes in good condition.

RF Generators—Transmitters, Induction Heaters, etc.

1. Use special multiple-shield enclosures.
2. Bypass and filter all lines entering or leaving the shield enclosures.
3. Use resonant traps for specific frequencies.

Ignition Noise (Internal Combustion Engine)

1. Place a resistor (10 kΩ) in high-voltage lead near the coil or use resistive ignition leads.
2. Use shielded internal resistor spark plugs.
3. Shield the coil.
4. Use shielded ignition wires.
5. Use bypass capacitors on the dc lines into the coil and distributor.

Static, Corona, High-Voltage Arc Discharges

1. Use conductive belts on machinery.
2. Use brush-to-ground drive shafts.
3. Electrically bond equipment together.
4. Eliminate sharp points and corners.
5. Use coatings and paint on rough surfaces.

6. Keep dry and free of contamination.
7. Pressurize high-voltage components or seal them in a hard vacuum.

34.13 Summary

The reduction of interference can be accomplished in many ways. However, in general, each method includes some form of physical placement, shielding, and filtering. The severity of the problem will determine the extent of use of these principles. For instance, the problem may be solved by moving a sensitive instrument a few meters away from a specific interfering source, or it may be necessary to place the whole measuring site in a remote area and use battery operation. Often, laboratories have special low-noise measurement areas utilizing specially constructed multiply-shielded rooms.

Chapter

35

Electrical Grounding

D. A. Burt and K. D. Baker

Department of Electrical Engineering
Utah State University, Logan, Utah

35.1 Introduction

Even experienced electronics engineers on occasion have been known to say that there is some "black art" to building and using electronic equipment that does not oscillate or have some unwanted noise signal impressed on it. In many cases, however, this "black art" is actually a good basic knowledge of how to ground all interconnecting systems properly.

Originally, an "electrical ground" was a conducting connection between an electric circuit and the earth; however, the word has been rather loosely applied in the electronics industry as a point or points which are used as a zero-voltage reference.

In a measurement system, grounds are conventionally separated into three types:

1. *Power grounds:* the return path for electric current that provides the power necessary to operate the instrument.
2. *Signal grounds:* the reference point and return path for all signal currents.
3. *Chassis and shield grounds:* usually the chassis and outer metal case of instruments and cable shields.

The basic grounding criterion is the requirement that each type of ground have the same potential throughout the system. This requirement, however, is seldom, if ever, met completely in any practical system. For this reason, this chapter is devoted to showing some of the sources of potentials appearing on the grounds and ways to minimize the problems caused by them.

35.2 AC Power Grounds

Among the largest sources of electrical interference are ac power-distribution systems. For this reason, it is important to have some understanding of the grounding practices in these distribution systems.

Power systems are normally separated into three groups:

1. Transmission lines which carry the power over long distances at high voltages, i.e., 34 kV and above.
2. Primary distribution lines which carry power to a test facility or a town in a relatively small area—ranging in voltages from 2.4 to 25 kV.
3. Facility distribution lines which normally operate at 120 to 240 V. In all cases these power lines are referenced to earth ground. This is done to prevent transient voltages due to arcing between grounds, to permit the use of lower insulation levels, and to aid in protective fusing.

Since these power lines are all referenced to earth ground, current will be flowing in the earth. This current will find and flow along the path of least resistance, thus producing potential drops in the earth and generating a magnetic flux. The potential drops make it impossible to connect to the earth at two different points and assume that they are and will remain at the same potential. The magnetic flux that is created will couple with a transformer action into long wires or loops of wire, thus generating low-impedance circulating currents which are difficult to eliminate.

High-voltage transmission lines are normally grounded at the generator end only. Over long lines, however, the capacitance of the wire to the earth can allow considerable current to flow through the earth. For primary distribution lines, the *National Electrical Safety Code* requires that the neutral conductor be connected to earth at least four times every mile. Under this system, the current will divide and flow into the earth and the wire in proportion to the conductivity of each.

In facility distribution, the neutral line is grounded at the source and should not be grounded at the load. Unless there is a fault somewhere, there should be very little earth current due to the facility distribution. Because of the extensive power systems throughout the world, it can be said generally that there will be ground currents almost everywhere. This is the 60-Hz signal that you see when you touch your finger to the input of an oscilloscope or the hum you hear when you touch the input of an audio amplifier. Since this signal is so prevalent, it must be considered in all interconnecting systems, especially if the systems are sensitive or widely separated.

35.3 Instrument Power Input

35.3.1 AC-powered instruments

Most quality instruments have a three-wire line cord. In the ac power line, one wire is the hot wire, the second is the common or power return, and the third is earth ground. Inside the instrument, the hot wire and the common go to a power transformer, while the earth ground connects directly to the metal case. The third wire provides personnel safety by maintaining the instrument case at earth potential and should not be circumvented. While this third wire or earth ground is fine for safety in preventing electric shock and in providing an overall grounded shield, it can cause problems when a number of instruments are all plugged into the power line and their cases are then all connected together to form a large loop. This loop then becomes a shorted turn, and the magnetic coupling to earth currents can produce large circulating currents. This is commonly referred to as a "ground loop."

Some inexpensive commercial electronic items such as clock radios and television sets do not use internal power supply transformers but derive the necessary voltages directly from the ac line. The chassis then may be connected to the ac common or to the high side of the line depending on the orientation of the plug with respect to the socket. Thus, if the plug is inserted such that the chassis is connected to the hot side of the ac power, the chassis is hot with respect to earth ground, and a person working on the equipment could receive a severe shock. If measurements are being made on this chassis with a quality instrument, a connection of the instrument ground to the hot chassis is a short circuit. Sparks, smoke, and a blown fuse will rapidly occur in that order. This type of equipment will always be completely encased in an insulating material and use an interlocking power cord to prevent shock to the user; however, a person working on the equipment must use due caution.

35.3.2 Battery-operated instruments

Rechargeable batteries and low-power-drain solid-state circuitry allow the instrument designer to design portable units that are completely isolated. Because of the isolation, this equipment offers a number of advantages in eliminating many of the problems associated with grounds. For instance, ground loops are readily eliminated, as is the problem of measuring a signal with respect to a point not at earth ground.

35.4 Instrument Grounds

35.4.1 Normal input grounds

Signal ground paths vary depending on the type of instrument; however, it is very important that no extraneous currents flow on them. A commonly used instrument is the measuring-type instrument, such as a voltmeter, oscilloscope, or chart recorder. These instruments take some input quantity and display it to the operator. In general, this type of instrument has a complete ground system, as shown in Fig. 35.1.

It should be noted that the input of these instruments is referenced to case ground, and as stated in the preceding section, this case ground is also tied to earth ground through the third wire of the ac power line. As long as the third wire is plugged in, the input ground should not be connected to any

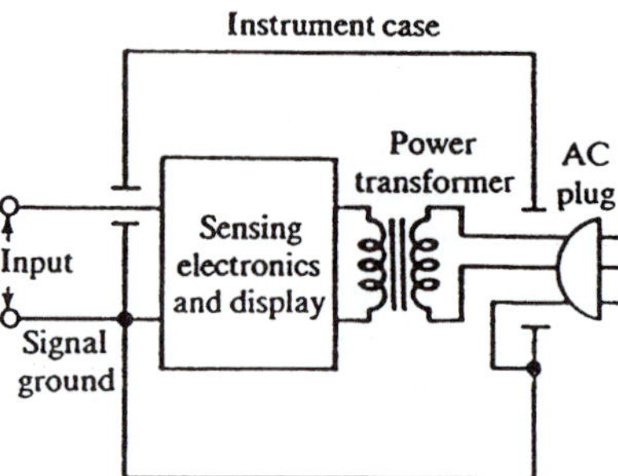

Figure 35.1 Most common input and power grounds of measuring instruments.

point that has a potential with respect to earth ground, or it could produce disastrous consequences.

35.4.2 Differential input

Some measuring instruments have a differential (floating) input (signal return isolated from case) and a case or shield ground, as shown in Fig. 35.2. In this type of instrument, the signal that is displayed is the voltage difference between the two input signals. For example, if a 5-V signal were applied to the positive input with respect to case ground and a 3-V signal on the negative input, the only portion that would be sensed is the difference voltage (differential voltage) of 2 V. The part of the voltage that is common to both inputs is termed "common-mode voltage" and is ideally not sensed. Since both input terminals are isolated, either terminal can be considered as signal ground. This reference terminal can be at an arbitrary voltage with respect to case ground as long as instrument voltage ratings are not exceeded.

35.4.3 Normal ground for output type of instrument

An output-type instrument is the signal source, such as a power supply or signal generator, which is to be connected to another instrument. Diagrams

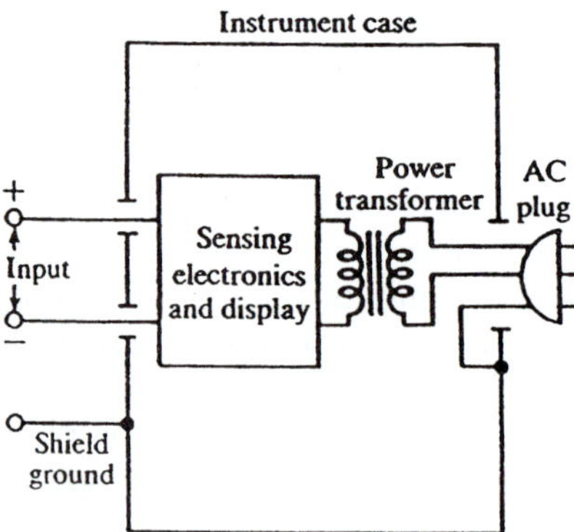

Figure 35.2 Grounding system for differential-input devices.

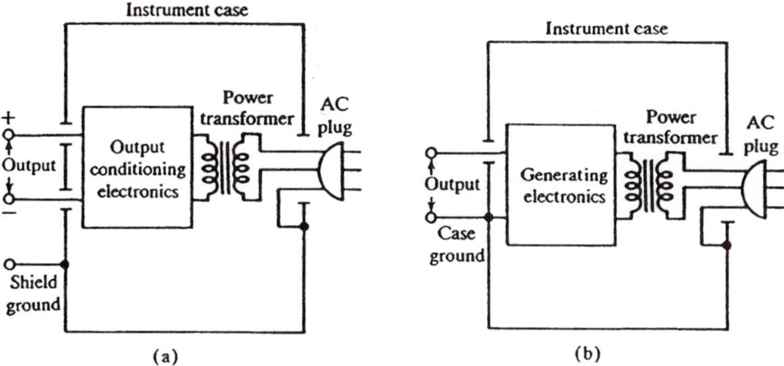

Figure 35.3 Typical ground system for output-type instruments. (a) Ground system of low-frequency and dc output instruments; (b) ground system of rf output instruments.

of these types of instruments are given in Fig. 35.3. For dc or low-frequency outputs, there are usually plus and minus output terminals with a separate case terminal that can be connected where desired, as shown in Fig. 35.3*a*. For rf generation or where there are high-frequency components, the output will be on an rf connector, where the outside or ground side of the connector is connected to the case, as shown in Fig. 35.3*b*.

35.4.4 Ground connections in measurement systems

An absolute general method of properly connecting the grounds of instruments together does not exist. Depending on the circumstances, there may be a number of ways that will produce satisfactory results. General guidelines for ground interconnections are

1. Connect all grounds so that shield, power, and signal ground currents cannot intermix but can only flow in their respective return paths.
2. Maintain short ground paths and use large conductors to minimize impedance between ground points.
3. Avoid multiple paths for ground currents.
4. Design each individual ground circuit in such a manner that high-level ground currents cannot flow in low-level input return circuits.

In most instances, it is not practical to adhere rigorously to all the foregoing rules at the same time. For example, all instruments should be connected to earth ground for safety purposes, and this requirement is often in conflict with the rule regarding multiple ground paths. In actual practice, the foregoing "rules" are used as a guide, and all grounding is done in such a manner as to minimize the effects of compromising these rules. The best way to illustrate good grounding techniques is by the use of a number of examples.

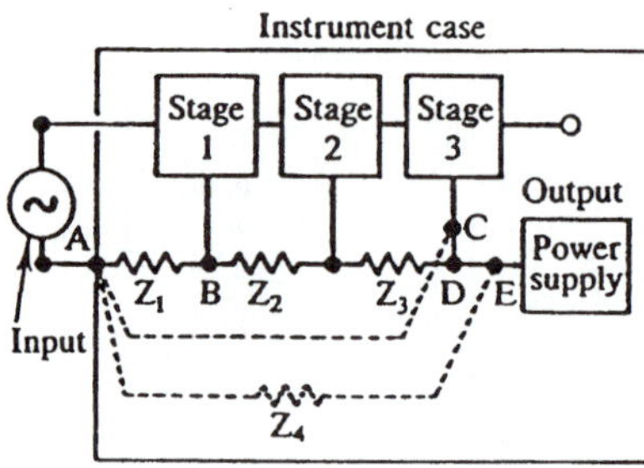

Figure 35.4 Ground system inside an instrument.

An example of the grounding system typical of the internal connections of instruments is shown in Fig. 35.4. In this case, the power and signal grounds are not isolated; however, care is taken to minimize the problems by proper ground design. The ground system runs in a continuous line from the high-level stage to the low-level stage in sequence with no doubling back. Even though a wire or chassis is designated a ground, it cannot be overemphasized that the same potential may not exist at all points, particularly at high frequencies. This is illustrated by the equivalent impedances, Z_1, Z_2, and Z_3, which are the inductance and resistance of the interconnecting wires. Although these impedances need to be minimized by choice of short paths and large conductors, the current that flows through these impedances may produce significant voltage drops, resulting in different potentials at the various ground points. Accordingly, a stage or instrument cannot be arbitrarily connected to any point considered ground. To illustrate the consequences of indiscriminate grounding, assume that stage 3 had been connected to point A instead of point D as the dotted line indicates. Now any signal and power currents that flow to stage 3 must pass through Z_1, Z_2, and Z_3. Often stage 3 is a high-level stage drawing large, varying currents. These currents will produce a voltage drop in Z_1 that will be directly in series with the input of the first stage, which often is a sensitive low-level amplifier. Potential drops also will appear across Z_2 and Z_3 in series with the inputs of stages 2 and 3. These extraneous signals on the input can result in many serious problems. Depending on what the instrument may be, the system may oscillate, have a dc offset, or show spikes or many other spurious responses.

Note also that the power-supply ground should be connected first to the high-level stage as shown. If the power supply were connected to point A, as shown by the dotted line and Z_4, again, the high-level current of stage 3 would be forced to flow through Z_1 and produce spurious responses. The low-level current from stage 1 will flow through Z_2 and Z_3 in the proper connection; however, this current is of such a low magnitude that the resulting signals in stages 2 and 3 are probably insignificant.

A classic illustration of the fallacy in assuming that ground is the same everywhere is encountered in the design of a simple dc power supply, as

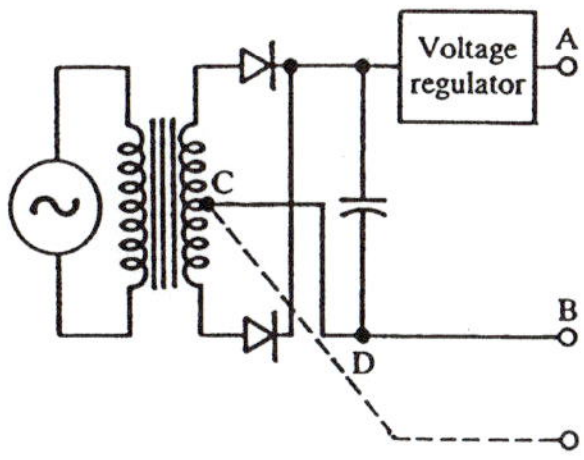

Figure 35.5 Example of the use of proper and improper ground reference for a simple dc power supply.

shown in Fig. 35.5. Normally, in a well-designed and well-constructed power supply there will be very little ripple and noise on the output between *A* and *B*. However, if by some chance it were more convenient to use point *C* as the ground reference rather than point *D*, considerable ripple will appear in the output. This occurs because the current that flows through the rectifiers to charge the filter capacitor flows only for a small part of a cycle, producing short pulses of current many times greater than the direct output current of the supply. Since these current pulses will usually be at least several amperes, it takes very little resistance between *C* and *D* to produce a considerable voltage pulse in series with the output.

Many problems associated with measurement systems are a result of ground loops. The best definition of a ground loop is the inability to provide the same potential at two different ground points. This can be due to multiple-point grounding, magnetic pickup on a large wire loop, or electric pickup on a long wire. Whenever 60-Hz hum appears in a system, it usually can be traced to a ground loop. To illustrate the problems associated with a ground loop, refer to the system shown in Fig. 35.6. Proper procedures dictate that the power-supply return connection should be connected only to point *B*. If instead it were connected to both points *A* and *B* (dotted line), the high-level currents from stage 3 would split and flow through both paths from point *B*. The amount of current flowing through each path would be proportional to the conductivity of each path. Again, the relative conductivities could be such that the system would oscillate or produce erroneous results. The ground loop may produce additional interference in the presence of a changing magnetic

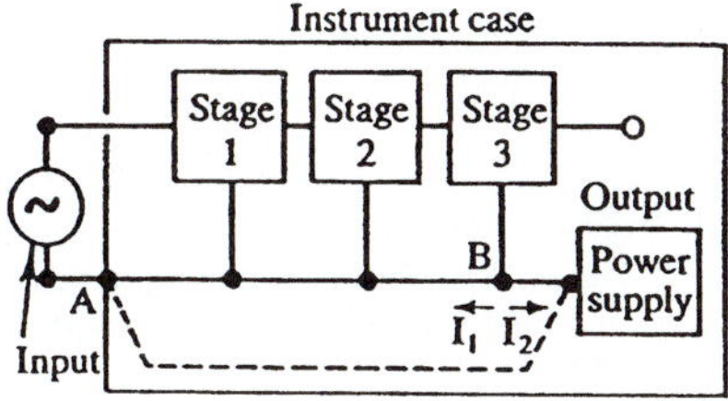

Figure 35.6 Ground loop inside an instrument.

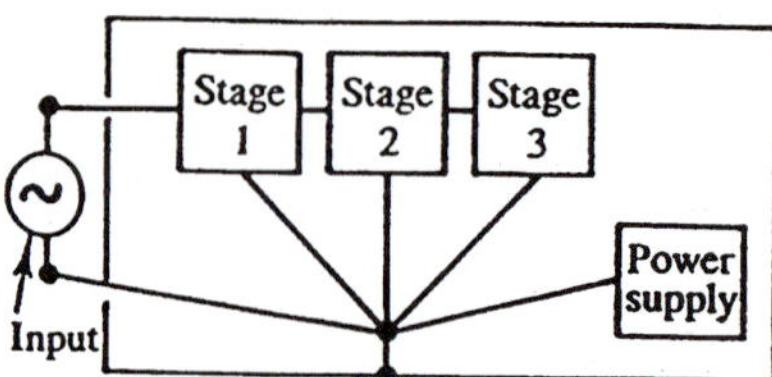

Figure 35.7 Instrument stages connected to a common power supply with single-point grounding.

field by inducing circulating currents in the loop. When this interference appears, because of long cables and widely separated ground points, it will probably appear at the signal ground and the signal input level at the same time. This is termed "common-mode interference" and requires special attention. This particular situation is discussed in Chap. 34. A method of grounding known as "single-point grounding" is shown in Fig. 35.7. In this example, the only ground impedance that is common to any of the stages is the output impedance of the power supply including the conductor to the common point. This usually can be made small enough to be insignificant.

The method of properly grounding a system consisting of an electronic instrument and an external power supply is illustrated in Fig. 35.8. Most good power supplies have three terminals: positive and negative terminals and a shield ground (the instrument case). When connecting this power supply to other equipment, the plus and minus outputs are connected to the necessary power inputs, and the shield ground is connected to the shield of the instrument as shown. This provides a continuous shield system for minimizing interference pickup from external sources.

An example of the complete grounding system for a typical test setup is shown in Fig. 35.9. The electronic circuit under test is powered by a dc power supply and excited by an audio oscillator, and the output is observed with an oscilloscope. Multiple ground paths have been held to a minimum, and the shields have been connected together to eliminate external interference. This is relatively simple for this illustration, since most of the instruments have isolated inputs and outputs. There is one ground loop through the instrument cases and the earth ground pin on the power input. Under normal circum-

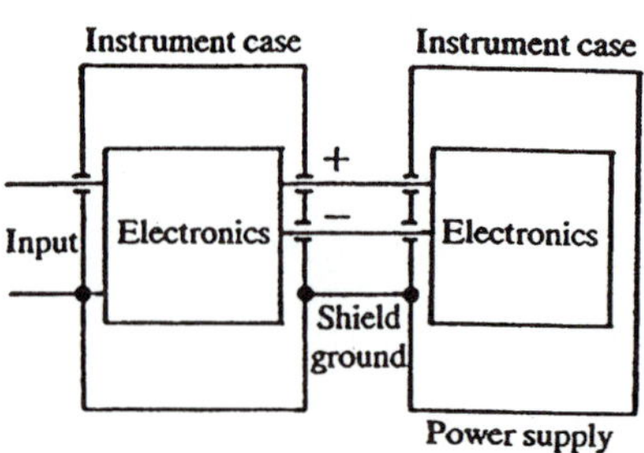

Figure 35.8 Proper ground system for connecting external dc power supply to separate electronic instrument.

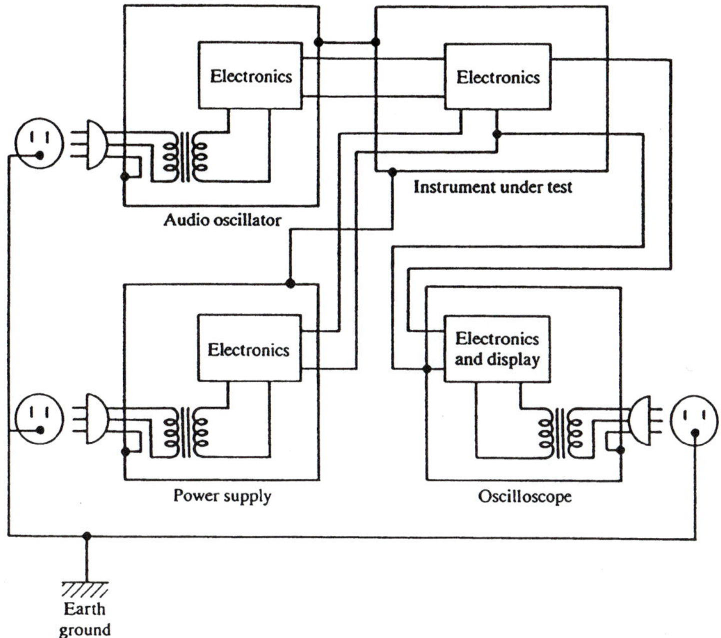

Figure 35.9 Proper ground system of measurement setup consisting of an instrument being excited by an audio oscillator and observed by an oscilloscope.

stances, since the shields are isolated from the other grounds, there are no significant currents flowing, and this ground loop should have negligible effect as long as the circulating currents are not large enough to produce strong magnetic fields.

Another interesting grounding system is a spacecraft payload system shown in an idealized case in Fig. 35.10. All power grounds are tied together at one point, as are the positive power leads, so that no power leads are common between the instruments. Thus, only the battery is common between instruments. A dc-to-dc converter is used on each instrument so that power ground is completely isolated from signal or case ground, preventing the flow of power currents in the supporting structure or signal grounds. Each instrument has its own reference to shield ground at its input, and the signal and signal ground are run via a shielded twisted pair. The shield may be grounded to the supporting structure at many points. Thus the twisted pair decreases magnetic coupling, and the grounded shield decreases any electrostatic coupling that may come from any radiating source.

Power-supply noise rejection can be further enhanced by operating all the converters at the same frequency and synchronizing them together so that they not only operate at the same frequency but also are actually phase-locked together. This prevents different supply frequencies from beat-

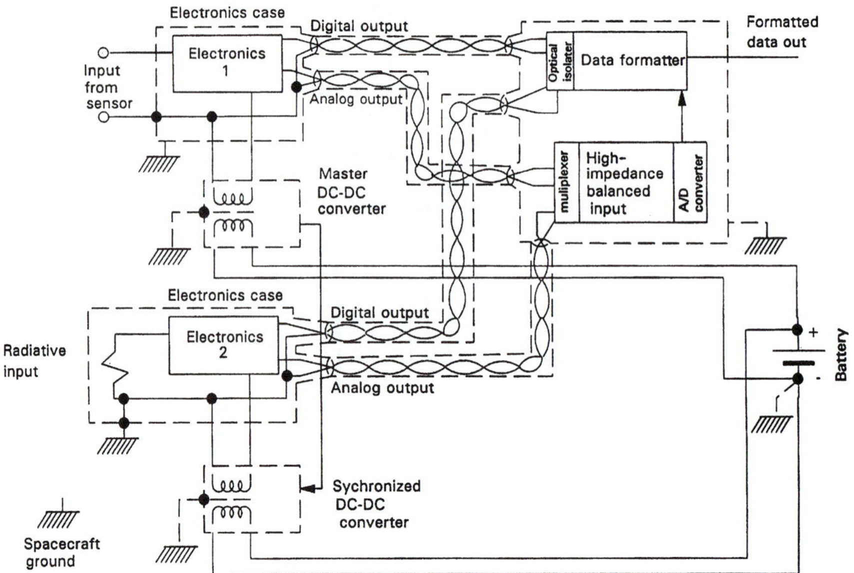

Figure 35.10 Idealized connections for a remote measurement device used on a spacecraft payload system.

ing together and producing spurious interference throughout the spectrum. Thus, in the locked system, the interference-sensitive circuits have to reject only the common fundamental power-supply frequency and some harmonics. An extension of this power-supply synchronization is to use only one dc-to-dc converter and provide individual isolation by separate supplies from different taps on the power transformer. Most output signals are converted to digital signals for transmission. The best isolation for these outputs is to use optical isolators. If the digital data rate is too high for optical isolators, there are several balanced matched digital line drivers and receivers that can be used.

If analog signals need to be multiplexed at a single point, this also can be accomplished by an analog optical isolator. Another method is to multiplex both the signal and signal ground at the same time with a break-before-make multiplexer so that the grounds between the two instruments are never connected together. The output of the multiplexer should then be fed into a high-impedance balanced-to-ground amplifier, as shown in Fig. 35.11.

35.4.5 RF and shield grounds

RF grounding is difficult to accomplish because of the distributed inductance of any grounding system used. A brute-force method employing large metal grounding areas and short grounding wires usually provides the best system.

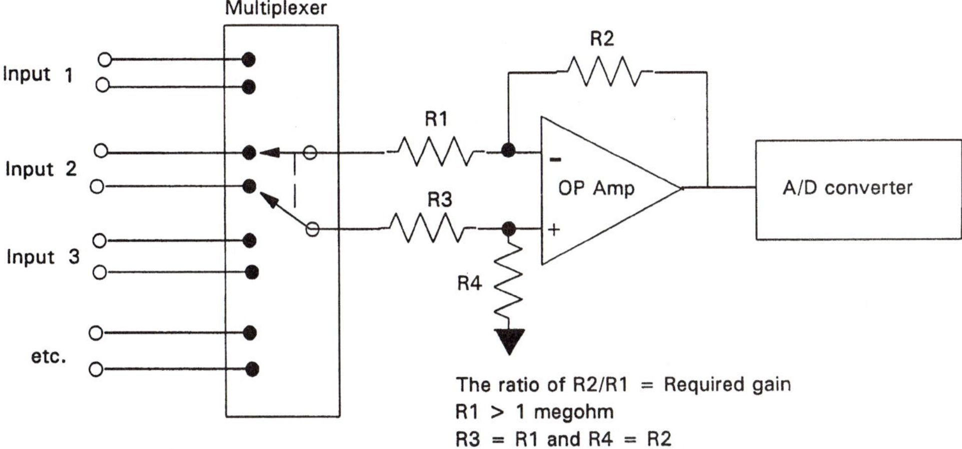

Figure 35.11 Analog multiplexed input to provide ground isolation.

The output and input terminals of most rf instruments are referenced directly to the metal case (shield ground) of the instruments. Separate instruments should be connected together with properly matched rf shielded cable, primarily to prevent standing waves on the lines, cable radiation, and induced signals (see Chaps. 33 and 34).

Cable shields are basically an extension of the cases around the instruments being connected together. If you have done all the rest of the power and signal isolation and grounding correctly, there will be no current flow on these shields due to the instruments. Their primary purpose is to guard the cable and minimize conducted interferences into the circuits from external radiation sources or from being a source of radiated interference.

When grounding cable shields, the necessary type of grounding depends on the frequencies involved, the length of the line, and the sensitivity of the instrument. In general, however, a shield is not of much value unless it is grounded at least every 0.15 wavelength of the pertinent signals. This type of multiple-point grounding is not effective if the ground points used have high impedances between them because of the multiple grounding currents that will be induced in the ground loops. The effect that these currents will have on the instruments should be evaluated. In many cases, the instruments used will have no sensitivity to the induced frequencies, and hence there is no problem.

However, in the case of wide-band circuits such as video circuits, there will invariably be some hum pickup. There are a few methods to reduce this hum, e.g., use of a diode clamping circuit as employed in the television industry to restore the dc level and hum cancellation. This cancellation is accomplished by adding the proper level out-of-phase signal to the system to cancel the interfering hum as much as possible.

35.5 Summary

All the preceding examples have shown sophisticated methods of grounding; however, it should be noted that a good brute-force method of large ground busbars or the instrument chassis should not be underrated. A typical example of this concept is the common practice used in large equipment racks such as a radar base or telemetry station. In these instances, all the instruments are bolted together in metal consoles as individual units with large bonding straps connecting the consoles. Shield and signal grounds are tied together; however, these currents are normally small enough not to produce serious problems. The input power is isolated because each instrument has its own power transformer.

In many instruments, the chassis is used as the ground, with each ground element connected with a short lead to the chassis near the component. This is paralleled in printed circuits, where one side of the printed board is a solid copper plate used as a ground.

Chapter

36

Distributed Parameters and Component Considerations

K. D. Baker and D. A. Burt

Department of Electrical Engineering

Utah State University, Logan, Utah

36.1 Introduction

In dc and low-frequency circuits, the elements of resistance, inductance, and capacitance are treated as lumped-circuit elements. In reality, the parameters of wires and other elements are distributed over some volume, and at higher frequencies, their distributed characteristics become more and more important. Distributed parameters are probably one of the largest contributors to the discrepancies between drawing-board design and actual performance of working equipment. For this reason, this chapter is devoted primarily to showing the magnitude of these parameters and some examples of the troubles that can occur because of them.

36.2 DC Resistance

Dc resistance is the best known and most thoroughly tabulated electrical parameter of materials. The formula for the resistance of a body of uniform cross-sectional area is simply

$$R = \frac{\rho l}{A} \tag{36.1}$$

where R = resistance, Ω

ρ = volume resistivity of the material

l = length of the conductor

A = cross-sectional area of the conductor

The values of ρ for many materials have been tabulated in reference books and hence will not be duplicated here.

The resistivity of all materials is a function of temperature. In general, the resistance of semiconductors and insulators will decrease with increasing temperature, whereas the resistance of most metals will increase with increasing temperatures. As an example, the resistivity of copper will increase approximately 0.4 percent per degree Celsius rise in temperature.

There is a simple rule of thumb that may be convenient for approximating the resistance of copper wire. The resistance of a copper wire 0.001 in in diameter* and 1 ft long (a circular mil-foot) is approximately 10 Ω. Thus, if a known wire has a cross-sectional area of 100 circular mils (cmil) and a length of 1000 ft, the approximate resistance will be

$$R \approx \frac{10 \times 1000}{100} = 100\Omega$$

Wire sizes have been standardized according to American (B and S) gauges (the smaller the gauge, the larger the wire diameter). Number 40 wire has close to a 10-cmil cross-sectional area, and for every three wire sizes, that area will approximately double. Accordingly, no. 37 wire will have a cross-sectional area of approximately 20 cmil; no. 34 wire, 40 cmil; etc. Thus the total resistance per foot will halve for each decrease of three wire sizes.

The effect of dc resistance with short wires will normally not be important if the wire is large enough to handle the current. However, for the transmission of power at longer distances, the wire resistance must be taken into account.

A classic example of the effects of wire resistance is the transmission of electric power used throughout the world.† For instance, assume that a small load of 10 A is required at 120 V at a distance of 1 mi from the generator. If no. 6 wire were used, it would present a total resistance of approximately 4 Ω. Thus 10 A flowing through the lines would then require a generator voltage of 160 V and produce a 400-W power loss in the wire just to deliver 1.2 kW at the load. However, if the voltage at the generator could be increased to 12,000 V and then stepped back down with a transformer to 120 V at the load, the line current would be reduced to 0.1 A to supply the same power down the line. Under these conditions, there would only be a 0.04-W power loss in the line. Anytime power transmission is required at low voltage, the wire resistance becomes very important.

---

* It is common to express the cross-sectional area of wires in circular mils, which is numerically equal to the square of the wire diameter in thousandths of an inch.

† Although the discussion has been strictly for dc resistance, this resistance is applicable for audio frequencies and below. The more complicated case of ac resistance at higher frequencies is discussed in the next section.

As another example, it may be necessary to transmit 10 A at 6 V to operate a rocket payload on the launch pad 1000 ft away. If no. 6 wire were used, the resistance for a round trip (2000 ft of wire) would be approximately 0.8 Ω; the 10-A current would produce an 8-V drop in the wire. It would then be necessary to supply 14 V at the input to obtain the required 6 V at the payload. Under some conditions, this may be allowable; however, consider what happens if the payload should require a load that varies from 5 to 10 A. If 14 V were applied to provide 6 V at 10 A, the voltage at the payload would rise to 10 V under the 5-A load. Some equipment could, of course, be damaged by such an overvoltage.

On occasion, it may be desirable to design a component that uses the wire resistance for current limiting, such as a coil to actuate a relay. In this case, it must be kept in mind that power is being dissipated in the windings where the wire is densely wound. The coil design then must be able to dissipate the heat generated at a high enough rate to maintain the coil temperature within the temperature bounds of the insulation used on the wire.

In summary, it should be stated that the dc resistance must be taken into account anytime large currents are being transmitted, long-distance wire runs are necessary, or the power dissipated in the wire is appreciable for the given configuration.

36.3 AC Resistance

At frequencies above the audio range, conductors present an effective resistance that is significantly higher than the simple dc resistance discussed above. This occurs because the magnetic flux produced by the current in the wire does not all encircle the wire; there will be some internal flux in the conductor. As a result, the impedance at the center of the wire is greater than that near the surface. This effect is more pronounced at higher frequencies. At higher frequencies, then, the current density in the conductor will be nonuniform, and the effective resistance increases accordingly. A round wire will have a uniform distribution of current around the cylinder; however, the current density will be greatest at the surface and decrease toward the center. This phenomenon has been termed "skin effect." The "skin depth" is defined as the depth at which the current density decreases to a value of $1/e$ or 37 percent of the surface density. When this skin thickness is one-third or less than the radius of the conductor, the resistance is nearly the same as with a uniform current distribution in a hollow cylinder with wall thickness equal to the skin depth. This skin depth can be calculated from the following equation:

$$\delta = \sqrt{\frac{\rho}{\pi f \mu_0 \mu_r}} \tag{36.2}$$

where δ = skin depth, m
$\mu_0 = 4\pi \times 10^{-7}$ H/m (permeability of freespace)
μ_r = relative permeability of conductor material ($\mu_r = 1$ for copper and other nonmagnetic materials)
ρ = resistivity of conductor material, $\Omega \cdot$ m
f = frequency, Hz

The ac resistance of a round wire can be approximated as

$$R_{ac} \approx \frac{l\rho}{\delta \pi D} \tag{36.3}$$

where ρ = conductivity of the wire
R_{ac} = ac resistance of the wire
l = the wire length
δ = skin depth
D = wire diameter

This approximation is reasonably accurate for ratios of diameter to skin depth greater than 3. If the conductor is copper and the length is given in feet and the diameter in inches, this equation numerically reduces to

$$R_{ac} = 0.996 \times 10^{-6} \frac{\sqrt{f}}{\mathrm{D}} \Omega/\mathrm{ft} \tag{36.4}$$

Example 36.1: Calculation of the ac resistance of a wire. What is the ac resistance per foot of a no. 18 cylindrical copper wire at 10 MHz? Compare this with the dc resistance. Assume that the wire is at a temperature of 20°C.

$$R_{ac} = 0.996 \times \frac{10^{-6}}{0.0403} \sqrt{10^7} = 0.078 \Omega/\mathrm{ft}$$

Taking the value of the dc resistance from a standard wire table, i.e.,

$$R_{dc} = 0.0063 \Omega/\mathrm{ft}$$

In this specific example, due to skin effect, the ac resistance is a factor of 12 greater than the dc resistance.

As illustrated by the example, the effective resistance for radio frequencies is much higher than for direct current. For this reason, the chassis and the shields of rf instruments often will be plated with a higher-conductivity metal such as silver. Because of the skin effect, the current will flow primarily in the thin layer of silver plating, thus reducing the effective resistance.

Another problem area occurs when trying to wind low-loss (high-Q) coils. The increased ac resistance greatly increases the loss of the coil. A special wire, termed "litz wire," is often used at lower frequencies to decrease this resistance. Litz wire consists of multiple strands of tiny wires bundled together but insulated from each other. These strands are transposed such that any single conductor, when averaged over a reasonable length, will have as much distance in the center of the bundle as on the outside. Thus each wire will present the same impedance to the current, and the current will divide uniformly between the wires and effectively increase the total surface area. The effectiveness of this technique decreases at frequencies above several hundred kilohertz because of wire irregularities and capacitance between strands; hence the wire is seldom used above a few megahertz.

36.4 Leakages

A problem occurs when making very low level measurements due to the leakage currents that occur in materials that are normally considered insulators. These leakages result from finite resistance to ground accumulated throughout the system, i.e., wire insulation, etc.

Most instruments that measure very low currents require a high input impedance to develop a voltage that can be detected without requiring tremendous amplification. This type of equipment may require all insulation to have greater than 10^{15} Ω resistance. For this reason, all leads which must present a high resistance must be kept short, clean, and dry. Another method of reducing the effects of dc leakages is the use of a guarded system, as described in Sec. 36.4.3

36.4.1 High-voltage leakages

At high voltages, air or gases which surround a conductor will begin to ionize and produce a corona leakage. If the voltage gradient is high enough, the gases can break down completely (arc) and cause a low-resistance path due to the ionized gases. The voltage gradient depends not only on the voltage and spacing of the conductors but also on their shape. For example, in dry air at sea level, an electrode gap of 0.2 in for 1-in spheres will arc at voltages above 20 kV, whereas two sharp points with the same gap will arc at approximately 6 kV.

As the gas pressure is decreased, the voltage required to initiate an arc for a given electrode spacing is decreased until a certain critical pressure is reached. As the pressure is decreased past this point, the arcing potential starts to increase. The minimum arcing voltage at the critical pressure is independent of the electrode spacing, although this critical pressure will depend on the electrode configuration. This means that any exposed electrodes with voltages greater than about 300 V can be expected to arc at some pressure between sea-level pressure and an absolute vacuum. For a gap of 0.2 in, the critical pressure will be about 1 mmHg or in the vicinity of 1/1000 atm. Since the arcing voltage increases rapidly for pressures less than the critical pressure, an ultrahigh vacuum is often utilized for a high-voltage dielectric to prevent arcing in electric circuits.

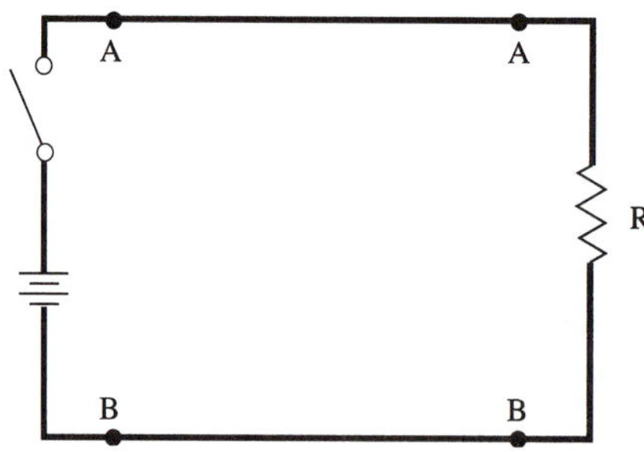

Figure 36.1 Example of a distributed capacitance causing an electric arc on a high-voltage switching circuit.

This variation of arcing with pressure is usually the reason that some equipment can operate only to altitudes of 50,000 to 70,000 ft. The same equipment, however, might be very satisfactory if it were to be used at a much greater altitude.

The stray parameters discussed in this chapter can cause high-voltage breakdown due to switching transients. To illustrate this, consider the schematic given in Fig. 36.1. When this circuit is actually made, there will be wires between *AA* and *BB*. These wires will have both series inductance and shunt capacitance, as discussed in Secs. 36.4.2 and 36.4.5. These stray parameters will cause the wiring to act as a transmission line with a characteristic impedance of

$$Z = \sqrt{\frac{L}{C}}$$

where Z = characteristic impedance, Ω
L = inductance per unit length, H/m
C = capacitance per unit length, F/m

When the switch is closed, the initial current that flows on the wiring is determined by the output impedance of the voltage source and this characteristic impedance. If the voltage source impedance is low and the final load impedance is much higher than the characteristic impedance, excess current will travel down the wiring until it reaches the load. Since the load cannot absorb the excess current, the energy is reflected back to the relay, thus effectively doubling the voltage on the lines. This short high-voltage spike can initiate ionization in the air surrounding the lines, and an arc discharge ensues, shorting out the whole system.

High-voltage leakages can be minimized in a number of ways:

1. Conductors must have adequate spacing.
2. Sharp edges or points on conductors should be avoided.
3. All exposed surfaces can be insulated; for instance, high-voltage power transformers are filled with transformer oil. Silastics and other potting

compounds are often used; however, care must be taken so that there is no contamination between the conducting surface and the potting compound, and there can be no voids in the potting.

4. All high voltages may be sealed in an airtight chamber and either pressurized or pumped to a high vacuum.
5. Analyze all high-voltage switching circuits for voltage transients amplified by circuit-distributed parameters.

36.4.2 Stray capacitance

The conductors of an electrical system will have capacitive coupling to each other through what is termed "stray capacitance." The magnitudes of these stray capacitances are difficult to calculate; however, a simple example will illustrate the stray-capacitance problem. The capacitance between two long parallel wires is given in the following equation:

$$C = \frac{3.677}{\log_{10}(2D/d)} \text{pF/ft} \qquad (36.5)$$

where C = capacitance per unit length
D = spacing between the wires
d = diameter of the wires (d is assumed to be much less than D)

For a 2-in spacing between two no. 24 wires, this gives a capacitance of 0.33 pF/in. While at first glance this may not seem to be such a large amount, consider the capacitive reactance between two wires 2 in long and spaced 2 in apart at a frequency of 1 MHz.

$$X_c = \frac{1}{\omega C} = \frac{1}{(2\pi \times 10^6)(0.66 \times 10^{-12})} \approx 240\text{k}\Omega$$

If one wire had a 5-V, 1-MHz signal impressed on it and the other wire were the input of an amplifier with a 240-kΩ impedance, a 2.5-V signal would be coupled to the amplifier. High-input-impedance amplifiers will have other shunt capacitances. When these shunt capacitances are taken into account, the problem is decreased. For instance, if the same amplifier had a 10-pF shunt capacitance on the input, a voltage of approximately 0.3 V would be impressed on the input of the amplifier.

Usually, the capacitance problems are worse than the simple case given because of capacitances to other large metallic surfaces. The effect of other large surfaces can only increase the capacitance unless the surface is positioned as a shield, which then effectively shorts extraneous signals to ground. The effects of leakage capacitance can be reduced by short wires with wide spacing, shielding, and guarding, as described next.

36.4.3 Cancellation of leakage effects by guarding

The basic principle of canceling leakage effects by the guarding technique is to require the conductors to be at the same potential. If this is the case, there will be no current between conductors, regardless of the impedance between them. The application of this concept is best illustrated by an example.

Example 36.2: Measurement limitations due to leakage and shunt capacitance. It is desired to measure the voltampere characteristics of a high-impedance nonlinear resistance R_m, which is a short distance from the measuring device, by applying a slowly varying sweep voltage to it and detecting the current. Assume that the impedance may vary from 10^{13} to 10^{14} Ω over a variation of voltage from 0 to 100 V. Find the effects due to a 3-ft length of low-capacitance cable (6.5 pF/ft) connecting the instrument to R_m.

Solution. The critical problems that will occur are due to the dc leakage and the shunt capacitance between leads which connect the unknown resistance and the instrument. If it were desired to make the measurement in 1 s, the displacement or capacitive current due to the shunt capacitance of the cable would be

$$i = C\frac{dv}{dt}$$

where i = capacitive or displacement current
C = total cable capacitance
dv/dt = time rate of change of the voltage (in this linear case, dv/dt = 100 V/s)

Thus

$$i = (19.5 \times 10^{-12})(10^2) = 19.5 \times 10^{-10}\text{A}$$

The current due to the resistance R_m measured would be less than 10^{-11} A and so would be masked by the displacement current. Even if the voltage were swept at such a slow rate that displacement currents became negligible, dc leakage currents in the cable could be on the same order of magnitude as the measured current. These problems can be eliminated by driving the shield of the cable at the same potential as the center conductor and allowing the current return on a third wire, as shown in Fig. 36.2. Thus there will be no current flow from the center conductor to the shield because there is no potential difference. Any current that flows from the shield to the third wire is not measured, as can be seen in Fig. 36.2.

36.4.4 Neutralization

Another method to eliminate the effects of leakage currents is by a neutralization scheme. Leakage currents are canceled by applying a second current

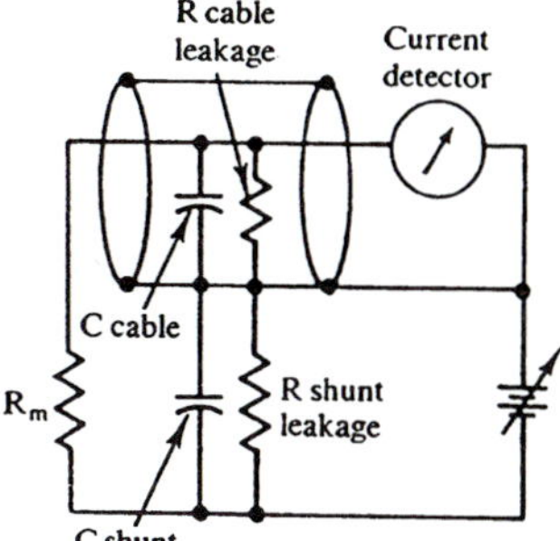

Figure 36.2 Guarded system to eliminate leakage.

to the input of the amplifier which has the same magnitude as the leakage component but with the opposite polarity such that the total net current is zero. This is done as shown in Fig. 36.3 for the same measurement situation as discussed in Fig. 36.2. In this case, an opposite-polarity that tracks the measuring voltage is applied to the input through R_{se} and C_{se}, which are made equivalent to the leakage resistance and capacitance of cable R_s and C_s This technique, of course, requires a sensitive balance that is difficult to maintain.

36.4.5 Stray inductance

A magnetic field will exist about any conductor that carries a current. The coupling of this field to the conductor will always produce an inductance. This inductance is a function of the conductor configuration, where it is placed with respect to all other components, and to some extent the frequency of operation. It is difficult to calculate the exact inductance for a particular wire in a circuit. However, the following equations for a straight, round, nonmagnetic wire should give the reader an idea of the magnitude of the inductance involved. At low frequencies, this self-inductance of a wire is given by

$$L = 0.005081(2.303 \log_{10} 4\frac{l}{d} - 0.75)\mu\text{H} \tag{36.6}$$

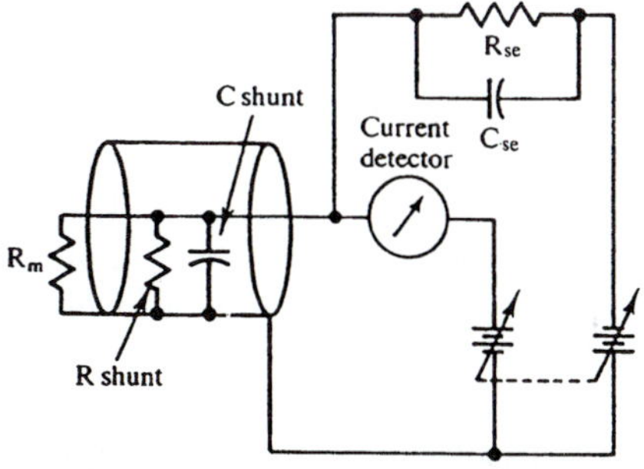

Figure 36.3 Neutralization to reduce leakages.

where L = inductance of the wires
l = length of wire, in
d = diameter of wire, in

At high frequencies, the self-inductance decreases slightly because of skin effects:

$$L = 0.005081(2.303 \log_{10} 4\frac{l}{d} - 1)\mu\text{H} \tag{36.7}$$

For a 5-in-long no. 20 wire, the high-frequency inductance will be approximately 0.14 μH (the resistance would be less than 0.01 Ω). At 10 MHz, this wire would present an inductive reactance of approximately 9 Ω. It is obvious that for any length of wire at very high frequencies, the wire impedance will become very high.

The inductance of interconnecting leads is one of the serious high-frequency limitations in measurement situations. At frequencies up to a few megahertz, short, heavy wires may be satisfactory; however, at higher frequencies, a transmission line may be the only effective means of connection. At microwave frequencies, strip-line and waveguide techniques are means of connection.

36.5 Component Stray Parameters

Mistakes are often made by assuming that capacitors, resistors, and inductors exhibit the same capacitance, resistance, and inductances, respectively, at all frequencies, voltages, and currents. Some of the changes that occur in these components due to distributed parameters are given in the following subsections.

36.5.1 Resistors

The characteristics of resistors depend on their physical construction. Most resistors are made either from resistive granules such as carbon or from some continuous resistive filament. In granular construction, there will be a small capacitance between the granules; thus, at high frequencies, this capacitance may shunt the resistive component to the extent that the resistor will appear as a leaky capacitor. As the frequency increases, the lead inductance becomes dominant. At high voltages (hundreds of volts), the resistance of common carbon resistors will vary with voltage. This is caused by tiny arcs that occur between the granules, again shunting out some of the resistance.

Filament resistors are made in many different ways. They can be wound from resistive wire, produced by carbon deposits, metal-etched, etc. Wirewound resistors require many turns of the filament on a form. This type of resistor will then have an inductance that is significant even at low frequencies. Some filament-wound resistors are designed to have as low an inductance as possible. This is accomplished by winding two filaments and connect-

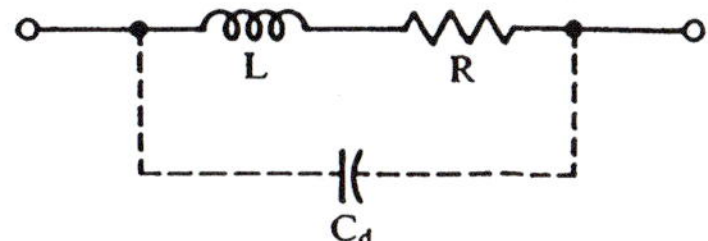

Figure 36.4 Equivalent circuit of an inductor (L = inductance; R = resistance resulting from losses due to wire resistance, hysteresis loss, and eddy current in the core; C_d = equivalent distributed capacitance between winding).

ing them in series so that the inductive field cancels, thus extending the frequency limit at which these resistors can operate. It is impossible to obtain perfect coupling between the two windings; therefore, there will appear some leakage inductance that will eventually limit the upper frequency at which the resistors can be used.

36.5.2 Inductors

Inductors consist of many turns of wire wound on a core. In various designs this core may be anywhere from air to a magnetic material of high permeability. All inductors will exhibit a shunt capacitance and series resistance as shown in the simplified lumped-parameter equivalent circuit of Fig. 36.4.

As the frequency is increased, two things will happen:

1. The resistance will increase because of skin effect, hysteresis, and eddy-current losses.
2. The distributed capacitance will determine a frequency at which the inductor will be self-resonant. At frequencies above this self-resonance, the inductor will appear capacitive.

When a magnetic material is used for the core, the inductance is also a function of the current through the coil because the magnetic permeability will vary with the current. Sometimes an air gap is used in the core material to increase the range over which linearity can be achieved. This, of course, decreases the inductance per turn that can be achieved in any given inductor.

36.5.3 Capacitors

Probably more varieties of capacitors are available than any other component. To describe the advantages and disadvantages of each is beyond the scope of this book. In general, however, all capacitors can be represented ap-

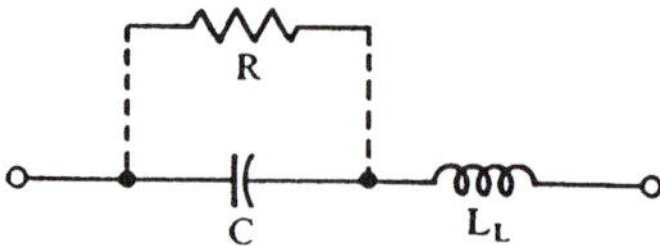

Figure 36.5 Equivalent circuit of a capacitor (C = capacitance; L_L = lead inductance; R = resistance due to leakage and dielectric loss).

proximately by the equivalent circuit shown in Fig. 36.5, and a few general comments are in order. The resistive loss of a given capacitor depends on the types of dielectric in the capacitor, the frequency of operation, and the applied voltage. Many dielectrics such as those in electrolytic capacitors are very lossy at high frequencies. As the voltage increases, the capacitance may vary significantly with some dielectrics, and eventually a voltage breakdown will occur, limiting the amount of voltage that can be applied safely to a capacitor. All capacitors will have some internal inductance due to leads and connections that will limit the discharge rate and current flow. Special low-inductance capacitors are built for high discharge rates.

At microwave frequencies, almost all components will appear inductive because of the connecting-lead inductance. At these frequencies, special techniques such as strip lines, cavities, and waveguides must be used rather than lumped-circuit components.

Chapter

37

Digital Interface Issues

D. A. Burt and K. D. Baker
Department of Electrical Engineering
Utah State University, Logan, Utah

37.1 Introduction

The advent of digital electronics and instruments, computer-controlled instruments, and computer networks has spawned many different types of digital interface protocols to allow the transfer of data back and forth. Each standard was developed for a specific purpose, such as speed, reliability, cost, politics, copyrights, etc. An exhaustive description of all these standards is beyond the scope of this chapter; however, they all have similarities that we will try to describe. The practical considerations associated with the interchange of digital data between measuring instruments and between these instruments and the rest of the world will be briefly presented. These explanations will include voltage levels, limited bandwidth, jitter, noise effects, sensitivity, pulse rise time, data rate, and other errors.

37.2 Asynchronous Systems

When this type of transmission is used to transfer data in the form of a serial bit stream, the receiving system has no way of knowing when a transmission is going to occur or even when a data word is going to be transmitted. Usually, the bit rate is known, but each word has to carry a code that tells the receiver where the word begins and ends. This type of transmission is used for many applications. It is required where only one path is available, and the data must be sent in a serial manner. Examples of this are FAX machines, modems, teletypes, and serial computer peripherals. Theoretically, asynchronous transfer could be accomplished on parallel bus systems, but if wires are available to allow parallel bit transfer, it is logical to assume that there are sufficient wires available for synchronous "handshaking" wires. The conversion of parallel data to a serial bit stream for transmission and the reverse process at the receiver are done by a special integrated circuit (IC) called a "universal asynchronous receiver/transmitter" (UART).

As stated previously, each word in a serial data stream has to stand alone, with everything required contained within the bit pattern for the receiver to recognize start, stop, and data bits. In addition, it is usual to include some type of data verification. For instance, each word may consist of 10 bits; one bit would indicate the start of the word, which usually occurs after a "long" off time, and eight bits would carry the data, with one bit used as parity to indicate the validity of the received data. A common scheme is to determine the parity bit by adding the bits in the data word and then sending a high-parity bit when the total is odd. This is called "odd parity." For an even-parity system, the parity bit is set high when the total of the high data bits is even. A "parity error" can either tell the receiver to ignore that data word or request retransmission.

These types of data-transfer systems work well for modest data rates when the signals are relatively noise-free. Unfortunately, this is not always the case, especially when the signal path is large, such as over telephone lines or over rf transmission links. Several methods have been devised to decrease the deleterious effects of noise. One method that is used often enough to warrant mentioning in this brief description is the "phase-locked loop." The phase-locked loop basically locks an internal oscillator to the bit rate of the incoming signal and thus virtually regenerates the bit clock that originally generated the data bit stream. This regenerated bit clock can then be used to sample the data signal at precisely the time a bit is supposed to occur. This requires that the transmitted frequency be stable so that the phase-locked loop can acquire and maintain lock. The word rate does not need to be fixed. The use of a phase-locked loop basically emulates some of the noise immunity inherent in synchronous systems, to be described next.

37.3 Synchronous Systems

There are significant noise immunity advantages over the asynchronous system by using synchronous digital data-transfer techniques. The asynchronous system requires that the receiver be open all the time to receive data any time the sender is ready. Thus any noise will be interpreted as a data word, and the receiver will try to act on it or use some sophisticated error-detection scheme to reject it. In contrast, the synchronous system will accept information only for the very narrow time slot that it specifies. Generally, this acceptance time is narrowed down to a very short time during a single bit transfer. This is accomplished by providing "handshaking" capability between the systems. For a serial data system, the handshake generally consists of data-ready, bit-clock, word-gate, and data-signal lines. When the transmitter has data that are ready to be sent, it raises a flag on the data-ready bus. The receiver monitors that flag, and when the flag is detected, the receiver will send back a word gate that will only allow a certain length for a word. Another signal is then sent, called the "data clock," which will only allow each bit to be placed on the data line at a specific time. The receiver will be set up to accept the data only for a certain short time of the bit clock. The synchronous handshake system works equally well on serial or parallel data transfer. Of course, for a parallel word, a word gate is not needed. Sometimes a block transfer may be made, which requires a frame gate. The various gate names are given here only for reference. Virtually everyone who works with these interfaces has their own favorite names, but they all do basically the same thing. There may be several additional gates or signals used on any given system. For instance, a printer will usually have a paper-out flag, etc.

37.4 Interface Drivers

No matter what type of digital system is used, some connecting medium is required between the transmitter and receiver such as wires, optical fibers, an rf link, etc. This connection needs to be free from loading effects and as noise-free as possible. In addition to incorporating methods to minimize the effects of common-mode and conducted noise, protection must be provided for the instruments that are interconnected. For example, a computer may have many terminals or instruments connected to it. If not properly designed and implemented, someone can scuff their feet on the rug and touch the terminal, resulting in a high-voltage discharge that will wipe out the host computer.

37.4.1 Optical isolators

Optical isolators can be installed between the two digital devices so that there is no actual conducting path between them. This is highly desirable, although it is not currently general practice throughout the computer and electronics industry. This type of isolation will provide voltage protection up to the capability of the optical isolator, which is usually rated at 1500 V. The optical isolator also provides a very high common-mode impedance, resulting in significant common-mode interference rejection. A practical limitation of the presently available optical isolators is the limited bit rate without the use of expensive special devices.

37.4.2 Balanced drivers/receivers

Integrated circuits are now commonly available which are built to drive signals over a line to another matched integrated circuit receiver at the end of the line. Numerous types are available. Some drive the signals with respect to a common ground (single-ended); others are balanced to ground (double-ended). Various devices have different data-rate capabilities, input and output impedances, transmission distances, etc. The selection of a specific device will depend on the application. For instance, a computer networking system is presently in use and being improved to send signals all over the world. The network card that is being installed in each computer uses specially designed transmitters and receivers based on rf techniques. The inputs and outputs are balun transformers matched to a 50-Ω transmission line. This provides the same common-mode isolation as an optical isolator and prevents one careless person on the line from wiping out computers throughout the world.

37.4.3 Internal logic levels

In the preceding sections, the general discussion has related to communication between senders and receivers in separate instruments over significant distances. Within any given piece of equipment, digital signals are transferred from one subsystem to another. As long as the subsystems use the same power supply voltages, the logic levels will be transferred from chip to chip using the same technologies: TTL to TTL, ECL to ECL, etc. In some cases it is necessary or advantageous to change from one technology to another. Again, there are several special devices designed to accomplish whatever voltage-level shifting or drive power required to drive the different families of devices.

37.5 Addressing

If the communications link is over a dedicated bus with a single sender and a single receiver, there is no need for an addressing capability. Often, however, to provide maximum efficiency, a digital interface will interconnect several different devices. When this occurs, a scheme is required that allows the transmitter to send the information to the desired on-line receiver. This is accomplished by sending the data in packets, with each packet carrying an address on the first few bits of the data. This address may consist of just one or two bits in some cases or numerous bytes depending on the number of separate addressable units that are planned to be connected to the bus. A prime example of this is the present computer networking system, which has more than sufficient capacity to have a distinct address for each individual person who is living on this earth. A receiver connected to such a system is designed to ignore any data not containing its unique address in the bit stream of data on the bus.

37.6 Data Compression

A technology that is rapidly expanding is the ability to increase information transfer rates by compressing the digital data without losing any information or at least losing very little. Data-compression people like to talk about the entropy of the signal. A "high entropy" simply means that a signal has a lot of detail; "low entropy," a low amount of detail. Consider a printed page; most of the page is still blank or white. Data compression for a FAX machine takes advantage of this by looking for a series of bit levels that are the same, counting them, and sending out a digital word that tells the signal level and the number of bytes that are all identical. Very significant reductions are thus effected in the bit rate being sent for the data transfer. This is one of the main innovations that has made the FAX machines really operational. If the entropy is high, different methods are required, but significant data-rate reductions can be obtained by sophisticated techniques, especially if some limited degradation can be tolerated.

37.7 Operational Concerns

Several operational concerns have already been expressed in this chapter. A listing in one place is included here for your consideration. This is by no means an exhaustive list, and if you are having trouble, sometimes it will seem that your particular problem has never been thought of before. Generally, however, the problem will be due to wrong logic levels or because of one or more of the following considerations.

37.7.1 Timing

The most critical parameter in a digital interface is to make sure that the bit being sampled is available when the receiver is looking for it. This is important for both parallel and serial interfaces. In the case of a serial interface, both systems need to operate at the same bit rate. Some smart systems, such as a FAX machine, will determine which of several standard bit rates is being sent and set the receive rate properly. Of course, the sender must use a standard bit rate within a certain tolerance, and that rate must be stable.

37.7.2 Jitter

The spacing between pulses in digital signals may vary slightly in time. This effect is called "jitter." The jitter results from additional noise, instability of the bit-rate generator, limited bandwidth, and in most systems because some bits may come through different circuits. In systems that are pushing the maximum data rate, jitter will introduce errors as the bits are misinterpreted. Any good design must be made to tolerate a certain amount of this jitter.

37.7.3 Noise and interference

Noise and interference on any digital interface can cause false triggers and signals, especially if the noise occurs during the sample time. Interference is generated many different ways and can be coupled into your circuit by radiation and conduction. Conducted interference can be single-ended and/or common-mode. This topic is covered more thoroughly in Chap. 34. A special type of interference in high-speed digital circuitry is that caused by improperly terminated wiring. A short, fast rise-time pulse sent down a line with an improper impedance match can echo back and forth several times and virtually cause numerous pulses to appear at the receiver even though only one was sent.

37.7.4 Rise time and bandwidth

Some digital systems are triggered by the rising and falling edges of the signals. In these circuits, the rise times must be maintained to provide proper triggering. The rise and fall times are determined by the bandwidth of the system. Achieving short transition times requires very large bandwidth. Again, it is important that the proper impedance match be used in any given transmission line. Often the signals may be transmitted over long distances, and significant degradation of the rise and fall times will occur, resulting in smearing out of the pulses. These can be restored quite well by using phase-locked loop techniques. The ultimate rate of signaling is set by the limit that the pulse being sent is too short to distinguish it distinctly from the next pulse due to spreading of the pulse as a result of the limited bandwidth. The limit to being able to distinguish these pulses uniquely is that the bandwidth must be at least twice the pulse rate. In a real system degraded with noise, as this limit is approached, the error rate becomes significant.

37.7.5 Bit error rates

"Bit-error" rates are defined as the number of errors that occur for a given number of transmitted bits, such as 1 error in 10^6 for a reasonably good system. The error rate that can be tolerated, of course, depends on the application. For instance, if you are transmitting a critical program, no error at all can be tolerated. For this reason, most communications systems utilize some sort of error detection, and some will even have an error-correction system. If only error detection is used, the system will have a method to request that the data be retransmitted when an error is detected.

37.8 Interface Standards

Many instruments are now computer-controlled and/or data collection at their output is in a digital format for analysis. Accordingly, several interface standards or data protocols have emerged for transfer of digital data between instruments. The most important of these will be summarized here starting with the interface used to connect computer equipment with programmable instruments.

37.8.1 IEEE-488/HP-IB interface for programmable instruments

The IEEE-488 interface emerged from the HP interface bus (HP-IB), also called the "general-purpose interface bus" (GPIB) first developed in the 1960s

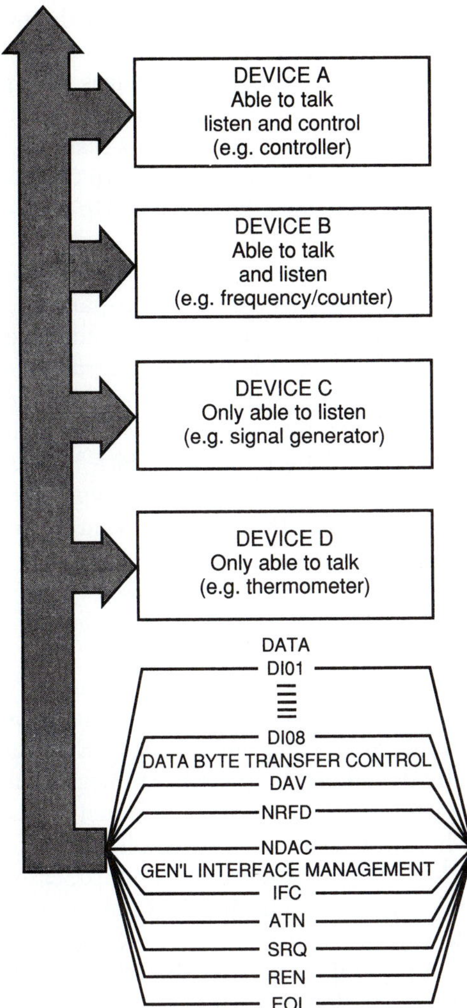

Figure 37.1 Interface connections and bus structure.

by Hewlett-Packard. This system, which evolved into the IEEE-488 interface, allows programmable instruments to be interconnected in a standard fashion for automated measurements. In addition to the IEEE, the International Electrotechnical Commission adopted this standard in IEC Publication 625-1, and the American National Standards Institute adopted and published it initially as ANSI Standard MC1.1. The standardized interface for programmable instruments is now widely accepted internationally.

The general interface and bus structure of the HP-IB/IEEE-488 is illustrated in Fig. 37.1. The system links several IEEE-488 devices that are listeners, talkers, or controllers together on a bus that transfers data in a parallel fashion. A "listener" is an instrument such as a printer or programmable power supply that can receive data from other devices via the bus. A "talker" is an instrument such as a voltmeter or frequency counter capable of transmitting data to other devices via the bus. Some devices can perform both roles. For example, a programmable instrument can "listen" to receive its control instructions and "talk" to send its measurement results. A "controller" manages the bus system by designating which devices are to send and receive data, and it can also command other specific actions. Up to as many as 15 devices can be connected to the bus in parallel.

The interface definition prescribes 16 signal lines in a cable using a 24-pin connector. Eight of these signal lines carry data in bit-parallel format, with the rest of the lines for data-byte transfer control (3) and general interface management (5). The signal lines carry addresses, program data, measurement data, universal commands, and status bytes to and from the devices interconnected in a system. The data lines are held at +5 V for a binary 0 and pulled to ground for a binary 1. The maximum data rate is 1 megabyte per second (Mbyte/s) over short distances or more typically 250 to 500 kbytes/s over a full transmission path that ranges from 2 to 20 m under various conditions.

37.8.2 RS-232C, RS-422A, RS-449, and RS-530

The RS-232C serial interface is widely used on PCs and other computers to interface to modems, mouse pointing devices, and other peripherals and instruments. This interface standard is also known as the "EIA standard interface" because it was originally developed by the Electronics Industries Association (EIA) in conjunction with the Bell system and industries that produced computers and modems. The International Telegraph and Telephone Consultive Committee (CCITT) has adopted a similar standard, called V.24. Also, military standard MIL-188C is similar. The RS-232C interface standard defines the electrical and data-interchange characteristics for transfer of data to and from a computer in bit-serial format. Although the standard states that this interface will be used for distances less than 15 m and data rates less than 9.6 kbits/s, the interface is useful under some conditions for data rates up to 20 kbits/s for cable lengths up to about 15 m or longer with the use of twisted pair cable. The total loading capacitance must be kept below about 2500 pF. The voltage levels are $-V$ for a binary 1 (mark) and $+V$ for a binary 0 (space), where V is 3 V or greater. More typically, the levels used will be ± 10 V or more. The voltage levels are with respect to a common signal ground lead, so-called single-ended signaling. Most RS-232 applications utilize asynchronous mode. Although not formally part of the RS-232C standard, most systems utilize a 25-pin D-type connector (DB25).

To increase the data rates, a new standard RS-422A (V.11) was developed for using balanced signaling with two wires per circuit. Since this technique is a differential system not using a common return path, it has reduced interference and crosstalk and hence allows higher signal rates and longer transmission paths. The balanced signal allows the use of smaller voltage levels than the single-ended system. Positive and negative levels from a few tenths of a volt to 25 V are used. Data rates using the RS-422A system can be up to 100 kbits/s at distances up to 1500 m or 40 Mbits/s at 15 m. The RS-422A interface uses a DB37 and a DB9 for connectors.

The RS-449 (V.35) interface is intended to succeed the RS-232C interface to allow for higher signaling rates and more flexibility. This system uses the same connectors as the RS-422A standard, but the system can be used single-ended with the RS-232C levels or balanced with those of the RS-422A. The RS-449 used in the balanced-signaling mode can conservatively provide signal rates up to 10 Mbits/s at 10 m or 100 kbits/s at 1000 m. Under the best conditions, the system can be used at rates of 2 Mbits/s at distances up to about 250 m. Because of the relatively high data rates, these systems use a synchronous mode of transmission. The RS-530 interface is basically the same as the RS-449 except using a DB25 connector.

37.8.3 Current-loop interface (20-mA current loop)

One of the earliest interfaces for the transfer of digital data, originally used to drive mechanical teleprinters, is the “current-loop system.” In this system, two or more devices are connected in a series loop, and the sender device is a source of a specified current through the loop, usually 20 mA. The current is interrupted to signify a space and reestablished for a mark to create the binary-coded bit stream. The interruption of the current is detected at the receiving devices to reproduce the binary signal. The noise immunity of the current-loop interface is much better than that of the voltage-driven interface of the RS-232C system because external noise picked up by the two-wire system has little effect on the current-sensitive receiver at the 20-mA level. As a result, the 20-mA current loop is particularly good for long distances up to about 1 km at modest bit rates due to the limitations of the current switches and receivers. Many devices provide two separate outputs, a usual-voltage-level RS-232C and a 20-mA current loop. The user can then choose either output depending on device separation and the signaling rate.

37.8.4 Centronics parallel interface

Most printers used with PCs employ a Centronics parallel interface with a 37-pin Centronics connector at the printer end of the interface cable. The parallel-port connector on the computer is usually a DB25. Eight bits of data are sent in parallel on eight data lines that are sequentially strobed out on command from a strobe pulse. The data levels are unbalanced at TTL logic levels of 0 and +5 V, nominally. This parallel interface is sometimes used to transfer data between the computer and other devices rather than just a printer.

37.8.5 Other interfaces

Because of the rapid proliferation of computer-controlled instruments and computer networks, there are a number of evolving interfaces in addition to those described in earlier subsections. A few of these will be mentioned here. The SCSI (small computer system interface) is a parallel interface standard for connecting high-performance peripherals to microcomputers. It is an 8-bit parallel cable interface with handshakes and protocols for handling multiple hosts and multiple peripherals. SCSI supports data rates up to 1.5 Mbytes/s in asynchronous mode or 4 Mbytes/s in synchronous mode with cable lengths up to 6 m (single-ended) or 25 m (differential). A 16-bit interface bus known as an "intelligent peripheral interface" (IPI), which offers higher data transfer rates, is being designed into many new systems. Interface standards also have been developed for use of data transfer in the packet mode (X.25), for advanced military systems (MIL STD-1553), and for computer local area networks (LANs). The entire area of data communications is a dynamic, rapidly evolving field; consequently, many innovations and changes can be expected in the near future.

Chapter

38

Instrument Systems

James M. McGillivary

Agilent Technologies
Rockaway, New Jersey

38.1 Introduction

The term *instrument system* can be used to describe a broad spectrum of systems, such as a voltmeter and a printer to many instruments and computers interlinked into an automated manufacturing environment. The basic reason that instruments are placed into systems is to achieve capabilities that the individual instruments do not have. If an instrument could do all the measurements that the user required, automated them, and compiled the data, there would be no need for a system. In this case, the system design and integration were performed by the instrument manufacturer. The likelihood of this occurring is very small because there are too many different types of measurements, and each device being tested uses its own particular subset of them.

38.2 Instrument System Hardware Architecture

A typical block diagram is shown in Fig. 38.1. General descriptions of the major elements follow the block diagram.

38.3 Instruments

Instruments are the devices used to take the actual measurements. These can be standard "off the shelf" instruments or custom instruments designed for a special purpose. A system designer has to specify which instruments are required in the system to make all of the correct measurements. The four main

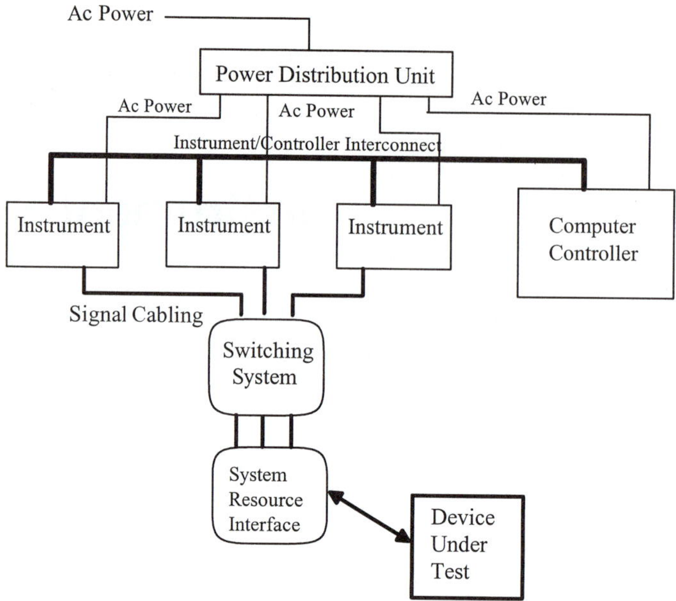

Figure 38.1 A typical hardware block diagram.

types of instruments are:

- Measurement instruments, which measure the output of the DUT.
- Stimulus instruments, which supply the required input signal to the DUT.
- Switching instruments, which route the measurement and stimulus signals to the DUT.
- Distributed instruments, which can be located far from the instrument controller.

To determine what instruments are required for the system, make a list of all the required tests for the device(s) under test (see Table 38.1). For each test, write down the stimulus signals and the expected response. Remember to include the power supply voltage and current levels. Then determine which instrument can provide the stimulus and measure the response. This will yield the list of instruments that are required for the system.

Take this list of instruments and try to reduce the number of instruments required. This will lower the cost of the systems and decrease the cable lengths required to connect to the DUT. The following techniques can help reduce the list.

- *Look for duplicates.* One particular instrument can be used in multiple tests. For example, a voltmeter might be needed for 10 tests, but only one per test. Just buy one. If a test requires two voltmeters at the same time, put two voltmeters in the system.

TABLE 38.1 A Sample Worksheet

Test name	Required stimulus	Expected response	Stimulus instrument	Response instrument
Power on test	12.0 V +/− 5% @ 1 amp	5.0 V +/− 1%	Power Supply	Digital Multimeter
Resistor test	None	10 kΩ +/− 5%	None	Digital Multimeter
Bandwidth test	5 V p-p, 10 MHz Sine wave	5 V p-p 10 MHz Sine wave	Function Generator	General Purpose Oscilloscope

- *Look for instruments that fill two requirements.* If the system requires two power supplies, one 5 V and one 12 V, one supply that covers the 5-V and 12-V range might fill the need (if both the 5 V and 12 V are not needed at the same time).
- *Look for multipurpose instruments.* Many instruments have multiple functions. Using one of these instruments can save the expense and size of multiple instruments.

If your instrument system will be distributed into different locations, be sure that you go through this exercise for each location.

38.4 Switching System

Switching is used to route the stimulus and response instruments to the DUT. Switches are mostly used in an instrument system to lower costs and to save space. Using one voltmeter and a 16-channel relay card is much less expensive and more compact than using 16 voltmeters.

In distributed systems, it is very difficult to use switches to economize the instrumentation. The measurement points are just too far away from each other to share instruments. This forces each measurement node to be self contained and forces the manufacturer of the node to find other ways to economize. This is accomplished by putting all of the node's functionality into one chip or using much less expensive instrumentation.

To design the switching system, follow the following steps:

- Define the switching topology.
- Select the type of relay.
- Determine the quantity of switches required.

38.4.1 Switching topology

The first decision on the switching topology (this defines the physical layout of the switches) is crucial. The two ways to design a switching system are: multiplexer topology or matrix topology. Figure 38.2 shows the difference.

The matrix switching topology can switch any of its inputs to any of the outputs. This provides the user of the system with a great deal of flexibility and does not require much up-front planning to design the system. If the system has 5 instruments and 50 test points, all that is required is a 5 × 50 matrix switch.

The problem with matrix switch systems is that they have a limited bandwidth and are prone to picking up and generating electrical noise. When a connection is made in the matrix, some unterminated lines are left connected to the circuit that are called *RF stubs.* These stubs reflect and radiate any signal that is passed through the matrix. This problem gets worse with bigger matrix-switch systems. Most matrix switch systems have this problem around 1 MHz and greater. The problem tends to be proportional to the size of the smallest part of the matrix. So, a 5 × 50 matrix would have approximately the same bandwidth limit as a 5 × 100 matrix. A 10 × 10 matrix would be worse than a 5 × 50.

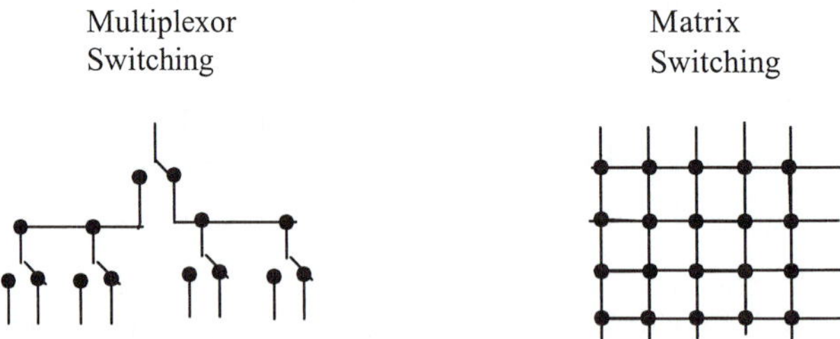

Figure 38.2 Multiplexer vs. matrix switching topology.

Multiplexer switching systems do not have the "stub" problems that matrix switching have. Multiplexer switch systems can have very high-frequency and low-noise signal characteristics. The drawback here is in the flexibility. A multiplexer switch has much less flexibility than a matrix switch; therefore, more planning is required when using a multiplexer switch.

38.4.2 Selecting the type of relay

Many types of relays can be used to switch signals in the instrument system. When selecting the relays, consider the following:

- Current requirements
- Voltage requirements
- Relay speed requirements
- Frequency content

Current requirements. Low current is in the milliampere range. High current is in the ampere range. Reed relays are recommended for low current. Armature relays are recommended for high current.

Voltage requirements. Reed relays are suitable for most voltages, but special contacts might be required for high voltages. Extremely low voltages (submillivolt range) also require special contacts—the contacts of the relays will generate some small thermal offset voltages that can interfere with the measurements.

Relay speed requirements. Armature relays are limited to 30 to 40 closures per second. Reed relays can do up to 500 closures per second. Requirements for faster speeds would need electronic relays. Electronic relays are typically based on a *field-effect transistor* (*FET*). This electronic device is capable of changing from a very low resistance (relay closed) to a very high resistance (relay open) very quickly, compared with mechanical switches. More care must be taken when designing a switching system with electronic switches because of parasitic transistors, diodes, leakage currents, and voltage restrictions that accompany electronic switches.

Frequency content. Reed relays can handle up to about 10 MHz. RF relays go to 1 GHz and beyond. Matrix relay systems have a limit of about 1 MHz.

Determining the quantity of switches required. If the system is going to use a matrix switch topology, it is very easy to determine the quantity of switches in the instrument system. Take the number of instrument signals that are going to be switched and multiply it by the number of pins that they could be connected to on the DUT. If five signals will be switched to 50 pins on the DUT, use a 5×50 matrix switch.

If the system is going to use a multiplexer switch topology, determining the number of switches is not as simple. The best way to do this is to make a schematic drawing of the instrumentation and the power and signal pins of the DUT. Create one drawing for each instrument in the system. Put the instrument on the left side of the paper and the pins for the DUT on the right. If the system is going to be used for multiple devices, repeat this for all the devices (or at least the worst case). Walk through the tests one by one to verify the switching system.

38.5 Computer Controller

The computer controller controls the instrumentation, gathers the data, and formats it for further analysis. Many different computers can be used, depending on the requirements for:

- *The interface to the instrument.* Some computers do not support all of the interfaces to instruments and can affect your system design.
- *Real-time measurement requirements.* Certain applications of instrument systems require real-time responses from the computer. This might require a specialized computer system.
- *Availability of software tools.* Software tools that can dramatically speed the development of automatic tests might only be available on certain computers or require certain computer configurations.
- *Environmental concerns.* Some applications might require that the computer be located in a harsh environment, whereas most computers are designed for a typical office environment.
- *Conformance to standards.* Some instrument systems might be required to meet corporate or end-user specified standards for the computer system.

38.6 Device Under Test

The *device under test* (*DUT*) is the product or component that is being tested. This can be anything electrical, from an integrated circuit to an entire airplane. In a distributed instrument system, the DUT might be an entire manufacturing plant or a large network.

38.7 Instrument-To-Controller Interconnect

The *instrument-to-controller interconnect* is the part of the system that interconnects the computer to the instrumentation. This interconnect point controls the instruments and retrieves the measurement data. At one time, there was very little choice on how to do this. A system that used a computer to control instruments used the IEEE-488 bus. With advances in technology, standardization of computers, and the need for higher-speed data transfer, a number of different interconnect options have become available.

38.7.1 IEEE-488 bus

The *IEEE-488 bus* has been the original standard for connecting computers to instruments since 1975. It is still the most popular and flexible interconnect scheme available. Almost every type of computer system has an IEEE-488 card and drivers available and most instrument manufacturers support this interface on their instruments.

This system consists of an interface card that plugs into the computer and cables that connect to the instruments. It is a bus structure that can connect 14 instruments to one interface card. They are all connected in parallel by chaining the cable from instrument to instrument. The maximum data transfer is just under 1 Mbytes/Sec. The system has heavily shielded cables and does a good job of isolating the computer noise from the instruments.

One big problem with IEEE-488 is that the specification only covers the electrical characteristics and some of the low-level messaging. This has made it difficult to write consistent software for this standard. In 1988, the IEEE came out with IEEE-488.2, which further defined the lower-level messaging. Shortly after this, instrument manufacturers further completed the specification with a standard called *SCPI* (*standard command programmable instrument*). SCPI standardized the higher-level commands and the error-reporting methods.

IEEE-488 does not address the instrument on a card technology, an increasing need for higher data-transfer rates or distributed instrument requirements. (See References 1 to 3 for further information).

38.7.2 VXIbus

A very high performance instrument-to-controller interconnect is the *VXIbus* (see Fig. 38.3) (*VMEbus extensions for instrumentation*). VXI has its roots in the VME backplane. *VME* (*VersaBus Modular European*) is a standardized backplane that is primarily focused on computer systems. The original intent of VME was to help engineers standardize their computer system designs. Before the VME standard, each product that required multiple printed circuit cards had a custom-designed backplane to connect the cards and allow the signals to pass back and forth between the cards. This usually ended up being a major part of the product design. With each product having its own backplane, it was impossible to leverage much of the design from one product to another. With the arrival of the VME standard backplane, designers were able to eliminate the time it used to take to design the backplane and leverage cards from previous designs. The VME standard specifies the following:

- Printed circuit card size. (VME calls them *A* and *B size,* where A is 6.99″ × 3.937″ and B is 6.299″ × 9.187″)

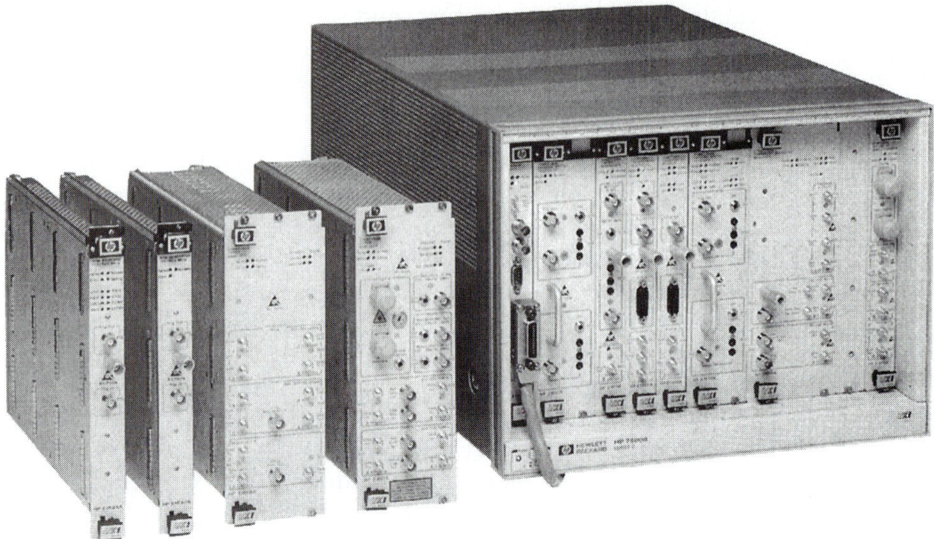

Figure 38.3 VXI card cage.

- Placement and type of connectors. (VME calls them *P1* and *P2 connectors*)
- Location of signals on the connectors
- Bandwidth and timing of signals
- Placement of power pins on the connectors
- Voltage levels and current rating of power pins
- Spacing between the printed circuit cards (0.8 inches, slot to slot)

VXI is an extension of the VME specification for instrumentation. VXI is a card cage, typically containing 13 slots (VME can have 21), with at least one of the slots being a controller card (called *slot 0 card*). This controller card is either a local VXI command interpreter or a fully functional computer. The VXI specification (reference 4) adds the following extensions to the VME standard:

- Additional printed circuit sizes (adds C and D size)
- Additional connector (P3)
- Additional signals on the connectors for instrumentation
- More power pins
- Additional spacing between the C- and D-size cards (1.2 inches slot to slot)
- More electromagnetic shielding
- Slot 0 card for timing, system controller functions, and data-communication ports

TABLE 38.2 The Four Configurations

Controller	Programming	Characteristics
Command module	Message based	Acts like typical IEEE-488 instrument (allows for access to latest controller technology)
Command module	Register based	Typical IEEE-488 instrument with less message overhead (fewer characters sent)
Embedded controller	Message based	Fast communications—almost backplane speed (message overhead)
Embedded controller	Register based	Fastest communications—backplane speed and low message overhead

The VXI systems have two different controller configurations and two methods to program them. Table 38.2 shows the four configurations.

38.7.3 PCbus

The standard *ISA* (*Industry Standard Architecture*) or *EISA* (*Extended Industry Standard Architecture*) personal computer bus can be used as an instrument-to-controller interface. To do this, a standard personal computer is selected as the controller. The instruments are "instruments on a card" that plug directly into the PC backplane. This is generally a very low-cost system solution caused by the huge volumes of personal computers and the low cost of fabricating cards for them. This approach has similar tradeoffs, as the VXI system, except that the PCbus was never intended to be used for instruments. Also, the card form factor is even smaller than VXI with more severe power limitations. The best application for this: medium-speed digital functional testing or data-acquisition.

38.7.4 LAN connections

With the ever-increasing pace of computer advances, new computer systems have lifetimes less than one year, but instruments generally last 10 years or more. Only a very small percentage of all computers are sold for use in instrument systems.

This means that it will become increasingly difficult to get instrument interface cards like IEEE-488 supported on the newer computer systems. One solution to this problem is to use the LAN connection (available on all new computers) to control instrumentation. LAN traffic is very fast and many software tools are available to support its development.

Some instruments support LAN control. Typically, these instruments are very intelligent (frequently, they have their own embedded computer) and have very data-intensive operations to perform. An example of this is a logic analyzer that might have many megabytes of test data to transfer to the test controller. An added benefit of using LAN as an interconnect is that it is not required to have the computer controller very close to the instrument because of the long lengths that LAN can support. This allows a "distributed system."

38.8 Power Conditioning/Distribution Unit

A power conditioning/distribution unit provides clean, controlled ac power for all the instruments and the controller. A central, dedicated power conditioning/distribution unit allows the system designer to add:

- A single on/off switch for the entire instrument system. This is better than requiring the user to hunt for all the different power switches (some of them are mounted in the back of an instrument). If the user misses a switch and runs a test, the test might fail or damage the DUT.
- An emergency power off switch to remove power quickly to the entire system in an emergency and prevent damage to personnel and equipment.
- Power-line conditioning to protect sensitive instruments from noise, glitches, and other powerline problems.
- Provide uninterruptable power to the system controller. These devices are essentially batteries that "keep the controller alive" during a power failure. They can be programmed to save current test data in case of a power failure or hold up the system during a short power-outage period. The power of these devices is limited, so it is unlikely that it can keep the entire system operating during the outage.

For distributed systems, the power conditioning can be a challenge because of the lack of a centralized power point. Each of the nodes will have to deal with the power issues by themselves. Another challenge is that the distributed systems are frequently deployed in areas where clean ac power is not conveniently available.

38.9 System Resource Interface

The system resource interface is the place in the instrument system where all the signals from the instruments are available. This is typically a group of connectors with cables that go back to the instruments in the system. These signal connectors are all in one central place to allow for easy access to the DUT.

38.9.1 Test fixtures

The connection to the DUT is called a *test fixture.* The several different types of fixtures are:

- Manual test fixtures
- Patch panels

- Commercial “quick disconnect” panels
- “Bed of nails”
- Booms

Manual test fixtures. This can be as simple as some cables coming off the front of the instruments and connected by hand to the DUT. These types of fixtures are generally reserved for systems with very short lifetimes, such as design verification systems and performance test systems. This saves the overhead of designing and building a more-sophisticated fixture, but makes the test less repeatable and the system difficult to document and troubleshoot.

Patch panel. The patch panel is a more-sophisticated version of the manual test fixture. It has a distinct central point where all the instruments are connected. The DUT is then connected to this central patch panel with its own set of cables. This is the same idea as manual test fixturing, except that a patch panel is between the DUT and the instruments. It is easier to reconfigure and will yield more repeatable tests. The connecting and disconnecting of the DUT will be much less error prone.

Commercial “quick disconnect” panel. The commercial “quick disconnect” panel is an automated patch panel where all the connections to the panel are made at one time using a mechanical connection system. This system is typically mounted in a 19-inch rack and has several types of connector blocks, where the instrument signals are connected. Generally, the panels can carry anything from 30 Amps to 20 GHz through them. On the other side of the panel is a removable fixture that makes the panel-to-DUT connection. These systems are even more reliable than the patch panel system because the user only has to connect one thing to the system. This makes the instrument system suitable for a manufacturing environment. This system is preferred in the environment test, functional test, and field test applications.

“Bed of nails.” This is a special fixture used in the in-circuit test systems. The fixture is a flat horizontal plate with sharp spring-loaded pins pointing upward. The instruments in the system are connected through cables to the pins. The pins connect to the DUT (almost exclusively a printed circuit card) directly. Alignment is crucial and tolerances are precise. There are two major types of fixtures available: fixtures for “through-hole” printed circuit cards and fixtures for “surface-mount” PC cards. The “surface-mount” type of fixture is more expensive because of the smaller and tighter tolerances of “surface-mount” cards. These fixtures are also available in a “clam-shell” type to handle double-sided “surface-mount” cards. Each printed circuit card tested on the in-circuit tester has its own unique fixture. Many companies specialize in custom building these fixtures.

Booms. Frequently in field test systems, the tester must be connected to a device located inside a piece of equipment. A good example of this is testing the electronics in an automobile. The electronics are deeply embedded inside the car and must be left in the car. In this case, the best system resource interface is a mechanical boom with long cables attached. The cables connect to the instruments and the electronics inside the car. Frequently, it has an intermediate connection point, similar to a patch panel, where the cable is broken in two. This is to enable the replacement of broken cables (the DUT end) without rewiring the entire tester.

38.9.2 Cabling to the system resource interface

When cabling from the instruments to the system resource interface (SRI), consider several factors:

- *Raw cable selection.* Use coaxial, twisted pair, individual wires, or special microwave cables. See the section on signal cabling for more detailed information.
- *Routing the cables.* Keep the high-power signal cables away from sensitive signals. Digital signals should be kept away from the analog signals. This will help keep the noise from digital and high-power signals from interfering with the sensitive signals.
- *Remember to keep the cable lengths as short as possible, but leave enough room for service loops.* These service loops are handy when it is time to service the instruments. One or two extra feet in the cable makes it possible to pull the instrument out of the rack without disconnecting the cables.
- *When fabricating the cable, be sure to design some mechanical strain relief into the cables.* In big systems, the cables can quickly become interwined and moving any cable will pull on others. This will cause cables without proper strain relief to become disconnected. Repairing a disconnected cable in a large system can be very frustrating and time consuming.

38.9.3 Specifications at the system resource Interface

Frequently in an instrument system, the users will ask for the specifications at the system resource interface. This is not a simple task. The specifications are typically the instrument specifications, less the cable and connector degradation. The difficulty is factoring in the interaction of all the other instruments on a given signal. For instance, the system might have a 70-dB noise floor, except when a 30-amp relay contact is closed. When specifying signal characteristics in a system, be sure to specify that all other parts of the system are not active. The same is true when designing the tests: use the specifications, but remember that other parts of the system might be active and distort the signal.

The following are useful techniques to compensate for system degradation or interaction:

- *Always use shielded cables—even if the signal is not crucial.* For example, a high-voltage dc power cable should be shielded—even if it is not sensitive to noise. It is shielded to prevent noise generation.
- *Use the power-supply sense lines.* These are lines on most dc power supplies that effectively measure the voltage of the supply at the load and adjust the supply output to compensate for voltage drops in the cabling. To work correctly, they must be connected to the load out at the DUT. Remember to include these in the switching system; they must be switched with the power-supply wires.
- *Use four-wire resistance measurements.* A four-wire resistance measurement is very similar to a power-supply sense line. The extra two wires null out the resistance of the cables connecting the instrument (ohmmeter) to the DUT. These should also be connected directly to the DUT and included in the switching system.
- *Software techniques can also be used to calculate the losses in the cabling and used to null out the errors.* The general technique is to measure a known value and calculate the error. This error is then subtracted from all subsequent measurements. An easy example is an ohmmeter. A four-wire measurement can be very expensive when measuring a large number of test points; it doubles the number of cables and relays used by the test. A less expensive method is to start the test by measuring a short circuit as close to the DUT as possible. The resulting value is the resistance error in the test cables. This value is then subtracted from the rest of the measurements. This technique is less accurate (but less expensive) than using four-wire measurements.

38.9.4 System resource interface layout

When designing the layout of the system resource interface, keep the following factors in mind:

- Keep high-power pins away from sensitive signals.
- Keep digital signals away from analog signals.
- Bring power-supply sense lines to the panel. To be used correctly, they must connect directly to the load (DUT).
- Bring all four wires of the ohmmeter to the panel. They also must be connected to the DUT.
- Leave room for expansion.
- When using "quick disconnect" panels, be sure that the mechanical load is balanced. Connector insertion forces can vary from a few ounces to several

pounds. If the load is not balanced throughout the panel, the fixture could warp and fail prematurely.

38.9.5 Safety issues

Many safety issues must be considered when designing the system resource interface. The following is a list of the major concerns:

Hazardous voltage. Many instruments are capable of generating voltages that are hazardous to humans. When these signals are brought to a system resource interface, be sure that these are inaccessible to people. There are several methods to do this, ranging from recessing the pin to safety interlocks. Check the local electrical codes.

High current. High-current signals are considered a safety issue because of the high amount of energy that they can deliver. A good example is a car battery. It can only deliver 12 volts, but many hundreds of amps. The hazard occurs when it is short-circuited; the leads weld together and generate heat that can cause fires or burn people. The same thing can occur with high-power instrumentation. These leads are protected in much the same way as hazardous voltages.

Ergonomic. The panel should be located at a height that allows the system operator easy access. A generally accepted height range is 42 to 54 inches from the floor. Also, be careful to limit the weight of the test fixture.

Power/panel interlocks. Occasionally, a system designer will design the system resource interface ("quick disconnect" type) so that when it is opened, it trips a power breaker and disables instruments that are capable of hazardous voltage or current.

38.9.6 Adapter identification

When using a "quick disconnect" panel in production, multiple test fixtures and test programs are frequently used. This can create a dangerous situation when running the wrong combination of test programs and test fixtures. This can cause damage to the DUT and the instrument system. The typical method of avoiding this is to have each test program begin by looking for some unique identification on the test fixture. Two of the most-common methods are a unique resistor value in the fixture or a set of jumpers and opens that are read as a digital code. If the resistor or code does not match, the program is aborted and the user is notified of the mismatch.

38.10 Signal Cabling

One of the most often misunderstood parts of an instrument system is the design of the signal cabling between the instruments and the DUT. The cabling typically goes to a system resource interface, but for this example, it is assumed that it goes directly from the instrument to the DUT. The system resource interface can be thought of as a big connector in between the two. A poor design of this cabling can introduce noise and measurement inaccuracies into the system (reference 5).

The typical design of signal cabling is to use a wire list and assemble it with point-to-point wiring. This method typically yields improper grounding and poor transmission lines resulting in noise problems and signal degradation.

38.10.1 Signal cable design philosophy

The whole design philosophy for signal cabling can be summarized into two rules:

- Define the single-point ground at the DUT.
- Connect the instruments in the system to the DUT with the best transmission line possible.

Each of these two rules is covered in detail in the following sections.

Single-point ground at the DUT. One of the first decisions made by an electrical engineer is to define the ground reference point. Usually, this is defined as ground plane, a power-supply terminal, or the ac power ground. Instrument systems are a very different environment because of the long distances between the signal sources, power supplies, measurement devices, and the DUT. If the ac power input to the instrument system is chosen, the instruments are 10, 20, or even 30 feet from the single-point ground. This causes each instrument to operate from different ground references. These differences will be "seen" by the DUT as noise and will interfere with the measurements. This effect can be reduced slightly by using heavier-gauge wire, but this just addresses the resistance of the wire, not the inductance. The inductance is the major contributor of different ground references—especially at higher frequencies (at 10 kHz, a #22 wire has more inductive reactance than resistance).

The "single-point ground" should be located at the DUT. Whatever the DUT considers as ground, the tester should also consider to be ground. After all, the tester is really just an attempt to emulate the system to which the DUT usually connects. To an automotive module, the tester should appear to be an automobile; to an avionics module, the tester should appear to be an airplane; etc.

A common grounding problem. Some instrument systems are designed with "system ground" pins at the *system resource interface* (*SRI*) panel. These pins are frequently used to bring the ac power ground from the power-distribution unit to the SRI. They are then incorrectly connected to the DUT ground in the hope that this will lower system noise. Even though this is not the best connection to make, a decrease in noise is sometimes apparent. The reason

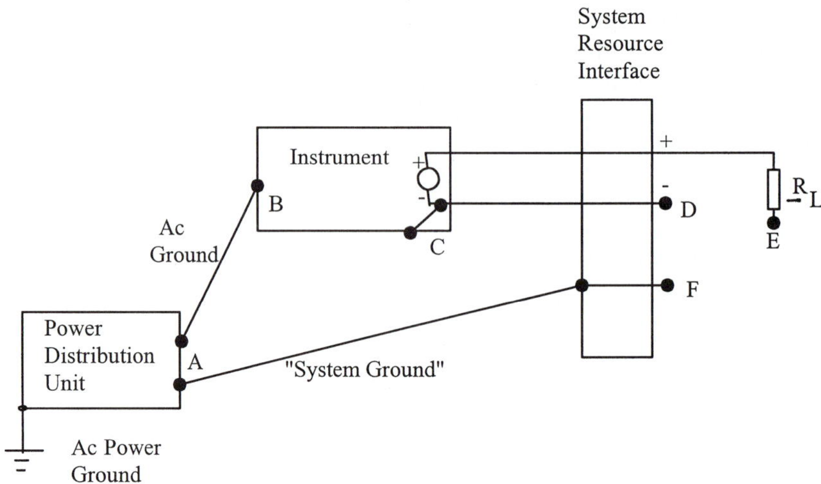

Figure 38.4 A common grounding problem.

for the decrease in noise is that a signal return for a given signal was not connected correctly. This causes the return current to flow back to the signal generator through chassis ground. The additional ac ground connection will just give the current a better return path. Even though the noise was reduced, this solution is just masking over the larger problem of not connecting the correct signal return. (Figure 38.4 shows a common grounding problem.)

Point A is the ac power ground. Point B is the ac ground on the instrument. Point C is the point where the instrument output is referenced to ground (usually through a BNC connector tied to the chassis). Point D is the return wire for the instrument output at the interface panel (this point is called the *shield* for instruments with coaxial connectors). Point E is the ground point of the DUT load and point F is the system ground pin at the interface panel (not all systems have this point). The real grounding question that all test engineers must answer is, "Where should I connect point E?" If the ac power ground is the system single-point ground, and ground loops are to be avoided, point E is often connected to point F. This is the worst connection to make because the current lacks a controlled impedance path back from the load. The current travels through a large loop. The lack of a controlled impedance path yields a poor quality signal because of the impedance mismatch. This causes ringing, reflections, and a decrease in signal level. The large loop area that the current has to travel causes noise generation and makes the circuit prone to noise susceptibility. The amount of noise is proportional to the loop area, so the larger the loop, the greater the noise.

If the DUT ground is the single-point ground (this sets point E as the single-point ground), one would connect point E to point D. This setup yields the minimum loop area and provides the signal with a controlled impedance (assuming that the cable used has a controlled impedance). This setup results in the lowest noise and maximum signal fidelity.

A third possibility is to connect point E to D and F. This is not as bad as just connecting E to F, but it is not as good as connecting D to E. This setup creates a ground loop from A to B, C, D, E, F, and back to A. This will pick up some low-frequency and instrument line-filter noise, and will be coupled into the load via the impedance between points C and D.

Use of transmission lines. Every signal in the tester should be connected to the DUT with a transmission line. This generally results in a very low noise, highly accurate signal at the DUT. Typical values depend upon the type of cable used, cable lengths, ambient noise, and instrument capabilities. Some typical characteristics to expect:

- *Coaxial cable.* Less than 1 mV of noise and a bandwidth of 500 MHz.
- *Shielded twisted pair.* Less than 5 mV of noise and a bandwidth of 50 MHz.
- *Twisted pair.* Less than 10 mV of noise and a bandwidth of 10 MHz.

Most noise problems are introduced in the signal cabling when system designers do not use a good transmission line or do not connect the signal returns correctly. A very good way of auditing the system is to trace the current flows for all signals in the test. The current from the instrument output comes out of the instrument, through the cable to the system resource interface panel, and into the DUT, but how does it go back to the instrument in the system? If it must return through the station ground, the system will have noise problems. The transmission line is broken so that the characteristic impedance is gone. The signal cable is now a large antenna that will radiate and receive noise.

Definition of a transmission line. Many system designs do not ensure that all signals are carried on a transmission line. A bad transmission line makes an excellent antenna. An antenna is equally good at receiving noise as it is at generating it, so a bad transmission line will not only have poor signal quality, but will generate noise that affects other signals.

Transmission lines are modeled as two wires with series inductance and shunt capacitance. The inductance is from the inductance of the wire and the capacitance is from two wires being next to each other. The magic of transmission lines is that the reactive component of the inductance and capacitance cancel each other out when the line is terminated in its characteristic impedance (50-Ω resistor for a 50-Ω line). In this case, the transmission line is lossless. The only problem with this model is that resistance has been left out. It is the resistance within the wires and between the wires that cause the losses in actual operation. Figure 38.5 is a diagram of a transmission line.

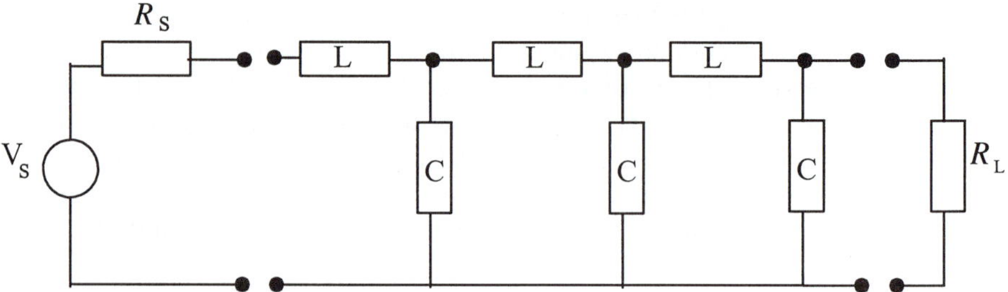

Figure 38.5 Transmission line diagram.

Misconceptions of transmission lines. It is important to recognize the distinction between resistance and impedance. Do not assume that a 50-Ω transmission line has 50 Ω of resistance. This is not true. The impedance of the line is calculated by taking the square root of the inductance divided by the capacitance. A 50-Ω line might have dc resistances that range from 50 to 10 MΩ dcx resistance.

Another common problem is computing the rise time and capacitive loading of a transmission line in a circuit. The wrong way is to add up the capacitance and neglect the inductance. For example, what is the rise time of a signal through 100 feet of 50-Ω RG-174 cable (RG stands for *Rated Gauge*) with 30-pF/ft capacitance into a 50-Ω load? Using the model that only takes into account the capacitance, the rise time is: $R \times C = 50 \times (30 \times 100) = 150$ ns. The measured rise time is much less than the predicted 150 ns (typically around 15 ns). The model did not take into account the nulling effect of the inductance on the capacitance.

When a good transmission line is used, most of the signal travels from the source to the load. Very little of it is lost via radiation (noise generation). One of the nice "rules" of electromagnetic radiation is that if a line does not generate any noise, it does not pickup any noise. Thus, a good transmission line does not pickup much noise.

Transmission lines and instrument systems. The system designer should be aware of which type of transmission line is being used for each signal. Use the best type of transmission lines possible. The problem is that the best transmission lines are usually the most expensive, so cost could be a limiting factor in the design. The best type is a coaxial cable, followed by shielded twisted pair, twisted pair, and ribbon cable. Point-to-point wiring is not a transmission line and should be avoided at all costs.

Instruments with coaxial signal connectors should use coaxial cable. The system resource interface should have coaxial connectors. Connect the shield to the DUT so that the current in the center conductor returns through the shield.

Instruments without coaxial connectors should use twisted pair as a minimum, shielded twisted pair, if possible, and coaxial cable if noise is crucial. Again, the signal current should return through the same transmission line; for twisted pair, the current on each pair should be equal and opposite. When using shielded twisted pair in the fixture, connect the shield to the signal-return wire at both ends: this makes the twisted pair function much like a coaxial cable.

A common problem that occurs in an instrument system is when one signal has to be brought to two different places. Instead of going through relays, the test engineer "tees" the line and connects it to two places at once. This breaks the transmission line and creates a "stub." This stub is an unterminated transmission line. When a transmission line is not terminated, the "magic" of the transmission line (the reactances cancel) stops working. The transmission line now acts like a large capacitor. Also, at higher frequencies, the signal starts to reflect off the unterminated stub and begins to build standing waves in the cable, which degrades the measurements.

Practical examples of transmission lines in instrument systems

Coaxial instruments. Coaxial instruments or switches should use coaxial cable. The recommended coaxial cable is a double-shielded 50-Ω RG-174 cable, where one shield is a foil shield for 100% coverage. If very long cable lengths (>3 feet) are required, use a short piece of RG-174 to connect to the coaxial pin. Connect RG-58 to the RG-174 with standard coaxial connectors to go the rest of the distance. Use RG-174 to mate with the coaxial pins in the panel and to keep the cable bulk down (RG-174 is only 0.1 inches in diameter).

If the system-resource interface does not have coaxial pins, use two standard pins to bring the signal through the interface panel. Keep using coaxial cable, but separate it into two pins. Be sure to keep the stripped-back part of the cable as short as possible. The amount of noise that this cable will pick up is directly proportional to this length.

General-purpose switching. General-purpose switching with double-pole double-throw switches (see Fig. 38.6) should use twisted-pair cable. For crucial signals, use shielded twisted-pair cable. Be sure that the cable has 50-Ω impedance and sufficient wire gauge to carry the maximum current. The twisted pair should always be connected to the high and the low relay contacts. The

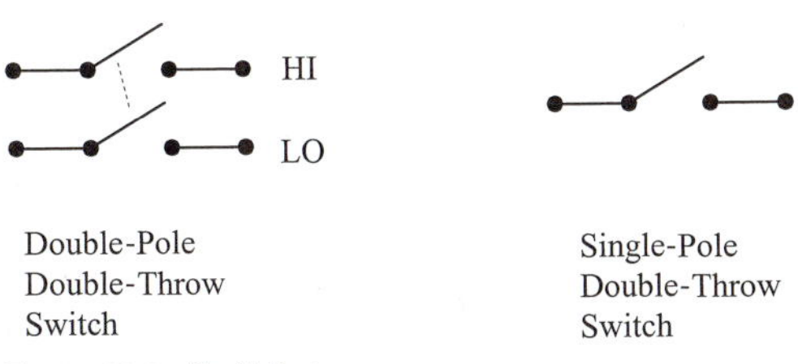

Figure 38.6 Switch types.

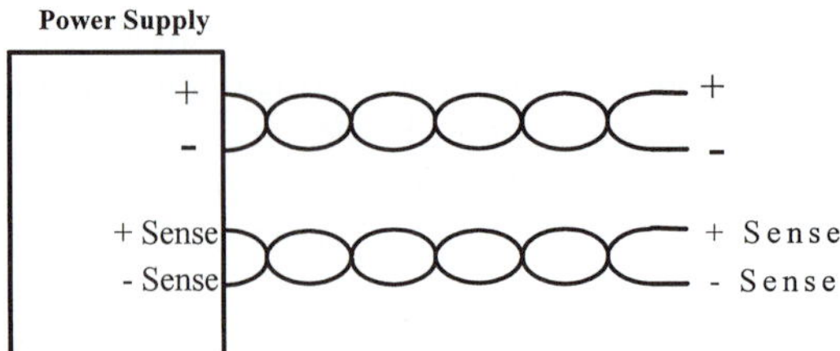

Figure 38.7 Power supply cabling.

shield should be connected to a guard pin on the connector. At the DUT, tie the shield to the DUT ground.

For switching that has only single-pole, double-throw switches, twisted pair should still be used. The biggest mistake is not carrying the signal return with the signal. The two options to switch these are: switch both the signal and the signal return with two relays, or use one relay for the signal and do not switch the return. If the return is not switched, jumper it to the cable on the other side of the relay in the fixture.

Digital multimeters. Digital multimeters have five terminals. These terminals are: Hi, Lo, Hi sense, Lo sense, and guard. The cable should be two shielded twisted pairs. The shields should be tied to the guard pin. The Hi and Lo pins should share a twisted pair and the Hi sense and Lo sense should share a twisted pair. For best measurement results, connect the guard terminal to the low side of the measurement source.

Power supplies. Power supplies have four terminals. Use two twisted pairs with "+" paired with the "−" and the "+ sense" paired with the "− sense." See Fig. 38.7

Digital testers. Digital testers typically use 100-Ω cable. The best cable for this is 100-Ω coaxial cable, the next best is 100-Ω twisted pair. It is very important to use good transmission lines with digital signals because of the fast rise times. These fast edges contain a lot of high-frequency energy. If the digital signal is running at 1 kHz–3*n* second rise times will raise the frequency content to the 100-MHz region, where transmission lines become very important.

Most digital testers should have a matching return line located close to the signal line. If using a coaxial cable, attach the center conductor to the signal pin and attach the shield to the signal-return pin. At the DUT, the shield should be attached to the DUT digital ground. Using twisted pair, one of the pair should be attached to the signal pin and the other to the signal return. At the DUT, the signal return should be attached to the DUT digital ground.

38.11 Putting It All Together: The Rack Layout

Most instrument systems are physically connected together in a rack. This keeps the instruments together and allows the user to move the entire system around. The standard rack is 19 inches wide, has holes for screws at regular intervals of ⅝ inch, ⅝ inch, and ½ inch. This allows for instrument placement at 1¾-inch intervals (reference 6).

A wide variety of cabinets is available from many cabinet manufacturers. Cabinet selection should be based on the harshness of the environment and the size of the test system. For large systems, choose tall cabinets for a narrow system or lower cabinets for a wider system. Some environments require expensive hardened cabinets, such as NEMA 12 (National Electrical Manufactures Association) enclosures. Some general recommendations are:

- Be sure that the cabinets have extender legs to prevent tipping over.
- 56″ maximum height (any higher and some users may not be able to reach all of the instruments).
- Rear doors for instrument serviceability.
- Accommodations for additional cooling (some cabinets have knock-out panels for additional fans).
- Accommodations for additional power wiring.
- Wheels or some other method to move them around.

To plan out the system, start with a blank rack diagram, such as shown in Fig. 38.8.

Now place the instruments, switching systems, system resource interface, and controller in the available rack space. To do this, use the following criteria:

- Size—be sure that everything fits and minimizes unused space.
- Crucial instruments should be mounted close to the system resource interface to minimize the cable lengths.
- Heavy instruments should be at the bottom of the cabinet to help keep the center of gravity low.
- An instrument that a technician can adjust or read should be at eye level. (This can vary if they are standing or seated).
- Try to balance the weight distribution between multiple racks of equipment. This makes the system more mobile and safer to ship.
- Design for heat dissipation—some instruments need additional cooling space around them to work correctly.
- Be sure that the heat in each cabinet does not exceed the cabinet's ability to dissipate it.
- Draw air in from the lower back of the cabinet (intakes in the front tend to suck in hair), exhaust air out the upper back (users do not like hot air blowing on them).

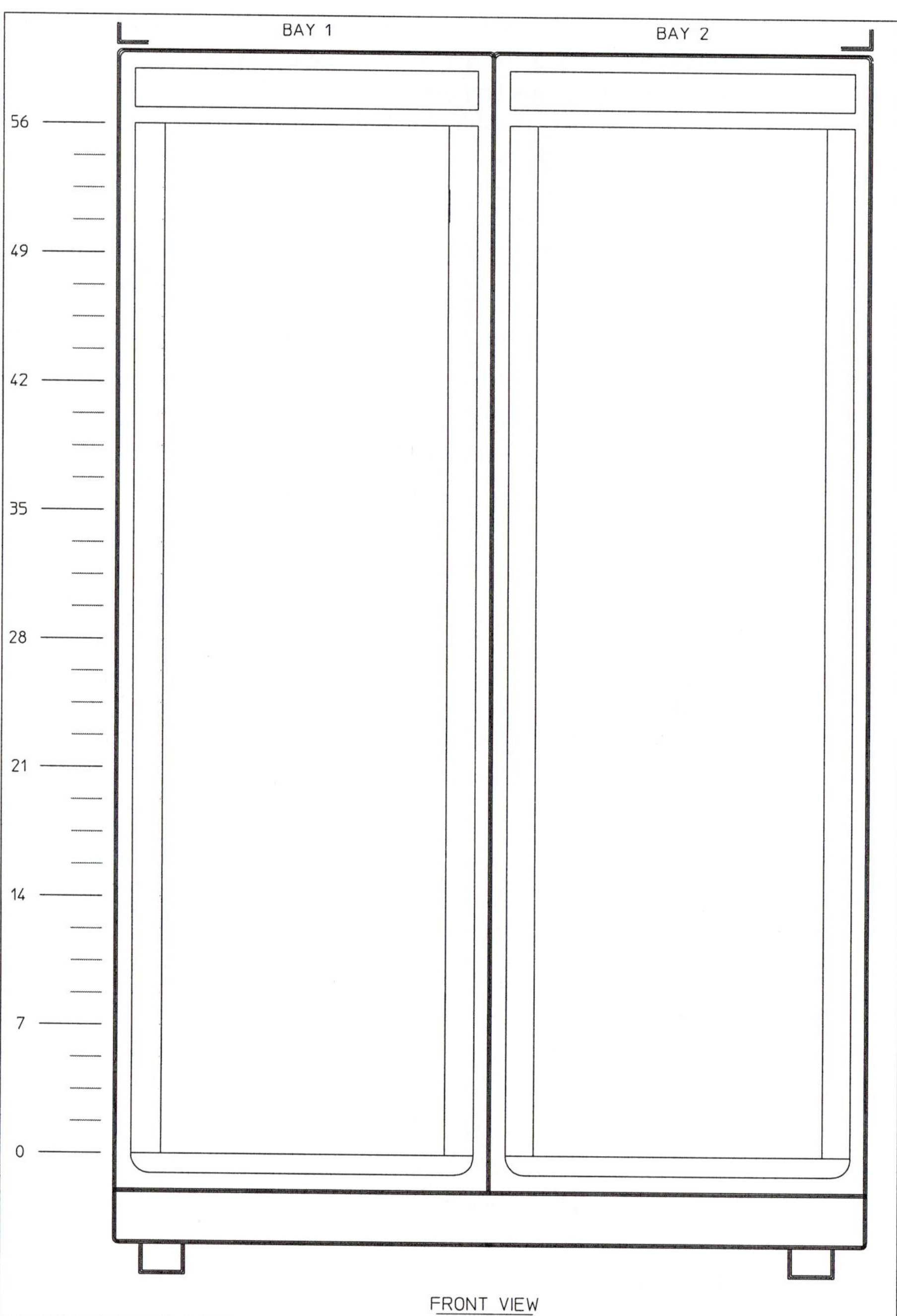

Figure 38.8 A blank rack diagram.

- Put instruments that require frequent servicing on rack slides so that they can be moved easily.
- Keep the ac power requirements balanced—too much power draw on one phase will cause excessive neutral line current.

38.12 Software

Software in the instrument system is used to automate the system. The best way to understand this software is to break it into different levels that correspond to the hardware that it controls. In Fig. 38.9, each software level has its own corresponding hardware component. The connection is shown with the arrowed lines. For example, the instrument-control drivers control the instrument and are connected in the diagram with the arrowed line.

An example of how to read the diagram is to see what happens when a test program programs an instrument. The test program sends a command to the instrument-control driver. The instrument-control driver formats it for the particular instrument interconnect bus. The instrument interconnect bus driver sends it to the operating system. The operating system converts it to an electrical signal and directs it to the correct instrument interconnect bus hardware (usually an interface card). This command is then sent to the instrument, where it is executed.

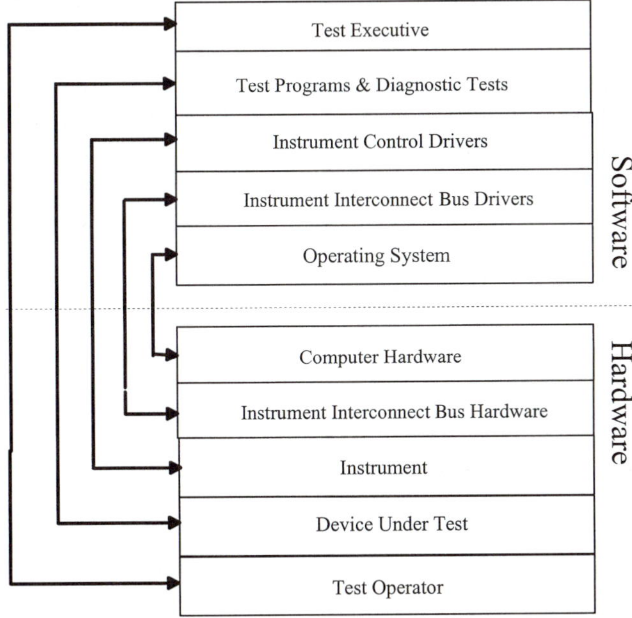

Figure 38.9 Instrument system software architecture.

38.12.1 Operating system

The *operating system* is the layer of software that controls the computer hardware. This includes the disk drives, printers, etc. Typically, the operating system is written for specific computer hardware. This is because of the tight coupling between this software and the hardware.

38.12.2 Instrument interconnect bus drivers

This is the software (drivers) that controls the instrument interconnect bus hardware. These drivers send the data and commands across the interconnect bus. For example, when an IEEE-488 bus is used, special drivers come with the IEEE-488 hardware. These drivers handle the reading, writing, etc. on the hardware.

Instrument control drivers. Instrument control drivers are software that control the instruments. These drivers send data and commands to the instrument interconnect bus drivers to be sent through the operating system and on to the instrument. Each instrument has its own set of instrument control drivers.

Test programs. Test programs are software that control (test) the DUT. These programs use the instrument control drivers to make the instruments stimulate and perform measurements on the DUT. The program then interprets the measurements and decides if the DUT has passed or failed.

Diagnostic tests. These are very similar to the test programs, except that they are used to find out what is wrong with a failing unit. Typically, "test programs" are limited to pass/fail operations. This keeps the tests fast and simple to write. Diagnostic tests are usually much more complicated to write and take longer to run.

Test executive. This is the software with which the test operator interacts. It generally does the following tasks:

- User login control
- Help the operator select the correct test to run for a particular DUT
- Logs the test data to selected databases
- Fault diagnostics
- Provide simple operator interfaces, such as touch screens, bar-code readers, etc.
- Report generation

38.13 Applications of Instrument Systems

Instrument systems are used in many different areas. The three most common ways to classify them are: research and development, manufacturing, and field repair systems. Each has its own particular characteristics. Some systems are highly automated and are used for many years. Other systems are operated manually and are used only for a few hours.

38.13.1 Research and development systems

Research and development systems are used for design verfication, performance testing, and environmental testing. These systems check the soundness of the design and typically are used by engineers.

Design verification systems. These systems are built by engineers to verify their designs. They generally have a very short lifetime (several months, at most) and are assembled and disassembled very quickly. There is generally very little automation because of the overhead of automating a test that has such a short lifetime.

The device under test is usually hand built and has internal test points or even extra circuitry built in to help with testing. These test points and circuitry are generally removed when the design moves into production.

A design verification test is performed with very precise and accurate instrumentation. The designer is trying to see exactly how the design is performing. For example, a very fast oscilloscope could be used to measure circuit delays to ensure that enough margin is in the design so that, when in production, with normal part variances, it will still function correctly. Sometimes this is also called *margin testing*. The result of this test is the confidence that the design can meet the design goals in mass production.

Automation in this system can save time if the engineer will be testing many different design variants. A good example of this would be designing an amplifier. The designer might have several different amplifier designs or even different types of components within a chosen design. The ability to run several tests on each design or component combination can speed the time required to make a sound engineering tradeoff.

Performance test systems. This type of testing is performed in the development phase of a project by the design engineer or by a highly trained technician. It is very similar to the design verification testing, except that the device being tested usually is very close to the final product. The test program and system can be seen as early prototypes of a production system. The performance test is concerned with how all the subcircuits will work together and what type of variance exists between a limited number of prototypes.

During the performance test, the tester will also stress the design to find the design limits. This is very valuable information—it gives the engineer the confidence that the design will still meet the design specification when it is mass produced. Typically, at this phase of the project, a limited number of PC cards are in existence. When the design is stressed beyond its limits (i.e., a 1-MHz amplifier tester to 10 MHz), the designer gets a good idea of how the design will perform in mass production or how it could fail if misused in the field.

Environmental test systems. These tests are used to see if the design can operate during certain environmental changes. This includes temperature, humidity, corrosives, electromagnetic interference, etc. These tests are either specified by company or government regulations, based on the type of equipment. Generally, an environmental lab will be set up to run the standard tests using a test system built specifically for that purpose. Some large companies have one central test group set up to service many different research and development projects. Smaller companies will use outside sources that specialize in this type of testing.

Typically, these instrument systems are highly automated because they are testing to rigid test specifications (these might even be specified by law) and the tests are almost exactly the same for each device being tested.

Some of these environmental tests can be destructive and the tester might need to monitor the device under test for some catastrophic failures and take the appropriate steps to prevent further damage. This can be as simple as turning off power or as complicated as operating a fire-extinguisher system. This type of instrument system must also record and present a great deal of data. This data must be stored for a fair amount of time because the tests can take a long time (some temperature tests can run for weeks). The security of this data is very important—if the data is lost after a two-week thermal test, the test might have to be run again, adding two weeks to the development schedule! The large amount of data must also be condensed into a format that is easily understandable. In the case of the two-week-long thermal test, the tester might be taking a measurement every second. This test would generate more than one million data points. The instrument system needs to find ways to make this understandable and presentable. Frequently, this data is condensed into some type of graph.

Instrument systems used for government regulation testing (FCC radiated emissions tests) might need to be certified by some government agency. This certification might require the tester to have a security system that detects changes in the hardware and software of the system. This is commonly performed with dated seals on the doors of the cabinets on the system.

38.13.2 Manufacturing systems

Manufacturing systems are used to ensure that the product being produced has been built correctly. These systems are built by engineers and are operated by production workers. The two main types of instrument systems in manufacturing are in-circuit testing systems and functional test systems.

In-circuit test systems. An in-circuit test is essentially a test of all the individual components on a PC card after they have been soldered into place. It verifies that the correct component has been loaded and is functioning correctly. It generally does not check for the interaction between multiple components.

An in-circuit test system is a collection of instruments and a computer that are connected to a test fixture called a *bed of nails.* This "bed of nails" is a flat horizontal panel with numerous sharp probes (nails) pointing upwards. A fully loaded PC card is placed on top of this and locked into place. Each PC card has its own unique test fixture because of the varied placement of components. Each "nail" is connected to a component lead (sometimes called *nodes*). The tester now has electrical access to all the components on the card.

The computer then instructs the instrument to measure each component on the card. For example, the first step might be to have an ohmmeter connect through the "nails" to all the resistors, measure the resistance, and verify that they are within specification. This is then repeated for all the other components on the card. When the test is complete, all components on the card are identified as being good or bad. The bad ones can then readily be identified and replaced.

This sounds like a perfect tester because it can measure every component on the card and isolate the failure to a component. The problem is that in the real world, it does not always have access to all the components, adjacent components will affect the tested component, and sometimes a card full of good components does not meet specifications. Also, few products are simply a printed circuit card. The in-circuit test system is usually used as a process check to verify the component loading and soldering process on PC card assembly, especially in the fabrication of surface-mount PC cards, which have a high amount of automation. The in-circuit test system can very quickly determine if any of the pick and place machines (that load and attach the components) are not functioning correctly or if the soldering process is not working correctly.

The test software for this system is relatively easy to write. The engineer writing the software uses standard routines, called *test libraries,* to test the common parts. Such parts as resistors, capacitors, and diodes have prewritten tests that require only the expected value of the part. More complicated digital parts can also have standard test routines.

Functional test systems. A *functional test system* is typically used at the very last stage of the production system. This test system is used to verify that the product being produced functions correctly. The tester typically emulates the environment where the product will operate. For example, the functional tester will emulate the personal computer if the device under test is a PC card for a personal computer. The tests consist of connecting the card into the tester and verifying that all the functions of the card work correctly.

This type of tester is expensive to build and maintain. The test fixtures are difficult to make and the test programs are usually complicated. When the product fails the test, it is also very difficult to use the tester to repair it. Unlike the in-circuit tester, the tests on a functional tester do not usually identify the failing component. Typically, a trained technician with additional test equipment must be brought in to diagnose the failure. Functional testers are frequently referred to as a *screening device,* keeping bad product from reaching the customer. In a well-tuned manufacturing environment, the in-circuit tester will find a majority of the failures and the functional tester will pick up almost all of the rest.

38.13.3 Field systems

A *field system* is an instrument system that is used after the product has been delivered to the customer. The primary purpose of these systems is to repair failing products. These systems are usually located close to the customer, rather than in a central place. They support a large number of products and frequently need to be portable. Cost is also a big factor because of the large number of systems in the field.

Because of this cost and size constraint, these systems are typically custom instruments, and use very little standard instrumentation. They are usually highly automated to keep training costs low. Their testing goal is to quickly find the failing subassembly so that the product can be returned to service. The failing subassembly is then either thrown away or sent back to a central place to be repaired and refurbished.

References

1. *IEEE Standard Digital Interface for Programmable Instrumentation*, ANSI/IEEE Std. 488.1, 1987.
2. *IEEE Standard Codes, Formats, Protocols and Common Commands*, ANSI/IEEE Std. 488.2, 1987.
3. Eppler, Barry, *A Beginner's Guide to SCPI*, Addison-Wesley, Reading, MA, 1991.
4. VITA-VMEbus International Trade Association, *VXIbus Specification Manual*, Scottsdale, AZ, or VXIbus, ANSI/IEEE Std 1155, 1992.
5. Ott, Henry W., *Noise Reduction Techniques in Electronics Systems,* Wiley-Interscience, New York, NY, 1988.
6. *Electronic Industry Association*, 310C Specification from American National Standards, 1977.

Chapter

39

Switches in Automated Test Systems

Calvin Erickson
Agilent Technologies
Loveland, Colorado

39.1 Introduction

Switching is a frequently overlooked but vital component of automated test systems. Test systems typically consist of a computer, test software, instrument hardware, switching, and a communication bus. Figure 39.1 shows a simple system block diagram. Switching provides the interface between the test system and the device under test (DUT). The switching block routes the test signals, power, and often the control lines to and from the DUT.

39.1.1 Automatic switching advantages

Switching provides test system flexibility and expandability. Signals may be redirected automatically to different DUT points or to different instruments. This allows a system to be reconfigured regularly on a high-mix production line or to be expanded in the future for the next-generation DUT.

Switching allows the set of expensive instrumentation to be minimized. This is especially important as density increases. One voltmeter, for example, can be used to measure many different points sequentially on the DUT. It is also possible to make many different measurements on the same set of points.

Switching simplifies the task of system calibration, self-test, and diagnostics. Source instrumentation (such as a pulse generator) can be routed back to sensor instrumentation (such as a counter). This arrangement may be used to check the operation of the instruments and the switches, as well as to monitor the effects of cabling and interconnect.

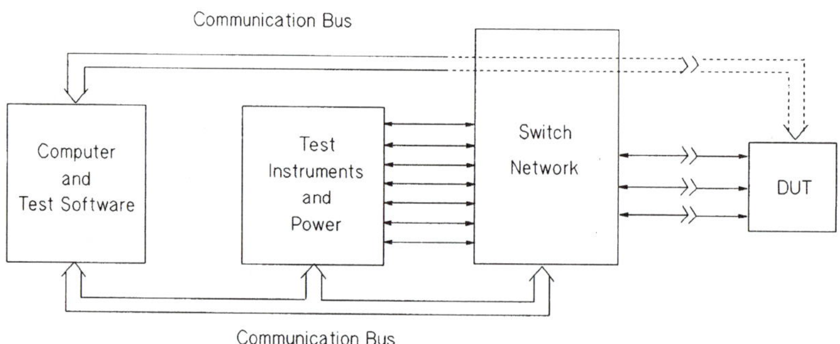

Figure 39.1 Test system block diagram.

39.1.2 Automatic switching disadvantages

Switching equipment degrades signals by introducing path resistance, thermal offsets, inductance, and capacitance (both signal-to-signal and signal-to-ground). This may produce unacceptable crosstalk between high-frequency signals or between power lines and low-level signals. This degradation is magnified by the cabling and interconnects present in most systems. The system designer should plan on minimizing this degradation early in the design.

Switches and connectors have a limited lifetime and are prone to failure. Some method of isolating faults should be built into the system to maximize uptime. Switches take time to operate. Depending on the switch element, this time may range from a few microseconds to a few hundred milliseconds. This time will decrease system throughput and should be accounted for by the system programmer. Switches also cost money. However, this should be more than offset by the savings in other test equipment, reduction in number of systems, and system versatility.

39.2 Switch Topologies

This section lists a few of the most widely used switch topologies. Most are commercially available for general-purpose use. In some cases, it may be necessary to design a custom topology specific to the application.

39.2.1 Simple switch configurations

Simple switch configurations are the basic building blocks of switching networks. They may be used individually or combined into topologies such as multiplexers or matrices. The most common simple configurations are forms *A, B,* and *C*. Forms *A* and *B* are also known as "single-pole, single-throw

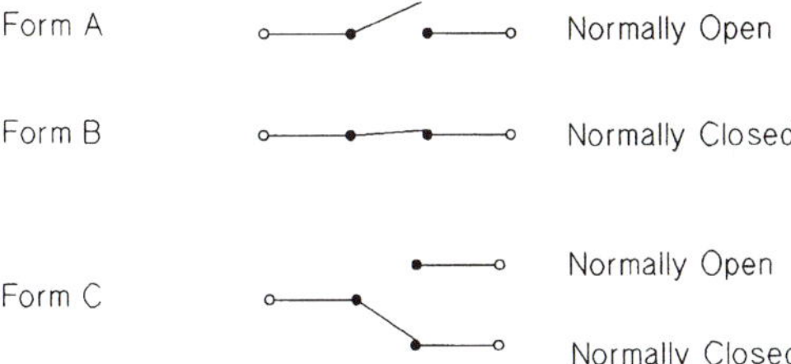

Figure 39.2 Simple switch configurations.

(SPST) switches." Form *C* is also known as a "single-pole, double-throw (SPDT) switch" (Fig. 39.2).

In many test applications, these simple forms are arranged into two-wire or three-wire (double-pole or triple-pole) configurations (Fig. 39.3). The multiple-wire configurations place multiple switch elements into a single package under control of a single line. This has three advantages. The first two are size reduction and simplified programming. If a certain measurement requires the connection of two lines, it is simpler to close a single device. The third is reduction of errors such as thermal emf. Since both elements are in a single package, they are kept at the same temperature. Typical applications include single-wire switches for higher frequency, two-wire switches for floating measurements, three-wire switches for guarded measurements, and four-wire switches for four-wire resistance measurements.

39.2.2 Binary ladder

The "binary ladder" is made up of form *C* switches (Fig. 39.4). It satisfies a major safety issue by ensuring that only one device is connected at a time. It also isolates the capacitance and reflection of all unused signal paths, making it a good choice for high-frequency applications.

39.2.3 Multiplexers

Multiplexers are used primarily to connect multiple signals to a single device. In most cases, only one signal may be connected to the device at any one time

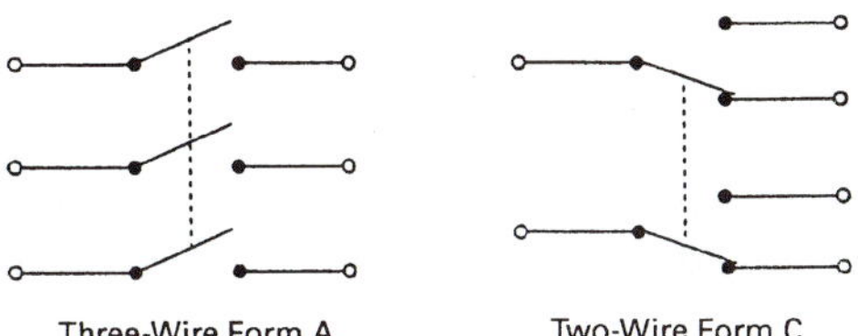

Figure 39.3 Multiple-wire switches.

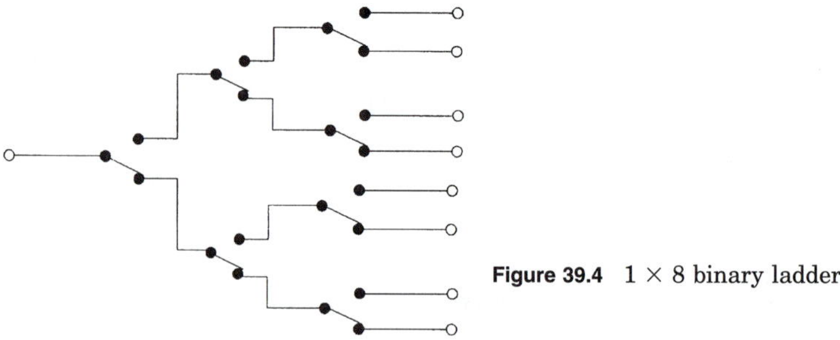

Figure 39.4 1 × 8 binary ladder.

(Fig. 39.5). Their primary advantages are the minimization of switch elements in the signal path and low cost.

39.2.4 Matrices

The full matrix allows any point to be connected to any other point. As such, it offers the most flexibility of any switch topology (Fig. 39.6). However, it also has a few important disadvantages. The first is loss of signal integrity. Capacitance, crosstalk, reflections, and impedance mismatch all increase dramatically with the size of a matrix. Special configurations such a stubless matrix can minimize these effects. Nonetheless, bandwidth is typically reduced when compared with a multiplexed system. The second disadvantage is the need for more relays. A full matrix requires a number of relays equal to the number of inputs multiplied by the number of outputs. Each relay not only adds cost but also increases the probability of failure. The third disadvantage is safety. A matrix allows any point to be connected to any other point. For example, a high-voltage line may be accidentally connected to a sensitive circuit, or worse yet, a point may be exposed to the system operator.

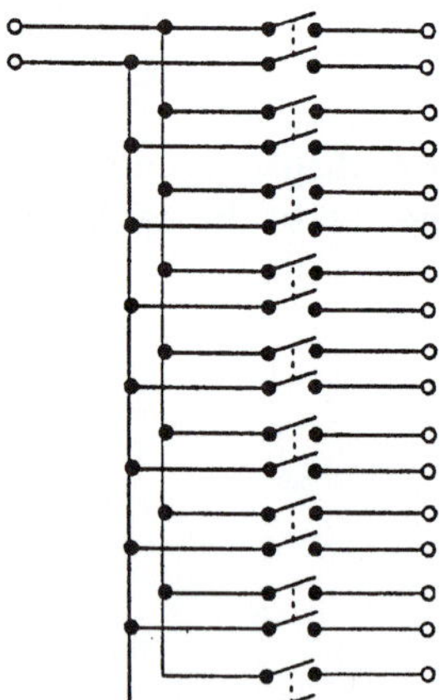

Figure 39.5 An eight-channel two-wire multiplexer.

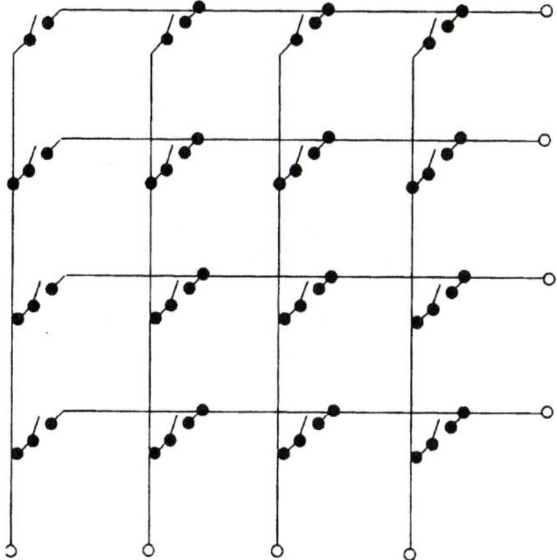

Figure 39.6 A 4 × 4 full matrix.

This places greater responsibility on the system designer to prevent unacceptable conditions.

In most applications, it is usually possible to minimize the size of a matrix or eliminate it altogether. For example, if a system requires 8 connections to system instruments and 32 connections to a DUT, the obvious solution would be an 8 × 32 matrix. However, if a maximum of 4 connections are required at any one time, 1 to each quarter of the DUT, then the matrix could be substantially reduced (Fig. 39.7). The new configuration is somewhat less

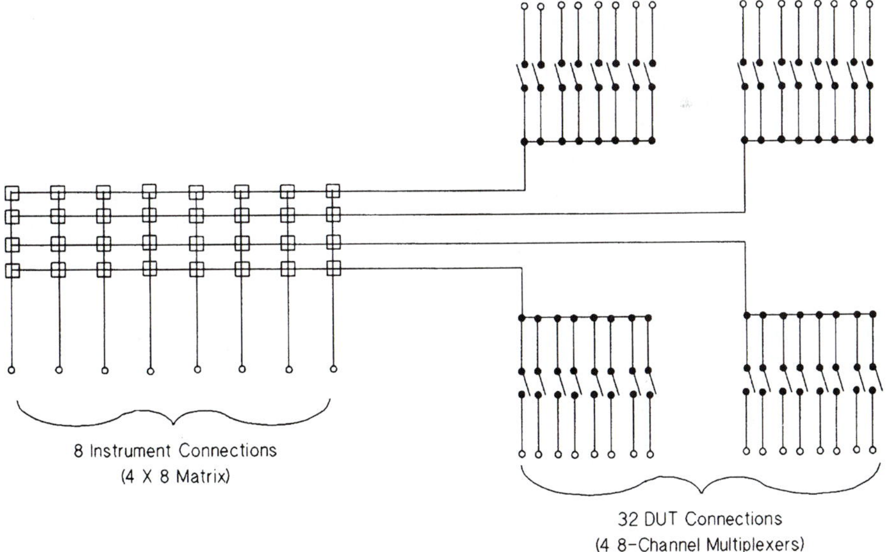

Figure 39.7 Matrix reduction. An 8 × 32 matrix has been reduced to a 4 × 8 matrix and four eight-channel multiplexers.

flexible, but it has much improved signal characteristics, lower cost, higher reliability, and better safety.

39.3 Switch-Element Technologies

Individual switching elements can be electromechanical relays or solid-state relays. Just as there are many kinds of signals to be switched (low and high frequency, low and high current, low and high voltage, etc.), there are many kinds of switch elements. Each kind of element has its own set of strengths and weaknesses. This section discusses a few of the most popular switch-element technologies. Table 39.1 presents the key differences. Note that ranges are given for most of the specifications. These should be used as guidelines only. Different designs and products may vary considerably.

39.3.1 Electromechanical armature relays

Armature relays operate with electromagnetic force. Current passes through a coil, causing a metallic contact to physically open or close. Armatures are the most widely used switch element. They are available in forms *A, B,* and *C* and are produced with a wide variety of specifications. They typically have low resistance and can be made with low thermal emf. Power versions, also known as "actuators" or "contactors," can handle the highest currents and voltages of any switch-element technology. The primary problem with armatures is a relatively low lifetime. Contact degradation due to electrical arcing and to mechanical wear-out both contribute to failures.

39.3.2 Electromechanical reed relays

Reed relays also operate with an energized coil, but they are much lighter in construction. The moving contacts are thin, flexible strips, or "reeds," housed in a sealed glass tube. Reed relays retain the low contact resistance of armatures, but they are able to operate much faster over a longer lifetime and have a slightly better thermal emf. They are not able to handle the higher current or voltage of an armature. Reeds are typically available only in forms *A* and *B*.

TABLE 39.1 Switch-Element Comparisons

	Armature	Power armature	Dry reed	Hg reed	FET	Power FET
Max speed, Hz	50–250	10–50	100–500	100–1000	500–100,000	100–250
Max voltage, V	100–250	250–1000	100–200	100–200	5–50	100–250
Max current, A	1–2	5–15	0.05–0.1	1–2	0.001–0.2	1–5
Contact R, Ω	0.01–2	0.01–2	0.01–2	0.005–0.075	5–100	1–3
Insulation R, Ω	10^7–10^{10}	10^9–10^{10}	10^{10}–10^{12}	10^8–10^{11}	10^7–10^{10}	10^7–10^{10}
Contact potential, μV	1–30	4–30	1–2	10–70	1–15	2–20
Life, cycles	10^5–10^7	10^5–10^7	10^7–10^8	10^7–10^8	$>10^9$	$>10^9$

An important variant of the reed relay is the mercury-wetted version. These relays have a small amount of mercury inside the glass tube. The mercury greatly reduces contact resistance and allows currents similar to those of small armature relays. Insulation resistance is slightly degraded, thermal emf is much higher, and many are sensitive to physical orientation (they cannot operate upside down).

39.3.3 Solid-state field-effect transistors (FETs)

Solid-state relays typically consist of an optically isolated input which activates a solid-state FET switch. They have the advantages of high-speed operation and a very long life. If a device is carefully kept within its specified operating conditions, life is practically infinite. Disadvantages include poor isolation, leakage currents, relatively high impedance when the switch is on, and sensitivity of key specifications to temperature. They are typically available only in form *A*. As these disadvantages are overcome, FETs will become more attractive due to their inherent speed, reliability, and life.

39.4 Reducing Electronic Noise in Switch Network Design

Combining switch elements and topologies into a network can generate or amplify electronic noise. In switch networks, electronic noise may be generated internally by the drive circuits, by the thermal imbalance between two switches, or by coupling among signal paths. Noise also may be generated outside the network and conducted or coupled in. Although noise problems apply to the entire system, they can become especially acute for switching. Switch networks contain a high concentration of signals, which magnify the errors. Most problems with electronic noise can be traced to improper grounding and shielding. This section recommends basic practices for reducing such errors. Because of the complexity of the subject, however, it cannot be a thorough discussion. Each system will present its own unique set of challenges. The system designer should seek additional information, including several other chapters in this handbook.

39.4.1 Grounding techniques

Test system grounding has two primary objectives: (1) to avoid ground loops susceptible to differences in ground potential and to magnetic fields, and (2) to minimize noise created by return currents from multiple circuits flowing through a common ground impedance.

In general, for frequencies below 1 MHz or for low-level signals, use single-

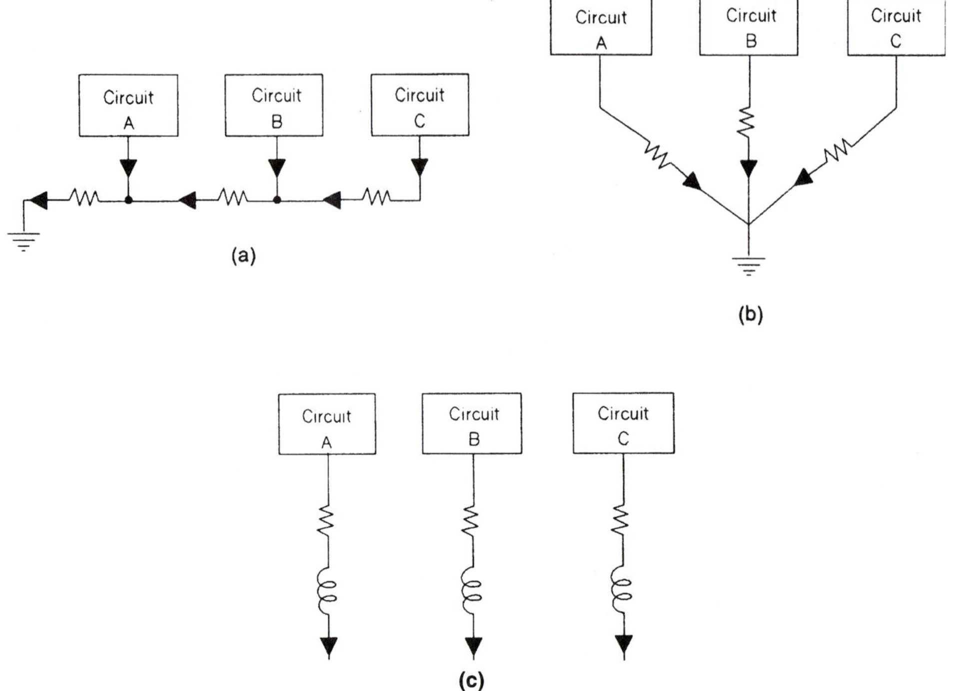

Figure 39.8 Ground systems. (*a*) A series single-point ground system; (*b*) a parallel single-point ground system; (*c*) a multipoint ground system.

point grounding (Fig. 39.8). The parallel configuration is superior but is also more expensive and difficult to wire. If the series configuration is adequate, the most critical points should be positioned near the primary ground point. For frequencies above 10 MHz, use the multipoint grounding system. For signals between 1 and 10 MHz, a single-point system may be used if the longest ground return path is kept to less than 1/20 of a wavelength. In all cases, return-path resistance and inductance should be minimized.

Most systems should have at least three separate ground returns. The first is for signals. This may be further subdivided into low-level signals and digital signals. The second is for noisy hardware such as relays, motors, and high-power equipment. The third is for chassis, racks, and cabinets. The ac power ground generally should be connected to this third ground.

Common-mode noise is often avoided by breaking the ground loop of the return with a floating input. The low measurement terminal is not connected to a common low but directly to the source low. Both are allowed to float above the chassis. In switch networks, this is most conveniently implemented with multiwire switches.

Ground loops also may be broken by isolating them with transformers or optical isolators. If it is necessary to maintain a dc path, chokes may be used to reduce high-frequency noise.

39.4.2 Shielding techniques

Shielding against noise must involve both capacitive (electric) and inductive (magnetic) coupling. The addition of a grounded shield around the conductor is highly effective against capacitive coupling. In switch networks, this takes the form of coaxial cables and connectors, stripline design on the printed circuit board, and special coaxial relays. These measures are recommended for frequencies above a few megahertz.

Reducing loop area is the most effective method against inductive coupling. Below a few hundred kilohertz, twisted pairs may be used against inductive coupling, unshielded for magnetic immunity and shielded for immunity from magnetic and capacitive pickup. For maximum protection below 1 MHz, the shield should not be one of the signal conductors.

39.4.3 Separation of high-level and low-level signals

High-level and low-level signals should be separated as much as possible. The entire signal path should be examined: cabling, connector pin-outs, even traces inside switch modules. All unused lines should be grounded and placed between sensitive signal paths.

39.4.4 Reduction of capacitance with tree switching

Tree switching is often used to isolate the capacitance of unused relays. This may be accomplished by placing switches between banks of multiplexers, for example. When a bank is not used, the tree switches open to isolate it from the rest of the network (Fig. 39.9).

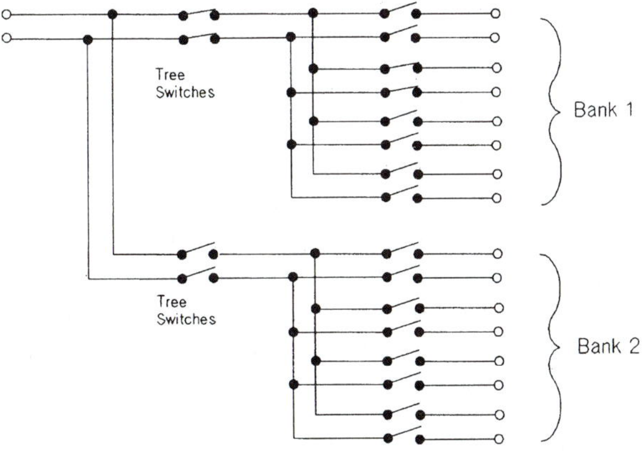

Figure 39.9 Tree switching. Closed tree switches allow a measurement in bank 1. Open tree switches protect the measurement from all bank 2 capacitance.

39.4.5 Compatibility of signals with switch elements and topologies

It is often expedient or necessary to use a switch product that is not ideally suited for the application. For example, suppose a new switch must be compatible with an existing switch network. It must have the same form factor, electrical interface, and programming language, and it must be available off the shelf. This greatly limits the choice of products. In this case, it may be necessary to use an existing 4×16 matrix when only a 3×10 is required. The additional crosstalk on the larger matrix degrades the signals. To minimize the effects of such compromises, the system designer should continually ask a number of questions. Are the switch specifications suitable for the application? Has impedance mismatch been minimized? Is contact resistance acceptable? Is there sufficient matching of thermal emf between signal high and low? All these may contribute to signal noise.

39.4.6 Cables and connectors

Switch networks typically include a concentration of cables and connectors. Cables add capacitance as a function of length. They should be kept as short as possible, therefore, while still complying with maintenance and workspace requirements. If necessary, it may be possible to reduce the overall cabling capacitance with a passive compensating network. Matched cable lengths may be required for time-critical measurements. To reduce noise, high-level and low-level signals should not share the same cable harness, unused connector pins should be grounded, and cable shields should be tied to their own connector pins.

39.5 Integration Procedure for Test System Switching

A systematic approach is essential for dealing with the many complexities and tradeoffs of switch network design. Switching is an integral part of the system and must be designed as part of the overall system. This section presents a generalized procedure applicable to most systems.

39.5.1 Determine system requirements

The first step is to define the system requirements. These requirements are generally global, independent of the detailed DUT requirements. They will influence the remainder of the integration procedure. Table 39.2 shows a few examples of the effects of system requirements.

What computer and software platforms are required? What is the desired communication bus? These may influence the choice of switch products based on compatibility and the availability of software drivers.

How many different DUTs must be tested by the system? What is the expected volume of each? How often will it be necessary to change among them?

TABLE 39.2 Example of System Requirements

System requirements	Effects on switch network
System is controlled by a PC via IEEE-488.	Switch network has a 488 interface and is able to respond to ASCII commands.
System will be programmed in C under an existing test software package.	Switch drivers already exist; command set is ideally consistent with existing and future switch products.
Only 1 type of DUT; high volume (2500/day); uptime is critical.	Self-test will be performed daily; switch failures are easy to isolate and repair; switch life is high, redundant signal paths exist, and/or scheduled replacement is acceptable; switch speed is high for throughput.
System will be used for a future revision of the DUT; should be similar but is still undefined.	Switch products, cabling, and programming are modular and/or expandable; a small number of additional channels (~20%) may easily be added.
System must fit in a single rack; will operate on a light manufacturing floor.	Switch network fits in less than 300 mm of rack space; switches comply with environmental standard MIL 28800 or equivalent.
DUTs are manually placed in fixtures.	No power applied to fixture when empty; complies with all local safety and regulatory requirements.
System must be ready in 6 months; development must cost <$200K; each system must cost <$100K.	All hardware is commercially available; switch hardware costs <$10K; cabling, fixtures, and self-test hardware costs <$14K; switch software engineering is <4 months; switch hardware engineering is <3 months; switch technician support is <5 months.

Is the system expected to support future, currently undefined DUTs? These factors may influence the versatility and expandability of the switching network and the design of a connector interface system and fixturing.

What are the requirements for system uptime? Is it necessary to diagnose problems and exchange a switch product quickly? These may affect the modularity of switch products, the design of self-test and diagnostic features, and the design of cabling.

What is the expected environment for the test system? This should include conditions such as temperature, humidity, vibration, and electric and magnetic fields. How will the system be used? Are there special operator restrictions? What are the safety requirements? What are the schedule and cost requirements?

39.5.2 Determine the required tests and specifications for each DUT

List all inputs and outputs for each DUT, along with important timing and signal characteristics such as voltage and current ranges and bandwidth. Then list all the tests required for the DUT. For example, one test may be to measure the propagation delay between signals on points 8 and 14. Both signals are at TTL levels. An acceptable (PASS) result will range from 100 to 190 μs. All other results are unacceptable (FAIL). A typical DUT will have many such tests.

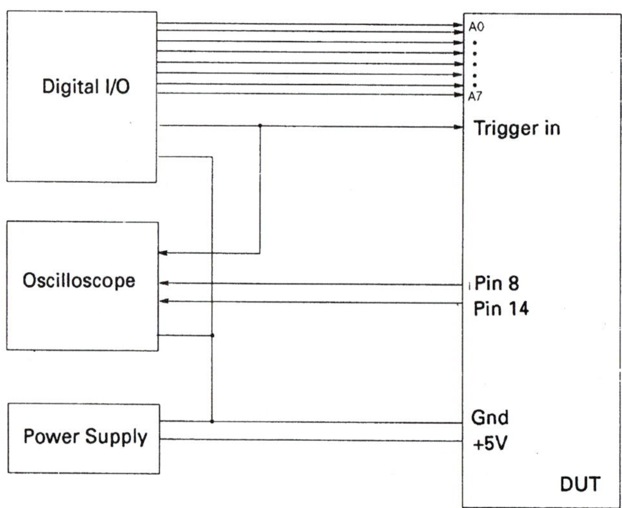

Figure 39.10 Test block diagram.

It is important to thoroughly understand the test requirements at this point. Overtesting can be time-consuming and costly. Underestimating the requirements can result in a large investment unable to perform the task.

39.5.3 Select instrumentation, source, and sense for each test

Based on the tests listed above, choose the instrumentation for each test. This step will be somewhat iterative, starting with choosing the ideal instrument and then minimizing the set based on factors such as cost goals and availability. It is important to include margin in the instrument specifications. Remember that test requirements have been specified at the DUT and that instruments are usually specified at their front panel. The signal degradation caused by switching and cabling is not yet quantifiable. Margin also may be valuable in meeting future, as yet undefined, requirements.

39.5.4 Specify setup for each test without switching

Draw a simple block diagram for each test. Do not include switching, but do show necessary functions such as isolation and load resistors. If possible, set up the test on the bench. This will allow problems to be identified and the preceding steps refined. For the propagation delay test example discussed above, the block diagram is shown in Fig. 39.10. A digital I/O is used to set the address lines and supply a trigger signal. An oscilloscope is used to measure the propagation delay between points 8 and 14.

In addition to the DUT tests, specify setups for system requirements such as self-test, diagnostics, fixture identification, and expandability. Self-test usually consists of routing source outputs (such as a power supply) back into sensor inputs (such as a voltmeter). Once a failure is identified by self-test, it will be necessary to isolate it. For switching, this may require scanning each signal path with a voltmeter to check impedance and to check for opens and shorts. Fixtures may be identified with a specific value resistor or identified digitally with a few bits shorted to high or ground. For expandability, it is worthwhile to imagine possible future requirements and identify how the system could accommodate them.

39.5.5 Determine switch network requirements

Based on the setups, determine the switch network requirements. The first task is to group all the switched lines into signal classes. For instance, classes may include high-frequency 50-Ω signals, low-level sensitive signals, and power signals. Next, for each signal class, determine the switch topology requirements. For example, of 12 power-line outputs (including returns) from the system, a maximum of 6 may be connected to the DUT at any one time. It is also a good idea to identify any limitations. For example, if one power-line voltage is connected, another may be prohibited for safety reasons. Or if one is present, another may always be present (e.g., −5.2 V and −2 V for ECL logic). These limitations will aid in the minimization of the switch network.

39.5.6 Combine switch elements and topologies into the desired network

This step will typically involve minimizing the network, reviewing the available switch products, and fitting them into the desired topologies. This process is highly iterative and should involve a thorough knowledge of the test system requirements listed in Sec. 39.5.1. Although a large number of products are available, variables such as signal integrity, topology, cost, density, ease of programming, form factor, versatility, and maintenance may make the choice difficult. Choosing the best products for an application will require compromise. In most cases, a commercially available product is preferred. Reasons include lower cost, faster delivery, built-in command sets, conformance to industry standards, features such as self-test, and a warranty for performance and reliability. In some cases, however, an acceptable switch is not available. It may then be necessary to design and build a custom switch. The key advantage is obtaining a switch optimized for the application. Disadvantages include higher cost and longer delivery.

The greatest problem at this point in the process is often signal degradation. Combining switch elements and topologies into larger networks will increase the errors discussed in Sec. 39.4. In particular, the amount of cabling, interconnect, and full cross-point matrices should be carefully monitored. Another problem is switch-element life. An estimate can be made of the number of cycles the switch will make per day, and the result compared with the requirements. It may be necessary to replace switch modules on a regular basis. If system uptime is especially important, it may be desirable to include redundant switches on parallel signal paths.

39.5.7 Specify system layout

Specify the location of each system component in the cabinet. A key consideration is the minimization of cabling. This generally means that switching should be centrally located, as close as possible to the DUT.

39.5.8 Confirm that system requirements are satisfied

It is now possible to estimate the effects of switching and interconnect on system requirements. One of the most important requirements is signal integrity. Given the complexity, signal integrity is best checked by setting up a few bench tests and taking measurements. These tests may be identical to those in Sec. 39.5.4 with the addition of switches and cables. The choice of tests usually includes all the most critical and a sampling of the others. If equipment is unavailable, it may be possible to adequately determine some of the effects on paper. Other requirements also should be reviewed. For example, do all products and parts meet the goals for programming, ambient environment, and cost? This step will often result in changes to some of the preceding definitions.

39.5.9 Obtain the switch products and cabling

Product availability and delivery are often overlooked. Delivery for standard commercial switch products often ranges from 4 to 10 weeks. Designing and fabricating a custom switch may take 12 weeks or more. This may seriously delay the overall system schedule. The delay can be minimized by using this time for programming, fixture design, cable harness fabrication, etc.

39.5.10 Integrate the system

Hardware integration consists of many individual tasks. The first is usually installation of the components into the cabinet(s). Next, the communication bus is connected, and each component is configured appropriately. This allows everything to be powered up and checked for basic operation. Finally, the signal cabling is attached and dressed to the DUT interface.

Test programming usually proceeds in parallel with the hardware integration. The programmer should consider a number of issues particular to switching. One issue is safety. It is usually necessary to break one connection before making another to avoid damage to the DUT or test system or injury to the operator. Another is the tradeoff between switch life and throughput. By removing power before activating a switch (cold switching), it may be possible to substantially increase switch life. The extra operations take time and come at the expense of throughput.

39.5.11 Verify system operation and specifications

System operation is normally verified by using a known good DUT or by using a self-test or calibration fixture in place of the DUT. If a problem is suspected, it may then be necessary to install a different fixture that provides more visibility into the individual signal paths. Faults are typically isolated by dividing the problem into smaller and smaller units. This final step of system development should be consistent with the long-term system self-test and maintenance plans.

Chapter

40

Standards-Based Modular Instruments

Dave Richey
Agilent Technologies
Loveland, Colorado

40.1 Overview of Modular Instruments

Modular instruments use a frame (Fig. 40.1), into which different types, or a varying number, of functional cards can be plugged. This is so that the instrument can accommodate a range of input/output channels or tailor its measurement capability according to the specific application being addressed.

40.1.1 Modular instrument types

The modular instruments described in this chapter support an industry standard. The modular instrument standards include:

- VME standard
- VXI standard
- Personal computer plug-ins
- CompactPCI standard.

PC plug-ins are not part of a formal standard. The ubiquity of the personal computer, however, has made the PC motherboard I/O bus a defacto standard for instruments.

Although all these standards are used for instrument systems, only VXI and a derivative of CompactPCI (called *PXI*) were developed expressly for instrumentation. For general-purpose instrumentation, VXI has the most products. PXI is emerging in the market, and it generally has the same features as VXI. More attention is given to VXI in this section.

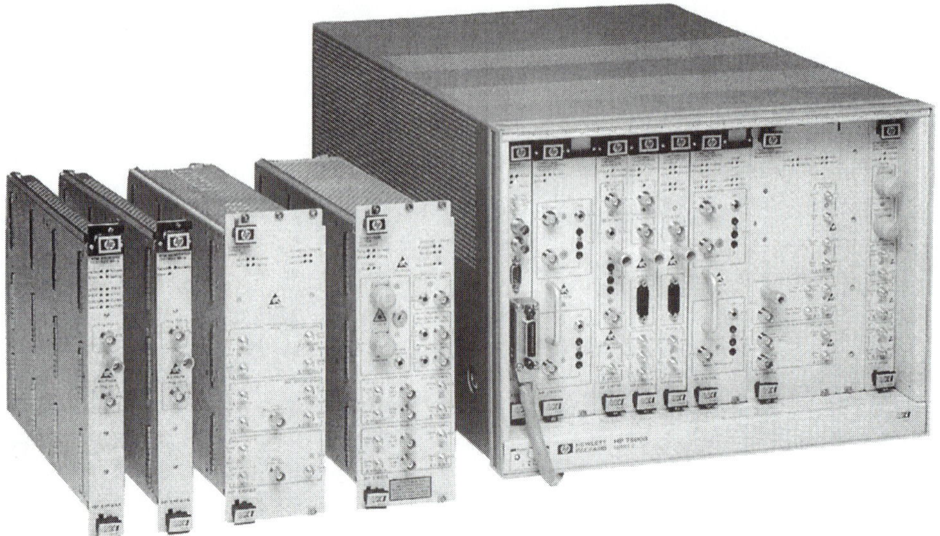

Figure 40.1 A typical modular instrument: VXI modules standing alone and installed in a VXI mainframe. Photo courtesy of Hewlett-Packard Co.

Standards-based modular instruments can accept products from many different vendors, as well as user-defined and constructed modules.

Modular instruments generally use a computer user interface instead of displays and controls embedded in the instrument's frame or package. Because they do not have their own user interface, modular instruments are often called *faceless instruments*. By sharing a computer display and keyboard/mouse (the "face" of the instrument), modular instruments can save the expense of multiple front-panel interfaces.

Without the traditional front panel, the most common approach to using modular instruments involves writing a test program, which configures the instruments, conducts the measurements, and reports results. For this reason, modular instruments typically are supplied with programmatic software interfaces, called *drivers*, to ease the job of communicating with an instrument module from a programming language.

40.1.2 Advantages of modular instruments

Modular instruments are an excellent choice for high-channel-count measurements, automated test systems, applications where space is at a premium, or complex measurements where several instruments need to be coordinated.

High channel count. High-channel-count measurements, such as data acquisition, really benefit from the standards-based modular form factor. Module size is relatively big, so a large number of channels can be included. Because no operating controls are on the front panel, the space can be devoted to connector space for a large number of wires/cables. Data acquisition usually requires many types of channels (i.e., different sampling rates and different resolution). Not everything can be purchased from one vendor. The mix and match capability of modular instruments allows the system to be easily configured from multiple vendors.

Automated test systems. Modular instruments address the needs of automated test systems very well. Automated test systems typically involve many types of instruments. The instruments are driven by software, not directly manipulated by a user. Packaging density is a concern, and, in fact, the logistics of racking and cabling are as important to system design as making measurements. With modular instruments, modules can be acquired from multiple vendors. The physical configuration of mainframe/module lends itself well to racking and cabling. The tight integration with computers makes it easier to program.

Space. For many instrument applications in the aerospace/defense sectors, space is at a premium (e.g., on ships and aircraft). The mainframe/module configuration is intrinsically high density. Several of the modular standards offer backplanes and subracks that can be directly integrated with existing racks or enclosures housing other equipment. For a given volume and multiple instruments, modular instruments usually create the smallest, most compact solution.

Element coordination. In a system where signal sources, signal routing via switches, and measurements must be carefully coordinated, modular instruments are a very good choice. Just being on the same backplane can facilitate a lot of coordination. System-level clocks, interrupts, and reset/start/stop signals provide a hardware mechanism for tight coordination. Some of the standards (such as VXI) also include instrument triggers and local buses between modules for the highest level of coordination.

Other advantages. Other intrinsic advantages of modular instruments include speed and cost. Communication between modules can be very fast because the instrument modules share a common system backplane, which is usually a high-speed parallel computer bus. This can enable fast measurement throughput for test systems or very fast data transfer in data-acquisition systems.

The cost of the frame, which houses the modules, the backplane communication electronics, and the power supply can all be shared across the instrument modules.

40.1.3 Disadvantages of modular instruments

Modular, standards-based instruments are not appropriate for all applications. They tend to be bulky and awkward, so (at best) they are transportable, but not portable. Even a small mainframe with the smallest possible module size is larger than a single-purpose, traditionally packaged instrument. A laptop computer, or screen and keyboard, usually must be brought along. That adds to the load.

For R&D and service, there is simplicity in single-purpose, traditionally packaged instruments. That makes them easier to operate. It also makes them easier to share, which can minimize spending.

The biggest reason not to use modular instruments is that, because they must be programmed, direct manipulation of instrument controls is difficult when performing experiments or troubleshooting. Graphical programming languages can be useful in making small real-time adjustments to instrument parameters in such situations. But it is not the same as viewing an oscilloscope screen and twisting the knobs to frame the desired signal.

40.1.4 Comparison of different modular instrument standards

All of the modular standard instruments evolved from high-speed parallel backplanes developed for the computing industry. All except PC plug-ins use the Eurocard form factor. For these reasons, all of the modular standards have similar appearance, configuration, and bus performance.

VME is the oldest of the modular standards. For applications supported by VME, it offers the widest range of vendor support. CompactPCI is relatively new and emerging.

Applications that are dominated by computing or analysis, with a small amount of instrumentation, will be best served by VME or CompactPCI. These modular standards offer a broader range of CPUs, more computer peripherals, and more choice in operating systems.

The simple PC plug-in configuration, using a conventional (nonindustrialized) PC and instrument cards, is the simplest, lowest-cost approach. It is limited to smaller, lower-performance applications. However, as an application grows, it might evolve to a passive backplane configuration (with higher cost and complexity).

Two of the modular instrument standards were designed specifically to meet the needs of higher-performance instrumentation and complex system building. These are the VXI and PXI modular instruments. In addition to special measurement features, these standards address system software and system environment. If the application involves general-purpose instrumentation and measurement switching, these modular instrument standards are the best choices. At this point, VXI has the largest number of vendors and the widest range of products.

Outside of general-purpose instrumentation, most modular instrument applications involve data acquisition and industrial control. VME, PC plug-ins,

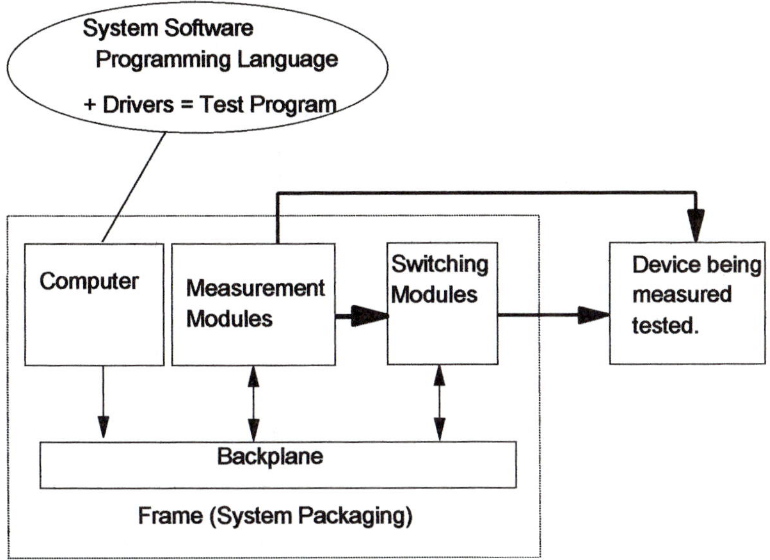

Figure 40.2 Key elements of modular standard-based instruments.

and CompactPCI instruments are all good choices. Demanding, noise-sensitive, or dynamic data-acquisition applications are probably best served with a VXI instrument.

40.2 Elements of Modular Instruments

As shown in Fig. 40.2, key elements of modular, standard-based instruments include a frame that contains a backplane with computer, measurement modules, and switching modules, all working with the *Device Under Test* (*DUT*). This is controlled by system software, including the programming language and drivers that make up the test program.

40.3 The System Backplane

At the center of modular instruments is the system backplane, or set of buses, which connect system modules together. These buses are high-speed parallel buses. All of the modular standards have a data-transfer bus, a master/slave arbitration bus (except ISA PC plug-ins), an interrupt bus, and a bus of special functions. The VME (32-bit) backplane is used as an example.

40.3.1 VME backplane example

The VME data-transfer bus consists of 31 address lines, 3 data-sizing lines, 6 address modifier lines, 32 data lines, and 4 control lines (Table 40.1).

TABLE 40.1 VME Transfer Bus

Address lines	Data lines	Address modifiers	Data sizing lines	Control lines
A01–A31	D00–D31	AM0–AM5	DS0	AS
			DS1	BERR
			LWORD	DTACK
				WRITE

Address modifiers. The address modifier lines indicate whether the bus cycle is a program fetch, a data read or write, or a block transfer. The modifier lines also indicate whether the bus is undergoing a supervisory cycle or a nonprivileged cycle, whether it was a standard or extended cycle. The meaning of these different cycles is defined in the VME specification.

Data-sizing lines. The data-sizing lines select whether a 32-bit (long word), 16-bit (word), or 8-bit (byte) transfer is in progress. In addition, they determine which word or byte out of the full 32 bits of data is addressed.

Control lines. The control lines determine the timing of the bus cycle (i.e., when it starts and stops, and when data is valid and ready to be accepted). The BERR signal indicates when an invalid cycle has occurred so that recovery can be attempted.

Bus mastering. At any moment, only one module on the VME bus can control data transfers. Usually, the system processor controls these bus cycles. It is increasingly common for a system to have multiple processors or to have another device (e.g., a disk drive) that needs to periodically control the bus. Control of the bus (Table 40.2) can be passed from module to module using the master/slave arbitration bus. The VME arbitration bus uses 10 lines: six lines that are in common with every module and four that are daisy chained from module to module. The daisy-chained signals are called *Bus Grant IN* when they enter a module and *Bus Grant OUT* when they leave a module.

TABLE 40.2 VME Arbitration Bus

Daisy chained signals	Common signals
BG0OUT to BG0IN	BBSY
BG1OUT to BG1IN	BCLR
BG2OUT to BG2IN	BR0
BG3OUT to BG3IN	BR1
	BR2
	BR3

TABLE 40.3 VME Interrupt Bus

IRQ1	Lowest priority
IRQ2-6	
IRQ7	Highest priority
IACK	Bussed acknowledge
IACKOUT to IACKIN	Daisy-chained acknowledge

The BBSY signal indicates whether a bus request is in progress. If a request is not in progress then one can be started using one of the BR0-BR3 signals. The bus is granted using the BG0OUT/IN-BG3OUT/IN signals. Although the system is clocked, the request and grant of the bus is asynchronous. The protocol of using the 10 signals ensures that orderly handoffs are done.

Interrupt bus. The VME interrupt bus has nine signals. The bus defines a priority interrupt scheme for seven inputs. There are two interrupt acknowledge signals. One is bused to all modules. One is daisy chained (See Table 40.3).

To request an interrupt one of the signals, IRQ1 through IRQ7 (depending on priority), is asserted. *IACK* indicates that the system is acknowledging the request. *IACK* also initiates the assertion of the daisy-chained *IACKOUT/IACKIN* signals. The reason for the daisy chain is so that more than one module can simultaneously assert an IRQ line. The arbitration of which module actually gets acknowledged is done with the daisy chain.

Utility bus. The VME utility bus (Table 40.4) provides signals to control the timing and start-up of the system, and to assist in troubleshooting when there is a problem. The utility bus includes six signals.

40.4 Form Factors

One place to start in understanding the similarities and differences between standards is board size.

TABLE 40.4 VME Utility Bus

SYSCLK	16-MHz 50% duty-cycle system clock
SERCLK	Serial clock driver conforming to IEEE P1132 spec
SERDAT	Serial data
ACFAIL	Indicates loss of system power, start power down
SYSRESET	System reset signal initiated by power up
SYSFAIL	Indicates various system failures

TABLE 40.5 EUROCARD Sizes

Eurocard sizes	VME name	VXI name	Compact PCI name
10 × 16 cm	Single height	A	3U
23 × 16 cm	Double height	B	6U
23 × 34 cm	—	C	—
36 × 34 cm	—	D	—

40.4.1 Board size

To ensure interchangeability between vendors, standardization of the instrument module's board size and spacing are important. The board size also determines, to a large degree, the complexity and capability of a given modular standard (i.e., more channels, more measurement capability). More performance, and higher resolution requires more space. With the exception of the PC plug-ins, all the other modular forms use Eurocard board sizes (see Table 40.5). These board sizes were standardized as IEEE 1101.1, ANSI 310-C, and IEC 297.

Four sizes in common use are as follows:

1. 10 × 16 cm
2. 23 × 16 cm
3. 23 × 34 cm
4. 36 × 34 cm

40.4.2 VME and VXI standards

The VME standard uses sizes 1 and 2, referring to them as *single-height* and *double-height boards*. The VXI standard uses all four sizes and refers to them as sizes *A, B, C,* and *D*, respectively. Most VXI products have adopted the B and C sizes. Compact PCI uses sizes 1 and 2, referring to the Eurocard names *3U* and *6U*.

40.4.3 PC plug-in modules

PC plug-in modules use the full- and half-card sizes adopted in the original IBM PC.

The board size is approximately 12.2 × 33.5 cm for the full-size card and 12.2 × 18.3 cm for the half-size card.

40.5 VME (VMEbus) Specification

The VME standard is more properly known as the *VMEbus specification*. The specification was developed for microcomputer systems utilizing single or multiple microprocessors. The specification was not directed at instrumentation, but the concept of computer functions integrated with measurement and control functions on a single backplane was certainly a factor in its creation. The specification was released by the VMEbus International Trade Association in August 1982. It was approved as *IEEE 1014–87* in March 1987.

In 1994, the VMEbus specification was amended to include 64-bit buses. This is known as *VME64*. Additional backplane connectors are involved with VME64, so care in choosing system components is required.

40.5.1 Common functions and applications

The most common applications are data acquisition, industrial control, and custom electronic-control systems. The number of VME modules for sale is dominated by CPU, computer storage, and computer I/O functions versus instrument functions.

40.5.2 VME standard

The VME standard defines the backplane functionality, electrical specifications, and mechanical specifications. The functionality is a derivation of the signals found on the Motorola 68000 microprocessor. No signals are specifically defined for instrumentation. The single-height board has no user-defined pins available. The double-height board has 64 user-defined signals. These are sometimes used for instrumentation needs. Again, there is no standardization.

40.5.3 System components

A VME system is usually formed with one or more frames, an embedded computer module, optionally various computer storage and I/O modules (e.g., LAN), and various measurement modules. Many choices are available when selecting and configuring a VME system. That comes with a lot of detailed work to integrate a system together. There is not a common programming model or instrument driver standard. As with other devices on a computer bus, programs must read/write device registers. In some cases, vendors supply software functions or subroutines to help with the programming.

VME frames. *Frames* refer to the backplane, power supply, and packaging that encloses a system (Fig. 40.3). System builders can choose a solution at many different levels from the component level (i.e., a backplane to complete powered desktop enclosures). Backplanes, subracks, and enclosures are available from many manufacturers, in single-, double-, and mixed-height sizes. A 19-inch rack will typically accommodate 21 cards. Power supplies are available from 150 W to 1500 W.

Figure 40.3 VME backplanes and subracks are available in many different configurations. Photo courtesy of APW.

VME embedded computers. VME is very good for building single- and multiple-processor computer systems. A large number of embedded processors are available. The bus was patterned after the Motorola 68000, so it is no surprise that Motorola processors are available for the entire family, including 68060 and 683XX. Products are also available for the PowerPC, Intel-compatible processors, SPARC, Digital Equipment Corporation, Hewlett-Packard, MIPS, and many *digital signal processors* (*DSPs*), such as from Texas Instruments.

VME measurement modules. Instrument functions in VME are primarily digital I/O modules and modules containing *Analog-to-Digital* (*ADC*) and *Digital-to-Analog* (*DAC*) converters. For these functions, there are many suppliers.

Typical digital I/O boards would have 16, 32, or 64 channels. Voltage range is typically 5 to 24 V. Some products are opto-isolated. Single-ended or differential channels are available.

ADC and DAC modules are available from 16 to 96 channels, depending on whether they are single-ended or differential, multiplexed, or parallel channels. Sampling rates range from 50 kHz to 150 MHz. Resolution ranges from 8 to 16 bits.

There are very few traditional instruments, like voltmeters, oscilloscopes, waveform/pulse generators, spectrum analyzers, etc. in VME.

VME switching. Many measurement systems require switching between measurement channels and signals from the device being tested. These switches would typically be multiplexers or matrix switches. There are several simple relay cards available in VME, but essentially no multiplexers or matrix switches.

VME software. Many VME systems are developed with real time operating systems. Software is available for VxWorks, pSOS+, OS-9, and various other operating systems. Source code for the C programming language is the most common. There is no standardized driver software.

40.6 VXI (VMEbus Extensions for Instrumentation)

The VXI standard is a derivative of the VME standard. In fact, *VXI* refers to *VMEbus eXtensions for Instrumentation*. It was driven by a U.S. Air Force program in the 1980s to design a single *Instrument-on-A-Card* (*IAC*) standard to substantially reduce the size of electronic test and operational systems.

In April 1987, five companies, Hewlett-Packard, Tektronix, Colorado Data Systems, Racal-Dana, and Wavetek, started discussions on creating an IAC standard that would benefit both the commercial and military test communities. The VXIbus Consortium was formed, and the VXIbus System Specification was released in July of 1987. The IEEE adopted it as *IEEE 1155* in 1992. After the basic VXIbus specification was adopted, it was recognized that the software aspect of test system development should be addressed. In 1993, the VXIplug&play Systems Alliance was formed. Several system "frameworks" were defined, based on established industry-standard operating systems. The Alliance released Revision 1.0 of the VXIplug&play specification in 1994.

40.6.1 Common VXI functions and applications

VXI is used for most instrument applications, including:

- Automated test systems. With its aerospace/defense roots, some of the specific applications have been for weapons and command/control systems, and as the foundation for operational systems, such as an artillery firing-control system.
- Manufacturing test systems. Examples include cellular phone testing and testers for automotive engine-control modules.
- Data acquisition, particularly for complex and sophisticated situations, such as satellite thermal vacuum testing.

40.6.2 VXI standard

The VXI standard including VXIplug&play is a comprehensive system-level specification. It defines backplane functionality, electrical specifications, mechanical specifications, power management, electromagnetic compatibility (EMC), system cooling, and programming model. The backplane functionality has features specifically for instrumentation.

40.6.3 VXI system hardware components

VXI systems can be configured with either an external computer connected to a VXI mainframe, or with an embedded computer. External computers are the more common configuration because the cost is lower and they allow the user to choose the latest and fastest PC available. Embedded computers are used when the highest test speed or throughput is required. They are also used when high-density packaging or a single package solution is needed.

A VXI system in the external computer configuration consists of a PC, PC I/O, a slot 0 controller, a mainframe, various measurement modules, various switching modules, and the supporting software (including VXI Plug-and-Play drivers).

A VXI system configured with an embedded computer is formed with an embedded computer module, a mainframe, various measurement modules, various switching modules and the supporting software, including VXI Plug-and-Play drivers.

Mainframes. VXI system builders have a choice of backplanes or complete mainframes. Mainframes are the most common choice (Fig. 40.4). They in-

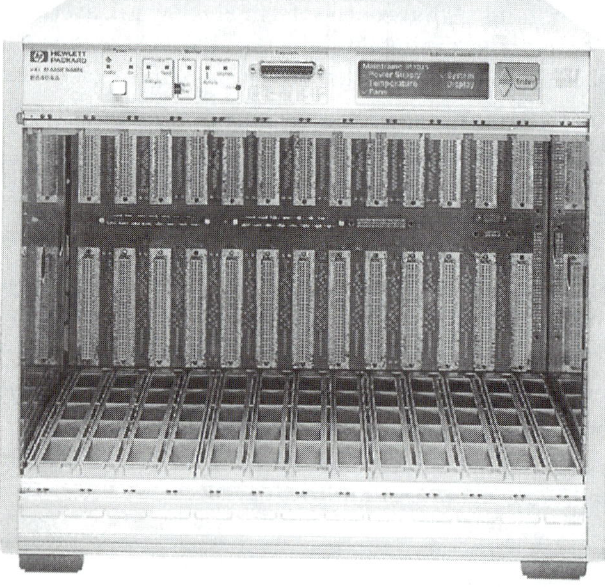

Figure 40.4 A 13-slot C-size VXI mainframe. Notice the system monitoring displays, which are unique, relative to other modular instrument types. Photo courtesy of Hewlett-Packard Co.

clude backplane, power supply, and enclosure. Mainframes for C-size modules are available with 4, 5, 6, or 13 slots. D-size mainframes are available with 5 or 13 slots. Power capabilities range from 550 W to 1500 W. B-size mainframes are typically available with 9 or 20 slots. These mainframes typically offer from 120 W to 400 W. Backplanes are available with basically the same slot choices as mainframes.

VXI external connection to computers. A VXI module is required to connect and communicate with an external computer.

Two choices are available: Slot 0 controllers and bus extenders. The most common bus extender is called *MXIbus*.

Slot 0 controllers and bus extenders take commands from the external computer, interpret them, and direct them to the various measurement and switch modules. Slot 0 controllers and bus extenders must be placed in a unique slot of the mainframe. Both provide special VXI functions that include the Resource Manager. Bus extenders translate a personal computer's I/O bus to the VXI backplane using a unique high-speed parallel bus. Cable lengths must be relatively short. Transactions over a bus extender have a register-to-register bit level flavor. The advantage of bus extenders is high speed.

Slot 0 controllers support these register level transactions, but also support message-based transactions. These transactions are ASCII commands following the *SCPI* (*Standard Commands for Programmable Instrumentation*) standard that was accepted by the instrument industry in 1990. SCPI commands are very readable, although slower to interpret. The advantage of message-based transactions is easier, quicker program development. The introduction of VXI Plug-and-Play drivers has superseded SCPI for easier, quicker programming.

Slot 0 controllers require a relatively high-speed communication interface to communicate with an external computer. GP-IB, known as *IEEE-488*, and Firewire, known as *IEEE-1394*, are the standard interfaces for Slot 0 controllers. The external computer must have a matching communication interface to the Slot 0 controller. Firewire interfaces are a computer-industry standard and are being built into PCs as standard equipment. Plug-in Firewire interface boards are used widely for older PCs. Basic Firewire software driver support was included in Windows 98 from Microsoft. GP-IB has been an instrument communication standard since 1975. Because it is not a computer standard, a plug-in interface board must be added to the external PC for connection to the Slot 0 controller. GP-IB interfaces are about five times slower than Firewire.

VXI embedded computers. Embedded controllers are available for C-size mainframes. Products with Intel-architecture processors running Windows 3.11/95/98/NT are available from several vendors. One problem for vendors and users of embedded computers is the rapid change of the PC industry. Users in industry want product availability and support for a given model to last 5 to 10 years. Lifetimes for PC microprocessors, chip sets, and peripherals is typically three to six months. Vendors of embedded computers take special steps to stockpile parts and selectively support microprocessors to ensure a longer lifetime. This generally means that embedded computers do not offer the latest technology.

A few other computer architectures and operating systems are available. Hewlett-Packard offers its HP-PA architecture running HP-UX.

VXI measurement modules. VXI offers the broadest and most complete line of instrumentation modules (see Fig. 40.5 for an example of a VXI measurement module). The standard was designed for this purpose so that is not a big surprise. Like the other modular standards, VXI offers a full line of digital I/O modules and modules containing ADCs and DACs. These functions are commonly used for data-acquisition and industrial-control applications.

Typical digital I/O boards would have 16, 32, or 64 channels. The voltage range is typically 5 to 24 V. Some products are opto-isolated. Single-ended or differential channels are available.

ADC and DAC modules are available from 16 to 96 channels, depending on whether they are single-ended or differential, or multiplexed or parallel channels. Sampling rates range from 50 kHz to 150 MHz and the resolution ranges from 8 to 16 bits.

A large and full line of traditional instruments, such as digital multimeters, oscilloscopes, waveform/pulse generators, spectrum analyzers, etc., are available.

Digital multimeters are available from 4½ to 6½ digits. They are capable of a full range of ac/dc voltage, current, and resistance measurements. Some are scanning voltmeters with up to 64 input channels. Some are configurable to optimize for measurement throughput.

Digital counters are available from three to eight channels, which can operate from 100 kHz to 23 GHz. Rubidium frequency standards are also available.

Oscilloscopes and waveform analyzers are available in several configurations. Products offer 2 or 4 channels, from 10 Msample/s to 5 Gsamples/s; bandwidths range from 100 MHz to 1 GHz.

The wide range of signal generators include arbitrary waveform generators (ARBs), pulse generators, and RF and microwave signal sources. Typical ARBs are 12 bits, 40 to 500 Msample/s, with AM, FM, PM, FSK, PSK modulation, and some frequency-hopping capabilities. Typical pulse generators offer 2 to 8 channels, running from 20 to 300 MHz. RF sources range from 800 MHz to 20 GHz.

RF/microwave measurement includes spectrum analyzers, power and modulation meters, receivers, downconverters, amplifiers, and attenuators.

The wide range of digital test products include stimulus/response, bound-

Figure 40.5 A VXI measurement module: C-size arbitrary waveform generator (RFI covers removed). Photo courtesy of Hewlett-Packard Co.

ary scan, and digital signal processors. High-speed pattern generators typically work up to 50 MHz. Channel count and memory depth varies widely. Channel count ranges from 8 to 192 per module. These can often be ganged together to form a single pattern of over 1000 channels. Memory depth ranges from 64 k to 1 M. The type, number, and range of VXI instruments continues to grow each year.

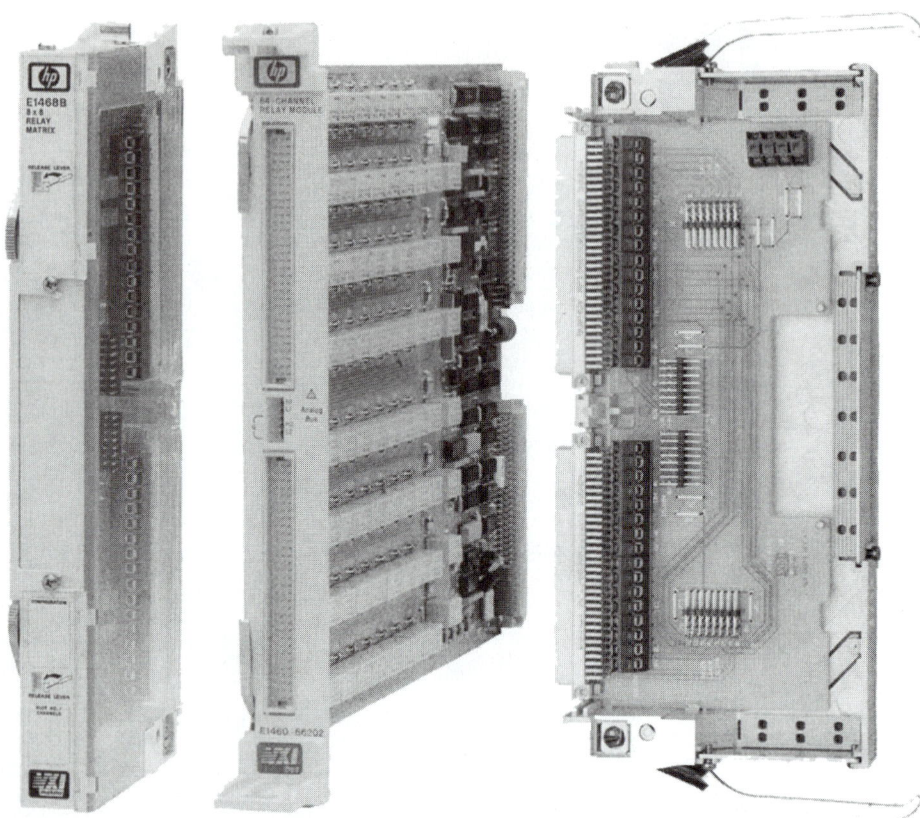

Figure 40.6 VXI switch module and termination blocks for wiring, C size. Photo courtesy of Hewlett-Packard Co.

Switching modules. VXI has the broadest line of switching modules among the modular instruments (see Fig. 40.6 for an example of a switching module). Switching can be used in many different ways. Switching products include simple relays, multiplexers and matrices, RF/microwave switches, and analog/FET switches. VXI switches are available for almost all applications from multiple vendors.

Simple relays (e.g., SPDT switches) are used to control the test environment. They are used to apply power to the device under test. They control motors, turn on heaters or lights, and control safety interlock systems. Armature relays are typically used. Maximum current and voltage ratings are important parameters. Maximum voltage is typically in the range of 150 V to 250 V. Maximum current ranges from 1 A to 10 A. Both of these parameters vary with the number of switches on the module. The number of switches on a C-size module ranges from 20 to 80. One product on the market offers 40 10-A SPST switches. Another product has 64 1-A SPDT switches.

Multiplexers and matrix switches are generally used to route signals from a sensor, or from a device under test, to one or more instruments. For example, a data-acquisition system can be built with a multiplexer and a voltmeter. Sensors, such as thermocouples, are routed one at a time through the multiplexer to the voltmeter. As another example, a manufacturing test system would need to connect signals from a device under test to various signal generators, loads, and instruments. The connections change dynamically as the test progresses. At first, the signal generator and instrument would be connected to one set of inputs and outputs. Later, they would be connected to another set.

These switch types are usually constructed with reed relays. Reed relays are faster than armature relays, which is important for the applications that use multiplexers and matrixes. Maximum current is usually in the 1-A to 2-A range. The important switch parameters vary with application. Data-acquisition switches might need to have low thermal offset. Matched path delays and bandwidth are important for a manufacturing test system. 10 MHz to 30 MHz of bandwidth is typical. Flexibility in configuration is usually the most important consideration. Multiplexers from 8-to-1 to 256-to-1 are widely available in C-size modules. Matrixes are available in many configurations: 4 × 16, 4 × 32, 8 × 8, 8 × 32, 48 × 48, 64 × 4, and 256 × 256.

RF/microwave switches are a special class of multiplexer and matrix switch. As the name implies, they are used to switch very high-frequency signals. Bandwidth and path loss are usually the crucial parameters. Coaxial switches, instead of reed relays, must be used. RF switches are usually available in 50-Ω or 75-Ω impedances. Typical bandwidth ranges from 200 MHz to 18 GHz. One C-size product offers a 60-to-1 multiplexer at 500 MHz. Another product offers eight 4-to-1 multiplexers at 3 GHz.

Analog/FET switches are another special class of multiplexer and matrix switch. Instead of mechanical relays, semiconductor devices are used to route the signals. This allows much higher-speed switching, but at the cost of lower voltage and current ranges, and usually a high-resistance connection (5 Ω to 100 Ω). Just a few products are available: 16-to-1, 96-to-1, and 48 × 48.

40.6.4 VXI software

The VXI standard has emphasized system software as an important element to improve the interoperability of modules from multiple vendors, and to provide system developers a head start in developing their software.

The VXIplug&play Systems Alliance developed a series of system frameworks encompassing both hardware and software.

Six frameworks have been defined:

- Windows 3.1
- Windows 95
- Windows NT

- HP-UX
- Sun
- GWIN

In common with all of the frameworks is a specification for basic instrument communication called *Virtual Instrument Software Architecture* (*VISA*). VISA is a common communications interface, regardless of whether the physical interconnect is GP-IB, Ethernet, or Firewire. Also in common among the frameworks is a specification for instrument drivers. Instrument drivers are functions, which are called from programming languages, to control instruments. VXIplug&play drivers must include four features:

- C function library files
- Interactive soft front panel
- Knowledge base file
- Help file

The C function library files must include a dynamic link library (.DLL or .SL), ANSI C source code, and a function panel file (.FP). The function must use the VISA I/O library for all I/O functions.

The interactive soft front panel is a graphical user interface for directly interacting with a VXI instrument module. Some front panels closely resemble the front panel of a traditional box instrument. This tool is meant to assist in the initial turn-on of a system and to help troubleshoot integration problems.

The knowledge base file is an ASCII description of all the instrument module's specifications. The help file provides information on the C function library, on the soft front panel, and on the instrument itself. Programming examples should also be included.

The VXIbus specification contains important software elements. These include a standard definition of module registers, and communication protocols for sending commands, checking on status, etc. The VXIbus specification also calls for a resource manager. The resource manager, at power up, will identify all VXI devices on the bus, manage the system's self test and diagnostic sequence, configure addressing and interrupts, and initiate normal system operation. This function greatly eases the system software developer's task, particularly if VXI modules have been selected from multiple vendors.

40.7 VXI Standard Specifications

Because the intent of the VXI standard was to create a standard specifically for instruments, several extensions were made to the VME specification. These extensions include additions to the backplane bus, and power, cooling, and RFI specifications.

40.7.1 Unique VXI signals

The A-board size has only one connector (P1) and has no backplane extensions over VME. The B-, C-, and D-size boards have at least a second connector (P2). On P2 are the following additions:

- Additional supply voltages to support analog circuits: −5.2 V, −2 V, +24 V, and −24 V.
- Additional pins were also added for a +5 V increase in the maximum current capacity.
- 10 MHz differential clock.
- Two parallel ECL trigger lines.
- Eight parallel TTL trigger lines.
- A module identification signal.
- A 12-line local bus that connects adjacent modules. The manufacturer of the module defines the functionality of these lines.
- An analog summing bus terminated in 50 Ω.

The 10-MHz clock is differential ECL. Its purpose is for synchronizing the timing of several modules.

40.7.2 Trigger lines

The TTL and ECL trigger lines are open collector lines used between modules for trigger, handshake, clock, or logic state communications. Several standard protocols have been defined for these lines, including synchronous (SYNC), asynchronous (ASYNC), and start/stop (STST). For some measurements, several instruments might need to be simultaneously started or stopped. One instrument might need to wait for another instrument to complete before starting. It might be desirable to understand the relative timing between instruments. These situations are the purpose for the trigger lines. Although software coordination can sometimes work, nanosecond speed is required for some measurements. The trigger lines provide that kind of speed.

Slot identification. The module identification signal allows a specific physical slot to be identified. This is used by the resource manager to identify modules, regardless of their physical location, and to assign addresses. This avoids the need for configuration switches, which are frequently set in error.

The local bus connects two adjacent slots. It does not extend across the entire backplane, but can be daisy chained through a module to reach several modules. The bus has been used to connect proprietary analog signals between modules. It has also been used as a very high speed parallel bus. An example might be to send high-speed data from a digitizer to a DSP. The local bus is typically used between two modules from a single manufacturer.

40.7.3 Module power

The VXIbus specification has set standards for mainframe and module power, mainframe cooling, and electromagnetic compatibility between modules. This ensures that products from multiple vendors will operate together. It substantially reduces the burden on a system designer to deal with these issues.

The standard for mainframe power includes maximum allowed variations in voltage, dc load ripple/noise, and induced ripple/noise. Mainframes must specify peak and dynamic current for all voltages supplied. Modules must specify their requirements for all voltages and currents. This makes it an easy calculation for system designers to manage their power budget.

Conducted emissions. The VXIbus specification sets the maximum conducted emission level for all modules, and has set up a test method to ensure compliance. Likewise, the specification has set the conducted susceptibility levels (continued operation in the presence of noise) and established a test method to ensure compliance. Susceptibility includes electrostatic discharge through a module faceplate.

Radiated emissions. The VXIbus specification for radiated emissions is consistent with FCC, VDE, and MIL specifications. The specification applies to a completely filled mainframe (i.e., a mainframe with 13 modules installed). Each module is expected to radiate only its fair share, $\frac{1}{13}$ of the allowable radiation. In addition to far-field measurements, the specification also sets limits on close-field emissions (i.e., radiation between modules).

As with conducted emissions, the specification has far-field and close-field limits for radiated susceptibility. A test method is recommended for all cases.

40.7.4 Cooling specification

The mainframe and module-cooling specification focuses on the test method for determining whether proper cooling will be available in a system.

40.8 Personal Computer Plug-Ins (PCPIs)

Since the IBM PC was introduced in the mid-1970s, a huge number of companies have been designing products to plug into the open slots of the PC motherboard. Three standards have defined those open slots. They are the ISA (Industry Standard Architecture), EISA (Extended Industry Standard Architecture), and PCI (Peripheral Component Interconnect) standards. EISA is an extension of ISA. All three were defined by the computer industry and none have any special support for instrumentation.

In 1994, a group of industrial computer vendors formed a consortium to develop specifications for systems and boards used in industrial computing applications. They called themselves *PICMG* (*PCI Industrial Computer Manufacturers Group*). PICMG was formed by seven companies: Digital Equipment,

GESPAC, I-Bus, Pro-Log, Teknor, Hybricon, and Ziatech. Today, it includes more than 350 vendors. The group's first specification defined passive backplane computers. PICMG 1.0 "PCI-ISA Passive Backplane" was adopted in October 1994. A second specification, PICMG 1.1 "PCI-PCI Bridge Board," was adopted in May 1995. It was a crucial specification for slot-hungry industrial computers that enabled systems larger than eight slots.

40.8.1 PCPI common functions and applications

The most common applications are data-acquisition, industrial-control, and custom electronic-control systems.

40.8.2 PCPI system components

Two different configurations are common with PC plug-in measurement systems. The first configuration consists simply of a personal computer and a few measurement modules. The second configuration consists of a passive backplane, a single-board computer, and several measurement modules. This latter approach can also be referred to as an *industrialized PC*. In both cases, the measurement modules are the same (Fig. 40.7).

PCPI frames. In the simple and most common configuration, the personal computer is the backplane, power supply, and packaging for the measurement system. System builders can choose a variety of PC form factors—from desktops to server towers. PC backplanes usually have seven or eight slots. After the installation of standard PC peripherals, only a couple of slots are open for instrumentation. Extender frames are available, but not common.

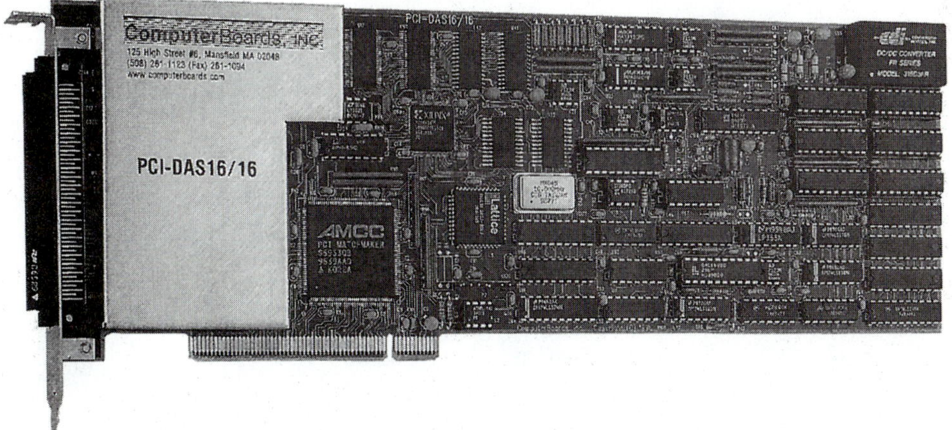

Figure 40.7 PC plug-in: Multi-function data-acquisition module, 16 analog inputs, 2 analog outputs, 24 digital I/O. Photo courtesy of ComputerBoards Inc.

This type of PC plug-in system tends to be small, where only a few instrument functions are needed.

For the passive backplane configuration, a frame solution can be built up from backplanes and chassis from many manufacturers (See Fig. 40.8). A 19-inch rack will have up to 20 slots. PCI-ISA passive backplanes usually contain a mixture of ISA and PCI slots. In addition, a backplane has one or two dedicated CPU slots for a single-board computer. Backplanes are available from 4 to 19 slots. For example, one product is a 19-slot backplane. It has one dedicated CPU slot, eight ISA slots, and 10 PCI slots. Because of the PCI specification, two of the PCI slots are primary slots, and eight are secondary slots. Primary and secondary slots are joined by a PCI-PCI bridge chip, which is included on the backplane. The chassis come in many configurations—from desktop to rackmount. Chassis power supplies range from 200 W to 570 W.

PCPI single board computers. For the passive backplane configuration, a single-board computer must be used. Like desktop PCs, these single-board computers use Intel architecture processors. One exception is a DEC Alpha processor product. A few other processor boards are used in conjunction with an Intel architecture processor to offload the main CPU. For example, one product uses an Intel 960 processor to offload high-performance I/O tasks. These single-board computers also contain hard-disk and floppy controllers, commu-

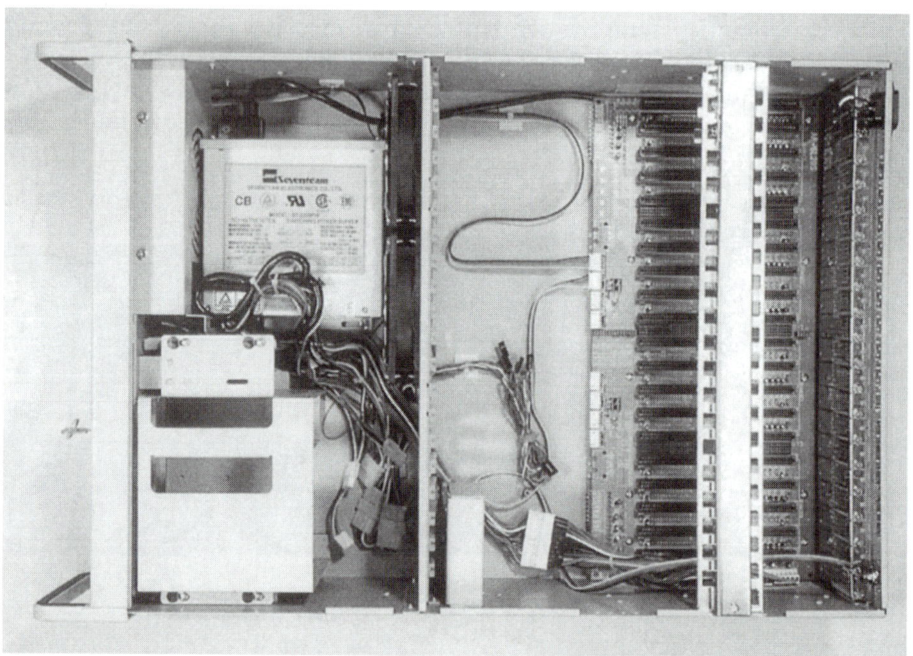

Figure 40.8 A passive backplane configuration: backplane, power supplies, and frame. Photo courtesy of ADAC Corp.

nication ports both LAN and serial, keyboard port, and (optionally) a video port.

PCPI measurement modules. PC plug-in instrument functions are primarily digital I/O modules and modules containing ADCs and DACs. In addition, there is a small number of basic instruments including oscilloscopes, digital multimeters, and pulse/function generators.

Digital I/O boards have 16, 32, or up to 96 channels, depending on whether they are simple TTL lines, opto-isolated, etc. Voltage range is typically 5 to 24 V.

ADC and DAC modules are available from 8 to 64 channels, depending on whether they are single-ended or differential. Sampling rates are available from 60 kHz to 10 MHz, resolution from 12 to 16 bits. These modules are sometimes multifunction modules also containing 8 to 16 channels of digital and counter/timers.

Traditional instrument functions include: a 2-channel, 15-MHz bandwidth, 20-megasample/sec, 8-bit oscilloscope, a 5½ digit digital multimeter, a 40 megasample/sec, 12-bit arbitrary waveform generator, and a 2-channel 10-mega-bit/sec serial data analyzer.

40.9 CompactPCI

CompactPCI is a derivative of the *Peripheral Component Interconnect* (*PCI*) specification from the personal-computer industry. CompactPCI was developed for industrial and embedded applications, including real-time data acquisition and instrumentation. The specification is driven and controlled by PICMG, the same group covered in PC plug-ins. The CompactPCI specification was released in November of 1995. To support the needs of general-purpose instrumentation, a PICMG subgroup developed a CompactPCI instrumentation specification, called *PXI* (*PCI eXtended for Instrumentation*). This specification was authored and copyrighted by National Instruments Corporation. The first public revision of PXI was released August 1997. The PXI specification continues to be developed and maintained by the *PXI Systems Alliance* (*PXISA*), a separate industry group.

40.9.1 CompactPCI specification

Similar to VME, the CompactPCI specification defines backplane functionality, electrical specifications, and mechanical specifications. The functionality is the same as the PCI bus, as defined by the PCI local bus specification, but with some additions. These additions provide pushbutton reset, power-supply status, system slot identification, and legacy IDE interrupt features. A unique feature of CompactPCI is *Hot Swap*, the ability to insert or remove modules while power is applied (an extension to the core specification).

The smaller 3U-size board has one 220-pin connector, split into two halves, called *J1* and *J2*. The larger 6U board can have additional connectors referred to as *J3* through *J5*. The 3U board has no user-defined pins available, but there are several reserved pins that are the subject of additional PICMG subgroup work. The 6U board has 315 pins that are also the subject of subgroup work. One of those subgroups is PXI. The additional signals are essentially the same as those extra bus signals defined by the VXI standard to enhance the VME bus.

40.9.2 CompactPCI system components

A CompactPCI measurement system usually consists of one or more frames, an embedded computer module, and various measurement modules.

Frames. *Frames* refer to the backplane, power supply, and packaging that encloses a system. System builders can choose a solution at many different levels from the component level (i.e., a backplane to complete powered desktop enclosures). A CompactPCI backplane has up to eight slots. They are available with two, four, six, or eight slots. It is possible to go beyond eight slots using a bridge chip on the frame, or a bridge card. Most instrumentation systems will probably not go beyond eight. An embedded computer or a bridge card occupies one of the eight slots. Backplanes, subracks, and enclosures are available from many manufacturers. Power supplies are available in many sizes.

Embedded computers. The PCI bus is commonly used in many computers—from personal computers to high-end workstations. For that reason, a large number of CompactPCI embedded processors are available. Intel architecture processors, such as the Pentium II, are available from several vendors. Embedded processors are also available with PowerPC, SPARC, MIPS, and many DSPs, such as from Texas Instruments.

The list of operating systems includes (but is not limited to) Windows 3.11/95/98/NT, LINUX, and VxWorks. Software support for instrumentation modules is generally limited to Windows. PXI modules are required by that specification to support the VISA standard (WIN framework) and to supply initialization files for proper system configuration.

Measurement modules. Instrument functions in CompactPCI are primarily digital I/O modules and modules containing ADCs and DACs. In addition, there is a small but growing number of traditional instruments, including oscilloscopes, digital multimeters, and serial data analyzers. These are specifically designed for PXI (Fig. 40.9), but, of course, can operate with the PXI extensions turned off in a general CompactPCI system.

Digital I/O boards have 16, 32, or up to 96 channels, depending on whether they are simple TTL lines, optoisolated, etc. The voltage range is typically 5 to 24 V.

ADC and DAC modules are available from 8 to 64 channels, depending on

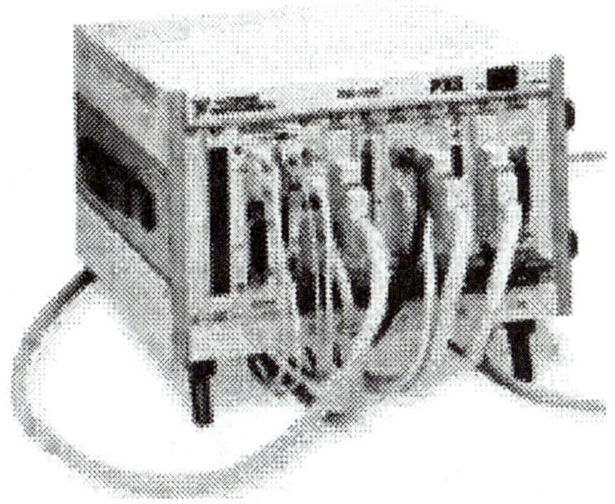

Figure 40.9 PXI mainframe with CPU and measurement modules. Photo courtesy of National Instruments Corp., Austin, Texas.

whether they are single-ended or differential. Sampling rates are available from 60 kHz to 10 MHz. Resolution ranges from 12 to 16 bits. These modules are sometimes multifunction modules also containing 8 to 16 channels of digital and counter/timers.

Traditional instrument functions include: a 2-channel, 15-MHz bandwidth, 20-Msample/s, 8-bit oscilloscope, a 5-digit digital multimeter, and a 2-channel 10 Mb/s serial data analyzer.

Switching modules. CompactPCI, specifically PXI, has a good range of switching products, including simple relays, multiplexers and matrices, RF switches, and FET switches. Simple relays include 8 and 16 SPDT, 5-A to 10-A products. Multiplexers are available up to 24 channels. Matrixes include 2×16, 4×8, and 4×40. RF multiplexers for 50 Ω and 75 Ω are available as four 4-to-1 switches with 200 MHz bandwidth.

Compact PCI software. PXI adopted as part of its specification many of the features of VXIplug&play software. It adopted software frameworks for Windows 95 and Windows NT. These frameworks are required to support the VISA I/O standard.

Chapter

41

Software and Instrumentation

Phil Christ and Bonnie Stahlin
Agilent Technologies
Loveland, Colorado

41.1 Software Role in Instrumentation

Three technology areas have had a profound influence on how instruments and computers are used together:

- software
- computing technologies
- input/output (I/O)

The convergence of these three technologies in instrumentation has created new capabilities and new ways of using test and measurement equipment. Here is a brief overview of these technologies and how they influence the test and measurement process.

41.1.1 Software as a driving technology

Having the right hardware in place is not sufficient for many testing and measurement applications. A large amount of software is necessary. Indeed, software has become an equal partner to hardware in providing instrument functionality. This chapter focuses on three key software application areas:

- Moving test data from instrumentation to computer applications for analysis and documentation
- Automating complex or repetitive measurements
- Remote monitoring and control of instrumentation

For a discussion of software-defined instruments, see Chapter 46.

Software advances are simplifying these tasks so that engineers focus less on hardware and software technology and more on making measurements and interpreting measurement information. The pervasive nature of the personal computer (PC) has driven huge levels of investment in development of new software technologies and applications. It has also resulted in increasing standardization of the applications and tools used in offices, laboratories, and production environments. Widespread usage of so-called "applications suites," for example, allows instrument vendors to focus their software development investment on simply providing easy connectivity to these applications. The pervasiveness of these applications also allows easier sharing of measurement information from one engineer to another.

Internet software technologies are another driving influence. The Internet has become the standard approach for remote monitoring and control of instrumentation. The Internet browser is the preferred user interface for these applications, both for its familiarity and its ease-of-use benefits.

The use of PCs has also defined what software tools are most commonly used for connecting instrumentation with computer applications. Technologies, such as Microsoft ActiveX and COM/DCOM, as described in the following paragraphs, are the most widely used. Instrument drivers based on these technologies have the benefits of working with a variety of programming environments and applications and having greater longevity because of their increased standardization.

The growing use of computer software connectivity standards in testing and measurement also allows instruments to integrate more readily with other computer devices not normally associated directly with instrumentation. Examples are transfer of test and measurement data to network-based file servers and *Structured Query Language* (*SQL*) relational databases.

41.1.2 Computing technology

Personal computers are widely used in test and measurement. Engineers use them for a broad range of tasks, including product design, report writing, data analysis, Internet-based research, etc. Because of the PC's broad range of capabilities and widespread usage, engineers naturally expect to use them with instrumentation. As PCs continue to grow in their capability, they will be a driving technology in testing and measurement for the foreseeable future.

41.1.3 Input/Output (I/O) technologies

I/O is the physical connection (and closely affiliated software layers) between instrument and computer and hence is the basic enabler for instrument-to-computer connectivity.

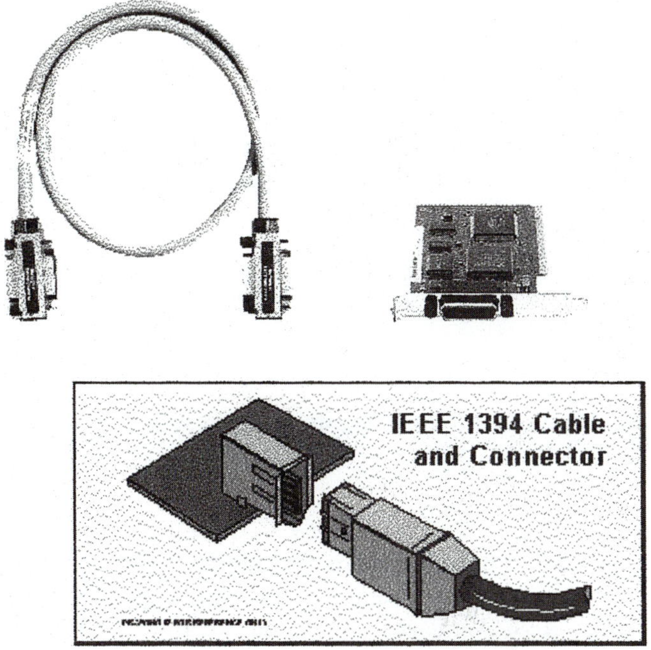

Figure 41.1 Test and measurement-specific I/O connector and PC plug-in card (GPIB, upper) and computer-standard I/O cable with connector built into PC (IEEE 1394, lower).

Test and measurement-specific I/O. High-performance I/O buses specifically designed for testing and measurement applications have been available for many years. The most popular has been the IEEE 488 bus, also known as *GPIB* (see Fig. 41.1). Computer-standard I/O has advantages of lower cost and ease of configuration, and, as a result, has grown in usage.

Computer-standard I/O. Computer industry-standard I/O has increased in performance and flexibility to the point where it is suitable for any testing and measurement application. It has the advantage of being readily available as a standard feature on PCs, which greatly simplifies the instrument-to-computer connection. The cables required are simple, inexpensive, and widely available. An example is IEEE 1394, also known as *Firewire* (see Fig. 41.1). Increasingly, connection of instruments to PCs is as simple as connecting computer peripheral devices, such as disk drives and printers. Computer-standard I/O is a crucially important driving force in instrument-to-computer connectivity.

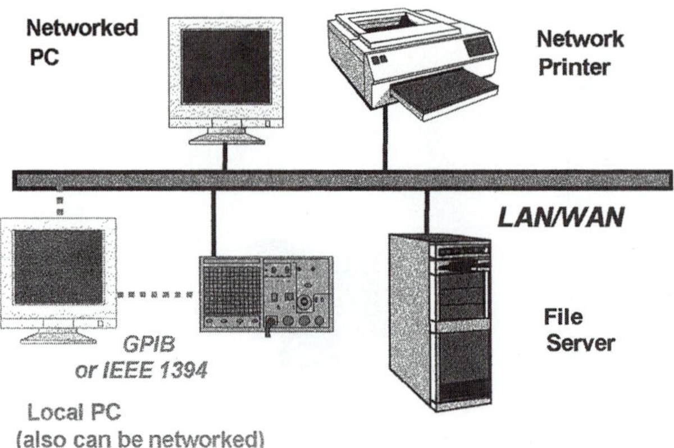

Figure 41.2 Many instruments can be directly connected to a LAN, like other networked devices.

Network connection. The I/O examples described previously require a direct connection between the instrument and computer. *Local-Area Networks* (*LAN*) and *Wide-Area Networks* (*WAN*) do not have this limitation. The instrument simply needs to be connected to the network. Once this is achieved, the instrument is able to form a software connection with practically any other device on the network. This includes geographically remote computers that could be in another building or another part of the world. Networking brings dramatic flexibility to instrument connectivity (see Fig. 41.2).

The most-prevalent networking standard is IEEE 802.3 ("Ethernet"). Networking provides significant performance benefits. Ethernet is widely available in 10-megabit/sec (10BaseT or 10Base2) and 100-megabit/sec (100BaseT) implementations, and a 1-gigabit/sec implementation (Gigabit Ethernet) implementation has rapidly grown in use.

41.2 Software Used by Instrumentation

Many types of software can be considered in the field of instrumentation. Included are:

- Software embedded within the instrument for making measurements.
- Computer-based software for making measurements.
- Software for connecting instruments to computers.
- Computer-based applications for analysis and documentation of instrument data.
- Programming languages for automating instruments.

- Software for emulating the instrument front-panel user interface.
- Other software user interfaces to instrument functions.

Issues in each of these topic areas will be briefly described.

41.2.1 Embedded software

This is the software within the instrument that intercepts the physical measurement and makes it available for collecting data, transforming data into measurements, displaying measurements, and running all the basic operations of the instrument. This software is only accessible to the user through the instrument front panel or user interface. The user is not able to change or manipulate the software beyond the functions available on the front panel, other than to install software updates from the manufacturer. Most instruments are built on an embedded *real-time operating system* (*RTOS*). Instrument designers evaluate the strengths and limitations of the RTOS when they are designing the instrument. Embedded software is covered in detail in Chapter 10.

41.2.2 Computer-based software for making measurements

Computer-based software can be used to provide additional capability that enhances what can be done with stand-alone instruments. Some of the things that can be done with computer-based software are:

- Groups of instruments can work in concert to make measurements that individual instruments cannot make alone. For instance, a source and an analyzer can be combined to both stimulate the device under test and measure its response.
- Measurement functions can be customized to a specific domain. For instance, a general-purpose spectrum analyzer can be used in a software environment to determine if a hybrid filter meets its specifications. Although the spectrum analyzer is capable of the basic measurements, such as power bandwidth, the system software applies the knowledge about the type of filter and its specifications.

41.2.3 Software for connecting instruments to computers

Many useful applications can result from connecting instruments to computers. Because of the variety of instruments available and because instrument and computer architectures often differ significantly, the software required to establish a connection between an instrument and a computer is not trivial. Two key software layers required are:

- I/O libraries
- Instrument drivers

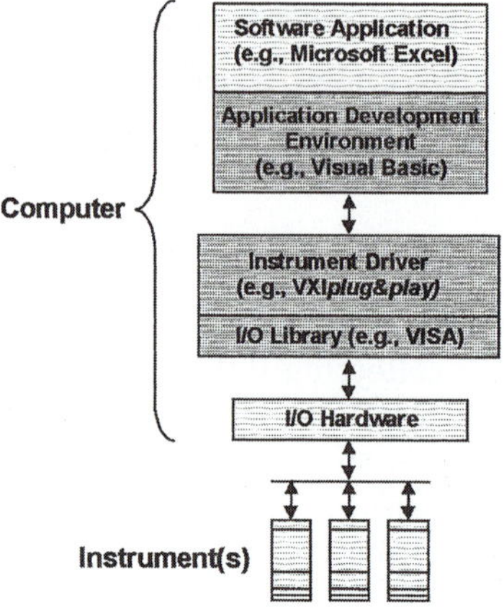

Figure 41.3 Communication between a computer application and an instrument requires I/O hardware and the following software layers: an I/O library (a set of standardized instrument functions that communicates to the instrument in character strings), an instrument driver (a software module, which allows instrument functions to be called using a software subroutine approach), and an Application Development Environment (the programming language in which the computer application is written).

These are illustrated in Fig. 41.3. These software layers are transparent to users who only use programs written by someone else (programs purchased from a vendor or programs written by programmers within the user's company). Users interested in developing their own applications for instrument automation must become familiar with how these layers operate.

An I/O library is a set of functions that can be called from a program that carries messages to and from an instrument in ASCII character strings. It takes care of all of the details of the I/O connection and allows the programmer to focus on the messages sent to the instrument. Standard sets of functions exist, such as the *Virtual Instruments Systems Architecture* (*VISA*) standard.

An instrument driver is an instrument-specific software module that gives the programmer access to the instrument without needing to bother directly with ASCII strings. Instead, the programmer uses a familiar subroutine approach. *VXI*plug&play is one example of a standard for instrument drivers.

An *application development environment* (*ADE*) is also required. An example ADE is Microsoft Visual Basic. If the application is commercially available, the existence of the ADE might be inherent in the application (e.g., Visual Basic within Microsoft Excel).

41.2.4 Computer applications for analysis and documentation

Technically oriented applications. Many personal computer applications are useful to the user of instrumentation. These include technically oriented applications, such as MATLAB and MathCAD. MATLAB is an application that combines numeric computation, advanced graphics and visualization, and a high-level programming language. It can be used for:

- Data analysis and visualization
- Numeric and symbolic computation

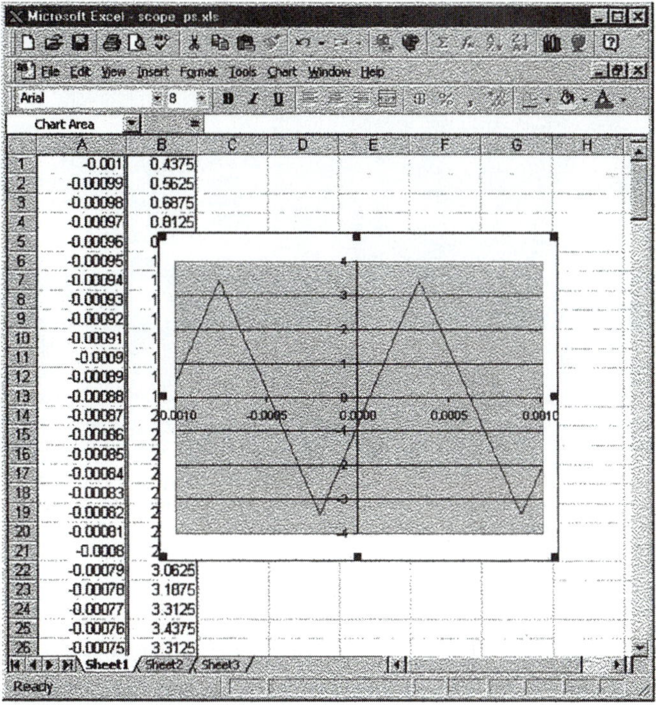

Figure 41.4 Spreadsheet with oscilloscope data in columns and overlaid graph of data.

- Engineering and scientific graphics
- Modeling, simulation, and prototyping

MathCAD provides similar functions, with a worksheet-style user interface.

Technical documentation applications, such as FrameMaker, are also used by engineers to document the results of their work.

Business-oriented applications. More frequently, standard business-oriented applications, such as Microsoft Excel and Word, are used for analysis and documentation respectively. Figure 41.4 illustrates data collection and analysis using Microsoft Excel.

41.2.5 Programming languages for automating instruments

Some users need to automate complex or repetitive test sequences. An *Application Development Environment* (*ADE*) is required, which could be a graphical programming language written specifically for test and measurement applications, such as VEE from Hewlett-Packard and LabVIEW from National Instruments. As with analysis and documentation applications, industry-

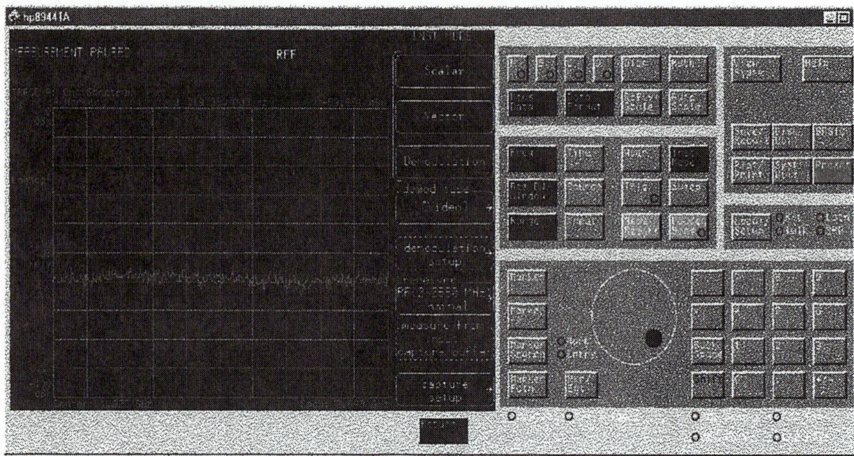

Figure 41.5 Virtual front panel for vector signal analyzer as it appears on computer monitor. The controls can be activated and manipulated with a mouse to perform the same functions they are used for on an actual instrument.

standard programming languages, such as Microsoft Visual BASIC, are frequently preferred over the graphical programming languages mentioned above. This is because more programmers are familiar with standard programming languages than test and measurement-specific languages, and a larger amount of documentation, debugging tools, etc., are available for these languages.

41.2.6 Software for emulating the instrument front-panel user interface

It is possible to emulate the entire front-panel user interface of the instrument in software. The user is able to access all the instrument's functions from a personal computer, with a very similar look and feel to the instrument itself (Fig. 41.5).

These "virtual front-panels" seldom appear in isolation, but are typically incorporated into another software component, such as an instrument driver or a remote monitoring and control application.

41.2.7 Other user interfaces to instrument functions

Frequently, the user requires only a subset of the instrument's functions in a computer-based user interface. A simple user interface for displaying selected instrument parameters might be sufficient. These can be developed as ActiveX controls. An ActiveX control is a reusable software object, based on the Microsoft *Common Object Model* (*COM*) standard. It has its own user inter-

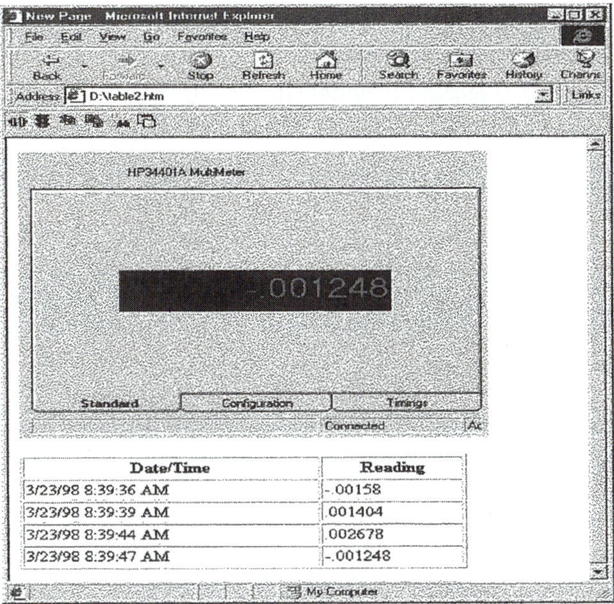

Figure 41.6 Multimeter ActiveX control viewed inside a browser

face and it can be easily inserted into other programs. Figure 41.6 shows a simple ActiveX control displaying the readout from a multimeter, inserted into an Internet browser.

See Chapter 44 for a further discussion on graphical user interfaces.

41.3 Connecting Models: Instruments and Software

In describing how software and instrumentation work together, it is helpful to organize the text based on the type of connection that exists between the instrument and the computer running the software. The connection models considered are:

- No connection model
- Local connection model

  Software for analysis and documentation
  Software for instrument automation
- Network connection model

  Accessing network resources
  Software for remote analysis/documentation or automation

- Internet "client-server" connection

  Instrument as Internet server model
  Instrument as Internet client model

41.3.1 No connection model

Instruments are frequently used as "stand-alone" devices with no connection to a computer or computer network. In many cases, the data available from the instrument's front panel is sufficient for the task at hand. If additional work is required to analyze or document the measurement data, and the data is not too voluminous, it might be acceptable for the user to manually enter the data into a computer disconnected from the instrument. If the instrument has its own built-in floppy disk drive or other common data device, the amount of data that can be practically transferred is considerably greater. The user simply moves the floppy disk from the instrument to the PC. Under the "no connection model," more complex or repetitive measurements are also done manually.

While appropriate for some applications, the lack of a computer connection limits the user to doing only simple tasks with the software. Manual intervention for data transfer and completing complex or repetitive tasks gets to be very tedious as the measurement task gets more complicated.

41.3.2 Local connection model

The instrument is connected directly to the computer through either test and measurement-specific or computer-standard I/O.

Software for analysis and documentation. Direct connection of the instrument to the computer simplifies the task of getting data into applications for analysis and documentation.

To perform these tasks, the user needs to have access to an instrument-specific driver and the appropriate ADE. This approach, however, is not cost-effective because ADEs can easily cost more than the target spreadsheet or word-processing application. It also has the disadvantage of requiring the user to be familiar with how to use the ADE. Learning the ADE might not be practical for the nonprogramming engineer. Finally, it might require the user to go through many steps just to achieve a simple data transfer.

A simpler and more economical approach is to use an ActiveX-based toolbar, which can be loaded into the target application. The toolbar includes the underlying driver software required to make the instrument connection. The user is able to access basic instrument functions (e.g., initiate a scan and upload resulting data) directly from the computer application. For example, a user can fill columns of a spreadsheet with data from the instrument with

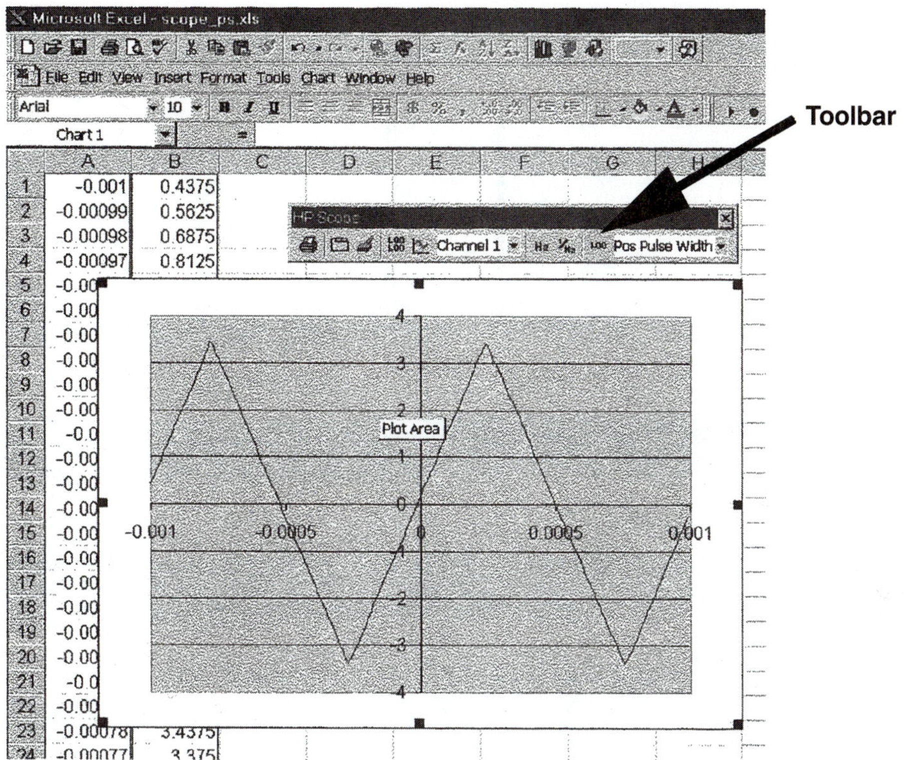

Figure 41.7 ActiveX toolbar for an oscilloscope embedded in an Excel spreadsheet. The toolbar provides direct access from Excel to the connected scope's configuration, measurements, and data. The icons on the toolbar specifically provide the following functions: establish the I/O connection to the scope, save and restore scope configurations, upload scope waveform data, capture a "screen shot" from the scope, capture a "single" measurement, and access the scope's help system.

one mouse-click on the spreadsheet toolbar. An example spreadsheet toolbar is shown in Fig. 41.7.

From a word-processor toolbar, an engineer can upload instrument parameter data and an image from the instrument screen into the word processor for documentation purposes. The word processor becomes a powerful electronic substitute to the engineer's conventional lab notebook.

Software for instrument automation. Programming languages allow automation of repetitive measurements and/or complex sequences of measurements. Many engineers do not have the time or desire to become computer programming professionals to complete these tasks. Advances in computer-standard ADEs (such as Visual Basic) give engineers the ability to perform sophisticated automation programming without this requirement.

Graphical programming languages, such as HP VEE and National Instru-

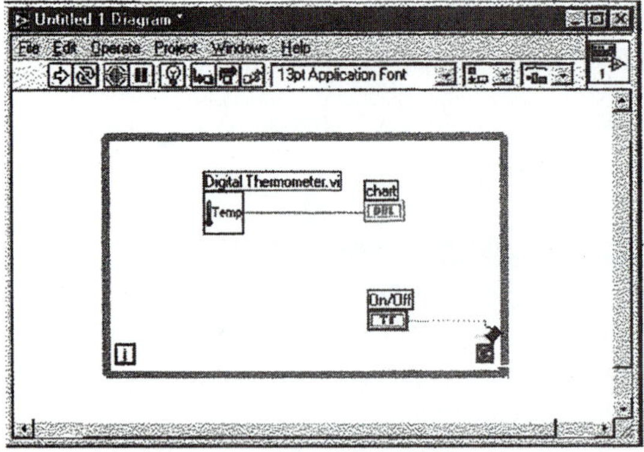

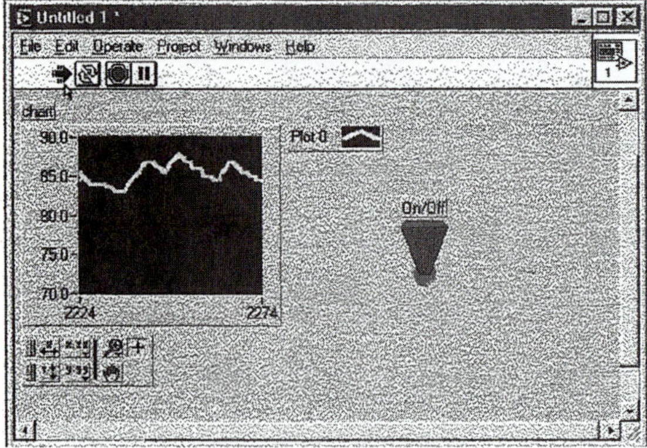

Figure 41.8 Programming with LabVIEW. Upper is a block diagram with a "while loop" (surrounding gray arrow), a digital thermometer object with its output connected to a waveform graphing object, and an on/off toggle switch object. The lower shows the resulting "software front panel" output. While the on/off switch is in the ON position, the program will read and display a temperature value every time the loop executes. (Courtesy of National Instrument Corporation, reprinted by permission.)

ments LabVIEW, allowed engineers to assemble graphical software objects into an automation sequence. These languages include instrument user interfaces, control objects, analysis objects, and driver libraries for connecting programs to a large number of instruments. Sequencing of measurements is accomplished through a form of flowcharting between objects.

Here is a typical sequence of automating with a graphical programming language (see Fig. 41.8):

1. The programmer develops a software "front panel" from a selection of graphical software controls and data displays.

2. To program the front panel, the user graphically selects icon-based functions and connects them together with lines, instead of writing textual programs. The functions can range from simple arithmetic operators to advanced acquisition and analysis routines. The execution order is determined by the flow of data between functions.

3. The programmer can choose to optimize the execution speed of the program by running compilers and optimization tools built into the ADE.

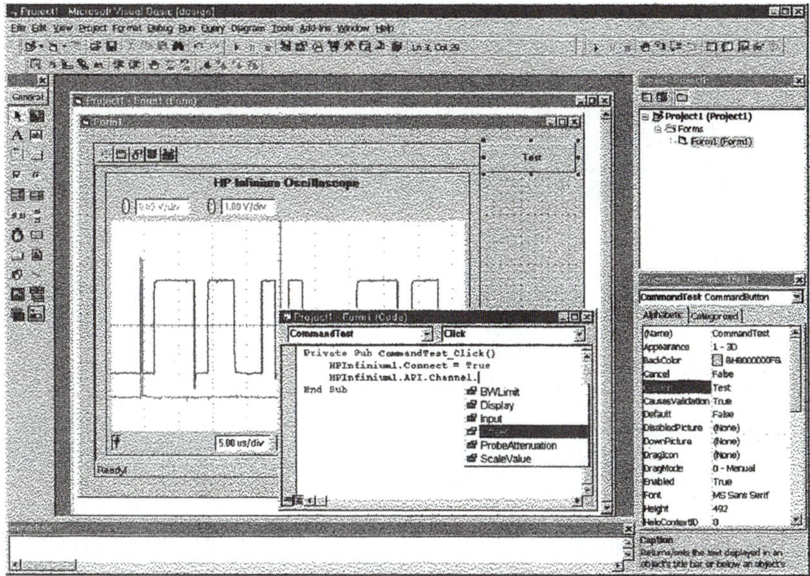

Figure 41.9 Visual BASIC programming with insertion of an ActiveX control for an oscilloscope.

4. The user returns to the front panel and begins the execution of the program by clicking on the appropriate "start" button.

Other programming languages are used, such as Microsoft Visual Basic. Visual Basic has many ease-of-programming benefits. Test and measurement-specific ActiveX controls are available, which can be used in conjunction with Visual Basic or other industry-standard programming languages. With Visual Basic (VB) (see Fig. 41.9), these ActiveX controls bring twofold benefits of:

1. making VB appear more graphical
2. making VB have more of a test and measurement "feel"

The process for developing an automation program with VB and ActiveX controls is virtually identical to the four-step process described for a graphical ADE. For example, the user creates a software front panel by placing ActiveX controls on a "form." Special editing tools are used to graphically manipulate the controls.

There is an important difference in step 2 of the process, however. Instead of connecting objects together graphically with lines, the user must use program commands in VB to achieve this. This has the disadvantage of being less graphical and slightly more complex. The VB commands are simple, how-

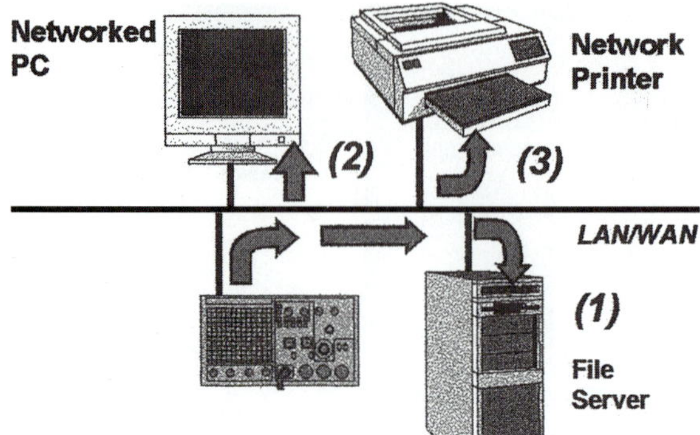

Figure 41.10 Example applications of a networked instrument: (1) Saving instrument data on a network file server, (2) directing instrument data to a networked PC, and (3) directing instrument output to a network printer.

ever, resembling the BASIC programming language that has been around for many years. Using VB also has the following advantages:

1. VB has more sophisticated functions available than graphical ADEs.
2. VB is more economical.
3. Because the user community for VB is much larger, a greater selection of training material and supplemental programming tools is available.

The combination of VB with test and measurement-specific ActiveX controls makes powerful automation capabilities readily available to the novice programmer.

Engineers who are willing to develop greater programming expertise and/or whose complex automation needs exceed even the capabilities of VB might prefer to use a purely textual programming language, such as C or C++, to achieve these tasks (see Fig. 41.10).

41.3.3 Network connection model

LAN/WAN connectivity for instruments makes additional use models and applications available to instrument users.

Accessing network resources. Once an instrument is on the network, other network resources become available. For example, file servers on the network can be used to store instrument data. The *Network File System* (*NFS*) is software technology that allows users to "mount" the file-server disk drive to the instrument so that it behaves just like the instrument's own local drive. Similarly, network-connected instruments are able to easily send data to other networked PCs for data display and/or storage.

Many or all of the printers available to the user can be network-based. Data from a network-connected instrument can be directed to the network printer for paper output.

Software for remote analysis/documentation or automation. ActiveX and COM/DCOM are based on a distributed software architecture. When these technologies are used for connectivity to analysis/documentation applications or an ADE, the same connection models described previously can be applied from a computer physically removed from the instrument. The engineer can be running an analysis/documentation application or automation program on a computer in his/her office or home, collecting data from a networked instrument based in a remote laboratory or production-floor environment.

41.4 Internet "Client-Server" Connection

Client-server is the dominant model for corporate computing. "Client" computers are usually PCs on users' desktops, connected through the corporate network. Local applications can be run independently on the PCs, but the PCs are also able to connect to server computers running larger server-based software applications. A small part of the application can be resident on the client PC, primarily as a user interface. Server computers provide other services to client computers, such as file sharing and network printing. Internet technology has further revolutionized client-server computing. Using a combination of Web-server software and Web-browser client software, a user has ready access to vast amounts of information and powerful applications. Using Internet technology in this way does not assume that the user is accessing the public Internet. The same Internet technology tools, however (e.g., browsers, Web servers, and Web pages), can be used internally to an enterprise as the foundation for client-server applications. Internet-based client-server computing further enhances the ways computer users interact remotely with instruments.

41.4.1 Instrument as internet server model

Under this model, the instrument "serves" up information about its status to a computer (Fig. 41.11).

One of the most important applications is remote monitoring of instrument parameters. This enables the engineer to check on the status of an instru-

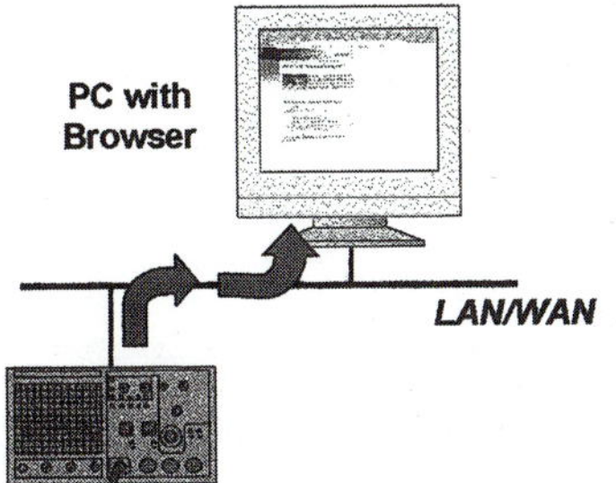

Figure 41.11 Instrument as Internet server. A remote browser on the PC can access web pages served up by the instrument. Web pages may contain documentation, access to "real-time" parameters and functions, hyperlinks to external web pages, etc.

ment that is physically distant. The laboratory or production line can be in another building, removed from the engineer's office. The engineer can set up a test and monitor it from home in the evening or over a weekend. This is especially useful, as the engineer might be waiting for an infrequent event to occur in the device under test. It might not be practical or desirable to stay and watch. The instrument can even be programmed to send a message via e-mail or to the user's pager to notify him of the event.

If the instrument comes equipped with a virtual front panel, as in Fig. 41.5, the engineer has access to all available instrument parameters for remote monitoring and control. The engineer can remotely view the virtual front panel in a Web browser. She can also exercise control of most of the instrument's functions. In many cases, control is not required, however. It might not even be desirable because hazards can occur from changing instrument conditions without an operator physically present. The number of instrument parameters required for successful monitoring might also be limited. In such cases, a full virtual front panel is not required. A simplified user interface, as in Fig. 41.6, might be sufficient. This interface might emulate a portion of the panel, as in the upper half of Fig. 41.6, or it might simply display instrument parameters in an HTML table, as in the lower half of Figure 42.6. HTML (Hypertext Markup Language) is the computer language understood by web browsers. The simplicity of set-up and the familiarity of the browser

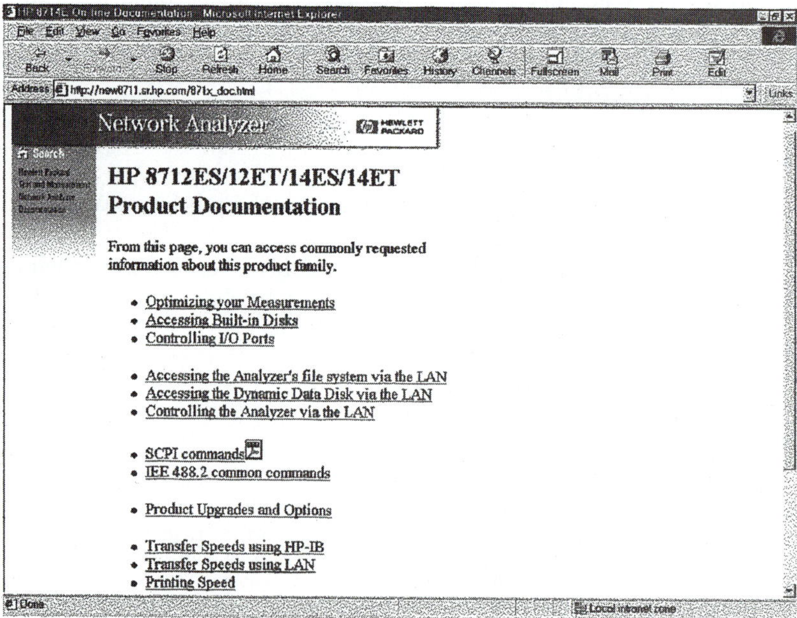

Figure 41.12 Product documentation from network analyzer, viewed in a PC-based browser.

user interface make using Internet technology attractive for remote monitoring applications.

The Internet server in the instrument may also serve up other useful information, including calibration and warranty status, updated regulatory information, and instrument documentation. Figure 41.12 shows an example of instrument documentation viewed from a browser.

The instrument as Internet server model enables still another testing and measurement application: sharing of an instrument by multiple users. Some instruments are quite expensive and, as result, might not be owned by an individual user. The instrument might be a shared resource. As such, scheduling the use of the instrument and segregating the data of different users could be problematic. Internet technology makes it straightforward to create separate areas for data and configuration information for different users. Internet security tools ensure that only authorized users have access to the information. Use of the instrument can even be scheduled by a calendar application running on the instrument, accessible to remote users of the instrument through an Internet browser.

41.4.2 Instrument as internet client model

Finally, the instrument may act as an Internet client. Under this model, the user may operate a browser directly from the instrument user interface. The model is useful for "downloading" information from a remote website. Exam-

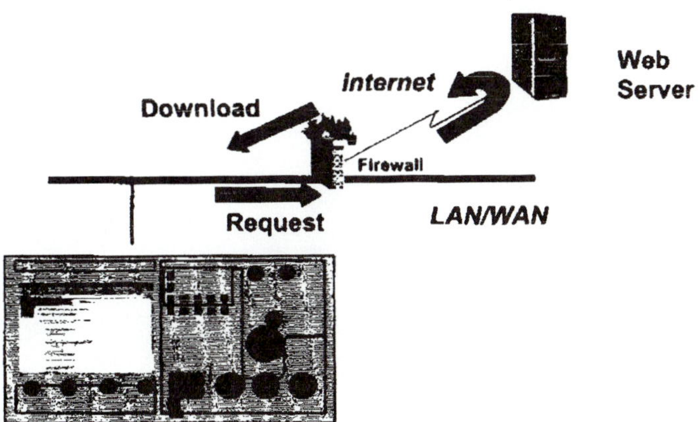

Figure 41.13 Instrument as Internet client. A web browser on the instrument is able to view pages from an external website, through the corporate "firewall," software which provides security by limiting corporate information that can be accessed externally. The user may request a download of information from the website, such as software updates for the instrument.

ples are downloading firmware updates, new software personalities for the instrument, and other add-on software capabilities, such as instrument drivers. In this way the user has access to many more software capabilities than are originally available in the instrument. This model is illustrated in Figure 41.13.

Chapter

42

Networked Connectivity for Instruments

Tim Tillson
Agilent Technologies
Loveland, Colorado

The development of digital technologies in instrumentation, along with the pervasive use of computers and networks, changed the way we think of and use both electronic instruments and measurements themselves. Major drivers, such as those listed, are having the effect of distributing the traditional notions of instrument and measurement.

- The growth of networks along with the numbers and kinds of devices attached to them.
- The doubling of computational power, roughly every 18 months ("Moore's Law"), affects both computers and the speed measurements can be made.
- The pervasive use of Windows operating systems and applications on the desktop.

Consider a scenario where many inexpensive temperature sensors are publishing real-time data on a network. A small computer might be "listening" to all the reports, deciding which ones are of interest, and reducing the data by some level. This information could be passed on to a larger personal computer, where an engineer is writing a program that interacts with a software interface on a still larger computer to further analyze the data. The output of this program might go to a large enterprise database somewhere else on the network.

In this example, the "instrument" is smeared across the dimensions of time and space, in both hardware and software. Likewise, the "measurement" is a hierarchy of increasing data abstractions. A person at the sensor level and a person at the enterprise data level are likely to think of the measurement differently.

These changes have introduced alternative models for instruments, which are covered in the next section. In these models, it is important to think of instruments from a point of view that is rooted in computers and networks, as servers of data. From this point of view, an instrument is a specialized computer, which can be thought of abstractly as a measurement server.

42.1 Instrument Models

42.1.1 Standalone model

Standalone instrument models involve an engineer or technician interacting directly with an instrument, performing setups, and making measurements, where both the device under test and the instrument are on the desk or workbench. Control is affected by direct manipulation of the (hard) front panel's switches, dials, rotary pulse generators, etc. Often, conclusions and decisions can be made immediately after viewing the instrument's output, for example, determining maximum and minimum values. Capture and preservation of data might involve writing entries in a notebook, printing results and pasting them in a notebook, or perhaps saving data on a floppy, where that floppy is hand carried to a computer, where further analysis could be performed.

42.1.2 Personal computer connected model

Instruments often have the ability to connect to personal computers via an I/O cable. This is a local connectivity model because the computer has a "hard-wired" (usually high performance) I/O connection to the instrument. Table 42.1 describes common personal computer I/O interfaces.

In this model, an operator or engineer might be interacting both with the instrument's hard front panel and the PC. Roles for the personal computer (PC) in testing and measurement are outlined in Table 42.2.

42.1.3 Network connected model

A professional working in a medium- to large-size corporation is almost certain to be in a networked computing environment, where devices of all kinds are interconnected by various kinds of I/O, and where data can flow onto a Local Area Network (LAN) to the wider corporate network (Intranet). Perhaps it will even be accessible from outside of the enterprise on the Internet. Figure 42.1 depicts a typical networked work environment.

Once an instrument is "on the network," either because it has a built-in network interface or because it is connected to a local PC, which is itself on the network, it can be an important addition to the other resources on the network, such as computers, printers, plotters, etc. It can:

- Be a measurement server that provides data either continuously or on demand to other resources on the network.

TABLE 42.1 Common Personal Computer I/O Interfaces

Interface	Description
IEEE 488.1, Hewlett-Packard Instrument Bus (HP-IB), General Purpose Instrument Bus (GPIB)	The IEEE 488 interface has been one of the primary instrument interfaces. It allows (electrically) for up to 15 devices up to 2 meters apart (each). It can achieve 1 Mbyte/sec transfer rates.
RS-232, RS-422A, RS-449, RS-530, CCITT V.24, IEEE 1174, MIL-188C	RS-232C is the venerable serial interface. Almost all computers have some form of RS-232 interface. An ongoing challenge with RS-232 is configuration, which is not defined. This configuration normally includes data rate, number of stop bits, connector size (9 pin, 15 pin), connector style (male or female), hardware handshake settings, which pin has transmitted data and which pin has received data (pin 2 and 3 problems). The other interfaces on the left are variations of the RS-232 standard with variations for data rates, configuration, and connectors. Data rates for RS-232 are around under 56 kbit; the variants can get above 10 Mbit/second.
Centronics	This is a standard-output (one-direction) parallel interface used for connecting computers and instruments to printers or other peripherals. It has also been used as an instrument interface (connecting the computer to the instrument). The original interface has been extended to the ECP (the Extended Capabilities Port) and EPP (the Enhanced Parallel Port). These extensions are both bidirectional interfaces that are about 10 times faster.
Universal Serial Bus	USB is an interface between a PC and devices (targeted at consumer peripherals). The USB peripheral bus is a multi-company industry standard. USB supports a data speed of 12 megabits per second.
FireWire (IEEE 1394)	FireWire (Apple's version) or IEEE 1394 High Performance Serial Bus is another interface between a PC and devices. It provides a single plug-and-socket connection, on which up to 63 devices can be attached with data-transfer speeds up to 400 Mbps (and eventually more). It is intended for high-performance devices.
Ethernet or LAN (IEEE 802.3)	Ethernet is a local-area network (LAN) protocol and hardware specification developed in 1976 for communication between computers. It supports transfer rates of 10 Mbps. Although originally on medium to large computers, it is available on PCs and is also implemented in several instruments. 100Base-T is one of several variations of Ethernet that happens to support 100 Mbps.

- Be a remotely controlled device.
- Be a self-aware device that is proactively interacting with other resources on the network: triggering alerts, uploading status, sending mail, and assuming new capabilities and personalities as it downloads new functions from remote computers.

42.2 Introduction to Computer Networks

Networks have grown from being the backbone of computing (which has engendered marketing slogans such as "The Network is the Computer") to being

TABLE 42.2 Roles for the Personal Computer

Role of the PC	Comment
Data Capture	Numeric and visual measurements, represented as binary or ASCII data, can be stored as files.
Operator Interface	Some of the capabilities of the instrument are available via a visual interface on the PC. This can take the form of a literal representation of the instrument's front panel, where every button and knob has a corresponding visual element on the screen, or as a more abstract operator interface, where the instrument's function is captured in a hierarchy of menu picks and dialog boxes. The interface might follow the standard look and feel conventions of Windows applications. In fact, it might offer the operator *more* functionality than is available through the instrument's front panel.
Programmatic Interface	The PC can run a program that configures and controls the instrument and captures measurements at appropriate times. The program might be using Standard Commands for Programmable Instruments (SCPI) or a language proprietary to the instrument to control its function.
Network Interface	Most computers are inherently more useful if they are connected to a network, as opposed to being islands of computing resource. The network connection allows transmission and receipt of data from other machines on the network, ability to execute programs downloaded from the network, or the possibility of collaborating in the execution of a larger distributed program that might be running on many networked machines simultaneously.

the fundamental enabler for almost all information workers. Modern offices and laboratories grind to a halt when the network goes down. Networks are so pervasive that they are changing the paradigms of work itself: consider how much office work involves either sending, receiving, or processing data from a network. In this context, it is not surprising that instruments are more useful if they are measurement servers (and clients) on a network.

To set the stage for information concerning various models for networked instruments, it is helpful to better understand computer networks themselves.

42.2.1 Circuit versus packet switched networks

A computer network implements agreements about how data is sent and received between all the devices that are connected to it. The telephone system is probably the most familiar network to most people. Its fundamental characteristic is that it is a circuit-switched network; once a call is connected, the user has a dedicated circuit from point A to point B for the duration of the call. Because the circuit is dedicated, it is possible to transmit silence (which can be very important information, depending on the nature of the conversation!). Computer networks, on the other hand, are packet switched. Information (including possibly digitized voice, music, or video) is broken up into small packets of binary data, each of which could conceivably take a separate route from its source to destination! At the receiving end, all packets that are

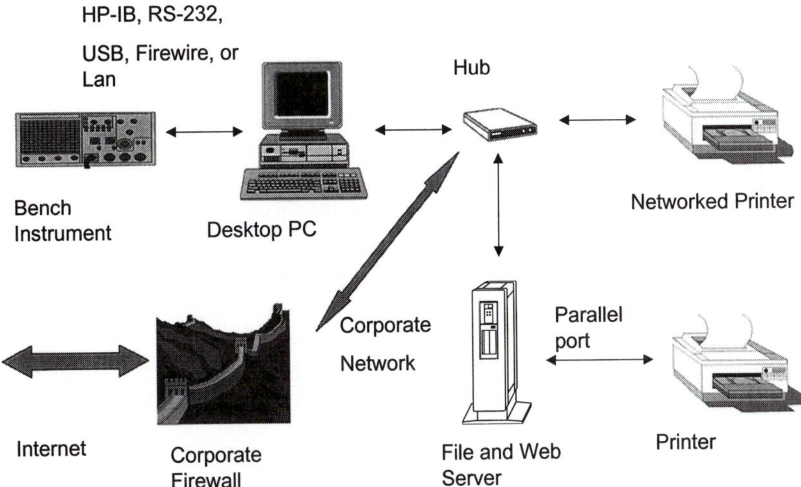

Figure 42.1 In the model shown, data flows from a bench instrument to a locally connected PC for analysis. The user might decide to print some data on the networked printer and/or forward it to the file and Web server. The data, in summary form, can be published on the file and Web server for access by others in the department or company. Remote control of the instrument can be enabled through a Web page on the server. With appropriate permissions, the file and Web server can also be accessed through the corporate firewall, and over the Internet, which can also facilitate remote support by the instrument vendor. If desired, the data could be made available in an unrestricted fashion to the general public on the Internet.

part of this "conversation" must be ordered by sending time and reassembled to reconstruct the message.

The difference between these two types of networks is illustrated by how access to an instrument via a modem on a dedicated phone line can differ from that on a local area network. In the former case, the user is only limited by the bandwidth of the modem connection and will have deterministic performance, subject to that constraint. In the latter, performance might be nondeterministic because of other traffic on the network, and the user will have to use the more complex network protocols covered in the following section.

42.2.2 Open Systems Interconnect (OSI) network protocol model

The International Standards Organization has developed the *Open Systems Interconnect* (*OSI*) model to facilitate standards to enable the interconnection of various computing devices. It is often referred to as the *Seven-Layer Model,* as shown in Table 42.3. At every layer, the OSI model specifies the rules that devices must follow to successfully talk to other devices on the network. These protocols, which are agreed-upon methods by which devices communi-

TABLE 42.3 The OSI Seven-Layer Model

Level	Layer	Comment
7	Application	Network services for a specific application (e.g., file transfer between systems and electronic mail).
6	Presentation	Syntax and semantics of information, including encryption and decryption.
5	Session	Establishes sessions on different machines and provides synchronization and mutual exclusion services for programs that must run to conclusion.
4	Transport	Integrity of the data from the origin to the termination of the link.
3	Network	Switching and routing of information and the establishment of logical associations between local and remote devices.
2	Data Link	Responsible for an error-free connection between network elements.
1	Physical	Transmission of bits over the physical communications link.

cate, are implemented in software or firmware, depending on the layer in question. It's important to realize the complexity of even a simple operation, such as sending electronic mail from one computer to another over the network. Figure 42.2 shows how a piece of mail starts at the application layer, is processed step by step through the presentation and lower layers until it is a stream of packets. These packets travel to the physical layer of its desti-

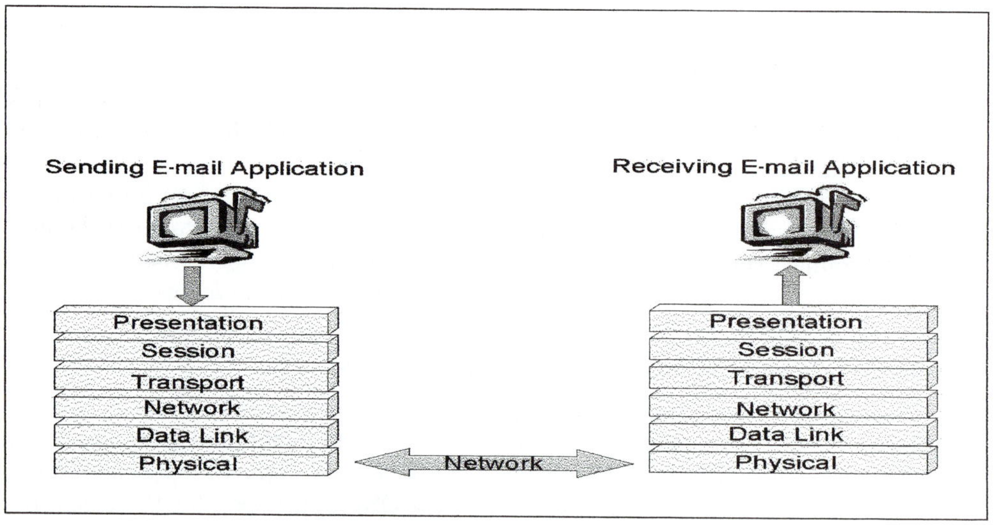

Figure 42.2 Electronic mail passes through many layers of processing described by the Open Systems Interconnect model from its source to destination.

nation computer, and are processed upward by the various protocols defined by the OSI layers until they are received by the e-mail application on the destination computer.

42.2.3 Computer networks and the Internet

Development of the Internet. The U.S. Advanced Research Projects Agency (ARPA) is generally regarded as the developer of the Internet, when, in 1969, lines were installed to link it to the sites of military suppliers. The ARPAnet grew by linking to other networks, including USENET (a network for users of the Unix operating system), BITNET in 1981, and NFSNET, created by the National Science Foundation in 1986 to link five super computer sites in the U.S. In 1991, the Commercial Internet Exchange was tied in after the NSF allowed commercial use of the network.

Browsers. The growth of ARPAnet and the lifting of noncommercial restrictions set the stage for the explosive growth of the Internet. But early users of the Internet were computer professionals, often Unix programmers, who were not put off by typing long, cryptic commands. What was needed was an ease-of-use breakthrough. This was provided by the first *browser,* a program that could navigate links between computer files, regardless of type or location, by clicking a mouse. The first browser became publicly available in 1994; this spark triggered the Internet explosion.

Firewalls. The Internet is often thought of as a "network of networks" that spans the planet. One of the strengths of the Internet is that it is largely unregulated by governments; in most countries, a company or individual can publish almost any material it likes. Precisely this open nature requires most corporations with connections to the Internet to erect firewalls, which are systems that comprise the connection point between the internal corporate network and the external Internet. Firewalls are configured to block certain traffic coming in from the outside, and to allow certain types from the inside to the outside. Typically, e-mail traffic and public Web site access is allowed inward, but logins on corporate systems from the outside are blocked.

42.3 Network Protocols and Software Standards

The success of corporate intranets and the Internet is because of the degree of protocol standardization that has been achieved. Users take it for granted that e-mail with attachments (such as embedded spreadsheets, applications, or pictures) will arrive successfully at any destination for which they have a valid e-mail address. This is truly remarkable when considering all of the other activity simultaneously occurring on the Internet, and numerous national boundaries that the e-mail might have to cross! The protocol standards that help make this happen are as follows:

- Internet Protocol (IP)
- Transport Control Protocol over Internet Protocol (TCP/IP)
- Sockets
- Telenet
- File Transfer Protocol (FTP)
- Networked File System (FTS)
- HyperText Transfer Protocol (HTTP)
- Universal Resource Locator (URL)
- VXI-11
- JetSend

42.3.1 IP (Internet Protocol)

The *Internet Protocol* (*IP*) layer is responsible for getting packets of data to a physical network interface device, which involves mapping the IP address in the packet header to a physical address. Devices that receive an IP packet forward it on to the next closer device to the destination specified by its address. At this level, the protocol is said to be *unreliable* because packets can be discarded if problems occur.

42.3.2 TCP/IP (Transport Control Protocol over Internet Protocol)

The TCP layer provides a connection from source to a destination, which looks like a sequential stream of characters to the programmer. As such, it has to deal with packet dropouts, as well as potential reordering, based on time stamps.

Most commercial real-time operating systems include "networking stacks," which implement the various layers of protocol so that the embedded system can be a networked device. These stacks typically include some of the applications covered in the following section, such as Telnet and FTP, and are usually additional-cost items. They are also almost always present in Unix and Windows operating systems.

42.3.3 Sockets

A *socket* is a pipe that works over a network, where a pipe is the operating system's way of connecting the output of one program to the input of another. This could be done by the first program writing a temporary file and the second program reading it; however, Windows and Unix provide pipes built into the operating system, which eliminate the need for temporary files and are more efficient. Using sockets, a program on one computer can communicate with a program on another computer on the LAN much as two programs on the same computer-use pipes.

42.3.4 Telnet

Telnet is a TCP/IP application that allows a user on a host computer to log into a remote computer or device as if he or she were on a terminal attached to that computer or device. A Telnet user is either in command mode, issuing commands that affect the Telnet session, or input mode, where the characters go straight into the input processing of the device.

42.3.5 FTP (File Transfer Protocol)

FTP is an application on top of TCP/IP, which allows a user to interactively perform remote file operations, such as sending or receiving text or binary files, deleting, and renaming them. It is a TCP/IP application in the "client-server" model, where the user is the client and the remote computer addressed by FTP is the server. FTP works by creating two connections, one for commands and one for data.

42.3.6 NFS (Networked File System)

NFS enables mounting a remote file system, making it appear to be a local disk drive. Thus, a client node is able to perform transparent file access over the network. By using NFS, a client node operates on files that reside on a variety of servers and server architectures, and across a variety of operating systems. File access calls on the client (such as read requests) are converted to NFS protocol requests and sent to the server system over the network. The server receives the request, performs the actual file system operation, and sends a response back to the client.

42.3.7 HTTP (HyperText Transfer Protocol)

The HyperText Transfer Protocol is a mechanism designed to allow and promote file sharing. It does not require a lot of code to implement, and is fast enough to support information sharing among workgroups, which might be widely distributed across a network. It is a generic, stateless protocol, which can be extended by adding methods and data representations.

On the Internet, HTTP is run over TCP/IP. An HTTP session starts with a connection by a client to an HTTP server; the client sends request messages to which the server responds. Then the connection is closed by either or both participants. The protocol in detail defines the formats for requests and responses.

42.3.8 URL (Uniform Resource Locator)

A client request is typically in the form of an address in string form called a *Uniform Resource Locator* (*URL*). It defines the location of a page the client would like to access. An example is **http: // www.hp.com,** the URL for the Hewlett-Packard corporation. Note that the first part of the URL through the // denotes a protocol such as HTTP. Other protocols are allowed in URLs, such as **ftp:** //.

42.3.9 VXI-11

VXI-11 is a testing and measurement industry standard that allows ASCII-based communications between a controller and a device over the LAN, using IEEE 488 (GPIB) style access.

42.3.10 JetSend

JetSend is a device-to-device communications protocol developed by Hewlett-Packard, which allows devices to negotiate information exchange intelligently. The protocol allows two devices to connect, negotiate the best possible data type, provide device status, and exchange information, without user intervention. JetSend acts as on-board intelligence, allowing devices to communicate directly with one another without the need for a server or detailed information (such as a device driver) to communicate with another JetSend device or PC. A standardized device-to-device protocol helps to ensure the interoperability of devices from diverse manufacturers. When JetSend is designed into instruments, it enables them to communicate with network devices, such as printers, without an intervening PC.

42.4 Network Connectivity Models

By using one or more of the following models, an instrument can expand its role from measurement device to measurement server.

42.4.1 Instrument as file server

Some network-connected instruments support and document FTP as a way to get data (in the form of files) out of the instrument. From a terminal window on a computer, which can address the desired instrument, a typical user session might look like the FTP session outlined in Table 42.4.

42.4.2 Time capture

Some instruments have an important feature called *time capture.* It allows raw data from the instrument's ADC and DSP hardware to be stored, along with current calibration settings, before processing by the instrument's DSP software. These files are often many megabytes in size, but they can be

TABLE 42.4 An FTP Session. This Sequence of Commands Uploads a Data File from the Instrument to the Current Directory on the Operator's Computer

Operating system command	FTP command	Comment
ftp <address of instrument>		Invoke the ftp program
	bin	Put ftp in binary mode so that all bytes will be uploaded without change
	cd <file system path to desired data file>	Change ftp to the desired directory
	get <name of data file>	Tell ftp to upload the file
	quit	End this ftp session

quickly and easily transported via LAN to a computer for storage. Once there, they can be loaded back onto the same or different instruments for analysis at a different time.[1]

42.4.3 Instrument as remote file system

The Network File Server protocol allows a file system on a computer or device to become visible to a remote computer, as if it were local to that computer. On Unix machines, the entire file system appears as a hierarchical tree with the root directory at the top (see Fig. 42.3). The "mount" operating system command is used to attach a remote file system somewhere in the local file hierarchy. On Windows machines, local hard disk drives are assigned letters, such as C. Remote file systems can also be assigned drive letters (usually later in the alphabet, such as J), making them appear as if they were additional hard drives on the local machine. Users can mount the disk drive in their instrument on their PC using this mechanism. Data appears in the form of files, which can be read or copied onto any computer on the network.

42.4.4 Remotely controllable instruments

I/O standards. Products based on I/O standards, such as the *Standard Instrument Control Library* (*SICL*) and *Virtual Instrument Software Architecture* (*VISA*) enable programmers to control instruments from PCs. The programmer uses the libraries to open a communications channel to an instrument, and then uses languages, such as SCPI, to control instrument behavior.

---

[1] Developed from "LAN in an Instrument: Two Years of Practical Use," HP internal document, by Bob Cutler.

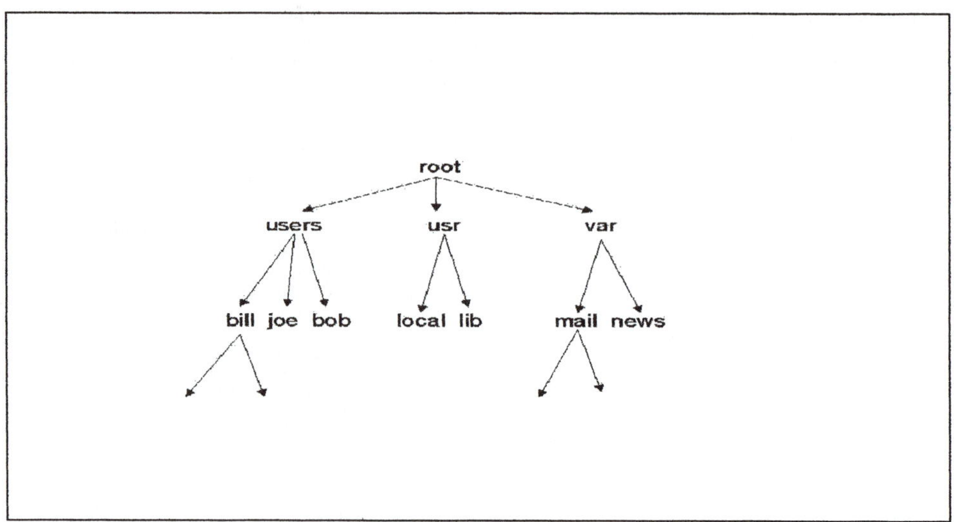

Figure 42.3 The file system on Unix operating systems is a tree or logical hierarchy, which has the root directory (/) at the top.

Portal box devices. Portal box devices have been used to put instruments on the LAN—even if they did not have LAN drivers/connection built in. A portal box will have both a LAN connection and a local I/O connection, such as GP-IB or RS-232. Because a portal box is a device on the LAN, it has its own Internet Protocol address, which can be assigned by the user.

A remote user can then Telnet to the portal box, which creates a session, wherein commands to the portal can be issued. One command will cause entry into the instrument's command loop, at which point the remote operator or programmer can begin issuing ASCII commands to the instrument itself.

Embedded LAN drivers. One major technique to enable remote control of instrumentation (either by a remote operator or program) is to embed LAN driver software in instrument firmware itself.

Some products built on the Unix operating system already have a LAN connection. Figure 42.4 is a block architecture diagram that illustrates various methods by which this can be achieved. A programmer might use testing and measurement graphical programming language environments, such as HP VEE or LabView from National Instruments, or use a general computer program development environment, such as Microsoft Developer Studio, to create a controller program. This program will issue SCPI commands that flow through I/O libraries or directly into a socket connection to get onto the network. A live operator could control the instrument by issuing SCPI commands in a terminal window (Telnet session).

On the instrument side of the connection, LAN driver code will typically provide multiple interfaces to the network, such as support for the VXI-11

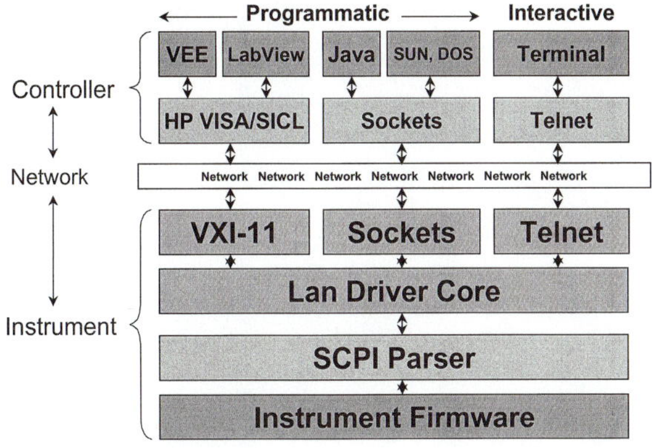

Figure 42.4 LAN driver software in instruments enables remote clients who might be programs or live operators.

standard, in addition to direct socket and Telnet interfaces for the interactive case. The LAN driver gets the SCPI commands into the SCPI parser loop of the firmware, which interprets the ASCII commands into calls to the actual firmware routines that make the measurement.

Often, significant performance gains can be realized by using socket connections, as opposed to VXI-144. In general command, query, and read and write operations using sockets are about twice as fast as those using VXI-144.

From a use-model point of view, LAN drivers supported in the instrument firmware provide full remote transparency to a programmer in environments, such as VEE from Hewlett-Packard, LabView from National Instruments, or Developer Studio. Instead of requiring the controlled instrument to have a direct connection to the PC where these environments are running, the instrument can be on the LAN anywhere in the world, but look to these environments as if they were local.

42.4.5 Instrument as web server

Once an instrument is connected to the LAN, it is natural to want the same control and access capabilities from anywhere on the LAN that one has when sitting in front of the device. Given that almost all PC users are familiar with navigating the Internet using browsers, Web pages make for a very convenient, easy-to-use remote user interface to networked instruments. The code that makes Web pages available for browsing is called a *Web* or *HTTP server.*

Web servers are not necessarily large programs. Code for HTTP servers written in ANSI C is in the public domain; it can be found using Web search engines, such as AltaVista, Yahoo, etc. HTTP 44.1 servers are small, on the order of perhaps 10 pages of source code and less than 10 Kb of binary code.

One important factor for instrument developers in embedding an HTTP

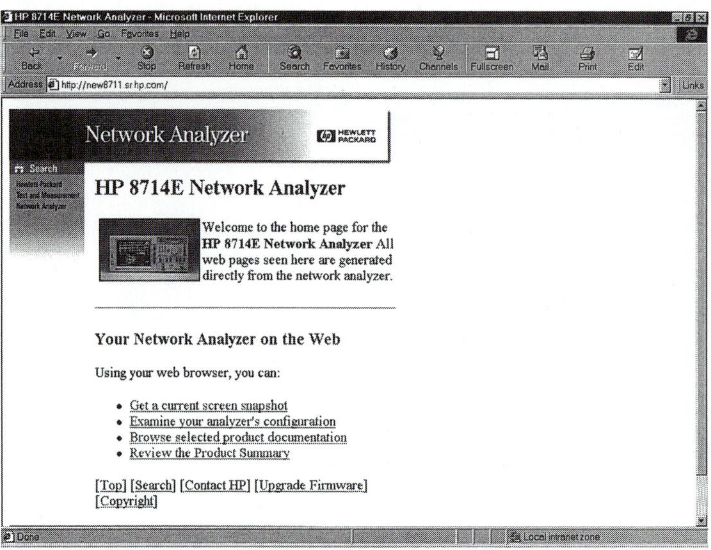

Figure 42.5 Home page of a network analyzer with an embedded Web server, as displayed in a browser. From a browser, the user can get a current screen snapshot, examine configuration information, browse selected product documentation, and read a product summary.

server is whether or not that instrument supports a file system internally. The HTTP protocol assumes a file system. In an embedded situation where a file system does not exist, URLs must be mapped to method or function calls in the firmware, rather than to locations of files that contain HTTP commands.

Instrument Web interfaces typically contain the following elements:

- *Home page.* Figure 42.5 shows the home page of a Network Analyzer. Note the product number and type are very prominent. Links are provided to search, not only for other pages on this instrument, but also to search central Web sites provided by the instrument vendor. These links can be very helpful in assisting customers find more information than is shipped with the product itself, such as related products, etc.
- *CRT screen access.* Here are a wide range of possibilities. A static snapshot of the screen can be taken at the time a user browses a hyperlink. This is easy to provide by writing image files, but will not satisfy applications or users that want an approximation of real-time screen updates. Real-time CRT updates over the LAN, in general, are not possible, unless the LAN is private, with bounds on other competing traffic. Even for approximate real time, care must be taken in transmitting visual data over a network (sending all the pixels on a continuous basis is much too inefficient). Schemes where the client code has some inherent graphics capability and can interpret changes from the last picture frame are more effective and are much more compact in their demands on network bandwidth.

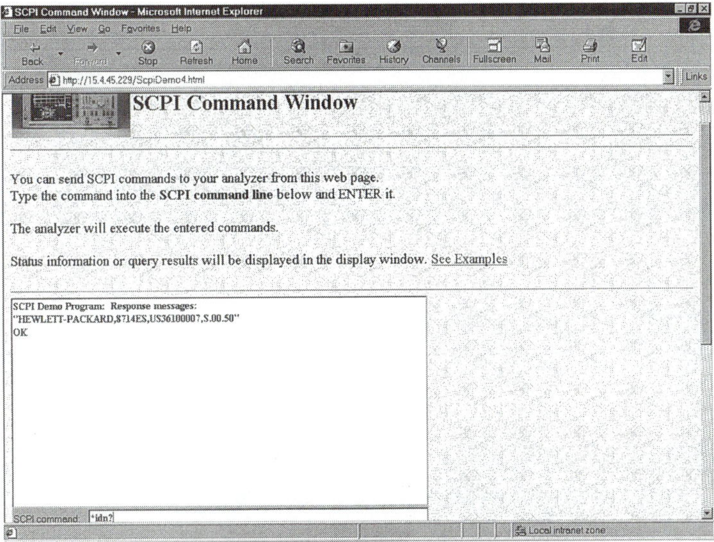

Figure 42.6 If an instrument supports SCPI, remote control and monitoring can easily be performed via a command window embedded in a Web page.

- *Configuration.* Information (such as firmware and boot ROM versions, serial number, installed options, types of memory installed, disk size, and other instrument parameters) are easy to provide and often very helpful, particularly when a customer calls in for support. Just knowing whether or not an instrument is under warranty can often save many minutes or hours of research time on the part of the vendor.
- *Documentation.* Inclusion of product manuals, help files, and product summaries makes the product more usable. It is also easy to include links on the documentation pages to further documentation and help that is resident on other servers on the network, providing the user with a (hopefully) seamless flow from information provided by the instrument to more information provided by the instrument's vendor.
- *Remote control via SCPI.* Figure 42.6 shows a Web page with an embedded SCPI command window. The user can send SCPI commands to the instrument from this window, with automatic logging and history of the replies.
- *Complete remote display.* Some instruments implicitly contain entire windowing/display systems, such as products which are built on a Microsoft Windows 98 or NT operating system framework, or those built on an X11/Unix operating system framework. In these cases, code can be added to the instrument, which, from the instrument's point of view, "looks" like an expected or added display, but actually remotes the entire display over the network to client code running in a browser. This provides the (remote) user with

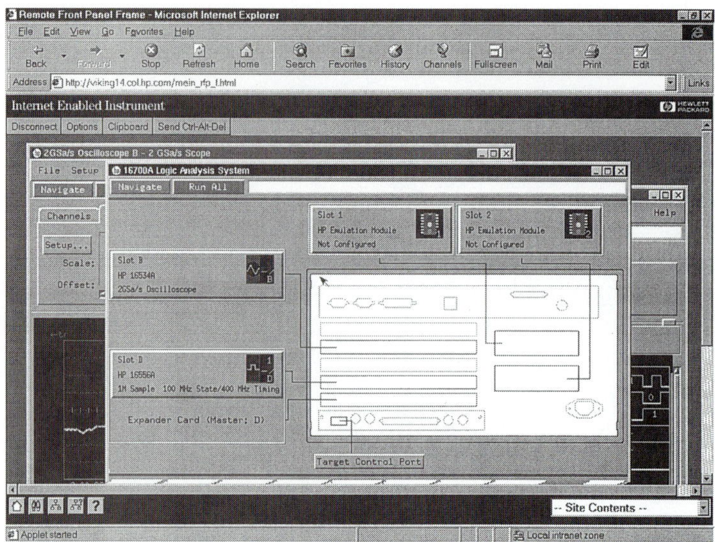

Figure 42.7 When instruments are built on top of the Unix or Windows operating systems, it is possible to add servers that can remotely display their multi-windowed interfaces in a browser.

the same level of control over the product as if he were sitting in front of it. Figure 42.7 shows a native Unix/X Windows Logic Analyzer instrument interface that is completely accessible through a browser. This instrument can be controlled from a browser anywhere in the world, as long as the user has permission.

- *Data export.* Mechanisms exist for easily getting data produced by the instrument exported into files with a common format (text, spreadsheet) on a remote machine, into desktop facilities (such as cut buffers), or directly into desktop applications, such as Microsoft Word or Excel. One way to do this is via dynamic data disk URLs. The user is presented with a page of possible files of different types, which capture appropriate data. A click on a particular file causes data to be written to it in a type-appropriate format at that moment. The files can then be copied using FTP to the client computer.
- *Server for ActiveX/COM components.* Web pages can be used to conveniently install COM and/or ActiveX components (defined in the following section), which then run inside a user's browser or in other ActiveX containers, such as Microsoft Word or Excel.

When choosing an instrument that will need to have remote access and control facilities, look for:

- Completeness of remote displays, including facilities for multiple users separated across the LAN to view a display simultaneously, and collaborate on interpreting a measurement.

- Site maps that can identify all instrumentation in a lab on the LAN.
- Built-in capabilities for remote support.
- Demo programs and code samples.
- Usage and command histories, including searching and indexing capabilities.
- Personal user workspaces, allowing a user to operate in an environment that is unique and customized to her needs. This is analogous to each user on a computer having a login that brings up a customized desktop and tools.
- Lab notebooks, which allow a user to paste in setups, configurations, screen savers, notes, probes used, etc.

42.4.6 Instrument as distributed software objects: Microsoft Common Object Model and ActiveX

Microsoft, through its Common Object Model (COM) and ActiveX standards (defined in the following section), has created what will likely become the next major local connectivity model for instruments. It is important to understand the local case first because it turns out that COM is an inherently distributed model, able to be used remotely over a network.

Microsoft stabilized the core specification for COM in 1993. It is best thought of as a way to package independently developed pieces of software so that they can interact in a consistent, predictable, object-oriented fashion.

COM objects are defined by a binary specification in a language-independent way. The specification defines how objects communicate, not what they do. Popular computer languages for implementing COM objects include C++, Java, Visual Basic, and Delphi.

Common Object Model (COM). COM brings the software world much closer to hardware paradigms. A software COM component can now be treated much like an integrated circuit, that is, it can be bought out of a catalog (on the World Wide Web!) and relied upon to work correctly in a larger program, where it interacts with other COM components from other suppliers. COM solves the fundamental problem of integrating diverse pieces of software that were implemented at different times by different suppliers in different languages. Independent software vendors can build their "software ICs" in any language they wish and sell them piecemeal without requiring central coordination. Software Developers can make intelligent build versus buy decisions. Large products can come to market faster because they are built from trusted, tested components.

With Windows NT 4.0 in 1996, Microsoft introduced Distributed COM (DCOM). A program or application running on a computer typically consists of a number of COM objects, which are communicating among themselves. Some of those objects might reside in the program's address space, but others might be outside of the program (in fact, be shared by multiple programs),

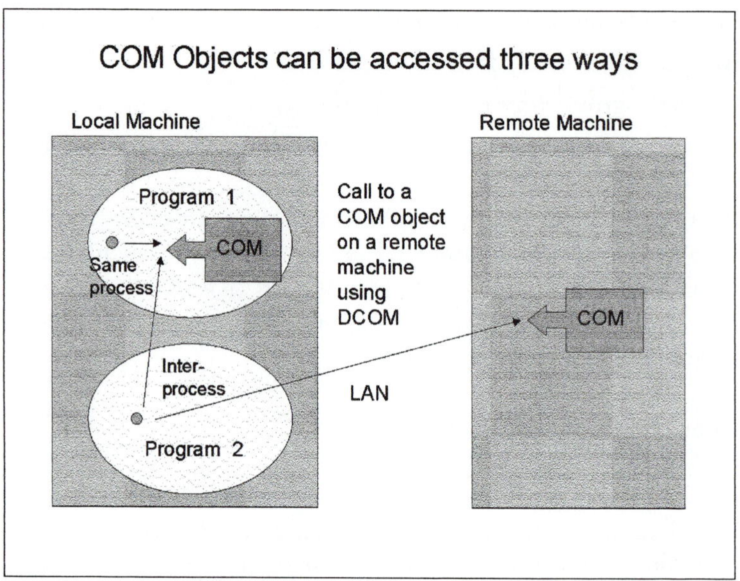

Figure 42.8 A COM object can be called by a client in the same process (program), by a client in a different program on the same computer, or by a client that is on a remote computer. The COM object infrastructure makes the COM object appear to be local to the caller's program, even when it is not.

but running on the same computer as the application. DCOM completed the architectural goal of location transparency for COM objects by allowing a COM object to appear as if it is running on the same computer as the application, but, in fact, be remote, running on a different computer accessible to the first over the network. Figure 42.8 shows these three cases. A COM object can be called by a client:

1. In the same process (program)
2. In a different program on the same computer
3. On a remote computer

The COM object infrastructure makes the COM object appear to be local to the caller's program, even when it is not.

DCOM solves an enormous set of problems for the computer industry. Its largest limitation is that COM objects require layers of Windows software beneath them to run.

Thus, they are language independent, but not platform independent, as contrasted with the goals of Java and Java Beans, covered in the following sections.

ActiveX. *ActiveX* is another layer of component standardization that Microsoft has defined on top of COM (see Fig. 42.9). They can be thought of as

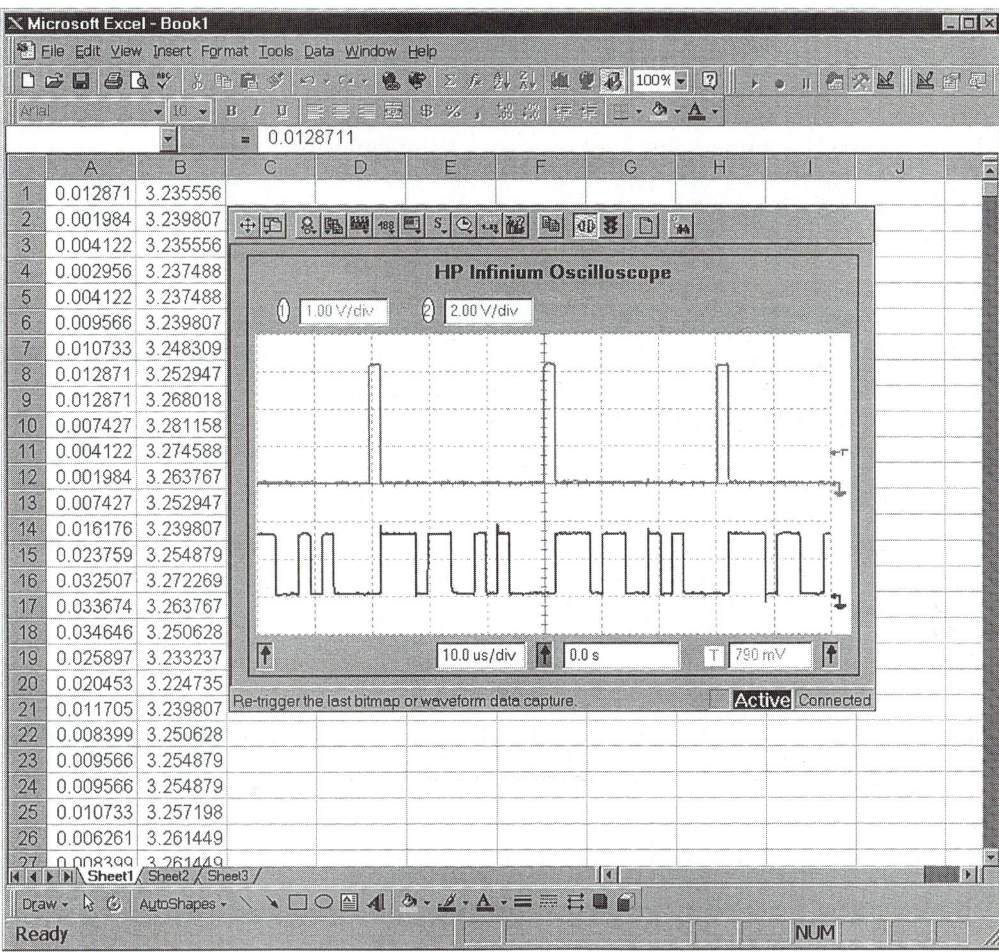

Figure 42.9 Microsoft Excel contains an ActiveX control, which is an operator interface to an oscilloscope. Not only are the traces for the selected input channels visible, but numerical trace data can be exported directly into Excel, as shown in columns A and B.

downloadable components (called *controls*) with additional capabilities, including a design-time user interface. They also have built-in persistence, which saves the state of the object, such as configuration information and user-defined settings.

Most Windows applications (such as Microsoft Excel or Word) are themselves built out of COM objects and ActiveX controls. They can also be containers that hold independently developed ActiveX controls. Likewise, modern software development environments, such as Microsoft Developer Studio or Delphi, allow programmers to use ActiveX controls provided by third parties in the user interfaces of the programs that they are developing. These development environments are also ActiveX containers.

This ActiveX control is running inside Excel as a container, and uses I/O libraries to communicate with the instrument over the network. If the instrument had a "hard" I/O connection to the PC using GP-IB or IEEE 1394, the behavior of the control would be the same. The flexibility of COM objects to behave the same, whether they are in the same program, on the same machine shared by multiple programs, or on a remote machine, but appearing to be local, is the key to seamless flow between local and network-connected use models.

42.4.7 Instrument as Java applet versus COM/ActiveX

A Java applet is a Java program that can be included in an *Hypertext Markup Language* (*HTML*) page, much like images are put on the page by inclusion of image files. When a page containing a Java applet is accessed by a Java-capable browser, the applet code is transparently downloaded and executed by the browser. For security reasons, Java applets have limitations, compared to standalone Java applications, which do not require a browser to run. For example, generic Java applets do not have access to the local file system at all.

In principle, instrument interfaces could be built as Java applets with comparable appearance and behavior to ActiveX controls. One advantage of this approach is that Java applets would run on any Java-capable browser, whether the browser were running on a Windows, *Real-Time Operating System* (*RTOS*), palmtop, Unix, or future OS platform. ActiveX controls, on the other hand, require 32-bit Windows run-time support layers (typically only found in Windows operating systems) to run. A disadvantage to the applet approach is that Windows environments tend to provide more robust user interface components.

42.4.8 Downloadable personalities/user programmable instruments

Many instruments support some level of user programmability, sometimes with *Instrument Basic* (*IBASIC*). Higher-end products support capability upgrades, often via floppy disk. Markets where upgrade capabilities are particularly important include telecommunications, where protocol analyzers of various types often support new protocols via this mechanism.

The combination of the Java programming language and the virtual computer that executes it (called the *Java Virtual Machine, JVM*) offers a platform where downloadable personalities are easier to access, and faster and more convenient to install. Java as a user instrument programming language offers many improvements over IBASIC, providing the programmer with a full object-oriented language, which, among other things, does its own memory management, freeing the programmer from the details of memory allocation and de-allocation.

Figure 42.10 shows an internal instrument architecture that supports both user programming in Java and more convenient and dynamic downloadable personality features. The instrument needs to be LAN capable, with LAN

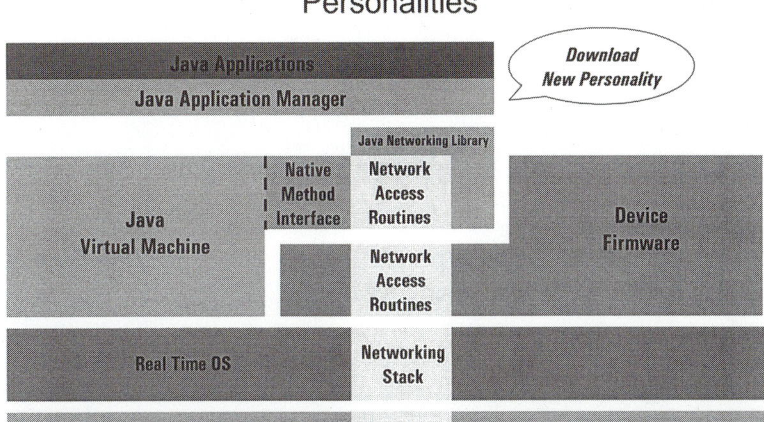

Figure 42.10 A Java instrument architecture to support downloadable personalities.

hardware and running a Real-Time Operating System, such as PSOS+ or VxWorks, which supports a TCP/IP networking stack.

The Java Virtual Machine runs on top of the RTOS. A typical Java Virtual Machine might be 500K bytes in size, and require two mega-bytes of RAM. The Java run-time environment needs to include the Java networking class library which makes calls on the RTOS networking stack. This enables Java applications to access the network; one possible Java application might be an HTTP/Web server, which provides capabilities such as described in the *Instrument as Web Server* section.

A Java Application Manager layer facilitates how new functionality is downloaded to the device and how it is managed once it is running in the device. The new functionality would be implemented in the form of Java class files, which are Java sources compiled into Java byte codes. The user might browse to a corporate Web page that contains lists of new personalities or functionality upgrades, then pick (and pay for) the ones he desires. The Application Manager layer would download the bits, ensure that currently running applications are stopped gracefully, load the new classes dynamically into the Java application environment, and restart. It also would ensure that this is done, given the memory constraints of the device.[2]

[2] From "Patent Application for Java Application Manager," Hewlett-Packard, Frank Judge and Chia-Chu Dorland, May 6, 1998.

Chapter

43

Computer Connectivity for Instruments

Joseph E. Mueller
Agilent Technologies
Loveland, Colorado

43.1 Introduction

In many test and measurement applications, it is important to connect instruments to computers. The applications vary from complex, high-speed automation to simply transferring measurement data from the instrument into an application on the computer for analysis or presentation. The actual implementation that provides the connection is composed of:

- The physical I/O connection
- The I/O library that runs on the computer
- Instrument interface software on the computer (also known as *instrument drivers*)

In addition to these implementation layers, a key part of the total system is the actual messages that are transmitted between an instrument and computer. Figure 43.1 shows the implementation layers for I/O on the computer, and Fig. 43.2 shows a typical instrument-computer interchange. This

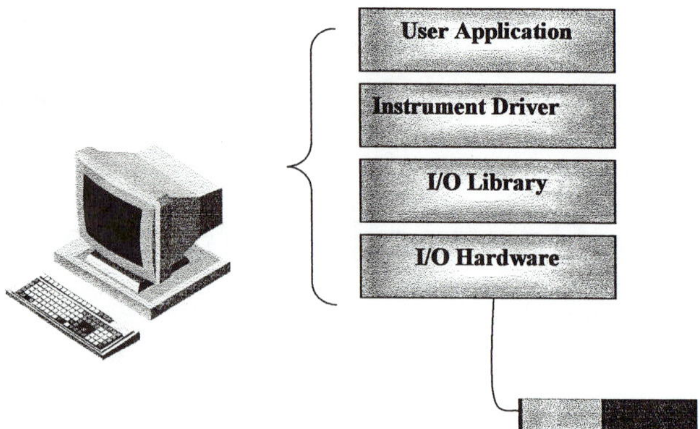

Figure 43.1 Implementation layers for I/O on the computer.

chapter considers each of these aspects of the computer to instrument connection.

43.1.1 Overview

Originally, instruments were stand-alone devices that helped engineers design and test their products. With the advent of low-cost computers in the '70s, people began to look for ways to automate their measurement tasks by connecting instruments to computers. This was particularly important in manufacturing, where product testing required lots of repetition and the time required to do the test was a key part of the manufacturing cost of the product. Automation was also crucial where the test might require many hours or weeks, for instance, when monitoring the processes in a facility or verifying an extremely complex product, such as a satellite.

43.1.2 IEEE 488

As these needs grew several ways of connecting instruments to computers evolved. In 1975, the IEEE adopted IEEE 488, a high-speed bus that can be used to control and read values back from instrumentation. IEEE 488 is still a popular way to connect instruments to computers because it is high-speed, reliable, and supported in a consistent manner by many instrument manufac-

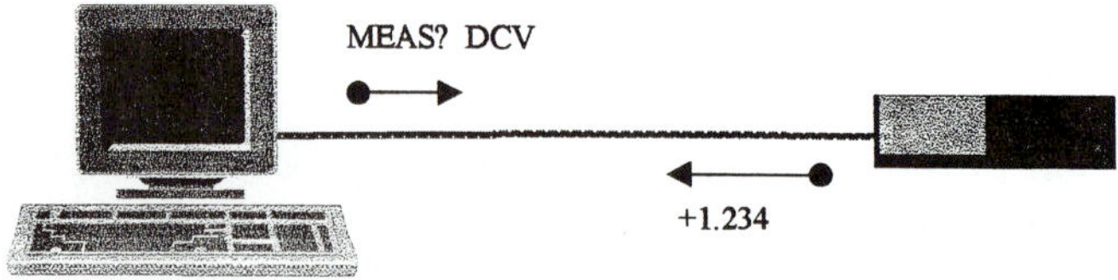

Figure 43.2 I/O traffic between a computer and an instrument.

turers. Another factor in the success of IEEE 488 is that interfaces are available for virtually every computer.

With the adoption of IEEE 488, virtually all instruments were designed to be capable of computer connection at some point. The historical automation tasks are still an excellent application example; however, instruments are also connected to computers for other purposes. For instance, virtually all engineers use word processors to write reports and computer-based analysis programs to do numerical analysis on measurement data. For these engineers, a connection between the instrument and the computer transfers measurement data directly into the appropriate application program (Fig. 43.2).

Input/output. The most obvious portion of the connectivity solution is the physical I/O. This is the physical hardware, and lowest levels of software that deal with sending 1s and 0s between the computer and the instrument. Several alternatives are currently available, including IEEE 488, RS-232, IEEE 1394, and the network.

Software. The software that provides the connection between the computer and the instrument varies somewhat with the application. For automation applications, a software developer writes a program that will accomplish some task. In this case, the developer needs to send messages to each instrument in the system to configure it for the test and then read a value back from a measuring instrument. The developer of the program uses either an I/O library or an instrument driver to control the instrument.

I/O libraries. The I/O library sends the data to and receives it from the instrument. This allows the programmer to work in terms of the actual messages that are exchanged with the instrument, without having to deal with the mechanics of transmitting the data. The messages are ASCII strings that are interpreted by the instrument. From the perspective of the developer, the manipulation of these strings is very similar to printing data to a console or a printer. The difference is that the messages, instead of being intended for a human reader, instruct the instrument what actions to take. Figure 43.3 shows a typical example.

```
10 DIM Dmm As New HPIOGPIB
11 Dmm.Connect("GPIB1::22::INSTR")
12 Dmm.Clear()
13 Dmm.Output("FUNCTION DCV")
14 Dmm.Output("READ?")
15 Dmm.Enter(Result)
16 Dmm.Close()
17 END
```

Figure 43.3 Making a simple measurement with an I/O library.

```
10 Dim Dmm as New HP34401
11 Dmm.Connect("GPIB1::22::INSTR")
12 Dmm.Clear()
13 Dmm.Function = DMM_DCV
14 Dmm.Read(Result)
15 Dmm.Close
16 END
```

Figure 43.4 Making a simple measurement using an instrument driver.

In this example program, line 10 creates the DMM object that will be responsible for the I/O occurring with the DMM (digital multimeter). Line 11 indicates that the DMM I/O library object should connect with the GPIB instrument with address 22. The first step in the program is to clear the communication channel (line 12). Then, two strings are sent to the instrument. The first string sent to the instrument (line 13) instructs it to configure to make a DC voltage measurement. The second string (line 14) initiates the voltage measurement and tells the instrument to transmit the result of the measurement back to the computer. Next, line 15 tells the I/O library to read the result back from the instruments and places it into the numeric variable *Result*. Presumably, for this application, the value received in result will be used to make subsequent decisions about the device under test. Finally, line 16 terminates the session with the instrument by doing a close. This does not result in any transactions with the instrument, but indicates to the I/O library that no further I/O operations will be done with this device. The I/O library provides the implementation of the commands *Connect*, *Clear*, *Output*, *Enter*, and *Close*. The actual strings sent to the instrument (FUNCTION DCV and READ?) are recognized and parsed by the instrument and are known as the *instrument language*. The *Standard Commands for Programmable Instruments* (*SCPI*) standard, along with the IEEE 488.2 standard, defines these strings.

When the developer controls the instrument via the instrument language, the developer uses functions and features from the I/O library that send, receive, and manipulate the strings. Some developers prefer to use an "instrument driver" to control the instrument. An instrument driver encapsulates some of the functions of the instrument and provides an interface to the instrument that is a subroutine call in the programming language that the developer is using. Generally, the same basic steps are taken, but the developer uses a familiar subroutine metaphor to control the instrument, instead of dealing with the I/O library directly. The instrument driver encapsulates the necessary I/O library calls and the instrument language. Figure 43.4 shows how operations done in Fig. 43.5 could be done with an instrument driver.

43.1.3 Instrument drivers

The term *instrument driver* is often used to describe the piece of software that allows a particular instrument to be used with certain test and

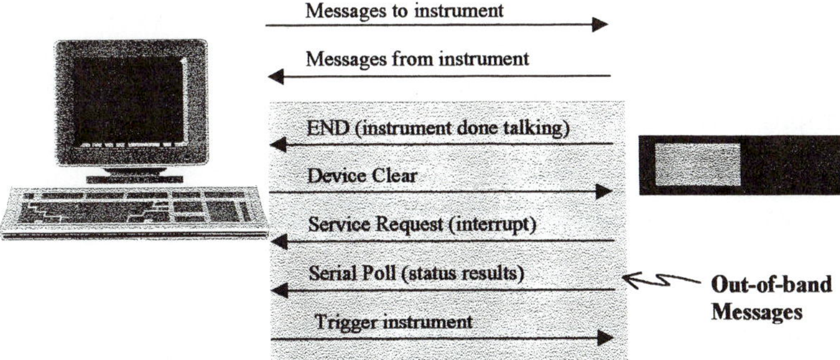

Figure 43.5 Types of messages sent between a computer and an instrument.

measurement-specific programming languages (for instance, LabView). Usually, these test and measurement-specific environments include an I/O library, but it is rather awkward to use within the environment. In addition, in these environments, the instrument driver usually provides a graphical interface to the instrument that is unique to the development environment.

Neither the direct use of an I/O library nor the use of an instrument driver works well for applications where data needs to be sent from an instrument directly into a PC application. A typical example of this is moving measurement results directly into a spreadsheet for plotting or storage. Neither an I/O library nor an instrument driver is helpful for this user because both of these tools is focused on simplifying programming instead of simplifying this manual task. This application requires a software module that understands how to communicate with the instrument and also how to deposit data into the spreadsheet.

The following sections of this chapter describe these layers in more detail and compare some of the different mechanisms available. The final section explores the performance of the computer-to-instrument connection.

43.2 Physical Input/Output Layer

The physical I/O layer provides the basic transfer of data between the instrument and the computer. Normally, the data is thought of as being organized into streams of bytes that flow between the instrument and the computer. The data stream flowing to the instrument is the instrument language. The flow into the computer is the responses (usually measurements) from the instrument. These byte streams can potentially flow simultaneously between the instrument and the computer so that the instrument is providing results while receiving new commands from the computer.

43.2.1 Out-of-band messages

In addition to the byte streams flowing between an instrument and a computer, additional messages are very useful in an instrument system. These messages can occur while the data is flowing to or from the instrument; therefore, they are known as *out-of-band messages*. The out-of-band messages commonly available in instrument systems are described in the following paragraphs.

Device clear. *Device Clear* is an indication from the computer to the instrument that the current communications processes should be aborted. This is useful if the computer and instrument lose synchronization (possibly through a programming error on the computer). Device Clear is also important if some external event occurs that indicates to the computer that the current operation should be aborted in favor of some higher-precedence operation.

Device Clear is carried as an out-of-band message because the primary communication stream might be occupied with some operation that can not be interrupted. For instance, a large block of data could be in the process of being transferred. Without an out-of-band indication, the instrument will not abort the data transfer until it is complete.

Service request. *Service Request* is an indication from the instrument to the computer that some condition requiring the computer's attention has occurred. When the instrument is initially configured, the computer specifies under what conditions the instrument should assert a service request. Some possible reasons for the service request might be:

- Something is potentially wrong with the data being collected
- An external limit has been reached
- Some error condition has occurred

Service Request is an out-of-band because these conditions need to be indicated to the computer without interfering in the primary communication stream. Service Request also needs to be delivered to the computer, independent of any actions that the computer is taking. For instance, the computer could be engaged in I/O to one device and then a separate device might assert a service request. Notice that for the I/O library to support this function, it must be able to start an additional process or thread on the computer in response to an instrument requesting service.

In the case of GPIB, a single physical line is used by all of the instrument to indicate the service request. As a result, when the computer detects the service request, it needs to determine which device actually asserted the service request. The computer also needs to determine the reason for the service request. The computer accomplishes this by reading the status of each instrument in the system. This operation is known as *Serial Poll*. Serial Poll also must be an out-of-band operation because the computer could ultimately decide to schedule the response to the service request as a lower priority than the currently executing tasks. Therefore, the serial poll must not interfere in whatever operations might have been underway.

Serial poll. As noted, serial poll provides an indication to the computer of the reason that a service request occurred. Normally upon receiving a service request, the computer interrogates the device for the reason it requested service by using Serial Poll. Some I/O systems allow the device to provide this status information when it requests service, eliminating the need for a serial function.

End. *End* is an out-of-band indication that a data transfer is complete. End is used in situations where it is impossible to determine the length of the transmission before it begins. One common example of this is when the instrument is configured to transmit data until some external condition is met. In this situation, it is impossible for the computer (or the instrument) to determine a priority when the data transfer will be complete. Therefore, an out-of-band end indication is required.

For some applications, a similar situation can exist when the computer is sending data to an instrument. An example of this is a directly digital synthesis source that is continuously downloading arbitrary data. For the computer to exit this mode requires some out-of-band indication. However, in this scenario, a device clear could serve the purpose.

Trigger. Some interfaces provide an out-of-band trigger indication. The configuration of the instrument will determine the exact interpretation of this trigger. For most system architectures this does not occur while data is being transferred, so it is not crucial that Trigger be transferred out-of-band.

Some physical I/O alternatives do not provide any or all of these out-of-band messages. Where these out-of-band messages are not available, it is frequently possible to get by without the associated function. For instance, a source is unlikely to need an out-of-band end message because it probably does not need to send an arbitrary amount of information back to the computer. GPIB and other interfaces designed specifically for testing and measurement include each of these functions.

43.2.2 Transmitting out-of-band messages

Where they do exist, it appears to the user that the out-of-band messages are actually carried by a mechanism that is independent of the hardware that actually transfers the data. In fact, this might not be the case. For instance, several interfaces are physically only a single wire. In these cases, rules, known as *protocols*, describe exactly how messages should be composed and sent over the wire. For instance, the protocol might state that whenever the instrument begins to communicate, it will send 4096 bytes of information (even if it is sent one bit at a time). This block of 4096 bytes is known as a *packet*. The protocol can then specify that the first byte of the packet indicates if the packet contains an out-of-band message or an in-band message. With this sort of protocol, layers above that fundamental layer pretend that two channels of information exist between the computer and the instrument. In actuality, the protocols used for I/O are much more complex than this, but they do use this sort of conceptual layering.

43.2.3 Physical I/O alternatives

Several I/O alternatives are used for instrumentation. In some cases, long-lived test and measurement-specific solutions are used. In many other cases, instruments are able to use standard computer I/O mechanisms.

At first thought, it might seem that test and measurement I/O solutions should be replaced by computer standards. Computer standards have the advantage of providing excellent bandwidth with very low cost cables and interfaces. However, the various test and measurement I/O solutions provide key capabilities for test and measurement applications. For instance, test and measurement I/O solutions, such as VXI, provide all of the out-of-band messages described previously. VXI provides high-bandwidth, low-latency transfers that are difficult to achieve with conventional computer I/O. (See Chapter 40 for a detailed description of VXI.) Perhaps one of the most crucial aspects of test and measurement I/O solutions is their longevity. Test and measurement products typically remain current for 5 to 10 years; however, once purchased, the user frequently expects to use the product for more than 10 years. This requires that I/O solutions include both the connection of new instruments into existing systems and the connection of older instruments into new systems.

General-purpose interface bus (GPIB) IEEE 488. *GPIB*, also known as the *Hewlett-Packard Interface Bus* (*HP-IB*) or *IEEE 488*, remains the most common test and measurement I/O solution. It has been an IEEE standard since 1975, with many implementations available before then. It provides performance up to one megabyte per second and excellent electrical characteristics. It allows as many as 15 devices to be connected in parallel on an 8-bit bus. The bus provides additional lines that are used to deliver the out-of-band messages.

The greatest problem with GPIB is the difficulty of getting the GPIB port onto the computer. Because GPIB is a test and measurement-specific interface, it is necessary for the user to purchase an interface card that plugs into the backplane of a computer. This is not only inconvenient because of the hardware installation, but it is extremely difficult to locate free resources on the computer to dedicate to GPIB (for instance, interrupt request lines and I/O addresses).

Some solutions avoid the configuration difficulties of installing a GPIB interface card by placing a GPIB adapter on some existing interface (for instance a network connection). In this case, all of the GPIB traffic is sent over the network connection and then bridged over to GPIB. This helps considerably with the complexity of the hardware installation, but pays some price in performance because all outbound traffic must travel first to the adapter and then over GPIB to the instrument. Similarly, traffic bound from GPIB to the computer must travel over both interfaces.

Using computer-standard I/O. There are several advantages to using computer-standard I/O with instruments. Specifically:

- Because the software and hardware are a standard part of the computer, no additional effort is required to install and configure the hardware and software.
- Because the computer industry has tremendous volumes, the cost of I/O cards, cables, and software are much lower than for an instrument specific solution.
- Because of the computer industry's tremendous volumes, the performance advantages to computer-standard solutions are significant.

As a result of these facts, most instrument systems strive to use computer standards wherever possible. Two general approaches are used for instruments to use computer standards for I/O. The first is to have the instrument pose as a conventional computer peripheral and directly use the supported protocols. In this scenario, no special software or configuration is necessary because the standard computer mechanisms are used. A typical example of this would be an instrument impersonating as a disk drive. In this way, moving measurement data is as simple as copying a file. Another example is to use a protocol, such as TCP/IP, which transports byte streams between devices, to carry the testing and measurement traffic.

The difficulty with computer-standard protocols is that they generally do not meet 100 percent of the needs of test and measurement applications. For instance, if the instrument is impersonating a disk drive, there is a convenient way to transfer the data, but it is extremely difficult to configure the instrument or trigger a measurement. In the example of using TCP/IP to transfer a byte stream, there is no way to provide the necessary out-of-band messages. In general, these problems can be dealt with by defining an additional set of rules that provide the missing mechanisms. This set of rules is known as a *protocol*. The goal is to select the computer-standard protocol that does the best job of meeting the testing and measurement needs, then creating protocols to fully solve the problem.

Consider the case of using TCP/IP to send data between a computer and an instrument. Because this provides a byte stream, it is a good replacement for conventional instrument communication; however, it does not support out-of-band messages. In this case, because TCP/IP supports multiple byte streams between two devices, the test and measurement application can open two separate TCP/IP connections. One connection is used for data, and the messages sent on the second connection provide the out-of-band messages.

In summary, it is important to use computer-standard protocols wherever possible because of the benefits in cost and performance. However, when using a computer-standard mechanism, it is occasionally necessary to define an additional protocol that defines how some of the extraordinary needs of the testing and measurement application are met.

To achieve consistency among instruments, several standards specify protocols for how test and measurement applications should use computer standard I/O. Three common standards are:

IEEE 1174	For use with RS-232
VXI–11 TCP/IP Instrument protocol	For use over a network based on TCP/IP
1394ta IIWG	For use with IEEE 1394

Computer-standard I/O falls into two different categories: desktop I/O and networking. Each of these is important for test and measurement, but each excels at slightly different applications.

Desktop I/O. *Desktop I/O* refers to those I/O connections that are built into standard computers for accessing conventional computer peripherals. Historically, this has been primarily RS-232 and parallel; however, modern computers use Universal Serial Bus (USB) and IEEE 1394 (also known as *Firewire*). There are also wireless, infrared, and RF mechanisms.

The primary difference between desktop I/O and network I/O is that desktop I/O is used to access peripherals private to a particular computer, whereas network I/O is used to communicate between computers and to access shared peripherals. Because desktop I/O is targeted at individual PC users, it is normally extremely easy for any individual to configure and use. Because network I/O is targeted at the larger enterprise, it might be more difficult to configure and might require the intervention of information technology (IT) personnel, who are responsible for smooth operation of the intracomputer communications.

Some of the characteristics of desktop I/O are:

- Low-cost, widely available cables
- The available I/O bandwidth is shared among peripherals attached to the same PC.
- The equipment on the desktop I/O bus is private to the computer.
- The extreme ease of configuration, which usually supports hot swapping of devices

Most PCs support both IEEE 1394 and USB for desktop I/O connections. They are both serial interfaces and have several architectural similarities. Both provide high-bandwidth isochronous unreliable transfers and reliable transfers. IEEE 1394 is available in several bandwidths from 400 Mbits/s to 3.2 Gbits/s. USB is somewhat slower at 12 Mbits/s; however, this data rate is competitive with both GPIB and slower network solutions.

Next to GPIB, RS-232 is the most common test and measurement interface. Despite the fact that it is notoriously difficult to properly configure the handshaking and cable type, this is still the interface of choice for many applications. RS-232 continues to be important for testing and measurement applications because:

- It is available on virtually any computer (any age, any manufacturer).
- It provides remote access. Using a modem, an RS-232 instrument can be accessed anywhere there is a phone line.

Network I/O. Network connections have been included on instruments since the late 1980s. They are popular for applications where:

- The instrument is part of a distributed measurement system that can be connected via a network—especially where measurements are made in geographically diverse areas that must be coordinated to produce a useful result. An example is monitoring radio transmissions and coordinating the measurements to locate various transmitters.
- The instrument can utilize a conventional network protocol to provide convenient access. For instance, a logic analyzer can benefit from emulating a file system. In this application, the user can copy a link map to the instrument using the file system. The instrument can then display its results, complete with all of the information provided by the link map.
- The instrument is convenient to a network. For example, a LAN analyzer, by its very nature, is already on the network, so it can most easily transmit its results over the network.

Several physical layers are available for network designs; however, the most popular ones are 10BaseT, 10Base2, and 100BaseT. 10BaseT and 10Base2 are both 10-megabit/second implementations and 100BaseT is a 100-megabit/second implementation. In a 10Base2 system, every device is connected in parallel on a coaxial cable. This is generally convenient for a new system, but the coaxial cable is rather costly. Both 10BaseT and 100BaseT are run on twisted-pair lines with a modular connector. In this case, each device has a separate connection run to it. The various connections are joined together with a hub. The hub connects several devices on one end; on the other end, it connects to the shared network (possibly another 10BaseT or 100BaseT). Higher-speed connects, such as Gigabit/second network connections, are available, as well.

Network connections are characterized by:

- Computer-to-computer connections or computer-to-shared-resource connections. This differs from desktop I/O, where the connection is to a local system.
- Bandwidth is shared with other users on the network.

- Some network connections require the intervention of an IT department, which "owns" the network.
- Network connections have a tremendous wealth of protocols from which to draw. Standard protocols exist for GPIB emulation (TCP/IP Instrument Protocol), basic reliable data transfer (TCP/IP), remote procedure calls (ONC RPC), remote human interfaces (http and HTML), and others.

43.3 Network Protocols for Test and Measurement

Several network protocols are particularly useful for testing and measurement applications. Some of the key protocols are:

- *Hypertext transfer protocol* (*http*) allows any network device to provide Web pages to any other device that is running a browser. This protocol is used with instruments that need to support remote connections.
- *Distributed object protocols* (primarily Microsoft Distributed Component Object Model, DCOM). These protocols allow a users program to invoke a function on a remote device in the same way that a software object on the local system is accessed. This capability allows the instrument driver to be delivered in the firmware of the instrument.
- The *TCP/IP instrument protocol* (also known as the *VXI 11 standard*) provides all of the same functions as GPIB for network instruments. In this case, users must have an appropriate library installed on their computer (e.g., HP VISA) to access the instrument because these protocols are not included in standard computer network-support packages.

The network has captured the imagination of many users and presents diverse opportunities for instrument connectivity. Networks are intriguing both because they are physically present in many facilities and because of the wealth of protocols available over the network. Using a network connection, a user can do everything from conventional GPIB operations to transferring files or browsing instrument-generated Web pages. (See Chapter 44 for a more detailed discussion of "Networked Connectivity for Instruments.")

43.4 Messages Sent Between Computers and Instruments

To fully understand I/O in a test and measurement application, it is important to further consider the actual messages that are transported between an instrument and a computer (see Fig. 43.5). This traffic is usually based on the *American Standard Code for Information Interchange* (*ASCII*). ASCII is the basic standard that establishes how computers represent alphabetic and numeric characters.

		Desktop I/O Alternatives			Network I/O	T&M Specific I/O	
		IEEE 1394	USB	RS-232	100BaseT : 10Base2	GPIB	VXI
Performance✦	Xfer rate Bytes/sec	20 MB/s	40 KB/s	3 KB/s	8 MB/s : 800KB/s	200 KB/s	50 MB/s
	Latency	~150 μs	1 ms	1 ms	100 μs	100 μs	10 μs
Inexpensive cables		Yes'	Yes	Yes	Yes	No	No
Computer SW installation required?		Yes	Yes	No	Protocol dependent	Yes	Yes
Ease of configuration		Good	Good	Weak	Weak see ③	Bad	Bad
Bandwidth shared with:		Desktop peripherals	Desktop Peripherals	Private	Computers and shared peripherals	Other instruments	Other instruments
Communication architecture		Peer-Peer	Master-Slave	Master-Slave	Client-Server	Master-Slave	Master-Slave and Peer-Peer
Other advantages		②	②④	④⑤	③④⑥⑦	⑤	
Other disadvantages				①			

✦ Performance numbers are ***estimates*** of achievable throughput in an actual system. Theoretical limits are somewhat higher.
① Cabling and setup is difficult.
② Devices can be powered from bus.
③ Numerous protocols to select from including TCP/IP Instrument Protocol, which provides T&M functions
④ Broadly available on standard computing platforms (although RS-232 is probably weakening and network not in many low-cost PCs).
⑤ Significant installed base across testing and measurement.
⑥ Most necessary SW is standard on the OS although specific T&M functions require additional software installation.
⑦ Natural match for distributed system applications for example distributed data acquisition.

Figure 43.6 A comparison of popular I/O mechanisms for T&M applications.

The actual messages sent between instruments and computers are described by the *Standard Commands for Programmable Instruments* (*SCPI*) standard. This standard is basically a dictionary that lists the various commands used to control an instrument. The SCPI commands are based on a standard instrument model (shown in Fig. 43.6).

It might seem that ASCII messages are inefficient, but when transferring the short messages used to setup an instrument, the overhead of ASCII is negligible. Actually, many network protocols, such as *ftp* (*file transfer protocol*) and http, are also based on ASCII messages for similar reasons. Whenever a large block of data is sent to an instrument or read from an instrument, the representation does affect the system throughput. An SCPI instrument always transmits or receives these blocks of data in binary. IEEE 488.2 specifies how the blocks are encapsulated.

The software interface to the instrument is largely independent of the hardware connection. Most hardware connections provide an eight-bit data path to the instrument. The instrument is controlled by commands and queries sent over this data path. Normally, the commands and queries are ASCII strings. IEEE 488.2 and *Standard Commands for Programmable Instruments* (*SCPI*) are two important standards that describe the software characteristics of the instrument interface. IEEE 488.2 describes the syntax that is accepted by the instrument. SCPI describes the actual meaning of the commands that

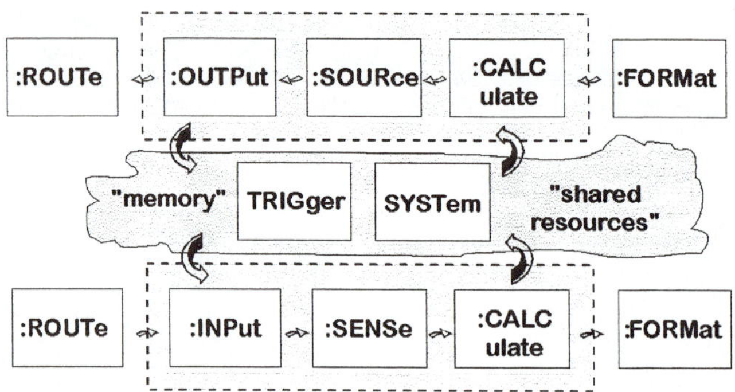

Figure 43.7 The SCPI instrument model.

are formed using the IEEE 488.2 synta45 (see Fig 43.7 for the SCPI instrument model).

For instance, the following ASCII string is the SCPI query that requests that a voltmeter measure a voltage and return the result:

MEAS:VOLT? 17,.001

To clarify the domain of IEEE 488.2 and SCPI, consider this example. The IEEE 488.2 rules indicate that the program mnemonic is MEAS:VOLT. The first parameter is 17 and the second parameter is .001. IEEE 488.2 specifies how the program mnemonic is formed and what characters are legal. IEEE 488.2 also contains the rules that describe exactly what syntax must be accepted by the instrument for numbers (for instance, 17, +1.7E1, and 17.0 must all be treated identically). IEEE 488.2 requires that the program mnemonic be followed by a question mark when the operation will return a result to the computer (such an operation is called a *query*).

The SCPI standard indicates that the MEAS:VOLT? program mnemonic shall return a measurement of the voltage at the input to the voltmeter. Further, it states that the first parameter indicates an estimate of the input (allowing the instrument to select the correct range) and that the second parameter specifies the resolution to which the conversion should be made (in this case, the conversion is made to within a factor of 0.001, which is 0.1 percent).

43.5 I/O Libraries

When an instrument is programmed by sending ASCII strings, it is useful to have a library to take care of the details of sending data between the com-

```
10 DIM Dmm as New HPIOGPIB
11 DIM Source As New HPIOGPIB
12 DIM Response(11) As Double
13 DIM I As Integer
14 Dmm.Connect("GPIB1::22::INSTR")
15 Source.Connect("GPIB1::26::INSTR")
16 Dmm.Clear()
17 Source.Clear()
18 Source.Output("*RST")
19 Dmm.Output("*RST;FUNCTION DCV")
20 I=1
21 FOR Stimulus = 1 To 2 Step .1
22     Source.Output("VOLTAGE " & Stimulus)
23     Dmm.Output("READ?")
24     Dmm.Enter(Response(I))
25     I=I+1
26 NEXT Stimulus
27 Dmm.Close()
28 Stimulus.Close()
29 END
```

Figure 43.8 Using an I/O library to control an instrument.

puter and the instrument. Figure 43.8 shows a typical example of controlling an instrument with an I/O library. Numerous I/O libraries are available; however, this example shows a simple example as it might appear in a Visual Basic application.

In this simple example, a source is swept from one volt to two volts in 100-millivolt steps and a voltage measurement is made at each step. The I/O library is first initialized in lines 14 and 15. The syntax indicates to the I/O library that both the DMM and the source are on the first GPIB interface (GPIB1), they are at addresses on the bus of 22 and 26, respectively, and each has a conventional instrument interface.

Lines 16 and 17 issue Device Clear to each instrument. Although this might be unnecessary, it guarantees that the communication channel is clear and the instruments will receive the subsequent commands.

Notice that this is the only out-of-band message used in this example. However, sending this out-of-band message is trivial using the I/O library.

In lines 18 and 19, the I/O library is used to send simple strings to the instruments, resetting them and configuring them for the test. Notice that these strings are defined by the SCPI and IEEE 488.2 standards. Line 22 shows how the I/O library is used to send a variable to the instrument. The Visual Basic interpreter recognizes the syntax within the parenthesis and converts the variable stimulus to a string and concatenates it with the string "VOLTAGE" to form the appropriate SCPI syntax to send to the instrument. Notice the trailing space in the string; this is necessary to separate the parameter from the command. This technique uses the tools familiar to the programmer for working with strings and printing results. The only difference is that the string is sent to the instrument instead of being printed. Line 23 is similar to lines 18 and 19 because a fixed string is again sent to the

DMM; however, in this case, the DMM is instructed to make a measurement and send a string back corresponding to the voltage that it measured. The question mark in the string indicates that a value should be sent back.

Finally, line 24 reads the response back from the instrument and converts it into a number that is placed into the array Response.

In summary, the I/O library takes strings and numbers from within the program and sends them to the instrument as strings. Similarly, the I/O library takes strings from the instrument and builds them into appropriate numbers that can be used in the program. The I/O library also has functions that can be invoked to send the out-of-band messages to the instrument.

43.6 Instrument Drivers

The term *instrument driver* is rather confusing because it is frequently applied to a lot of different things. Most generally, an instrument driver is a layer of software that allows the user to work with an instrument without directly accessing an I/O library.

Figure 43.9 illustrates how the program in Fig. 43.8 might look if instrument drivers were used to access the two instruments.

Notice that the I/O library calls have been replaced with calls to a library that has specific knowledge about the instruments (the DMM object is specific to the HP34401 digital multimeter and the source object is specific to the HP33120 source).

As was the case with the I/O library, the first step here is to initialize the driver. This occurs on lines 13 and 14. This syntax is almost identical to the one used with the I/O library because the function is exactly the same. That is, these lines indicate to the underlying library what physical instrument is to be used. This syntax indicates that these are GPIB instruments addressed as 22 and 26, respectively.

```
10  Dim Dmm as New HP34401
11  Dim Source as New HP33120
12  Dim I as Integer
13  Dmm.Connect("GPIB1::22::INSTR")
14  Source.Connect("GPIB1:26::INSTR")
15  Dmm.Clear()
16  Source.Clear()
17  Dmm.Function = DMM_DCV
18  I = 1
19  FOR Stimulus = 1 To 2 Step .1
20        Source.Voltage = Stimulus
21        Dmm.Read(Response(I))
22        I = I+1
23  NEXT Stimulus
24  Dmm.Close
25  Stimulus.Close
26  END
```

Figure 43.9 Controlling instruments with instrument drivers.

Similarly, the step of clearing the communication channel (lines 15 and 16) to the instrument is much the same if a driver is used or not. However, when using a driver, it is likely that this step actually occurs when the connection is first opened, so it is probably redundant here.

Line 17 demonstrates the first major difference between working with a driver and controlling the instrument with the I/O library. This line uses the conventional syntax of the programming environment to control the instrument. So, although the meaning is the same as the I/O library example (*Dmm. Output("FUNCTION:DCV")*), the user is not required to understand the syntax or rules for dealing with the instrument. Nor does the user need to understand any of the mechanics of using the I/O library.

This same simplification is evident in lines 20 and 21. Notice that the actual abstraction of what is happening is the same in the two examples; however, with the instrument driver, the user remains in the paradigm of the programming language.

Another advantage of drivers is that they provide a software layer that can be modified to support different instruments. For instance, in the previous example, suppose that some other instrument is used in place of the HP 34401. In the case where strings are sent directly to the instrument, it would be necessary for the two instruments to accept the same instrument language (for instance, if both instruments are based on the SCPI standard). If the new instrument is different, the control program must be updated to the new instrument language.

With the instrument driver, to use a different multimeter would only require that line 10 be changed. Then the instrument driver can be made to behave similarly between the two applications.

Unfortunately, this example does not entirely provide the ability to change instrument drivers. The difficulty comes if other characteristics of the instruments are not specified by the driver. For instance, in the previous example, suppose that the measurement required a high input impedance. In the case of the first instrument, the programmer might have included a line to set the instrument to the 10-volt range, where it has its highest input impedance:

Dmm.DCVoltage.Range = 10

The difficulty occurs if the new instrument, although capable of making the measurement, has its high input impedance on the 1-volt range. Now, although the two drivers each support the syntax to set the range, it can be set in the same fashion, and each instrument is capable of making the measurement, the program will not work with the new instrument unless line 17.1 is changed in addition to line 10. The solution to this problem is to add a measurement layer on top of the instrument drivers. The instrument drivers are still valuable to the developer, but an additional layer that is aware of the actual measurement that is occurring (in this example, it would be aware that a high input impedance were required for the measurement) needs to be included in the system architecture. Then, to insert a new instrument, the developer can check each measurement that is used by the system, instead of each instrument function.

43.7 Performance and System Throughput

Several important measures of the performance of an instrument relate to the connection to the computer. These figures of merit help to show how fast the instrument will perform in a test system and the resulting system throughput.

43.7.1 Block transfers

The most common specification quoted for I/O is how fast a block of data can be transferred between an instrument and an arbitrarily fast computer. This specification is important when large amounts of data will be transferred between the instrument and the computer. An example is when numerous oscilloscope traces or blocks of data from a logic analyzer are transferred. Typically, transfer rates vary between 10 kilobytes/second and 100 megabyte/second.

When considering these transfer rates, it is crucial to remember that both the instrument and the computer contribute to the total time. So, a computer that can transfer data at 1 megabyte/second can be connected to an instrument that is capable of 1 megabyte/second, but the system throughput could easily drop to 500 kilobytes/second. To see why this happens, consider the time required to transfer each byte. If the computer consumes 1 μsec (corresponding to a 1 megabyte/second rate) and the instrument consumes 1 μsec (corresponding to its 1 megabyte/second rate), the total time will be 2 μsec if the time required by the instrument and computer does not overlap in any way (2 μsec implies a net rate of 500 kilobytes/second). Normally, some overlap will exist between the two times, yielding rates from 600 kilobytes/second to 800 kilobytes/second for this type of scenario.

To see the affect of block transfer rates on system throughput, consider an instrument that is transferring a trace of 500 elements. Suppose that each element in the trace is transferred as an ASCII number in scientific notation with a six-digit mantissa and a two-digit exponent. The IEEE 488.2 standard form for the list includes 13 characters per number, as shown by the following example:

$$+1.23456E+04$$

Therefore, a 500-element trace would occupy 6500 bytes. At 1-Mb/s, this yields a transfer time of 13 μs per number and a total transfer time of 6.5 ms. At 10 Kb/s, this yields a transfer time of 1.30 ms per number and a total transfer time of 650 ms. The slower case could significantly impact system throughput. However, one consideration is more important.

In this scenario, the instrument is probably representing the number internally, as either an integer or an IEEE 754 floating-point number. The computer's internal representation is probably similar. Therefore, to transfer the data in this fashion requires two format conversions: one in the instrument before the transmission and another in the computer after the transmission. These format conversions are very computationally intensive, and require approximately the same amount of time as the transmission. Depending on the system, each conversion will vary between 100 μs and 1 ms. As a result, even at the typical transfer rate of 100 Kb/s, the throughput is dominated by the type conversions, not the transfer time.

Transferring the data in binary form easily solves this problem. IEEE 488.2 provides syntax for transmitting binary data. This syntax begins with a short header, stating how long the transmission will be, followed by the raw data. For a number of the range used in the previous example, only four bytes will be transferred per number for total transfer of 2000 bytes. This yields a transfer time approximately three times faster (because four bytes are transferred instead of 13) with the improvement that format conversions have been avoided, which is even more significant.

In some cases, format conversions will be necessary when transferring binary data. For instance, the instrument might transfer 16-bit two's complement integer data, which must be converted to 64-bit floating point in the computer. This conversion does not present the same performance problem as conversion to ASCII because both numbers are in binary.

43.7.2 Short transfers

The block transfer rates provide a good measure of the maximum speed of the interface and the ability of the computer and instrument to transfer large blocks of data. However, short transfers dominate the throughput of many systems. This is because:

- Many instruments only transfer short blocks (such as power meters and frequency counters).
- For trace-oriented instruments, the time to configure the instrument might dominate over the time to transfer the data.
- Many devices, although trace oriented from the front panel, provide scalar measurements. For instance, oscilloscopes can typically return a scalar measurement of rise time.

Short transfers are slower for both computers and instruments. Depending on the hardware interface, the time to reverse the direction of transfer (that is to switch between sending the instrument a query and receiving the result) can be significant in itself.

There are two chief measures of how fast an instrument and computer can transfer short blocks. One is how fast the instrument can receive a command, parse it, and act on it. Because most instruments buffer requests, this measurement must be made in a way that measures the time to actually parse and act on the request, not just accept the data into its input buffer. Typically, this operation ranges from 1 to 10 ms. The second key measure is *turn-around time*. The turn-around time is how long it requires the instrument to receive a short query and return the response. Although the computer contributes to this, the instrument typically dominates this figure. Typically, this will be about twice as long as the time to receive and act on a command. Any value less than 10 ms is good for this specification, with virtually no instruments able to do this in less than one ms.

For either of these benchmarks, it is crucial to consider the action that is being taken by the instrument. In many cases, relays are closed or analog control loops are changed. The time constants for these types of operations can be as great as one second. In such cases, these benchmarks are meaningless and frequently misleading. On the other hand, many systems instruments go to great expense of optimizing the set-up and configuration time by including fast analog hardware. Even for these products, it is important to consider the time required by the instrument to accurately predict system performance.

Chapter

44

Graphical User Interfaces for Instruments

Jan Ryles

Agilent Technologies
Loveland, Colorado

44.1 Introduction

The user interface is the medium by which the user makes use of the underlying technology described in many of the chapters of this book. Literally, as the interface to the user, it can be presented in a manner that facilitates the user's task success or inhibits this activity. It can serve to simplify the presentation of complicated technology or expose it in excruciating detail. The instrument user often interacts with a combination of hardware interfaces (as described in Chapter 12) and graphical user interfaces, both of which vary across instrument types and instrument vendors. Users of multiple instruments are challenged to learn a variety of interface styles, which increases their learning time.

As personal computer (PC) standards dominate the workplace, they also make their way onto the workbench and the factory environment. In many cases, text-based user interfaces have been replaced by graphical user interfaces and graphical user interfaces replaced by Windows-based interfaces, providing a more-standardized appearance and behavior. PC-based functionality can be used to augment and/or replace test and measurement functionality available in the instrument box. These changes have altered the way that users perceive their instruments and systems and accomplish their tasks.

The goal of this chapter is to describe these changes as they affect instrument configurations, the evolution of graphical user interfaces and the subsequent impact on the instrument users' experience. The final section of the chapter includes criteria for assessing the usability of the graphical user in-

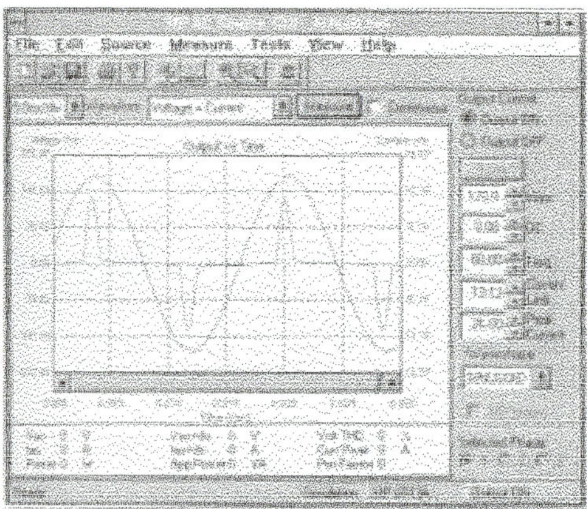

Figure 44.1 A graphical user interface. A measurement interface can take on the appearance of a more standard graphical user interface. Note the menubar and toolbar at the top of the interface.

terface of the instrument. A brief usability checklist is provided to assist users in evaluating instruments prior to purchase.

44.2 Introduction to Graphical User Interfaces

Graphical user interfaces (GUIs) (see Fig. 44.1) can be defined as a visual software interface, in which the user interacts with windows, icons, menus, and a pointing device (sometimes called a *WIMP interface*), in order to achieve certain tasks. They have replaced most character-based interfaces on computers, although many conventional instruments still rely on text-based interfaces and pointing devices embedded in the instrument front panel, such as arrow keys. The user operates the graphical user interface by using a pointing device, typically a mouse, to point at objects on the screen and perform an action by pushing a button, doubleclicking on the item, dragging one object onto another, and other similar direct-manipulation actions. Functionality is typically arranged based on menus, buttons, and dialog boxes. This typical arrangement allows for users to move from one interface to another with some pre-knowledge of how the interface should work.

Many computer-controlled test systems, as well as some standalone instruments, have migrated to graphical user interfaces. As instruments and computers become more compatible, the world of graphical user interfaces is also forging ahead with new design paradigms, which should provide ever greater ease of use. Users of instruments will be impacted by these changes, requiring a shift in their interaction styles.

44.3 Instrument Configurations

Before looking at the changes in user interfaces, it is important to look first at the alternatives in instrument configurations. The conventional configurations are handheld, standalone instruments; rack and stack instruments; and faceless and distributed instruments, as covered in Chapter 12 on Instrument-User Interfaces. With the infusion of the PC into the equation, some additional configurations can be considered, which, by their very nature, include a graphical user interface. These PC/instrument configurations impact the users' experience with and perception of the instrument/system. The following are some of the predominant configurations, which have evolved because of the increasing prevalence of PCs in the workplace.

44.3.1 PC embedded in instrument

In this configuration, the PC operating system is embedded within the instrument. This makes the Windows interface available for the graphical user interface part of the product. The instrument hardware front panel still retains some of the familiar hardware controls mentioned in Chapter 12. However, the ratio of software to hardware controls has shifted so that the software controls are a superset of the hardware controls. The user can interact with the Windows interface via a mouse, touch screen, or pointing device of their choice.

This configuration can retain the appearance of the instrument, as in the oscilloscope in Fig. 44.2A, or it can appear more like a personal computer. In

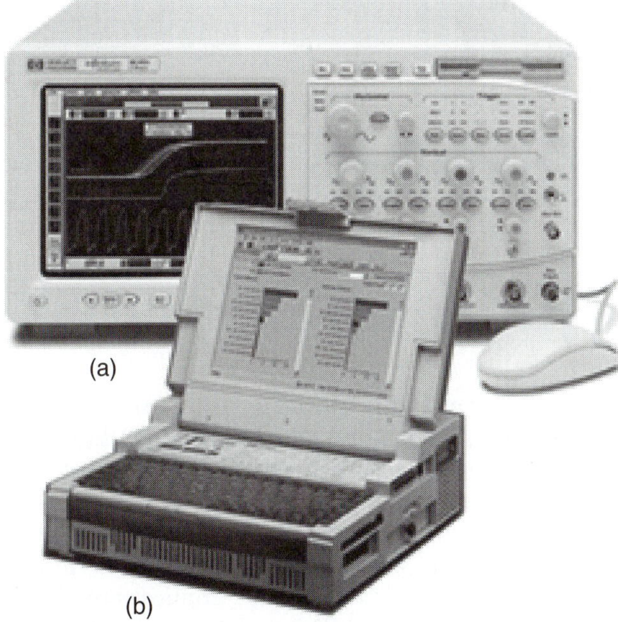

Figure 44.2 PC is embedded in the instrument. Notice that the instrument (a) retains the appearance of an instrument; in (b), it looks similar to a laptop computer.

Figure 44.3 Companion software. Software running on a PC can be used to supplement the functionality of the instrument. In this case, the user can capture screen shots from the instruments and export them to standard PC applications.

Fig. 44.2B, a network advisory instrument appears to look like a laptop. The instrument is actually located in the lower portion of the laptop casing.

In some cases, the user might have access to the Windows desktop and other applications via the instrument and in other cases, the user might be "locked out" from interacting with the desktop and underlying file system. Start-up and shutdown might become issues if the Windows desktop is hidden. For more information on this configuration, see Chapter 10 on "Embedded Computers in Electronic Instruments."

44.3.2 PC companion software

In this configuration, (see Fig. 44.3) the computer software is run on the user's desktop computer, which is connected to the instrument via standard I/O, such as GPIB, RS-232, or Ethernet. The software typically adds functionality, which enhances the instrument's capabilities. For example, the software can be used to capture data and/or screen snapshots, which can be imported into a desktop application, such as Excel or Word. The companion software can also be used to automate monitoring of measurements and compile the data into a summary report. Companion software might range from simple add-ins to a standard spreadsheet to a sophisticated standalone statistics package. See Chapter 41 on Software and Instrumentation for additional examples.

44.3.3 Virtual front panels

The rack and stack instruments or card-cage configurations are not new, nor are the virtual front panels that provide the interface to the user. A virtual front panel can be defined as a software-based user interface that allows a user to access instrument functionality without directly interacting with the instrument itself. In many cases, the interfaces formerly performed using Unix interface standards have migrated to Windows-style interfaces. In addition, the potential for integration with other applications and use of the Internet can greatly enhance the capabilities of the previous configurations. The virtual front panel on the factory floor might be a simple touch screen running on a PC platform, but a supervisor might view a summary of the tests

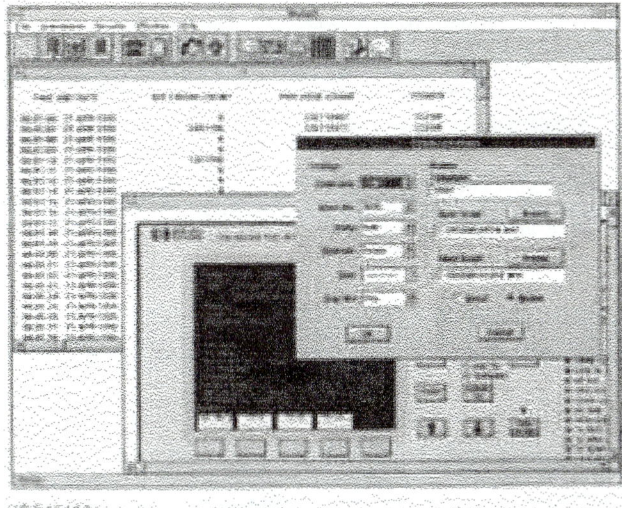

Figure 44.4 Virtual front panel. Test sequences are displayed, in addition to test results, that can be exported into standard PC applications for analysis and reporting.

being run via a Web interface. In another distributed model, an engineer at a central control site can view the same graphical user interface as technicians in the field, similar to the one shown in Fig. 44.4. See Chapter 45 on "Virtual Instruments" for further information.

44.3.4 Remote front panels

Users often need remote access to their instruments and systems. This can be for the purposes of monitoring, troubleshooting and in some cases, for remote control. With the advent of Web capabilities, it is possible to display test and measurement information via a browser, which can then be viewed by multiple users in remote locations. This can be a replacement for, or in addition to, the virtual front panel local to the instrument/system. Web browser interfaces have a different navigation paradigm than a typical Windows application, which can lead to some design challenges. In some cases, the Windows interface itself is exported remotely, as a mirror of the primary application. Users can become confused as to what functionality is available remotely, so the designer must make that very clear. For example, sometimes remote monitoring is available, but remote control is not—because of potential safety issues. In Fig. 44.4, the same information is available to different users in the same format. Sometimes just a subset of information is available for quick troubleshooting, as in Fig. 44.5. The subset of information might appear in a Web browser or a Windows interface.

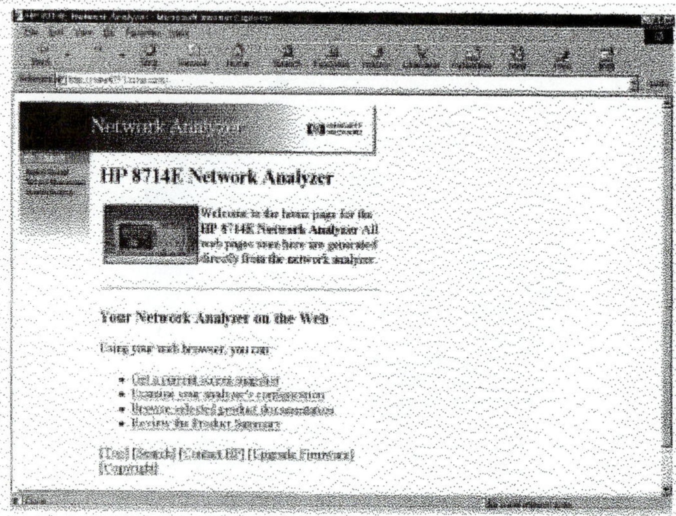

Figure 44.5 Remote access. From this Web page, the user can get a screen snapshot from the instrument, as well as access online documentation.

44.3.5 Web for help and documentation

Web browsers are used increasingly to display documentation that was previously printed in large manuals. These manuals are often outdated and/or lost in the user's workplace. Internet technology allows for easy updates and immediate access to documentation. In some cases, a Web server is placed inside of the actual instrument, which allows the user to bring up the documentation on the instrument itself or export it to their desktop computer. Via the Internet, the user can contact the vendor with support issues and/or read comments and hints from other users. In Fig. 44.5, the user not only has access to the instrument documentation, but also to the company's external Website. See Chapter 42 on "Network Connectivity for Instruments" for further information concerning Internet capabilities.

44.4 Changes in User's Overall Experience

The additional configurations described previously can change the user's overall experience in performing their test and measurement tasks. Three key aspects of the user's experience can vary, based on these additional configurations:

- Focus of attention
- Transfer of learning
- Mental model

44.4.1 Focus of attention

In the past, the engineers' *focus of attention* has been on the instrument, with the exception of rack and stack instruments, in which the users' attention is on a centralized front panel. When the PC is embedded in the instrument, the user's attention stays focused on the instrument, yet the interaction might more closely resemble interaction with a PC. With the PC being used as companion software, the user's attention could be split between the instrument and the PC, depending on the user's task. With remote access via a Web browser, the user's attention will shift back and forth from the instrument to the Web browser to access documentation required to complete the task.

This potential for divided attention impacts the user's interaction styles and causes users to develop strategies for moving efficiently between the available interfaces.

44.4.2 Transfer of learning

Transfer of learning refers to how much of what a user already knows impacts what they need to learn. Transfer of learning can be positive or negative. A positive transfer of learning means that the users can make use of much of what they had previously learned, in order to learn something new. With the evolution of user interfaces, some users will prefer to see interfaces, which are very similar to what they have been used to. Other users might want to see instrument functionality translated into user interfaces that look more like Microsoft applications. This becomes a challenge for designers and users alike to balance the competing transfer of learning issues.

44.4.3 Conceptual model

Users form a *mental model* or conceptual picture of whatever they interact with. Sometimes this model is consistent with what the designer had in mind and sometimes it is not. The conceptual model is the picture the user forms that drives their style of interaction with the device.

Users typically attempt to match this picture with something else familiar in their environment. As PCs are added into the configuration equation, they might change the user's conceptual model of the instrument and subsequent style of interaction.

Table 44.1 shows how the user's primary attention, transfer of learning, and mental models could vary across different configurations. It is important to understand that the choice of configuration can change the user's experience.

It is obvious from the table that the user's experience varies across the different configurations and designers of these configurations should be cognizant of the impact on the users' experience.

TABLE 44.1 Changes in User's Experience

Configuration	Focus of attention	Transfer of learning from	Conceptual model
Embedded PC	Instrument	Instrument/PC	Specialized box
Companion software	Instrument/PC	Both	PC helper for instrument
Virtual Front Panel (faceless)	PC front panel	PC	PC-driven system
Virtual Front Panel (rack and stack)	PC/some on instrument	PC/instruments	PC front end
Web remote access to subset of functionality	PC/Web	Web browsers	Web page
Remote front panel mimic of instrument	PC remote front panel	Instrument	Mirror of instrument
Web-based documentation	PC/Web	Web browsers	Web page
Distributed	PC front panel	PC	System

44.5 Drivers for the Evolution of Graphical User Interfaces

In addition to evolving configurations, users are also being confronted with evolving user interface models.

With the increasing prevalence of the PC in the instrument world, instrument users are more likely to encounter user interface models that resemble desktop applications. However, these user interface models are rapidly evolving and it will be important for the instrument user to have some understanding of what drives these models and how it might impact their instrument usage.

A number of drivers contribute to the direction and evolution of user interfaces in general, which also impact instruments and systems. The key drivers include:

- Change in workforce skills
- Lower cost of PCs and PC boards
- Upgrades in operating systems
- Graphical user interface standards
- Multimedia capabilities
- Increasing use of networking and the Internet
- Change in user expectations
- Move from separate functions to task components

44.5.1 Change in workforce skills

The new graduating classes of engineers grew up with computers. They feel comfortable with computers, graphical user interface, and desktop applications. In addition, not everyone who interacts with an instrument/system is necessarily a measurement expert. This implies that it is increasingly necessary to leverage what users already know from the computer world to facilitate their learning in the instrument world.

44.5.2 Lower cost of PCs and PC boards

The cost of PCs and PC boards is low enough to allow engineers to have not only a desktop system, but also a PC on their workbench. PC boards can now be considered for use as embedded microprocessors. See Chapter 10 on "Embedded Computers." The availability of PC technology enables the use of graphical user interfaces for both embedded instruments and virtual front panels.

44.5.3 Upgrades in operating systems

Often with operating system changes come changes in the desktop appearance. This impacts subsequent application interactions and user interface standards.

44.5.4 Graphical user interface standards

Because Microsoft Windows has become the standard desktop environment for many users, it sets an expectation for the behavior and appearance of other PC applications. This standardization allows users to transfer learning from one application to another. User interface style guides dictate usage and layout of screen components to achieve a common "look and feel" across applications.

44.5.5 Multimedia capabilities

GUIs can be augmented with multimedia capabilities, including audio, video, and animation. The application can actually interact with the user via agents and voice recognition. These enhanced capabilities can be used to effectively incorporate training into the instrument.

44.5.6 Increasing use of networking and the internet

The exponential growth of the Internet and Web browsing has changed the way many people do their work. This is also impacting the world of engineering. See Chapter 44 on "Network Connectivity."

Many companies now use Websites to provide software updates, documentation, and support. Some instruments are incorporating Web servers to enable the instrument to serve up Web pages.

44.5.7 Change in user expectations

Users are now familiar with the evolving ease of use of PC standard applications and have come to expect the transfer of learning advantage they experience across applications. This expectation can now be applied to instrument interfaces. In addition, users expect that their needs will be taken into account as products are developed. Increasingly, designers are adopting user-centered design processes to take into account users' input into product design.

44.5.8 Move from separate functions to task components

User interfaces are increasingly becoming task-focused, as opposed to function-focused. A *task focus* means that the interface is presented in a manner that supports what the user needs to accomplish and the steps toward achieving that goal. A *function focus* implies that the interface presents functions or features that the instrument is capable of performing without regard to the user's real task goals. In a function-focused interface, it is up to the user to link together the functions, usually via menu items and dialogs to perform the task goal. In a task-focused interface, the tasks can be represented as objects, which allows the user to select one object, for example "perform measurement." When tasks are intelligently encapsulated into objects, both usability and productivity can be improved.

44.6 The Evolution of Instrument User Interfaces

Although many handheld instruments have not changed significantly in appearance (except to move from analog to digital display), the above drivers for change have heavily influenced other instrument configurations. This has resulted in the following key changes for instrument/system users.

44.6.1 Software versus hardware controls

Some instrument graphical front panels include a one-to-one mapping of the hardware components in the software. The larger CRT display of the computer allows for more information presentation options than the smaller displays available in most stand-alone instruments. The virtual front panel can be configured to take on different user interface presentations (personalities), as appropriate for different types of measurements. These "personalities" might contain a relevant subset of the overall available functionality.

In PC embedded instruments, the graphical user interface might contain a superset of the functionality represented by the hardware controls. Advanced functions or less frequently used functions can be accessed via pull-down menus.

The general trend is for fewer hardware controls (knobs, dials, and buttons), with more functionality represented in the software. This allows for software updates, online help capabilities, and message feedback, which cannot be performed in hardware controls. The exception is, of course, emergency controls, which require immediate user access and hardwiring. Controls that are frequently accessed are likely to be duplicated in hardware and software controls.

44.6.2 Interaction with desktop

Virtual front panels for instruments/systems can now be launched from a Windows desktop. Users can more readily interact with other applications, such as spreadsheets and statistics packages. File system management can be handled via the desktop utilities. Users can get quick access to the instrument application via the taskbar, toolbar, start menu, or desktop icon. The embedded PC/instrument model provides the user with automatic ready access to the desktop, unless steps are taken to lock out access. This could have negative implications for the user, if they load multiple applications, thereby slowing the performance of the instrument itself.

44.6.3 Interaction with other applications

Because of application standards, it is now possible to "drag and drop" files and/or data from one application onto another. It is possible for set-up files from the instrument to be saved to the user's desktop PC and re-imported into the instrument, when needed. With some formatting specifications, data can be sent from the instrument into a spreadsheet or similar application for later statistical analysis. The results of this analysis can be exported into a word processor in order to create a report. Thus, other applications can serve to extend the capability of the instrument.

44.6.4 Interaction with web browsers

From the instrument or system, which includes an embedded PC or PC front end, users can access a Web browser. This can give the user access to documentation, reference information, vendor updates, etc. Users can potentially interact with vendors via "frequently asked questions" and/or converse with other users. This expands the users' world and could enhance productivity. In some configurations, the Windows graphical user interface can be exported remotely and viewed through a Web browser. See Chapter 42 on "Network Connectivity for Instruments" for additional information.

44.6.5 Changes in online help and learning paradigms

The evolving user interfaces have also included improvements in methods of providing user guidance. Online help includes task-based and context-sensitive (help on an item) options. Wizards (sequenced dialogs) are commonplace as a mechanism to provide guidance through sequenced tasks. These could be used to guide users through instrument set-up tasks. Multimedia tutorials are becoming commonplace to teach users about a new application. There is less emphasis on paper manuals and more emphasis on online user guidance. This places the information closer to the context of the task. In many cases, the user can print out a paper document and/or pages from the help system, if so desired.

44.6.6 Menu access styles

Many conventional instruments used front-panel buttons (function keys) to provide hierarchical menu structures. The menus were often used for setting options, but sometimes included advanced features. The standard windows menubar is now used in many systems applications—especially test executives, which drive many rack-and-stack systems. Instruments with windows embedded can also make use of the menubar. The toolbar, although originally intended to provide shortcuts to menu operations, is being used as a general

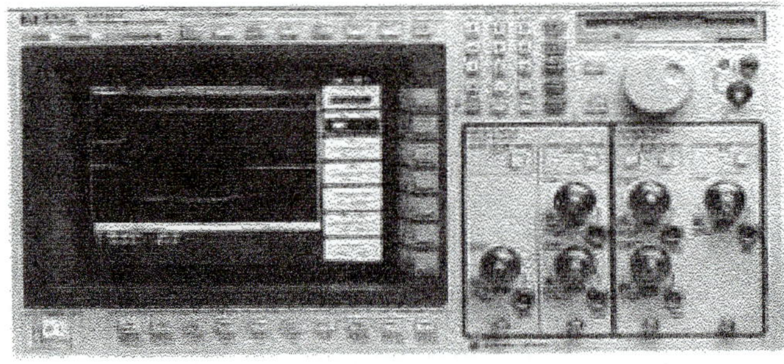

(a)

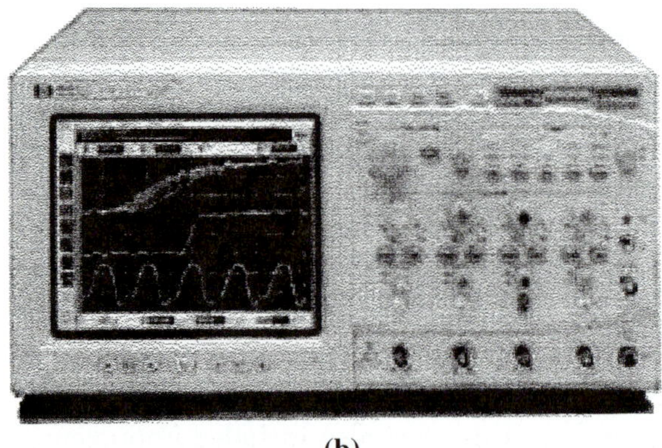

(b)

Figure 44.6 Conventional vs. Windows-based instrument. The menus are accessed via the function keys (a) and the menus are accessed via the menubar (b).

launch area for actions, dialog access, and shortcuts. Some applications include a taskbar on the left side of the application, which provides for shortcuts to specific task areas. Icons are often used on taskbars and toolbars to represent an item or action. Measurement-specific icons are being developed, although the lack of standardization in this area can present some challenges for users. (See Fig. 44.6)

44.6.7 Pointing devices

Graphical user interfaces are referred to as *point-and-click interfaces,* meaning that the user points the cursor at an item to select it and clicks or doubleclicks to take action on the selection. The mouse has been the preferred pointing device, but others are available. These include glide paths, trackballs, embedded eraser points, stylus, and even touch screens. Voice-recognition and voice-activation technology is improving to the degree that it might be incorporated into some instrument interfaces.

44.6.8 Printing

Printing from an instrument is a common user requirement. Historically, this had been accomplished via printers connected by GPIB and/or RS-232. With the growing popularity of desktop printers, users often prefer to print from the instrument to their desktop printer. This can pose a problem when a legacy instrument does not support a newer printer driver. Some users resort to sneaker net (floppy disks) and others go through the computer to print data from the instrument. PC companion software (as described earlier in this chapter) can facilitate this method. The inclusion of the newer I/O interfaces, built into instruments, such as FireWire® and USB, can provide direct connection to the newer printers (see Chapter 43 on "Instrument to Computer Connection").

44.6.9 Changes toward task-focused models

As described in the previous section on drivers for user interface changes, the move towards task-based interfaces is responsible for subtle, but significant changes in user interfaces. There is an increasing focus on support for users' primary tasks. Conventional interfaces can present all the functionality available to the user without regard for design based on frequency or primacy of user tasks. The task-based interface facilitates the user to both find the desired task and execute it efficiently. The encapsulation of common measurements and tests into task objects could enhance both usability and productivity for the user.

Table 44.2 shows a comparison between conventional instruments and instruments with Windows user interfaces in some of the areas described.

Figure 44.6 shows a visual comparison of a conventional oscilloscope compared to an oscilloscope with a Windows-based graphical user interface. Al-

TABLE 44.2 Summary of Major User Interface Changes

User-interface component	Conventional instruments	Windows GUI instruments
Ratio hard controls/soft controls	More hard controls	More soft controls
Interaction with desktop	None	File system, application launch
Interaction with other apps	Minimal interactivity	Data transfer (cut and paste)
Interaction with web browsers	Not available	Readily available
Online help	Text, not context-sensitive	Context-sensitive, wizards, Web
Menu access	Function Keys	Menubar, Toolbar
Pointing devices	Hardware arrow keys	Mouse, touchscreen, trackball
Printing	Via Parallel or RS232 ports	Print via PC or PC standard I/O
Function vs. Task Focus	More function focus	More task focus

though at first glance they appear similar, Table 44.2 summarizes some of the differences.

44.7 Evaluating the Instrument User Interface

Users are placing more demands on designers for "user-friendly" applications. The instrument world is no exception in articulating this requirement. As productivity demands increase, users do not have time to spend with time-consuming trial-and-error learning styles to learn how to make a measurement. Usability is increasing in importance as a criterion for equipment purchase. Although the evolving user interfaces have generally resulted in progress toward usability, some challenges are still ahead for instrument users. When making purchase decisions, it would benefit users to consider the following key areas:

- Usefulness
- Learnability
- Usability
- Data manipulation
- Hardware-related aspects
- The vendor's customer focus

Section 44.8 of this chapter provides a checklist that maps to the areas included in this section.

44.7.1 Usefulness

The instrument is said to be *useful* if it provides the functions required to perform the task. Whereas, *usability* refers to how easy it is for the user to perform the functions (ease of use).

Some instruments are very specialized and handle only certain types of measurements; others are more generalized, such as mixed-signal oscilloscopes, which perform both scope functions and logic analyzer functions. Each model has both pros and cons. With a specialized instrument, the user might need to have access to additional instruments to successfully complete their task goal, for example, troubleshooting a board failure. However, a more generalized instrument often presents more functionality to the user than they need at any given time, which can result in usability problems. Consider your own task flow when deciding if you should purchase a specialized or more generalized piece of equipment.

An instrument can be useful, but have poor usability. Once the user has determined that the instrument is "useful," the user might then want to assess usability. The following are some key aspects of usability for the user to consider. The first aspect of usability is termed *learnability* or ease of the initial learning of a product.

44.7.2 Learnability

The intermittent use of some instruments, such as logic analyzers, means that the user must essentially relearn the interface each time they interact with it. In assessing learnability, the user should look for interfaces that comply with common standards, are task-based, and provide adequate learning tools, such as tutorials or quick start guides. Users might also want to consider transfer of learning from previous products and/or desktop applications and their ability to form a coherent conceptual model of the instrument. These two important aspects of learnability are elaborated on, as follows.

Transfer of learning. Users generally get used to a particular instrument and how it works. Because instruments typically have a much longer shelf life than PCs, it might be even harder for a user to give up an existing instrument for a new one and risk a loss of productivity. Although the use of Windows standards should facilitate common look and feel, there is still room for variability across instruments. This can be frustrating for users. Because user interfaces are evolving, the users might need to relearn an interface as new standards are established. All of these factors could negatively impact the transfer of learning. A delicate balance exists between innovation and standardization, which needs to be considered in product design.

Look for the product that gives you the most transfer of learning. If the users are avid desktop PC users, then the instrument that incorporates PC use models and GUI standards will provide them with some automatic transfer of learning.

Conceptual models. By interacting with a device, users form a conceptual model or mental model of how it works and how they can most efficiently interact with it. Users might also have preconceived conceptual models of how something "should" work, based on previous experiences with that device or a similar one. A conceptual model is a picture that the user forms, but rarely articulates, about the overall behavior of a device. Good designs incorporate an explicit conceptual model into the design that serves to pull the parts of the design together into a meaningful whole. The user can then use the conceptual design to predict the behavior of a device, which generally leads to greater efficiency. If designers are not conscious of this phenomenon, then it is likely to result in an incoherent design. Users' attempts to form a conceptual model will result in frustration and inaccurate predictions regarding the behavior of the device.

As you evaluate an instrument, can you predict how it will behave? Does it behave in a manner that is consistent with your conceptual model? Does it behave consistently within itself?

44.7.3 Usability

Usability is defined as the ease of using the product to perform the functions that allow the user to successfully reach their task goals. Several key areas of usability related to instruments are covered in this section.

Task-based design. It is difficult to overemphasize the importance of task-based design to facilitate usability. A task is defined as the steps the user needs to go through to accomplish their ultimate goal. For example, measurement setup is a task. Task-based design has been covered with regard to graphical user interfaces; however, it must be extended to include the overall interface. If task completion requires that a user complete numerous measurement set-ups within graphical user interface dialog boxes and then push a hardware button to initiate the measurement, then this task sequence needs to be facilitated via the overall design. Frequently performed task actions might be available to the user via front-panel hardware controls, grouped together according to task function. This is especially useful if one setup is used to make multiple measurements. These same actions might also be available as menu choices within the GUI so that the user can maintain attention within the same interface, rather than shifting back and forth between hard and soft controls. Regardless, the terminology should be identical so that the user knows that there are two ways to achieve the same result. Look for instruments that intuitively support the tasks that you most frequently perform.

Menu organization/terminology. The optimal organization of menus can contribute to the usability of the instrument. Top-level menu items should correspond to Windows standards, whenever possible. Beyond that, top-level menus should represent the primary functional groupings of objects that the user is interested in. The subsequent menu items should be arranged in task sequence and/or functionally grouped. The terminology should match the users' own terms. The users should be able to readily locate the desired item in the menu hierarchy without engaging in trial-and-error behavior.

Instrument set-up. In many cases, the set up of the instrument is more challenging and time consuming than the actual taking of the measurement or running of a test. Many opportunities are available for improvement via the use of online wizards, templates, defaults, examples, and/or task guidance. With the PC in the equation, the user could save the set-up files for later reuse. Look for GUIs that provide set-up guidance online and capabilities for saving and importing set-up files.

Although the usability of instrument user interfaces is constantly improving, for some instruments, probing continues to be problematic. Probes could fall off, break, or have poor connections, which affects the outcome of the measurement. Comparably, for systems, fixtures, which are essentially multiple probe points, present an opportunity for usability improvements. Look for graphical user interfaces that support probing tasks, such as feedback on good and bad probe points and/or facilitating probe assignment.

Feedback and troubleshooting. Feedback, in terms of a graphical user interface, includes dialog boxes that indicate errors, warnings, or notices, as well as status indicators. It is important that feedback dialogs are informative and "user friendly." Feedback that appears to "scold" the user and provides no information about what to do next is frustrating and impedes the user's progress. If a setup, test, and/or measurement fails, it is important that the GUI provide troubleshooting assistance. This can be located within the online help system and/or can be provided within the context of the task problem. Look for a GUI that provides informative feedback messages.

User expertise. Users have different levels of expertise, both in domain knowledge and general PC knowledge. It is difficult for designers to accommodate all levels of knowledge. If the designer assumes too much knowledge, the user will find it difficult to use the interface. If the designer assumes too little knowledge, then the user might be inundated with unnecessary helpful features and extra task steps. Standard approaches to this problem have included the use of modes (expert and novice modes), optional wizards, and/or reliance on online help. Some instruments include options to personalize the user interface, based on the users' characteristics, including such variables as experience level, frequency of use, and key tasks performed. Consider your level of expertise and whether or not the instrument accommodates your needs. If the instrument is shared among several users, does it accommodate the varied levels of expertise?

44.7.4 Data manipulation

Instruments provide different levels of data manipulation. PC companion software can facilitate the export of data into a spreadsheet and/or statistical package. Virtual front panels might have direct linkages to related data applications. Test systems might require links to databases. In the simplest case, users just want to print a snapshot of the screen, for example, printing a waveform.

Determine what your data-manipulation needs are and check to see if the instrument and/or PC companion software can meet these needs adequately. See Chapter 41 on "Software and Instrumentation" for information on software-enhanced data manipulation.

44.7.5 Hardware-related aspects of usability

When assessing the usability of an instrument, it is difficult to separate out hardware controls from the GUI-related controls. It is the combination of the hardware components, described in Chapter 12, "Instrument Hardware User Interfaces," and the evolving graphical user interfaces described in this chapter, which together form the *gestalt* (the whole is more than just the sum of the parts). The related hardware aspects are included here.

Relationship of hardware and software controls. As previously mentioned, the GUI often includes a superset of the functionality found on the hardware front panel. The hardware front panel should be utilized for functions that are frequently accessed and/or require immediate user action. Many start/stop actions are located on the front panel, as well as in the GUI. Identify which functions you need immediate access to and check to see that they are in a convenient location. Review whether or not the terminology is consistent between the hard and soft controls.

Ergonomics. Ergonomics is traditionally defined as the study of "man" and "machine" interactions. Good ergonomic design, as it relates to instruments, would focus on the location, shape, and relative position of hardware controls on the instrument box. Numerous ergonomic guidelines must be considered, beyond the scope of this chapter. Chapter 12 covered some of the aspects of hardware ergonomics. Overall ergonomic considerations would include the relative position of the CRT screen area (that displays the GUI) and the hardware front-panel components. Generally speaking, the hardware controls should not be placed in a position that requires the user to reach across the screen area, thereby, blocking the view of the screen. Related controls should be grouped together. Important controls should be visually distinguishable from less-important controls.

Pointing devices. It is especially important to consider the use of the instrument and the environment when evaluating the pointing devices available with the instrument. Rack-mounted instruments often do not have a surface area available for the use of a mouse. However, some types of vertically embedded pointing devices on the front panel might create ergonomic strain for users. Consider the environment in which you will use the instrument and your personal preferences to determine if the pointing devices provided with the instrument are acceptable.

Screen real estate. Instruments typically have less screen real estate than the average desktop monitor. This can lead to ease-of-use problems because screen objects are usually smaller, closer together. Therefore, they are more difficult targets in terms of fine motor behavior. Even in the case of PCs embedded in instruments, the advantages of the desktop-like interface might be compromised by the small-screen real estate. Some instrument configurations allow for use of a desktop monitor. Consider whether or not the screen real estate on the instrument is adequate for your needs.

44.8 Evaluation Checklist

There are a number of criteria the potential user can employ, in order to select an instrument that is easy to learn and operate. The previous section described key areas to consider. Table 44.3 provides a checklist as a summary of those areas and can be used as a starting point for evaluating the instrument user interface.

TABLE 44.3 Evaluation Checklist for Instrument Graphical User Interfaces

Usefulness		The instrument supports the tasks I want to perform.
		The functions I frequently use are easy to find.
		The interface is task-focused.
		The instrument is generalized or specialized based on my needs.
Learnability		The instrument GUI includes task-based help.
		The instrument GUI includes context-sensitive online help.
		The instrument GUI includes wizards and/or set-up guidance.
		The instrument GUI provides assistance with probing/connections.
		I can transfer learning from my previous experience with PCs.
		I can transfer learning from my previous experience with other instruments.
		I can experience success within 15 minutes of using the interface.
		The instrument model matches my conceptual model.
		I can predict the behavior of the GUI.
		Levels of user expertise are accommodated.

TABLE 44.3 Evaluation Checklist for Instrument Graphical User Interfaces (*Continued*)

Usability		The menus are organized so that I can find items of interest.
		The icons are recognizable (or include tool tips).
		The terminology used is familiar to me.
		The terminology is used consistently across the application.
		The primary tasks are easy to locate.
		The primary tasks are easy to execute.
		Task dependencies are evident.
		Measurement set-up is facilitated.
		Appropriate feedback is provided to indicate progress.
		Error messages are easy to understand.
		Error messages include recommendations.
		Troubleshooting assistance is available.
Data Manipulation		It is easy to access other relevant applications.
		It is easy to export data to other applications.
		It is easy to import data from other applications.
		It is easy to record measurement results.
		It is easy to annotate measurement results.
		It is easy to connect the instrument/system to a printer.
		It is easy to print measurement results.
Hardware-Related Usability		The frequently accessed functions are available on the front panel.
		Any emergency controls are available on the front panel.
		The GUI includes a superset of the functionality on the front panel.
		The front panel layout follows ergonomic principles.
		The pointing device is appropriate for the environment.
		The pointing device is familiar.
		The screen real estate is adequate for my tasks.
Vendor's Customer Focus		The vendor is customer-focused.
		The vendor considers customer input.
		The vendor follows customer-centered design principles.
		The products has been designed from the user's point of view.

Bibliography

Nielsen, Jakob, *Usability Engineering.* New York, Academic Press, 1993.

Norman, Donald A., *The Invisible Computer.* Cambridge, Massachusetts, 1998.

Chapter 45

Virtual Instruments

Darren Kwock
Agilent Technologies
Loveland, Colorado

45.1 Introduction

Virtual instruments are devices configured from independent hardware components by software to perform custom instrumentation functions. Because they are software defined, they can be reconfigured to perform different measurement and instrumentation functions when needed. This differs from traditional electronic instruments, which are hard wired into one configuration and used to perform one specific task.

Traditionally, electronic instruments have been easily identifiable, box-shaped objects with a front panel. The front panel provides a user interface that allows the operator to interact and control the instrument. Typically, a front panel contains knobs, buttons, and possibly a display, as shown in Fig. 45.1. Advances in electronics have provided great improvements in accuracy, functionality, and reliability for instrumentation. At the same time, the rise of digital technology has led to data buses that connect instruments to computers efficiently. This provides the computer with the ability to control and exchange data with the instrument in an effective and timely manner.

45.1.1 Computer technology in test and measurement applications

Computers have evolved at a very high rate. Low-cost, high-performance computers are very common tools for engineers and technicians. This pervasiveness of computer technology has had a profound impact on traditional instruments. A computer, along with advanced software and graphical user interfaces, allows users to emulate and often surpass the capabilities of traditional instrumentation. This is also assisted by modular instruments, which provide basic measurement functionality on printed circuit boards.

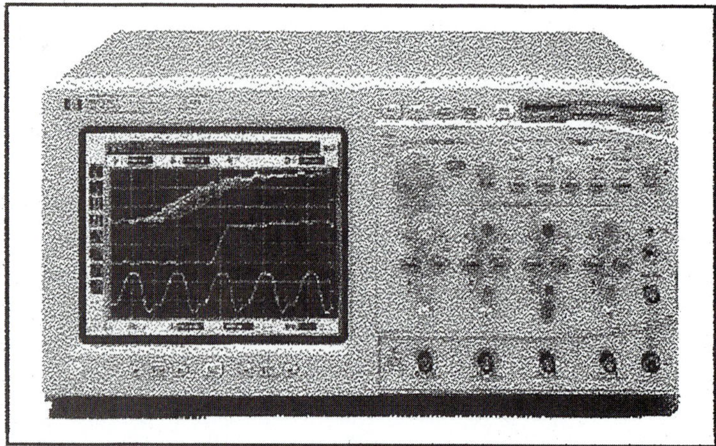

Figure 45.1 A traditional "box" instrument with front panel.

45.1.2 Modular instrumentation technology

Two examples of modular instruments are PC plug-in cards and VXI instruments (see Chapter 40 for a detailed description on modular instruments). PC plug-in cards are breadboards that follow computer bus standards and follow conventions used for computer peripherals. VXI is a standard for modular instrumentation that extends the VME computer architecture to support instrumentation. These modular instruments can be placed directly into a computer, as in PC plug-in cards; or placed in a separate enclosure, called a *mainframe* (such as in VXI).

Modular instruments often provide just the basic measurement functionality. They do not have front panels, and thus do not have knobs, buttons, or other methods to physically control the instrument (refer to Fig. 45.2A and Fig. 45.2B for a PC plug-in card and a VXI instrument). Instead, modular instruments are dependent on software to provide control capabilities. The combination of measurement capabilities and computers created a new type of instrument called a *virtual instrument.*

45.2 Instrument Models

Virtual instrumentation combines measurement, computers, and software to solve a particular measurement problem. Virtual instruments provide an alternative paradigm to stand-alone instruments, which contain all the functionality in a single device. Virtual instruments allow people to customize an instrument's functionality and control how it is presented to the user through software. Software is the key component of a virtual instrumentation. Software provides virtual instruments with "personality" and transforms a collection of measurement hardware and computers into customized instrumentation.

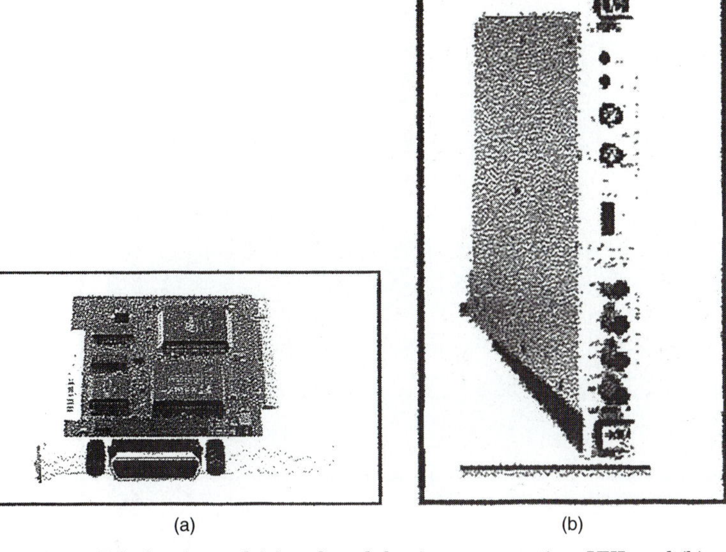

(a) (b)

Figure 45.2 PC plug-in card (a) and modular instrumentation: VXI card (b).

Virtual instrumentation came about as computers and instruments became more closely intertwined. This section describes four models that developed as computers and instruments evolved.

- Stand-alone model
- Computer-assisted model
- Modular instrument model
- Network instrument model

45.2.1 Stand-alone model

Prior to the 1970s, electronic instruments were stand-alone devices that people manually controlled. They contained knobs and buttons and displayed results on a front-panel display. All measurements needed to be recorded by hand. This can be called the *stand-alone instrument-use model.* For example, many common hand-held and traditional "box" instruments are used in this manner.

45.2.2 Computer-assisted model

The 1970s marked the first time that computers became generally available. Computers could also be used to control instruments because of the development of *GPIB* (*General-Purpose Instrument Bus,* also known as *IEEE 488.2*).

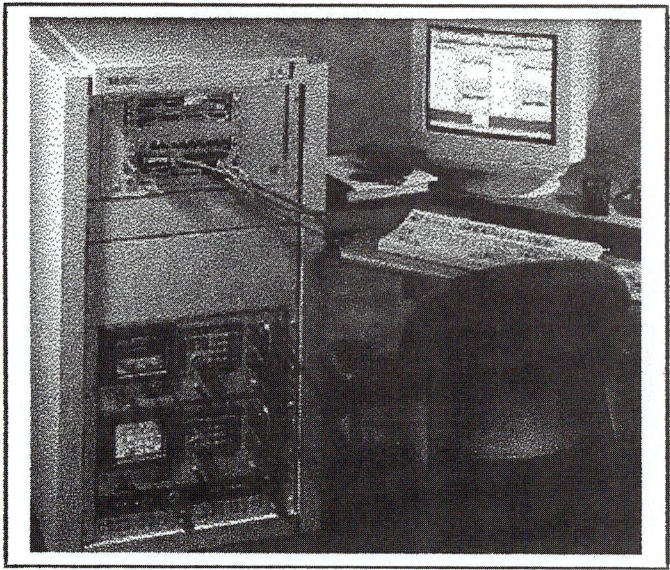

Figure 45.3 Computer-assisted model with a rack of GPIB instruments.

Typically, each GPIB instrument performed a specific measurement and users would "rack and stack" these instruments together to form a measurement system, as shown in Fig. 45.3. The earliest computer users essentially followed the same procedures that were developed with stand-alone instruments, but used the computer to help automate some of these tasks. This usage can be described as the computer-assisted-use model. The computer allows measurements to be completed quickly and repeatedly, as compared to the stand-alone instruments, which was a manual process. This system was also flexible because different rack-and-stack instruments could be combined and controlled by the computer to perform more customized measurements.

45.2.3 Modular instrument model

The 1980s saw the development of modular instrumentation systems. These systems were the first devices designed to be completely computer controlled. A major distinction is that modular instruments do not have a user interface built into the device. They were dependent on the host computer to provide a mechanism for people to interact with the instruments. Modular instruments, combined with a computer and software, provided expanded flexibility. These measurement capabilities are defined by the software, which allows the user to control how the measurements were completed and presented. The concept can also be applied to traditional stand-alone instrumentation as well. This combination of measurement, computation, and software formed the virtual

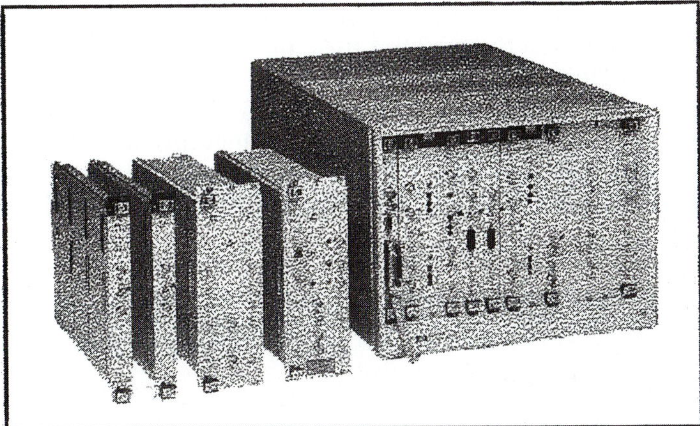

Figure 45.4 A modular instrument model.

instrumentation concept. An example of a modular instrumentation system is shown in Fig. 45.4.

45.2.4 Network instrument model

The 1990s gave rise to networking and the explosive growth of the Internet. This provided the technology to further expand virtual instruments to span large geographic distances. Previously, instrumentation systems were connected to a local computer. Networks allowed multiple computers to be connected and thus allowed instrumentation, which was connected to the computers, to also be distributed across a wide area, as shown in Fig. 45.5.

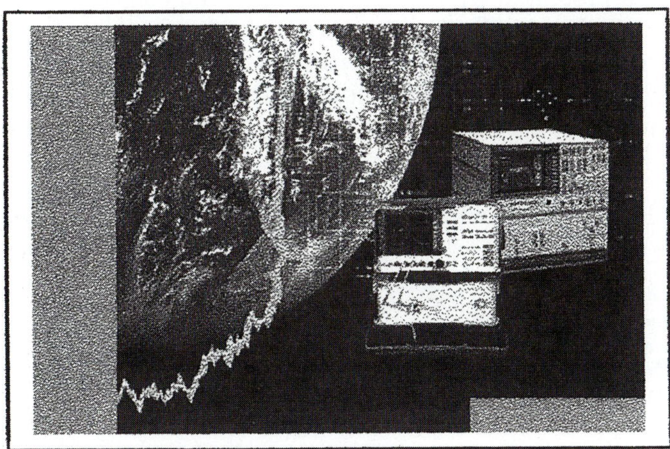

Figure 45.5 A network instrument model.

People using instruments can apply virtual instrumentation to the computer-assisted model, the modular instrument model, and the network instrument model. This occurs because people can choose different configurations of measurement hardware and software in order to construct an instrument system that is optimized to perform their measurement task. From the instrument user's perspective, virtual instrumentation does not apply to the stand-alone model because the instrument is used manually and does not involve a computer. However, from the instrument developer's perspective, virtual instrument concepts can also apply to the construction of complex box instruments that are developed around PCs.

45.3 Instrumentation Components

At a high level, all instruments generally have the following basic components:

- Measurement
- Computation
- User interface

Other items that are less visible, but are also important, are:

- Software
- Communication infrastructure

A hardware-oriented block diagram of the measurement, computation, and user-interface elements of a spectrum analyzer is shown in Fig. 45.6. Test and measurement components include both sensing and sourcing operations. The sensing components convert physical world signals into digital signals that can be processed by the computation components. The results of this operation can then be displayed on the user interface.

Software and communication infrastructure elements are also important items that are part of an instrument. *Software* is the program that controls the computational hardware. The communication infrastructure connects the

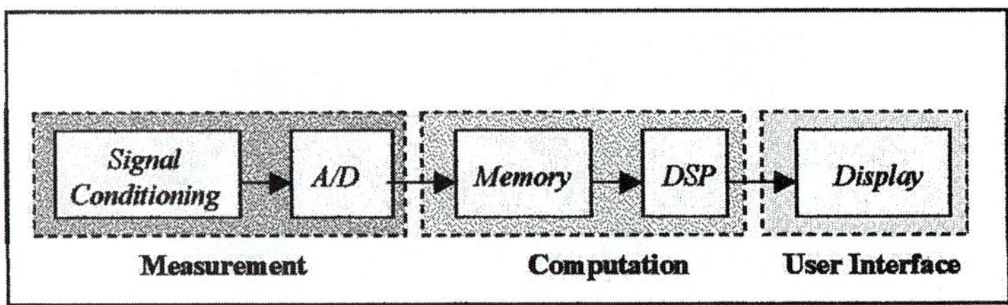

Figure 45.6 A block diagram of a complex instrument, showing the measurement, computation, and user-interface components.

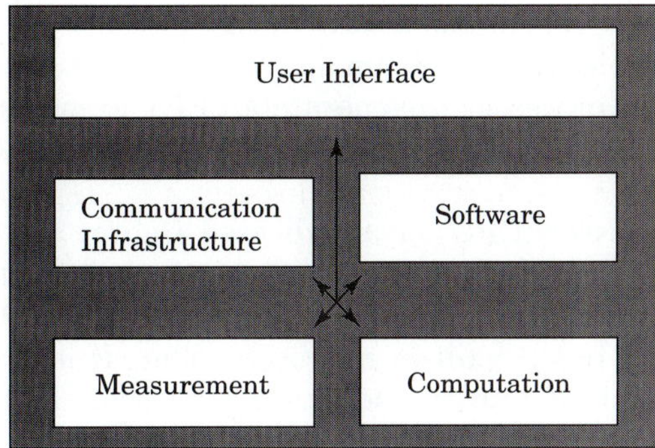

Figure 45.7 Instrument components consist of measurement, computation, user interface, and are connected by the communication infrastructure and controlled by software.

measurement, computation, and user-interface components together to form a single system. Typically, the communication infrastructure is defined in terms of a standardized connection called a *bus*.

In traditional or stand-alone instruments all of these functions are combined into a single device. Virtual instruments have the same components, but they are not constrained to be in a single device. All of these instrument components are described in Fig. 45.7.

45.3.1 Measurement

Measurement components are the part of the instrument that actually performs the primary instrumentation function. For purposes of this section, *measurement* applies to both stimulus (sourcing) devices, as well as measurement (sensing) devices. Measurements involve physical processes that are typically analog in nature. In sensing devices, the measurement portion converts the physical (analog) world into a digital format that can be processed using computational resources, such as a digital signal processor. In stimulus devices, the measurement portion performs the reverse operation and converts digital information back into an analog signal that can be used. Typically, devices called *Analog-to-Digital Converters* (*ADCs*) and *Digital-to-Analog Converters* (*DACs*) perform these operations.

Real-world signals typically can not be directly inputted to an A/D or used directly from a DAC. This occurs because these signals typically are outside the operating parameters of the ADC and DAC. Signal conditioning is used to modify the signals to make them valid signals that can be used with ADCs and DACs. Signal conditioning typically involves amplifiers or attenuators, as well as a wide variety of filters that are used to condition the signal in to the desired waveform.

45.3.2 Computation

Computation deals with calculations and managing information. This is needed to manipulate and process signals that are sensed from ADCs. It is also needed to generate digital signals that are sent to DACs to create desired analog signals. For example, in a spectrum analyzer, the computation component uses a digital signal processor to generate an FFT (Fast Fourier Transform) on the ADC generated time-domain signal to convert it to the frequency domain.

Generally, a microprocessor (CPU) or a dedicated *Digital Signal Processor* (*DSP*) accomplishes the computational tasks. Computation can take place in one or more locations at the same time. For example, a traditional stand-alone instrument usually has a built-in microcontroller. If it is connected to a host computer, the host computer also has a processor. In a distributed measurement environment, several computers and/or instruments could be involved in the measurement process. Each one of these devices can have its own processor and can contribute to the measurement task. However, their operation and coordination must be controlled by software developed for this purpose.

Computation also involves all aspects of managing information, including memory and storage mechanisms. These functions can occur in an instrument or in the attached computer. Generally, the host computer has more resources available than the instrument. However, moving information between the instrument and computer can have an impact on performance.

45.3.3 User interface

The user interface is the part of the instrument with which the operator interacts. The two types of interfaces are: physical interfaces, which are associated with traditional instruments, and computer-controlled interfaces, such as graphical user interfaces. Computer user interfaces are typically associated with modular and distributed instruments, but can also be used with computer-assisted instruments, as well.

Physical interfaces. In traditional stand-alone instruments, the front panel of the instrument consists of physical knobs, buttons, and displays, which allow the instrument user to easily adjust the various controls and easily view the CRT display to view the results. The user interface in stand-alone instruments are optimized for human control. The front panel of a typical stand-alone instrument is shown in Fig. 45.8A.

Computer-controlled interfaces. Modular instruments, in comparison, are optimized for computer control and do not have physical interfaces, such as knobs or buttons. At best, modular instruments have LEDs that light to indicate their current operational status. A modular instrument is shown in Fig. 45.8B. People interact with modular instruments through the computer. This interaction can occur using a graphical user interface, if operator control is required. If operator interaction is not required, the computer can control the instrumentation without providing a user interface.

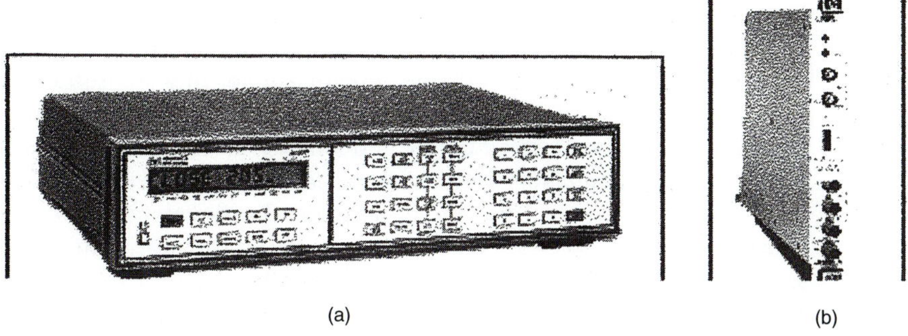

(a) (b)

Figure 45.8 A comparison of the front panel of a stand-alone instrument (a) compared to a VXI instrument (b). Notice that the VXI instrument lacks a means of physically interacting with the instrument, as compared to the knobs and display screen on the front panel of the stand-alone instrument.

A graphical user interface provides a software-defined environment for interacting with the instrument. A graphical user interface can simulate a front panel by providing software "widgets," which represent physical knobs, buttons, and displays. This allows people who are familiar with traditional instruments to easily use virtual instruments—even though the underlying instrument hardware is different from traditional instruments. An example of a simulated front panel is shown in Fig. 45.9A.

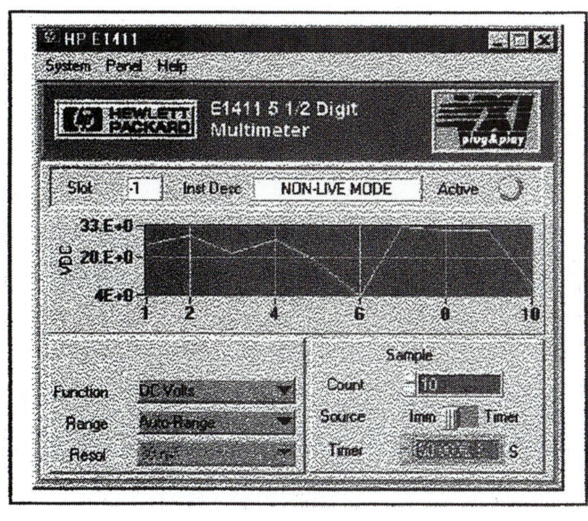

(a)

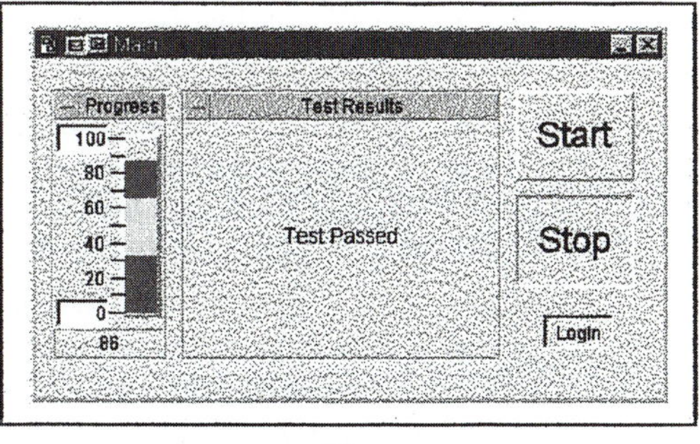

(b)

Figure 45.9 A comparison of simulated front panels and operator interfaces. A simulated front panel (a) and an operator interface (b). Simulated front panels emulate the physical controls used in traditional instruments. Operator interfaces are simplified views of the information from an instrument system.

The graphical user interface is not restricted to simulating an instrument's front panel. In complex instrument systems, it can be desirable to have a simple user interface with a few buttons and a simple graphical display to indicate the status of the system. For example, in a manufacturing test environment, a non-skilled operator is typically running the instrumentation system. For this type of operator, it is desirable to have as few controls as possible to make the measurement system easy to use. An example of an operator interface is shown in Fig. 45.9B. During the development and debugging of such a system, a test engineer can use simulated front panels and other more-sophisticated tools to access the instruments.

In virtual instruments, the user interface is the only mechanism in which people interact. The user interface hides the implementation from the user. The "instrument" could consist of a single box instrument or it could be composed of many instruments working together to cross a network and provide the desired outcome. The user interface hides this fact from the user.

45.3.4 Communication infrastructure

Communication infrastructure is important because it provides the means by which instrument components are connected together. In traditional standalone instruments, all components were inside the box and communication was not an issue. Instrument-to-computer communication became important as computers became more involved in instrument systems.

Two communication approaches are:

- Bus structures
- Network structures

Bus structures. The most common types of buses are the computer bus standards, such as ISA, PCI, or VME. *ISA* (*Industry Standard Association*) and *PCI* (*Peripheral Components Interconnect*) are standard computer buses used in personal computers. Many standard computer peripherals (such as modems, sound cards, and video cards) are designed to work in the ISA and PCI bus. *VME* (*Versa Modular European*) is used in industrial computers. Additional buses are the VXI bus and the MXI bus. VXI is based on VME and adds extensions for instrumentation. MXI is an instrumentation bus developed by National Instruments. It has many uses, such as extending the VXI bus or connecting the VXI bus to the PCI or ISA bus.

Other common buses include RS-232 and GPIB. *RS-232* is a serial protocol used in a wide variety of computers and devices. The IEEE has standardized RS-232 for instrumentation under IEEE 1176. *GPIB* (*General-Purpose Instrument Bus*) is the most popular instrument bus.

Network structures. Computer-to-computer communications have become more important as networking becomes more pervasive. The most popular are the Internet and Local-Area Networks (LANs).

45.3.5 Software

Software provides the instrument with personality. It embodies the procedures and algorithms that control the analysis and presentation (user interface) of information. The several types of software are:

- Firmware
- Instrument drivers
- Application development environments
- Test programs
- Test executives

Software for a particular purpose can be obtained from many sources, including being bundled by the instrument manufacturer, developed by the end user, or part of a third-party software package.

Software is important because it provides the instrument's personality and is a unifying factor. It also unites all the parts into a single system and makes everything appear as one cohesive unit. It also allows the same type of hardware to be used for different purposes. For example, a general-purpose cellular telephone test instrument can be customized with software to be a specialized *Code-Division Multiple Access* (*CDMA*) tester for a specific vendor's cellular telephone.

Software also hides the underlying measurement hardware and provides a higher level of abstraction that is easier for people to use and interact with.

Firmware. *Firmware* is a specialized type of software that is permanently fixed in a hardware element, such as a *Read-Only Memory* (*ROM*). Typically, the instrument manufacturer provides firmware. Firmware defines low-level "intrinsic" functionality and is not modifiable by end users. An example of firmware is the software that defines an instrument's programming language, such as *Standard Commands for Programmable Instruments* (*SCPI*). SCPI is a standard for an instrument's language and provides standard syntax and semantics for communicating with instruments.

Instrument drivers. Instrument drivers are instrument specific pieces of software that provide easy to use control of the instrument in a standard fashion. Instrument drivers provide a standardized *Application Programming Interface* (*API*) to an instrument. The standardized API allows the instrument user to program all instruments in an easy-to-use and consistent method. It allows the instrument's physical interface (e.g., VXI, PC plug-in, GPIB, etc.) to be abstracted, and the instrument drivers make all components appear similar from a software programming point of view. Typically, an instrument driver is provided by the instrument manufacturer and shipped with the instrument. Periodic updates are usually available from the vendor and are easily accessible from their Web site. If an instrument driver is not available, people can write their own instrument driver or use third parties to create an instrument driver. Some instruments also provide a graphical user interface associated specifically with the instrument hardware that provides easy use of the instrument's specific features. This is particularly important with modular instruments because they do not have a front panel.

Application development environments. *Application development environments (ADEs)* are specialized environments used to develop software. Generally, they include support for a programming language, such as C or BASIC, as well as support tools that make programming easier. ADEs are primarily used to create and modify instrument drivers and test programs.

Test programs. *Test programs* control all the instruments in a test system. It controls the sequencing, information processing, and display of all the instruments in the system. Test programs are usually written by the instrument user and are customized to perform their specific tasks. Test programs are easier to write and maintain by making calls to instrument drivers to perform the desired functionality.

Test executives. *Test executives* are additional software tools that make creating, managing, and executing test programs easier. It provides a standardized method for controlling tests and provides built-in capabilities for managing test programs. Some standardized functionality includes test sequencing, standardized test results, connections to databases, and other features. Test executives are typically used by organizations with many test programs.

45.4 Virtual Instrument Classes

The basic components of all instruments include measurement, computation, user interface, communication infrastructure, and software. These components can be combined in different ways to create different types of virtual instruments. Combining components in different ways can form different classes or categories of virtual instruments. Some virtual instrument classes are:

- Virtual front panels
- Software-defined instruments
- Distributed measurement systems

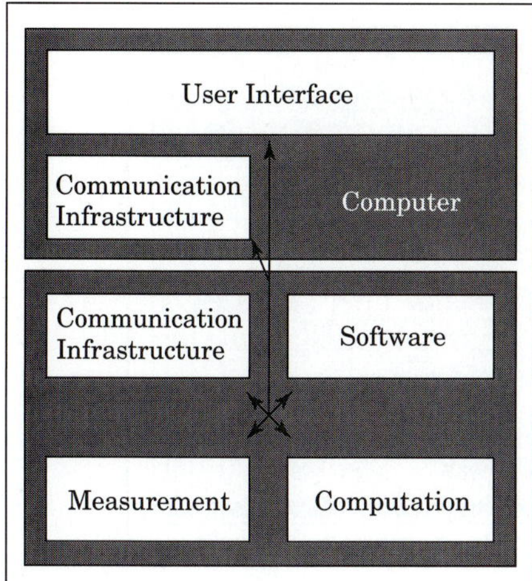

Figure 45.10 Components in a virtual front panel.

45.4.1 Virtual front panels

Virtual front panels are the simplest type of virtual instrument. They consist of a traditional stand-alone instrument connected to a computer. A generalized block diagram of the instrument components in a virtual front panel is shown in Fig. 45.10.

A graphical user interface is provided for the instrument that runs on the host computer. The graphical user interface usually duplicates the functionality of a traditional instrument's physical front panel. This is sometimes called a *soft front panel.* It typically consists of buttons and knobs that emulate interacting with a traditional front panel.

Software front panels are particularly valuable when running with modular instruments. Modular instruments were designed for computer control and hence do not have a physical front panel. All interactions with the instruments depend on the host computer. A software front panel is required for interactive use. A software front panel is shown in Fig. 45.11.

45.4.2 Software-defined instrumentation

Software-defined instruments consist of a computer connected to instruments and instrument components. However, some of the computational tasks and software tasks are run on the controller, in addition to the instruments. This occurs because the physical instruments are not capable of completing the desired measurement by themselves. They require software to combine and coordinate the actions of multiple pieces of measurement hardware to perform the desired measurement. A block diagram of instrument components for a software-defined instrument is shown in Fig. 45.12.

A traditional stand-alone instrument has all of the instrument components (measurement, computation, user interface, software, and communication in-

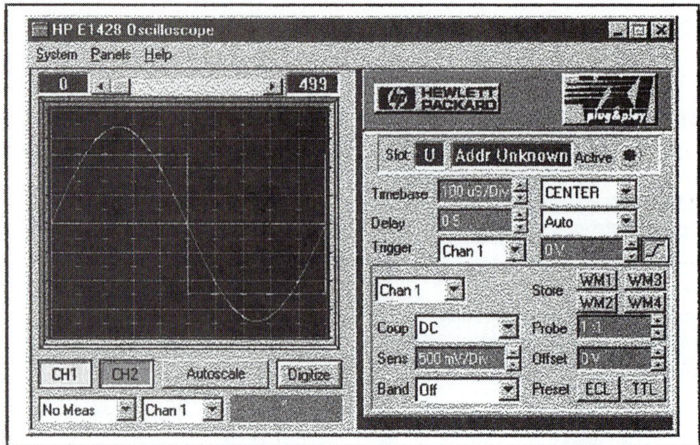

Figure 45.11 A software front panel.

frastructure) combined in a single unit. A software-defined instrument, in contrast, does not have all the functionality and instrument components assigned to a single self-contained unit. The software and communication infrastructure allows multiple instrument hardware to be coordinated and reconfigured to perform custom measurement tasks.

Software-defined instruments are used extensively with modular instruments, such as PC plug-in cards or VXI instruments. The software is used to provide personality to the instrument system. It allows basic measurement

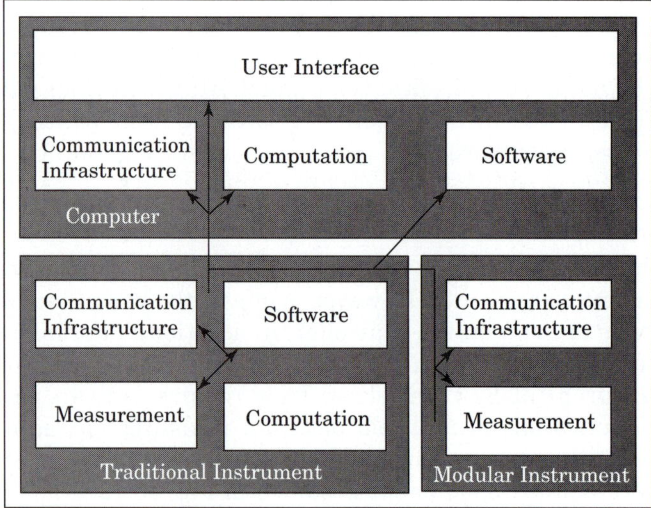

Figure 45.12 Instrument components in a software-defined instrument.

blocks to be combined in an optimal solution. Software-defined instruments are more versatile than traditional instruments because the capabilities and features are not defined solely by the instrument vendor. However, software-defined instruments require specific personality software running on a computer that is connected to the measurement hardware.

45.4.3 Distributed-measurement systems

Distributed-measurement systems extend software-defined instruments to span a network. This allows measurements and computational resources to be geographically spread apart. In contrast, software-defined instrument systems are typically only connected to a local computer. A block diagram of a distributed measurement system is shown in Fig. 45.13.

Distributed instruments allow for new applications. Such applications include measurement systems that span large geographic regions. An example includes telephone traffic monitors, which are connected to key locations across an entire telephone network. Each site monitors network traffic and provides measurement information as part of single virtual instrument for monitoring the entire network.

45.5 Implementing Virtual Instrument Systems

This section describes issues associated with creating and maintaining virtual instrument systems.

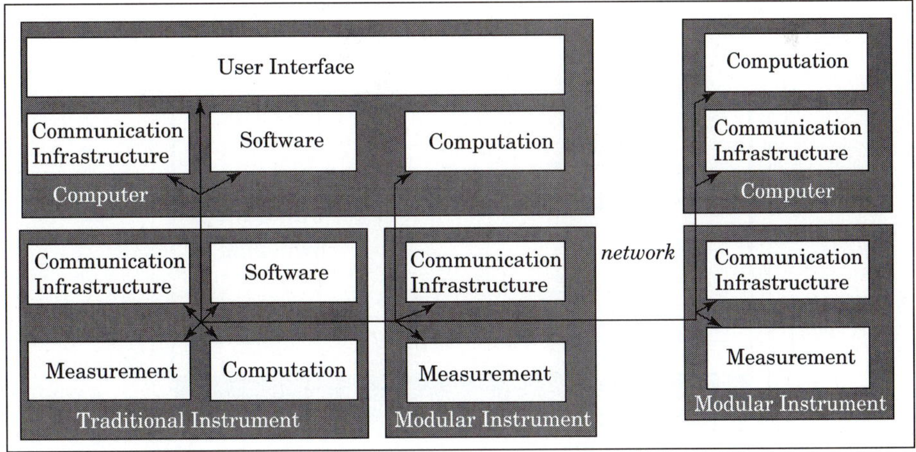

Figure 45.13 Distributed instrument using two computers.

45.5.1 Partitioning

Virtual instrumentation combines physical instruments with computers. Partitioning involves selecting which type of instruments to use and how each instrument component responsibility is to be divided between the various instruments and the computer.

Partitioning can be demonstrated by using an example of creating a scanning voltmeter virtual instrument. A scanning voltmeter takes voltage measurements over several channels. A scanning voltmeter could exist as a stand-alone instrument. This instrument would be connected to a computer using *General Purpose Instrument Bus* (*GPIB*) and programmed to take voltage reading over several channels. Software tasks are divided between the computer and the instrument. The instrument driver provides a consistent programming model for programming the instrument. It uses the I/O library software on the computer to control the GPIB interface. Instrument-specific commands (e.g., Standard Commands for Programmable Instruments, SCPI) are sent to the instrument. The instrument's firmware then converts the SCPI into actions that control the voltmeter and built-in switches to perform the scanning-measurement operation. Partitioning of a scanning voltmeter using a stand-alone instrument is shown in Fig. 45.14.

An alternate approach would be to create a scanning voltmeter using the VXI modular instrument standard. This could be constructed using a VXI voltmeter and VXI switches. VXI requires a control device called a *GPIB-to-VXI Slot-zero controller,* which allows the computer to control the VXI instruments using GPIB. Certain GPIB-to-VXI slot-zero controllers are capable of running downloadable "firmware," which makes the VXI voltmeter and switches look like a single virtual instrument. This type of partitioning of a VXI-based scanning voltmeter is shown in Fig. 45.15. Because both alternatives perform the same function, why is it important to know how they were partitioned? It is important because partitioning can have a large impact on many factors, including expandability, performance, cost, maintenance, and

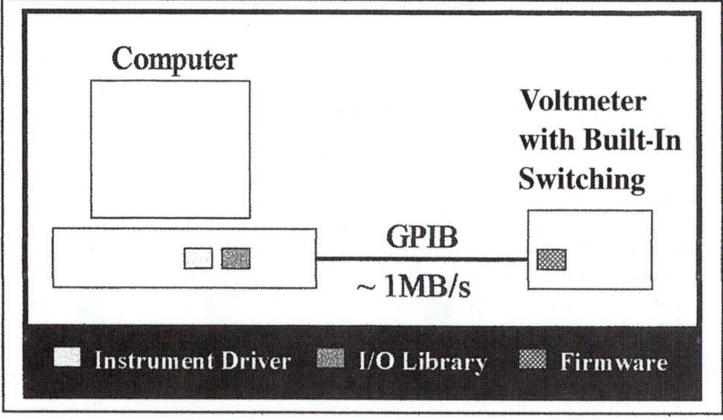

Figure 45.14 Traditional instrument scanning voltmeter.

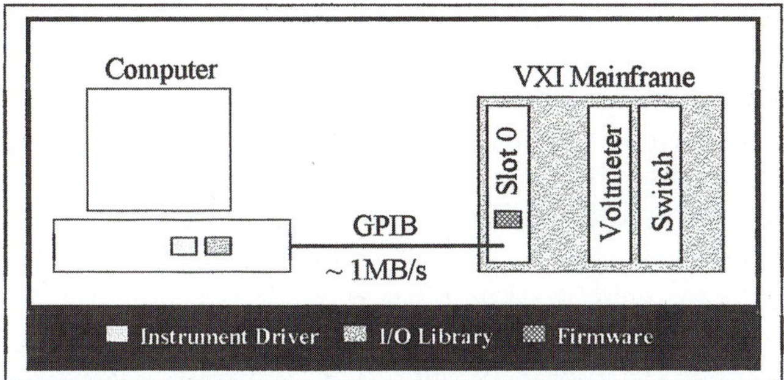

Figure 45.15 A GPIB-connected VXI scanning voltmeter—the standalone instrument can be replaced by a modular VXI system, which provides more expansion and customization.

software. In the first example, the stand-alone instrument has a fixed number of channels. It is not as expandable as the VXI configuration, which is composed of modular instrumentation and can be easily expanded using higher-density switches. Cost also has an impact on partitioning. The stand-alone instrument typically costs less than the VXI configuration.

An alternate VXI implementation can be constructed by replacing the GPIB-to-VXI slot-zero controller with a MXI interface. The MXI interface is a high-performance parallel bus that replaces the GPIB interface in connecting the computer to the VXI instruments. The original VXI system depended on software that was running in the GPIB-to-VXI slot-zero controller. Because the MXI system does not have a microcontroller, it is necessary to have this software available on the controlling computer. Virtual instrumentation allows the components to be repartitioned. Software that was formerly in the GPIB-to-VXI slot-zero controller can be ported to run on the controlling computer. This new partitioning is shown in Fig. 45.16.

Partitioning has a big impact on performance. For a large number of channels, the stand-alone instrument and the GPIB-connected VXI instruments

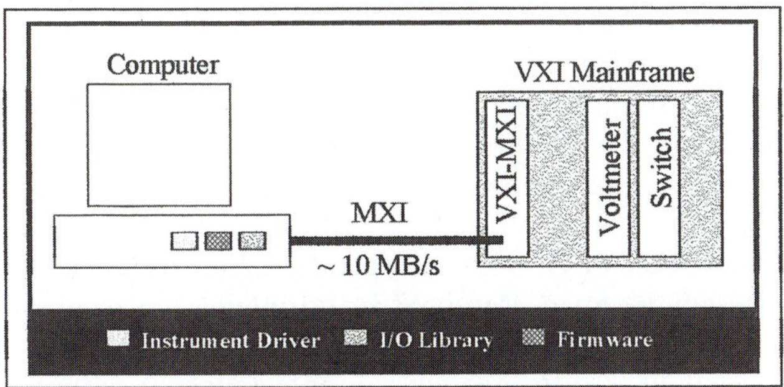

Figure 45.16 MXI-connected VXI scanning voltmeter.

can have similar performance because GPIB is the limiting factor. The MXI-connected solution provided faster performance because MXI is a much-faster communication interface and it can transfer data much faster. The increased performance also came with increased costs. The stand-alone instrument system is less expensive than the GPIB-connected VXI solution, which, in turn, is less expensive than the MXI solution.

45.5.2 Software and development environments

Software is the most important part of a virtual instrument. Software controls how the various instrument components are combined and provides the instrument with personality. In large test systems, software investments often are much greater than the instrumentation hardware costs. Hardware has standardized categories: voltmeters, spectrum analyzers, and oscilloscopes. Software does not have the standardized equivalent. This prevents large amounts of re-use and generally causes escalating software development and maintenance costs. Software does have the benefits of being very easy to customize. This allows people to have tremendous flexibility, and existing applications can often be adapted to suit a particular purpose.

To develop virtual instrument systems, a software-development environment needs to be selected. Software-development environments can be categorized into standard "textual" programming environments or "graphical" programming environments. *Textual programming environments* include Microsoft Visual C/C++ and Microsoft Visual BASIC. *Graphical programming environments* include National Instrument's LabVIEW and Hewlett-Packard's VEE. When selecting a programming environment, be sure that it supports instrument-specific software, such as instrument drivers or measurement routines.

Graphical programming. Graphical programming involves using icons (pictures) and widgets, which represent basic blocks of functionality. The first step in graphical program is to locate the icon that performs the action or measurement that suits your needs. A common icon is to use an instrument-driver icon. The *instrument driver* is typically provided by the instrument manufacturer and it allows a person to easily utilize the functionality of the instrument hardware. Icons are visually connected together through software "programming" to create custom measurements. Connecting icons is typically done through "wiring," which involves using "software wires" to connect icons together. The wires allow control to pass from one icon to the next in a sequential fashion.

A graphical software environment typically has a wide variety of icons and widgets. These can perform standard instrumentation components, such as user interface components, communication infrastructure, and computational components. Input user interface components (such as software knobs, software sliders, software dials, etc.) are readily available.

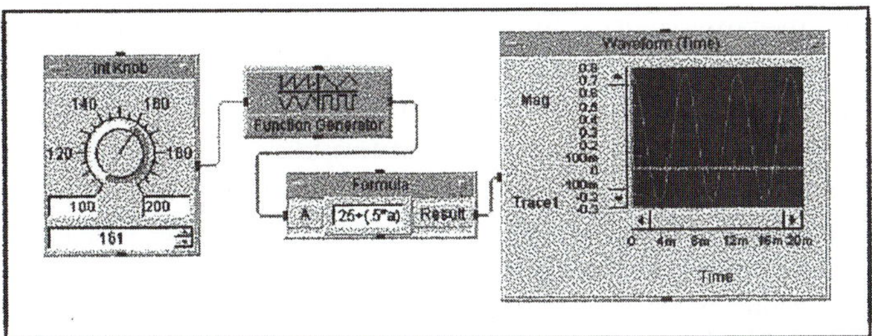

Figure 45.17 Graphical programming—visually connecting icons (pictures) to coordinate measurement operations.

Similarly, output user interface components (such as software displays, software simulated LEDs, and graphs) are also available. In addition, computational elements are also available to perform mathematical operations on measurement data. After the measurement signal is mathematically manipulated, the new signal can be displayed using output user interface components, such as graphs, simulated LEDs, etc.

Grouping a set of basic icons together and assigning them to a new icon can create custom measurements or functions. Every time you want to perform the custom measurement, you can use the new icon. When the custom measurement is complete, the physical-measurement components can be reconfigured to perform a different measurement.

Figure 45.17 shows an example of graphical programming with icons. In Fig. 45.17, a simulated knob is connected or "wired" to a function generator, which is programmed to generate a cosine wave with an amplitude of one. The knob controls the frequency of the generated cosine wave. The output of the function generator is mathematically altered using the formula box. The output is offset by 0.25 and the amplitude of the cosine wave is decreased by half. Finally, the output is displayed as a waveform, so the user can see the result.

When the program is started, control starts at the first icon or widget. In Fig. 45.17, the first widget is the knob. When the user adjusts the knob, the new value of the knob is passed to the function generator. This causes the function generator to change its frequency to correspond to the value that was passed from the knob. From the function generator, the output signal is passed to the formula box, which performs the desired calculation. The final output signal is then displayed on the waveform display.

Textual programming. Textual programming involves programming instruments using a standard programming language, such as BASIC or C. Conceptually, textual programming is very similar to graphical programming. However, instead of icons, textual languages use "functions" to perform basic blocks of functionality. Functions are typically found in instrument drivers or I/O libraries. Instrument drivers are collections of "functions" that control a particular instrument. I/O libraries are collections of "functions" that control the communication infrastructure.

Calling instrument driver functions and I/O library functions in the desired sequence creates custom measurements. Instead of assigning this custom measurement to a new icon, textual languages create a new "function." Whenever the desired custom measurement is required, the function can be called to setup the measurement hardware and perform the measurement. After the function completes, the measurement hardware is available to be reconfigured for a different measurement.

Graphical programming compared to textual programming. Graphical programming is a visually oriented approach to programming. Proponents of graphical programming claim that graphical programming is easier and more intuitive to use than traditional textual programming.

Textual programming requires the programmer to be reasonably proficient in the programming language. Proponents of graphical programming claim that nonprogrammers have an easier time learning the graphical approach and, thus, are more productive in a shorter amount of time. However, the advantage of textual languages (especially C) is that they tend to have faster execution times and better performance than comparable programs developed with graphical programming. Textual programming environments are typically used in demanding high throughput virtual instrumentation systems, such as manufacturing test systems. Textual programming environments are popular and many engineers are trained to use these standardized tools. Graphical programming environments are "easier to use" and require less expertise than textual programming environments. Graphical environments are better for "nonprogrammers" and are useful for quickly developing virtual instruments that are needed quickly and/or need to be reconfigured rapidly.

Virtual instruments can also exist without a programming environment. Many people use instruments directly with standard analysis or documentation software, such as Microsoft Excel (spreadsheet) or Microsoft Word (word processor). This typically occurs in R&D design characterization or in a troubleshooting situation. The most-important task is to understand what is going on as quickly and as accurately as possible. In these situations, the standard analysis package can directly input data from the instruments and can be used to analyze, store, and present the information in a useful format.

45.5.3 Instrument drivers and measurement libraries

Large virtual instrumentation systems are typically composed of many instruments that are combined together through software. Instrument drivers can aid the construction of virtual instrument systems significantly. The instrument manufacturer typically provides instrument drivers, which are a software interface to control the instrument. Customized test programs can be developed using calls to instrument drivers. This reduces the amount of new software that needs to be written and can reduce development time.

Advanced virtual instrumentation systems typically have another layer of software, called a *measurement library.* A measurement library's purpose is to abstract the desired measurement functionality from a specific instrument driver. Instrument drivers are developed for a specific instrument. They highlight the instrument's features and unique capabilities. Measurement libraries, in contrast, focus on common capabilities that are task (measurement) focused, instead of hardware (instrument) focused. This abstraction allows for easier instrument interchangability.

45.5.4 Maintenance

Maintaining a virtual instrument system can be complicated. This occurs because computers (and even instruments) can have a shorter lifetime than the virtual instrument system. Another problem is that operating systems tend to be updated rather frequently and new and improved functionality can also be desired or required. A well-designed system can help to minimize these problems.

Computers continue to evolve with new versions appearing routinely. After a virtual instrumentation system is deployed for several years, it might be necessary to replace the existing computer. Typically, it is impossible to find a similar computer to the one built into the test system. In this situation, a new computer is used and the existing software is installed on the new computer. This seemingly harmless change can adversely affect the virtual instrumentation system because the timing of key software routines will have changed.

Operating systems are also in a state of constant improvement. The choice of operating systems might not be under the user's control. For example, a site *Information Technology* (*IT*) department might control all operating systems on the site and only support one operating system. They might force a virtual instrumentation system to upgrade to the latest operating system. In this situation, it is important that all the software components be available on the new platform. This includes the software-development environment, instrument drivers, and measurement libraries. If they are not available, then some mechanism must be available to port the code to the new environment.

Porting software is important in virtual instrumentation because the software provides the personality of the instrument. The software controls how the various measurement components interact with each other. If the software is not ported to a new environment, then the virtual instrument is not able to operate properly and coordinate the needed activities to perform the desired measurement.

Instruments usually have long lifetimes compared to the computer products, but even instruments might not outlive the virtual instrument system.

If this situation is anticipated, the virtual instrumentation software should be developed using measurement libraries as much as possible, instead of only making calls to instrument drivers.

45.6 Computer Industry Impact on Virtual Instrumentation

Advances in the computer industry have major impacts on Virtual Instrumentation. This section describes some of the trends and impact that computer-industry advances will have on virtual instrumentation.

45.6.1 Abundant computation resources

There is a historic trend for increasing computational power at lower costs. This is manifested in desktop computers' ever increasing performance at reduced costs. This has the result of making computers more pervasive and more likely to be found next to an instrument.

Increasingly powerful desktop computers allow many analysis and presentation tasks to be repartitioned on the computer, instead of in the instrument. This allows virtual instruments to take advantage of the additional resources provided by the computer, including large amounts of memory and permanent storage capabilities. In addition, shared resources, such as file servers or network printers, can also be accessed.

However, just because additional tasks can be partitioned on the computer, it does not mean that all tasks should be partitioned on a computer. Complex systems, composed of many instruments, might experience throughput limitations if all information is routed through the computer. This occurs because most PC operating systems are not designed for the heavy barrage of transactions that occur when many instruments are sending information to the PC. Modern operating systems time slice between multiple threads of execution. Each transaction interrupts the normal flow and requires attention. In this situation, it would be better to have the instrument perform the transaction and store the information in its local buffer or to have the instrument perform an analysis or other data-reduction activities. Fortunately, advances in computer technology are also present in embedded microcontrollers for instruments. This provides instruments with additional capabilities, such as expanded memory or permanent storage, as included in CD-ROM drives and hard disks.

Traditional instrumentation is also experiencing a trend toward the PC architecture. These instruments are built around traditional PCs that are embedded into the instrument. From the outside, they look like traditional stand-alone instruments, however, internally they are PC based and have many of the benefits of virtual instrumentation. This allows instrument manufacturers to take advantages of many of the advances in the computer industry.

45.6.2 Computer I/O standards

The computer industry is providing new and improved standards for communication. Some of these computer standards include RS-232, Universal Serial Bus (USB), IEEE 1394 (Firewire), and 10/100BaseT LAN. Instrument manufacturers are adopting many of these interfaces on their instruments.

These interfaces are attractive because they are built into the computer. Traditional instrument interfaces, such as GPIB, require additional I/O hardware to allow instruments to be connected to the computer. This leads to additional installation and configuration efforts. Easier I/O connections make it easier to connect instruments to computers and will increase the usage of virtual instruments.

Serial bus. Computer I/O also affects partitioning. Computer I/O is trending toward serial buses, such as USB or IEEE 1394. Serial I/O works better with large blocks of information, as compared to small, frequent packets of information. This means that better performance can be achieved by having more computational tasks on the instrumentation side. This allows the instrumentation hardware to perform more processing and control and just the results can be sent back to the computer. This reduces the amount of information that needs to be sent across the serial connection and tends to be more block oriented. Instrument hardware that requires many small transactions and constant attention will not perform as well in serial I/O.

Internet. Another trend is the growing use of the Internet and the trend toward computer to computer linkages. This trend makes distributed measurements practical. This will lead to more uses of virtual instruments as new distributed-measurement applications are created. Networking will also become more pervasive in certain types of instruments hardware. Instruments that are very expensive or those that are typically shared can benefit from network connections because many people can access the same resource over the network.

45.7 Summary

Virtual instruments are devices that are configured from independent hardware components by software to perform custom instrumentation functions. This provides a different paradigm for instrumentation, compared to traditional “box” instruments, which are independent monolithic devices.

Instrument components consist of:

- Measurement
- Computation
- User interface

- Software
- Communication infrastructure

These components can be combined (partitioned) together to form different virtual instrument classes, including:

- Virtual front panels
- Software defined instruments
- Distributed measurements systems

Creating virtual instrument systems involves understanding partitioning and the impact that it can have on key attributes, such as performance, maintenance, and cost. Software and software-development environments play a crucial role in developing virtual instrument systems. Software-development environments can be classified as using graphical programming or using textual programming.

Chapter

46

Distributed Measurement Systems

Geri Georg
Agilent Technologies
Loveland, Colorado

46.1 Introduction to Distributed Measurement Systems

Developments in the technologies of testing and measurement, computation, and networks have allowed them to converge and create a new capability in *distributed measurement.* This is the ability to have testing and measurement elements that are physically remote from each other and the user, organized, and controlled by computers to perform measurement and/or control functions. The networks that provide the communications between the testing and measurement elements and the computer can be of any sort, such as *Local Area Networks* (*LANs*) or public *Wide Area Networks* (*WANs*).

The issues associated with this capability include:

- The availability of testing and measurement devices, such as "smart sensors" that allow information or action, at a distance, to be controlled over a network by computers, combined with "instrument drivers" that provide the interface between an instrument and a computer.
- The availability of computers and software that can control the instrumentation through appropriate interfaces and languages.
- The availability of networks with sufficient bandwidth (information transmission capacity) to allow the timely transmission and reception of commands and information to and from remote locations.

These issues are described in greater detail in separate chapters of this book. This chapter describes the combined capability, which is defined as *distributed measurement systems.*

46.2 Distributed Computation System Characteristics

Distributed computation systems represent the general case of capabilities where groups of computing elements, including one or more computers and remote input/output devices, work together to solve problems. These systems can be *tightly coupled* or loosely coupled. In addition, either type of system can be structured as *client/server, peer to peer,* or *tiered.*

46.2.1 Tightly coupled systems

Tightly coupled means that the computers share computing, memory, input/output (or other resources) and communicating through the shared memory. Each computer might also have its own local high-speed buffer cache. These systems can consist of several, hundreds, or even thousands of processors.

In tightly coupled systems (Fig. 46.1), the resources must be well-coordinated or the results will be chaotic. This is usually done via clock synchronization and special hardware/software that keeps the local memories consistent. Even in loosely coupled groups of computers, the resources that are

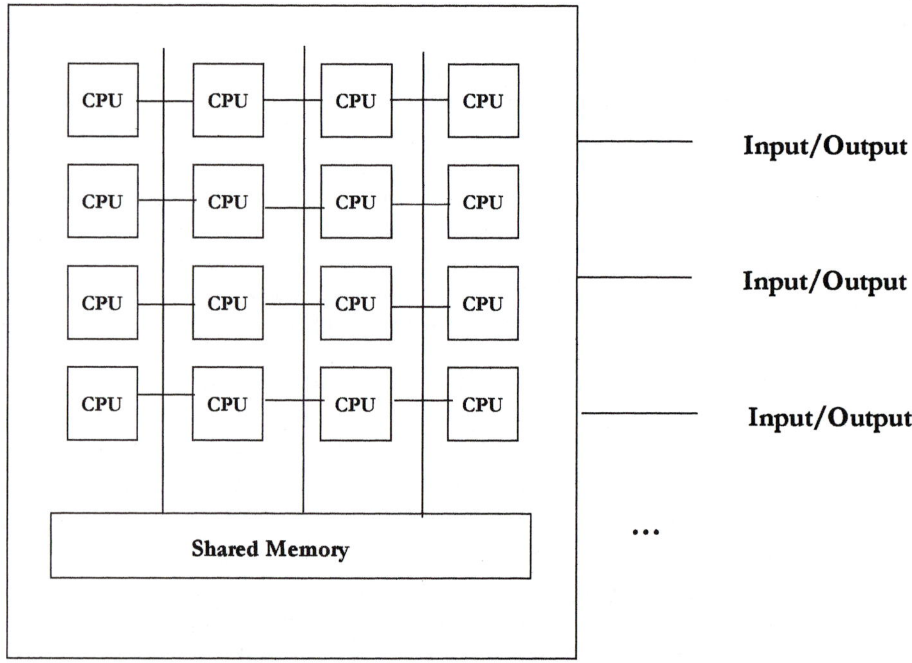

Figure 46.1 A tightly coupled computing system consisting of 16 processing elements that communicate with each other via shared memory. All other resources, such as input/output to user interfaces, networks, and archive media, are shared. CPUs in these kinds of systems often have their own local cache memories to decrease the amount of accesses they need to make to the shared memory.

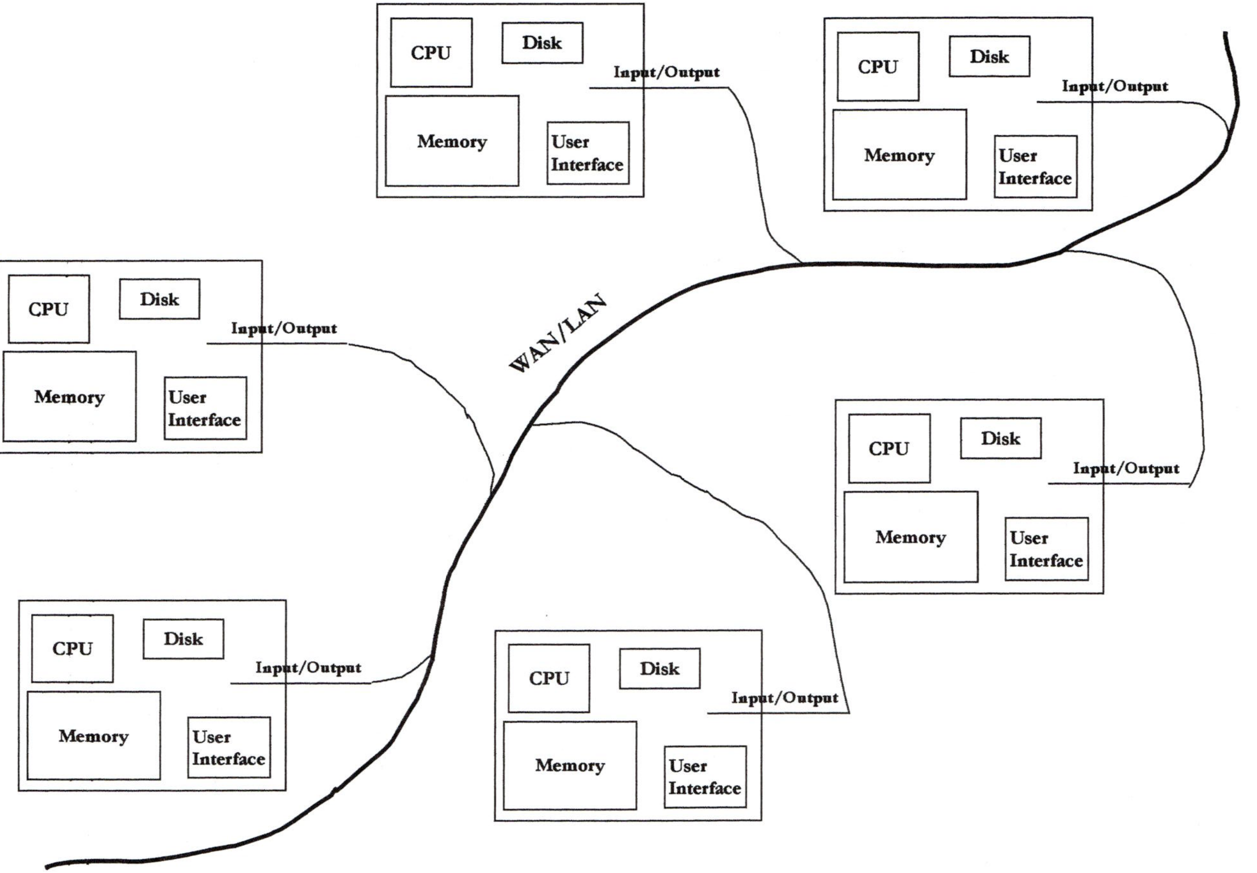

Figure 46.2 A loosely coupled computing system consisting of six computers that communicate with each other via message passing. Each computer has its own local memory, archive, user interface, and input/output capabilities.

shared (input or intermediate results, for example), must be coordinated or the final outcome will not be meaningful.

46.2.2 Loosely coupled systems

Loosely coupled means a set of computers (Fig. 46.2) communicating by messages; with each computer having its own local memory and input/output capabilities. These systems typically consist of several to hundreds of computers. In either case, loosely or tightly coupled, if the distributed system is well designed, it can be tailored so that the resources (time, equipment, and human) available closely match those needed to solve the problem. Well-designed systems are maintainable and their capabilities are extendable to meet future needs.

46.2.3 Client/server, peer-to-peer, and tiered operation

Applications running on either tightly or loosely coupled systems can be structured as client/server or peer-to-peer and can additionally exhibit tiers.

In a *client / server structure,* one computer acts as a server for data or services that another, client computer requests. Computers can act as servers for some requests and also as clients when they make requests.

In *peer-to-peer computing,* applications elements work cooperatively on multiple computers, perhaps passing data back and forth for various computations before a result is finally achieved. In this case, no part of the application really acts as a client and no part really acts as a server; they cooperate.

In *tiered structures,* an application is partitioned in logical parts across multiple computers. The most-common partitioning is to put the user interface on one computer and the rest of the application on another. This is called two-tier partitioning. Three-tier partitioning occurs if the user interface, data, and application reside on different computers.

46.3 Distributed Measurement Systems

Although distributed measurement systems have a different look than distributed computer systems, the ideas of coupling and coordination are similar; they can be either "tightly coupled" or "loosely coupled." In addition, however, they include test and measurement elements and capabilities that are remote from the user and contain the ability to sense a local situation, perform some local computation functions, and communicate over a network, creating "virtual instruments."

If the distributed measurement system has been well designed to meet the needs of the problem domain, the issues specific to that domain and the use of the system, much more can be accomplished than by using a single measurement system. Quite often, a reduction of resources and the number of locations where they are needed can also be achieved. This gain must be balanced, however, by the extra effort required to understand the problem domain, intended use, long-term goals, and the extra design, implementation,

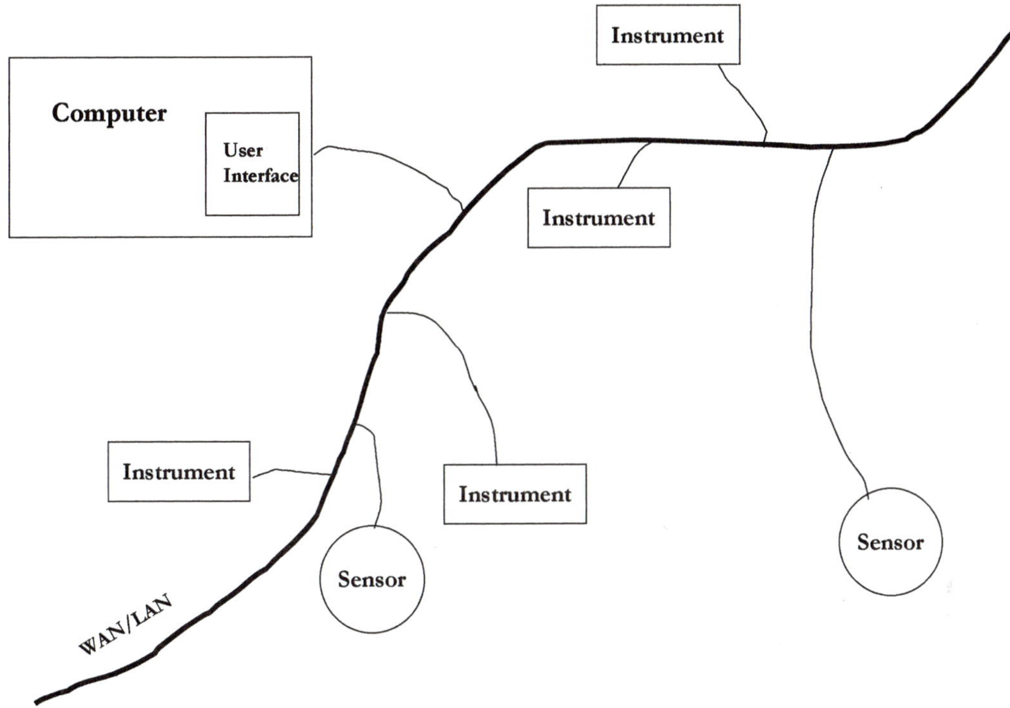

Figure 46.3 The computer is acting as the controller and also contains the user interface for this system. Instruments and sensors can be configured in and out of the system, but their role is fixed; they are the measuring devices.

deployment, and maintenance efforts needed to create and sustain the distributed measurement system. Many tradeoffs are involved; some of the more common ones are described later in this chapter.

46.3.1 Tightly coupled and loosely coupled distributed measurement systems

Distributed measurement systems are usually one of two types:

- *Type 1.* A tightly coupled system of distributed measurements, where a single computer is controlling several instruments to coordinate a measurement. The configuration of this type of distributed system is fairly rigid. That is, the controller is always the same, the user interface is always at the same location. Although instruments can be moved around, their role in the system is fixed (Fig. 46.3). The software for this kind of distributed measurement system is the easiest to build because many of the issues described later in this chapter are not relevant to it, such as a common user environment (one controller and one user environment already exist). (See section 46.9 for more details on software issues. It is the system of choice when a single controller is acceptable.)

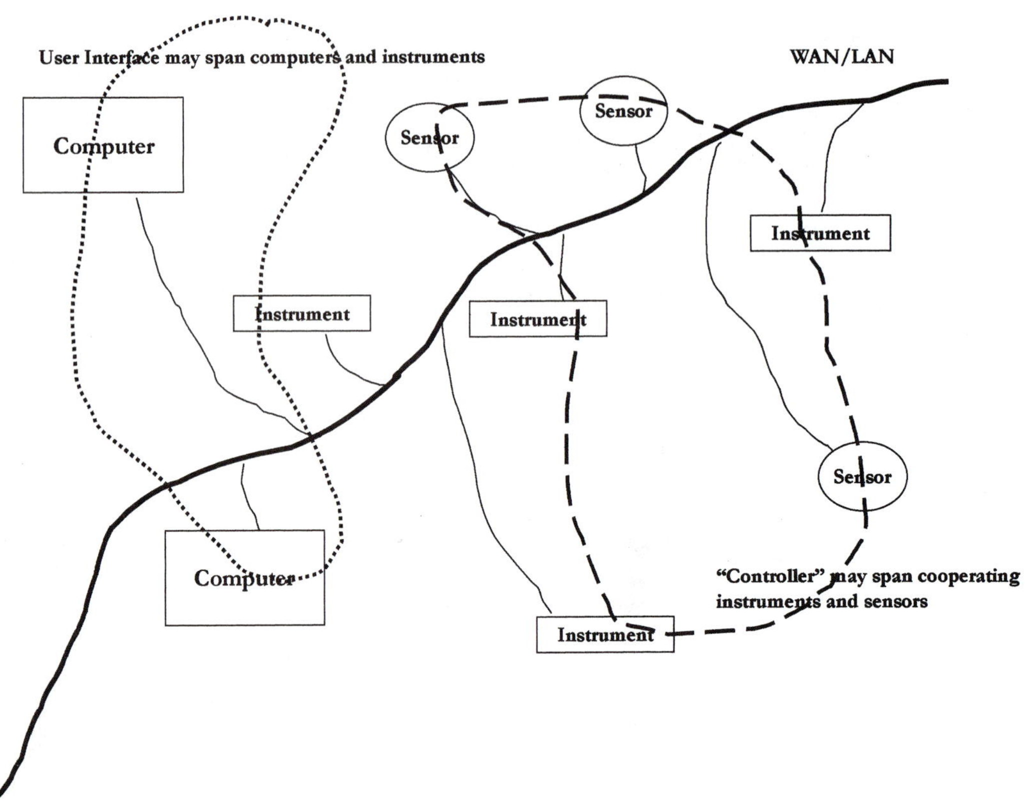

Figure 46.4 In this figure, multiple computers can act as the controllers and also contain the user interface for this system; they tend to change roles as needed, and the roles can be distributed across them. This is shown by the dotted line. Instruments and sensors can be configured in and out of the system; they not only take measurements but can also execute parts of cooperating programs that make up the "controller" of these measurements. This is indicated by the dashed line. Together, this spanning of roles and possible reconfiguration create a set of "virtual instruments" that work together in the system.

- *Type 2.* A loosely coupled distributed measurement system, where "virtual instruments" are accessible from multiple locations. A virtual instrument in this situation is defined as physically dispersed and independent elements of an instrument, such as interface, controller, sensor(s), algorithm/analysis, storage, and communication that can be configured, through appropriate software, to accomplish a given test or control task. This dispersion also allows reconfiguration at will and across different virtual instruments to perform specific measurements that would otherwise necessitate more equipment, time, or human resources.

These systems are very fluid, with the controller, user interface, and instruments changing not only location, but also functionality often (Fig. 46.4). This kind of distributed measurement system is more versatile than a tightly coupled measurement system, but the software needed is much more difficult to develop and maintain. (See the description of software issues in sections 46.9 and 46.10.) These kinds of systems are found where the additional versatility over a tightly coupled system is needed and where the tradeoff in effort to develop the software is warranted.

In the case of Type 1, the instruments or sensors are located some distance away from the computer, usually connected by a network (field bus, LAN, or WAN). Although the instruments or sensors might have their own local controllers, the computer is the overall controller for the measurement. (See the description in section 46.11.1 for an example of the decision process leading to a tightly coupled measurement system.)

In the case of Type 2, the measurement interface, controller, instrument, sensor, and computer are all located at different geographical points, again usually over some sort of network. It is often useful to think of this kind of system as one in which a number of "virtual instruments" are coordinated to take a measurement.

46.3.2 Smart sensors

Smart sensors are devices that transmit not only raw data from the transducer, but also configuration information, such as calibration information, its location, and how to interpret the raw data.

46.3.3 Virtual instrument

A virtual instrument can be thought of as physically dispersed elements of an instrument: interface, controller, sensor(s), algorithm/analysis, storage, and communication, working together. This dispersion allows reconfiguration at will and across different virtual instruments to perform specific measurements that would otherwise necessitate more equipment, time, or human resources.

46.4 Instrument/Computer Communication Standards

Many standards that ease communication among computers and instruments have also been developed; these work by restricting the ways communication and inter-operation occur to a few well-understood methods. For example, VXI specifies some hardware aspects of VME-based instrumentation, including card size and use of lines not included in VME, as well as protocols for instrument communication. IEEE 488 primarily specifies the way that instruments communicate with controllers, while *Universal Serial Bus* (*USB*) specifies more general communication between a computer and its peripherals. IEEE 1451.1 and 1451.2 specify the way that smart sensors identify and describe themselves as well as the way they communicate their information. *Transmission Control Protocol* (*TCP*) and *Internet Protocol* (*IP*) specify the way computers can communicate over a network.

Along with these standards have come different ways to program them together into systems, such as using graphical languages (e.g., VEE for Hewlett-

Packard and LabView for National Instruments), or object-oriented program composition with technologies, such as *Distributed Component Object Model* (*DCOM*) and *Common Object Request Broker Architecture* (*CORBA*).

46.5 System Planning and Development

Unfortunately, these technologies and programming methods are not enough in themselves to ensure that the resulting distributed measurement system will perform to expectations. Very often, these systems fail their requirements of such factors as equipment utilization, performance (time), flexibility, manageability, and/or correctness. This chapter addresses this missing "glue": the knowledge and analysis needed to create a distributed measurement system that lives up to its expectations and potential.

The outline of the rest of this chapter is as follows. First, examples of distributed measurement systems are shown, along with their context. These examples also show what kinds of things can be distributed in such a system. Next, some of the reasons for creating distributed measurement systems are explored, along with basic issues in their development. Distributed measurement systems have a number of inherent issues that affect their success in solving the intended problem. These issues and their manifestations are covered next. The final section presents a high-level methodology that can be followed to develop a distributed system, using the components and ideas covered in the rest of the chapter.

46.6 Typical Distributed Measurement System Topology

Figure 46.5 shows typical kinds of distributed measurement systems that are also used as examples in the rest of this chapter. The figure shows a hypothetical, greatly simplified, business that has several different locations and a variety of functions that occur at those locations. Each location is surrounded by a box. Network connections of various kinds (i.e., LAN and WAN) occur between locations.

46.6.1 Enterprise system

Figure 46.6 shows the Enterprise computing system

The kinds of tasks that occur at the Enterprise building include:

- Analyzing, managing, and storing Enterprise-level business data.
- Determining what should be built tomorrow and transmitting instructions to the manufacturing computers (C and D in the manufacturing building), based on manufacturing data.
- Detecting fraud and initiating action (with field computer B at the field), based on field data.

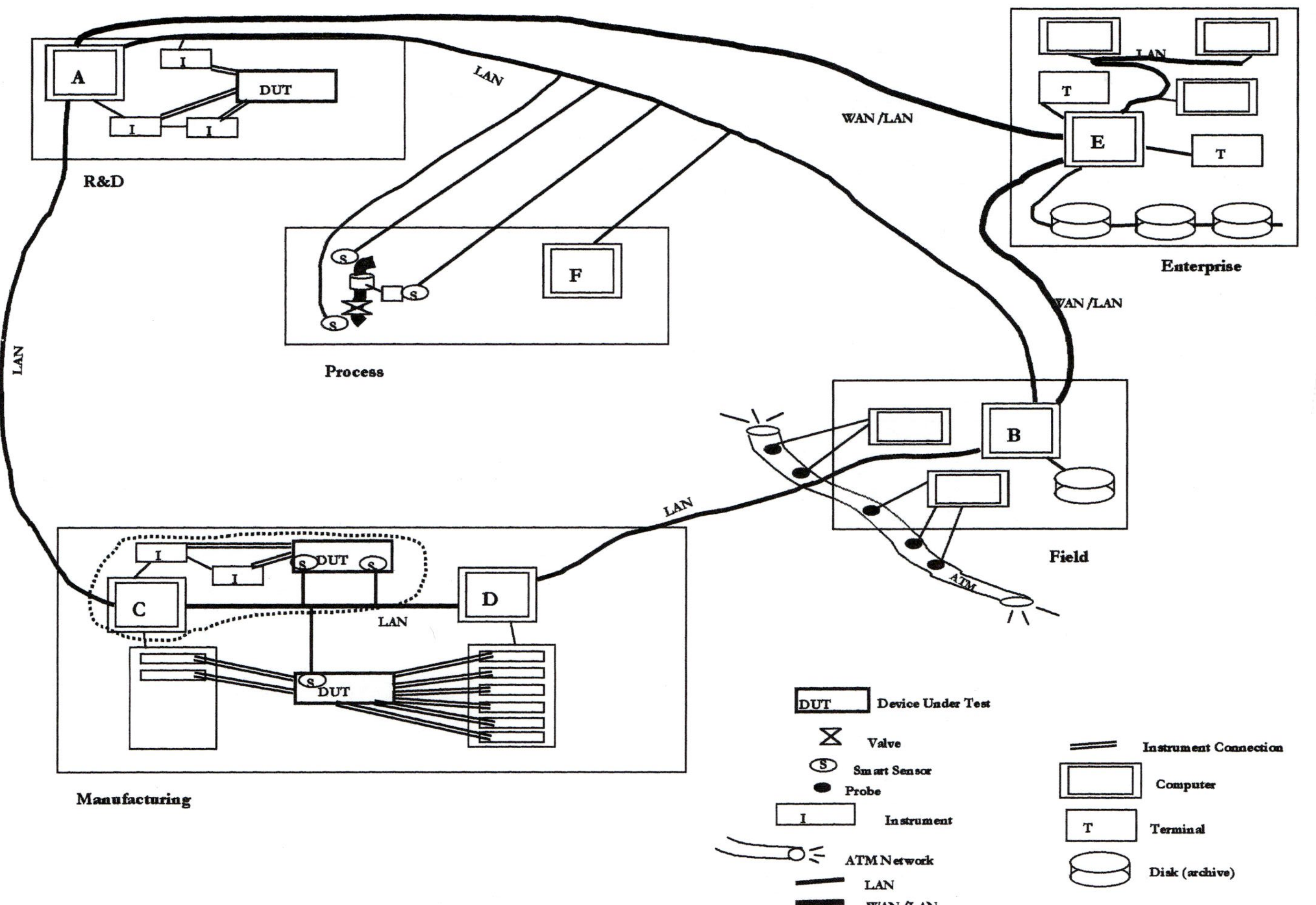

Figure 46.5 An example of a distributed computing system is shown, along with examples of tightly coupled and loosely coupled distributed measurement systems. Computer A is in the R&D building controlling a tightly coupled measurement system. Computer B is at the field, working in a loosely coupled distributed measurement system. Computers C and D are in the manufacturing building; C is controlling a tightly coupled measurement system, as well as participating with computer D in a loosely coupled distributed measurement system. Computer E is in the Enterprise building, participating in both client/server and peer-to-peer applications with other workstations in the building. Computer F is in the process building, participating in a loosely coupled distributed measurement system.

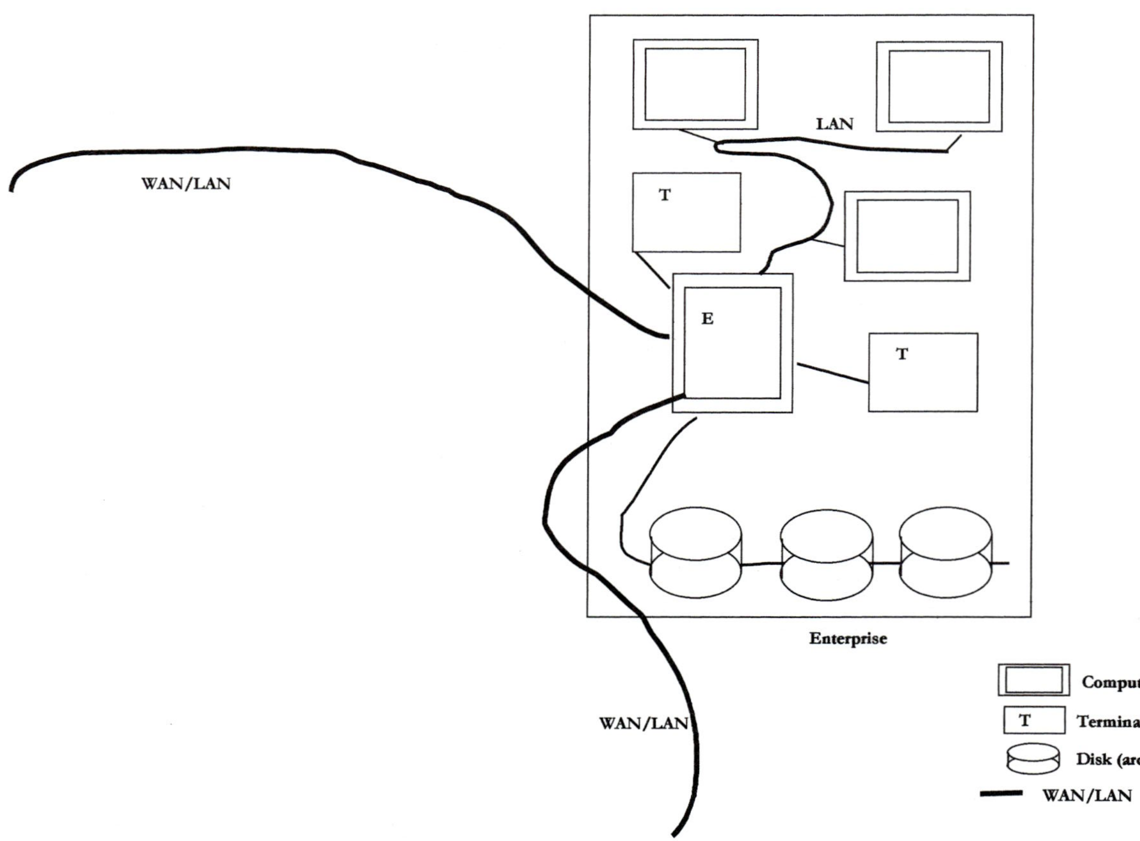

Figure 46.6 Company headquarters is located at the Enterprise building, and computer E is the main company computer. Computer E has terminals connected directly to it, where operators can control it directly, as well as write, test, maintain business applications. A local area network (LAN) is through the building so that users of personal computers can access data or business applications on computer E. Additionally, computer E controls a set of secondary storage disks.

- Transmitting instructions for overall management (with process computer F in the process building), based on the process data.

The group of computers at the Enterprise building demonstrates distributed computing, as follows. Business data resides on the disks attached to computer E, and personal computers can access that data, copying it to their own disks, manipulating it, and sending the results back to computer E. In this case, computer E is acting as a server for data, which the client personal computers request. If, in addition, an application is running on computer E that requires manipulated data and that can request it from the personal computers that generate it, then the personal computers are acting as servers and computer E is acting as a client.

46.6.2 Research and development system

Figure 46.7 shows the *Research and Development* (*R&D*) building of this company. The kinds of tasks that can occur at the R&D building are:

- Locally testing a device under test (DUT) as part of design/development.
- Troubleshooting a manufacturing problem or a DUT on the manufacturing floor.
- Sending data to the Enterprise.

The ideas of client/server and peer-to-peer system structuring seen in the Enterprise computing system can also be seen in the measurement world. In the R&D example, a user of computer A causes it to send commands (requests) to some combination of all three instruments to coordinate some measurement (response) of the DUT. Here, computer A is acting as a client and the instruments are acting as servers. This is an example of a tightly coupled distributed measurement system.

46.6.3 Manufacturing system

Another example of a tightly coupled distributed measurement system (where a single computer controls a set of instruments) is shown in Fig. 46.8.

The typical tasks that occur in the manufacturing building are all related to test executives controlling instruments, which are measuring attributes of manufactured products.

Figure 46.9 shows an example of a loosely coupled distributed measurement system that also exists in the manufacturing building.

It might be that computer D always controls the actions of the instruments in the rack attached to it, computer C always controls the actions of the rack connected to it, and one or the other of them always controls the smart sensor, but, in most cases, either computer will have to control instruments in its own rack, as well as the other computer's rack and the sensor. The bounds of the virtual instrument are constantly changing.

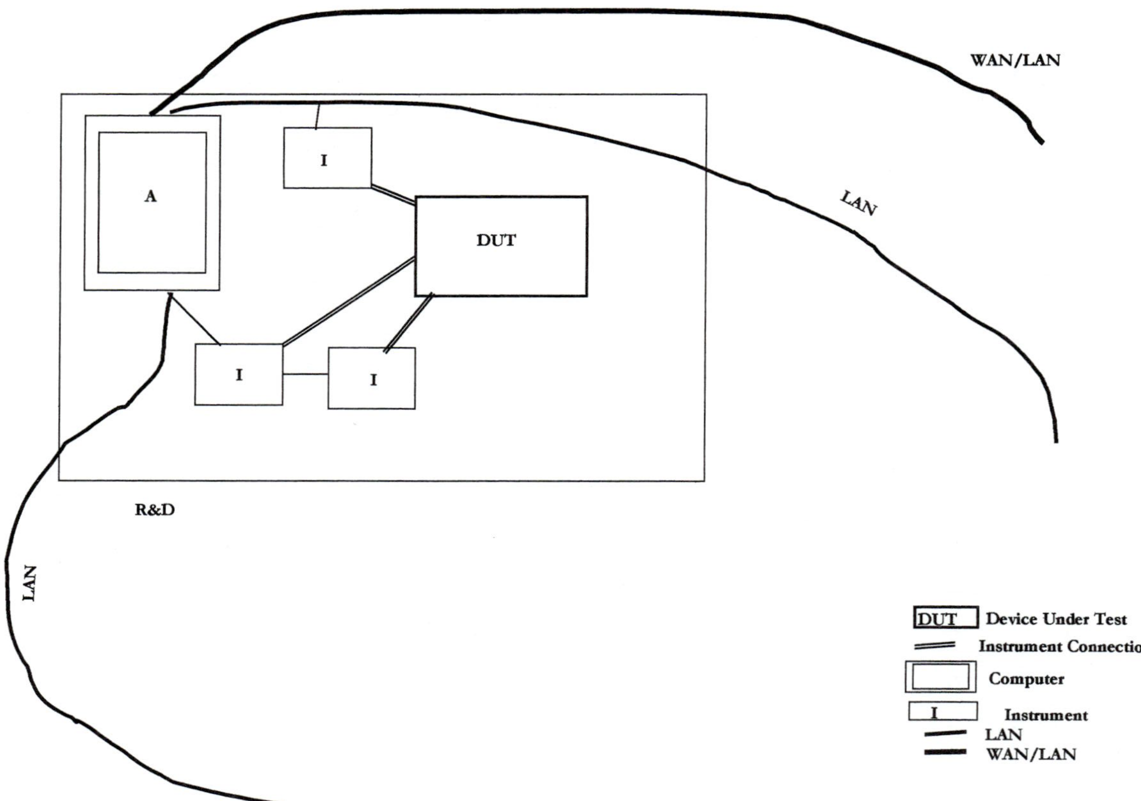

Figure 46.7 In the research and development (R&D) building of this company, computer A is controlling several instruments on an engineer's bench, which are testing a device that will soon be in production. Two of the instruments are connected directly to computer A, and a third is connected to the LAN in the building so that multiple computers can access it.

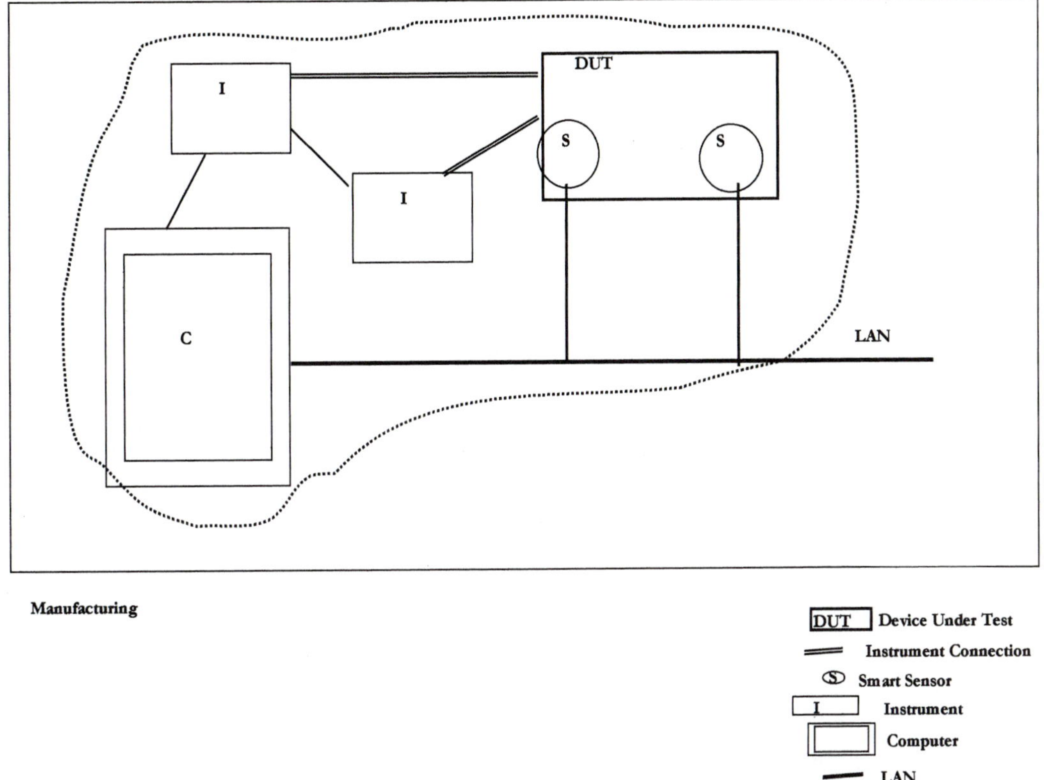

Figure 46.8 In the manufacturing building, in the dotted line enclosing computer C, two directly connected instruments, two smart sensors connected to the LAN and a DUT demonstrate testing on the manufacturing line.

The software in this system must be able to coordinate two computers, their access to a pool of instruments, and a smart sensor connected to the LAN. In a satellite test, the DUT is so complicated that multiple parts of it are tested at the same time, requiring further coordination. Also, if a test fails, someone must troubleshoot the system to fix it. Usually, that person is not on the manufacturing floor; in this example, they might be in the R&D building, working on computer A.

In this scenario, the distributed measurement system must now be able to add computer A to its configuration, and allow computer A to coordinate efforts of all of the instruments and the smart sensor to perform ad-hoc troubleshooting tests. One way to accomplish this is to allow the interface to the testing system to be distributed to computer A, while computers C and D still direct the actual measurements (a two-tier application again, but this time in the measurement world). Another way is to allow part of the coordination software to run on computer A, resulting in a group of cooperating programs that control the instruments and sensor in a peer-to-peer configuration.

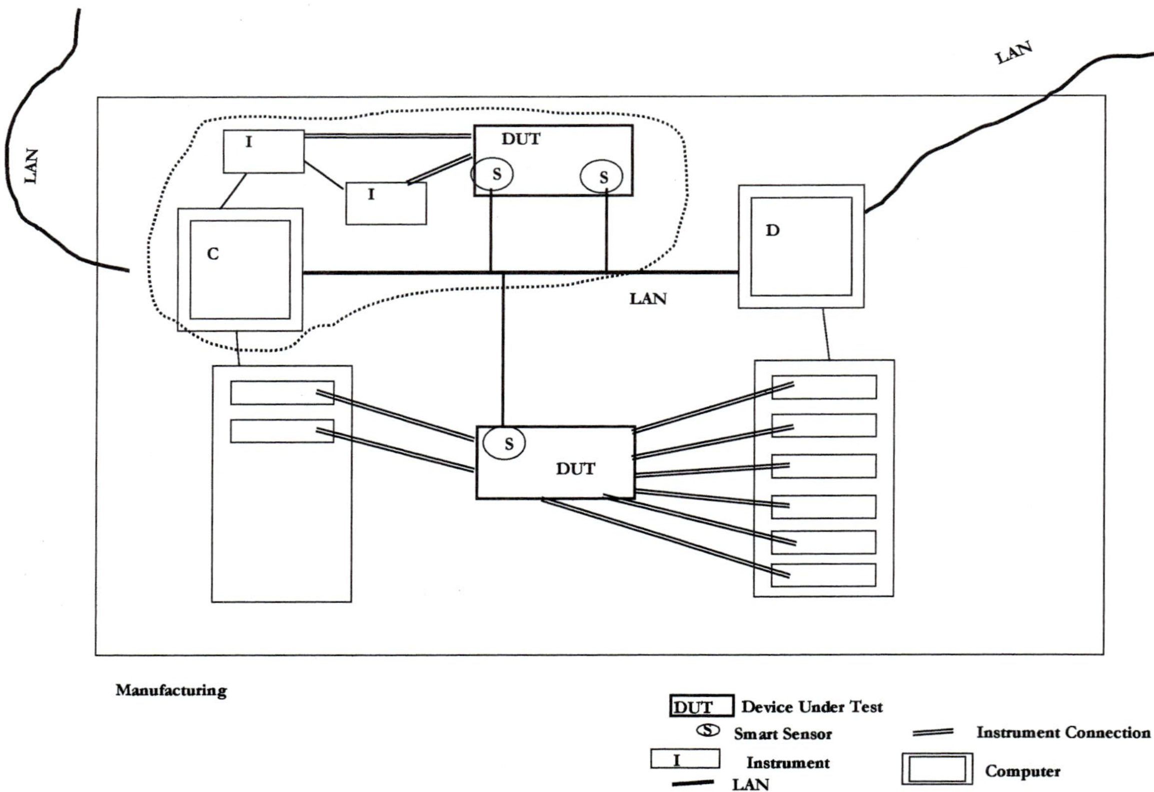

Figure 46.9 Computer C is cooperating with computer D, multiple racks of instruments, and a sensor connected to the LAN to test a DUT. In this case, the DUT is a satellite, and it takes the coordinated efforts of multiple racks of instruments to perform measurements and test it.

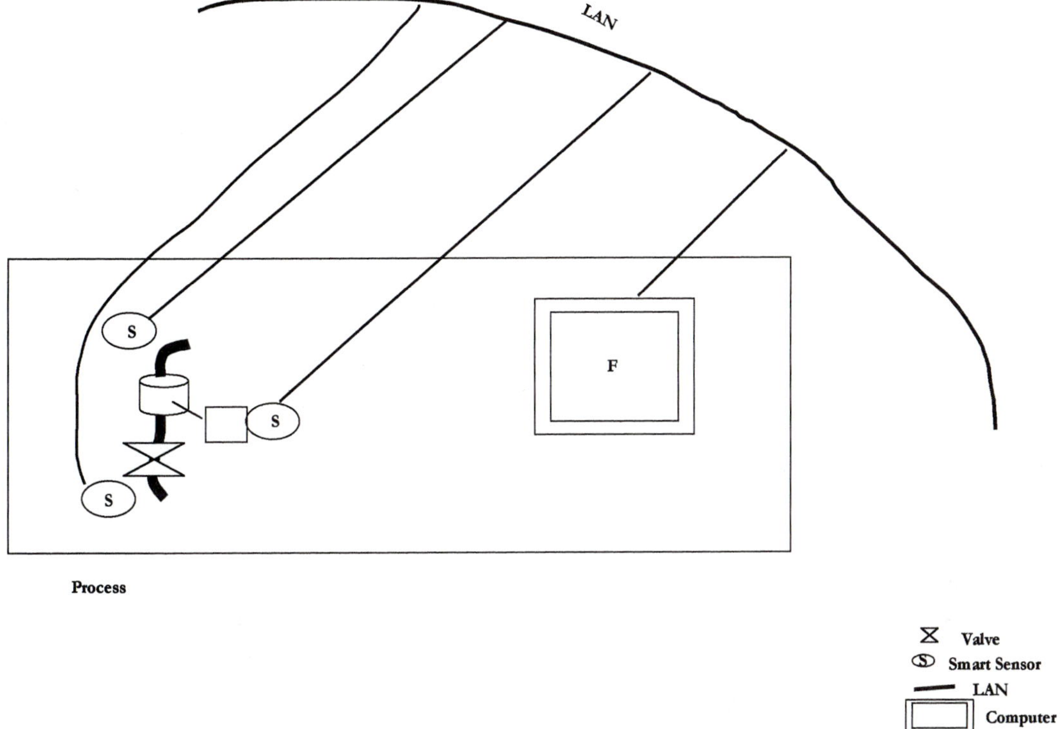

Figure 46.10 In the process building, computer F coordinates the reading of three smart sensors connected to the LAN, to determine what is happening with the process. Based on results from the sensors, computer F can adjust how they take measurements, how often, and of what, and it might also do something like control the valve.

46.6.4 Process system

Another example of a loosely coupled distributed measurement system is in the process building, shown in Fig. 46.10.

Tasks that occur in this area include:

- Monitoring process and control when needed.
- Getting overall management instructions from the enterprise (via computer B, then computer E).
- Making process adjustments based on instructions from the enterprise.

This is another example of a loosely coupled distributed measurement system. Each smart sensor has a computer built into it, which controls the transducer and can perform computation prior to sending results out on a network. The overall result is a loosely coupled set of processors taking measurements and communicating with each other and the Computer F to monitor and control the process.

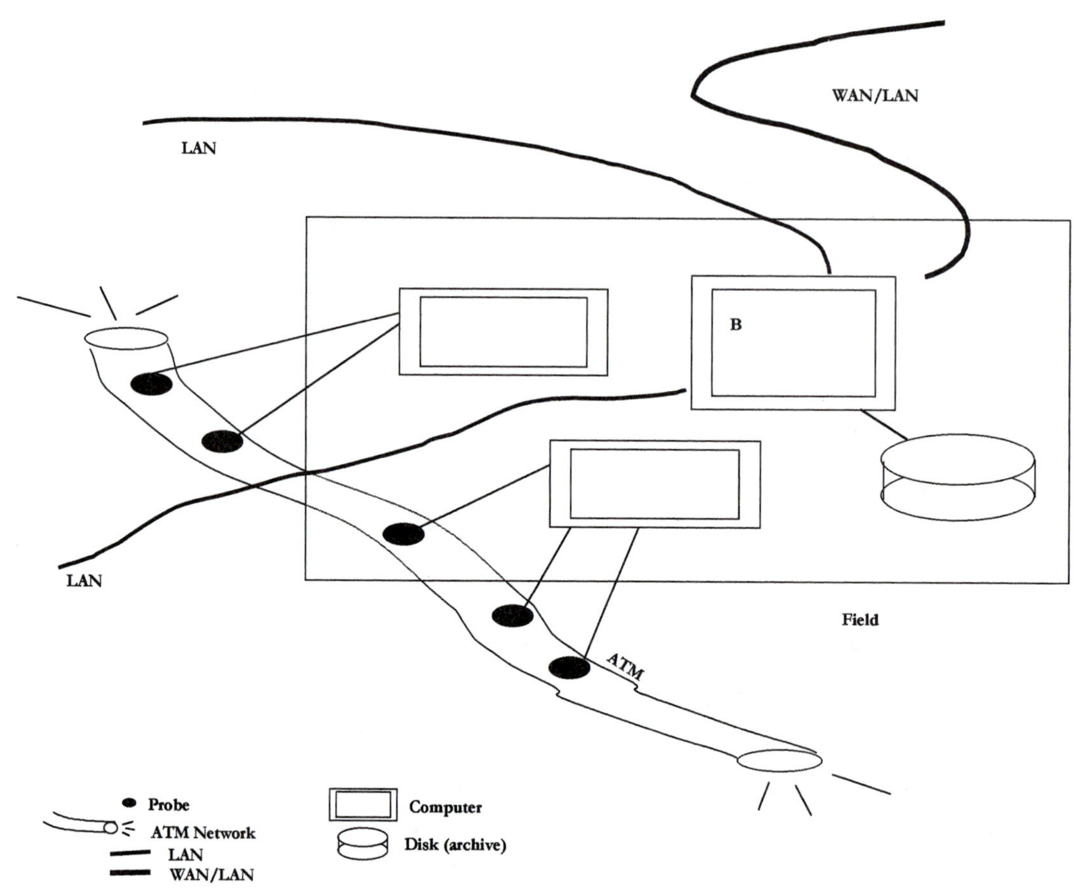

Figure 46.11 In the field, Computer B is working with other computers which have attached probes that are measuring an Asynchronous Transfer Mode (ATM) network. The probes simply report data packets crossing the network, doing minimal processing to extract header information. The computers attached to them reduce data from the probes, looking for patterns. Computers like computer B further coordinate and reduce data, and can send it to the enterprise (e.g., computer E) for further analysis.

46.6.5 Field system

The system shown in the field with Computer B is also an example of a loosely coupled distributed measurement system. It is shown in Fig. 46.11.

Tasks that occur at the field are:

- Looking for patterns and reporting to the enterprise.
- Adjusting collectors based on input from the enterprise.

One application of this kind of system is in fraud detection. As soon as a possible fraudulent situation is detected, the system might need to make adjustments on the patterns it is looking for, or in the analysis it needs to perform. Thus, multiple computers and probes could be reconfiguring and exchanging information frequently to coordinate the work.

46.7 Distributable Functions

The kinds of things that can be distributed in a measurement system include:

- Users, or user interfaces
- Computations
- Measurements
- Test executives
- Archives
- DUTs

Note that some of the distributable functions do not imply a loosely coupled distributed measurement system. For example, geographically distributed measurements can be done with either a tightly coupled distributed measurement system or a loosely coupled distributed measurement system. Archiving results to another location can also be done fairly simply, without a loosely coupled system. And finally, simultaneous testing of a DUT can be performed in a system of distributed measurement, using parallelism in the controller, rather than a loosely coupled system.

46.7.1 User interfaces

The user interface is by far the most common item distributed across resources, in both computer and measurement systems. In Fig. 46.5, the user/user interface is distributed in the computer sense at the Enterprise, when personal computers act as the interface to business applications running on computer E. In the measurement sense, this case occurs when the user/user interface exists on computer A and drives ad-hoc testing of a satellite by coordinating computers C and D and their associated instruments.

46.7.2 Computations

Another item easily distributed; measurements can be taken in one location, then sent to another computer for analysis or aggregation. In Fig. 46.5 this is most easily seen in the field, where computers connected to the probes aggregate data and computer B further aggregates and analyzes it.

46.7.3 Measurements

This form of distribution is most easily seen in systems of distributed measurements, where a single computer coordinates the actions of various instruments or sensors to perform a measurement task. An example in Fig. 46.5 is in R&D, where computer A coordinates several instruments to perform measurements on a DUT.

46.7.4 Test executives

A *test executive* is a coordination program that allocates measurement resources and executes a sequence of instructions to perform a test. An example of distributing a test executive is in the satellite test in manufacturing, between computers C and D. Coordinating programs (test executive) must be running on both computers to allow each computer coordinated/structured access to all the instruments in both racks, as well as the smart sensor on the LAN.

46.7.5 Archives

Archives include storing raw data, as well as the sequence of measurements used to generate them, usually on a hard disk. Although not mentioned in the previous example, in any number of points in Fig. 46.5, data and results can be sent from the originating system to another system for archival purposes. Probably the easiest to see is that the set of disks at the Enterprise could be used to archive measurement data from the Process, ATM data from the field, or manufacturing data.

46.7.6 DUT

In the example of satellite tests, the device under test itself is *distributed* in the sense that multiple tests are occurring simultaneously on different parts of the DUT. Clearly, measurements taken on different parts can be used in sets of different measurements or tests at the same time.

46.7.7 Time component

There is clearly a time component to distribution as well, for both systems of distributed measurements and for loosely coupled systems. For example:

- distribution can occur before a measurement is taken; in this case, it probably takes the form of distributed resource allocation. (This kind of distribution occurs in loosely coupled systems; it isn't necessary in systems of distributed measurements.)
- distribution can occur during a measurement; data is gathered from multiple sensors at the same time.
- distribution can also occur after the measurement, in the form of distributed analysis or archiving.

The latter two cases can be seen in systems of distributed measurements, as well as loosely coupled systems.

46.8 Reasons for Distribution

- Scarce resources
- Expected longevity of the system
- Work partitioning

Usually, the decision to build a distributed measurement system is made in order to take full advantage of some scarce resources. More than likely, these scarce resources are computational or geographical in nature.

46.8.1 Scarce resources

An example of a scarcity of computation resources occurs when a complicated measurement is needed, perhaps involving taking several sensor readings and performing an analysis on them, then creating the final result. If this must occur very quickly, it might be easier to partition the analysis so that multiple computation components can work simultaneously to perform the whole task more quickly. Thus, a loosely coupled distributed measurement system might make sense.

An example of a geographical scarcity is when a single instrument is available to take measurements, and results of its measurements are needed in several locations. Again, a loosely coupled system might make sense. However, it is also quite possible that a simpler tightly coupled distributed measurement system could be adequate.

Examples of scarce resources (computational and geographical) that sometimes drive the decision to build a distributed system are:

- Instrument (e.g., one instrument for multiple people to use)
- Network bandwidth (computed/integrated data versus raw data or large-grained control versus fine-grained control)
- Disk (e.g., archival space in another location, not where the data is generated)
- Input/output bandwidth or devices
- Processing elements (e.g., computing power exists at some location other than where the measurement is being taken)
- Time (e.g., measurement/analysis must occur within a specific timeframe)
- DUT (e.g., multiple measurements must be taken at the same time in order to meet time constraints)

46.8.2 Expected system longevity

Even if the system consists of scarce resources, secondary reasons also affect the decision to create a distributed measurement system. Most of these have to do with the expected longevity of the system. One of the major tradeoffs that must be understood when developing distributed measurement systems is that of longevity; if the system will only be used for a small amount of time, there is probably no point in going through the extra work needed to build a loosely coupled distributed system. However, if the system is expected to be used for a long period of time, requires loose coupling, or flexibility and maintainability, then the tradeoff is probably worthwhile in order to realize the additional capabilities of a loosely coupled distributed system.

46.8.3 Work partitioning

The final reasons to consider developing a loosely coupled distributed system surround the actual development itself. If the system is to be developed by multiple teams, it will quite often end up being loosely coupled as a result of the work partitioning, where such issues as complexity management and the size of the system come into play.

46.9 Software for Distributed Measurement Systems

If a distributed measurement system will be developed, two other items must be taken into account. These are:

1. How much effort will be spent on deciding the software structure itself
2. The effort and complexity of developing the resulting system.

The software in a distributed measurement system is responsible for the basic measurement control; sending commands to instruments, getting measurement data back, perhaps performing computation on raw data to generate results, and sending the data where it is needed. In a distributed measurement system, the software is also responsible for coordinating the efforts of multiple sets of instruments and perhaps computers; taking care of synchronization across all these components, and coordinating all the efforts so the system as a whole is coherent.

In a tightly coupled distributed measurement system, only the efforts of all the instruments must be coordinated by software running on the single controller (and perhaps the instruments if they are capable of executing user-supplied application software). However, in a loosely coupled distributed measurement system, the efforts of multiple computers must be coordinated as well, and the software resides and must operate cooperatively on all of these computers and instruments. In both types of systems, a software application developer must develop this software and somehow transport executable copies to all of the computers (and perhaps instruments) in the system and get it running properly, in a coordinated fashion in all of those locations. These are some of the issues involved in developing distributed measurement system software. These and others are covered in more detail in section 46.10.

46.9.1 Deciding software structure

The expected longevity of the system and the way that work is partitioned can drive the software structure. The three basic possibilities are:

- *Software architecture.* This is the structure of the system at the highest level of abstraction. It provides a way to discuss the software, in the broadest sense, and make the highest level decisions about how it will work.
- *Platform.* This is a less abstract structure; it provides software for the common parts of the system that several versions will use.
- *Instance.* This is the most concrete structure; it is a full implementation of all of the software in the system.

46.9.2 Software architecture

If a system is expected to last a long time, through many modifications, then developing a software architecture that demonstrates the required system properties and flexibility is probably worthwhile. (The specific properties important to distributed measurement systems are covered in the next section.) Components can then be developed that will be interchangeable with future components because they conform to the architecture. Similarly, if the work is partitioned so that many developers are involved, an architecture that specifies how all the pieces communicate and fit together will help the integration, bring-up, and maintenance proceed smoothly.

46.9.3 Software platform

If a system is expected to have a short life span, but many similar, derivative systems will be developed later, then developing a software platform to exploit the similarities of these systems is probably the best choice. The platform developers can then supply common components that each system will use, thereby increasing the quality of the systems (the assumption being that re-using components leads to better tested and higher-quality components). Part of such a platform is a complete specification of the interfaces and properties of the parts of the system that are not part of the platform. (Specific properties are covered in the next section.)

46.9.4 Software instance

If a system is not expected to be used a long time, and if it is unique, then even if it is complex and the work is partitioned across several groups of developers, a completely specified design and associated software implementation is probably adequate. The design must specify the system properties that all components need to exhibit (see the next section for specific properties), as well as interfaces that must be met between components. Such a design will allow system software integration, bring-up, and maintenance to flow smoothly.

46.9.5 Effort and complexity

The effort and complexity involved to develop any distributed system, much less a distributed measurement system, are greater than that needed for a non-distributed system. (In addition, a tightly coupled distributed measurement system is easier to build than a loosely coupled one.) This greater work manifests itself in all areas of software development: design, implementation, testing, deployment, and maintenance. And experience has shown that the effort in design and implementation are always much, much greater than ever expected (no matter what is expected).

Similarly, the effort and complexity of developing a platform (a set of common software components usable in a group of similar systems) is much greater than that of developing a single system. The effort and complexity of developing an architecture (a complete high-level specification of how the system and potential derivatives will be structured and their components interact) is much greater than that of developing a platform. Much of this increased effort comes in the role of communication, both between developers and between the users of the platform or architecture.

So, if the decision is made to create a distributed measurement system, this increase in effort and complexity must be planned for, with additional increases resulting from the development of a loosely coupled system, and structurally of a platform or an architecture.

46.10 Software Issues Inherent in Distributed Measurement Systems

All distributed measurement systems must deal with the issues described in the following paragraphs to some extent. Naturally, systems designed for different measurements and environments will experience them to differing degrees (e.g., a tightly coupled system generally has fewer issues it needs to deal with than a loosely coupled one). However, issues that are often not thought important at first product design often become major stumbling blocks as these systems are extended and redesigned over time. (If the system is tightly coupled, some of these issues are not relevant. These are noted in the summary section.)

To develop the right software to solve the intended problem, the problem domain and information model must first be understood. From there, deciding which distributed measurement system issues are relevant to the software helps direct development effort.

46.10.1 Problem domain and information model

Probably the most important issue associated with developing a distributed measurement system is that of understanding the problem domain. This is not only the problem itself, but also the context in which it needs to be solved; how and where people need to solve it by using the distributed measurement system. The problem domain can include specific ways the system will be used (use cases) and the various "states" that items in it will pass through.

From this understanding of the problem domain, the things that are important to that domain and must be included or modeled in the software system can be identified. These items constitute an information model; a model that relates important items in the real world to software items or objects in the distributed system. For example, in an ATM network monitoring system, an aspect of the real world is the call record/header that accompanies packets across the network. A representation of this call record must also exist in the software developed for the monitoring system because it is the only way that portions of calls can be uniquely identified.

Once a thorough understanding of the problem domain and an accurate information model has been grasped, it is much easier to decide which of the other distributed systems issues are really important and, therefore, where design and development effort is best spent.

46.10.2 Distributed measurement system issues*

The major issues listed in this section are associated with distributed measurement systems. They are all addressed through the system software that is developed to run on the various computers and instruments in the distributed measurement system.

- Controlling system behavior
- Enterprise communication
- User interaction with the system
- System characteristics

Controlling system behavior. Controlling system behavior includes turning the system on and off, installing new software/firmware revisions, configuring functionality, detecting/logging faults and correcting them, and monitoring and adjusting the system performance and resource utilization. Uniformity and consistency are key to this issue. Users at any location, running any portion of the system, must be able to accomplish things consistent with their roles. Roles can be broad (such as developer, user, or maintainer of the system) or more focused on use cases (such as test operator, test developer, R&D engineer, or administrator).

For any large distributed measurement system (tightly or loosely coupled), turning the system on and off can be a problem; for example, how do you turn off an instrument in a remote geographic location (perhaps you just reset it

---

* For more information, see "An Approach to Architecting Enterprise Solutions," by Robert Seliger, *Hewlett-Packard Journal,* Feb. 1999.

to a known state and leave the power on)? In a large loosely coupled system, the issue of getting the right version of software components to the right machines can be very difficult (it usually involves sending software over a network, so the network addresses of each location must be specified). These are typical software issues for both tightly coupled and loosely coupled distributed measurement systems.

Enterprise communication. Enterprise communication is the ability to exchange information with other systems, based on formats and protocols important in the domain. Although this issue includes basic communication (e.g., serial or network protocols) and data formats (e.g., ASCII or byte-swapped binary), there is much more to it. Usually, the programs specific to a problem domain carry not only their own data formats, but also consistency and integrity rules that must be met by that data. Enterprise communication in a distributed system must address all of these aspects.

User interaction with the system. Users must be able to interact with the system using a consistent method that governs all interactions with the system and all its applications. Although uniformity and consistency are related to this issue (see the controlling system behavior section), it also means presenting the system so that users familiar with the real-world problem domain can make sense of the computerized solution to the problem, which is embodied in the distributed measurement system.

System characteristics. System characteristics are concepts perceived about the system as a whole, but must be adequately supported by all the pieces that make up that whole. Interactions between those pieces must be understood well enough that achieving the goal is feasible. Examples of system characteristics are:

- *Functionality.* Consistent data models and common services available from all parts of the system, scalability of the system, performance of the system, its real-time characterization, etc.
- *Trust.* Quite often, this term is used to mean *security.* In a distributed system, it also means the reliability of the system as a whole, and its availability; for example, what happens when part of the system goes down (as often happens with a network) and then comes back up?
- *Synchronization.* One of the startling (though obvious) aspects of a distributed measurement system is that it truly exhibits simultaneous behavior. The only control over sequence of events is through synchronization.

An example occurs when two computations complete (in different locations of a distributed system) and their data is needed by a third computation (located in yet another location). In this case, both inputs must be present for the third computation to start. However, without some sort of synchroniza-

TABLE 46.1 The Relative Importance of Distributed System Software Issues to Loosely and Tightly Coupled Distributed Measurement Systems

Issue	Relevance to loosely coupled distributed measurement system	Relevance to tightly coupled distributed measurement system
Problem domain and information model	HIGH	HIGH
Controlling system behavior	HIGH	LOW. Because there is one controller, this aspect is uniform by definition.
Enterprise communication	Might be high: problem domain specific.	Might be high: problem domain specific.
User interaction with the system	HIGH	LOW: only one interface because there is only one controller.
System characteristics: ■ functionality ■ trust ■ synchronization	HIGH	Might be high for functionality and trust: scalability, performance, and real-time characteristics are likely high, as is security. LOW for synchronization; one controller can handle this aspect easily.

This table shows the relative importance of specific software issues for loosely coupled and tightly coupled distributed measurement systems. It is not feasible to address all of these issues fully in every distributed system; however, picking the few initial areas to concentrate effort can make the difference between a successful software implementation and one that never finishes, or never fulfills its expectations. Therefore once the development of a loosely coupled or tightly coupled system has been chosen, this table can be used, along with the previous detailed discussion of these issues, to decide where development efforts needs to be directed. (See section 46.6 for a discussion of how to choose whether to develop a loosely coupled or tightly coupled system.)

tion, none of the computations can "know" when the others have finished—especially if they execute several times. So, some mechanism is necessary to let the third computation know when it can start its processing; when all the data it needs to start is present.

This issue can be approached by using the idea of a lock-step mechanism, where the ordering of some events is strict, or a looser model can be used that is not as deterministic, but still yields coherency at particular points (e.g., time points or data points). In either case, message passing is the technique most often used to achieve synchronization.

46.10.3 Issue analysis

A complete understanding of the problem domain and an accurate information model help the designer decide which of these issues are really worth the focus and effort in order to avoid system-specific stumbling blocks that are created as the system is extended and redesigned over time.

Table 46.1 summarizes the issues relevant to a tightly coupled distributed measurement system and to a loosely coupled distributed measurement system.

46.10.4 Dealing with distributed measurement system issues

Knowledge of the problem domain and information model allows the identification of the relevant issues for the distributed measurement system. From there, it is necessary to use tools such as use cases (a detailed sequence of steps that a user performs to accomplish a particular task, such as configuring a set of instruments to perform a measurement), which identify complete, specific user tasks, to decide how to address these issues.

The following sections show how these use cases can be used to identify the important aspects of the distributed measurement system software, and, to some extent, how they can be addressed. In the following example, refer to Fig. 46.5. This example covers the case where a tightly coupled distributed measurement system (the R&D computer A and of its associated instruments and network connected smart sensors) is being developed. The engineer in R&D is the person developing the software, all of which will reside on Computer A. (Notice that the smart sensors might execute measurement applications, too, associated with the kind of measurements that they are making. This software is most likely supplied by the smart sensor vendor.) The assumption is further made that the software being developed will have to work with software running on the Enterprise Computer E. Software on Computer E cannot be changed by the R&D engineer.

Example use case. As an example from Fig. 46.5, suppose that an R&D engineer needs to generate measurements on a product under development, create a report, and upload it to a research progress application running on the Enterprise computer. For simplicity, assume that another use case has covered the way that the engineer creates tests or code to perform the measurements. Then the high-level use case for this task might be:

1. Connect the necessary instruments to the device under test, power up the system (computer, instruments, DUT), invoke the tests/measurements, and gather results into some location. Repeat this loop until all the necessary measurements are complete.
2. Identify the kind of report that needs to be sent to the Enterprise, extract the appropriate data from the raw measurements, analyze it to create the data expected by the report generator, and generate the report.
3. Connect to the Enterprise system and application of interest. Transfer data to the Enterprise application.
4. Close the connection to the Enterprise.
5. Shut down the test/measurement system, including instruments (so that somebody else can use them) and the DUT.

Suppose that the Enterprise application expects ASCII data and that this data will be stored in a database at the Enterprise, where it can be correlated from many locations. Transactions into the database are only allowed if all of the data for a record exists. For an R&D entry, assume that the record is expecting a product number, project code name, date in the format mm/dd/yyyy, test identification number, and analyzed test results. If the test identification number does not exist in the database, there must be an addition of the number and a text description of the purpose of the test and expected results.

Analyzing the use case for Enterprise Communication issues. From the use case described in the last section, one can determine what aspects of Enterprise Communication are important and need to be solved in the system software. (Note that in this example, the analysis will be valid for either a tightly coupled distributed measurement system or a loosely coupled distributed measurement system.) The following items must be provided by the measurement system:

- Primary connection to the Enterprise system.
- Connection to the Enterprise application; in this case, a database-entry application.
- Determination of what data the application needs; in this case, the product number, project code name, date, test identification number, analyzed test results. As a subset of this action, it is also required to determine if the test identification number exists in the database and, if not, interact with the R&D engineer to create the proper data (description of test purpose and expected results) and upload it to the database. More than likely, if this branch operation needs to be taken, some sort of security/role checking will determine if this R&D engineer is authorized to add test descriptions to the database.
- Packaging of the database record and transferring it to the Enterprise.
- Waiting for acknowledgement that the transaction completed. If the acknowledgement is not received, the error must be handled, either by fixing it in the application, logging, or reporting the failure.
- Disconnection from the database and the Enterprise.

This list is one example where a use case can be used to drive the identification of specific issues in the area of Enterprise communication. Notice that in this example, data format (e.g., ASCII) was important, as well as data consistency (the test identification number must match information already in the database) and data integrity (a complete record must be transmitted at once). The other use cases important to this system also need to be analyzed for their Enterprise communication aspects; then the overall requirements of this issue are identified.

Analyzing the same use case for system behavioral control issues. The same use case described in the last section can also be used to identify specific issues related to controlling system behavior that the software in the measurement system must address. For example, turning the system on and off are the main things related to controlling system behavior for this use case. This is probably fairly simple, but can include being sure that the instruments needed for a measurement are actually available and set up correctly. Because they are all locally connected to the computer, and there is only one user at a time, this is straightforward.

As in the Enterprise communication aspect, the other use cases that will be met by this system need to be analyzed for their system behavioral control components. Once this is complete, the system behavioral control requirements have been defined.

46.11 Developing Distributed Measurement Systems

The real key to developing a distributed measurement system is a thorough understanding of the problem domain and the problems that the system will be solving (including its context). Knowledge of the problem domain will allow a reasoned analysis of the many issues involved in a distributed system and the choice of the issues that really matter for the application. Knowledge of the context and problem to be solved will also help drive the actual decision of whether or not to go to the effort of building a distributed system. Here is how such a decision might be made.

46.11.1 Development steps

1. Identify the problem domain and the information model that will be used for the system. From these, develop the relevant-use cases that must be met by the system.
2. Look at the reasons to distribute a system. Identify the scarce resources, if there are any. Realistically determine the life of the system and what will be important during its lifetime. Analyze the work partitioning and complexity of the system.
3. From these analyses, decide whether the system needs to be loosely coupled or if a tightly coupled system will suffice.
4. Next, determine which distributed system issues are important for this problem. (Use Table 46.1 to help determine which issues are important for a tightly coupled versus a loosely coupled system, and use-case analysis described in section 46.10.4 to obtain the detailed necessary requirements for the important issues.)

At this point, all the relevant information is known in order to provide focus for design and software development. The next step is to implement the system.

46.11.2 Examples of tightly coupled and loosely coupled distributed measurement systems

The following examples are taken from the systems shown in Fig. 46.5. Each example shows how the type of distributed measurement system (tightly or loosely coupled) was chosen, as well as an analysis of the issues that need to be solved in the system software.

Tightly coupled system: R&D. Consider the R&D building in Fig. 46.12.

The engineer using computer A must share one of the instruments commonly used with four other engineers. That instrument is connected to the network for easy access by any engineer from their own computer. Computer A's user could build a distributed system to allow coherent, cooperative access to the instrument. Here is how the decision might be made.

As a first step, the engineer needs to look at reasons to build a distributed system, as outlined previously. When this is done, clearly, the instrument is a scarce resource. It is also the only scarce resource in this scenario. However, the context in which it will be used is one that requires extreme flexibility (R&D bench testing), but not over long periods of time; that is, longevity is not an issue. Furthermore, any DUT being tested by any of the engineers must be connected manually to the instrument. Thus, something as simple as a note indicating the current user might be sufficient to ensure coherency in its use. Therefore, a loosely coupled system is not necessary; a simpler tightly coupled distributed measurement system will suffice.

Loosely coupled distributed measurement systems. The following examples are all loosely coupled distributed measurement systems. They are taken from Fig. 46.5 and, therefore, are generalizations of systems found in the real world.

Manufacturing test. For another example, consider the satellite test system in use in the Manufacturing building of Fig. 46.13.

In this case, test developers create a suite of complicated tests for the preflight testing of a satellite. Test operators run the tests and results are fed to other computers for analysis, then transmittal to the enterprise. When failures occur, R&D engineers are often needed to troubleshoot the failing parts, which could be interacting in complicated and subtle ways. Pre-flight test results are kept for comparison purposes when the satellite is in actual operation and possibly for troubleshooting once it is operational. Testing occurs over very long periods of time. Each satellite tested varies to some extent from the previous one, although some components are the same on each one.

Decision to build a loosely coupled system. In this case, the decision of whether to build a distributed system is more complicated than in the case of the R&D example. Here is how it might be made, using the steps outlined previously.

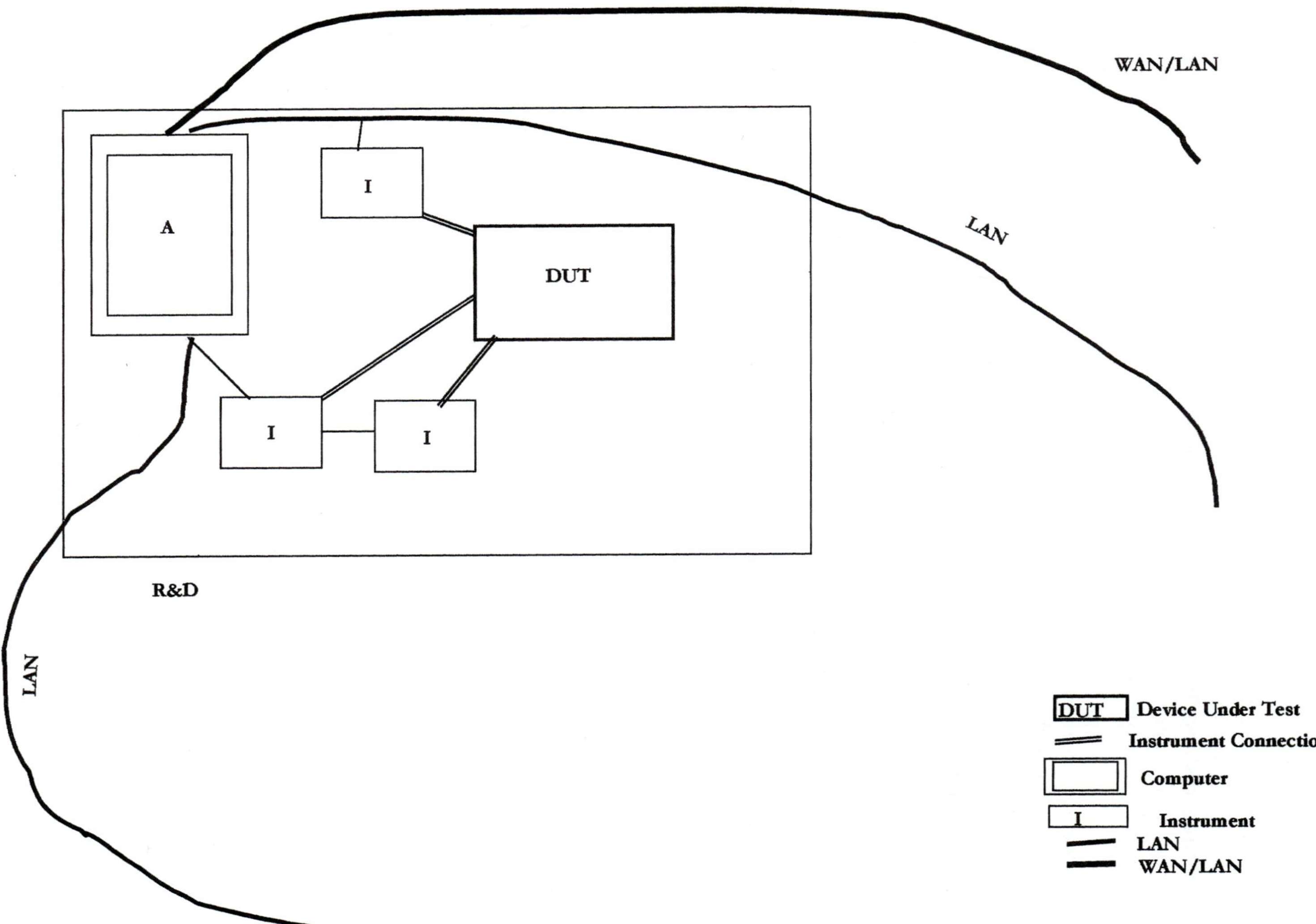

Figure 46.12 Computer A is controlling several instruments on an engineer's bench, which are testing a device that will soon be in production. Two of the instruments are connected directly to computer A and a third is connected to the LAN in the building so that multiple users can access it.

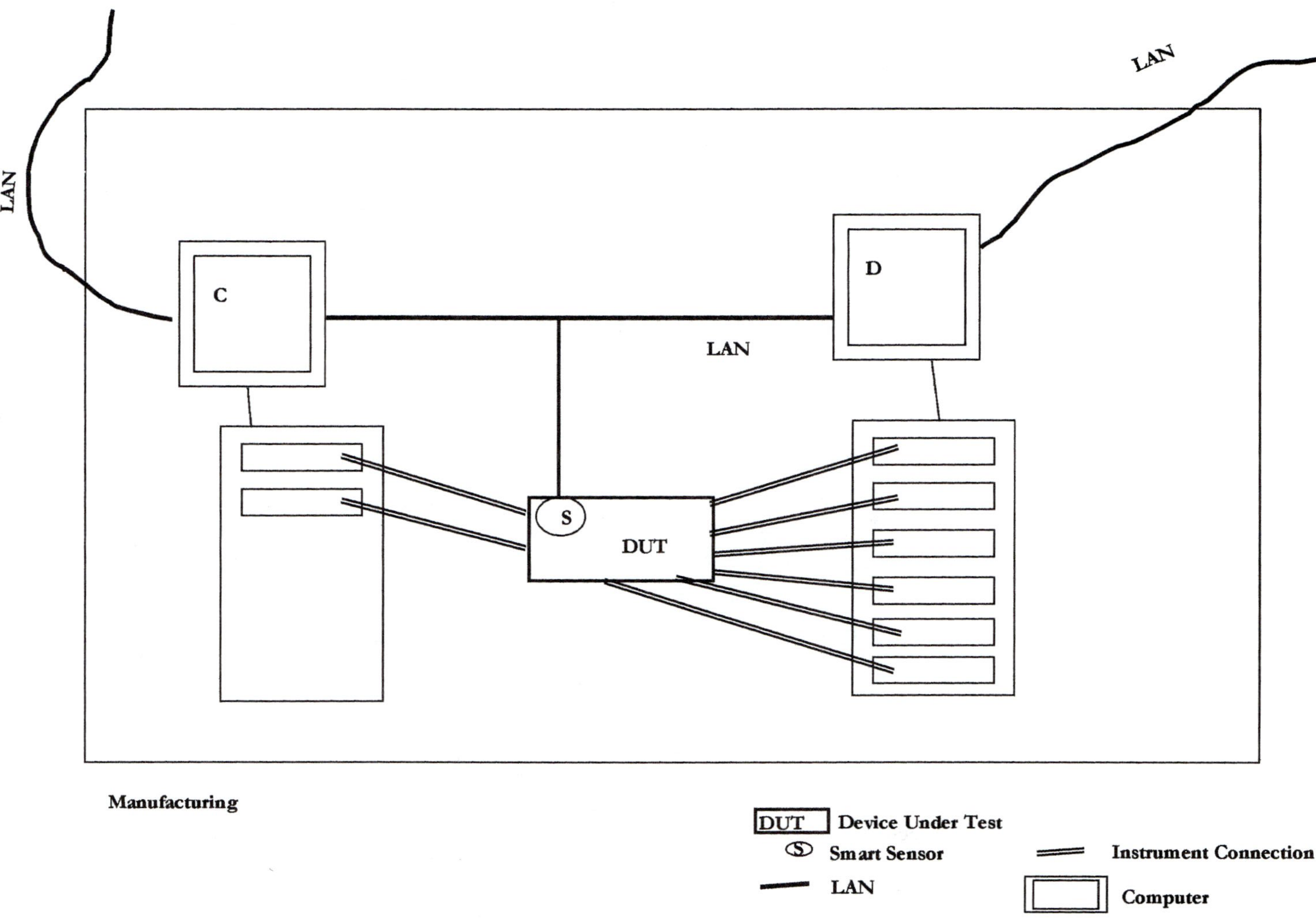

Figure 46.13 Computer C is cooperating with computer D, multiple racks of instruments, and a sensor connected to the LAN to test a DUT. In this case, the DUT is a satellite; the coordinated efforts of multiple racks of instruments are required to perform measurements and test it.

1. An analysis of scarce resources shows the following:
   - The DUT itself is a scarce resource, as is time. The satellite needs to be tested and launched as quickly as possible to generate revenue for the company. However, parts must be tested over long periods of time, so simultaneous testing will be needed.
   - More than likely, the instruments needed to test the various parts of the satellite are scarce resources, as well. If the instruments can be used to test different parts of the satellite using careful coordination, then fewer duplicates, usually doing nothing, are needed.
   - Computational resources are also scarce. The measurements that need to be taken are very complicated and require a great deal of computation. Multiple computers will be needed; it will be more economical to borrow computational resources from an idle computer than to invest in additional computers that, again, could be idle a great deal of the time. In addition, more control resources are necessary to coordinate the operation of all the instruments that will be needed to test the satellite.
   - Finally, the expert user (i.e., R&D engineer) is also a scarce resource. If something fails on the satellite, it must be found and fixed as quickly as possible. R&D engineers might be on call to work with test operators and test developers, and they might have to physically visit the manufacturing site on occasion.
2. An analysis of the probable life of the system shows the following:
   - The system is very loosely coupled. Instruments will be reconfigured often and different tests will need to be run simultaneously; occasionally remote users will need to access the system for various operations.
   - The system must also be very flexible, in terms of instrument and control configuration.
   - If the company will be building and testing satellites, it is probably in that business for a long time. A testing system will have a long longevity. The expectation must be that satellites will change a great deal over that time and tests themselves will evolve, as well. Another aspect of longevity is that of the tests themselves. Although individual tests might run for a relatively long period of time, each complete satellite test has a definite beginning and end.
3. Finally, it should be noted that although the basic framework of the test system could be developed by a single group of people, those doing the test development are most certainly a different group, as are those experts who develop ad-hoc tests for troubleshooting.

   Based on this reasoning, the decision is made to develop a loosely coupled distributed measurement system to do the satellite test. Further, based on the longevity of the system and the fact that satellites do vary in their testing needs, the loosely coupled system needs to provide a basic structure into which differing parts can be added, as needed, to test different items. The ability to extend the scope and capability of the system will also be important because it is impossible to know completely how it will be used over its lifetime.

Relevant issues in manufacturing test example. Given this development and use environment and knowledge of the physical problem, the distributed systems issues can be analyzed for their relative importance. Although some of these issues are quite generic, they tend to exhibit slightly different characteristics, depending on the specific measurement system being developed. (For example, monitoring system performance is a generic issue; in the manufacturing system monitoring needs to be able to occur around resource utilization in order to answer questions, such as "are all instruments being used to their full capacity?" This necessitates capabilities in the system-management area for the software that drives individual measurements.)

- *Controlling system behavior.* (See section 46.10 for a detailed definition of this issue). Controlling system behavior is a key issue for this measurement system. The software needed to perform complicated measurements will be created by different measurement experts, and, unless great care is taken, each measurement will be probably controlled differently (for example, in terms of configuring the software, as well as in terms of updating and monitoring it). Another point of this issue is that it is very difficult to go back and add these kinds of capabilities to existing code. Unless the capabilities are built into the software initially, things like detecting and logging faults (for monitoring or notification purposes) are almost impossible. Software probes and measurement capabilities in the system itself are needed to be able to check the system against predictions.
- *Enterprise communication.* Enterprise applications probably take the point of overall analysis of test results, comparison with in-flight operation, and potentially drawing conclusions across results. If standards exist across these applications, then the system must ensure their rules and formats, as well.
- *User interaction with the system.* Complicated measurements will most likely be developed by (different) measurement experts. Unless some care is taken, they will not necessarily be presented in a format that necessarily makes sense to the expert or test operator.
- *Functionality.* (See section 46.10 for a detailed definition of this issue.) If complicated measurements are created by different experts, consistent models and services are probably important for configuration and set-up purposes. Data movement can be very important for something as complicated as satellite test because measurements can be used in multiple tests of different parts of the satellite at the same time.

  Scalability, performance, and real-time aspects are probably bounded by this kind of application. That is, the system probably does not have a wide variation on how large or small it needs to be. In addition, its performance must be consistent with the length of time that the tests need to be run; if a test requires a week to perform, the real performance issue is to be consistent with the concurrency of the other tests that need to run in that time. Finally, data from the measurements must be handled at the rate it arrives, thus bounding the real-time aspects of the system. The ability to interrogate the system for various operation characteristics (which are problem dependent) is also needed. One example of such a characteristic is the percent of time an instrument or rack is being utilized. Based on this resource utilization, the system might need to be modified to increase the measurements that can be performed at the same time so that the instrument resources are fully utilized and overall testing takes less time.

- *Trust.* (See section 46.10 for a detailed definition of this issue.) The system is fairly well contained in the manufacturing building. Security of data is probably important to the enterprise, and availability of the remote access for use by the expert is necessary. Clearly, reliability of the measurements themselves are crucial.
- *Synchronization.* This is a big, very difficult issue for this application. Multiple computers driving a shared set of instruments and sensors to simultaneously test different parts of the same DUT must be carefully coordinated. Resource allocation is a crucial synchronization point, as is final computed measurement. Appropriate locks must be available so that when resources are allocated, they cannot be used by other tests and/or measurements, but (at the same time) too strict a locking will disallow concurrency. A tradeoff must be found that ensures that neither deadlock or live-lock (the continual inability of some part of the system to proceed as a result of lack of resources) of the system occurs.

Field network monitoring. Consider the ATM digital communications network monitoring in the field, shown in Fig. 46.14.

Decision to build a loosely coupled system

1. An analysis of scarce resources shows the following:
   - The DUT itself is not a single entity, but a very geographically dispersed network.
   - Probes needed to test the network are not scarce; they reside at each testing point on the network.
   - Computational resources are scarce; the amount of data gathered by the probes needs to be reduced as close to the source as possible so that the available computational resources can deal with it. A related scarce resource is network bandwidth, another reason for data reduction as close to the source as possible.
2. An analysis of the probable life of the system shows the following:
   - The system will be running all the time.
   - Because the system has no clear beginning or end, there must be some way to check its "state," and save and restore it, if necessary.

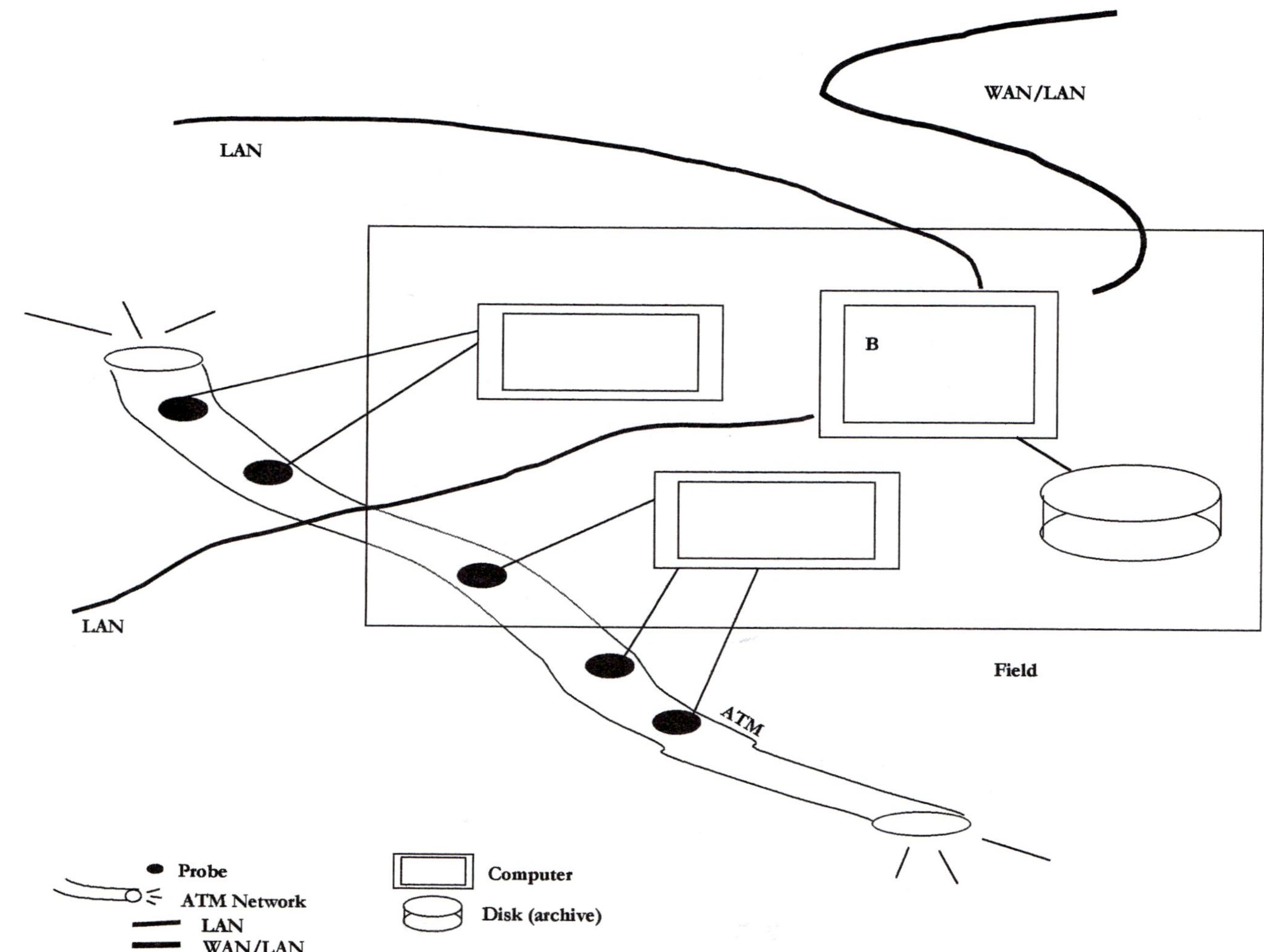

Figure 46.14 Computer B is working with other computers, which have attached probes that are measuring an ATM digital communications network. The probes simply report data packets crossing the network, performing minimal processing to extract header information. The computers attached to them reduce data from the probes, looking for patterns. Such computers as computer B further coordinate and reduce data, and send it to the enterprise for further analysis.

- Clearly, probes and networks will be coming and going in the system all of the time, so flexibility in configuration and control will be important.
- Because parts of the system might not be working at any particular point in time, reliability of those parts that are operating will be important.

These reasons all drive the decision to build a loosely coupled system to solve this monitoring problem. The ability to extend the scope and capability of the system will also be necessary because it is not possible to determine what kinds of monitoring will be of interest over a long period of time.

Relevant issues for ATM digital communications network monitoring example

- *Controlling system behavior.* (See section 46.10 for a detailed definition of this issue.) For this application, turning the system on and off doesn't really occur, except at system initialization; the system is constantly being re-configured as applications and probes and computers and networks constantly change. System initialization is a large problem, as is the way that these configurations are specified and handled. Another issue is the deployment of the binaries (executable code) and versions of binaries to the applications running on various computers. Still another issue is getting the right programs (or parts of programs) up and running at the right times on the various computers. (For example, an application might need to have one version of a service-providing program running on one of several computers in order to work.) So, the service-providing program must be up before the application attempts to execute. Controlling system behavior thus takes on a time aspect, as well as a location and content aspect. Clearly, in such a diverse application, detecting and logging and correcting faults is also an issue. Given that the distributed system is huge, such actions must have a granularity that is fairly large or else the system will be overrun with logging data. Resource utilization is probably an issue, as is the ability to adjust it in order to affect system performance. This, again, is caused by the huge scale of the system and its profusion of networks.
- *Enterprise communication.* Enterprise applications probably take the point of overall analysis of patterns, and perhaps direct attention toward the specific part of the network being analyzed. Also, some enterprise applications, such as billing, must be used; in this case, specific data formats and models might be necessary.
- *User interaction with the system.* This is an issue mostly because of the immense scale of the system. An operator or administrator must be able to consistently deal with all parts of the system.
- *Functionality.* (See section 46.10 for a detailed definition of this issue.) If several applications can run on the system at one time, using data gathered by the probes for various reasons, consistent models and services are probably important. Moving partially analyzed data from application to ap-

plication might also be expedient. Scalability is a crucial issue for this system, as is real-time characterization, because probes are generating large amounts of data, which must quickly be reduced, all of the time.

- *Trust.* (See section 46.10 for a detailed definition of this issue.) Security in this system is a large issue because the patterns being analyzed may contain confidential information. Reliability of the system overall may be somewhat alleviated simply by the fact that so many probes and computers are involved (e.g., some redundancy). Probably the bigger issue here is availability; what happens to the application when part of the network goes down, when a computer goes down, or when a probe goes down, etc.

- *Synchronization.* This is a big issue for this application, mostly because of the sheer size of the system. Thousands of probes generating data for hundreds of computers, which all contribute to a single application, require precise synchronization in order to generate expected results. And, at the same time, using messages to synchronize all these components is not feasible because of the enormous quantity of network traffic that would be generated. Synchronization of the system in some sort of hierarchy that preserves data and computation locality is more feasible.

Process monitoring. Next, consider the process monitoring and control in the process area, shown in Fig. 46.15.

Decision to build a loosely coupled system

1. An analysis of the scarce resources in the system shows the following:
   - The smart sensors used in this system can be viewed as scarce resources in that they are very powerful instrument devices, capable of many different operations. Using them in multiple roles, remotely, could mean that fewer are needed for the monitoring job.
   - This system is geographically dispersed; smart sensors are monitoring different areas of the process.
   - Each smart sensor has its own controller, which can cooperate with other smart-sensor controllers and computers to deliver a coordinated measurement application.
   - Time is a scarce resource; if a process changes, the results must be reported in real time and dealt with very quickly.

2. An analysis of the probable life of the system reveals that, although it is expected to be in use a long time, the overall system functionality is not expected to change very much.

   Because smart sensors are involved, the broadcast of data by the sensors and the use of events to trigger action in computers is present. Many controllers are involved just because smart sensors are being used. These reasons all drive the decision to develop a loosely coupled system.

LAN

S

S

S

F

Process

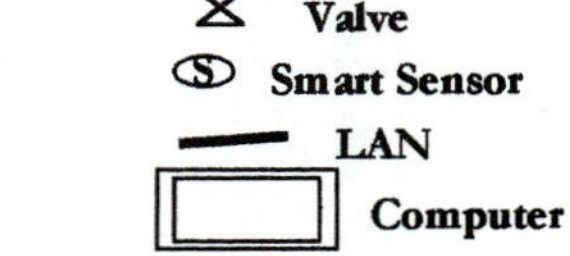

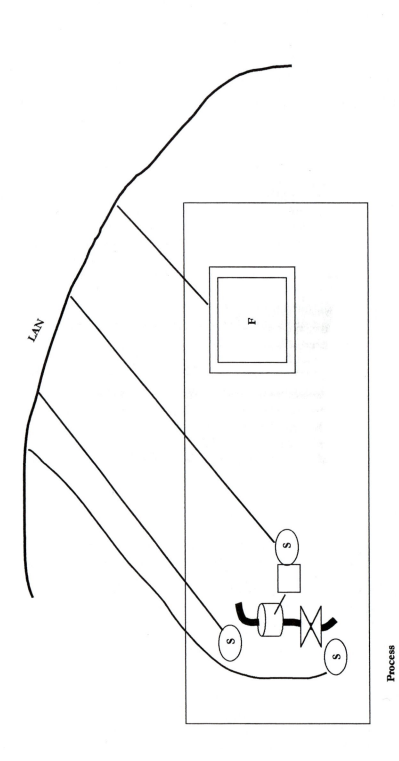

Figure 46.15 Computer F coordinates the reading of three smart sensors connected to the LAN, to determine what is happening with the process. Based on results from the sensors, computer F can adjust how they take measurements, how often, and of what, and it can also do something like control the valve.

Relevant issues for process-monitoring example

- *Controlling system behavior.* (See section 46.10 for a detailed definition of this issue.) In this system, each of these items is important, and, like the ATM system, the idea of portions of the system coming and going because of network problems also exists. Smart sensors have the ability to run applications, so the deployment problem (getting the right binaries to the right points) grows beyond the computers in the system, all the way to the sensors. As monitoring and control change through the system, controlling the behavior of all of the components must be consistent.
- *Enterprise communication.* After intermediate aggregation and analysis by system computers, Enterprise applications further manipulate results. These applications typically have their own data formats and models, which must be used.
- *User interaction with the system.* This is an issue for writers of smart-sensor applications, as well as those working with the applications running on intermediate computers in the system.
- *Functionality.* (See section 46.10 for a detailed definition of this issue.) Consistent models and services are necessary for intermediate computers and smart sensors. These need to interoperate easily. Scalability is a crucial issue for this system because it real-time characterization; sensors are generating data constantly, and intermediate computers must quickly decide which events are triggers for their own application execution.
- *Trust.* (See section 46.10 for a detailed definition of this issue.) Security in this system might be an issue, depending on what is being monitored. Reliability, as in the case of the ATM system, may be somewhat alleviated simply by the fact that multiple sensors and computers are involved (e.g., some redundancy). And, also, as in the ATM case, availability is an issue because of the networks.
- *Synchronization.* Synchronization is important for this system, given that intermediate computers are dealing with data from multiple sources. Smart sensors tend to send their data out when available, for processing by whatever computers have registered an interest. Thus, the existence of such an event is in itself a synchronization mechanism at this low level. Synchronization across intermediate computers must be handled, as well.

Chapter

47

Smart Transducers (Sensors or Actuators), Interfaces, and Networks*

Kang Lee**

National Institute of Standards and Technology
Gaithersburg, Maryland

47.1 Introduction

The word *smart* has been added as prefix to many things that are perceived to possess some form of intelligence. The term *smart sensor* was adopted in the mid-1980s in the sensor fields to differentiate this class of sensors from conventional sensors.[1] A conventional sensor measures a physical, biological, or chemical parameters, such as displacement, acceleration, pressure, temperature, humidity, oxygen, or carbon monoxide content, and converts them into an electrical signal, either voltage or current. However, a smart sensor with some form of intelligence, provided by an additional microcontroller unit or microprocessor, can convert this raw signal into a level or form which makes it more convenient to use. This might include signal amplification, conditioning, processing, or conversion. In addition, over time, smart functions were not only built into sensors, but applied to actuators as well. Therefore, the term *smart transducers* as used in this chapter refers to smart sen-

* Contribution of National Institute of Standards and Technology. Gaithersburg, MD 20899-8200. Not subject to copyright.

** Additional material provided by Stan Woods of Hewlett-Packard Laboratories and Dr. Janice Bryzek of Maxim.

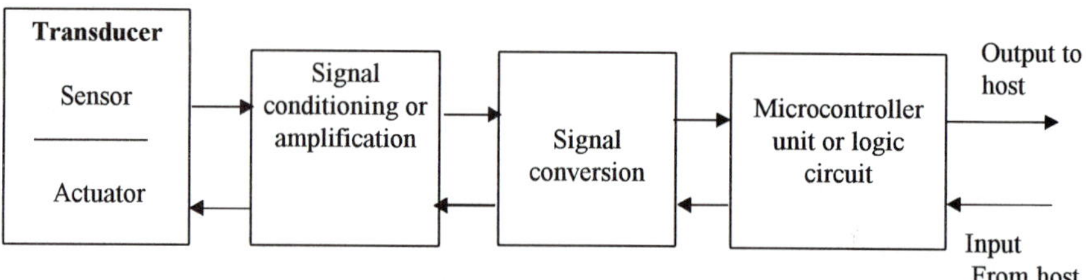

Figure 47.1 A smart transducer is either a sensor or an actuator that is instrumented or integrated with signal conditioning and conversion and a microcontroller or microprocessor to provide intelligent functions. Its output is migrating from an analog to a digital format for added capability to communicate with a host or a network.

sors or smart actuators. Figure 47.1 illustrates the partitioning of a smart transducer's functions.

A smart transducer provides value-added functions to a sensor or actuator and does more than just pass the raw sensor signal through or to a host in a digital form. In addition, it operates on the signal and increases the value of information by processing. It might also be able to perform such functions as self-identification, self-calibration, and self-testing, as well as communicate the data and information to other smart transducers, or with a host in a digital mode. With these enhanced capabilities, smart transducers can respond to industry's need for:

- Remote or telemonitoring, access, and control, utilizing a *Local-Area Network* (*LAN*) or *Wide-Area Network* (*WAN*), such as the Internet, and its communications infrastructure.
- Higher bandwidth and more relevant information on processes for making intelligent decisions for better control in manufacturing, laboratory, or production operation.

47.1.1 Definitions of smart transducer terms

The specific definitions of terms with regard to transducers and sensors, including connecting them to networks are defined in the IEEE Standard 1451.2-1997:

- *Smart sensor.* A sensor version of a smart transducer.
- *Smart transducer.* A transducer that provides functions beyond those necessary for generating a correct representation of a sensed or controlled quantity. This functionality typically simplifies the integration of the transducer into applications in a networked environment.
- *Smart Transducer Interface Module* (*STIM*). A module that contains the transducer electronic data sheet, logic to implement the transducer interface, the transducer(s), and any signal conversion or signal conditioning.

This standard expressly requires that no operating mode of the Smart Transducer Interface Module ever permits those components to be physically separated. They may, however, be separated during manufacturing and repair.

- *Transducer.* A device converting energy from one domain to another. The device might either be a sensor or an actuator.
- *Transducer Electronic Data Sheet (TEDS).* A data sheet describing a transducer stored in some form of electronically readable memory.
- *Transducer Independent Interface (TII).* The digital interface used to connect a Smart Transducer Interface Module to a Network Capable Application Processor.
- *Transducer interface.* The physical connection by which a transducer communicates with the control or data systems that it is a member of, including the physical connector, the signal wires used and the rules by which information is passed across the connection.

47.1.2 Examples of smart transducers

An example of an earlier smart transducer design is shown in Figs. 47.2 and 47.3. Figure 47.3 shows a Motorola smart air pressure transducer with ampli-

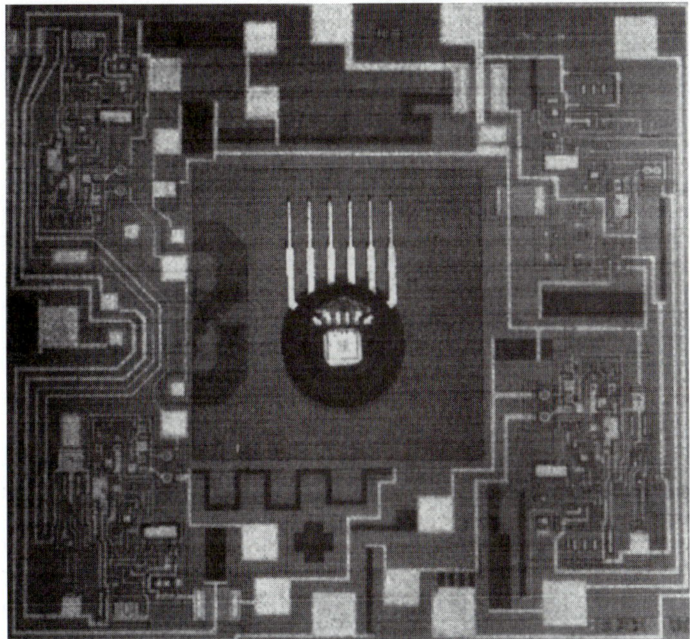

Figure 47.2 Motorola air-pressure smart transducer. This generation of a Motorola air-pressure smart transducer included the pressure transducer element (front center) and the amplification signal conditioning and circuitry (background). Courtesy of Motorola. Used by permission.

Figure 47.3 Motorola air-pressure smart transducer system designs. This smart pressure transducer design included a physical transducer element (right) and the microcontroller (middle) with digital display. Courtesy of Motorola. Used by permission.

fication and signal conditioning circuitry, and it produced an analog signal output. Figure 47.3 shows Motorola's two-chip smart transducer consisting of the integrated pressure sensors and a microcontroller with an onboard A/D converter and EEPROM that produces calibrated output.

47.1.3 Transducers, smart transducers, and instruments compared

Figure 47.4 illustrates the different features of transducers, smart transducers, and instruments. Basic transducers or signal-conditioned transducers have analog outputs. They still need some interface circuitry for connecting to a data-acquisition card, a host, or an instrument. Normally, only a transducer signal is sent to the host. On the other hand, instruments are fully equipped with signal conditioning and conversion capability, as well as computer interfaces for data collection and analysis. Nevertheless, instruments are relatively bulky, are usually designed for use in a laboratory, and are relatively expensive for just reading transducer outputs. However, smart transducers are somewhere between transducers and instruments in terms

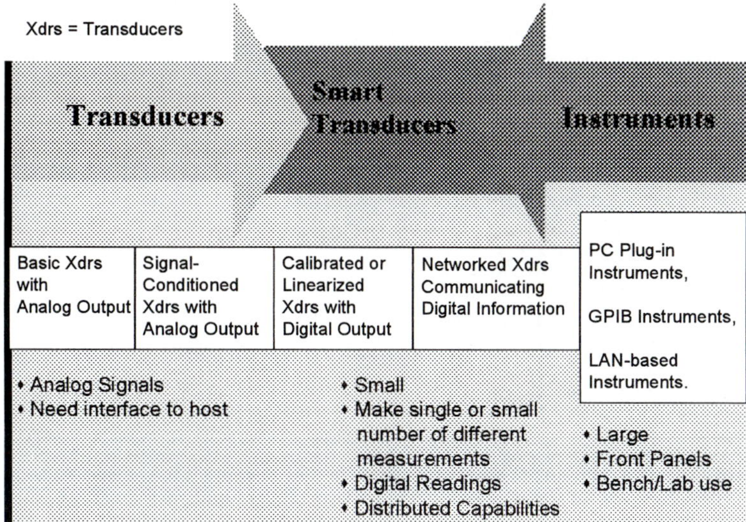

Figure 47.4 Comparison of features of transducers, smart transducers, and instruments. Smart transducers have many levels of built-in capability, similar to those possessed by instruments. They are, therefore, much easier to apply than basic transducers. A transducer-based application can be taken from concept to functional state in a relatively short amount of time.

of capability and cost. They have various degrees of sophistication, ranging from those with calibrated and digital outputs to those with network-ready connectivity. In addition to the transducer output signal, smart transducers can communicate other information such as its identity, its health, and operating status. Equipped with these capabilities, smart transducers are very small and easy to use. Those equipped with a standard interface can also plug-and-play with different networks.

A diagram depicting the domain of the smart transducers and instruments in terms of communication speed, measurement speed, computation, and cost is shown in Fig. 47.5. The radar plot diagram is used mainly for illustration purposes. Smart transducers stay in the domain of low cost, simple computation, and low measurement and communication speed. However, this paradigm is changing as the price of microchips decreases, speed of microprocessors increases, and demand for measurement speed and smart transducer size grows. This trend will push the capability of smart transducer further into the domain of instruments. This blurring of the distinction between smart transducers and instruments leads to using smart transducers as special-purpose instruments, or for networking them into a virtual instrument, as described elsewhere in this book.

47.1.4 Desirable capabilities and features of smart transducers

The following is a set of capabilities that are desirable for smart transducers to possess, and criteria that can be used to qualify smart transducers. In

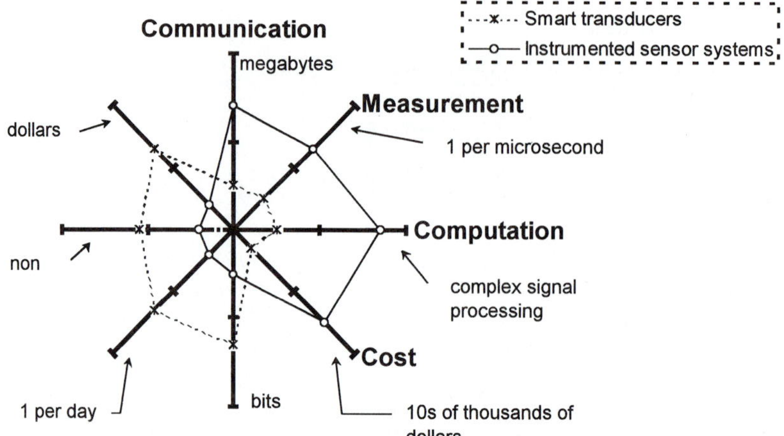

Figure 47.5 The domains of smart transducers and instrumented sensor systems. A "radar plot" shows the domains of smart transducers and instrumented smart sensors in terms of communication speed, measurement rate, computational capability, and unit cost. The smart transducers are less complex, cost less, and are simpler to use, but instrumented smart sensor systems are more complex, usually possess higher performance, and thus cost more.

conventional sensor systems, as a sensor is excited by a physical stimulus, it generates a low-level signal. This raw signal is then amplified and fed to a data-acquisition unit before being applied to an instrumentation system or computer host. Therefore, in addition to the sensor(s), a number of separate modules need to be connected to form the sensor system. However, it is expected that smart transducers will provide a more integrated, efficient solution. Hence, besides providing conditioned and formatted output signals, the smart transducer provides additional manufacture-related information for the transducers. This information, attached to the smart transducer, is used for self-identification purposes.

Hence, some desirable capabilities that smart transducers have or be able to do are:

- Perform self-identification or self-describing to the system upon power-up or upon inquiry
- Perform self-test or self-diagnostic and report status to the system
- Perform self-check or self-calibration to maintain sensor accuracy
- Convert transducer data into digital form and engineering units
- Have multivariable sensing capability
- Perform such information processing or software functions as data logging, filtering, averaging, signal processing, virtual sensing by fusing multiple measured variables into a derived variable

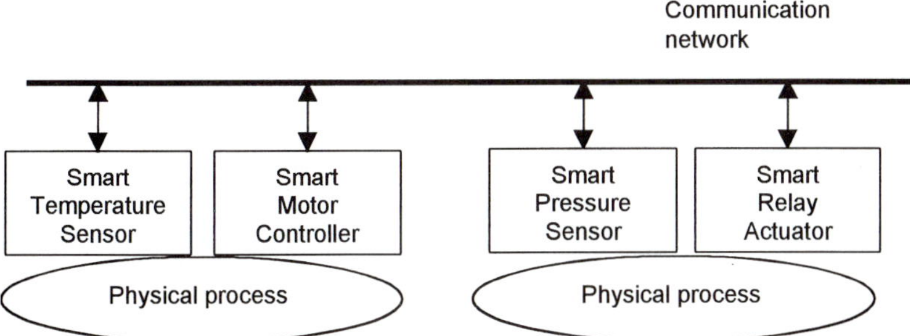

Figure 47.6 A distributed measurement and control system consists of smart transducers connected to a network. Each smart transducer can pass information to others. A group of smart transducers consisting of sensor(s) and actuator(s) can form a closed-loop control system. Clusters of control applications can be formed anywhere in the network. This distributed system approach allows a great deal of flexibility to form control systems.

- Have knowledge of time and can maintain and associate time with transducer(s) data
- Adhere to a standardized messaging and communication protocol

If smart transducers possess these capabilities, they will improve the quality and integrity of the measurements. Equipped with some of these integrated functionalities, smart transducers will pave a path for easier deployment in distributed measurement and control (DMC) applications. An example DMC system is shown in Fig. 47.6.

47.2 The Evolution of Smart Transducers

The advance of silicon integrated circuit and microprocessor technologies has enabled the integration of functions, such as signal conversion and processing (e.g., linearization and filtering) into sensors. As a result, new generations of smart sensors and actuators in the form of monolithic microsensors and *microelectromechanical systems* (*MEMS*), respectively, have emerged. MEMS technology, for example, has made significant contribution in producing *manifold absolute pressure* (*MAP*) *sensors* for use in automotive electronic engine control since 1976 and silicon accelerometers in crash sensing and airbag deployment since early 1990s. The result is that many automobiles contain one or more of these types of devices. In addition, the availability of high-volume, low-cost microcontrollers and memory devices can enhance vehicle diagnostics with smart sensors of high reliability.

The concept of smart transducers evolved from sensors with analog output to sensors with digital output because of the need to improve the accuracy of sensors and the availability of microelectronics technology. In the late 1960s, the concept of smart sensors began to surface and the accuracy of silicon sensors was enhanced through intelligent temperature compensation. Ac-

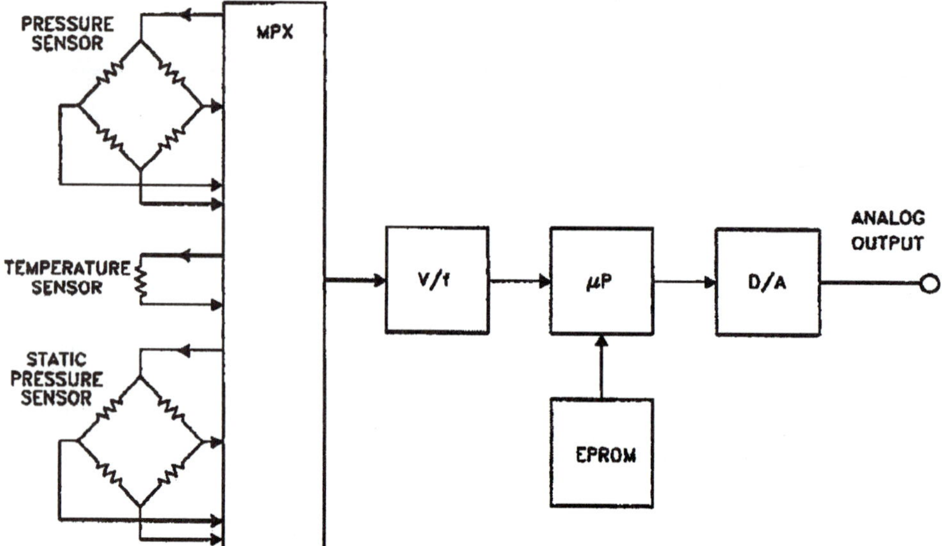

Figure 47.7 First smart processor-control transmitter multiplexes three sensors to the output. The three analog sensor signals were multiplexed and converted to pulses through a voltage-to-frequency converter (VFC). A microprocessor-based digital signal processor (DSP) read in the pulse trains, processed the signals in digital form, and then outputted the pressure signal to the digital-to-analog converter (DAC), which converted the signal into analog form.

cording to Janusz Bryzek* the first implementation of such a design was a silicon pressure sensor built by Honeywell for the air data system on McDonnell Douglas DC-9 airplanes.[2] To perform compensation of temperature sensitivity and normalization, an on-board computer and a temperature sensor were used. The pressure value was determined from both the pressure and temperature signals. The accuracy was claimed to be in the range of 0.01%, which was an order of magnitude better than any other silicon sensors built in the 1970s.

In the early 1980s, Honeywell also made the next-generation smart transducers. Two designs were released. One design was for process control applications and the other was for aerospace applications. The schematic of the design is shown in Fig. 47.7. The process-control transmitter output was based on three input signals of differential pressure, temperature, and static pressure. The three analog sensor signals were multiplexed and converted to pulses through a *voltage-to-frequency converter* (*VFC*). A microprocessor-based digital signal processor (DSP) received the pulse trains, processed the signals in digital form, and then output the pressure signal to the *digital-to-analog*

* Janusz Bryzek is a "Sensor Magazine" Lifetime Achievement Award recipient.

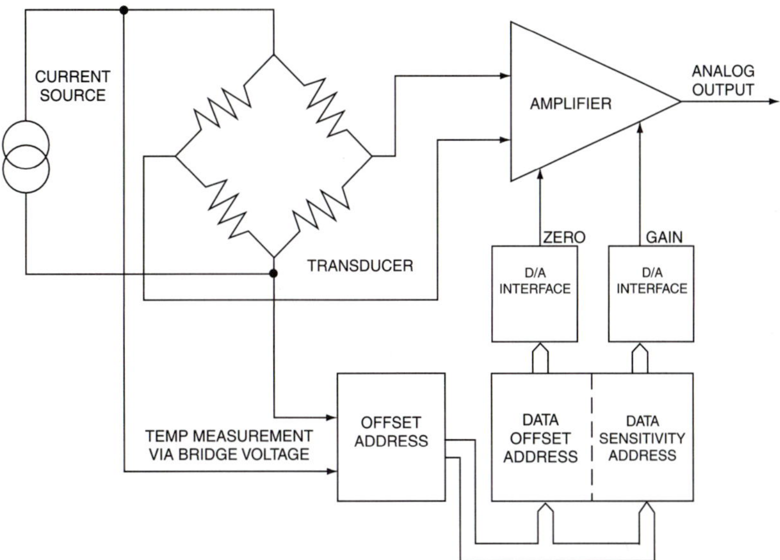

Figure 47.8 The first digitally controlled analog signal processor. The digital value of the temperature signal is used to control the address for two locations in an EEPROM, where the offset and full-scale output adjustment values for a given temperature are stored. These values, converted by two DACs, are used to set zero and gain on the programmable amplifier to produce the desired analog transducer output.

converter (*DAC*), which converted the signal into analog form. A software algorithm in the DSP performed static pressure compensation which was required for process control functions. The key advanced digital features of the process-control transmitter were remote calibration over a 400:1 range, addressability, and diagnostics. The aerospace transducer version used dual absolute pressure sensors and a temperature sensor. The VFC was upgraded over time to a 21-bit resolution, which yielded a 1-PPM transducer output resolution. The transducers were rated at 0.02% total error band over the entire operating pressure range, temperature range, and three-year time period. The cost of transducers with these capabilities was high, on the order of thousands of dollars.

The first digitally controlled and compensated analog sensor signal processor for smart transducers was introduced by Keller.[3] The design, which is shown in Fig. 47.8, was based on a digitally programmable amplifier with look-up table corrections. The transducer signal was generated by temperature-dependent bridge circuitry. In operation, an ADC digitized the temperature of the bridge. The digital value of this signal controlled the address for two locations in an *electrically erasable programmable read-only memory* (*EEPROM*), where the offset and *full-scale output* (*FSO*) adjustment values for a given temperature were stored. These values, converted by two DACs, were used to set zero and gain on the programmable amplifier to produce the

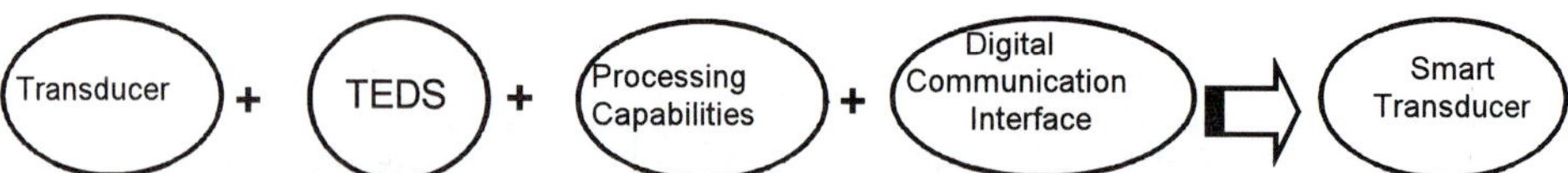

Figure 47.9 A transducer can be either a sensor or actuator. A smart transducer is a transducer integrated with a TEDS, has microprocessing capability to perform some intelligent functions, and has a digital communication interface for outputting (Per IEEE 1451.2).

desired output signal. Applying this compensation technique, transducer offset variations were reduced from 25 millivolts (mV) to 0.1 mV and FSO variations from 25% to 0.1%. All the components in this system were implemented in a single ASIC including the on-chip EEPROM.

Hence, smart transducers are transducers that have some built-in enhanced functions. The degree of added functions varies considerably. According to the industry standard, IEEE Std. 1451.2-1997, *A Smart Transducer Interface for Sensors and Actuators,*[4] *a smart transducer* is a transducer that provides functions beyond those necessary for generating a correct representation of a sensed or controlled quantity. This functionality typically simplifies the integration of the transducer into any applications in a networked environment. Figure 47.9 illustrates how a transducer is transformed into a smart transducer based on this definition.

1451.2-based smart transducers have a *transducer electronic data sheet* (*TEDS*) and digital interface for communication. The TEDS, which contains standardize manufacture-related data (such as manufacturer, name, type of transducer, serial number, and calibration data) enables smart transducer self-identification. The digital interface allows the communication of the TEDS data and transducer output to a network, an instrument, or a host microcomputer.

Smart transducers have come a long way from simple temperature-compensated transducers, through the digital computer revolution, to transducers that possess ability for self-identification and digital communication. The evolution of smart transducers is outlined in a chronological order in Fig. 47.10.

47.3 Capabilities of a Smart Transducer

The added functions of smart transducers could be implemented in hardware electronics or software algorithms that would improve the performance of smart transducers. Hardware implementations can include memory devices, *field-programmable arrays* (*FPGAs*), *application-specific integrated circuits* (*ASICs*), or microprocessing units. On the other hand, software implementations might include compensation algorithms, virtual sensing, signal processing, and other sophisticated functions. The functions in consideration are as follows:

- Physical memory for data storage

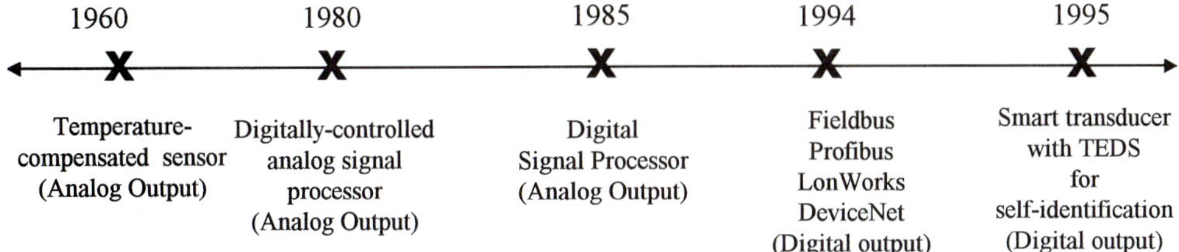

Figure 47.10 Smart transducers evolved from simple temperature-compensated sensors through digital signal processing enhancement, to transducers having abilities for self-identification and digital communication.

- Computational capability
- Communication (input/output) capability

47.3.1 Physical memory for data storage

The output of a raw sensor element is not a perfectly linear function. Its output often needs to be linearized or calibrated. The degree of linearization depends on the degree of accuracy required for an application. It is very desirable to have the calibration data for the sensor stored with the smart sensor. In addition, any manufacture-related information and data about the smart sensor (such as manufacturer name, type of sensor, and serial number) will also be advantageous to the users. These calibration and manufacturers' data can be stored in a memory chip. As a matter of fact, the ability to store and modify these parameters and data in an EEPROM will also allow modification and updating of the calibration data as needed. The calibration function will probably be performed at the manufacturer's site.

47.3.2 Computational capability

In the smart-sensor functional partitioning diagram of Fig. 47.1, the calibration function in a smart transducer can either be implemented in the signal conditioning or amplification stage (analog signal domain), signal conversion stage (analog signal-processing domain), or microprocessor stage (digital signal domain). The calibration done in the first domain is performed during the sensor manufacturing process to produce a standardized analog output signal, as an example, by laser trimming the resistive elements in a bridge circuit. This method is efficient, but could be expensive. It is highly dependent on production volume because it is time consuming to set up a production line to automate the process. Calibration can be achieved in the microprocessor stage by storing the correction data and correction algorithm in the microprocessor memory or EEPROM. The microprocessor will execute the correction algorithm through table lookup or by using parametric polynomial solutions. Besides improving linearization for the smart sensor, the microprocessor can convert the measurement data to display different units, such as kPascal or pounds per square inch, as required by the users. It can also perform autozero or autoranging, as well as self-testing. Therefore, the powerful computation ability of the microprocessor can process raw sensor data, incorporate digital compensation algorithms, and perform signals processing and information processing.

47.3.3 Communication (input/output) capability

Smart sensors should have the ability to easily interface with a host or system device. They should be able to communicate and share measurement data with the host in a commonly acceptable or standardized format. This need stems from the increased interest in communicating transducer information in *distributed measurement and control* (*DMC*) systems. In DMC systems, data from various sensors are sent to a network host. Based on the sensors' information and the control algorithm, the host determines a course of action and manipulates the actuator accordingly. The communications between the smart transducers and the host devices are more than just raw signals. For example, high-level, conditioned, linearized, scaled, and formatted ready-to-use sensor data may be passed. It will be easier for the system device to interpret and digest. In addition, a standardized data format is easier to achieve among numerous sensor and host devices and, therefore, interoperability is attainable.

Various methods for communicating system information and measurement data exist in industry. They range from analog data transmission to serial and parallel digital techniques. The term used in industry to describe a set of rules for communication is called a *protocol.* The ISA's 4- to 20-mA standard is a 20-year-old analog data communication protocol used widely in process-control instrumentation systems. Rosemount's *highway addressable remote transducer* (*HART*) is a mixed analog and digital communication protocol, where digital information is transmitted simultaneously and on top of the 4- to 20-mA current-loop analog signal. *Controller area network* (*CAN*), developed by Robert Bosch GmbH, is a serial digital communication protocol designed specifically for high-speed data communication for distributed real-time control applications for automotive systems.

47.4 Interfaces and Networks for Smart Transducers

Transducers serve a wide variety of industry's needs (e.g., manufacturing, industrial control, aerospace, automotive, biomedicine, building, and process control). The demands for automation systems are continually growing, as well as those for transducers. The need for transducers has created a good opportunity for transducer manufacturers, but it has also created some challenges for them as well. Because the transducer market is very diverse, and, as transducer usage increases, sensor manufacturers are seeking ways to build low-cost, smart transducers to serve the various industries. With the emergence of control networking technology, many transducer buses, transducer control networks, and fieldbus implementations have mushroomed in the marketplace. Transducer manufacturers and users alike are finding ways to apply the networking technology to their transducers for distributed measurement and control applications.[5]

47.4.1 Buses and networks

Numerous buses and field networks are on the market. These buses and networks were initially conceived to solve or address a specific need of a particular sector of the manufacturing automation and process-control industry. They can basically be divided into three categories:

- Sensor buses
- Device level-control networks
- Field networks

Sensor buses. The sensor buses are basic level buses for connecting sensors or actuators and transferring the sensor readings to a controller or a host. They primarily pass the sensor data straight through and seldom process the data.

Device level-control networks. The device level-control networks are higher-level buses that format and perform some basic processing of the data at each network node before passing it to the host computer upon request.

Field networks. The field-level networks were designed to take care of all levels of process-control operations from sensors and devices all the way to the controlling of the processes.

Protocols. The following is a list of transducer bus, device control network, and field network protocols that attempt to facilitate sensor and actuator connectivity.

ARCNET (Attached Resource Computer Network). Developed in 1977 by Datapoint. Maximum data rate of 2.5 Mbps.

ASI (Actuator Sensor Interface). Developed in Germany by a consortium of sensor suppliers. A low-cost, bit-level system designed to handle four bits per message for binary devices in a master/slave structure operating in distances up to 100 meters.

BACnet (Building Automation Control Network). An American *Association of Heating, Ventilation, Refrigeration & Air Conditioning Engineers (ASHRAE)* standard developed by HVAC system suppliers. It supports networking options—ARCNET, Ethernet, master/slave token passing (MS/TP) network based on RS-485 protocol, and LONWORKS.

Bitbus. Developed in 1984 by Intel around the 8044 microprocessor. Features multitasking with a master/slave structure using RS-485 serial linking.

CAN (Control Area Network) Bus. Developed in Germany by Robert Bosch GmbH with Intel and Philips in the early 1980s for automotive in-vehicle networking. A peer-to-peer *Carrier Sense Multiple Access* (*CSMA*) system. Selectable baud rates up to 1 Mbps, and twisted-pair, fiber, coax, and RF media supported. CAN is ISO Standard #11898, approved for passenger vehicle applications. CAN-based systems were approved by SAE as Standard J1850 for American passenger cars and Standard J1939 for trucks and large vehicles. A *CAN in Automation* (*CIA*) group has been formed in Germany to work on application issues.

CEbus (Consumer Electronic Bus). This is being developed by a consortium of consumer electronics manufacturers through the *Electronic Industries Association* (*EIA*). It is primarily used for home automation, and supports coax, RF, and power-line media.

DeviceNet. A version of CAN developed by Allen-Bradley. It features use of object-oriented software and is used primarily in industrial control systems. It uses a four-wire (signal pair and power pair) shielded cable and can support up to 64 nodes per network at speeds up to 500 kbps at 100 m and 125 kbps at 500 m. An *Open DeviceNet Vendors Association* (*ODVA*) has been formed.

Ethernet. 10-Mbps local-area network primarily used for office and enterprise automation. It is recently becoming a very strong contender for control and field network for manufacturing because of its speed, real-time control capability,[6] and readily available Ethernet cabling infrastructure. Ethernet communication speed has increased to 100 Mbps and is expected to achieve wider commercial use in 1-ps arena.

Foundation Fieldbus. Formed from the merging of components of specifications by WorldFIP and Profibus supporters. This was formed to test/demonstrate fieldbus components to support an eventual single, universal fieldbus standard.

GPIB (General-Purpose Interface Bus). GPIB became the IEEE-488 standard in 1978. More of a data-acquisition system with limited node capabilities. It is used mainly in laboratories and industrial instrument systems.

HART (Highway Addressable Remote Transducer). A network produced by Rosemount. Provides two-way digital communication atop traditional 4- to 20-mA loops. A HART organization has been formed.

Interbus S. An open system developed by Phoenix Contact. A fast sensor/actuator data-ring-type bus. Utilizes RS-422 transceiver technology and handles analog via separate I/O modules. Up to 256 drops per network and up to 4096 digital I/Os can be supported. The network is deterministic with data throughput in the low milliseconds.

ISA SP50. Organized in 1985 to develop a digital signal-based standard to complement the traditional 4- to 20-mA standard of the process industries. Beset by individual company interests, as well as Profibus/WorldFIP polarizations, SP50 has had a long, tough trail in pursuing an acceptable fieldbus standard. However, progress is encouraging with the constructive support of the Fieldbus Foundation. Both Profibus and WorldFIP offer eventual migration paths to any forthcoming IEC 1158/SP50 world standard.

J1850. An SAE standard for passenger cars, covering mid-speed data rates optimized at 10.4 and 41.6 kbps—rates used by GM and Ford.

LONWORKS (Local Operating Network). A distributed control network developed by Echelon Company. It uses custom Neuron chips implementing ISO/OSI seven-layer stack protocol. It supports such media as twisted pair, coax, fiber optic, RF, infrared, and power line with data rates up to 1.25 Mbps for distances up to 500 meters and 78 kbps at 2000 meters.

SDS (Smart Distributed System). Developed by Honeywell Micro-Switch. An "open" CAN-based system using a four-wire cable (two twisted pairs; signal and power). It can support up to 128 nodes at speeds up to 1.25 Mbps interfacing with PLCs and PCs for industrial control applications.

SERCOS (serial, real-time communication system). A bus developed in Europe for motors and motion control applications.

SERIPLEX-developed by Automation Process Control (APC) Company. An ASIC-based multiplexing system which offers both peer-to-peer and master/slave communications.

Profibus (Process Field Bus). Developed in Germany and strongly supported by Siemens. It is German DIN Standard 19245. In three parts, Parts 1 and 2 are designated Profibus-FMS and cover automation in general. Part 3, Profibus-DP is a faster system for factory automation. A fourth Profibus-PA Part is in preparation for process control. A number of installations are operating, covering various industries. Chips and tools are available.

WorldFIP (Factory Information Protocol). A French National Fieldbus Standard, based on the three OSI control-related layers 1, 2, and 7. There are a number of installations primarily in France and Italy. Chips and products are available. *FICOMP* (*Fieldbus Consortium*), begun in late 1992, is developing board-level products and software in accordance with IEC, Fieldbus Foundation, and WorldFIP specs.

Table 47.1 shows a summary comparing the performance characteristics of some of the popular ones.[7]

The buses and networks in the listing are primarily developed in the private sector. Other interfaces and networks were developed in university laboratories. These solutions have addressed some particular needs and have not been explored commercially. They are briefly described:

Parallel bus for sensor-driven systems. The *Michigan Parallel Standard* (*MPS*) is an "interface between one host computer and up to 256 sensor/actuator nodes employing an 8-bit parallel binary data interchange" designed

TABLE 47.1 A Comparison of the Performance Characteristics and Capability of some of the Popular Buses and Field Networks

Characteristics \ Buses	ASI	Seriplex	Interbus-S	CAN	DeviceNet/SDS	LonWorks	Profibus	Fieldbus Foundation
OSI Layers	1,2,7	1,2,7	1,2,7	1,2	1,2–4,7	1,2,3,4,5,6,7	1,2,7	1,2,7
Media Access	Master/ slave	Slotted M/S or peer-to-peer	Slatted ring with master	CSMA/CR	CSMA/CR	CSMA/CA CR	Master/slave token	M/S token
Media	UTP, flat cable 2-wire LP	4-wire TP	TP, LP for local loops + others	TP, Fiber	TP, 4-wire LP	TP, RF, PL, Fiber, Coax, IR, TP/IS	TP, LP	TP, TP/IS, RF, Fiber, Coax
Max Bit Rate	167kbps	500kbps	500kbps@40m	1Mbps@40m	500kbps@100m	1.25Mbps @130m/ segment	500kbps @200m, 12Mbps (new)	2.5Mbps
Max Distance	100m	2,000m @100kbps	12.8km@400m segments	1km@50kbps	500m@125kbps	1,400m/ segment @78kbps	4km@93kbps, 1.2km segments	1.5km@ 31.25 kbps
Max Nodes	31 w/4-bit data	256	Limited by delay	CAN 1:211 msg IDs CAN 1:229 msg IDs	64/128	32 nodes/ domain 32k domains	128	240/ segment

for sensor/actuator-driven systems. This is a centralized master-slave system. Nodes are not allowed to communicate with each other. They are only allowed to respond to messages from the host and are not allowed to call the host. Each node has an unique address for up to 256 addresses. The MPS uses a half-duplex asynchronous communication with hardware handshaking. It has 16 lines, including three ground and power lines. It has eight-bit bidirectional parallel lines and one parity line for single-bit error checking. It has a special purpose control (SP) line, which makes it compatible with different applications.[8] The speed of the bus is limited only by the drivers, the receivers, and the physical characteristics of the bus. It seems that a maximum communication rate of 1 Mbps is quite achievable.

Integrated Smart-Sensor (IS2) bus. The Delft University of Technology. The IS2 bus is a mixed analog/digital two-line serial bus interface.[9]

47.5 Software for Networked Smart Transducers

When smart transducers are networked together, a single or multiple transducer might be connected to a network node. Besides the smart transducer hardware (such as the transducer element, signal conditioning and converting circuits, and microprocessor and communication transceiver), software plays a key role in coordinating smart transducer operation. The operating system is the central nervous system for the smart transducers in a network node. Software (such as device driver) allows the operating system and application software module in a node to obtain information about the transducer and data from the transducer for configuration, diagnostic, data

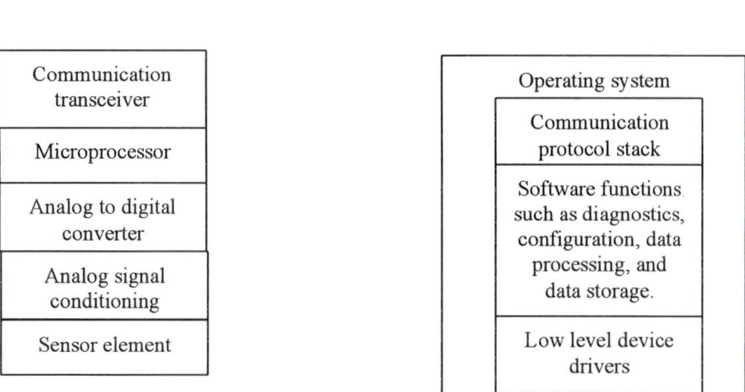

Figure 47.11 Basic hardware and software block diagrams for a networked smart transducer. The diagrams show some of the functionality of the hardware and software modules that provide the capability for the networked smart transducers.

processing, and control functions. Part of the software functions is to interface with the transducer hardware and communicate with the node in the network. Figure 47.11 shows basic hardware and software block diagram for a smart transducer that has networking capability.

47.6 A Smart Transducer Interface Standard: IEEE 1451

Due to the emergence of numerous transducer buses and networks on the market, transducer manufacturers and users alike have encountered both economical and practical barriers in interfacing to these networks. The *National Institute of Standards and Technology* (*NIST*), with the cooperation of the *Institute of Electrical and Electronics Engineers* (*IEEE*), has hosted a series of workshops to provide an open forum for holding discussions on smart-sensor communication interfaces. The first one was held in March 1994 with the specific purpose to explore the possibility of developing a standard interface that would simplify connectivity of smart sensors to control networks. In the workshops participants favored the establishment of a hardware-independent communication interface standard for connecting smart sensors to control networks leveraging existing and emerging sensor-networking technologies.[10] Later that year, the Committee on Sensor Technology in the Instrumentation and Measurement Society of IEEE sponsored a project to pursue the development of a smart transducer communication interface standard, IEEE P 1451. Draft Standards for Smart Transducer Interface for Sensors and Actuators. The main objective was to define a set of common communication interfaces for connecting smart transducers to control networks to facilitate transducer/network interoperability. Four working groups were formed to develop a family of four standards to address the needs of different sectors of the transducer industry:

- 1451.1 Network Capable Application Processor Information Model
- 1451.2 Transducer to Microprocessor Communications Protocols and Transducer Electronic Data Sheet (TEDS) Formats
- 1451.3 Digital Communication and Transducer Electronic Data Sheet (TEDS) Formats for Distributed Multidrop Systems
- 1451.4 Mixed-mode Communication Protocols and Transducer Electronic Data Sheet (TEDS) Formats

47.6.1 Aspects of standardizing connectivity

During the working group discussions, it was determined that standardizing the connectivity of smart transducers to networks involved two aspects:

- The physical interfacing of smart transducers to control networks.
- The modeling of the behavior of smart transducers to the networks.

The first aspect entails the design of a hardware interface between the smart transducers and a microprocessor that would reside in a network node and the design of a set of communication protocols and data formats. The latter entails the development of an information model and the appropriate logical interfaces interconnecting the components of the model. Standard mappings from this model onto various network protocols make it easier for sensor and actuator manufacturers to support multiple networks. The IEEE P1451 smart transducer interface standard's objective is to define a uniform approach for connecting sensors and actuators to field networks. The implementation of the various interfaces of the standards is intended to facilitate sensor and actuator interoperability across different networks and thus achieve network independence.[11]

The IEEE P1451 standard is explained in detail, including how it addresses the following problems in the sensor, control, and field networks industries.

- Simplify the connectivity of transducers to control and field networks.
- Enable transducer vendors to support multiple networks.
- Enable system integrators and users alike to select from a wide variety of transducers and networks based on merit.

47.6.2 Relationship of the family of IEEE P1451 standards

The relationships among the IEEE P1451 standards are illustrated in Fig. 47.12. This diagram is mainly used as a systematic tool to describe the con-

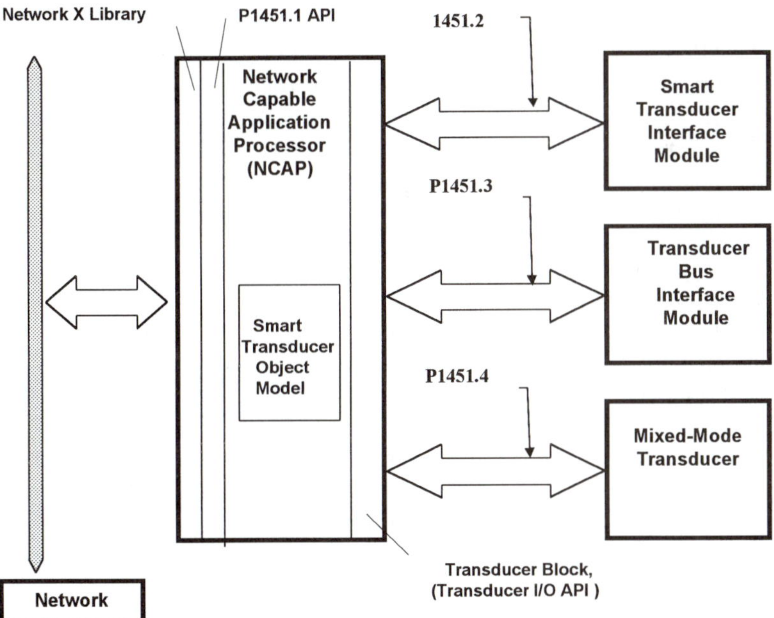

Figure 47.12 IEEE P1451 family relationship.

ceptual idea, it is left to the transducer, interface electronics, and network producers to design products adhering to the 1451 standards. It is the intention of the working groups that all the standards in the IEEE P1451 family can be used either standalone or with each other. Each of the 1451.2, P1451.3, and P1451.4 interfaces communicates with the P1451.1 NCAP through the 1451.2, P1451.3, and P1451.4 transducer blocks, respectively. The transducer blocks are transducer input/output *application programming interfaces (APIs)*.

IEEE P1451.1. The P1451.1 uses object modeling to describe the smart transducer. The smart transducer model provides a simple, yet extensible, systematic procedure to support the basic requirements of the smart transducer. The model will support various transducers by way of a series of transducer blocks, which are the input/output (I/O) driver abstraction of hardware. Application software, which might reside in the NCAP or system, can access the transducer(s) through the application programming interface. The API is a set of standardized software function routines, such as IO_read to request specific type of TEDS data and IO_Control to change sensor sampling rate. When the P1451.1 common model is deployed, it enables smart devices to easily connect to different networks.

IEEE 1451.2. The 1451.2 standard can be implemented independently or can function in concert with the P1451.1 to specify a standard environment for networked smart transducers. In addition to control networks, STIMs can be used with microprocessors in a variety of applications, such as instrumentation systems, data-acquisition cards, and host computers. Because of the STIM's unique capability to provide a self-describing feature for sensors and actuators, many vendors and users express interest in using 1451.2-compatible devices for ease of integration and maintenance. Some interesting applications have emerged, such as monitoring and control of independent wheel motion in very large earth-moving machines deployed in mining operation, high-altitude air-temperature and quality monitoring, and water-quality monitoring in water-treatment plants.

Transducer electronic data sheets (TEDS). *Transducer Electronic Data Sheets (TEDS)*, stored in EEPROMs, contain fields that fully describe the type, operation, calibration, and attributes of the transducer. The TEDS included is scalable to any size the sensor manufacturer desires. One of the key requirements in the 1451.2 standard is that the TEDS must be physically connected to the transducer. This integrated transducer-TEDS functional boundary definition provides one distinct feature that makes self-identification of transducers to the system or network possible.

47.6.3 IEEE P1451.1, network-capable application processor (NCAP) information model

A *Network-Capable Application Processor (NCAP)* is a microprocessor that resides in a node of a control network. The microprocessor has built-in hard-

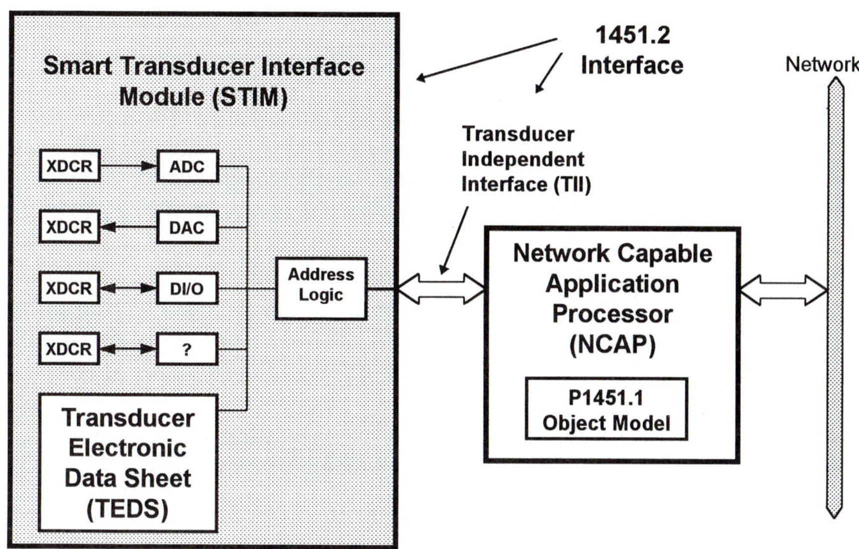

Figure 47.13 System block diagram illustrating the P1451.1 and 1451.2 interfaces. This diagram shows where the P1451.1 and 1451.2 interfaces are defined. The P1451.1 defines a common object model that describes the behavior of smart transducers. The 1451.2 defines the TEDS, the STIM, the transducer independent interface, and the protocols for the STIM to communicate with the NCAP.

ware (communicating medium) and software (protocols) to support connectivity to a network. It also has operating software for controlling the node activities, as well as programming memory set aside and made available for user applications. The aim of the P1451.1 standard is to utilize existing control networking technology and develop standardized connection methods for smart transducers to control networks. As a result, little or no changes will be required to use different methods of *analog-to-digital* (*A/D*) conversion, different microprocessors, or different network protocols and transceivers. This objective is achieved through the definition of a common object model for the components of a networked smart transducer, together with interface specifications to these components. The IEEE P1451.1 smart transducer object model resides in an NCAP. It is shown in the P1451 system block diagram in Fig. 47.13. An NCAP can be manufacturer specific. It can communicate with a *Smart Transducer Interface Module* (*STIM*) as long as it has implemented the IEEE 1451.2 interface.

Smart transducer object model interfaces. The smart transducer object model provides two key interfaces:

1) The interface to the transducer block encapsulates the details of the transducer hardware implementation within a simple programming model. This makes the sensor or actuator hardware interface look like an I/O (input-output) driver.

2) The interface to the NCAP block and ports together encapsulate the details of the different network protocol implementations behind a small set of communications methods.

These two logical models and interface specifications are used by the smart transducer object model in defining a data model for the smart transducer that is supported by the network. Application-specific behavior is modeled by function blocks. To produce the desired behavior the function blocks communicate with other blocks both on and off the smart transducer. Function blocks are supported by a range of components, such as files, parameters, actions, and services that include event generation, communication and synchronization to provide standardized behaviors. This common network-independent application model has the following key advantages:

- Establishes a high degree of interoperability between sensors/actuators and networks, thus enabling "plug-and-play" capability.
- Simplification of the support of multiple sensor/actuator control network protocols.

47.6.4 IEEE 1451.2 Transducer to microprocessor communication protocols and transducer electronic data sheet (TEDS) formats

The IEEE 1451.2 specifications allow the "plug and play" of transducers to a network at the smart transducer module level and the P1451.1 smart transducer information model enables "plug and play" of transducers at the network level. A system block diagram of the 1451.2 standard interface and the proposed P1451.1 interface is depicted in Fig. 47.13. The 1451.2 standard defines the specifications for the *smart transducer interface module* (*STIM*) and the transducer independent interface (TII). The STIM contains the TEDS, transducers (XDCR), any necessary signal conditioning and conversion, such as *analog-to-digital converter* (*ADC*), *digital-to-analog converter* (*DAC*), or simple discrete *digital input/output* (*DI/O*), and logic circuitry to facilitate digital communication between the STIM and NCAP. Through the TII 10-wire digital communication interface, the communication protocols between the smart transducers and microprocessor, the NCAP or host can initiate sensor readings or manipulate actuator actions, as well as request TEDS data.[12,13] The 10-wire interface is shown in Table 47.2. Some of the benefits of the IEEE 1451.2 TEDS areas are:

- Enables self-configuration transducers to NCAPs
- Human error is reduced
- Long-term self-documentation is available
- Extensible TEDS

TABLE 47.2 IEEE 1451.2 Interface

Line	Logic	Driven by	Function
DIN	positive logic	NCAP	Address and data transport from NCAP to STIM
DOUT	positive logic	STIM	Data transport from STIM to NCAP
DCLK	positive edge	NCAP	Positive-going edge latches data on both DIN and DOUT
NIOE	active low	NCAP	Signals that the data transport is active and delimits data transport framing
NTRIG	negative edge	NCAP	Performs triggering function
NACK	negative edge	STIM	Serves two functions: ▪ trigger acknowledge ▪ data transport acknowledge
NINT	negative edge	STIM	Used by the STIM to request service from the NCAP
NSDET	active low	STIM	Used by the NCAP to detect the presence of a STIM
POWER	N/A	NCAP	+5 V power supply
COMMON	N/A	NCAP	Signal common or ground

The normal procedure to configure a non-1451.2 transducer to an NCAP is to:

1. Read paper data sheet and decide whether signal conditioning is necessary.
2. If necessary, build signal conditioning.
3. Calibrate the transducer or enter calibration information by hand from a paper data sheet into the software in the data-acquisition system.
4. Connect the transducer.

In this situation, human errors can be introduced in reading the calibration data sheet and entering the data. In this case, *documentation* means keeping the calibration data sheet for record. Paper can often be misplaced and even get lost.

The method to configure a transducer with the IEEE 1451.2 TEDS is just plug in the STIM. The STIM will automatically send the manufacture-related data, including the calibration data, to a host.

Another advantage is that if the transducer is moved to a new location, the TEDS moves with the transducer. Thus, replacing or upgrading transducers with more-accurate ones will be simply "plug and play." Manual entering of sensor information and human errors will be eliminated. IEEE Std 1451.2-1997 has been approved and published by IEEE as a full-use standard.[4]

47.6.5 IEEE P1451.3, digital communication and transducer electronic data sheet (TEDS) formats for distributed multidrop systems

In some applications, transducers are located in harsh environments that don't allow the TEDS to be physically attached to the sensor. In other situations, the transducers must be distributed across an area in which it is undesirable to install a NCAP for each transducer. This is because it could complicate the logic design at each measurement site and increase the total power requirements. Thus, the P1451.3 project is to develop a standard digital interface for connecting large arrays of physically separated transducers in a multidrop configuration that satisfies the following requirements:

- Need to synchronously acquire large arrays of transducer readings.
- Need to accommodate an aggregate of 100s Mbps from 100s of transducers.
- Transferring of data on a power cable—minimum number of wire.
- Need self-configuring dynamically controlled network.

47.6.6 IEEE P1451.4, mixed-mode communication protocols and transducer electronic data sheet (TEDS) formats

In another segment of the transducer industry (such as the piezoelectric transducers, piezoresistive transducers, and strain gauges, etc.), companies embrace the idea of having small TEDSs physically attached to their analog transducers. However, they would also like to share the communication of the digital TEDS data with the analog signal from the transducer on a minimum set of wires (two wires), which is far less than the 10-digital communication scheme will first allow the transmission of the digital TEDS data for self-describing the transducer and then switch into another mode of operation—analog signal output from the transducer.[14]

47.7 Advantages of Networking Smart Transducers

There are many advantages and benefits of networking smart transducers. In a networked system, a single network cable can replace many point-to-point connections. This reduces a large amount of parallel cabling. A transducer-based control system includes sensors, equipment, and system-integration costs. In general, system (hardware and software) integration is a lot more expensive than equipment cost. Networking smart transducers will basically use "plug-and-play" transducers, and therefore will require minimum development effort. This will effectively lower the total system integration expense. This scenario applies both to transducer maintenance and upgrade. From another perspective, the smartness of the transducer-based system can be placed in the transducer or at the system level. The latter requires system software development and is a lot more costly. A smart transducer will certainly simplify the system development for smartness. Because smart transducers can accommodate application programs, they have a high degree of flexibility in measurement capability that can be programmed or adjusted to meet the need for speed, linearity, and accuracy. The standardized TEDS data format and communication protocols realize these benefits. The advantages are summarized in the following:

- Usage of standardized data formats, physical units specification, and communication protocol
- Distributed smartness at the transducer level
- Enhanced measurement capabilities: speed, linearity, and accuracy
- Ease of system integration
- Ease of use/maintenance/replacement/upgrade by "plug and play" of transducers
- Networked transducer solutions offer different levels of smartness
- Minimize wiring cost

References

1. Frank, R., "Understanding Smart Sensors," Artech House, Norwood, MA, 1996.
2. Bryzek, J., Evolution of Smart Electronics for Smart Transducers. SPIE Short courses, San Diego, CA, March 1995.
3. Zias, A.; Keller, H.; and van Ritter, M., "PromComp, Digital Signal Conditioning for Pressure Transducers," Test and Transducer Conference, Wembley, UK, October 1985.
4. "IEEE Std 1451.2-1997, Standard for a Smart Transducer Interface for Sensors and Actuators—Transducer to Microprocessor Communication Protocols and Transducer Electronic Data Sheet (TEDS) Formats." Institute of Electrical and Electronics Engineers, Inc., Piscataway, NJ, September 26, 1997.
5. Lee, K. and Schneeman, R., "Distributed Measurement and Control Based on the IEEE 1451 Smart Transducer Interface Standards," Proceedings of the Instrumentation and Measurement Conference (IMTC) '99, Venice, Italy. May 24–27, 1999.
6. Eidson, J. and Cole, W., "Closed Loop Control Using Ethernet as the Fieldbus." ISA Tech/97, Anaheim, CA, October 7–9, 1997.
7. Madan, P., "Introduction to Control Network, An Overview of Sensor/Actuator Network," Proceedings of the Fifth IEEE TC-9/NIST Workshop on Smart Transducer Interface Standards—IEEE P1451. Gaithersburg, MD, November 15–17, 1995, pp. 41–55.
8. Riedijk, F. and Huijsng, J., "Sensor Interface Environment based on Sigma-Delta Conversion and Serial Bus Interface," Proceedings of Sensors Conference, 1993.
9. Van der Horn, G., Huijsing, J., "Integrated Smart Sensors: Design and Calibration," Kluwer Academic Publishers, Dordrecht, The Netherlands, 1998.
10. Bryzek, J., "Summary Report," Proceedings of the IEEE/NIST First Smart Sensor Interface Standard Workshop, NIST, Gaithersburg, MD, pp. 5–12, March 31, 1994.
11. Lee, K., "The Proposed Smart Transducer Interface Standard," Proceedings of the Instrumentation Measurement Conference (IMTC) '98, St. Paul, MN, May 18–21, 1998, Vol. 1, pp. 129–133A.
12. Woods, S.; Bryzek, J.; Chen, S.; Cranmer, J.; El-Kareh, E.; Geipel, M.; Gen-Kuong, F.; Hould-

sworth, J.; LeComte, N.; Lee, K.; Mattes, M.; and Rasmussen, D., "IEEE-P1451.2 Smart Transducer Interface Module," Proceedings of SENSORS Conference, Philadelphia, PA, October 22–24, 1996.

13. Woods, S.; Lee, K.; and Bryzek, J., "An Overview of the IEEE-P1451.2 Smart Transducer Interface Module," Smart Sensor Interfaces, Analog Integrated Circuits and Signal Processing, *An International Journal,* Kluwer Academic Publishers, Vol. 14, No. 3, November 1997, pp. 3–15.
14. Chen, S.; and Lee, K., "IEEE P1451.4–A Proposed Smart Transducer Interface for Mixed-Mode Communication." *Sound and Vibration Magazine,* April 1998, pp. 24–27.

ACRONYMS AND ABBREVIATIONS

A	Ampere (electric current)
A-O	Acoustic-optic transducers
A/D	Analog to digital
ADC	Analog-to-digital converter
AFM	Atomic force microscope
AKC	Automatic level control
AM	Amplitude modulation
ANSI	American National Standards Institute
ASCII	American Standard Code for Information Interchange
ASIC	Application specific integrated circuit
ASTTL	Advanced Schottky Transistor transistor logic
ATE	Automatic test equipment
ATN	Attention (used by controllers)
B	Susceptance
BER	Bit error rate
BERT	Bit error rated testing
BOB	Break out box
C	Capacitance
CCITT	International Telephone and Telegraph Committee
CCOTDR	Complementary code OTDR
cd	Candela (luminous intensity)
CF	Crest factor
CGA	Color graphics adapter
ChemFET	Chemical field effect field-effect transistor
CMR	Common mode rejection
CPU	Central processing unit
CPW	Coplanar waveguide
CRC	Cyclic redundancy checksum
CRT	Cathode ray tube
CW	Continuous wave
D	Dissipation factor
D/A	Digital to analog
DAC	Digital-to-analog converter
dBm	Decibels referred to 1 mW of power
DCE	Data communications equipment
DDS	Direct digital synthesis
DDS	Direct digital signal
DAV	Data valid
DFB	Distributed feedback laser
DFT	Direct Fourier transform
DH	Double heterostructure
DIP	Dual inline package
DMA	Direct memory access
DMM	Digital multimeter
DMs	Degraded minutes
DNL	Differential nonlinearity
DNL	Differential Nonlinearity of ADC
DNRZ	Delayed not return to zero
DOP	Degree of polarization
DSP	Digital signal processing
DTE	Data terminal equipment
DUT	Device under test
DVM	Digital voltmeter
E/O	Electical signal to optical signal transform
ECL	Emitter coupled logic
ECMR	Effective common mode rejection
EDFA	Erbium-doped fiber amplifiers
EGA	Enhanced graphics adapter
EISA	Extended industry standard architecture
EMC	Electromagnetic compatibility
EMI	Electromagnetic interference
EOI	End of identity
ES	Errored second
EW	Electronic warfare
F-P	Fabry-Perot (Laser diodes)
FCC	Federal Communications Commission
FET	Field effect transistor
FFT	Fast Fourier transform
FIR	Finite-impulse-response
FM	Frequency modulation
FMG	Force/measure generator
FO	Fiber optic
FSK	Frequency shift keying
FT	Fourier Transform

A.2 Acronyms and Abbreviations

FTIA — Frequency and time interval frequency
FWHM — Full width at half maximum
G — Conductance
GaAs — Gallium arsenide
GFI — General format identifier
GPIB — General purpose interface bus
GPS — Global positioning system
GUI — Graphic use interface
HET — Hall effect device
HPIB — Hewlett-Packard Interface Bus
ICT — IC temperature transducer
IDT — Interdigital transducer
IEC — International Electrotechnical Commission
IEEE — Institute for Electrical and Electronic Engineers
IF — Intermediate frequency
IFC — Interface clear
IFET — Ion-sensitive field-effect transistor
IIR — Infinite-impulse-response
INL — Integral nonlinearity of ADC
ISA — Industry standard architecture
ISDN — Integrated services digital network
j — Imaginary unit
K — Kelvin (temperature)
kg — kilogram (mass)
L — Inductance
LAN — Local area network
LAP — Line access procedure
LC — Linear coefficient
LCD — Liquid crystal display
LCN — Logical channel number
LD — Laser diode
LE — Load enable
LED — Light emitting diode
LGCN — Logical channel group number
LSB — Least significant bit
LSD — Least significant digit
LVDT — Linear variable differential transformer
LVT — Linear velocity transducers
m — Meter (length)
M&TE — Measurement and test equipment
MAC — Media access and control
MCM — Multichip module
MCXO — Microprocessor compensated quartz crystal
MDA — Monochrome display adapter
MMS — Modular measurement system
MOSFET — Metal oxide semiconductor field-effect transistor
MTBF — Mean time between failures
MUX — Microwave multiplexer
NA — Numerical aperture
NCO — Numerically controlled oscillator
NDAC — No data accepted
NDT — Nondestructive testing
NIST — National Institute for Science and Technology
NMR — Normal mode noise rejection
NRFD — Not ready for data
NRZ — Not return to zero
O/E — Optical signal to electrical signal transform
O/O — Optical to optical
OCP — Overcurrent protection
OCXO — Oven compensated quartz crystal
OSA — Optical spectrum analyzer
OTDR — Optical time domain reflectometer
OVP — Overvoltage protection
P(R) — Packets received
P(S) — Packets sent
PC — Personal computer
PCO2 — Partial pressure of carbon dioxide
PD — Power divider
PDL — Polarization-dependent loss
PE — Digital pin electronics
pF — Picofarad
PGA — Pin grid array component package
PIN — *p*-type and *n*-type semiconductor materials separated by an intrinsic layer
PLL — Phase lock loop
PMD — Polarization-mode dispersion
PMMC — Permanent magnet moving coil
PMU — Parametric measurement unit
PO_2 — Partial pressure of oxygen
ppm — Parts per million
PRBS — Pseudo-random bit sequence
PRF — Pulse repetition frequency
PRTD — Platinum resistance temperature detectors
PSD — Pulse sensitive detector
PTI — Packet type identifier
PVDG — Polyvinylidene diflouoride
PW — Pulsed wave
PZT — Ferroelectric ceramic
Q — Measure of energy stored in a resonant circuit

QAM	Quadrature amplitude modulation
QPSK	Quadrature phase shift keying
R	Resistance
rad	Radian (phase angle)
RAM	Random access memory
REN	Remote enable
RG	Rated gauge (transmission line)
RIN	Relative intensity noise
RISC	Reduced instruction set computers
RMS	Root mean square
ROM	Read only memory
RPG	Rotary pulse generator
RSS	Root sum squared
RTD	Resistance temperature detectors
S (parameters)	Scattering parameters
s	Second (time)
SA	Selective availability
SABM	Set asynchronous balanced mode
SAW	Surface acoustic wave
SCC	Serial communications controller
SCPI	Standard commands for programmable instruments
SCR	Silicon controlled rectifier
SDH	Synchronous digital hierarchy
SEM	Standard reference materials
SES	Severely errored second
SMC	Smith chart
SMU	Source monitor unit
SNDR	Signal-to-noise-and-distortion ratio
SNR	Signal-to-noise ratio
SONET	Synchronous optical network
SOP	State of polarization
SPDT	Single pole double throw
SPE	Surface profiling transducers
SPST	Single pole single throw switch
SPT	Surface profiling transducer
sr	Steradian (solid angle)
SRI	System resource interface
SRQ	Service request
STM	Scanning tunneling microscope
T/R	Transmit/receive
TC	Teminal count
TC	Thermocouple
TDM	Time domain multiplexed
TDR	Time domain reflectometer
TE	Transverse electric modes
TEM	Transverse electromagnetic mode
TIA	Time interval analyzers
TIR	Total internal reflection
TM	Transverse magnetic modes
TMSL	Test and measurement systems language
TP	Triple point
TRF	Tuned radio frequency
TRL	Through/reflection/line
TTC	Transition time converters
TTL	Transistor transistor logic
TXCO	Temperature controlled crystal oscillator
UART	Universal asynchronous receiver/transmitter
UNSO	United States Naval Observatory
UT	Universal time
UTC	Universal time coordinated
VA	Volt-ampere
VAR	Volt-ampere reactive
V_{BD}	Reverse breakdown voltage
VCO	Voltage controlled oscillator
V_F	Forward voltage
VGA	Video graphics array
VL	Longitudinal mode transducer
VLSI	Very large scale integrated circuit
VME	VersaBus Modular European
VNA	Vector network analyzer
V_S	Velocity of propagation
VSWR	Voltage standing wave ratio
VXI	VMEbus extensions for instrumentation
WAN	Wide area network
WDM	Wavelength division multiplexing
WPT	Wave-propagation transducers
X	Reactance
Y	Admittance
YIG	Yttrium-iron garnet
YTM	YIG tuned multiplier
Z	Impedance
Z_0	Characteristic impedance
Z_S	Acoustic impedance

Index

ABOUT THE EDITOR

Clyde F. Coombs Jr. is one of the best-selling editors in professional publishing. He developed and edited four editions of the *Printed Circuits Handbook* and the *Communications Network Test and Measurement Handbook*, two of McGraw-Hill's top-selling technical books. An experienced electronics engineer and manager, he is currently semi-retired from Hewlett-Packard. He resides in Los Altos, California.